Springer Undergraduate Texts in Mathematics and Technology (SUMAT) publishes textbooks aimed primarily at the undergraduate. Each text is designed principally for students who are considering careers either in the mathematical sciences or in technology-based areas such as engineering, finance, information technology and computer science, bioscience and medicine, optimization or industry. Texts aim to be accessible introductions to a wide range of core mathematical disciplines and their practical, real-world applications; and are fashioned both for course use and for independent study.

Jeff Calder • Peter J. Olver

Linear Algebra, Data Science, and Machine Learning

 Springer

Jeff Calder
School of Mathematics
University of Minnesota
Minneapolis, USA

Peter J. Olver 🆔
School of Mathematics
University of Minnesota
Minneapolis, MN, USA

ISSN 1867-5506 ISSN 1867-5514 (electronic)
Springer Undergraduate Texts in Mathematics and Technology
ISBN 978-3-031-93763-7 ISBN 978-3-031-93764-4 (eBook)
https://doi.org/10.1007/978-3-031-93764-4

This Springer imprint is published by the registered company Springer Nature Switzerland AG
The registered company address is: Gewerbestrasse 11, 6330 Cham, Switzerland

If disposing of this product, please recycle the paper.

Jeff: *To my amazing wife Hayley, and our wonderful children Violet, Jack, and Nora, who fill my life with inspiration, love, and joy.*

Peter: *To my wife, Chehrzad Shakiban — to fifty years of love, family, and memories.*

Contents

Preface

Introduction to Data Science and Machine Learning

The purpose of this text is to provide a self-contained and mathematically rigorous introduction to modern methods of machine learning and data analysis at the advanced undergraduate/beginning graduate student level. The underlying mathematics — linear algebra, optimization, elementary probability and statistics, graph theory — is developed in detail with a minimum of prerequisites, relying only on very basic calculus, as described below.

The book takes a mathematical approach to the subject, with a focus on understanding how algorithms work, using a range of linear algebraic tools, that are presented ab initio, combined with some elementary differential calculus, the latter primarily applied to optimization. In particular, this is not a book on statistical machine learning, and we do not make heavy use of probabilistic concepts or interpretations (although probability does make several appearances). We have done this in an effort to make the book accessible to as broad a range of readers and students as possible, while still providing mathematical depth to the material we have chosen to cover.

While the overarching goal is to introduce readers to a broad range of modern machine learning methods and algorithms, enabling them to apply such techniques to real-world problems, we do not shy away from stating theorems and writing out proofs, especially when they lead to insight into the underlying mathematics and an understanding as to when algorithms work well and when they work suboptimally, if not fail outright. In our approach, applications go hand in hand with theory, each reinforcing and inspiring the other. In this way, the reader will be well prepared when confronting recalcitrant practical problems, as well as being able to understand, or even actively contribute to, future developments in the field.

Linear Algebra

The first five chapters develop, from the ground up, a broad range of ideas and techniques coming from linear algebra — meaning the theory and application of vectors and matrices — in a concrete and comprehensive form for direct use in data science and machine learning. These materials are inspired by the second author's text, *Applied Linear Algebra* [181], coauthored with his wife, Chehrzad Shakiban. While there is some overlap in the exposition and the selection of examples and exercises, the material has been extensively rearranged and rewritten. Indeed, this book takes an unusual and, in many ways, unique approach to introductory linear algebra that differs substantially from all existing introductory texts, including [181]. It begins with real vectors, reviewing their basic properties, including a detailed development of the key tools of the trade: bases, inner products, norms, and orthogonality, with a particular emphasis on the utility of orthonormal bases. Matrices appear afterwards, building on their interrelationships with vectors and linear algebraic systems of equations. However, the standard solution method for linear systems, namely Gaussian elimination, is *not* covered or used. This is because it exhibits numerical instabilities on challenging (also known as ill-conditioned) systems and, like all direct solution methods, scales poorly with an increase in dimension. Consequently, Gaussian elimination is unsuitable for practical computations involving the large linear systems arising in the applications under consideration. For us, the direct solution method of choice for linear systems (including least squares solutions) relies on a generalized version of the QR factorization of a matrix based on the Gram–Schmidt orthogonalization process, which is less prone to such numerical difficulties. On the other hand, when confronted with extremely large systems, especially those involving sparse matrices, meaning those with lots of zero entries, we appeal to powerful indirect iterative solution techniques. Unfortunately, space considerations compel us to refer to the literature — for example, [56, 88, 105, 245] — for the detailed development of these more advanced numerical methods. Computer implementations can be found in many general purpose numerical software packages, including Python; see also, for instance, [230].

Despite the text's unconventional pedagogical approach, the linear algebra covered in the first part is completely self-contained (modulo high school algebra) and, at least in principle, can be learned by a student who is previously unfamiliar with the subject. On the other hand, the reader will be well served by their experience in a first course in the subject, coupled with a significant level of mathematical maturity, including the ability to handle abstraction when required. That said, we will avoid introducing abstraction for its own sake, but, rather, only when necessary, as a tool to aid in understanding the required constructions, while unifying in a common framework and increasing the breadth and depth of the many applications of linear algebra to data science, machine learning, and beyond.

Calculus and Optimization

While the edifice of modern machine learning is founded on linear algebra, the essential task is optimization, meaning finding — or at least well approximating — the minimum (or maximum) of a certain real-valued nonlinear function, known in this context as the loss function, or objective function, or cost function, or entropy, or energy, etc., that measures, in some sense, the performance of the algorithm. For example, in supervised learning with labeled data, the loss function is optimized so as to impose, to the extent possible, fidelity to the training data labels, thus, hopefully, giving good results when applied to new data, including those whose labels are to be predicted. While a loss function can depend on a huge number of parameters that are to be optimally adjusted when training the algorithm, it is typically built up from fairly elementary components, namely linear (or affine) functions combined with surprisingly simple nonlinearities.

The optimization task will rely on some basic results from differential calculus, which are stated without proof; for details, the reader can consult any introductory text in the subject, including [4, 158]. These include elementary functions of one and several variables, continuity, limits, differentiability, computation of ordinary and partial derivatives, and, particularly, the chain rule. (One of the key algorithms in the subject, automatic differentiation, is a streamlined approach to computing the derivatives of complicated loss functions via the chain rule.) Gradients are essential, along with, at times, second order derivative Hessian matrices; both are introduced from first principles and in a general form that relies on a choice of underlying inner product. The method of gradient descent is the fundamental tool used to approximate the minimizer (or at least one of them if there are several) of the loss function, and we devote significant effort to studying it in detail, rigorously establishing rates of convergence under a variety of assumptions, and also developing several enhancements. We will occasionally refer to Taylor's formula for functions of both one and several variables, but (almost) exclusively the first and second order versions. Some familiarity with very basic ordinary differential equations, particularly linear equations, as well as simple numerical solution methods, is assumed later on; a good reference is [30]. Elementary integration of scalar functions, including basic numerical approximations, appears occasionally.

We also assume a little familiarity with very basic point set topology, but only in the context of Euclidean space. This includes the notions of open, closed, and compact (closed and bounded) subsets, their interior and boundary, and their (pathwise) connectedness. Basic references include [1, 202]. One crucial result, again stated without proof, is the existence of a maximum and minimum of a continuous real-valued function on a compact subset, the applied goal being how to find them. The method of proof by induction will appear often, again without elaboration. On the other hand, we make no assumptions on the reader's familiarity with probability and statistics, graph theory, or complex numbers, and these will be developed from scratch when required.

Comments on Individual Chapters

Chapter 1 introduces vectors, starting with their basic arithmetical operations — addition and scalar multiplication. As noted above, with rare exceptions, only real vectors in finite dimensional Euclidean space are considered throughout. The remainder of the chapter covers the absolutely fundamental concepts of subspace, span, linear independence, basis, and dimension within this context. While they may, upon first encounter, strike the novice as unnecessarily abstract, further success in the subject rests on one's ability to fully assimilate and confidently utilize these concepts, both practically and theoretically.

Chapter 2 reviews the standard dot product and Euclidean norm in order to motivate introducing general inner products and their associated norms. The ability to work in different norms and inner products becomes important when developing and refining machine learning algorithms. The fundamental Cauchy–Schwarz inequality is easily derived in this framework, and the more familiar triangle inequality, for inner product-based norms, is a simple consequence. The orthogonality of vectors and, subsequently, subspaces, under an inner product leads to the notion of an orthonormal basis, of fundamental importance for both theoretical developments and computational algorithms. (Indeed, without some form of orthonormality, many problems arising in machine learning applications would be computationally infeasible, even on supercomputers.) We then develop a couple of versions of the Gram–Schmidt algorithm for converting an arbitrary basis — of Euclidean space or a subspace — into an orthonormal basis. We also show how orthonormal bases are used to construct the orthogonal projection of a vector onto a subspace, thereby solving the closest point problem, which we subsequently apply to produce least squares solutions to incompatible linear systems. The chapter concludes by introducing norms that do not arise from inner products, some of which will play very useful roles in later developments.

Chapter 3 finally introduces matrices, beginning with their basic arithmetical operations — matrix addition along with scalar and matrix multiplication — and how they act on vectors. The elementary transpose operation serves to introduce the important class of symmetric matrices. The connection between matrices and linear algebraic systems of equations motivates the introduction of two of the fundamental subspaces associated with a matrix — its image and kernel — whose respective dimensions are its rank and nullity. Following a brief discussion of superposition principles for linear systems is a section on matrix inverses, which includes determinants of 2×2 matrices. Because our goal is to learn algorithms and techniques used in real world applications, the latter section is short; indeed, while at times useful for theoretical formulas and proofs, there is no practical need to ever compute the inverse or determinant of even a moderately large matrix. The chapter concludes by introducing linear and affine functions, also of importance in geometry, as local approximations of more general nonlinear functions, and a key building block for many loss functions of importance in machine learning.

Chapter 4 investigates how inner products and norms impact matrices. Classification of general inner products on Euclidean space requires symmetric positive definite matrices, which appear in many other contexts. Gram matrices, whose entries are inner products of a finite collection of vectors, are a particularly fruitful source of positive definite and semidefinite matrices. The transpose of a matrix is seen to be a particular case (for the dot product) of the adjoint of the matrix with respect to a general inner product, thereby preacribing self-adjoint and positive definite matrices that generalize the symmetric case. Again, all of these constructions are developed in anticipation of the development of machine learning applications, in particular preconditioning to enhance their utility and speed. The image and kernel of its adjoint, both of which depend on the choice of inner product, are a matrix's other two fundamental subspaces — known as its coimage and cokernel — which, by the Fundamental Theorem of Linear Algebra, satisfy important orthogonality relations with its image and kernel. These four subspaces serve to fully describe the remarkable geometry underlying matrix multiplication and the solution of linear algebraic systems. The following section introduces matrices that preserve inner product-based norms, concentrating on orthogonal matrices whose columns form an orthonormal basis, and which represent rigid rotations and reflections, thereby of importance not only in geometry but also mechanics, robotics, molecular and protein dynamics, computer graphics and gaming, and beyond. The Gram–Schmidt orthonormalization process is reinterpreted as the QR factorization of a matrix into the product of an orthogonal and an upper triangular matrix, which is here extended to include matrices of nonmaximal rank and rectangular matrices. The QR factorization leads to a useful direct method for solving linear systems of equations or, in the incompatible case, producing their least squares solutions, efficiently bypassing the more standard normal equations. The chapter closes by developing the concept of a matrix norm associated with a norm on Euclidean space, as well as the Frobenius inner product and norm on matrices.

Chapter 5 is devoted to the final essential topic in linear algebra: eigenvalues and eigenvectors. Apart from small illustrative examples, computing eigenvalues and eigenvectors is best left to computer software packages designed for this purpose, and so we do not dwell on this aspect at the outset. Instead, we focus on developing and understanding their key properties and many ramifications. Following terminology introduced in [181], we concentrate on what we call complete matrices, meaning those that possess an eigenvector basis of Euclidean space and are hence (real) diagonalizable. Not all matrices are complete (even if we were to expand our scope to include complex eigenvectors) but the incomplete ones play essentially no role in our applications and only serve as cautionary examples. As we will show, a matrix is complete if and only if it is self-adjoint with respect to some inner product if and only if it possesses an orthonormal eigenvector basis; indeed, this is how orthonormal bases of importance arise. The spectral theorem, which is the finite-dimensional version of a fundamental result in quantum physics, formalizes the diagonalization of symmetric and, more generally, self-adjoint matrices. We then show how their eigenvalues can be characterized by optimization principles involving a certain suitably constrained quadratic function. Basic practical methods for computing eigenvalues and eigenvectors rely on iteratively multiplying the matrix by an initial non-zero vector; such iterative schemes also define the basic probabilistic notion of a Markov process. Here, we cover the power method and orthogonal iteration for efficiently computing some or all of the eigenvalues and eigenvectors of a self-adjoint matrix. The final section covers the singular value decomposition, of fundamental importance in modern statistical analysis and data science, forming the basis of principal component analysis. The section ends by introducing the condition number of a matrix, which quantifies how difficult it is to numerically construct the solution to an associated linear system.

The minimization of what is referred to as a loss function, which can depend on a potentially huge number of variables, lies at the heart of most machine learning algorithms, and the development of practical algorithms for minimization is of central importance. Chapter 6 develops several basic strategies for optimizing nonlinear functions. It begins with the simplest case, namely a quadratic function, whose minima are characterized as the solutions to an associated linear system with positive (semi)definite coefficient matrix, followed by an extension where they are subject to linear constraints. Critical points, where the gradient vanishes, include (local) maxima, minima, and saddle points, and the Hessian matrix can often be used to test their character. We next introduce the all-important method of gradient descent for finding — or, rather, successively approximating — minima and minimizers, both local and global. In order to account for preconditioning, the gradient and the Hessian are defined intrinsically with respect to a general inner product, and many results continue to hold in this general setting. Refinements include proximal gradient descent and the method of conjugate gradients. Basic convergence results for gradient descent, using the notions of Lipschitz continuity, convexity, and extensions, are presented with complete proofs. The chapter concludes with a brief discussion of the classical Newton method, that often converges faster, but which, however, is of lesser importance in large scale problems owing to its higher computational costs. More advanced optimization techniques are deferred until the final chapter of the book.

Chapter 7 introduces the basics of data science and machine learning that underlie the in-depth study of fundamental algorithms in this chapter and its successors. We begin with a discussion of how data, which includes measurements, signals, images, etc., is assembled to form the data matrix. Basic quantities including mean, variance, and covariance, and the notion of labeled data are presented. The three main types of machine learning — fully supervised, unsupervised, and semi-supervised — are introduced; practical algorithms for handling each appear throughout the remainder of the text. When applying machine learning algorithms, the importance of properly splitting data into training, testing, and, possibly, validation subsets is emphasized, particularly since the misuse of the proper protocols can lead to misleading if not false claims concerning their effectiveness and utility. Basic algorithms covered in this chapter include linear, ridge, and lasso regression, support vector machines, k nearest neighbor classification, and k means clustering. The final section introduces kernel methods, which enable one to significantly extend the range of applicability of these and other algorithms.

Chapter 8 is devoted to principal component analysis (PCA), which applies the singular value decomposition of a matrix in order to simplify and visualize data. The chapter starts with a brief introduction to statistical data analysis. After introducing the basic ideas behind PCA, we provide a proof of its optimality for linearly approximating a data set by a low dimensional affine subspace. We then cover robust versions of PCA that are better able to handle outliers in noisy data sets. We also study other linear dimension reduction algorithms related to PCA, including kernel PCA, linear discriminant analysis (LDA), and multidimensional scaling (MDS).

Chapter 9 is devoted to graph theory and its ramifications and utility for data science and machine learning. We begin by introducing graphs and directed graphs, also known as digraphs, which are combinatorial objects consisting of nodes connected by edges; the edges may carry weights characterizing their importance. Data is often endowed with a graph-theoretic structure which aids in the design of machine learning algorithms. The associated weight and degree matrices are used to construct the graph Laplacian matrix, which comes in several flavors. Spectral graph theory refers to the application of the spectrum, meaning the eigenvalues and eigenvectors, of the graph Laplacian(s). We use it to develop algorithms for clustering and community detection, including spectral clustering and modularity optimization. Next up are various notions of distance between nodes (data points) in graphs, leading to a graph-based adaptation of MDS called ISOMAP. One method for prescribing internodal distances is based on the notion of diffusion on graphs and digraphs, which leads to Google's PageRank internet search engine, as well as diffusion map embeddings and multiclass spectral clustering methods. We then introduce the t-SNE algorithm, which is a widely used graph-based data visualization technique that improves linear techniques like PCA, ISOMAP, and spectral embedding, to be followed by some graph-theoretic semi-supervised learning algorithms. The final section surveys an important application to contemporary signal and image processing: the discrete Fourier representation of a sampled function, which, in fact, is a particular instance of spectral theory in the case of a cyclic graph; this section culminates in the justly famous and widely employed fast Fourier transform (FFT).

Chapter 10 covers neural networks and deep learning. We begin by introducing the mathematical framework of a fully connected neural network, consisting of the iterated composition of affine functions and simple nonlinearities, which can be regarded as simple, mathematically idealized neurons. A key complication is the potentially huge number of parameters appearing in such networks, and the computation of the required gradients for optimization in order to train the network relies on adapted chain rule techniques including automatic differentiation and backpropagation. Of fundamental importance in applications is the use of neural network architectures that are designed for the type of learning data — graphs, images, language, etc. We cover convolutional neural networks, designed for images and video, graph convolutional neural networks, designed for graph-based learning, and the transformer neural network architecture that is now widely used in natural language processing and powers the recent stunning advances in large language models such as Chat-GPT. The final section of this chapter expounds on the issue of universal approximation, meaning the ability of a prescribed class of functions, including polynomials, trigonometric (Fourier) polynomials, continuous piecewise affine functions, and a variety of neural networks, to closely approximate any (reasonable) function.

Chapter 11 returns to further study optimization, presenting some of the more sophisticated algorithms that are utilized in the large scale and challenging problems, including those that arise in the training of deep neural networks. Following further analysis of the convergence of gradient descent, we introduce momentum-based algorithms, including the heavy ball and Nesterov's accelerated methods, for attaining improved rates of convergence. We also study iterative Krylov subspace methods for solving linear systems with (sparse) positive definite coefficient matrices, and show that the conjugate gradient method is, in a sense, the optimal one. Stochastic gradient descent (SGD) can be used to accelerate the convergence of standard gradient descent when confronted with very large scale problems; the analysis requires a brief review of conditional probability. The penultimate section further analyzes gradient-based optimization algorithms by treating their continuum limits, which are certain ordinary differential equations that are amenable to basic analytical tools, to thereby better understand convergence issues and results in the discrete setting. In the final section, we study the problem of optimizing neural networks.

Exercises

Exercise sets appear at the end of every section. The exercises come in a variety of flavors. Typically, the set begins with straightforward problems testing comprehension of the new techniques and the required computational skills. We advocate solving some of the less challenging exercises by hand before resorting to software. These are followed by less routine exercises, which can range over proofs that were not supplied in the text, additional practical and theoretical results of interest, further developments in the subject, computational problems, at times making use of publicly available data sets, and beyond. Some are quite routine, while others will challenge even the most advanced reader. Larger scale computational problems require use of suitably powerful software, and, when appropriate, include links to Python notebooks.

Advice to instructors: Consider assigning only a couple of parts of a multi-part exercise. We have found the True/False exercises to be a particularly useful indicator of a student's level of understanding. Emphasize to the students that a full answer is not merely a T or F, but must include a detailed justification for the chosen answer, e.g., a proof, a counterexample, a reference to a result in the text, or the like.

A *Students' Solutions Manual*, containing the solutions to roughly a third of the exercises, is available to anyone at the text's Github website: `https://github.com/jwcalder/LAML`. An *Instructors' Solutions Manual*, that includes all the solutions in the students' manual along with additional solutions, is available to registered instructors at the text's Springer website: `https://link.springer.com/book/9783031937637`. Since solutions tend to unavoidably leak out onto the internet, roughly a third of the exercises do not have posted solutions in either manual. The authors will consider assisting a reader seeking advice on exercises with unposted solutions.

Software

For the computational activities associated with this text, access to a reasonably powerful computer (a decent laptop will suffice) and the internet is assumed. We rely on the increasingly popular open source programming language Python. Any student who has some computer programming experience can easily ramp up to speed in Python by working through the notebooks listed below. Additional Python notebooks appear throughout the text, and are all publicly available on a GitHub website (`https://github.com/jwcalder/LAML`). The easiest way to access them is to click on the corresponding link in the green "Python Notebook" box in the ebook, or to use the associated QR code[1] in the printed version. For example, the following notebook provides an introduction to basic aspects of the Python programming language that will be used in subsequent notebooks.

Python Notebook: Intro to Python (.ipynb)

Clicking on the link above, or using the QR code, will open the Python notebook from GitHub in Google Colab (`https://colab.research.google.com/`) which is a free cloud-based Python notebook environment hosted by Google. Colab uses Jupyter notebooks, which offer a way to interleave text (including mathematics in LaTeX) with Python code. Running Python code in a Google Colab notebook requires only a web browser and internet access, and in particular does not require the user to install Python, or any Python packages, on their own computer, which often causes difficulties for beginners. On the other hand, advanced users can certainly install and run Python on their own computers, and Python notebooks can be downloaded to run and modify locally.

One reason Python has become widely used in a variety of applications is the availability of high quality third party Python packages for tasks such as numerical analysis, data analysis, scientific computation, deep learning, etc. We will make extensive use of several packages in this text, including `numpy`, `scipy`, `sklearn`, `pandas`, `pytorch`, and `graphlearning`, the

---

[1]Which has nothing in common with the aforementioned QR algorithm!

last of which was created by the first author. Many of these packages are introduced by the way of examples in the accompanying notebooks. We will assume the reader is eventually able to achieve familiarity with the `numpy` and `pandas` packages, via the Python notebooks listed below. In addition, there is an introduction to `pytorch` in a notebook at the start of Chapter 10.

The `numpy` package provides support for multi-dimensional arrays and linear algebraic operations on them, and is one of the most useful packages for implementing linear algebra in Python. The following introductory notebook provides an introduction to `numpy.`, and includes the basics of how to define vectors and matrices (i.e., arrays) in `numpy` and how to operate on them.

Python Notebook: Intro to Numpy (.ipynb)

Writing efficient code that makes use of vectorization requires some of the more advanced aspects of `numpy`, which can be found in the following notebook.

Python Notebook: Advanced Numpy (.ipynb)

Both of the preceding notebooks contain a number of exercises, and we recommend the reader complete some or, better, all of these before proceeding. Given some familiarity with `numpy`, the next notebook overviews basic approaches to solving linear systems and computing eigenvectors and singular value decompositions. Some of this material requires that the reader be familiar with the basic material in Chapter 5.

Python Notebook: Numpy Linear Algebra (.ipynb)

Finally, let us mention that the `pandas` Python package is useful for loading, storing, and manipulating data. Readers are encouraged to explore the following introductory `pandas` notebook.

Python Notebook: Intro to Pandas (.ipynb)

Course Outlines

The material in this textbook is currently being used for two semester long courses on the mathematics of machine learning and data analysis at the University of Minnesota. The two courses divide the book chapters roughly as follows:

First Course:

1. Basics of optimization (Chapter 6).

2. A basic introduction to machine learning and data (Chapter 7).

3. Principal component analysis and related algorithms (Chapter 8).

Second Course:

1. Graph theory and graph-based learning (Chapter 9).

2. Neural networks and deep learning (Chapter 10).

3. Advanced topics in optimization (Chapter 11).

Provided there is time, the instructor often spends a couple of lectures introducing neural networks and deep learning in the first course. The meeting pattern for each semester is 2 lectures per week for 14 weeks, with each lecture lasting about 2 hours (which includes time for the instructor to lecture and time for students to work on mathematics or Python programming exercises during class). Roughly speaking, each section in the book is covered in a single lecture, though some longer sections require more time, and vice versa for shorter ones. In lieu of a final exam, both courses utilize a final Python project in which students work together in groups to apply machine learning algorithms to real data sets.

The first five chapters of the book offer a self-contained development of the necessary concepts in linear algebra, and can be used as review material in either course. The amount of review necessary depends on the expected level of familiarity with linear algebra among incoming students. The courses taught at the University of Minnesota require students to have taken a basic linear algebra course, which includes some (but certainly not many or all) of the topics in the first 4 chapters of the book, and some limited exposure to eigenvectors and eigenvalues (though very likely no experience with singular values). Thus, a majority of the time spent reviewing linear algebra is focused on eigenvalues and singular values (Chapter 5).

While the two courses can be taught in the order in which the chapters appear in the textbook, with the linear algebra material reviewed as needed, we have found that the first course can be implemented more effectively by rearranging the material slightly in order to get to the machine learning applications earlier on. A suggested order of topics for the first course is given below.

Suggested Order of Topics in a First Course:

1. Review of vectors, matrices, inner products, norms, and orthogonality, including orthonormal bases, as needed from the first four chapters.

2. Basic introduction to optimization (Section 6.1 through Section 6.4)

3. Introduction to basic machine learning algorithms (Chapter 7)

4. Review of eigenvalues and eigenvectors for self-adjoint matrices, singular value decomposition, and the spectral decomposition and norm of a matrix (Chapter 5).

5. Convergence results for gradient descent (Section 6.4 through Section 6.10).

6. Principal component analysis and related methods (Chapter 8).

7. Beginning of Chapter 10 on neural networks and deep learning, as time permits.

The logic behind this structure is that the vast majority of Chapter 7 requires only the basics of gradient descent and optimizing quadratic functions, which are covered in the first half of Chapter 6. The few topics in Chapter 7 that require knowledge of eigenvectors, eigenvalues, singular value decompositions, and strong convexity would have to be skipped on a first pass through the chapter (namely, some parts of the analysis of ridge regression, and lasso regression). The instructor can return to these topics after reviewing Chapter 5 and completing the remainder of Chapter 6. It is important to note that Chapter 8 relies heavily on the material from Chapter 5.

We suggest to follow the order of the remaining chapters in the book for the second course. Here, it is important to point out some of the dependencies between the final three chapters. A majority of the graph-based learning topics in Chapter 9 depend heavily on knowledge of eigenvector decompositions for self-adjoint matrices and (to a lesser extent) the singular value decomposition, both from Chapter 5, and on many of the topics introduced in Chapter 8. Subsequently, our coverage of convolutional neural networks, graph neural networks, and transformers in Chapter 10 requires many of the topics from Chapter 9, such as the spectral theory for graph Laplacians and an understanding of how to define convolution on graphs.

The courses outlined above are simply our suggestions based on the experience of teaching from a working copy of this textbook over several years. We fully expect that instructors may find other ways to construct courses using the material contained in the text, and we certainly would appreciate hearing about any such developments.

Some Final Remarks

To the student: You are about to learn the fundamental mathematical foundations along with a broad range of current techniques and algorithms that underlie modern machine learning and data science. No matter how the subject evolves in the years to come, we are confident that this material will continue to form the essential foundation that will enable one to maintain proficiency with any and all future developments. We hope you enjoy the experience, and profit from it in your studies and your career.

To the instructor: Thank you for adopting our text! We hope you enjoy teaching from it as much as we enjoyed writing it. Whatever your experience, we want to hear from you. Let us know which parts you liked and which you didn't. Which sections worked and which were less successful. Which parts your students enjoyed, which parts they struggled with, and which parts they disliked. Were enough examples included? Were the exercises of sufficient variety and at an appropriate level to enable your students to learn and use the material? How can we improve the text in future editions?

To all readers: Please send us your comments and suggestions for improving the exposition, the mathematical developments, the statements of results and their proofs, etc., as well as suggestions for additional topics that should be covered in a text at this level. Like every author, we sincerely hope that we have written an error-free text. On the other hand, in our experience, no matter how many times you proofread, mistakes still manage to sneak through. If you spot one, please notify us. Known errors, typos, and corrections will be posted when found on the text's Github website (`https://github.com/jwcalder/LAML`).

Numbering and Referencing Conventions

Theorems, Lemmas, Propositions, Definitions, and Examples are numbered consecutively within each chapter, using a common index. Thus, in Chapter 1, Lemma 1.2 follows Definition 1.1, and precedes Theorem 1.3 and Example 1.4. We find this numbering system to be the most conducive for navigating the material. Equations are also numbered consecutively within chapters, so that, for example, (3.12) refers to the twelfth numbered equation in Chapter 3. Tables and figures are included in a separate common numbering scheme. All tables, figures, and images are due to the authors, except for those that include or are based on images in referenced public data bases.

Exercises appear the end of each section, and are indicated by section and exercise numbers, followed, as necessary, by part. References to exercises within the chapter just include these numbers, while those in a different chapter are so indicated. Bibliographic references are listed alphabetically at the end of the text, and are referred to by number. For example, [181] refers to the first author's linear algebra text. Clickable links are provided in the ebook version.

The end of a proof is indicated by the symbol ■.

The end of an Example or a Remark is indicated by the symbol ▲.

An exercise or part thereof whose solution appears in the both the Students' and Instructors' Solutions Manuals is indicated by the symbol ♡.

An exercise or part thereof whose solution appears in only the Instructors' Solutions Manual is indicated by the symbol ◇.

Notation

Here we review some basic notations that are used throughout the book.

$\mathbb{Z}, \mathbb{N}, \mathbb{Q}, \mathbb{R}, \mathbb{C}$ denote, respectively, the integers, the natural numbers (nonnegative integers), the rational numbers, the real numbers, and the complex numbers, where we use i to denote the imaginary unit, i.e., one of the two square roots of -1, the other being $-i$. Since almost everything takes place in the n-dimensional real Euclidean space $\mathbb{R}^n$, complex numbers only appear in a couple of sections, and can mostly be ignored. As usual $e = 2.71828182845904\ldots$ denotes the base of the natural logarithm, while $\pi = 3.14159265358979\ldots$ is the area of a circle of unit radius. Modular arithmetic is indicated by $j \equiv k \bmod n$, for $j, k \in \mathbb{Z}$ and $0 < n \in \mathbb{N}$, meaning that $j - k$ is divisible by n. An equals sign with a colon is occasionally used to define a quantity; thus, $x := y + 1$ serves to define x. Sometimes this is written in reverse: $y + 1 =: x$. The notation $x \simeq y$ means that the objects on the left and right hand side can be identified in some prescribed manner.

The absolute value of a real number x is denoted by $|x|$. We use the standard notations $e^x = \exp(x)$ to denote the exponential function. We always use $\log x$ for its inverse, i.e., the natural, meaning base e, logarithm, while $\log_a x = \log x / \log a$ is used for logarithms with base a. Angles are always measured in radians (although occasionally degrees will be used in descriptive sentences), and all trigonometric functions, $\cos, \sin, \tan$, etc., are evaluated on radian arguments. We write $x \approx y$ to mean that x is approximately equal to y, usually without precisely stating how close they are. Conversely, we write $x \gg y$, which is equivalent to writing $y \ll x$, if x is much greater than y, again without saying precisely how much greater.

The standard notations

$$\sum_{i=1}^{n} a_i = a_1 + a_2 + \cdots + a_n, \qquad \prod_{i=1}^{n} a_i = a_1 a_2 \cdots a_n,$$

are used for the sum and product of the quantities $a_1, \ldots, a_n$. The binomial coefficients, that arise as the coefficients of the monomials $x^i y^{n-i}$ in the expansion of $(x + y)^n$ for $0 \leq i \leq n$, are denoted by $\binom{n}{i} = \dfrac{n!}{i!\,(n-i)!}$, the exclamation mark indicating the factorial of a natural number, whereby $n! = n\,(n-1)\,(n-2)\cdots 3 \cdot 2 \cdot 1$ when $0 < n \in \mathbb{N}$, while $0! = 1$ by convention.

We use $S = \{\, F \mid C \,\}$ to denote a set, where F is a formula for the members of the set and C is a list of conditions; when clear from context, one or the other may be omitted. For example, $\{\, x \mid 0 \leq x \leq 1,\ x \in \mathbb{R} \,\} = \{\, 0 \leq x \leq 1 \,\}$ means the closed unit interval between 0 and 1, also denoted $[0,1]$, while $\{0\}$ is the set consisting only of the number 0. More generally, $[a,b] = \{\, a \leq x \leq b \,\}$ denotes a closed interval, while $(a,b) = \{\, a < x < b \,\}$ is the corresponding open interval. In the latter case a could be $-\infty$ and/or b could be $+\infty$; thus, $(-\infty,\infty) = \mathbb{R}$. Half open intervals are denoted similarly, e.g., $(a,b] = \{\, a < x \leq b \,\}$.

We write $x \in S$ to indicate that x is an element of the set S, while $y \notin S$ says that y is not an element. The empty set is denoted by the symbol $\varnothing$. The cardinality of a set S, which is number of elements therein, which may be infinite, is denoted by $\#S$. The union and intersection of the sets A, B are respectively denoted by $A \cup B$ and $A \cap B$. The subset notation $A \subset B$ or, equivalently, $B \supset A$, meaning that every element of A is an element of B, includes the possibility that the sets might be equal, although for emphasis we sometimes write $A \subseteq B$, while $A \subsetneq B$ specifically implies that $A \neq B$. We use $B \setminus A = \{\, x \mid x \in B,\ x \notin A \,\}$ to denote set-theoretic difference, meaning all elements of B that do not belong to A, which need not be a subset of B for this to make sense. We use $A \times B$ to denote the Cartesian product of two sets, which is the set of all ordered pairs (a,b) where $a \in A$ and $b \in B$. Similarly for iterated Cartesian products; for example $A \times B \times C$ is the set of ordered triples (a,b,c) with $a \in A$, $b \in B$, $c \in C$. Given $0 < n \in \mathbb{N}$, we will write[2] $S^n = S \times \cdots \times S$ for the n fold Cartesian product of a set S with itself; we remark that this notation is consistent with our notation $\mathbb{R}^n$ for n-dimensional real Euclidean space, so if $S \subset \mathbb{R}$, then $S^n \subset \mathbb{R}^n$.

We use $\min S$ and $\max S$ to denote the minimum and maximum, respectively, of a closed subset $S \subset \mathbb{R}$; if the set is unbounded these can be $-\infty$ and/or ∞. This notation is also sometimes (sloppily) used even when the set is not closed, since our goal is to convey the basic idea without undue technicalities. To be terminologically precise, we should replace them by the more formal terms infimum, denoted $\inf S$, for greatest lower bound and supremum, denoted $\sup S$, for least upper bound. Readers familiar with the latter can readily make the substitutions when required.

An arrow $\to$ is used in two senses: first, to indicate convergence of a sequence to a limit: $x_n \to x^\star$ as $n \to \infty$, which means that $\lim_{n \to \infty} x_n = x^\star$; second, to indicate a function, so $F\colon X \to Y$ means that F defines a function from the domain set X to the codomain[3] set Y, written $y = f(x) \in Y$ for $x \in X$; this is sometimes abbreviated as $x \mapsto y$. Note that the image of the function, namely $F(X) := \{\, f(x) \mid x \in X \,\} \subset Y$, is only required to be a subset of the codomain. Composition of functions is denoted $F \circ G$, so that $F \circ G(x) = F\big[G(x)\big]$, which requires that the codomain of G be a subset of the domain of F. Given a real-valued function F defined on a set S, we write $\min F$ and $\max F$ for its minimum (or infimum) and maximum (or supremum); the underlying domain set is explicitly indicated as required, e.g., $\max_{x \in S} F(x)$. Similarly, $\operatorname{argmin} F$ and $\operatorname{argmax} F$ will denote, respectively, a value of x that minimizes or maximizes $F(x)$, if such exists.

Given a function F with domain $\mathbb{R}$ and codomain either $\mathbb{R}$, i.e., scalar-valued, or a higher-dimensional Euclidean space $\mathbb{R}^n$, i.e., vector-valued, we will use the usual notation $\dfrac{dF}{dx}$ for its derivative (when it exists) with respect to $x \in \mathbb{R}$, sometimes also denoted by a prime: $F'(x)$, and similarly for higher order derivatives, e.g., $\dfrac{d^2 F}{dx^2} = F''(x)$. We also employ the standard notations $\dfrac{\partial F}{\partial x}, \dfrac{\partial^2 F}{\partial x^2}, \dfrac{\partial^2 F}{\partial x\, \partial y}$, etc., for partial derivatives of a function F depending on several

---

[2] Of course, if a is a real or complex number, or even a matrix, then a^n denotes its n-th power.

[3] We prefer "codomain" to "range", which has several different meanings in linear algebra, depending on which text one consults.

variables $x, y, \ldots$. We use $\lim\limits_{n \to \infty} a_n$ and $\lim\limits_{x \to a} F(x)$ to denote the usual limits of a sequence a_n and a function F with domain $\mathbb{R}$, where the limit point a can be ∞ or $-\infty$. In particular, $F(a^+)$ and $F(a^-)$ denote the left and right hand limits at the point $x = a$, respectively. The definite integral of the function $F(x)$ on the interval $a \leq x \leq b$ is denoted by $\int_a^b F(x)\, dx$.

We find it sometimes convenient to employ the "big O" notation to describe the "rate of convergence" (or "divergence") of a function, a sequence, or an algorithm. Given two functions F, G, we write $F = \mathrm{O}(G)$ if there exists a constant $C \geq 0$ such that $|F(x)| \leq C\, |G(x)|$ for all sufficiently large or all sufficiently small x (depending on the context) lying in their common domain. The big O notation can also be similarly employed when dealing with sequences of real numbers; see [92] for more details.

We consistently use boldface lowercase letters, e.g., $\mathbf{v}, \mathbf{x}, \mathbf{a}$, to denote vectors (almost always column vectors), whose entries are the corresponding non-bold subscripted letter: v_1, x_i, a_n, etc. Matrices are denoted by ordinary capital letters, e.g., A, C, K, M — but not all such letters refer to matrices. The entries of a matrix, say A, are indicated by the corresponding subscripted lowercase letters, a_{ij} being the entry in its i-th row and j-th column.

Acknowledgments

Thanks to Chehrzad Shakiban for generously allowing us to adapt a significant amount of the linear algebra material from the book [181]. Thanks to Joseph Malkoun for showing us how to use the Schur–Horn inequalities (5.64) to simplify the proof of von Neumann's trace inequality (5.65), and to Alexander Heaton for sharing his proof of Theorem 9.13. Thanks also to Nick Higham and Darij Grinberg for discussions on the generalized QR factorization; to Linda Ness for input and additional references on machine learning; to Sheehan Olver for initial discussions concerning automatic differentiation; to Marc Paolella for suggestions on early drafts of the material; to Will Traves for feedback on using it in a course he taught at the Naval Academy; and to our colleague Scot Adams for pedagogical suggestions. Thanks to Daniela Beckelhymer for providing feedback on Chapter 7. We are grateful to Andrea L. Bertozzi, Leon Bungert, William Leeb, Gilad Lerman, Kevin Miller, Dejan Slepčev, Matthew Thorpe, and Nicolás García Trillos for discussions that contributed to the development of this book. We must particularly thank Katrina Yezzi–Woodley for ongoing collaborations on the applications of machine learning to the study of broken bones in paleoanthropology, which served to inspire us to envision writing this text and then to carry it out. We thank Loretta Bartolini for initially encouraging us to publish with Springer, and, subsequently, Elizabeth Loew at Springer for her continual support, help, and enthusiasm during the at times arduous writing and publication process. We finally thank all the students who took Math 5465/5466 at the University of Minnesota in the past two years while this textbook was being developed and written — their feedback has positively influenced the presentation and topics in the final version. And of course, we offer a profound thanks to our families for their patience and forbearance while we devoted so many hours to its completion.

Chapter 1

Vectors

In this chapter we introduce our first main protagonist — vectors — and present some of their fundamental properties. Throughout almost all of this text, we will only need to deal with vectors that have real entries, and this restriction will help to streamline and focus the exposition. Vectors with a prescribed number of entries fill out what is known as Euclidean space, since it forms the realm of ordinary Euclidean geometry. The entries of a vector can be viewed as Cartesian coordinates, and their number indicates the underlying dimension.

In many applications where machine learning plays a role, e.g., image processing, the dimension of the underlying space can be extremely large — thousands or millions or even more. Thus, the need for systematic and efficient computational tools is essential, and forms the underlying philosophy of our approach to linear algebra, which takes a novel tack. In this chapter, following an introduction to the basic arithmetic properties of vector addition and scalar multiplication, we introduce the fundamental notion of a subspace of Euclidean space, and then develop the all-important concepts of span, linear independence/dependence, and basis, which form the foundations of all that follows.

1.1 Vectors

A *column vector* consists of a finite number of real numbers, known as its *entries*, arranged in a vertical column. Given a positive integer $n = 1, 2, 3, \ldots$, the set of all vectors with n entries is denoted by $\mathbb{R}^n$, where the symbol $\mathbb{R}$ is used to denote the field of real numbers. For example, here are some vectors in $\mathbb{R}^3$:

$$
\begin{pmatrix} 1 \\ 0 \\ 3 \end{pmatrix}, \qquad
\begin{pmatrix} \pi \\ \sqrt{2} \\ -\frac{4}{7} \end{pmatrix}, \qquad
\begin{pmatrix} 3.14 \\ 1.41 \\ -.57 \end{pmatrix}, \qquad
\begin{pmatrix} 0 \\ 0 \\ 0 \end{pmatrix}.
$$

Thus, in general, a vector $\mathbf{v} \in \mathbb{R}^n$ has the form

$$
\mathbf{v} = \begin{pmatrix} v_1 \\ v_2 \\ \vdots \\ v_n \end{pmatrix}, \qquad \text{where} \qquad v_1, \ldots, v_n \in \mathbb{R}. \tag{1.1}
$$

Two vectors are equal, $\mathbf{v} = \mathbf{w}$, if and only if they have the same number of entries, so $\mathbf{v}, \mathbf{w} \in \mathbb{R}^n$ for some $0 < n \in \mathbb{N}$, and *all* their entries are equal: $v_i = w_i$, $i = 1, \ldots, n$. In particular, when $n = 1$, a column vector $\mathbf{v} = (v_1) \in \mathbb{R}^1$ has but a single entry. Such a vector can be uniquely identified with the corresponding real number $v_1 \in \mathbb{R}$, and so $\mathbb{R}^1 \simeq \mathbb{R}$. In linear algebra, the real numbers are often referred to as *scalars*, so as to distinguish them from more general vectors.

Remark. The set $\mathbb{R}^n$ is known as *n-dimensional Euclidean space*, which forms the basic setting for Euclidean geometry. Thus, for example, $\mathbb{R}^1 \simeq \mathbb{R}$ can be identified as the real line; $\mathbb{R}^2$ is the two-dimensional Euclidean plane; $\mathbb{R}^3$ can be identified with three-dimensional space; and so on. (A linear algebraic formulation of the notion of dimension will appear below.) A vector $\mathbf{v} \in \mathbb{R}^n$ can be regarded as a directed line segment, indicating both direction and magnitude; see Figure 1.1 for examples. Placing the start of the vector at the origin, we can identify its end as a point in n-dimensional space, and its entries are identified with the end point's Cartesian coordinates. For example the three entries of a vector in $\mathbb{R}^3$ can be viewed as the x, y, and z coordinates of its end point in three-dimensional Euclidean space. We will not dwell on nitpicking distinctions between points and vectors in $\mathbb{R}^n$, and identify them without further comment throughout. ▲

Remark. One can also consider vectors whose entries are other objects, e.g., complex numbers, functions, etc. However, throughout this text, we will almost exclusively use real vectors, and thus never need to specify the precise nature of their entries. ▲

Similarly, a *row vector* contains a finite number of real numbers arranged in a horizontal row. It is important, for mathematical reasons, to distinguish between row and column vectors. And, as we shall see, column vectors are the more important of the two, and so the term "vector" without qualification will always mean "column vector". Thus, writing $\mathbf{v} \in \mathbb{R}^n$ means that $\mathbf{v}$ is a column vector with n entries.

The operation of converting a column vector into a row vector, and vice versa, is known as the *transpose*, and denoted with a T superscript. Thus,

$$\begin{pmatrix} v_1 \\ v_2 \\ \vdots \\ v_n \end{pmatrix}^T = (v_1, v_2, \ldots, v_n), \qquad \text{while} \qquad (v_1, v_2, \ldots, v_n)^T = \begin{pmatrix} v_1 \\ v_2 \\ \vdots \\ v_n \end{pmatrix}. \tag{1.2}$$

Note that transposing twice takes you back to where you started: $(\mathbf{v}^T)^T = \mathbf{v}$. To conserve vertical space in the typeset text, we will often use the transpose notation, as in the second equation in (1.2), as a compact way of writing column vectors.

Of especial importance is the *zero vector*, all of whose entries are zero, denoted by a bold face $\mathbf{0} = (0, 0, \ldots, 0)^T$. Technically, we should also indicate the number of entries, since each $\mathbb{R}^n$ contains a different zero vector, but this extra notation is almost always superfluous, being clear from context. Similarly, we denote the vector in $\mathbb{R}^n$ all of whose entries are one by the bold face symbol $\mathbf{1} = (1, 1, \ldots, 1)^T$. We often use the *standard basis vectors*, all of whose entries are zero except for a single 1. (The term "basis" will be officially defined below.) In machine learning, these are also known as the *one-hot vectors*. Thus, in $\mathbb{R}^n$, there are n standard basis or one-hot vectors:

$$\mathbf{e}_1 = \begin{pmatrix} 1 \\ 0 \\ 0 \\ \vdots \\ 0 \\ 0 \end{pmatrix}, \qquad \mathbf{e}_2 = \begin{pmatrix} 0 \\ 1 \\ 0 \\ \vdots \\ 0 \\ 0 \end{pmatrix}, \qquad \ldots \qquad \mathbf{e}_n = \begin{pmatrix} 0 \\ 0 \\ 0 \\ \vdots \\ 0 \\ 1 \end{pmatrix}, \tag{1.3}$$

so that $\mathbf{e}_i$ is the vector with 1 in its i-th entry and 0's elsewhere. Again, to streamline the notation, we do not attach an extra index indicating the number of entries in each $\mathbf{e}_i \in \mathbb{R}^n$.

There are two important arithmetical operations on vectors. The first is *vector addition*. Given two vectors $\mathbf{v}, \mathbf{w} \in \mathbb{R}^n$, their sum $\mathbf{v} + \mathbf{w}$ is obtained by adding each entry of $\mathbf{v}$ to the corresponding entry of $\mathbf{w}$, so

$$
\begin{pmatrix} v_1 \\ v_2 \\ \vdots \\ v_n \end{pmatrix} + \begin{pmatrix} w_1 \\ w_2 \\ \vdots \\ w_n \end{pmatrix} = \begin{pmatrix} v_1 + w_1 \\ v_2 + w_2 \\ \vdots \\ v_n + w_n \end{pmatrix}.
$$

Addition between vectors $\mathbf{v} \in \mathbb{R}^n$ and $\mathbf{w} \in \mathbb{R}^m$ when $n \neq m$ is not allowed. The second operation is known as *scalar multiplication*. Given a scalar $c \in \mathbb{R}$ and a vector $\mathbf{v} \in \mathbb{R}^n$, the scalar product $c\mathbf{v}$ is the vector obtained by multiplying all the entries of $\mathbf{v}$ by c, so

$$
c \begin{pmatrix} v_1 \\ v_2 \\ \vdots \\ v_n \end{pmatrix} = \begin{pmatrix} cv_1 \\ cv_2 \\ \vdots \\ cv_n \end{pmatrix}.
$$

These two operations are illustrated in Figure 1.1; the sum of two vectors is, geometrically, the diagonal of the parallelogram they form;[1] scalar multiplication amounts to stretching (or shrinking) the vector by a factor $|c|$, and, when $c < 0$, reversing its direction. In particular, $-\mathbf{v} = (-1)\mathbf{v}$ is the vector obtained by reversing the signs of all entries, and is geometrically realized by reflecting the vector through the origin.

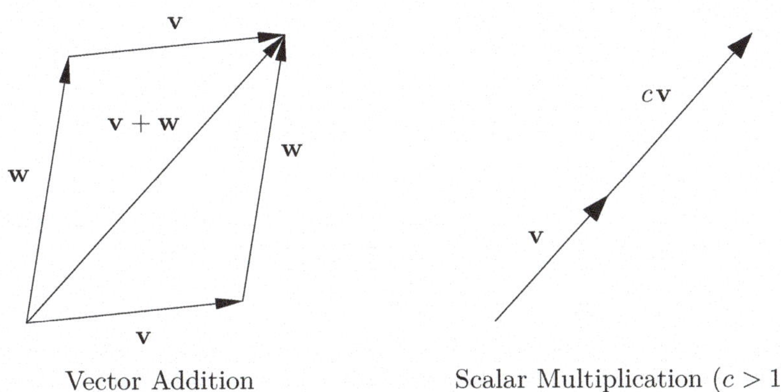

Vector Addition   Scalar Multiplication ($c > 1$)

Figure 1.1: Vector Space Operations in $\mathbb{R}^n$

Using the basic properties of real arithmetic, one easily verifies that vector addition and scalar multiplication obey the following properties, valid for all vectors $\mathbf{u}, \mathbf{v}, \mathbf{w} \in \mathbb{R}^n$, and all scalars $c, d \in \mathbb{R}$:

(a) *Commutativity of Addition*: $\mathbf{v} + \mathbf{w} = \mathbf{w} + \mathbf{v}$.

(b) *Associativity of Addition*: $\mathbf{u} + (\mathbf{v} + \mathbf{w}) = (\mathbf{u} + \mathbf{v}) + \mathbf{w}$.

---

[1] The geometric construction of vector addition based on parallelograms dates back to Newton's formulation of how to combine two forces that act a body; see [12] for extensive historical details.

    (c) *Distributivity of Scalar Multiplication*:
$$(c+d)\,\mathbf{v} = (c\,\mathbf{v}) + (d\,\mathbf{v}), \text{ and } c\,(\mathbf{v}+\mathbf{w}) = (c\,\mathbf{v}) + (c\,\mathbf{w}).$$

    (d) *Associativity of Scalar Multiplication*: $c\,(d\,\mathbf{v}) = (c\,d)\,\mathbf{v}$.

    (e) *Unit for Scalar Multiplication*: $1\,\mathbf{v} = \mathbf{v}$.

    (f) *Additive Identities*: $\mathbf{v} + \mathbf{0} = \mathbf{v} = \mathbf{0} + \mathbf{v}, \quad \mathbf{v} + (-\mathbf{v}) = \mathbf{0} = (-\mathbf{v}) + \mathbf{v}.$
$$\text{In general, if } \mathbf{u} = \mathbf{v} + \mathbf{w}, \text{ then } \mathbf{w} = \mathbf{u} - \mathbf{v} = \mathbf{u} + (-\mathbf{v}).$$

    (g) *Null Properties*: $0\,\mathbf{v} = \mathbf{0}, \; c\,\mathbf{0} = \mathbf{0},$ and if $c\,\mathbf{v} = \mathbf{0},$ then either $c = 0$ or $\mathbf{v} = \mathbf{0}.$

These properties endow $\mathbb{R}^n$ with a mathematical structure known as a (real) *vector space*. While vector spaces can be much more general, [181], in this text the only vector spaces we will encounter are $\mathbb{R}^n$ and subspaces thereof, as we next discuss. While the set of row vectors[2] also satisfies all these properties, we will always be focused on column vectors and $\mathbb{R}^n$.

    We will at times consider ordered pairs of vectors $(\mathbf{v}, \mathbf{w})$, where $\mathbf{v} \in \mathbb{R}^m$ and $\mathbf{w} \in \mathbb{R}^n$, say, where m is not necessarily equal to n. The set of all such pairs is known as the *Cartesian product* of $\mathbb{R}^m$ and $\mathbb{R}^n$, and denoted by $\mathbb{R}^m \times \mathbb{R}^n$. The Cartesian product space can be identified with $\mathbb{R}^{m+n}$, by identifying the pair $(\mathbf{v}, \mathbf{w})$, where $\mathbf{v} = (v_1, \ldots, v_m)^T$, $\mathbf{w} = (w_1, \ldots, w_n)^T$, with the vector $(v_1, \ldots, v_m, w_1, \ldots, w_n)^T \in \mathbb{R}^{m+n}$. However, it is sometimes useful to use the Cartesian product notation to remind us what we are dealing with. One can clearly extend this construction to ordered k-tuples of vectors in the evident manner. For example, ordered triples $(\mathbf{u}, \mathbf{v}, \mathbf{w}) \in \mathbb{R}^l \times \mathbb{R}^m \times \mathbb{R}^n$, with $\mathbf{u} \in \mathbb{R}^l$, $\mathbf{v} \in \mathbb{R}^m$, $\mathbf{w} \in \mathbb{R}^n$, can be identified with vectors in $\mathbb{R}^{l+m+n}$.

Exercises

1.1. Plot the following vectors in $\mathbb{R}^2$.
    $(a)\,\heartsuit\;(-2, 2)^T, \quad (b)\,\heartsuit\;(0, -1)^T, \quad (c)\,\diamondsuit\;3\,(1, 1)^T, \quad (d)\;(-2, 3)^T - (-5, 3)^T.$

1.2. Suppose $\mathbf{v} = (1, 2, -1)^T$ and $\mathbf{w} = (0, -1, 2)^T$. Determine the following vectors:
    $(a)\,\heartsuit\;-\mathbf{v}, \quad (b)\;3\,\mathbf{v}, \quad (c)\,\heartsuit\;-5\,\mathbf{w}, \quad (d)\,\diamondsuit\;\mathbf{v}+\mathbf{w}, \quad (e)\;\mathbf{v}-\mathbf{w}, \quad (f)\;2\,\mathbf{v}-3\,\mathbf{w}.$

1.3. Prove the arithmetic properties $(a)\,\heartsuit, (b)\,\diamondsuit, (c)\,\heartsuit, (d), (e), (f)\,\heartsuit, (g)$ for vectors in $\mathbb{R}^n$.

1.2 Subspaces

In linear algebra, the most important subsets of $\mathbb{R}^n$ are those that closed under the operations of vector addition and scalar multiplication. They serve to generalize the geometric notions of point, line, and plane in two- and three-dimensional space. More precisely:

> **Definition 1.1.** A *subspace* of $\mathbb{R}^n$ is a nonempty subset $\varnothing \neq V \subseteq \mathbb{R}^n$ that satisfies
>
>   (a) for every $\mathbf{v}, \mathbf{w} \in V$, the sum $\mathbf{v} + \mathbf{w} \in V$, and
>
>   (b) for every $\mathbf{v} \in V$ and every $c \in \mathbb{R}$, the scalar product $c\,\mathbf{v} \in V$.

    In particular, a subspace *must* contain the zero vector $\mathbf{0} \in V$. Indeed, if $\mathbf{v} \in V$ is any vector, then $0\,\mathbf{v} = \mathbf{0}$ must also lie in V by closure under scalar multiplication. It is sometimes

---

[2]In more theoretical treatments of the subject, the space of row vectors is identified as the "dual vector space" to $\mathbb{R}^n$; see, e.g., [181].

convenient to combine the two closure conditions. Thus, to prove that V is a subspace, it suffices to check that $c\mathbf{v} + d\mathbf{w} \in V$ for all $\mathbf{v}, \mathbf{w} \in V$ and $c, d \in \mathbb{R}$.

Example 1.2. Let us list some examples of subspaces of the three-dimensional Euclidean space $\mathbb{R}^3$.

(a) The trivial subspace $V = \{\mathbf{0}\}$. Demonstrating closure is easy: since there is only one vector $\mathbf{0}$ in V, we just need to check that $\mathbf{0} + \mathbf{0} = \mathbf{0} \in V$ and $c\mathbf{0} = \mathbf{0} \in V$ for every scalar c.

(b) The entire space $V = \mathbb{R}^3$. Here closure is immediate.

(c) The set of all vectors of the form $(x, y, 0)^T$, i.e., the xy coordinate plane. To prove closure, we check that all sums $(x, y, 0)^T + (\widehat{x}, \widehat{y}, 0)^T = (x + \widehat{x}, y + \widehat{y}, 0)^T$ and scalar multiples $c(x, y, 0)^T = (cx, cy, 0)^T$ of vectors in the xy-plane remain in the plane.

(d) The set of solutions $(x, y, z)^T$ to the homogeneous linear equation

$$3x + 2y - z = 0. \tag{1.4}$$

Indeed, if $\mathbf{x} = (x, y, z)^T$ is a solution, then so is every scalar multiple $c\mathbf{x} = (cx, cy, cz)^T$ since $3(cx) + 2(cy) - (cz) = c(3x + 2y - z) = 0$. Moreover, if $\widehat{\mathbf{x}} = (\widehat{x}, \widehat{y}, \widehat{z})^T$ is a second solution, so $3\widehat{x} + 2\widehat{y} - \widehat{z} = 0$, their sum $\mathbf{x} + \widehat{\mathbf{x}} = (x + \widehat{x}, y + \widehat{y}, z + \widehat{z})^T$ is also a solution, since

$$3(x + \widehat{x}) + 2(y + \widehat{y}) - (z + \widehat{z}) = (3x + 2y - z) + (3\widehat{x} + 2\widehat{y} - \widehat{z}) = 0.$$

The solution space to (1.4) can be identified as the two-dimensional plane passing through the origin with normal vector $(3, 2, -1)^T$.

(e) The set of all vectors lying in the plane spanned by the vectors $\mathbf{v}_1 = (2, -3, 0)^T$ and $\mathbf{v}_2 = (1, 0, 3)^T$. In other words, we consider all vectors of the form

$$\mathbf{v} = a\mathbf{v}_1 + b\mathbf{v}_2 = a \begin{pmatrix} 2 \\ -3 \\ 0 \end{pmatrix} + b \begin{pmatrix} 1 \\ 0 \\ 3 \end{pmatrix} = \begin{pmatrix} 2a + b \\ -3a \\ 3b \end{pmatrix},$$

where $a, b \in \mathbb{R}$ are arbitrary scalars. If $\mathbf{v} = a\mathbf{v}_1 + b\mathbf{v}_2$ and $\mathbf{w} = \widehat{a}\,\mathbf{v}_1 + \widehat{b}\,\mathbf{v}_2$ are any two vectors in the span, then so is

$$c\mathbf{v} + d\mathbf{w} = c(a\mathbf{v}_1 + b\mathbf{v}_2) + d(\widehat{a}\,\mathbf{v}_1 + \widehat{b}\,\mathbf{v}_2) = (ac + \widehat{a}\,d)\mathbf{v}_1 + (bc + \widehat{b}\,d)\mathbf{v}_2 = \widetilde{a}\,\mathbf{v}_1 + \widetilde{b}\,\mathbf{v}_2,$$

where $\widetilde{a} = ac + \widehat{a}\,d$, $\widetilde{b} = bc + \widehat{b}\,d$. This demonstrates that the span is a subspace of $\mathbb{R}^3$. The reader may have already noticed that this subspace is the same plane defined by (1.4).   ▲

Example 1.3. The following subsets of $\mathbb{R}^3$ are *not* subspaces.

(a) The set A of all vectors of the form $(x, y, 1)^T$, i.e., the plane parallel to the xy coordinate plane passing through $(0, 0, 1)^T$. Indeed, $(0, 0, 0)^T \notin A$, which is the most basic requirement for a subspace. In fact, neither of the closure axioms hold for this subset.

(b) The nonnegative orthant $\mathcal{O}^+ = \{x \geq 0,\ y \geq 0,\ z \geq 0\}$. Although $\mathbf{0} \in \mathcal{O}^+$, and the sum of two vectors in $\mathcal{O}^+$ also belongs to $\mathcal{O}^+$, multiplying by negative scalars takes us outside the orthant, violating closure under scalar multiplication.

(c) The unit sphere $S_1 = \{\, x^2 + y^2 + z^2 = 1 \,\}$. Again, $\mathbf{0} \notin S_1$. More generally, curved surfaces, such as the paraboloid $P = \{\, z = x^2 + y^2 \,\}$, are not subspaces. Although $\mathbf{0} \in P$, most scalar multiples of vectors in P do not belong to P. For example, $(\,1,1,2\,)^T \in P$, but $2\,(\,1,1,2\,)^T = (\,2,2,4\,)^T \notin P$.   ▲

In fact, there are only four fundamentally different types of subspaces of three-dimensional Euclidean space:

$$(i) \quad \text{a point — the trivial subspace } V = \{\mathbf{0}\},$$

$$(ii) \quad \text{a line passing through the origin,}$$

$$(iii) \quad \text{a plane passing through the origin,}$$

$$(iv) \quad \text{the entire three-dimensional space } V = \mathbb{R}^3.$$

We can establish this observation by the following argument. If $V = \{\mathbf{0}\}$ contains only the zero vector, then we are in case (iv). Otherwise, $V \subset \mathbb{R}^3$ contains a nonzero vector $\mathbf{0} \neq \mathbf{v}_1 \in V$. But since V must contain all scalar multiples $c\,\mathbf{v}_1$, it includes the entire line in the direction of $\mathbf{v}_1$. If V contains another vector $\mathbf{v}_2$ that does not lie in the line through $\mathbf{v}_1$, then it must contain the entire plane $\{c\,\mathbf{v}_1 + d\,\mathbf{v}_2\}$ spanned by $\mathbf{v}_1, \mathbf{v}_2$. Finally, if there is a third vector $\mathbf{v}_3$ not contained in this plane, then we claim that $V = \mathbb{R}^3$. This final fact will be an immediate consequence of general results in this chapter, although the interested reader might try to prove it directly before proceeding.

Exercises

2.1. ♡ (a) Prove that the set of all vectors $(\,x, y, z\,)^T$ such that $x - y + 4z = 0$ forms a subspace of $\mathbb{R}^3$. (b) Explain why the set of all vectors that satisfy $x - y + 4z = 1$ does not form a subspace.

2.2. Which of the following are subspaces of $\mathbb{R}^3$? Justify your answers! (a) ♡ The set of all vectors $(\,x, y, z\,)^T$ satisfying $x + y + z + 1 = 0$. (b) ◊ The set of vectors of the form $(\,t, -t, 0\,)^T$ for $t \in \mathbb{R}$. (c) ♡ The set of vectors of the form $(\,r - s, r + 2\,s, -s\,)^T$ for $r, s \in \mathbb{R}$. (d) The set of vectors whose first component equals 0. (e) The set of vectors whose last component equals 1. (f) ♡ The set of all vectors $(\,x, y, z\,)^T$ with $x \geq y \geq z$. (g) ♡ The set of all solutions to the equation $z = x - y$. (h) ◊ The set of all solutions to the equation $z = xy$. (i) The set of all solutions to the equation $x^2 + y^2 + z^2 = 0$. (j) The set of all solutions to the system $xy = yz = xz$.

2.3. Determine which of the following sets of vectors $\mathbf{x} = (\,x_1, x_2, \ldots, x_n\,)^T$ are subspaces of $\mathbb{R}^n$: (a) ♡ all equal entries $x_1 = \cdots = x_n$; (b) ♡ all positive entries: $x_i \geq 0$; (c) ◊ first and last entries equal to zero: $x_1 = x_n = 0$; (d) the entries add up to zero: $x_1 + \cdots + x_n = 0$; (e) first and last entries differ by one: $x_1 - x_n = 1$.

2.4. Show that if $W \subset \mathbb{R}^2$ is a subspace containing the vectors $(\,1, -1\,)^T$, $(\,1, 1\,)^T$, then $W = \mathbb{R}^2$.

2.5. ◊ (a) Can you construct an example of a subset $S \subset \mathbb{R}^2$ with the property that $c\mathbf{v} \in S$ for all $c \in \mathbb{R}$, $\mathbf{v} \in S$, and yet S is not a subspace?   (b) What about an example in which $\mathbf{v} + \mathbf{w} \in S$ for every $\mathbf{v}, \mathbf{w} \in S$, and yet S is not a subspace?

2.6. Show that if V and W are subspaces of $\mathbb{R}^n$, then (a) ♡ their *intersection* $V \cap W$ is a subspace; (b) their *sum* $V + W = \{\, \mathbf{v} + \mathbf{w} \mid \mathbf{v} \in V,\ \mathbf{w} \in W \,\}$ is a subspace; but (c) ◊ their *union* $V \cup W$ is not a subspace, unless $V \subset W$ or $W \subset V$.

2.7. Let $V \subset \mathbb{R}^n$ be a subspace. A subset of the form $W = V + \mathbf{b} = \{\mathbf{v} + \mathbf{b} \mid \mathbf{v} \in V\}$, where $\mathbf{b} \in V$ is a fixed vector, is known as an *affine subspace* of $\mathbb{R}^n$. (a) Show that an affine subspace $W \subset \mathbb{R}^n$ is a genuine subspace if and only if $\mathbf{b} \in V$. (b) Draw the affine subspaces $W \subset \mathbb{R}^2$ when (i) V is the x-axis and $\mathbf{b} = (2,1)^T$, (ii) V is the line $y = \frac{3}{2}x$ and $\mathbf{b} = (1,1)^T$, (iii) V is the line $\{(t,-t)^T \mid t \in \mathbb{R}\}$, and $\mathbf{b} = (2,-2)^T$. (c) Show that the line $x - 2y = 1$ is an affine subspace of $\mathbb{R}^2$.

2.8. $\diamond$ A *line* in the plane is a subset of the form $L = \{\mathbf{a} + t\mathbf{v} \mid t \in \mathbb{R}\} \subset \mathbb{R}^2$ where $\mathbf{a}$ and $\mathbf{0} \neq \mathbf{v}$ are vectors in $\mathbb{R}^2$; thus, in the language of Exercise 2.7 they are affine subspaces. Two lines are *parallel* if and only if the corresponding vectors $\mathbf{v}$ are parallel, meaning that they are nonzero scalar multiple of each other. (a) Express the y axis in the above form, and find all lines that are parallel to it. (b) More generally, express the line $ax + by = c$ with a, b not both 0 in the above form, and find all lines that are parallel to it. (c) Show that two different lines in the plane are parallel if and only if they do not intersect. (d) The definition of parallel lines extends, as written, to $\mathbb{R}^3$. Answer part (a) in this case, but show that part (c) is no longer true.

1.3 Span and Linear Independence

The definition of the span of a collection of vectors generalizes, in a natural fashion, the geometric notion of two vectors spanning a plane in $\mathbb{R}^3$. As such, it describes the first of two general methods for constructing subspaces.

Definition 1.4. Let $\mathbf{v}_1, \ldots, \mathbf{v}_k \in \mathbb{R}^n$. A sum of the form

$$c_1 \mathbf{v}_1 + c_2 \mathbf{v}_2 + \cdots + c_k \mathbf{v}_k = \sum_{i=1}^{k} c_i \mathbf{v}_i, \tag{1.5}$$

where the coefficients $c_1, c_2, \ldots, c_k \in \mathbb{R}$ are any scalars, is known as a *linear combination* of the vectors $\mathbf{v}_1, \ldots, \mathbf{v}_k$.

For instance,

$$3\mathbf{v}_1 + \mathbf{v}_2 - 2\mathbf{v}_3, \qquad\qquad 8\mathbf{v}_1 - \tfrac{1}{3}\mathbf{v}_3 = 8\mathbf{v}_1 + 0\mathbf{v}_2 - \tfrac{1}{3}\mathbf{v}_3,$$
$$\mathbf{v}_2 = 0\mathbf{v}_1 + 1\mathbf{v}_2 + 0\mathbf{v}_3, \qquad\qquad \mathbf{0} = 0\mathbf{v}_1 + 0\mathbf{v}_2 + 0\mathbf{v}_3,$$

are four different linear combinations of the three vectors $\mathbf{v}_1, \mathbf{v}_2, \mathbf{v}_3$.

By repeatedly applying the closure conditions, one easily sees that if $V \subset \mathbb{R}^n$ is a subspace and $\mathbf{v}_1, \ldots, \mathbf{v}_k \in V$, then any linear combination (1.5) also belongs to V.

Definition 1.5. The *span* of a finite collection of vectors $\mathbf{v}_1, \ldots, \mathbf{v}_k \in \mathbb{R}^n$ is the subset $V = \mathrm{span}\,\{\mathbf{v}_1, \ldots, \mathbf{v}_k\} \subset \mathbb{R}^n$ consisting of all possible linear combinations (1.5) for $c_1, \ldots, c_k \in \mathbb{R}$.

A key observation is that the span always forms a subspace.

Proposition 1.6. *The span of any collection of vectors in $\mathbb{R}^n$ is a subspace.*

Proof. We need to show that if

$$\mathbf{v} = c_1\mathbf{v}_1 + \cdots + c_k\mathbf{v}_k \qquad \text{and} \qquad \widehat{\mathbf{v}} = \widehat{c}_1\mathbf{v}_1 + \cdots + \widehat{c}_k\mathbf{v}_k$$

are any two linear combinations, then their sum is also a linear combination, since

$$\mathbf{v} + \widehat{\mathbf{v}} = (c_1 + \widehat{c}_1)\mathbf{v}_1 + \cdots + (c_k + \widehat{c}_k)\mathbf{v}_k = \widetilde{c}_1\mathbf{v}_1 + \cdots + \widetilde{c}_k\mathbf{v}_k,$$

where $\widetilde{c}_i = c_i + \widehat{c}_i$. Similarly, for any scalar multiple,

$$a\,\mathbf{v} = (a\,c_1)\mathbf{v}_1 + \cdots + (a\,c_k)\mathbf{v}_k = \overline{c}_1\mathbf{v}_1 + \cdots + \overline{c}_k\mathbf{v}_k,$$

where $\overline{c}_i = a\,c_i$, which completes the proof. ∎

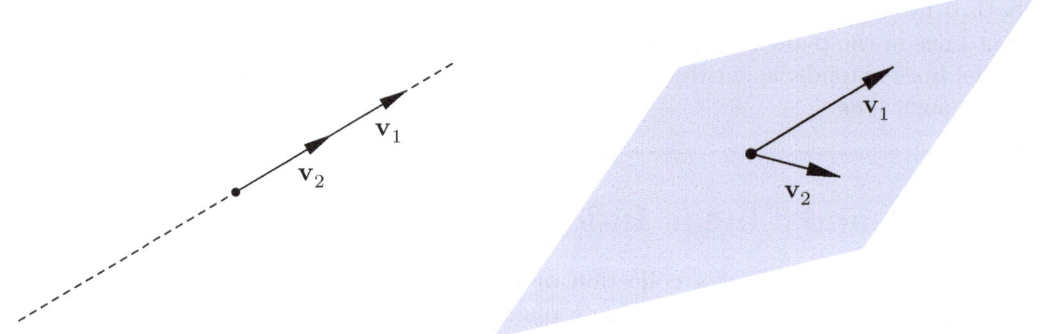

Figure 1.2: Line and Plane Spanned by Two Vectors

Example 1.7. *Examples of subspaces spanned by vectors in* $\mathbb{R}^3$:

(i) If $\mathbf{v}_1 \neq \mathbf{0}$ is any nonzero vector in $\mathbb{R}^3$, then span $\{\mathbf{v}_1\}$ is the line $\{c\,\mathbf{v}_1 \mid c \in \mathbb{R}\}$ consisting of all vectors *parallel* to $\mathbf{v}_1$. If $\mathbf{v}_1 = \mathbf{0}$, then its span just contains the origin, i.e., it is a point.

(ii) If $\mathbf{v}_1$ and $\mathbf{v}_2$ are any two vectors in $\mathbb{R}^3$, then span $\{\mathbf{v}_1, \mathbf{v}_2\}$ is the set of all vectors of the form $c_1\mathbf{v}_1 + c_2\mathbf{v}_2$. Typically, such a span prescribes a plane passing through the origin. However, if $\mathbf{v}_1$ and $\mathbf{v}_2$ are parallel, then their span is just a line. The most degenerate case occurs when $\mathbf{v}_1 = \mathbf{v}_2 = \mathbf{0}$, where the span is just a point — the origin.

(iii) If we are given three non-coplanar vectors $\mathbf{v}_1, \mathbf{v}_2, \mathbf{v}_3$, then their span is all of $\mathbb{R}^3$, as we shall prove below. If they all lie in a plane, then their span is the plane — unless they are all parallel, in which case their span is a line — or, in the completely degenerate situation $\mathbf{v}_1 = \mathbf{v}_2 = \mathbf{v}_3 = \mathbf{0}$, a single point. ▲

Thus, every subspace of $\mathbb{R}^3$ can be realized as the span of some set of vectors. One can consider subspaces spanned by four or more vectors in $\mathbb{R}^3$, but these continue to be limited to being either a point (the origin), a line, a plane, or the entire three-dimensional space.

Example 1.8. Let $W \subset \mathbb{R}^3$ be the plane spanned by the vectors $\mathbf{v}_1 = (1, -2, 1)^T$ and $\mathbf{v}_2 = (2, -3, 1)^T$. Question: Does the vector $\mathbf{v} = (0, 1, -1)^T$ belong to W? To answer, we need to see whether we can find scalars c_1, c_2 such that

$$\mathbf{v} = c_1\,\mathbf{v}_1 + c_2\,\mathbf{v}_2; \quad \text{that is,} \quad \begin{pmatrix} 0 \\ 1 \\ -1 \end{pmatrix} = c_1\begin{pmatrix} 1 \\ -2 \\ 1 \end{pmatrix} + c_2\begin{pmatrix} 2 \\ -3 \\ 1 \end{pmatrix} = \begin{pmatrix} c_1 + 2c_2 \\ -2c_1 - 3c_2 \\ c_1 + c_2 \end{pmatrix}.$$

Thus, c_1, c_2 must satisfy the linear algebraic system

$$c_1 + 2c_2 = 0, \qquad -2c_1 - 3c_2 = 1, \qquad c_1 + c_2 = -1,$$

which has the evident solution $c_1 = -2$, $c_2 = 1$. We conclude that $\mathbf{v} = -2\mathbf{v}_1 + \mathbf{v}_2$ does belong to the span. On the other hand, $\widetilde{\mathbf{v}} = (1, 0, 0)^T$ does not belong to W. Indeed, there are no scalars c_1, c_2 such that $\widetilde{\mathbf{v}} = c_1\mathbf{v}_1 + c_2\mathbf{v}_2$, because the corresponding linear system has no solutions. ▲

Warning: It is entirely possible for different sets of vectors to span the *same* subspace. For instance, $\mathbf{e}_1 = (1, 0, 0)^T$ and $\mathbf{e}_2 = (0, 1, 0)^T$ span the xy-plane in $\mathbb{R}^3$, as do the three coplanar vectors $\mathbf{v}_1 = (1, -1, 0)^T$, $\mathbf{v}_2 = (-1, 2, 0)^T$, $\mathbf{v}_3 = (2, 1, 0)^T$.

Often, all of the vectors used to form a span are essential. For example, we cannot use fewer than two vectors to span a plane in $\mathbb{R}^3$, since the span of a single vector is at most a line. However, in degenerate situations, some of the spanning vectors may be redundant. For instance, if the two vectors are parallel, then their span is a line, but only one of the vectors is really needed to prescribe the line. The elimination of such superfluous spanning vectors is encapsulated in the following important definition.

Definition 1.9. The vectors $\mathbf{v}_1, \ldots, \mathbf{v}_k \in \mathbb{R}^n$ are called *linearly dependent* if there exist scalars $c_1, \ldots, c_k \in \mathbb{R}$, *not all zero*, such that

$$c_1\mathbf{v}_1 + \cdots + c_k\mathbf{v}_k = \mathbf{0}. \tag{1.6}$$

Vectors that are not linearly dependent are called *linearly independent*.

The restriction that not all the c_i's are zero is essential: if $c_1 = \cdots = c_k = 0$, then the linear combination (1.6) is automatically zero. Thus, to check linear independence, one needs to show that the *only* linear combination that produces the zero vector (1.6) is this trivial one; in other words, $c_1 = \cdots = c_k = 0$ is the *one and only* solution to the vector equation (1.6). Observe that if $\mathbf{v}_1, \ldots, \mathbf{v}_k$ are linearly independent, so is any subset thereof, e.g., $\mathbf{v}_1, \ldots, \mathbf{v}_j$ for $j < k$. But this is *not* valid for linear dependence: $\mathbf{v}_1, \mathbf{v}_2$ might be linearly dependent, but if $\mathbf{v}_1 \neq \mathbf{0}$, on its own it forms a linearly independent set.

Example 1.10. Some examples of linear independence and dependence:

(a) The vectors

$$\mathbf{v}_1 = \begin{pmatrix} 1 \\ 2 \\ -1 \end{pmatrix}, \qquad \mathbf{v}_2 = \begin{pmatrix} 0 \\ 3 \\ 1 \end{pmatrix}, \qquad \mathbf{v}_3 = \begin{pmatrix} -1 \\ 4 \\ 3 \end{pmatrix},$$

are linearly dependent, because

$$\mathbf{v}_1 - 2\mathbf{v}_2 + \mathbf{v}_3 = \mathbf{0}.$$

On the other hand, the first two vectors $\mathbf{v}_1, \mathbf{v}_2$ are linearly independent. To see this, suppose that

$$c_1\mathbf{v}_1 + c_2\mathbf{v}_2 = \begin{pmatrix} c_1 \\ 2c_1 + 3c_2 \\ -c_1 + c_2 \end{pmatrix} = \begin{pmatrix} 0 \\ 0 \\ 0 \end{pmatrix}.$$

For this to happen, c_1, c_2 must satisfy the homogeneous linear system

$$c_1 = 0, \qquad 2c_1 + 3c_2 = 0, \qquad -c_1 + c_2 = 0,$$

which, as you can check, has only the trivial solution $c_1 = c_2 = 0$.

(b) In general, any collection $\mathbf{v}_1, \ldots, \mathbf{v}_k$ that includes the zero vector, say $\mathbf{v}_1 = \mathbf{0}$, is automatically linearly dependent, since $1\,\mathbf{0} + 0\,\mathbf{v}_2 + \cdots + 0\,\mathbf{v}_k = \mathbf{0}$ is a nontrivial linear combination that adds up to $\mathbf{0}$.

(c) Two vectors $\mathbf{v}, \mathbf{w} \in V$ are linearly dependent if and only if they are *parallel*, meaning that one is a scalar multiple of the other. Indeed, if $\mathbf{v} = a\mathbf{w}$, then $\mathbf{v} - a\mathbf{w} = \mathbf{0}$ is a nontrivial linear combination summing to zero. Conversely, if $c\mathbf{v} + d\mathbf{w} = \mathbf{0}$ and $c \neq 0$, then $\mathbf{v} = -(d/c)\mathbf{w}$, while if $c = 0$ but $d \neq 0$, then $\mathbf{w} = \mathbf{0} = 0\,\mathbf{v}$.   ▲

Lemma 1.11. *Let $\mathbf{v}_1, \ldots, \mathbf{v}_k$ be linearly independent. If $\mathbf{v}_{k+1} \notin \operatorname{span}\{\mathbf{v}_1, \ldots, \mathbf{v}_k\}$, then $\mathbf{v}_1, \ldots, \mathbf{v}_{k+1}$ are also linearly independent.*

Proof. Suppose

$$c_1\mathbf{v}_1 + \cdots + c_k\mathbf{v}_k + c_{k+1}\mathbf{v}_{k+1} = \mathbf{0}. \tag{1.7}$$

If $c_{k+1} = 0$, then, by the linear independence of $\mathbf{v}_1, \ldots, \mathbf{v}_k$, this implies $c_1{}^{\backprime} = \cdots = c_k = 0$. On the other hand, if $c_{k+1} \neq 0$, then we can divide (1.7) by c_{k+1} and rewrite the resulting equation in the form

$$\mathbf{v}_{k+1} = (-c_1/c_{k+1})\,\mathbf{v}_1 + \cdots + (-c_k/c_{k+1})\,\mathbf{v}_k,$$

which implies $\mathbf{v}_{k+1} \in \operatorname{span}\{\mathbf{v}_1, \ldots, \mathbf{v}_k\}$, in contradiction to our hypothesis. Thus, the only linear combination satisfying (1.7) is the trivial one $c_1{}^{\backprime} = \cdots = c_k = c_{k+1} = 0$.   ∎

Exercises

3.1. ♡ Show that $\begin{pmatrix} -1 \\ 2 \\ 3 \end{pmatrix}$ belongs to the subspace of $\mathbb{R}^3$ spanned by $\begin{pmatrix} 2 \\ -1 \\ 2 \end{pmatrix}, \begin{pmatrix} 5 \\ -4 \\ 1 \end{pmatrix}$ by writing it as a linear combination of the spanning vectors.

3.2. (a) Determine whether $\begin{pmatrix} 1 \\ -1 \end{pmatrix}$ is in the span of $\begin{pmatrix} 1 \\ 1 \end{pmatrix}$ and $\begin{pmatrix} 2 \\ 1 \end{pmatrix}$.

(b) Are $\begin{pmatrix} 1 \\ 0 \\ -2 \end{pmatrix}$ and $\begin{pmatrix} 1 \\ -2 \\ -3 \end{pmatrix}$ in the span of $\begin{pmatrix} 1 \\ 1 \\ 0 \end{pmatrix}$ and $\begin{pmatrix} 0 \\ 1 \\ 1 \end{pmatrix}$?

3.3. Which of the following sets of vectors span all of $\mathbb{R}^2$? (a)♡ $\begin{pmatrix} 1 \\ -1 \end{pmatrix}$; (b)♡ $\begin{pmatrix} 2 \\ -1 \end{pmatrix}, \begin{pmatrix} 1 \\ 3 \end{pmatrix}$; (c)♢ $\begin{pmatrix} 6 \\ -9 \end{pmatrix}, \begin{pmatrix} -4 \\ 6 \end{pmatrix}$; (d) $\begin{pmatrix} 2 \\ -1 \end{pmatrix}, \begin{pmatrix} -1 \\ 2 \end{pmatrix}$; (e)♡ $\begin{pmatrix} 1 \\ 2 \end{pmatrix}, \begin{pmatrix} 2 \\ 4 \end{pmatrix}, \begin{pmatrix} 4 \\ 8 \end{pmatrix}$; (f) $\begin{pmatrix} 0 \\ 0 \end{pmatrix}, \begin{pmatrix} 1 \\ 2 \end{pmatrix}, \begin{pmatrix} 3 \\ 4 \end{pmatrix}$.

3.4. Determine whether the given vectors are linearly independent or linearly dependent:

(a)♡ $\begin{pmatrix} 1 \\ 2 \end{pmatrix}, \begin{pmatrix} 2 \\ 1 \end{pmatrix}$, (b)♡ $\begin{pmatrix} 1 \\ 3 \end{pmatrix}, \begin{pmatrix} -2 \\ -6 \end{pmatrix}$, (c) $\begin{pmatrix} 2 \\ 1 \end{pmatrix}, \begin{pmatrix} -1 \\ 3 \end{pmatrix}, \begin{pmatrix} 5 \\ 2 \end{pmatrix}$, (d)♡ $\begin{pmatrix} 1 \\ 3 \\ -2 \end{pmatrix}, \begin{pmatrix} 0 \\ 2 \\ -1 \end{pmatrix}$,

$$(e) \diamond \begin{pmatrix} 0 \\ 1 \\ 1 \end{pmatrix}, \begin{pmatrix} 1 \\ -1 \\ 0 \end{pmatrix}, \begin{pmatrix} 1 \\ 1 \\ 2 \end{pmatrix}, \quad (f) \diamond \begin{pmatrix} 1 \\ 1 \\ 0 \end{pmatrix}, \begin{pmatrix} 1 \\ 0 \\ 1 \end{pmatrix}, \begin{pmatrix} 0 \\ 1 \\ 1 \end{pmatrix}, \quad (g) \begin{pmatrix} 4 \\ 2 \\ 0 \\ -6 \end{pmatrix}, \begin{pmatrix} -6 \\ -3 \\ 0 \\ 9 \end{pmatrix}.$$

3.5. Prove or give a counter-example: if $\mathbf{z}$ is a linear combination of $\mathbf{u}, \mathbf{v}, \mathbf{w}$, then $\mathbf{w}$ is a linear combination of $\mathbf{u}, \mathbf{v}, \mathbf{z}$.

3.6. $\diamond$ *True or false*: A set of vectors is linearly dependent if the zero vector belongs to their span.

3.7. $\heartsuit$ Prove or give a counterexample to the following statement: If $\mathbf{v}_1, \ldots, \mathbf{v}_k$ do not span $\mathbb{R}^n$, then $\mathbf{v}_1, \ldots, \mathbf{v}_k$ are linearly independent.

3.8. $\diamond$ Suppose $\mathbf{v}_1, \ldots, \mathbf{v}_k$ span the subspace $V \subset \mathbb{R}^n$. Let $\mathbf{v}_{k+1}, \ldots, \mathbf{v}_m \in V$ be any other vectors. Prove that the combined collection $\mathbf{v}_1, \ldots, \mathbf{v}_m$ also spans V.

3.9. (a) Prove that if $\mathbf{v}_1, \ldots, \mathbf{v}_m$ are linearly independent, then every subset $\mathbf{v}_{i_1}, \ldots, \mathbf{v}_{i_k}$ with $1 \leq k < m$, is also linearly independent. (b) Does the same hold true for linearly dependent vectors?

1.4 Basis and Dimension

In order to span a subspace, we must employ a sufficient number of distinct vectors. On the other hand, including too many vectors in the spanning set will violate linear independence, and cause redundancies. The optimal spanning sets are those that are also linearly independent. By combining the properties of span and linear independence, we arrive at the all-important concept of a basis.

> **Definition 1.12.** A *basis* of a subspace $V \subseteq \mathbb{R}^n$ is a finite set of vectors $\mathbf{v}_1, \ldots, \mathbf{v}_k \in V$ that (a) spans V, and (b) is linearly independent.

Example 1.13. As we already noted, the *standard basis* of $\mathbb{R}^n$ consists of the n vectors (1.3). They clearly span $\mathbb{R}^n$, since we can write any vector

$$\mathbf{x} = \begin{pmatrix} x_1 \\ x_2 \\ \vdots \\ x_n \end{pmatrix} = x_1 \mathbf{e}_1 + x_2 \mathbf{e}_2 + \cdots + x_n \mathbf{e}_n \tag{1.8}$$

as a linear combination, whose coefficients are its entries. Moreover, the only linear combination that yields the zero vector $\mathbf{x} = \mathbf{0}$ is the trivial one $x_1 = \cdots = x_n = 0$, which shows that $\mathbf{e}_1, \ldots, \mathbf{e}_n$ are linearly independent. We remark that this is but one of many possible bases for $\mathbb{R}^n$. ▲

A key fact is that every basis of a subspace $V \subset \mathbb{R}^n$ contains the same number of vectors. This result serves to motivate a linear algebraic characterization of dimension. In particular, every basis of $\mathbb{R}^n$ consists of exactly n vectors, and hence, as stated earlier, $\mathbb{R}^n$ has dimension n. Bear in mind that not every set of n vectors $\mathbf{v}_1, \ldots, \mathbf{v}_n \in \mathbb{R}^n$ forms a basis; indeed, they may be linearly dependent.

Theorem 1.14. *Suppose the subspace $V \subset \mathbb{R}^n$ has a basis $\mathbf{v}_1, \ldots, \mathbf{v}_k \in V$. Then every other basis of V has the same number, k, of vectors in it. This number is called the* dimension *of V, and written $\dim V = k$. In particular, $0 < k \leq n$.*

A proof of this result will appear below. The only subspace that has no basis is the trivial subspace $V = \{\mathbf{0}\}$, which by convention has dimension 0. To see that every other subspace has a basis, we can proceed as follows. First choose any nonzero vector $\mathbf{v}_1 \in V$. If every other vector $\mathbf{v} \in V$ is a scalar multiple of $\mathbf{v}_1$, then the basis consists of the single vector $\mathbf{v}_1$ and the subspace has dimension equal to 1, i.e., it is a line. Otherwise let $\mathbf{v}_2 \in V$ be any vector that is not a scalar multiple of $\mathbf{v}_1$, and hence $\mathbf{v}_1, \mathbf{v}_2$ are linearly independent. If they span V, then they form a basis, which has dimension $= 2$. Otherwise, we can find $\mathbf{v}_3$ which does not belong to their span. Lemma 1.11 implies that $\mathbf{v}_1, \mathbf{v}_2, \mathbf{v}_3$ are linearly independent. We proceed iteratively; at step k we have linearly independent vectors $\mathbf{v}_1, \ldots, \mathbf{v}_k \in V$. If they span V, they form a basis, and hence $\dim V = k$. Otherwise, we can choose $\mathbf{v}_{k+1} \in V$ such that $\mathbf{v}_{k+1} \notin \operatorname{span}\{\mathbf{v}_1, \ldots, \mathbf{v}_k\}$, and, again by Lemma 1.11, $\mathbf{v}_1, \ldots, \mathbf{v}_{k+1}$ are linearly independent vectors that belong to V. The process terminates when the number of vectors chosen equals the dimension of V, which must be less than or equal to n. Thus the origin — a point — has dimension 0, lines have dimension 1, planes have dimension 2, and so on. A subspace $V \subset \mathbb{R}^n$ of submaximal dimension $n - 1$ is known as a *hyperplane*. Again, all subspaces must pass through the origin.

As a consequence of the above argument, we have established the following result.

Lemma 1.15. *Suppose $V \subset \mathbb{R}^n$ is a subspace with $\dim V = k > 0$. Suppose $\mathbf{v}_1, \ldots, \mathbf{v}_j \in V$ are linearly independent vectors in the subspace with $1 \leq j < k$. Then there exist linearly independent vectors $\mathbf{v}_{j+1}, \ldots, \mathbf{v}_k \in V$ such that $\mathbf{v}_1, \ldots, \mathbf{v}_k$ form a basis for V.*

Remark. As we have seen, determining whether a set of vectors is linearly independent or linearly dependent, or determining whether a vector lies in their span, requires solving a linear system of algebraic equations. Systematic techniques for effecting this rely on matrices, and will be developed in Chapters 3 and 4.   ▲

The proof of Theorem 1.14 rests on the following lemma. Consider a *homogeneous linear system* of equations

$$
\begin{aligned}
a_{11} x_1 + a_{12} x_2 + \cdots + a_{1n} x_n &= 0, \\
a_{21} x_1 + a_{22} x_2 + \cdots + a_{2n} x_n &= 0, \\
\vdots \qquad\qquad \vdots \qquad\qquad \vdots \quad\;\; & \\
a_{m1} x_1 + a_{m2} x_2 + \cdots + a_{mn} x_n &= 0.
\end{aligned}
\tag{1.9}
$$

consisting of m equations in the n unknowns $x_1, \ldots, x_n$, with right hand sides all zero. Clearly setting all the unknowns to zero, $x_1 = x_2 = \cdots = x_n = 0$, solves the system; we call this the *trivial solution*. Any other solution (which may or may not exist), when at least one of the x_j is nonzero, is called *nontrivial*.

We next note that a homogeneous linear system that has more unknowns than equations always has a nontrivial solution. On the other hand, if the number of unknowns is less than or equal to the number of equations, this may or may not be the case, i.e., depending on the system, there may only be the trivial solution.

Lemma 1.16. *If $n > m$, the homogeneous linear system (1.9) has a nontrivial solution.*

Proof. We prove this result by induction on the number of equations. The initial case, with $m = 1$, so we have one equation in 2 or more unknowns, is left as an exercise for the reader. Now, if all the coefficients of x_1 in (1.9) vanish, so $a_{11} = a_{21} = \cdots = a_{m1} = 0$, then we can take $x_1 = 1$, $x_2 = \cdots = x_n = 0$ as our nontrivial solution. Otherwise, at least one of these coefficients is nonzero, and we can assume, by relabeling the equations if necessary, that $a_{11} \neq 0$. For each $j = 2, \ldots, m$, we then subtract a_{j1}/a_{11} times the first equation from the j-th equation in order to eliminate x_1 from it. The resulting linear system has the same solutions, and consists of an initial equation that involves all n unknowns, followed by $m - 1$ equations involving only the $n - 1$ unknowns $x_2, \ldots, x_n$. Since $n - 1 > m - 1$, we can use the induction hypothesis that says that the latter system has a nontrivial solution, meaning not all $x_2, \ldots, x_n$ are zero. Given this solution, we use the initial equation to solve for x_1 (which may be 0 but that doesn't matter) and the result forms a nontrivial solution to the original system. ∎

Lemma 1.17. *Suppose $\mathbf{v}_1, \ldots, \mathbf{v}_k$ span a subspace $V \subset \mathbb{R}^n$. Then every set of $m > k$ vectors $\mathbf{w}_1, \ldots, \mathbf{w}_m \in V$ is linearly dependent.*

Proof. Let us write each vector

$$\mathbf{w}_j = \sum_{i=1}^{k} a_{ij} \mathbf{v}_i, \qquad j = 1, \ldots, m,$$

as a linear combination of the spanning set. Then

$$c_1 \mathbf{w}_1 + \cdots + c_m \mathbf{w}_m = \sum_{i=1}^{k} \sum_{j=1}^{m} a_{ij} c_j \mathbf{v}_i. \tag{1.10}$$

This linear combination will be zero whenever $c_1, \ldots, c_m$ solves the homogeneous linear system

$$\sum_{j=1}^{m} a_{ij} c_j = 0, \qquad i = 1, \ldots, k,$$

consisting of k equations in $m > k$ unknowns. Lemma 1.16 guarantees that this system has a nontrivial solution, with not all c_j being 0. For this choice of $c_1, \ldots, c_m$, the right hand side of (1.10) is the zero vector, while the left hand side is a nontrivial linear combination. This then implies that $\mathbf{w}_1, \ldots, \mathbf{w}_m$ are linearly dependent. ∎

Proof of Theorem 1.14: Suppose we have two bases containing a different number of vectors. By definition, the smaller basis spans the subspace. But then Lemma 1.17 tell us that the vectors in the larger purported basis must be linearly dependent, which contradicts our initial assumption that the latter is a basis. ∎

Proposition 1.18. *The vectors* $\mathbf{v}_1, \ldots, \mathbf{v}_k$ *form a basis of a subspace* $V \subset \mathbb{R}^n$ *if and only if every* $\mathbf{x} \in V$ *can be written* uniquely *as a linear combination of the basis vectors:*

$$\mathbf{x} = c_1 \mathbf{v}_1 + \cdots + c_k \mathbf{v}_k = \sum_{i=1}^{k} c_i \mathbf{v}_i. \tag{1.11}$$

Proof. Suppose first that $\mathbf{v}_1, \ldots, \mathbf{v}_k$ form a basis of V. The fact that they span V implies that every $\mathbf{x} \in V$ can be written as some linear combination of the basis vectors. Suppose we can write a vector

$$\mathbf{x} = c_1 \mathbf{v}_1 + \cdots + c_k \mathbf{v}_k = \widehat{c}_1 \mathbf{v}_1 + \cdots + \widehat{c}_k \mathbf{v}_k \tag{1.12}$$

as two different combinations. Subtracting one from the other, we obtain

$$(c_1 - \widehat{c}_1) \mathbf{v}_1 + \cdots + (c_k - \widehat{c}_k) \mathbf{v}_n = \mathbf{0}.$$

The left-hand side is a linear combination of the linearly independent basis vectors, and hence vanishes if and only if all its coefficients $c_i - \widehat{c}_i = 0$, meaning that the two linear combinations (1.12) are one and the same.

On the other hand, if $\mathbf{v}_1, \ldots, \mathbf{v}_k$ are not a basis, then either they do not span V, which means that some vectors cannot be expressed as a linear combination of them, or they are linearly dependent, in which case there is a nontrivial linear combination which equals zero, and hence the zero vector in particular can be written as more than one linear combination. $\blacksquare$

One sometimes refers to the coefficients $c_1, \ldots, c_k$ in (1.11) as the *coordinates* of the vector $\mathbf{x}$ with respect to the given basis. For the standard basis (1.3) of $\mathbb{R}^n$, according to (1.8), the coordinates of a vector $\mathbf{x} = (x_1, \ldots, x_n)^T = x_1 \mathbf{e}_1 + \cdots + x_n \mathbf{e}_n$ are its entries, i.e., its usual Cartesian coordinates.

As a direct consequence of the preceding developments, we can now give a precise meaning to the optimality of bases. We state the result for $\mathbb{R}^n$, but a similar result holds for subspaces thereof, in which we replace n by $\dim V$.

Theorem 1.19.

(a) *Every set of more than n vectors in $\mathbb{R}^n$ is linearly dependent.*

(b) *No set of fewer than n vectors spans $\mathbb{R}^n$.*

(c) *A set of n vectors forms a basis if and only if it spans $\mathbb{R}^n$.*

(d) *A set of n vectors forms a basis if and only if it is linearly independent.*

Thus, to check $\mathbf{v}_1, \ldots, \mathbf{v}_n$ forms a basis of $\mathbb{R}^n$ one only needs to check either that they are linearly independent or that they span all of $\mathbb{R}^n$; the second fact then follows automatically. More generally, if $\mathbf{v}_1, \ldots, \mathbf{v}_k \in \mathbb{R}^n$ are linearly independent, then they form a basis for their span $V = \mathrm{span}\,\{\mathbf{v}_1, \ldots, \mathbf{v}_k\} \subset \mathbb{R}^n$.

Example 1.20. *A Wavelet Basis.* The vectors

$$\mathbf{v}_1 = \begin{pmatrix} 1 \\ 1 \\ 1 \\ 1 \end{pmatrix}, \qquad \mathbf{v}_2 = \begin{pmatrix} 1 \\ 1 \\ -1 \\ -1 \end{pmatrix}, \qquad \mathbf{v}_3 = \begin{pmatrix} 1 \\ -1 \\ 0 \\ 0 \end{pmatrix}, \qquad \mathbf{v}_4 = \begin{pmatrix} 0 \\ 0 \\ 1 \\ -1 \end{pmatrix}, \tag{1.13}$$

form a basis of $\mathbb{R}^4$, and is an example of a *wavelet basis*. Wavelets play an increasingly central role in modern signal and digital image processing, [181, 242], and this constitutes a very simple example.

How do we find the coordinates of a vector, say $\mathbf{x} = (\, 4, -2, 1, 5\,)^T$, relative to the wavelet basis? We need to find the coefficients c_1, c_2, c_3, c_4 such that

$$\mathbf{x} = c_1\,\mathbf{v}_1 + c_2\,\mathbf{v}_2 + c_3\,\mathbf{v}_3 + c_4\,\mathbf{v}_4.$$

The individual entries of this vector equation,

$$c_1 + c_2 + c_3 = 4, \qquad c_1 + c_2 - c_3 = -2, \qquad c_1 - c_2 + c_4 = 1, \qquad c_1 - c_2 - c_4 = 5,$$

form a linear system of 4 equations for c_1, c_2, c_3, c_4. The solution[3]

$$c_1 = 2, \qquad c_2 = -1, \qquad c_3 = 3, \qquad c_4 = -2,$$

gives the coordinates of

$$\mathbf{x} = \begin{pmatrix} 4 \\ -2 \\ 1 \\ 5 \end{pmatrix} = 2\,\mathbf{v}_1 - \mathbf{v}_2 + 3\,\mathbf{v}_3 - 2\,\mathbf{v}_4 = 2\begin{pmatrix} 1 \\ 1 \\ 1 \\ 1 \end{pmatrix} - \begin{pmatrix} 1 \\ 1 \\ -1 \\ -1 \end{pmatrix} + 3\begin{pmatrix} 1 \\ -1 \\ 0 \\ 0 \end{pmatrix} - 2\begin{pmatrix} 0 \\ 0 \\ 1 \\ -1 \end{pmatrix}.$$

in the wavelet basis. ▲

Why would one want to employ a different basis? The answer is *simplification* and *speed* — many computations and formulas become much easier, and hence faster, to perform in a basis that is adapted to the problem at hand. In signal processing, wavelet bases are particularly appropriate for denoising, compression, and efficient storage of signals, including audio, still images, videos, and so on. These processes would be quite time-consuming — if not impossible in large data regimes like video and three-dimensional image processing — to accomplish in the standard basis. Later, we will see many such examples that arise in machine learning.

The proof of the next result is left to Exercise 4.7.

Proposition 1.21. *If $W \subseteq V \subseteq \mathbb{R}^n$ are subspaces, then $0 \leq \dim W \leq \dim V \leq n$. Moreover, $\dim V = \dim W$ if and only if $V = W$.*

Let $V, W \subset \mathbb{R}^n$ be subspaces. According to Exercise 2.6, their intersection $V \cap W$ and their sum $V + W = \{\,\mathbf{v} + \mathbf{w} \mid \mathbf{v} \in V,\ \mathbf{w} \in W\,\}$ are also subspaces. Our final result in this chapter relates the dimensions of these four subspaces.

Proposition 1.22. *Let $V, W \subset \mathbb{R}^n$ be subspaces. Then*

$$\dim(V \cap W) + \dim(V + W) = \dim V + \dim W. \tag{1.14}$$

Proof. We leave the cases when $V \cap W = \{\mathbf{0}\}$ for the reader. Assume $V \cap W$ has dimension $1 \leq i = \dim(V \cap W)$, with basis $\mathbf{u}_1, \ldots, \mathbf{u}_i$. Suppose $\dim V = j$ and $\dim W = k$. Since $\mathbf{u}_1, \ldots, \mathbf{u}_i \in V$ are linearly independent, Lemma 1.15 implies we can find $\mathbf{v}_1, \ldots, \mathbf{v}_{j-i} \in V$ such that $\mathbf{u}_1, \ldots, \mathbf{u}_i, \mathbf{v}_1, \ldots, \mathbf{v}_{j-i}$ form a basis for V. Similarly, we can find $\mathbf{w}_1, \ldots, \mathbf{w}_{k-i} \in W$

---

[3] It is not hard to solve the system by hand, but a much simpler method will appear in Example 2.20.

such that $\mathbf{u}_1, \ldots, \mathbf{u}_i, \mathbf{w}_1, \ldots, \mathbf{w}_{k-i}$ form a basis for W. We claim that $\mathbf{u}_1, \ldots, \mathbf{u}_i, \mathbf{v}_1, \ldots, \mathbf{v}_{j-i}$, $\mathbf{w}_1, \ldots, \mathbf{w}_{k-i}$ form a basis for $V + W$, and therefore

$$\dim(V + W) = i + (j - i) + (k - i) = j + k - i = \dim V + \dim W - \dim(V \cap W),$$

which establishes (1.14).

To prove the claim, we first note that given any $\mathbf{v} + \mathbf{w} \in V + W$, with $\mathbf{v} \in V$ and $\mathbf{w} \in W$, then we can express

$$\mathbf{v} = a_1 \mathbf{u}_1 + \cdots + a_i \mathbf{u}_i + b_1 \mathbf{v}_1 + \cdots + b_{j-i} \mathbf{v}_{j-i},$$
$$\mathbf{w} = c_1 \mathbf{u}_1 + \cdots + c_i \mathbf{u}_i + d_1 \mathbf{w}_1 + \cdots + d_{k-i} \mathbf{w}_{k-i},$$

as linear combinations of their respective bases, and hence

$$\mathbf{v} + \mathbf{w} = (a_1 + c_1)\mathbf{u}_1 + \cdots + (a_i + c_i)\mathbf{u}_i$$
$$+ b_1 \mathbf{v}_1 + \cdots + b_{j-i} \mathbf{v}_{j-i} + d_1 \mathbf{w}_1 + \cdots + d_{k-i} \mathbf{w}_{k-i},$$

is a linear combination of the vectors $\mathbf{u}_1, \ldots, \mathbf{u}_i, \mathbf{v}_1, \ldots, \mathbf{v}_{j-i}, \mathbf{w}_1, \ldots, \mathbf{w}_{k-i}$, which proves that they span $V + W$. The only remaining step is to show that these vectors are linearly independent. Suppose

$$a_1 \mathbf{u}_1 + \cdots + a_i \mathbf{u}_i + b_1 \mathbf{v}_1 + \cdots + b_{j-i} \mathbf{v}_{j-i} + d_1 \mathbf{w}_1 + \cdots + d_{k-i} \mathbf{w}_{k-i} = \mathbf{0}.$$

Then,

$$a_1 \mathbf{u}_1 + \cdots + a_i \mathbf{u}_i + b_1 \mathbf{v}_1 + \cdots + b_{j-i} \mathbf{v}_{j-i} = -d_1 \mathbf{w}_1 - \cdots - d_{k-i} \mathbf{w}_{k-i}. \tag{1.15}$$

The left hand side belongs to V, whereas the right hand side belongs to W, and hence they both belong to $V \cap W$. This implies we can write them as a linear combination of the basis $\mathbf{u}_1, \ldots, \mathbf{u}_i$, so

$$-d_1 \mathbf{w}_1 - \cdots - d_{k-i} \mathbf{w}_{k-i} = c_1 \mathbf{u}_1 + \cdots + c_i \mathbf{u}_i,$$

or, equivalently,

$$c_1 \mathbf{u}_1 + \cdots + c_i \mathbf{u}_i + d_1 \mathbf{w}_1 + \cdots + d_{k-i} \mathbf{w}_{k-i} = \mathbf{0}.$$

Linear independence of $\mathbf{u}_1, \ldots, \mathbf{u}_i, \mathbf{w}_1, \ldots, \mathbf{w}_{k-i}$ implies that all coefficients are 0; in particular

$$d_1 = \cdots = d_{k-i} = 0,$$

and hence the right hand side of (1.15) is $\mathbf{0}$. Equating this to the left hand side and using linear independence of $\mathbf{u}_1, \ldots, \mathbf{u}_i, \mathbf{v}_1, \ldots, \mathbf{v}_{j-i}$, we deduce that

$$a_1 = \cdots = a_i = b_1 = \cdots = b_{j-i} = 0$$

also, which proves the desired linear independence. ∎

Since $\dim(V + W) \leq n$, as an immediate consequence of (1.14), we have the following useful inequality,

$$\max\{j + k - n, 0\} \leq \dim(V \cap W) \leq \min\{j, k\}, \qquad \text{where} \qquad \begin{aligned} \dim V &= j, \\ \dim W &= k, \end{aligned} \tag{1.16}$$

the upper bound following from Proposition 1.21, since $V \cap W \subset V$ and $V \cap W \subset W$.

Exercises

4.1. Determine which of the following sets of vectors are bases of $\mathbb{R}^2$: $(a) \heartsuit$ $\begin{pmatrix} 2 \\ 1 \end{pmatrix}$;

$(b) \heartsuit$ $\begin{pmatrix} 1 \\ -1 \end{pmatrix}, \begin{pmatrix} -1 \\ 1 \end{pmatrix}$; $(c) \diamondsuit$ $\begin{pmatrix} 1 \\ 2 \end{pmatrix}, \begin{pmatrix} 2 \\ 1 \end{pmatrix}$; $(d) \heartsuit$ $\begin{pmatrix} 3 \\ 5 \end{pmatrix}, \begin{pmatrix} 0 \\ 0 \end{pmatrix}$; (e) $\begin{pmatrix} 2 \\ 0 \end{pmatrix}, \begin{pmatrix} -1 \\ 2 \end{pmatrix}, \begin{pmatrix} 0 \\ -1 \end{pmatrix}$.

4.2. Determine which of the following are bases of $\mathbb{R}^3$: $(a) \heartsuit$ $\begin{pmatrix} 2 \\ 1 \\ 5 \end{pmatrix}, \begin{pmatrix} 1 \\ 5 \\ 2 \end{pmatrix}$; $(b) \heartsuit$ $\begin{pmatrix} 0 \\ 1 \\ -5 \end{pmatrix},$

$\begin{pmatrix} -1 \\ 3 \\ 0 \end{pmatrix}, \begin{pmatrix} 1 \\ 3 \\ 0 \end{pmatrix}$; $(c) \diamondsuit$ $\begin{pmatrix} -1 \\ 0 \\ 1 \end{pmatrix}, \begin{pmatrix} 0 \\ 4 \\ -1 \end{pmatrix}, \begin{pmatrix} 1 \\ -4 \\ 0 \end{pmatrix}$; (d) $\begin{pmatrix} 2 \\ 0 \\ -2 \end{pmatrix}, \begin{pmatrix} -1 \\ 2 \\ -1 \end{pmatrix}, \begin{pmatrix} 0 \\ -1 \\ 0 \end{pmatrix}, \begin{pmatrix} -1 \\ 2 \\ 1 \end{pmatrix}$.

4.3. Let $\mathbf{v}_1 = \begin{pmatrix} 1 \\ 2 \end{pmatrix}$, $\mathbf{v}_2 = \begin{pmatrix} 3 \\ 1 \end{pmatrix}$, $\mathbf{v}_3 = \begin{pmatrix} 2 \\ -1 \end{pmatrix}$. (a) Do $\mathbf{v}_1, \mathbf{v}_2, \mathbf{v}_3$ span $\mathbb{R}^2$? (b) Are $\mathbf{v}_1, \mathbf{v}_2, \mathbf{v}_3$ linearly independent? (c) Do $\mathbf{v}_1, \mathbf{v}_2, \mathbf{v}_3$ form a basis for $\mathbb{R}^2$? If not, is it possible to choose some subset that is a basis?

4.4. Find a basis for the following planes in $\mathbb{R}^3$:
$(a) \heartsuit$ the xy plane; $\quad$ (b) $z - 2y = 0$; $\quad$ $(c) \diamondsuit$ $4x + 3y - z = 0$.

4.5. $\heartsuit$ Show, by computing an example, how the uniqueness result in Proposition 1.18 fails if one has a linearly dependent set of vectors.

4.6. Show that if $\mathbf{v}_1, \ldots, \mathbf{v}_k$ span the subspace $\{\mathbf{0}\} \neq V \subset \mathbb{R}^n$, then one can choose a subset $\mathbf{v}_{i_1}, \ldots, \mathbf{v}_{i_j}$ that forms a basis of V, and hence $j = \dim V \leq k$. Under what conditions is $\dim V = k$?

4.7. $\diamondsuit$ Prove Proposition 1.21.

Chapter 2

Inner Product, Orthogonality, Norm

The geometry of Euclidean space is founded on the familiar properties of length and angle. In Euclidean geometry, distance between points is measured by the length of the difference between the corresponding vectors, while angle relies on their dot product. The dot product is formalized by the more general concept of an inner product. Other types of inner product arise naturally in statistics, data analysis, and elsewhere. Each inner product has an associated norm, which is used to measure lengths of vectors. Inner products and norms lie at the heart of linear (and nonlinear) analysis, including machine learning.

Mathematical analysis relies on the exploitation of inequalities. The most fundamental is the Cauchy–Schwarz inequality, which is valid for every inner product. The more familiar triangle inequality for the associated norm is then derived as a simple consequence. Not every norm comes from an inner product, and, in such cases, the triangle inequality becomes part of the general definition.

Orthogonality is the mathematical formalization of the geometrical property of perpendicularity, and is a remarkably powerful tool that appears throughout the manifold applications of linear algebra. Two vectors are said to be orthogonal if their inner product vanishes. The orthogonal projection of a vector onto a subspace coincides with the closest point on the subspace, and thus has applications to data analysis through the method of least squares. Bases consisting of mutually orthogonal elements that each have unit norm play an essential role in both practical and theoretical developments, throughout applications of linear algebra, and in the design of practical numerical algorithms. Indeed, computations become dramatically simpler and less prone to numerical inaccuracies when performed in suitably adapted orthonormal coordinate systems. Most large scale modern applications, ranging over machine learning, data analysis, signal and image processing, and elsewhere, would be impractical, if not completely infeasible, were it not for the dramatic simplifying power of orthonormality. In Section 2.5, we develop a general version of the Gram–Schmidt Process that produces an orthonormal basis for the subspace spanned by an arbitrary collection of vectors, thereby demonstrating that every subspace has an orthonormal basis.

J. Calder, P. J. Olver, *Linear Algebra, Data Science, and Machine Learning*, Springer Undergraduate Texts in Mathematics and Technology, https://doi.org/10.1007/978-3-031-93764-4_2

2.1   Inner Products

Our starting point is the familiar *dot product*

$$\mathbf{v} \cdot \mathbf{w} = v_1 w_1 + v_2 w_2 + \cdots + v_n w_n = \sum_{i=1}^{n} v_i w_i \tag{2.1}$$

between (column) vectors $\mathbf{v} = (v_1, v_2, \ldots, v_n)^T$, $\mathbf{w} = (w_1, w_2, \ldots, w_n)^T \in \mathbb{R}^n$. A key fact is that the dot product of a vector with itself,

$$\mathbf{v} \cdot \mathbf{v} = v_1^2 + v_2^2 + \cdots + v_n^2,$$

is the sum of the squares of its entries, and hence, by the classical Pythagorean Theorem, equals the square of its length. Consequently, the *Euclidean norm* or *length* of a vector is found by taking the square root:

$$\|\mathbf{v}\| = \sqrt{\mathbf{v} \cdot \mathbf{v}} = \sqrt{v_1^2 + v_2^2 + \cdots + v_n^2}. \tag{2.2}$$

Figure 2.1 shows the two- and three-dimensional versions. Note that every vector $\mathbf{v} \in \mathbb{R}^n$, has nonnegative Euclidean norm: $\|\mathbf{v}\| \geq 0$; moreover, only the zero vector has zero norm: $\|\mathbf{v}\| = 0$ if and only if $\mathbf{v} = \mathbf{0}$.

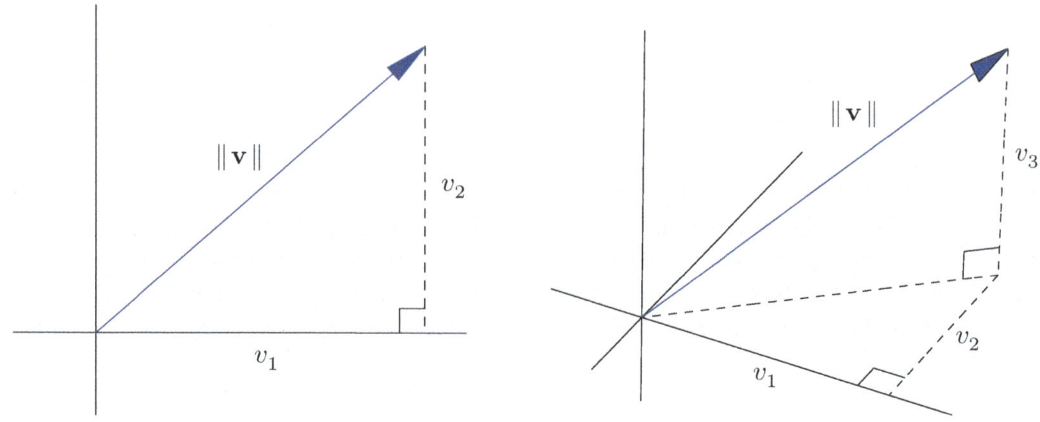

Figure 2.1: The Euclidean Norm in $\mathbb{R}^2$ and $\mathbb{R}^3$

The elementary properties of dot product and Euclidean norm serve to inspire the definition of general inner products.

Definition 2.1. An *inner product* is a pairing that takes two vectors $\mathbf{v}, \mathbf{w} \in \mathbb{R}^n$ and produces a real number $\langle \mathbf{v}, \mathbf{w} \rangle \in \mathbb{R}$. The inner product is required to satisfy the following three axioms for all $\mathbf{u}, \mathbf{v}, \mathbf{w} \in \mathbb{R}^n$, and scalars $c, d \in \mathbb{R}$.

(*i*) *Bilinearity*:

$$\begin{aligned} \langle c\,\mathbf{u} + d\,\mathbf{v}, \mathbf{w} \rangle &= c\,\langle \mathbf{u}, \mathbf{w} \rangle + d\,\langle \mathbf{v}, \mathbf{w} \rangle, \\ \langle \mathbf{u}, c\,\mathbf{v} + d\,\mathbf{w} \rangle &= c\,\langle \mathbf{u}, \mathbf{v} \rangle + d\,\langle \mathbf{u}, \mathbf{w} \rangle. \end{aligned} \tag{2.3}$$

(*ii*) *Symmetry*:

$$\langle \mathbf{v}, \mathbf{w} \rangle = \langle \mathbf{w}, \mathbf{v} \rangle. \tag{2.4}$$

(*iii*) *Positivity*:

$$\langle \mathbf{v}, \mathbf{v} \rangle > 0 \quad \text{whenever} \quad \mathbf{v} \neq \mathbf{0}, \quad \text{while} \quad \langle \mathbf{0}, \mathbf{0} \rangle = 0. \tag{2.5}$$

Verification of the inner product axioms (2.3)–(2.5) for the Euclidean dot product is straightforward, and left as an exercise for the reader.

Given an inner product, the associated *norm* of a vector $\mathbf{v} \in \mathbb{R}^n$ is, in analogy with (2.2), defined as the positive square root of the inner product of the vector with itself:

$$\|\mathbf{v}\| = \sqrt{\langle \mathbf{v}, \mathbf{v} \rangle}. \tag{2.6}$$

The positivity axiom implies that $\|\mathbf{v}\| \geq 0$ is real and nonnegative, and equals 0 if and only if $\mathbf{v} = \mathbf{0}$ is the zero vector. Note also that if $c \in \mathbb{R}$ is any scalar, then, by bilinearity of the inner product, the norm satisfies the following *homogeneity* property:

$$\|c\mathbf{v}\| = \sqrt{\langle c\mathbf{v}, c\mathbf{v} \rangle} = \sqrt{c^2 \langle \mathbf{v}, \mathbf{v} \rangle} = |c| \, \|\mathbf{v}\|, \tag{2.7}$$

where $|c|$ denotes the absolute value of c. In particular, $\|-\mathbf{v}\| = \|\mathbf{v}\|$.

Example 2.2. While certainly the most common inner product on $\mathbb{R}^2$, the dot product

$$\langle \mathbf{v}, \mathbf{w} \rangle = \mathbf{v} \cdot \mathbf{w} = v_1 w_1 + v_2 w_2, \qquad \mathbf{v} = \begin{pmatrix} v_1 \\ v_2 \end{pmatrix}, \qquad \mathbf{w} = \begin{pmatrix} w_1 \\ w_2 \end{pmatrix},$$

is by no means the only possibility. A simple example is provided by the weighted inner product

$$\langle \mathbf{v}, \mathbf{w} \rangle = 2 v_1 w_1 + 5 v_2 w_2. \tag{2.8}$$

Let us verify that this formula does indeed define an inner product. The symmetry axiom (2.4) is immediate. Moreover,

$$\begin{aligned}
\langle c\mathbf{u} + d\mathbf{v}, \mathbf{w} \rangle &= 2(cu_1 + dv_1)w_1 + 5(cu_2 + dv_2)w_2 \\
&= c(2u_1 w_1 + 5u_2 w_2) + d(2v_1 w_1 + 5v_2 w_2) = c\langle \mathbf{u}, \mathbf{w} \rangle + d\langle \mathbf{v}, \mathbf{w} \rangle,
\end{aligned}$$

which verifies the first bilinearity condition; the second follows by a very similar computation.[1] Moreover, $\langle \mathbf{0}, \mathbf{0} \rangle = 0$, while $\langle \mathbf{v}, \mathbf{v} \rangle = 2v_1^2 + 5v_2^2 > 0$ whenever $\mathbf{v} \neq \mathbf{0}$, since at least one of the summands is strictly positive. This establishes (2.8) as a legitimate inner product on $\mathbb{R}^2$. The associated weighted norm

$$\|\mathbf{v}\| = \sqrt{2v_1^2 + 5v_2^2} \tag{2.9}$$

defines an alternative, "non-Pythagorean" notion of length for vectors in $\mathbb{R}^2$.

A less evident example of an inner product on $\mathbb{R}^2$ is provided by the expression

$$\langle \mathbf{v}, \mathbf{w} \rangle = v_1 w_1 - v_1 w_2 - v_2 w_1 + 4 v_2 w_2. \tag{2.10}$$

Bilinearity is verified in the same manner as before, and symmetry is immediate. Positivity is ensured by noticing that the expression

$$\langle \mathbf{v}, \mathbf{v} \rangle = v_1^2 - 2v_1 v_2 + 4v_2^2 = (v_1 - v_2)^2 + 3v_2^2 \geq 0$$

is always nonnegative, and, moreover, is equal to zero if and only if $v_1 - v_2 = 0$, $v_2 = 0$, i.e., only when $v_1 = v_2 = 0$ and so $\mathbf{v} = \mathbf{0}$. We conclude that (2.10) defines yet another inner product on $\mathbb{R}^2$, with associated norm

$$\|\mathbf{v}\| = \sqrt{\langle \mathbf{v}, \mathbf{v} \rangle} = \sqrt{v_1^2 - 2v_1 v_2 + 4v_2^2}.$$

---

[1] Alternatively, it is not hard to see that symmetry and the first bilinearity condition implies that the second bilinearity condition is satisfied.

On the other hand, despite having all positive coefficients, the expression

$$\langle\, \mathbf{v}, \mathbf{w}\,\rangle = v_1\, w_1 + 2\, v_1\, w_2 + 2\, v_2\, w_1 + v_2\, w_2,$$

does *not* define an inner product. It evidently satisfies the bilinearity and symmetry requirements. However, setting $\mathbf{v} = \mathbf{w}$,

$$q(v_1, v_2) = \langle\, \mathbf{v}, \mathbf{v}\,\rangle = v_1^2 + 4\, v_1\, v_2 + v_2^2 \qquad \text{satisfies} \qquad q(1, -1) = -2,$$

and hence fails the positivity requirement.   ▲

The second example (2.8) is a particular case of a general class of inner products.

Example 2.3. Let $c_1, \ldots, c_n > 0$ be a set of *positive* numbers. The corresponding *weighted inner product* and *weighted norm* on $\mathbb{R}^n$ are defined by

$$\langle\, \mathbf{v}, \mathbf{w}\,\rangle = \sum_{i=1}^{n} c_i\, v_i\, w_i, \qquad\qquad \|\,\mathbf{v}\,\| = \sqrt{\langle\, \mathbf{v}, \mathbf{v}\,\rangle} = \sqrt{\sum_{i=1}^{n} c_i\, v_i^2}\ . \tag{2.11}$$

The numbers $c_i > 0$ are the *weights*. Observe that the larger the weight c_i, the more the i-th coordinate of $\mathbf{v}$ contributes to the norm. Weighted norms are particularly relevant in statistics and data fitting, [110, 241], when one wants to emphasize the importance of certain measurements and de-emphasize others; this is done by assigning appropriate weights to the different components of the data vector $\mathbf{v}$.   ▲

Let us now try to determine the most general inner product that can be placed on $\mathbb{R}^n$. We begin by noting that, by iterating the bilinearity condition (2.3), we find

$$\begin{aligned}
\langle\, c_1\mathbf{v}_1 + \cdots + c_k\mathbf{v}_k, \mathbf{w}\,\rangle &= c_1\langle\, \mathbf{v}_1, \mathbf{w}\,\rangle + \cdots + c_k\langle\, \mathbf{v}_k, \mathbf{w}\,\rangle, \\
\langle\, \mathbf{v}, c_1\mathbf{w}_1 + \cdots + c_k\mathbf{w}_k\,\rangle &= c_1\langle\, \mathbf{v}, \mathbf{w}_1\,\rangle + \cdots + c_k\langle\, \mathbf{v}, \mathbf{w}_k\,\rangle,
\end{aligned} \tag{2.12}$$

for any vectors $\mathbf{v}_1, \ldots, \mathbf{v}_k, \mathbf{v}, \mathbf{w}_1, \ldots, \mathbf{w}_k, \mathbf{w}$ and scalars $c_1, \ldots, c_k$. Thus, writing the vectors

$$\mathbf{v} = \begin{pmatrix} v_1 \\ v_2 \\ \vdots \\ v_n \end{pmatrix} = v_1\,\mathbf{e}_1 + \cdots + v_n\,\mathbf{e}_n, \qquad\qquad \mathbf{w} = \begin{pmatrix} w_1 \\ w_2 \\ \vdots \\ w_n \end{pmatrix} = w_1\,\mathbf{e}_1 + \cdots + w_n\,\mathbf{e}_n, \tag{2.13}$$

as linear combinations of the standard basis vectors (1.3), we can successively apply the identities in (2.12) to expand their inner product as follows:

$$\langle\, \mathbf{v}, \mathbf{w}\,\rangle = \left\langle\, \sum_{i=1}^{n} v_i\,\mathbf{e}_i, \sum_{j=1}^{n} w_j\,\mathbf{e}_j\, \right\rangle = \sum_{i,j=1}^{n} v_i\, w_j\langle\, \mathbf{e}_i, \mathbf{e}_j\,\rangle = \sum_{i,j=1}^{n} c_{ij}\, v_i\, w_j, \tag{2.14}$$

where

$$c_{ij} = \langle\, \mathbf{e}_i, \mathbf{e}_j\,\rangle, \qquad i, j = 1, \ldots, n. \tag{2.15}$$

We conclude that any inner product must be expressed in the general *bilinear form* (2.14). The two remaining inner product axioms will impose certain constraints on the coefficients (2.15). Symmetry implies that

$$c_{ij} = \langle\, \mathbf{e}_i, \mathbf{e}_j\,\rangle = \langle\, \mathbf{e}_j, \mathbf{e}_i\,\rangle = c_{ji}, \qquad i, j = 1, \ldots, n. \tag{2.16}$$

The final condition is positivity, which requires that

$$q(\mathbf{v}) = \langle \mathbf{v}, \mathbf{v} \rangle = \sum_{i,j=1}^{n} c_{ij}\, v_i\, v_j = \sum_{i=1}^{n} c_{ii}\, v_i^2 + 2 \sum_{i<j} c_{ij}\, v_i\, v_j > 0 \quad \text{for all} \quad \mathbf{0} \neq \mathbf{v} \in \mathbb{R}^n, \qquad (2.17)$$

where we used (2.16) when writing the second expression. The function (2.17) is a homogeneous quadratic polynomial depending on $\mathbf{v} = (v_1, \ldots, v_n)^T$, also known as a *quadratic form*. The precise implications of this positivity condition are not so immediately evident. As we saw in Example 2.2, positivity of all the coefficients does not imply (2.17), while (2.17) does not imply that all $c_{ij} > 0$.

Example 2.4. Let us first investigate the two-dimensional case, and classify all inner products on $\mathbb{R}^2$. According to (2.14), (2.16), they assume the bilinear form

$$\langle \mathbf{v}, \mathbf{w} \rangle = a\, v_1 w_1 + b(v_1 w_2 + v_2 w_1) + c\, v_2 w_2, \quad \text{where} \quad \mathbf{v} = (v_1, v_2)^T, \quad \mathbf{w} = (w_1, w_2)^T,$$

where

$$a = c_{11} = \langle \mathbf{e}_1, \mathbf{e}_1 \rangle, \qquad b = c_{12} = c_{21} = \langle \mathbf{e}_1, \mathbf{e}_2 \rangle, \qquad c = c_{22} = \langle \mathbf{e}_2, \mathbf{e}_2 \rangle.$$

The positivity condition (2.17) requires

$$q(v_1, v_2) = a\, v_1^2 + 2b\, v_1 v_2 + c\, v_2^2 > 0 \quad \text{for all} \quad \mathbf{0} \neq \mathbf{v} = (v_1, v_2)^T. \qquad (2.18)$$

The implied requirements on a, b, c can be determined by recalling the algebraic technique known as "completing the square". First, we note that $0 < q(1,0) = a$ and so the initial coefficient must be positive. We then write

$$q(v_1, v_2) = \left(\sqrt{a}\, v_1 + \frac{b}{\sqrt{a}}\, v_2 \right)^2 + \frac{ac - b^2}{a}\, v_2^2.$$

The first term is ≥ 0; moreover, $0 < q(-b/a, 1) = (ac - b^2)/a$. We deduce that the coefficients a, b, c must satisfy

$$a > 0, \qquad \frac{ac - b^2}{a} > 0. \qquad (2.19)$$

It is not hard to see that, conversely, if (2.19) holds, then $q(v_1, v_2) > 0$ unless $v_1 = v_2 = 0$. Thus conditions (2.19) are necessary and sufficient for the quadratic form (2.18) to be positive definite. ▲

Proving positivity of a quadratic form (2.17) in $n > 2$ variables is accomplished, iteratively, by a similar argument. We first note that positivity requires

$$c_{jj} = \langle \mathbf{e}_j, \mathbf{e}_j \rangle > 0.$$

Thus if any one of these coefficients is ≤ 0, we immediately conclude that the form does not satisfy the positivity requirement. (On the other hand, as we saw above, strict positivity of these coefficients is not sufficient to establish positivity of the quadratic form.) We then complete the square by combining all the terms in $q(\mathbf{v})$ that involve v_1 into a square, at the expense of introducing extra terms involving only the other variables; that is, we write

$$q(v_1, \ldots, v_n) = (b_{11} v_1 + b_{12} v_2 + \cdots + b_{1n} v_n)^2 + \widetilde{q}(v_2, \ldots, v_n), \qquad (2.20)$$

where $\widetilde{q}$ is a quadratic form that does not depend on v_1. Comparing with (2.17), this requires

$$c_{11} = b_{11}^2, \quad c_{1j} = b_{11} b_{1j}, \quad \text{and hence} \quad b_{11} = \sqrt{c_{11}} > 0, \quad b_{1j} = \frac{c_{1j}}{\sqrt{c_{11}}}, \quad j = 2, \ldots, n. \qquad (2.21)$$

We claim that q is positive if and only if $\widetilde{q}$ is positive. Indeed, $\widetilde{q}(v_2, \ldots, v_n) > 0$ for all $(v_2, \ldots, v_n) \neq \mathbf{0}$. Let $(v_1, \ldots, v_n) \neq \mathbf{0}$. If $v_1 \neq 0$, then the first term on the right hand side of (2.20) is > 0 and hence $q(v_1, \ldots, v_n) > 0$. Otherwise, if $v_1 = 0$, then $(v_2, \ldots, v_n) \neq \mathbf{0}$ and again formula (2.20) implies $q(v_1, \ldots, v_n) > 0$. On the other hand, if $\widetilde{q}(v_2, \ldots, v_n) \leq 0$ for some $(v_2, \ldots, v_n) \neq \mathbf{0}$, and we set $v_1 = -(b_{12} v_2 + \cdots + b_{1n} v_n)/b_{11}$, then $(v_1, \ldots, v_n) \neq \mathbf{0}$ and $q(v_1, \ldots, v_n) \leq 0$, thus violating positivity.

The quadratic form

$$\widetilde{q}(v_2, \ldots, v_n) = \sum_{i,j=2}^{n} \widetilde{c}_{ij}\, v_i\, v_j$$

depends upon one fewer variable and hence we can inductively apply the preceding algorithm to it. In particular, positivity requires that all $\widetilde{c}_{jj} > 0$, $j = 2, \ldots, n$. In the next step, we complete the square for the terms involing v_2 in $\widetilde{q}$ and thereby produce a quadratic form depending on only $v_3, \ldots, v_n$, which must be positive if $\widetilde{q}$ and hence q are to be positive. And so on. If the algorithm succeeds all the way to the end, the original quadratic form is positive, and the final result is to re-express it as a sum of squares of the form

$$q(\mathbf{v}) = y_1^2 + y_2^2 + \cdots + y_n^2, \qquad \text{where} \qquad y_i = \sum_{j=i}^{n} b_{ij} v_j, \qquad (2.22)$$

for certain coefficients b_{ij} for $j \geq i$, with $b_{ii} > 0$, so that each y_i depends linearly on $v_i, v_{i+1}, \ldots, v_n$ only. This inductive procedure, based on successive completions of squares, provides us with a practical algorithm for determining whether of not a given quadratic form is positive definite, and hence whether or not a given set of symmetric coefficients $c_{ij} = c_{ji}$ defines an inner product (2.14).

Example 2.5. Let us determine whether

$$\langle\, \mathbf{v}, \mathbf{w} \,\rangle = v_1 w_1 + 2\,(v_1 w_2 + v_2 w_1) - (v_1 w_3 + v_3 w_1) + 6\,v_2 w_2 + 9\,v_3 w_3 \qquad (2.23)$$

determines an inner product on $\mathbb{R}^3$. According to the above reasoning, we need only chack positivity of the associated quadratic form

$$q(\mathbf{v}) = \langle\, \mathbf{v}, \mathbf{v} \,\rangle = v_1^2 + 4\,v_1\,v_2 - 2\,v_1\,v_3 + 6\,v_2^2 + 9\,v_3^2.$$

We begin by completing the square for the terms involving v_1, writing

$$v_1^2 + 4\,v_1\,v_2 - 2\,v_1\,v_3 = (v_1 + 2\,v_2 - v_3)^2 - 4\,v_2^2 + 4\,v_2\,v_3 - v_3^2.$$

Therefore,

$$q(\mathbf{v}) = (v_1 + 2\,v_2 - v_3)^2 + 2\,v_2^2 + 4\,v_2\,v_3 + 8\,v_3^2 = (v_1 + 2\,v_2 - v_3)^2 + \widetilde{q}(v_2, v_3),$$

where

$$\widetilde{q}(v_2, v_3) = 2\,v_2^2 + 4\,v_2\,v_3 + 8\,v_3^2$$

is a quadratic form that involves only v_2, v_3. We then repeat the process, combining all the terms involving v_2 in the remaining quadratic form into a square, writing

$$\widetilde{q}(v_2, v_3) = \left(\, \sqrt{2}\,v_2 + \sqrt{2}\,v_3 \,\right)^2 + 6\,v_3^2.$$

This gives the final form

$$q(\mathbf{v}) = y_1^2 + y_2^2 + y_3^2 = (v_1 + 2\,v_2 - v_3)^2 + \left(\, \sqrt{2}\,v_2 + \sqrt{2}\,v_3 \,\right)^2 + \left(\, \sqrt{6}\,v_3 \,\right)^2.$$

Since it is a sum of squares, we have $q(\mathbf{v}) \geq 0$ for all $\mathbf{v} \in \mathbb{R}^3$. Moreover, $q(\mathbf{v}) = 0$ if and only if all three squares vanish:

$$v_1 + 2v_2 - v_3 = \sqrt{2}\,v_2 + v_3/\sqrt{2} = \sqrt{6}\,v_3 = 0,$$

which clearly requires $v_1 = v_2 = v_3 = 0$. We conclude that (2.23) does define an inner product on $\mathbb{R}^3$. ▲

Given an inner product and associated norm, the vectors $\mathbf{u} \in \mathbb{R}^n$ that have unit norm, $\|\mathbf{u}\| = 1$, play a special role, and are known as *unit vectors*. The following lemma shows how to construct a unit vector pointing in the same direction as any given nonzero vector.

Lemma 2.6. *If $\mathbf{v} \neq \mathbf{0}$ is any nonzero vector, then the vector $\mathbf{u} = \mathbf{v}/\|\mathbf{v}\|$ obtained by dividing $\mathbf{v}$ by its norm is a unit vector parallel to $\mathbf{v}$.*

Proof. We write $\mathbf{u} = c\mathbf{v}$ where $c = 1/\|\mathbf{v}\|$ is a scalar. Making use of the homogeneity property (2.7) of the norm, we find

$$\|\mathbf{u}\| = \|c\mathbf{v}\| = c\|\mathbf{v}\| = \frac{\|\mathbf{v}\|}{\|\mathbf{v}\|} = 1. \qquad \blacksquare$$

Example 2.7. The vector $\mathbf{v} = (-1, 2)^T$ has length $\|\mathbf{v}\| = \sqrt{5}$ with respect to the standard Euclidean norm. Therefore, the unit vector pointing in the same direction is

$$\mathbf{u} = \frac{\mathbf{v}}{\|\mathbf{v}\|} = \frac{1}{\sqrt{5}}\begin{pmatrix} -2 \\ 1 \end{pmatrix} = \begin{pmatrix} -\frac{2}{\sqrt{5}} \\ \frac{1}{\sqrt{5}} \end{pmatrix}.$$

On the other hand, for the weighted norm (2.9), $\|\mathbf{v}\| = \sqrt{13}$, and so

$$\mathbf{u} = \frac{\mathbf{v}}{\|\mathbf{v}\|} = \frac{1}{\sqrt{13}}\begin{pmatrix} -2 \\ 1 \end{pmatrix} = \begin{pmatrix} -\frac{2}{\sqrt{13}} \\ \frac{1}{\sqrt{13}} \end{pmatrix}$$

is the unit vector parallel to $\mathbf{v}$. Thus, the notion of unit vector will depend upon which norm is being used. ▲

Exercises

1.1. Which of the following formulas for $\langle \mathbf{v}, \mathbf{w} \rangle$ define inner products on $\mathbb{R}^2$?
 (a)♡ $2v_1 w_1 + 3v_2 w_2$,   (b)♡ $v_1 w_2 + v_2 w_1$,   (c) $(v_1 + v_2)(w_1 + w_2)$,   (d) $v_1^2 w_1^2 + v_2^2 w_2^2$,
 (e)◇ $2v_1 w_1 + (v_1 - v_2)(w_1 - w_2)$,   (f) $4v_1 w_1 - 2v_1 w_2 - 2v_2 w_1 + 4v_2 w_2$.

1.2. For which values of b does the formula $\langle \mathbf{v}, \mathbf{w} \rangle = v_1 w_1 - v_1 w_2 - v_2 w_1 + b v_2 w_2$ define an inner product on $\mathbb{R}^2$?

1.3. Prove that each of the following formulas for $\langle \mathbf{v}, \mathbf{w} \rangle$ defines an inner product on $\mathbb{R}^3$. Verify all the inner product axioms in careful detail:
 (a)♡ $v_1 w_1 + 2v_2 w_2 + 3v_3 w_3$,   (b) $4v_1 w_1 + 2v_1 w_2 + 2v_2 w_1 + 4v_2 w_2 + v_3 w_3$,
 (c)◇ $2v_1 w_1 - 2v_1 w_2 - 2v_2 w_1 + 3v_2 w_2 - v_2 w_3 - v_3 w_2 + 2v_3 w_3$.

1.4. Prove that the following quadratic forms on $\mathbb{R}^3$ are positive definite by writing each as a sum of squares. Then write down the corresponding inner product.
$\quad$ $(a) \heartsuit$ $x^2 + 4xz + 3y^2 + 5z^2,$ $\quad$ $(b) \diamondsuit$ $x^2 + 3xy + 3y^2 - 2xz + 8z^2,$
$\quad$ (c) $2x_1^2 + x_1 x_2 - 2x_1 x_3 + 2x_2^2 - 2x_2 x_3 + 2x_3^2.$

1.5. Prove that the second bilinearity formula (2.3) is a consequence of the first and the other two inner product axioms.

1.6. $(a) \heartsuit$ Prove that $\langle \mathbf{x}, \mathbf{v} \rangle = 0$ for all $\mathbf{v} \in \mathbb{R}^n$ if and only if $\mathbf{x} = \mathbf{0}$. $(b) \diamondsuit$ Prove that $\langle \mathbf{x}, \mathbf{v} \rangle = \langle \mathbf{y}, \mathbf{v} \rangle$ for all $\mathbf{v} \in \mathbb{R}^n$ if and only if $\mathbf{x} = \mathbf{y}$. (c) Let $\mathbf{v}_1, \ldots, \mathbf{v}_n$ be a basis for $\mathbb{R}^n$. Prove that $\langle \mathbf{x}, \mathbf{v}_i \rangle = \langle \mathbf{y}, \mathbf{v}_i \rangle$ for all $i = 1, \ldots, n$ if and only if $\mathbf{x} = \mathbf{y}$.

1.7. Let $\langle \cdot, \cdot \rangle$ be an inner product on $\mathbb{R}^n$ and let $\|\cdot\|$ be the induced norm.
$\quad (a) \heartsuit$ Show that the norm satisfies the *parallelogram identity*

$$\|\mathbf{v} + \mathbf{w}\|^2 + \|\mathbf{v} - \mathbf{w}\|^2 = 2\|\mathbf{v}\|^2 + 2\|\mathbf{w}\|^2 \quad \text{for all} \quad \mathbf{v}, \mathbf{w} \in \mathbb{R}^n. \tag{2.24}$$

$(b) \diamondsuit$ Prove the identity

$$\langle \mathbf{v}, \mathbf{w} \rangle = \tfrac{1}{4}\left(\|\mathbf{v} + \mathbf{w}\|^2 - \|\mathbf{v} - \mathbf{w}\|^2\right), \tag{2.25}$$

which allows one to reconstruct an inner product from its norm.
$\quad (c)$ Use (2.25) to find the inner product on $\mathbb{R}^2$ corresponding to the norm

$$\|\mathbf{v}\| = \sqrt{v_1^2 - 3v_1 v_2 + 5v_2^2}\,.$$

1.8. Suppose $\langle \mathbf{v}, \mathbf{w} \rangle_1$ and $\langle \mathbf{v}, \mathbf{w} \rangle_2$ are two inner products on $\mathbb{R}^n$. For which $\alpha, \beta \in \mathbb{R}$ is the linear combination $\langle \mathbf{v}, \mathbf{w} \rangle = \alpha \langle \mathbf{v}, \mathbf{w} \rangle_1 + \beta \langle \mathbf{v}, \mathbf{w} \rangle_2$ a legitimate inner product? *Hint:* The case $\alpha, \beta \geq 0$ is easy. However, some negative values are also permitted, and your task is to decide which.

2.2   Inequalities

There are two fundamental inequalities that are valid for *any* inner product. The first, which is named after the nineteenth-century mathematicians Augustin Cauchy and Herman Schwarz, is inspired by the geometric interpretation of the dot product on Euclidean space in terms of the angle between vectors.[2] The more familiar triangle inequality, that the length of any side of a triangle is bounded by the sum of the lengths of the other two sides, is, in fact, an immediate consequence of the Cauchy–Schwarz inequality, and hence also valid for any norm based on an inner product.

2.2.1   The Cauchy–Schwarz Inequality

In Euclidean geometry, the dot product between two vectors $\mathbf{v}, \mathbf{w} \in \mathbb{R}^n$ can be geometrically characterized by the equation

$$\mathbf{v} \cdot \mathbf{w} = \|\mathbf{v}\|\,\|\mathbf{w}\|\cos\theta, \tag{2.26}$$

---

[2]Russians also give credit for its discovery to their compatriot Viktor Bunyakovsky, and, indeed, some authors append his name to the inequality.

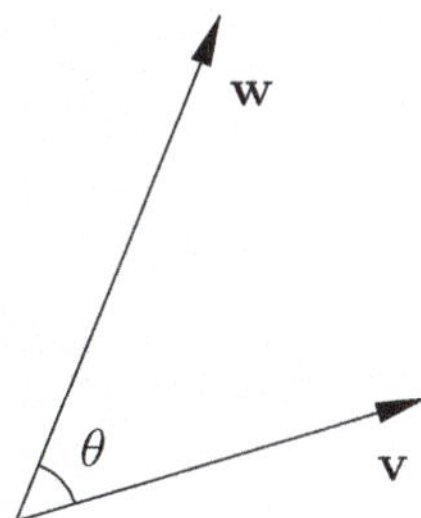

Figure 2.2: Angle Between Two Vectors

where $\theta = \sphericalangle(\mathbf{v}, \mathbf{w})$ measures the angle between the two vectors, as illustrated in Figure 2.2. Since $-1 \le \cos\theta \le 1$, the dot product between two vectors is bounded by the product of their lengths:

$$-\|\mathbf{v}\|\,\|\mathbf{w}\| \;\le\; \mathbf{v}\cdot\mathbf{w} \;\le\; \|\mathbf{v}\|\,\|\mathbf{w}\|, \qquad \text{or, equivalently,} \qquad |\,\mathbf{v}\cdot\mathbf{w}\,| \;\le\; \|\mathbf{v}\|\,\|\mathbf{w}\|,$$

where $|\cdot|$ denotes the absolute value of a real number. This is the simplest form of the general *Cauchy–Schwarz inequality*. We present a direct algebraic proof that does not rely on the geometrical notions of length and angle, and thus demonstrates its universal validity for *any* inner product.

Theorem 2.8. *Every inner product satisfies the Cauchy–Schwarz inequality*

$$|\langle \mathbf{v}, \mathbf{w} \rangle| \;\le\; \|\mathbf{v}\|\,\|\mathbf{w}\|, \qquad \text{for all} \qquad \mathbf{v}, \mathbf{w} \in \mathbb{R}^n, \tag{2.27}$$

where $\|\cdot\|$ is the associated norm. Equality holds in (2.27) if and only if $\mathbf{v}$ and $\mathbf{w}$ are parallel vectors, i.e., $\mathbf{v} = \lambda\mathbf{w}$ for some scalar λ.

Proof. If either $\mathbf{v}$ or $\mathbf{w}$ is the zero vector, the inequality is trivial, since both sides are equal to 0; moreover the zero vector is parallel to any other vector. Thus, we will assume $\mathbf{v}, \mathbf{w} \neq \mathbf{0}$. Dividing both sides of (2.27) by the product $\|\mathbf{v}\|\,\|\mathbf{w}\|$ reduces it to

$$|\langle \widetilde{\mathbf{v}}, \widetilde{\mathbf{w}} \rangle| \;\le\; 1, \qquad \text{where} \qquad \widetilde{\mathbf{v}} = \frac{\mathbf{v}}{\|\mathbf{v}\|}, \qquad \widetilde{\mathbf{w}} = \frac{\mathbf{w}}{\|\mathbf{w}\|}. \tag{2.28}$$

Lemma 2.6 tells us that $\widetilde{\mathbf{v}}, \widetilde{\mathbf{w}}$ are unit vectors, i.e., $\|\widetilde{\mathbf{v}}\| = \|\widetilde{\mathbf{w}}\| = 1$. Let us next expand the squared norms of their sum and difference:

$$0 \le \|\widetilde{\mathbf{v}} \pm \widetilde{\mathbf{w}}\|^2 = \|\widetilde{\mathbf{v}}\|^2 \pm 2\langle \widetilde{\mathbf{v}}, \widetilde{\mathbf{w}} \rangle + \|\widetilde{\mathbf{w}}\|^2 = 2\big(1 \pm \langle \widetilde{\mathbf{v}}, \widetilde{\mathbf{w}} \rangle\big),$$

which implies $\pm\langle \widetilde{\mathbf{v}}, \widetilde{\mathbf{w}} \rangle \le 1$, and hence (2.28) holds, which implies the general inequality. Moreover, we have equality of (2.28) at $+1$ if and only if $\widetilde{\mathbf{v}} = \widetilde{\mathbf{w}}$ and at -1 if and only if $\widetilde{\mathbf{v}} = -\widetilde{\mathbf{w}}$, either of which means the original vectors $\mathbf{v}, \mathbf{w}$ are parallel. ∎

Remark 2.9. Since $-|a| \le a \le |a|$ for any $a \in \mathbb{R}$, the Cauchy–Schwarz inequality implies

$$-\|\mathbf{v}\|\,\|\mathbf{w}\| \;\le\; \langle \mathbf{v}, \mathbf{w} \rangle \;\le\; \|\mathbf{v}\|\,\|\mathbf{w}\|, \qquad \text{for all} \qquad \mathbf{v}, \mathbf{w} \in \mathbb{R}^n. \tag{2.29}$$

Moreover, $\langle \mathbf{v}, \mathbf{w} \rangle = \|\mathbf{v}\|\,\|\mathbf{w}\|$ if and only if $\mathbf{v}$ and $\mathbf{w}$ are parallel vectors pointing in the same direction, so $\langle \mathbf{v}, \mathbf{w} \rangle \ge 0$, while $\langle \mathbf{v}, \mathbf{w} \rangle = -\|\mathbf{v}\|\,\|\mathbf{w}\|$ if and only if they are parallel and point in opposite directions. ▲

2.2.2   The Triangle Inequality

The familiar triangle inequality states that the length of one side of a triangle is at most equal to the sum of the lengths of the other two sides. Referring to Figure 2.3, if the first two sides are represented by vectors $\mathbf{v}$ and $\mathbf{w}$, then the third corresponds to their sum $\mathbf{v} + \mathbf{w}$. The triangle inequality turns out to be an elementary consequence of the Cauchy–Schwarz inequality (2.27), and hence is valid for *every* norm based on an inner product.

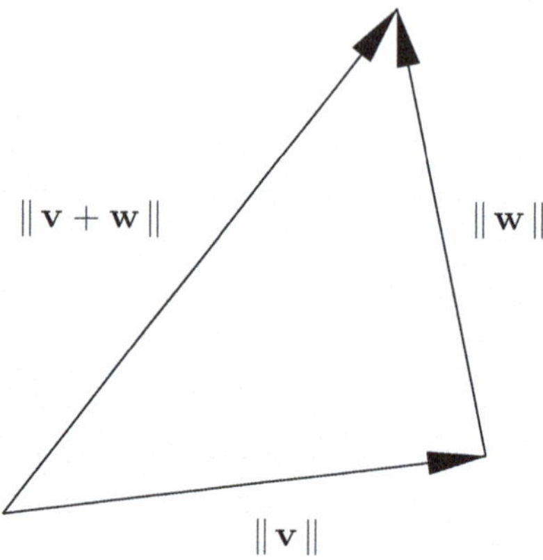

Figure 2.3: Triangle Inequality

Theorem 2.10. *The norm associated with an inner product satisfies the* triangle *inequality*

$$\| \mathbf{v} + \mathbf{w} \| \;\leq\; \| \mathbf{v} \| + \| \mathbf{w} \| \qquad for\ all \qquad \mathbf{v}, \mathbf{w} \in \mathbb{R}^n. \tag{2.30}$$

Equality holds if and only if $\mathbf{v}$ and $\mathbf{w}$ are parallel vectors that point in the same direction, i.e., $\mathbf{v} = c\,\mathbf{w}$ for some nonnegative scalar $c \geq 0$.

Proof. We compute

$$\| \mathbf{v} + \mathbf{w} \|^2 = \langle \mathbf{v} + \mathbf{w}, \mathbf{v} + \mathbf{w} \rangle = \| \mathbf{v} \|^2 + 2\,\langle \mathbf{v}, \mathbf{w} \rangle + \| \mathbf{w} \|^2$$

$$\leq \| \mathbf{v} \|^2 + 2\,\| \mathbf{v} \|\,\| \mathbf{w} \| + \| \mathbf{w} \|^2 = \big(\| \mathbf{v} \| + \| \mathbf{w} \| \big)^2,$$

where the middle inequality is a consequence of (2.29). Taking square roots of both sides and using the fact that the resulting expressions are both positive completes the proof of the triangle inequality. The fact that equality holds under the stated conditions follows from Remark 2.9.   ∎

Example 2.11. The vectors $\mathbf{v} = \begin{pmatrix} 3 \\ 1 \end{pmatrix}$ and $\mathbf{w} = \begin{pmatrix} -1 \\ 2 \end{pmatrix}$ sum to $\mathbf{v} + \mathbf{w} = \begin{pmatrix} 2 \\ 3 \end{pmatrix}$. Their Euclidean norms are $\| \mathbf{v} \| = \sqrt{10}$ and $\| \mathbf{w} \| = \sqrt{5}$, while $\| \mathbf{v} + \mathbf{w} \| = \sqrt{13}$. The triangle inequality (2.30) in this case says $\sqrt{13} \leq \sqrt{10} + \sqrt{5}$, which is true. On the other hand, if we use the weighted norm (2.9), the triangle inequality becomes

$$\| \mathbf{v} + \mathbf{w} \| = \sqrt{53} \leq \sqrt{23} + \sqrt{22} = \| \mathbf{v} \| + \| \mathbf{w} \|. \qquad \blacktriangle$$

Exercises

2.1. Verify the Cauchy–Schwarz and triangle inequalities for the vectors $\mathbf{v} = (1, 2)^T$ and $\mathbf{w} = (1, -3)^T$ using $(a)\,\heartsuit$ the dot product; $(b)\,\diamondsuit$ the weighted inner product $\langle \mathbf{v}, \mathbf{w} \rangle = v_1 w_1 + 2 v_2 w_2$; (c) the inner product (2.10).

2.2. Verify the Cauchy–Schwarz and triangle inequalities for each of the following pairs of vectors $\mathbf{v}, \mathbf{w}$, using the standard dot product, and then determine the angle between them:
$(a)\,\heartsuit\ (1, 2)^T, (-1, 2)^T,\ \ (b)\,\diamondsuit\ (1, -1, 0)^T, (-1, 0, 1)^T,\ \ (c)\ (1, -1, 1, 0)^T, (-2, 0, -1, 1)^T.$

2.3. Prove that the points $(0, 0, 0), (1, 1, 0), (1, 0, 1), (0, 1, 1)$ form the vertices of a regular tetrahedron, meaning that all sides have the same length. What is the common Euclidean angle between the edges? What is the angle between any two rays going from the center $\left(\frac{1}{2}, \frac{1}{2}, \frac{1}{2}\right)$ to the vertices? *Remark*: Methane molecules assume this geometric configuration, and the angle influences their chemistry.

2.4. $\heartsuit$ Given an inner product on $\mathbb{R}^n$, define the corresponding (non-Euclidean) *angle* θ between two nonzero vectors $\mathbf{0} \neq \mathbf{v}, \mathbf{w} \in \mathbb{R}^n$ by the formula $\langle \mathbf{v}, \mathbf{w} \rangle = \|\mathbf{v}\|\,\|\mathbf{w}\|\,\cos\theta$. Prove that the *Law of Cosines* holds in general:

$$\|\mathbf{v} - \mathbf{w}\|^2 = \|\mathbf{v}\|^2 + \|\mathbf{w}\|^2 - 2\,\|\mathbf{v}\|\,\|\mathbf{w}\|\,\cos\theta. \tag{2.31}$$

2.5. Let $t > 0$. Prove the inequality $|\langle \mathbf{v}, \mathbf{w} \rangle| \leq t\,\|\mathbf{v}\|^2 + \dfrac{1}{4t}\,\|\mathbf{w}\|^2$.

2.3 Orthogonal Vectors and Orthogonal Bases

In Euclidean geometry, a particularly noteworthy configuration occurs when two vectors are *perpendicular*, meaning that they meet at a right angle, so $\theta = \sphericalangle(\mathbf{v}, \mathbf{w}) = \frac{1}{2}\pi$ or $\frac{3}{2}\pi$, and hence $\cos\theta = 0$. The angle formula (2.26) implies that the vectors $\mathbf{v}, \mathbf{w}$ are perpendicular if and only if their dot product vanishes: $\mathbf{v} \cdot \mathbf{w} = 0$. Perpendicularity is of similar importance for general inner products, but, for historical reasons, has been given a more suggestive name.

> **Definition 2.12.** Two vectors $\mathbf{v}, \mathbf{w} \in \mathbb{R}^n$ are called *orthogonal* if their inner product vanishes: $\langle \mathbf{v}, \mathbf{w} \rangle = 0$.

In particular, the zero vector is orthogonal to all other vectors: $\langle \mathbf{0}, \mathbf{v} \rangle = 0$ for all $\mathbf{v} \in \mathbb{R}^n$, and is the only vector with this property, since $\mathbf{v}$ is orthogonal to itself, so $\langle \mathbf{v}, \mathbf{v} \rangle = 0$, if and only if $\mathbf{v} = \mathbf{0}$.

Example 2.13. The vectors $\mathbf{v} = (1, 2)^T$ and $\mathbf{w} = (6, -3)^T$ are orthogonal with respect to the Euclidean dot product in $\mathbb{R}^2$, since $\mathbf{v} \cdot \mathbf{w} = 1 \cdot 6 + 2 \cdot (-3) = 0$. We deduce that they meet at a right angle. However, these vectors are *not* orthogonal with respect to the weighted inner product (2.8):

$$\langle \mathbf{v}, \mathbf{w} \rangle = \left\langle \begin{pmatrix} 1 \\ 2 \end{pmatrix}, \begin{pmatrix} 6 \\ -3 \end{pmatrix} \right\rangle = 2 \cdot 1 \cdot 6 + 5 \cdot 2 \cdot (-3) = -18 \neq 0.$$

Thus, orthogonality depends upon which inner product is being used. ▲

As we will see, calculations involving bases are considerably simplified when their elements are mutually orthogonal unit vectors.

Definition 2.14. A basis $\mathbf{u}_1, \ldots, \mathbf{u}_k$ of a k-dimensional subspace $V \subseteq \mathbb{R}^n$ is called *orthogonal* if $\langle \mathbf{u}_i, \mathbf{u}_j \rangle = 0$ for all $i \neq j$. The basis is called *orthonormal* if, in addition, each vector has unit length: $\|\mathbf{u}_i\| = 1$, for all $i = 1, \ldots, k$.

For the Euclidean space $\mathbb{R}^n$ equipped with the standard dot product, the simplest example of an orthonormal basis is the standard basis $\mathbf{e}_1, \ldots, \mathbf{e}_n$, as given in (1.3). Orthogonality follows because $\mathbf{e}_i \cdot \mathbf{e}_j = 0$, for $i \neq j$, while $\|\mathbf{e}_i\| = 1$ implies normality.

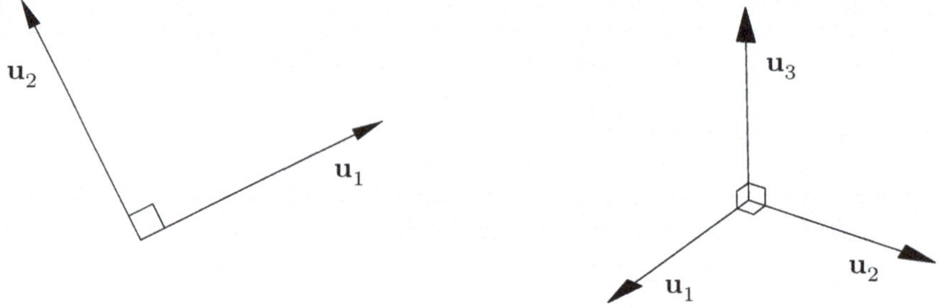

Figure 2.4: Orthonormal Bases in $\mathbb{R}^2$ and $\mathbb{R}^3$

Example 2.15. The vectors

$$\mathbf{v}_1 = \begin{pmatrix} 1 \\ 2 \\ -1 \end{pmatrix}, \qquad \mathbf{v}_2 = \begin{pmatrix} 0 \\ 1 \\ 2 \end{pmatrix}, \qquad \mathbf{v}_3 = \begin{pmatrix} 5 \\ -2 \\ 1 \end{pmatrix},$$

are mutually orthogonal under the dot product: $\mathbf{v}_1 \cdot \mathbf{v}_2 = \mathbf{v}_1 \cdot \mathbf{v}_3 = \mathbf{v}_2 \cdot \mathbf{v}_3 = 0$. Theorem 2.18 implies that they form an orthogonal basis of $\mathbb{R}^3$. When we divide each orthogonal basis vector by its Euclidean length, the result is the orthonormal basis

$$\mathbf{u}_1 = \frac{1}{\sqrt{6}} \begin{pmatrix} 1 \\ 2 \\ -1 \end{pmatrix} = \begin{pmatrix} \frac{1}{\sqrt{6}} \\ \frac{2}{\sqrt{6}} \\ -\frac{1}{\sqrt{6}} \end{pmatrix}, \quad \mathbf{u}_2 = \frac{1}{\sqrt{5}} \begin{pmatrix} 0 \\ 1 \\ 2 \end{pmatrix} = \begin{pmatrix} 0 \\ \frac{1}{\sqrt{5}} \\ \frac{2}{\sqrt{5}} \end{pmatrix}, \quad \mathbf{u}_3 = \frac{1}{\sqrt{30}} \begin{pmatrix} 5 \\ -2 \\ 1 \end{pmatrix} = \begin{pmatrix} \frac{5}{\sqrt{30}} \\ -\frac{2}{\sqrt{30}} \\ \frac{1}{\sqrt{30}} \end{pmatrix},$$

satisfying $\mathbf{u}_1 \cdot \mathbf{u}_2 = \mathbf{u}_1 \cdot \mathbf{u}_3 = \mathbf{u}_2 \cdot \mathbf{u}_3 = 0$ and $\|\mathbf{u}_1\| = \|\mathbf{u}_2\| = \|\mathbf{u}_3\| = 1$. The appearance of square roots in the entries of orthonormal basis vectors is fairly typical.   ▲

Example 2.16. Let us find all the orthonormal bases of $\mathbb{R}^2$. Since every unit vector must lie on the unit circle, which is parametrized by $x = \cos\theta$, $y = \sin\theta$, the first basis vector has the form $\mathbf{u}_1 = (\cos\theta, \sin\theta)^T$ for some angle $0 \leq \theta < 2\pi$. It is not hard to see that there are only two unit vectors that are orthogonal to $\mathbf{u}_1$, namely $\mathbf{u}_2 = (-\sin\theta, \cos\theta)^T$ or $\mathbf{u}_2 = (\sin\theta, -\cos\theta)^T$. Thus, every orthonormal basis of $\mathbb{R}^2$ has one of the following two forms:

$$\mathbf{u}_1 = \begin{pmatrix} \cos\theta \\ \sin\theta \end{pmatrix}, \qquad \mathbf{u}_2 = \begin{pmatrix} -\sin\theta \\ \cos\theta \end{pmatrix}$$

$$\mathbf{u}_1 = \begin{pmatrix} \cos\theta \\ \sin\theta \end{pmatrix}, \qquad \mathbf{u}_2 = \begin{pmatrix} \sin\theta \\ -\cos\theta \end{pmatrix} \qquad \text{for} \qquad 0 \leq \theta < 2\pi. \qquad (2.32)$$

▲

Since a basis cannot contain the zero vector, there is an easy way to convert an orthogonal basis to an orthonormal basis. Namely, we replace each basis vector $\mathbf{v}_i$ by the unit vector $\mathbf{u}_i = \mathbf{v}_i / \|\mathbf{v}_i\|$ pointing in the same direction, as in Lemma 2.6.

Lemma 2.17. *If $\mathbf{v}_1, \ldots, \mathbf{v}_k$ is an orthogonal basis of a subspace V, then the normalized vectors $\mathbf{u}_i = \mathbf{v}_i / \|\mathbf{v}_i\|$, $i = 1, \ldots, k$, form an orthonormal basis.*

A useful observation is that every orthogonal collection of nonzero vectors is automatically linearly independent.

Theorem 2.18. *Let $\mathbf{v}_1, \ldots, \mathbf{v}_k \in \mathbb{R}^n$ be nonzero and mutually orthogonal, so $\mathbf{v}_i \neq \mathbf{0}$ and $\langle \mathbf{v}_i, \mathbf{v}_j \rangle = 0$ for all $i \neq j$. Then $\mathbf{v}_1, \ldots, \mathbf{v}_k$ are linearly independent, and hence form an orthogonal basis for $V = \operatorname{span}\{\mathbf{v}_1, \ldots, \mathbf{v}_k\} \subset \mathbb{R}^n$, which is therefore a subspace of dimension $k = \dim V \leq n$.*

Proof. Suppose

$$c_1 \mathbf{v}_1 + \cdots + c_k \mathbf{v}_k = \mathbf{0}.$$

Let us take the inner product of this equation with any $\mathbf{v}_i$. Using bilinearity of the inner product and orthogonality, we compute

$$0 = \langle \mathbf{v}_i, c_1 \mathbf{v}_1 + \cdots + c_k \mathbf{v}_k \rangle = c_1 \langle \mathbf{v}_i, \mathbf{v}_1 \rangle + \cdots + c_k \langle \mathbf{v}_i, \mathbf{v}_k \rangle = c_i \langle \mathbf{v}_i, \mathbf{v}_i \rangle = c_i \|\mathbf{v}_i\|^2.$$

Therefore, given that $\mathbf{v}_i \neq \mathbf{0}$, we conclude that $c_i = 0$. Since this holds for all $i = 1, \ldots, k$, the linear independence of $\mathbf{v}_1, \ldots, \mathbf{v}_k$ follows. $\blacksquare$

What are the advantages of orthogonal and orthonormal bases? Once one is in possession of a basis of a subspace V, a key issue is how to express other vectors $\mathbf{v} \in V$ as linear combinations of the basis vectors — that is, to find their *coordinates* in the prescribed basis. In general, this is not so easy, since it requires solving a system of linear equations. In high-dimensional situations arising in applications, computing the required coordinates may require a considerable, if not infeasible, amount of time and effort. However, if the basis is orthonormal, then the computation requires almost no work, and, moreover, is is not nearly as prone to numerical errors and instabilities.

Theorem 2.19. *Let $\mathbf{u}_1, \ldots, \mathbf{u}_k$ be an orthonormal basis for a k-dimensional subspace $V \subseteq \mathbb{R}^n$. Then one can write any vector $\mathbf{v} \in V$ as a linear combination*

$$\mathbf{v} = c_1 \mathbf{u}_1 + \cdots + c_k \mathbf{u}_k, \tag{2.33}$$

in which its coordinates

$$c_i = \langle \mathbf{u}_i, \mathbf{v} \rangle, \qquad i = 1, \ldots, k, \tag{2.34}$$

are explicitly given as inner products. Moreover, its norm is given by the Pythagorean formula

$$\|\mathbf{v}\| = \sqrt{c_1^2 + \cdots + c_k^2} = \sqrt{\sum_{i=1}^{k} \langle \mathbf{u}_i, \mathbf{v} \rangle^2}, \tag{2.35}$$

namely, the square root of the sum of the squares of its orthonormal basis coordinates.

Proof. Let us compute the inner product of the element (2.33) with one of the basis vectors. Using the orthonormality conditions

$$\langle\, \mathbf{u}_i, \mathbf{u}_j \,\rangle = \begin{cases} 0 & i \neq j, \\ 1 & i = j, \end{cases} \tag{2.36}$$

and bilinearity of the inner product, we obtain

$$\langle\, \mathbf{u}_i, \mathbf{v} \,\rangle = \left\langle\, \mathbf{u}_i, \sum_{j=1}^{k} c_j\, \mathbf{u}_j \,\right\rangle = \sum_{j=1}^{k} c_j \,\langle\, \mathbf{u}_i, \mathbf{u}_j \,\rangle = c_i\, \|\, \mathbf{u}_i \,\|^2 = c_i.$$

To prove formula (2.35), we similarly expand

$$\|\, \mathbf{v} \,\|^2 = \langle\, \mathbf{v}, \mathbf{v} \,\rangle = \left\langle\, \sum_{i=1}^{k} c_i\, \mathbf{u}_i, \sum_{j=1}^{k} c_j\, \mathbf{u}_j \,\right\rangle = \sum_{i,j=1}^{k} c_i\, c_j \,\langle\, \mathbf{u}_i, \mathbf{u}_j \,\rangle = \sum_{i=1}^{k} c_i^2,$$

again making use of orthonormality of the basis elements. $\blacksquare$

Example 2.20. The wavelet basis

$$\mathbf{v}_1 = \begin{pmatrix} 1 \\ 1 \\ 1 \\ 1 \end{pmatrix}, \qquad \mathbf{v}_2 = \begin{pmatrix} 1 \\ 1 \\ -1 \\ -1 \end{pmatrix}, \qquad \mathbf{v}_3 = \begin{pmatrix} 1 \\ -1 \\ 0 \\ 0 \end{pmatrix}, \qquad \mathbf{v}_4 = \begin{pmatrix} 0 \\ 0 \\ 1 \\ -1 \end{pmatrix}, \tag{2.37}$$

introduced in Example 1.20 is, in fact, an orthogonal basis of $\mathbb{R}^4$ under the dot product, meaning that $\mathbf{v}_i \cdot \mathbf{v}_j = 0$ for $i \neq j$. Their Euclidean norms are

$$\|\, \mathbf{v}_1 \,\| = 2, \qquad \|\, \mathbf{v}_2 \,\| = 2, \qquad \|\, \mathbf{v}_3 \,\| = \sqrt{2}, \qquad \|\, \mathbf{v}_4 \,\| = \sqrt{2},$$

and hence the corresponding orthonormal wavelet basis is

$$\mathbf{u}_1 = \begin{pmatrix} \frac{1}{2} \\ \frac{1}{2} \\ \frac{1}{2} \\ \frac{1}{2} \end{pmatrix}, \qquad \mathbf{u}_2 = \begin{pmatrix} \frac{1}{2} \\ \frac{1}{2} \\ -\frac{1}{2} \\ -\frac{1}{2} \end{pmatrix}, \qquad \mathbf{u}_3 = \begin{pmatrix} \frac{1}{\sqrt{2}} \\ -\frac{1}{\sqrt{2}} \\ 0 \\ 0 \end{pmatrix}, \qquad \mathbf{u}_4 = \begin{pmatrix} 0 \\ 0 \\ \frac{1}{\sqrt{2}} \\ -\frac{1}{\sqrt{2}} \end{pmatrix}.$$

Therefore, using (2.34), we can readily express any vector as a linear combination of the orthonormal wavelet basis vectors. For example,

$$\mathbf{v} = \begin{pmatrix} 4 \\ -2 \\ 1 \\ 5 \end{pmatrix} = 4\, \mathbf{u}_1 - 2\, \mathbf{u}_2 + 3\sqrt{2}\, \mathbf{u}_3 - 2\sqrt{2}\, \mathbf{u}_4 = 2\, \mathbf{v}_1 - \mathbf{v}_2 + 3\, \mathbf{v}_3 - 2\, \mathbf{v}_4,$$

where the orthonormal wavelet basis coordinates are computed directly by taking dot products:

$$\mathbf{u}_1 \cdot \mathbf{v} = 4, \qquad \mathbf{u}_2 \cdot \mathbf{v} = -2, \qquad \mathbf{u}_3 \cdot \mathbf{v} = 3\sqrt{2}, \qquad \mathbf{u}_4 \cdot \mathbf{v} = -2\sqrt{2},$$

thereby reproducing the result in Example 1.20 without the need to solve any equations. We also note that $46 = \|\, \mathbf{v} \,\|^2 = 4^2 + (-2)^2 + (3\sqrt{2})^2 + (-2\sqrt{2})^2$, in conformity with the Pythagorean formula (2.35). $\blacktriangle$

Exercises

Note: Unless stated otherwise, the inner product is the standard dot product on $\mathbb{R}^n$.

3.1. ♡ (a) Find $a \in \mathbb{R}$ such that $(2, a, -3)^T$ is orthogonal to $(-1, 3, -2)^T$.

(b) Is there any value of a for which $(2, a, -3)^T$ is parallel to $(-1, 3, -2)^T$?

3.2. ♡ Find all vectors in $\mathbb{R}^3$ that are orthogonal to both $(1, 2, 3)^T$ and $(-2, 0, 1)^T$.

3.3. Answer Exercises 3.1 and 3.2 using the weighted inner product
$$\langle \mathbf{v}, \mathbf{w} \rangle = 3 v_1 w_1 + 2 v_2 w_2 + v_3 w_3.$$

3.4. (a) Prove that $\mathbf{v}_1 = \left(\frac{3}{5}, 0, \frac{4}{5}\right)^T$, $\mathbf{v}_2 = \left(-\frac{4}{13}, \frac{12}{13}, \frac{3}{13}\right)^T$, $\mathbf{v}_3 = \left(-\frac{48}{65}, -\frac{5}{13}, \frac{36}{65}\right)^T$, form an orthonormal basis for $\mathbb{R}^3$. (b) Find the coordinates of $\mathbf{v} = (1, 1, 1)^T$ relative to this basis. (c) Verify the Pythagorean formula (2.35) in this particular case.

3.5. Using the dot product, classify the following pairs of vectors in $\mathbb{R}^2$ as
(*i*) basis, (*ii*) orthogonal basis, and/or (*iii*) orthonormal basis:

(a)♡ $\begin{pmatrix} -1 \\ 2 \end{pmatrix}, \begin{pmatrix} 2 \\ 1 \end{pmatrix}$; (b)◇ $\begin{pmatrix} \frac{1}{\sqrt{2}} \\ \frac{1}{\sqrt{2}} \end{pmatrix}, \begin{pmatrix} -\frac{1}{\sqrt{2}} \\ \frac{1}{\sqrt{2}} \end{pmatrix}$; (c) $\begin{pmatrix} -1 \\ -1 \end{pmatrix}, \begin{pmatrix} 2 \\ 2 \end{pmatrix}$; (d)♡ $\begin{pmatrix} 2 \\ 3 \end{pmatrix}, \begin{pmatrix} 1 \\ -6 \end{pmatrix}$;

(e)◇ $\begin{pmatrix} -1 \\ 0 \end{pmatrix}, \begin{pmatrix} 0 \\ 3 \end{pmatrix}$; (f) $\begin{pmatrix} \frac{3}{5} \\ \frac{4}{5} \end{pmatrix}, \begin{pmatrix} -\frac{4}{5} \\ \frac{3}{5} \end{pmatrix}$.

3.6. Repeat Exercise 3.5, but use the weighted inner product $\langle \mathbf{v}, \mathbf{w} \rangle = v_1 w_1 + \frac{1}{9} v_2 w_2$ instead of the dot product.

3.7. ◇ Prove that if $\mathbf{u}, \mathbf{v}$ are both unit vectors, then $\mathbf{u} + \mathbf{v}$ and $\mathbf{u} - \mathbf{v}$ are orthogonal. Are they also unit vectors?

3.8. ♡ Suppose that $\mathbf{u}_1, \ldots, \mathbf{u}_n$ form an orthonormal basis of $\mathbb{R}^n$. Prove that the inner product between two vectors $\mathbf{v} = c_1 \mathbf{u}_1 + \cdots + c_n \mathbf{u}_n$ and $\mathbf{w} = d_1 \mathbf{u}_1 + \cdots + d_n \mathbf{u}_n$ is equal to the dot product of their coordinates: $\langle \mathbf{v}, \mathbf{w} \rangle = c_1 d_1 + \cdots + c_n d_n$.

3.9. Prove that the triangle inequality is an equality, $\| \mathbf{v} + \mathbf{w} \|^2 = \| \mathbf{v} \|^2 + \| \mathbf{w} \|^2$, if and only if $\mathbf{v}, \mathbf{w}$ are orthogonal. Explain why this formula can be viewed as the generalization of the Pythagorean Theorem.

2.4 Orthogonal Projection and the Closest Point

Throughout this section, $V \subsetneq \mathbb{R}^n$ will be a subspace of dimension $0 < k < n$, and we fix an inner product on $\mathbb{R}^n$. To facilitate your geometric intuition, you may initially want to concentrate on the ordinary dot product, which is both the simplest and the most important case.

Definition 2.21. A vector $\mathbf{q} \in \mathbb{R}^n$ is said to be *orthogonal* to the subspace $V \subset \mathbb{R}^n$ if it is orthogonal to every vector in V, so $\langle \mathbf{v}, \mathbf{q} \rangle = 0$ for all $\mathbf{v} \in V$.

Lemma 2.22. *If $\mathbf{q} \in V$, then $\mathbf{q}$ is orthogonal to V if and only if $\mathbf{q} = \mathbf{0}$.*

Proof. Since $\mathbf{q}$ is required to be orthogonal to every vector in V, it must, in particular, be orthogonal to itself, and so $0 = \langle \mathbf{q}, \mathbf{q} \rangle = \| \mathbf{q} \|^2$, which implies $\mathbf{q} = \mathbf{0}$. ∎

If $\mathbf{v}_1, \ldots, \mathbf{v}_k$ span the subspace V, e.g., they form a basis, then $\mathbf{q}$ is orthogonal to V if and only if it is orthogonal to each basis vector: $\langle \mathbf{v}_i, \mathbf{q} \rangle = 0$ for $i = 1, \ldots, k$. Indeed, any other vector in V has the form $\mathbf{v} = c_1 \mathbf{v}_1 + \cdots + c_k \mathbf{v}_k$, and hence, by linearity, $\langle \mathbf{v}, \mathbf{q} \rangle = c_1 \langle \mathbf{v}_1, \mathbf{q} \rangle + \cdots + c_k \langle \mathbf{v}_k, \mathbf{q} \rangle = 0$, as required.

> **Definition 2.23.** The *orthogonal projection* of a vector $\mathbf{b} \in \mathbb{R}^n$ onto the subspace V is the element $\mathbf{p} \in V$ that makes the difference $\mathbf{q} = \mathbf{b} - \mathbf{p}$ orthogonal to V.

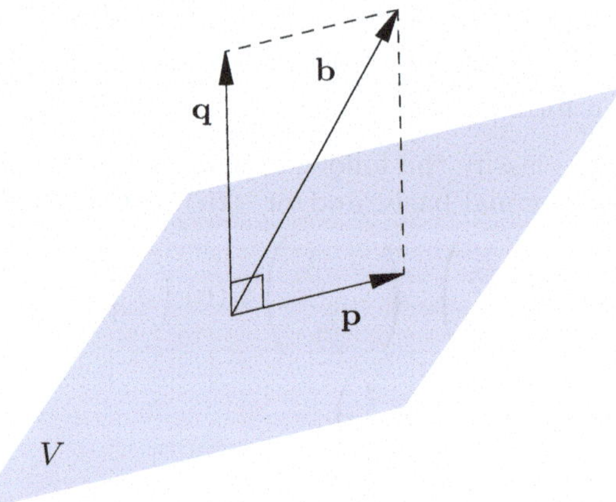

Figure 2.5: The Orthogonal Projection of a Vector onto a Subspace

The geometric configuration underlying orthogonal projection is sketched in Figure 2.5. We note that both $\mathbf{p}$ and $\mathbf{q}$ are uniquely determined, as a consequence of Theorem 2.24 below. Also, since $\langle \mathbf{v}, \mathbf{q} \rangle = 0$ when $\mathbf{v} \in V$, we have

$$\langle \mathbf{v}, \mathbf{b} \rangle = \langle \mathbf{v}, \mathbf{p} \rangle \quad \text{for all} \quad \mathbf{v} \in V. \tag{2.38}$$

The explicit construction of the orthogonal projection is greatly simplified by taking an orthonormal basis of the subspace. (A proof of the existence of such an orthonormal basis, which is, in fact, based on this construction, appears below.)

> **Theorem 2.24.** *Let $\mathbf{u}_1, \ldots, \mathbf{u}_k$ be an orthonormal basis for the subspace $V \subset \mathbb{R}^n$. Then the orthogonal projection of $\mathbf{b} \in \mathbb{R}^n$ onto V is given by*
>
> $$\mathbf{p} = c_1 \mathbf{u}_1 + \cdots + c_k \mathbf{u}_k, \quad \text{where} \quad c_i = \langle \mathbf{u}_i, \mathbf{b} \rangle, \quad i = 1, \ldots, k. \tag{2.39}$$
>
> *Moreover, the projection has norm*
>
> $$\| \mathbf{p} \| = \sqrt{c_1^2 + \cdots + c_k^2} = \sqrt{\sum_{i=1}^{k} \langle \mathbf{u}_i, \mathbf{b} \rangle^2}. \tag{2.40}$$

Proof. First, since $\mathbf{u}_1, \ldots, \mathbf{u}_k$ form a basis of the subspace, the orthogonally projected vector must be some linear combination thereof: $\mathbf{p} = c_1 \mathbf{u}_1 + \cdots + c_k \mathbf{u}_k$. Definition 2.23 requires that the difference $\mathbf{q} = \mathbf{b} - \mathbf{p}$ be orthogonal to V, and, as noted above, it suffices to check orthogonality to the basis vectors. By our orthonormality assumption, for each $i = 1, \ldots, n$,

$$0 = \langle \mathbf{u}_i, \mathbf{q} \rangle = \langle \mathbf{u}_i, \mathbf{b} - \mathbf{p} \rangle = \langle \mathbf{u}_i, \mathbf{b} - c_1 \mathbf{u}_1 - \cdots - c_k \mathbf{u}_k \rangle$$
$$= \langle \mathbf{u}_i, \mathbf{b} \rangle - c_1 \langle \mathbf{u}_i, \mathbf{u}_1 \rangle - \cdots - c_k \langle \mathbf{u}_i, \mathbf{u}_k \rangle = \langle \mathbf{u}_i, \mathbf{b} \rangle - c_i,$$

which establishes formula (2.39). The proof of (2.40) follows in the same manner as (2.35). ∎

An intriguing observation is that the formula for the coefficients in the orthogonal projection formula (2.39) coincides with the formula (2.34) for writing a vector in terms of an orthonormal basis. Indeed, if $\mathbf{b}$ were an element of V, then it would coincide with its orthogonal projection, $\mathbf{p} = \mathbf{b}$. (Why?) Consequently, the orthogonal projection formula includes the orthogonal basis formula as a special case.

Orthogonal projection also solves the problem of finding the closest point on a subspace $V \subset \mathbb{R}^n$ to a given vector $\mathbf{b} \in \mathbb{R}^n$. In other words, we seek $\mathbf{v} \in V$ that minimizes the distance $\mathrm{dist}(\mathbf{b}, \mathbf{v}) = \| \mathbf{b} - \mathbf{v} \|$ from $\mathbf{b}$ to $\mathbf{v}$.

Theorem 2.25. *Let $V \subset \mathbb{R}^n$ be a subspace, and suppose $\mathbf{b} \in \mathbb{R}^n$. Then, the vector $\mathbf{v} \in V$ that is closest in distance to $\mathbf{b}$ is its orthogonal projection $\mathbf{p} \in V$.*

Proof. Let $\mathbf{v} \in V$ be any vector in the subspace. Using the decomposition (2.55), its squared distance to $\mathbf{b} = \mathbf{p} + \mathbf{q}$ is given by

$$\mathrm{dist}(\mathbf{b}, \mathbf{v})^2 = \| \mathbf{b} - \mathbf{v} \|^2 = \| \mathbf{b} \|^2 - 2 \langle \mathbf{b}, \mathbf{v} \rangle + \| \mathbf{v} \|^2$$
$$= \| \mathbf{b} \|^2 - 2 \langle \mathbf{p}, \mathbf{v} \rangle + \| \mathbf{v} \|^2 = \| \mathbf{b} \|^2 - \| \mathbf{p} \|^2 + \| \mathbf{v} - \mathbf{p} \|^2,$$

where we used (2.38) for the third equality, and then completed the square. Now, the first two terms in the final formula do not depend on $\mathbf{v}$. Thus, its minimum value as $\mathbf{v} \in V$ varies is obtained when the final term vanishes, which requires $\mathbf{v} = \mathbf{p}$, with

$$\mathrm{dist}(\mathbf{b}, V) = \| \mathbf{q} \| = \| \mathbf{b} - \mathbf{p} \| = \sqrt{\| \mathbf{b} \|^2 - \| \mathbf{p} \|^2} \tag{2.41}$$

being the minimum distance to V. ∎

Example 2.26. In this example, we use the dot product on $\mathbb{R}^3$. Consider the plane $V \subset \mathbb{R}^3$ spanned by the orthogonal vectors

$$\mathbf{v}_1 = \begin{pmatrix} 1 \\ -2 \\ 1 \end{pmatrix}, \qquad \mathbf{v}_2 = \begin{pmatrix} 1 \\ 1 \\ 1 \end{pmatrix}.$$

We first replace $\mathbf{v}_1, \mathbf{v}_2$ by the orthonormal basis

$$\mathbf{u}_1 = \frac{\mathbf{v}_1}{\| \mathbf{v}_1 \|} = \begin{pmatrix} \frac{1}{\sqrt{6}} \\ -\frac{2}{\sqrt{6}} \\ \frac{1}{\sqrt{6}} \end{pmatrix}, \qquad \mathbf{u}_2 = \frac{\mathbf{v}_2}{\| \mathbf{v}_2 \|} = \begin{pmatrix} \frac{1}{\sqrt{3}} \\ \frac{1}{\sqrt{3}} \\ \frac{1}{\sqrt{3}} \end{pmatrix}.$$

Then, using (2.39), the orthogonal projection of $\mathbf{b} = (1, 0, 0)^T$ onto V is the vector

$$\mathbf{p} = \langle \mathbf{u}_1, \mathbf{b} \rangle \mathbf{u}_1 + \langle \mathbf{u}_2, \mathbf{b} \rangle \mathbf{u}_2 = \frac{1}{\sqrt{6}} \begin{pmatrix} \frac{1}{\sqrt{6}} \\ -\frac{2}{\sqrt{6}} \\ \frac{1}{\sqrt{6}} \end{pmatrix} + \frac{1}{\sqrt{3}} \begin{pmatrix} \frac{1}{\sqrt{3}} \\ \frac{1}{\sqrt{3}} \\ \frac{1}{\sqrt{3}} \end{pmatrix} = \begin{pmatrix} \frac{1}{2} \\ 0 \\ \frac{1}{2} \end{pmatrix},$$

while

$$\mathbf{q} = \mathbf{b} - \mathbf{p} = \begin{pmatrix} 1 \\ 0 \\ 0 \end{pmatrix} - \begin{pmatrix} \frac{1}{2} \\ 0 \\ \frac{1}{2} \end{pmatrix} = \begin{pmatrix} \frac{1}{2} \\ 0 \\ -\frac{1}{2} \end{pmatrix}$$

is orthogonal to both $\mathbf{v}_1$ and $\mathbf{v}_2$, and hence orthogonal to the subspace V. We deduce that the distance from $\mathbf{b} = (1, 0, 0)^T$ to the plane V is $\|\mathbf{q}\| = 1/\sqrt{2}$. $\blacktriangle$

Exercises

4.1. Using the dot product on $\mathbb{R}^3$, given $\mathbf{v} = (1, 1, 1)^T$ find its orthogonal projection onto and distance to the following subspaces: $(a)\,\heartsuit$ the line in the direction $\left(-\frac{1}{\sqrt{3}}, \frac{1}{\sqrt{3}}, \frac{1}{\sqrt{3}} \right)^T$; (b) the line spanned by $(2, -1, 3)^T$; $(c)\,\diamondsuit$ the plane spanned by $(1, 1, 0)^T, (-2, 2, 1)^T$.

4.2. Redo Exercise 4.1 using the weighted inner product $\langle \mathbf{v}, \mathbf{w} \rangle = 2 v_1 w_1 + 2 v_2 w_2 + 3 v_3 w_3$.

4.3. Using the weighted norm $\|\mathbf{v}\|^2 = 4 v_1^2 + 3 v_2^2 + 2 v_3^2 + v_4^2$, find the closest point on the subspace spanned by $(1, -1, 2, 5)^T$ and $(2, 1, 0, -1)^T$ to the vector $\mathbf{v} = (1, 2, -1, 2)^T$.

4.4. $\diamondsuit$ Let $V, W \subset \mathbb{R}^n$ be subspaces. *True or false*: If the orthogonal projection of a vector $\mathbf{v} \in V$ onto W is the vector $\mathbf{w} \in W$, then the orthogonal projection of $\mathbf{w}$ onto V is $\mathbf{v}$.

2.5 The Gram–Schmidt Process

In this section we show how to explicitly construct orthonormal bases of subspaces with respect to a prescribed inner product. The resulting algorithm is an extension of what is known as the *Gram–Schmidt process*, in honor of the nineteenth/twentieth-century Danish and German mathematicians Jørgen Gram and Erhard Schmidt, although it was apparently first formulated in the eighteenth century by the famous French mathematician Pierre–Simon Laplace.

The starting point is the orthonormal basis formula contained in Theorem 2.19. Suppose that $\mathbf{u}_1, \ldots, \mathbf{u}_k$ form an orthonormal basis for a k-dimensional subspace

$$V_k = \operatorname{span} \{\mathbf{u}_1, \ldots, \mathbf{u}_k\} \subseteq \mathbb{R}^m.$$

Then, in view of Theorem 2.19, any vector $\mathbf{v} \in V_k$ can be written as

$$\mathbf{v} = c_1 \mathbf{u}_1 + c_2 \mathbf{u}_2 + \cdots + c_{k-1} \mathbf{u}_{k-1} + c_k \mathbf{u}_k, \quad \text{where} \quad c_i = \langle \mathbf{u}_i, \mathbf{v} \rangle, \quad i = 1, \ldots, k. \tag{2.42}$$

According to Theorem 2.24, we can interpret the first $k-1$ terms in the sum as the orthogonal projection of $\mathbf{v}$ onto the $(k-1)$-dimensional subspace $V_{k-1} = \operatorname{span} \{\mathbf{u}_1, \ldots, \mathbf{u}_{k-1}\} \subset V_k$ because the last term, $c_k \mathbf{u}_k$, is orthogonal to the subspace V_{k-1} since it is orthogonal to its basis vectors $\mathbf{u}_1, \ldots, \mathbf{u}_{k-1}$. Note also that $\mathbf{v} \in V_{k-1}$ if and only if $c_k = 0$.

Vice versa, if we are given $\mathbf{u}_1, \ldots, \mathbf{u}_{k-1}$ and a vector $\mathbf{v} \notin V_{k-1} = \operatorname{span} \{\mathbf{u}_1, \ldots, \mathbf{u}_{k-1}\}$, then we can use formula (2.42) to construct a unit vector $\mathbf{u}_k$ that is orthogonal to V_{k-1}. Namely, we first compute $c_1, \ldots, c_{k-1}$ using the inner product formulas in (2.42). The final coefficient c_k can be obtained by rewriting (2.42) as $c_k \, \mathbf{u}_k = \mathbf{v} - c_1 \, \mathbf{u}_1 - \cdots - c_{k-1} \, \mathbf{u}_{k-1}$ and then taking norms of both sides:

$$c_k = \| \, \mathbf{v} - c_1 \, \mathbf{u}_1 - \cdots - c_{k-1} \, \mathbf{u}_{k-1} \, \|. \tag{2.43}$$

With this in hand, we can then solve (2.42) for

$$\mathbf{u}_k = \frac{1}{c_k} \bigl(\mathbf{v} - c_1 \, \mathbf{u}_1 - \cdots - c_{k-1} \, \mathbf{u}_{k-1} \bigr), \tag{2.44}$$

where we are assured that $c_k \neq 0$ by our assumption that $\mathbf{v} \notin V_{k-1}$. The combined vectors $\mathbf{u}_1, \ldots, \mathbf{u}_k$ form an orthonormal basis of $V_k = \operatorname{span} \{ \mathbf{u}_1, \ldots, \mathbf{u}_k \} = \operatorname{span} \{ \mathbf{u}_1, \ldots, \mathbf{u}_{k-1}, \mathbf{v} \}$. This construction can be recast into a recursive algorithm that will effectively construct orthonormal bases of subspaces, and thus prove that every subspace has one.

To wit, suppose $\mathbf{v}_1, \ldots, \mathbf{v}_n \in \mathbb{R}^m$, endowed with a prescribed inner product and norm. Note that we do not make any assumptions on the vectors; in particular, they are allowed to be linearly dependent. For each $k = 1, \ldots, n$, let $V_k \subset \mathbb{R}^m$ be the subspace spanned by $\mathbf{v}_1, \ldots, \mathbf{v}_k$, so that

$$\{ \mathbf{0} \} \subseteq V_1 \subseteq V_2 \subseteq \cdots \subseteq V_{n-1} \subseteq V_n \subseteq \mathbb{R}^m. \tag{2.45}$$

Let $s_k = \dim V_k$, so that $0 \leq s_1 \leq s_2 \leq \cdots \leq s_n \leq m$, and, moreover, $s_k \leq k$. Indeed, for $1 \leq k \leq n-1$, either $s_{k+1} = s_k$, which occurs when $V_{k+1} = V_k$ and so $\mathbf{v}_{k+1} \in V_k$, or $s_{k+1} = s_k + 1$, which occurs when $V_{k+1} \supsetneq V_k$ and so $\mathbf{v}_{k+1} \notin V_k$.

The first step is to construct an orthonormal basis of $V_1 = \operatorname{span} \{ \mathbf{v}_1 \}$. If $\mathbf{v}_1 = \mathbf{0}$, there is nothing to do, since the trivial subspace $V_1 = \{ \mathbf{0} \}$ has no basis. Otherwise we set

$$\mathbf{u}_1 = \frac{1}{r_{11}} \, \mathbf{v}_1, \qquad \text{where} \qquad r_{11} = \| \, \mathbf{v}_1 \, \| > 0.$$

Proceeding to the second step, there are three possibilities. If $\mathbf{v}_1 = \mathbf{v}_2 = \mathbf{0}$, so $V_2 = \{ \mathbf{0} \}$, there is still nothing to do. If $\mathbf{v}_1 = \mathbf{0}$ but $\mathbf{v}_2 \neq \mathbf{0}$, then $V_2 = \operatorname{span} \{ \mathbf{v}_1, \mathbf{v}_2 \} = \operatorname{span} \{ \mathbf{v}_2 \}$ has dimension $s_2 = 1$, and we can choose the unit basis vector

$$\mathbf{u}_1 = \frac{1}{r_{12}} \, \mathbf{v}_2, \qquad \text{where} \qquad r_{12} = \| \, \mathbf{v}_2 \, \| > 0.$$

Finally, when $\mathbf{v}_1 \neq \mathbf{0}$, there are two further subcases. If $\mathbf{v}_2 \in V_1$, which means that $\mathbf{v}_2$ is a scalar multiple of $\mathbf{v}_1$, then

$$\mathbf{v}_2 = r_{12} \, \mathbf{u}_1, \qquad \text{where} \qquad r_{12} = \langle \, \mathbf{u}_1, \mathbf{v}_2 \, \rangle.$$

Thus, $V_2 = V_1$ has dimension $s_2 = s_1 = 1$, and its orthonormal basis is also given by $\mathbf{u}_1$. Otherwise, $\mathbf{v}_1, \mathbf{v}_2$ are linearly independent, and hence $\dim V_2 = 2$. Using the preceding formulas, we can write

$$\mathbf{v}_2 = r_{12} \, \mathbf{u}_1 + r_{22} \, \mathbf{u}_2, \qquad \text{where} \qquad r_{12} = \langle \, \mathbf{u}_1, \mathbf{v}_2 \, \rangle,$$

while, using (2.43),

$$r_{22} = \| \, \mathbf{v}_2 - r_{12} \, \mathbf{u}_1 \, \|, \qquad \text{and then} \qquad \mathbf{u}_2 = \frac{1}{r_{22}} \bigl(\mathbf{v} - r_{12} \, \mathbf{u}_1 \bigr).$$

We conclude that $\mathbf{u}_1, \mathbf{u}_2$ form an orthonormal basis of V_2. As noted above, $r_{22} \neq 0$, as otherwise we would have $\mathbf{v}_2 \in V_1$, and be back in the first subcase.

In the general recursive step, we assume that we have constructed an orthonormal basis $\mathbf{u}_1, \ldots, \mathbf{u}_{s_{k-1}}$ of V_{k-1}, which may be empty if $\mathbf{v}_1 = \cdots = \mathbf{v}_{k-1} = \mathbf{0}$ and hence $V_1 = \cdots = V_{k-1} = \{\mathbf{0}\}$. Now, consider V_k, which is spanned by $\mathbf{v}_1, \ldots, \mathbf{v}_k$. There are two possibilities: either $V_{k-1} = V_k$ or $V_{k-1} \subsetneq V_k$. Let us set

$$\mathbf{w}_k = \mathbf{v}_k - r_{1,k}\,\mathbf{u}_1 - \cdots - r_{s_{k-1},k}\,\mathbf{u}_{s_{k-1}}, \quad \text{where} \quad r_{i,k} = \langle \mathbf{u}_i, \mathbf{v}_k \rangle, \quad i = 1, \ldots, s_{k-1}.$$

Note that, by orthonormality of the constructed basis vectors,

$$\langle \mathbf{u}_i, \mathbf{w}_k \rangle = \langle \mathbf{u}_i, \mathbf{v}_k \rangle - r_{ik} = 0, \qquad i = 1, \ldots, s_{k-1},$$

and hence $\mathbf{w}_k$ is orthogonal to $\mathbf{u}_1, \ldots, \mathbf{u}_{s_{k-1}}$ and hence to the subspace V_{k-1}. If $\mathbf{w}_k = \mathbf{0}$, then $\mathbf{v}_k \in V_{k-1}$, and we are in the first case, with

$$s_k = \dim V_k = \dim V_{k-1} = s_{k-1},$$

and where $\mathbf{u}_1, \ldots, \mathbf{u}_{s_{k-1}}$ continue to form an orthonormal basis for V_k. On the other hand, if $\mathbf{w}_k \neq \mathbf{0}$, then $\mathbf{v}_k \notin V_{k-1}$, and hence

$$s_k = \dim V_k = \dim V_{k-1} + 1 = s_{k-1} + 1.$$

We are therefore in need of one more unit vector in order to form an orthonormal basis of V_k, which, by the preceding constructions, is obtained by setting

$$r_{s_k,k} = \| \mathbf{w}_k \| > 0, \qquad \mathbf{u}_{s_k} = \frac{\mathbf{w}_k}{r_{s_k,k}}. \tag{2.46}$$

We continue this process until we reach the final subspace V_n, which then has orthonormal basis $\mathbf{u}_1, \ldots, \mathbf{u}_{s_n}$, where $s_n = \dim V_n$. Observe that we can express each

$$\mathbf{v}_k = r_{1,k}\,\mathbf{u}_1 + \cdots + r_{s_k,k}\,\mathbf{u}_{s_k} \tag{2.47}$$

in terms of the orthonormal basis $\mathbf{u}_1, \ldots, \mathbf{u}_{s_k}$ of V_k using the preceding formulas for the coefficients.

We call the preceding algorithm the (*general*) *Gram–Schmidt process*. The classical version corresponds to the case when $\mathbf{v}_1, \ldots, \mathbf{v}_n$ are linearly independent, which implies $\dim V_k = k$ for $k = 1, \ldots, n$, and at each step of the algorithm we append a new orthonormal basis vector $\mathbf{u}_{k+1}$ using (2.46).

Example 2.27. Here is a simple example that illustrates the algorithm. Let us, for simplicity, use the dot product. Consider the vectors

$$\mathbf{v}_1 = \begin{pmatrix} 1 \\ 2 \\ -2 \end{pmatrix}, \quad \mathbf{v}_2 = \begin{pmatrix} 3 \\ 6 \\ -6 \end{pmatrix}, \quad \mathbf{v}_3 = \begin{pmatrix} 3 \\ 1 \\ 1 \end{pmatrix}, \quad \mathbf{v}_4 = \begin{pmatrix} 2 \\ -1 \\ 3 \end{pmatrix}.$$

Since $\mathbf{v}_1 \neq \mathbf{0}$, the process starts by setting

$$r_{11} = \| \mathbf{v}_1 \| = 3, \qquad \mathbf{u}_1 = \frac{\mathbf{v}_1}{r_{11}} = \begin{pmatrix} \frac{1}{3} \\ \frac{2}{3} \\ -\frac{2}{3} \end{pmatrix}.$$

Next,

$$r_{12} = \mathbf{u}_1 \cdot \mathbf{v}_2 = 9.$$

Since $\mathbf{v}_2 = r_{12}\mathbf{u}_1$, the first two vectors $\mathbf{v}_1, \mathbf{v}_2$ are linearly dependent, and the subspace $V_2 = \operatorname{span}\{\mathbf{v}_1, \mathbf{v}_2\} = V_1$ is one-dimensional with orthonormal basis just consisting of $\mathbf{u}_1$. Next,

$$r_{13} = \mathbf{u}_1 \cdot \mathbf{v}_3 = 1, \qquad r_{23} = \|\mathbf{v}_3 - r_{13}\mathbf{u}_1\| = \sqrt{10}, \qquad \mathbf{u}_2 = \frac{\mathbf{v}_3 - r_{13}\mathbf{u}_1}{r_{23}} = \begin{pmatrix} \frac{8}{3\sqrt{10}} \\ \frac{1}{3\sqrt{10}} \\ \frac{5}{3\sqrt{10}} \end{pmatrix},$$

and so $V_3 = \operatorname{span}\{\mathbf{v}_1, \mathbf{v}_2, \mathbf{v}_3\}$ is two-dimensional with orthonormal basis $\mathbf{u}_1, \mathbf{u}_2$. Finally,

$$r_{14} = \mathbf{u}_1 \cdot \mathbf{v}_4 = -2, \qquad r_{24} = \mathbf{u}_2 \cdot \mathbf{v}_4 = \sqrt{10}.$$

Since $\mathbf{v}_4 = r_{14}\mathbf{u}_1 + r_{24}\mathbf{u}_2$, the subspace $V_4 = \operatorname{span}\{\mathbf{v}_1, \mathbf{v}_2, \mathbf{v}_3, \mathbf{v}_4\}$ is also two-dimensional, with orthonormal basis $\mathbf{u}_1, \mathbf{u}_2$, which is the final output of the Gram–Schmidt process. $\blacktriangle$

It turns out that in practical, large-scale computations, the Gram–Schmidt process as formulated above may be subject to numerical instabilities, and accumulating round-off errors can corrupt the computations, leading to inaccurate, non-orthonormal vectors. Fortunately, there is a simple rearrangement of the calculation that ameliorates this difficulty and leads to the numerically robust algorithm that is most often used in practice; see [56, 88, 105, 230] for full details. The key idea is to treat the vectors simultaneously rather than sequentially, making full use of the orthonormal basis vectors as they arise.

The first source of potential numerical instability is that, at step k, the algorithm introduces a new orthonormal basis vector, as in (2.46), whenever $\|\mathbf{w}_k\| > 0$ and only when $\|\mathbf{w}_k\| = 0$ is this not done. However, accumulating numerical errors may well turn a zero value of $\|\mathbf{w}_k\|$ into a small nonzero quantity. In this case, computing the next orthonormal basis vector by dividing $\mathbf{w}_k$ by its norm will produce a spurious result that should not be used. To avoid this issue, we introduce a suitably small threshold $\varepsilon > 0$, which is related to the machine precision being used, and deem that when

$$\|\mathbf{w}_k\| < \varepsilon, \tag{2.48}$$

one regards the subspace $V_k = V_{k-1}$ (modulo numerical error), and so it does not include a new orthonormal basis vector.

With this in hand, a further potential problem could arise if $\mathbf{v}_k$ has very large entries, so that the criterion (2.48) is not satisfied, but nevertheless makes a very small angle with the subspace V_{k-1} and hence should be viewed as (approximately) lying therein. Vice versa, multiplying $\mathbf{v}_k$ by a very small scalar would satisfy the threshold criterion (2.48) even though it makes a large angle with the preceding subspace and should be viewed as independent of the preceding vectors. Both issues can be effectively avoided by "preconditioning" by dividing each vector by its norm, producing all unit (but not orthogonal) vectors $\widetilde{\mathbf{v}}_k = \mathbf{v}_k / \|\mathbf{v}_k\|$. On the other hand, if $\|\mathbf{v}_k\|$ is very small, it may be a better idea to set $\widetilde{\mathbf{v}}_k = \mathbf{0}$ or, equivalently, just omit $\mathbf{v}_k$ from the computation as zero vectors do not affect the final outcome. From here on, we revert to $\mathbf{v}_1, \ldots, \mathbf{v}_k$ to denote the resulting initial vectors.

To avoid the second mode of numerical instability, the algorithm begins as before — assuming $\mathbf{v}_1 \neq \mathbf{0}$, we take $\mathbf{u}_1 = \mathbf{v}_1 / \|\mathbf{v}_1\|$. We then subtract off the appropriate multiples of $\mathbf{u}_1$ from *all* of the remaining vectors before proceeding, which is accomplished by setting

$$\widehat{\mathbf{v}}_j = \mathbf{v}_j - r_{1j}\mathbf{u}_1, \quad \text{where} \quad r_{1j} = \langle \mathbf{u}_1, \mathbf{v}_j \rangle \quad \text{for} \quad j = 2, \ldots, n.$$

Once a second orthonormal basis vector $\mathbf{u}_2$ is found we similarly modify the as yet unused $\widehat{\mathbf{v}}_j$, and continue the process until all the orthonormal basis vectors have been found.

More explicitly, given a threshold $\varepsilon > 0$, the modified Gram–Schmidt algorithm starts with the initial basis vectors $\mathbf{v}_j^{(0)} = \mathbf{v}_j$ for all $j = 1, \ldots, n$. More generally, we can precondition by either setting $\mathbf{v}_j^{(0)} = \mathbf{v}_j / \|\mathbf{v}_j\|$, or, if $\|\mathbf{v}_j\|$ is very small, either setting $\mathbf{v}_j^{(0)} = \mathbf{0}$ or just discarding it from consideration. Let us further initialize by setting $V_0 = \{\mathbf{0}\}$ and hence $s_0 = \dim V_0 = 0$.

At each step $1 \le k \le n$, we have already determined $\mathbf{u}_1, \ldots, \mathbf{u}_{s_{k-1}}$, the orthonormal basis of V_{k-1}, where $s_{k-1} = \dim V_{k-1}$. (If $s_{k-1} = 0$, there are no basis vectors as yet.) If

$$\| \mathbf{v}_k^{(s_{k-1})} \| < \varepsilon, \tag{2.49}$$

then we set $s_k = s_{k-1}$, and there is nothing further to do at this step. Otherwise, $s_k = s_{k-1}+1$, and we define

$$r_{s_k,k} = \| \mathbf{v}_k^{(s_{k-1})} \|, \qquad \mathbf{u}_{s_k} = \frac{\mathbf{v}_k^{(s_{k-1})}}{r_{s_k,k}}. \tag{2.50}$$

Finally, if $k < n$, we update the remaining vectors by setting

$$\mathbf{v}_j^{(s_k)} = \mathbf{v}_j^{(s_{k-1})} - r_{s_k,j}\,\mathbf{u}_{s_k}, \quad \text{where} \quad r_{s_k,j} = \langle\, \mathbf{u}_{s_k}, \mathbf{v}_j^{(s_{k-1})} \,\rangle, \quad \text{for} \quad j = k+1, \ldots, n, \tag{2.51}$$

while when $k = n$, the recursion terminates. The resulting algorithm is a numerically stable computation of the *same* orthonormal basis vectors $\mathbf{u}_1, \ldots, \mathbf{u}_{s_n}$ that were produced earlier; see [56, 88, 105] for a detailed analysis.

Example 2.28. Let us apply the modified Gram–Schmidt process to the vectors

$$\mathbf{v}_1^{(0)} = \mathbf{v}_1 = \begin{pmatrix} 1 \\ 1 \\ -1 \end{pmatrix}, \qquad \mathbf{v}_2^{(0)} = \mathbf{v}_2 = \begin{pmatrix} 1 \\ 0 \\ 2 \end{pmatrix}, \qquad \mathbf{v}_3^{(0)} = \mathbf{v}_3 = \begin{pmatrix} 2 \\ -2 \\ 3 \end{pmatrix},$$

using the dot product and Euclidean norm on $\mathbb{R}^3$. Starting at $k = 1$, we compute

$$r_{11} = \| \mathbf{v}_1^{(0)} \| = \sqrt{3}, \qquad \text{and so} \qquad \mathbf{u}_1 = \frac{\mathbf{v}_1^{(0)}}{r_{11}} = \begin{pmatrix} \frac{1}{\sqrt{3}} \\ \frac{1}{\sqrt{3}} \\ -\frac{1}{\sqrt{3}} \end{pmatrix},$$

is the first orthonormal basis vector, with $s_1 = \dim V_1 = 1$ (since r_{11} is not small). Next, we compute

$$r_{12} = \mathbf{u}_1 \cdot \mathbf{v}_2^{(0)} = -\frac{1}{\sqrt{3}}, \qquad\qquad r_{13} = \mathbf{u}_1 \cdot \mathbf{v}_3^{(0)} = -\sqrt{3},$$

$$\mathbf{v}_2^{(1)} = \mathbf{v}_2^{(0)} - r_{12}\,\mathbf{u}_1 = \begin{pmatrix} \frac{4}{3} \\ \frac{1}{3} \\ \frac{5}{3} \end{pmatrix}, \qquad\qquad \mathbf{v}_3^{(1)} = \mathbf{v}_3^{(0)} - r_{13}\,\mathbf{u}_1 = \begin{pmatrix} 3 \\ 1 \\ -2 \end{pmatrix},$$

which completes the first step. Moving on to $k = 2$, we have

$$r_{22} = \| \mathbf{v}_2^{(1)} \| = \sqrt{\frac{14}{3}}, \qquad \mathbf{u}_2 = \frac{\mathbf{v}_2^{(1)}}{r_{22}} = \begin{pmatrix} \frac{4}{\sqrt{42}} \\ \frac{1}{\sqrt{42}} \\ \frac{5}{\sqrt{42}} \end{pmatrix},$$

which is the second orthonormal basis vector, and so $s_2 = \dim V_2 = 2$. Further,

$$r_{23} = \mathbf{u}_2 \cdot \mathbf{v}_3^{(1)} = \sqrt{\frac{21}{2}}, \qquad \mathbf{v}_3^{(2)} = \mathbf{v}_3^{(1)} - r_{23}\,\mathbf{u}_2 = \begin{pmatrix} 1 \\ -\frac{3}{2} \\ -\frac{1}{2} \end{pmatrix}.$$

Setting $k = 3$, we finally produce

$$r_{33} = \|\mathbf{v}_3^{(2)}\| = \sqrt{\frac{7}{2}}, \qquad \mathbf{u}_3 = \frac{\mathbf{v}_3^{(2)}}{r_{33}} = \begin{pmatrix} \frac{2}{\sqrt{14}} \\ -\frac{3}{\sqrt{14}} \\ -\frac{1}{\sqrt{14}} \end{pmatrix},$$

which finishes the process. The resulting vectors $\mathbf{u}_1, \mathbf{u}_2, \mathbf{u}_3$ form the desired orthonormal basis, and hence the original vectors $\mathbf{v}_1, \mathbf{v}_2, \mathbf{v}_3$ form a basis for $V_3 = \mathbb{R}^3$, with $s_3 = \dim V_3 = 3$. ▲

Exercises

5.1. Use the first version of the Gram–Schmidt process to determine an orthonormal basis for $\mathbb{R}^3$ with the dot product starting with the following sets of vectors:

$$(a)\,\heartsuit\ \begin{pmatrix} 1 \\ 0 \\ 1 \end{pmatrix}, \begin{pmatrix} 1 \\ 1 \\ 1 \end{pmatrix}, \begin{pmatrix} -1 \\ 2 \\ 1 \end{pmatrix}; \quad (b)\,\heartsuit\ \begin{pmatrix} 1 \\ 1 \\ 0 \end{pmatrix}, \begin{pmatrix} 0 \\ 1 \\ -1 \end{pmatrix}, \begin{pmatrix} 1 \\ 0 \\ -1 \end{pmatrix}; \quad (c)\,\diamondsuit\ \begin{pmatrix} 1 \\ 2 \\ 3 \end{pmatrix}, \begin{pmatrix} 4 \\ 5 \\ 0 \end{pmatrix}, \begin{pmatrix} 2 \\ 3 \\ -1 \end{pmatrix}.$$

5.2. Apply the Gram–Schmidt process to the following sets of vectors using the dot product on $\mathbb{R}^4$. Which produce an orthonormal basis?

$$(a)\,\heartsuit\ (1,0,1,0)^T,(0,1,0,-1)^T,(1,0,0,1)^T,(1,1,1,1)^T;$$
$$(b)\ (1,0,0,1)^T,(4,1,0,0)^T,(1,0,2,1)^T,(0,2,0,1)^T;$$
$$(c)\,\diamondsuit\ (1,-1,0,1)^T,(0,-1,1,2)^T,(2,-1,-1,0)^T,(2,2,-2,1)^T.$$

5.3. Redo Exercises 5.1 and 5.2 by implementing the numerically stable Gram–Schmidt process (2.50), (2.51), and verify that you end up with the same orthonormal bases.

5.4. Use the Gram–Schmidt process to construct an orthonormal basis under the dot product for the following subspaces of $\mathbb{R}^3$: $(a)\,\heartsuit$ the plane spanned by $(0,2,1)^T,(1,-2,-1)^T$; $(b)\,\diamondsuit$ the plane defined by the equation $2x - y + 3z = 0$; (c) the set of all vectors orthogonal to $(1,-1,-2)^T$.

5.5. Redo Exercises 5.1 and 5.4 using the weighted inner product
$$\langle \mathbf{v}, \mathbf{w} \rangle = 3v_1 w_1 + 2v_2 w_2 + v_3 w_3.$$

5.6.$\,\heartsuit$ Using the dot product on $\mathbb{R}^3$, find the orthogonal projection of the vector $(1,3,-1)^T$ onto the plane spanned by $(-1,2,1)^T,(2,1,-3)^T$ by first using the Gram–Schmidt process to construct an orthonormal basis.

5.7. (a) Show that one can alternatively compute $r_{s_k,k} = \sqrt{\|\mathbf{v}_k\|^2 - r_{1,k}^2 - \cdots - r_{s_k-1,k}^2}$ in the Gram–Schmidt formula (2.46). (b) Explain why this formula, while valid when using exact arithmetic, can potentially cause numerical difficulties.

2.6   Orthogonal Subspaces and Complements

We now extend the notion of orthogonality from individual elements to subspaces. We begin by studying the set containing all vectors that are orthogonal to a given vector.

Definition 2.29. The *orthogonal complement* of a vector $\mathbf{v} \in \mathbb{R}^n$ is the subspace

$$\mathbf{v}^\perp = \{\, \mathbf{q} \in \mathbb{R}^n \mid \langle \mathbf{v}, \mathbf{q} \rangle = 0 \,\}. \tag{2.52}$$

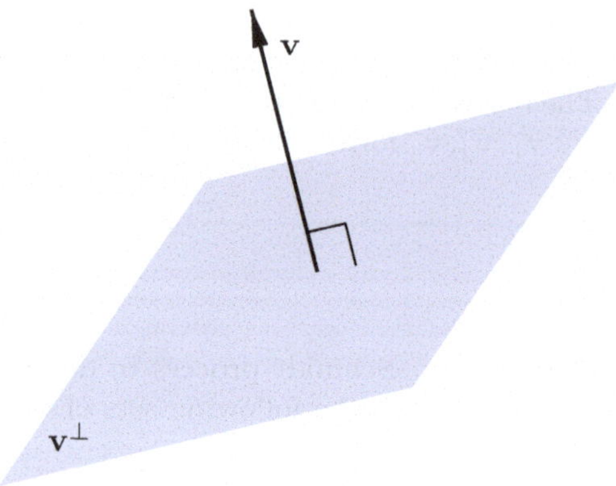

Figure 2.6: Orthogonal Complement to a Vector

Given $\mathbf{x}, \mathbf{y} \in \mathbf{v}^\perp$, and $c, d \in \mathbb{R}$, then

$$\langle \mathbf{v}, (c\mathbf{x} + d\mathbf{y}) \rangle = c\langle \mathbf{v}, \mathbf{x} \rangle + d\langle \mathbf{v}, \mathbf{y} \rangle = 0,$$

and hence $c\mathbf{x} + d\mathbf{y} \in \mathbf{v}^\perp$, which proves that $\mathbf{v}^\perp \subset \mathbb{R}^n$ is indeed a subspace. In particular, $\mathbf{0}^\perp = \mathbb{R}^n$ because every vector is orthogonal to the zero vector.

Example 2.30. Let $\mathbf{v} = (2, -1, 3)^T \in \mathbb{R}^3$. Then, when using the dot product, a vector $\mathbf{q} = (x, y, z)^T$ belongs to its orthogonal complement if and only if $\mathbf{v} \cdot \mathbf{q} = 2x - y + 3z = 0$. Thus, $\mathbf{v}^\perp \subset \mathbb{R}^3$ is the plane passing through the origin with normal vector $\mathbf{v}$. On the other hand, for the weighted inner product $\langle \mathbf{v}, \mathbf{w} \rangle = 3 v_1 w_1 + 2 v_2 w_2 + v_3 w_3$, the orthogonal complement is the plane $\langle \mathbf{v}, \mathbf{q} \rangle = 6x - 2y + 3z = 0$.   ▲

Two subspaces $V, W \subset \mathbb{R}^n$ are called *orthogonal* (with respect to the given inner product) if every vector in V is orthogonal to every vector in W. The most important configuration is as follows.

Definition 2.31. The *orthogonal complement* of a subspace $V \subset \mathbb{R}^n$ is defined as the set of all vectors that are orthogonal to V:

$$V^\perp = \{\, \mathbf{x} \in \mathbb{R}^n \mid \langle \mathbf{v}, \mathbf{x} \rangle = 0 \ \text{ for all } \ \mathbf{v} \in V \,\}. \tag{2.53}$$

In particular, $\mathbf{v}^\perp$ is the orthogonal complement to the one-dimensional subspace (line) spanned by $\mathbf{v}$. One easily checks that the orthogonal complement (2.53) is also a subspace. Moreover, the only vector that belongs to both V and $V^\perp$ is the zero vector (since it must be orthogonal to itself) and hence $V \cap V^\perp = \{\mathbf{0}\}$.

Theorem 2.32. *Let $V \subset \mathbb{R}^n$ be a subspace. Then its orthogonal complement $V^\perp \subset \mathbb{R}^n$ is also a subspace, and its orthogonal complement is $V = (V^\perp)^\perp$. Moreover,*

$$\dim V + \dim V^\perp = n. \tag{2.54}$$

Every vector $\mathbf{b} \in \mathbb{R}^n$ can be uniquely decomposed into

$$\mathbf{b} = \mathbf{p} + \mathbf{q}, \tag{2.55}$$

where $\mathbf{p} \in V$ is the orthogonal projection of $\mathbf{b}$ onto V and $\mathbf{q} \in V^\perp$ is the orthogonal projection of $\mathbf{b}$ onto $V^\perp$. Moreover,

$$\|\mathbf{b}\|^2 = \|\mathbf{p}\|^2 + \|\mathbf{q}\|^2. \tag{2.56}$$

Proof. Given $\mathbf{b} \in \mathbb{R}^n$, let $\mathbf{p} \in V$ be its orthogonal projection onto V, so that $\mathbf{q} = \mathbf{b} - \mathbf{p}$ is orthogonal to V and hence $\mathbf{q} \in V^\perp$. On the other hand, $\mathbf{p} \in V$ is orthogonal to $V^\perp$, and hence $\mathbf{q}$ is the orthogonal projection of $\mathbf{b}$ onto $V^\perp$. Note that $\mathbf{b} \in (V^\perp)^\perp$ if and only if $\mathbf{q} = \mathbf{0}$ and hence $\mathbf{b} \in V$, which proves that $V = (V^\perp)^\perp$. Furthermore,

$$\|\mathbf{b}\|^2 = \|\mathbf{p} + \mathbf{q}\|^2 = \|\mathbf{p}\|^2 + 2\langle \mathbf{p}, \mathbf{q} \rangle + \|\mathbf{q}\|^2 = \|\mathbf{p}\|^2 + \|\mathbf{q}\|^2,$$

since $\mathbf{p}, \mathbf{q}$ are orthogonal, thus proving (2.56). In particular,

$$\|\mathbf{p}\|^2 \leq \|\mathbf{b}\|^2, \tag{2.57}$$

with equality if and only if $\mathbf{q} = \mathbf{0}$ and hence $\mathbf{p} = \mathbf{b} \in V$. Moreover, (2.56) implies that, given $\mathbf{p} \in V$, $\mathbf{q} \in V^\perp$, their sum $\mathbf{p} + \mathbf{q} = \mathbf{0}$ if and only if $\mathbf{p} = \mathbf{q} = \mathbf{0}$.

Finally, if $\mathbf{v}_1, \ldots, \mathbf{v}_k$ is a basis for V and $\mathbf{w}_1, \ldots, \mathbf{w}_\ell$ is a basis for $V^\perp$, then we claim they combine to form a basis for $\mathbb{R}^n$, which implies $\dim V + \dim V^\perp = k + \ell = n$. Indeed, the combined bases span $\mathbb{R}^n$ since, given $\mathbf{b} = \mathbf{p} + \mathbf{q}$, the vector $\mathbf{p} \in V$ can be written as a linear combination of $\mathbf{v}_1, \ldots, \mathbf{v}_k$, while $\mathbf{q} \in V^\perp$ can be written as a linear combination of $\mathbf{w}_1, \ldots, \mathbf{w}_\ell$, and thus $\mathbf{b} = \mathbf{p} + \mathbf{q}$ is a linear combination of $\mathbf{v}_1, \ldots, \mathbf{v}_k, \mathbf{w}_1, \ldots, \mathbf{w}_\ell$. To prove linear independence of the combined set, if

$$\mathbf{p} = \sum_{i=1}^{k} c_i \mathbf{v}_i, \qquad \mathbf{q} = \sum_{j=1}^{\ell} d_j \mathbf{w}_j \qquad \text{satisfy} \qquad \mathbf{p} + \mathbf{q} = \sum_{i=1}^{k} c_i \mathbf{v}_i + \sum_{j=1}^{\ell} d_j \mathbf{w}_j = \mathbf{0},$$

then, by the preceding remarks, $\mathbf{p} = \mathbf{q} = \mathbf{0}$, and hence, given that $\mathbf{v}_i$ and $\mathbf{w}_j$ are bases of their respective subspaces, $c_1 = \cdots = c_k = d_1 = \cdots = d_\ell = 0$, as required. Note, furthermore, that if $\mathbf{v}_1, \ldots, \mathbf{v}_k$ and $\mathbf{w}_1, \ldots, \mathbf{w}_\ell$ form orthonormal bases of V and $V^\perp$, respectively, then they combine to form an orthonormal basis of $\mathbb{R}^n$. $\blacksquare$

Remark. Observe that, according to (2.41), the distance from $\mathbf{b}$ to the subspace V equals the norm of its orthogonal projection $\mathbf{q}$ onto the orthogonal complementary subspace $V^\perp$. If $\dim V = k$, so $\dim V^\perp = n - k$, and we introduce a orthonormal basis $\mathbf{w}_1, \ldots, \mathbf{w}_{n-k}$ of $V^\perp$, then we can use formula (2.40) to compute the norm of the projection $\mathbf{q}$ and hence

$$\mathrm{dist}(\mathbf{b}, V) = \|\mathbf{q}\| = \sqrt{\sum_{j=1}^{n-k} \langle \mathbf{w}_j, \mathbf{b} \rangle^2} \tag{2.58}$$

is the distance. Vice versa, the distance from $\mathbf{b}$ to $V^\perp$ is given by $\mathrm{dist}(\mathbf{b}, V^\perp) = \|\mathbf{p}\|$. $\blacktriangle$

Example 2.33. Let $V \subset \mathbb{R}^4$ be the two-dimensional subspace spanned by the linearly independent vectors $\mathbf{v}_1 = (1, 0, 1, 0)^T$, $\mathbf{v}_2 = (0, 1, -1, 1)^T$. Under the dot product, its orthogonal complement $V^\perp$ consists of all vectors $\mathbf{x} = (x_1, x_2, x_3, x_4)^T$ that are orthogonal to both $\mathbf{v}_1$ and $\mathbf{v}_2$, and hence satisfy the two linear equations

$$\mathbf{x} \cdot \mathbf{v}_1 = x_1 + x_3 = 0, \qquad \mathbf{x} \cdot \mathbf{v}_2 = x_2 - x_3 + x_4 = 0.$$

Thus, the solution

$$\mathbf{x} = \begin{pmatrix} -x_3 \\ x_3 - x_4 \\ x_3 \\ x_4 \end{pmatrix} = x_3 \begin{pmatrix} -1 \\ 1 \\ 1 \\ 0 \end{pmatrix} + x_4 \begin{pmatrix} 0 \\ -1 \\ 0 \\ 1 \end{pmatrix}$$

belongs to the two-dimensional subspace spanned by the indicated vectors on the right hand side, which thus form a basis of $V^\perp$. Note that $\dim V = 2$ and so $\dim V^\perp = 4 - 2 = 2$ also.

To orthogonally project vectors in $\mathbb{R}^4$ onto these two subspaces, we apply the Gram–Schmidt process to determine orthonormal bases:

$$\mathbf{u}_1 = \begin{pmatrix} \frac{1}{\sqrt{2}} \\ 0 \\ \frac{1}{\sqrt{2}} \\ 0 \end{pmatrix}, \quad \mathbf{u}_2 = \begin{pmatrix} \frac{1}{\sqrt{10}} \\ \frac{\sqrt{2}}{\sqrt{5}} \\ -\frac{1}{\sqrt{10}} \\ \frac{\sqrt{2}}{\sqrt{5}} \end{pmatrix} \in V, \qquad \mathbf{w}_1 = \begin{pmatrix} -\frac{1}{\sqrt{3}} \\ \frac{1}{\sqrt{3}} \\ \frac{1}{\sqrt{3}} \\ 0 \end{pmatrix}, \quad \mathbf{w}_2 = \begin{pmatrix} \frac{1}{\sqrt{15}} \\ -\frac{2}{\sqrt{15}} \\ -\frac{1}{\sqrt{15}} \\ \frac{\sqrt{3}}{\sqrt{5}} \end{pmatrix} \in V^\perp.$$

Thus, the orthogonal projections of, say, $\mathbf{b} = (1, 1, 1, 1)^T$ onto the two subspaces are the vectors

$$\mathbf{p} = (\mathbf{u}_1 \cdot \mathbf{b})\mathbf{u}_1 + (\mathbf{u}_2 \cdot \mathbf{b})\mathbf{u}_2 = \begin{pmatrix} \frac{7}{5} \\ \frac{4}{5} \\ \frac{3}{5} \\ \frac{4}{5} \end{pmatrix} \in V, \qquad \mathbf{q} = (\mathbf{w}_1 \cdot \mathbf{b})\mathbf{w}_1 + (\mathbf{w}_2 \cdot \mathbf{b})\mathbf{w}_2 = \begin{pmatrix} \frac{7}{5} \\ \frac{4}{5} \\ \frac{3}{5} \\ \frac{4}{5} \end{pmatrix} \in V^\perp,$$

noting that $\mathbf{p} + \mathbf{q} = \mathbf{b}$ and, furthermore, $\mathbf{p} \cdot \mathbf{q} = 0$. We conclude that the distances from the vector $\mathbf{b}$ to these two subspaces are

$$\operatorname{dist}(\mathbf{b}, V) = \|\mathbf{q}\| = 3\sqrt{\frac{2}{5}}, \qquad \operatorname{dist}(\mathbf{b}, V^\perp) = \|\mathbf{p}\| = \sqrt{\frac{2}{5}},$$

which, moreover, satisfy the Pythagorean formula $\|\mathbf{p}\|^2 + \|\mathbf{q}\|^2 = \|\mathbf{b}\|^2 = 4$.   ▲

Exercises

6.1. Using the dot product on $\mathbb{R}^3$, find the orthogonal complement $V^\perp$ of the subspaces $V \subset \mathbb{R}^3$ spanned by the indicated vectors. What is the dimension of $V^\perp$ in each case?

$(a)\,\heartsuit\ \begin{pmatrix} 3 \\ -1 \\ 1 \end{pmatrix}, \quad (b)\,\heartsuit\ \begin{pmatrix} 1 \\ 2 \\ 3 \end{pmatrix}, \begin{pmatrix} 2 \\ 0 \\ 1 \end{pmatrix}, \quad (c)\ \begin{pmatrix} 1 \\ 2 \\ 3 \end{pmatrix}, \begin{pmatrix} 2 \\ 4 \\ 6 \end{pmatrix}, \quad (d)\,\diamondsuit\ \begin{pmatrix} 1 \\ 1 \\ 0 \end{pmatrix}, \begin{pmatrix} 1 \\ 0 \\ 1 \end{pmatrix}, \begin{pmatrix} 0 \\ 1 \\ 1 \end{pmatrix}.$

6.2. Use the dot product to decompose each of the following vectors with respect to the indicated subspace as $\mathbf{b} = \mathbf{p} + \mathbf{q}$, where $\mathbf{p} \in V$, $\mathbf{q} \in V^{\perp}$.

(a) $\heartsuit$ $\mathbf{b} = \begin{pmatrix} 0 \\ 1 \end{pmatrix}$, $V = \left\{ \begin{pmatrix} x \\ y \end{pmatrix} \,\middle|\, 3x + 2y = 0 \right\}$; (b) $\diamond$ $\mathbf{b} = \begin{pmatrix} 1 \\ 2 \end{pmatrix}$, $V = \text{span}\left\{ \begin{pmatrix} -3 \\ 1 \end{pmatrix} \right\}$;

(c) $\mathbf{b} = \begin{pmatrix} 1 \\ 0 \\ 0 \end{pmatrix}$, $V = \{ x - y + z = 0 \}$; (d) $\mathbf{b} = \begin{pmatrix} 1 \\ 2 \\ 1 \end{pmatrix}$, $V = \text{span}\left\{ \begin{pmatrix} 2 \\ 2 \\ 1 \end{pmatrix}, \begin{pmatrix} 1 \\ 0 \\ 1 \end{pmatrix} \right\}$.

6.3. Find an orthonormal basis under the dot product for the orthogonal complement of the following subspaces of $\mathbb{R}^3$: (a) $\heartsuit$ the plane $3x + 4y - 5z = 0$; (b) the plane spanned by $(1, -1, 3)^T$, $(2, 0, -1)^T$; (c) $\diamond$ the line in the direction $(-2, 1, 3)^T$.

6.4. Redo Exercises 6.1 and 6.3 using the weighted inner product
$$\langle \mathbf{v}, \mathbf{w} \rangle = v_1 w_1 + 2 v_2 w_2 + 3 v_3 w_3.$$

6.5. $\heartsuit$ Prove that if $V_1 \subset V_2 \subset \mathbb{R}^n$ are subspaces, then $V_1^{\perp} \supset V_2^{\perp}$.

6.6. Let $V \subset \mathbb{R}^n$ have dimension $1 \leq k < n$. Suppose $\mathbf{u}_1, \ldots, \mathbf{u}_k$ is an orthonormal basis for V and $\mathbf{u}_{k+1}, \ldots, \mathbf{u}_n$ is an orthonormal basis for $V^{\perp}$. (a) Prove that the combination $\mathbf{u}_1, \ldots, \mathbf{u}_n$ forms an orthonormal basis of $\mathbb{R}^n$. (b) Show that if $\mathbf{v} = c_1 \mathbf{u}_1 + \cdots + c_n \mathbf{u}_n$ is any vector in $\mathbb{R}^n$, then its orthogonal decomposition is given by $\mathbf{v} = \mathbf{p} + \mathbf{q}$, where $\mathbf{p} = c_1 \mathbf{u}_1 + \cdots + c_k \mathbf{u}_k \in V$ and $\mathbf{q} = c_{k+1} \mathbf{u}_{k+1} + \cdots + c_n \mathbf{u}_n \in V^{\perp}$.

2.7 Norms

Not every norm that is useful for applications arises from an inner product. To define a general norm, we will extract those properties that do not directly rely on the inner product structure.

Definition 2.34. A *norm* on $\mathbb{R}^n$ assigns a nonnegative real number $\| \mathbf{v} \|$ to each vector $\mathbf{v} \in \mathbb{R}^n$, subject to the following axioms, valid for every $\mathbf{v}, \mathbf{w} \in \mathbb{R}^n$ and $c \in \mathbb{R}$:

(*i*) *Positivity*: $\quad \| \mathbf{v} \| \geq 0$, with $\| \mathbf{v} \| = 0$ if and only if $\mathbf{v} = \mathbf{0}$.

(*ii*) *Homogeneity*: $\quad \| c \mathbf{v} \| = | c | \, \| \mathbf{v} \|$.

(*iii*) *Triangle inequality*: $\quad \| \mathbf{v} + \mathbf{w} \| \leq \| \mathbf{v} \| + \| \mathbf{w} \|$.

Every inner product gives rise to a norm satisfying the preceding properties. Indeed, positivity of the norm is one of the inner product axioms. The homogeneity property was proved in (2.7), while the triangle inequality for an inner product norm was established in Theorem 2.10.

2.7.1 Basic Examples

Let us introduce the most important examples of norms that do not come from inner products. The 1 *norm* of a vector $\mathbf{v} = (v_1, v_2, \ldots, v_n)^T \in \mathbb{R}^n$ is defined as the sum of the absolute values of its entries:
$$\| \mathbf{v} \|_1 = | v_1 | + | v_2 | + \cdots + | v_n |. \tag{2.59}$$
This is sometimes referred to as the *Manhattan* or *city block* or *taxicab norm*, since it represents the (minimal) distance traveled by a car on city streets arranged in a rectangular grid, with

travel restricted to east/west/north/south. The *max* or ∞ *norm* of a vector is equal to its maximal entry (in absolute value):

$$\| \mathbf{v} \|_\infty = \max \left\{ \, |v_1|, |v_2|, \ \ldots \ , |v_n| \, \right\}. \tag{2.60}$$

Verification of the positivity and homogeneity properties for these two norms is straightforward; the triangle inequality is a direct consequence of the elementary inequality

$$|a + b| \le |a| + |b|, \qquad a, b \in \mathbb{R}, \tag{2.61}$$

for absolute values.

The Euclidean norm, 1 norm, and ∞ norm on $\mathbb{R}^n$ are just three instances of the general *p norm*

$$\| \mathbf{v} \|_p \ = \ \sqrt[p]{\, |v_1|^p + |v_2|^p + \ \cdots \ + |v_n|^p \,} . \tag{2.62}$$

This quantity defines a norm for all $1 \le p < \infty$, and the ∞ norm is a limiting case of (2.62) as $p \to \infty$. Note that the Euclidean norm (2.2) is the *2 norm*, and is often designated as such; it is the only p norm which comes from an inner product. The positivity and homogeneity properties of the p norm are not hard to establish. However, when $p \ne 1, 2, \infty$, the triangle inequality is not trivial; in detail, it reads

$$\sqrt[p]{\sum_{i=1}^{n} |v_i + w_i|^p} \ \le \ \sqrt[p]{\sum_{i=1}^{n} |v_i|^p} \ + \ \sqrt[p]{\sum_{i=1}^{n} |w_i|^p} \, , \tag{2.63}$$

and is known as *Minkowski's inequality*, named after the early twentieth century Lithuanian-German mathematician Hermann Minkowski, whose proof follows Theorem 6.46. .

2.7.2 Spheres and Balls

According to Lemma 2.6, which applies as stated to any norm, if $\mathbf{v} \ne \mathbf{0}$, then $\mathbf{u} = \mathbf{v}/\| \mathbf{v} \|$ is a *unit vector*, $\| \mathbf{u} \| = 1$, pointing in the same direction as $\mathbf{v}$. The *unit sphere* for a given norm is defined as the set of all unit vectors

$$S_1 = \left\{ \, \| \mathbf{u} \| = 1 \, \right\} \subset \mathbb{R}^n, \tag{2.64}$$

while the *unit ball*

$$B_1 = \left\{ \, \| \mathbf{v} \| \le 1 \, \right\} \subset \mathbb{R}^n \tag{2.65}$$

consists of all vectors of norm less than or equal to 1, and has the unit sphere as its boundary. Note that $\mathbf{0} \in B_1$, but $\mathbf{0} \notin S_1$. More generally, the sphere and ball of size (or "radius") $r \ge 0$ are defined as

$$S_r = \left\{ \, \| \mathbf{u} \| = r \, \right\}, \qquad B_r = \left\{ \, \| \mathbf{v} \| \le r \, \right\}, \tag{2.66}$$

and are obtained by scaling the unit sphere and ball by the factor r. Note that $B_r \subset B_R$ whenever $r \le R$. In particular, $S_0 = B_0 = \{ \mathbf{0} \}$.

The unit sphere for the Euclidean norm on $\mathbb{R}^n$ is the usual round sphere with unit radius:

$$S_1^{(2)} = \left\{ \, \| \mathbf{x} \|^2 = x_1^2 + \ \cdots \ + x_n^2 = 1 \, \right\}. \tag{2.67}$$

In two dimensions, the unit sphere is the circle of radius 1 and the unit ball is the disk of radius 1.

The unit sphere for the ∞ norm is the surface of a unit cube (or square in two dimensions):

$$S_1^{(\infty)} = \left\{ \mathbf{x} \in \mathbb{R}^n \ \middle| \ \begin{array}{l} |x_i| \leq 1, \ i = 1,\ldots,n, \quad \text{and either} \\ x_1 = \pm 1 \ \text{ or } \ \ldots \ \text{ or } \ x_n = \pm 1 \end{array} \right\}. \tag{2.68}$$

For the 1 norm,

$$S_1^{(1)} = \{ \mathbf{x} \in \mathbb{R}^n \mid |x_1| + \cdots + |x_n| = 1 \} \tag{2.69}$$

is the unit diamond in two dimensions, the unit octahedron in three dimensions, and the unit cross polytope in general. See Figure 2.7 for the two-dimensional pictures. In one dimension, these all coincide with the interval $[-1,1]$.

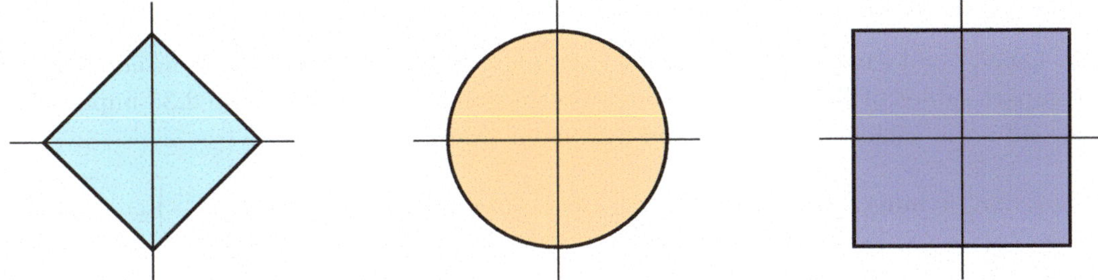

Figure 2.7: Unit Balls and Spheres for 1, 2, and ∞ Norms in $\mathbb{R}^2$

It is not hard to prove, using the triangle inequality, that a norm defines a continuous real-valued function $F(\mathbf{x}) = \|\mathbf{x}\|$ on $\mathbb{R}^n$. This result is used to prove that the unit sphere and unit ball for any norm on $\mathbb{R}^n$ are *compact*, meaning closed and bounded, subsets, cf. [1, 202].

We will often make use of the following fundamental result, which can also be found in the preceding references.

Theorem 2.35. *Let $S \subset \mathbb{R}^n$ be a compact subset. Let $F \colon S \to \mathbb{R}$ be a continuous function. Then F is bounded and, moreover, achieves its maximum and minimum values on S. In other words, there exists at least one $\mathbf{x}_{min}, \mathbf{x}_{max} \in S$ such that*

$$F(\mathbf{x}_{min}) = m = \min \{ F(\mathbf{x}) \mid \mathbf{x} \in S \}, \qquad F(\mathbf{x}_{max}) = M = \max \{ F(\mathbf{x}) \mid \mathbf{x} \in S \}, \tag{2.70}$$

and hence $m \leq F(\mathbf{x}) \leq M$ for all $\mathbf{x} \in S$.

In particular, any continuous function on the unit sphere or unit ball is bounded and achieves its minimum and maximum values. Note that, in contrast to Theorem 2.35, functions defined on noncompact subsets, e.g., the entire space, or an open subset (either bounded or unbounded), need not have any maxima or minima, simple examples being the scalar functions $F(x) = x$, e^x, and $\arctan x$ defined on $S = \mathbb{R}$, the latter being bounded between $\pm \frac{1}{2}\pi$ but nowhere achieving these values.

2.7.3 Equivalence of Norms

While there are many different types of norms on $\mathbb{R}^n$, they are all more or less equivalent. "Equivalence" does not mean that they assume the same values, but rather that they are, in a certain sense, always relatively close to one another, and so, for many analytical purposes, may be used interchangeably. As a consequence, we may be able to simplify the analysis of a problem by choosing a suitably adapted norm.

Theorem 2.36. *Let $\|\cdot\|_a$ and $\|\cdot\|_b$ be any two norms on $\mathbb{R}^n$. Then there exist positive constants $0 < r^\star \le R^\star$ such that*

$$r^\star \|\mathbf{v}\|_a \le \|\mathbf{v}\|_b \le R^\star \|\mathbf{v}\|_a \qquad \textit{for every} \qquad \mathbf{v} \in \mathbb{R}^n. \tag{2.71}$$

Remark. If we take $\mathbf{v}$ to have $\|\mathbf{v}\|_a \le 1$, and so $\mathbf{v} \in B_1^{(a)}$, then the inequalities (2.71) tell us that $\|\mathbf{v}\|_b \le R^\star$, and hence $\mathbf{v} \in B_{R^\star}^{(b)}$. Thus, the unit ball for the a norm lies inside the ball of radius $R^\star$ for the b norm. By a similar reasoning, the ball of radius $r^\star$ for the b norm lies inside the unit ball for the a norm. More generally, each ball in one norm is contained in and also contains a ball in the other norm of a suitable radius. ▲

Proof. Let $S_1^{(a)} = \{\, \|\mathbf{u}\|_a = 1 \,\}$ denote the unit sphere of the first norm which, as noted above, is a compact subset of $\mathbb{R}^n$. Since norms are continuous functions, Theorem 2.35 implies that the second norm achieves minimum and maximum values on $S_1^{(a)}$:

$$r^\star = \min\left\{\, \|\mathbf{u}\|_b \,\middle|\, \mathbf{u} \in S_1^{(a)} \,\right\}, \qquad R^\star = \max\left\{\, \|\mathbf{u}\|_b \,\middle|\, \mathbf{u} \in S_1^{(a)} \,\right\}. \tag{2.72}$$

Moreover, since the minimum and maximum values are achieved at one or more points on $S_1^{(a)}$, we have $0 < r^\star \le R^\star < \infty$, with equality holding if and only if the norms are the identical. The minimum and maximum (2.72) will serve as the constants in the desired inequalities (2.71). Indeed, by definition,

$$r^\star \le \|\mathbf{u}\|_b \le R^\star \qquad \text{when} \qquad \|\mathbf{u}\|_a = 1, \tag{2.73}$$

which proves that (2.71) is valid for all unit vectors $\mathbf{v} = \mathbf{u} \in S_1^{(a)}$. To prove the inequalities in general, assume $\mathbf{v} \ne \mathbf{0}$. (The case $\mathbf{v} = \mathbf{0}$ is trivial.) Lemma 2.6 says that $\mathbf{u} = \mathbf{v}/\|\mathbf{v}\|_a$ is a unit vector in the first norm: $\|\mathbf{u}\|_a = 1$, and hence $\mathbf{u} \in S_1^{(a)}$. Moreover, by the homogeneity property of the norm, $\|\mathbf{u}\|_b = \|\mathbf{v}\|_b/\|\mathbf{v}\|_a$. Substituting into (2.73) and clearing denominators completes the proof of (2.71). ∎

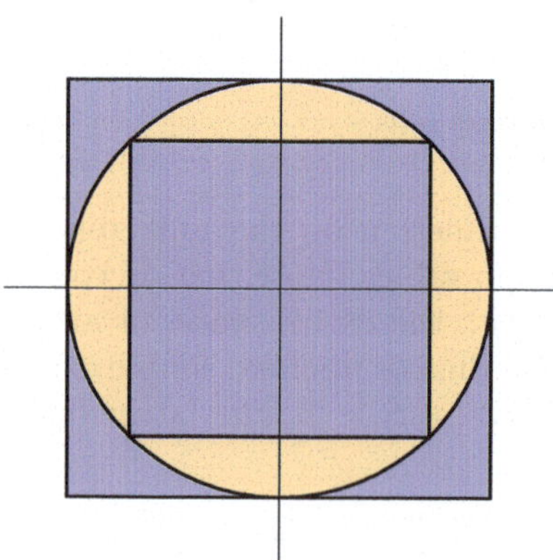

Figure 2.8: Equivalence of the ∞ and 2 Norms in $\mathbb{R}^2$

Example 2.37. Consider the Euclidean norm $\|\cdot\|_2$ and the max norm $\|\cdot\|_\infty$ on $\mathbb{R}^n$. According to (2.72), the bounding constants are found by minimizing and maximizing the max norm $\|\mathbf{u}\|_\infty = \max\{\, |u_1|, \ldots, |u_n| \,\}$ over all vectors $\mathbf{u}$ with unit Euclidean norm, $\|\mathbf{u}\|_2 = 1$, thus lying on the round unit sphere. The maximal value is achieved at the poles $\pm \mathbf{e}_k$, with $R^\star = \| \pm \mathbf{e}_k \|_\infty = 1$. The minimal value is attained at the points $\left(\pm 1/\sqrt{n}, \ldots, \pm 1/\sqrt{n} \right)^T$, whereby $r^\star = 1/\sqrt{n}$. Therefore,

$$\frac{1}{\sqrt{n}} \, \| \mathbf{v} \|_2 \leq \| \mathbf{v} \|_\infty \leq \| \mathbf{v} \|_2. \tag{2.74}$$

We can interpret these inequalities as follows. Suppose $\mathbf{v}$ is a vector lying on the unit sphere in the Euclidean norm, so $\| \mathbf{v} \|_2 = 1$. Then (2.74) tells us that its ∞ norm is bounded from above and below by $1/\sqrt{n} \leq \| \mathbf{v} \|_\infty \leq 1$. Thus, the Euclidean unit sphere sits inside the ∞ norm unit sphere (cube) and outside the ∞ norm sphere (cube) of size $1/\sqrt{n}$. Figure 2.8 plots the two-dimensional case: the unit circle is inside the unit square, and contains the square of size $1/\sqrt{2}$. Interestingly, the size of the inner cube goes to 0 as the dimension $n \to \infty$. ▲

One consequence of Theorem 2.36 is that all norms on $\mathbb{R}^n$ induce the same topology — convergence of sequences, notions of open and closed sets, and so on — and hence analysis on $\mathbb{R}^n$ is essentially independent of the choice of norm. Further details can be found, for instance, in [1, 202].

One final remark: There are many optimization problems arising in applications, including data analysis and machine learning, that involve norms. An example is the closest point problem we analyzed in Section 2.4, which makes sense for any norm. Typically, optimization that involves norms coming from inner products can be solved by linear algebra, whereas optimization involving other types of norms requires calculus and the associated analytical methods to be developed in Chapters 6 and 11. Thus, the solution to the closest point problem for an inner product norm leads to the linear algebraic method of least squares — see Section 6.2 — whereas solving this problem for other norms requires calculus-based techniques such as gradient descent.

2.7.4 Metrics and Distance

The mathematical concept of a metric space is predicated on a notion of distance between points therein.

> **Definition 2.38.** A set S is called a *metric space* if there is a real-valued *distance function* $\mathrm{dist} \colon S \times S \to \mathbb{R}^+$, satisfying the following axioms for all $x, y, z \in S$:
>
> (a) *Symmetry*: $\mathrm{dist}(x, y) = \mathrm{dist}(y, x)$;
>
> (b) *Positivity*: $\mathrm{dist}(x, y) \geq 0$ and $\mathrm{dist}(x, y) = 0$ if and only if $x = y$;
>
> (c) *Triangle inequality*: $\mathrm{dist}(x, z) \leq \mathrm{dist}(x, y) + \mathrm{dist}(y, z)$.

The distance function is sometimes referred to as a *metric*. Two points $x, y \in S$ in a metric space are considered to be close if their distance is, in some sense, small: $\mathrm{dist}(x, y) \ll 1$. The fundamental example is Euclidean space. Every norm on $\mathbb{R}^n$ defines a distance between vectors, namely

$$\mathrm{dist}(\mathbf{v}, \mathbf{w}) = \| \mathbf{v} - \mathbf{w} \|, \qquad \mathbf{v}, \mathbf{w} \in \mathbb{R}^n, \tag{2.75}$$

which satisfies the above axioms. The first two axioms are immediate, and to establish the third we apply the triangle inequality (2.30) for the norm:

$$\text{dist}(\mathbf{v}, \mathbf{z}) = \|\mathbf{v} - \mathbf{z}\| = \|(\mathbf{v} - \mathbf{w}) + (\mathbf{w} - \mathbf{z})\|$$
$$\leq \|\mathbf{v} - \mathbf{w}\| + \|\mathbf{w} - \mathbf{z}\| = \text{dist}(\mathbf{v}, \mathbf{w}) + \text{dist}(\mathbf{w}, \mathbf{z}).$$

For the standard Euclidean norm, we recover the usual notion of distance between points (i.e., endpoints of vectors that are based at the origin) in Euclidean space. Other norms produce alternative and at times quite useful distances. Any subset $S \subset \mathbb{R}^n$ of Euclidean space is clearly a metric space, under the chosen norm-based distance (2.75). Later we will encounter other metric spaces, in particular we will construct a distance function on graphs; see Section 9.5. Another example is the *Hamming distance* which simply measures how many entries of two vectors are different; it originally arose in information theory, and an important application is to coding, [156, 198]. Given a metric space S, a key question is whether it can be *isometrically embedded* into Euclidean space with some prescribed distance function (2.75): does there exist a map $\varphi \colon S \to \mathbb{R}^n$ such that $\text{dist}(\varphi(x), \varphi(y)) = \text{dist}(x, y)$ for all $x, y \in S$? We will discuss this problem in Sections 8.5 and 9.5.

There are yet more general notions of distance that arise in applications. Sometimes one or more of the defining conditions are weakened so that the notion of closeness of two points varies. For example, one may not require $\text{dist}(x, y) = 0$ only when $x = y$. Similarly, the triangle inequality may be relaxed to say that if x is close to y and y is close to z, then x is, in some sense, close to z.

In Section 6.7, we will introduce the relative entropy or Kullback–Leibler divergence between points, of importance in information theory, statistics, and finance, [134]. Another measure of closeness used in machine learning is the *cosine distance*, which is defined as

$$d_{\cos}(\mathbf{v}, \mathbf{w}) = 1 - \frac{\mathbf{v} \cdot \mathbf{w}}{\|\mathbf{v}\| \, \|\mathbf{w}\|} := 1 - \cos\theta, \qquad \mathbf{0} \neq \mathbf{v}, \mathbf{w} \in \mathbb{R}^n. \tag{2.76}$$

In view of (2.26), the cosine distance serves to measure the cosine of the angle $\theta = \sphericalangle(\mathbf{v}, \mathbf{w})$ between vectors, but does not depend on their norms. As a consequence of (2.29) and the subsequent remarks, $0 \leq d_{\cos}(\mathbf{v}, \mathbf{w}) \leq 2$. In particular, $d_{\cos}(\mathbf{v}, \mathbf{w}) = 0$ if and only if $\mathbf{v}$ and $\mathbf{w}$ point in the same direction, so $\mathbf{v} = c\mathbf{w}$ for some $c > 0$, while $d_{\cos}(\mathbf{v}, \mathbf{w}) = 2$ if and only if $\mathbf{v}$ and $\mathbf{w}$ point in opposite directions, so $\mathbf{v} = c\mathbf{w}$ for some $c < 0$. Moreover, $d_{\cos}(\mathbf{v}, \mathbf{w}) = 1$ if and only if $\mathbf{v}$ and $\mathbf{w}$ are orthogonal. According to Exercise 7.18,

$$d_{\cos}(\mathbf{v}, \mathbf{w}) = \frac{1}{2} \left\| \frac{\mathbf{v}}{\|\mathbf{v}\|} - \frac{\mathbf{w}}{\|\mathbf{w}\|} \right\|^2. \tag{2.77}$$

Thus, the cosine distance is equivalent to normalizing the vectors to have unit Euclidean norm, and then computing the Euclidean distance between the normalized vectors. Moreover, while the cosine distance does not satisfy the triangle inequality, its square root does; that is, as a consequence of Exercise 7.18,

$$\sqrt{d_{\cos}(\mathbf{v}, \mathbf{z})} \leq \sqrt{d_{\cos}(\mathbf{v}, \mathbf{w})} + \sqrt{d_{\cos}(\mathbf{w}, \mathbf{z})} \quad \text{for all} \quad \mathbf{v}, \mathbf{w}, \mathbf{z} \in \mathbb{R}^n. \tag{2.78}$$

The cosine distance is not a true distance on $\mathbb{R}^n$, since it does not satisfy the triangle inequality. Moreover, its square root is not a distance since $d_{\cos}(\mathbf{v}, \mathbf{w}) = 0$ when $\mathbf{v} \neq \mathbf{w}$ point in the same direction. Nevertheless, the cosine distance is a useful way to compare the similarity of vectors, especially in high dimensions, where it focuses on the larger values of the vector and ignores the smaller ones, which may be noise. The reader should note that the cosine distance (2.76) can be defined for *any* inner product in place of the dot product and using the induced norm.

Exercises

7.1. Compute the 1, 2, 3, and ∞ norms of the following vectors, and then verify the triangle inequality in each case.

$$(a)\,\heartsuit\ \begin{pmatrix} 1 \\ 0 \end{pmatrix},\begin{pmatrix} 0 \\ 1 \end{pmatrix}; \quad (b)\ \begin{pmatrix} 2 \\ -1 \end{pmatrix},\begin{pmatrix} 1 \\ -2 \end{pmatrix}; \quad (c)\ \begin{pmatrix} 1 \\ 0 \\ -1 \end{pmatrix},\begin{pmatrix} -1 \\ 1 \\ 0 \end{pmatrix}; \quad (d)\,\diamondsuit\ \begin{pmatrix} 1 \\ -2 \\ -1 \end{pmatrix},\begin{pmatrix} 2 \\ -1 \\ -3 \end{pmatrix}.$$

7.2. Find a unit vector in the same direction as $\mathbf{v} = (1, 2, -3)^T$ for $(a)\,\heartsuit$ the Euclidean norm, $(b)\,\diamondsuit$ the weighted norm $\|\mathbf{v}\|^2 = 2v_1^2 + v_2^2 + \frac{1}{3} v_3^2$, $(c)\,\heartsuit$ the 1 norm, (d) the ∞ norm.

7.3. Which two of the vectors $\mathbf{u} = (-2, 2, 1)^T$, $\mathbf{v} = (1, 4, 1)^T$, $\mathbf{w} = (0, 0, -1)^T$ are closest in distance for $(a)\,\heartsuit$ the Euclidean norm? $(b)\,\diamondsuit$ the 1 norm? (c) the ∞ norm?

7.4. Carefully prove that $\|(x, y)^T\| = |x| + 2\,|x - y|$ defines a norm on $\mathbb{R}^2$.

7.5. Prove that the following formulas define norms on $\mathbb{R}^2$:
$(a)\,\heartsuit\ \|\mathbf{v}\| = \sqrt{2v_1^2 + 3v_2^2}$, $\quad (b)\ \|\mathbf{v}\| = \sqrt{2v_1^2 - v_1 v_2 + 2v_2^2}$, $\quad (c)\,\heartsuit\ \|\mathbf{v}\| = 2\,|v_1| + |v_2|$,
$(d)\ \|\mathbf{v}\| = \max\{\,2\,|v_1|, |v_2|\,\}$, $\quad (e)\,\diamondsuit\ \|\mathbf{v}\| = \max\{\,|v_1 - v_2|, |v_1 + v_2|\,\}$.

7.6. Which of the following formulas define norms on $\mathbb{R}^3$? $(a)\,\heartsuit\ \|\mathbf{v}\| = \sqrt{2v_1^2 + v_2^2 + 3v_3^2}$,
$(b)\,\heartsuit\ \|\mathbf{v}\| = \sqrt{v_1^2 + 2v_1 v_2 + v_2^2 + v_3^2}$, $\quad (c)\,\diamondsuit\ \|\mathbf{v}\| = \max\{|v_1|, |v_2|, |v_3|\}$,
$(d)\ \|\mathbf{v}\| = |v_1 - v_2| + |v_2 - v_3| + |v_3 - v_1|$, $\quad (e)\ \|\mathbf{v}\| = |v_1| + \max\{|v_2|, |v_3|\}$.

7.7. $\heartsuit$ Prove that any norm on $\mathbb{R}^n$ satisfies the *reverse triangle inequality*

$$\|\mathbf{x} + \mathbf{y}\| \geq \big|\,\|\mathbf{x}\| - \|\mathbf{y}\|\,\big| \qquad \text{for all} \qquad \mathbf{x}, \mathbf{y} \in \mathbb{R}^n. \tag{2.79}$$

7.8. Let $a > 0$. Let $\|\cdot\|$ be any norm on $\mathbb{R}^n$. Prove that $\|\mathbf{v}\|_a = a\,\|\mathbf{v}\|$ also defines a norm.

7.9. $\diamondsuit$ Prove that two parallel vectors $\mathbf{v}$ and $\mathbf{w}$ have the same norm if and only if $\mathbf{v} = \pm\mathbf{w}$.

7.10. $\heartsuit$ *True or false*: If $\|\mathbf{v} + \mathbf{w}\| = \|\mathbf{v}\| + \|\mathbf{w}\|$, then $\mathbf{v}, \mathbf{w}$ are parallel vectors.

7.11. $\heartsuit$ How many unit vectors are parallel to a given vector $\mathbf{v} \neq \mathbf{0}$? (a) 0, (b) 1, (c) 2, (d) 3, $(e)\ \infty$, (f) depends on the norm. Explain your answer.

7.12. Let $\|\cdot\|$ be a norm on $\mathbb{R}^n$. Prove that there is a constant $C > 0$ such that the entries of every $\mathbf{v} = (v_1, \ldots, v_n)^T \in \mathbb{R}^n$ are all bounded, in absolute value, by $|v_i| \leq C\,\|\mathbf{v}\|$.

7.13. $\diamondsuit$ Prove that the ∞ norm on $\mathbb{R}^2$ does not come from an inner product.
$\qquad$ *Hint*: Look at Exercise 1.7.

7.14. Check the validity of the inequalities (2.74) for the particular vectors

$$(a)\,\heartsuit\ (1, -1)^T, \quad (b)\,\diamondsuit\ (1, 2, 3)^T, \quad (c)\ (1, 1, 1, 1)^T.$$

7.15. Show the equivalence of the Euclidean norm and the 1 norm on $\mathbb{R}^n$ by proving

$$\|\mathbf{v}\|_2 \leq \|\mathbf{v}\|_1 \leq \sqrt{n}\,\|\mathbf{v}\|_2.$$

Then verify that the vectors in Exercise 7.14 satisfy both inequalities.

7.16. $\diamondsuit$ Let $\mathbf{v}^{(1)}, \mathbf{v}^{(2)}, \mathbf{v}^{(3)}, \ldots \in \mathbb{R}^n$ be a series of vectors such one or more of their entries satisfy $|v_i^{(k)}| \to \infty$ as $k \to \infty$. Prove that $\|\mathbf{v}^{(k)}\| \to \infty$ as $k \to \infty$ for any norm on $\mathbb{R}^n$.

7.17. Compute the cosine distance between the pairs of vectors in Exercise 7.1.

7.18. Show that formulae (2.77) and (2.78) hold.

Chapter 3

Matrices

This chapter introduces and begins the study of our second main protagonist — matrices. As in our treatment of vectors, we will only need to consider matrices with real entries in this text. We first discuss the basic properties of and arithmetic operations on matrices — addition, scalar multiplication, multiplication of matrices, and transposes, the latter leading to the important class of symmetric matrices. One basic use of matrices is to represent and help solve linear systems of algebraic equations. Here we only discuss the basic connections; practical solution methods for the large systems arising in machine learning and data science will be developed in subsequent chapters. Each matrix possesses two fundamental subspaces, its image and its kernel, which leads to a definition of its most important numerical invariant: its rank. We also briefly discuss the matrix inverse, which we only employ in theoretical arguments, and the determinant, but only of 2×2 matrices. Finally we show how matrices serve to define linear and affine functions on Euclidean space.

Further basic properties of matrices, namely how they relate to norms and inner products, will be covered in the following chapter.

3.1 Matrices and Matrix Arithmetic

A *matrix* is a rectangular array of real numbers.[1] Thus,

$$
\begin{pmatrix} 1 & 0 & 3 \\ -2 & 4 & 1 \end{pmatrix}, \qquad
\begin{pmatrix} \pi & 0 \\ e & \frac{1}{2} \\ -1 & .83 \\ \sqrt{5} & -\frac{4}{7} \end{pmatrix}, \qquad
(.2 \quad -1.6 \quad .32), \qquad
\begin{pmatrix} 0 \\ 0 \end{pmatrix}, \qquad
\begin{pmatrix} 1 & 3 \\ -2 & 5 \end{pmatrix},
$$

are all examples of matrices. We use the notation

$$
A = \begin{pmatrix} a_{11} & a_{12} & \cdots & a_{1n} \\ a_{21} & a_{22} & \cdots & a_{2n} \\ \vdots & \vdots & \ddots & \vdots \\ a_{m1} & a_{m2} & \cdots & a_{mn} \end{pmatrix} \tag{3.1}
$$

for a general matrix of size $m \times n$ (read "m by n"), where m denotes the number of *rows* and n the number of *columns*. Thus, the preceding examples of matrices have respective sizes 2×3, 4×2, 1×3, 2×1, and 2×2. A matrix is *square* if $m = n$, i.e., it has the same number

---

[1] One can, of course, consider matrices whose entries are allowed to be arbitrary objects, including complex numbers, functions, etc.; however, in this book, only matrices with real entries need be considered.

J. Calder, P. J. Olver, *Linear Algebra, Data Science, and Machine Learning*, Springer Undergraduate Texts in Mathematics and Technology, https://doi.org/10.1007/978-3-031-93764-4_3

of rows as columns. A *column vector* is an $m \times 1$ matrix, while a *row vector* is a $1 \times n$ matrix. Thus, an $m \times n$ matrix contains m column vectors in $\mathbb{R}^n$ and n row vectors having m entries each. A 1×1 matrix is both a column and a row vector, and, as before, can be identified with its single scalar entry.

Notation: We will consistently use bold face lower case letters to denote column vectors, and ordinary capital letters to denote matrices.

The number that lies in the i-th row and the j-th column of A is called the (i, j) *entry* of A, and is denoted by a_{ij} or, sometimes with a separating comma, $a_{i,j}$, in order to avoid ambiguities. The row index always appears first and the column index second. Two matrices are equal, $A = B$, if and only if they have the same size, say $m \times n$, and *all* their entries are the same: $a_{ij} = b_{ij}$ for $i = 1, \ldots, m$ and $j = 1, \ldots, n$.

An important example is provided by a standard rectangular gray scale digital image, which is composed of a grid of pixels, each possessing a level of grayness, $0 \leq a_{ij} \leq 1$, with 0 corresponding to black and 1 corresponding to white. The collection of pixel levels can be thus be identified with the entries of a matrix that represents the image. Alternatively, we can view a matrix or gray scale image as a function

$$F \colon \mathbb{N}_m \times \mathbb{N}_n \longrightarrow \mathbb{R}, \qquad \text{where} \qquad \mathbb{N}_m = \{1, 2, \ldots, m\}, \quad \mathbb{N}_n = \{1, 2, \ldots, n\}, \qquad (3.2)$$

so that $a_{ij} = F(i, j)$. Color images require multiple numbers at each pixel, and hence correspond to functions $F \colon \mathbb{N}_m \times \mathbb{N}_n \longrightarrow \mathbb{R}^d$, where d is the number of channels, usually 3 (e.g., RGB) or 4 (e.g., CMYK). Such functions/images can be identified with a collection of d matrices, each of size $m \times n$.

Matrix arithmetic involves three basic operations: matrix addition, scalar multiplication, and matrix multiplication. One is allowed to add two matrices if and only if they are of the *same size*, and *matrix addition*, like vector addition, is performed entry by entry. For example,

$$\begin{pmatrix} 1 & 2 \\ -1 & 0 \end{pmatrix} + \begin{pmatrix} 3 & -5 \\ 2 & 1 \end{pmatrix} = \begin{pmatrix} 4 & -3 \\ 1 & 1 \end{pmatrix}.$$

Therefore, if A and B are $m \times n$ matrices, their sum $C = A + B$ is the $m \times n$ matrix whose entries are given by $c_{ij} = a_{ij} + b_{ij}$ for $i = 1, \ldots, m$ and $j = 1, \ldots, n$.

Scalar multiplication takes a scalar $c \in \mathbb{R}$ and an $m \times n$ matrix A and computes the $m \times n$ matrix $B = c\,A$ by multiplying each entry of A by c. For example,

$$3 \begin{pmatrix} 1 & 2 \\ -1 & 0 \end{pmatrix} = \begin{pmatrix} 3 & 6 \\ -3 & 0 \end{pmatrix}.$$

In general, $b_{ij} = c\,a_{ij}$ for $i = 1, \ldots, m$ and $j = 1, \ldots, n$.

Finally, we define *matrix multiplication*. First, the product of a row vector $\mathbf{v}^T$ and a column vector $\mathbf{w}$ having the *same* number of entries is the *scalar* or 1×1 matrix defined by the following rule:

$$\mathbf{v}^T \mathbf{w} = (v_1, v_2, \ldots, v_n) \begin{pmatrix} w_1 \\ w_2 \\ \vdots \\ w_n \end{pmatrix} = v_1 w_1 + v_2 w_2 + \cdots + v_n w_n = \sum_{i=1}^{n} v_i w_i. \qquad (3.3)$$

A key observation is that the matrix product of a row and column vector is the same as the dot product (2.1) between the corresponding column vectors

$$\mathbf{v} \cdot \mathbf{w} = \mathbf{v}^T \mathbf{w} = \mathbf{w}^T \mathbf{v} = \mathbf{w} \cdot \mathbf{v}. \qquad (3.4)$$

It should be emphasized that the matrix product between two column vectors $\mathbf{v}, \mathbf{w} \in \mathbb{R}^n$ is *not* defined, except in the scalar case $n = 1$ when it coincides with multiplication in $\mathbb{R}$.

More generally, if A is an $m \times n$ matrix and B is an $n \times p$ matrix, so that the number of *columns* in A equals the number of *rows* in B, then the matrix product $C = AB$ is defined as the $m \times p$ matrix whose (i, j) entry equals the product of the i-th row of A and the j-th column of B. Therefore,

$$c_{ij} = \sum_{k=1}^{n} a_{ik}\, b_{kj}, \qquad i = 1, \ldots, m, \qquad j = 1, \ldots, n. \tag{3.5}$$

Note that our restriction on the sizes of A and B guarantees that the relevant row and column vectors will have the same number of entries, and so their product is defined.

The bad news is that matrix multiplication is *not* commutative — that is, BA is not necessarily equal to AB. For example, BA may not be defined even when AB is due to their sizes. Even if both are defined, they may be different sized matrices. For example the product $c = \mathbf{v}^T \mathbf{w}$ of a row vector $\mathbf{v}^T$, a $1 \times n$ matrix, and a column vector $\mathbf{w}$, an $n \times 1$ matrix with the same number of entries, is a 1×1 matrix, i.e., a scalar, whereas the reversed product $C = \mathbf{w}\,\mathbf{v}^T$ is an $n \times n$ matrix. For instance,

$$(1 \quad 2) \begin{pmatrix} 3 \\ 0 \end{pmatrix} = 3, \qquad \begin{pmatrix} 3 \\ 0 \end{pmatrix} (1 \quad 2) = \begin{pmatrix} 3 & 6 \\ 0 & 0 \end{pmatrix}.$$

In computing the latter product, don't forget that we multiply the *rows* of the first matrix by the *columns* of the second, each of which has but a single entry. Moreover, even if the matrix products AB and BA have the same size, which requires both A and B to be square matrices, we may still have $AB \neq BA$. For example,

$$\begin{pmatrix} 1 & 2 \\ 3 & 4 \end{pmatrix} \begin{pmatrix} 0 & 1 \\ -1 & 2 \end{pmatrix} = \begin{pmatrix} -2 & 5 \\ -4 & 11 \end{pmatrix} \neq \begin{pmatrix} 3 & 4 \\ 5 & 6 \end{pmatrix} = \begin{pmatrix} 0 & 1 \\ -1 & 2 \end{pmatrix} \begin{pmatrix} 1 & 2 \\ 3 & 4 \end{pmatrix}.$$

Fortunately, matrix multiplication is associative, so

$$A(BC) = (AB)C \tag{3.6}$$

whenever A has size $m \times n$, B has size $n \times p$, and C has size $p \times q$; the result is a matrix of size $m \times q$. The proof of associativity is a tedious computation based on the definition of matrix multiplication that, for brevity, we omit. Matrix multiplication is also distributive over matrix addition:

$$A(B + C) = AB + AC, \qquad (A + B)C = AC + BC, \tag{3.7}$$

for matrices of the appropriate size. Consequently, the one difference between matrix algebra and ordinary algebra is that you need to be careful not to change the order of multiplicative factors without proper justification.

Since matrix multiplication acts by multiplying rows by columns, one can compute the columns in a matrix product AB by multiplying the matrix A and the individual columns of B. For example, the two columns of the matrix product

$$\begin{pmatrix} 1 & -1 & 2 \\ 2 & 0 & -2 \end{pmatrix} \begin{pmatrix} 3 & 4 \\ 0 & 2 \\ -1 & 1 \end{pmatrix} = \begin{pmatrix} 1 & 4 \\ 8 & 6 \end{pmatrix}$$

are obtained by multiplying the first matrix with the individual columns of the second:

$$\begin{pmatrix} 1 & -1 & 2 \\ 2 & 0 & -2 \end{pmatrix}\begin{pmatrix} 3 \\ 0 \\ -1 \end{pmatrix} = \begin{pmatrix} 1 \\ 8 \end{pmatrix}, \qquad \begin{pmatrix} 1 & -1 & 2 \\ 2 & 0 & -2 \end{pmatrix}\begin{pmatrix} 4 \\ 2 \\ 1 \end{pmatrix} = \begin{pmatrix} 4 \\ 6 \end{pmatrix}.$$

In general, if we use $\mathbf{b}_k$ to denote the k-th column of B, then

$$AB = A\left(\mathbf{b}_1\ \mathbf{b}_2\ \ldots\ \mathbf{b}_p\right) = \left(A\mathbf{b}_1\ A\mathbf{b}_2\ \ldots\ A\mathbf{b}_p\right), \tag{3.8}$$

indicating that the k-th column of their matrix product is $A\mathbf{b}_k$.

We also note that multiplying an $m \times n$ matrix A by the standard basis vector $\mathbf{e}_j \in \mathbb{R}^n$ produces the j-th column $\mathbf{v}_j = A\mathbf{e}_j$ of A. Thus, the individual entries of a matrix A can be obtained by multiplying it on the left and the right by the standard basis vectors:

$$a_{ij} = \mathbf{e}_i^T A \mathbf{e}_j = \mathbf{e}_i \cdot (A\mathbf{e}_j), \tag{3.9}$$

keeping in mind that, in this formula, $\mathbf{e}_i \in \mathbb{R}^m$ while $\mathbf{e}_j \in \mathbb{R}^n$.

Although matrix multiplication AB is defined by multiplying rows of A by columns of B, if you suitably interpret the operation, you can also compute the product by multiplying columns of A by rows of B! Suppose that A is an $m \times n$ matrix with columns $\mathbf{v}_1, \ldots, \mathbf{v}_n \in \mathbb{R}^m$. Suppose B is an $n \times p$ matrix with rows $\mathbf{w}_1^T, \ldots, \mathbf{w}_n^T$, where $\mathbf{w}_1, \ldots, \mathbf{w}_n \in \mathbb{R}^p$. Then we claim that

$$AB = \mathbf{v}_1 \mathbf{w}_1^T + \mathbf{v}_2 \mathbf{w}_2^T + \cdots + \mathbf{v}_n \mathbf{w}_n^T, \tag{3.10}$$

where each summand is a matrix of size $m \times p$. For example,

$$\begin{pmatrix} 1 & 2 \\ 3 & 4 \end{pmatrix}\begin{pmatrix} 0 & -1 \\ 2 & 3 \end{pmatrix} = \begin{pmatrix} 1 \\ 3 \end{pmatrix}(0\ \ -1) + \begin{pmatrix} 2 \\ 4 \end{pmatrix}(2\ \ 3) = \begin{pmatrix} 0 & -1 \\ 0 & -3 \end{pmatrix} + \begin{pmatrix} 4 & 6 \\ 8 & 12 \end{pmatrix} = \begin{pmatrix} 4 & 5 \\ 8 & 9 \end{pmatrix},$$

which, as you can check, agrees with the usual method for computing the matrix product. Equation (3.10) is straightforwardly justified by writing out the formulas for the individual entries of both sides.

There are two important special matrices. The first is the *zero matrix*, all of whose entries are 0. We use $O_{m \times n}$ to denote the $m \times n$ zero matrix, almost always written as just O because its size will usually be clear from the context. The zero matrix is the additive unit, so $A + O = A = O + A$ when O has the same size as A. In particular, the zero vector $\mathbf{0} \in \mathbb{R}^n$ is the same as the $n \times 1$ zero matrix: $\mathbf{0} = O_{n \times 1}$.

The role of the multiplicative unit is played by the square *identity matrix*

$$I = I_n = \begin{pmatrix} 1 & 0 & 0 & \cdots & 0 & 0 \\ 0 & 1 & 0 & \cdots & 0 & 0 \\ 0 & 0 & 1 & \cdots & 0 & 0 \\ \vdots & \vdots & \vdots & \ddots & \vdots & \vdots \\ 0 & 0 & 0 & \cdots & 1 & 0 \\ 0 & 0 & 0 & \cdots & 0 & 1 \end{pmatrix}$$

of size $n \times n$. The entries along the *main diagonal* — which runs from top left to bottom right — are equal to 1, while the *off-diagonal* entries are all 0. Note that the columns of I are the standard basis vectors (1.3) of $\mathbb{R}^n$. As you can check, if A is any $m \times n$ matrix, then $I_m A = A = A I_n$. We will generally write the preceding equation as just $I A = A = A I$, since each matrix product is well-defined for exactly one size of identity matrix.

The identity matrix is a particular example of a diagonal matrix. In general, a square matrix A is said to be *diagonal* if all its off-diagonal entries are zero: $a_{ij} = 0$ for all $i \neq j$. We will write

$$D = \operatorname{diag}(c_1, \ldots, c_n) = \operatorname{diag} \mathbf{c}, \qquad \text{where} \qquad \mathbf{c} = (c_1, \ldots, c_n)^T \in \mathbb{R}^n \tag{3.11}$$

for the $n \times n$ diagonal matrix with diagonal entries $d_{ii} = c_i$. For example, $\operatorname{diag}(1, 0, 3)$ refers to the diagonal matrix $\begin{pmatrix} 1 & 0 & 0 \\ 0 & 0 & 0 \\ 0 & 0 & 3 \end{pmatrix}$. Thus, the $n \times n$ identity matrix $\mathrm{I} = \operatorname{diag} \mathbf{1}$ is the diagonal matrix associated with the all ones vector $\mathbf{1} = (1, \ldots, 1)^T \in \mathbb{R}^n$.

A square matrix is said to be *upper triangular* if all its entries above the main diagonal vanish. Similarly, it is *lower triangular* if all its entries below the main diagonal vanish. Thus,

$$U = \begin{pmatrix} 1 & 0 & -3 \\ 0 & 4 & 1 \\ 0 & 0 & -2 \end{pmatrix} \quad \text{and} \quad L = \begin{pmatrix} -3 & 0 & 0 \\ 4 & 0 & 2 \\ 0 & 1 & 2 \end{pmatrix} \quad \text{are, respectively, upper and lower triangular}$$

3×3 matrices. A matrix is both upper and lower triangular if and only if it is diagonal.

Let us summarize the basic properties of matrix arithmetic. In the following table, A, B, C are matrices; c, d are scalars; O is a zero matrix; and I is an identity matrix. All matrices are assumed to have the proper sizes in order that the indicated operations are defined. Proofs are left as exercises for the reader.

Basic Matrix Arithmetic

Matrix Addition:	Commutativity	$A + B = B + A$
	Associativity	$(A + B) + C = A + (B + C)$
	Zero Matrix	$A + O = A = O + A$
	Additive Inverse	$A + (-A) = O, \quad -A = (-1)A$
Scalar Multiplication:	Associativity	$c(dA) = (cd)A$
	Distributivity	$c(A + B) = (cA) + (cB)$ $(c + d)A = (cA) + (dA)$
	Unit Scalar	$1A = A$
	Zero Scalar	$0A = O$
Matrix Multiplication:	Associativity	$(AB)C = A(BC)$
	Distributivity	$A(B + C) = AB + AC,$ $(A + B)C = AC + BC,$
	Compatibility	$c(AB) - (cA)B - A(cB)$
	Identity Matrix	$A\,\mathrm{I} = A = \mathrm{I}\,A$
	Zero Matrix	$A\,O = O, \quad O\,A = O$

Let $\mathcal{M}_{m \times n}$ denote the space of all real matrices of size $m \times n$. We are able to identify $\mathcal{M}_{m \times n} \simeq \mathbb{R}^{mn}$, i.e., the Euclidean space consisting of column vectors with mn entries, whose dimension is the total number of entries in an $m \times n$ matrix. One way to do this is to stack the n columns of a matrix on top of each other, i.e., write out the entries as a single column vector. For example, we identify the 2×3 matrix $A = \begin{pmatrix} 1 & 2 & 3 \\ 4 & 5 & 6 \end{pmatrix}$ with the column vector $(1,4,2,5,3,6)^T \in \mathbb{R}^6$. In particular, when subjecting two-dimensional images to data analysis, each image matrix can be identified with the corresponding vector, which is interpreted as a data point in a high dimensional Euclidean space. This identification of matrices with vectors coincides with how they are stored in computer memory. The operations of matrix addition and scalar multiplication correspond to the operations of vector addition and scalar multiplication. However, matrix multiplication has no vectorial counterpart. Moreover, the identification $\mathcal{M}_{m \times n} \simeq \mathbb{R}^{mn}$ enables us to talk about the linear independence and span of a collection of matrices, matrix bases, subspaces of matrices, their dimension, and so on.

Exercises

1.1. Let $A = \begin{pmatrix} -2 & 0 & 1 & 3 \\ -1 & 2 & 7 & -5 \\ 6 & -6 & -3 & 4 \end{pmatrix}$. (a) What is the size of A? (b) What is its $(2,3)$ entry? (c) $(3,1)$ entry? (d) 1-st row? (e) 2-nd column?

1.2. Let $A = \begin{pmatrix} 1 & -1 & 3 \\ -1 & 4 & -2 \\ 3 & 0 & 6 \end{pmatrix}$, $B = \begin{pmatrix} -6 & 0 & 3 \\ 4 & 2 & -1 \end{pmatrix}$, $C = \begin{pmatrix} 2 & 3 \\ -3 & -4 \\ 1 & 2 \end{pmatrix}$.

Compute the indicated combinations where possible. $(a) \heartsuit \, 3A - B$, $(b) \, AB$, $(c) \heartsuit \, BA$, $(d) \heartsuit \, (A + B)C$, $(e) \, A + BC$, $(f) \diamondsuit \, A + 2CB$, $(g) \, A^2 - 3A + I$, $(h) \, (B - I)(C + I)$.

1.3. Which of the following pairs of matrices commute under matrix multiplication?

$(a) \heartsuit \, \begin{pmatrix} -1 \\ 1 \end{pmatrix}, (4 \quad 3)$, $(b) \, \begin{pmatrix} 1 & 2 \\ -2 & 1 \end{pmatrix}, \begin{pmatrix} 2 & 3 \\ 5 & 0 \end{pmatrix}$, $(c) \heartsuit \, \begin{pmatrix} 1 & 2 \\ 2 & 1 \end{pmatrix}, \begin{pmatrix} 3 & -2 \\ -2 & 3 \end{pmatrix}$,

$(d) \diamondsuit \, \begin{pmatrix} 3 & -1 \\ 0 & 2 \\ 1 & 4 \end{pmatrix}, \begin{pmatrix} 4 & 2 & -2 \\ 5 & 2 & 4 \end{pmatrix}$, $(e) \, \begin{pmatrix} 3 & 0 & -1 \\ -2 & -1 & 2 \\ 2 & 0 & 0 \end{pmatrix}, \begin{pmatrix} 2 & 0 & -1 \\ 1 & 1 & -1 \\ 2 & 0 & -1 \end{pmatrix}$.

1.4. Let A be an $m \times n$ matrix. What are the permissible sizes for the zero matrices appearing in the identities $A\,\mathrm{O} = \mathrm{O}$ and $\mathrm{O}\,A = \mathrm{O}$?

1.5. $\diamondsuit$ Let A be an $m \times n$ matrix and let c be a scalar. Show that if $cA = \mathrm{O}$, then either $c = 0$ or $A = \mathrm{O}$.

1.6. $\heartsuit$ Find a nonzero matrix $A \neq \mathrm{O}$ such that $A^2 = \mathrm{O}$.

1.7. Let A have a row all of whose entries are zero. (a) Explain why the product AB also has a zero row. (b) Find an example where BA does not have a zero row.

1.8. $(a)\,\heartsuit$ Let A be an $m \times n$ matrix. Let $\mathbf{e}_j \in \mathbb{R}^n$ denote the j-th standard basis vector. Explain why the product $A\,\mathbf{e}_j$ equals the j-th column of A. $(b)\,\diamondsuit$ Similarly, let $\widehat{\mathbf{e}}_i \in \mathbb{R}^m$ be the i-th standard basis vector. Explain why the triple product $\widehat{\mathbf{e}}_i^T A\,\mathbf{e}_j = a_{ij}$ equals the (i,j) entry of the matrix A.

1.9. $\heartsuit$ Prove that $A\mathbf{v} = \mathbf{0}$ for every vector $\mathbf{v}$ (with the appropriate number of entries) if and only if $A = \mathrm{O}$ is the zero matrix.

1.10. Let A and B be $m \times n$ matrices. (a) $\diamond$ Suppose that $\mathbf{v}^T A \mathbf{w} = \mathbf{v}^T B \mathbf{w}$ for all vectors $\mathbf{v}, \mathbf{w}$. Prove that $A = B$. (b) Give an example of two matrices such that $\mathbf{v}^T A \mathbf{v} = \mathbf{v}^T B \mathbf{v}$ for all vectors $\mathbf{v}$, but $A \neq B$.

1.11. Show that if the matrices A and B commute, then they necessarily are both square and the same size.

1.12. Prove that matrix multiplication is associative: $A(BC) = (AB)C$ when defined.

1.13. Write out the following diagonal matrices: (a) $\heartsuit$ $\operatorname{diag}(1, 0, -1)$, (b) $\operatorname{diag}(2, -2, 3, -3)$.

1.14. *True or false*: (a) The sum of two diagonal matrices of the same size is a diagonal matrix. (b) The product is also diagonal.

1.15. The *trace* of a square matrix $A \in \mathcal{M}_{n \times n}$ is defined to be the sum of its diagonal entries:

$$\operatorname{tr} A = a_{11} + a_{22} + \cdots + a_{nn}. \tag{3.12}$$

Let A, B, C be $n \times n$ matrices. Prove that the trace satisfies the following identities: (a) $\heartsuit$ $\operatorname{tr}(A + B) = \operatorname{tr} A + \operatorname{tr} B$; (b) $\heartsuit$ $\operatorname{tr}(AB) = \operatorname{tr}(BA)$; (c) $\diamond$ $\operatorname{tr}(ABC) = \operatorname{tr}(CAB) = \operatorname{tr}(BCA)$. On the other hand, find an example where $\operatorname{tr}(ABC) \neq \operatorname{tr}(ACB)$. (d) Is part (b) valid if A has size $m \times n$ and B has size $n \times m$?

1.16. A *block matrix* has the form $M = \begin{pmatrix} A & B \\ C & D \end{pmatrix}$ in which A, B, C, D are matrices with respective sizes $i \times k$, $i \times l$, $j \times k$, $j \times l$. (a) What is the size of M? (b) Write out the block matrix M when $A = \begin{pmatrix} 1 \\ 3 \end{pmatrix}$, $B = \begin{pmatrix} 1 & -1 \\ 0 & 1 \end{pmatrix}$, $C = \begin{pmatrix} 1 \\ -2 \\ 1 \end{pmatrix}$, $D = \begin{pmatrix} 1 & 3 \\ 2 & 0 \\ 1 & -1 \end{pmatrix}$.

(c) Show that if $N = \begin{pmatrix} P & Q \\ R & S \end{pmatrix}$ is a block matrix whose blocks have the same size as those of M, then $M + N = \begin{pmatrix} A + P & B + Q \\ C + R & D + S \end{pmatrix}$, i.e., matrix addition can be done in blocks.

(d) Show that if $P = \begin{pmatrix} X & Y \\ Z & W \end{pmatrix}$ has blocks of a compatible size, the matrix product is

$$MP = \begin{pmatrix} AX + BZ & AY + BW \\ CX + DZ & CY + DW \end{pmatrix},$$ in analogy with multiplication of 2×2 matrices. Explain

what "compatible" means. (e) Write down a compatible block matrix P for the matrix M in part (b), and validate the block matrix product identity of part (d) for your chosen matrices.

1.17. The naïve way to "multiply" matrices is known as the *Hadamard product*, and is occasionally useful. More specifically, given two $m \times n$ matrices A, B, necessarily of the *same* size, their Hadamard product is the $m \times n$ matrix $C = A \circ B$ whose (i, j) entry is merely the product of the (i, j) entries of A and B, so $c_{ij} = a_{ij} b_{ij}$.

(a) $\heartsuit$ Prove that the Hadamard product is commutative: $A \circ B = B \circ A$.
(b) Which of the matrix arithmetic properties does the Hadamard product satisfy?
(c) $\heartsuit$ What is the multiplicative identity for the Hadamard product?
(d) Let $D = \operatorname{diag} \mathbf{d}$ be a diagonal matrix. Show that $D\mathbf{x} = \mathbf{d} \circ \mathbf{x}$.
(e) $\diamond$ Let $\mathbf{x}, \mathbf{y}, \mathbf{z} \in \mathbb{R}^n$. Prove the Hadamard product vector identities

$$(i)\ (\mathbf{x} \circ \mathbf{y}) \cdot \mathbf{z} = (\mathbf{x} \circ \mathbf{z}) \cdot \mathbf{y}, \qquad (ii)\ (\mathbf{x}\mathbf{x}^T) \circ (\mathbf{y}\mathbf{y}^T) = (\mathbf{x} \circ \mathbf{y})(\mathbf{x} \circ \mathbf{y})^T.$$

3.2   Transposes and Symmetric Matrices

Another basic operation on a matrix is to interchange its rows and columns, generalizing the transpose operation (1.2) on vectors. If A is an $m \times n$ matrix, then its *transpose*, denoted by A^T, is the $n \times m$ matrix whose (i, j) entry equals the (j, i) entry of A; thus

$$B = A^T \quad \text{means that} \quad b_{ij} = a_{ji}, \quad i = 1, \ldots, m, \quad j = 1, \ldots, n.$$

For example, if

$$A = \begin{pmatrix} 1 & 2 & 3 \\ 4 & 5 & 6 \end{pmatrix}, \quad \text{then} \quad A^T = \begin{pmatrix} 1 & 4 \\ 2 & 5 \\ 3 & 6 \end{pmatrix}.$$

Observe that the rows of A become the columns of A^T and vice versa. In particular, if A is a 1×1 matrix, i.e., a scalar, then $A^T = A$. In general, transposing twice returns you to where you started:

$$(A^T)^T = A. \tag{3.13}$$

Note that, in particular, the transpose of an upper triangular matrix is lower triangular and vice versa. The transpose operation does not alter a diagonal matrix.

Transposition is compatible with matrix addition and scalar multiplication:

$$(A + B)^T = A^T + B^T, \qquad (cA)^T = cA^T. \tag{3.14}$$

It is also compatible with matrix multiplication, but *reverses* the order:

$$(AB)^T = B^T A^T. \tag{3.15}$$

Indeed, if A has size $m \times n$ and B has size $n \times p$, so they can be multiplied, then A^T has size $n \times m$ and B^T has size $p \times n$, and so, in general, one has no choice but to multiply $B^T A^T$ in that order. Formula (3.15) is a straightforward consequence of the basic laws of matrix multiplication, and its proof is delegated to the reader as Exercise 2.5. More generally, an inductive argument shows

$$(A_1 A_2 \, \cdots \, A_{k-1} A_k)^T = A_k^T A_{k-1}^T \, \cdots \, A_2^T A_1^T, \tag{3.16}$$

when defined. An important special case is the product of a row vector $\mathbf{v}^T$ and a column vector $\mathbf{w}$ with the same number of entries. In this case,

$$\mathbf{v}^T \mathbf{w} = (\mathbf{v}^T \mathbf{w})^T = \mathbf{w}^T \mathbf{v}, \qquad \mathbf{v}, \mathbf{w} \in \mathbb{R}^n, \tag{3.17}$$

because their product is a scalar, namely the dot product $\mathbf{v} \cdot \mathbf{w}$, and so equals its own transpose.

A particularly important class of square matrices is those that are unchanged by the transpose operation.

Definition 3.1. A matrix S is called *symmetric* if it equals its own transpose: $S = S^T$.

Clearly, a symmetric matrix must be square. Thus, S is symmetric if and only if it is square and its entries satisfy $s_{ji} = s_{ij}$ for all i, j. In other words, entries lying in "mirror image" positions relative to the main diagonal must be equal. For example, the most general symmetric 3×3 matrix has the form

$$S = \begin{pmatrix} a & b & c \\ b & d & e \\ c & e & f \end{pmatrix}.$$

Note that all diagonal matrices, including the identity, are symmetric.

> **Lemma 3.2.** *An $n \times n$ matrix S is symmetric if and only if*
>
> $$\mathbf{x} \cdot (S\mathbf{y}) = (S\mathbf{x}) \cdot \mathbf{y} \quad \text{for all} \quad \mathbf{x}, \mathbf{y} \in \mathbb{R}^n. \tag{3.18}$$

Proof. First, if $S = S^T$, then, using (3.4) and (3.15),

$$\mathbf{x} \cdot (S\mathbf{y}) = \mathbf{x}^T S \mathbf{y} = \mathbf{x}^T S^T \mathbf{y} = (S\mathbf{x})^T \mathbf{y} = (S\mathbf{x}) \cdot \mathbf{y}.$$

Conversely, evaluating (3.18) on the standard basis vectors and using (3.9) produces

$$s_{ij} = \mathbf{e}_i \cdot (S\mathbf{e}_j) = (S\mathbf{e}_i) \cdot \mathbf{e}_j = s_{ji},$$

proving symmetry of S.   ∎

Exercises

2.1. Write down the transpose of the following matrices:

$$(a)\, \heartsuit \begin{pmatrix} 1 \\ 5 \end{pmatrix}, \quad (b)\, \begin{pmatrix} 1 & 2 \\ 2 & 1 \end{pmatrix}, \quad (c)\, \heartsuit \begin{pmatrix} 1 & 2 & -1 \\ 2 & 0 & 2 \end{pmatrix}, \quad (d)\, \begin{pmatrix} 1 & 2 \\ 3 & 4 \\ 5 & 6 \end{pmatrix}, \quad (e)\, \diamondsuit \begin{pmatrix} 1 & 2 & -1 \\ 0 & 3 & 2 \\ 1 & 1 & 5 \end{pmatrix}.$$

2.2. Let $A = \begin{pmatrix} 3 & -1 & -1 \\ 1 & 2 & 1 \end{pmatrix}$, $B = \begin{pmatrix} -1 & 2 \\ 2 & 0 \\ -3 & 4 \end{pmatrix}$. Compute A^T and B^T. Then compute $(AB)^T$ and $(BA)^T$ without first computing AB or BA.

2.3. $\diamondsuit$ *True or false*: Every square matrix A commutes with its transpose A^T.

2.4. $\heartsuit$ Let A be a square matrix. Prove that $A + A^T$ is symmetric.

2.5. Prove formula (3.15).

2.6. If $\mathbf{v}, \mathbf{w}$ are column vectors with the same number of entries, does
$$(a)\, \heartsuit\ \mathbf{v}^T \mathbf{w} = \mathbf{w}^T \mathbf{v}? \qquad (b)\, \diamondsuit\ \mathbf{v} \mathbf{w}^T = \mathbf{w} \mathbf{v}^T?$$

2.7. Let A be an arbitrary matrix. Prove that the matrix product $A^T A$ is well defined and symmetric. Write out a couple of examples to verify this result.

2.8. $\heartsuit$ Let $A = (\,\mathbf{v}_1 \ \ldots \ \mathbf{v}_n\,)$ be an $m \times n$ matrix with the indicated columns. Prove that the trace (see Exercise 1.15) of the symmetric matrix $A^T A$ equals the sum of the squared Euclidean norms of the columns of A, i.e., $\operatorname{tr}(A^T A) = \| \mathbf{v}_1 \|_2^2 + \cdots + \| \mathbf{v}_n \|_2^2$.

2.9. Suppose R, S are symmetric matrices. Prove that $(a)\, \heartsuit$ their sum $R + S$ is symmetric; $(b)\, \diamondsuit$ their product RS is symmetric if and only if R and S commute: $RS = SR$.

3.3 Linear Systems and Vectors

If A is an $m \times n$ matrix, and $\mathbf{x}$ is a column vector in $\mathbb{R}^n$, then the product $A\mathbf{x}$ is a column vector in $\mathbb{R}^m$. Let $\mathbf{b} \in \mathbb{R}^m$ be another vector. Performing the indicated multiplication, we find that the vector equation

$$A\mathbf{x} = \mathbf{b} \tag{3.19}$$

is equivalent to a system

$$
\begin{aligned}
a_{11}x_1 + a_{12}x_2 + \cdots + a_{1n}x_n &= b_1, \\
a_{21}x_1 + a_{22}x_2 + \cdots + a_{2n}x_n &= b_2, \\
\vdots \qquad\qquad \vdots \qquad\qquad \vdots \qquad & \\
a_{m1}x_1 + a_{m2}x_2 + \cdots + a_{mn}x_n &= b_m.
\end{aligned}
\tag{3.20}
$$

consisting of m linear algebraic equations in n unknowns, in which A, with entries a_{ij}, is the *coefficient matrix*, $\mathbf{x} = (x_1, x_2, \ldots, x_n)^T$ is a column vector containing the unknowns, while $\mathbf{b} = (b_1, b_2, \ldots, b_m)^T$ is the column vector containing the right-hand sides. This correspondence is one of the principal reasons for the definition of matrix multiplication. In particular, the *homogeneous linear system*, in which the right hand sides are all 0, can be written in vectorial form as $A\mathbf{x} = \mathbf{0}$, where $\mathbf{0}$ is the zero vector in $\mathbb{R}^m$.

Example 3.3. For the linear system

$$x + 3y + 2z - w = 0, \qquad 6y + z + 4w = 3, \qquad -x - 3z + 2w = 1,$$

the coefficient matrix, vector of unknowns, and right hand side are

$$
A = \begin{pmatrix} 1 & 3 & 2 & -1 \\ 0 & 6 & 1 & 4 \\ -1 & 0 & -3 & 2 \end{pmatrix}, \qquad
\mathbf{x} = \begin{pmatrix} x \\ y \\ z \\ w \end{pmatrix}, \qquad
\mathbf{b} = \begin{pmatrix} 0 \\ 3 \\ 1 \end{pmatrix}. \qquad\qquad \blacktriangle
$$

Let us now connect such linear algebraic systems with the basic vectorial concepts of span, linear (in)dependence, and basis. Given a set of vectors $\mathbf{v}_1, \ldots, \mathbf{v}_n \in \mathbb{R}^m$, one can form an $m \times n$ matrix $A = (\mathbf{v}_1 \ \ldots \ \mathbf{v}_n)$ with the indicated columns. To this end, we note the useful formula

$$
A\mathbf{x} = x_1 \mathbf{v}_1 + x_2 \mathbf{v}_2 + \cdots + x_n \mathbf{v}_n, \qquad \text{where} \qquad
\mathbf{x} = \begin{pmatrix} x_1 \\ x_2 \\ \vdots \\ x_n \end{pmatrix},
\tag{3.21}
$$

that expresses any linear combination of the vectors in terms of matrix multiplication. For example,

$$
\begin{pmatrix} 1 & 3 & 0 \\ -1 & 2 & 1 \\ 4 & -1 & -2 \end{pmatrix}
\begin{pmatrix} x_1 \\ x_2 \\ x_3 \end{pmatrix}
= \begin{pmatrix} x_1 + 3x_2 \\ -x_1 + 2x_2 + x_3 \\ 4x_1 - x_2 - 2x_3 \end{pmatrix}
= x_1 \begin{pmatrix} 1 \\ -1 \\ 4 \end{pmatrix}
+ x_2 \begin{pmatrix} 3 \\ 2 \\ -1 \end{pmatrix}
+ x_3 \begin{pmatrix} 0 \\ 1 \\ -2 \end{pmatrix}.
$$

Indeed, (3.21) is a special case of the alternative matrix multiplication formula (3.10) when the second matrix is a column vector. The key result is the following:

Theorem 3.4. *Let $\mathbf{v}_1, \ldots, \mathbf{v}_n \in \mathbb{R}^m$, and let $A = (\mathbf{v}_1 \ldots \mathbf{v}_n)$ be the corresponding $m \times n$ matrix.*

   (a) *The vectors are linearly dependent if and only if there is a nonzero solution $\mathbf{x} \neq \mathbf{0}$ to the homogeneous linear system $A\mathbf{x} = \mathbf{0}$.*

   (b) *The vectors are linearly independent if and only if the only solution to the homogeneous system $A\mathbf{x} = \mathbf{0}$ is the trivial one, $\mathbf{x} = \mathbf{0}$.*

   (c) *A vector $\mathbf{b}$ lies in the span of $\mathbf{v}_1, \ldots, \mathbf{v}_n$ if and only if the linear system $A\mathbf{x} = \mathbf{b}$ has a solution.*

Proof. We prove the first statement, leaving the other two as exercises for the reader. The condition that $\mathbf{v}_1, \ldots, \mathbf{v}_n$ be linearly dependent is that there exists a nonzero vector

$$\mathbf{x} = (x_1, \ldots, x_n)^T \neq \mathbf{0} \qquad \text{such that} \qquad A\mathbf{x} = x_1 \mathbf{v}_1 + \cdots + x_n \mathbf{v}_n = \mathbf{0}.$$

Therefore, linear dependence requires the existence of a nontrivial solution to the homogeneous linear system $A\mathbf{x} = \mathbf{0}$.   ∎

Example 3.5. Given the vectors $\mathbf{v}_1 = \begin{pmatrix} 1 \\ 0 \\ 1 \end{pmatrix}$, $\mathbf{v}_2 = \begin{pmatrix} 0 \\ 1 \\ -2 \end{pmatrix}$, $\mathbf{v}_3 = \begin{pmatrix} -1 \\ -2 \\ 3 \end{pmatrix}$, the correspond-

ing matrix is $A = \begin{pmatrix} 1 & 0 & -1 \\ 0 & 1 & 2 \\ 1 & -2 & 3 \end{pmatrix}$. Setting $\mathbf{x} = \begin{pmatrix} x_1 \\ x_2 \\ x_3 \end{pmatrix}$ and $\mathbf{b} = \begin{pmatrix} b_1 \\ b_2 \\ b_3 \end{pmatrix}$, the linear system

$A\mathbf{x} = \mathbf{b}$ is

$$x_1 - x_3 = b_1, \qquad x_2 - 2x_3 = b_2, \qquad x_1 - 2x_2 + 3x_3 = b_3. \tag{3.22}$$

The first two equations are readily solved for

$$x_1 = x_3 + b_1, \qquad x_2 = 2x_3 + b_2. \tag{3.23}$$

Substituting these expressions into the third equation produces

$$b_1 - 2b_2 + b_3 = 0, \tag{3.24}$$

which is a compatibility condition that needs to be imposed on the right hand side of the system in order that there be a solution. In view of part (c) of Theorem 3.4, we deduce that $\mathbf{b} \in \operatorname{span}\{\mathbf{v}_1, \mathbf{v}_2, \mathbf{v}_3\}$ if and only if it satisfies the compatibility condition (3.24); in other words, the span is the plane in $\mathbb{R}^3$ defined by the equation (3.24). Moreover, setting $b_1 = b_2 = b_3 = 0$, the solution to the homogeneous system $A\mathbf{x} = \mathbf{0}$ is $x_1 = x_3$, $x_2 = 2x_3$, where x_3 is a "free variable" that can assume any value. Thus, the homogeneous system admits nonzero solutions, implying that the vectors $\mathbf{v}_1, \mathbf{v}_2, \mathbf{v}_3$ are linearly dependent.   ▲

One of the first things one normally learns in a basic course in linear algebra, [181, 224], is how to solve general linear systems (3.20). Not every linear system has a solution; in general the right hand sides for which a solution exists are subject to certain *compatibility conditions*, which we will characterize below. On the other hand, every homogeneous linear system has at least one solution, namely when all the unknowns are equal to zero.

However, in this text, we will not cover the standard solution technique known as *Gaussian elimination*. The reason is that, even with modifications such as pivoting, [181], Gaussian elimination is unable to accurately and efficiently handle many of the large linear systems that arise in applications to data science and machine learning. Instead we will develop an alternative direct solution technique based on the so-called QR factorization of the coefficient matrix; see Section 4.7 for details. We refer the reader to [56, 105, 181] for alternative iterative methods for (approximately) solving large linear systems with suitably structured coefficient matrices. To solve any challenging linear systems that arise in the exercises, the reader is advised to either use the Python notebooks that were referenced in the Preface, or other suitable software.

Exercises

3.1. For each of the following linear systems, write down the coefficient matrix A and the vectors $\mathbf{x}$ and $\mathbf{b}$.

$(a)\,\heartsuit\quad \begin{aligned} x - y &= 7, \\ x + 2y &= 3; \end{aligned}$
$(b)\quad \begin{aligned} 6u + v &= 5, \\ 3u - 2v &= 5; \end{aligned}$
$(c)\,\diamondsuit\quad \begin{aligned} p + 3q - 3r &= 0, \\ q - r &= 1, \\ 2p - q + 3r &= 3, \\ 2p - 5r &= -1; \end{aligned}$
$(d)\,\heartsuit\quad \begin{aligned} 2u - v + 2w &= 2, \\ -u - v &= 1, \\ 3u - 2w &= 1. \end{aligned}$

3.2. Write out and solve the linear systems corresponding to the indicated matrix, vector of unknowns, and right-hand side. $(a)\,\heartsuit\ A = \begin{pmatrix} 1 & -1 \\ 2 & 3 \end{pmatrix}, \ \mathbf{x} = \begin{pmatrix} x \\ y \end{pmatrix}, \ \mathbf{b} = \begin{pmatrix} -1 \\ -3 \end{pmatrix};$

$(b)\,\diamondsuit\ A = \begin{pmatrix} 0 & -4 \\ 5 & 1 \end{pmatrix}, \ \mathbf{x} = \begin{pmatrix} c \\ d \end{pmatrix}, \ \mathbf{b} = \begin{pmatrix} 4 \\ 1 \end{pmatrix}; \quad (c)\,\heartsuit\ A = \begin{pmatrix} 1 & 0 & 1 \\ 1 & 1 & 0 \\ 0 & 1 & 1 \end{pmatrix}, \ \mathbf{x} = \begin{pmatrix} u \\ v \\ w \end{pmatrix},$

$\mathbf{b} = \begin{pmatrix} -1 \\ -1 \\ 2 \end{pmatrix}; \quad (d)\ A = \begin{pmatrix} 3 & 0 & -1 \\ -2 & -1 & 0 \\ 0 & -3 & 0 \end{pmatrix}, \ \mathbf{x} = \begin{pmatrix} x_1 \\ x_2 \\ x_3 \end{pmatrix}, \ \mathbf{b} = \begin{pmatrix} 1 \\ 0 \\ 1 \end{pmatrix}.$

3.3. Write out the linear system that determines whether the following sets of vectors are linearly independent or dependent. Then determine which of the two possibilities holds.

$(a)\,\heartsuit\ \begin{pmatrix} 1 \\ 2 \end{pmatrix}, \begin{pmatrix} 2 \\ 1 \end{pmatrix}, \quad (b)\ \begin{pmatrix} 1 \\ 3 \end{pmatrix}, \begin{pmatrix} -2 \\ -6 \end{pmatrix}, \quad (c)\,\heartsuit\ \begin{pmatrix} 2 \\ 1 \end{pmatrix}, \begin{pmatrix} -1 \\ 3 \end{pmatrix}, \begin{pmatrix} 5 \\ 2 \end{pmatrix}, \quad (d)\ \begin{pmatrix} 1 \\ 3 \\ -2 \end{pmatrix}, \begin{pmatrix} 0 \\ 2 \\ -1 \end{pmatrix},$

$(e)\,\diamondsuit\ \begin{pmatrix} 0 \\ 1 \\ 1 \end{pmatrix}, \begin{pmatrix} 1 \\ -1 \\ 0 \end{pmatrix}, \begin{pmatrix} 3 \\ -1 \\ 2 \end{pmatrix}, \quad (f)\ \begin{pmatrix} 2 \\ 1 \\ 3 \end{pmatrix}, \begin{pmatrix} 1 \\ -2 \\ 1 \end{pmatrix}, \begin{pmatrix} 2 \\ -3 \\ 0 \end{pmatrix}, \begin{pmatrix} 0 \\ -1 \\ 4 \end{pmatrix}.$

3.4. For each of the corresponding sets of vectors in Exercise 3.3, write out the linear system that determines whether the indicated vector lies in their span. Then determine whether or not this holds.

$(a)\,\heartsuit\ \begin{pmatrix} 1 \\ 0 \end{pmatrix}, \quad (b)\ \begin{pmatrix} 2 \\ 1 \end{pmatrix}, \quad (c)\ \begin{pmatrix} -1 \\ -1 \end{pmatrix}, \quad (d)\,\heartsuit\ \begin{pmatrix} 1 \\ 0 \\ 0 \end{pmatrix}, \quad (e)\,\diamondsuit\ \begin{pmatrix} 1 \\ 1 \\ 1 \end{pmatrix}, \quad (f)\ \begin{pmatrix} 2 \\ -2 \\ 4 \end{pmatrix}.$

3.4   Image, Kernel, Rank, Nullity

Let $A = (\,\mathbf{v}_1 \ \ldots \ \mathbf{v}_n\,)$ be an $m \times n$ matrix, whose columns $\mathbf{v}_1, \ldots, \mathbf{v}_n$ form a set of n vectors in $\mathbb{R}^m$. The subspace spanned by its column vectors is known as the *image*[2] of A, and denoted

$$\operatorname{img} A = \operatorname{span}\{\mathbf{v}_1, \ldots, \mathbf{v}_n\} \subset \mathbb{R}^m. \tag{3.25}$$

Alternative names appearing in the literature include *column space* and *range*.

By definition, a vector $\mathbf{b} \in \mathbb{R}^m$ belongs to $\operatorname{img} A$ if can be written as a linear combination,

$$\mathbf{b} = x_1 \mathbf{v}_1 + \ \cdots \ + x_n \mathbf{v}_n,$$

of the columns. By our basic matrix multiplication formula (3.21), the right-hand side of this equation equals the product $A\mathbf{x}$ of the matrix A with the column vector $\mathbf{x} = (\,x_1, \ldots, x_n\,)^T$, and hence $\mathbf{b} = A\mathbf{x}$ for some $\mathbf{x} \in \mathbb{R}^n$. Thus,

$$\operatorname{img} A = \{\,A\mathbf{x} \mid \mathbf{x} \in \mathbb{R}^n\,\} \subset \mathbb{R}^m. \tag{3.26}$$

We conclude that *a vector $\mathbf{b}$ lies in the image of A if and only if the linear system $A\mathbf{x} = \mathbf{b}$ has a solution.*

The dimension of the image subspace provides an important numerical quantity associated with any matrix.

Definition 3.6. The *rank* of a matrix A is the dimension of its image:

$$\operatorname{rank} A = \dim \operatorname{img} A. \tag{3.27}$$

Note that since $\operatorname{img} A \subset \mathbb{R}^n$, we have $0 \le \operatorname{rank} A \le n$. The only matrix of rank 0 is the zero matrix: $\operatorname{rank} O = 0$, with $\operatorname{img} O = \{\mathbf{0}\}$. We will develop an algorithm for computing the rank of a general matrix in Section 4.7.

Proposition 3.7. *An $m \times n$ matrix A has $\operatorname{rank} A = 1$ if and only if there are nonzero vectors $\mathbf{0} \ne \mathbf{v} \in \mathbb{R}^m$, $\mathbf{0} \ne \mathbf{w} \in \mathbb{R}^n$, such that $A = \mathbf{v}\mathbf{w}^T$. More generally, we have $\operatorname{rank} A = r > 0$ if and only if there exist linearly independent vectors $\mathbf{v}_1, \ldots, \mathbf{v}_r \in \mathbb{R}^m$, $\mathbf{w}_1, \ldots, \mathbf{w}_r \in \mathbb{R}^n$ such that A can be expressed as a sum of r rank one matrices,*

$$A = \mathbf{v}_1 \mathbf{w}_1^T + \ \cdots \ + \mathbf{v}_r \mathbf{w}_r^T = V W^T, \tag{3.28}$$

where the matrices $V = (\,\mathbf{v}_1 \ \ldots \ \mathbf{v}_r\,)^T \in \mathcal{M}_{m \times r}$, $W = (\,\mathbf{w}_1 \ \ldots \ \mathbf{w}_r\,)^T \in \mathcal{M}_{n \times r}$ with the indicated columns both have rank r. Moreover, the vectors $\mathbf{v}_1, \ldots, \mathbf{v}_r$ form a basis for $\operatorname{img} A = \operatorname{img} V$.

Proof. If $\operatorname{rank} A = \dim \operatorname{img} A = 1$, the image of A consists of scalar multiples of a single nonzero vector $\mathbf{0} \ne \mathbf{v} \in \mathbb{R}^m$. In particular, the j-th column $\mathbf{v}_j$ of A is in the image, and hence $\mathbf{v}_j = w_j \mathbf{v}$ for some $w_j \in \mathbb{R}$. Setting $\mathbf{w} = (\,w_1, \ldots, w_n\,)^T$, we deduce that $A = \mathbf{v}\mathbf{w}^T$ as claimed. Note that $\mathbf{w} \ne \mathbf{0}$, as otherwise $A = \mathbf{v}\,\mathbf{0}^T = O$ would be the zero matrix, which has rank 0.

---

[2] The term "image" comes from the interpretation of a matrix as a linear function; see Section 3.7 and [181] for details.

Now, suppose rank $A = \dim \operatorname{img} A = r$. Note first that the j-th column of (3.28) expresses the j-th column $\mathbf{a}_j$ of A as a linear combination of $\mathbf{v}_1, \ldots, \mathbf{v}_r$:

$$\mathbf{a}_j = w_{1j}\mathbf{v}_1 + \cdots + w_{rj}\mathbf{v}_r, \qquad j = 1, \ldots, n. \tag{3.29}$$

Consequently, the vectors $\mathbf{v}_1, \ldots, \mathbf{v}_r$ span $\operatorname{img} A = \operatorname{span}\{\mathbf{a}_1, \ldots, \mathbf{a}_n\}$ which, by Theorem 1.19 (with $\operatorname{img} A$ replacing $\mathbb{R}^n$) implies that they form a basis. Now, given an $m \times n$ matrix A of rank r, let $\mathbf{v}_1, \ldots, \mathbf{v}_r \in \mathbb{R}^m$ form a basis of $\operatorname{img} A$, whereby each column of A can be written as a linear combination thereof, as in (3.29), for certain coefficients w_{ij}. But this system of vector equations is just a rewritten form of the matrix equation (3.28).   ∎

A second important subspace consists of all vectors in $\mathbb{R}^n$ that are annihilated, i.e., sent to zero, when multiplied by A. It is known as the *kernel* or, alternatively, *null space* of A and denoted by

$$\ker A = \{\mathbf{z} \in \mathbb{R}^n \mid A\mathbf{z} = \mathbf{0}\} \subset \mathbb{R}^n. \tag{3.30}$$

The kernel is the set of solutions $\mathbf{z}$ to the homogeneous linear system $A\mathbf{z} = \mathbf{0}$. The proof that $\ker A$ is a subspace requires us to verify the usual closure conditions: Suppose that $\mathbf{z}, \mathbf{w} \in \ker A$, so that $A\mathbf{z} = \mathbf{0} = A\mathbf{w}$. Then, by the compatibility of scalar and matrix multiplication, $A(c\mathbf{z} + d\mathbf{w}) = cA\mathbf{z} + dA\mathbf{w} = \mathbf{0}$ for any scalars c, d, which implies that $c\mathbf{z} + d\mathbf{w} \in \ker A$. We will develop a computational algorithm for determining the kernel of a matrix in Section 4.7.

Definition 3.8. The *nullity* of a matrix A is the dimension of its kernel:

$$\operatorname{nullity} A = \dim \ker A. \tag{3.31}$$

The rank and nullity are directly related by the following important formula.

Theorem 3.9. *Let A be an $m \times n$ matrix. Then*

$$\operatorname{rank} A + \operatorname{nullity} A = n. \tag{3.32}$$

Proof. Let $r = \operatorname{rank} A = \dim \operatorname{img} A$. Let $\mathbf{v}_1, \ldots, \mathbf{v}_r \in \mathbb{R}^n$ be such that the image vectors $\mathbf{b}_1 = A\mathbf{v}_1, \ldots, \mathbf{b}_r = A\mathbf{v}_r$ form a basis for $\operatorname{img} A$. Let $\mathbf{z}_1, \ldots, \mathbf{z}_s$ be a basis for $\ker A$, so that $s = \operatorname{nullity} A$. We claim that, when combined, $\mathbf{v}_1, \ldots, \mathbf{v}_r, \mathbf{z}_1, \ldots, \mathbf{z}_s$ form a basis for $\mathbb{R}^n$. From this, the rank-nullity formula (3.32) follows immediately from the fact that every basis of $\mathbb{R}^n$ has exactly $n = r + s$ vectors.

To prove the claim, let us first show that these vectors are linearly independent. Suppose

$$c_1\mathbf{v}_1 + \cdots + c_r\mathbf{v}_r + d_1\mathbf{z}_1 + \cdots + d_s\mathbf{z}_s = \mathbf{0} \tag{3.33}$$

for some $c_i, d_j \in \mathbb{R}$. Multiplying by A produces

$$\mathbf{0} = c_1 A\mathbf{v}_1 + \cdots + c_r A\mathbf{v}_r + d_1 A\mathbf{z}_1 + \cdots + d_s A\mathbf{z}_s = c_1\mathbf{b}_1 + \cdots + c_r\mathbf{b}_r,$$

because $\mathbf{z}_j \in \ker A$ and hence $A\mathbf{z}_j = \mathbf{0}$. Linear independence of $\mathbf{b}_1, \ldots, \mathbf{b}_r$ implies that $c_1 = \cdots = c_r = 0$. Substituting this back into (3.33) produces $d_1\mathbf{z}_1 + \cdots + d_s\mathbf{z}_s = \mathbf{0}$, which, by the linear independence of $\mathbf{z}_1, \ldots, \mathbf{z}_s$, implies $d_1 = \cdots = d_s = 0$. Thus, the only linear combination that vanishes, as in (3.33), is the trivial one, which establishes the linear independence of the full set $\mathbf{v}_1, \ldots, \mathbf{v}_r, \mathbf{z}_1, \ldots, \mathbf{z}_s$.

Second, to show that they span $\mathbb{R}^n$, suppose $\mathbf{x} \in \mathbb{R}^n$. Then $A\mathbf{x} \in \operatorname{img} A$, and hence we can write

$$A\mathbf{x} = c_1 \mathbf{b}_1 + \cdots + c_r \mathbf{b}_r = A(c_1 \mathbf{v}_1 + \cdots + c_r \mathbf{v}_r)$$

for some $c_1, \ldots, c_r \in \mathbb{R}$. This in turn implies that

$$A(\mathbf{x} - c_1 \mathbf{v}_1 - \cdots - c_r \mathbf{v}_r) = \mathbf{0} \qquad \text{and hence} \qquad \mathbf{x} - c_1 \mathbf{v}_1 - \cdots - c_r \mathbf{v}_r \in \ker A,$$

from which we conclude

$$\mathbf{x} - c_1 \mathbf{v}_1 - \cdots - c_r \mathbf{v}_r = d_1 \mathbf{z}_1 + \cdots + d_s \mathbf{z}_s$$

for some $d_1, \ldots, d_s \in \mathbb{R}$. Rearranging the final equation produces

$$\mathbf{x} = c_1 \mathbf{v}_1 + \cdots + c_r \mathbf{v}_r + d_1 \mathbf{z}_1 + \cdots + d_s \mathbf{z}_s.$$

Since $\mathbf{x} \in \mathbb{R}^n$ was arbitrary, we deduce that the vectors span $\mathbb{R}^n$. $\blacksquare$

Example 3.10. Consider the 2×3 matrix

$$A = \begin{pmatrix} 1 & 0 & -1 \\ 0 & 1 & -2 \end{pmatrix}.$$

The image is spanned by its three columns, which is easily seen to be all of $\mathbb{R}^2 = \operatorname{img} A$. We deduce that $\operatorname{rank} A = \dim \operatorname{img} A = 2$.

On the other hand, the homogeneous system, $A\mathbf{x} = \mathbf{0}$ with $\mathbf{x} = (x_1, x_2, x_3)^T$ takes the form

$$x_1 - x_3 = 0, \qquad x_2 - 2x_3 = 0,$$

whose general solution is given by $x_1 = x_3$, $x_2 = 2x_3$, where x_3 is a free variable that can assume any value. Thus, the general element of the kernel has the form $\mathbf{x} = x_3 (1, 2, 1)^T$. We deduce that $\ker A \subset \mathbb{R}^3$ is a one-dimensional line, with basis $(1, 2, 1)^T$, whence nullity $A = 1$. Thus, equation (3.32) is verified: $\operatorname{rank} A + \operatorname{nullity} A = 2 + 1 = 3$. $\blacktriangle$

The most important subcase is that of a square matrix, when the associated linear system has the same number of equations as unknowns.

Definition 3.11. Let A be a square $n \times n$ matrix. Then A is said to be *nonsingular* if its rank is maximal, namely $\operatorname{rank} A = n$.

The next result is an immediate consequence of (3.32) and the preceding constructions.

Theorem 3.12. *Let A be a square $n \times n$ matrix. Then the following are equivalent:*
(a) A is nonsingular; (b) $\operatorname{rank} A = n$; (c) $\operatorname{nullity} A = 0$; (d) $\operatorname{img} A = \mathbb{R}^n$;
(e) $\ker A = \{\mathbf{0}\}$.

For example, the $n \times n$ identity matrix $I = I_n$ is nonsingular because its columns are the standard basis vectors (1.3) which span all of $\mathbb{R}^n$, and hence $\operatorname{img} I = \mathbb{R}^n$. Thus, $\operatorname{rank} I = n$, while $\ker I = \{\mathbf{0}\}$. On the other hand, the $n \times n$ zero matrix $O = O_{n \times n}$ is singular since $\operatorname{rank} O = 0$, its columns spanning the 0-dimensional subspace $\operatorname{img} O = \{\mathbf{0}\}$, while $\ker O = \mathbb{R}^n$ and so $\operatorname{nullity} O = n$.

If A, B are matrices, of respective sizes $m \times n$ and $n \times k$, so that the matrix product AB is defined, then

$$\operatorname{img}(AB) \subseteq \operatorname{img} A, \qquad \ker(AB) \supseteq \ker B. \tag{3.34}$$

Indeed, every vector $\mathbf{w} = AB\mathbf{x} \in \operatorname{img}(AB)$ satisfies $\mathbf{w} = A\mathbf{v}$ for $\mathbf{v} = B\mathbf{x}$, and hence $\mathbf{w} \in \operatorname{img} A$. Similarly, if $\mathbf{z} \in \ker B$, then $B\mathbf{z} = \mathbf{0}$, so $AB\mathbf{z} = \mathbf{0}$, and hence $\mathbf{z} \in \ker(AB)$. As a consequence of (3.34) and Proposition 1.21, we thus deduce

$$\operatorname{rank}(AB) \le \operatorname{rank} A, \qquad \operatorname{nullity}(AB) \ge \operatorname{nullity} B. \tag{3.35}$$

A fundamental result is that a matrix and its transpose have the same rank. This is remarkable, because the rank of A is the dimension of the subspace of $\mathbb{R}^m$ spanned by its columns, whereas the rank of A^T is the dimension of the subspace of $\mathbb{R}^n$ spanned by the transposes of the rows of A. The fact that these subspaces, which, if $m \ne n$, are not even in the same Euclidean space, have the same dimension is far from obvious.

Theorem 3.13. *Let A be a matrix. Then*

$$\operatorname{rank} A = \operatorname{rank} A^T. \tag{3.36}$$

We will establish formula (3.36) as a consequence of Theorem 4.24 below. Applying our earlier remark to A^T, we obtain the more precise rank inequality

$$0 \le \operatorname{rank} A \le \min\{m, n\} \tag{3.37}$$

for an $m \times n$ matrix A. It is not hard to find matrices whose rank achieves the upper bound, cf. Exercise 4.7.

Example 3.14. The transpose of the matrix A considered in Example 3.10 is

$$A^T = \begin{pmatrix} 1 & 0 \\ 0 & 1 \\ -1 & -2 \end{pmatrix}.$$

The subspace $\operatorname{img} A^T \subset \mathbb{R}^3$ is spanned by its two columns, which, since they are linearly independent, forms a two-dimensional plane. Therefore, in accordance with Theorem 3.13, $\operatorname{rank} A^T = \dim \operatorname{img} A^T = 2 = \operatorname{rank} A.$ ▲

Corollary 3.15. *If A is a nonsingular square matrix, so is A^T.*

Indeed, if A is nonsingular, then $\operatorname{rank} A = n = \operatorname{rank} A^T$, which implies that A^T is also nonsingular.

Exercises

4.1. Find a basis, if it exists, of the image and the kernel of the following matrices:

$(a)\,\heartsuit\ \begin{pmatrix} 2 & -1 & 5 \end{pmatrix}, \quad (b)\ \begin{pmatrix} 8 & -4 \\ -6 & 3 \end{pmatrix}, \quad (c)\,\heartsuit\ \begin{pmatrix} 1 & -1 & 2 \\ -2 & 2 & -4 \end{pmatrix}, \quad (d)\,\diamond\ \begin{pmatrix} 1 & 2 & 3 \\ 0 & 4 & 5 \\ 0 & 0 & 6 \end{pmatrix}.$

4.2. Prove that the average of all the entries in each row of A is 0 if and only if $\mathbf{1} \in \ker A$.

4.3. $\diamondsuit$ Prove that $\ker A \subseteq \ker A^2$. More generally, prove $\ker A \subseteq \ker(BA)$ for every compatible matrix B.

4.4. Prove that $\operatorname{img} A \supseteq \operatorname{img} A^2$. More generally, prove $\operatorname{img} A \supseteq \operatorname{img}(AB)$ for every compatible matrix B.

4.5. $\heartsuit$ *True or false*: If A is a square matrix, then $\ker A \cap \operatorname{img} A = \{\mathbf{0}\}$.

4.6. A matrix P is called *idempotent* if it satisfies $P^2 = P$. (*a*) Explain why P must be square. (*b*) Find all 2×2 idempotent matrices. (*c*) Prove that $\mathbf{w} \in \operatorname{img} P$ if and only if $P\mathbf{w} = \mathbf{w}$, and hence $\mathbf{w} \in \ker(P - \mathrm{I})$. (*d*) Show that every $\mathbf{v} \in \mathbb{R}^n$ can be uniquely written as $\mathbf{v} = \mathbf{w} + \mathbf{z}$ where $\mathbf{w} \in \operatorname{img} P$, $\mathbf{z} \in \ker P$.

4.7. $\diamondsuit$ Given $m, n \geq 1$, construct an $m \times n$ matrix A such that $\operatorname{rank} A = \min\{m, n\}$.

4.8. $\heartsuit$ *True or false*: If $\ker A = \ker B$, then $\operatorname{rank} A = \operatorname{rank} B$.

4.9. Referring to Proposition 3.7, show that $\mathbf{w}_1, \ldots, \mathbf{w}_r$ in the decomposition (3.28) form a basis for $\operatorname{img} A^T$.

4.10. (*a*) $\diamondsuit$ Let $\mathbf{v}_1, \ldots, \mathbf{v}_r \in \mathbb{R}^n$ be linearly independent vectors. Prove that the matrix
$$S = \mathbf{v}_1\,\mathbf{v}_1^T + \cdots + \mathbf{v}_r\,\mathbf{v}_r^T$$
is symmetric and has rank r.

(*b*) *True or false*: Every symmetric rank r matrix can be written in this form.

3.5 Superposition Principles for Linear Systems

The principle of superposition lies at the heart of linearity. For homogeneous systems, superposition allows one to generate new solutions by combining known solutions. For inhomogeneous systems, one form of superposition rests on combining the solution to the corresponding homogeneous system with a particular solution. Another superposition mechanism is to combine the solutions corresponding to different inhomogeneities. Superposition is the reason why linear systems are so much easier to solve, since one only needs to find relatively few solutions in order to construct the general solution.

As before, A denotes an $m \times n$ matrix, and we set $r = \operatorname{rank} A$. We consider linear systems of the form $A\mathbf{x} = \mathbf{b}$ for various right hand sides. The system is said to be compatible if it has at least one solution. For example, the homogeneous system $A\mathbf{x} = \mathbf{0}$ is always compatible since $\mathbf{x} = \mathbf{0}$ is a solution.

Let us start with the fact that the kernel of A forms a subspace, which can be re-expressed as a *superposition principle* for solutions to a homogeneous system of linear equations.

Theorem 3.16. *If $\mathbf{z}_1, \ldots, \mathbf{z}_k$ are individual solutions to the same homogeneous linear system $A\mathbf{z} = \mathbf{0}$, then so is every linear combination $c_1\mathbf{z}_1 + \cdots + c_k\mathbf{z}_k$.*

In particular, if $\mathbf{z}_1, \ldots, \mathbf{z}_{n-r}$ form a basis for $\ker A$, which, according to Theorem 3.9, has dimension $n - r$, then the general solution to the homogeneous linear system $A\mathbf{z} = \mathbf{0}$ is a linear combination or *superposition* of the individual basis solutions:

$$\mathbf{z} = c_1\mathbf{z}_1 + \cdots + c_{n-r}\mathbf{z}_{n-r} \qquad \text{for arbitrary} \qquad c_1, \ldots, c_{n-r} \in \mathbb{R}. \tag{3.38}$$

The next result characterizes the general structure of solutions to inhomogeneous linear systems of algebraic equations.

> **Theorem 3.17.** *Let A be an $m \times n$ matrix. Then the linear system $A\mathbf{x} = \mathbf{b}$ for $\mathbf{b} \in \mathbb{R}^m$ is compatible, and so has a solution $\mathbf{x}^\star \in \mathbb{R}^n$ if and only if $\mathbf{b} \in \operatorname{img} A$. Moreover, the general solution to the system is given by $\mathbf{x} = \mathbf{x}^\star + \mathbf{z}$ where $\mathbf{z} \in \ker A$ is an arbitrary element of the kernel of the coefficient matrix.*

Proof. The first part was already noted as a consequence of the definition of the image. As for the second, given that $A\mathbf{x}^\star = \mathbf{b}$, we have $A(\mathbf{x} - \mathbf{x}^\star) = \mathbf{b} - \mathbf{b} = \mathbf{0}$, which implies that $\mathbf{z} = \mathbf{x} - \mathbf{x}^\star \in \ker A$. ∎

In Theorem 3.17, the solution $\mathbf{x}^\star$ is often referred to as a *particular solution* and the theorem says that the most general solution to the linear system is obtained by adding to the particular solution any solution to the homogeneous system $A\mathbf{z} = \mathbf{0}$. In view of (3.38), we can thus write the general solution in the form

$$\mathbf{x} = \mathbf{x}^\star + c_1\mathbf{z}_1 + \cdots + c_{n-r}\mathbf{z}_{n-r}, \tag{3.39}$$

where $c_1, \ldots, c_{n-r} \in \mathbb{R}$ are arbitrary. Note that any of the vectors that appear on the right hand side of (3.39) could equally well serve as the particular solution. Thus, the solution to a linear system $A\mathbf{x} = \mathbf{b}$, when it exists, depends on $n - r = \operatorname{nullity} A$ arbitrary constants. In particular, the solution to the system is unique if and only if $\operatorname{nullity} A = 0$, or, equivalently, $\operatorname{rank} A = n$. We have therefore proved the following result characterizing the possible number of solutions to a *linear system*:

> **Theorem 3.18.** *A system $A\mathbf{x} = \mathbf{b}$ has either*
>
> (a) *exactly one solution, when $\mathbf{b} \in \operatorname{img} A$ and $\operatorname{nullity} A = 0$, or*
>
> (b) *infinitely many solutions, when $\mathbf{b} \in \operatorname{img} A$ and $\operatorname{nullity} A > 0$, or*
>
> (c) *no solutions when $\mathbf{b} \notin \operatorname{img} A$.*

Thus, a linear system can *never* have a finite number — other than 0 or 1 — of solutions. As a consequence, any linear system that admits two or more solutions automatically has infinitely many!

Example 3.19. Consider the homogeneous linear system

$$x - y + z = 1, \qquad y + 2z = 3,$$

with coefficient matrix $A = \begin{pmatrix} 1 & -1 & 1 \\ 0 & 1 & 2 \end{pmatrix}$. An evident particular solution is $x^\star = 4$, $y^\star = 3$, $z^\star = 0$. The general solution to the homogeneous system, where the right hand sides are zero, is $x = -3z$, $y = -2z$, where z is a free variable which can assume any value. Thus, the kernel of the coefficient matrix consists of all vectors

$$\mathbf{z} = \begin{pmatrix} -3z \\ -2z \\ z \end{pmatrix} = z\begin{pmatrix} -3 \\ -2 \\ 1 \end{pmatrix}, \qquad \text{so that} \qquad \mathbf{z}_1 = \begin{pmatrix} -3 \\ -2 \\ 1 \end{pmatrix}$$

forms a basis for the one-dimensional $\ker A$. The general solution to the preceding inhomogeneous system is

$$\begin{pmatrix} x \\ y \\ z \end{pmatrix} = \mathbf{x} = \mathbf{x}^\star + c\,\mathbf{z}_1 = \begin{pmatrix} 4 - 3c \\ 3 - 2c \\ c \end{pmatrix} \qquad \text{for any} \qquad c \in \mathbb{R}. \qquad \blacktriangle$$

Next, suppose we know particular solutions $\mathbf{x}_1^\star$ and $\mathbf{x}_2^\star$ to two inhomogeneous linear systems

$$A\mathbf{x} = \mathbf{b}_1, \qquad A\mathbf{x} = \mathbf{b}_2,$$

that have the *same* coefficient matrix A but different right hand sides. Consider the system

$$A\mathbf{x} = c_1\mathbf{b}_1 + c_2\mathbf{b}_2,$$

whose right-hand side is a linear combination, or *superposition*, of the previous two. Then a particular solution to the combined system is given by the *same* superposition of the previous solutions:

$$\mathbf{x}^\star = c_1\mathbf{x}_1^\star + c_2\mathbf{x}_2^\star.$$

The proof is immediate:

$$A\mathbf{x}^\star = A(c_1\mathbf{x}_1^\star + c_2\mathbf{x}_2^\star) = c_1 A\mathbf{x}_1^\star + c_2 A\mathbf{x}_2^\star = c_1\mathbf{b}_1 + c_2\mathbf{b}_2.$$

In physical applications, the inhomogeneities $\mathbf{b}_1, \mathbf{b}_2$ typically represent external forces, and the solutions $\mathbf{x}_1^\star, \mathbf{x}_2^\star$ represent the respective responses of the physical apparatus. The linear superposition principle says that if we know how the system responds to the individual forces, we immediately know its response to any combination thereof. The precise details of the system are irrelevant — all that is required is its linearity.

The preceding construction is easily extended to several inhomogeneities, and the result is the general *superposition principle* for inhomogeneous linear systems.

Theorem 3.20. *Suppose that* $\mathbf{x}_1^\star, \ldots, \mathbf{x}_k^\star$ *are particular solutions to each of the inhomogeneous linear systems*

$$A\mathbf{x} = \mathbf{b}_1, \qquad A\mathbf{x} = \mathbf{b}_2, \qquad \ldots \qquad A\mathbf{x} = \mathbf{b}_k, \tag{3.40}$$

all having the same coefficient matrix, and where $\mathbf{b}_1, \ldots, \mathbf{b}_k \in \mathrm{img}\, A$. *Then, for any choice of scalars* $c_1, \ldots, c_k$, *a particular solution to the combined system*

$$A\mathbf{x} = c_1\mathbf{b}_1 + \cdots + c_k\mathbf{b}_k \tag{3.41}$$

is the corresponding superposition

$$\mathbf{x}^\star = c_1\mathbf{x}_1^\star + \cdots + c_k\mathbf{x}_k^\star \tag{3.42}$$

of individual solutions. The general solution to (3.41) *is*

$$\mathbf{x} = \mathbf{x}^\star + \mathbf{z} = c_1\mathbf{x}_1^\star + \cdots + c_k\mathbf{x}_k^\star + \mathbf{z}, \tag{3.43}$$

where $\mathbf{z} \in \ker A$ *is the general solution to the homogeneous system* $A\mathbf{z} = \mathbf{0}$.

Example 3.21. The system

$$\begin{pmatrix} 4 & 1 \\ 1 & 4 \end{pmatrix} \begin{pmatrix} x_1 \\ x_2 \end{pmatrix} = \begin{pmatrix} f_1 \\ f_2 \end{pmatrix}$$

models the mechanical response of a pair of masses connected by a spring, and subject to external forcing represented by the right hand side. The solution $\mathbf{x} = (x_1, x_2)^T$ represents the displacements of the masses, while the entries of the right-hand side $\mathbf{f} = (f_1, f_2)^T$ are the applied forces. We can directly determine the response of the system $\mathbf{x}_1^\star = \left(\frac{4}{15}, -\frac{1}{15}\right)^T$

to a unit force $\mathbf{e}_1 = (1,0)^T$ on the first mass, and the response $\mathbf{x}_2^\star = \left(-\frac{1}{15}, \frac{4}{15}\right)^T$ to a unit force $\mathbf{e}_2 = (0,1)^T$ on the second mass. Superposition gives the response of the system to a general force, since we can write

$$\mathbf{f} = \begin{pmatrix} f_1 \\ f_2 \end{pmatrix} = f_1\,\mathbf{e}_1 + f_2\,\mathbf{e}_2 = f_1\begin{pmatrix} 1 \\ 0 \end{pmatrix} + f_2\begin{pmatrix} 0 \\ 1 \end{pmatrix},$$

and hence

$$\mathbf{x} = f_1\,\mathbf{x}_1^\star + f_2\,\mathbf{x}_2^\star = f_1\begin{pmatrix} \frac{4}{15} \\ -\frac{1}{15} \end{pmatrix} + f_2\begin{pmatrix} -\frac{1}{15} \\ \frac{4}{15} \end{pmatrix} = \begin{pmatrix} \frac{4}{15}f_1 - \frac{1}{15}f_2 \\ -\frac{1}{15}f_1 + \frac{4}{15}f_2 \end{pmatrix}. \qquad \blacktriangle$$

Generalizing Example 3.21, if we know particular solutions $\mathbf{x}_1^\star, \dots, \mathbf{x}_m^\star$ to

$$A\,\mathbf{x} = \mathbf{e}_i, \qquad \text{for each} \qquad i = 1, \dots, m, \tag{3.44}$$

where $\mathbf{e}_1, \dots, \mathbf{e}_m$ are the standard basis vectors of $\mathbb{R}^m$, then we can reconstruct a particular solution $\mathbf{x}^\star$ to the general linear system $A\,\mathbf{x} = \mathbf{b}$ by first writing

$$\mathbf{b} = b_1\,\mathbf{e}_1 + \cdots + b_m\,\mathbf{e}_m$$

as a linear combination of the basis vectors, and then using superposition to form

$$\mathbf{x}^\star = b_1\,\mathbf{x}_1^\star + \cdots + b_m\,\mathbf{x}_m^\star. \tag{3.45}$$

This idea will be developed further in the next section.

Exercises

5.1. Find the solution $\mathbf{x}_1^\star$ to the system $\begin{pmatrix} 1 & 2 \\ -3 & -4 \end{pmatrix}\begin{pmatrix} x \\ y \end{pmatrix} = \begin{pmatrix} 1 \\ 0 \end{pmatrix}$, and the solution $\mathbf{x}_2^\star$ to $\begin{pmatrix} 1 & 2 \\ -3 & -4 \end{pmatrix}\begin{pmatrix} x \\ y \end{pmatrix} = \begin{pmatrix} 0 \\ 1 \end{pmatrix}$. Then express the solution to $\begin{pmatrix} 1 & 2 \\ -3 & -4 \end{pmatrix}\begin{pmatrix} x \\ y \end{pmatrix} = \begin{pmatrix} 1 \\ 4 \end{pmatrix}$ as a linear combination of $\mathbf{x}_1^\star$ and $\mathbf{x}_2^\star$.

5.2. ♡ Let $A = \begin{pmatrix} 1 & 2 & -1 \\ 2 & 5 & -1 \\ 1 & 3 & 2 \end{pmatrix}$. Given that $\mathbf{x}_1^\star = \begin{pmatrix} 5 \\ -1 \\ 2 \end{pmatrix}$ solves $A\,\mathbf{x} = \mathbf{b}_1 = \begin{pmatrix} 1 \\ 3 \\ 6 \end{pmatrix}$ and $\mathbf{x}_2^\star = \begin{pmatrix} -11 \\ 5 \\ -1 \end{pmatrix}$ solves $A\,\mathbf{x} = \mathbf{b}_2 = \begin{pmatrix} 0 \\ 4 \\ 2 \end{pmatrix}$, find a solution to $A\,\mathbf{x} = 2\,\mathbf{b}_1 + \mathbf{b}_2 = \begin{pmatrix} 2 \\ 10 \\ 14 \end{pmatrix}$.

5.3. ◇ Applying a unit external force in the horizontal direction moves a mass 3 units to the right, while applying a unit force in the vertical direction moves it up 2 units. Assuming linearity, where will the mass move under the applied force $\mathbf{f} = (2, -3)^T$?

5.4. *True or false*: If A, B are matrices of the same size, $\mathbf{x}_1^\star$ solves $A\,\mathbf{x} = \mathbf{c}$, and $\mathbf{x}_2^\star$ solves $B\,\mathbf{x} = \mathbf{d}$, then $\mathbf{x}^\star = \mathbf{x}_1^\star + \mathbf{x}_2^\star$ solves $(A + B)\,\mathbf{x} = \mathbf{c} + \mathbf{d}$.

3.6 Matrix Inverses

Let us revisit the superposition construction we saw at the end of Section 3.5, focusing on the case when the coefficient matrix A is square and nonsingular, of size $n \times n$. Under this assumption, $\operatorname{img} A = \mathbb{R}^n$ and hence we can uniquely solve the linear system $A\mathbf{x} = \mathbf{b}$ for any right hand side $\mathbf{b} \in \mathbb{R}^n$. In particular, there are uniquely defined vectors $\mathbf{x}_1, \ldots, \mathbf{x}_n$ that satisfy the linear systems

$$A\mathbf{x}_1 = \mathbf{e}_1, \qquad \ldots \qquad A\mathbf{x}_n = \mathbf{e}_n. \tag{3.46}$$

Writing $\mathbf{b} = b_1\mathbf{e}_1 + \cdots + b_n\mathbf{e}_n$, according to the remarks at the end of the preceding section, the solution to $A\mathbf{x} = \mathbf{b}$ is given by

$$\mathbf{x} = b_1\mathbf{x}_1 + \cdots + b_n\mathbf{x}_n. \tag{3.47}$$

Let us rewrite the preceding formulas in matrix form. We assemble the solution vectors into an $n \times n$ matrix $X = (\mathbf{x}_1 \ldots \mathbf{x}_n)$, while the right hand sides in (3.46) form the identity matrix $\mathrm{I} = (\mathbf{e}_1 \ldots \mathbf{e}_n)$. Then the column-wise matrix multiplication formula (3.8) implies that the n vector equations (3.46) are equivalent to the single matrix equation

$$AX = \mathrm{I}. \tag{3.48}$$

The resulting matrix X is known as the (right) *inverse* of the matrix A, and commonly denoted by $X = A^{-1}$. Thus, using (3.21), we can write the solution formula (3.47) as $\mathbf{x} = X\mathbf{b} = A^{-1}\mathbf{b}$. We have thus proved:

Theorem 3.22. *If A is square and nonsingular, then the linear system $A\mathbf{x} = \mathbf{b}$ has a unique solution given by $\mathbf{x} = A^{-1}\mathbf{b}$.*

For this reason, the term "invertible" is often used as a synonym for "nonsingular". Although an elegant result, and of great theoretical significance, the practical value of this solution formula is rather limited because the computation of the inverse matrix is usually too difficult and time-consuming once its size n is even moderately large.

Theorem 3.23. *An $n \times n$ matrix has an inverse if and only if it is nonsingular or, equivalently, $\operatorname{rank} A = n$ or, equivalently, $\operatorname{nullity} A = 0$.*

Indeed, if A has an inverse $X = A^{-1}$ satisfying (3.48), then $\mathbf{x} = A^{-1}\mathbf{b}$ satisfies $A\mathbf{x} = AA^{-1}\mathbf{b} = \mathbf{b}$, which implies every $\mathbf{b} \in \mathbb{R}^n$ belongs to $\operatorname{img} A$, and hence $\operatorname{rank} A = n$, so A is nonsingular.

Example 3.24. Since

$$\begin{pmatrix} 1 & 2 & -1 \\ -3 & 1 & 2 \\ -2 & 2 & 1 \end{pmatrix} \begin{pmatrix} 3 & 4 & -5 \\ 1 & 1 & -1 \\ 4 & 6 & -7 \end{pmatrix} = \begin{pmatrix} 1 & 0 & 0 \\ 0 & 1 & 0 \\ 0 & 0 & 1 \end{pmatrix},$$

we conclude that when $A = \begin{pmatrix} 1 & 2 & -1 \\ -3 & 1 & 2 \\ -2 & 2 & 1 \end{pmatrix}$, then $A^{-1} = \begin{pmatrix} 3 & 4 & -5 \\ 1 & 1 & -1 \\ 4 & 6 & -7 \end{pmatrix}$. Observe that there is no obvious way to anticipate the entries of A^{-1} from the entries of A. ▲

Example 3.25. Let us compute the inverse $X = \begin{pmatrix} x & y \\ z & w \end{pmatrix}$, when it exists, of a general 2×2 matrix $A = \begin{pmatrix} a & b \\ c & d \end{pmatrix}$. The inverse condition (3.48), namely

$$A X = \begin{pmatrix} a\,x + b\,z & a\,y + b\,w \\ c\,x + d\,z & c\,y + d\,w \end{pmatrix} = \begin{pmatrix} 1 & 0 \\ 0 & 1 \end{pmatrix} = I,$$

holds if and only if x, y, z, w satisfy the linear system

$$\begin{aligned} a\,x + b\,z &= 1, & a\,y + b\,w &= 0, \\ c\,x + d\,z &= 0, & c\,y + d\,w &= 1. \end{aligned}$$

Solving by standard techniques, we find

$$x = \frac{d}{a\,d - b\,c}, \qquad y = -\,\frac{b}{a\,d - b\,c}, \qquad z = -\,\frac{c}{a\,d - b\,c}, \qquad w = \frac{a}{a\,d - b\,c},$$

provided the denominator $a\,d - b\,c \neq 0$ does not vanish. Therefore, the matrix

$$X = A^{-1} = \frac{1}{a\,d - b\,c} \begin{pmatrix} d & -b \\ -c & a \end{pmatrix} \tag{3.49}$$

forms the inverse to A.

The denominator appearing in the preceding formulas has a special name; it is called the *determinant* of the 2×2 matrix A, and denoted by

$$\det \begin{pmatrix} a & b \\ c & d \end{pmatrix} = a\,d - b\,c. \tag{3.50}$$

Thus, the 2×2 matrix A is nonsingular if and only if $\det A \neq 0$. ▲

Remark. As you may already know, there is a generalization of the notion of determinant to an arbitrary square matrix, cf. [181, 224]. There is also a quite complicated formula for the inverse of an $n \times n$ matrix involving determinants, [224]. However, other than in this simple 2×2 case, there is no valid reason to ever compute a determinant or a matrix inverse; all algorithms for large scale linear systems and matrices rely on alternative, more efficient, and more accurate, algorithms. ▲

Proposition 3.26. *If A and B are nonsingular matrices of the same size, then their product, $A B$, is also nonsingular, and*

$$(A B)^{-1} = B^{-1} A^{-1}. \tag{3.51}$$

Note that, as with transposes, the order of the factors is reversed under inversion.

Proof. By associativity,

$$(A B)(B^{-1} A^{-1}) = A B B^{-1} A^{-1} = A\,I\,A^{-1} = A A^{-1} = I. \qquad \blacksquare$$

Warning: In general, $(A + B)^{-1} \neq A^{-1} + B^{-1}$. Indeed, this equation is not even true for scalars (1×1 matrices)!

Similarly, according to Corollary 3.15, if A is nonsingular, its transpose A^T is also nonsingular, and hence we can construct an $n \times n$ matrix $Y = (A^T)^{-1}$ satisfying

$$A^T Y = I.$$

Let us take the transpose of the latter equation:

$$(A^T Y)^T = Y^T A = I, \qquad \text{hence} \qquad Y^T = Y^T I = Y^T A A^{-1} = I A^{-1} = A^{-1}. \tag{3.52}$$

Thus, we have proved that transposing a matrix and then inverting yields the same result as first inverting and then transposing.

Proposition 3.27. *If A is a nonsingular matrix, so is A^T, and its inverse is denoted by*

$$A^{-T} = (A^T)^{-1} = (A^{-1})^T. \tag{3.53}$$

Replacing $Y^T = A^{-1}$ in the first equation in (3.52) implies that

$$A^{-1} A = I = A A^{-1}. \tag{3.54}$$

In other words, for square matrices, a right inverse is also a left inverse. Equation (3.54) also shows that inverting a matrix twice brings us back to where we started.

Proposition 3.28. *If A is nonsingular, then A^{-1} is nonsingular and $(A^{-1})^{-1} = A$.*

Exercises

6.1. Verify by direct multiplication that the following matrices are inverses:

$$(a) \; \heartsuit \; \begin{pmatrix} 2 & 3 \\ -1 & -1 \end{pmatrix}, \; \begin{pmatrix} -1 & -3 \\ 1 & 2 \end{pmatrix}; \qquad (b) \; \begin{pmatrix} 2 & 1 & 1 \\ 3 & 2 & 1 \\ 2 & 1 & 2 \end{pmatrix}, \; \begin{pmatrix} 3 & -1 & -1 \\ -4 & 2 & 1 \\ -1 & 0 & 1 \end{pmatrix}.$$

6.2. Show that the inverse of $L = \begin{pmatrix} 1 & 0 & 0 \\ a & 1 & 0 \\ b & 0 & 1 \end{pmatrix}$ is $L^{-1} = \begin{pmatrix} 1 & 0 & 0 \\ -a & 1 & 0 \\ -b & 0 & 1 \end{pmatrix}$. However, the

inverse of $M = \begin{pmatrix} 1 & 0 & 0 \\ a & 1 & 0 \\ b & c & 1 \end{pmatrix}$ is *not* $\begin{pmatrix} 1 & 0 & 0 \\ -a & 1 & 0 \\ -b & -c & 1 \end{pmatrix}$. What is M^{-1}?

6.3. $\diamondsuit$ Find all real 2×2 matrices that are their own inverses: $A^{-1} = A$.

6.4. Show that if A is a nonsingular matrix, so is every power A^n.

6.5. $\heartsuit$ Prove that a diagonal matrix $D = \mathrm{diag}\,(d_1, \dots, d_n)$ is invertible if and only if all its diagonal entries are nonzero, in which case $D^{-1} = \mathrm{diag}\,(1/d_1, \dots, 1/d_n)$.

6.6. Prove that an upper triangular matrix U is nonsingular if and only if all its diagonal entries are nonzero, $u_{ii} \neq 0$, in which case U^{-1} is also upper triangular with diagonal entries $1/u_{ii}$. Does the same hold for lower triangular matrices?

6.7. ♡ (a) Prove that the inverse transpose operation (3.53) respects matrix multiplication:
$$(AB)^{-T} = A^{-T}B^{-T}. \quad \text{(b) Verify this identity for } A = \begin{pmatrix} 1 & -1 \\ 1 & 0 \end{pmatrix}, \quad B = \begin{pmatrix} 2 & 1 \\ 1 & 1 \end{pmatrix}.$$

3.7   Linear and Affine Functions

Among the multitude of functions of one or more variables, the simplest are the linear and affine functions. These basic functions must be thoroughly understood before venturing into the vast nonlinear wilderness. For example, in calculus, one often approximates a (sufficiently smooth) nonlinear function near a point by the tangent space to its graph, which is, in general, the graph of an affine function, namely its first order Taylor polynomial, [4, 158].

In this section, we will first show how every linear function on Euclidean space can be characterized by multiplication by a matrix, and thereby reinterpret a matrix as the coordinate representation of a linear function. We then define an affine function by supplementing a linear function by the addition of a fixed vector. Important examples will appear in the exercises and the following chapter.

3.7.1   Linear Functions

A function between Euclidean spaces is said to be linear if it respects the operations of vector addition and scalar multiplication.

Definition 3.29. A function $L : \mathbb{R}^n \to \mathbb{R}^m$ is called *linear* if it satisfies

$$L[\mathbf{v} + \mathbf{w}] = L[\mathbf{v}] + L[\mathbf{w}], \qquad L[c\,\mathbf{v}] = c\,L[\mathbf{v}], \qquad (3.55)$$

for all $\mathbf{v}, \mathbf{w} \in \mathbb{R}^n$ and all scalars $c \in \mathbb{R}$.

Remark. Given any function $F : \mathbb{R}^n \to \mathbb{R}^m$, we will refer to $\mathbb{R}^n$ as its *domain* and $\mathbb{R}^m$ as its *codomain*.[3]   ▲

In particular, setting $c = 0$ in the second condition implies that a linear function always maps the zero vector $\mathbf{0} \in \mathbb{R}^n$ to the zero vector $\mathbf{0} \in \mathbb{R}^m$, so

$$L[\mathbf{0}] = \mathbf{0}. \qquad (3.56)$$

We can readily combine the two defining conditions (3.55) into a single rule

$$L[c\,\mathbf{v} + d\,\mathbf{w}] = c\,L[\mathbf{v}] + d\,L[\mathbf{w}], \qquad \text{for all} \qquad \mathbf{v}, \mathbf{w} \in V, \quad c, d \in \mathbb{R}, \qquad (3.57)$$

that characterizes linearity of a function L. An easy induction proves that a linear function respects linear combinations, so

$$L[c_1\mathbf{v}_1 + \cdots + c_k\mathbf{v}_k] = c_1\,L[\mathbf{v}_1] + \cdots + c_k\,L[\mathbf{v}_k] \qquad (3.58)$$

for all $c_1, \ldots, c_k \in \mathbb{R}$ and $\mathbf{v}_1, \ldots, \mathbf{v}_k \in V$.

---

[3] The terms "range" and "target" are also sometimes used for the codomain. However, some authors use "range" to mean the image of L, and so the term is potentially confusing. An alternative name for domain is "source".

Example 3.30. The simplest linear function is the zero function $O[\mathbf{v}] \equiv \mathbf{0}$, which maps every element $\mathbf{v} \in \mathbb{R}^n$ to the zero vector in $\mathbb{R}^m$. Note that, in view of (3.56), this is the *only* constant linear function; a nonzero constant function is *not*, despite its evident simplicity, linear. Another simple but important linear function is the identity function $I \colon \mathbb{R}^n \to \mathbb{R}^n$, which leaves every vector unchanged: $I[\mathbf{v}] = \mathbf{v}$. Slightly more generally, the operation of scalar multiplication $M_a[\mathbf{v}] = a\,\mathbf{v}$ by a scalar $a \in \mathbb{R}$ defines a linear function from $\mathbb{R}^n$ to itself, with $M_0 = O$, the zero function, and $M_1 = I$, the identity function, appearing as special cases. $\blacktriangle$

Example 3.31. We claim that every linear function $L \colon \mathbb{R} \to \mathbb{R}$ has the form

$$y = L[x] = a\,x,$$

for some constant a. Therefore, the only scalar linear functions are those whose graph is a straight line passing through the origin. To prove this, we write $x \in \mathbb{R}$ as a scalar product $x = x\,1$. Then, by the second property in (3.55),

$$L[x] = L[x\,1] = x\,L[1] = a\,x, \qquad \text{where} \qquad a = L[1],$$

as claimed. $\blacktriangle$

Warning: Even though the graph of the function

$$y = a\,x + b, \tag{3.59}$$

is a straight line, it is *not* a linear function — unless $b = 0$, so the line goes through the origin. The proper mathematical name for a function of the form (3.59) is an *affine function*, which will be the subject of the following subsection.

Example 3.32. Let A be an $m \times n$ matrix. Then the function $L[\mathbf{v}] = A\mathbf{v}$ given by matrix multiplication is easily seen to be linear. Indeed, the requirements (3.55) reduce to the basic distributivity and scalar multiplication properties of matrix multiplication:

$$A(\mathbf{v} + \mathbf{w}) = A\mathbf{v} + A\mathbf{w}, \qquad A(c\,\mathbf{v}) = c\,A\mathbf{v}, \qquad \text{for all} \qquad \mathbf{v}, \mathbf{w} \in \mathbb{R}^n, \quad c \in \mathbb{R}.$$

In particular, if $m = n$, and $A = I$ is the identity matrix, then $L[\mathbf{v}] = \mathbf{v}$ is the identity function. $\blacktriangle$

In fact, *every* linear function on Euclidean space has this form.

Theorem 3.33. *Every linear function $L \colon \mathbb{R}^n \to \mathbb{R}^m$ is given by matrix multiplication:* $L[\mathbf{v}] = A\mathbf{v}$, *where A is an $m \times n$ matrix.*

Warning: Pay attention to the order of m and n. While A has size $m \times n$, the linear function L goes *from $\mathbb{R}^n$ to $\mathbb{R}^m$*.

Proof. The key idea is to look at what the linear function does to the basis vectors. Let $\mathbf{e}_1, \ldots, \mathbf{e}_n$ be the standard basis of $\mathbb{R}^n$, as in (1.3), and let $\widehat{\mathbf{e}}_1, \ldots, \widehat{\mathbf{e}}_m$ be the standard basis of $\mathbb{R}^m$. (We temporarily place hats on the latter to avoid confusing the two.) Since $L[\mathbf{e}_j] \in \mathbb{R}^m$, we can write it as a linear combination of the latter basis vectors:

$$L[\mathbf{e}_j] = \mathbf{a}_j = \begin{pmatrix} a_{1j} \\ a_{2j} \\ \vdots \\ a_{mj} \end{pmatrix} = a_{1j}\,\widehat{\mathbf{e}}_1 + a_{2j}\,\widehat{\mathbf{e}}_2 + \cdots + a_{mj}\,\widehat{\mathbf{e}}_m, \qquad j = 1, \ldots, n. \tag{3.60}$$

Let us construct the $m \times n$ matrix

$$A = (\, \mathbf{a}_1 \ \mathbf{a}_2 \ \ldots \ \mathbf{a}_n \,) = \begin{pmatrix} a_{11} & a_{12} & \cdots & a_{1n} \\ a_{21} & a_{22} & \cdots & a_{2n} \\ \vdots & \vdots & \ddots & \vdots \\ a_{m1} & a_{m2} & \cdots & a_{mn} \end{pmatrix} \tag{3.61}$$

whose columns are the image vectors (3.60). Using (3.58), we then compute the effect of L on a general vector $\mathbf{v} = (\, v_1, \ldots, v_n \,)^T \in \mathbb{R}^n$:

$$L[\mathbf{v}] = L[v_1\, \mathbf{e}_1 + \cdots + v_n\, \mathbf{e}_n] = v_1\, L[\mathbf{e}_1] + \cdots + v_n\, L[\mathbf{e}_n] = v_1\, \mathbf{a}_1 + \cdots + v_n\, \mathbf{a}_n = A\,\mathbf{v}.$$

The final equality follows from our basic formula (3.21) connecting matrix multiplication and linear combinations. We conclude that the vector $L[\mathbf{v}]$ coincides with the vector $A\,\mathbf{v}$ obtained by multiplying $\mathbf{v}$ by the coefficient matrix A. $\blacksquare$

The proof of Theorem 3.33 shows us how to construct the matrix representative of a given linear function $L : \mathbb{R}^n \to \mathbb{R}^m$. We merely assemble the image column vectors, namely $\mathbf{a}_1 = L[\mathbf{e}_1], \ \ldots, \mathbf{a}_n = L[\mathbf{e}_n]$, into an $m \times n$ matrix A.

The composition of two linear functions is again a linear function.

Proposition 3.34. *If $L : \mathbb{R}^n \to \mathbb{R}^m$ and $M : \mathbb{R}^m \to \mathbb{R}^k$ are linear functions, then the composite function $M \circ L : \mathbb{R}^n \to \mathbb{R}^k$, defined by $(M \circ L)[\mathbf{v}] = M[L[\mathbf{v}]]$, is also linear.*

Proof. This is straightforward:

$$\begin{aligned}(M \circ L)[c\,\mathbf{v} + d\,\mathbf{w}] &= M[L[c\,\mathbf{v} + d\,\mathbf{w}]] = M[c\,L[\mathbf{v}] + d\,L[\mathbf{w}]] \\ &= c\,M[L[\mathbf{v}]] + d\,M[L[\mathbf{w}]] = c\,(M \circ L)[\mathbf{v}] + d\,(M \circ L)[\mathbf{w}],\end{aligned}$$

where we used, successively, the linearity of L and then of M. $\blacksquare$

According to Theorem 3.33, $L[\mathbf{v}] = A\,\mathbf{v}$ for some $m \times n$ matrix A, while $M[\mathbf{w}] = B\,\mathbf{w}$ for some $k \times m$ matrix B. Their composition $M \circ L : \mathbb{R}^n \to \mathbb{R}^k$ is given by

$$(M \circ L)[\mathbf{v}] = M[L[\mathbf{v}]] = B(A\,\mathbf{v}) = (BA)\,\mathbf{v},$$

and hence corresponds to the $k \times n$ product matrix BA. In other words, *composition of linear functions on Euclidean space is the same as matrix multiplication*, which is another reason for the original definition of matrix multiplication. And, like matrix multiplication, composition of functions, including linear functions, is not, in general, commutative. Further, this identification gives a simple proof of the associativity of matrix multiplication, which follows immediately from the (easily proved) associativity of the composition of functions.

Finally, we note that the inverse of a linear function, when it exists, is also linear and is prescribed by the inverse matrix. Details are left to the reader; see Exercise 7.6.

Proposition 3.35. *If the function $L : \mathbb{R}^n \to \mathbb{R}^n$ is linear and invertible, then its inverse $L^{-1} : \mathbb{R}^n \to \mathbb{R}^n$ is also a linear function.*

3.7.2 Affine Functions

Of course, not every elementary function of importance in applications is linear. A simple example is a *translation*, whereby all the points in $\mathbb{R}^m$ are moved in the same direction by a common distance. The function $T: \mathbb{R}^m \to \mathbb{R}^m$ that accomplishes this is

$$T[\mathbf{x}] = \mathbf{x} + \mathbf{b}, \qquad \mathbf{x} \in \mathbb{R}^m, \tag{3.62}$$

where $\mathbf{b} \in \mathbb{R}^n$ determines the direction and the distance that the points are translated. Except in the trivial case $\mathbf{b} = \mathbf{0}$, the translation T is *not* a linear function because

$$T[\mathbf{x}+\mathbf{y}] = \mathbf{x}+\mathbf{y}+\mathbf{b} \neq T[\mathbf{x}]+T[\mathbf{y}] = (\mathbf{x}+\mathbf{b}) + (\mathbf{y}+\mathbf{b}) = \mathbf{x}+\mathbf{y}+2\mathbf{b}.$$

Or, more simply, we note that $T[\mathbf{0}] = \mathbf{b}$, which must be $\mathbf{0}$ if T is to be linear.

Combining translations and linear functions leads us to a more general important class of functions. The word "affine" comes from the Latin "affinus", meaning "related", because such functions preserve the relation of parallelism between lines; see Exercise 7.5.

> **Definition 3.36.** A function $F: \mathbb{R}^n \to \mathbb{R}^m$ of the form
>
> $$F[\mathbf{x}] = A\,\mathbf{x} + \mathbf{b}, \tag{3.63}$$
>
> where A is an $m \times n$ matrix and $\mathbf{b} \in \mathbb{R}^m$, is called an *affine function*.

For example, every affine function from $\mathbb{R}$ to $\mathbb{R}$ has the form (3.59). In general, $F[\mathbf{x}]$ is an affine function if and only if $L[\mathbf{x}] = F[\mathbf{x}] - F[\mathbf{0}]$ is a linear function. In the particular case (3.63), $F[\mathbf{0}] = \mathbf{b}$, and so $L[\mathbf{x}] = A\,\mathbf{x}$. If $A = O$ is the zero matrix, then $F[\mathbf{x}] = \mathbf{b}$ is a constant function, so every constant function is affine, and is linear if and only if $\mathbf{b} = \mathbf{0}$.

Observe that the affine function (3.63) can be constructed by composing a linear function $L[\mathbf{x}] = A\,\mathbf{x}$ and a translation $T[\mathbf{x}] = \mathbf{x} + \mathbf{b}$, so

$$F[\mathbf{x}] = T \circ L[\mathbf{x}] = T[L[\mathbf{x}]] = T[A\,\mathbf{x}] = A\,\mathbf{x} + \mathbf{b}.$$

More generally, the composition of any two affine functions is again an affine function. Specifically, given

$$F[\mathbf{x}] = A\,\mathbf{x} + \mathbf{a}, \qquad G[\mathbf{y}] = B\,\mathbf{y} + \mathbf{b},$$

then

$$(G \circ F)[\mathbf{x}] = G[F[\mathbf{x}]] = G[A\,\mathbf{x}+\mathbf{a}] = B\,(A\,\mathbf{x}+\mathbf{a}) + \mathbf{b} = C\,\mathbf{x} + \mathbf{c}, \tag{3.64}$$
$$\text{where} \quad C = BA, \quad \mathbf{c} = B\,\mathbf{a} + \mathbf{b}.$$

Note that the coefficient matrix of the composition is the product of the coefficient matrices, but the resulting vector of translation is *not* the sum of the two translation vectors.

Exercises

7.1. ♡ (a) Show that the function $R\colon \mathbb{R}^2 \to \mathbb{R}^2$ that rotates vectors in the plane by $90°$ is linear and find its matrix representative.

(b) Answer the same question for rotation by a specified angle θ.

7.2. (a) Show that the function $T\colon \mathbb{R}^2 \to \mathbb{R}^2$ that reflects vectors through the x axis is linear and find its matrix representative.

(b) Answer the same question for the reflection through the line $x = y$.

7.3. ◇ Let $t \in \mathbb{R}$. The function $S_t\colon \mathbb{R}^3 \to \mathbb{R}^3$ defined by

$$S_t(x, y, z) = (x \cos t - y \sin t, x \sin t + y \cos t, z + t)$$

is called a *screw motion* in the direction of the z axis. Is S_t linear? affine? Describe in geometrical terms what happens to a point $\mathbf{x} = (x, y, z)^T \in \mathbb{R}^3$.

7.4. Let $F\colon \mathbb{R}^n \to \mathbb{R}^m$, $G\colon \mathbb{R}^m \to \mathbb{R}^l$, $H\colon \mathbb{R}^l \to \mathbb{R}^k$ be linear functions with respective matrix representatives A, B, C. What are the sizes of these matrices? What is the matrix representative of the composition function $H \circ G \circ F\colon \mathbb{R}^n \to \mathbb{R}^k$? Use your answer to prove that matrix multiplication is associative.

7.5. ♡ Prove that an affine function maps parallel lines to parallel lines.

7.6. If $F\colon \mathbb{R}^n \to \mathbb{R}^n$ is any function, its *inverse*, if it exists, is the function $F^{-1}\colon \mathbb{R}^n \to \mathbb{R}^n$ such that the composite functions $F \circ F^{-1} = \mathrm{I} = F^{-1} \circ F$ are the identity function: $\mathrm{I}\,[\mathbf{x}] = \mathbf{x}$ for all $\mathbf{x} \in \mathbb{R}^n$. (a) ◇ Under what conditions does a linear function $L\colon \mathbb{R}^n \to \mathbb{R}^n$ with matrix form $L[\mathbf{x}] = A\mathbf{x}$ have an inverse? Show that, when it exists, L^{-1} is also linear and find its matrix representative. (b) Answer the same question for an affine function $F\colon \mathbb{R}^n \to \mathbb{R}^n$ with $F[\mathbf{x}] = A\mathbf{x} + \mathbf{b}$. Is the inverse necessarily affine? (c) ◇ What is the inverse of a rotation, as in Exercise 7.1? (d) What are the inverses of the reflections in Exercise 7.2? (e) What is the inverse of a screw motion in Exercise 7.3?

7.7. Suppose we identify $\mathcal{M}_{m \times n} \simeq \mathbb{R}^{mn}$. (a) ♡ Show that, for a fixed $k \times m$ matrix B, matrix multiplication $L[A] = BA$ defines a linear function $L\colon \mathbb{R}^{mn} \to \mathbb{R}^{km}$. (b) Show that, similarly, the trace $\operatorname{tr} A$ of a square matrix $A \in \mathcal{M}_{n \times n}$ defines a linear function $\operatorname{tr}\colon \mathbb{R}^{n^2} \to \mathbb{R}$.

Chapter 4

How Matrices Interact with Inner Products and Norms

In this chapter we discuss how matrices interact with inner products and their induced norms and, occasionally, with more general norms. Our first task is to determine the most general inner product that can be placed on Euclidean space; this will lead us to the important notion of a symmetric positive definite matrix. We next discuss the Gram matrix construction, which enables one to readily construct positive definite matrices. According to Section 3.7, an $m \times n$ matrix serves to define a linear function from $\mathbb{R}^n$ to $\mathbb{R}^m$. If we endow each of these spaces with an inner product, we can define the adjoint matrix and adjoint linear function, which goes in the reverse direction, and generalizes the ordinary transpose of a matrix. Self-adjoint and self-adjoint positive definite matrices form important generalizations of symmetric and symmetric positive definite matrices. This construction leads to the other two fundamental subspaces associated with a matrix, the cokernel and coimage, and the Fundamental Theorem of Linear Algebra that codifies the dimensions and interrelationships between the four fundamental matrix subspaces. We then turn to a study or orthogonal matrices and, more generally, norm-preserving matrices. Geometrically, orthogonal matrices correspond to rotations and reflections of Euclidean space. This is followed by a brief discussion of the matrices that induce orthogonal projections. The QR factorization of an arbitrary nonzero matrix, which is based on the general Gram–Schmidt Process presented in the preceding chapter, will be employed as our preferred method for direct solution of linear algebraic systems, as well as finding least squares solutions to incompatible systems. Finally we discuss norms on the space of matrices of a fixed size and, in particular, how norms on Euclidean space induce natural matrix norms.

4.1 Symmetric Positive Definite Matrices

Let us now return to the study of inner products. Our starting point is the general formula (2.14) for an inner product on $\mathbb{R}^n$, which we can rewrite in matrix form as follows

$$\langle \mathbf{v}, \mathbf{w} \rangle = \sum_{i,j=1}^{n} c_{ij}\, v_i\, w_j = \mathbf{v}^T C \mathbf{w}. \tag{4.1}$$

Here C is the $n \times n$ matrix whose entries are the coefficients $c_{ij} = \langle \mathbf{e}_i, \mathbf{e}_j \rangle$ that prescribe the inner product. Note that the symmetry requirement (2.16) implies that $C = C^T$ is a symmetric matrix. As we observed in Section 2.1, not every set of coefficients defines an inner product; those that do lead to the following very important class of matrices.

Definition 4.1. A symmetric $n \times n$ matrix C is called *positive definite* if it defines an inner product via the formula (4.1).

Note that, for all symmetric matrices C, formula (4.1) automatically satisfies the bilinearity and symmetry conditions of the Definition 2.1 of an inner product. Thus, we deduce that C is positive definite if and only if

$$\mathbf{x}^T C \mathbf{x} > 0 \qquad \text{for all} \qquad \mathbf{0} \neq \mathbf{x} \in \mathbb{R}^n. \tag{4.2}$$

We will sometimes write $C > 0$ to mean that the symmetric matrix C is positive definite.

Warning: The condition $C > 0$ does *not* mean that all the entries of C are positive. There are many positive definite matrices that have some negative entries. Conversely, many symmetric matrices with all positive entries are not positive definite!

Example 4.2. According to Example 2.4, although the symmetric matrix $C = \begin{pmatrix} 2 & -1 \\ -1 & 4 \end{pmatrix}$ has two negative entries, it is, nevertheless, a positive definite matrix. Indeed, as we saw,

$$\langle \mathbf{x}, \mathbf{y} \rangle = \mathbf{x}^T C \mathbf{y} = x_1 y_1 - x_1 y_2 - x_1 y_2 + 4 x_2 y_2$$

is a bona fide inner product on $\mathbb{R}^2$. On the other hand, since

$$\mathbf{x}^T C \mathbf{y} = x_1 y_1 + 2 x_1 y_2 + 2 x_1 y_2 + x_2 y_2$$

is not an inner product, the associated coefficient matrix $C = \begin{pmatrix} 1 & 2 \\ 2 & 1 \end{pmatrix}$ is not positive definite, despite it having all positive entries. ▲

Our preliminary analysis has resulted in the following general characterization of inner products on $\mathbb{R}^n$.

Theorem 4.3. *Every inner product on $\mathbb{R}^n$ is given by*

$$\langle \mathbf{x}, \mathbf{y} \rangle = \mathbf{x}^T C \mathbf{y} \qquad \text{for} \qquad \mathbf{x}, \mathbf{y} \in \mathbb{R}^n, \tag{4.3}$$

where C is a symmetric, positive definite $n \times n$ matrix.

Remark. When there is a need to explicitly indicate which inner product is being used, we will add a subscript, and so write $\langle \mathbf{x}, \mathbf{y} \rangle_C = \mathbf{x}^T C \mathbf{y}$; similarly for the associated norm $\|\mathbf{x}\|_C = \sqrt{\langle \mathbf{x}, \mathbf{x} \rangle_C} = \sqrt{\mathbf{x}^T C \mathbf{x}}$. In particular, if $C = I$ is the identity matrix, the inner product reduces to the dot product and the norm is the Euclidean norm. In this case, because it coincides with the p norm (2.62) when $p = 2$, we will use the subscript 2 instead of I, so

$$\langle \mathbf{x}, \mathbf{y} \rangle_2 = \mathbf{x}^T \mathbf{y} = \mathbf{x} \cdot \mathbf{y}, \qquad \|\mathbf{x}\|_2 = \sqrt{\mathbf{x} \cdot \mathbf{x}}.$$

The subscript will be omitted when the choice of inner product is clear from the context. ▲

In order to determine whether or not a symmetric matrix is positive definite, we use the algorithm at the end of Section 2.1 for testing positivity of the associated quadratic form

$\mathbf{x}^T C \mathbf{x}$. Recall the condition (2.22) that says that C defines an inner product if and only if the quadratic form can be written as a sum of squares:

$$\mathbf{x}^T C \mathbf{x} = y_1^2 + y_2^2 + \cdots + y_n^2, \qquad \text{where} \qquad y_i = \sum_{j=i}^{n} b_{ij}\, x_j.$$

This can be reformulated in purely matrix form as follows. Let $B = (b_{ij})$ be the $n \times n$ matrix of coefficients, so that the preceding equation can be written in matrix form

$$\mathbf{x}^T C \mathbf{x} = \mathbf{y}^T \mathbf{y} = \|\mathbf{y}\|_2^2, \qquad \text{where} \qquad \mathbf{y} = B\mathbf{x}. \tag{4.4}$$

Thus,

$$\mathbf{x}^T C \mathbf{x} = (B\mathbf{x})^T (B\mathbf{x}) = \mathbf{x}^T (B^T B)\, \mathbf{x}.$$

Since both C and $B^T B$ are symmetric, and this holds for all $\mathbf{x}$ — see Lemma 4.21 — we deduce the matrix factorization

$$C = B^T B. \tag{4.5}$$

The matrix B is *positive upper triangular*, meaning that all its entries below the diagonal are zero, so $b_{ij} = 0$ for $i > j$, while its diagonal entries are positive, $b_{ii} > 0$, while B^T is similarly positive lower triangular. Equation (4.5) is known as the *Cholesky factorization* of the matrix C, first proposed by the early twentieth-century French geographer André-Louis Cholesky for solving problems in geodetic surveying.

> **Theorem 4.4.** *A symmetric matrix C is positive definite if and only if it has a Cholesky factorization $C = B^T B$ where B is positive upper triangular.*

Example 4.5. Referring back to the inner product in Example 2.5, we deduce the Cholesky factorization of its coefficient matrix:

$$C = \begin{pmatrix} 1 & 2 & -1 \\ 2 & 6 & 0 \\ -1 & 0 & 9 \end{pmatrix} = \begin{pmatrix} 1 & 0 & 0 \\ 2 & \sqrt{2} & 0 \\ -1 & \sqrt{2} & \sqrt{6} \end{pmatrix} \begin{pmatrix} 1 & 2 & -1 \\ 0 & \sqrt{2} & \sqrt{2} \\ 0 & 0 & \sqrt{6} \end{pmatrix} = B^T B,$$

reconfirming its positive definiteness.   ▲

One can reformulate the algorithm in Section 2.1 for checking positive definiteness based on completing the square into a purely matrix form in which one constructs the Cholesky factorization (4.5) by inductively computing the successive rows of the upper triangular matrix B. For example, the entries in the first row can be computed by noting that the factorization requires, in particular,

$$b_{11} b_{1j} = c_{1j}, \qquad \text{and hence} \qquad b_{11} = \sqrt{c_{11}} > 0, \quad b_{1j} = \frac{c_{1j}}{\sqrt{c_{11}}}, \quad j = 2, \ldots, n, \tag{4.6}$$

which is the same as (2.21). One then subtracts the rank 1 matrix $B_1 = \mathbf{b}_1 \mathbf{b}_1^T$, where $\mathbf{b}_1^T = (b_{11}, \ldots, b_{1n})$ is the first row of B (and hence $\mathbf{b}_1$ is the first column of B^T). In view of (4.6), the resulting matrix $C_1 = C - \mathbf{b}_1 \mathbf{b}_1^T$ has all zeros in its first row and first column. One then repeats the algorithm on the $(n-1) \times (n-1)$ matrix $\widetilde{C}_1$ obtained by deleting these zero entries. After performing n steps of the algorithm, making sure the corresponding diagonal entries, and hence the required square roots, are always positive — as otherwise C is not positive definite — the result is a decomposition

$$C = \mathbf{b}_1 \mathbf{b}_1^T + \cdots + \mathbf{b}_n \mathbf{b}_n^T = B^T B \tag{4.7}$$

that is equivalent to the Cholesky factorization of C.

Slightly more generally, a quadratic form and its associated symmetric coefficient matrix are called *positive semidefinite* if

$$q(\mathbf{x}) = \mathbf{x}^T C \mathbf{x} \geq 0 \qquad \text{for all} \qquad \mathbf{x} \in \mathbb{R}^n, \tag{4.8}$$

in which case we write $C \geq 0$. A positive semidefinite matrix that is not positive definite will have *null directions*, meaning nonzero vectors $\mathbf{z} \neq \mathbf{0}$ such that $q(\mathbf{z}) = \mathbf{z}^T C \mathbf{z} = \mathbf{0}$. Clearly, every nonzero vector $\mathbf{z} \in \ker C$ defines a null direction. On the other hand, a positive definite matrix is not allowed to have null directions, and so $\ker C = \{\mathbf{0}\}$. Recalling Proposition 3.12, we deduce that all positive definite matrices are nonsingular.

Proposition 4.6. *If a matrix is positive definite, then it is nonsingular.*

The converse, however, is certainly *not* valid; many symmetric, nonsingular matrices fail to be positive definite.

Example 4.7. The matrix $C = \begin{pmatrix} 1 & -1 \\ -1 & 1 \end{pmatrix}$ is positive semidefinite, but not positive definite.

Indeed, the associated quadratic form

$$q(\mathbf{x}) = \mathbf{x}^T C \mathbf{x} = x_1^2 - 2\,x_1 x_2 + x_2^2 = (x_1 - x_2)^2 \geq 0$$

is a perfect square, and so clearly nonnegative. However, the elements of $\ker C$, namely the scalar multiples of the vector $(\,1,1\,)^T$, define null directions: $q(c, c) = 0$.   ▲

In a similar fashion, a quadratic form $q(\mathbf{x}) = \mathbf{x}^T C \mathbf{x}$ and its associated symmetric matrix C are called *negative semidefinite* if $q(\mathbf{x}) \leq 0$ for all $\mathbf{x}$ and *negative definite* if $q(\mathbf{x}) < 0$ for all $\mathbf{x} \neq \mathbf{0}$. Note that C is negative (semi)definite if and only if $-C$ is positive (semi)definite. A quadratic form is called *indefinite* if it is neither positive nor negative semidefinite, equivalently, if there exist vectors $\mathbf{x}_+$ where $q(\mathbf{x}_+) > 0$ and vectors $\mathbf{x}_-$ where $q(\mathbf{x}_-) < 0$; see, for instance, the second matrix in Example 4.2.

A slight extension to the proof of (4.7) produces the following characterization of positive semidefinite matrices; details are left to the reader to complete in Exercise 1.10. The reader may compare this result with Proposition 3.7 and Exercise 4.10.

Theorem 4.8. *An $n \times n$ matrix C is symmetric, positive semidefinite if and only if there exist linearly independent vectors $\mathbf{v}_1, \ldots, \mathbf{v}_r \in \mathbb{R}^n$ such that*

$$C = \mathbf{v}_1 \mathbf{v}_1^T + \cdots + \mathbf{v}_r \mathbf{v}_r^T = V V^T, \tag{4.9}$$

where $V = (\,\mathbf{v}_1 \ \ldots \ \mathbf{v}_r\,)^T$ has the indicated columns. Moreover, $\mathrm{rank}\, C = r$, and $\mathbf{v}_1, \ldots, \mathbf{v}_r$ form a basis for $\mathrm{img}\, C = \mathrm{img}\, V$. Finally, C is positive definite if and only if $r = n$.

Corollary 4.9. *If C is symmetric, positive semidefinite, then $\mathbf{z}$ is a null direction, and so $\mathbf{z}^T C \mathbf{z} = 0$, if and only if $\mathbf{z} \in \ker C$. In particular, C is positive definite if and only if $\ker C = \{\mathbf{0}\}$.*

Remark. Only positive definite matrices define *bona fide* inner products. However, indefinite matrices play a fundamental role in Einstein's theory of special relativity, [169]. In particular, the quadratic form associated with the matrix

$$C = \begin{pmatrix} c^2 & 0 & 0 & 0 \\ 0 & -1 & 0 & 0 \\ 0 & 0 & -1 & 0 \\ 0 & 0 & 0 & -1 \end{pmatrix}, \qquad \begin{aligned} &\text{namely} \quad q(\mathbf{x}) = \mathbf{x}^T C \mathbf{x} = c^2 t^2 - x^2 - y^2 - z^2, \\ &\text{where} \quad \mathbf{x} = (\, t, x, y, z\,)^T, \end{aligned} \qquad (4.10)$$

with c representing the speed of light, is the so-called Minkowski "metric" on four-dimensional relativistic space-time $\mathbb{R}^4$. The set of null directions, i.e., $\mathcal{N} = \{\, \mathbf{z} \mid q(\mathbf{z}) = 0\,\}$, forms the relativistic light cone. In this case, $\ker C = \{\mathbf{0}\}$, so an indefinite matrix can be nonsingular, and yet possess nonzero null directions, so Corollary 4.9 is *not* valid for indefinite matrices. ▲

Exercises

1.1. Are the following matrices are positive definite? In the positive definite cases, write down its Cholesky factorization and the formula for the associated inner product.

$$(a)\,\heartsuit \ \begin{pmatrix} 1 & 0 \\ 0 & 2 \end{pmatrix}, \quad (b)\,\heartsuit \ \begin{pmatrix} 1 & 1 \\ 1 & 1 \end{pmatrix}, \quad (c)\,\diamondsuit \ \begin{pmatrix} 1 & -1 \\ -1 & 3 \end{pmatrix}, \quad (d)\ \begin{pmatrix} 1 & 1 & 2 \\ 1 & 2 & 1 \\ 2 & 1 & 1 \end{pmatrix},$$

$$(e)\,\heartsuit \ \begin{pmatrix} 1 & 1 & 1 \\ 1 & 2 & -2 \\ 1 & -2 & 4 \end{pmatrix}, \quad (f)\ \begin{pmatrix} 2 & 1 & 1 & 1 \\ 1 & 2 & 1 & 1 \\ 1 & 1 & 2 & 1 \\ 1 & 1 & 1 & 2 \end{pmatrix}, \quad (g)\,\diamondsuit \ \begin{pmatrix} -1 & 1 & 1 & 1 \\ 1 & -1 & 1 & 1 \\ 1 & 1 & -1 & 1 \\ 1 & 1 & 1 & -1 \end{pmatrix}.$$

1.2. (a) For which values of c is the matrix $A = \begin{pmatrix} 1 & 1 & 0 \\ 1 & c & 1 \\ 0 & 1 & 1 \end{pmatrix}$ positive definite? (b) For the particular value $c = 3$, find its Cholesky factorization. (c) Use your result from part (b) to rewrite $q(x, y, z) = x^2 + 2\,x\,y + 3\,y^2 + 2\,y\,z + z^2$ as a sum of squares.

1.3. $\heartsuit$ Let $C = \begin{pmatrix} 1 & 2 \\ 2 & 3 \end{pmatrix}$. Prove that the associated quadratic form $q(\mathbf{x}) = \mathbf{x}^T C \mathbf{x}$ is indefinite by finding a point $\mathbf{x}^+$ where $q(\mathbf{x}^+) > 0$ and a point $\mathbf{x}^-$ where $q(\mathbf{x}^-) < 0$.

1.4. (a) Prove that an $n \times n$ diagonal matrix $D = \mathrm{diag}\,(c_1, c_2, \ldots, c_n)$ is positive definite if and only if $c_i > 0$ for all $i = 1, \ldots, n$.
 (b) Write down and identify the associated inner product.

1.5. (a) $\heartsuit$ Prove that the sum of two positive definite matrices is positive definite.
 (b) More generally, prove that the sum of a positive definite matrix and a positive semidefinite matrix is positive definite.
 (c) $\diamondsuit$ Can the sum of two positive semidefinite matrices be positive definite?
 (d) Give an example of two matrices that are not positive definite or semidefinite, but whose sum is positive definite.

1.6. Prove that if C is positive definite and $a > 0$, then $a\,C$ is also positive definite.

1.7. ◇ Prove that if C is positive semidefinite and $\alpha > 0$, then $C + \alpha\, I$ is positive definite.

1.8. Suppose H and K are both positive semidefinite. Prove that $H + K$ is also positive semidefinite. Moreover, $\ker(H + K) = \ker H \cap \ker K$, and hence $H + K$ is positive definite if and only if H and K have no common nonzero null directions.

1.9. ♡ (a) Show that every diagonal entry of a positive definite matrix must be strictly positive. (b) Write down a symmetric matrix with all positive diagonal entries that is not positive definite. (c) Find a nonzero matrix with one or more zero diagonal entries that is positive semidefinite.

1.10. (a) Show that if C is a positive semidefinite $n \times n$ matrix then every diagonal entry $c_{jj} \geq 0$. (b) Show that if $c_{jj} = 0$, then $c_{ij} = 0$ for all $i = 1, \ldots, n$. (c) Use part (b) to complete the proof of Theorem 4.8 by adapting the complete the squares algorithm at the end of Section 2.1. (d) Is every positive semidefinite quadratic form a sum of squares? If so, how many squares are required?

1.11. ◇ Find two positive definite matrices H and K whose product HK is not positive definite.

1.12. ◇ Let C be a nonsingular symmetric matrix. (a) Show that $\mathbf{x}^T C^{-1} \mathbf{x} = \mathbf{y}^T C \mathbf{y}$, where $C\mathbf{y} = \mathbf{x}$. (b) Prove that if C is positive definite, then so is C^{-1}.

1.13. ♡ Let A be an $n \times n$ matrix. Prove that $\mathbf{x}^T A \mathbf{x} = \mathbf{x}^T S \mathbf{x}$, where $S = \frac{1}{2}(A + A^T)$ is a symmetric matrix. Therefore, we do not lose any generality by restricting our discussion to quadratic forms that are constructed from symmetric matrices.

1.14. ◇ (a) Let R and S be symmetric $n \times n$ matrices. Prove that $\mathbf{x}^T R \mathbf{x} = \mathbf{x}^T S \mathbf{x}$ for all $\mathbf{x} \in \mathbb{R}^n$ if and only if $R = S$. (b) Find an example of two non-symmetric matrices $R \neq S$ such that $\mathbf{x}^T R \mathbf{x} = \mathbf{x}^T S \mathbf{x}$ for all $\mathbf{x} \in \mathbb{R}^n$.

1.15. Let $S(t)$ be a one-parameter family of symmetric matrices depending continuously on $t \in \mathbb{R}$. (a) Prove that if $S(t_0)$ is positive definite for some t_0, then $S(t)$ is positive definite for all t sufficiently close to t_0. (b) Explain why this is not necessarily true for positive semidefinite matrices.

1.16. Suppose $C = \begin{pmatrix} C_1 & C_2 \\ C_2^T & C_3 \end{pmatrix}$ is an $n \times n$ symmetric matrix, written in block form where C_1, C_2, C_3 have respective sizes $p \times p$, $p \times (n - p)$, $(n - p) \times (n - p)$. (a) ◇ Prove that if C is positive definite, then both C_1 and C_3 are positive definite. (b) Prove that if $C_2 = O$ and C_1 and C_3 are positive definite, then C is positive definite. (c) ◇ Give an example with $n = 4$ and $p = 2$ where C_1 and C_3 are positive definite, but C is not positive definite.

4.2 Gram Matrices

Symmetric matrices whose entries are given by inner products of vectors will appear throughout this text. They are named after the nineteenth-century Danish mathematician Jørgen Gram, whom we already met. We endow $\mathbb{R}^n$ with an inner product, of which the most important case is, as always, the dot product.

Definition 4.10. Let $\mathbf{v}_1, \ldots, \mathbf{v}_k \in \mathbb{R}^n$. The associated *Gram matrix*

$$
G = \begin{pmatrix}
\langle \mathbf{v}_1, \mathbf{v}_1 \rangle & \langle \mathbf{v}_1, \mathbf{v}_2 \rangle & \cdots & \langle \mathbf{v}_1, \mathbf{v}_k \rangle \\
\langle \mathbf{v}_2, \mathbf{v}_1 \rangle & \langle \mathbf{v}_2, \mathbf{v}_2 \rangle & \cdots & \langle \mathbf{v}_2, \mathbf{v}_k \rangle \\
\vdots & \vdots & \ddots & \vdots \\
\langle \mathbf{v}_k, \mathbf{v}_1 \rangle & \langle \mathbf{v}_k, \mathbf{v}_2 \rangle & \cdots & \langle \mathbf{v}_k, \mathbf{v}_k \rangle
\end{pmatrix}
\tag{4.11}
$$

is the $k \times k$ matrix whose entries are the inner products between the selected vectors.

Symmetry of the inner product implies symmetry of the Gram matrix:

$$
g_{ij} = \langle \mathbf{v}_i, \mathbf{v}_j \rangle = \langle \mathbf{v}_j, \mathbf{v}_i \rangle = g_{ji}, \qquad \text{and hence} \qquad G^T = G.
\tag{4.12}
$$

One example of a Gram matrix is the positive definite matrix C in the inner product formula (4.1), whose entries are given by the inner products of the standard basis vectors.

Example 4.11. Consider the vectors $\mathbf{v}_1 = \begin{pmatrix} 1 \\ 2 \\ -1 \end{pmatrix}$, $\mathbf{v}_2 = \begin{pmatrix} 3 \\ 0 \\ 6 \end{pmatrix}$. For the standard Euclidean dot product on $\mathbb{R}^3$, the Gram matrix is

$$
G = \begin{pmatrix} \mathbf{v}_1 \cdot \mathbf{v}_1 & \mathbf{v}_1 \cdot \mathbf{v}_2 \\ \mathbf{v}_2 \cdot \mathbf{v}_1 & \mathbf{v}_2 \cdot \mathbf{v}_2 \end{pmatrix} = \begin{pmatrix} 6 & -3 \\ -3 & 45 \end{pmatrix}.
$$

On the other hand, for the weighted inner product

$$
\langle \mathbf{v}, \mathbf{w} \rangle = 3\, v_1\, w_1 + 2\, v_2\, w_2 + 5\, v_3\, w_3,
\tag{4.13}
$$

the corresponding Gram matrix is

$$
\widetilde{G} = \begin{pmatrix} \langle \mathbf{v}_1, \mathbf{v}_1 \rangle & \langle \mathbf{v}_1, \mathbf{v}_2 \rangle \\ \langle \mathbf{v}_2, \mathbf{v}_1 \rangle & \langle \mathbf{v}_2, \mathbf{v}_2 \rangle \end{pmatrix} = \begin{pmatrix} 16 & -21 \\ -21 & 207 \end{pmatrix}.
\tag{4.14}
$$

▲

Theorem 4.12. *All Gram matrices are positive semidefinite. The Gram matrix (4.11) is positive definite if and only if $\mathbf{v}_1, \ldots, \mathbf{v}_k$ are linearly independent.*

Thus, since $\mathbf{v}_1, \mathbf{v}_2$ are evidently linearly independent, both matrices in Example 4.11 are positive definite. In fact, the simplest and most common method for producing positive definite and semidefinite matrices is through the Gram matrix construction.

Proof. To prove positive (semi)definiteness of G, we examine the associated quadratic form

$$
q(\mathbf{x}) = \mathbf{x}^T G \mathbf{x} = \sum_{i,j=1}^{k} g_{ij}\, x_i\, x_j.
$$

Substituting the values (4.12) for the matrix entries, and then invoking the bilinearity of the inner product, we deduce

$$
q(\mathbf{x}) = \sum_{i,j=1}^{k} \langle \mathbf{v}_i, \mathbf{v}_j \rangle\, x_i\, x_j = \left\langle \sum_{i=1}^{k} x_i \mathbf{v}_i,\ \sum_{j=1}^{k} x_j \mathbf{v}_j \right\rangle = \langle \mathbf{v}, \mathbf{v} \rangle = \|\mathbf{v}\|^2 \geq 0,
$$

where $\mathbf{v} = x_1 \mathbf{v}_1 + \cdots + x_k \mathbf{v}_k$ lies in the subspace V spanned by the given vectors. This immediately proves that G is positive semidefinite.

Moreover, $q(\mathbf{x}) = \|\mathbf{v}\|^2 > 0$ as long as $\mathbf{v} \neq \mathbf{0}$. If $\mathbf{v}_1, \ldots, \mathbf{v}_k$ are linearly independent, then

$$\mathbf{v} = x_1 \mathbf{v}_1 + \cdots + x_k \mathbf{v}_k = \mathbf{0} \qquad \text{if and only if} \qquad x_1 = \cdots = x_k = 0,$$

and hence $q(\mathbf{x}) = 0$ if and only if $\mathbf{x} = \mathbf{0}$. This implies that, in this situation, $q(\mathbf{x})$ and hence G are positive definite. ∎

In the case of the Euclidean dot product, the construction of the Gram matrix G can be directly implemented as follows. Given column vectors $\mathbf{v}_1, \ldots, \mathbf{v}_k \in \mathbb{R}^n$, let us form the $n \times k$ matrix $A = (\, \mathbf{v}_1 \, \ldots \, \mathbf{v}_k \,)$. In view of the identification (3.4) between the dot product and multiplication of row and column vectors, the (i, j) entry of G is given as the product

$$g_{ij} = \mathbf{v}_i \cdot \mathbf{v}_j = \mathbf{v}_i^T \mathbf{v}_j$$

of the i-th row of the transpose A^T and the j-th column of A. In other words, the Gram matrix can be evaluated as a matrix product:

$$G = A^T A. \tag{4.15}$$

For the preceding Example 4.11,

$$A = \begin{pmatrix} 1 & 3 \\ 2 & 0 \\ -1 & 6 \end{pmatrix}, \quad \text{and so} \quad G = A^T A = \begin{pmatrix} 1 & 2 & -1 \\ 3 & 0 & 6 \end{pmatrix} \begin{pmatrix} 1 & 3 \\ 2 & 0 \\ -1 & 6 \end{pmatrix} = \begin{pmatrix} 6 & -3 \\ -3 & 45 \end{pmatrix}.$$

Changing the underlying inner product will, of course, change the Gram matrix. As noted in Theorem 4.3, every inner product on $\mathbb{R}^n$ has the form

$$\langle \mathbf{v}, \mathbf{w} \rangle = \mathbf{v}^T K \mathbf{w} \qquad \text{for} \qquad \mathbf{v}, \mathbf{w} \in \mathbb{R}^n, \tag{4.16}$$

where K is a symmetric, positive definite $m \times m$ matrix. Thus, given k vectors $\mathbf{v}_1, \ldots, \mathbf{v}_k \in \mathbb{R}^n$, the entries of the $k \times k$ Gram matrix with respect to this inner product are

$$g_{ij} = \langle \mathbf{v}_i, \mathbf{v}_j \rangle = \mathbf{v}_i^T K \mathbf{v}_j.$$

If, as above, we assemble the column vectors into an $n \times k$ matrix $A = (\, \mathbf{v}_1 \, \ldots \, \mathbf{v}_k \,)$, then the Gram matrix entry g_{ij} is obtained by multiplying the i-th row of A^T by the j-th column of the product matrix $K A$. Therefore, the Gram matrix based on the alternative inner product (4.16) is given by

$$G = A^T K A. \tag{4.17}$$

The Gram matrices constructed in (4.17) arise in a wide variety of applications, including least squares approximation theory, mechanical structures, and electrical circuits, cf. [181]. They will also play an essential role in our machine learning algorithms.

Theorem 4.13. *Suppose A is an $n \times k$ matrix. If K is any positive definite $n \times n$ matrix, then the Gram matrix $G = A^T K A$ is a positive semidefinite $k \times k$ matrix, and is positive definite if and only if $\operatorname{rank} A = k$. In general, $\ker G = \ker A$, and hence $\operatorname{rank} G = \operatorname{rank} A$.*

Proof. The first part follows immediately from Theorem 4.12 and the fact that the columns of A are linearly independent if and only if it has rank k. If $A\mathbf{x} = \mathbf{0}$, then $G\mathbf{x} = A^T K A\mathbf{x} = \mathbf{0}$, and hence $\ker A \subset \ker G$. Conversely, if $G\mathbf{x} = \mathbf{0}$, then

$$0 = \mathbf{x}^T G\mathbf{x} = \mathbf{x}^T A^T K A\mathbf{x} = \mathbf{y}^T K\mathbf{y}, \qquad \text{where} \qquad \mathbf{y} = A\mathbf{x}.$$

Since $K > 0$, this occurs if and only if $\mathbf{y} = \mathbf{0}$, and so $\mathbf{x} \in \ker A$. Finally, by Theorem 3.9,

$$\operatorname{rank} G = n - \dim \ker G = n - \dim \ker A = \operatorname{rank} A. \qquad \blacksquare$$

Example 4.14. Returning to the situation of Example 4.11, the weighted inner product (4.13) corresponds to the diagonal positive definite matrix $K = \begin{pmatrix} 3 & 0 & 0 \\ 0 & 2 & 0 \\ 0 & 0 & 5 \end{pmatrix}$. Therefore, the weighted Gram matrix (4.17) based on the vectors $\mathbf{v}_1 = \begin{pmatrix} 1 \\ 2 \\ -1 \end{pmatrix}$, $\mathbf{v}_2 = \begin{pmatrix} 3 \\ 0 \\ 6 \end{pmatrix}$, is

$$G = A^T K A = \begin{pmatrix} 1 & 2 & -1 \\ 3 & 0 & 6 \end{pmatrix} \begin{pmatrix} 3 & 0 & 0 \\ 0 & 2 & 0 \\ 0 & 0 & 5 \end{pmatrix} \begin{pmatrix} 1 & 3 \\ 2 & 0 \\ -1 & 6 \end{pmatrix} = \begin{pmatrix} 16 & -21 \\ -21 & 207 \end{pmatrix},$$

thereby reproducing (4.14). $\blacktriangle$

Finally, we observe that the Cholesky factorization (4.5) implies that every positive definite matrix can be realized as the Gram matrix of a collection of vectors under the dot product, namely the columns of the upper triangular matrix B.

Exercises

2.1. Find the Gram matrix corresponding to each of the following sets of vectors using the Euclidean dot product on $\mathbb{R}^n$. Which are positive definite? $(a)\,\heartsuit$ $\begin{pmatrix} -1 \\ 3 \end{pmatrix}$, $\begin{pmatrix} 0 \\ 2 \end{pmatrix}$,

$(b)\,\diamondsuit$ $\begin{pmatrix} 1 \\ 2 \end{pmatrix}$, $\begin{pmatrix} -2 \\ 3 \end{pmatrix}$, $\begin{pmatrix} -1 \\ -1 \end{pmatrix}$, $(c)\,\heartsuit$ $\begin{pmatrix} 2 \\ 1 \\ -1 \end{pmatrix}$, $\begin{pmatrix} -3 \\ 0 \\ 2 \end{pmatrix}$, $(d)\,\diamondsuit$ $\begin{pmatrix} 1 \\ 1 \\ 0 \end{pmatrix}$, $\begin{pmatrix} 1 \\ 0 \\ 1 \end{pmatrix}$, $\begin{pmatrix} 0 \\ 1 \\ 1 \end{pmatrix}$,

$(e)\,\heartsuit$ $\begin{pmatrix} 1 \\ -2 \\ 2 \end{pmatrix}$, $\begin{pmatrix} 2 \\ -1 \\ 1 \end{pmatrix}$, $\begin{pmatrix} -1 \\ -1 \\ 1 \end{pmatrix}$, (f) $\begin{pmatrix} 1 \\ 0 \\ -1 \\ 0 \end{pmatrix}$, $\begin{pmatrix} -1 \\ 1 \\ 0 \\ 1 \end{pmatrix}$, (g) $\begin{pmatrix} 1 \\ 2 \\ 3 \\ 4 \end{pmatrix}$, $\begin{pmatrix} -2 \\ 1 \\ -4 \\ 3 \end{pmatrix}$, $\begin{pmatrix} -1 \\ 3 \\ -1 \\ -2 \end{pmatrix}$.

2.2. Recompute the Gram matrices for cases $(c–e)$ in the previous exercise using the weighted inner product $\langle \mathbf{x}, \mathbf{y} \rangle = x_1 y_1 + 2\,x_2 y_2 + 3 x_3 y_3$. Does this change their positive definiteness?

2.3. Express the following as Gram matrices or explain why this is not possible.

$(a)\,\heartsuit$ $\begin{pmatrix} 2 & 3 \\ 3 & 4 \end{pmatrix}$, $(b)\,\heartsuit$ $\begin{pmatrix} 4 & -1 \\ -1 & 1 \end{pmatrix}$, $(c)\,\diamondsuit$ $\begin{pmatrix} 3 & 2 \\ 1 & 4 \end{pmatrix}$, (d) $\begin{pmatrix} 1 & 1 & 1 \\ 1 & 0 & 1 \\ 1 & 1 & 1 \end{pmatrix}$, $(e)\,\diamondsuit$ $\begin{pmatrix} 9 & 3 & 3 \\ 3 & 2 & 2 \\ 3 & 2 & 6 \end{pmatrix}$.

2.4. Suppose $\mathbf{v}_1, \ldots, \mathbf{v}_k \in \mathbb{R}^n$ are nonzero mutually orthogonal elements. Write down their Gram matrix. Why is it nonsingular?

2.5. ♡ (*a*) Prove that if C is a positive definite matrix, then C^2 is also positive definite.
(*b*) More generally, if S is symmetric and nonsingular, then S^2 is positive definite.

2.6. (*a*) ◇ Find an example of two matrices A, K with K not positive definite and $\ker A = \{0\}$ such that the matrix $G = A^T K A$ is positive definite. Thus, the requirement that K be positive definite is not necessary in order that the matrix product G be positive definite.
(*b*) Show that if K is negative definite, then $A^T K A$ cannot be positive definite.

2.7. ◇ Is every positive semidefinite matrix a Gram matrix?

4.3   Adjoints

The adjoint of a matrix, which relies on a choice of inner products, generalizes its transpose, and appears in many applications, as well as naturally extending to more general types of linear functions when there is no obvious way to define the transpose. In particular, a self-adjoint matrix generalizes the notion of a symmetric matrix. We then extend the class of positive definite matrices to include self-adjoint matrices that satisfy a suitable positivity requirement.

Let A be an $m \times n$ matrix. According to Section 3.7, we can view multiplication of vectors by A as defining a linear function $L : \mathbb{R}^n \to \mathbb{R}^m$, where $L[\mathbf{x}] = A\mathbf{x} \in \mathbb{R}^m$ for $\mathbf{x} \in \mathbb{R}^n$. Suppose we place inner products on the domain and codomain spaces. To keep track of which is which, we will use subscripts to denote the inner products and associated norms, whereby

$$\langle \mathbf{x}, \widetilde{\mathbf{x}} \rangle_C = \mathbf{x}^T C \widetilde{\mathbf{x}}, \qquad \|\mathbf{x}\|_C = \sqrt{\mathbf{x}^T C \mathbf{x}}, \qquad \mathbf{x}, \widetilde{\mathbf{x}} \in \mathbb{R}^n,$$
$$\langle \mathbf{y}, \widetilde{\mathbf{y}} \rangle_K = \mathbf{y}^T K \widetilde{\mathbf{y}}, \qquad \|\mathbf{y}\|_K = \sqrt{\mathbf{y}^T K \mathbf{y}}, \qquad \mathbf{y}, \widetilde{\mathbf{y}} \in \mathbb{R}^m. \tag{4.18}$$

Here C, K are symmetric positive definite matrices of respective sizes $n \times n$ and $m \times m$. We allow the possibility of using different inner products on the domain and codomain spaces even when $m = n$. With these in hand, we make the following definition.

Definition 4.15. The *adjoint* of an $m \times n$ matrix A is the $n \times m$ matrix A^* that satisfies

$$\langle \mathbf{x}, A^* \mathbf{y} \rangle_C = \langle A\mathbf{x}, \mathbf{y} \rangle_K \qquad \text{for all} \qquad \mathbf{x} \in \mathbb{R}^n, \quad \mathbf{y} \in \mathbb{R}^m. \tag{4.19}$$

In order to determine a formula for the adjoint, let us write out the condition (4.19) using the formulas for the inner products. We find

$$\langle \mathbf{x}, A^* \mathbf{y} \rangle_C = \mathbf{x}^T C A^* \mathbf{y}, \qquad \langle A\mathbf{x}, \mathbf{y} \rangle_K = (A\mathbf{x})^T K \mathbf{y} = \mathbf{x}^T A^T K \mathbf{y}.$$

Equating these two expressions, and noting that the resulting equation holds for all $\mathbf{x}, \mathbf{y}$, we conclude that

$$C A^* = A^T K, \qquad \text{or, equivalently,} \qquad A^* = C^{-1} A^T K. \tag{4.20}$$

keeping in mind that C is positive definite and hence nonsingular. Equation (4.20) provides a general formula for the adjoint of A. In particular, if both inner products are the dot product, so C, K are identity matrices (of the appropriate sizes), then the adjoint reduces to the transpose: $A^* = A^T$. Thus, the transpose should be viewed as a particular case of the adjoint operation when both inner products are the dot product.

Example 4.16. Suppose $A = \begin{pmatrix} 1 & -3 & 1 \\ 0 & 2 & 0 \end{pmatrix}$. If we use the dot product on both $\mathbb{R}^2$ and $\mathbb{R}^3$, then the adjoint of A is its transpose: $A^* = A^T = \begin{pmatrix} 1 & 0 \\ -3 & 2 \\ 1 & 0 \end{pmatrix}$.

However, if we use the weighted inner products

$$\langle \mathbf{x}, \widetilde{\mathbf{x}} \rangle = 3x_1\widetilde{x}_1 + 2x_2\widetilde{x}_2 + x_3\widetilde{x}_3, \qquad \langle \mathbf{y}, \widetilde{\mathbf{y}} \rangle = 6y_1\widetilde{y}_1 + 8y_2\widetilde{y}_2,$$

then, using (4.20),

$$A^* = \begin{pmatrix} \frac{1}{3} & 0 & 0 \\ 0 & \frac{1}{2} & 0 \\ 0 & 0 & 1 \end{pmatrix} \begin{pmatrix} 1 & 0 \\ -3 & 2 \\ 1 & 0 \end{pmatrix} \begin{pmatrix} 6 & 0 \\ 0 & 8 \end{pmatrix} = \begin{pmatrix} 2 & 0 \\ -9 & 8 \\ 6 & 0 \end{pmatrix}. \qquad \blacktriangle$$

Everything that we learned about transposes can be reinterpreted in the more general language of adjoints. First, applying the adjoint operation twice returns you to where you began; this is an immediate consequence of the defining equation (4.19).

Proposition 4.17. *The adjoint of the adjoint of A is just $A = (A^*)^*$.*

We also note that the adjoint of the sum of two matrices is the sum of their adjoints and similarly for scalar multiples. Furthermore, the adjoint of the product is the product of the adjoints but in the reverse order:

$$(A + B)^* = A^* + B^*, \qquad (cA)^* = cA^*, \qquad (AB)^* = B^*A^*. \tag{4.21}$$

Proofs of these facts are relegated to the exercises.

4.3.1 Self-Adjoint and Positive Definite Matrices

We now specialize to square matrices of size $n \times n$, which serve to define linear functions $L\colon \mathbb{R}^n \to \mathbb{R}^n$. While we could impose different inner products on the two copies of $\mathbb{R}^n$, corresponding to the domain and codomain of the linear function L, for simplicity we will now assume that these are the same inner product, namely

$$\langle \mathbf{x}, \widetilde{\mathbf{x}} \rangle_C = \mathbf{x}^T C \widetilde{\mathbf{x}}, \tag{4.22}$$

where C is a symmetric positive definite $n \times n$ matrix.

Definition 4.18. An $n \times n$ matrix H is called *self-adjoint* if it equals its adjoint, $H^* = H$, meaning that

$$\langle \mathbf{x}, H\mathbf{y} \rangle_C = \langle H\mathbf{x}, \mathbf{y} \rangle_C \qquad \text{for all} \qquad \mathbf{x}, \mathbf{y} \in \mathbb{R}^n. \tag{4.23}$$

Applying (4.20) with $A \mapsto H$ and $K \mapsto C$, we see that H is self-adjoint with respect to the inner product defined by C provided

$$H = C^{-1}H^T C, \qquad \text{or, equivalently,} \qquad CH = H^T C. \tag{4.24}$$

In particular, if we use the dot product, so $C = I$, then the self-adjointness condition (4.24) requires that H itself be symmetric. Thus, one should view symmetric matrices as the special case of self-adjoint matrices when one uses the dot product. The second equation in (4.24) supplies us with a criterion for self-adjointness.

Proposition 4.19. *A matrix H is self-adjoint with respect to the inner product defined by the symmetric positive definite matrix C if and only if $H = C^{-1}S$, where $S = CH = H^T C = S^T$ is a symmetric matrix.*

Example 4.20. The non-symmetric matrix $H = \begin{pmatrix} 1 & 2 \\ 1 & 3 \end{pmatrix}$ is self-adjoint with respect to the inner product

$$\langle \mathbf{x}, \mathbf{y} \rangle = x_1 y_1 + 2 x_2 y_2$$

on $\mathbb{R}^2$. Indeed, note that

$$\langle \mathbf{x}, H\mathbf{y} \rangle = \left\langle \begin{pmatrix} x_1 \\ x_2 \end{pmatrix}, \begin{pmatrix} y_1 + 2 y_2 \\ y_1 + 3 y_2 \end{pmatrix} \right\rangle = x_1 y_1 + 2 x_1 y_2 + 2 x_2 y_1 + 6 x_2 y_2.$$

The expression on the right remains unchanged when swapping x_i for y_i, hence (4.23) holds. Alternatively, we can note that $S = CH = \begin{pmatrix} 1 & 0 \\ 0 & 2 \end{pmatrix} \begin{pmatrix} 1 & 2 \\ 1 & 3 \end{pmatrix} = \begin{pmatrix} 1 & 2 \\ 2 & 6 \end{pmatrix}$ is symmetric.   ▲

The question then arises as to which matrices are self-adjoint with respect to a suitable choise of inner product on $\mathbb{R}^n$. This will be answered in Theorem 5.31 below.

Lemma 4.21. *Suppose H is an $n \times n$ self-adjoint matrix and $\langle \mathbf{x}, H\mathbf{x} \rangle_C = 0$ for all $\mathbf{x} \in \mathbb{R}^n$. Then $H = O$.*

Proof. Suppose $\mathbf{x}, \mathbf{y} \in \mathbb{R}^n$. Then

$$\begin{aligned} 0 &= \langle \mathbf{x} + \mathbf{y}, H(\mathbf{x} + \mathbf{y}) \rangle_C \\ &= \langle \mathbf{x}, H\mathbf{x} \rangle_C + \langle \mathbf{y}, H\mathbf{x} \rangle_C + \langle \mathbf{x}, H\mathbf{y} \rangle_C + \langle \mathbf{y}, H\mathbf{y} \rangle_C = 2\langle \mathbf{y}, H\mathbf{x} \rangle_C, \end{aligned}$$

since H is self-adjoint. Thus, $\langle \mathbf{y}, H\mathbf{x} \rangle_C = 0$ for all $\mathbf{x}$ and $\mathbf{y}$. We conclude that $H\mathbf{x} = \mathbf{0}$ for all $\mathbf{x} \in \mathbb{R}^n$, and hence $H = O$.   ■

We now formulate a more general definition of a positive definite matrix.

Definition 4.22. Let H be a self-adjoint $n \times n$ matrix with respect to the inner product defined by the symmetric positive definite matrix C. Then H is called *positive definite* if

$$\langle \mathbf{x}, H\mathbf{x} \rangle_C > 0 \qquad \text{for all} \qquad \mathbf{0} \neq \mathbf{x} \in \mathbb{R}^n. \tag{4.25}$$

More generally, H is *positive semidefinite* if

$$\langle \mathbf{x}, H\mathbf{x} \rangle_C \geq 0 \qquad \text{for all} \qquad \mathbf{x} \in \mathbb{R}^n. \tag{4.26}$$

In particular, if we use the dot product, then H must be symmetric, and (4.25) reduces to our earlier positivity requirement (4.2). Keep in mind that, for more general inner products, H need not be symmetric. As a counterpart to Proposition 4.19, we have the following characterization of general positive definite matrices.

> **Proposition 4.23.** *A matrix H is positive definite with respect to the inner product defined by the symmetric positive definite matrix C if and only if $S = CH$ is symmetric and positive definite.*

Proof. As above, symmetry of S implies self-adjointness of H. Moreover,

$$\langle \mathbf{x}, H\mathbf{x} \rangle_C = \mathbf{x}^T C H \mathbf{x} = \mathbf{x}^T S \mathbf{x} > 0 \qquad \text{for all} \qquad \mathbf{x} \neq \mathbf{0}$$

if and only if S is positive definite, which as before, can be checked by establishing the existence of a Cholesky factorization: $S = CH = B^T B$, where B is positive upper triangular. $\blacksquare$

Exercises

3.1. Choose one from the following list of inner products on $\mathbb{R}^3$ for both the domain and codomain, and find the adjoint of $A = \begin{pmatrix} 1 & 1 & 0 \\ -1 & 0 & 1 \\ 0 & -1 & 2 \end{pmatrix}$: $(a)\,\heartsuit$ the Euclidean dot product;

$(b)\,\heartsuit$ the weighted inner product $\langle \mathbf{v}, \mathbf{w} \rangle = v_1 w_1 + 2 v_2 w_2 + 3 v_3 w_3$; $(c)\,\diamondsuit$ the inner product $\langle \mathbf{v}, \mathbf{w} \rangle = \mathbf{v}^T C \mathbf{w}$ defined by the positive definite matrix $C = \begin{pmatrix} 2 & 1 & 0 \\ 1 & 2 & 1 \\ 0 & 1 & 2 \end{pmatrix}$.

3.2. From the list in Exercise 3.1, choose different inner products on the domain and codomain, and then compute the adjoint of the matrix A.

3.3. $\diamondsuit$ Prove that $A = \begin{pmatrix} 6 & -3 \\ -2 & 4 \end{pmatrix}$ is self-adjoint with respect to the weighted inner product $\langle \mathbf{v}, \mathbf{w} \rangle = 2 v_1 w_1 + 3 v_2 w_2$. Is A positive definite?

3.4. $\heartsuit$ Consider the weighted inner product $\langle \mathbf{v}, \mathbf{w} \rangle = v_1 w_1 + \frac{1}{2} v_2 w_2 + \frac{1}{3} v_3 w_3$ on $\mathbb{R}^3$.
(a) What are the conditions on the entries of a 3×3 matrix A in order that it be self-adjoint?
(b) Write down an example of a non-diagonal self-adjoint matrix.

3.5. Answer Exercise 3.4 for the inner product based on $C = \begin{pmatrix} 2 & -1 & 0 \\ -1 & 2 & -1 \\ 0 & -1 & 2 \end{pmatrix}$.

3.6. Prove the following adjoint identities: $(a)\,\heartsuit$ $(A+B)^* = A^* + B^*$, $(b)\,\diamondsuit$ $(AB)^* = B^* A^*$, (c) $(cA)^* = cA^*$ for $c \in \mathbb{R}$, $(d)\,\heartsuit$ $(A^*)^* = A$, $(e)\,\diamondsuit$ $(A^{-1})^* = (A^*)^{-1}$.

3.7. Is $I^* = I$?

3.8. $\heartsuit$ Let C, K be positive definite matrices defining inner products on $\mathbb{R}^n$ and $\mathbb{R}^m$, respectively. Let A be an $m \times n$ matrix with adjoint A^*. Prove that $\mathbf{x}$ solves the inhomogeneous linear system $A\mathbf{x} = \mathbf{b}$ if and only if

$$\langle \mathbf{x}, A^* \mathbf{y} \rangle_C = \langle \mathbf{b}, \mathbf{y} \rangle_K \qquad \text{for all} \qquad \mathbf{y} \in \mathbb{R}^m. \tag{4.27}$$

Remark: Equation (4.27) is known as the *weak formulation* of the linear system. Its generalizations plays an essential role in the analysis of differential equations and their numerical approximations, [180, 192, 225].

4.4　The Fundamental Matrix Subspaces

In this section, we introduce the remaining two of the four fundamental subspaces associated with a matrix, and establish important orthogonality relations. Recall that the image and kernel of an $m \times n$ matrix A are subspaces of, respectively, $\mathbb{R}^m$ and $\mathbb{R}^n$. The other two subspaces rely on introducing inner products on $\mathbb{R}^m$ and $\mathbb{R}^n$, which serve to specify the adjoint A^*, of size $n \times m$. The image of A^* is called the *coimage* of A, and so

$$\operatorname{coimg} A = \operatorname{img} A^* \subset \mathbb{R}^n. \tag{4.28}$$

Similarly, the kernel of A^* is called the *cokernel* of A:

$$\operatorname{coker} A = \ker A^* \subset \mathbb{R}^m. \tag{4.29}$$

These four,

$$\operatorname{img} A, \ \operatorname{coker} A \subset \mathbb{R}^m, \qquad \operatorname{coimg} A, \ \ker A \subset \mathbb{R}^n, \tag{4.30}$$

are known as the *fundamental subspaces* associated with a matrix or, equivalently, the associated linear function between inner product spaces. In particular, if we use the dot products on both $\mathbb{R}^n$ and $\mathbb{R}^m$, then the coimage and cokernel become the image and kernel of the transpose matrix A^T. In this case, the coimage is also known as the *row space* of A because it is, by definition, the span of the columns of A^T, which are the transposed rows of A. This is the standard case; however, extending the arguments to more general inner products is straightforward, as we will now show.

It turns out that the relevant pairs of subspaces are, in fact, orthogonal complements under the imposed inner products on $\mathbb{R}^m$ and $\mathbb{R}^n$. Moreover, their dimensions are prescribed by the common rank of A and its adjoint. This important result is known as the *Fundamental Theorem of Linear Algebra*.

> **Theorem 4.24.** *Let A be an $m \times n$ matrix. Let $\langle \cdot, \cdot \rangle_C$ and $\langle \cdot, \cdot \rangle_K$ be inner products on $\mathbb{R}^n$ and $\mathbb{R}^m$, respectively. Then the kernel and coimage of A are orthogonal complementary subspaces of $\mathbb{R}^n$, while its cokernel and image are orthogonal complementary subspaces of $\mathbb{R}^m$:*
>
> $$\operatorname{img} A = (\operatorname{coker} A)^\perp \subset \mathbb{R}^m, \qquad \operatorname{coimg} A = (\ker A)^\perp \subset \mathbb{R}^n. \tag{4.31}$$
>
> *The dimensions of these subspaces are*
>
> $$\dim \operatorname{img} A = \dim \operatorname{coimg} A = r, \quad \dim \ker A = n - r, \quad \dim \operatorname{coker} A = m - r, \tag{4.32}$$
>
> *where*
>
> $$r = \operatorname{rank} A = \operatorname{rank} A^*. \tag{4.33}$$

Proof. By definition, a vector $\mathbf{b} \in \mathbb{R}^m$ lies in $\operatorname{img} A$ if and only if there is $\mathbf{x} \in \mathbb{R}^n$ such that $\mathbf{b} = A\mathbf{x}$. On the other hand, a vector $\mathbf{y} \in \mathbb{R}^m$ lies in $\operatorname{coker} A = \ker A^*$ if and only if $A^*\mathbf{y} = \mathbf{0}$. Thus, if $\mathbf{b} \in \operatorname{img} A$ and $\mathbf{y} \in \operatorname{coker} A$ then

$$\langle \mathbf{y}, \mathbf{b} \rangle_K = \langle \mathbf{y}, A\mathbf{x} \rangle_K = \langle A^*\mathbf{y}, \mathbf{x} \rangle_C = 0,$$

which shows that $\operatorname{img} A \subset (\operatorname{coker} A)^\perp$, or, equivalently[1], that $\operatorname{coker} A \subset (\operatorname{img} A)^\perp$. Now, if $\mathbf{y} \in (\operatorname{img} A)^\perp$ then

$$0 = \langle \mathbf{y}, A\mathbf{x} \rangle_K = \langle A^*\mathbf{y}, \mathbf{x} \rangle_C \quad \text{for all} \quad \mathbf{x} \in \mathbb{R}^n,$$

---

[1] Here, we use that $V \subset W$ implies $W^\perp \subset V^\perp$ and $(V^\perp)^\perp = V$; see Theorem 2.32 and Exercise 6.5.

and hence $A^*\mathbf{y} = \mathbf{0}$, i.e., $\mathbf{y} \in \ker A^* = \operatorname{coker} A$. Therefore $(\operatorname{img} A)^\perp \subset \operatorname{coker} A$, from which the opposite inclusion $(\operatorname{coker} A)^\perp \subset \operatorname{img} A$ follows, and hence $\operatorname{img} A = (\operatorname{coker} A)^\perp$. Orthogonality of the coimage and kernel follows by the same argument applied to the adjoint matrix A^*.

Finally, since $r = \operatorname{rank} A = \dim \operatorname{img} A$, as a consequence of the formula (2.56) for the dimensions of orthogonal complements and Theorem 3.9, we deduce

$$\dim \operatorname{coker} A = m - \dim \operatorname{img} A = m - r,$$
$$\dim \operatorname{coimg} A = n - \dim \ker A = n - (n - r) = r = \dim \operatorname{img} A.$$

Note that the second formula implies

$$\operatorname{rank} A = \dim \operatorname{img} A = r = \dim \operatorname{coimg} A = \dim \operatorname{img} A^* = \operatorname{rank} A^*,$$

proving that the matrix and its adjoint have the same rank. In the particular case of the dot product, this establishes our previously stated Theorem 3.13. $\blacksquare$

4.4.1 Applications to Self-Adjoint Matrices

One method for producing positive definite and semidefinite self-adjoint matrices is modeled on the Gram matrix construction (4.15). The Fundamental Theorem 4.24 enables us to determine their ranks and fundamental subspaces.

Theorem 4.25. *Let A be an $m \times n$ matrix with adjoint A^* relative to inner products on $\mathbb{R}^m$ and $\mathbb{R}^n$. Then the $n \times n$ matrix $H = A^*A$ is self-adjoint with respect to the inner product on $\mathbb{R}^n$, while the $m \times m$ matrix $J = AA^*$ is self-adjoint with respect to the inner product on $\mathbb{R}^m$. Both H and J are positive semidefinite and have the same rank $r = \operatorname{rank} A$; furthermore,*

$$\ker H = \ker A, \quad \operatorname{img} H = \operatorname{coimg} A, \quad \ker J = \operatorname{coker} A, \quad \operatorname{img} J = \operatorname{img} A. \quad (4.34)$$

Moreover, H is positive definite if and only if $\operatorname{rank} A = n$, while J is positive definite if and only if $\operatorname{rank} A = m$.

Proof. Using the defining equation (4.19) for the adjoint and Proposition 4.17, we have

$$\langle \mathbf{x}, H\widetilde{\mathbf{x}} \rangle_C = \langle \mathbf{x}, A^*A\widetilde{\mathbf{x}} \rangle_C = \langle A\mathbf{x}, A\widetilde{\mathbf{x}} \rangle_K = \langle A^*A\mathbf{x}, \widetilde{\mathbf{x}} \rangle_C = \langle H\mathbf{x}, \widetilde{\mathbf{x}} \rangle_C,$$

for all $\mathbf{x}, \widetilde{\mathbf{x}} \in \mathbb{R}^n$, proving self-adjointness. Moreover, setting $\widetilde{\mathbf{x}} = \mathbf{x}$ in the above formula, we find

$$\langle \mathbf{x}, H\mathbf{x} \rangle_C = \langle A\mathbf{x}, A\mathbf{x} \rangle_K = \| A\mathbf{x} \|_K^2 \geq 0,$$

and hence H satisfies the positive semidefinite requirement (4.26). Moreover, H is positive definite if and only if $\ker A = \{\mathbf{0}\}$, which is equivalent to the rank condition. The proof for J is identical, replacing A by A^*. The proof of (4.34) is left to the reader as Exercise 4.4. $\blacksquare$

Note that, in view of (4.20), the matrices in Theorem 4.25 are explicitly given by

$$H = A^*A = C^{-1}A^T K A, \qquad J = AA^* = A C^{-1}A^T K. \qquad (4.35)$$

In particular, if $C = I$, then $A^*A = A^T K A$ takes the form of a Gram matrix (4.17) with respect to the inner product induced by K, and Theorem 4.25 reduces to Theorem 4.13, but with additional information concerning the images. Furthermore, if both $C = I$ and $K = I$, so that both inner products are the dot product, then $H = A^T A$ and $J = AA^T$ are both Gram matrices, as in (4.15), corresponding, respectively, to the columns and rows of A.

Example 4.26. Consider the 2×3 rank 2 matrix A in Example 4.16 whose adjoints with respect to the dot products as well as a pair of weighted inner products were determined. Theorem 4.25 implies that both of the product matrices

$$
A^T A = \begin{pmatrix} 1 & 0 \\ -3 & 2 \\ 1 & 0 \end{pmatrix} \begin{pmatrix} 1 & -3 & 1 \\ 0 & 2 & 0 \end{pmatrix} = \begin{pmatrix} 1 & -3 & 1 \\ -3 & 13 & -3 \\ 1 & -3 & 1 \end{pmatrix},
$$

$$
A^* A = \begin{pmatrix} 2 & 0 \\ -9 & 8 \\ 6 & 0 \end{pmatrix} \begin{pmatrix} 1 & -3 & 1 \\ 0 & 2 & 0 \end{pmatrix} = \begin{pmatrix} 2 & -6 & 2 \\ -9 & 43 & -9 \\ 6 & -18 & 6 \end{pmatrix},
$$

have rank 2 and are positive semidefinite and self-adjoint with respect to the relevant inner product, while

$$
A A^T = \begin{pmatrix} 1 & -3 & 1 \\ 0 & 2 & 0 \end{pmatrix} \begin{pmatrix} 1 & 0 \\ -3 & 2 \\ 1 & 0 \end{pmatrix} = \begin{pmatrix} 11 & -6 \\ -6 & 4 \end{pmatrix},
$$

$$
A A^* = \begin{pmatrix} 1 & -3 & 1 \\ 0 & 2 & 0 \end{pmatrix} \begin{pmatrix} 2 & 0 \\ -9 & 8 \\ 6 & 0 \end{pmatrix} = \begin{pmatrix} 35 & -24 \\ -18 & 16 \end{pmatrix},
$$

also have rank 2 and are positive definite and self-adjoint with respect to the relevant inner product, as can be checked directly.   ▲

4.4.2   Applications to Linear Systems

One important consequence of the Fundamental Theorem 4.24 is the following characterization of compatible linear systems. As we know, when the coefficient matrix has positive nullity, the solution to a compatible linear system is not unique. One can, in fact, single out one particular solution by the property of its belonging to the coimage of the coefficient matrix; moreover, this solution is distinguished as having minimal norm among all solutions.

> **Theorem 4.27.** *A linear system $A\mathbf{x} = \mathbf{b}$ has a solution if and only if $\mathbf{b}$ is orthogonal to the cokernel of A, so $\mathbf{b} \in \operatorname{img} A = (\operatorname{coker} A)^\perp$. In this case, the system has a unique solution $\mathbf{p} \in \operatorname{coimg} A = (\ker A)^\perp$ satisfying $A\mathbf{p} = \mathbf{b}$. The general solution is $\mathbf{x} = \mathbf{p} + \mathbf{q}$, where $\mathbf{q} \in \ker A$, and thus $\mathbf{p}$ is the common orthogonal projection of all the solutions $\mathbf{x}$ onto the coimage of the coefficient matrix. Moreover, the particular solution $\mathbf{p} \in \operatorname{coimg} A$ has the smallest norm of all possible solutions: $\|\mathbf{p}\| \leq \|\mathbf{x}\|$ whenever $A\mathbf{x} = \mathbf{b}$.*

Proof. Indeed, the system has a solution if and only if the right-hand side belongs to the image of the coefficient matrix, $\mathbf{b} \in \operatorname{img} A$, which, by (4.31), requires that $\mathbf{b}$ be orthogonal to its cokernel. Thus, the compatibility conditions for the linear system $A\mathbf{x} = \mathbf{b}$ can be expressed in the form

$$
\langle \mathbf{y}, \mathbf{b} \rangle_K = 0 \qquad \text{for every } \mathbf{y} \text{ satisfying} \qquad A^* \mathbf{y} = \mathbf{0}. \tag{4.36}
$$

In practice, one only needs to check orthogonality of $\mathbf{b}$ with respect to a basis $\mathbf{y}_1, \ldots, \mathbf{y}_{m-r}$ of the cokernel, leading to a system of $m - r$ compatibility constraints

$$
\langle \mathbf{y}_i, \mathbf{b} \rangle_K = 0, \qquad i = 1, \ldots, m - r, \tag{4.37}
$$

which ensure the orthogonality of $\mathbf{b}$ to the entire cokernel of A. The compatibility conditions (4.37) are known as the *Fredholm alternative*, named after the Swedish mathematician Ivar Fredholm, who introduced them in his study of linear integral equations. Later, his compatibility criterion was recognized to be a general property of linear systems, including linear algebraic systems, linear differential equations, and so on.

To establish the second part of the Theorem, let $\mathbf{x}$ be any solution, and let $\mathbf{p}$ be the orthogonal projection of $\mathbf{x}$ onto $\operatorname{coimg} A$. Then $\mathbf{q} = \mathbf{x} - \mathbf{p} \in (\operatorname{coimg} A)^{\perp} = \ker A$, and thus $A\mathbf{p} = A\mathbf{x} - A\mathbf{q} = \mathbf{b}$, proving that $\mathbf{p}$ is a solution. To prove uniqueness, if $\widetilde{\mathbf{p}} \in \operatorname{coimg} A$ is another solution belonging to the coimage, then $\mathbf{p} - \widetilde{\mathbf{p}} \in \ker A$, but orthogonality implies $\ker A \cap \operatorname{coimg} A = \{\mathbf{0}\}$, and hence $\mathbf{p} - \widetilde{\mathbf{p}} = \mathbf{0}$. Finally, the norm of a general solution $\mathbf{x} = \mathbf{p} + \mathbf{q}$ is, using (2.56),

$$\|\mathbf{x}\|^2 = \|\mathbf{p} + \mathbf{q}\|^2 = \|\mathbf{p}\|^2 + \|\mathbf{q}\|^2 \geq \|\mathbf{p}\|^2,$$

with equality if and only if $\mathbf{q} = \mathbf{0}$. ∎

In summary, the linear system $A\mathbf{x} = \mathbf{b}$ has a solution if and only if $\mathbf{b} \in \operatorname{img} A$, or, equivalently, is orthogonal to every vector $\mathbf{y} \in \operatorname{coker} A$. If the Fredholm compatibility conditions (4.37) hold, then the system has a *unique* solution $\mathbf{p} \in \operatorname{coimg} A$, which is the solution of minimal norm. The general solution to the system is $\mathbf{x} = \mathbf{p} + \mathbf{z}$, where $\mathbf{p}$ is the particular solution belonging to the coimage, while $\mathbf{z} \in \ker A$ is an arbitrary element of the kernel.

Given any solution $\mathbf{x}$ to the system, to find the solution $\mathbf{p} \in \operatorname{coimg} A$ of minimal norm, we can either use the orthogonal projection formula (2.39) onto the coimage:

$$\mathbf{p} = \sum_{i=1}^{r} \langle \mathbf{u}_i, \mathbf{x} \rangle \mathbf{u}_i, \quad \text{where } \mathbf{u}_1, \ldots, \mathbf{u}_r \text{ form an orthonormal basis of } \operatorname{coimg} A, \qquad (4.38)$$

or, if we know the general solution, we can characterize $\mathbf{p}$ by requiring that it be orthogonal to the kernel:

$$\langle \mathbf{p}, \mathbf{z}_j \rangle = 0, \qquad j = 1, \ldots, n - r, \qquad \text{where } \mathbf{z}_1, \ldots, \mathbf{z}_{n-r} \text{ form a basis of } \ker A. \qquad (4.39)$$

The orthonormal basis basis of $\operatorname{coimg} A$ can be constructed by applying the Gram–Schmidt process to the columns of A^*.

Example 4.28. In this example, we use dot products, and hence the adjoint of a matrix is just its transpose. Consider the linear system

$$A\mathbf{x} = \begin{pmatrix} 1 & 0 & -1 \\ 0 & 1 & -1 \\ 2 & -1 & -1 \end{pmatrix} \begin{pmatrix} x_1 \\ x_2 \\ x_3 \end{pmatrix} = \begin{pmatrix} 2 \\ 1 \\ 3 \end{pmatrix}. \qquad (4.40)$$

We easily solve for

$$\mathbf{x} = \begin{pmatrix} x_1 \\ x_2 \\ x_3 \end{pmatrix} = \begin{pmatrix} 2 + t \\ -1 + t \\ t \end{pmatrix} = \begin{pmatrix} 2 \\ -1 \\ 0 \end{pmatrix} + \begin{pmatrix} t \\ t \\ t \end{pmatrix}, \qquad (4.41)$$

where $t \in \mathbb{R}$ is arbitrary. In the last expression, referring back to Theorem 3.17, the first vector is a particular solution, $\mathbf{x}^{\star} = (2, -1, 0)^T$, while the second is the general element of the kernel, which is hence one-dimensional with basis $\mathbf{z} = (1, 1, 1)^T$. Thus nullity $A = 1$ and hence, by (3.32), rank $A = 2$.

We next compute the cokernel by solving

$$A^T\mathbf{y} = \begin{pmatrix} 1 & 0 & 2 \\ 0 & 1 & -1 \\ -1 & -1 & -1 \end{pmatrix} \begin{pmatrix} u \\ v \\ w \end{pmatrix} = \begin{pmatrix} 0 \\ 0 \\ 0 \end{pmatrix},$$

leading to $\mathbf{y} = s\,\mathbf{v}$ where $s \in \mathbb{R}$ and $\mathbf{v} = (-2, 1, 1)^T$. Thus the cokernel is one-dimensional, again in accordance with Theorem 4.24, with basis $\mathbf{v}$. Moreover, the Fredholm conditions in Theorem 4.27, says the system $A\mathbf{x} = \mathbf{b} = (b_1, b_2, b_3)^T$ is compatible if and only if

$$\mathbf{v} \cdot \mathbf{b} = -2b_1 + b_2 + b_3 = 0,$$

which is satisfied in the above particular case.

To find the solution $\mathbf{p}$ of minimum Euclidean norm, we must determine the value of t such that the solution (4.41) belongs to the coimage of A, or, equivalently, is orthogonal to the kernel of A, so

$$0 = \mathbf{x} \cdot \mathbf{z} = \begin{pmatrix} 2+t \\ -1+t \\ t \end{pmatrix} \cdot \begin{pmatrix} 1 \\ 1 \\ 1 \end{pmatrix} = 1 + 3t, \quad \text{hence} \quad t = -\tfrac{1}{3} \quad \text{and} \quad \mathbf{p} = \begin{pmatrix} \tfrac{5}{3} \\ -\tfrac{4}{3} \\ -\tfrac{1}{3} \end{pmatrix}.$$

Let us check that its norm is indeed the smallest among all solutions to the original system:

$$\|\mathbf{p}\| = \sqrt{\frac{14}{3}} \le \|\mathbf{x}\| = \|(2+t, -1+t, t)^T\| = \sqrt{3t^2 + 2t + 5},$$

where the quadratic function inside the square root achieves its minimum at $t = -\tfrac{1}{3}$. ▲

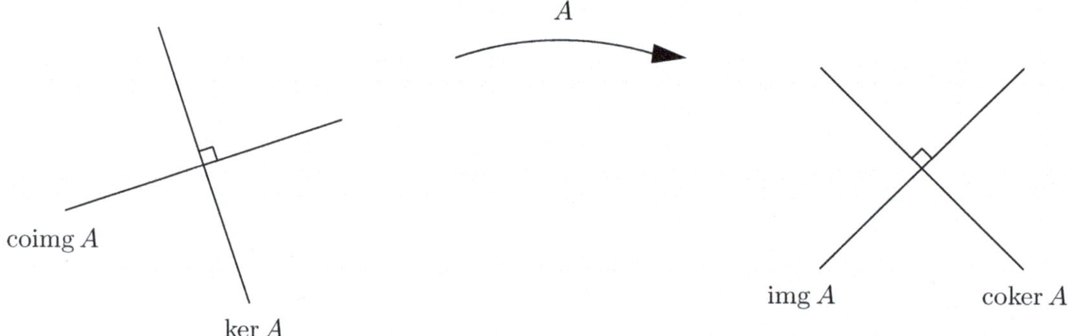

Figure 4.1: The Fundamental Matrix Subspaces

Referring to Figure 4.1, we have now attained a full understanding of the fascinating geometry that lurks behind the simple algebraic operation of multiplying a vector $\mathbf{x} \in \mathbb{R}^n$ by an $m \times n$ matrix, resulting in a vector $\mathbf{b} = A\mathbf{x} \in \mathbb{R}^m$ or, equivalently, the associated linear system $A\mathbf{x} = \mathbf{b}$. Since the kernel and coimage of A are orthogonal complements in the domain space $\mathbb{R}^n$, Theorem 2.32 tells us that we can uniquely decompose $\mathbf{x} = \mathbf{p} + \mathbf{z}$, where $\mathbf{p} \in \operatorname{coimg} A$, while $\mathbf{z} \in \ker A$. Since $A\mathbf{z} = \mathbf{0}$, we have

$$\mathbf{b} = A\mathbf{x} = A(\mathbf{p} + \mathbf{z}) = A\mathbf{p}.$$

Therefore, we can regard multiplication by A as a combination of two operations:

(i) The first is an orthogonal projection onto the coimage of A taking $\mathbf{x}$ to $\mathbf{p}$.

(ii) The second maps a vector in $\operatorname{coimg} A \subset \mathbb{R}^n$ to a vector in $\operatorname{img} A \subset \mathbb{R}^m$, taking the orthogonal projection $\mathbf{p}$ to the image vector $\mathbf{b} = A\mathbf{p} = A\mathbf{x}$.

Moreover, if A has rank r, then both $\operatorname{img} A$ and $\operatorname{coimg} A$ are r-dimensional subspaces, albeit of different vector spaces. Each vector $\mathbf{b} \in \operatorname{img} A$ corresponds to a *unique* vector $\mathbf{p} \in \operatorname{coimg} A$. Indeed, if $\mathbf{p}, \widetilde{\mathbf{p}} \in \operatorname{coimg} A$ satisfy $\mathbf{b} = A\mathbf{p} = A\widetilde{\mathbf{p}}$, then $A(\mathbf{p}-\widetilde{\mathbf{p}}) = \mathbf{0}$, hence $\mathbf{p}-\widetilde{\mathbf{p}} \in \ker A$. But, since the kernel and the coimage are orthogonal complements, the only vector that belongs to both is the zero vector, and thus $\mathbf{p} = \widetilde{\mathbf{p}}$. In this manner, we have proved the first part of the following result; the second is left as Exercise 4.6.

Corollary 4.29. *Multiplication by an $m \times n$ matrix A of rank r defines a one-to-one correspondence between the r-dimensional subspaces $\operatorname{coimg} A \subset \mathbb{R}^n$ and $\operatorname{img} A \subset \mathbb{R}^m$. Moreover, if $\mathbf{v}_1, \ldots, \mathbf{v}_r$ forms a basis of $\operatorname{coimg} A$ then their images $A\mathbf{v}_1, \ldots, A\mathbf{v}_r$ form a basis for $\operatorname{img} A$.*

The preceding results enable us to bound the rank of the product of two matrices.

Proposition 4.30. *Suppose $A \in \mathcal{M}_{m \times n}$, $B \in \mathcal{M}_{n \times p}$, and $\operatorname{rank} A = r$, $\operatorname{rank} B = s$. Then the rank of the product matrix $AB \in \mathcal{M}_{m \times p}$ can be bounded by the Sylvester inequalities*

$$\max\{r + s - n, 0\} \le \operatorname{rank}(AB) \le \min\{r, s\}. \tag{4.42}$$

Proof. By Corollary 4.29, multiplication by A defines a 1-1 map from $\operatorname{coimg} A$ to $\operatorname{img} A$. Thus,

$$\operatorname{rank}(AB) = \dim \operatorname{img}(AB) = \dim(\operatorname{coimg} A \cap \operatorname{img} B),$$

where the subspaces $\operatorname{coimg} A, \operatorname{img} B \subset \mathbb{R}^n$ have respective dimensions r, s. Equation (1.16) implies that their intersection satisfies the Sylvester inequalities. $\blacksquare$

Exercises

Note: Unless stated otherwise, the inner product is the standard dot product on $\mathbb{R}^n$.

4.1. For each of the following matrices find bases (when they exist) for the
(i) image, (ii) coimage, (iii) kernel, and (iv) cokernel.

$(a)\,\heartsuit\; \begin{pmatrix} 1 & -3 \\ 2 & -6 \end{pmatrix}$, $\quad (b)\,\heartsuit\; \begin{pmatrix} 1 & -3 & 1 \\ 0 & 2 & 0 \end{pmatrix}$, $\quad (c)\,\diamondsuit\; \begin{pmatrix} 0 & 0 & -8 \\ 1 & 2 & -1 \\ 2 & 4 & 6 \end{pmatrix}$, $\quad (d)\; \begin{pmatrix} 1 & 1 & 3 & 1 \\ 1 & 1 & 0 & 1 \\ 0 & 0 & 3 & 0 \end{pmatrix}$.

4.2. Find bases for the image and coimage of $\begin{pmatrix} 1 & -3 & 0 \\ 2 & -6 & 4 \\ -3 & 9 & 1 \end{pmatrix}$. Make sure they have the same number of elements. Then write each row and column as a linear combination of the appropriate basis vectors.

4.3. Find bases for the coimage and cokernel of the matrix $\begin{pmatrix} 1 & 1 & 0 \\ 1 & 0 & -1 \\ 0 & 1 & 1 \end{pmatrix}$ using

$(a)\,\heartsuit$ the dot product; $(b)\,\diamondsuit$ the weighted inner product $\langle \mathbf{v}, \mathbf{w} \rangle = v_1 w_1 + 2 v_2 w_2 + 3 v_3 w_3$. Make sure that the dimensions satisfy the formulas in the Fundamental Theorem 4.24.

4.4. $\diamondsuit$ Prove the equations in (4.34).

4.5.♡ *True or false*: nullity A = nullity A^*.

4.6. Prove that if $\mathbf{v}_1, \ldots, \mathbf{v}_r$ are a basis of coimg A, then their images $A\mathbf{v}_1, \ldots, A\mathbf{v}_r$ are a basis for img A.

4.5   Orthogonal and Norm–Preserving Matrices

In this section, we continue to use the inner products and norms on $\mathbb{R}^n$ and $\mathbb{R}^m$ as written in (4.18). Even when $m = n$, we can allow different norms on the two copies of $\mathbb{R}^n$, although usually they will be the same. In the later parts of this section, we restrict our attention to the dot product for simplicity.

Definition 4.31. An $m \times n$ matrix Q is called *norm-preserving* if it satisfies

$$\| Q\mathbf{x} \|_K = \| \mathbf{x} \|_C \qquad \text{for all} \qquad \mathbf{x} \in \mathbb{R}^n. \tag{4.43}$$

Remark. We restrict to norms defined by inner products because for most other norms there are very few norm-preserving matrices. Indeed, only signed permutation matrices, which act on vectors by permuting and, possibly, changing some of the signs of their entries, preserve the p norms (2.62) on $\mathbb{R}^n$ when $p \neq 2$; see [144]. ▲

Theorem 4.32. *A matrix Q is norm-preserving if and only*

$$Q^* Q = \mathrm{I}, \tag{4.44}$$

i.e., its adjoint with respect to the inner products forms a left inverse.

Proof. We square the left hand side of (4.43) and use the adjoint equation (4.19):

$$\| Q\mathbf{x} \|_K^2 = \langle Q\mathbf{x}, Q\mathbf{x} \rangle_K = \langle \mathbf{x}, Q^* Q\mathbf{x} \rangle_C.$$

Equating this to $\| \mathbf{x} \|_C^2 = \langle \mathbf{x}, \mathbf{x} \rangle_C$ yields

$$\langle \mathbf{x}, (Q^* Q - \mathrm{I})\mathbf{x} \rangle_C = 0 \qquad \text{for all} \qquad \mathbf{x} \in \mathbb{R}^n.$$

The matrix $Q^* Q - \mathrm{I}$ is self-adjoint, and hence Lemma 4.21 implies (4.44). ∎

Corollary 4.33. *There are no norm-preserving matrices of size $m \times n$ if $m < n$.*

Proof. According to Theorem 4.25, $\mathrm{rank}(Q^* Q) = \mathrm{rank}\, Q \leq m$, while the $n \times n$ identity matrix has rank $\mathrm{I} = n$. Thus, if $m < n$, the ranks are not the same, and so the two matrices cannot be equal. Hence, no matrix Q of this size can satisfy (4.44). ∎

In other words, viewing multiplication by Q as defining a linear map $\mathbf{x} \mapsto Q\mathbf{x}$ from $\mathbb{R}^n$ to $\mathbb{R}^m$, the corollary tells us that it is not possible to preserve the norms of all vectors when mapping to a lower dimensional space.

In view of formula (4.20), the norm-preserving condition (4.44) is, explicitly,

$$Q^T K Q = C. \tag{4.45}$$

Let $\mathbf{q}_1, \ldots, \mathbf{q}_n \in \mathbb{R}^m$ denote the columns of Q, where, by Corollary 4.33, $m \geq n$. Then the (i, j) entry of (4.45) is

$$\langle \mathbf{q}_i, \mathbf{q}_j \rangle_K = c_{ij} = \langle \mathbf{e}_i, \mathbf{e}_j \rangle_C. \tag{4.46}$$

In other words, the n columns of Q have the same inner products as the standard basis vectors of $\mathbb{R}^n$. In particular, we deduce:

Proposition 4.34. *Suppose we set $C = \mathrm{I}$, so that $\mathbb{R}^n$ has the Euclidean norm. Then the $m \times n$ matrix Q is norm-preserving, meaning $\|Q\mathbf{x}\|_K = \|\mathbf{x}\|_2$ for all $\mathbf{x} \in \mathbb{R}^n$, if and only if its columns are orthonormal. In this case $Q^* = Q^T K$, and (4.44) reduces to*

$$Q^* Q = Q^T K Q = \mathrm{I}. \tag{4.47}$$

For the remainder of this section, we will concentrate our attention on $n \times n$ matrices Q, and the dot product and Euclidean norm on $\mathbb{R}^n$, so that $C = K = \mathrm{I}$. In this case, the norm-preserving condition (4.45) becomes

$$Q^T Q = \mathrm{I}, \quad \text{and hence} \quad Q^{-1} = Q^T, \quad \text{so} \quad Q Q^T = \mathrm{I}. \tag{4.48}$$

Thus, (4.46) takes the form

$$\mathbf{q}_i \cdot \mathbf{q}_j = \mathbf{q}_i^T \mathbf{q}_j = \begin{cases} 1, & i = j, \\ 0, & i \neq j, \end{cases} \tag{4.49}$$

and so, in accordance with Proposition 4.34, the columns $\mathbf{q}_1, \ldots, \mathbf{q}_n$ of Q form an orthonormal basis of $\mathbb{R}^n$ under the dot product. A matrix Q satisfying (4.48), or, equivalently, (4.49) is known as an *orthogonal matrix*. In other words, a matrix is Euclidean norm-preserving if and only if it is an orthogonal matrix, whose columns form an orthonormal basis. In particular, the $n \times n$ identity matrix I, whose columns are the standard orthonormal basis $\mathbf{e}_1, \ldots, \mathbf{e}_n$ of $\mathbb{R}^n$, is orthogonal. The last equation in (4.48) implies that if Q is orthogonal, so is $Q^{-1} = Q^T$, whose columns, which are the transposed rows of Q, form another, usually different, orthonormal basis of $\mathbb{R}^n$.

Remark. Technically, since its columns for an orthonormal basis, Q should be called an "orthonormal" matrix, not an "orthogonal" matrix. But the terminology is so standard throughout mathematics and physics that we have no choice but to adopt it here. There is no commonly accepted name for a matrix whose columns form an orthogonal but not orthonormal basis. ▲

Example 4.35. The vectors

$$\mathbf{v}_1 = \begin{pmatrix} -1 \\ 1 \\ 1 \end{pmatrix}, \qquad \mathbf{v}_2 = \begin{pmatrix} 2 \\ 1 \\ 1 \end{pmatrix}, \qquad \mathbf{v}_3 = \begin{pmatrix} 0 \\ 1 \\ -1 \end{pmatrix},$$

are mutually orthogonal, and hence, by Theorem 2.18, form a basis of $\mathbb{R}^3$. An orthonormal basis is obtained by dividing each by its length $\|\mathbf{v}_1\| = \sqrt{3}$, $\|\mathbf{v}_2\| = \sqrt{6}$, $\|\mathbf{v}_3\| = \sqrt{2}$, which produces the corresponding orthonormal basis vectors

$$\mathbf{q}_1 = \begin{pmatrix} -\frac{1}{\sqrt{3}} \\ \frac{1}{\sqrt{3}} \\ \frac{1}{\sqrt{3}} \end{pmatrix}, \qquad \mathbf{q}_2 = \begin{pmatrix} \frac{2}{\sqrt{6}} \\ \frac{1}{\sqrt{6}} \\ \frac{1}{\sqrt{6}} \end{pmatrix}, \qquad \mathbf{q}_3 = \begin{pmatrix} 0 \\ \frac{1}{\sqrt{2}} \\ -\frac{1}{\sqrt{2}} \end{pmatrix}.$$

These form the columns of a 3×3 orthogonal matrix

$$Q = (\,\mathbf{q}_1\ \mathbf{q}_2\ \mathbf{q}_3\,) = \begin{pmatrix} -\frac{1}{\sqrt{3}} & \frac{2}{\sqrt{6}} & 0 \\[2mm] \frac{1}{\sqrt{3}} & \frac{1}{\sqrt{6}} & \frac{1}{\sqrt{2}} \\[2mm] \frac{1}{\sqrt{3}} & \frac{1}{\sqrt{6}} & -\frac{1}{\sqrt{2}} \end{pmatrix}, \tag{4.50}$$

which, as the reader can check, satisfies (4.48). The three columns and three (transposed) rows of Q form two different orthonormal bases of $\mathbb{R}^3$. ▲

Example 4.36. The orthonormal bases of $\mathbb{R}^2$ were completely classified in Example 2.16. The columns of an orthogonal 2×2 matrix must be an orthonormal basis, and hence every 2×2 orthogonal matrix has one of two possible forms

$$\begin{pmatrix} \cos\theta & -\sin\theta \\ \sin\theta & \cos\theta \end{pmatrix} \quad \text{or} \quad \begin{pmatrix} \cos\theta & \sin\theta \\ \sin\theta & -\cos\theta \end{pmatrix}, \quad \text{where} \quad 0 \le \theta < 2\pi. \tag{4.51}$$

Those in the first class have determinant $+1$, and, given a vector $\mathbf{v} \in \mathbb{R}^2$, the transformed vector $\mathbf{w} = Q\mathbf{v}$ is obtained by rotating $\mathbf{v}$ through an angle θ. Those in the second class have determinant -1, and $\mathbf{w} = Q\mathbf{v}$ is obtained by reflecting $\mathbf{v}$ through a line that makes an angle $\frac{1}{2}\theta$ with the horizontal axis. Thus, geometrically, each 2×2 orthogonal matrix defines either a rotation or a reflection. ▲

Remark. As in the two-dimensional case discussed in Example 4.36, the set of $n \times n$ orthogonal matrices can be split into rotations and reflections. In three-dimensional space, a rotation can be physically realized, whereas a reflection takes you to Alice's mirror image world. It appears that the only way to distinguish between the two case is via the determinant, which is $+1$ for a rotation and -1 for a reflection. Since we do not cover determinants in this book, we defer to other linear algebra texts, e.g., [181, 224], for details. ▲

Proposition 4.37. *The inverse and the transpose of an orthogonal matrix are both orthogonal matrices. The product of two orthogonal matrices is also orthogonal.*

Proof. The first statement was already established. As for the second, if $Q_1^T Q_1 = I = Q_2^T Q_2$, then $(Q_1 Q_2)^T (Q_1 Q_2) = Q_2^T Q_1^T Q_1 Q_2 = Q_2^T Q_2 = I$, and so $Q_1 Q_2$ is also orthogonal. ■

The two properties in Proposition 4.37 tell us that the set of all orthogonal matrices forms a *group*.[2] The *orthogonal group* lies at the foundation of everyday Euclidean geometry, as well as computer graphics, animation, and gaming, [24], atomic structure and chemistry, [85], crystallography, [123], rigid body mechanics, [87], including robots, spacecraft, satellites, airplanes, drones, and underwater vehicles, and many diverse areas of mathematics.

Proposition 4.38. *Let $\mathbf{u}_1, \ldots, \mathbf{u}_n$ and $\mathbf{v}_1, \ldots, \mathbf{v}_n$ be orthonormal bases of $\mathbb{R}^n$. Then there exists an $n \times n$ orthogonal matrix Q such that $\mathbf{v}_i = Q\mathbf{u}_i$ for $i = 1, \ldots, n$.*

Proof. Let $U = (\,\mathbf{u}_1\ \ldots\ \mathbf{u}_n\,)$ and $V = (\,\mathbf{v}_1\ \ldots\ \mathbf{v}_n\,)$ be the corresponding orthogonal matrices. Let $Q = V U^T = V U^{-1}$, which is orthogonal by Proposition 4.37. Moreover, $V = QU$, and the columns of the latter matrix equation are the desired relations. ■

---

[2] Although they will not play a significant role in this text, groups underlie the mathematical formalization of symmetry and, as such, form one of the most fundamental concepts in advanced mathematics and its applications, particularly quantum mechanics and modern theoretical physics, [167]. Moreover, according to the mathematician Felix Klein, cf. [252], all geometry is based on group theory.

4.5.1 Rigid Motions

In this section, we will investigate functions that preserve distance, as measured by the norm of the difference between vectors in $\mathbb{R}^n$, cf. (2.75).

Definition 4.39. A function $F\colon \mathbb{R}^n \to \mathbb{R}^n$ is called an *isometry* if it preserves distance, meaning

$$d\big(F[\mathbf{v}], F[\mathbf{w}]\big) = d(\mathbf{v}, \mathbf{w}) \qquad \text{for all} \qquad \mathbf{v}, \mathbf{w} \in \mathbb{R}^n. \tag{4.52}$$

The mathematical term *metric* refers to an underlying notion of distance; thus, "isometry" translates as "distance-preserving function". Since the distance between points is just the norm of their difference, $d(\mathbf{v}, \mathbf{w}) = \|\mathbf{v} - \mathbf{w}\|$, the isometry condition (4.52) can be restated as

$$\big\| F[\mathbf{v}] - F[\mathbf{w}] \big\| = \|\mathbf{v} - \mathbf{w}\| \qquad \text{for all} \qquad \mathbf{v}, \mathbf{w} \in \mathbb{R}^n. \tag{4.53}$$

Clearly, any translation (3.62) defines an isometry, since

$$T[\mathbf{v}] - T[\mathbf{w}] = (\mathbf{v} + \mathbf{b}) - (\mathbf{w} + \mathbf{b}) = \mathbf{v} - \mathbf{w}.$$

Let us focus on the ordinary Euclidean distance induced by the Euclidean norm. Functions that preserve Euclidean distance, i.e., Euclidean isometries, are known as *rigid motions*, since they "move" objects in space without deforming them. It can be proved, [253], that the most general Euclidean isometry of $\mathbb{R}^n$ is an affine function, and hence of the form $F[\mathbf{x}] = Q\,\mathbf{x} + \mathbf{b}$, where Q is an $n \times n$ matrix. The isometry condition (4.53) becomes

$$\|Q\mathbf{v} - Q\mathbf{w}\| = \|Q(\mathbf{v} - \mathbf{w})\| = \|\mathbf{v} - \mathbf{w}\|.$$

Writing $\mathbf{x} = \mathbf{v} - \mathbf{w}$, this requires $\|Q\mathbf{x}\|^2 = \|\mathbf{x}\|^2$ for all $\mathbf{x} \in \mathbb{R}^n$. Thus, Q is Euclidean norm-preserving, and hence an orthogonal matrix.

Proposition 4.40. *An affine function* $L[\mathbf{x}] = Q\,\mathbf{x} + \mathbf{b}$ *defines a Euclidean isometry of* $\mathbb{R}^n$ *if and only if* Q *is an orthogonal matrix.*

The linear part $\mathbf{x} \longmapsto Q\,\mathbf{x}$ of an affine isometry represents a rotation or a reflection, and hence every Euclidean rigid motion is a combination of translations, rotations, and reflections. The isometries of $\mathbb{R}^2$ and $\mathbb{R}^3$ are indispensable for understanding of how physical objects move in three-dimensional space. Basic computer graphics and animation require efficient implementation of rigid isometries and their compositions in three-dimensional space — coupled with appropriate perspective maps prescribing the projection of three-dimensional objects onto a two-dimensional viewing screen, [34, 206].

Exercises

5.1. Determine which of the following are orthogonal matrices: $\quad (a)\,\heartsuit \begin{pmatrix} 1 & 1 \\ -1 & 1 \end{pmatrix},$

$(b) \begin{pmatrix} \frac{12}{13} & \frac{5}{13} \\ -\frac{5}{13} & \frac{12}{13} \end{pmatrix}, \quad (c)\,\heartsuit \begin{pmatrix} 0 & 1 & 0 \\ -1 & 0 & 0 \\ 0 & 0 & -1 \end{pmatrix}, \quad (d) \begin{pmatrix} -\frac{1}{3} & \frac{2}{3} & \frac{2}{3} \\ \frac{2}{3} & -\frac{1}{3} & \frac{2}{3} \\ \frac{2}{3} & \frac{2}{3} & -\frac{1}{3} \end{pmatrix}, \quad (e)\,\diamondsuit \begin{pmatrix} \frac{1}{2} & \frac{1}{3} & \frac{1}{4} \\ \frac{1}{3} & \frac{1}{4} & \frac{1}{5} \\ \frac{1}{4} & \frac{1}{5} & \frac{1}{6} \end{pmatrix}.$

5.2. Write down all diagonal $n \times n$ orthogonal matrices. How many are there?

5.3.$\diamondsuit$ Prove that every orthogonal upper triangular matrix is necessarily a diagonal matrix. What diagonal entries are possible?

5.4. *True or false*:
   (a)$\heartsuit$ A matrix whose columns form an orthogonal basis of $\mathbb{R}^n$ is an orthogonal matrix.
   (b)$\diamondsuit$ A matrix whose rows form an orthonormal basis of $\mathbb{R}^n$ is an orthogonal matrix.
   (c) An orthogonal matrix is symmetric if and only if it is a diagonal matrix.

5.5. Which of the indicated maps define isometries of the Euclidean plane?
$$(a)\heartsuit \begin{pmatrix} y \\ -x \end{pmatrix}, \quad (b)\diamondsuit \begin{pmatrix} x - y + 1 \\ x + 2 \end{pmatrix}, \quad (c)\heartsuit \frac{1}{\sqrt{2}} \begin{pmatrix} x + y - 3 \\ x + y - 2 \end{pmatrix}, \quad (d) \frac{1}{5} \begin{pmatrix} 3x + 4y \\ -4x + 3y + 1 \end{pmatrix}.$$

5.6. Which of the following matrices are Euclidean norm-preserving?
$$(a)\heartsuit \begin{pmatrix} 1 & 0 & 0 \\ 0 & 1 & 0 \end{pmatrix}, \quad (b)\heartsuit \begin{pmatrix} 1 & 0 \\ 0 & 1 \\ 0 & 0 \end{pmatrix}, \quad (c)\diamondsuit \begin{pmatrix} \frac{1}{3} & \frac{1}{3} \\ \frac{1}{3} & -\frac{2}{3} \\ \frac{1}{3} & \frac{1}{3} \end{pmatrix}, \quad (d) \begin{pmatrix} \frac{2}{3} & \frac{1}{\sqrt{2}} \\ -\frac{1}{3} & 0 \\ -\frac{2}{3} & \frac{1}{\sqrt{2}} \end{pmatrix}.$$

5.7. *True or false*: There are no norm-preserving linear maps from $\mathbb{R}^2$ to $\mathbb{R}^3$ when $\mathbb{R}^2$ has the norm induced by the inner product corresponding to the matrix $C = \begin{pmatrix} 4 & -1 \\ -1 & 4 \end{pmatrix}$ and $\mathbb{R}^3$ has the Euclidean norm. If true, explain why not. If false, explain how to construct such norm-preserving maps and write down at least one explicit example.

5.8.$\diamondsuit$ Let $\mathbf{v}_1, \dots, \mathbf{v}_n$ and $\mathbf{w}_1, \dots, \mathbf{w}_n$ be two sets of linearly independent vectors in $\mathbb{R}^n$. Show that all their dot products are the same, so $\mathbf{v}_i \cdot \mathbf{v}_j = \mathbf{w}_i \cdot \mathbf{w}_j$ for all $i, j = 1, \dots, n$, if and only if there is an orthogonal matrix Q such that $\mathbf{w}_i = Q\mathbf{v}_i$ for all $i = 1, \dots, n$.

5.9. A set of $n + 1$ points $\mathbf{a}_0, \dots, \mathbf{a}_n \in \mathbb{R}^n$ is said to be *in general position* if the differences $\mathbf{a}_i - \mathbf{a}_j$ for $0 \le i < j \le n$ span $\mathbb{R}^n$. (a) Show that the points are in general position if and only if they do not all lie in a proper affine subspace $W \subsetneq \mathbb{R}^n$, as defined in Exercise 2.7. (b) Let $\mathbf{a}_0, \dots, \mathbf{a}_n$ and $\mathbf{b}_0, \dots, \mathbf{b}_n$ be two sets in general position. Show that there is an isometry $F \colon \mathbb{R}^n \to \mathbb{R}^n$ such that $F[\mathbf{a}_i] = \mathbf{b}_i$ for all $i = 0, \dots, n$, if and only if their interpoint distances agree: $\| \mathbf{a}_i - \mathbf{a}_j \| = \| \mathbf{b}_i - \mathbf{b}_j \|$ for all $0 \le i < j \le n$. *Hint*: Use Exercise 5.8.

4.6 Projection Matrices

In this section, we show how orthogonal projection of a vector onto a subspace, as introduced in Section 2.4, can be realized by matrix multiplication. To begin with, we restrict our attention to the dot product, where the formulas are slightly simpler. At the end of the section we indicate how to modify the constructions for a more general inner product.

Suppose $V \subset \mathbb{R}^n$ is a k-dimensional subspace for some $0 < k < n$. Let $\mathbf{u}_1, \dots, \mathbf{u}_k$ be an orthonormal basis for V. We form the $n \times k$ matrix $U = (\mathbf{u}_1 \ \dots \ \mathbf{u}_k)$ whose columns are the basis vectors. Orthonormality implies that $U^T U = I$ is a $k \times k$ identity matrix. On the other hand, because U is not square, and hence not an orthogonal matrix, the $n \times n$ symmetric matrix
$$P = U U^T \tag{4.54}$$
is *not* necessarily an identity matrix. In fact, P is the matrix that produces the orthogonal projection of vectors onto the subspace V.

Indeed, our orthogonal projection formula (2.39) can be recast into an equivalent matrix form, so that the orthogonal projection of $\mathbf{b} \in \mathbb{R}^n$ onto the subspace V is given by

$$\mathbf{p} = \sum_{i=1}^{k} (\mathbf{u}_i \cdot \mathbf{b})\,\mathbf{u}_i = \sum_{i=1}^{k} \mathbf{u}_i\,(\mathbf{u}_i^T \mathbf{b}) = U\,U^T \mathbf{b} = P\,\mathbf{b}, \tag{4.55}$$

where the third equality follows from the alternative matrix multiplication formula (3.10). Note that, by the properties of orthogonal projection,

$$\operatorname{img} P = V, \qquad \ker P = V^{\perp}, \tag{4.56}$$

the latter being the orthogonal complement to V. The *projection matrix* P satisfies

$$P^2 = U\,U^T U\,U^T = U\,U^T = P, \tag{4.57}$$

and hence P is idempotent, as in Exercise 4.6. This is a restatement of the fact orthogonal projection does not affect a vector that is already in the subspace, and hence reprojecting a projected vector does nothing further to it. In the particular case when Q is an $n \times n$ orthogonal matrix whose columns form an orthonormal basis of $\mathbb{R}^n$, the projection matrix $Q\,Q^T = \mathrm{I}$ is the identity matrix, reflecting the fact that orthogonal projection of a vector $\mathbf{v} \in \mathbb{R}^n$ onto $\mathbb{R}^n$ is simply $\mathbf{v}$ itself, and (4.55) reduces to the orthonormal basis formula (2.33).

Vice versa, the $n \times n$ symmetric matrix

$$R = \mathrm{I} - P = \mathrm{I} - U\,U^T \tag{4.58}$$

corresponds to orthogonal projection onto the orthogonal complementary subspace $V^{\perp}$. If $\mathbf{w}_1, \ldots, \mathbf{w}_{n-k}$ are an orthonormal basis for $V^{\perp}$ and $W = (\,\mathbf{w}_1 \ \ldots \ \mathbf{w}_{n-k}\,)$ is the corresponding $n \times (n - k)$ matrix, so that $W^T W = \mathrm{I}$, then

$$R = W\,W^T = \mathrm{I} - U\,U^T = \mathrm{I} - P, \qquad \operatorname{img} R = V^{\perp}, \qquad \ker R = V. \tag{4.59}$$

The orthogonal decomposition formula in Theorem 2.32 can thus be written as

$$\mathbf{b} = \mathbf{p} + \mathbf{q}, \qquad \text{where} \quad \mathbf{p} = P\,\mathbf{b} \in V \quad \text{and} \quad \mathbf{q} = R\,\mathbf{b} \in V^{\perp}. \tag{4.60}$$

Note that the orthonormal bases of V and $V^{\perp}$ can be combined to form an orthonormal basis $\mathbf{u}_1, \ldots, \mathbf{u}_k, \mathbf{w}_1, \ldots, \mathbf{w}_{n-k}$ of $\mathbb{R}^n$. Equation (4.59) is thereby equivalent to the condition that the corresponding $n \times n$ orthogonal matrix $Q = (U\,W) = (\,\mathbf{u}_1 \ldots \mathbf{u}_k\,\mathbf{w}_1 \ldots \mathbf{w}_{n-k}\,)$ obtained by combining U and W satisfies $Q\,Q^T = \mathrm{I} = Q^T Q$.

Example 4.41. Consider the line

$$V = \left\{ (\,t, 0, -t\,)^T \ \middle| \ t \in \mathbb{R} \right\} \subset \mathbb{R}^3$$

in the direction of the vector $\mathbf{v} = (\,1, 0, -1\,)^T$. An orthonormal basis of V is obtained by dividing $\mathbf{v}$ by its length, producing $\mathbf{u} = \left(\frac{1}{\sqrt{2}}, 0, -\frac{1}{\sqrt{2}} \right)^T$. The corresponding projection matrix is

$$P = \mathbf{u}\,\mathbf{u}^T = \begin{pmatrix} \frac{1}{\sqrt{2}} \\ 0 \\ -\frac{1}{\sqrt{2}} \end{pmatrix} \left(\frac{1}{\sqrt{2}}, 0, -\frac{1}{\sqrt{2}} \right) = \begin{pmatrix} \frac{1}{2} & 0 & -\frac{1}{2} \\ 0 & 0 & 0 \\ -\frac{1}{2} & 0 & \frac{1}{2} \end{pmatrix}.$$

Given

$$\mathbf{b} = \begin{pmatrix} b_1 \\ b_2 \\ b_3 \end{pmatrix} \in \mathbb{R}^3, \qquad \text{then} \qquad P\mathbf{b} = \begin{pmatrix} \frac{1}{2}\,(b_1 - b_3) \\ 0 \\ \frac{1}{2}\,(b_3 - b_1) \end{pmatrix} \in V$$

is the orthogonal projection of $\mathbf{b}$ onto the line V. Furthermore, the matrix

$$R = \mathrm{I} - P = \begin{pmatrix} \frac{1}{2} & 0 & \frac{1}{2} \\ 0 & 1 & 0 \\ \frac{1}{2} & 0 & 1 - \frac{1}{2} \end{pmatrix}$$

projects $\mathbf{b}$ onto the line's orthogonal complement $V^\perp$, so

$$R\,\mathbf{b} = \begin{pmatrix} \frac{1}{2}\,(b_1 + b_3) \\ b_2 \\ \frac{1}{2}\,(b_1 + b_3) \end{pmatrix} \in V^\perp = \mathbf{u}^\perp.$$

Note that the line's orthogonal complement

$$V^\perp = \{\, x_1 - x_3 = 0 \,\} \subset \mathbb{R}^3$$

is the plane with normal vector $\mathbf{v} \in V$. An orthonormal basis for $V^\perp$ is provided by the vectors $\mathbf{w}_1 = \left(\frac{1}{\sqrt{2}}, 0, \frac{1}{\sqrt{2}} \right)^T$, $\mathbf{w}_2 = (0, 1, 0)^T$. Setting

$$W = (\mathbf{w}_1 \; \mathbf{w}_2) = \begin{pmatrix} \frac{1}{\sqrt{2}} & 0 \\ 0 & 1 \\ \frac{1}{\sqrt{2}} & 0 \end{pmatrix},$$

one easily sees that the corresponding projection matrix coincides with

$$R = W W^T = \begin{pmatrix} \frac{1}{\sqrt{2}} & 0 \\ 0 & 1 \\ \frac{1}{\sqrt{2}} & 0 \end{pmatrix} \begin{pmatrix} \frac{1}{\sqrt{2}} & 0 & \frac{1}{\sqrt{2}} \\ 0 & 1 & 0 \end{pmatrix} = \begin{pmatrix} \frac{1}{2} & 0 & \frac{1}{2} \\ 0 & 1 & 0 \\ \frac{1}{2} & 0 & 1 - \frac{1}{2} \end{pmatrix},$$

thereby confirming formula (4.59) in this example. ▲

Finally, let us explain how to modify the above formulas in the case of an inner product $\langle \mathbf{x}, \mathbf{y} \rangle_C = \mathbf{x}^T C \mathbf{y}$ on $\mathbb{R}^n$ provided by the symmetric positive definite matrix C. In this case, the orthonormality of the basis $\mathbf{u}_1, \ldots, \mathbf{u}_k$ of V implies that the $n \times k$ matrix $U = (\mathbf{u}_1 \; \ldots \; \mathbf{u}_k)$ satisfies $U^T C U = \mathrm{I}$. Let us continue to use the dot product on the domain space $\mathbb{R}^k$ of U, so that, referring back to (4.47) (with C replacing K),

$$U^* = U^T C, \qquad \text{and hence} \qquad U^* U = U^T C U = \mathrm{I}. \tag{4.61}$$

The orthogonal projection formula (4.55) becomes

$$\mathbf{p} = \sum_{i=1}^{k} \langle \mathbf{u}_i, \mathbf{b} \rangle \, \mathbf{u}_i = \sum_{i=1}^{k} \mathbf{u}_i \, (\mathbf{u}_i^T C \mathbf{b}) = U U^T C \mathbf{b} = P\mathbf{b}, \tag{4.62}$$

so that

$$P = U U^T C = U U^* \tag{4.63}$$

is the orthogonal projection matrix onto the subspace V under the prescribed inner product. Idempotency of the projection matrix, $P^2 = P$, is a straightforward consequence of (4.61). Finally, setting $W = (\mathbf{w}_1 \ \dots \ \mathbf{w}_{n-k})$ to be the $n \times (n-k)$ matrix whose columns form an orthonormal basis for $V^\perp$, so that $W^T C W = \mathrm{I}$, and $R = W W^T C = \mathrm{I} - U U^T C$ is the projection matrix onto $V^\perp$.

Exercises

6.1. Using the dot product, write out the projection matrix corresponding to the subspaces spanned by

$$(a)\ \heartsuit\ \begin{pmatrix} \frac{1}{\sqrt{2}} \\ \frac{1}{\sqrt{2}} \end{pmatrix},\quad (b)\ \begin{pmatrix} \frac{2}{3} \\ -\frac{2}{3} \\ \frac{1}{3} \end{pmatrix},\quad (c)\ \heartsuit\ \begin{pmatrix} \frac{1}{\sqrt{6}} \\ -\frac{2}{\sqrt{6}} \\ \frac{1}{\sqrt{6}} \end{pmatrix},\quad \begin{pmatrix} \frac{1}{\sqrt{3}} \\ \frac{1}{\sqrt{3}} \\ \frac{1}{\sqrt{3}} \end{pmatrix},\quad (d)\ \diamondsuit\ \begin{pmatrix} \frac{1}{2} \\ \frac{1}{2} \\ \frac{1}{2} \\ -\frac{1}{2} \end{pmatrix},\ \begin{pmatrix} \frac{1}{2} \\ -\frac{1}{2} \\ \frac{1}{2} \\ \frac{1}{2} \end{pmatrix},\ \begin{pmatrix} \frac{1}{2} \\ \frac{1}{2} \\ -\frac{1}{2} \\ \frac{1}{2} \end{pmatrix}.$$

6.2. Write out the projection matrices onto the orthogonal complements of the subspaces in Exercise 6.1.

6.3. $\heartsuit$ Prove that a projection matrix is positive semidefinite. When is it positive definite?

6.4. *True or false*: Given a linearly independent set of vectors, the associated Gram matrix is a projection matrix.

6.5. $\diamondsuit$ Let $V, \widetilde{V} \subset \mathbb{R}^n$ be two subspaces of $\mathbb{R}^n$, equipped with the dot product. Let $P, \widetilde{P}$ be the corresponding projection matrices. *True or false*: (a) The product matrix $Q = \widetilde{P} P$ is a projection matrix. If true, which subspace does Q project onto? If false, describe the effect of Q on a vector in $\mathbb{R}^n$. (b) The matrices P and $\widetilde{P}$ commute.

6.6. Let $\mathbb{R}^n$ be equipped with a given inner product. (a) Write out a definition of the orthogonal projection of a point $\mathbf{c} \in \mathbb{R}^n$ onto the affine subspace $W = \mathbf{b} + V \subset \mathbb{R}^n$, as defined in Exercise 2.7 in Chapter 1. (b) Find a formula for the projection you defined in part (a). (c) Test your formula on the affine subspaces (i) $\{x - 2y = 1\} \subset \mathbb{R}^2$, (ii) $\{x + y + z = 1\} \subset \mathbb{R}^3$, using the dot product. (d) Choose a weighted inner product on the indicated Euclidean space, and redo part (c).

4.7 The General QR Factorization and the Solution of Linear Systems

Python Notebook: QR Factorization (.ipynb)

Until now, we have not seriously discussed how to actually solve a linear system of algebraic equations. Readers who are already familiar with Gaussian elimination, as presented in almost all linear algebra texts, including [181, 224], already know an algorithm that can, at least in principle, systematically solve linear systems, checking that the right hand side belongs to the image of the coefficient matrix, i.e., the required compatibility conditions are satisfied. However, as we argued above, when dealing with the large linear systems arising in machine learning, data analysis, and elsewhere, Gaussian elimination exhibits significant weaknesses, and there is a need for a better, more numerically robust algorithm.

In this section we develop one such algorithm, that is based on the so-called QR factorization of the coefficient matrix. This is simply a matrix reformulation of the general Gram–Schmidt algorithm that was developed in Section 2.5 for constructing orthonormal bases of subspaces spanned by a collection of vectors, in this case the columns of the coefficient matrix, which span its image. We will only use the dot product and Euclidean norm when constructing our orthonormal bases here, although the methods can be straightforwardly adapted to more general inner products; see Exercise 3.13 in Chapter 5.

4.7.1　The QR Factorization of a Matrix

As usual, we write our linear system in vectorial form

$$A\mathbf{x} = \mathbf{b}, \tag{4.64}$$

where the coefficient matrix A has size $m \times n$, the right hand side $\mathbf{b} \in \mathbb{R}^m$, and the desired solution (if it exists) $\mathbf{x} \in \mathbb{R}^n$. We assume $A \neq O$, in order to avoid a meaningless system of equations. Let $\mathbf{v}_1, \ldots, \mathbf{v}_n \in \mathbb{R}^m$ denote the columns of $A = (\mathbf{v}_1 \ \ldots \ \mathbf{v}_n)$. We define $V_k = \mathrm{span}\,\{\mathbf{v}_1, \ldots, \mathbf{v}_k\}$ to be the subspace of $\mathbb{R}^m$ spanned by the first k columns; in particular, $V_n = \mathrm{img}\,A$. We set $s_k = \dim V_k$, and hence, by (3.27), $s_n = s = \mathrm{rank}\,A > 0$. The general Gram–Schmidt algorithm presented in Section 2.5 recursively constructs orthonormal bases $\mathbf{u}_1, \ldots, \mathbf{u}_{s_k}$ of each subspace V_k; in particular, $\mathbf{u}_1, \ldots, \mathbf{u}_s$ form an orthonormal basis for the image (column space) of A.

Now, copying (2.47), one can express each column

$$\mathbf{v}_k = r_{1,k}\mathbf{u}_1 + \cdots + r_{s_k,k}\mathbf{u}_{s_k}, \qquad k = 1, \ldots, n, \tag{4.65}$$

as a linear combination of the basis vectors of V_k. Since we are using the dot product here, the formulas for the coefficients are found by the usual orthonormal basis formula, $r_{i,k} = \mathbf{u}_i \cdot \mathbf{v}_k$; see Theorem 2.19. In practice, one should use the numerically stable version of the Gram–Schmidt algorithm to effect the computations.

We can rewrite the Gram–Schmidt equations in an equivalent matrix form to produce the (*general*[3]) QR *factorization*

$$A = QR, \tag{4.66}$$

of the matrix A, so that (4.65) is simply the k-th column of this matrix equation. Here $Q = (\mathbf{u}_1 \ \ldots \ \mathbf{u}_s)$ is the $m \times s$ matrix containing the orthonormal basis vectors of $V_n = \mathrm{img}\,A$. The $s \times n$ matrix R is in what is called *row echelon form*, [181, 224], meaning that it has the

---

[3]The classical QR factorization, [181], is the particular case when A is a nonsingular square matrix.

following "staircase" structure[4]:

$$
\begin{pmatrix}
\circledast & * & \cdots & * & * & * & \cdots & * & * & \cdots & \cdots & * & * & * & \cdots & * \\
0 & 0 & \cdots & 0 & \circledast & * & \cdots & * & * & \cdots & \cdots & * & * & * & \cdots & * \\
0 & 0 & \cdots & 0 & 0 & 0 & \cdots & 0 & \circledast & \cdots & \cdots & * & * & * & \cdots & * \\
0 & 0 & \cdots & 0 & 0 & \cdots & 0 & 0 & 0 & \cdots & \cdots & * & * & * & \cdots & * \\
\vdots & \vdots & \ddots & \vdots & \vdots & \vdots & \ddots & \vdots & \vdots & \ddots & \ddots & \vdots & \vdots & \vdots & \ddots & \vdots \\
0 & 0 & \cdots & 0 & 0 & \cdots & 0 & 0 & 0 & \cdots & \cdots & 0 & \circledast & * & \cdots & *
\end{pmatrix}
\tag{4.67}
$$

There may also be one or more all zero initial columns, which would result from initial zero columns of A. (Although for linear systems, this possibility effectively never arises since it would mean that one or more variables do not appear anywhere in the system and can thus be ignored.) The starred entries are the coefficients $r_{i,j}$ appearing in (4.65). The s circled stars, $\circledast$, (one per row) are the nonzero entries $r_{s_k,k} > 0$, that arise when the dimension of the subspaces increase, and are given in (2.46). We will call these entries the QR *pivots*. Each row of R contains exactly one pivot that is either its first entry, or follows one or more initial entries of 0; moreover all successive pivots appear in columns to its right. The entries below the "staircase", indicated by the solid line, are all zero, while the non-pivot entries above the staircase, indicated by uncircled stars, can be either zero or nonzero. For later reference, let

$$
1 \le \ell_1 < \ell_2 < \cdots < \ell_s \le n
\tag{4.68}
$$

index the columns of R containing the pivots. The pivot columns are easily seen to be linearly independent vectors (see Exercise 7.6), and hence form a basis for $\mathbb{R}^s = \operatorname{img} R$, which implies that $\operatorname{rank} R = s = \operatorname{rank} A$.

In the classical version of this algorithm, cf. [181], A is a square, nonsingular $n \times n$ matrix, so that $\operatorname{rank} A = n$, and hence Q is an orthogonal $n \times n$ matrix whose columns are the orthonormal basis $\mathbf{u}_1, \ldots, \mathbf{u}_n$ of $\operatorname{img} A = \mathbb{R}^n$ obtained by applying the Gram–Schmidt process to the columns of A, while R is a nonsingular $n \times n$ upper triangular matrix with strictly positive diagonal entries, i.e., a positive upper triangular matrix.

Example 4.42. The columns of the nonsingular matrix $A = \begin{pmatrix} 1 & 1 & 2 \\ 1 & 0 & -2 \\ -1 & 2 & 3 \end{pmatrix}$ are the vectors considered in Example 2.28. The orthonormal basis $\mathbf{u}_1, \mathbf{u}_2, \mathbf{u}_3$ and coefficients r_{ij} constructed using the Gram–Schmidt algorithm lead to the orthogonal and upper triangular matrices

$$
Q = \begin{pmatrix}
\frac{1}{\sqrt{3}} & \frac{4}{\sqrt{42}} & \frac{2}{\sqrt{14}} \\
\frac{1}{\sqrt{3}} & \frac{1}{\sqrt{42}} & -\frac{3}{\sqrt{14}} \\
-\frac{1}{\sqrt{3}} & \frac{5}{\sqrt{42}} & -\frac{1}{\sqrt{14}}
\end{pmatrix},
\qquad
R = \begin{pmatrix}
\sqrt{3} & -\frac{1}{\sqrt{3}} & -\sqrt{3} \\
0 & \frac{\sqrt{14}}{\sqrt{3}} & \frac{\sqrt{21}}{\sqrt{2}} \\
0 & 0 & \frac{\sqrt{7}}{\sqrt{2}}
\end{pmatrix}.
\tag{4.69}
$$

The reader may wish to verify that, indeed, $A = QR$. ▲

---

[4]Unlike the row echelon matrices appearing in Gaussian elimination, R contains no all zero rows.

Example 4.43. The vectors in the Gram–Schmidt Example 2.27 correspond to the factorization $A = QR$, where

$$A = \begin{pmatrix} 1 & 3 & 3 & 2 \\ 2 & 6 & 1 & -1 \\ -2 & -6 & 1 & 3 \end{pmatrix}, \quad Q = \begin{pmatrix} \frac{1}{3} & \frac{8}{3\sqrt{10}} \\ \frac{2}{3} & \frac{1}{3\sqrt{10}} \\ -\frac{2}{3} & \frac{5}{3\sqrt{10}} \end{pmatrix}, \quad R = \begin{pmatrix} 3 & 9 & 1 & -2 \\ 0 & 0 & \sqrt{10} & \sqrt{10} \end{pmatrix}. \quad (4.70)$$

Observe that $\operatorname{rank} A = \operatorname{rank} R = 2$, and that $A = QR$. ▲

We will later need to know the uniqueness of the QR factorization.

Proposition 4.44. *Let $A \neq O$. Suppose $A = QR = \widehat{Q}\widehat{R}$, where Q and $\widehat{Q}$ have orthonormal columns, while R and $\widehat{R}$ are in row echelon form with positive pivots. Then $Q = \widehat{Q}$ and $R = \widehat{R}$.*

Proof. Let A have size $m \times n$. Let $Q = (\mathbf{u}_1 \ \ldots \ \mathbf{u}_r)$ and $Q = (\widehat{\mathbf{u}}_1 \ \ldots \ \widehat{\mathbf{u}}_r)$ where, owing to the form of the factorization, $r = \operatorname{rank} A$. Let k be the smaller of the indices of the columns that contain the first pivot of R or $\widehat{R}$. (Initial zero columns of both do not produce any restrictions.) Then the k-th column of the product matrix $QR = \widehat{Q}\widehat{R}$ is

$$r_{1k}\mathbf{u}_1 = \widehat{r}_{1k}\widehat{\mathbf{u}}_1.$$

Taking norms of both sides and using the fact that the columns of $Q, \widehat{Q}$ are unit vectors, we deduce that $|r_{1k}| = |\widehat{r}_{1k}|$. Thus both $r_{1k}, \widehat{r}_{1k}$ are nonzero and must be the first pivot of their respective matrices, so $k = \ell_1$. Since we assume that they are positive, this implies $r_{1k} = \widehat{r}_{1k}$ and hence $\mathbf{u}_1 = \widehat{\mathbf{u}}_1$.

We now proceed inductively to consider the following columns. For column $\ell_{\nu-1} < k \leq \ell_\nu$, where $1 < \nu \leq r$, the inductive hypothesis is that $\mathbf{u}_1 = \widehat{\mathbf{u}}_1, \ \ldots \ , \mathbf{u}_{\ell_{\nu-1}} = \widehat{\mathbf{u}}_{\ell_{\nu-1}}$, and $r_{ij} = \widehat{r}_{ij}$ for $i = 1, \ldots, m$ and $j = 1, \ldots, k-1$. If $k < \ell_\nu$, the k-th column of $QR = \widehat{Q}\widehat{R}$ is

$$r_{1k}\mathbf{u}_1 + \cdots + r_{\ell_{\nu-1}k}\mathbf{u}_{\ell_{\nu-1}} = \widehat{r}_{1k}\widehat{\mathbf{u}}_1 + \cdots + \widehat{r}_{\ell_{\nu-1}k}\widehat{\mathbf{u}}_{\ell_{\nu-1}}.$$

Taking the dot product with each $\mathbf{u}_i = \widehat{\mathbf{u}}_i$ for $i = 1, \ldots, \ell_{\nu-1}$ produces $r_{ik} = \widehat{r}_{ik}$. On the other hand, if $k = \ell_\nu$, then the k-th column is

$$r_{1k}\mathbf{u}_1 + \cdots + r_{\ell_{\nu-1}k}\mathbf{u}_{\ell_{\nu-1}} + r_{\ell_\nu k}\mathbf{u}_{\ell_\nu} = \widehat{r}_{1k}\widehat{\mathbf{u}}_1 + \cdots + \widehat{r}_{\ell_{\nu-1}k}\widehat{\mathbf{u}}_{\ell_{\nu-1}} + r_{\ell_\nu k}\mathbf{u}_{\ell_\nu}.$$

As before, taking the dot product with each $\mathbf{u}_i = \widehat{\mathbf{u}}_i$ for $i = 1, \ldots, \ell_{\nu-1}$ produces $r_{ik} = \widehat{r}_{ik}$ for $1 \leq i \leq \ell_{\nu-1}$. Taking the squared Euclidean norms of both sides of the preceding equation and using orthonormality, we find

$$r_{1k}^2 + \cdots + r_{\ell_{\nu-1}k}^2 + r_{\ell_\nu k}^2 = \widehat{r}_{1k}^2 + \cdots + \widehat{r}_{\ell_{\nu-1}k}^2 + r_{\ell_\nu k}^2,$$

which, by what we already showed, implies $r_{\ell_\nu k}^2 = r_{\ell_\nu k}^2$ and hence, by positivity, $r_{\ell_\nu k} = r_{\ell_\nu k}$, which completes the inductive step. ■

4.7.2 Solutions to Linear Systems and Least Squares

Let us see how we can use the general $A = QR$ factorization (4.66) to solve the linear system (4.64). Replacing the coefficient matrix A produces the equivalent system

$$QR\mathbf{x} = \mathbf{b}. \tag{4.71}$$

Since the columns of Q are orthonormal, we have $Q^T Q = I$, bearing in mind that, in general, QQ^T is not an identity matrix; see (4.73) below. Thus, after multiplying by Q^T, the linear system (4.71) becomes

$$R\mathbf{x} = \mathbf{y}, \qquad \text{where} \qquad \mathbf{y} = Q^T \mathbf{b}. \tag{4.72}$$

Suppose that $\mathbf{b} \in \operatorname{img} A$, and hence the system (4.71) is compatible and, furthermore, haa one or more solutions $\mathbf{x}$. The reduced "row echelon" system (4.72) can then be solved by *back substitution*. The variables x_i assigned to columns of R that do *not* contain a pivot are called *free variables*, and can assume any value. If $s = \operatorname{rank} A = n$, then there are no free variables, and the solution is unique. The remaining *pivot variables* associated with the pivot columns, namely $x_{\ell_1}, x_{\ell_2}, \ldots, x_{\ell_s}$, are then expressed in terms of the free variables by working backwards, as follows.

Owing to the staircase structure of R, and using the indexing (4.68) of the pivot columns, the last equation only involves $x_{\ell_s}, \ldots, x_n$, and the coefficient of the last pivot variable x_{ℓ_s} is the nonzero pivot r_{s,ℓ_s}. We can thus solve for x_{ℓ_s} in terms of the subsequent free variables (if any), namely $x_{\ell_s+1}, \ldots, x_n$. The penultimate equation only involves $x_{\ell_{s-1}}, \ldots, x_n$, and again the coefficient of the pivot variable $x_{\ell_{s-1}}$ is the nonzero pivot. Thus, we can solve this equation for $x_{\ell_{s-1}}$ in terms of $x_{\ell_{s-1}+1}, \ldots, x_n$. Of these, x_{ℓ_s} has already been determined, and the remainder are all free, and hence $x_{\ell_{s-1}}$ can be expressed in terms of the free variables only. And so on until the first equation has been solved for the first pivot variable x_{ℓ_1} in terms of the remaining variables, and hence in terms of the free variables. In this way, the pivot variables are all expressed in terms of the free variables. Unless $\ker R = \{\mathbf{0}\}$, which implies $s = n$ and R is upper triangular, the solution depends on $n - s$ free variables and is not unique. In particular, if the right hand side $\mathbf{b} = \mathbf{0}$, we obtain a formula for the general element of $\ker R = \ker A$, which, by the Fundamental Theorem 4.24 is a subspace of dimension $n - s$, written as a linear combination of $n - s$ basis vectors whose coefficients are the free variables.

On the other hand, suppose $\mathbf{b} \notin \operatorname{img} A$. The row echelon system (4.72) can still be solved by back substitution as above, but the resulting solution $\mathbf{x}$ no longer solves the original linear system (4.64), which, after all, is incompatible. We claim that $\mathbf{x}$ is, in fact, the least squares solution to the incompatible system (4.64), in accordance with the following definition:

Definition 4.45. A *least squares solution* to a linear system of equations $A\mathbf{x} = \mathbf{b}$ is a vector $\mathbf{x}^\star \in \mathbb{R}^n$ that minimizes the Euclidean norm $\| A\mathbf{x} - \mathbf{b} \|$ over all $\mathbf{x} \in \mathbb{R}^n$.

If the system actually has a solution $\mathbf{x}^\star$, so $A\mathbf{x}^\star = \mathbf{b} \in \operatorname{img} A$, then it is automatically the least squares solution, since $\| A\mathbf{x}^\star - \mathbf{b} \| = 0$, which is clearly the minimum value. The concept of least squares solution is new only when the system does not have a solution, i.e., $\mathbf{b}$ does not lie in the image of A, and so the minimum is strictly positive.

Lemma 4.46. *A vector* $\mathbf{x}^\star \in \mathbb{R}^n$ *is a least squares solution to the linear system* $A\mathbf{x} = \mathbf{b}$ *if and only if* $A\mathbf{x}^\star = \mathbf{p}$ *is the orthogonal projection of* $\mathbf{b}$ *onto* $\operatorname{img} A$.

Proof. According to Theorem 2.25, the orthogonal projection $\mathbf{p}$ of $\mathbf{b}$ onto img A is the closest point to $\mathbf{b}$ in img A, meaning the vector $\mathbf{v} = \mathbf{p}$ that minimizes $\|\mathbf{v} - \mathbf{b}\|$ over all vectors $\mathbf{v} = A\mathbf{x} \in$ img A. This implies that $\mathbf{p} = A\mathbf{x}^{\star}$ minimizes $\|A\mathbf{x} - \mathbf{b}\|$, in accordance with Definition 4.45. $\blacksquare$

> **Theorem 4.47.** *Let $A = QR$ be a nonzero $m \times n$ matrix. Given $\mathbf{b} \in \mathbb{R}^n$, let us set $\mathbf{y} = Q^T\mathbf{b} \in \mathbb{R}^s$, where $s = \operatorname{rank} A$. Then every solution $\mathbf{x}$ to the row echelon system $R\mathbf{x} = \mathbf{y}$ is a least squares solution to the linear system $A\mathbf{x} = \mathbf{b}$. In particular, if $\mathbf{b} \in$ img A, then $\mathbf{x}$ solves the linear system.*

Proof. Given that the columns $\mathbf{u}_1, \dots, \mathbf{u}_s$ of Q form an orthonormal basis of img A, according to (4.55) (replacing U by Q), the orthogonal projection of $\mathbf{b}$ onto img A is given by

$$\mathbf{p} = QQ^T\mathbf{b}, \tag{4.73}$$

i.e., QQ^T is the projection matrix onto img A. On the other hand, multiplying (4.72) by Q, we see that

$$QR\mathbf{x} = A\mathbf{x} = Q\mathbf{y} = QQ^T\mathbf{b} = \mathbf{p},$$

and hence Lemma 4.46 tells us that $\mathbf{x}$ is a least squares solution. The least squares solution is unique if and only if $\operatorname{rank} A = n$; otherwise we can add in any element of $\ker A$ without affecting its status as a least squares solution. $\blacksquare$

The method of least squares is of great importance in linear analysis, and has a vast range of applications in data fitting, statistics, approximation theory, and beyond. We refer the reader to [181, 222, 224] for details. We remark that the usual approach to least squares requires construction of the so-called normal equations, which we will cover in Section 6.2.2. Strikingly, the present method, based on the general QR factorization of the coefficient matrix, completely bypasses the less direct normal equations.

Example 4.48. Let us use the factorization constructed in (4.70) to construct the solution $\mathbf{x} = (x, y, z, w)^T$ to a couple of linear systems of the form $A\mathbf{x} = \mathbf{b}$, i.e.,

$$x + 3y + 3z + 2w = b_1, \qquad 2x + 6y + z - w = b_2, \qquad -2x - 6y + z + 3w = b_3.$$

. First, if $\mathbf{b} = (1, -3, 5)^T$, the row echelon system

$$R\mathbf{x} = Q^T\mathbf{b} = \begin{pmatrix} -5 \\ \sqrt{10} \end{pmatrix} \qquad \text{is} \qquad \begin{aligned} 3x + 9y + z - 2w &= -5, \\ \sqrt{10}\,z + \sqrt{10}\,w &= \sqrt{10}, \end{aligned}$$

and its general solution is easily found by back substitution. The basic variables x, z correspond to the columns in R with pivots, while y, w are the free variables, of which there are two since $\dim \ker A = 2 = 4 - \operatorname{rank} A$. We first solve the second equation for the basic variable $z = 1 - w$. Substituting this into the first equation we solve for $x = -2 - 3y + w$. Observe that the resulting solution

$$\mathbf{x} = \begin{pmatrix} -2 - 3y + w \\ y \\ 1 - w \\ w \end{pmatrix} = \begin{pmatrix} -2 \\ 0 \\ 1 \\ 0 \end{pmatrix} + y\begin{pmatrix} -3 \\ 1 \\ 0 \\ 0 \end{pmatrix} + w\begin{pmatrix} 1 \\ 0 \\ -1 \\ 1 \end{pmatrix}$$

satisfies $A\mathbf{x} = \mathbf{b}$, implying that $\mathbf{b} \in \operatorname{img} A$ and $\mathbf{x}$ is a bona fide solution. Also, the terms involving y and w form the general solution to the homogeneous system $A\mathbf{x} = 0$, and the vectors they multiply form a basis for $\ker A$. An orthonormal basis can be constructed using the Gram–Schmidt process, producing

$$\mathbf{u}_1 = \left(-\frac{3}{\sqrt{10}}, \frac{1}{\sqrt{10}}, 0, 0\right)^T, \qquad \mathbf{u}_2 = \left(\frac{1}{\sqrt{210}}, \frac{3}{\sqrt{210}}, -\frac{10}{\sqrt{210}}, \frac{10}{\sqrt{210}}\right)^T.$$

According to Theorem 4.27, the unique solution $\mathbf{x}^\star$ of minimal Euclidean norm is obtained by projecting any solution to $\operatorname{coimg} A = (\ker A)^\perp$, whence

$$\mathbf{x}^\star = \mathbf{x} - (\mathbf{x} \cdot \mathbf{u}_1)\mathbf{u}_1 - (\mathbf{x} \cdot \mathbf{u}_2)\mathbf{u}_2 = \left(-\tfrac{1}{7}, -\tfrac{3}{7}, \tfrac{3}{7}, \tfrac{4}{7}\right)^T.$$

When performing this computation, one can set $y = w = 0$ in the above solution formula since the kernel vectors project to zero. One can check that it minimizes the Euclidean norm $\|\mathbf{x}\|$ among all possible solutions.

Second, if $\mathbf{b} = (1, 0, -1)^T$, the general solution to the row echelon system

$$R\mathbf{x} = Q^T\mathbf{b} = \begin{pmatrix} 1 \\ \frac{1}{\sqrt{10}} \end{pmatrix}, \qquad \text{or, in detail,} \qquad \begin{aligned} 3x + 9y + z - 2w &= 1, \\ \sqrt{10}\,z + \sqrt{10}\,w &= \frac{1}{\sqrt{10}}, \end{aligned}$$

is $x = \tfrac{3}{10} - 3y + w$, $z = \tfrac{1}{10} - w$. In this case,

$$\mathbf{x} = \begin{pmatrix} \frac{3}{10} - 3y + w \\ y \\ \frac{1}{10} - w \\ w \end{pmatrix} = \begin{pmatrix} \frac{3}{10} \\ 0 \\ \frac{1}{10} \\ 0 \end{pmatrix} + y\begin{pmatrix} -3 \\ 1 \\ 0 \\ 0 \end{pmatrix} + w\begin{pmatrix} 1 \\ 0 \\ -1 \\ 1 \end{pmatrix}.$$

Since $A\mathbf{x} = \left(\tfrac{3}{5}, \tfrac{7}{10}, -\tfrac{1}{2}\right)^T \neq \mathbf{b}$, the system does not have an exact solution, but $\mathbf{x}$ is the general least squares solution. As before, one can obtain the unique least squares solution of minimal norm by projecting to $\operatorname{coimg} A = (\ker A)^\perp$, which produces

$$\mathbf{x}^\star = \mathbf{x} - (\mathbf{x} \cdot \mathbf{u}_1)\mathbf{u}_1 - (\mathbf{x} \cdot \mathbf{u}_2)\mathbf{u}_2 = \left(\tfrac{1}{30}, \tfrac{1}{10}, \tfrac{1}{15}, \tfrac{1}{30}\right)^T. \qquad \blacktriangle$$

Remark. If we precondition by omitting small or zero columns of the coefficient matrix, the effect is to omit the corresponding variable x_i, which is effectively a free variable for the original linear system, but its value does not affect the values of any of the other variables. If we further scale some or all of the columns of A to make them of unit norm (or at least not too large or small in norm), the effect is to replace the coefficient matrix by $\widetilde{A} = AD$, where D is the diagonal matrix containing the scaling factors. Solving $\widetilde{A}\widetilde{\mathbf{x}} = \mathbf{b}$ by the above algorithm, we recover the corresponding values of the original variables $\mathbf{x} = D^{-1}\widetilde{\mathbf{x}}$ by dividing the entries of $\widetilde{\mathbf{x}}$ by the corresponding scaling factors. $\qquad \blacktriangle$

Exercises

7.1. Find the QR factorization of the following matrices: $(a)\,\heartsuit$ $\begin{pmatrix} 1 & -3 \\ 2 & 1 \end{pmatrix}$, (b) $\begin{pmatrix} 4 & 3 \\ 3 & 2 \end{pmatrix}$,

$(c)\,\heartsuit$ $\begin{pmatrix} 2 & 1 & -1 \\ 0 & 1 & 3 \\ -1 & -1 & 1 \end{pmatrix}$, (d) $\begin{pmatrix} 0 & 1 & 2 \\ -1 & 1 & 1 \\ -1 & 1 & 3 \end{pmatrix}$, $(e)\,\diamondsuit$ $\begin{pmatrix} 0 & 0 & 2 \\ 0 & 4 & 1 \\ -1 & 0 & 1 \end{pmatrix}$, (f) $\begin{pmatrix} 1 & 1 & 1 & 1 \\ 1 & 2 & 1 & 0 \\ 1 & 1 & 2 & 1 \\ 1 & 0 & 1 & 1 \end{pmatrix}$.

7.2. For each of the following linear systems, find the QR factorization of the coefficient matrix, and then use your factorization to solve the system: $(a)\,\heartsuit$ $\begin{pmatrix} 1 & 2 \\ -1 & 3 \end{pmatrix} \begin{pmatrix} x \\ y \end{pmatrix} = \begin{pmatrix} -1 \\ 2 \end{pmatrix}$, $(b)\,\diamondsuit$ $\begin{pmatrix} 2 & 1 & -1 \\ 1 & 0 & 2 \\ 2 & -1 & 3 \end{pmatrix} \begin{pmatrix} x \\ y \\ z \end{pmatrix} = \begin{pmatrix} 2 \\ -1 \\ 0 \end{pmatrix}$, (c) $\begin{pmatrix} 1 & 1 & 0 \\ -1 & 0 & 1 \\ 0 & -1 & 1 \end{pmatrix} \begin{pmatrix} x \\ y \\ z \end{pmatrix} = \begin{pmatrix} 0 \\ 1 \\ 0 \end{pmatrix}$.

7.3. Determine the rank of the following matrices using the extended QR method:

$(a)\,\heartsuit$ $\begin{pmatrix} 1 & 2 \\ -2 & -4 \end{pmatrix}$, (b) $\begin{pmatrix} 3 & -1 & -2 \\ -6 & 2 & 1 \end{pmatrix}$, $(c)\,\heartsuit$ $\begin{pmatrix} 2 & -5 & -1 \\ 1 & -6 & -4 \\ 3 & -4 & 2 \end{pmatrix}$, $(d)\,\diamondsuit$ $\begin{pmatrix} 1 & 2 & 3 & 4 \\ 3 & 6 & 4 & 7 \\ 1 & 2 & 2 & 3 \\ 3 & 6 & 5 & 8 \end{pmatrix}$.

7.4. Use the QR method to compute the least squares solution to the linear system $A\mathbf{x} = \mathbf{b}$ when $(a)\,\heartsuit$ $A = \begin{pmatrix} 1 \\ 2 \\ 1 \end{pmatrix}$, $\mathbf{b} = \begin{pmatrix} 1 \\ 1 \\ 0 \end{pmatrix}$; $(b)\,\diamondsuit$ $A = \begin{pmatrix} 1 & 0 \\ 2 & -1 \\ 3 & 2 \end{pmatrix}$, $\mathbf{b} = \begin{pmatrix} 1 \\ 1 \\ 0 \end{pmatrix}$;

$(c)\,\heartsuit$ $A = \begin{pmatrix} 2 & 1 & -1 \\ 1 & -2 & 0 \\ 3 & -1 & -1 \end{pmatrix}$, $\mathbf{b} = \begin{pmatrix} 1 \\ 0 \\ -1 \end{pmatrix}$; (d) $A = \begin{pmatrix} 2 & 1 & 1 \\ 1 & -2 & -1 \\ 1 & 0 & -1 \\ 5 & 0 & 1 \end{pmatrix}$, $\mathbf{b} = \begin{pmatrix} 0 \\ 0 \\ 1 \\ 0 \end{pmatrix}$.

7.5. Use the numerically stable version of the Gram–Schmidt process to find the QR factorizations of the $3 \times 3, 4 \times 4$ and 5×5 versions of the tridiagonal matrix that has 4's along the diagonal and 1's on the sub- and super-diagonals.

7.6. Let R be a row echelon matrix of the form (4.67). Prove that the columns of R containing the pivots are linearly independent.

7.7.$\,\diamondsuit$ Implement QR factorization via Gram–Schmidt in Python. Test the method on small toy examples, on a large random matrix, and on a poorly conditioned matrix. A classic example of a poorly conditioned matrix is the $n \times n$ *Hilbert matrix*

$$H_n = \begin{pmatrix} 1 & \frac{1}{2} & \frac{1}{3} & \cdots & \frac{1}{n} \\ \frac{1}{2} & \frac{1}{3} & \frac{1}{4} & \cdots & \frac{1}{n+1} \\ \frac{1}{3} & \frac{1}{4} & \frac{1}{5} & \cdots & \frac{1}{n+2} \\ \vdots & \vdots & \vdots & \ddots & \vdots \\ \frac{1}{n} & \frac{1}{n+1} & \frac{1}{n+2} & \cdots & \frac{1}{2n-1} \end{pmatrix}. \tag{4.74}$$

It is known, cf. [181], that the Hilbert matrix is positive definite and hence nonsingular. What is the largest Hilbert matrix for which your code produces correct results, up to 5 decimal places? *Hint*: Check the orthogonality of the matrix Q that is produced. The notebook from this section is a good place to start.

7.8. Find an implementation of the QR algorithm in the **numpy** Python package and try it out on the Hilbert matrix (4.74). Does it work? Also compare the run-time of the **numpy** version with your code from Exercise 7.9. Can you find out what algorithm is used by **numpy**?

7.9. $\diamond$ Another approach to address the numerical instability of Gram–Schmidt for QR factorization is the *reorthogonalization trick*, which essentially just repeats the orthogonalization step in the Gram–Schmidt algorithm a second time. Let $\mathbf{a}_1, \mathbf{a}_2, \ldots, \mathbf{a}_m \in \mathbb{R}^n$ denote the columns of the $n \times m$ matrix A. We initialize $\mathbf{q}_1 = \mathbf{a}_1 / r_{11}$, where $r_{11} = \|\mathbf{a}_1\|$ and repeat the following three steps for $k = 2$ through $k = m$.

1. Compute $s_{jk} = \mathbf{q}_j \cdot \mathbf{a}_k$ for $j \leq k - 1$, and set $\mathbf{v} = \mathbf{a}_k - \sum_{j=1}^{k-1} s_{jk} \mathbf{q}_j$.

2. Compute $t_{jk} = \mathbf{q}_j \cdot \mathbf{v}$ for $j \leq k - 1$, and set $\mathbf{x}_k = \mathbf{v} - \sum_{j=1}^{k-1} t_{jk} \mathbf{q}_j$.

3. Set $r_{jk} = s_{jk} + t_{jk}$ for $j \leq k - 1$, and then set $r_{kk} = \|\mathbf{x}_k\|$ and $\mathbf{q}_k = \mathbf{x}_k / r_{kk}$.

Step 1 is the first orthogonalization, like in Gram–Schmidt, while step 2 is the second one. In exact arithmetic we have $t_{jk} = 0$ and step 2 does nothing. In in-exact floating point arithmetic, step 2 corrects for a loss of orthogonality in the computation of $\mathbf{v}$. Implement the Gram–Schmidt method with re-orthogonalization in Python. Test the method on large random matrices and on a large Hilbert matrix (4.74). The Python notebook from this section will be helpful.

4.8 Matrix Norms

In this section, we investigate norms on spaces of matrices. Of course, since we can identify $\mathcal{M}_{m \times n} \simeq \mathbb{R}^{mn}$, any norm on the latter Euclidean space induces a norm on the space of $m \times n$ matrices. However, such norms tend not to be of use unless they behave well under matrix multiplication.

4.8.1 Natural Matrix Norms

We begin by fixing a norm $\|\cdot\|$ on $\mathbb{R}^n$, which will naturally induce a norm on the space $\mathcal{M}_{n \times n}$ of all $n \times n$ matrices. The original norm may or may not come from an inner product — this is irrelevant as far as the construction goes.

Theorem 4.49. *If $\|\cdot\|$ is any norm on $\mathbb{R}^n$, then the quantity*

$$\|A\| = \max\{\|A\mathbf{u}\| \mid \|\mathbf{u}\| = 1\} \tag{4.75}$$

defines the norm of a matrix $A \in \mathcal{M}_{n \times n}$, called the associated natural matrix norm.

Proof. First note that $\|A\| < \infty$, since the function $F(\mathbf{x}) = \|A\mathbf{x}\|$ for $\mathbf{x} \in \mathbb{R}^n$ is continuous, and the maximum is taken on a compact subset, namely the unit sphere $S_1 = \{\|\mathbf{u}\| = 1\}$ for the given norm, cf. Theorem 2.35.

To show that (4.75) defines a norm, we need to verify the three basic axioms of Definition 2.34. Positivity $\|A\| \geq 0$, is immediate. Suppose $\|A\| = 0$. This means that, for every unit vector, $\|A\mathbf{u}\| = 0$, and hence $A\mathbf{u} = \mathbf{0}$ whenever $\|\mathbf{u}\| = 1$. If $\mathbf{0} \neq \mathbf{v} \in \mathbb{R}^n$ is any nonzero vector, then $\mathbf{u} = \mathbf{v}/r$, where $r = \|\mathbf{v}\|$, is a unit vector, so

$$A\mathbf{v} = A(r\mathbf{u}) = rA\mathbf{u} = \mathbf{0}. \tag{4.76}$$

Therefore, $A\mathbf{v} = \mathbf{0}$ for every $\mathbf{v} \in \mathbb{R}^n$, which implies that $A = O$ is the zero matrix. This serves to prove the positivity property: $\|A\| = 0$ if and only if $A = O$.

As for homogeneity, if $c \in \mathbb{R}$ is any scalar, then

$$\|cA\| = \max\{\|cA\mathbf{u}\|\} = \max\{|c|\,\|A\mathbf{u}\|\} = |c|\max\{\|A\mathbf{u}\|\} = |c|\,\|A\|.$$

Finally, to prove the triangle inequality, we use the fact that the maximum of the sum of quantities is bounded by the sum of their individual maxima. Therefore, since the norm on $\mathbb{R}^n$ satisfies the triangle inequality,

$$\|A + B\| = \max\{\|A\mathbf{u} + B\mathbf{u}\|\} \le \max\{\|A\mathbf{u}\| + \|B\mathbf{u}\|\}$$
$$\le \max\{\|A\mathbf{u}\|\} + \max\{\|B\mathbf{u}\|\} = \|A\| + \|B\|. \qquad \blacksquare$$

Example 4.50. For any natural matrix norm, the identity matrix has norm $1 = \|I\|$.   ▲

Remark. The matrix norm formula (4.75) can be readily extended to rectangular matrices. If $A \in \mathcal{M}_{m \times n}$, then $\mathbf{u} \in \mathbb{R}^n$, while $A\mathbf{u} \in \mathbb{R}^m$, and one can employ *any* pair of norms on $\mathbb{R}^n$ and on $\mathbb{R}^m$. However, in what follows, we concentrate on the case of square matrices.   ▲

The property that distinguishes a matrix norm from a generic norm on the space of matrices is the fact that it also obeys two very useful *product inequalities*.

Theorem 4.51. *A natural matrix norm satisfies*

$$\|A\mathbf{v}\| \le \|A\|\,\|\mathbf{v}\|, \qquad \text{for all} \qquad A \in \mathcal{M}_{n \times n}, \quad \mathbf{v} \in \mathbb{R}^n. \qquad (4.77)$$

Furthermore,

$$\|AB\| \le \|A\|\,\|B\|, \qquad \text{for all} \qquad A, B \in \mathcal{M}_{n \times n}. \qquad (4.78)$$

Proof. Note first that, by definition $\|A\mathbf{u}\| \le \|A\|$ for all unit vectors $\|\mathbf{u}\| = 1$. Then, as in (4.76), letting $\mathbf{v} = r\,\mathbf{u}$ where $\mathbf{u}$ is a unit vector and $r = \|\mathbf{v}\| \ge 0$, we have

$$\|A\mathbf{v}\| = \|A(r\,\mathbf{u})\| = r\,\|A\mathbf{u}\| \le r\,\|A\| = \|\mathbf{v}\|\,\|A\|,$$

proving the first inequality. To prove the second, we apply the first, replacing $\mathbf{v}$ by $B\,\mathbf{u}$:

$$\|AB\| = \max\{\|AB\mathbf{u}\|\} = \max\{\|A(B\mathbf{u})\|\}$$
$$\le \max\{\|A\|\,\|B\mathbf{u}\|\} = \|A\|\max\{\|B\mathbf{u}\|\} = \|A\|\,\|B\|. \qquad \blacksquare$$

The multiplicative inequality (4.78) implies, in particular, that $\|A^2\| \le \|A\|^2$; keep in mind that equality is not necessarily valid. More generally:

Definition 4.52. A matrix A is called *convergent* if its powers converge to the zero matrix, $A^k \to O$, meaning that all the entries of A^k go to 0 as $k \to \infty$.

Proposition 4.53. *If A is a square matrix, then $\|A^k\| \le \|A\|^k$. Thus, if $\|A\| < 1$, then A is a convergent matrix.*

While having matrix norm strictly less than one guarantees that the matrix is convergent, there are matrices that have norm ≥ 1 which are nevertheless convergent; see Example 5.26.

Let us determine the explicit formula for the matrix norm induced by the ∞ norm

$$\| \mathbf{v} \|_\infty = \max \{ \, | v_1 |, \, \ldots, | v_n | \, \}.$$

Definition 4.54. The i-th *absolute row sum* of an $m \times n$ matrix A is the sum of the absolute values of the entries in its i-th row:

$$s_i = | a_{i1} | + \cdots + | a_{in} | = \sum_{j=1}^{n} | a_{ij} |. \tag{4.79}$$

Proposition 4.55. *The ∞ matrix norm of an $n \times n$ matrix A is equal to its maximum absolute row sum:*

$$\| A \|_\infty = \max\{s_1, \ldots, s_n\} = \max \left\{ \left. \sum_{j=1}^{n} | a_{ij} | \; \right| \; 1 \le i \le n \right\}. \tag{4.80}$$

Proof. Let $s = \max\{s_1, \ldots, s_n\}$ denote the right-hand side of (4.80). Given any $\mathbf{v} \in \mathbb{R}^n$, we compute the ∞ norm of the image vector $A\mathbf{v}$:

$$\| A\mathbf{v} \|_\infty = \max \left\{ \left| \sum_{j=1}^{n} a_{ij} v_j \right| \right\} \le \max \left\{ \sum_{j=1}^{n} | a_{ij} v_j | \right\}$$

$$\le \max \left\{ \sum_{j=1}^{n} | a_{ij} | \right\} \max \{ \, | v_j | \, \} = s \, \| \mathbf{v} \|_\infty.$$

In particular, by specializing $\mathbf{v} = \mathbf{u}$ to a unit vector, with $\| \mathbf{u} \|_\infty = 1$, we deduce that $\| A \|_\infty \le s$. On the other hand, suppose the maximal absolute row sum occurs at row i, so

$$s_i = \sum_{j=1}^{n} | a_{ij} | = s. \tag{4.81}$$

Let $\mathbf{u} \in \mathbb{R}^n$ be the vector with the following entries: $u_j = +1$ if $a_{ij} \ge 0$, while $u_j = -1$ if $a_{ij} < 0$, so $\| \mathbf{u} \|_\infty = 1$. Moreover, since $a_{ij} u_j = | a_{ij} |$, the i-th entry of $A\mathbf{u}$ is equal to the i-th absolute row sum (4.81). This implies that $\| A \|_\infty \ge \| A\mathbf{u} \|_\infty \ge s$. $\blacksquare$

Example 4.56. Consider the symmetric matrix $A = \begin{pmatrix} \frac{1}{2} & -\frac{1}{3} \\ -\frac{1}{3} & \frac{1}{4} \end{pmatrix}$. Its two absolute row sums are $\left| \frac{1}{2} \right| + \left| -\frac{1}{3} \right| = \frac{5}{6}$, $\left| -\frac{1}{3} \right| + \left| \frac{1}{4} \right| = \frac{7}{12}$, so $\| A \|_\infty = \max \left\{ \frac{5}{6}, \frac{7}{12} \right\} = \frac{5}{6}$. This implies that $\| A\mathbf{v} \|_\infty \le \frac{5}{6} \| \mathbf{v} \|_\infty$ for any vector $\mathbf{v} \in \mathbb{R}^2$. $\blacktriangle$

Proposition 4.57. *The 1 matrix norm is the maximum absolute column sum:*

$$\| A \|_1 = \max \left\{ \left. \sum_{i=1}^{n} | a_{ij} | \; \right| \; 1 \le j \le n \right\}. \tag{4.82}$$

Thus, $\| A \|_1 = \| A^T \|_\infty.$

The proof of Proposition 4.57 is left to the reader as Exercise 8.4.

Remark. The reader may have noticed that we have not written down a formula for the Euclidean matrix norm $\|A\|_2$ or, more generally, the matrix norm based on an inner product on $\mathbb{R}^n$. This is because we need additional tools, and so defer their explicit expressions until Section 5.7.2. Unfortunately, there is no good explicit formula for the matrix p norm when $p \neq 1, 2, \infty$, cf. [102]. ▲

4.8.2 The Frobenius Inner Product and Norm

Besides the matrix norms coming from norms on $\mathbb{R}^n$, there is another matrix norm that can can be extended to the space of rectangular matrices and plays an important role in some of our later applications. First, let us define the *trace* of a square matrix $M \in \mathcal{M}_{n \times n}$ to be the sum of its diagonal entries:

$$\operatorname{tr} M = m_{11} + m_{22} + \cdots + m_{nn}. \tag{4.83}$$

Basic properties of the trace can be found in Exercise 1.15.

Now suppose that $A, B \in \mathcal{M}_{m \times n}$ are matrices of size $m \times n$. Observe that the product matrix $M = A^T B$ is square of size $n \times n$. The following inner product on the space of matrices is named after the influential German algebraist Georg Frobenius.

> **Definition 4.58.** The *Frobenius inner product* on the space of all real matrices of size $m \times n$ is defined as
>
> $$\langle A, B \rangle_F = \operatorname{tr}(A^T B) = \sum_{i=1}^{m} \sum_{j=1}^{n} a_{ij} b_{ij} = \operatorname{tr}(A B^T) = \langle A^T, B^T \rangle_F, \tag{4.84}$$
>
> for $A, B \in \mathcal{M}_{m \times n}$.

The corresponding *Frobenius norm* is

$$\|A\|_F = \sqrt{\operatorname{tr}(A^T A)} = \sqrt{\sum_{i=1}^{m} \sum_{j=1}^{n} a_{ij}^2} = \sqrt{\operatorname{tr}(A A^T)} = \|A^T\|_F, \qquad A \in \mathcal{M}_{m \times n}. \tag{4.85}$$

Note that if we identify $\mathcal{M}_{m \times n} \simeq \mathbb{R}^{mn}$, then the Frobenius inner product becomes the usual dot product between vectors, and the Frobenius norm is the usual Euclidean norm. If $\mathbf{r}_1^T, \ldots, \mathbf{r}_m^T$ and $\mathbf{a}_1, \ldots, \mathbf{a}_n$ denote the rows and columns of A, respectively, (so that $\mathbf{r}_1, \ldots, \mathbf{r}_m$ are the columns of A^T) while $\mathbf{s}_1^T, \ldots, \mathbf{s}_m^T$ and $\mathbf{b}_1, \ldots, \mathbf{b}_n$ are the rows and columns of B, then

$$\langle A, B \rangle_F = \sum_{i=1}^{m} \mathbf{r}_i \cdot \mathbf{s}_i = \sum_{j=1}^{n} \mathbf{a}_j \cdot \mathbf{b}_j \tag{4.86}$$

is the sum of their respective dot products. Thus the squared Frobenius norm of a matrix A can be written as the sum of the squared Euclidean norms of either its row vectors or its column vectors:

$$\|A\|_F^2 = \sum_{i=1}^{m} \|\mathbf{r}_i\|^2 = \sum_{j=1}^{n} \|\mathbf{a}_j\|^2. \tag{4.87}$$

Note that the Frobenius norm of the $n \times n$ identity matrix is $\|I\|_F = \sqrt{n}$, and hence Example 4.50 implies that it is not a natural matrix norm on the space of $n \times n$ matrices when $n > 1$. On the other hand, the Frobenius norm does satisfy the inequality (4.78) for multiplication of general matrices, and so is said to define a (non-natural) *matrix norm.*.

Proposition 4.59. *Let A, B be, respectively, $m \times n$ and $n \times p$ matrices. Then*

$$\| AB \|_F \leq \| A \|_F \| B \|_F. \tag{4.88}$$

Proof. Let $\mathbf{r}_1^T, \ldots, \mathbf{r}_m^T$ be the rows of A, and $\mathbf{b}_1, \ldots, \mathbf{b}_p$ the columns of B. Then the (i, j) entry of AB is $\mathbf{r}_i^T \mathbf{b}_j = \mathbf{r}_i \cdot \mathbf{b}_j$. Thus, using the Cauchy–Schwarz inequality (2.27) on these dot products,

$$\| AB \|_F^2 = \sum_{i=1}^{m} \sum_{j=1}^{p} (\mathbf{r}_i \cdot \mathbf{b}_j)^2 \leq \sum_{i=1}^{m} \sum_{j=1}^{p} \| \mathbf{r}_i \|^2 \| \mathbf{b}_j \|^2$$

$$= \left(\sum_{i=1}^{m} \| \mathbf{r}_i \|^2 \right) \left(\sum_{j=1}^{p} \| \mathbf{b}_j \|^2 \right) = \| A \|_F^2 \| B \|_F^2,$$

where the final equality follows from (4.87). $\blacksquare$

Example 4.60. Let $a, b, c, d \in \mathbb{R}$, and consider the 2×2 matrices

$$A = \begin{pmatrix} a & b \\ c & d \end{pmatrix}, \quad B = \begin{pmatrix} d & -b \\ -c & a \end{pmatrix}, \quad \text{with product} \quad AB = \begin{pmatrix} ad - bc & 0 \\ 0 & ad - bc \end{pmatrix},$$

so that $B = (ad - bc) A^{-1}$, where the prefactor is the determinant of A; see Example 3.25. Their respective Frobenius norms are

$$\| A \|_F = \| B \|_F = \sqrt{a^2 + b^2 + c^2 + d^2}, \quad \| AB \|_F = \sqrt{2} \, |ad - bc|.$$

Thus the multiplicative inequality (4.88) implies the following inequality:

$$\sqrt{2} \, |ad - bc| \leq a^2 + b^2 + c^2 + d^2,$$

valid for any real numbers a, b, c, d. The reader may enjoy finding a direct proof. $\blacktriangle$

More generally, given a weighted inner product $\langle \mathbf{v}, \mathbf{w} \rangle = \mathbf{v}^T C \mathbf{w}$ on $\mathbb{R}^m$, where C is symmetric, positive definite, the corresponding *weighted Frobenius inner product* and *norm* are given by

$$\langle A, B \rangle_C = \sum_{j=1}^{n} \langle \mathbf{a}_j, \mathbf{b}_j \rangle = \sum_{j=1}^{n} \mathbf{a}_j^T C \mathbf{b}_j - \operatorname{tr}(A^T C B),$$

$$\| A \|_C = \sqrt{\sum_{j=1}^{n} \| \mathbf{a}_j \|^2} = \sqrt{\sum_{j=1}^{n} \mathbf{a}_j^T C \mathbf{a}_j} = \sqrt{\operatorname{tr}(A^T C A)}. \tag{4.89}$$

If we identify $\mathcal{M}_{m \times n} \simeq \mathbb{R}^{mn}$, then this corresponds to the weighted inner product associated with the positive definite block diagonal matrix that has n copies of C along the diagonal. In this case, there is no corresponding multiplicative inequality.

Exercises

8.1. Compute (i) the 1 matrix norm, (ii) the ∞ matrix norm, and (iii) the Frobenius norm of the following matrices:

$(a)\heartsuit$ $\begin{pmatrix} 2 & -2 \\ -3 & 5 \end{pmatrix}$, $\quad (b)\diamondsuit$ $\begin{pmatrix} \frac{5}{3} & \frac{4}{3} \\ -\frac{7}{6} & -\frac{5}{6} \end{pmatrix}$, $\quad (c)\heartsuit$ $\begin{pmatrix} 1 & -2 & 3 \\ -1 & 0 & 1 \\ 2 & -1 & 1 \end{pmatrix}$, $\quad (d)$ $\begin{pmatrix} 0 & .2 & .8 \\ -.3 & 0 & .1 \\ -.4 & .1 & 0 \end{pmatrix}$.

8.2. Let $A = \begin{pmatrix} 1 & 1 \\ 1 & -2 \end{pmatrix}$. Compute the natural matrix norm $\| A \|$ using the following norms

on $\mathbb{R}^2$: $(a)\heartsuit$ the 1 norm; $(b)\diamondsuit$ the ∞ norm; $(c)\heartsuit$ the weighted 1 norm $\| \mathbf{v} \| = 2 \, | \, v_1 \, | + 3 \, | \, v_2 \, |$; (d) the weighted ∞ norm $\| \mathbf{v} \| = \max \{ \, 2 \, | \, v_1 \, |, 3 \, | \, v_2 \, | \, \}$.

8.3. $\heartsuit$ Find a matrix A such that $\| A^2 \|_\infty \neq \| A \|_\infty^2$.

8.4. $\diamondsuit$ Prove formula (4.82) for the 1 matrix norm.

8.5. $\diamondsuit$ Explain why $\| A \| = \max | \, a_{ij} \, |$ defines a norm on the space of $n \times n$ matrices. Show by example that this is *not* a matrix norm, i.e., (4.78) is not necessarily valid.

8.6. Let Q be an $n \times n$ orthogonal matrix. Prove that $\| Q \, A \, Q^T \|_F = \| A \|_F$ for any $n \times n$ matrix A.

8.7. $\heartsuit$ *True or false*: If $\| A \|_F < 1$, then $A^k \to O$ as $k \to \infty$.

8.8. For $A \in \mathcal{M}_{n \times n}$, set $\| A \| = n^{-1/2} \| A \|_F$. Prove that $\| A \|$ defines a norm and, further, $\| I \| = 1$, but it does not satisfy the multiplicative inequality (4.78) when $n > 1$, and hence is also not a natural matrix norm.

Chapter 5

Eigenvalues and Singular Values

Each square matrix possesses a collection of one or more distinguished scalars, called eigenvalues, each associated with certain distinguished vectors known as eigenvectors. From a geometrical viewpoint, when the matrix acts on vectors via matrix multiplication, the eigenvectors specify the directions of pure scaling and the eigenvalues the extent the eigenvector is scaled. Eigenvalues and eigenvectors are of absolutely fundamental importance, and assume an essential role in a broad range of applications, including machine learning and data analysis, dynamical systems, both continuous and discrete, statistics, and many more.

In this text, we will exclusively deal with what we call complete (also known as real diagonalizable) matrices, meaning those whose eigenvectors form a basis of the underlying Euclidean space. Fortunately almost all matrices, including all symmetric matrices, and almost every matrix appearing in machine learning applications, are complete and so we will not lose much by ignoring the incomplete ones. For the latter, which are more technically complicated, we refer the reader to comprehensive linear algebra texts, e.g., [181,224]. Another complication is that, even though the matrix is always assumed to be real, its eigenvalues and associated eigenvectors may be complex. This is one of the few places in the text where we must deal, briefly, with complex numbers. (See Section 9.10.1 for a basic introduction to complex numbers and vectors.) On the other hand, almost all matrices appearing in our applications have only real eigenvalues and eigenvectors, and so we will not go into much depth in the complex case. In particular, every symmetric matrix is complete, has only real eigenvalues, and its eigenvectors always form an orthogonal basis of $\mathbb{R}^n$; in fact, this is how orthogonal and orthonormal bases most naturally appear.

A non-square matrix does not possess eigenvalues. In their place, one studies the eigenvalues of the associated square Gram matrix, the square roots of which are known as the singular values of the original matrix and its eigenvectors are the associated singular vectors. Singular values and vectors underlie the powerful method of statistical data analysis known as principal component analysis (PCA), and are of immense importance in an increasingly broad range of contemporary applications, including image processing, semantics, language and speech recognition, and machine learning.

Most of the widely used solution methods, for both linear and nonlinear systems, rely on some form of iteration, meaning the repeated application of a function or process. One begins with an approximation (or guess) to the desired solution, and then, in favorable circumstances, the iterations lead to successively closer and closer approximations. In this chapter, we concentrate on linear and affine iterative systems that are based on repeated multiplication of an initial vector by a square matrix, possibly supplemented by addition of a fixed vector. Iterative methods are particularly effective for solving the very large systems arising in ma-

© The Author(s), under exclusive license to Springer Nature Switzerland AG 2025
J. Calder, P. J. Olver, *Linear Algebra, Data Science, and Machine Learning*, Springer Undergraduate Texts in Mathematics and Technology, https://doi.org/10.1007/978-3-031-93764-4_5

chine learning, as well as in the numerical solution of both ordinary and partial differential equations, [32, 225]. All practical methods for computing eigenvalues and eigenvectors rely on some form of iteration. A detailed historical development of iterative methods for solving linear systems and eigenvalue problems can be found in the recent survey paper [227].

Remark: Except in very low dimensions, the accurate numerical computation of eigenvalues, eigenvectors, singular values, and singular vectors of matrices is a challenge. Consequently, solving the more substantial computational problems in this chapter will require access to suitable computer software.

5.1   Eigenvalues and Eigenvectors

We inaugurate our discussion by stating the basic definition. Its importance will become manifestly evident as we proceed.

Definition 5.1. Let A be a square matrix. A scalar λ is called an *eigenvalue* of A if there is a *nonzero* vector $\mathbf{v} \neq \mathbf{0}$, called an *eigenvector*, such that

$$A\mathbf{v} = \lambda\mathbf{v}. \tag{5.1}$$

In geometric terms, the matrix A scales (stretches) the eigenvector $\mathbf{v}$ by an amount specified by the eigenvalue λ. The requirement that the eigenvector $\mathbf{v}$ be nonzero is important, since $\mathbf{v} = \mathbf{0}$ is a trivial solution to the eigenvalue equation (5.1) for *every* scalar λ.

Remark. The odd-looking terms "eigenvalue" and "eigenvector" are hybrid German–English words. In the original German, they are *Eigenwert* and *Eigenvektor*, which can be fully translated as "proper value" and "proper vector". For some reason, the half-translated terms have acquired a certain charm, and are now standard. The alternative English terms *characteristic value* and *characteristic vector* can be found in some (mostly older) texts. ▲

The eigenvalue equation (5.1) is a system of linear equations for the entries of the eigenvector $\mathbf{v}$ — provided that the eigenvalue λ is specified in advance — but is "mildly" nonlinear as a combined system for λ and $\mathbf{v}$. Let us rewrite the equation in the form[1]

$$(A - \lambda\,\mathrm{I})\,\mathbf{v} = \mathbf{0}, \tag{5.2}$$

where I is the identity matrix of the correct size, so $\lambda\,\mathrm{I}\,\mathbf{v} = \lambda\mathbf{v}$. Now, for given λ, equation (5.2) is a homogeneous linear system for $\mathbf{v}$, and always has the trivial zero solution $\mathbf{v} = \mathbf{0}$, but we are specifically seeking a nonzero solution! According to Theorem 1.16, a homogeneous linear system has a nonzero solution $\mathbf{v} \neq \mathbf{0}$ if and only if its coefficient matrix, which in this case is $A - \lambda\,\mathrm{I}$, is singular. This observation is the key to resolving the eigenvector equation.

Theorem 5.2. *A scalar λ is an eigenvalue of the $n \times n$ matrix A if and only if the matrix $A - \lambda\,\mathrm{I}$ is singular, i.e.,* $\mathrm{rank}(A - \lambda\,\mathrm{I}) < n$ *or, equivalently,* $\mathrm{nullity}(A - \lambda\,\mathrm{I}) > 0$. *The corresponding eigenvectors are all the nonzero solutions to the eigenvalue equation* (5.2).

For a fixed scalar λ, we will call the subspace

$$V_\lambda = \ker(A - \lambda\,\mathrm{I}). \tag{5.3}$$

---

[1]Note that it is not legal to write (5.2) in the form $(A - \lambda)\mathbf{v} = \mathbf{0}$ since we do not know how to subtract a scalar λ from a matrix A. Worse, if you type $A - \lambda$ in some common software packages including Python, the result will be to subtract λ from *all* the entries of A, which is *not* what we are after!

i.e., the set of solutions to the eigenvalue equation (5.2), the associated *eigenspace*. Thus, λ is an eigenvalue if and only if the eigenspace is nontrivial, $V_\lambda \neq \{\mathbf{0}\}$, in which case the eigenvectors are all its nonzero elements: $\mathbf{0} \neq \mathbf{v} \in V_\lambda$. The dimension of V_λ will be called the *multiplicity*[2] of the eigenvalue. Thus, recalling the definition of the nullity of a matrix, and invoking Theorem 3.9,

$$\text{multiplicity } \lambda = \dim V_\lambda = \text{nullity}(A - \lambda\,\mathrm{I}) = n - \text{rank}(A - \lambda\,\mathrm{I}), \tag{5.4}$$

when A has size $n \times n$, so that λ is an eigenvalue if and only if its multiplicity, as defined by (5.4), is ≥ 1. In particular, $\lambda = 0$ is an eigenvalue if and only if its eigenspace, which coincides with the kernel of the matrix A, is nontrivial, $V_0 = \ker A \neq \{\mathbf{0}\}$, and hence A is a singular matrix. The nonzero vectors $\mathbf{0} \neq \mathbf{v} \in \ker A$ are known as *null eigenvectors*.

Proposition 5.3. *A matrix is singular if and only if it has a zero eigenvalue.*

Example 5.4. Let's consider the simplest[3] case in detail. Let

$$A = \begin{pmatrix} a & b \\ c & d \end{pmatrix} \tag{5.5}$$

be a general 2×2 real matrix with the indicated entries $a, b, c, d \in \mathbb{R}$. A scalar λ will be an eigenvalue if and only if the matrix

$$A - \lambda\,\mathrm{I} = \begin{pmatrix} a - \lambda & b \\ c & d - \lambda \end{pmatrix}$$

is singular. As in Example 3.25, this is the case if and only if its determinant vanishes:

$$\det(A - \lambda\,\mathrm{I}) = (a - \lambda)(d - \lambda) - bc = \lambda^2 - (a + d)\lambda + (ad - bc) = 0. \tag{5.6}$$

Thus, the eigenvalues are the solutions to a certain quadratic polynomial equation, known as the *characteristic equation* associated with the matrix (5.5). The characteristic equation can be immediately solved using the quadratic formula. As such, there are three possibilities, which can be characterized by the sign of the *discriminant* of the quadratic equation (5.6):

$$\Delta = (a - d)^2 + 4bc. \tag{5.7}$$

(a) $\Delta > 0$: The characteristic equation has two different real roots $\lambda_1 \neq \lambda_2$. In this case, A has two distinct eigenvalues. Moreover, it is not hard to show that $\dim V_{\lambda_i} = \dim \ker(A - \lambda_i\,\mathrm{I}) = 1$ for $i = 1, 2$, and hence each eigenvalue has multiplicity 1.

(b) $\Delta = 0$: The characteristic equation has a single real root λ_1, and so A has only one eigenvalue. Its multiplicity can be either 1 or 2.

(c) $\Delta < 0$: The characteristic equation has complex conjugate roots $\lambda_\pm = \mu \pm i\nu$, where $i = \sqrt{-1}$ is the imaginary unit. In this case, A has two *complex eigenvalues*, and the associated eigenvectors have complex entries.

---

[2] In linear algebra, this is often referred to as the *geometric multiplicity* in order to distinguish it from the, possibly different, *algebraic multiplicity*. The latter will not concern us, although the interested reader can consult [181] for details.

[3] Well, technically the second simplest. The case of a 1×1 matrix is delegated as an easy exercise for the reader.

Here are representative examples of each case: The matrix

$$A = \begin{pmatrix} 3 & 1 \\ 1 & 3 \end{pmatrix} \quad \text{has characteristic equation} \quad \lambda^2 - 6\lambda + 8 = (\lambda - 2)(\lambda - 4) = 0,$$

and hence has two real eigenvalues: $\lambda_1 = 2$ and $\lambda_2 = 4$. Solving the corresponding eigenvector equations (5.2) associated with each eigenvalue produces the corresponding eigenvectors

$$\lambda_1 = 2, \quad \begin{pmatrix} c \\ -c \end{pmatrix} = c\begin{pmatrix} 1 \\ -1 \end{pmatrix} = c\,\mathbf{v}_1, \qquad \lambda_2 = 4, \quad \begin{pmatrix} c \\ c \end{pmatrix} = c\begin{pmatrix} 1 \\ 1 \end{pmatrix} = c\,\mathbf{v}_2,$$

where $c \neq 0$ is any nonzero scalar.

Remark 5.5. In general, if $\mathbf{v}$ is an eigenvector of A for the eigenvalue λ, then so is every nonzero scalar multiple of $\mathbf{v}$. In practice, we distinguish only linearly independent eigenvectors. Thus, as in this example, we shall say, somewhat loosely, "$\mathbf{v}_1 = (1, -1)^T$ is *the* eigenvector corresponding to the eigenvalue $\lambda_1 = 2$", when we really mean that the set of eigenvectors for $\lambda_1 = 2$ consists of all nonzero scalar multiples of $\mathbf{v}_1$.　　　　　　▲

The matrix

$$A = \begin{pmatrix} 2 & 0 \\ 0 & 2 \end{pmatrix} \quad \text{has characteristic equation} \quad \lambda^2 - 4 = 0,$$

and hence has only a single eigenvalue: $\lambda_1 = 2$. Every nonzero vector $\mathbf{0} \neq \mathbf{v} \in \mathbb{R}^2$ is an eigenvector, and hence the eigenvalue has multiplicity $2 = \dim \ker(A - 2\,I) = \mathbb{R}^2$.

On the other hand, the matrix

$$A = \begin{pmatrix} 2 & 1 \\ 0 & 2 \end{pmatrix} \quad \text{has the same characteristic equation} \quad \lambda^2 - 4 = 0, \tag{5.8}$$

and hence also has just one eigenvalue: $\lambda_1 = 2$. However, the solutions to the eigenvector equation $A\mathbf{v} = 2\mathbf{v}$ are just $(c, 0)^T = c(1, 0)^T$ for $c \in \mathbb{R}$, and hence the eigenvalue only has multiplicity 1 with a single (up to scalar multiple) eigenvector $\mathbf{v}_1 = (1, 0)^T$.

Finally, the matrix

$$A = \begin{pmatrix} 1 & 4 \\ -1 & 1 \end{pmatrix} \quad \text{has characteristic equation} \quad \lambda^2 - 2\lambda + 5 = 0. \tag{5.9}$$

In this case, the characteristic equation has no real roots, and hence A has no real eigenvalues. The two complex conjugate roots $\lambda_{\pm} = 1 \pm 2\,i$, where $i = \sqrt{-1}$, are viewed as *complex eigenvalues*. They correspond to complex conjugate eigenvectors

$$\mathbf{z}_+ = \begin{pmatrix} -2\,i \\ 1 \end{pmatrix} = \begin{pmatrix} 0 \\ 1 \end{pmatrix} + i\begin{pmatrix} -2 \\ 0 \end{pmatrix}, \qquad \mathbf{z}_- = \begin{pmatrix} 2\,i \\ 1 \end{pmatrix} = \begin{pmatrix} 0 \\ 1 \end{pmatrix} - i\begin{pmatrix} -2 \\ 0 \end{pmatrix},$$

that satisfy the respective eigenvector equations

$$A\,\mathbf{z}_+ = \lambda_+\,\mathbf{z}_+ = (1 + 2\,i)\,\mathbf{z}_+, \qquad A\,\mathbf{z}_- = \lambda_-\,\mathbf{z}_- = (1 - 2\,i)\,\mathbf{z}_-.$$

Writing the complex eigenvalues $\lambda_{\pm} = \mu \pm i\,\nu$ and eigenvectors $\mathbf{z}_{\pm} = \mathbf{u} \pm i\,\mathbf{v}$ in terms of their real and imaginary parts, so

$$\mu = 1, \qquad \nu = 2, \qquad \mathbf{u} = \begin{pmatrix} 0 \\ 1 \end{pmatrix}, \qquad \mathbf{v} = \begin{pmatrix} -2 \\ 0 \end{pmatrix},$$

we find the preceding eigenvector equations are equivalent to the pair of real vector equations

$$A\mathbf{u} = \mu\mathbf{u} - \nu\mathbf{v} = \mathbf{u} - 2\mathbf{v}, \qquad A\mathbf{v} = \nu\mathbf{u} + \mu\mathbf{v} = 2\mathbf{u} + \mathbf{v}. \tag{5.10}$$

Note that $\mathbf{u}, \mathbf{v}$ are *not* eigenvectors of A. ▲

Remark. Although we promised at the outset of this text to restrict our attention to real vectors and real matrices, the latter may have complex eigenvalues and hence the associated eigenvectors will necessarily have complex entries. The complex eigenvalues and eigenvectors can be converted into real numbers and vectors by taking their real and imaginary parts, which satisfy a system similar to that in (5.10). Thus, one could, even in this situation, remain entirely within the real domain. Fortunately, as noted above, essentially all the matrices of importance in machine learning and data analysis have only real eigenvalues and eigenvectors. Nevertheless, complex eigenvalues do play an important role in many other applications, particularly dynamical systems, both continuous (differential equations) and discrete (Markov processes and the like). Details can be found in comprehensive introductions to linear algebra, such as [181, 224]. ▲

Just as the determinant can be generalized to $n \times n$ matrices, so can the *characteristic equation*, which is given by

$$p_A(\lambda) = \det(A - \lambda I) = 0, \tag{5.11}$$

and based on the (complicated) determinant function, [181]. The left hand side turns out to be a polynomial of degree n, known as the *characteristic polynomial*, and every solution to the characteristic polynomial equation is an eigenvalue of the matrix A. According to the fundamental theorem of algebra, cf. [74], every (complex) polynomial of degree $n \geq 1$ can be completely factored, and so we can write the characteristic polynomial in factored form:

$$p_A(\lambda) = (-1)^n (\lambda - \lambda_1)(\lambda - \lambda_2) \cdots (\lambda - \lambda_n). \tag{5.12}$$

The complex numbers $\lambda_1, \ldots, \lambda_n$, some of which may be repeated, are the *roots* of the characteristic equation (5.11), and hence the eigenvalues of the matrix A.

Therefore, we immediately conclude:

Theorem 5.6. *An $n \times n$ matrix possesses at least one and at most n distinct complex eigenvalues.*

While of importance for certain theoretical developments, as soon as the size of the matrix is moderately large, say $n \geq 4$, constructing and solving the characteristic equation is a terrible method for practically computing eigenvalues and eigenvectors, and so, unlike most introductions to the subject, we shall not dwell on this approach. Some basic numerical algorithms for computing eigenvalues and eigenvectors can be found in Section 5.6. △

Proposition 5.7. *If A and B are square matrices of the same size, then AB and BA have the same eigenvalues.*

Proof. Let $\mathbf{u}$ be an eigenvector of AB with eigenvalue λ, so $AB\mathbf{u} = \lambda\mathbf{u}$. Let $\mathbf{w} = B\mathbf{u}$. Then $BA\mathbf{w} = BAB\mathbf{u} = B(\lambda\mathbf{u}) = \lambda B\mathbf{u} = \lambda\mathbf{w}$. Thus, if $\mathbf{w} \neq \mathbf{0}$, then it is an eigenvector of BA with the same eigenvalue λ. On the other hand, if $\mathbf{w} = \mathbf{0}$ then $\mathbf{0} \neq \mathbf{u} \in \ker B$ and hence $\lambda = 0$. This means both AB and BA are also singular, and both have $\lambda = 0$ as an eigenvalue. ∎

Corollary 5.8. *Let A be a square matrix and V a nonsingular matrix of the same size. Then A and the similar matrix $V^{-1}AV$ have the same eigenvalues.*

The final result of this section relies on the introduction of an inner product on $\mathbb{R}^n$, with the consequential Definition 4.15 of the adjoint matrix. As above, the matrix A is square, of size $n \times n$, and, for simplicity, we impose the same inner product on its domain and codomain, which are both $\mathbb{R}^n$. As always, the most important case is the dot product, for which the adjoint coincides with the transpose A^T.

Proposition 5.9. *A square matrix A and its adjoint A^* have the same eigenvalues with the same multiplicities.*

Proof. First, since we are using a single inner product on $\mathbb{R}^n$, we have $I^* = I$. Thus, $(A - \lambda I)^* = A^* - \lambda I$. Equation (4.33) implies that $A - \lambda I$ and $A^* - \lambda I$ have the same rank, and so the result immediately follows from (5.4). ∎

Remark. While A^* has the same eigenvalues as A, its eigenvectors and eigenspaces are, in general, different. In particular, an eigenvector $\mathbf{v}$ of the transpose A^T, which satisfies $A^T \mathbf{v} = \lambda \mathbf{v}$, is sometimes referred to as a *left eigenvector* or *co-eigenvector* of A, since its transpose satisfies $\mathbf{v}^T A = \lambda \mathbf{v}^T$. ▲

Exercises

1.1. Find the eigenvalues and eigenvectors of the following 2×2 matrices:

$$(a)\,\heartsuit \begin{pmatrix} 0 & 1 \\ 1 & 0 \end{pmatrix}, \quad (b)\,\diamondsuit \begin{pmatrix} 1 & -2 \\ -2 & 1 \end{pmatrix}, \quad (c) \begin{pmatrix} 1 & -\frac{2}{3} \\ \frac{1}{2} & -\frac{1}{6} \end{pmatrix}, \quad (d)\,\heartsuit \begin{pmatrix} 3 & 1 \\ -1 & 1 \end{pmatrix}.$$

1.2. Write down $(a)\,\heartsuit$ a 2×2 matrix that has 0 as one of its eigenvalues and $(1,2)^T$ as a corresponding eigenvector; (b) a 3×3 matrix that has $(1,2,3)^T$ as an eigenvector for the eigenvalue -1; $(c)\,\heartsuit$ a 4×4 matrix that has -4 as an eigenvalue with multiplicity 2.

1.3. Find all eigenvalues and eigenvectors of $(a)\,\heartsuit$ the $n \times n$ zero matrix O; (b) the $n \times n$ identity matrix I; $(c)\,\diamondsuit$ the $n \times n$ matrix $E = \mathbf{1}\mathbf{1}^T$ with every entry equal to 1..

1.4. Prove that an $n \times n$ matrix has a eigenvalue of multiplicity n if and only if it is a scalar multiple of the identity matrix.

1.5. $\diamondsuit$ A matrix A is called *nilpotent* if $A^k = O$ for some $k \geq 1$. (a) Prove that a nilpotent matrix has only 0 as an eigenvalue. (b) Write down a nonzero nilpotent matrix $A \neq O$.

1.6. Let A be a square matrix. $(a)\,\heartsuit$ Explain in detail why every nonzero scalar multiple of an eigenvector of A is also an eigenvector. $(b)\,\heartsuit$ Show that every nonzero linear combination of two eigenvectors $\mathbf{v}, \mathbf{w}$ corresponding to the *same* eigenvalue is also an eigenvector. $(c)\,\diamondsuit$ Prove that a linear combination $c\mathbf{v} + d\mathbf{w}$, with $c, d \neq 0$, of two eigenvectors corresponding to *different* eigenvalues is never an eigenvector.

1.7. Suppose that λ is an eigenvalue of A, and $b, c \in \mathbb{R}$. (a) Prove that $c\lambda$ is an eigenvalue of the scalar multiple $c\,A$. (b) Prove that $\lambda + b$ is an eigenvalue of $A + b\,I$. (c) More generally, $c\lambda + b$ is an eigenvalue of $B = c\,A + b\,I$ for scalars c, b.

1.8. ♡ (a) Show that if λ is an eigenvalue of A, then λ^2 is an eigenvalue of A^2. (b) Is the converse valid: if μ is an eigenvalue of A^2, then $\sqrt{\mu}$ is an eigenvalue of A?

1.9. (a) Prove that if $\lambda \neq 0$ is a nonzero eigenvalue of the nonsingular matrix A, then $1/\lambda$ is an eigenvalue of A^{-1}. (b) What happens if A has 0 as an eigenvalue?

1.10. ◇ Does Proposition 5.7 hold when A, B are rectangular matrices?

1.11. *True or false*:
    (a) ♡ If λ is an eigenvalue of both A and B, then it is an eigenvalue of the sum $A + B$.
    (b) ◇ If $\mathbf{v}$ is an eigenvector of both A and B, then it is an eigenvector of $A + B$.
    (c) If λ is an eigenvalue of A and μ is an eigenvalue of B, then $\lambda\mu$ is an eigenvalue of the matrix product $C = AB$.

1.12. *Deflation*: Suppose A has eigenvalue λ and corresponding eigenvector $\mathbf{v}$. (a) Let $\mathbf{b}$ be any vector. Prove that the matrix $B = A - \mathbf{v}\mathbf{b}^T$ also has $\mathbf{v}$ as an eigenvector, now with eigenvalue $\lambda - \beta$, where $\beta = \mathbf{v} \cdot \mathbf{b}$. (b) Prove that if $\mu \neq \lambda - \beta$ is any other eigenvalue of A, then it is also an eigenvalue of B. *Hint*: Look for an eigenvector of the form $\mathbf{w} + c\mathbf{v}$, where $\mathbf{w}$ is an eigenvector of A. (c) Given a nonsingular matrix A with eigenvalues $\lambda_1, \lambda_2, \ldots, \lambda_n$ and $\lambda_1 \neq \lambda_j$ for all $j \geq 2$, explain how to construct a *deflated matrix* B whose eigenvalues are $0, \lambda_2, \ldots, \lambda_n$. (d) Try out your method on the matrix $\begin{pmatrix} 3 & 3 \\ 1 & 5 \end{pmatrix}$.

5.2   Eigenvector Bases

Most of the bases of $\mathbb{R}^n$ that play a distinguished role in applications are assembled from the eigenvectors of some $n \times n$ matrix. However, not every square matrix has an eigenvector basis; elementary examples include the 2×2 matrices (5.8) and (5.9). However, the vast majority of matrices of importance in machine learning, including, as we will see, all symmetric and self-adjoint matrices, do possess real eigenvector bases, and thus we do not lose much by focusing our attention on them from here on.

The first task is to show that eigenvectors corresponding to distinct eigenvalues are automatically linearly independent.

Proposition 5.10. *If $\lambda_1, \ldots, \lambda_k$ are distinct eigenvalues of a matrix A, so $\lambda_i \neq \lambda_j$ when $i \neq j$, then any set of associated eigenvectors $\mathbf{v}_1, \ldots, \mathbf{v}_k$ is linearly independent.*

Proof. The result is proved by induction on the number of eigenvalues. The case $k = 1$ is immediate, since an eigenvector cannot be zero. Assume that we know that the result is valid for $k - 1$ eigenvalues. Suppose we have a vanishing linear combination:

$$c_1 \mathbf{v}_1 + \cdots + c_{k-1} \mathbf{v}_{k-1} + c_k \mathbf{v}_k = \mathbf{0}. \tag{5.13}$$

Let us multiply this equation by the matrix A:

$$A(c_1 \mathbf{v}_1 + \cdots + c_{k-1} \mathbf{v}_{k-1} + c_k \mathbf{v}_k) = c_1 A\mathbf{v}_1 + \cdots + c_{k-1} A\mathbf{v}_{k-1} + c_k A\mathbf{v}_k$$
$$= c_1 \lambda_1 \mathbf{v}_1 + \cdots + c_{k-1} \lambda_{k-1} \mathbf{v}_{k-1} + c_k \lambda_k \mathbf{v}_k = \mathbf{0}.$$

On the other hand, if we multiply the original equation (5.13) by λ_k, we also have

$$c_1 \lambda_k \mathbf{v}_1 + \cdots + c_{k-1} \lambda_k \mathbf{v}_{k-1} + c_k \lambda_k \mathbf{v}_k = \mathbf{0}.$$

Subtracting this from the previous equation, the final terms cancel, and we are left with the equation

$$c_1(\lambda_1 - \lambda_k)\mathbf{v}_1 + \cdots + c_{k-1}(\lambda_{k-1} - \lambda_k)\mathbf{v}_{k-1} = \mathbf{0}.$$

This is a vanishing linear combination of the first $k-1$ eigenvectors, and so, by our induction hypothesis, can happen only if all the coefficients are zero:

$$c_1(\lambda_1 - \lambda_k) = 0, \qquad \cdots \qquad c_{k-1}(\lambda_{k-1} - \lambda_k) = 0.$$

The eigenvalues were assumed to be distinct, and consequently $c_1 = \cdots = c_{k-1} = 0$. Substituting these values back into (5.13), we find that $c_k\mathbf{v}_k = \mathbf{0}$, and so $c_k = 0$ also, since the eigenvector $\mathbf{v}_k \neq \mathbf{0}$. Thus we have proved that (5.13) holds if and only if $c_1 = \cdots = c_k = 0$, which implies the linear independence of the eigenvectors $\mathbf{v}_1, \ldots, \mathbf{v}_k$. This completes the induction step. $\blacksquare$

The most important consequence of this result concerns when a matrix has the maximum allotment of eigenvalues.

Corollary 5.11. *If the $n \times n$ matrix A has n distinct real eigenvalues $\lambda_1, \ldots, \lambda_n$, then the corresponding real eigenvectors $\mathbf{v}_1, \ldots, \mathbf{v}_n$ form a basis of $\mathbb{R}^n$.*

Example 5.12. As we saw earlier, the matrix

$$A = \begin{pmatrix} 3 & 1 \\ 1 & 3 \end{pmatrix} \qquad \text{has eigenvectors} \qquad \mathbf{v}_1 = \begin{pmatrix} 1 \\ 1 \end{pmatrix}, \quad \mathbf{v}_2 = \begin{pmatrix} 1 \\ -1 \end{pmatrix},$$

corresponding to its eigenvalues: $\lambda_1 = 4$, $\lambda_2 = 2$. Since $\lambda_1 \neq \lambda_2$, the eigenvectors $\mathbf{v}_1, \mathbf{v}_2$ necessarily form a basis of $\mathbb{R}^2$, as can be checked. On the other hand, the matrix $A = \begin{pmatrix} 2 & 0 \\ 0 & 2 \end{pmatrix}$ has only one eigenvalue: $\lambda_1 = 2$, and every nonzero vector $\mathbf{0} \neq \mathbf{v} \in \mathbb{R}^2$ is an eigenvector. In this case, any basis of $\mathbb{R}^2$ serves as an eigenvector basis. $\blacktriangle$

The following slightly non-standard terminology is taken from [181].

Definition 5.13. An $n \times n$ matrix A is called *complete* if there exists a basis $\mathbf{v}_1, \ldots, \mathbf{v}_n$ of $\mathbb{R}^n$ consisting of eigenvectors of A.

Any real matrix with n distinct real eigenvalues is automatically complete, while those with fewer may or may not be complete. In view of Theorem 5.15 below, complete matrices are also often called (real) *diagonalizable*, although this term could be misinterpreted, since we require the matrices appearing in the diagonalization equation (5.14) to be real, whereas most texts allow them to have complex entries.

Lemma 5.14. *If A is a complete $n \times n$ matrix, with eigenvector basis $\mathbf{v}_1, \ldots, \mathbf{v}_n$ and corresponding eigenvalues $\lambda_1, \ldots, \lambda_n$, some of which may be equal, then these constitute all the eigenvalues of A.*

Proof. Suppose $A\mathbf{w} = \mu\mathbf{w}$. We express $\mathbf{w}$ as a linear combination of the eigenvector basis, so

$$\mathbf{w} = c_1\mathbf{v}_1 + \cdots + c_n\mathbf{v}_n.$$

Then, on the one hand,

$$A\mathbf{w} = c_1 A\mathbf{v}_1 + \cdots + c_n A\mathbf{v}_n = c_1 \lambda_1 \mathbf{v}_1 + \cdots + c_n \lambda_n \mathbf{v}_n,$$

while, on the other hand,

$$A\mathbf{w} = \mu \mathbf{w} = c_1 \mu \mathbf{v}_1 + \cdots + c_n \mu \mathbf{v}_n.$$

Equating these two linear combinations of the eigenvector basis vectors, we deduce that their coefficients must be equal: $c_i(\lambda_i - \mu) = 0$, which implies that either $\mu = \lambda_i$ or $c_i = 0$. If μ is not equal to any of the eigenvalues, then all the coefficients $c_1 = \cdots = c_n = 0$, which implies $\mathbf{w} = \mathbf{0}$ and is hence not an eigenvector, contrary to our hypothesis. ∎

Remark. Real matrices possessing complex eigenvalues (which necessarily come in complex conjugate pairs) can also be deemed complete provided the real and imaginary parts of their complex eigenvectors, along with their real eigenvectors, if any, can be used to form a basis of $\mathbb{R}^n$; see [181] for details. However, in this text *complete* always means that all the eigenvalues are real and that $\mathbb{R}^n$ has an eigenvector basis. Incomplete matrices are more painful to deal with, and, if confronted with one (which, fortunately, almost never happens in machine learning), one is advised to consult a more comprehensive linear algebra text, e.g., [181]. ▲

Let us now state a result establishing the diagonalizability of complete matrices.

Theorem 5.15. *A square matrix A is complete if and only if there exists a nonsingular matrix V and a diagonal matrix Λ such that*

$$V^{-1}AV = \Lambda, \qquad or, \ equivalently, \qquad A = V\Lambda V^{-1}. \tag{5.14}$$

Proof. We rewrite (5.14) in the equivalent form

$$AV = V\Lambda. \tag{5.15}$$

Using the columnwise action (3.8) of matrix multiplication, one easily sees that the j-th column of the matrix equation (5.14) is given by $A\mathbf{v}_j = \lambda_j \mathbf{v}_j$, where $\mathbf{v}_j$ denotes the j-th column of V and λ_j the j-th diagonal entry of Λ. Therefore, the columns of V are necessarily eigenvectors, and the diagonal entries of Λ are the corresponding eigenvalues. (Repeated eigenvalues appear as many times as their multiplicity.) And, as a result, (5.14) requires that A have n linearly independent eigenvectors, i.e., an eigenvector basis, to form the columns of the nonsingular matrix V. ∎

Corollary 5.16. *If A is a complete matrix of size $n \times n$, then the sum of its eigenvalues equals its trace, i.e., the sum of its diagonal entries:*

$$\sum_{i=1}^{n} \lambda_i = \operatorname{tr} A = \sum_{i=1}^{n} a_{ii}. \tag{5.16}$$

Proof. We take the trace of the diagonalization equation (5.14), and use the property in Exercise 1.15(e) in Chapter 3:

$$\operatorname{tr} A = \operatorname{tr}(V\Lambda V^{-1}) = \operatorname{tr}(V^{-1}V\Lambda) = \operatorname{tr}\Lambda = \sum_{i=1}^{n} \lambda_i. \qquad ∎$$

Example 5.17. The matrix $A = \begin{pmatrix} 1 & -1 \\ 2 & 4 \end{pmatrix}$ has eigenvalues $\lambda_1 = 2$, $\lambda_2 = 3$ with correspond-

ing eigenvectors $\mathbf{v}_1 = \begin{pmatrix} 1 \\ -1 \end{pmatrix}$, $\mathbf{v}_2 = \begin{pmatrix} 1 \\ -2 \end{pmatrix}$. Thus, in the diagonalization equation (5.14),

$V = \begin{pmatrix} 1 & 1 \\ -1 & -2 \end{pmatrix}$, $\Lambda = \begin{pmatrix} 2 & 0 \\ 0 & 3 \end{pmatrix}$, whereby

$$V^{-1}AV = \begin{pmatrix} 2 & 1 \\ -1 & -1 \end{pmatrix} \begin{pmatrix} 1 & -1 \\ 2 & 4 \end{pmatrix} \begin{pmatrix} 1 & 1 \\ -1 & -2 \end{pmatrix} = \begin{pmatrix} 2 & 0 \\ 0 & 3 \end{pmatrix} = \Lambda.$$

Observe that $\operatorname{tr} A = 5 = \lambda_1 + \lambda_2$, in accordance with (5.16).   ▲

Remark. When summing the eigenvalues in (5.16), repeated eigenvalues must be summed in accordance with their multiplicity. Corollary 5.16 remains valid when the matrix has complex eigenvalues and is complex complete. There is a version that is valid for incomplete matrices, but the multiplicities used are not the same as those defined above; see [181] for details. For example, the matrix $A = \begin{pmatrix} 2 & 1 \\ 0 & 2 \end{pmatrix}$ has a single eigenvalue $\lambda_1 = 2$ with multiplicity 1, but $\operatorname{tr} A = 4 = 2 + 2$.   ▲

There is also a notion of completeness for pairs of matrices.

Definition 5.18. A pair of $n \times n$ matrices A, B is called *simultaneously complete*, also known as *simultaneously diagonalizable*, if there exists a basis $\mathbf{v}_1, \ldots, \mathbf{v}_n$ of $\mathbb{R}^n$ consisting of common eigenvectors of both A and B.

In other words, the common eigenvector basis of A, B satisfies

$$A\mathbf{v}_i = \lambda_i \mathbf{v}_i, \qquad B\mathbf{v}_i = \mu_i \mathbf{v}_i, \qquad i = 1, \ldots, n, \tag{5.17}$$

where $\lambda_1, \ldots, \lambda_n, \mu_1, \ldots, \mu_n \in \mathbb{R}$, some of which may be equal, are the respective eigenvalues of A, B. Alternatively, (5.17) is equivalent to simultaneous diagonalization of the two matrices:

$$V^{-1}AV = \Lambda, \quad V^{-1}BV = M, \quad \text{or, equivalently,} \quad A = V\Lambda V^{-1}, \quad B = VMV^{-1}, \tag{5.18}$$

where $\Lambda = \operatorname{diag}(\lambda_1, \ldots, \lambda_n)$ and $M = \operatorname{diag}(\mu_1, \ldots, \mu_n)$ are the diagonal eigenvalue matrices, and $V = (\mathbf{v}_1 \ \ldots \ \mathbf{v}_n)$ is the common eigenvector matrix. There is also an evident notion of simultaneous completeness for matrices with complex eigenvalues, but this extension will not be required in this text.

Theorem 5.19. *Let A, B be $n \times n$ matrices. If A, B are simultaneously complete, then they commute: $AB = BA$. Conversely, if A, B commute and either (i) A has n distinct real eigenvalues, or (ii) B has n distinct real eigenvalues, or (iii) both A and B are complete, then A, B are simultaneously complete, and so have a common eigenvector basis.*

Proof. Since diagonal matrices always commute, if A, B satisfy (5.18), then

$$AB = (V\Lambda V^{-1})(VMV^{-1}) = V\Lambda MV^{-1} = VM\Lambda V^{-1} = (VMV^{-1})(V\Lambda V^{-1}) = BA,$$

and hence they commute.

To prove the converse, we first note that if A, B commute and $\mathbf{v}$ is an eigenvector of A with associated eigenvalue λ, then $\mathbf{w} = B\mathbf{v}$, if nonzero, is also an eigenvector of A with the same eigenvalue. Indeed, $A\mathbf{w} = AB\mathbf{v} = BA\mathbf{v} = \lambda B\mathbf{v} = \lambda \mathbf{w}$, which establishes the result.

Now, if A has n distinct real eigenvalues with corresponding eigenvector basis $\mathbf{v}_1, \ldots, \mathbf{v}_n$, then each eigenspace is one-dimensional. Thus, by the preceding result, each $B\mathbf{v}_i$ must be a scalar multiple of $\mathbf{v}_i$, and so $B\mathbf{v}_i = \mu_i \mathbf{v}_i$ for some $\mu_i \in \mathbb{R}$, which may be 0. Thus, $\mathbf{v}_i$ is also an eigenvector of B, proving simultaneous completeness. The same goes for when B has n distinct real eigenvalues.

The third case is a little trickier. According to the first result, B preserves the eigenspaces of A, namely if $\mathbf{v} \in V_i = \ker(A - \lambda_i I)$, then $B\mathbf{v} \in V_i$, and hence, in the terminology of Exercise 2.7, V_i is an invariant subspace for B (as well as A). Thus, because we are assuming B is complete, there must be a basis of V_i consisting of eigenvectors of B, which are also automatically eigenvectors of A because they all belong to V_i. Assembling all these eigenspace bases together produces the required common eigenvectors of A and B.   ∎

Remark. Part (iii) of the theorem is not valid for general matrices. For example, the matrices $A = \begin{pmatrix} 1 & 0 \\ 0 & 1 \end{pmatrix}$ and $B = \begin{pmatrix} 1 & 1 \\ 0 & 1 \end{pmatrix}$ commute — indeed the identity matrix commutes with any matrix — and A is complete, but B does not have an eigenvector basis.   ▲

Corollary 5.20. *Two symmetric matrices R, S commute if and only if they are have a common eigenvector basis. The same holds for a pair of matrices that are self-adjoint with respect to the same inner product on $\mathbb{R}^n$.*

5.2.1 Powers of Matrices and the Spectral Radius

Another application of the diagonalization formula (5.14) is that it enables us to easily compute powers of complete matrices.

Proposition 5.21. *Let A be an $n \times n$ complete matrix satisfying (5.14). Then, for any integer $k \geq 0$,*

$$A^k = V \Lambda^k V^{-1}, \qquad \text{where} \qquad \Lambda^k = \mathrm{diag}\,(\lambda_1^k, \ldots, \lambda_n^k). \tag{5.19}$$

Proof. First, using (5.14),

$$A^2 = (V \Lambda V^{-1})(V \Lambda V^{-1}) = V \Lambda^2 V^{-1},$$

proving (5.19) for $k = 2$. Iterating this argument proves the formula for all $k > 2$, while for $k = 0$ we have $I = A^0 = V \Lambda^0 V^{-1} = V I V^{-1}$.   ∎

Observe that the powers of the eigenvalues λ_j^k are the eigenvalues of the power matrix A^k, and the columns $\mathbf{v}_j$ of V are the corresponding eigenvectors, which are the same as the eigenvectors of A. If A is regular and nonsingular, and hence does not have a zero eigenvalue, then formula (5.19) also applies to negative powers, and hence is valid for all $k \in \mathbb{Z}$. Indeed,

$$(V \Lambda^{-k} V^{-1})(V \Lambda^k V^{-1}) = V \Lambda^{-k} \Lambda^k V^{-1} = V V^{-1} = I,$$

and hence $V \Lambda^{-k} V^{-1} = (V \Lambda^k V^{-1})^{-1} = (A^k)^{-1} = A^{-k}$.

Example 5.22. Return to the matrix $A = \begin{pmatrix} 1 & -1 \\ 2 & 4 \end{pmatrix}$ considered in Example 5.17. According to our calculations, the k-th power of A is given by

$$A^k = V \Lambda^k V^{-1} = \begin{pmatrix} 1 & 1 \\ -1 & -2 \end{pmatrix} \begin{pmatrix} 2^k & 0 \\ 0 & 3^k \end{pmatrix} \begin{pmatrix} 2 & 1 \\ -1 & -1 \end{pmatrix} = \begin{pmatrix} 2^{k+1} - 3^k & 2^k - 3^k \\ 2 \cdot 3^k - 2^{k+1} & 2 \cdot 3^k - 2^k \end{pmatrix}. \quad \blacktriangle$$

The behavior of the k-th powers of a matrix will thus depend on its eigenvalues. Recall first that if $a \in \mathbb{R}$ is any real number, then, as $k \to \infty$, its powers $a^k \to 0$ if $|a| < 1$, but are exponentially unbounded if $|a| > 1$. If $|a| = 1$, so $a = \pm 1$, the powers a^k remain bounded but do not go to 0. The same holds for complex numbers $a \in \mathbb{C}$, where $|a|$ denotes its modulus; see (9.162). Similarly, in view of formula (5.19), if A is complete and has one or more (real or complex) eigenvalues satisfying $|\lambda_j| > 1$, then the powers A^k become unbounded, meaning that one or more entries become arbitrarily large in absolute value. On the other hand, if *all* its eigenvalues satisfy $|\lambda_j| < 1$, then the powers $A^k \to O$ converge to the zero matrix as $k \to \infty$. Finally, if all its eigenvalues satisfy $|\lambda_j| \leq 1$, then its powers A^k remain bounded. The first two statements also hold for incomplete matrices, [181], but the third is not necessarily true. For example, the only eigenvalue of the incomplete matrix $A = \begin{pmatrix} 1 & 1 \\ 0 & 1 \end{pmatrix}$ is $\lambda = 1$, but $A^k = \begin{pmatrix} 1 & k \\ 0 & 1 \end{pmatrix}$ becomes unbounded as $k \to \infty$, albeit at a polynomial rather than exponential rate.

These remarks motivate the following important definition.

Definition 5.23. The *spectral radius* of a matrix A is defined as the maximal modulus of all of its real and complex eigenvalues: $\rho(A) = \max \{ |\lambda_1|, \ldots, |\lambda_k| \}$.

Note that the only complete matrix with zero spectral radius is the zero matrix, although there are nonzero incomplete matrices with this property, e.g., $A = \begin{pmatrix} 0 & 1 \\ 0 & 0 \end{pmatrix}$.

Recall that a square matrix A is called *convergent* if its powers $A^k \to O$ as $k \to \infty$; see Definition 4.52. Convergent matrices are characterized by their spectral radius.

Theorem 5.24. *The matrix A is convergent if and only if its spectral radius is strictly less than one: $\rho(A) < 1$. On the other hand, if $\rho(A) > 1$, then, for any norm on the space of $n \times n$ matrices, $\| A^k \| \to \infty$ as $k \to \infty$.*

Proof. When A is complete, the first part of the theorem is a direct consequence of the power formula (5.19), using the fact that each diagonal entry of Λ^k satisfies $\lambda_j^k \to 0$ since $|\lambda_j| \leq \rho(A) < 1$. On the other hand, if $\rho(A) > 1$, then at least one eigenvalue of A satisfies $|\lambda_j| > 1$, and hence $|\lambda_j|^k \to \infty$. This implies $\| \Lambda^k \| \to \infty$; see Exercise 7.16 in Chapter 2. Therefore, as a consequence of the diagonalization equation (5.19),

$$\| \Lambda^k \| = \| V^{-1} A^k V \| \leq \| V^{-1} \| \, \| A^k \| \, \| V \|.$$

Since $\| V \| \neq 0$, this implies $\| A^k \| \to \infty$.

With a little extra work, this proof can be adapted to matrices with complex eigenvalues that are diagonalizable. The proof for incomplete matrices relies on the Jordan canonical form, and we refer the reader to [181] for the details. $\blacksquare$

Remark. The converse to the second part of the theorem is not always valid. If $\rho(A) = 1$, then $\| A^k \|$ remains bounded when A is (complex) diagonalizable, but not necessarily when A is incomplete. An example of the former is $A = I$, and of the latter is the matrix $A = \begin{pmatrix} 1 & 1 \\ 0 & 1 \end{pmatrix}$ that appears immediately before Definition 5.23. $\blacktriangle$

5.2.2 Connections with Matrix Norms

Let $\| A \|$ denote a natural matrix norm, as defined in Theorem 4.49. According to Proposition 4.53, if $\| A \| < 1$ then A is convergent; the converse to this statement is, however, not valid.

Theorem 5.25. *The spectral radius of a matrix is bounded by its matrix norm:*

$$\rho(A) \le \| A \|. \tag{5.20}$$

Proof. If λ is a real eigenvalue, and $\mathbf{u}$ a corresponding unit eigenvector, so that $A\mathbf{u} = \lambda \mathbf{u}$ with $\| \mathbf{u} \| = 1$, then

$$\| A\mathbf{u} \| = \| \lambda \mathbf{u} \| = |\lambda| \, \| \mathbf{u} \| = |\lambda|. \tag{5.21}$$

Since $\| A \|$ is the maximum of $\| A\mathbf{u} \|$ over all possible unit vectors, this implies that

$$|\lambda| \le \| A \|. \tag{5.22}$$

If all the eigenvalues of A are real, then the spectral radius is the maximum of their absolute values, and so it too is bounded by $\| A \|$, proving (5.20). The proof when A has complex eigenvalues is a bit trickier and can be found in [181, Theorem 9.21]. $\blacksquare$

Example 5.26. Consider the matrix $A = \begin{pmatrix} \frac{1}{2} & -\frac{1}{3} \\ -\frac{1}{3} & \frac{1}{4} \end{pmatrix}$. Since $\| A \|_\infty = \frac{5}{6} \approx .8333$ is less than 1, A is a convergent matrix. Indeed, its eigenvalues are $\dfrac{9 \pm \sqrt{73}}{24}$, and hence its spectral radius is $\rho(A) = \dfrac{9 + \sqrt{73}}{24} \approx .7310$, which is slightly smaller than its ∞ norm. The matrix $A = \begin{pmatrix} \frac{1}{2} & -\frac{3}{5} \\ -\frac{3}{5} & \frac{1}{4} \end{pmatrix}$ has matrix norm $\| A \|_\infty = \frac{11}{10} > 1$. On the other hand, its eigenvalues are $\dfrac{15 \pm \sqrt{601}}{40}$ and hence its spectral radius is $\rho(A) = \dfrac{15 + \sqrt{601}}{40} \approx .9879$, which implies that A is (just barely) convergent, even though its ∞ matrix norm is larger than 1. $\blacktriangle$

Based on the accumulated evidence, one might be tempted to speculate that the spectral radius itself defines a matrix norm. Unfortunately, this is not the case. Indeed, as we already noted, the nonzero matrix $A = \begin{pmatrix} 0 & 1 \\ 0 & 0 \end{pmatrix}$ has zero spectral radius, $\rho(A) = 0$, in violation of a basic norm axiom.

Nevertheless there is an intimate connection between matrix norms and the spectral radius, which is best encapsulated in the following result, known as *Gel'fand's formula* in honor of the influential twentieth century Russian mathematician Israel Gel'fand.

Theorem 5.27. *If A is any square matrix and $\|A\|$ its norm in any natural matrix norm, then*

$$\lim_{k\to\infty} \|A^k\|^{1/k} = \rho(A). \tag{5.23}$$

Proof. Let us assume $\rho(A) > 0$. Let $0 < \varepsilon < \rho(A)$, and define

$$A_+ = \frac{A}{\rho(A) + \varepsilon}, \qquad A_- = \frac{A}{\rho(A) - \varepsilon}.$$

Since $\rho(cA) = c\rho(A)$ for any positive scalar $c > 0$, we have

$$\rho(A_+) = \frac{\rho(A)}{\rho(A) + \varepsilon} < 1, \qquad \rho(A_-) = \frac{\rho(A)}{\rho(A) - \varepsilon} > 1.$$

Thus, according to Theorem 5.24, A_+ is a convergent matrix, and hence there exists n_+ such that $\|A_+^k\| < 1$ for all $k \geq n_+$. This implies that

$$\|A^k\| = \big(\rho(A) + \varepsilon\big)^k \|A_+^k\| < \big(\rho(A) + \varepsilon\big)^k, \quad \text{and hence} \quad \|A^k\|^{1/k} < \rho(A) + \varepsilon$$

whenever $k \geq n_+$. Similarly, again by Theorem 5.24, A_- satisfies $\|A_-^k\| \to \infty$, and hence there exists n_- such that $\|A_-^k\| > 1$ for all $k \geq n_-$. This implies that

$$\|A^k\| = \big(\rho(A) - \varepsilon\big)^k \|A_-^k\| > \big(\rho(A) - \varepsilon\big)^k, \quad \text{and hence} \quad \|A^k\|^{1/k} > \rho(A) - \varepsilon.$$

whenever $k \geq n_-$. Thus, when $n \geq \max\{n_+, n_-\}$, we have

$$\rho(A) - \varepsilon < \|A^k\|^{1/k} < \rho(A) + \varepsilon.$$

Since this holds for all sufficiently small $\varepsilon > 0$, it implies the limiting formula (5.23).

The remaining case, when $\rho(A) = 0$, is trivial if A is complete, since then $A = O$ is the zero matrix. The incomplete case relies on the fact that A is nilpotent, meaning that $A^k = O$ for all sufficiently large k; see [181] for a proof. ∎

As an immediate application of Gel'fand's formula (5.23), we obtain the following useful estimate.

Corollary 5.28. *Let A be an $n \times n$ matrix. Let $\|A\|$ be its natural matrix norm based on a norm $\|\cdot\|$ on $\mathbb{R}^n$. Let $\varepsilon > 0$. Then there exists a positive integer N, depending on ε, such that, for all $k \geq N$,*

$$\|A^k\| \leq \big(\rho(A) + \varepsilon\big)^k, \tag{5.24}$$

and hence

$$\|A^k \mathbf{v}\| \leq \big(\rho(A) + \varepsilon\big)^k \|\mathbf{v}\| \qquad \text{for any} \qquad \mathbf{v} \in \mathbb{R}^n.$$

Exercises

2.1. Which of the following matrices are complete? For those that are, exhibit an eigenvector basis of $\mathbb{R}^2$. For those that are not, what is the dimension of the subspace of $\mathbb{R}^2$ spanned by the eigenvectors? $(a)\,\heartsuit\;\begin{pmatrix} 1 & 3 \\ 3 & 1 \end{pmatrix}$, $(b)\,\diamondsuit\;\begin{pmatrix} 1 & 3 \\ -3 & 1 \end{pmatrix}$, $(c)\,\heartsuit\;\begin{pmatrix} 1 & 3 \\ 0 & 1 \end{pmatrix}$, $(d)\;\begin{pmatrix} 3 & 3 \\ 1 & 5 \end{pmatrix}$.

2.2. Find the spectral radius of the following matrices. Which are convergent?

$(a)\,\heartsuit\;\begin{pmatrix} .3 & -.4 \\ -.2 & .6 \end{pmatrix}$, $(b)\,\diamondsuit\;\begin{pmatrix} 0 & \frac{4}{5} \\ \frac{3}{5} & \frac{2}{3} \end{pmatrix}$, $(c)\,\heartsuit\;\begin{pmatrix} .3 & 2.2 & -1.7 \\ 0 & -.6 & 4.1 \\ 0 & 0 & .5 \end{pmatrix}$, $(d)\;\begin{pmatrix} 0 & 1 & -1 \\ 0 & 1 & 0 \\ 1 & 0 & 2 \end{pmatrix}$.

2.3. Use (5.19) to write down an explicit formula for the k-th power of the following matrices:

$(a)\,\heartsuit\;\begin{pmatrix} 5 & 2 \\ 2 & 2 \end{pmatrix}$, $(b)\;\begin{pmatrix} 4 & 1 \\ -2 & 1 \end{pmatrix}$, $(c)\,\heartsuit\;\begin{pmatrix} 1 & -1 & 0 \\ 0 & 0 & 1 \\ 0 & 0 & -1 \end{pmatrix}$, $(d)\,\diamondsuit\;\begin{pmatrix} 1 & 1 & 2 \\ 1 & 2 & 1 \\ 2 & 1 & 1 \end{pmatrix}$.

2.4. *True or false*: (a) Every diagonal matrix is complete. (b) Every upper triangular matrix is complete.

2.5. $\heartsuit$ Prove that if A is a complete matrix, then so is $cA + dI$, where c, d are any scalars.

2.6. (a) Prove that if A is complete, then so is A^2. (b) Give an example of an incomplete matrix A such that A^2 is complete.

2.7. $\diamondsuit$ Let A be an $n \times n$ matrix. A nontrivial subspace $\{\mathbf{0}\} \neq V \subset \mathbb{R}^n$ is called an *invariant subspace* for A if $A\mathbf{v} \in V$ whenever $\mathbf{v} \in V$. (a) Prove that every eigenspace of A is an invariant subspace. (b) Prove that if A is complete, then every invariant subspace is spanned by one or more of its eigenvectors. *Hint*: Adapt the method used to prove Proposition 5.10.

2.8. Suppose A has spectral radius $\rho(A)$. Can you predict the spectral radius of $cA + dI$, where c, d are scalars? If not, what additional information do you need?

2.9. *True or false*: $(a)\,\heartsuit\;\rho(cA) = c\,\rho(A)$, $(b)\;\rho(V^{-1}AV) = \rho(A)$, $(c)\,\heartsuit\;\rho(A^2) = \rho(A)^2$, $(d)\,\heartsuit\;\rho(A^{-1}) = 1/\rho(A)$, $(e)\,\diamondsuit\;\rho(A + B) = \rho(A) + \rho(B)$, $(f)\;\rho(AB) = \rho(A)\,\rho(B)$.

2.10. Prove that if A is any square matrix, then there exists $c \neq 0$ such that the scalar multiple cA is convergent. Find a formula for the largest possible such c.

5.3   Eigenvalues of Self-Adjoint Matrices

Fortunately, the matrices that arise in applications to machine learning and data analysis are complete and, in fact, possess some additional structure that ameliorates the calculation of their eigenvalues and eigenvectors. The most important class consists of the self-adjoint matrices, which includes symmetric and positive definite matrices. In fact, not only are the eigenvalues of a self-adjoint matrix necessarily real, the eigenvectors *always* form an *orthogonal basis* of the underlying Euclidean space, enjoying all the wonderful properties we studied in Section 2.3. One can, of course, convert an orthogonal eigenvector basis into an orthonormal eigenvector basis by dividing each basis vector by its norm, which does not alter its status as an eigenvector. In fact, this is by far the most common way for orthonormal bases to appear

— as the eigenvector bases of particular self-adjoint matrices. Let us state this important result, but defer its proof until the following section.

Theorem 5.29. *Let $S = S^*$ be a self-adjoint $n \times n$ matrix with respect to an inner product on $\mathbb{R}^n$. Then*

    (a) *All the eigenvalues of S are real.*

    (b) *Eigenvectors corresponding to distinct eigenvalues are orthogonal.*

    (c) *There is an orthonormal basis of $\mathbb{R}^n$ consisting of n eigenvectors of S.*

In particular, all self-adjoint matrices are complete.

Note that if S has n distinct eigenvalues, then the orthonormal basis vectors are uniquely determined up to plus or minus signs. When the matrix has eigenvalues with higher multiplicity, there is more freedom in their specification, since they can contain any orthonormal basis of each eigenspace.

In this chapter, we will always sort the eigenvalues of a self-adjoint matrix S in *decreasing* order, so

$$\lambda_1 \geq \lambda_2 \geq \cdots \geq \lambda_n, \tag{5.25}$$

where repeated eigenvalues are listed as many times as their multiplicity. We will sometimes also write $\lambda_j(S)$ to denote the j-th eigenvalue of the matrix S, and $\lambda_{max} = \lambda_{max}(S) = \lambda_1$ and $\lambda_{min} = \lambda_{min}(S) = \lambda_n$ to denote the largest and smallest eigenvalues, respectively. Note that the spectral radius $\rho(S)$ is either $|\lambda_{max}(S)|$ or $|\lambda_{min}(S)|$, depending upon which one is larger is absolute value.

Example 5.30. The 2×2 matrix $S = \begin{pmatrix} 3 & 1 \\ 1 & 3 \end{pmatrix}$ in Example 5.12 is symmetric, and so has real eigenvalues: $\lambda_1 = \lambda_{max} = 4$ and $\lambda_2 = \lambda_{min} = 2$. Thus $\rho(S) = 4$. The corresponding eigenvectors $\mathbf{v}_1 = (1,1)^T$ and $\mathbf{v}_2 = (1,-1)^T$ are orthogonal under the dot product, i.e., $\mathbf{v}_1 \cdot \mathbf{v}_2 = 0$, and hence form an orthogonal basis of $\mathbb{R}^2$. The orthonormal eigenvector basis promised by Theorem 5.29 is obtained by dividing each eigenvector by its Euclidean norm:

$$\mathbf{u}_1 = \begin{pmatrix} \frac{1}{\sqrt{2}} \\ \frac{1}{\sqrt{2}} \end{pmatrix}, \qquad \mathbf{u}_2 = \begin{pmatrix} \frac{1}{\sqrt{2}} \\ -\frac{1}{\sqrt{2}} \end{pmatrix}. \qquad\qquad \blacktriangle$$

We are now in a position to characterize all self-adjoint matrices.

Theorem 5.31. *An $n \times n$ matrix S is self-adjoint with respect to some inner product on $\mathbb{R}^n$ if and only if it is complete.*

Proof. We have already established the direct statement. To prove the converse, given a complete matrix S, we need to construct an inner product that makes S self-adjoint. Let $V = (\mathbf{v}_1 \ \ldots \ \mathbf{v}_n)$ be the nonsingular matrix whose columns are the eigenvectors of S, so that, by (5.15),

$$S V = V \Lambda, \qquad \text{and hence} \qquad V^T S^T = \Lambda V^T, \tag{5.26}$$

where $\Lambda = \operatorname{diag}(\lambda_1, \ldots, \lambda_n)$ is the diagonal eigenvalue matrix. Then $\mathbf{v}_1, \ldots, \mathbf{v}_n$ will form an orthonormal eigenvector basis of S for the inner product determined by a positive definite symmetric matrix C provided

$$V^T C V = I, \qquad \text{and hence} \qquad C = V^{-T} V^{-1} = (V V^T)^{-1}. \tag{5.27}$$

Note that the resulting matrix C is symmetric and positive definite, as required, since it is the Gram matrix for the nonsingular matrix V^{-1}. Furthermore, using (5.26), we have

$$V V^T S^T = V \Lambda V^T = S V V^T,$$

and hence, using (5.27),

$$S^* = C^{-1} S^T C = V V^T S^T (V V^T)^{-1} = V \Lambda V^T (V V^T)^{-1} = S V V^T (V V^T)^{-1} = S,$$

thus establishing self-adjointness of S with respect to the inner product that is prescribed by the inverse eigenvector Gram matrix $C = (V V^T)^{-1}$.   ∎

The eigenvalues of a self-adjoint matrix can be used to test its positive definiteness.

Theorem 5.32. *A self-adjoint matrix $H = H^*$ is positive definite if and only if all of its eigenvalues are strictly positive.*

Proof. First, if H is positive definite, then, by definition, $\langle \mathbf{x}, H \mathbf{x} \rangle > 0$ for all nonzero vectors $\mathbf{0} \neq \mathbf{x} \in \mathbb{R}^n$. In particular, if $\mathbf{x} = \mathbf{v}$ is an eigenvector with (necessarily real) eigenvalue λ, then

$$0 < \langle \mathbf{v}, H \mathbf{v} \rangle = \langle \mathbf{v}, \lambda \mathbf{v} \rangle = \lambda \langle \mathbf{v}, \mathbf{v} \rangle = \lambda \| \mathbf{v} \|^2, \tag{5.28}$$

which immediately proves that $\lambda > 0$.

Conversely, suppose H has all positive eigenvalues. Let $\mathbf{u}_1, \ldots, \mathbf{u}_n$ be the orthonormal eigenvector basis guaranteed by Theorem 5.29, with $H \mathbf{u}_j = \lambda_j \mathbf{u}_j$ for $\lambda_j > 0$. Writing

$$\mathbf{x} = c_1 \mathbf{u}_1 + \cdots + c_n \mathbf{u}_n, \qquad \text{we obtain} \qquad H \mathbf{x} = c_1 \lambda_1 \mathbf{u}_1 + \cdots + c_n \lambda_n \mathbf{u}_n.$$

Therefore, using the orthonormality of the eigenvectors,

$$\langle \mathbf{x}, H \mathbf{x} \rangle = \left\langle \sum_{i=1}^{n} c_i \mathbf{u}_i, \sum_{j=1}^{n} c_j \lambda_j \mathbf{u}_j \right\rangle = \sum_{i,j=1}^{n} \lambda_j c_i c_j \langle \mathbf{u}_i, \mathbf{u}_j \rangle = \sum_{i=1}^{n} \lambda_i c_i^2 > 0$$

whenever $\mathbf{x} \neq \mathbf{0}$, since all $\lambda_i > 0$ and only $\mathbf{x} = \mathbf{0}$ has coordinates $c_1 = \cdots = c_n = 0$. This inequality establishes the positive definiteness of H.   ∎

The same proof shows that H is positive semidefinite if and only if all its eigenvalues satisfy $\lambda_j \geq 0$. A positive semidefinite matrix that is not positive definite admits a zero eigenvalue with eigenspace $V_0 = \ker H \neq \{\mathbf{0}\}$. In both cases, the spectral radius is the largest eigenvalue, $\rho(H) = \lambda_{max}(H) > 0$.

Finally, combining Theorems 5.31 and 5.32, we are able to characterize all matrices which are self adjoint and positive definite with respect to some inner product.

Theorem 5.33. *A matrix H is self-adjoint and positive definite with respect to some inner product if and only if is complete and has all strictly positive eigenvalues.*

Example 5.34. Consider the 3×3 matrix $A = \begin{pmatrix} -1 & -2 & -2 \\ 2 & 3 & 2 \\ 2 & 2 & 3 \end{pmatrix}$. It can be shown that A has two eigenvalues: $\lambda_1 = 3$ and $\lambda_2 = 1$. The first has, up to scalar multiple, a single eigenvector $\mathbf{v}_1 = (1, -1, -1)^T$; on the other hand, the second eigenvalue has a two-dimensional eigenspace, with basis $\mathbf{v}_2 = (1, -1, 0)^T$, $\mathbf{v}_3 = (1, 0, -1)^T$, and any nonzero linear combination of these two vectors is an eigenvector for λ_2. We conclude that A is complete, with all positive eigenvalues. Theorem 5.31 and Theorem 5.33 imply that A is self-adjoint and positive definite with respect to the inner product defined by the symmetric positive definite matrix

$$C = (VV^T)^{-1} = \begin{pmatrix} 3 & 2 & 2 \\ 2 & 2 & 1 \\ 2 & 1 & 2 \end{pmatrix}, \quad \text{where} \quad V = \begin{pmatrix} 1 & 1 & 1 \\ -1 & -1 & 0 \\ -1 & 0 & -1 \end{pmatrix} \quad \text{is the eigenvector matrix.}$$

Indeed, the product matrix $CA = \begin{pmatrix} 5 & 4 & 4 \\ 4 & 4 & 3 \\ 4 & 3 & 4 \end{pmatrix}$ is symmetric, verifying the self-adjointness

criterion in Proposition 4.19, and positive definite. Observe that having a negative entry on the main diagonal of A does not preclude its positive definiteness.   ▲

5.3.1   The Spectral Theorem

We have now established a key result known as the *Spectral Theorem*. The term "spectrum" refers to the set of eigenvalues of a matrix, or, more generally, a linear operator, [181]. The terminology is motivated by physics, where the spectral energy lines of atoms, molecules, nuclei, etc., are characterized as the eigenvalues of the governing quantum mechanical Schrödinger operator, [167]. The Spectral Theorem 5.35 is the finite-dimensional version of the decomposition of a quantum mechanical linear operator into its spectral eigenstates, cf. [192].

Theorem 5.35. *Let S be a self-adjoint $n \times n$ matrix with respect to the inner product defined by the symmetric positive definite matrix C. Let $Q = (\mathbf{u}_1 \ \ldots \ \mathbf{u}_n)$ be the matrix whose columns form an orthonormal basis of $\mathbb{R}^n$ consisting of eigenvectors of S, and let $\Lambda = \mathrm{diag}\,(\lambda_1, \ldots, \lambda_n)$ be the diagonal matrix containing the corresponding eigenvalues. Then*

$$S = Q\Lambda Q^{-1} = Q\Lambda Q^* = Q\Lambda Q^T C = \sum_{j=1}^{n} \lambda_j\, \mathbf{u}_j \mathbf{u}_j^T C. \tag{5.29}$$

Here, the adjoint of Q is computed using the dot product on its domain space and the inner product determined by C on its codomain.

Proof. As in (4.61), the condition that the columns of Q form an orthonormal basis of $\mathbb{R}^n$ for the inner product defined by C is

$$Q^* Q = Q^T C Q = \mathrm{I}, \qquad \text{and hence} \qquad Q^{-1} = Q^* = Q^T C. \tag{5.30}$$

The initial matrix equation in (5.29) is an immediate consequence of (5.14), while the second and third equations follow from (5.30). The third equation follows by applying the alternative rule (3.10) for matrix multiplication to the product $Q\Lambda Q^T$, noting that, since Λ is diagonal, the k-th row of ΛQ^T is $\lambda_k\, \mathbf{u}_k^T$. Alternatively, it can be established directly by noting that, for any $\mathbf{x} \in \mathbb{R}^n$, we can write

$$S\mathbf{x} = S \sum_{j=1}^{n} \langle \mathbf{u}_j, \mathbf{x} \rangle\, \mathbf{u}_j = \sum_{j=1}^{n} \langle \mathbf{u}_j, \mathbf{x} \rangle\, S\mathbf{u}_j = \sum_{j=1}^{n} (\mathbf{u}_j^T C \mathbf{x})\lambda_j \mathbf{u}_j = \left(\sum_{j=1}^{n} \lambda_j \mathbf{u}_j \mathbf{u}_j^T \right) C\mathbf{x}.$$

Since this holds for all $\mathbf{x}$, the spectral decomposition (5.29) follows.   ■

Remark. The traditional and most common form of the Spectral Theorem is the special case when $C = I$, corresponding to the dot product on $\mathbb{R}^n$. Then S is a symmetric matrix, Q is an orthogonal matrix, and the spectral factorization equation (5.29) becomes

$$S = Q\Lambda Q^{-1} = Q\Lambda Q^T = \sum_{j=1}^{n} \lambda_j \, \mathbf{u}_j \mathbf{u}_j^T. \tag{5.31}$$

If $\operatorname{rank} S = k$, then S has k nonzero eigenvalues (multiple eigenvalues counted accordingly). Let $\widetilde{\Lambda}$ denote the $k \times k$ diagonal matrix containing the nonzero eigenvalues, and let $\widetilde{Q}$ be the $n \times k$ matrix whose columns are the corresponding non-null orthonormal eigenvectors. Then formula (5.31) can be written in the reduced form

$$S = \widetilde{Q}\,\widetilde{\Lambda}\,\widetilde{Q}^T = \sum_{\lambda_j \neq 0} \lambda_j \, \mathbf{u}_j \mathbf{u}_j^T. \tag{5.32}$$

An analogous result holds in the more general self-adjoint case (5.29).   ▲

Example 5.36. Using the dot product on $\mathbb{R}^2$, the orthonormal eigenvectors of the 2×2 matrix considered in Example 5.30 can be assembled into the orthogonal matrix $Q = \begin{pmatrix} \frac{1}{\sqrt{2}} & \frac{1}{\sqrt{2}} \\ \frac{1}{\sqrt{2}} & -\frac{1}{\sqrt{2}} \end{pmatrix}$. The reader can validate the resulting spectral factorization:

$$\begin{pmatrix} 3 & 1 \\ 1 & 3 \end{pmatrix} = S = Q\Lambda Q^T = \begin{pmatrix} \frac{1}{\sqrt{2}} & \frac{1}{\sqrt{2}} \\ \frac{1}{\sqrt{2}} & -\frac{1}{\sqrt{2}} \end{pmatrix} \begin{pmatrix} 4 & 0 \\ 0 & 2 \end{pmatrix} \begin{pmatrix} \frac{1}{\sqrt{2}} & \frac{1}{\sqrt{2}} \\ \frac{1}{\sqrt{2}} & -\frac{1}{\sqrt{2}} \end{pmatrix}$$

$$= 4 \begin{pmatrix} \frac{1}{\sqrt{2}} \\ \frac{1}{\sqrt{2}} \end{pmatrix} \begin{pmatrix} \frac{1}{\sqrt{2}} & \frac{1}{\sqrt{2}} \end{pmatrix} + 2 \begin{pmatrix} \frac{1}{\sqrt{2}} \\ -\frac{1}{\sqrt{2}} \end{pmatrix} \begin{pmatrix} \frac{1}{\sqrt{2}} & -\frac{1}{\sqrt{2}} \end{pmatrix}. \qquad ▲$$

The Spectral Theorem 5.35 is a fundamental tool in the study of symmetric and self-adjoint matrices. Let us now present a few initial applications.

5.3.2   Powers of Self-Adjoint Matrices

Combining the formula (5.19) for the powers of a complete matrix with the Spectral Theorem 5.35 allows us to easily compute powers of a self-adjoint matrix $S = S^*$, namely

$$S^k = Q\Lambda^k Q^{-1} = Q\Lambda^k Q^* = Q\Lambda^k Q^T C = \sum_{j=1}^{n} \lambda_j^k \, \mathbf{u}_j \mathbf{u}_j^T C. \tag{5.33}$$

Here $\lambda_1, \ldots, \lambda_n$ are the eigenvalues of S (necessarily real), while $\mathbf{u}_1, \ldots, \mathbf{u}_n$ the corresponding orthonormal eigenvectors, and C is a symmetric positive definite matrix defining the underlying inner product on $\mathbb{R}^n$. The standard case corresponds to the dot product with $C = I$ and $S = S^T$ symmetric, for which (5.33) reduces to

$$S^k = Q\Lambda^k Q^T = \sum_{j=1}^{n} \lambda_j^k \, \mathbf{u}_j \mathbf{u}_j^T. \tag{5.34}$$

Formula (5.33) is valid for all nonnegative integers $k \geq 0$ and, if S is nonsingular and hence 0 is not an eigenvalue, for all $k \in \mathbb{Z}$. For example,

$$S^{-1} = Q\Lambda^{-1}Q^{-1} = Q\Lambda^{-1}Q^* = Q\Lambda^{-1}Q^T C = \left(\sum_{j=1}^{n} \frac{\mathbf{u}_j \mathbf{u}_j^T}{\lambda_j} \right) C. \tag{5.35}$$

If, in addition, S is positive definite, then all its eigenvalues $\lambda_j > 0$, and the right hand side of formula (5.33) makes sense when k is *any* real number. This motivates the following definition. For this, we make the convention that, for $p \in \mathbb{R}$, the p-th power of a positive real number $0 < a \in \mathbb{R}$ is defined to be $a^p = e^{p \log a} > 0$. Observe that $(a^p)^q = a^{p+q}$ for all $p, q \in \mathbb{R}$.

> **Definition 5.37.** Let S be a positive definite self-adjoint matrix. Then, using the above notations, for any $p \in \mathbb{R}$, the p-th power matrix S^p is defined as
>
> $$S^p = Q\,\Lambda^p\,Q^{-1} = Q\,\Lambda^p\,Q^* = Q\,\Lambda^p\,Q^T C = \sum_{j=1}^{n} \lambda_j^p\,\mathbf{u}_j\,\mathbf{u}_j^T C, \qquad (5.36)$$
>
> where $\Lambda^p = \operatorname{diag}(\lambda_1^p, \ldots, \lambda_n^p)$.

Observe that S^p is positive definite and self-adjoint for all $p \in \mathbb{R}$. Moreover, its eigenvalues are the p-th powers of the eigenvalues of S. The matrix power S^p satisfies the familiar rule that $S^p S^q = S^{p+q}$. Indeed, we have

$$S^p S^q = Q\,\Lambda^p Q^{-1} Q\Lambda^q Q^{-1} = Q\,\Lambda^p \Lambda^q\,Q^{-1} = Q\,\Lambda^{p+q}\,Q^{-1} = S^{p+q}.$$

If S is only positive semidefinite, then we can still define S^p in the same manner for any $p \geq 0$, but, since one or more eigenvalues vanish, we are not able to extend to negative exponents.

An important special case is when $p = \frac{1}{2}$; then

$$S^{1/2} = Q\,\Lambda^{1/2}\,Q^{-1} = Q\,\Lambda^{1/2}\,Q^* = Q\,\Lambda^{1/2}\,Q^T C \qquad (5.37)$$

is called the *matrix square root* of the self adjoint matrix S and often denoted $\sqrt{S}$, although we prefer the former notation. By the above property, we have $S^{1/2} S^{1/2} = (S^{1/2})^2 = S$. Note also that $S^{-1/2}$ is the inverse of $S^{1/2}$ when S, and hence $S^{1/2}$, is nonsingular. Practical methods of computing the matrix square root can be found in [104].

Now, given that C is a symmetric positive definite $n \times n$ matrix, it is self-adjoint with respect to the dot product on $\mathbb{R}^n$. Let Q now denote the orthogonal matrix whose columns are the orthonormal eigenvector basis for C, so that, by the classical spectral equation (5.31), $C = Q\Lambda Q^{-1} = Q\Lambda Q^T$, where Λ is the corresponding diagonal eigenvalue matrix. The matrix square root

$$C^{1/2} = Q\,\Lambda^{1/2}\,Q^T \qquad (5.38)$$

of C allows us to relate concepts involving the C inner product and norm to the ordinary dot product and Euclidean norm. We shall prove the first of the following interconnected statements, and leave the rest to Exercise 3.5.

(a) If $\mathbf{y} = C^{1/2}\mathbf{x}$, then $\|\mathbf{x}\|_C = \|\mathbf{y}\|_2$.

(b) A matrix $\widetilde{S}$ is self-adjoint under the C inner product if and only if $S = C^{1/2}\widetilde{S}C^{-1/2}$ is symmetric. Moreover, $\widetilde{S}$ is positive definite if and only if S is positive definite.

(c) An $n \times n$ matrix $\widetilde{Q}$ is C norm preserving if and only if $Q = C^{1/2}\widetilde{Q}C^{-1/2}$ is an orthogonal matrix.

(d) The columns of an $n \times n$ matrix $\widehat{Q}$ form an orthonormal basis for the C inner product if and only if $Q = C^{1/2}\widehat{Q}$ is an orthogonal matrix and hence its columns form an orthonormal basis under the dot product.

To prove part (a), since C is symmetric,

$$\|\mathbf{y}\|_2^2 = \mathbf{y}^T \mathbf{y} = (C^{1/2}\mathbf{x})^T (C^{1/2}\mathbf{x}) = \mathbf{x}^T C^{1/2} C^{1/2}\mathbf{x} = \mathbf{x}^T C\mathbf{x} = \|\mathbf{x}\|_C^2.$$

Remark. Although these properties enable us, in essence, to perform all calculations and constructions using only the dot product, the flexibility afforded by more general inner products and norms will prove to be of great utility when we discuss key machine learning algorithms, and it is, in our opinion, important to retain the general inner product constructions throughout, thereby providing a natural framework for what is known as preconditioning that will often serve to enhance the speed and accuracy of the required numerical computations. ▲

5.3.3 The Schur Product Theorem

Now, whereas the powers of positive definite matrices are also positive definite, the same cannot be said for the product of two different positive definite matrices. Indeed, the product of two symmetric matrices R, S is not necessarily symmetric, let alone positive definite; the same goes for more general self-adjoint matrices. Indeed, according to Exercise 2.9(b) in Chapter 3, $(RS)^T \neq RS$ unless R and S commute.

However, it turns out that, in the symmetric case, the alternative Hadamard matrix product, as defined in Exercise 1.17, does maintain symmetry and positive definiteness. Recall that the Hadamard product is the "wrong" way to multiply matrices. More specifically, given two $m \times n$ matrices A, B, necessarily of the same size, their *Hadamard product* is the $m \times n$ matrix $C = A \circ B$ whose (i,j) entry is merely the product of the (i,j) entries of A and B, so $c_{ij} = a_{ij} b_{ij}$. Unlike matrix multiplication, the Hadamard product is clearly commutative, that is, $A \circ B = B \circ A$, as well as satisfying all the usual properties of multiplication, e.g., associativity.

The *Schur Product Theorem* states that the Hadamard product of two symmetric positive definite matrices is also symmetric positive definite. *Warning*: This result is *not* true for more general self-adjoint positive definite matrices.

Theorem 5.38. *If R and S are symmetric positive (semi)definite matrices, then $R \circ S$ is symmetric positive (semi)definite.*

Proof. Let R and S be symmetric matrices. According to the spectral decomposition (5.31), we can write

$$R = \sum_{j=1}^{n} \lambda_j \mathbf{u}_j \mathbf{u}_j^T, \qquad S = \sum_{j=1}^{n} \mu_j \mathbf{v}_j \mathbf{v}_j^T,$$

where $\lambda_1, \ldots, \lambda_n$ are the eigenvalues and $\mathbf{u}_1, \ldots, \mathbf{u}_n$ the corresponding orthonormal eigenvectors of R, while $\mu_1, \ldots, \mu_n$ are the eigenvalues and $\mathbf{v}_1, \ldots, \mathbf{v}_n$ the corresponding orthonormal eigenvectors of S. Their Hadamard product is then given by

$$R \circ S = \sum_{i=1}^{n} \sum_{j=1}^{n} \lambda_i \mu_j (\mathbf{u}_j \mathbf{u}_j^T) \circ (\mathbf{v}_j \mathbf{v}_j^T) = \sum_{i=1}^{n} \sum_{j=1}^{n} \lambda_i \mu_j (\mathbf{u}_i \circ \mathbf{v}_j)(\mathbf{u}_i \circ \mathbf{v}_j)^T,$$

where the second equality follows from the second identity in Exercise 1.17(d) in Chapter 3. Thus, if $\mathbf{x} \in \mathbb{R}^n$,

$$\mathbf{x}^T (R \circ S) \mathbf{x} = \sum_{i=1}^{n} \sum_{j=1}^{n} \lambda_i \mu_j \left[\mathbf{x}^T (\mathbf{u}_i \circ \mathbf{v}_j) \right]^2 = \sum_{i=1}^{n} \sum_{j=1}^{n} \lambda_i \mu_j \left[(\mathbf{u}_i \circ \mathbf{v}_j) \cdot \mathbf{x} \right]^2.$$

If all $\lambda_i, \mu_j \geq 0$, the right hand side is clearly ≥ 0, proving the positive semidefinite case. As for positive definite symmetric matrices R, S, where all $\lambda_i, \mu_j > 0$, the right hand side can only vanish if

$$0 = (\mathbf{u}_i \circ \mathbf{v}_j) \cdot \mathbf{x} = (\mathbf{u}_i \circ \mathbf{x}) \cdot \mathbf{v}_j, \qquad \text{for all} \qquad i, j = 1, \dots, n, \tag{5.39}$$

where we used the first identity in Exercise 1.17(d) in Chapter 3. Since both $\mathbf{u}_1, \dots, \mathbf{u}_n$ and $\mathbf{v}_1, \dots, \mathbf{v}_n$ are bases of $\mathbb{R}^n$, it is readily seen that (5.39) holds if and only if $\mathbf{x} = \mathbf{0}$, thus establishing positive definiteness of the Hadamard product. ∎

Example 5.39. The matrices $R = \begin{pmatrix} 3 & -1 \\ -1 & 2 \end{pmatrix}$ and $S = \begin{pmatrix} 4 & 2 \\ 2 & 2 \end{pmatrix}$ are both symmetric and

positive definite. Their Hadamard product $R \circ S = \begin{pmatrix} 12 & -2 \\ -2 & 4 \end{pmatrix}$ is also positive definite, as

can easily be checked. ▲

5.3.4 Generalized Eigenvalues and Eigenvectors

In certain applications, it is useful to generalize the notion of eigenvalue by replacing the identity matrix that appears in (5.2) by a general square matrix.

Definition 5.40. Let A, B be $n \times n$ matrices. A scalar λ is called a *generalized eigenvalue* of the matrix pair A, B if there is a *nonzero* vector $\mathbf{v} \neq \mathbf{0}$, called a *generalized eigenvector*, such that

$$A\mathbf{v} = \lambda B \mathbf{v}. \tag{5.40}$$

Thus, λ is a generalized eigenvalue if and only if the matrix $A - \lambda B$ is singular, and the associated generalized eigenvectors are the nonzero elements of its kernel. The generalized eigenvalues are hence the roots of the *generalized characteristic equation*:

$$p_{A,B}(\lambda) = \det(A - \lambda B) = 0, \tag{5.41}$$

the left hand side of which, provided $B \neq O$, is a nonconstant polynomial in λ of degree $\leq n$. Consequently, every such matrix pair has at least one and at most n distinct complex generalized eigenvectors. If B is nonsingular, we can rewrite the generalized eigenvector equation (5.40) as

$$B^{-1}A\mathbf{v} = \lambda \mathbf{v}, \tag{5.42}$$

and hence generalized eigenvalues and eigenvectors of the pair A, B are ordinary eigenvalues and eigenvectors of the product matrix $B^{-1}A$.

We are particularly interested in the case when the matrices are symmetric and at least one is positive definite.

Theorem 5.41. *Let K, C be symmetric $n \times n$ matrices, with C positive definite. Then their generalized eigenvalues are all real. Moreover, they are complete in the sense that there exists a generalized eigenvector basis of $\mathbb{R}^n$, whose elements are orthonormal under the inner product defined by C. Moreover, K is also positive definite if and only if their generalized eigenvalues are all strictly positive.*

Proof. According to the above remarks, the generalized eigenvalues and eigenvectors are the ordinary eigenvalues and eigenvectors of the matrix $H = C^{-1}K$. Proposition 4.19 says that H is self-adjoint with respect to the inner product defined by C, and hence Theorem 5.41 follows immediately from Theorems 5.29 and 5.32. ∎

Example 5.42. Suppose $K = \begin{pmatrix} -2 & 1 \\ 1 & 3 \end{pmatrix}$, $C = \begin{pmatrix} 2 & -1 \\ -1 & 4 \end{pmatrix}$. The generalized characteristic equation is

$$0 = \det(K - \lambda C) = \det \begin{pmatrix} -2 - 2\lambda & 1 + \lambda \\ 1 + \lambda & 3 - 4\lambda \end{pmatrix} = 7\lambda^2 - 7,$$

and hence the generalized eigenvalues are $\lambda_1 = 1$, $\lambda_2 = -1$. The corresponding generalized eigenvectors are $\mathbf{v}_1 = (1, 0)^T$, $\mathbf{v}_2 = (1, 2)^T$, obtained by solving the homogeneous systems $(K - \lambda_i C)\mathbf{v}_i = \mathbf{0}$, $i = 1, 2$. As you can check, these are the ordinary eigenvalues and eigenvectors of the matrix $C^{-1}K = \begin{pmatrix} -1 & 1 \\ 0 & 1 \end{pmatrix}$. $\blacktriangle$

Remark. As with ordinary eigenvalues, aside from very small examples, one *never* computes generalized eigenvalue using the generalized characteristic equation (5.41). Furthermore, unless C has a simple inverse, it is better to compute them within the generalized eigenvalue framework instead of working with $H = C^{-1}K$. $\blacktriangle$

Exercises

3.1. Find the eigenvalues and an orthonormal eigenvector basis for the following symmetric matrices, and then write out their spectral factorization. Use this to determine which are positive definite. $(a) \heartsuit$ $\begin{pmatrix} 2 & 6 \\ 6 & -7 \end{pmatrix}$ (b) $\begin{pmatrix} 5 & -2 \\ -2 & 5 \end{pmatrix}$, $(c) \diamondsuit$ $\begin{pmatrix} 1 & 0 & 4 \\ 0 & 1 & 3 \\ 4 & 3 & 1 \end{pmatrix}$.

3.2. Construct a symmetric matrix that has the following eigenvalues and associated eigenvectors, or explain why none exists:
$(a) \heartsuit$ $\lambda_1 = -2$, $\mathbf{v}_1 = (1, -1)^T$, $\lambda_2 = 1$, $\mathbf{v}_2 = (1, 1)^T$, $(b) \diamondsuit$ $\lambda_1 = 3$, $\mathbf{v}_1 = (2, -1)^T$, $\lambda_2 = -1$, $\mathbf{v}_2 = (-1, 2)^T$, (c) $\lambda_1 = 2$, $\mathbf{v}_1 = (2, 1)^T$, $\lambda_2 = 2$, $\mathbf{v}_2 = (1, 2)^T$.

3.3. Find a symmetric positive definite matrix whose eigenvectors are the wavelet basis vectors (2.37), or explain why none exists.

3.4. How many orthonormal eigenvector bases does a symmetric $n \times n$ matrix have?

3.5. Prove the properties $(b) \heartsuit$, $(c) \diamondsuit$, (d) listed on page 140.

3.6. $\diamondsuit$ *True or false*: A matrix with a real eigenvector basis that is orthonormal under the dot product is symmetric.

3.7. $\diamondsuit$ Let S be symmetric positive definite, $\mathbf{b} \in \mathbb{R}^n$, and let $\mathbf{x} \in \mathbb{R}^n$ be the solution of the linear system $S\mathbf{x} = \mathbf{b}$. Let $\lambda_1 \geq \lambda_2 \geq \cdots \geq \lambda_n > 0$ be the eigenvalues of S and $\mathbf{u}_1, \ldots, \mathbf{u}_n$ the corresponding orthonormal eigenvectors. (a) Show that $\mathbf{x} = \sum_{i=1}^{n} \lambda_i^{-1}(\mathbf{b} \cdot \mathbf{u}_i)\mathbf{u}_i$.

(b) Given $2 \leq k \leq n$, show that the spectrally truncated approximate solution

$$\mathbf{x}_k = \sum_{i=k}^{n} \lambda_i^{-1}(\mathbf{b} \cdot \mathbf{u}_i)\mathbf{u}_i \qquad \text{satisfies} \qquad \|\mathbf{x}_k - \mathbf{x}\| \leq \lambda_{k-1}^{-1}\|\mathbf{b}\|. \tag{5.43}$$

Remark: The spectrally truncated approximate solution (5.43) can be an efficient way to approximately solve a linear system $A\mathbf{x} = \mathbf{b}$, provided one can choose k so that λ_{k+1}^{-1} is sufficiently small. This is particularly useful when the linear system $A\mathbf{x} = \mathbf{b}$ needs to be repeatedly solved for different values of $\mathbf{b}$, since the eigenvectors need only be computed once.

3.8. ♡ Given an inner product on $\mathbb{R}^n$, let $\mathbf{u}_1, \ldots, \mathbf{u}_n$ be an orthonormal basis. Prove that they form an eigenvector basis for some self-adjoint $n \times n$ matrix S. Can you characterize all such matrices? Under what conditions can you construct such an S that is positive definite?

3.9. ♡ Find a non-symmetric 2×2 matrix S with real eigenvalues that does not satisfy the inequalities (5.62).

3.10. *Orthogonal Deflation.* Let S be an $n \times n$ symmetric matrix with eigenvalues (not necessarily ordered) $\lambda_1, \ldots, \lambda_n$ and corresponding orthonormal eigenvectors $\mathbf{u}_1, \ldots, \mathbf{u}_n$.
(a) Let $P_1 = \mathrm{I} - \mathbf{u}_1 \mathbf{u}_1^T$ be the projection matrix onto the orthogonal complement to the first eigenvector $\mathbf{u}_1$, i.e., the subspace spanned by $\mathbf{u}_2, \ldots, \mathbf{u}_n$. prove that the matrix $B_1 = P_1 A$ has the same eigenvectors $\mathbf{u}_1, \ldots, \mathbf{u}_n$ and, furthermore, the corresponding eigenvalues are $\mu_1 = 0, \mu_2 = \lambda_2, \ldots, \mu_n = \lambda_n$. *Note:* B_1 is not necessarily a symmetric matrix.
(b) More generally, for $1 \le j < n$, let $P_j = \mathrm{I} - \mathbf{u}_1 \mathbf{u}_1^T - \cdots - \mathbf{u}_j \mathbf{u}_j^T$ be the projection matrix onto the subspace spanned by the last $n - j$ eigenvectors $\mathbf{u}_{j+1}, \ldots, \mathbf{u}_n$. Prove that the matrix $B_j = P_j A$ has the same eigenvectors $\mathbf{u}_1, \ldots, \mathbf{u}_n$, and the corresponding eigenvalues are $\mu_1 = \cdots = \mu_j = 0$, $\mu_{j+1} = \lambda_{j+1}, \ldots, \mu_n = \lambda_n$.
(c) Does this result extend to self-adjoint matrices? If so formulate it precisely.

3.11. ♡ Write down two self adjoint positive definite matrices whose Hadamard product is not positive definite.

3.12. Compute the generalized eigenvalues and eigenvectors for the following matrix pairs. Verify orthogonality of the eigenvectors under the appropriate inner product.

(a) ♡ $K = \begin{pmatrix} 3 & -1 \\ -1 & 2 \end{pmatrix}$, $\quad C = \begin{pmatrix} 2 & 0 \\ 0 & 3 \end{pmatrix}$; $\quad$ (b) $K = \begin{pmatrix} 3 & 1 \\ 1 & 1 \end{pmatrix}$, $\quad C = \begin{pmatrix} 2 & 0 \\ 0 & 1 \end{pmatrix}$;

(c) $K = \begin{pmatrix} 2 & -1 \\ -1 & 4 \end{pmatrix}$, $C = \begin{pmatrix} 2 & -1 \\ -1 & 1 \end{pmatrix}$; $\quad$ (d) ◇ $K = \begin{pmatrix} 1 & 2 & 0 \\ 2 & 8 & 2 \\ 0 & 2 & 1 \end{pmatrix}$, $C = \begin{pmatrix} 1 & 1 & 0 \\ 1 & 3 & 1 \\ 0 & 1 & 1 \end{pmatrix}$.

3.13. Suppose one performs the Gram–Schmidt process on vectors $\mathbf{v}_1, \ldots, \mathbf{v}_n \in \mathbb{R}^m$ using the alternative inner product $\langle \mathbf{v}, \mathbf{w} \rangle_C = \mathbf{v}^T C \mathbf{w}$ where C is symmetric positive definite, producing the orthonormal vectors $\mathbf{u}_1, \ldots, \mathbf{u}_s$. Let $A = (\mathbf{v}_1 \ \ldots \ \mathbf{v}_n)$ and $Q = (\mathbf{u}_1 \ \ldots \ \mathbf{u}_s)$, where $s = \mathrm{rank}\, A$. (a) Show that this is equivalent to the matrix factorization $A = QR$ where $Q^T C Q = \mathrm{I}$ and R is in row echelon form (4.67). (b) Show further that if we set $\widetilde{A} = C^{1/2} A$ and $\widetilde{Q} = C^{1/2} Q$, then $\widetilde{A} = \widetilde{Q} R$ is the ordinary QR factorization of $\widetilde{A}$.

5.4 Optimization Principles

As above, C will be a symmetric positive definite matrix determining an inner product on $\mathbb{R}^n$, with $C = \mathrm{I}$ corresponding to the dot product. The eigenvalues of a self-adjoint matrix $S = S^*$ can be characterized by an optimization principle based on the associated quadratic form

$$q(\mathbf{x}) = \langle \mathbf{x}, S\mathbf{x} \rangle = \mathbf{x}^T C S \mathbf{x} = \sum_{i,j,k=1}^{n} c_{ij} s_{jk} x_i x_k \tag{5.44}$$

that we used in our analysis of positive definiteness. (Here, S is not necessarily positive definite.) The first remark is that if $\mathbf{v}$ is an eigenvector, with $S\mathbf{v} = \lambda \mathbf{v}$, then, as in (5.28),

$$q(\mathbf{v}) = \langle \mathbf{v}, S\mathbf{v} \rangle = \langle \mathbf{v}, \lambda \mathbf{v} \rangle = \lambda \langle \mathbf{v}, \mathbf{v} \rangle = \lambda \| \mathbf{v} \|^2. \tag{5.45}$$

If $\mathbf{u}$ is a *unit eigenvector*, so $S\mathbf{u} = \lambda\mathbf{u}$ and $\|\mathbf{u}\|^2 = \mathbf{u}^T C\mathbf{u} = 1$, then the value of

$$q(\mathbf{u}) = \lambda \tag{5.46}$$

is the associated eigenvalue. In particular, the minimal value of $q(\mathbf{u})$ among all unit eigenvectors is the smallest of the eigenvalues of S; similarly for the largest value. It turns out that these optimization principles extend to all unit vectors.

Theorem 5.43. *Suppose S is a self-adjoint matrix for a given inner product. Suppose $\mathbf{u}$ is a unit vector that minimizes the quadratic function $q(\mathbf{x}) = \langle\mathbf{x}, S\mathbf{x}\rangle$ over all vectors with $\|\mathbf{x}\| = 1$. Then $\mathbf{u}$ is an eigenvector of S and the minimum value $\lambda = q(\mathbf{u})$ is the smallest real eigenvalue of S. Similarly, if $\widehat{\mathbf{u}}$ is a unit vector that maximizes $q(\mathbf{x})$ over all unit vectors with $\|\mathbf{x}\| = 1$, then $\widehat{\mathbf{u}}$ is an eigenvector and the maximum value $\widehat{\lambda} = q(\widehat{\mathbf{u}})$ is the largest real eigenvalue.*

Proof. If we assume the validity of Theorem 5.29, then the proof is relatively easy. Let $\mathbf{u}_1, \ldots, \mathbf{u}_n$ be the orthonormal eigenvector basis associated with S, so $S\mathbf{u}_j = \lambda_j\mathbf{u}_j$. Writing

$$\mathbf{x} = a_1\mathbf{u}_1 + \cdots + a_n\mathbf{u}_n, \quad \text{so that, by (2.35),} \quad a_1^2 + \cdots + a_n^2 = \|\mathbf{x}\|^2 = 1,$$

orthonormality implies

$$q(\mathbf{x}) = \langle\mathbf{x}, S\mathbf{x}\rangle = \left\langle \sum_{i=1}^{n} a_i\mathbf{u}_i, \sum_{j=1}^{n} a_j S\mathbf{u}_j \right\rangle = \left\langle \sum_{i=1}^{n} a_i\mathbf{u}_i, \sum_{j=1}^{n} \lambda_j a_j\mathbf{u}_j \right\rangle$$
$$= \sum_{i,j=1}^{n} \lambda_j a_i a_j \langle\mathbf{u}_i, \mathbf{u}_j\rangle = \sum_{i=1}^{n} \lambda_i a_i^2. \tag{5.47}$$

Assuming $\lambda_1 \geq \lambda_2 \geq \cdots \geq \lambda_n$, the latter sum can be bounded from below by

$$q(\mathbf{x}) = \lambda_1 a_1^2 + \cdots + \lambda_n a_n^2 \geq \lambda_n(a_1^2 + \cdots + a_n^2) = \lambda_n. \tag{5.48}$$

On the other hand, setting $\mathbf{x} = \mathbf{u}_n$, so $a_1 = \cdots = a_{n-1} = 0$, $a_n = 1$, we have $q(\mathbf{u}_n) = \lambda_n$, and we conclude that λ_n is the minimum value of $q(\mathbf{x})$ when $\mathbf{x}$ ranges over all unit vectors, with the minimum achieved when $\mathbf{x} = \mathbf{u}_n$. (The minimum is also achieved when $\mathbf{x} = -\mathbf{u}_n$ or, if λ_n happens to be a multiple eigenvalue, when $\mathbf{x}$ is any unit vector in the associated eigenspace.) The proof that the maximum value of $q(\mathbf{x})$ over all unit vectors is λ_1, achieved by $\mathbf{x} = \mathbf{u}_1$ is almost identical. Or one can replace S by $-S$ and use the minimizer result.

However, the proof of Theorem 5.29 that appears below relies on the optimization principle of Theorem 5.43, and hence we cannot use it here without leading to a circular argument. A proof that does not rely on Theorem 5.29 proceeds as follows.

First, because the unit sphere $S_1 = \{\|\mathbf{x}\| = 1\}$ is a compact subset of $\mathbb{R}^n$ and the quadratic function $q: S_1 \to \mathbb{R}$ is continuous, Theorem 2.35 assures us that $q(\mathbf{x})$ achieves its minimum value at some unit vector $\mathbf{u}$. Our task is to prove that $\mathbf{u}$ is an eigenvector.

According to Theorem 2.32, every vector in $\mathbb{R}^n$ can be decomposed into a sum of a multiple of $\mathbf{u}$ and a vector belonging to its orthogonal complement $\mathbf{u}^\perp = \{\mathbf{v} \mid \langle\mathbf{u}, \mathbf{v}\rangle = 0\}$. In particular, we express

$$S\mathbf{u} = \lambda\mathbf{u} + \alpha\mathbf{v} \quad \text{for some} \quad \lambda, \alpha \in \mathbb{R}, \tag{5.49}$$

where $\langle\mathbf{u}, \mathbf{v}\rangle = 0$, and we can assume $\|\mathbf{v}\| = 1$. If we can prove that $\alpha = 0$, then (5.49) implies that $\mathbf{u}$ is an eigenvector, as claimed, with corresponding eigenvalue λ.

To establish the claim, note first that, using the orthonormality of $\mathbf{u}, \mathbf{v}$, and (5.49),

$$q(\mathbf{u}) = \langle \mathbf{u}, S\mathbf{u} \rangle = \lambda, \qquad \langle \mathbf{v}, S\mathbf{u} \rangle = \alpha.$$

Now consider the one-parameter family of unit vectors

$$\mathbf{w}_\theta = (\cos \theta)\,\mathbf{u} + (\sin \theta)\,\mathbf{v} \quad \text{for} \quad \theta \in \mathbb{R},$$

noting that

$$\|\mathbf{w}_\theta\|^2 = (\cos^2 \theta)\,\|\mathbf{u}\|^2 + 2\,(\cos \theta \sin \theta)\,\langle \mathbf{u}, \mathbf{v} \rangle + (\sin^2 \theta)\,\|\mathbf{v}\|^2 = \cos^2 \theta + \sin^2 \theta = 1,$$

as required. Furthermore,

$$S\mathbf{w}_\theta = (\cos \theta)\,S\mathbf{u} + (\sin \theta)\,S\mathbf{v} = (\cos \theta)\,(\lambda\mathbf{u} + \alpha\mathbf{v}) + (\sin \theta)\,S\mathbf{v}.$$

Define the scalar function

$$\begin{aligned} g(\theta) = q(\mathbf{w}_\theta) = \langle \mathbf{w}_\theta, S\mathbf{w}_\theta \rangle &= \langle (\cos \theta)\,\mathbf{u} + (\sin \theta)\,\mathbf{v}, (\cos \theta)\,(\lambda\mathbf{u} + \alpha\mathbf{v}) + (\sin \theta)\,S\mathbf{v} \rangle \\ &= \lambda \cos^2 \theta + 2\alpha \cos \theta \sin \theta + \beta \sin^2 \theta, \quad \text{where} \quad \beta = q(\mathbf{v}) = \langle \mathbf{v}, S\mathbf{v} \rangle, \end{aligned}$$

and we used the fact that $\mathbf{u}, \mathbf{v}$ are orthogonal unit vectors. According to our hypothesis on $\mathbf{u}$, the function $g(\theta)$ must achieve a minimum at $\theta = 0$, with $g(0) = \lambda$. Calculus tells us that its derivative must vanish at a minimum, and so $0 = g'(0) = 2\alpha$, and hence $\alpha = 0$, as desired. Thus, we conclude that $\mathbf{u}$ is indeed an eigenvector. Finally, equation (5.46) combined with the fact that $\mathbf{u}$ minimizes q, proves that $\mathbf{u}$ belongs to the smallest eigenvalue of S. ∎

Example 5.44. The problem is to minimize/maximize the value of the quadratic form

$$q(x, y) = 3\,x^2 + 2\,xy + 3\,y^2$$

for all x, y lying on the unit circle $x^2 + y^2 = 1$. This optimization problem is precisely of the form in Theorem 5.43. Writing $q(x, y) = \mathbf{x} \cdot S\mathbf{x}$ using the dot product, with $\mathbf{x} = (x, y)^T$, the symmetric coefficient matrix is $S = \begin{pmatrix} 3 & 1 \\ 1 & 3 \end{pmatrix}$, whose eigenvalues were found to be $\lambda_1 = 2$ and $\lambda_2 = 4$. Theorem 5.43 implies that the minimum is the smallest eigenvalue, and hence equal to 2, while its maximum is the largest eigenvalue, and hence equal to 4. Thus, evaluating $q(x, y)$ on the unit eigenvectors, we conclude that

$$q\left(\tfrac{1}{\sqrt{2}}, -\tfrac{1}{\sqrt{2}} \right) = 2 \le q(x, y) \le 4 = q\left(\tfrac{1}{\sqrt{2}}, \tfrac{1}{\sqrt{2}} \right) \qquad \text{for all} \qquad x^2 + y^2 = 1. \qquad \blacktriangle$$

Remark. To solve such constrained optimization problems using calculus, one could appeal to the method of Lagrange multipliers, cf. [158]. The multiplier, in fact, turns out to be the eigenvalue. In the two-dimensional case, an easier strategy would be to parametrize the unit circle, setting $x = \cos t$, $y = \sin t$, and then use one variable calculus to minimize or maximize the scalar function $f(t) = q(\cos t, \sin t)$. In higher dimensions, the latter strategy is more tricky, requiring (generalized) spherical coordinates, cf. [180]. $\blacktriangle$

In practical applications, the restriction of the quadratic form to unit vectors may not be particularly convenient. We can, however, rephrase the eigenvalue optimization principles in a form that utilizes general nonzero vectors. If $\mathbf{v} \ne \mathbf{0}$, then $\mathbf{x} = \mathbf{v}/\|\mathbf{v}\|$ is a unit vector. Substituting this expression for $\mathbf{x}$ in the quadratic form (5.44) leads to the following optimization principles for the extreme eigenvalues of a self-adjoint matrix S, listed as in (5.25):

$$\begin{aligned} \lambda_{max} = \lambda_1 = \max_{\mathbf{u}} \left\{ \langle \mathbf{u}, S\mathbf{u} \rangle \mid \|\mathbf{u}\| = 1 \right\} = \max_{\mathbf{x}} \left\{ \frac{\langle \mathbf{x}, S\mathbf{x} \rangle}{\|\mathbf{x}\|^2} \,\bigg|\, \mathbf{x} \ne \mathbf{0} \right\}, \\ \lambda_{min} = \lambda_n = \min_{\mathbf{u}} \left\{ \langle \mathbf{u}, S\mathbf{u} \rangle \mid \|\mathbf{u}\| = 1 \right\} = \min_{\mathbf{x}} \left\{ \frac{\langle \mathbf{x}, S\mathbf{x} \rangle}{\|\mathbf{x}\|^2} \,\bigg|\, \mathbf{x} \ne \mathbf{0} \right\}. \end{aligned} \qquad (5.50)$$

Thus, we can replace optimization of a quadratic polynomial over the unit sphere by optimization of a rational function over all of $\mathbb{R}^n \setminus \{\mathbf{0}\}$. The rational function to be optimized is known as the *Rayleigh quotient*, named after Lord Rayleigh, a prominent nineteenth-century British scientist. As always, the most important case is when we use the dot product and Euclidean norm, so that S is a symmetric matrix. For instance, referring back to Example 5.44, the minimum value of

$$r(x, y) = \frac{3x^2 + 2xy + 3y^2}{x^2 + y^2} \qquad \text{for all} \qquad \begin{pmatrix} x \\ y \end{pmatrix} \neq \begin{pmatrix} 0 \\ 0 \end{pmatrix}$$

is equal to 2, the same minimal eigenvalue of the corresponding coefficient matrix.

There is an alternative, useful optimization principle for characterizing the dominant eigenvalue of positive (semi)definite matrices.

Theorem 5.45. *Let H be a self-adjoint positive semidefinite matrix with respect to an inner product. Then the dominant eigenvalue of H is given by*

$$\lambda_{max}(H) = \max_{\mathbf{u}} \left\{ \| H\mathbf{u} \| \mid \| \mathbf{u} \| = 1 \right\} = \max_{\mathbf{x}} \left\{ \frac{\| H\mathbf{x} \|}{\| \mathbf{x} \|} \;\middle|\; \mathbf{x} \neq \mathbf{0} \right\}. \tag{5.51}$$

Proof. Since H is self-adjoint, we can write

$$\| H\mathbf{u} \|^2 = \langle H\mathbf{u}, H\mathbf{u} \rangle = \langle \mathbf{u}, H^2\mathbf{u} \rangle. \tag{5.52}$$

Thus, by Theorem 5.43, the maximum of (5.52) of all unit vectors $\mathbf{u}$ is the dominant eigenvalue of H^2, which, by positive semidefiniteness, is the square of the dominant eigenvalue of H, i.e., $\lambda_{max}(H^2) = \lambda_{max}(H)^2$. The first part of (5.51) follows immediately, and the second follows as in our derivation of the Rayleigh quotient. $\blacksquare$

We further note that, in light of the proof of Theorem 5.41, we can similarly characterize the largest and smallest generalized eigenvalues of a pair of symmetric matrices.

Theorem 5.46. *Let K, C be symmetric $n \times n$ matrices, with C positive definite. Then their extreme generalized eigenvalues can be characterized by the following optimization principles*

$$\lambda_{max} = \lambda_1 = \max_{\mathbf{u}} \left\{ \mathbf{u}^T K\mathbf{u} \mid \mathbf{u}^T C\mathbf{u} = 1 \right\} = \max_{\mathbf{x}} \left\{ \frac{\mathbf{x}^T K\mathbf{x}}{\mathbf{x}^T C\mathbf{x}} \;\middle|\; \mathbf{x} \neq \mathbf{0} \right\},$$

$$\lambda_{min} = \lambda_n = \min_{\mathbf{u}} \left\{ \mathbf{u}^T K\mathbf{u} \mid \mathbf{u}^T C\mathbf{u} = 1 \right\} = \min_{\mathbf{x}} \left\{ \frac{\mathbf{x}^T K\mathbf{x}}{\mathbf{x}^T C\mathbf{x}} \;\middle|\; \mathbf{x} \neq \mathbf{0} \right\}. \tag{5.53}$$

5.4.1 Intermediate Eigenvalues and the Min-Max Theorem

What about characterizing one of the intermediate eigenvalues? Then we need to be a little more sophisticated in designing the optimization principle. The key observation is the orthogonality of the eigenvectors. Thus, if we seek to find the next largest eigenvector $\mathbf{u}_2$, we should minimize over unit vectors that are orthogonal to the first eigenvector $\mathbf{u}_1$. More generally, the following result can be established using an argument similar to the one used in the proof of Theorem 5.43; details are left to the reader.

Theorem 5.47. *Let S be a self-adjoint matrix with eigenvalues $\lambda_1 \geq \lambda_2 \geq \cdots \geq \lambda_n$ and corresponding orthonormal eigenvectors $\mathbf{u}_1, \ldots, \mathbf{u}_n$. Then the maximal value of the quadratic form $q(\mathbf{x}) = \langle \mathbf{x}, S\mathbf{x} \rangle$ over all unit vectors that are orthogonal to the first $k - 1$ eigenvectors is its k-th eigenvalue:*

$$\lambda_k = q(\mathbf{u}_k) = \max_{\mathbf{u}} \left\{ q(\mathbf{u}) \mid \|\mathbf{u}\| = 1, \ \langle \mathbf{u}, \mathbf{u}_1 \rangle = \cdots = \langle \mathbf{u}, \mathbf{u}_{k-1} \rangle = 0 \right\}. \tag{5.54}$$

A similar result holds for the minimal values:

$$\lambda_k = q(\mathbf{u}_k) = \min_{\mathbf{u}} \left\{ q(\mathbf{u}) \mid \|\mathbf{u}\| = 1, \ \langle \mathbf{u}, \mathbf{u}_{k+1} \rangle = \cdots = \langle \mathbf{u}, \mathbf{u}_n \rangle = 0 \right\}. \tag{5.55}$$

Thus, at least in principle, one can compute the eigenvalues and eigenvectors of a self-adjoint matrix by the following recursive procedure. First, find the largest eigenvalue λ_1 by the basic maximization principle in Theorem 5.43 using, say, the optimization methods developed in Chapters 6 and 11. The associated eigenvector $\mathbf{u}_1$ is found by solving the eigenvector system (5.2), e.g., by using the solution method based on the QR factorization. (Keep in mind that the coefficient matrix is singular, and one requires nonzero elements of its kernel.) The next smallest eigenvalue λ_2 is then characterized by the constrained maximization principle (5.55), and so on. Although of some theoretical interest, this algorithm is of somewhat limited practical value, and in Section 5.6 we will develop some more practical approaches to computing eigenvalues and eigenvectors.

An alternative formulation is based on the observation that to find the k-th eigenvalue λ_k, we can minimize the restriction of the quadratic form (or, equivalently, the Rayleigh quotient), to the k-dimensional subspace $V_k = \operatorname{span}\{\mathbf{u}_1, \ldots, \mathbf{u}_k\}$ spanned by the first k eigenvectors:

$$\lambda_k = \min_{\mathbf{u}} \left\{ \langle \mathbf{u}, S\mathbf{u} \rangle \mid \|\mathbf{u}\| = 1, \ \mathbf{u} \in V_k \right\} = \min_{\mathbf{x}} \left\{ \frac{\langle \mathbf{x}, S\mathbf{x} \rangle}{\|\mathbf{x}\|^2} \ \middle| \ \mathbf{0} \neq \mathbf{x} \in V_k \right\}. \tag{5.56}$$

This follows from applying the first proof of Theorem 5.43 to such vectors, so that the sums in (5.47), (5.48) only go from 1 to k. Now, it turns out that, if we replace V_k by another k-dimensional subspace, we cannot achieve a larger value for the corresponding minimum. This result is stated as follows.

Proposition 5.48. *Let S be self-adjoint with eigenvalues $\lambda_1 \geq \lambda_2 \geq \cdots \geq \lambda_n$. Let $V \subset \mathbb{R}^n$ be a k-dimensional subspace. Then there exist unit vectors $\mathbf{u}, \widehat{\mathbf{u}} \in V$ such that*

$$\langle \mathbf{u}, S\mathbf{u} \rangle \leq \lambda_k, \qquad \langle \widehat{\mathbf{u}}, S\widehat{\mathbf{u}} \rangle \geq \lambda_{n-k+1}. \tag{5.57}$$

Proof. Let $\widehat{V}_k = \operatorname{span}\{\mathbf{u}_k, \ldots, \mathbf{u}_n\}$ be the subspace spanned by the last $n-k+1$ eigenvectors. Since $\dim V + \dim \widehat{V}_k = k + (n - k + 1) > n$, (1.16) implies that the two subspaces have nontrivial intersection, $V \cap \widehat{V}_k \neq \{\mathbf{0}\}$. This implies that the intersection contains a unit vector $\mathbf{u} = x_k \mathbf{u}_k + \cdots + x_n \mathbf{u}_n \in V \cap \widehat{V}_k$. But then, by orthonormality,

$$\langle \mathbf{u}, S\mathbf{u} \rangle = \sum_{i=k}^{n} \lambda_i x_i^2 \leq \lambda_k \sum_{i=k}^{n} x_i^2 = \lambda_k.$$

The second inequality is proved in a similar manner, or by simply replacing S by $-S$. $\blacksquare$

As an immediate corollary, we deduce the following optimization principle characterizing the intermediate eigenvalues of a self-adjoint matrix, known as the *Min-Max Theorem*. The reason for the name is because, when extended to self-adjoint operators on infinite-dimensional function spaces arising in analysis and quantum mechanics, the spectrum is typically only bounded from below, [192], and hence only the second min-max principle applies.

Theorem 5.49. *Let S be self-adjoint with eigenvalues $\lambda_1 \geq \lambda_2 \geq \cdots \geq \lambda_n$. Then*

$$
\begin{aligned}
\lambda_k &= \max_V \left\{ \min_{\mathbf{u}} \left\{ \langle \mathbf{u}, S\mathbf{u} \rangle \mid \mathbf{u} \in V,\ \|\mathbf{u}\| = 1 \right\} \,\Big|\, \dim V = k \right\} \\
&= \max_V \left\{ \min_{\mathbf{x}} \left\{ \frac{\langle \mathbf{x}, S\mathbf{x} \rangle}{\|\mathbf{x}\|^2} \,\Big|\, \mathbf{0} \neq \mathbf{x} \in V \right\} \,\Big|\, \dim V = k \right\}, \\
&= \min_V \left\{ \max_{\mathbf{u}} \left\{ \langle \mathbf{u}, S\mathbf{u} \rangle \mid \mathbf{u} \in V,\ \|\mathbf{u}\| = 1 \right\} \,\Big|\, \dim V = n - k + 1 \right\}, \\
&= \min_V \left\{ \max_{\mathbf{x}} \left\{ \frac{\langle \mathbf{x}, S\mathbf{x} \rangle}{\|\mathbf{x}\|^2} \,\Big|\, \mathbf{0} \neq \mathbf{x} \in V \right\} \,\Big|\, \dim V = n - k + 1 \right\}.
\end{aligned}
\tag{5.58}
$$

Proof. Consider the first equation. The first inequality in (5.57) implies that, for any k-dimensional subspace $V \subset \mathbb{R}^n$, the indicated minimum is at most λ_k. On the other hand, choosing $V = V_k$ produces a minimum value equal to λ_k by (5.56). Thus, the maximum of all such minima is λ_k. The second equation follows by replacing $\mathbf{u} \longmapsto \mathbf{x}/\|\mathbf{x}\|$ as in our derivation of the Rayleigh quotient (5.50). The last two equations follow in a similar fashion from the second inequality in (5.57), switching $k \longleftrightarrow n - k + 1$. ∎

For completeness, the corresponding optimization (min-max) principles for the intermediate generalized eigenvalues of pairs of symmetric matrices are stated below.

Theorem 5.50. *Let K, C be symmetric $n \times n$ matrices, with C positive definite. Let $\lambda_1 \geq \lambda_2 \geq \cdots \geq \lambda_n$ be their generalized eigenvalues and $\mathbf{u}_1, \ldots, \mathbf{u}_n$ a corresponding orthonormal eigenvector basis of $\mathbb{R}^n$ for the inner product defined by C. Then*

$$
\begin{aligned}
\lambda_k &= \max_{\mathbf{u}} \left\{ \mathbf{u}^T K \mathbf{u} \mid \mathbf{u}^T C \mathbf{u} = 1,\ \mathbf{u}^T C \mathbf{u}_1 = \cdots = \mathbf{u}^T C \mathbf{u}_{k-1} = 0 \right\} \\
&= \max_{\mathbf{x}} \left\{ \frac{\mathbf{x}^T K \mathbf{x}}{\mathbf{x}^T C \mathbf{x}} \,\Big|\, \mathbf{x} \neq \mathbf{0},\ \mathbf{x}^T C \mathbf{u}_1 = \cdots = \mathbf{x}^T C \mathbf{u}_{k-1} = 0 \right\} \\
&= \max_V \left\{ \min_{\mathbf{u}} \left\{ \mathbf{u}^T K \mathbf{u} \mid \mathbf{u} \in V,\ \mathbf{u}^T C \mathbf{u} = 1 \right\} \,\Big|\, \dim V = k \right\} \\
&= \max_V \left\{ \min_{\mathbf{x}} \left\{ \frac{\mathbf{x}^T K \mathbf{x}}{\mathbf{x}^T C \mathbf{x}} \,\Big|\, \mathbf{0} \neq \mathbf{x} \in V \right\} \,\Big|\, \dim V = k \right\} \\
&= \min_{\mathbf{u}} \left\{ \mathbf{u}^T K \mathbf{u} \mid \mathbf{u}^T C \mathbf{u} = 1,\ \mathbf{u}^T C \mathbf{u}_{k+1} = \cdots = \mathbf{u}^T C \mathbf{u}_n = 0 \right\} \\
&= \min_{\mathbf{x}} \left\{ \frac{\mathbf{x}^T K \mathbf{x}}{\mathbf{x}^T C \mathbf{x}} \,\Big|\, \mathbf{x} \neq \mathbf{0},\ \mathbf{x}^T C \mathbf{u}_{k+1} = \cdots = \mathbf{x}^T C \mathbf{u}_n = 0 \right\} \\
&= \min_V \left\{ \max_{\mathbf{u}} \left\{ \mathbf{u}^T K \mathbf{u} \mid \mathbf{u} \in V,\ \mathbf{u}^T C \mathbf{u} = 1 \right\} \,\Big|\, \dim V = n - k + 1 \right\} \\
&= \min_V \left\{ \max_{\mathbf{x}} \left\{ \frac{\mathbf{x}^T K \mathbf{x}}{\mathbf{x}^T C \mathbf{x}} \,\Big|\, \mathbf{0} \neq \mathbf{x} \in V \right\} \,\Big|\, \dim V = n - k + 1 \right\}.
\end{aligned}
\tag{5.59}
$$

We close this subsection with one final proof.

Proof of Theorem 5.29:   We net present a proof of Theorem 5.29, which will be done by an induction on the size of the matrix. The case of a 1×1 matrix is completely trivial. In general, for an $n \times n$ matrix S, let $\mathbf{u}_n$ be the minimizer of the quadratic form $q(\mathbf{x})$ over the unit sphere guaranteed by Theorem 5.43, so that $\mathbf{u}_n$ is an eigenvector of S with (minimal) eigenvalue $\lambda = \lambda_n$. Let $\mathbf{u}_n^\perp = \{\, \mathbf{v} \,|\, \langle\, \mathbf{u}_n, \mathbf{v}\,\rangle = 0\,\} \subset \mathbb{R}^n$ denote its orthogonal complement. Note that if $\mathbf{v} \in \mathbf{u}_n^\perp$, then $S\mathbf{v} \in \mathbf{u}^\perp$. Indeed, by self-adjointness,

$$\langle\, \mathbf{u}_n, S\mathbf{v}\,\rangle = \langle\, S\mathbf{u}_n, \mathbf{v}\,\rangle = \lambda\,\langle\, \mathbf{u}_n, \mathbf{v}\,\rangle = 0 \qquad \text{whenever} \qquad \mathbf{v} \in \mathbf{u}_n^\perp.$$

Now, select any orthonormal basis $\mathbf{v}_1, \ldots, \mathbf{v}_{n-1}$ of the orthogonal complement $\mathbf{u}_n^\perp$. Since, for each $j = 1, \ldots, n-1$, the vector $S\mathbf{v}_j \in \mathbf{u}^\perp$, we can write

$$S\mathbf{v}_j = \sum_{j=1}^{n-1} b_{ij}\mathbf{v}_i, \qquad j = 1, \ldots, n-1. \tag{5.60}$$

The $(n-1) \times (n-1)$ matrix B with entries b_{ij} is symmetric because, by orthonormality of the chosen basis and symmetry of S,

$$b_{ij} = \langle\, \mathbf{v}_i, S\mathbf{v}_j\,\rangle = \langle\, S\mathbf{v}_i, \mathbf{v}_j\,\rangle = b_{ji}.$$

Thus, by our induction hypothesis, because B is self-adjoint under the dot product, it possesses $n-1$ orthonormal eigenvectors, which we write as $\mathbf{w}_j = \left(\, w_{1j}, \ldots, w_{n-1,j}\,\right)^T$, $j = 1, \ldots, n-1$, so that

$$B\mathbf{w}_j = \lambda_j\mathbf{w}_j, \qquad \text{or, in components} \qquad \sum_{k=1}^{n-1} b_{ik}w_{kj} = \lambda_j w_{ij}, \tag{5.61}$$

with λ_j the corresponding eigenvalue. Orthonormality implies

$$\mathbf{w}_i \cdot \mathbf{w}_j = \sum_{k=1}^{n-1} w_{ki}w_{kj} = \begin{cases} 0 & i \neq j, \\ 1 & i = j. \end{cases}$$

For each j, the corresponding vector

$$\mathbf{u}_j = \sum_{k=1}^{n-1} w_{kj}\mathbf{v}_k$$

is an eigenvector of S with the same eigenvalue λ_j; indeed, using (5.60) and then (5.61),

$$S\mathbf{u}_j = \sum_{k=1}^{n-1} w_{kj}S\mathbf{v}_k = \sum_{i,k=1}^{n-1} b_{ik}w_{kj}\mathbf{v}_i = \lambda_j \sum_{i=1}^{n-1} w_{ij}\mathbf{v}_i = \lambda_j\mathbf{u}_j.$$

Moreover, they form an orthonormal basis of $\mathbf{u}_n^\perp$ since

$$\langle\, \mathbf{u}_i, \mathbf{u}_j\,\rangle = \left\langle\, \sum_{k=1}^{n-1} w_{ki}\mathbf{v}_k, \sum_{l=1}^{n-1} w_{lj}\mathbf{v}_l\,\right\rangle = \sum_{k,l=1}^{n-1} w_{ki}w_{lj}\langle\, \mathbf{v}_k, \mathbf{v}_l\,\rangle$$

$$= \sum_{k=1}^{n-1} w_{ki}w_{kj} = \begin{cases} 0 & i \neq j, \\ 1 & i = j. \end{cases}$$

Since each $\mathbf{u}_j$, for $j = 1, \ldots, n-1$, is orthogonal to $\mathbf{u}_n$, the vectors $\mathbf{u}_1, \ldots, \mathbf{u}_{n-1}, \mathbf{u}_n$ form an orthonormal eigenvector basis of $\mathbb{R}^n$, as claimed. Finally, part (a) of the theorem is, in view of Lemma 5.14, an immediate consequence of the existence of a (real) eigenvector basis. ■

5.4.2 Eigenvalue Inequalities

In this subsection, we collect together some useful inequalities concerning eigenvalues of symmetric matrices. Keep in mind that these do not extend to general non-symmetric matrices, even those with only real eigenvalues. (On the other hand, extensions to self-adjoint matrices are possible, but left for the reader to investigate.)

The first result says that, roughly speaking, submatrices of symmetric matrices have smaller eigenvalues. More precisely:

Lemma 5.51. *Let S be a symmetric $n \times n$ matrix, and let $\widetilde{S}$ be the $(n-1) \times (n-1)$ matrix obtained by deleting the last row and column of S. We order each of their eigenvalues from largest to smallest. Then*

$$\lambda_k(\widetilde{S}) \leq \lambda_k(S), \qquad for \qquad k = 1, \ldots, n-1. \tag{5.62}$$

Proof. Let $P = (\mathrm{I}_{n-1} \ \mathbf{0})$ be the $(n-1) \times n$ matrix that projects $\mathbb{R}^n$ onto $\mathbb{R}^{n-1}$ by omitting the last coordinate. Its transpose P^T maps $\mathbb{R}^{n-1}$ to $\mathbb{R}^n$ by appending a zero as the n-th coordinate. Thus, $\widetilde{S} = P S P^T$. Moreover the product $P^T P$ is the $n \times n$ diagonal matrix with ones on the diagonals except for a zero in the bottom right corner, and acts on a vector $\mathbf{x} \in \mathbb{R}^n$ by simply replacing its last component x_n by zero. In other words, $P^T P$ is the projection matrix that maps $\mathbb{R}^n$ onto the orthogonal complement $\mathbf{e}_n^\perp$ of the last standard basis vector.

Now, let $V_{k-1} \subset \mathbb{R}^n$ be the subspace spanned by the top $k-1$ eigenvectors of S. By the optimization principle for eigenvalues in Theorem 5.47,

$$\begin{aligned}
\lambda_k(S) &= \max\left\{ \mathbf{x}^T S \mathbf{x} \ \middle|\ \mathbf{x} \in V_{k-1}^\perp,\ \|\mathbf{x}\| = 1 \right\} \\
&\geq \max\left\{ (P^T P \mathbf{w})^T S (P^T P \mathbf{w}) \ \middle|\ P^T P \mathbf{w} \in V_{k-1}^\perp,\ \|P^T P \mathbf{w}\| = 1 \right\} \\
&= \max\left\{ (P \mathbf{w})^T P S P^T (P \mathbf{w}) \ \middle|\ P \mathbf{w} \in (P V_{k-1})^\perp,\ \|P \mathbf{w}\| = 1 \right\} \\
&= \max\left\{ \mathbf{y}^T \widetilde{S} \mathbf{y} \ \middle|\ \mathbf{y} \in (P V_{k-1})^\perp,\ \|\mathbf{y}\| = 1 \right\},
\end{aligned} \tag{5.63}$$

where we use the notation $P V_{k-1} = \{ P \mathbf{v} \mid \mathbf{v} \in V_{k-1} \} \subset \mathbb{R}^{n-1}$. Since $\dim(P V_{k-1}) \leq k-1$, we have $\dim(P V_{k-1})^\perp \geq n-k$, and so, by the third statement of the Min-Max Theorem 5.49 (with n replaced by $n-1$), the final quantity in (5.63) is bounded from below by $\lambda_k(\widetilde{S})$. $\blacksquare$

Our next result contains the *Schur–Horn inequalities*, named after the twentieth century mathematicians Issai Schur and Alfred Horn.

Theorem 5.52. *Let S be a symmetric $n \times n$ matrix and let $\lambda_1 \geq \lambda_2 \geq \cdots \geq \lambda_n$ be its eigenvalues. Then, for all $k = 1, \ldots, n$, the sum of the first k diagonal entries of S is bounded by the sum of its first k largest eigenvalues:*

$$\sum_{i=1}^{k} s_{ii} \leq \sum_{i=1}^{k} \lambda_i. \tag{5.64}$$

In particular, when $k = n$, the inequality (5.64) is an equality.

Proof. We work by induction on the size n of S, the case $n = 1$ being trivial. Let $\widetilde{S}$ be the $(n-1) \times (n-1)$ submatrix introduced in Lemma 5.51. Then, by (5.62) and our induction hypothesis,

$$\sum_{i=1}^{k} \lambda_i(S) \geq \sum_{i=1}^{k} \lambda_i(\widetilde{S}) \geq \sum_{i=1}^{k} \widetilde{s}_{ii} = \sum_{i=1}^{k} s_{ii}, \qquad k = 1, \ldots, n-1,$$

since the diagonal entries $\widetilde{s}_{ii} = s_{ii}$ agree for $i = 1, \ldots, n-1$. This proves the result for $k \leq n-1$. On the other hand, as noted in the statement of the Theorem, the case $k = n$ is an equality, namely (5.16), and hence the induction step is established. $\blacksquare$

Our final result is known as *von Neumann's trace inequality*, named after the influential Hungarian–American mathematician and physicist John von Neumann, a key founding figure of modern scientific computing.

Theorem 5.53. *Let R, S be symmetric $n \times n$ matrices. Then*

$$\operatorname{tr}(RS) \ \leq\ \sum_{i=1}^{n} \lambda_i(R)\,\lambda_i(S), \tag{5.65}$$

where the eigenvalues of both matrices are ordered from largest to smallest.

Proof. Let us diagonalize $R = Q\Lambda Q^T$, and note that, using Exercise 1.15(c),

$$\operatorname{tr}(RS) = \operatorname{tr}(Q\Lambda Q^T S) = \operatorname{tr}(\Lambda Q^T S Q).$$

Moreover, $Q^T S Q$ is symmetric and has the same eigenvalues as S. Thus, it suffices to prove the inequality (5.65) in the setting where R is a diagonal matrix. Next note that

$$\begin{aligned}
\sum_{i=1}^{n} \lambda_i(R)\lambda_i(S) = {}& \bigl[\lambda_1(R) - \lambda_2(R)\bigr]\lambda_1(S) + \bigl[\lambda_2(R) - \lambda_3(R)\bigr]\bigl[\lambda_1(S) + \lambda_2(S)\bigr] \\
& + \bigl[\lambda_3(R) - \lambda_4(R)\bigr]\bigl[\lambda_1(S) + \lambda_2(S) + \lambda_3(S)\bigr] + \cdots \\
& + \bigl[\lambda_{n-1}(R) - \lambda_n(R)\bigr]\bigl[\lambda_1(S) + \lambda_2(S) + \cdots + \lambda_{n-1}(S)\bigr] \\
& + \lambda_n(R)\bigl[\lambda_1(S) + \lambda_2(S) + \cdots + \lambda_n(S)\bigr].
\end{aligned}$$

Since $\lambda_i(R) \geq \lambda_{i+1}(R)$ when $1 \leq i < n$, we use the Schur–Horn inequalities (5.64) to bound the first $n-1$ terms on the right hand side from below and then use the equality on the last term to conclude that, when R is diagonal with $\lambda_i(R) = r_{ii}$,

$$\begin{aligned}
\sum_{i=1}^{n} \lambda_i(R)\lambda_i(S) \geq {}& \bigl[\lambda_1(R) - \lambda_2(R)\bigr]s_{11} + \bigl[\lambda_2(R) - \lambda_3(R)\bigr](s_{11} + s_{22}) + \cdots \\
& + \bigl[\lambda_{n-1}(R) - \lambda_n(R)\bigr](s_{11} + s_{22} + \cdots + s_{n-1,n-1}) \\
& + \lambda_n(R)(s_{11} + s_{22} + \cdots + s_{nn}) \\
= {}& \sum_{i=1}^{n} \lambda_i(R)\, s_{ii} = \sum_{i=1}^{n} r_{ii}\, s_{ii} = \operatorname{tr}(RS),
\end{aligned}$$

since the diagonal entries of RS are $r_{ii} s_{ii}$. This establishes (5.65) when R is diagonal, and hence, by the above remark, in general. $\blacksquare$

Exercises

4.1. Find the minimum and maximum values of the quadratic form $5x^2 + 4xy + 5y^2$ where x, y are subject to the constraint $x^2 + y^2 = 1$.

4.2. Write down and solve optimization principles characterizing the largest and the smallest eigenvalues of the following positive definite matrices:

$$(a) \heartsuit \begin{pmatrix} 2 & -1 \\ -1 & 3 \end{pmatrix}, \quad (b) \diamondsuit \begin{pmatrix} 4 & 1 \\ 1 & 4 \end{pmatrix}, \quad (c) \heartsuit \begin{pmatrix} 3 & 0 & -1 \\ 0 & 3 & 0 \\ -1 & 0 & 3 \end{pmatrix}, \quad (d) \begin{pmatrix} 4 & -1 & -2 \\ -1 & 4 & -1 \\ -2 & -1 & 4 \end{pmatrix}.$$

4.3. Write down and solve a maximization principle that characterizes the middle eigenvalue of the matrices in parts (c) and (d) of Exercise 4.2.

4.4. $\diamondsuit$ Suppose S is a symmetric matrix. What is the maximum value of $q(\mathbf{x}) = \mathbf{x}^T S \mathbf{x}$ when $\mathbf{x}$ is constrained to a sphere of radius $\|\mathbf{x}\| = r$?

4.5. Suppose H is symmetric, positive definite. Prove the product formula

$$\max\left\{\, \mathbf{x}^T H \mathbf{x} \;\middle|\; \|\mathbf{x}\| = 1 \,\right\} \; \min\left\{\, \mathbf{x}^T H^{-1} \mathbf{x} \;\middle|\; \|\mathbf{x}\| = 1 \,\right\} = 1.$$

4.6. Write out optimization principles for the largest and smallest generalized eigenvalues of the matrix pairs in Exercise 3.12.

4.7. $\diamondsuit$ Write out the details in the proof of Theorem 5.47.

4.8. (a) Prove the generalized eigenvalue optimization principles in Theorem 5.46.
(b) Extend your proof to establish one of the principles in Theorem 5.50.

5.5 Linear Iterative Systems

Iteration — meaning the repeated application of a function or process — appears throughout mathematics. Iterative methods are particularly important for finding numerical approximations. In this section, we are interested in iteration of linear and affine functions, and begin with the basic definition of an iterative system of linear equations.

Definition 5.54. A *linear iterative system* takes the form

$$\mathbf{x}_{k+1} = A\mathbf{x}_k, \qquad \mathbf{x}_0 = \mathbf{b}, \qquad\qquad (5.66)$$

where the *coefficient matrix* A is square.

The *initial value* $\mathbf{b}$ and the successive *iterates* $\mathbf{x}_k$ for $k = 0, 1, 2, \ldots$ are vectors in $\mathbb{R}^n$, and so A has size $n \times n$. The solution to the iterative system (5.66) is immediate. Clearly,

$$\mathbf{x}_1 = A\mathbf{x}_0 = A\mathbf{b}, \qquad \mathbf{x}_2 = A\mathbf{x}_1 = A^2\mathbf{b}, \qquad \mathbf{x}_3 = A\mathbf{x}_2 = A^3\mathbf{b},$$

and, in general,

$$\mathbf{x}_k = A^k\mathbf{b}. \qquad\qquad (5.67)$$

Thus, the iterates are simply determined by multiplying the initial vector $\mathbf{b}$ by the successive powers of the coefficient matrix A.

We have already noted the connection between the powers of a matrix and its eigenvalues. While we could employ the diagonalization formula (5.19) to analyze (5.67), let us instead proceed directly. If $\mathbf{v}$ is an eigenvector of A with eigenvalue λ, then $A\mathbf{v} = \lambda\mathbf{v}$, and hence $A^k\mathbf{v} = \lambda^k\mathbf{v}$. Thus, if the initial vector is a linear combination of eigenvectors, so

$$\mathbf{x}_0 = \mathbf{b} = c_1\mathbf{v}_1 + \cdots + c_j\mathbf{v}_j,$$

then the k-th iterate is given by

$$\mathbf{x}_k = A^k\mathbf{b} = c_1 A^k\mathbf{v}_1 + \cdots + c_j A^k\mathbf{v}_j = c_1\lambda_1^k\mathbf{v}_1 + \cdots + c_j\lambda_j^k\mathbf{v}_j,$$

where λ_i is the eigenvalue associated with the eigenvector $\mathbf{v}_i$. If A is complete, then it admits an eigenvector basis $\mathbf{v}_1, \ldots, \mathbf{v}_n$, and any initial vector can be expressed as a linear combination thereof. In this manner, we construct a formula for the general solution to the system.

Theorem 5.55. *If the coefficient matrix A is complete, then the solution to the linear iterative system $\mathbf{x}_{k+1} = A\mathbf{x}_k$ with initial vector*

$$\mathbf{x}_0 = \mathbf{b} = c_1\mathbf{v}_1 + \cdots + c_n\mathbf{v}_n,$$

is given by

$$\mathbf{x}_k = c_1\lambda_1^k\mathbf{v}_1 + c_2\lambda_2^k\mathbf{v}_2 + \cdots + c_n\lambda_n^k\mathbf{v}_n, \tag{5.68}$$

where $\mathbf{v}_1, \ldots, \mathbf{v}_n$ are an eigenvector basis and $\lambda_1, \ldots, \lambda_n$ the corresponding eigenvalues.

Remark. In general, even when A is a real matrix, some or all of its eigenvalues and eigenvectors may be complex, and, in such situations, the solution (5.68) involves powers of the complex eigenvalues. As long as the coefficient matrix A and the initial vector $\mathbf{b}$ are real, the solution remains real, and can be alternatively characterized as the real part of (5.68), while its imaginary part is zero. However, as almost all the coefficient matrices we consider are complete and have only real eigenvalues — e.g., when A is symmetric or, more generally, self-adjoint — we will not develop this aspect in the present text. Solutions in the incomplete cases are even more complicated to write down. For details on both, we refer the reader to [181].   ▲

Example 5.56. Consider the iterative system

$$x_{k+1} = \tfrac{3}{5}\,x_k + \tfrac{1}{5}\,y_k, \qquad y_{k+1} = \tfrac{1}{5}\,x_k + \tfrac{3}{5}\,y_k, \tag{5.69}$$

with initial conditions

$$x_0 = a, \qquad y_0 = b. \tag{5.70}$$

The system can be rewritten in matrix form (5.66), with

$$A = \begin{pmatrix} .6 & .2 \\ .2 & .6 \end{pmatrix}, \qquad \mathbf{x}_k = \begin{pmatrix} x_k \\ y_k \end{pmatrix}, \qquad \mathbf{b} = \begin{pmatrix} a \\ b \end{pmatrix}.$$

The eigenvalues and eigenvectors are

$$\lambda_1 = .8, \qquad \mathbf{v}_1 = \begin{pmatrix} 1 \\ 1 \end{pmatrix}, \qquad \lambda_2 = .4, \qquad \mathbf{v}_2 = \begin{pmatrix} -1 \\ 1 \end{pmatrix}.$$

Theorem 5.55 tells us that the general solution is given as a linear combination of the basic eigensolutions:

$$\mathbf{x}_k = c_1\lambda_1^k\mathbf{v}_1 + c_2\lambda_2^k\mathbf{v}_2 = c_1(.8)^k\begin{pmatrix} 1 \\ 1 \end{pmatrix} + c_2(.4)^k\begin{pmatrix} -1 \\ 1 \end{pmatrix} = \begin{pmatrix} c_1(.8)^k - c_2(.4)^k \\ c_1(.8)^k + c_2(.4)^k \end{pmatrix},$$

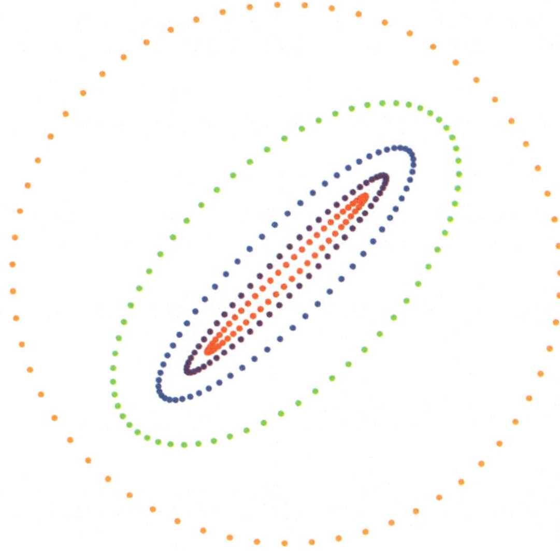

Figure 5.1: Stable Iterative System

where c_1, c_2 are determined by the initial conditions:

$$\mathbf{x}_0 = \begin{pmatrix} c_1 - c_2 \\ c_1 + c_2 \end{pmatrix} = \begin{pmatrix} a \\ b \end{pmatrix}, \qquad \text{and hence} \qquad c_1 = \frac{a+b}{2}, \qquad c_2 = \frac{b-a}{2}.$$

Therefore, the explicit formula for the solution to the initial value problem (5.69), (5.70) is

$$x_k = (.8)^k\,\frac{a+b}{2} + (.4)^k\,\frac{a-b}{2}, \qquad y_k = (.8)^k\,\frac{a+b}{2} + (.4)^k\,\frac{b-a}{2}.$$

In particular, as $k \to \infty$, the iterates $\mathbf{x}_k \to \mathbf{0}$ converge to zero at a rate governed by the dominant eigenvalue $\lambda_1 = .8$. Figure 5.1 illustrates the cumulative effect of the iteration; the initial data is colored orange, and successive iterates are colored green, blue, purple, red. The initial conditions consist of a large number of points on the unit circle $x^2 + y^2 = 1$, which are successively mapped to points on progressively smaller and flatter ellipses, whose semi-axes are in the directions of the two eigenvectors, that shrink down towards the origin.   ▲

This example motivates the definition of a convergent system.

Definition 5.57. A linear iterative system is called *convergent* if every solution goes to zero, i.e., $\mathbf{x}_k \to \mathbf{0}$ as $k \to \infty$.

Recalling the Definition 5.23 of the spectral radius and Theorem 5.24, we immediately deduce the basic convergence result for linear iterative systems:

Theorem 5.58. *A linear iterative system is convergent if and only if its coefficient matrix A is convergent, whence $\rho(A) < 1$.*

5.5.1 Affine Iterative Systems

An iterative system of the form

$$\mathbf{x}_{k+1} = A\mathbf{x}_k + \mathbf{c}, \qquad \mathbf{x}_0 = \mathbf{b}, \tag{5.71}$$

in which A is an $n \times n$ matrix and $\mathbf{c} \in \mathbb{R}^n$, is known as an *affine iterative system* since the right hand side is an affine function of $\mathbf{x}_k$. Suppose that the solutions converge: $\mathbf{x}_k \to \mathbf{x}^\star$ as $k \to \infty$. Then, by taking the limit of both sides of (5.71), we discover that the limit point $\mathbf{x}^\star$ solves the *fixed-point equation*

$$\mathbf{x}^\star = A\mathbf{x}^\star + \mathbf{c}, \qquad \text{or, equivalently,} \qquad (I - A)\mathbf{x}^\star = \mathbf{c}. \qquad (5.72)$$

As long as $I - A$ is nonsingular, or, equivalently, 1 is not an eigenvalue of A, the fixed point equation has a unique solution $\mathbf{x}^\star$.

The convergence of solutions to (5.71) to the fixed point $\mathbf{x}^\star$ is based on the behavior of the *error vectors*

$$\mathbf{y}_k := \mathbf{x}_k - \mathbf{x}^\star, \qquad (5.73)$$

which measure how close the iterates are to the true solution. Let us find out how the successive error vectors are related. We compute

$$\mathbf{y}_{k+1} = \mathbf{x}_{k+1} - \mathbf{x}^\star = (A\mathbf{x}_k + \mathbf{c}) - (A\mathbf{x}^\star + \mathbf{c}) = A(\mathbf{x}_k - \mathbf{x}^\star) = A\mathbf{y}_k,$$

showing that the error vectors satisfy a *linear* iterative system

$$\mathbf{y}_{k+1} = A\mathbf{y}_k, \qquad (5.74)$$

with the *same* coefficient matrix A. Therefore, the solutions to (5.71) converge to the fixed point, $\mathbf{x}_k \to \mathbf{x}^\star$, if and only if the error vectors converge to zero: $\mathbf{y}_k \to \mathbf{0}$ as $k \to \infty$. Our analysis of linear iterative systems, as summarized in Theorem 5.58, establishes the following basic convergence result.

Proposition 5.59. *The solutions to the affine iterative system (5.71) will all converge to the solution to the fixed point equation (5.72) if and only if A is a convergent matrix, or, equivalently, its spectral radius satisfies $\rho(A) < 1$.*

In particular, the spectral radius condition ensures that 1 is not an eigenvalue of A, and hence the fixed point equation (5.72) has a unique solution.

5.5.2   Markov Processes

A discrete probabilistic process in which the future state of a system depends only upon its current configuration is known as a *Markov process* or *Markov chain*, to honor the pioneering early twentieth century contributions of the Russian mathematician Andrei Markov. Markov processes are described by linear iterative systems whose coefficient matrices have a special form. They define the simplest examples of stochastic processes, [18, 66], which have many profound physical, biological, economic, and statistical applications, including networks, internet search engines, speech recognition, and routing.

To take a very simple (albeit slightly artificial) example, suppose you would like to predict the weather in your city. Consulting local weather records over the past decade, you determine that

(*a*) If today is sunny, there is a 70% chance that tomorrow will also be sunny,

(*b*) But, if today is cloudy, the chances are 80% that tomorrow will also be cloudy.

Question: given that today is sunny, what is the probability that next Saturday's weather will also be sunny?

To formulate this process mathematically, we let s_k denote the probability that day k is sunny and c_k the probability that it is cloudy. If we assume that these are the only possibilities, then the individual probabilities must sum to 1, so

$$s_k + c_k = 1.$$

According to our data, the probability that the next day is sunny or cloudy is expressed by the equations

$$s_{k+1} = .7\, s_k + .2\, c_k, \qquad c_{k+1} = .3\, s_k + .8\, c_k. \tag{5.75}$$

Indeed, day $k+1$ could be sunny either if day k was, with a 70% chance, or, if day k was cloudy, there is still a 20% chance of day $k+1$ being sunny. We rewrite (5.75) in a more convenient matrix form:

$$\mathbf{x}_{k+1} = A\mathbf{x}_k, \qquad \text{where} \qquad A = \begin{pmatrix} .7 & .2 \\ .3 & .8 \end{pmatrix}, \qquad \mathbf{x}_k = \begin{pmatrix} s_k \\ c_k \end{pmatrix}. \tag{5.76}$$

In a Markov process, the vector of probabilities $\mathbf{x}_k$ is known as the k-th *state vector* and the matrix A is known as the *transition matrix*, whose entries fix the transition probabilities between the various states.

By assumption, the initial state vector is $\mathbf{x}_0 = (1,0)^T$, since we know for certain that today is sunny. Rounded off to three decimal places, the subsequent state vectors are

$$\mathbf{x}_1 \simeq \begin{pmatrix} .7 \\ .3 \end{pmatrix}, \qquad \mathbf{x}_2 \simeq \begin{pmatrix} .55 \\ .45 \end{pmatrix}, \qquad \mathbf{x}_3 \simeq \begin{pmatrix} .475 \\ .525 \end{pmatrix}, \qquad \mathbf{x}_4 \simeq \begin{pmatrix} .438 \\ .563 \end{pmatrix},$$

$$\mathbf{x}_5 \simeq \begin{pmatrix} .419 \\ .581 \end{pmatrix}, \qquad \mathbf{x}_6 \simeq \begin{pmatrix} .410 \\ .591 \end{pmatrix}, \qquad \mathbf{x}_7 \simeq \begin{pmatrix} .405 \\ .595 \end{pmatrix}, \qquad \mathbf{x}_8 \simeq \begin{pmatrix} .402 \\ .598 \end{pmatrix}.$$

The iterates converge fairly rapidly to $(.4, .6)^T$, which is, in fact, a fixed point for the iterative system (5.76). Thus, in the long run, 40% of the days will be sunny and 60% will be cloudy. Let us explain why this happens.

Definition 5.60. A vector $\mathbf{x} = (x_1, \ldots, x_n)^T \in \mathbb{R}^n$ is called a *probability vector* if all its entries lie between 0 and 1, so that $0 \le x_i \le 1$ for $i = 1, \ldots, n$, and, moreover, their sum $x_1 + \cdots + x_n = 1$.

We interpret the entry x_i of a probability vector as the probability that the system is in state number i. The fact that the entries add up to 1 means that they represent a complete list of probabilities for the possible states of the system. The set of probability vectors defines a *simplex* in $\mathbb{R}^n$ that lies on the $(n-1)$-dimensional affine subspace $\{x_1 + \cdots + x_n = 1\}$. For example, the possible probability vectors in $\mathbb{R}^3$ fill the equilateral triangle plotted in Figure 5.2.

Remark. Every nonzero vector $\mathbf{0} \neq \mathbf{v} = (v_1, \ldots, v_n)^T$ with all nonnegative entries, $v_i \ge 0$ for $i = 1, \ldots, n$, can be converted into a parallel probability vector by dividing by the sum of its entries, or, equivalently, its 1 norm:

$$\mathbf{u} = \frac{\mathbf{v}}{v_1 + \cdots + v_n} = \frac{\mathbf{v}}{\|\mathbf{v}\|_1}. \tag{5.77}$$

For example, if $\mathbf{v} = (3, 2, 0, 1)^T$, then $\mathbf{u} = \left(\frac{1}{2}, \frac{1}{3}, 0, \frac{1}{6}\right)^T$ is the corresponding probability vector. ▲

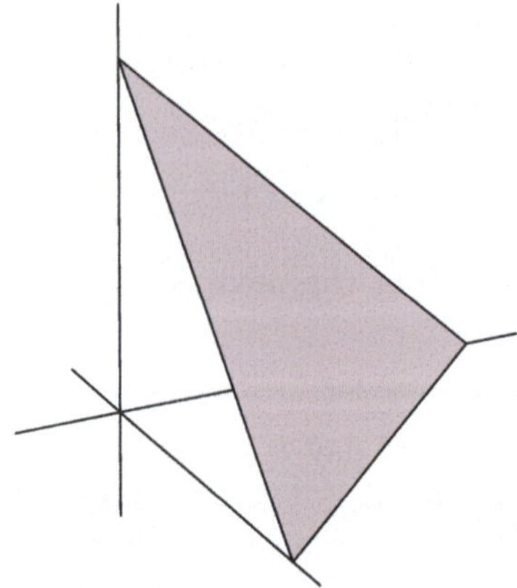

Figure 5.2: The Set of Probability Vectors in $\mathbb{R}^3$

In general, a *Markov process* is represented by a first order linear iterative system

$$\mathbf{x}_{k+1} = A\mathbf{x}_k, \tag{5.78}$$

whose initial state $\mathbf{x}_0$ is a probability vector. The entries of the *transition matrix* A must satisfy

$$0 \le a_{ij} \le 1, \qquad a_{1j} + \cdots + a_{nj} = 1. \tag{5.79}$$

The entry a_{ij} represents the *transitional probability* that the system will switch from state j to state i. (Note the reversal of indices.) Since this covers all possible transitions, the *column sums* of the transition matrix are all equal to 1, and hence each column of A is a probability vector, which is equivalent to condition (5.79), which can be restated as

$$\mathbf{1}^T A = \mathbf{1}^T, \tag{5.80}$$

where $\mathbf{1} \in \mathbb{R}^n$ is the all ones vector. In Exercise 5.13 you are asked to show that, under these assumptions, if $\mathbf{x}_k$ is a probability vector, then so is $\mathbf{x}_{k+1} = A\mathbf{x}_k$, and hence, given our assumption on the initial state, the solution $\mathbf{x}_k = A^k \mathbf{x}_0$ to the Markov process defines a sequence, or "chain", of probability vectors.

Let us now investigate the convergence of the Markov process. Not all Markov processes converge — see Exercise 5.9 for an example — and so we impose some additional mild restrictions on the transition matrix.

Definition 5.61. A transition matrix (5.79) is *regular* if some power A^k contains no zero entries. In particular, if A itself has no zero entries, then it is regular.

The entries of A^k describe the transition probabilities of getting from one state to another in k steps. Thus, regularity of the transition matrix means that there is a nonzero probability of getting from any state to any state (including the same state) in exactly k steps for some $k \ge 1$. A Markov process with a regular transition matrix is also called *aperiodic* since convergence precludes the periodic behavior observed in the irregular example contained in Exercise 5.9. The asymptotic behavior of an aperiodic Markov process is governed by the following fundamental result, originally due to the German mathematicians Oskar Perron and Georg Frobenius in the early part of the twentieth century. A proof can be found in [18].

Theorem 5.62. *If A is a regular transition matrix, then it admits a unique probability eigenvector $\mathbf{v}^\star$ with simple eigenvalue $\lambda_1 = 1$. All other eigenvalues — both real and complex — satisfy $|\lambda_j| < 1$. Moreover, all Markov processes with coefficient matrix A will converge to the probability eigenvector: $\mathbf{x}_k \to \mathbf{v}^\star$ as $k \to \infty$.*

Example 5.63. The eigenvalues and eigenvectors of the weather transition matrix (5.76) are

$$\lambda_1 = 1, \qquad \mathbf{v}_1 = \begin{pmatrix} \frac{2}{3} \\ 1 \end{pmatrix}, \qquad \lambda_2 = .5, \qquad \mathbf{v}_2 = \begin{pmatrix} -1 \\ 1 \end{pmatrix}.$$

The first eigenvector is then converted into a probability vector via formula (5.77):

$$\mathbf{v}^\star = \frac{1}{1 + \frac{2}{3}} \begin{pmatrix} \frac{2}{3} \\ 1 \end{pmatrix} = \begin{pmatrix} \frac{2}{5} \\ \frac{3}{5} \end{pmatrix}.$$

This distinguished probability eigenvector represents the final asymptotic state of the system after many iterations, *no matter what the initial state is.* Thus, our earlier observation that, in the long run, about 40% of the days will be sunny and 60% will be cloudy does not depend upon today's weather. ▲

Example 5.64. A taxi company in Minnesota serves the cities of Minneapolis and St. Paul, as well as the nearby suburbs. Records indicate that, on average, 10% of the customers taking a taxi in Minneapolis go to St. Paul and 30% go to the suburbs. Customers boarding in St. Paul have a 30% chance of going to Minneapolis and a 30% chance of going to the suburbs, while suburban customers choose Minneapolis 40% of the time and St. Paul 30% of the time. The owner of the taxi company is interested in knowing where, on average, the taxis will end up.

Let us write this as a Markov process. The entries of the state vector $\mathbf{x}_k$ tell what proportion of the taxi fleet is, respectively, in Minneapolis, St. Paul, and the suburbs, or, equivalently, the probability that an individual taxi will be in one of the three locations. Using the given data, we construct the relevant transition matrix

$$A = \begin{pmatrix} .6 & .3 & .4 \\ .1 & .4 & .3 \\ .3 & .3 & .3 \end{pmatrix}.$$

Note that A is regular since it has no zero entries. The probability eigenvector

$$\mathbf{v}^\star \simeq (\, .4714, \quad .2286, \quad .3 \,)^T$$

corresponding to the unit eigenvalue $\lambda_1 = 1$ is found by solving the linear system $(A - \mathrm{I})\mathbf{v} = 0$ and then converting the solution[4] $\mathbf{v}$ into a valid probability vector $\mathbf{v}^\star$ by use of formula (5.77). According to Theorem 5.62, no matter how the taxis are initially distributed, eventually about 47% of the taxis will be in Minneapolis, 23% in St. Paul, and 30% in the suburbs. This can be confirmed by running numerical experiments. Moreover, if the owner places this fraction of the taxis in the three locations, then they will more or less remain in such proportions forever. ▲

Remark. According to the general solution formula (5.68), the convergence rate of the Markov process to its steady state is governed by the size of the *subdominant eigenvalue* λ_2. The smaller $|\lambda_2|$ is, the faster the process converges. In the taxi example, $\lambda_2 = .3$ (and $\lambda_3 = 0$), and so the convergence to steady state is fairly rapid. ▲

---

[4]Theorem 5.62 guarantees that there is an eigenvector $\mathbf{v}$ with all nonnegative entries.

Exercises

5.1. Find the explicit formula for the solution to the following linear iterative systems:

(a) ♡ $x_{k+1} = x_k - 2\,y_k$, $y_{k+1} = -2\,x_k + y_k$, $x_0 = 1$, $y_0 = 0$;

(b) ◇ $x_{k+1} = x_k - \frac{2}{3}\,y_k$, $y_{k+1} = \frac{1}{2}\,x_k - \frac{1}{6}\,y_k$, $x_0 = -2$, $y_0 = 3$;

(c) $x_{k+1} = x_k - y_k$, $y_{k+1} = -x_k + 5\,y_k$, $x_0 = 1$, $y_0 = 0$.

5.2. Use your answers from Exercise 2.3 to solve the following iterative systems:

(a) ♡ $x_{k+1} = 5\,x_k + 2\,y_k$, $y_{k+1} = 2\,x_k + 2\,y_k$, $x_0 = 1$, $y_0 = -1$;

(b) $x_{k+1} = 4\,x_k + y_k$, $y_{k+1} = -2\,x_k + y_k$, $x_0 = 1$, $y_0 = -1$;

(c) ♡ $x_{k+1} = x_k - y_k$, $y_{k+1} = z_k$, $z_{k+1} = -z_k$, $\mathbf{x}_0 = 1$, $y_0 = 3$, $z_0 = 2$;

(d) ◇ $x_{k+1} = x_k + y_k + 2\,z_k$, $y_{k+1} = x_k + 2\,y_k + z_k$, $z_{k+1} = 2\,x_k + y_k + z_k$,

$\mathbf{x}_0 = 1$, $y_0 = 0$, $z_0 = 1$.

5.3. ♡ Explain why the j-th column $\mathbf{c}_j^{(k)}$ of the matrix power A^k satisfies the linear iterative system $\mathbf{c}_j^{(k+1)} = A\,\mathbf{c}_j^{(k)}$ with initial data $\mathbf{c}_j^{(0)} = \mathbf{e}_j$, the j-th standard basis vector.

5.4. Given a linear iterative system with complete but nonconvergent coefficient matrix, which solutions, if any, will converge to $\mathbf{0}$?

5.5. ◇ Suppose A is a complete matrix. Prove that every solution to the corresponding linear iterative system is bounded if and only if $\rho(A) \leq 1$.
Remark: If A is not complete, then this result does not hold when $\rho(A) = 1$.

5.6. *True or false*: (a) ♡ If A is convergent, then A^2 is convergent.

(b) ◇ If A is convergent, then $A^T A$ is convergent.

5.7. Prove that a matrix A with all integer entries is convergent if and only if it is nilpotent, i.e., $A^k = O$ for some $k \geq 0$. Give a nonzero example of such a matrix.

5.8. Determine if the following matrices are regular transition matrices. If so, find the associated probability eigenvector.

(a) ♡ $\begin{pmatrix} \frac{1}{2} & \frac{1}{3} \\ \frac{3}{4} & \frac{2}{3} \end{pmatrix}$, (b) ♡ $\begin{pmatrix} \frac{1}{4} & \frac{2}{3} \\ \frac{3}{4} & \frac{1}{3} \end{pmatrix}$, (c) $\begin{pmatrix} 0 & \frac{1}{5} \\ 1 & \frac{4}{5} \end{pmatrix}$, (d) ◇ $\begin{pmatrix} 0 & 1 & 0 \\ 1 & 0 & 0 \\ 0 & 0 & 1 \end{pmatrix}$, (e) $\begin{pmatrix} .3 & .5 & .2 \\ .3 & .2 & .5 \\ .4 & .3 & .3 \end{pmatrix}$.

5.9. ♡ Explain why the irregular Markov process with transition matrix $A = \begin{pmatrix} 0 & 1 \\ 1 & 0 \end{pmatrix}$ does not reach a steady state.

5.10. ♡ A certain plant species has either red, pink, or white flowers, depending on its genotype. If you cross a pink plant with any other plant, the probability distribution of the offspring is prescribed by the transition matrix $A = \begin{pmatrix} .5 & .25 & 0 \\ .5 & .5 & .5 \\ 0 & .25 & .5 \end{pmatrix}$. On average, if you continue crossing with only pink plants, what percentage of the three types of flowers would you expect to see in your garden?

5.11. ◇ The population of an island is divided into city and country residents. Each year, 5% of the residents of the city move to the country and 15% of the residents of the country move to the city. In 2023, 35,000 people live in the city and 25,000 in the country. Assuming no growth in the population, how many people will live in the city and how many will live in the country between the years 2024 and 2028? What is the eventual population distribution of the island?

5.12. A business executive is managing three branches, labeled A, B, and C, of a corporation. She never visits the same branch on consecutive days. If she visits branch A one day, she visits branch B the next day. If she visits either branch B or C that day, then the next day she is twice as likely to visit branch A as to visit branch B or C. Explain why the resulting transition matrix is regular. Which branch does she visit the most often in the long run?

5.13. Let A be an $n \times n$ transition matrix. Prove that if $\mathbf{x} \in \mathbb{R}^n$ is a probability vector, then so is $\mathbf{y} = A\mathbf{x}$.

5.14. $\diamond$ Let A be a regular transition matrix, so that all entries of T^k are strictly positive for some $k \geq 1$. (a) Prove that A cannot contain a row with all zero entries. (b) Show, by induction, that all entries of A^ℓ are strictly positive for all $\ell \geq k$.

5.15. Show that if A is an $n \times n$ transition matrix, then its 1 matrix norm is $\|A\|_1 = 1$, and hence $\|A\mathbf{x}\|_1 \leq \|\mathbf{x}\|_1$ for any $\mathbf{x} \in \mathbb{R}^n$.

5.6 Numerical Computation of Eigenvalues

Python Notebook: Numerical Computation of Eigenvalues (.ipynb)

In this section, we develop a couple of the most basic numerical algorithms for computing eigenvalues and eigenvectors. They are iterative in nature, and compute by successive approximation. They are both based on the aforementioned connections between the eigenvalues and the powers of a matrix. The power method, in the form developed here, requires that the $n \times n$ matrix A be complete, meaning that it admits an eigenvector basis of $\mathbb{R}^n$. It computes a single eigenvalue, namely the largest one (in absolute value) and its associated eigenvector, by repeatedly multiplying an initial vector by the matrix. Orthogonal iteration is a simple extension of the power method that computes several, or even all, of the eigenvalues and eigenvectors. It requires that A be symmetric, and relies on the consequential orthogonality properties of the eigenvectors. The method can be readily extended to self-adjoint matrices, or, equivalently, by Theorem 5.31, complete matrices.[5] We refer the reader to the literature, [88, 105, 181, 205, 245], for further details and extensions to more general matrices, along with the more advanced techniques that can be employed when numerically computing eigenvalues and eigenvectors of large recalcitrant matrices.

5.6.1 The Power Method

We have already noted the role played by the eigenvalues and eigenvectors in the solution to linear iterative systems. Now we are going to turn the tables, and use the iterative system as a mechanism for approximating one or more of the eigenvalues and eigenvectors of the coefficient matrix. The simplest of these computational procedures is the *power method*.

We assume, for simplicity, that A is a complete $n \times n$ matrix. Let $\mathbf{v}_1, \dots, \mathbf{v}_n$ denote its eigenvector basis, and $\lambda_1, \dots, \lambda_n$ the corresponding eigenvalues. As we have learned, the

---

[5] Although this latter remark is, in a sense, meaningless, since orthogonal iteration for computing the eigenvectors of a self-adjoint matrix requires knowing the underlying inner product, whereas Theorem 5.31 requires knowing the eigenvectors to determine the appropriate inner product.

solution to the linear iterative system

$$\mathbf{x}_{k+1} = A\mathbf{x}_k, \qquad \mathbf{x}_0 = \mathbf{b}, \tag{5.81}$$

is obtained by multiplying the initial vector $\mathbf{b}$ by the successive powers of the coefficient matrix: $\mathbf{x}_k = A^k \mathbf{b}$. If we write the initial vector in terms of the eigenvector basis

$$\mathbf{b} = c_1 \mathbf{v}_1 + \cdots + c_n \mathbf{v}_n, \tag{5.82}$$

then the solution takes the explicit form given in Theorem 5.55, namely

$$\mathbf{x}_k = A^k \mathbf{b} = c_1 \lambda_1^k \mathbf{v}_1 + \cdots + c_n \lambda_n^k \mathbf{v}_n. \tag{5.83}$$

Suppose further that A has a single *dominant real* eigenvalue[6] λ_1, that is larger than all others in magnitude, so

$$\rho(A) = |\lambda_1| > |\lambda_j| \qquad \text{for all} \qquad j > 1. \tag{5.84}$$

As its name implies, this eigenvalue will eventually dominate the iteration (5.83). Indeed, since

$$|\lambda_1|^k \gg |\lambda_j|^k \qquad \text{for all} \quad j > 1 \quad \text{and all} \quad k \gg 0,$$

the first term in the iterative formula (5.83) will eventually be much larger than the rest, and so, provided $c_1 \neq 0$,

$$\mathbf{x}_k \simeq c_1 \lambda_1^k \mathbf{v}_1 \qquad \text{for} \qquad k \gg 0.$$

Therefore, the solution to the iterative system (5.81) will, almost always, end up being a multiple of the dominant eigenvector of the coefficient matrix. We postone a rigorous statement of this to Theorem 5.67 below.

To compute the dominant eigenvalue, we note that the i-th entry of the iterate $\mathbf{x}_k$ is approximated by $x_{k,i} \simeq c_1 \lambda_1^k v_{1,i}$, where $v_{1,i}$ is the i-th entry of the eigenvector $\mathbf{v}_1$. Thus, as long as $v_{1,i} \neq 0$, we can recover the dominant eigenvalue by taking a ratio between selected components of successive iterates:

$$\lambda_1 \simeq \frac{x_{k,i}}{x_{k-1,i}} \qquad \text{provided that} \qquad x_{k-1,i} \neq 0. \tag{5.85}$$

The index i can be chosen as required, e.g., that of any entry of $\mathbf{x}_{k-1}$ that is not too small in absolute value.

Example 5.65. Consider the matrix $A = \begin{pmatrix} -1 & 2 & 2 \\ -1 & -4 & -2 \\ -3 & 9 & 7 \end{pmatrix}$. As you can check, its eigenvalues and eigenvectors are

$$\lambda_1 = 3, \quad \mathbf{v}_1 = \begin{pmatrix} 1 \\ -1 \\ 3 \end{pmatrix}, \quad \lambda_2 = 1, \quad \mathbf{v}_2 = \begin{pmatrix} -1 \\ 1 \\ -2 \end{pmatrix}, \quad \lambda_3 = -2, \quad \mathbf{v}_3 = \begin{pmatrix} 0 \\ 1 \\ -1 \end{pmatrix}.$$

Repeatedly multiplying the initial vector $\mathbf{a} = (1, 0, 0)^T$ by the matrix A results in the iterates $\mathbf{x}_k = A^k \mathbf{a}$ listed in the accompanying table.

---

[6]In terms of our prescribed ordering (5.25) of the eigenvalues in decreasing magnitude, if the smallest eigenvalue is large negative, so $|\lambda_n| > |\lambda_1|$, then replace λ_1 by λ_n in the following discussion. For positive definite and semidefinite matrices, this is not an issue since all their eigenvalues are nonnegative.

k		$\mathbf{x}_k$		λ
0	1	0	0	
1	−1	−1	−3	−1.
2	−7	11	−27	7.
3	−25	17	−69	3.5714
4	−79	95	−255	3.1600
5	−241	209	−693	3.0506
6	−727	791	−2247	3.0166
7	−2185	2057	−6429	3.0055
8	−6559	6815	−19935	3.0018
9	−19681	19169	−58533	3.0006
10	−59047	60071	−178167	3.0002
11	−177145	175097	−529389	3.0001
12	−531439	535535	−1598415	3.0000

The last column indicates the ratio $\lambda_k = x_{k,1}/x_{k-1,1}$ between the first components of successive iterates. (One could equally well use the second or third components.) The ratios are converging to the dominant eigenvalue $\lambda_1 = 3$, while the vectors $\mathbf{x}_k$ are converging to a very large multiple of the corresponding eigenvector $\mathbf{v}_1 = (1, -1, 3)^T$. ▲

Since the iterates of A are, typically, getting either very large — when $\rho(A) > 1$ — or very small — when $\rho(A) < 1$ — the iterated vectors will be increasingly subject to numerical overflow or underflow, and the method may break down before a reasonable approximation is achieved. One way to avoid this outcome is to restrict our attention to unit vectors relative to a given norm, e.g., the Euclidean norm or the ∞ norm, since their entries cannot be too large, and so are less likely to cause numerical errors in the computations. As usual, the unit vector $\mathbf{y}_k = \|\mathbf{x}_k\|^{-1}\mathbf{x}_k$ is obtained by dividing the iterate by its norm; it can be computed directly by the *renormalized power method*

$$\mathbf{y}_{k+1} = \frac{A\mathbf{y}_k}{\|A\mathbf{y}_k\|}, \qquad \mathbf{y}_0 = \frac{\mathbf{x}_0}{\|\mathbf{x}_0\|}. \tag{5.86}$$

If the dominant eigenvalue is positive, $\lambda_1 > 0$, then $\mathbf{y}_k \to \mathbf{u}_1$ will converge to one of the two dominant unit eigenvectors (the other is $-\mathbf{u}_1$). If $\lambda_1 < 0$, then the iterates will switch back and forth between the two eigenvectors, so $\mathbf{y}_k \simeq \pm\mathbf{u}_1$. In either case, the dominant eigenvalue λ_1 is obtained as a limiting ratio between nonzero entries of $A\mathbf{y}_k$ and $\mathbf{y}_k$. If some other sort of behavior is observed, it means that one of our assumptions is not valid; either A has more than one dominant eigenvalue of maximum modulus, e.g., it has a complex conjugate pair of eigenvalues of largest modulus, or it is not complete.

Example 5.66. For the matrix considered in Example 5.65, starting the iterative system (5.86) with $\mathbf{y}_0 = (1, 0, 0)^T$, the resulting unit vectors are tabulated below. The last column, being the ratio between the first components of $A\mathbf{y}_{k-1}$ and $\mathbf{y}_{k-1}$, again converges to the dominant eigenvalue $\lambda_1 = 3$. ▲

k	$\mathbf{u}^{(k)}$			λ
0	1	0	0	
1	$-.3015$	$-.3015$	$-.9045$	-1.0000
2	$-.2335$	$.3669$	$-.9005$	7.0000
3	$-.3319$	$.2257$	$-.9159$	3.5714
4	$-.2788$	$.3353$	$-.8999$	3.1600
5	$-.3159$	$.2740$	$-.9084$	3.0506
6	$-.2919$	$.3176$	$-.9022$	3.0166
7	$-.3080$	$.2899$	$-.9061$	3.0055
8	$-.2973$	$.3089$	$-.9035$	3.0018
9	$-.3044$	$.2965$	$-.9052$	3.0006
10	$-.2996$	$.3048$	$-.9041$	3.0002
11	$-.3028$	$.2993$	$-.9048$	3.0001
12	$-.3007$	$.3030$	$-.9043$	3.0000

Remark. The power method will continue to work even if A has complex eigenvalues, provided they are all of smaller in modulus than the dominant real eigenvalue λ_1. One can even drop the completeness (either real or complex) assumption, but this requires a more technical restriction on λ_1; see [181]. ▲

If the dominant eigenvalue is complex, then, because A is real, its complex conjugate is also an eigenvalue, so our underlying assumption does not hold. Moreover, starting with a real initial vector will only produce real iterates, and the above analysis does not work as stated. On the other hand, it is possible to modify the method to, in favorable situations, also compute the dominant complex eigenvalues and eigenvectors; the underlying idea is based on the real system (5.10). Variants of the power method for computing the other eigenvalues of the matrix are explored in the exercises.

We conclude this section by establishing the expected $(\lambda_2/\lambda_1)^k$ convergence rate for the power method. We state and prove the result for positive semidefinite matrices, though the extension to any self-adjoint (i.e., complete) matrix is straightforward; see Exercise 6.9.

Theorem 5.67. *Let $\langle \cdot , \cdot \rangle$ be an inner product on $\mathbb{R}^n$ with induced norm $\| \cdot \|$. Let A be self-adjoint positive semidefinite with eigenvalues $\lambda_1 > \lambda_2 \geq \cdots \geq \lambda_n \geq 0$. Let $\mathbf{u}_1, \ldots, \mathbf{u}_n$ be the corresponding orthonormal eigenvector basis, so that $\mathbf{u}_1$ is the dominant unit eigenvector. Let $\mathbf{y}_0$ be a unit vector such that $\langle \mathbf{y}_0, \mathbf{u}_1 \rangle > 0$ and let $\mathbf{y}_k$ denote the iterates (5.86) of the power method with initial vector $\mathbf{y}_0$. Then*

$$\| \mathbf{y}_k - \mathbf{u}_1 \| \leq \frac{\sqrt{2}}{|\mathbf{y}_0 \cdot \mathbf{u}_1|} \left(\frac{\lambda_2(A)}{\lambda_1(A)} \right)^k . \tag{5.87}$$

Proof. We can write

$$\mathbf{y}_k = \frac{\mathbf{x}_k}{\| \mathbf{x}_k \|}, \qquad \text{where} \qquad \mathbf{x}_k = A^k \mathbf{y}_0 = \sum_{i=1}^{n} c_i \lambda_i^k \mathbf{u}_i, \qquad c_i = \langle \mathbf{y}_0, \mathbf{u}_i \rangle .$$

Since $\mathbf{y}_k$ and $\mathbf{u}_1$ are unit vectors,

$$\|\mathbf{y}_k - \mathbf{u}_1\|^2 = 2\left(1 - \langle \mathbf{y}_k, \mathbf{u}_1 \rangle\right) \leq 2\left(1 - \langle \mathbf{y}_k, \mathbf{u}_1 \rangle^2\right) = 2\left(1 - \frac{c_1^2 \lambda_1^{2k}}{\|\mathbf{x}_k\|^2}\right), \tag{5.88}$$

where we used the inequality

$$0 \leq \langle \mathbf{y}_k, \mathbf{u}_1 \rangle = \frac{c_1 \lambda_1^k}{a_1} \leq \|\mathbf{y}_k\|\,\|\mathbf{u}_1\| = 1,$$

which is a consequence of Cauchy–Schwarz (2.27) and the assumption $c_1 = \langle \mathbf{y}_0, \mathbf{u}_1 \rangle > 0$. Since

$$\|\mathbf{x}_k\|^2 = \sum_{i=1}^n c_i^2 \lambda_i^{2k} \geq c_1^2 \lambda_1^{2k} = \langle \mathbf{y}_0, \mathbf{u}_1 \rangle^2 \lambda_1^{2k}, \qquad \text{and} \qquad 1 = \|\mathbf{y}_0\|^2 = \sum_{i=1}^n c_i^2,$$

we have

$$1 - \frac{c_1^2 \lambda_1^{2k}}{\|\mathbf{x}_k\|^2} = \frac{\|\mathbf{x}_k\|^2 - c_1^2 \lambda_1^{2k}}{\|\mathbf{x}_k\|^2} = \frac{1}{\|\mathbf{x}_k\|^2} \sum_{i=2}^n c_i^2 \lambda_i^{2k} \leq \frac{\lambda_2^{2k}}{\|\mathbf{x}_k\|^2} \sum_{i=2}^n c_i^2 \leq \frac{1}{\langle \mathbf{y}_0, \mathbf{u}_1 \rangle^2} \left(\frac{\lambda_2}{\lambda_1}\right)^{2k}.$$

Inserting this inequality into (5.88) and then taking square roots on both sides completes the proof. ∎

Remark 5.68. We see from Theorem 5.67 that convergence of the power method requires $\lambda_1(A) > \lambda_2(A)$, and the convergence rate is precisely the ratio $\lambda_2(A)/\lambda_1(A)$ between the subdominant and dominant eigenvalues. Thus, the farther the dominant eigenvalue lies away from the rest, the faster the power method converges. Since we can easily drop the positive semidefiniteness requirement — see Exercise 6.9 — we can prove convergence of the power method for any complete matrix, provided the eigenvalue λ_1 of largest absolute value is unique, meaning that it has a one-dimensional eigenspace and, in addition, $-\lambda_1$ is not an eigenvalue.

We also note that if $\langle \mathbf{y}_0, \mathbf{u}_1 \rangle < 0$ then the power method converges to $-\mathbf{u}_1$, and so the rate (5.87) holds with $\|\mathbf{y}_k + \mathbf{u}_1\|$ on the left hand side. However, if $\langle \mathbf{y}_0, \mathbf{u}_1 \rangle = 0$, and exact arithmetic is used, the method will not converge to the dominant eigenvector. As we do not know the eigenvectors in advance, it is not so easy to guarantee that this will not happen, although one must be quite unlucky to make such a poor choice of initial vector. Moreover, even if $\langle \mathbf{y}_0, \mathbf{u}_1 \rangle = 0$, numerical round-off error will typically come to one's rescue, since it will almost inevitably introduce a tiny component of the eigenvector $\mathbf{u}_1$ into some iterate, and this component will eventually dominate the computation, and the power method will converge at the same rate. The trick is to wait long enough for it to have the desired effect! ▲

5.6.2 Orthogonal Iteration

As stated, the power method produces only the dominant (largest in magnitude) eigenvalue of a matrix A. The inverse power method of Exercise 6.5 can be used to find the smallest eigenvalue. Additional eigenvalues can be found by using the shifted inverse power method of Exercise 6.6, or the deflation method of Exercises 1.12 and 3.10. However, if we need to know more than a couple of the eigenvalues, such piecemeal approaches are too time-consuming to be of much practical value. Here we present a simple modification of the power method that will enable us to simultaneously compute a specified number (including all of them if needed) of eigenvalues and the corresponding eigenvectors. We will restrict our attention to symmetric positive semidefinite matrices, A, since these are simpler, in that they have only real nonnegative eigenvalues and orthonormal eigenvector bases; moreover, these are the ones for which we will require such computational techniques. The methods described extend readily to self-adjoint positive semidefinite matrices, in which one merely replaces the dot product and Euclidean norm by the corresponding inner product and norm, and can be adapted to more general symmetric and self-adjoint matrices.

If we were to use the power method to capture several, say p, eigenvectors and eigenvalues of A, the first thought might be to try to perform it simultaneously on an initial collection $\mathbf{v}_1, \ldots, \mathbf{v}_p$ of linearly independent vectors instead of just one individual vector. The problem is that, for almost all vectors, the power iterates $\mathbf{v}_i^{(k)} = A^k \mathbf{v}_i$ all tend to a multiple of the dominant eigenvector $\mathbf{u}_1$. Normalizing the vectors at each step, as in (5.86), is not any better, since then they merely converge to one of the two dominant unit eigenvectors $\pm \mathbf{u}_1$. However, if, inspired by the form of the eigenvector basis, we *orthonormalize* the vectors at each step, then we effectively prevent them from all accumulating at the same dominant unit eigenvector, and so, with a bit of luck, the resulting vectors will converge to the required system of eigenvectors. The resulting method is known as orthogonal iteration.

Thus, let[7] $1 \leq p \leq n$. To initiate the method, we select p linearly independent vectors $\mathbf{v}_1, \ldots, \mathbf{v}_p \in \mathbb{R}^n$, which form the columns of an $n \times p$ matrix $V_0 = (\mathbf{v}_1 \ \ldots \ \mathbf{v}_p)$. For example, we can set $\mathbf{v}_i = \mathbf{e}_i$ to be the i-th standard basis vector; alternatively, we can choose $\mathbf{v}_1, \ldots, \mathbf{v}_p$ to be a random choice of linearly independent vectors. We apply the Gram–Schmidt process to orthonormalize the initial vectors, which is equivalent to factoring $V_0 = S_0 R_0$, where S_0 is an $n \times p$ matrix with orthonormal columns, so that $S_0^T S_0 = I$, and R_0 is a positive (i.e., with all positive entries along the diagonal) upper triangular $p \times p$ matrix. We then apply A to the orthonormal columns of S_0, and then orthonormalize the resulting vectors. This is equivalent to matrix multiplication and then factoring the resulting matrix, so $A S_0 = S_1 R_1$, where S_1 is an $n \times p$ matrix satisfying the orthonormality condition $S_1^T S_1 = I$ and R_1 is a positive upper triangular $p \times p$ matrix. *Orthogonal iteration* simply iterates this process:

$$A S_k = S_{k+1} R_{k+1}, \qquad S_0^T S_0 = I, \tag{5.89}$$

where the $n \times p$ matrix S_{k+1} has orthonormal columns, so $S_{k+1}^T S_{k+1} = I$, and R_{k+1} is positive upper triangular of size $p \times p$. As we will subsequently prove, subject to a certain technical condition, which is the analog of the power method convergence condition that the initial vector has a nonzero component in the direction of the dominant eigenvector, for most choices of initial matrix S_0, the resulting $p \times p$ matrices R_k converge to the diagonal matrix containing the largest p eigenvalues of A, ordered from largest to smallest, while the columns of the matrices S_k converge to the corresponding eigenvectors. In other words,

$$S_k \longrightarrow Q_p = (\mathbf{u}_1 \ \ldots \ \mathbf{u}_p), \qquad R_k \longrightarrow \Lambda_p = \operatorname{diag}(\lambda_1, \ldots, \lambda_p), \qquad k \to \infty. \tag{5.90}$$

Example 5.69. Consider the symmetric matrix $A = \begin{pmatrix} 2 & 1 & 0 \\ 1 & 3 & -1 \\ 0 & -1 & 6 \end{pmatrix}$. Let us apply orthogonal iteration to A, starting with $S_0 = I$. In the first step, we factorize $A S_0 = A = S_1 R_1$, where, to four decimal places,

$$S_1 \simeq \begin{pmatrix} .8944 & -.4082 & -.1826 \\ .4472 & .8165 & .3651 \\ 0 & -.4082 & .9129 \end{pmatrix}, \qquad R_1 \simeq \begin{pmatrix} 2.2361 & 2.2361 & -.4472 \\ 0 & 2.4495 & -3.2660 \\ 0 & 0 & 5.1121 \end{pmatrix}.$$

---

[7] The case $p = 1$ reduces to the power method as presented above.

We then factor $A S_1 = S_2 R_2$ to produce

$$S_2 \simeq \begin{pmatrix} .7001 & -.4400 & -.5623 \\ .7001 & .2686 & .6615 \\ -.1400 & -.8569 & .4962 \end{pmatrix}, \qquad R_2 \simeq \begin{pmatrix} 3.1937 & 2.1723 & -.7158 \\ 0 & 3.4565 & -4.3804 \\ 0 & 0 & 2.5364 \end{pmatrix}.$$

Continuing in this manner, after 10 iterations we have

$$S_{10} \simeq \begin{pmatrix} .0791 & -.5663 & -.8204 \\ .3179 & -.7657 & .5592 \\ -.9448 & -.3050 & .1195 \end{pmatrix}, \qquad R_{10} \simeq \begin{pmatrix} 6.3218 & .1218 & 0 \\ 0 & 3.3588 & -.0015 \\ 0 & 0 & 1.3187 \end{pmatrix}.$$

After 25 iterations, the process has completely settled down, and

$$S_{25} \simeq \begin{pmatrix} .0710 & -.5672 & -.8205 \\ .3069 & -.7702 & .5590 \\ -.9491 & -.2915 & .1194 \end{pmatrix}, \qquad R_{25} \simeq \begin{pmatrix} 6.3234 & 0 & 0 \\ 0 & 3.3579 & 0 \\ 0 & 0 & 1.3187 \end{pmatrix}.$$

The eigenvalues of A appear along the diagonal of R_{20}, while the columns of S_{20} are the corresponding orthonormal eigenvector basis, listed in the same order as the eigenvalues, both correct to 4 decimal places. ▲

Let us now investigate convergence of orthogonal iteration. The first observation connects it with the power method; namely, multiplying the the initial vectors by the k-th power of A produces the columns of the matrix

$$A^k S_0 = S_k T_k, \qquad \text{where} \qquad T_k = R_k T_{k-1} = R_k R_{k-1} \cdots R_2 R_1, \qquad T_0 = I, \qquad (5.91)$$

which is proved by induction. It trivially holds for $k = 0$. To justify the induction step, using (5.89),

$$A^{k+1} S_0 = A A^k S_0 = A S_k T_k = S_{k+1} R_{k+1} T_k = S_{k+1} T_{k+1},$$

where we use the fact that both $R_{k+1} T_k$ and T_{k+1} are positive upper triangular, and hence must be equal owing to the uniqueness of the QR factorization, as stated in Proposition 4.44.

Let $\Lambda = \operatorname{diag}(\lambda_1, \ldots, \lambda_n)$ be the diagonal eigenvalue matrix for A and $Q = (\mathbf{u}_1 \ \ldots \ \mathbf{u}_n)$ the corresponding $n \times n$ orthogonal eigenvector matrix. We substitute the spectral formula (5.34) for the powers of A into (5.91) to obtain

$$A^k S_0 = Q \Lambda^k Q^T S_0 = S_k T_k. \qquad (5.92)$$

To simplify the proof, let us assume that the largest $p + 1$ eigenvalues of A are distinct, so

$$\lambda_1 > \lambda_2 > \cdots \lambda_p > \lambda_{p+1} \geq \lambda_{p+2} \geq \cdots \geq \lambda_n \geq 0. \qquad (5.93)$$

The modification in the case of repeated eigenvalues will be indicated at the end of the section.

We now impose a *regularity* condition on the initial vectors. Let $V_i = \operatorname{span}\{\mathbf{v}_1, \ldots, \mathbf{v}_i\}$ be the i-dimensional subspace of $\mathbb{R}^n$ spanned by the first i initial vectors, and let $U_i = \operatorname{span}\{\mathbf{u}_1, \ldots, \mathbf{u}_i\}$ be the i-dimensional subspace spanned by the first i eigenvectors of A. We assume that

$$V_i \cap U_i^\perp = \{\mathbf{0}\}, \qquad \text{for all} \qquad i = 1, \ldots, p, \qquad (5.94)$$

meaning that there is no nonzero vector in V_i which is orthogonal to all the eigenvectors $\mathbf{u}_1, \ldots, \mathbf{u}_i$. (This is equivalent to the condition that the orthogonal projection of V_i onto U_i is a one-to-one map.) We assert that this condition on the initial vectors is generic, meaning that almost all choices of the initial vectors $\mathbf{v}_1, \ldots, \mathbf{v}_p$ will satisfy it. Indeed, for $i = 1$, (5.94) just requires that $\mathbf{v}_1$ be non-orthogonal to the eigenvector $\mathbf{u}_1$, which is equivalent to the generic condition $\mathbf{v}_1 \cdot \mathbf{u}_1 \neq 0$ required for the success of the power method. Next consider the case $i = 2$, and suppose, as a specific example, that U_2 is the xy plane in $\mathbb{R}^3$, so $U_2^{\perp}$ is the z axis. Almost all planes $\mathbf{0} \in V_2 \subset \mathbb{R}^3$ will not contain the z axis. This genericity can be readily extended to any plane $U_2 \subset \mathbb{R}^3$ and, more generally, any two-dimensional subspace $U_2 \subset \mathbb{R}^n$, and hence (5.94) for $i = 2$ is again generic. The general case is similar: almost all i-dimensional subspaces $V_i \subset \mathbb{R}^n$ will not contain a nonzero vector belonging to a fixed $(n-i)$-dimensional subspace $U_i^{\perp}$.

We now claim that our regularity condition (5.94) is equivalent to being able to factor the matrix[8]

$$Q^T S_0 = LU, \qquad \text{or, equivalently,} \qquad S_0 = QLU, \tag{5.95}$$

into the product of a lower triangular $n \times p$ matrix L and an upper triangular $p \times p$ matrix U both of which have nonzero entries along their main diagonals, so $l_{ij} = 0$ for $i < j$ while $l_{ii} \neq 0$, and $u_{ij} = 0$ for $i > j$ while $u_{ii} \neq 0$. Justification of the equivalence of (5.94) and (5.95) is the subject of Exercise 6.16. To continue, we work directly with the matrix factorization (5.95).

We first note that we can assume, without loss of generality, that the matrix U is positive upper triangular, since if its i-th diagonal entry is negative we can reverse its sign[9] by replacing $\mathbf{u}_i$ by $-\mathbf{u}_i$ in Q, which does not alter the status of Q as an orthogonal eigenvector matrix for A. Substituting (5.95) into (5.92) produces

$$Q\Lambda^k LU = S_k T_k, \qquad \text{or, equivalently,} \qquad Q\Lambda^k L = S_k T_k U^{-1}.$$

Multiplying the latter equation on the right by $\Lambda_p^{-k} = \text{diag}\,(\lambda_1^{-k}, \ldots, \lambda_p^{-k})$ yields

$$S\Lambda^k L\Lambda_p^{-k} = S_k Y_k, \qquad \text{where} \qquad Y_k = T_k U^{-1} \Lambda_p^{-k} \tag{5.96}$$

is also a positive upper triangular matrix, since T_k, U, Λ_p are all of that form. Let us now investigate what happens as $k \to \infty$. The entries of the $n \times p$ matrix $N_k = \Lambda^k L\Lambda_p^{-k}$ are readily computed:

$$n_{ij}^{(k)} = \begin{cases} l_{ij}\,(\lambda_i/\lambda_j)^k, & i > j, \\ 1, & i = j, \\ 0, & i < j, \end{cases} \qquad \text{for} \qquad \begin{array}{l} i = 1, \ldots, n, \\ j = 1, \ldots, p. \end{array}$$

In view of our assumption (5.93), $0 \leq \lambda_i < \lambda_j$ when $i > j$ and $j \leq p$, and hence $(\lambda_i/\lambda_j)^k \to 0$ as $k \to \infty$. Thus,

$$N_k = \Lambda^k L\Lambda_p^{-k} \longrightarrow E_p := \begin{pmatrix} \mathbf{e}_1 & \ldots & \mathbf{e}_p \end{pmatrix} = \begin{pmatrix} \mathrm{I} \\ \mathrm{O} \end{pmatrix}. \tag{5.97}$$

The rate of convergence is governed by the largest of the eigenvalue ratios $0 \leq \lambda_j/\lambda_i < 1$ for $1 \leq j < i \leq p+1$; thus the farther apart the first $p+1$ eigenvalues are, the faster the convergence. Substituting (5.97) back into (5.96), we conclude that

$$S_k Y_k \longrightarrow QE_p = \begin{pmatrix} \mathbf{u}_1 & \ldots & \mathbf{u}_p \end{pmatrix} =: Q_p.$$

We now appeal to the following lemma, whose proof can be found at the end of the section.

---

[8] A matrix that admits such a factorization is called "regular" in [181].

[9] This is analogous to the discussion of how to treat the sign of $\langle \mathbf{y}_0, \mathbf{u}_1 \rangle$ in the power method.

Lemma 5.70. *Let* $S_1, S_2, \ldots$ *and* Q_p *be* $n \times p$ *matrices with orthonormal columns, and let* $Y_1, Y_2, \ldots$ *be positive upper triangular* $p \times p$ *matrices. Then* $S_k Y_k \to Q_p$ *if and only if* $S_k \to Q_p$ *and* $Y_k \to I$, *as* $k \to \infty$.

Lemma 5.70 implies that, as claimed, the orthogonal matrices S_k do converge to the eigenvector matrix Q_p. Moreover, by (5.91) and (5.96),

$$R_k = T_k T_{k-1}^{-1} = \left(Y_k U^{-1} \Lambda_p^{-k} \right) \left(Y_{k-1} U^{-1} \Lambda_p^{1-k} \right)^{-1} = Y_k \Lambda_p Y_{k-1}^{-1}.$$

Since both Y_k and Y_{k-1} converge to the identity matrix, R_k converges to the diagonal eigenvalue matrix Λ_p, as claimed. We have thus proved the key convergence result for orthogonal iteration.

Theorem 5.71. *Suppose that* A *is a positive definite symmetric* $n \times n$ *matrix whose eigenvalues satisfy* (5.93). *Let* $Q = (\mathbf{u}_1 \ldots \mathbf{u}_n)$ *be the corresponding orthogonal eigenvector matrix and* $Q_p = (\mathbf{u}_1 \ldots \mathbf{u}_p)$ *the* $n \times p$ *matrix containing the top* p *eigenvectors. Suppose* S_0 *is an* $n \times p$ *matrix with orthonormal columns which satisfies the regularity condition* (5.95). *Then the matrices* $S_k \to S$ *and* $R_k \to \Lambda$ *appearing in the orthogonal iteration* (5.89) *converge to, respectively, the orthogonal eigenvector matrix* $Q_p = (\mathbf{u}_1 \ldots \mathbf{u}_p)$ *and the diagonal eigenvalue matrix* $\Lambda_p = \mathrm{diag}\,(\lambda_1, \ldots, \lambda_p)$.

An example that fails to satisfy the regularity condition can be found in Exercise 6.13; in this case taking a different initial condition for orthogonal iteration will almost certainly produce the correctly ordered eigenvalues and eigenvectors.

Remark. If A is symmetric and has distinct eigenvalues, then, for suitably large $\alpha \gg 0$, the *shifted matrix* $\widetilde{A} = A + \alpha\,I$ is positive definite, has the same eigenvectors as A, and has distinct shifted eigenvalues $\widetilde{\lambda}_k = \lambda_k + \alpha$. Thus, one can run the algorithm to determine the eigenvalues and eigenvectors of $\widetilde{A}$, and hence those of A by undoing the shift. ▲

The last remaining item is a proof of Lemma 5.70. We write

$$S = (\mathbf{u}_1 \ldots \mathbf{u}_n), \qquad S_k = \left(\mathbf{u}_1^{(k)} \ldots \mathbf{u}_n^{(k)} \right),$$

in columnar form. Let $y_{ij}^{(k)}$ denote the entries of the positive upper triangular matrix Y_k. The first column of the limiting equation $S_k Y_k \to S$ reads $y_{11}^{(k)} \mathbf{u}_1^{(k)} \to \mathbf{u}_1$. Since both $\mathbf{u}_1^{(k)}$ and $\mathbf{u}_1$ are unit vectors, and $y_{11}^{(k)} > 0$, it follows that

$$y_{11}^{(k)} = \| y_{11}^{(k)} \mathbf{u}_1^{(k)} \| \longrightarrow \| \mathbf{u}_1 \| = 1, \quad \text{and hence the first column} \quad \mathbf{u}_1^{(k)} \longrightarrow \mathbf{u}_1.$$

The second column reads

$$y_{12}^{(k)} \mathbf{u}_1^{(k)} + y_{22}^{(k)} \mathbf{u}_2^{(k)} \longrightarrow \mathbf{u}_2.$$

Taking the inner product with $\mathbf{u}_1^{(k)} \to \mathbf{u}_1$ and using orthonormality, we deduce $y_{12}^{(k)} \to 0$, and hence $y_{22}^{(k)} \mathbf{u}_2^{(k)} \to \mathbf{u}_2$, which, by the previous reasoning, implies that $y_{22}^{(k)} \to 1$ and $\mathbf{u}_2^{(k)} \to \mathbf{u}_2$. The proof is completed by working through the remaining columns, using a similar argument at each step. The details are left to the reader. △

Remark. If A has repeated eigenvalues, so (5.93), then the entries of N_k corresponding to equal eigenvalues will be constant, and so $N_k \to N$, which is a lower triangular matrix with

$n_{ii} = 1$, and $n_{ij} \neq 0$ for $i > j$ if and only if $\lambda_i = \lambda_j$. Then $S_k Y_k \to Q N = Z$ where the columns of Z are still eigenvectors since, due to the form of N, they are linear combinations of the eigenvectors that belong to the same eigenspace of A. We then orthonormalize the columns of Z, which amounts to replacing those that are in each eigenspace of dimension ≥ 2 by a corresponding orthonormal eigenspace basis, by performing a QR factorization: $Z = \widehat{Q}\,\widehat{R}$. The evident modification of Lemma 5.70 can then be used to complete the convergence proof as before; details are left to the motivated reader. $\blacktriangle$

Exercises

6.1. Use the power method to approximate the dominant eigenvalue and associated eigenvector of the following matrices. Write your code in Python and compare to the output of `numpy.linalg.eig`.

$$(a)\,\heartsuit \begin{pmatrix} -1 & -2 \\ 3 & 4 \end{pmatrix}, (b)\,\diamondsuit \begin{pmatrix} 3 & -1 & 0 \\ -1 & 2 & -1 \\ 0 & -1 & 3 \end{pmatrix}, (c)\,\heartsuit \begin{pmatrix} -2 & 0 & 1 \\ -3 & -2 & 0 \\ -2 & 5 & 4 \end{pmatrix}, (d) \begin{pmatrix} 2 & -1 & 0 & 0 \\ -1 & 2 & -1 & 0 \\ 0 & -1 & 2 & -1 \\ 0 & 0 & -1 & 2 \end{pmatrix}.$$

6.2. $\diamondsuit$ Write Python code to use the power method to compute the dominant eigenvector for a random $n \times n$ positive definite symmetric matrix A with a reasonably large value for n (e.g., $n \geq 100$). Compare your code against the output of `scipy.sparse.linalg.eigsh`. How many iterations are required? How quickly does your code run compared to `scipy`? How large can you take n? *Hint*: To construct a random positive definite matrix A, start with a random matrix B (which will almost certainly be of maximal rank) and construct the Gram matrix $A = B^T B$.

6.3. $\heartsuit$ Prove that, for the normalized iterative method (5.86), $\| A \mathbf{y}_k \| \to |\lambda_1|$. Assuming λ_1 is real, explain how to deduce its sign.

6.4. Discuss the asymptotic behavior of solutions to an iterative system that has two real eigenvalues of largest modulus: $\lambda_n = -\lambda_1$. How can you determine the eigenvalues and eigenvectors? *Remark*: With a bit more work, one can similarly treat the case when A is a real matrix with a complex conjugate pair of dominant eigenvalues, cf. [181].

6.5. $\heartsuit$ *The Inverse Power Method.* Let A be a nonsingular matrix. (i) Show that the eigenvalues of A^{-1} are the reciprocals $1/\lambda$ of the eigenvalues of A. How are the eigenvectors related? (ii) Show how to use the power method on A^{-1} to produce the smallest (in modulus) eigenvalue of A. (iii) What is the rate of convergence of the algorithm? (iv) Design a practical iterative algorithm based on the QR decomposition of A. (v) Apply your algorithm to find the smallest eigenvalues and associated eigenvectors of the matrices in Exercise 6.1.

6.6. *The Shifted Inverse Power Method.* Suppose that μ is *not* an eigenvalue of A. (i) Show that the iterative system $\mathbf{x}_{k+1} = (A - \mu\, \mathrm{I})^{-1}\mathbf{x}_k$ converges to the eigenvector of A corresponding to the eigenvalue $\lambda^\star$ that is *closest* to μ. (ii) Explain how to compute $\lambda^\star$. (iii) What is the rate of convergence of the algorithm? (iv) What happens if μ is an eigenvalue? (v) Apply the shifted inverse power method to the find the eigenvalue closest to $\mu = .5$ of the matrices in Exercise 6.1.

6.7. Let A be positive definite symmetric with a unique dominant eigenvector $\mathbf{u}_1$, i.e., $\lambda_1 > \lambda_2$, which we take to be a Euclidean unit vector, so $\|\mathbf{u}_1\| = 1$. Let $\|\cdot\|_*$ be another norm on $\mathbb{R}^n$. Let $\mathbf{x}_{k+1} = A\mathbf{x}_k / \| A\mathbf{x}_k \|_*$ be the iterations of the power method in this norm, with $\|\mathbf{x}_0\|_* = 1$. Show that if $\mathbf{x}_0 \cdot \mathbf{u}_1 > 0$ then $\mathbf{x}_k \to \mathbf{u}_1 / \|\mathbf{u}_1\|_*$ as $k \to \infty$.

6.8. $\diamondsuit$ Let A be a symmetric positive semidefinite matrix whose dominant eigenvalue has multiplicity $j \geq 2$. That is, its eigenvalues satisfy $\lambda_1 = \lambda_2 = \cdots = \lambda_j > \lambda_{j+1} \geq \cdots \geq \lambda_n \geq 0$. Let $\mathbf{u}_1, \ldots, \mathbf{u}_n$ denote the corresponding orthonormal eigenvectors, let $U = (\,\mathbf{u}_1 \ldots \mathbf{u}_k\,)$ and let $P = UU^T$ be the orthogonal projection matrix onto the dominant eigenspace. Consider the power method applied to A, so $\mathbf{x}_{k+1} = A\mathbf{x}_k / \|A\mathbf{x}_k\|$, starting from some initial unit vector $\mathbf{x}_0$. Show that if $P\mathbf{x}_0 \neq 0$, then $\|\mathbf{x}_k - P\mathbf{x}_k\| \leq \dfrac{\sqrt{2}}{\|P\mathbf{x}_0\|} \left(\dfrac{\lambda_{j+1}}{\lambda_1}\right)^k$.

6.9. Extend Theorem 5.67 to the setting where A is self-adjoint, not necessarily positive semidefinite, but has a unique dominant eigenvalue with largest absolute value.

6.10. Apply orthogonal iteration to the following symmetric matrices to find their eigenvalues and eigenvectors to 2 decimal places:

$$
(a)\,\heartsuit \begin{pmatrix} 1 & 2 \\ 2 & 6 \end{pmatrix}, \quad
(b) \begin{pmatrix} 3 & -1 \\ -1 & 5 \end{pmatrix}, \quad
(c)\,\heartsuit \begin{pmatrix} 2 & 1 & 0 \\ 1 & 2 & 3 \\ 0 & 3 & 1 \end{pmatrix}, \quad
(d)\,\diamondsuit \begin{pmatrix} 2 & 5 & 0 \\ 5 & 0 & -3 \\ 0 & -3 & 3 \end{pmatrix},
$$

$$
(e)\,\diamondsuit \begin{pmatrix} 3 & -1 & 0 & 0 \\ -1 & 3 & -1 & 0 \\ 0 & -1 & 3 & -1 \\ 0 & 0 & -1 & 3 \end{pmatrix}, \quad
(f) \begin{pmatrix} 6 & 1 & -1 & 0 \\ 1 & 8 & 1 & -1 \\ -1 & 1 & 4 & 1 \\ 0 & -1 & 1 & 3 \end{pmatrix}.
$$

6.11. $\diamondsuit$ Repeat Exercise 6.2, except use orthogonal iteration to compute the top k eigenvectors of A for some choice of $1 < k < n$.

6.12. Let A_n be the $n \times n$ matrix with all 2's on the diagonal and 1's on the the the sub- and super-diagonals. Use orthogonal iteration to compute the top 5 eigenvalues of A_n for $n = 10, 20$ and 50.

6.13. $\heartsuit$ Show that applying orthogonal iteration to the matrix $A = \begin{pmatrix} 4 & -1 & 1 \\ -1 & 7 & 2 \\ 1 & 2 & 7 \end{pmatrix}$, starting with the initial matrix $S_0 = I$, eventually results in a diagonal matrix with the eigenvalues on the diagonal, but not in decreasing order. Explain why. Try changing the initial condition S_0; does that produce the eigenvalues in the correct order?

6.14. $\heartsuit$ Assume that orthogonal iteration applied to a symmetric positive semidefinite matrix A converges to an $n \times k$ matrix Q, whose columns are orthonormal, and a $k \times k$ upper triangular matrix R, whose diagonal entries are positive. Then Q and R satisfy $AQ = QR$. Show that the columns of Q are eigenvectors of A, and R is a diagonal matrix containing the corresponding eigenvalues.

6.15. The QR *algorithm*, [78, 79, 133] for computing all the eigenvalues and eigenvectors of a symmetric positive semidefinite matrix A is the following iterative scheme:

$$
A = A_1 = Q_1 R_1, \qquad A_{k+1} = R_k Q_k = Q_{k+1} R_{k+1}, \qquad k = 1, 2, \ldots, \tag{5.98}
$$

where each Q_k is orthogonal and R_k is positive definite upper triangular. In other words, starting with the matrix $A = A_1$, one successively performs a QR factorization and then multiplies the factors in the wrong order to form the next matrix in the iteration. (a) Show that R_k are the same matrices that appear in the orthogonal iteration (5.89) with $p = n$ and $S_0 = I$, while $S_k = Q_1 Q_2 \cdots Q_k$. (b) Determine the appropriate regularity condition required for convergence, and then explain how to use the QR algorithm to compute the eigenvalues and eigenvectors of A.

6.16. Given the subspaces in (5.94), let π_i be the orthogonal projection map onto U_i.
(a) Explain why (5.94) is equivalent to the statement that $\pi_i \colon V_i \to U_i$ is a one-to-one map.
(b) Let $\mathbf{w}_i \in V_i$ be the unique vector such that $\pi_i(\mathbf{w}_i) = \mathbf{u}_i$, and let $W = \begin{pmatrix} \mathbf{w}_1 \ \ldots \ \mathbf{w}_p \end{pmatrix}$.
Prove that $W = S_0 Z = Q_p L$ where L is lower triangular with diagonal entries $l_{ii} = 1$, while
Z is upper triangular with nonzero diagonal entries. (c) Use part (b) to prove the equivalence
of condition (5.94) and the matrix factorization (5.95).

5.7   Singular Values

We have already indicated the central role played by the eigenvalues and eigenvectors of a
square matrix in both theory and applications. Alas, rectangular matrices do not have eigenvalues (why?), and so, at first glance, do not appear to possess any quantities of comparable
significance. However, if A is an $m \times n$ matrix, and we impose inner products on $\mathbb{R}^n$ and $\mathbb{R}^m$
as in Section 4.3 — e.g., the dot products — then the eigenvalues of the associated self-adjoint,
positive semidefinite *square* matrix $S = A^* A$ — which can be naturally formed even when A
is not square — play a comparably important role. Since they are not easily related to the
eigenvalues of A, which, in the non-square case, don't even exist, we shall endow them with
a new name. They were first systematically studied by the German mathematician Erhard
Schmidt in early days of the twentieth century, although intimations can be found a century
earlier in Carl Friedrich Gauss's work on rigid body dynamics.

> **Definition 5.72.** Given inner products on $\mathbb{R}^n$ and $\mathbb{R}^m$, the *singular values* $\sigma_1, \ldots, \sigma_r$
> of an $m \times n$ matrix A are the positive square roots, $\sigma_i = \sqrt{\lambda_i} > 0$, of the nonzero
> eigenvalues of the associated positive semidefinite self-adjoint matrix $S = A^* A$. The
> corresponding eigenvectors of S are known as the *singular vectors* of A.

Since Theorem 4.25 tells us that $S = A^* A$ is necessarily positive semidefinite, its eigenvalues are always nonnegative, $\lambda_i \geq 0$, independently of whether A itself has positive, negative,
or even complex eigenvalues, or is rectangular and has no eigenvalues at all. The nonzero
eigenvalues of S are thus the squares, $\lambda_i = \sigma_i^2 > 0$, of the singular values of A. We will follow
the standard convention, and label the singular values in decreasing order, so that

$$\sigma_1 \geq \sigma_2 \geq \ \cdots \ \geq \sigma_r > 0. \tag{5.99}$$

Thus, $\sigma_1 = \sigma_{max}(A)$ will always denote the largest, or *dominant*, singular value. If S has
repeated eigenvalues, the singular values of A are repeated with the same multiplicities. The
number r of singular values is equal to the common rank of A and S.

Warning: Some texts include the zero eigenvalues of S as singular values of A. We find this
to be less convenient, but you should be aware of the differences between the two conventions.
Later we will discuss what happens when A has one or more very small singular values.

As was the case with eigenvalues, we will sometimes also write $\sigma_i(A)$ to denote the i-th
singular value of the matrix A, and $\sigma_{max}(A) = \sigma_1$ and $\sigma_{min}(A) = \sigma_r$ to denote the largest
and smallest singular values. If $r = \operatorname{rank} S = \operatorname{rank} A < n$, then S also has a zero eigenvalue,
with multiplicity $n - r = \operatorname{nullity} S = \operatorname{nullity} A$.
According to (4.35),

$$S = A^* A = C^{-1} A^T K A,$$

where C, K are the symmetric positive definite matrices determining the inner products on $\mathbb{R}^n, \mathbb{R}^m$, respectively. The most important case is when we use dot products on both, whereby C, K are both identity matrices, and hence $S = A^T A$ is a basic Gram matrix; indeed, in most of the literature, only this case is used and the designation "singular value" only refers to its (nonzero) eigenvalues. Here, with an eye towards later applications, we find it convenient to retain the option of using alternative inner products, and hence, for us, the singular values of A will depend upon which inner products are used.

Example 5.73. Let $A = \begin{pmatrix} 3 & 5 \\ 4 & 0 \end{pmatrix}$. Using the dot product, the associated Gram matrix

$$S = A^T A = \begin{pmatrix} 3 & 4 \\ 5 & 0 \end{pmatrix} \begin{pmatrix} 3 & 5 \\ 4 & 0 \end{pmatrix} = \begin{pmatrix} 25 & 15 \\ 15 & 25 \end{pmatrix}$$

has eigenvalues $\lambda_1 = 40$, $\lambda_2 = 10$, with $\mathbf{v}_1 = \begin{pmatrix} 1 \\ 1 \end{pmatrix}$, $\mathbf{v}_2 = \begin{pmatrix} 1 \\ -1 \end{pmatrix}$ the corresponding eigenvectors. This implies that the singular values of A are $\sigma_1 = \sigma_{max} = \sqrt{40} \approx 6.3246$ and $\sigma_2 = \sigma_{min} = \sqrt{10} \approx 3.1623$, with $\mathbf{v}_1, \mathbf{v}_2$ being the singular vectors. Note that the singular values are *not* its eigenvalues, which are $\lambda_1 = \frac{1}{2}\left(3 + \sqrt{89}\right) \simeq 6.2170$ and $\lambda_2 = \frac{1}{2}\left(3 - \sqrt{89}\right) \simeq -3.2170$, nor are the singular vectors eigenvectors of A. ▲

Only in the special case of self-adjoint — in particular symmetric — matrices is there a direct connection between their singular values and their (necessarily real) eigenvalues.

Theorem 5.74. *If $A = A^*$ is a self-adjoint $n \times n$ matrix, then its singular values are the absolute values of its nonzero eigenvalues: $\sigma_i = |\lambda_i| > 0$, and its singular vectors coincide with its non-null eigenvectors. In particular, if A is positive definite, then $\sigma_i = \lambda_i$ for $i = 1, \ldots, n$.*

Proof. When A is self-adjoint, $S = A^* A = A^2$. So, if

$$A\mathbf{v} = \lambda \mathbf{v}, \qquad \text{then} \qquad S\mathbf{v} = A^2 \mathbf{v} = A(\lambda \mathbf{v}) = \lambda A \mathbf{v} = \lambda^2 \mathbf{v},$$

and hence every eigenvector $\mathbf{v}$ of A is also an eigenvector of S with eigenvalue λ^2. The eigenvector basis of A guaranteed by Theorem 5.29 is thus also an eigenvector basis for S, and hence the non-null eigenvectors form a complete system of singular vectors for A. ∎

Thus, if A is positive semidefinite, then its singular values are the same as its nonzero eigenvalues. In particular, if $A^T = A > 0$ is symmetric and positive definite, and hence self-adjoint with respect to the dot product, then Theorem 5.74 implies that its singular values — with respect to the dot product — are its eigenvalues. However, if we use the alternative inner product $\langle \mathbf{x}, \mathbf{y} \rangle = \mathbf{x}^T C \mathbf{y}$, then the corresponding singular values are the eigenvalues of the self-adjoint but non-symmetric matrix $C^{-1} A$, which are *not* the same as the eigenvalues of A. Note further that, as a consequence of Corollary 5.8, the eigenvalues of $C^{-1} A$ are the same as the eigenvalues of the similar symmetric positive definite matrix $C^{-1/2} A C^{-1/2} = C^{1/2}(C^{-1}A)C^{-1/2}$.

5.7.1 The Singular Value Decomposition

The generalization of the spectral factorization (5.31) to non-symmetric matrices is known as the *singular value decomposition*, commonly abbreviated SVD. Unlike the former, which applies only to square matrices, every nonzero matrix possesses a singular value decomposition. When computing adjoints, we will use the inner products defined by C and K on $\mathbb{R}^n$ and $\mathbb{R}^m$, respectively, and the dot product on $\mathbb{R}^r$, the latter dictated by the form of the matrices appearing in the decomposition. When stating this result, we recall Proposition 4.34 characterizing matrices with orthonormal columns.

Theorem 5.75. *A nonzero real $m \times n$ matrix $A \neq O$ of rank $r > 0$ can be factored,*

$$A = P\Sigma Q^* = P\Sigma Q^T C = \sum_{k=1}^{r} \sigma_k\, \mathbf{p}_k \mathbf{q}_k^T C, \tag{5.100}$$

into the product of an $m \times r$ matrix $P = (\mathbf{p}_1 \ \cdots \ \mathbf{p}_r)$ that has orthonormal columns, so $P^*P = P^T K P = I$, the $r \times r$ diagonal matrix $\Sigma = \mathrm{diag}\,(\sigma_1, \ldots, \sigma_r)$ that has the singular values of A as its diagonal entries, and the adjoint of an $n \times r$ matrix $Q = (\mathbf{q}_1 \ \cdots \ \mathbf{q}_r)$ that has orthonormal columns, so $Q^*Q = Q^T C Q = I$. Moreover, the columns $\mathbf{q}_1, \ldots, \mathbf{q}_r \in \mathbb{R}^n$ of Q form an orthonormal basis for $\mathrm{coimg}\,A$, while the columns $\mathbf{p}_1, \ldots, \mathbf{p}_r \in \mathbb{R}^m$ of P form an orthonormal basis for $\mathrm{img}\,A$.

Remark. In the classical case, when one only employs the dot product on all three spaces, the singular value decomposition (5.100) reduces to the standard form

$$A = P\Sigma Q^T = \sum_{k=1}^{r} \sigma_k\, \mathbf{p}_k \mathbf{q}_k^T, \tag{5.101}$$

where $P^T P = I$ and $Q^T Q = I$.   ▲

Proof. Let $\mathbf{q}_1, \ldots, \mathbf{q}_n \in \mathbb{R}^n$ be an orthonormal eigenvector basis of the self-adjoint matrix $S = A^*A$, where $\mathbf{q}_1, \ldots, \mathbf{q}_r$ are singular eigenvectors, corresponding to the nonzero eigenvalues, i.e., the squares of the singular values, so

$$S\mathbf{q}_i = A^*A\mathbf{q}_i = \sigma_i^2\, \mathbf{q}_i, \qquad i = 1, \ldots, r, \tag{5.102}$$

while $\mathbf{q}_{r+1}, \ldots, \mathbf{q}_n$ are null eigenvectors, so

$$A\mathbf{q}_j = \mathbf{0}, \qquad S\mathbf{q}_j = A^*A\mathbf{q}_j = \mathbf{0}, \qquad j = r+1, \ldots, n, \tag{5.103}$$

where the first equation follows from the fact that A and A^*A have the same kernel; see Theorem 4.25. Moreover, the singular vectors $\mathbf{q}_1, \ldots, \mathbf{q}_r$ form an orthonormal basis for $\mathrm{img}\,(A^*A) = \mathrm{coimg}\,A$.

Since $\mathbf{q}_1, \ldots, \mathbf{q}_n$ are an orthonormal basis, given $\mathbf{x} \in \mathbb{R}^n$, we have

$$A\mathbf{x} = A \sum_{k=1}^{n} \langle \mathbf{q}_k, \mathbf{x} \rangle_C\, \mathbf{q}_k = \sum_{k=1}^{n} (\mathbf{q}_k^T C\, \mathbf{x})\, A\mathbf{q}_k = \left(\sum_{k=1}^{r} (A\mathbf{q}_k)\, \mathbf{q}_k^T C \right) \mathbf{x},$$

where we used (5.103) to reduce the sum to $k = 1, \ldots, r$ in the last line. Since this holds for all $\mathbf{x} \in \mathbb{R}^n$, it follows, upon defining

$$\mathbf{p}_k := \frac{A\mathbf{q}_k}{\sigma_k}, \tag{5.104}$$

that

$$A = \sum_{k=1}^{r} (A\mathbf{q}_k)\,\mathbf{q}_k^T C = \sum_{k=1}^{r} \sigma_k \mathbf{p}_k \mathbf{q}_k^T C = P\Sigma Q^T C = P\Sigma Q^*,$$

proving (5.101).

It remains to show that the vectors $\mathbf{p}_1, \dots, \mathbf{p}_r$ are orthonormal. Indeed, by the definition (4.19) of the adjoint, the eigenvalue equation (5.102), and the orthonormality of $\mathbf{q}_1, \dots, \mathbf{q}_r$,

$$\langle \mathbf{p}_i, \mathbf{p}_j \rangle_K = \frac{\langle A\mathbf{q}_i, A\mathbf{q}_j \rangle_K}{\sigma_i \sigma_j} = \frac{\langle A^* A\mathbf{q}_i, \mathbf{q}_j \rangle_C}{\sigma_i \sigma_j} = \frac{\sigma_i^2 \langle \mathbf{q}_i, \mathbf{q}_j \rangle_C}{\sigma_i \sigma_j} = \begin{cases} 0, & i \ne j, \\ 1, & i = j. \end{cases}$$

Since they belong to $\operatorname{img} A$, which has dimension $r = \operatorname{rank} A$, they therefore form an orthonormal basis for the image. $\blacksquare$

Remark. If A has distinct singular values, its singular value decomposition (5.101) is almost unique, modulo simultaneously changing the signs of one or more of the corresponding columns of Q and P. Matrices with repeated singular values have more freedom, since one can use different orthonormal bases of each eigenspace of S. $\blacktriangle$

Observe that, taking the adjoint of (5.101) and noting that Σ is diagonal and hence self-adjoint with respect to the dot product on $\mathbb{R}^r$, so $\Sigma^* = \Sigma^T = \Sigma$, we obtain

$$A^* = Q\Sigma P^* = Q\Sigma P^T K, \tag{5.105}$$

which is a singular value decomposition of the adjoint matrix A^*. In particular, we obtain the following result:

Proposition 5.76. *A matrix A and its adjoint A^* have the same singular values.*

Note that their singular vectors are not the same; indeed, those of A are the orthonormal columns of Q, whereas those of A^* are the orthonormal columns of P, which are related by (5.104). Thus,

$$A^* \mathbf{p}_i = \sigma_i \mathbf{q}_i, \qquad i = 1, \dots, r, \tag{5.106}$$

which is also a consequence of (5.102).

Example 5.77. For the matrix $A = \begin{pmatrix} 3 & 5 \\ 4 & 0 \end{pmatrix}$ in Example 5.73, an orthonormal eigenvector basis of $S = A^T A = \begin{pmatrix} 25 & 15 \\ 15 & 25 \end{pmatrix}$ is given by the unit singular vectors $\mathbf{q}_1 = \begin{pmatrix} \frac{1}{\sqrt{2}} \\ \frac{1}{\sqrt{2}} \end{pmatrix}$ and $\mathbf{q}_2 = \begin{pmatrix} -\frac{1}{\sqrt{2}} \\ \frac{1}{\sqrt{2}} \end{pmatrix}$. Thus, $Q = \begin{pmatrix} \frac{1}{\sqrt{2}} & -\frac{1}{\sqrt{2}} \\ \frac{1}{\sqrt{2}} & \frac{1}{\sqrt{2}} \end{pmatrix}$. Next, according to (5.104),

$$\mathbf{p}_1 = \frac{A\mathbf{q}_1}{\sigma_1} = \frac{1}{\sqrt{40}}\begin{pmatrix} 4\sqrt{2} \\ 2\sqrt{2} \end{pmatrix} = \begin{pmatrix} \frac{2}{\sqrt{5}} \\ \frac{1}{\sqrt{5}} \end{pmatrix}, \qquad \mathbf{p}_2 = \frac{A\mathbf{q}_2}{\sigma_2} = \frac{1}{\sqrt{10}}\begin{pmatrix} \sqrt{2} \\ -2\sqrt{2} \end{pmatrix} = \begin{pmatrix} \frac{1}{\sqrt{5}} \\ -\frac{2}{\sqrt{5}} \end{pmatrix},$$

and thus $P = \begin{pmatrix} \frac{2}{\sqrt{5}} & \frac{1}{\sqrt{5}} \\ \frac{1}{\sqrt{5}} & -\frac{2}{\sqrt{5}} \end{pmatrix}$. You may wish to validate the resulting singular value factorization

$$A = \begin{pmatrix} 3 & 5 \\ 4 & 0 \end{pmatrix} = \begin{pmatrix} \frac{2}{\sqrt{5}} & \frac{1}{\sqrt{5}} \\ \frac{1}{\sqrt{5}} & -\frac{2}{\sqrt{5}} \end{pmatrix}\begin{pmatrix} \sqrt{40} & 0 \\ 0 & \sqrt{10} \end{pmatrix}\begin{pmatrix} \frac{1}{\sqrt{2}} & \frac{1}{\sqrt{2}} \\ -\frac{1}{\sqrt{2}} & \frac{1}{\sqrt{2}} \end{pmatrix} = P\Sigma Q^T. \qquad \blacktriangle$$

Example 5.78. Suppose the matrix A has only one singular value, so $\sigma_1 = \cdots = \sigma_r = \sigma$, where $r = \operatorname{rank} A$. Then $\Sigma = \sigma\,I$ is a multiple of the identity matrix, and hence $A = \sigma\,P\,Q^*$. Thus,[10]

$$A^*A = \sigma^2\,Q\,P^*\,P\,Q^* = \sigma^2\,Q\,Q^*, \qquad \text{where} \qquad Q^*Q = Q^T C\,Q = I.$$

According to (4.62) and identifying Q with U, this implies that A^*A is a multiple of the orthogonal projection matrix $Q\,Q^*$, for the inner product based on the matrix C, onto the subspace $\operatorname{img} Q = \operatorname{coimg} A$. In particular, this implies that, modulo a factor of σ^2, the matrix A^*A is idempotent: $(A^*A)^2 = \sigma^2 A^*A$. ▲

Finally, we note that practical numerical algorithms for computing singular values and the singular value decomposition can be found in [88, 230, 245].

5.7.2 The Euclidean Matrix Norm

Singular values allow us to finally write down a formula for the natural matrix norm induced by the Euclidean norm (or 2 norm) on $\mathbb{R}^n$, as defined in Theorem 4.49.

Theorem 5.79. *Let $\|\cdot\|_2$ denote the Euclidean norm on $\mathbb{R}^n$. Let $A \neq O$ be a nonzero $n \times n$ matrix. Then its Euclidean matrix norm equals its dominant (largest) singular value:*

$$\|A\|_2 = \max\left\{\|A\mathbf{u}\|_2 \mid \|\mathbf{u}\|_2 = 1\right\} = \sigma_{max}(A), \qquad \text{while} \qquad \|O\|_2 = 0. \qquad (5.107)$$

Proof. Observe that

$$\|A\mathbf{u}\|_2^2 = (A\mathbf{u})^T A\mathbf{u} = \mathbf{u}^T A^T A\mathbf{u} = \mathbf{u} \cdot (A^T A\mathbf{u}).$$

According to (5.50) (for the dot product), the maximum of the right hand side over all unit vectors, $\|\mathbf{u}\|_2 = 1$, is the maximal eigenvalue of the symmetric Gram matrix $S = A^T A$. Thus, provided $A \neq O$,

$$\begin{aligned}
\|A\|_2^2 &= \max\left\{\|A\mathbf{u}\|_2^2 \mid \|\mathbf{u}\|_2 = 1\right\} \\
&= \max\left\{\mathbf{u} \cdot (A^T A\mathbf{u}) \mid \|\mathbf{u}\|_2 = 1\right\} = \lambda_{max}(A^T A) = \sigma_{max}(A)^2,
\end{aligned}$$

by the definition of singular value. Taking the square roots of both sides completes the proof. ∎

As a consequence, we deduce the following inequality, valid for any $n \times n$ matrix $A \neq O$:

$$\|A\mathbf{v}\|_2 \leq \sigma_{max}(A)\,\|\mathbf{v}\|_2 \qquad \text{for all} \qquad \mathbf{v} \in \mathbb{R}^n. \qquad (5.108)$$

Example 5.80. Consider the matrix $A = \begin{pmatrix} 0 & -\frac{1}{3} & \frac{1}{3} \\ \frac{1}{4} & 0 & \frac{1}{2} \\ \frac{2}{5} & \frac{1}{5} & 0 \end{pmatrix} \simeq \begin{pmatrix} 0 & -.3333 & .3333 \\ .25 & 0 & .5 \\ .4 & .2 & 0 \end{pmatrix}$. The corresponding Gram matrix

$$A^T A \simeq \begin{pmatrix} .2225 & .0800 & .1250 \\ .0800 & .1511 & -.1111 \\ .1250 & -.1111 & .3611 \end{pmatrix},$$

---

[10] Keep in mind that the inner product on the range of Q is the dot product.

has eigenvalues $\lambda_1 \simeq .4472$, $\lambda_2 \simeq .2665$, $\lambda_3 \simeq .0210$, and hence the singular values of A are their square roots: $\sigma_1 \simeq .6687$, $\sigma_2 \simeq .5163$, $\sigma_3 \simeq .1448$. The Euclidean matrix norm of A is the largest singular value, and so $\|A\|_2 \simeq .6687$. $\blacktriangle$

Formula (5.107) relates the Euclidean matrix norm to the dominant singular value when the adjoint (i.e., transpose) is computed using the dot products. We can generalize this formula to when the singular values are computed using the adjoint with respect to other inner product norms, as above. The proof is left to the reader as Exercise 7.15.

Theorem 5.81. *Let $A \neq O$ be an $m \times n$ matrix. Using the inner products on $\mathbb{R}^n$ and $\mathbb{R}^m$ determined by the symmetric positive definite matrices C and K, respectively, the dominant singular value is given by*

$$\sigma_{max}(A) = \|A\|_{C,K} = \max\left\{\|A\mathbf{u}\|_K \mid \|\mathbf{u}\|_C = 1\right\}. \tag{5.109}$$

Given an inner product $\langle \cdot, \cdot \rangle_C$ determined by the symmetric positive definite matrix C, we can similarly characterize the corresponding natural matrix norm of a self-adjoint positive semidefinite matrix H. Referring to Theorem 5.45, we deduce

$$\|H\|_C = \max\left\{\|H\mathbf{u}\|_C \mid \|\mathbf{u}\|_C = 1\right\} = \lambda_{max}(H) = \sigma_{max}(H). \tag{5.110}$$

The details of the proof are left to the reader as Exercise 7.16.

5.7.3 Condition Number and Rank

Singular values play a key role in modern computational algorithms, and can be used to distinguish between well-behaved and what are known as ill-conditioned linear systems. Roughly speaking, the closer a nonsingular square matrix is to being singular, the harder it is to accurately solve the associated linear system $A\mathbf{x} = \mathbf{b}$. Those that are very close produce ill-conditioned linear systems that can be quite challenging to solve accurately on a computer due to the effects of numerical errors such as round-off. The method based on QR factorization introduced in Section 4.7 does a better job of treating mildly ill-conditioned system than more elementary solution algorithms such as Gaussian elimination. Iterative methods, [181], can perform even better when dealing with systems that have a particular form. Nevertheless, there always exist severely ill conditioned systems that can stymie even the best linear system solvers.

Recall that the number of singular values equals the rank, and so a nonsingular $n \times n$ matrix has n singular values. However, if one or more of these singular values is very small, the matrix is close to being of nonmaximal rank. This measurement of "closeness" can be quantified as follows.

Definition 5.82. The *condition number* of a nonsingular $n \times n$ matrix is the ratio between its largest and smallest singular values: $\kappa(A) = \sigma_1/\sigma_n$.

Remark 5.83. In particular, if H is positive definite, then, by Theorem 5.74, its condition number is the ratio between its largest and smallest eigenvalues: $\kappa(H) = \lambda_1/\lambda_n$. $\blacktriangle$

An $n \times n$ matrix with fewer than n singular values is singular, and is said to have condition number ∞. A nonsingular matrix with several very small singular value is close to being singular, which is indicated by its large condition number, and designated as *ill-conditioned*. In practical terms, ill-conditioning occurs when the condition number is larger than the reciprocal of the machine's precision, e.g., 10^7 for typical single-precision arithmetic.

Remark. Since the singular values of a matrix depend on the choice of inner products, so does its condition number. As we will see, an inspired choice may lead to better conditioning.   ▲

Example 5.84. A simple example of an ill-conditioned matrix is provided by

$$
A = \begin{pmatrix} 1.00001 & 1. & -1. \\ 2. & 2.00001 & -2. \\ 3. & 3. & -3.00001 \end{pmatrix},
$$

which has (dot product) singular values $\sigma_1 \approx 6.48075$, $\sigma_2 \approx \sigma_3 \approx .000001$, and hence has rank 3. On the other hand, it is very close to the singular rank 1 matrix $\widetilde{A} = \begin{pmatrix} 1 & 1 & -1 \\ 2 & 2 & -2 \\ 3 & 3 & -3 \end{pmatrix}$ obtained by rounding off its diagonal entries, which is a consequence of the smallness of its second and third singular values. Indeed, its condition number is quite large $\kappa(A) \approx 6.48 \times 10^6$, and so accurately solving any associated linear system requires some care.   ▲

This example serves to motivate an effective practical method for computing a good approximation for the rank of a matrix: first assign a threshold, e.g., 10^{-5}, and then treat any singular value lying below the threshold as if it were zero. In this way, the best guess for the actual rank of the matrix will be the number of singular values that are above the threshold. This idea will be justified by Theorem 8.15 appearing in Chapter 8.

Exercises

Note: Unless stated otherwise, the underlying inner product is the dot product on $\mathbb{R}^n$.

7.1. Find the singular values of the following matrices and then write out their singular value decomposition: $(a)\,\heartsuit\ (2, -1, 3)$, $\quad (b)\,\heartsuit\ \begin{pmatrix} 0 & 1 \\ -1 & 0 \end{pmatrix}$, $\quad (c)\ \begin{pmatrix} 1 & 1 \\ 0 & 2 \end{pmatrix}$, $\quad (d)\,\diamond\ \begin{pmatrix} 1 & -2 \\ -3 & 6 \end{pmatrix}$,

$(e)\,\heartsuit\ \begin{pmatrix} 2 & 0 & 0 \\ 0 & 3 & 0 \end{pmatrix}$, $\quad (f)\,\diamond\ \begin{pmatrix} 0 & 1 \\ 1 & -1 \\ -1 & 0 \\ 1 & -1 \end{pmatrix}$, $\quad (g)\,\diamond\ \begin{pmatrix} 2 & 1 & 0 & -1 \\ 0 & -1 & 1 & 1 \end{pmatrix}$, $\quad (h)\ \begin{pmatrix} 1 & -1 & 0 \\ -1 & 2 & -1 \\ 0 & -1 & 1 \end{pmatrix}$.

7.2. $\diamond$ What are the singular values of a $1 \times n$ matrix? Write down its singular value decomposition.

7.3. Prove that if the square matrix A is nonsingular, then the singular values of A^{-1} are the reciprocals of the singular values of A. How are their condition numbers related?

7.4. $\heartsuit$ *True or false*: If A is a symmetric matrix, then its singular values are the same as its eigenvalues.

7.5. *True or false*: The singular values of A^2 are the squares σ_i^2 of the singular values of A.

7.6. $\heartsuit$ Suppose Q is an orthogonal $n \times n$ matrix. What are its singular values?

7.7. $\diamond$ What can you say about a matrix whose singular values $\sigma_1 = \cdots = \sigma_r$ are all the same?

7.8. Let A be a square matrix. Prove that its maximal eigenvalue is smaller than its maximal singular value: $\max |\lambda_i| \le \max \sigma_i$.

7.9. Use the power method to find the largest singular value of the following matrices:

$$(a)\,\heartsuit \begin{pmatrix} 1 & 2 \\ -1 & 3 \end{pmatrix},\ (b)\,\diamondsuit \begin{pmatrix} 2 & 1 & -1 \\ -2 & 3 & 1 \end{pmatrix},\ (c)\,\heartsuit \begin{pmatrix} 2 & 2 & 1 & -1 \\ 1 & -2 & 0 & 1 \end{pmatrix},\ (d) \begin{pmatrix} 3 & 1 & -1 \\ 1 & -2 & 2 \\ 2 & -1 & 1 \end{pmatrix}.$$

7.10. Compute the Euclidean matrix norm of the following matrices.

$$(a)\,\heartsuit \begin{pmatrix} \frac{1}{2} & \frac{1}{4} \\ \frac{1}{3} & \frac{1}{6} \end{pmatrix},\quad (b)\,\diamondsuit \begin{pmatrix} \frac{5}{3} & \frac{4}{3} \\ -\frac{7}{6} & -\frac{5}{6} \end{pmatrix},\quad (c)\,\heartsuit \begin{pmatrix} \frac{2}{7} & -\frac{2}{7} \\ -\frac{2}{7} & \frac{6}{7} \end{pmatrix},\quad (d) \begin{pmatrix} \frac{1}{4} & \frac{3}{2} \\ -\frac{1}{2} & \frac{5}{4} \end{pmatrix}.$$

7.11. $\diamondsuit$ Find a matrix A whose Euclidean matrix norm satisfies $\|A^2\|_2 \neq \|A\|_2^2$.

7.12. $\heartsuit$ *True or false*: The minimum value of the quantity in (5.107) is the smallest singular value of A.

7.13. Let A be an $n \times n$ matrix with singular value vector $\boldsymbol{\sigma} = (\sigma_1, \ldots, \sigma_r)$. Prove that (a) $\|\boldsymbol{\sigma}\|_\infty = \|A\|_2$; (b) $\|\boldsymbol{\sigma}\|_2 = \|A\|_F$, the Frobenius norm. *Remark*: The 1 norm of the singular value vector $\|\boldsymbol{\sigma}\|_1$ also defines a useful matrix norm, the *Ky Fan norm*.

7.14. $\heartsuit$ Prove that the Euclidean matrix norm is bounded by the Frobenius norm, so that $\|A\|_2 \leq \|A\|_F$. When are they equal?

7.15. Prove formula (5.109).

7.16. Fill in the details of the proof of formula (5.110). What happens if H is not positive semidefinite?

7.17. Find the condition number of the following matrices. Which would you characterize as ill-conditioned?

$$(a)\,\heartsuit \begin{pmatrix} 2 & -1 \\ -3 & 1 \end{pmatrix},\ (b)\,\diamondsuit \begin{pmatrix} 1 & 2 \\ 1.001 & 1.9997 \end{pmatrix},\ (c)\,\heartsuit \begin{pmatrix} -1 & 3 & 4 \\ 2 & 10 & 6 \\ 1 & 2 & -3 \end{pmatrix},\ (d) \begin{pmatrix} 72 & 96 & 103 \\ 42 & 55 & 59 \\ 67 & 95 & 102 \end{pmatrix}.$$

7.18. Let A be a nonsingular square matrix. Prove the following formulas for its condition number:

$$(a)\,\diamondsuit\ \kappa(A) = \frac{\max\{\|A\mathbf{u}\| \mid \|\mathbf{u}\| = 1\}}{\min\{\|A\mathbf{u}\| \mid \|\mathbf{u}\| = 1\}},\qquad (b)\ \kappa(A) = \|A\|_2\,\|A^{-1}\|_2.$$

Chapter 6

Basics of Optimization

Optimization — that is, finding the minima and maxima of real-valued functions — is one of the most important problems throughout science and engineering. Minimization principles naturally arise in the fitting of data and in machine learning, where one seeks to minimize an appropriately chosen "loss function". The equilibrium solutions of systems of physical significance seek to minimize their potential energy. Engineering design is guided by a variety of optimization constraints, such as performance, longevity, safety, and cost. Additional applications naturally appear in economics and financial mathematics — one often wishes to minimize expenses or maximize profits — in biological and ecological systems, in pattern recognition and signal processing, in statistics, and many other fields.

In fact, since early human civilization, we have been concerned with optimization, such as finding the shortest path down a mountain. However, before mathematics was developed, we were constrained to crude simulations, e.g., roll a rock down the mountain, or follow a stream, [64]. The study of systematic mathematical solutions to optimization problems began with the French mathematician Pierre de Fermat in his work *Maxima and Minima* in 1636–1642 [54]. Fermat proposed to minimize a function by setting the gradient equal to zero and solving the resulting equation — a method which is still taught in every multivariable calculus class! In the nearly four centuries since Fermat's seminal work, the development of mathematical techniques and tools for optimization has been of intense mathematical and practical interest.

Fermat's techniques work well for simple functions (e.g., quadratic functions), but the equations become too difficult to solve (or even write down!) for more complicated problems. In this case, we often resort to iterative computational techniques that compute successively better approximations of the minimum, but may never exactly solve the problem. One of the most widely used methods, called *gradient descent*, was almost certainly known to Newton, but it appears that it was first formally proposed by the French mathematician and engineer Augustin-Louis Cauchy in 1847 [38]; the same ideas were arrived at independently by another French mathematician Jacques Hadamard in 1907 [48, 95]. The first mathematical results on the convergence of gradient descent are due to the American mathematician Haskell Curry in 1944, [50]. Gradient descent, and variations thereof, are the building blocks for training modern machine learning models, and understanding their ability to solve hard optimization problems is therefore of immense interest in mathematics.

In this chapter, we will describe some basic theoretical and numerical techniques used to solve, or closely approximate the solution to, fairly general optimization problems. The principal numerical technique is gradient descent, in which one determines the direction of the next iterate by the negative gradient of the objective function — the "downhill" direction of steepest decrease. While our analysis of gradient descent will be largely restricted to the setting of *convex* optimization problems, the method is often highly effective in the nonconvex setting — for example, in training of deep neural networks. We conclude the chapter by analyzing Newton's method, which makes use of the second derivative Hessian matrix of the objective function to, in favorable situations, speed up convergence. More advanced methods and results will be the focus of the subsequent Chapter 11.

Notation: Throughout this chapter, unless specifically noted otherwise, $\langle\,\cdot\,,\cdot\,\rangle$ denotes an inner product on $\mathbb{R}^n$, and $\|\cdot\|$ refers to the induced norm. In some places we will specialize results to the dot product $\mathbf{x}\cdot\mathbf{y} = \mathbf{x}^T\mathbf{y}$ and the Euclidean norm $\|\mathbf{x}\|_2 = \sqrt{\mathbf{x}\cdot\mathbf{x}}$.

6.1   The Objective Function

Throughout this chapter, the real-valued function $F(\mathbf{x}) = F(x_1,\ldots,x_n)$ to be optimized — the energy, entropy, work, cost, etc. — will be called the *objective function*. In machine learning, it is often referred to as the *loss function*, and serves to measure the performance of an algorithm. As such, F depends upon one or more variables $\mathbf{x} = (x_1, x_2, \ldots, x_n)^T$ that belong to a prescribed subset $\Omega \subset \mathbb{R}^n$. We will always assume that the objective function is at least continuous. Additional conditions, e.g., differentiability, will be imposed as needed.

Definition 6.1. A point $\mathbf{x}^\star \in \Omega$ is a *global minimizer* of the objective function $F(\mathbf{x})$ on the domain Ω if

$$F(\mathbf{x}^\star) \le F(\mathbf{x}) \qquad \text{for all} \qquad \mathbf{x} \in \Omega. \tag{6.1}$$

The minimizer is called *strict* if

$$F(\mathbf{x}^\star) < F(\mathbf{x}) \qquad \text{for} \qquad \mathbf{x}^\star \ne \mathbf{x} \in \Omega. \tag{6.2}$$

The point $\mathbf{x}^\star$ is called a (*strict*) *local minimizer* if the relevant inequality holds just for points $\mathbf{x} \in \Omega$ nearby $\mathbf{x}^\star$, i.e., satisfying $\|\mathbf{x} - \mathbf{x}^\star\| < \delta$ for some $\delta > 0$. A local minimizer is called *isolated* if there are no other local minimizers in a suitably small neighborhood; this requires the minimizer to be strict, but strictness in itself this does not suffice to show that the minimizer is isolated; see Exercise 1.3.

The definition of a *maximizer* — local or global — is the same, but with the reversed inequality: $F(\mathbf{x}^\star) \ge F(\mathbf{x})$ or, in the strict case, $F(\mathbf{x}^\star) > F(\mathbf{x})$. Alternatively, a maximizer of $F(\mathbf{x})$ is the same as a minimizer of the negative $-F(\mathbf{x})$. Therefore, every result that applies to minimization of a function can easily be translated into a result on maximization, which allows us to concentrate exclusively on the minimization problem without any loss of generality. We will use *extremizer* as a shorthand term for either a minimizer or a maximizer.

Remark. Any system of equations can be readily converted into a minimization principle. Given a system $G(\mathbf{x}) = \mathbf{0}$ consisting of m equations in n unknowns that are specified by a function $G\colon \mathbb{R}^n \to \mathbb{R}^m$, we introduce the scalar-valued objective function

$$F(\mathbf{x}) = \|G(\mathbf{x})\|. \tag{6.3}$$

By the basic properties of the norm, the minimum value is $F(\mathbf{x}) = 0$, and this is achieved if and only if $G(\mathbf{x}) = \mathbf{0}$, i.e., at a solution to the system. More generally, if the system does not have a solution, the minimizers of $F(\mathbf{x})$ will, in a sense, be the best attempt at solving the system. For example, if $G(\mathbf{x}) = A\mathbf{x} - \mathbf{b}$ is an affine function, then the minimizer of $F(\mathbf{x}) = \|A\mathbf{x} - \mathbf{b}\|$ is known as the *least squares solution* to the linear system $A\mathbf{x} = \mathbf{b}$, that we already encountered in Section 4.7.   ▲

In contrast to the rather difficult question of existence of solutions to systems of equations, as we noted in Theorem 2.35, a continuous function on a compact (i.e., closed and bounded) set is guaranteed to admit a minimizer. However, this existential result does not indicate how to go about finding it. Our goal, then, is to formulate practical algorithms that can accurately compute or at least closely approximate the minimizers of general nonlinear functions.

Let us first review the basic procedure for optimizing scalar functions that you learned in first and second year calculus. Throughout the remainder of this section, $f \colon I \to \mathbb{R}$ is a scalar objective function defined on an open interval $I \subset \mathbb{R}$, and the goal is the find its *extremizers*, that is its minimizers and maximizers, both local and global. At the very least, we will assume that $f \in \mathrm{C}^1$ is continuously differentiable on its domain.

Definition 6.2. Let $f \colon I \to \mathbb{R}$ be differentiable. If $f'(x^\star) = 0$, then $x^\star$ is called a *critical point* of f.

The *first derivative test* for extremizers says that they are necessarily critical points.

Theorem 6.3. *If $x^\star$ is a minimizer or maximizer, either local or global, then it is a critical point, so $f'(x^\star) = 0$.*

Remark. Thus, critical points can be minimizers or maximizers, but they might be neither. An example of the latter is the inflection point $x^\star = 0$ of the function $f(x) = x^3$. ▲

Proof. If $x^\star$ is a local minimizer, then for x sufficiently close but not equal to $x^\star$, the difference quotient

$$\frac{f(x) - f(x^\star)}{x - x^\star}$$

will be ≥ 0 when $x > x^\star$ and ≤ 0 when $x < x^\star$. Thus, its limit as $x \to x^\star$, which, by the definition of derivative, is $f'(x^\star)$, must be zero. ∎

If the domain of f is a closed interval, then one must also be concerned with *boundary minimizers* and *maximizers* which occur at its endpoints, and are not necessarily critical points. As important as they can be, in this book we will (mostly) ignore the boundary behavior of our objective functions.

The test for determining the nature of a critical point requires that the objective function have a continuous second order derivative.

Proposition 6.4. *Let $f \in \mathrm{C}^2$ be a twice continuous differentiable scalar function, and suppose that $x^\star$ a critical point: $f'(x^\star) = 0$. If $x^\star$ is a local minimizer, then $f''(x^\star) \geq 0$. Conversely, if $f''(x^\star) > 0$, then $x^\star$ is a strict local minimizer. Similarly, $f''(x^\star) \leq 0$ is required at a local maximizer, while $f''(x^\star) < 0$ implies that $x^\star$ is a strict local maximizer.*

Proof. As noted above, it suffices to prove the minimizer version of this result. The proof relies on the first order Taylor formula, [4],

$$f(x) = f(a) + f'(a)\,(x - a) + \tfrac{1}{2}\,f''(y)\,(x - a)^2, \tag{6.4}$$

which is valid for some y between x and a. In particular, setting $a = x^\star$ to be the critical point, the second term on the right hand side vanishes, so

$$f(x) = f(x^\star) + \tfrac{1}{2}\,f''(y)\,(x - x^\star)^2. \tag{6.5}$$

If $f''(x^\star) > 0$, then, by continuity, $f''(y) > 0$ for y sufficiently close to $x^\star$, whereby $f(x) > f(x^\star)$ for $x \neq x^\star$ sufficiently close. We conclude that $x^\star$ is a strict local minimizer. Conversely, if $f(x) \geq f(x^\star)$ then (6.5) implies $f''(y) \geq 0$. Letting $x \to x^\star$, which implies $y \to x^\star$ also, we conclude that, by continuity, $f''(x^\star) \geq 0$. $\blacksquare$

Remark. In the borderline case, when $f''(x^\star) = 0$, the second derivative test is inconclusive, and the point could be a maximizer or minimizer, perhaps strict, perhaps not, or neither of the two. In such cases, one must analyze the higher order terms in the Taylor expansion to try to resolve the status of the critical point. $\blacktriangle$

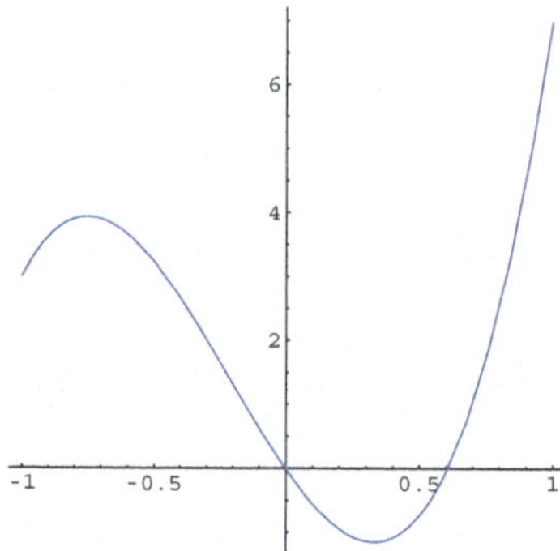

Figure 6.1: The function $8x^3 + 5x^2 - 6x$

Example 6.5. Let us optimize the scalar objective function

$$f(x) = 8x^3 + 5x^2 - 6x$$

on the domain $-1 \leq x \leq 1$. To locate the minimizer, the first step is to look at the critical points where the derivative vanishes:

$$f'(x) = 24x^2 + 10x - 6 = 0, \qquad \text{and hence} \qquad x = \tfrac{1}{3}, \ -\tfrac{3}{4}.$$

To ascertain the local nature of the two critical points, we apply the second derivative test. Since $f''(x) = 48x + 10$, we have

$$f''\left(\tfrac{1}{3}\right) = 26 > 0, \qquad \text{whereas} \qquad f''\left(-\tfrac{3}{4}\right) = -26 < 0.$$

We conclude that $\tfrac{1}{3}$ is a local minimizer, while $\tfrac{3}{4}$ is a local maximizer.

To find the global minimizer and maximizer on the interval $[-1, 1]$, we must also take into account the boundary points ± 1. Comparing the function values at the four points,

$$f(-1) = 3, \qquad f\left(\tfrac{1}{3}\right) = -\tfrac{31}{27} \approx -1.148, \qquad f\left(-\tfrac{3}{4}\right) = \tfrac{63}{16} = 3.9375, \qquad f(1) = 7,$$

we see that $\tfrac{1}{3}$ is the global minimizer, whereas 1 is the global maximizer — which occurs on the boundary of the interval. This is borne out by the graph of the function, which is displayed in Figure 6.1. $\blacktriangle$

While locating and characterizing the extremizers of scalar functions is usually relatively straightforward, the same cannot be said of functions of several variables, particularly when the dimension of their domain space is large. Interior local minimizers are easier to find and characterize, and, to keep the presentation simple, we shall focus our efforts on them. Indeed, unless otherwise indicated, we will assume that the objective function is defined on all of $\mathbb{R}^n$

and so we need not worry about boundary minimizers and maximizers. Moreover, most of our results remain valid when the domain is an open subset.

Exercises

1.1. Find all local and global extremizers on $\mathbb{R}$ of the following scalar functions:

$(a)\heartsuit\ x^3 - 2x + 1,\quad (b)\diamond\ \dfrac{x}{1+x^2},\quad (c)\heartsuit\ \dfrac{x^2 - 3x + 5}{x^2 + 1},\quad (d)\diamond\ e^{x^2 - 2x^4},\quad (e)\ \sin x + \tfrac{1}{2}\cos 2x.$

1.2. Minimize and maximize the following objective functions on the indicated domains:

$(a)\heartsuit\ x^3 - 2x^2 + x,\ -1 \le x \le 1;\quad (b)\diamond\ x^5 - 2x^3 + x - 3,\ 0 \le x \le 2;$

$(c)\heartsuit\ \dfrac{x^2 - x}{x^2 + 1},\ -3 \le x \le 3;\quad (d)\ \sin(x^2 + 1),\ 0 \le x \le 2.$

1.3. Prove that the scalar function $F(x) = \begin{cases} 2x^2 + x^2 \sin \dfrac{1}{x}, & x \ne 0, \\ 0, & x = 0, \end{cases}$ is continuous, has a

strict global minimizer at $x^\star = 0$, but there exist local minimizers arbitrarily close to $x^\star$, and hence $x^\star$ is not an isolated local minimizer.

1.4. Why can't you apply Theorem 2.35 to (6.3) and thereby prove the existence of solutions to the system $G(\mathbf{x}) = \mathbf{0}$?

1.5. $\diamond$ *True or false*: If $F(\mathbf{x}) \ge c$ is bounded from below for all $\mathbf{x} \in \mathbb{R}^n$, then F has a global minimizer.

6.2 Minimization of Quadratic Functions

The simplest algebraic equations are linear systems. As such, one must thoroughly understand them before venturing into the far more complicated nonlinear realm. For minimization problems, the starting point is a quadratic function. Nonconstant linear and affine functions do not have minimizers[1] — think of the function $F(x) = \alpha x + \beta$, whose graph is a straight line. In this section, we shall completely solve the problem of minimizing a general quadratic function of n variables using linear algebra. With this firmly in hand, the subsequent sections of this chapter and, later, Chapter 11 will develop methods for optimizing more general functions.

Let us begin by reviewing the very simplest example — minimizing a scalar quadratic polynomial

$$p(x) = \tfrac{1}{2}ax^2 + bx + c \tag{6.6}$$

over all possible values of $x \in \mathbb{R}$. If $a > 0$, then the graph of p is a parabola opening upwards, and so there exists a unique global minimizer. If $a < 0$, the parabola points downwards, and there is no minimizer, although there is a global maximizer. If $a = 0$, the graph is a straight line, and there is neither minimizer nor maximizer over all $x \in \mathbb{R}$ — except in the trivial case when $b = 0$ also, and the function $p(x) = c$ is constant, with every x qualifying as a minimizer and a maximizer. The three nontrivial possibilities are illustrated in Figure 6.2.

---

[1] Keep in mind that we are viewing the function defined on all of $\mathbb{R}^n$. Minimizing linear and affine functions over compact domains is an important and vast subject in its own right, that we do not have time to treat in this text. When the domain is a polyhedron, [52, 53] provide introductions to what is known as *linear programming*.

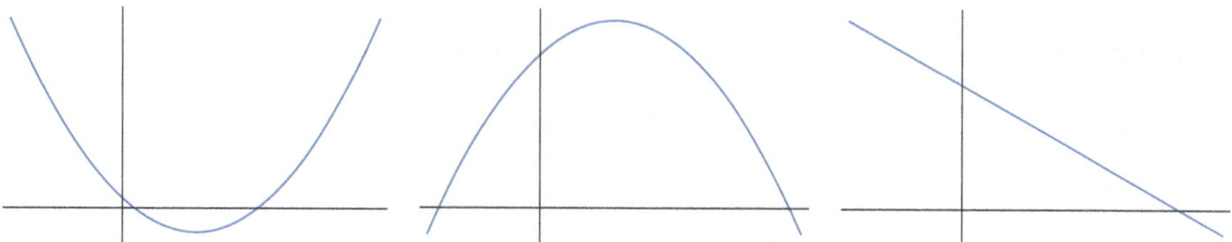

Figure 6.2: Parabolas

In the case $a > 0$, the minimizer can be found by calculus. The critical points are found by setting its derivative

$$p'(x) = a\,x + b = 0,$$

we conclude that the only possible minimum value occurs at

$$x^{\star} = -\frac{b}{a}, \qquad \text{where} \qquad p(x^{\star}) = c - \frac{b^2}{2\,a}. \tag{6.7}$$

Of course, one must check that this critical point is indeed a minimizer, and not a maximizer or inflection point. The second derivative test will show that $p''(x^{\star}) = a > 0$, and so $x^{\star}$ is at least a local minimizer.

A more instructive approach to this problem — and one that requires only elementary algebra — is to "complete the square", rewriting the polynomial in the form

$$p(x) = \frac{a}{2}\left(x + \frac{b}{a}\right)^2 + \left(c - \frac{b^2}{2\,a}\right). \tag{6.8}$$

If $a > 0$, then the first term is always ≥ 0, and, moreover, attains its minimum value 0 only at $x^{\star} = -b/a$. The second term is constant, and so is unaffected by the value of x. Thus, the *global minimizer* of $p(x)$ is at $x^{\star} = -b/a$. Moreover, its minimal value equals the constant term, $p(x^{\star}) = c - b^2/(2\,a)$, thereby reconfirming and strengthening the calculus result in (6.7). Indeed, if you go back to the proof of the calculus result, [4], it relies on the fact that one already knows how to minimize a quadratic function.

6.2.1　Unconstrained Minimization

Now that we have the one-variable case firmly in hand, let us turn our attention to the more substantial problem of minimizing quadratic functions of several variables. Thus, we seek to minimize a *quadratic polynomial*

$$P(\mathbf{x}) = P(x_1, \ldots, x_n) = \frac{1}{2}\sum_{i,j=1}^{n} h_{ij}\,x_i\,x_j - \sum_{i=1}^{n} f_i\,x_i + c, \tag{6.9}$$

depending on n variables $\mathbf{x} = (x_1, x_2, \ldots, x_n)^T \in \mathbb{R}^n$. The initial $\frac{1}{2}$ is included for later convenience. The coefficients h_{ij}, f_i, and c are all assumed to be real. (It does not make sense to talk about minimizers of complex-valued functions.) Moreover, we can assume, without loss of generality, that the coefficients of the quadratic terms are symmetric:[2] $h_{ij} = h_{ji}$. Note

---

[2]See Exercise 1.13 in Chapter 4 for a justification.

that $P(\mathbf{x})$ is more general than a quadratic form in that it also contains linear and constant terms. We seek a global minimizer, and so the variables $\mathbf{x}$ are allowed to vary over all of $\mathbb{R}^n$.

Let us begin by rewriting the *quadratic function* (6.9) in a more compact matrix notation:

$$P(\mathbf{x}) = \tfrac{1}{2}\mathbf{x}^T H \mathbf{x} - \mathbf{x}^T \mathbf{f} + c, \qquad \mathbf{x} \in \mathbb{R}^n, \tag{6.10}$$

in which $H = H^T$ is a symmetric $n \times n$ matrix with entries $h_{ij} = h_{ji}$, while $\mathbf{f} \in \mathbb{R}^n$ is a constant vector with entries f_i, and c is a constant scalar.

Example 6.6. Consider the quadratic polynomial

$$P(x_1, x_2) = 4x_1^2 - 2x_1 x_2 + 3x_2^2 + 3x_1 - 2x_2 + 1$$

depending on two real variables x_1, x_2. It can be written in the matrix form (6.10) as

$$P(x_1, x_2) = \tfrac{1}{2}(x_1 \ \ x_2)\begin{pmatrix} 8 & -2 \\ -2 & 6 \end{pmatrix}\begin{pmatrix} x_1 \\ x_2 \end{pmatrix} - (x_1 \ \ x_2)\begin{pmatrix} -3 \\ 2 \end{pmatrix} + 1, \tag{6.11}$$

whereby

$$\mathbf{x} = \begin{pmatrix} x_1 \\ x_2 \end{pmatrix}, \qquad H = \begin{pmatrix} 8 & -2 \\ -2 & 6 \end{pmatrix}, \qquad \mathbf{f} = \begin{pmatrix} -3 \\ 2 \end{pmatrix}, \qquad c = 1. \tag{6.12}$$

Pay attention to the symmetry of the coefficient matrix $H = H^T$.   ▲

We first note that in the simple scalar case (6.6), we needed to impose the condition that the quadratic coefficient a be *positive* in order to obtain a (unique) minimizer. The corresponding condition for the multivariable case is that the symmetric coefficient matrix H be *positive definite*, as formulated in Definition 4.1. This key assumption enables us to establish a general minimization criterion.

> **Theorem 6.7.** *If H is positive definite, then the quadratic function (6.10) has a unique global minimizer, which is the solution to the linear system*
>
> $$H\mathbf{x} = \mathbf{f}, \qquad namely \qquad \mathbf{x}^\star = H^{-1}\mathbf{f}. \tag{6.13}$$
>
> *The minimum value of $P(\mathbf{x})$ is equal to any of the following expressions:*
>
> $$P(\mathbf{x}^\star) = P(H^{-1}\mathbf{f}) = c - \tfrac{1}{2}\mathbf{f}^T H^{-1}\mathbf{f} = c - \tfrac{1}{2}\mathbf{f}^T\mathbf{x}^\star = c - \tfrac{1}{2}(\mathbf{x}^\star)^T H \mathbf{x}^\star. \tag{6.14}$$

Proof. First recall that positive definiteness implies that H is nonsingular — see Theorem 4.6 — and hence the linear system (6.13) has a unique solution $\mathbf{x}^\star = H^{-1}\mathbf{f}$. Since $\mathbf{f} = H\mathbf{x}^\star$, it follows that

$$\begin{aligned} P(\mathbf{x}) - P(\mathbf{x}^\star) &= \tfrac{1}{2}\mathbf{x}^T H \mathbf{x} - \mathbf{x}^T\mathbf{f} - \tfrac{1}{2}(\mathbf{x}^\star)^T H \mathbf{x}^\star + (\mathbf{x}^\star)^T\mathbf{f} \\ &= \tfrac{1}{2}\mathbf{x}^T H \mathbf{x} - \mathbf{x}^T H\mathbf{x}^\star + \tfrac{1}{2}(\mathbf{x}^\star)^T H \mathbf{x}^\star = \tfrac{1}{2}(\mathbf{x} - \mathbf{x}^\star)^T H(\mathbf{x} - \mathbf{x}^\star) \end{aligned} \tag{6.15}$$

for all $\mathbf{x} \in \mathbb{R}^n$, where we used the symmetry of $H = H^T$ to identify the scalar terms

$$\mathbf{x}^T H \mathbf{x}^\star = (\mathbf{x}^T H \mathbf{x}^\star)^T = (\mathbf{x}^\star)^T H^T \mathbf{x} = (\mathbf{x}^\star)^T H \mathbf{x}.$$

The final expression in (6.15) has the form $\tfrac{1}{2}\mathbf{y}^T H \mathbf{y}$, where $\mathbf{y} = \mathbf{x} - \mathbf{x}^\star$. Since we assumed that H is positive definite, we know that $\mathbf{y}^T H \mathbf{y} > 0$ for all $\mathbf{y} \neq \mathbf{0}$, i.e., for all $\mathbf{x} \neq \mathbf{x}^\star$. Thus, $P(\mathbf{x}) > P(\mathbf{x}^\star)$ whenever $\mathbf{x} \neq \mathbf{x}^\star$, which proves that $\mathbf{x}^\star$ is the unique global minimizer. The expressions in (6.14) for the minimum value follow from simple substitutions.   ■

Remark. In the preceding formulas and proof, we only use the inverse matrix as a convenient way to write the solution. In practice, one avoids ever computing H^{-1}, and more efficient computational techniques for solving the linear system (6.13) are employed, e.g., the QR method developed in Section 4.7.   ▲

Example 6.8. Let us minimize the quadratic function appearing in (6.11) above. According to Theorem 6.7, to find the minimizer we must solve the linear system $H\mathbf{x} = \mathbf{f}$, which, in this case, is

$$\begin{pmatrix} 8 & -2 \\ -2 & 6 \end{pmatrix} \begin{pmatrix} x_1 \\ x_2 \end{pmatrix} = \begin{pmatrix} -3 \\ 2 \end{pmatrix}.$$

One easily establishes that the coefficient matrix is positive definite, and hence $P(x_1, x_2)$ does have a unique minimizer, obtained by solving the preceding system:

$$\mathbf{x}^\star = \begin{pmatrix} x_1^\star \\ x_2^\star \end{pmatrix} = \begin{pmatrix} -\frac{7}{22} \\ \frac{5}{22} \end{pmatrix} \approx \begin{pmatrix} -.31818 \\ .22727 \end{pmatrix}.$$

The quickest way to compute the minimal value is to use the second formula in (6.14):

$$P(\mathbf{x}^\star) = P\left(-\tfrac{7}{22}, \tfrac{5}{22}\right) = 1 - \tfrac{1}{2}\,(-3, 2) \begin{pmatrix} -\frac{7}{22} \\ \frac{5}{22} \end{pmatrix} = \tfrac{13}{44} \approx .29546. \qquad ▲$$

Theorem 6.7 solves the general quadratic minimization problem when the coefficient matrix H is positive definite. Otherwise, the quadratic function (6.10) does not have a minimizer, apart from one exceptional situation.

> **Theorem 6.9.** *If the matrix H is positive definite, then the quadratic function (6.10) has a unique global minimizer $\mathbf{x}^\star$ satisfying $H\mathbf{x}^\star = \mathbf{f}$. If H is only positive semidefinite, and $\mathbf{f} \in \mathrm{img}\,H$, then every solution to the linear system $H\mathbf{x}^\star = \mathbf{f}$ is a global minimizer of $P(\mathbf{x})$, and vice-versa, but the minimizer is not unique, since $P(\mathbf{x}^\star + \mathbf{z}) = P(\mathbf{x}^\star)$ whenever $\mathbf{z} \in \ker H$. In all other cases, $P(\mathbf{x})$ has no global minimizer, and can assume arbitrarily large negative values.*

Proof. The first part is merely a restatement of Theorem 6.7. The second part is proved by a similar computation, by noting that (6.15) holds for any solution $\mathbf{x}^\star$ of $H\mathbf{x}^\star = \mathbf{f}$. Moreover, if $\mathbf{z} \in \ker H$, then

$$P(\mathbf{x}^\star + \mathbf{z}) = \tfrac{1}{2}\,(\mathbf{x}^\star + \mathbf{z})^T H\,(\mathbf{x}^\star + \mathbf{z}) - (\mathbf{x}^\star + \mathbf{z})^T H\mathbf{x}^\star + c = \tfrac{1}{2}\,\mathbf{x}^\star H\mathbf{x}^\star - \mathbf{x}^\star \mathbf{f} + c = P(\mathbf{x}^\star),$$

since $H\mathbf{z} = \mathbf{0}$ and H is a symmetric matrix.

If H is not positive semidefinite, then one can find a vector $\mathbf{y}$ such that $a = \mathbf{y}^T H\mathbf{y} < 0$. If we set $\mathbf{x} = t\mathbf{y}$, then $P(\mathbf{x}) = P(t\mathbf{y}) = \tfrac{1}{2}at^2 + bt + c$, with $b = \mathbf{y}^T \mathbf{f}$. Since $a < 0$, by choosing $|t| \gg 0$ sufficiently large, we can arrange that $P(t\mathbf{y}) \ll 0$ is arbitrarily large negative, and so P has no minimizer. The one remaining case — when H is positive semidefinite, but $\mathbf{f} \notin \mathrm{img}\,H$ — is the subject of Exercise 2.7.   ■

6.2.2   Least Squares

In Section 4.7, we encountered the method of least squares for solving, or, rather, coming as close to solving as possible, incompatible linear systems. In this section, we delve a bit deeper into this method, introducing the normal equations, and then showing how a simple regularization technique aids in their solution in ill-conditioned cases.

Let A be an $m \times n$ matrix, $\mathbf{b} \in \mathbb{R}^m$, and consider the linear system $A\mathbf{x} = \mathbf{b}$ that is to be solved for $\mathbf{x} \in \mathbb{R}^n$. Recall Definition 4.45, that states that a *least squares solution* to the system is a vector $\mathbf{x} \in \mathbb{R}^n$ that minimizes the Euclidean norm[3] $\|A\mathbf{x} - \mathbf{b}\|_2$ over all $\mathbf{x} \in \mathbb{R}^n$. All bona fide solutions are least squares solutions, since they have 0 as the minimum value. But when $\mathbf{b} \notin \operatorname{img} A$, the system is incompatible, so there is no ordinary solution, but there will always be a least squares solution.

Let us apply the minimization techniques developed in the preceding section to solve this problem. We begin by expanding the squared norm:

$$\begin{aligned}
\|A\mathbf{x} - \mathbf{b}\|_2^2 &= (A\mathbf{x} - \mathbf{b})^T(A\mathbf{x} - \mathbf{b}) \\
&= (A\mathbf{x})^T A\mathbf{x} - 2(A\mathbf{x})^T \mathbf{b} + \mathbf{b}^T \mathbf{b} = \mathbf{x}^T A^T A\mathbf{x} - 2\mathbf{x}^T A^T \mathbf{b} + \|\mathbf{b}\|_2^2.
\end{aligned} \tag{6.16}$$

The result is a quadratic function of the form (6.10), with $n \times n$ coefficient matrix $H = A^T A$, which is the Gram matrix (4.15) (with respect to the dot product) associated with the columns of A, while $\mathbf{f} = A^T \mathbf{b}$. According to Theorem 4.13, H is always positive semidefinite, and is positive definite and hence nonsingular if and only if the columns of A are linearly independent, or, equivalently, nullity $A = 0$. In particular, if $m < n$, so there are fewer equations than unknowns, the Gram matrix is inevitably singular.

Theorem 6.10. *The set of least squares solutions of $A\mathbf{x} = \mathbf{b}$ coincides with the set of solutions of the linear system*

$$A^T A\mathbf{x} = A^T \mathbf{b}. \tag{6.17}$$

Proof. Inspecting the expansion of the least squares objective $\|A\mathbf{x} - \mathbf{b}\|_2^2$ in (6.16), we see that $H = A^T A$ and $\mathbf{f} = A^T \mathbf{b}$. Theorem 4.25 tells us that $\operatorname{img}(A^T) = \operatorname{img}(A^T A)$, and hence $\mathbf{f} \in \operatorname{img} H$. The result thus follows directly from Theorem 6.9. ∎

The linear system (6.17), which consists of n equations in n unknowns, is known as the *normal equations* associated with the least squares problem. Solving the normal equations provides an alternative, and often used method for finding least squares solutions. In the positive definite case, where $A^T A > 0$, the least squares solution of $A\mathbf{x} = \mathbf{b}$, denoted $\mathbf{x}^\star$, is unique, and we can write

$$\mathbf{x}^\star = (A^T A)^{-1} A^T \mathbf{b}. \tag{6.18}$$

Note that if A is invertible — which necessitates $m = n$ so A is square — then so is A^T, and then formula (6.18) reduces to the standard solution formula $\mathbf{x}^\star = A^{-1}\mathbf{b}$. Of course, in practical situations, one would not invert $A^T A$ when solving the normal equations, and so the least squares solution formula (6.18) is primarily of theoretical interest. A direct solution of the normal equations would involve a QR factorization of the Gram matrix $A^T A$, which does not easily follow from the factorization $A = QR$ of the original matrix, although one can use the latter to slightly simplify the computation of $A^T A = R^T R$. For this reason, we advocate the QR method outlined in Theorem 4.47 as an efficient means of solving the least squares problem directly. Alternatively, one can employ a suitable iterative method, as discussed below, to determine the least squares solution.

---

[3] One can straightforwardly extend the ensuing analysis to any norm derived from an inner product, [181]. For other norms, the minimization problem is no longer quadratic, and hence its solution relies on the nonlinear minimization algorithms developed later in this chapter.

In the deficient case when nullity $A > 0$, the least squares solution is no longer uniquely specified. Here, one often singles out the particular solution that has minimal norm, as described in Theorem 4.27.

> **Theorem 6.11.** *Given a matrix A, there is a unique least squares solution $\mathbf{x}^\star \in \operatorname{img} A^T$ to the linear system $A\mathbf{x} = \mathbf{b}$ and the general least squares solution has the form $\mathbf{x} = \mathbf{x}^\star + \mathbf{z}$ where $\mathbf{z} \in \ker A$. Furthermore, $\mathbf{x}^\star$ is distinguished as the least squares solution with minimal Euclidean norm.*

Proof. The proof follows from Theorem 4.27, but is quite short and so we include the full version here. Since $\operatorname{img} A^T = (\ker A)^\perp$, any least squares solution can be written as $\mathbf{x} = \mathbf{x}^\star + \mathbf{z}$ where $\mathbf{x}^\star \in \operatorname{img}(A^T A) = \operatorname{img} A^T$ and $\mathbf{z} \in \ker(A^T A) = \ker A$. It follows that $\mathbf{x}^\star$ satisfies the normal equations (6.17); moreover, since $\ker(A^T A) \cap \operatorname{img}(A^T A) = \{\mathbf{0}\}$, the solution $\mathbf{x}^\star$ is unique. Writing $\|\mathbf{x}\|_2^2 = \|\mathbf{x}^\star\|_2^2 + \|\mathbf{z}\|_2^2$, which follows from the orthogonality of $\operatorname{coimg} A = \operatorname{img} A^T$ and $\ker A$, we deduce that the solution with minimal Euclidean norm is obtained by taking $\mathbf{z} = \mathbf{0}$. $\blacksquare$

6.2.3 Constrained Minimization

Let us next discuss the problem of minimizing the restriction of a quadratic function (6.10) to a nontrivial subspace $\{\mathbf{0}\} \neq V \subset \mathbb{R}^n$. We assume that the quadratic term coefficient matrix H is symmetric positive definite, although, as noted below, we are sometimes able to minimize even in the absence of this assumption.

Let $\mathbf{v}_1, \ldots, \mathbf{v}_p$ be a basis for V, and form the $n \times p$ matrix $A = (\mathbf{v}_1 \ \ldots \ \mathbf{v}_p)$ containing the basis vectors as its columns. The general element of V has the form

$$\mathbf{x} = y_1 \mathbf{v}_1 + \cdots + y_p \mathbf{v}_p = A\mathbf{y}, \qquad \text{where} \qquad \mathbf{y} = \left(y_1, \ldots, y_p\right)^T \in \mathbb{R}^p.$$

Thus, the restriction of $P(\mathbf{x})$ to V is also a quadratic function, taking the form

$$P(A\mathbf{y}) = \tfrac{1}{2}\mathbf{y}^T A^T H A \mathbf{y} + \mathbf{y}^T A^T \mathbf{f} + c. \tag{6.19}$$

Moreover, the $p \times p$ coefficient matrix $G = A^T H A$ of its quadratic terms has the form of a Gram matrix with respect to the inner product induced by H, cf. (4.17). Since $\operatorname{rank} A = p$, Theorem 4.13 implies that the symmetric matrix G is positive definite. Thus, we can apply our basic minimization Theorem 6.7 to (6.19) and produce a solution to the constrained minimization problem.

> **Theorem 6.12.** *Let $H \in \mathcal{M}_{n \times n}$ is symmetric positive definite, and let $A \in \mathcal{M}_{n \times k}$ have rank k. Then the restriction of the quadratic function (6.10) to the subspace $V = \operatorname{img} A$ has a unique minimizer $\mathbf{x}^\star = A\mathbf{y}^*$, where $\mathbf{y}^*$ is the solution to the linear system*
>
> $$A^T H A \mathbf{y} = A^T \mathbf{f}. \tag{6.20}$$

Thus, the minimizer $\mathbf{x}^\star = A\mathbf{y}^*$ satisfies

$$A^T(H\mathbf{x}^* - \mathbf{b}) = \mathbf{0}. \tag{6.21}$$

The entries of (6.21) are the dot products between the *residual vector* $\mathbf{r} = H\mathbf{x}^* - \mathbf{b}$ and the columns of A, i.e., the basis vectors of V. This implies the following characterization of the minimizer.

> **Corollary 6.13.** *The unique minimizer $\mathbf{x}^\star \in V$ of the quadratic function (6.10) on the subspace $V \subset \mathbb{R}^n$ can be characterized by the condition that its residual vector $\mathbf{r} = H\mathbf{x}^\star - \mathbf{b}$ be orthogonal to V.*

Remark. According to Exercise 2.6, it is not necessary that H be positive definite in order that the matrix $G = A^T H A$ be positive definite, even though it is then not a bona fide Gram matrix since H does not define an inner product. In other words, the restriction of an indefinite quadratic function to a subspace can at times produce a positive definite quadratic function, which thus has a minimizer of the same form as in Theorem 6.12. The reader may enjoy exploring the conditions that ensure that this is the case. ▲

More generally, we seek to minimize a quadratic function on a p-dimensional *affine subspace*[4]

$$W = V + \mathbf{b} = \{\, \mathbf{x} + \mathbf{b} \mid \mathbf{x} \in V \,\}, \tag{6.22}$$

where $V \subset \mathbb{R}^n$ is a p-dimensional subspace as above, and $\mathbf{b} \in \mathbb{R}^n$. For simplicity, we restrict attention to a quadratic form

$$Q(\mathbf{x}) = \tfrac{1}{2}\mathbf{x}^T H \mathbf{x}, \qquad \mathbf{x} \in \mathbb{R}^n, \tag{6.23}$$

where the coefficient matrix H is symmetric, positive definite. Extending our analysis to more general quadratic functions is straightforward, and is left for the reader to complete as Exercise 2.16.

As above, let $A = (\, \mathbf{v}_1 \ \ldots \ \mathbf{v}_p \,)$ be the $n \times p$ matrix whose columns form a basis of V. The general element of W has the form

$$y_1 \mathbf{v}_1 + \cdots + y_p \mathbf{v}_p + \mathbf{b} = A\mathbf{y} + \mathbf{b}, \qquad \text{where} \qquad \mathbf{y} = \left(y_1, \ldots, y_p\right)^T \in \mathbb{R}^p.$$

Thus, the restriction of Q to W takes the form of a quadratic function of $\mathbf{y}$:

$$R(\mathbf{y}) = Q(A\mathbf{y} + \mathbf{b}) = \tfrac{1}{2}\mathbf{y}^T A^T H A \mathbf{y} + \mathbf{y}^T A^T H \mathbf{b} + \tfrac{1}{2}\mathbf{b}^T H \mathbf{b}. \tag{6.24}$$

As noted above, $G = A^T H A$ is positive definite. Thus, we can apply our basic minimization Theorem 6.7 to $R(\mathbf{y})$ and produce a solution to the constrained minimization problem.

> **Theorem 6.14.** *Suppose $Q(\mathbf{x}) = \tfrac{1}{2}\mathbf{x}^T H \mathbf{x}$ is a positive definite quadratic function. Then its restriction to the affine subspace parametrized by $A\mathbf{y} + \mathbf{b}$ for $\mathbf{y} \in \mathbb{R}^p$, where A is an $n \times p$ matrix of rank p and $\mathbf{b} \in \mathbb{R}^n$, has a unique global minimizer $\mathbf{x}^\star = A\mathbf{y}^\star + \mathbf{b}$, where $\mathbf{y}^\star$ is the unique solution to the linear system*
>
> $$A^T H (A\mathbf{y} + \mathbf{b}) = A^T H A \mathbf{y} + A^T H \mathbf{b} = \mathbf{0}. \tag{6.25}$$

An important special case is when the affine subspace is prescribed by setting some of the variables x_i to constants. By possibly relabeling the coordinates, let us suppose that these are the last $n - p$ variables, and so the affine subspace is

$$W = \left\{\, \mathbf{x} = (x_1, x_2, \ldots, x_n)^T \;\middle|\; x_{p+1} = b_{p+1}, \ x_{p+2} = b_{p+2}, \ \ldots \ x_n = b_n \,\right\},$$

where $b_{p+1}, \ \ldots \ b_n \in \mathbb{R}$. In this case, we can write the general element of W in the form

$$\mathbf{x} = \begin{pmatrix} \mathbf{y} \\ \tilde{\mathbf{b}} \end{pmatrix} = A\mathbf{y} + \mathbf{b}, \text{ where } \mathbf{y} \in \mathbb{R}^p, \text{ and } A = \begin{pmatrix} \mathrm{I} \\ \mathrm{O} \end{pmatrix} \text{ consists of a } p \times p \text{ identity matrix on}$$

---

[4] See Exercise 2.7 for basic results on affine subspaces.

top of $n - p$ all zero rows, while $\mathbf{b} = \begin{pmatrix} \mathbf{0} \\ \widetilde{\mathbf{b}} \end{pmatrix} = \left(0, \dots, 0, b_{p+1}, \dots b_n \right)^T$. Writing the coefficient

matrix $H = \begin{pmatrix} H_1 & H_2 \\ H_2^T & H_3 \end{pmatrix}$ in block form (see Exercise 1.16), where H_1, H_2, H_3 have respective

sizes $p \times p$, $p \times (n - p)$, $(n - p) \times (n - p)$, then the linear system (6.25) takes the simple form

$$(H_1 \; H_2) \begin{pmatrix} \mathbf{y} \\ \widetilde{\mathbf{b}} \end{pmatrix} = H_1 \mathbf{y} + H_2 \widetilde{\mathbf{b}} = \mathbf{0}. \tag{6.26}$$

Whenever H_1 is positive definite, the solution $\mathbf{y}$ determines the global minimizer of the constrained minimization problem. In particular, if H itself is positive definite, this holds as a consequence of Exercise 1.16.

Exercises

2.1. For each of the following quadratic functions, determine whether there is a minimizer. If so, find the minimizer and the minimum value. (a) $\heartsuit$ $x^2 - 2xy + 4y^2 + x - 1$,
(b) $3x^2 + 3xy + 3y^2 - 2x - 2y + 4$, (c) $\heartsuit$ $x^2 + 5xy + 3y^2 + 2x - y$,
(d) $\heartsuit$ $x^2 + y^2 + yz + z^2 + x + y - z$, (e) $x^2 + xy - y^2 - yz + z^2 - 3$,
(f) $\Diamond$ $x^2 + 5xz + y^2 - 2yz + z^2 + 2x - z - 3$, (g) $x^2 + xy + y^2 + yz + z^2 + zw + w^2 - 2x - w$.

2.2. (a) For which numbers b (allowing both positive and negative numbers) is the matrix
$A = \begin{pmatrix} 1 & b \\ b & 4 \end{pmatrix}$ positive definite? (b) Find the minimum value (depending on b; it might be

finite or it might be $-\infty$) of the function $p(x, y) = x^2 + 2bxy + 4y^2 - 2y$.

2.3. For each matrix H, vector $\mathbf{f}$, and scalar c, write out the quadratic function $P(\mathbf{x})$ given by (6.10). Then either find the minimizer $\mathbf{x}^\star$ and minimum value $P(\mathbf{x}^\star)$, or explain why there is none.

(a) $\heartsuit$ $H = \begin{pmatrix} 4 & -12 \\ -12 & 45 \end{pmatrix}$, $\mathbf{f} = \begin{pmatrix} -\frac{1}{2} \\ 2 \end{pmatrix}$, $c = 3$; (b) $\heartsuit$ $H = \begin{pmatrix} 3 & 2 \\ 2 & 1 \end{pmatrix}$, $\mathbf{f} = \begin{pmatrix} 4 \\ 1 \end{pmatrix}$, $c = 0$;

(c) $\heartsuit$ $H = \begin{pmatrix} 3 & -1 & 1 \\ -1 & 2 & -1 \\ 1 & -1 & 3 \end{pmatrix}$, $\mathbf{f} = \begin{pmatrix} 0 \\ 4 \\ -4 \end{pmatrix}$, $c = 6$; (d) $\Diamond$ $H = \begin{pmatrix} 1 & 1 & 1 \\ 1 & 2 & -1 \\ 1 & -1 & 1 \end{pmatrix}$,

$\mathbf{f} = \begin{pmatrix} -3 \\ -1 \\ 2 \end{pmatrix}$, $c = 1$; (e) $H = \begin{pmatrix} 1 & 1 & 0 & 0 \\ 1 & 2 & 1 & 0 \\ 0 & 1 & 3 & 1 \\ 0 & 0 & 1 & 4 \end{pmatrix}$, $\mathbf{f} = \begin{pmatrix} -1 \\ 2 \\ -3 \\ 4 \end{pmatrix}$, $c = 0$.

2.4. Find the minimum value of the quadratic function

$$p(x_1, \dots, x_n) = 4 \sum_{i=1}^{n} x_i^2 - 2 \sum_{i=1}^{n-1} x_i x_{i+1} + \sum_{i=1}^{n} x_i \qquad \text{for} \qquad n = 2, 3, 4.$$

2.5. $\Diamond$ Let $H > 0$. Prove that a quadratic function $P(\mathbf{x}) = \frac{1}{2} \mathbf{x}^T H \mathbf{x} - \mathbf{x}^T \mathbf{f}$ without constant term has nonpositive minimum value: $P(\mathbf{x}^\star) \leq 0$. When is the minimum value zero?

2.6. $\heartsuit$ Show that the quadratic function $P(x, y) = x^2 + y$ has a positive semidefinite coefficient matrix, but no minimum.

2.7. $\diamond$ Prove that if H is a positive semidefinite matrix, and $\mathbf{f} \notin \operatorname{img} H$, then the quadratic function (6.10) has no minimum value. *Hint*: Look at what happens when $\mathbf{x} \in \ker H$.

2.8. Suppose H_1 and H_2 are symmetric, positive definite $n \times n$ matrices. Suppose that, for $i = 1, 2$, the minimizer of $P_i(\mathbf{x}) = \frac{1}{2}\mathbf{x}^T H_i \mathbf{x} - \mathbf{x}^T \mathbf{f}_i + c_i$, is $\mathbf{x}_i^\star$. Is the minimizer of $P(\mathbf{x}) = P_1(\mathbf{x}) + P_2(\mathbf{x})$ given by $\mathbf{x}^\star = \mathbf{x}_1^\star + \mathbf{x}_2^\star$? Prove or give a counterexample.

2.9. $\heartsuit$ Under what conditions does a quadratic function (6.10) have a finite global maximum? Explain how to find the maximizer and maximum value.

2.10. Find the maximum value of the quadratic functions
$$(a)\,\heartsuit\; -x^2 + 3xy - 5y^2 - x + 1, \quad (b)\; -2x^2 + 6xy - 3y^2 + 4x - 3y.$$

2.11. Use the normal equations to find the least squares solution to the linear systems in Exercise 7.4 of Chapter 4.

2.12. $\diamond$ Show that when AA^T is nonsingular, the least squares solution of $A\mathbf{x} = \mathbf{b}$ of minimum norm can be expressed as $\mathbf{x}^\star = A^T(AA^T)^{-1}\mathbf{b}$.

2.13. Find the minimizer and minimum value of the following quadratic functions when subject to the indicated constraint. $(a)\,\heartsuit\; x^2 - 2xy + 6y^2, \quad x + y = 1,$
$(b)\,\diamond\; x^2 + y^2 + 2yz + 4z^2, \quad x + 2y - z = 3, \quad (c)\; x^2 + xy - y^2 - yz + z^2, \quad x - y - z = 1.$

2.14. Let $P(x, y) = xy$. Show that P does not have a minimum on $\mathbb{R}^2$. However, the constrained minimization problem obtained by restricting $P(x, y)$ to the line $y = x$ does have a minimum. For which lines $y = ax + b$ does the restriction of $P(x, y)$ have a minimum? maximum? both? neither?

2.15. $\heartsuit$ Let H be a symmetric matrix. Suppose V is a subspace spanned by one or more eigenvectors of H having positive eigenvalues. Show that the restriction of the quadratic function (6.10) to V has a unique global minimum. Write down the linear system the minimum must satisfy.

2.16. Let $P(\mathbf{x})$ be a quadratic function as in (6.10) with symmetric positive definite coefficient matrix H. Let $W = V + \mathbf{b}$ be an affine subspace, as in (6.22). Explain when P has a unique minimizer when restricted to W, and show how to find it.

6.3 The Gradient and Critical Points

To study and compute the minimizers and maximizers of non-quadratic functions will require us to review some basic multivariable calculus. For details see, for instance, [4, 158].

6.3.1 The Gradient

As you learn in multivariable calculus, the (interior) minimizers and maximizers of a real-valued function $F(\mathbf{x}) = F(x_1, \ldots, x_n)$ are necessarily *critical points*, meaning places where its gradient vanishes. The *standard gradient*, also known as the *Euclidean gradient*, is the vector field whose entries are its first order partial derivatives:

$$\nabla F(\mathbf{x}) = \left(\frac{\partial F}{\partial x_1}, \; \cdots, \; \frac{\partial F}{\partial x_n} \right)^T. \tag{6.27}$$

A function $F(\mathbf{x})$ is said to be *continuously differentiable* if its gradient $\nabla F(\mathbf{x})$ is a continuously varying vector-valued function of $\mathbf{x}$. This is equivalent to the requirement that its first order partial derivatives $\partial F/\partial x_i$ are all continuous. From here on, all objective functions are assumed to be continuously differentiable on their domain of definition.

Let us reformulate the definition of the gradient in a more intrinsic manner.

Lemma 6.15. *The gradient of a real-valued function $F\colon \mathbb{R}^n \to \mathbb{R}$ at a point $\mathbf{x} \in \mathbb{R}^n$ is the vector $\nabla F(\mathbf{x}) \in \mathbb{R}^n$ that satisfies*

$$\nabla F(\mathbf{x}) \cdot \mathbf{y} = \frac{d}{dt} F(\mathbf{x} + t\,\mathbf{y}) \bigg|_{t=0} \qquad \text{for all} \qquad \mathbf{y} \in \mathbb{R}^n. \tag{6.28}$$

Remark. The quantity displayed in formula (6.28) is known as the *directional derivative* of F with respect to $\mathbf{y} \in V$, and typically denoted by $\partial F/\partial \mathbf{y}$. It measures the rate of change of F in the direction of the vector $\mathbf{y}$, scaled in proportion to its length, and equals the dot product between the gradient of the function and the direction vector $\mathbf{y}$. ▲

Proof of Lemma 6.15. We use the chain rule to compute

$$\frac{d}{dt} F(\mathbf{x} + t\,\mathbf{y}) = \frac{\partial F}{\partial x_1}(\mathbf{x} + t\,\mathbf{y})\, y_1 \; + \; \cdots \; + \; \frac{\partial F}{\partial x_n}(\mathbf{x} + t\,\mathbf{y})\, y_n. \tag{6.29}$$

Setting $t = 0$, the right hand side reduces to

$$\frac{d}{dt} F(\mathbf{x} + t\,\mathbf{y}) \bigg|_{t=0} = \frac{\partial F}{\partial x_1}(\mathbf{x})\, y_1 \; + \; \cdots \; + \; \frac{\partial F}{\partial x_n}(\mathbf{x})\, y_n = \nabla F(\mathbf{x}) \cdot \mathbf{y}.$$

Conversely, any $\mathbf{v} \in \mathbb{R}^n$ that satisfies (6.28) in place of $\nabla F(\mathbf{x})$ would necessarily satisfy $\mathbf{v} \cdot \mathbf{y} = \nabla F(\mathbf{x}) \cdot \mathbf{y}$ for all $\mathbf{y} \in \mathbb{R}^n$, and so $\mathbf{v} = \nabla F(\mathbf{x})$. ■

Example 6.16. Consider the quadratic function

$$P(\mathbf{x}) = \tfrac{1}{2}\,\mathbf{x}^T H \mathbf{x} - \mathbf{x}^T \mathbf{f} + c, \qquad \mathbf{x} \in \mathbb{R}^n, \tag{6.30}$$

that we analyzed in Section 6.2. To determine its gradient, we compute

$$\frac{d}{dt} P(\mathbf{x} + t\,\mathbf{y}) = \frac{d}{dt}\left[P(\mathbf{x}) + t\,\mathbf{y}^T (H\mathbf{x} - \mathbf{f}) + \tfrac{1}{2} t^2\, \mathbf{y}^T H \mathbf{y} \right] = \mathbf{y}^T (H\mathbf{x} - \mathbf{f}) + t\,\mathbf{y}^T H \mathbf{y}.$$

Setting $t = 0$, we find

$$\frac{d}{dt} P(\mathbf{x} + t\,\mathbf{y}) \bigg|_{t=0} = \mathbf{y}^T (H\mathbf{x} - \mathbf{f}) = (H\mathbf{x} - \mathbf{f}) \cdot \mathbf{y},$$

and hence

$$\nabla P(\mathbf{x}) = H\mathbf{x} - \mathbf{f} \tag{6.31}$$

determines its gradient. ▲

If $\mathbf{x}(t)$ represents a parametrized curve contained within the domain of definition of $F(\mathbf{x})$, then a similar chain rule computation shows that the instantaneous rate of change in the scalar quantity F as we move along the curve is given by

$$\frac{d}{dt} F(\mathbf{x}(t)) = \nabla F(\mathbf{x}(t)) \cdot \frac{d\mathbf{x}}{dt}, \tag{6.32}$$

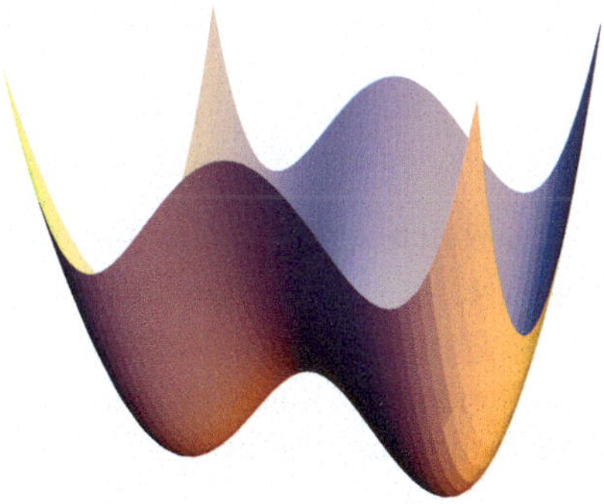

Figure 6.3: The Function $x^4 - 2x^2 + y^2$

which is the directional derivative of F with respect to the velocity or tangent vector $\mathbf{v} = d\mathbf{x}/dt$ to the curve. For instance, suppose $F(x_1, x_2)$ represents the elevation of a mountain range at position $\mathbf{x} = (x_1, x_2)^T$. If we travel through the mountains along the path whose horizontal coordinates are $\mathbf{x}(t) = (x_1(t), x_2(t))^T$, then our instantaneous rate of ascent or descent (6.32) is equal to the dot product of our velocity vector $d\mathbf{x}/dt$ with the gradient of the elevation function. This observation leads to an important interpretation of the gradient vector.

Theorem 6.17. *The gradient $\nabla F(\mathbf{x})$ of a scalar function $F(\mathbf{x})$ points in the direction of its steepest increase at the point $\mathbf{x}$. The negative gradient, $-\nabla F(\mathbf{x})$, which points in the opposite direction, indicates the direction of steepest decrease.*

Thus, when F represents elevation, ∇F tells us the direction that is steepest uphill, while $-\nabla F$ points directly downhill — the direction water will flow. Similarly, if F represents the temperature of a solid body, then ∇F tells us the direction in which it is heating up the quickest. Heat energy (like water) will flow in the opposite, coldest direction, namely that of the negative gradient vector $-\nabla F$.

But you need to be careful in how you interpret Proposition 6.17. Clearly, the faster you move along a curve, the faster the function $F(\mathbf{x})$ will vary, and one needs to take this into account when comparing the rates of change along different curves. The easiest way to effect the comparison is to assume that the tangent vector $\mathbf{u} = d\mathbf{x}/dt$ has unit Euclidean norm, so $\|\mathbf{u}\|_2 = 1$, which means that we are passing through the point $\mathbf{x}(t)$ with unit speed. Once this is done, Proposition 6.17 is an immediate consequence of the Cauchy–Schwarz inequality (2.27). Indeed,

$$\left| \frac{\partial F}{\partial \mathbf{u}} \right| = |\nabla F \cdot \mathbf{u}| \le \|\nabla F\|_2 \|\mathbf{u}\|_2 = \|\nabla F\|_2, \qquad \text{when} \qquad \|\mathbf{u}\|_2 = 1,$$

with equality if and only if $\mathbf{u}$ points in the same direction as the gradient. Therefore, assuming $\nabla F \ne \mathbf{0}$, the maximum rate of change is when $\mathbf{u} = \nabla F / \|\nabla F\|_2$ is the unit vector in the gradient direction, while the minimum is achieved when $\mathbf{u} = -\nabla F / \|\nabla F\|_2$ points in the opposite direction.

Example 6.18. Consider the function

$$F(x, y) = x^4 - 2x^2 + y^2,$$

which is defined and continuously differentiable on all of $\mathbb{R}^2$; see Figure 6.3. Its gradient is readily computed:

$$\nabla F(x, y) = \begin{pmatrix} \partial F/\partial x \\ \partial F/\partial y \end{pmatrix} = \begin{pmatrix} 4x^3 - 4x \\ 2y \end{pmatrix}.$$

For instance, $\nabla F(1, 1) = (-4, 2)^T$, which specifies the direction of steepest increase in F at the point $(1, 1)$, while its negative, $(4, -2)^T$ is the direction of steepest decrease.   ▲

An important but subtle point is that the gradient vector (6.27) is based upon on the Euclidean dot product on $\mathbb{R}^n$; changing the inner product will change the formula for the gradient. Lemma 6.15 suggests the following definition.

Definition 6.19. The gradient of $F\colon \mathbb{R}^n \to \mathbb{R}$ at the point $\mathbf{x} \in \mathbb{R}^n$ with respect to the inner product $\langle \cdot, \cdot \rangle$, denoted again by $\nabla F(\mathbf{x})$, is defined by the identity

$$\langle \nabla F(\mathbf{x}), \mathbf{y} \rangle = \frac{d}{dt} F(\mathbf{x} + t\,\mathbf{y}) \bigg|_{t=0} \qquad \text{for all} \qquad \mathbf{y} \in \mathbb{R}^n. \tag{6.33}$$

For the remainder of this chapter, we will fix an inner product $\langle \cdot, \cdot \rangle$ on $\mathbb{R}^n$, and write ∇F for the gradient of F with respect to this inner product, as defined in Definition 6.19. When the specific choice of inner product is important, we will write $\nabla_C F$, where

$$\langle \mathbf{v}, \mathbf{w} \rangle = \langle \mathbf{v}, \mathbf{w} \rangle_C = \mathbf{v}^T C \mathbf{w}, \qquad \mathbf{v}, \mathbf{w} \in \mathbb{R}^n, \tag{6.34}$$

with C a symmetric positive definite matrix. We will, when required, denote the standard or Euclidean gradient (6.27), obtained by choosing the dot product, so $C = I$, by $\nabla_2 F$ rather than $\nabla_I F$.

Let us next derive a formula for the gradient with respect to the alternative inner product (6.34). Our computation in (6.29) becomes

$$\frac{d}{dt} F(\mathbf{x} + t\,\mathbf{y}) \bigg|_{t=0} = \nabla_2 F(\mathbf{x})^T \mathbf{y} = \nabla F(\mathbf{x})^T C^{-1} C \mathbf{y}$$

$$= \big[C^{-1} \nabla_2 F(\mathbf{x}) \big]^T C \mathbf{y} = \langle C^{-1} \nabla_2 F(\mathbf{x}), \mathbf{y} \rangle_C, \tag{6.35}$$

where we use the fact that C is symmetric. Therefore, by Definition 6.19 the gradient $\nabla_C F$ satisfies

$$\nabla_C F(\mathbf{x}) = C^{-1} \nabla_2 F(\mathbf{x}). \tag{6.36}$$

Thus, every alternative gradient is obtained by multiplication of the Euclidean gradient by a positive definite matrix — in this case C^{-1}. Changing the inner product in the definition of the gradient is useful for *preconditioning* in optimization, where the matrix C^{-1} is known as the *preconditioner*. Preconditioning in the context of gradient descent will be discussed later in this chapter.

Example 6.20. Let us return to Example 6.16 to note that the gradient of the quadratic function (6.30) with respect to the inner product (6.34) is given by

$$\nabla_C P(\mathbf{x}) = C^{-1} (H\mathbf{x} - \mathbf{f}).$$

An important special case is when $H = C$ and $\mathbf{f} = C\mathbf{b}$, so that

$$P(\mathbf{x}) = \tfrac{1}{2}\mathbf{x}^T C \mathbf{x} - \mathbf{x}^T C \mathbf{b} + c = \tfrac{1}{2} \|\mathbf{x}\|_C^2 - \langle \mathbf{x}, \mathbf{b} \rangle_C + c.$$

In this case $\nabla_C P(\mathbf{x}) = \mathbf{x} - \mathbf{b}$, in direct analogy with the case of the Euclidean gradient.   ▲

6.3.2 Critical Points

The only points at which the gradient ∇F fails to indicate directions of increase/decrease of the objective function are where it vanishes. Such points play a critical role in the analysis of extrema, whence the following definition.

> **Definition 6.21.** A point $\mathbf{x}^\star \in \mathbb{R}^n$ is called a *critical point* of the objective function $F\colon \mathbb{R}^n \to \mathbb{R}$ if
> $$\nabla F(\mathbf{x}^\star) = \mathbf{0}. \tag{6.37}$$

Remark. Although the gradient depends on the underlying inner product, in view of equation (6.36), the condition (6.37) that $\mathbf{x}^\star$ be a critical point does not. ▲

Let us prove that all local minimizers are indeed critical points.

> **Theorem 6.22.** *Every local minimizer* $\mathbf{x}^\star$ *of a continuously differentiable function* $F(\mathbf{x})$ *is a critical point*: $\nabla F(\mathbf{x}^\star) = \mathbf{0}$.

Proof. Let $\mathbf{0} \neq \mathbf{y} \in \mathbb{R}^n$ be any vector. Consider the scalar function

$$g(t) = F(\mathbf{x}^\star + t\,\mathbf{y}),$$

which measures the values of F along the straight line passing through $\mathbf{x}^\star$ in the direction prescribed by $\mathbf{y}$. Since $\mathbf{x}^\star$ is a local minimizer,

$$F(\mathbf{x}^\star) \le F(\mathbf{x}^\star + t\,\mathbf{y}), \qquad \text{and hence} \qquad g(0) \le g(t)$$

for all t sufficiently close to zero. In other words, $g(t)$, as a function of the single variable t, has a local minimum at $t = 0$, and hence $g'(0) = 0$. Therefore, by the Definition 6.19 of the gradient,

$$0 = g'(0) = \frac{d}{dt} F(\mathbf{x}^\star + t\,\mathbf{y}) \bigg|_{t=0} = \langle\, \nabla F(\mathbf{x}^\star), \mathbf{y} \,\rangle.$$

We conclude that the gradient vector $\nabla F(\mathbf{x}^\star)$ at the critical point must be orthogonal to *every* vector $\mathbf{y} \in \mathbb{R}^n$, which is only possible if $\nabla F(\mathbf{x}^\star) = \mathbf{0}$. ∎

Thus, provided the objective function is continuously differentiable, every minimizer, both local and global, is necessarily a critical point. The converse is not true — critical points can also be saddle points, or have other degeneracies. By the same reasoning, every maximizer is also a critical point. An evident analytical method for determining the (interior) minimizers of a given function is to first find all its critical points by solving the system of equations (6.37). Each critical point then needs to be examined more closely — as it could be either a (local) minimizer, maximizer, or neither. These can often be distinguished through the second derivative test; see below. Later in this chapter we will develop better methods for effecting this task.

Example 6.23. Consider the function $F(x, y) = x^4 - 2x^2 + y^2$ introduced in Example 6.18. In view of the formula for its gradient, its critical points are obtained by solving the pair of equations

$$4x^3 - 4x = 0, \qquad 2y = 0.$$

The solutions to the first equation are $x = 0, \pm 1$, while the second equation requires $y = 0$. Therefore, F has three critical points:

$$\mathbf{x}_1^\star = \begin{pmatrix} 0 \\ 0 \end{pmatrix}, \qquad \mathbf{x}_2^\star = \begin{pmatrix} 1 \\ 0 \end{pmatrix}, \qquad \mathbf{x}_3^\star = \begin{pmatrix} -1 \\ 0 \end{pmatrix}. \tag{6.38}$$

Inspecting its graph in Figure 6.3, we suspect that the first critical point $\mathbf{x}_1^\star$ is a saddle point, whereas the other two appear to be local minimizers, having the same value $F(\mathbf{x}_2^\star) = F(\mathbf{x}_3^\star) = -1$. This will be confirmed once we learn how to analytically distinguish critical points. ▲

Example 6.24. For the quadratic function (6.30) whose gradient is given in (6.31), the critical points are at solutions to the linear system $H\mathbf{x} = \mathbf{f}$. If H is positive definite, as we know, there is a unique critical point $\mathbf{x}^\star = H^{-1}\mathbf{f}$ which is a global minimizer. If H is negative definite, the point $\mathbf{x}^\star$ is a global maximizer. For H indefinite and nonsingular, $\mathbf{x}^\star$ is a saddle point. Finally if H is singular, then there are either no critical points, when $\mathbf{f} \notin \operatorname{img} H$, or an entire affine subspace of critical points when $\mathbf{f} \in \operatorname{img} H$, namely $\ker H + \mathbf{b} = \{\, \mathbf{z} + \mathbf{b} \mid \mathbf{z} \in \ker H \,\}$, where $\mathbf{x} = \mathbf{b}$ is any particular solution to $H\mathbf{x} = \mathbf{f}$. In particular, if H is positive semidefinite and $\mathbf{f} \in \operatorname{img} H$, the affine subspace contains all global minimizers of $P(\mathbf{x})$. ▲

The reader should also pay attention to the distinction between local minimizers and global minimizers. In the absence of theoretical justification, one practical method for determining whether or not a minimizer is global is to find all the different local minimizers, including those on the boundary, and see which one gives the smallest value. If the domain is unbounded, one must also worry about the asymptotic behavior of the objective function as $\|\mathbf{x}\| \to \infty$.

Exercises

3.1. Find the standard gradient, where it exists, of the following functions.:
   $(a)\,\heartsuit\ x_1 x_2^2,$ $(b)\,\diamondsuit\ \log(x_1^2 + x_2^2),$ $(c)\,\heartsuit\ e^{x_1 - 2x_2},$ $(d)\ \tan^{-1}(x_1/x_2).$

3.2. Repeat Exercise 3.1 using the inner products
   $(i)\ \langle \mathbf{x}, \mathbf{y} \rangle = 3x_1 y_1 + 2x_2 y_2;$ $(ii)\ \langle \mathbf{x}, \mathbf{y} \rangle = x_1 y_1 - x_1 y_2 - x_2 y_1 + 4x_2 y_2.$

3.3. Find the critical points of the following objective functions:
   $(a)\,\heartsuit\ x^4 + y^4 - 4xy,$ $(b)\,\heartsuit\ xy(1 - x - y),$ $(c)\,\diamondsuit\ xy\,e^{-2x^2 - 2y^2},$ $(d)\ (x - y)\cos y.$

3.4. Find the critical points of the following objective functions:
   $(a)\,\diamondsuit\ x^2 + y^2 + \tfrac{1}{3}z^3 - yz - x,$ $(b)\ \cos(x^2 + y^2 + z^2),$ $(c)\,\diamondsuit\ y/z - x/y.$

3.5. $\diamondsuit$ Show that when the function $F(x, y) = 2x^4 - 4x^2 y + y^2$ is restricted to any line through the origin, so $(x, y) = t\,(a, b)$, its minimizer occurs when $x = y = 0$. Show that, however, $\mathbf{0}$ is *not* a local minimizer for $F(x, y)$. Thus, one cannot conclusively detect minimizers by looking along lines.

3.6. Let $y = f(x)$ and $z = g(y)$ be continuously differentiable scalar functions, and let $h(x) = g \circ f(x)$ denote their composition. *True or false:*
   $(a)\,\heartsuit$ A critical point of $f(x)$ is a critical point of $h(x)$.
   $(b)\,\diamondsuit$ A local minimizer of $f(x)$ is a local minimizer of $h(x)$.
   $(c)\,\heartsuit$ A critical point of $h(x)$ is a critical point of $f(x)$.
   (d) A local minimizer of $h(x)$ is a local minimizer of $f(x)$.

3.7. Suppose that the continuously differentiable scalar function $f(x)$ has only one critical point $x_\star$. (a) Show that if $x_\star$ is a local extremizer, then it is a global extremizer for f. (b) Is $x_\star$ necessarily an extremizer?

6.4 Gradient Descent

Python Notebook: Gradient descent (.ipynb)

Let us now turn our attention to multi-dimensional non-quadratic optimization problems. Our overall goal is, given a real-valued objective function $F \colon \mathbb{R}^n \to \mathbb{R}$, whereby $F(\mathbf{x}) = F(x_1, \ldots, x_n)$ depends on n variables, to find, or at least closely approximate its minimum, if such exists. We usually seek the global minimum, but local minima are easier to find, and, in the appropriate circumstances, sufficient for many applications. For simplicity, we assume throughout that the domain of F is all of $\mathbb{R}^n$, although much of what we say applies when F is only defined on an open subset thereof. As noted above, we will not investigate boundary minima here. As we discussed in the preceding section, we work in this chapter with a general inner product $\langle \, \cdot \, , \cdot \, \rangle$ and the induced norm $\| \cdot \|$ and gradient ∇F.

According to Theorem 6.17, which can be easily extended to a general inner product, at any point $\mathbf{x} \in \mathbb{R}^n$, the negative gradient vector $-\nabla F(\mathbf{x})$, if nonzero, points in the direction of steepest decrease in F. (Bear in mind that, since the gradient depends on the choice of inner product, so will the direction of steepest decrease.) Thus, to minimize F, an evident strategy is to "walk downhill", and, to be efficient, walk downhill as rapidly as possible, namely in the direction $-\nabla F(\mathbf{x})$. After walking in this direction for a little while, we recompute the gradient, and this tells us the new direction to head downhill. With luck, we will eventually end up at the bottom of the valley, i.e., at a (local) minimizer of the objective function. And the nearer we are when we start off, the faster we anticipate converging.

This simple idea forms the basis of the method of *gradient descent* for minimizing the objective function $F(\mathbf{x})$. In a numerical implementation, we start the iterative procedure with an initial guess $\mathbf{x}_0$, and let $\mathbf{x}_k$ denote the k-th approximation to the minimizer $\mathbf{x}^\star$. To compute the next approximation, we set out from $\mathbf{x}_k$ in the direction of the negative gradient there, and set

$$\mathbf{x}_{k+1} = \mathbf{x}_k - \alpha_k \, \nabla F(\mathbf{x}_k). \tag{6.39}$$

for some positive scalar $\alpha_k > 0$, that indicates how far we travel in the negative gradient direction. The scalar α_k is known as the k-th *time step* (thinking of the iterations being computed at successive "times"), or *learning rate* when using gradient descent to train machine learning algorithms. We are free to adjust α_k so as to optimize our descent path, and this is the key to the success of the method. Often, one chooses a uniform time step $\alpha_k = \alpha$, or only varies it occasionally during the course of the computations.

If $\nabla F(\mathbf{x}_k) \neq \mathbf{0}$, then, at least when $\alpha_k > 0$ is sufficiently small,

$$F(\mathbf{x}_{k+1}) < F(\mathbf{x}_k), \tag{6.40}$$

and so $\mathbf{x}_{k+1}$ is, presumably, a better approximation to the desired minimizer. Clearly, we cannot choose α_k too large or we run the risk of overshooting the minimizer and reversing the inequality (6.40). Think of walking downhill in the Swiss Alps. If you are not paying attention and walk too far in a straight line, which is what happens as α_k increases, then you might very well miss the valley and end up higher than you began — not a good strategy for descending to the bottom! On the other hand, if we choose α_k too small, taking very tiny steps, then the method may end up converging to the minimizer much too slowly to be of practical use.

How should we choose an optimal value for the time step α_k? Keep in mind that the goal is to minimize $F(\mathbf{x})$. Thus, a good strategy would be to set α_k equal to the value of $t > 0$

that minimizes the scalar objective function

$$g(t) = F\big(\mathbf{x}_k - t\,\nabla F(\mathbf{x}_k)\big) \tag{6.41}$$

obtained by restricting $F(\mathbf{x})$ to the ray emanating from $\mathbf{x}_k$ that lies in the negative gradient direction. Physically, this corresponds to setting off in a straight line in the direction of steepest decrease, and continuing on until we cannot go down any further. Barring luck, we will not have reached the actual bottom of the valley, but must then readjust our direction and continue on down the hill in a series of straight line paths. In practice, one can rarely compute the minimizing value $t^\star$ of (6.41) exactly, although it is not hard to approximate; see Exercise 10.8.

It is important to point out that gradient descent (6.39) depends on our choice of inner product on $\mathbb{R}^n$, since the notion of gradient ∇F depends on this choice. The most common choice is the dot product, which yields *Euclidean gradient descent*

$$\mathbf{x}_{k+1} = \mathbf{x}_k - \alpha_k\,\nabla_2 F(\mathbf{x}_k), \tag{6.42}$$

where we recall $\nabla_2 F$ is the usual gradient (6.27).

If we instead choose an inner product $\langle \mathbf{x}, \mathbf{y}\rangle = \mathbf{x}^T C_k\,\mathbf{y}$ defined by a positive definite matrix C_k, which is allowed to change with each iteration, then, courtesy of (6.36), the gradient descent iteration becomes

$$\mathbf{x}_{k+1} = \mathbf{x}_k - \alpha_k\,C_k^{-1}\nabla_2 F(\mathbf{x}_k). \tag{6.43}$$

This is referred to as *preconditioned gradient descent*, and the matrix C_k is called the *preconditioner*. This illustrates the advantage of working with general inner products on $\mathbb{R}^n$; our analysis applies equally well to standard gradient descent as well as preconditioned gradient descent. As we shall see in the examples and results that follow, a good choice of preconditioner C_k can substantially accelerate convergence.

Example 6.25. Consider the quadratic objective function

$$P(\mathbf{x}) = \tfrac{1}{2}\mathbf{x}^T H\mathbf{x} - \mathbf{x}^T \mathbf{b} + c, \qquad \mathbf{x} \in \mathbb{R}^n, \tag{6.44}$$

that we analyzed in Section 6.2. We assume $H > 0$ is positive definite, and hence there is a unique global minimum, at $\mathbf{x}^\star = H^{-1}\mathbf{b}$. At first, let us choose the dot product as our inner product. According to (6.31), the Euclidean gradient of P is

$$\nabla_2 P(\mathbf{x}) = H\mathbf{x} - \mathbf{b}, \tag{6.45}$$

which vanishes at $\mathbf{x}^\star$, as it must. Gradient descent (6.42) with a constant time step α thus takes the form

$$\mathbf{x}_{k+1} = \mathbf{x}_k - \alpha\,(H\mathbf{x}_k - \mathbf{b}) = (\mathrm{I} - \alpha H)\mathbf{x}_k + \alpha\mathbf{b}. \tag{6.46}$$

Convergence of the iterates to the minimizer $\mathbf{x}^\star$ can be determined by subtracting $\mathbf{x}^\star$ from both sides of (6.46), and using the fact that $\mathbf{b} = H\mathbf{x}^\star$, whence

$$\mathbf{x}_{k+1} - \mathbf{x}^\star = (\mathrm{I} - \alpha H)\,(\mathbf{x}_k - \mathbf{x}^\star).$$

Taking the Euclidean norm on both sides we have

$$\|\mathbf{x}_{k+1} - \mathbf{x}^\star\|_2 = \|(\mathrm{I} - \alpha H)\,(\mathbf{x}_k - \mathbf{x}^\star)\|_2 \le \|\mathrm{I} - \alpha H\|_2 \,\|\mathbf{x}_k - \mathbf{x}^\star\|_2, \tag{6.47}$$

where the first term in the final expression is the Euclidean matrix norm of $\mathrm{I} - \alpha H$, which is determined by Theorem 5.79.

Let us now make the restriction $0 < \alpha \leq 1/\lambda_{max}(H)$, which ensures that $I - \alpha H$ is positive semidefinite and so Theorem 5.45 implies that

$$\| I - \alpha H \|_2 = \lambda_{max}(I - \alpha H) = 1 - \alpha \lambda_{min}(H) \in (0, 1).$$

This ensures that the next iterate $\mathbf{x}_{k+1}$ will be closer to $\mathbf{x}^\star$ by the factor $1 - \alpha \lambda_{min}(H)$, which is less than 1. Inserting this into (6.47) and iterating k times yields

$$\| \mathbf{x}_k - \mathbf{x}^\star \|_2 \leq \left(1 - \alpha \lambda_{min}(H) \right)^k \| \mathbf{x}_0 - \mathbf{x}^\star \|_2.$$

Taking the largest allowable time step $\alpha = 1/\lambda_{max}(H)$ yields

$$\| \mathbf{x}_k - \mathbf{x}^\star \|_2 \leq \left(1 - \kappa(H)^{-1} \right)^k \| \mathbf{x}_0 - \mathbf{x}^\star \|_2, \tag{6.48}$$

where $\kappa(H) = \lambda_{max}(H)/\lambda_{min}(H)$ is the *condition number* of the positive definite matrix H; see Definition 5.82. Since $0 \leq 1 - \kappa(H)^{-1} < 1$, this proves that the iterates $\mathbf{x}_k$ converge to $\mathbf{x}^\star$ as $k \to \infty$

A convergence inequality of the form (6.48) is known as *linear convergence*, because the error $\| \mathbf{x}_k - \mathbf{x}^\star \|$ decreases at a constant rate $\beta = 1 - \kappa(H)^{-1}$ at each iteration. Notice how the rate of convergence depends on the condition number $\kappa(H)$ of the matrix H. Clearly, the smaller $\kappa(H)$ is, the smaller β is and the faster the convergence rate. A matrix that is well-conditioned, meaning that $\kappa(H)$ is close to one, exhibits faster convergence than an ill-conditioned matrix, where $\kappa(H)$ is very large. If $\kappa(H) = 1$, which requires $H = cI$ and so $\alpha = 1/c$, then $\beta = 0$, and convergence is immediate after only one step. We also mention that we can combine the elementary inequality[5] $1 - x \leq e^{-x} = \exp(-x)$ with (6.48) to deduce that

$$\| \mathbf{x}_k - \mathbf{x}^\star \|_2 \leq \| \mathbf{x}_0 - \mathbf{x}^\star \|_2 \exp\left(-\kappa(H)^{-1} k \right). \tag{6.49}$$

Thus, linear convergence corresponds to the error decreasing at an *exponential* rate of $\kappa(H)^{-1}$.

We now consider the role of preconditioning by changing the inner product to $\langle \mathbf{x}, \mathbf{y} \rangle = \mathbf{x}^T C \mathbf{y}$, where C is symmetric, positive definite. The *preconditioned gradient descent* iterations analogous to (6.46) are

$$\mathbf{x}_{k+1} = \mathbf{x}_k - \alpha_k C^{-1}(H \mathbf{x}_k - \mathbf{b}). \tag{6.50}$$

The iteration matrix $C^{-1}H$ is not symmetric, but it is self-adjoint for the chosen inner product; see Proposition 4.19. Thus, we can repeat the preceding analysis verbatim to find that, for $\alpha = 1/\lambda_{max}(C^{-1}H)$,

$$\| \mathbf{x}_k - \mathbf{x}^\star \|_C \leq \left(1 - \kappa^{-1} \right)^k \| \mathbf{x}_0 - \mathbf{x}^\star \|_C \leq e^{-k/\kappa} \| \mathbf{x}_0 - \mathbf{x}^\star \|_C, \tag{6.51}$$

where

$$\kappa = \frac{\lambda_{max}(C^{-1}H)}{\lambda_{min}(C^{-1}H)}$$

is the "preconditioned condition number", i.e., the condition number of $C^{-1}H$ with respect to the inner product defined by C, as formulated in Definition 5.82.

We conclude that the convergence rate for preconditioned gradient descent, at least for quadratic functions, depends on the spectrum of the preconditioned matrix $C^{-1}H$. Thus, the goal is to choose the preconditioning matrix C so that $C^{-1}H$ is well-conditioned. Of course, the optimal choice is simply $C = H$, in which case $C^{-1}H = I$, $\kappa = 1$, and convergence takes place in one iteration! However, this requires computing H^{-1}, or solving the linear system

---

[5] The former is the tangent line of the latter at $x = 0$, which is convex; see Section 6.7.

$H\mathbf{x} = \mathbf{b}$, which obviates the need for iteration to approximate the solution! Thus, the trick of preconditioning is to find a matrix K that is a good approximation of H^{-1}, and use K in place of C^{-1}. One way to do this is to solve another optimization problem

$$\min\left\{\, \| \,\mathrm{I} - KH\,\|_2 \mid K \in V \,\right\}, \tag{6.52}$$

where $V \subset \mathcal{M}_{n \times n} \simeq \mathbb{R}^{n^2}$ is an adroitly chosen subspace of the space of $n \times n$ matrices. If $V = \mathcal{M}_{n \times n}$, then $K = H^{-1}$, which, as noted above, defeats the point of the method. The trick is to choose a sufficiently small subspace — for example V could contain only certain types of sparse matrices, e.g., diagonal matrices — so that (6.52) is computationally tractable, and its minimal value is relatively small. We refer to Exercise 4.7 for more details.

The linear convergence rates in (6.48) and (6.51) can be extended to more general non-quadratic functions that are *strongly convex*, however, the linear rate does not hold in general for all convex functions; see Section 6.9.   ▲

Example 6.26. Let us extend the preceding example by constructing gradient descent for a constrained quadratic form, whose minimization was the subject of Theorem 6.14. In this case, we apply the unconstrained gradient descent of Example 6.25 to the quadratic polynomial (6.24). Assuming a uniform time step, in terms of $\mathbf{y} \in \mathbb{R}^p$, this produces the iteration

$$\mathbf{y}_{k+1} = \mathbf{y}_k - \alpha\, A^T H\,(A\,\mathbf{y}_k + \mathbf{b}). \tag{6.53}$$

The corresponding points on the affine subspace $\mathbf{x}_k = A\,\mathbf{y}_k + \mathbf{b}$ are given by

$$\begin{aligned}
\mathbf{x}_{k+1} = A\,\mathbf{y}_{k+1} + \mathbf{b} &= A\,\mathbf{y}_k + \mathbf{b} - \alpha\,A A^T H\,(A\,\mathbf{y}_k + \mathbf{b}) \\
&= \mathbf{x}_k - \alpha\,A A^T H \mathbf{x}_k = (\mathrm{I} - \alpha\,A A^T H)\,\mathbf{x}_k.
\end{aligned} \tag{6.54}$$

Assuming there is no numerical inaccuracies, this ensures that the iterates remain on the affine subspace $\mathbf{x} = A\,\mathbf{y} + \mathbf{b}$. On the other hand, numerical errors caused by floating point round off and the like will cause the iterates to drift off the subspace, and one will need to move them back onto it by, say, applying orthogonal projection either at each step, or perhaps just occasionally when the accumulated error becomes too large.

A particularly important case arises when the affine subspace is specified by setting the last $n - p$ entries of $\mathbf{x}$ to constants, so we require

$$x_{p+1} = b_{p+1}, \quad \dots \quad x_n = b_n. \tag{6.55}$$

Then, as we noted at the end of Section 6.2, the corresponding matrix is $A = \begin{pmatrix} \mathrm{I} \\ \mathrm{O} \end{pmatrix}$, where $\mathrm{I} = \mathrm{I}_p$ denotes the $p \times p$ identity matrix, and hence (6.54) reduces to

$$\mathbf{x}_{k+1} = (\mathrm{I} - \alpha\, P H)\,\mathbf{x}_k = P\,(\mathrm{I} - \alpha H)\,\mathbf{x}_k + \mathbf{b}, \tag{6.56}$$

where

$$P = A A^T = \begin{pmatrix} \mathrm{I} \\ \mathrm{O} \end{pmatrix} (\,\mathrm{I}\ \mathrm{O}\,) = \begin{pmatrix} \mathrm{I} & \mathrm{O} \\ \mathrm{O} & \mathrm{O} \end{pmatrix}, \quad \text{and, as before,} \quad \mathbf{b} = \left(0, \dots, 0, b_{p+1}, \dots b_n\right)^T.$$

Note that for any vector $\mathbf{x} \in \mathbb{R}^n$,

$$P\mathbf{x} + \mathbf{b} = \left(x_1, \dots, x_p, b_{p+1}, \dots, b_n\right)^T,$$

i.e., this operation is a projection onto the affine subspace that replaces the last $n - p$ components of $\mathbf{x}$ by the constant values (6.55). We conclude that the constrained gradient descent algorithm (6.56) can be recast in the following straightforward form. At each time step, perform the usual unconstrained gradient descent using the coefficient matrix H:

$$\widehat{\mathbf{x}}_{k+1} = (\mathrm{I} - \alpha H)\,\mathbf{x}_k. \tag{6.57}$$

Then, to obtain $\mathbf{x}_{k+1}$, replace the last $n - p$ components of $\widehat{\mathbf{x}}_{k+1}$ with their required constant values (6.55), i.e., set

$$\mathbf{x}_{k+1} = P\,\widehat{\mathbf{x}}_{k+1} + \mathbf{b}. \tag{6.58}$$

In practice, there is thus no need to calculate the last $n - p$ entries of $\widehat{\mathbf{x}}_{k+1}$, and so one can streamline the algorithm by writing $H = \begin{pmatrix} H_1 & H_2 \\ H_2^T & H_3 \end{pmatrix}$ in block form, as at the end of Section 6.2. Setting $\mathbf{y} = \left(x_1, x_2, \ldots, x_p \right)^T \in \mathbb{R}^p$, the algorithm reduces to calculating

$$\mathbf{y}_{k+1} = (\mathrm{I} - \alpha H_1)\,\mathbf{y}_k + \mathbf{c}, \quad \text{where} \quad \mathbf{c} = H_2\widetilde{\mathbf{b}}, \quad \text{with} \quad \widetilde{\mathbf{b}} = \left(b_{p+1}, \ldots, b_n \right)^T. \tag{6.59}$$

The corresponding point $\mathbf{x}_{k+1} = \begin{pmatrix} \mathbf{y}_{k+1} \\ \widetilde{\mathbf{b}} \end{pmatrix}$ on the affine subspace is simply obtained by appending the values $\widetilde{\mathbf{b}}$ to the preceding iterate. Here we no longer need to worry about numerical error taking us off the affine subspace. Note that, by suitably relabeling, the algorithm is easily adapted to setting any p of the entries of $\mathbf{x}$ to constants. ▲

6.4.1 Proximal Gradient Descent

In order to apply gradient descent, the objective function F must be differentiable, so that we can compute its gradient $\nabla F(\mathbf{x})$ in order to specify a descent direction. In many important applications, the objective function is nondifferentiable, and in such cases, additional techniques are required. In general, it is very hard to optimize nondifferentiable functions, but there are some special cases that can be handled with extensions of gradient descent. One such scenario is when the objective function has the form

$$F(\mathbf{x}) = G(\mathbf{x}) + \lambda\,H(\mathbf{x}),$$

where $\lambda \geq 0$, G is a continuously differentiable function, and H may be nondifferentiable. An important example of this form is the Lasso regression problem studied later in Chapter 7 where $H(\mathbf{x}) = \|\mathbf{x}\|_1$ is the one norm (2.59). In this setting, proximal gradient descent can sometimes be extremely effective.

The starting point for *proximal gradient descent* is the observation that each step of gradient descent on a differentiable objective function F can be interpreted as solving an optimization problem of the form

$$\mathbf{x}_{k+1} = \underset{\mathbf{x}}{\operatorname{argmin}} \left\{ F(\mathbf{x}_k) + \langle\, \nabla F(\mathbf{x}_k), \mathbf{x} - \mathbf{x}_k\,\rangle + \frac{1}{2\alpha_k}\,\|\mathbf{x} - \mathbf{x}_k\|^2 \right\}, \tag{6.60}$$

where we recall that argmin refers to the vector $\mathbf{x} \in \mathbb{R}^n$ that minimizes the objective function, and not the minimal value. The solution of (6.60) is simply the gradient descent step, namely $\mathbf{x}_{k+1} = \mathbf{x}_k - \alpha_k \nabla F(\mathbf{x}_k)$; we leave the verification of this to the reader in Exercise 4.6. We can think of the optimization problem in (6.60) as minimizing a local approximation to F near the point $\mathbf{x}_k$; indeed, quantity on the right hand side is the tangent plane approximation, i.e., the first order Taylor expansion, of F at the point $\mathbf{x}_k$, as discussed in Section 6.8, plus a quadratic penalty term to keep the solution from drifting too far away from $\mathbf{x}_k$, where the tangent plane approximation is invalid.

In proximal gradient descent, we apply the same idea, except that we perform the Taylor expansion only in the differentiable part of the loss, namely G. That is, a single step of proximal gradient descent is given by

$$\mathbf{x}_{k+1} = \underset{\mathbf{x}}{\operatorname{argmin}} \left\{ \langle \nabla G(\mathbf{x}_k), \mathbf{x} - \mathbf{x}_k \rangle + \frac{1}{2\alpha_k} \|\mathbf{x} - \mathbf{x}_k\|^2 + \lambda H(\mathbf{x}) \right\},$$

where we dropped $G(\mathbf{x}_k)$ since it does not affect the minimizer. We can complete the square and divide by λ to simplify this to read

$$\mathbf{x}_{k+1} = \underset{\mathbf{x} \in \mathbb{R}^n}{\operatorname{argmin}} \left\{ \frac{1}{2\alpha_k \lambda} \|\mathbf{x} - \mathbf{y}_k\|^2 + H(\mathbf{x}) \right\}, \quad \text{where} \quad \mathbf{y}_k = \mathbf{x}_k - \alpha_k \nabla G(\mathbf{x}_k), \qquad (6.61)$$

and we again dropped a term involving $\|\nabla G(\mathbf{x}_k)\|^2$, since it does not involve $\mathbf{x}$.

The minimization problem in (6.61) is called the *proximal operator* of H with step size $\alpha_k \lambda$ applied to $\mathbf{y}_k$. Thus, proximal gradient descent involves taking a step of gradient descent on the smooth part G, i.e., compute $\mathbf{y}_k$, followed by an application of the proximal operator for the nonsmooth part H applied to $\mathbf{y}_k$ with step size $\alpha_k \lambda$. In situations where the proximal operator for H can be computed efficiently (e.g., for Lasso in Section 7.2), optimization by proximal gradient descent can be quite effective. In more complicated situations, the solution of the proximal problem (6.61) may be no easier than solving the original optimization problem for F.

Exercises

4.1. ♡ Write Python code to implement gradient descent on the functions $F_1(x, y) = x^2 + 2y^2$, $F_2(x, y) = x^2 + 10y^2$ and $F_3(x, y) = \sin x \, \sin y$ and numerically investigate the rates of convergence. You will need to choose the time step α by hand in each case to get the fastest convergence rate. For which function does gradient descent converge the most quickly?

4.2. Show that gradient descent for minimizing a function $F(\mathbf{x})$ subject to the constraint $\mathbf{x} = A\mathbf{z} + \mathbf{b}$ is given by the iterations

$$\mathbf{z}_{k+1} = \mathbf{z}_k - \alpha A^T \nabla F(A\mathbf{z}_k + \mathbf{b}) \quad \text{and} \quad \mathbf{x}_{k+1} = \mathbf{x}_k - \alpha A A^T \nabla F(\mathbf{x}_k).$$

4.3. ◊ Repeat Exercise 4.1 where each optimization problem is subject to the constraint $x + y = 2\pi$, using Exercise 4.2.

4.4. ♡ Prove that, provided $\nabla F(\mathbf{x}_k) \neq \mathbf{0}$, the inequality (6.40) holds when $\alpha_k > 0$ is sufficiently small.

4.5. ♡ (a) Show that the system $x^2 + y^2 = 1$, $x + y = 2$, does not have a solution.

(b) Use gradient descent to construct a "least squares solution" by minimizing the scalar valued function $F(x, y) = (x^2 + y^2 - 1)^2 + (x + y - 2)^2$.

4.6. Verify that $\mathbf{x}_{k+1} = \mathbf{x}_k - \alpha_k \nabla F(\mathbf{x}_k)$ solves the minimization problem (6.60).

4.7. ◊ Let A be a square matrix with $\|I - A\|_2 = \varepsilon < 1$. Show that $0 < \dfrac{\lambda_{max}(A)}{\lambda_{min}(A)} \leq \dfrac{1 + \varepsilon}{1 - \varepsilon}$.

4.8. $\diamond$ In this exercise, we consider the problem of how to choose the time step α_k when minimizing the quadratic function $F(\mathbf{x}) = \frac{1}{2}\mathbf{x}^T H \mathbf{x} - \mathbf{b}^T \mathbf{x} + c$ in the gradient descent iteration $\mathbf{x}_{k+1} = \mathbf{x}_k - \alpha_k (A\mathbf{x}_k - \mathbf{b})$.

(a) Derive an expression for α_k that minimizes $F(\mathbf{x}_{k+1})$ over all choices of α_k. *Hint:* Write out $F(\mathbf{x}_{k+1}) = F\big(\mathbf{x}_k - \alpha_k(H\mathbf{x}_k - \mathbf{b})\big)$ using the definition of F and note that the resulting expression is a quadratic function of α_k. It may be helpful to write your choices of the time step in terms of the residual $\mathbf{r}_k = A\mathbf{x}_k - \mathbf{b}$.

(b) Derive an expression for α_k that minimizes $\|H\mathbf{x}_{k+1} - \mathbf{b}\|_2$ over all choices of α_k.

4.9. $\diamond$ Implement parts (a) and (b) from Exercise 4.8 in Python and compare against the choice $\alpha_k = 1/\lambda_{max}(H)$ from Example 6.25. Which method converges more quickly? As in Exercise 6.2 in Chapter 5, to generate a random symmetric positive definite matrix H, generate a random square matrix A and set $H = A^T A$.

4.10. Assume that H is continuously differentiable and that the proximal operator of H defined in (6.61) admits a minimizer. Show that

$$\mathbf{x}_{k+1} = \mathbf{y}_k - \alpha_k \lambda \nabla H(\mathbf{x}_{k+1}). \tag{6.62}$$

Thus, the proximal operator can be viewed as a version of *implicit* gradient descent, where the gradient ∇H is evaluated at the next iterate $\mathbf{x}_{k+1}$ (of course, solving (6.62) for $\mathbf{x}_{k+1}$ is not always straightforward).

4.11. Let $H(\mathbf{x}) = \frac{1}{2}\|\mathbf{x}\|^2$. Show that the proximal operator of H given in (6.61) can be explicitly solved and is given by $\mathbf{x}_{k+1} = (1 + \alpha_k \lambda)^{-1} \mathbf{y}_k$.

4.12. Repeat Exercise 4.11 for $H(\mathbf{x}) = \frac{1}{2}\langle S\mathbf{x}, \mathbf{x}\rangle$, where S is self-adjoint. Give an explicit formula for the proximal update step in (6.61).

6.5 The Conjugate Gradient Method

Gradient descent is a reasonable algorithm, and is guaranteed to converge to a global minimizer when applied to convex functions, as we shall subsequently see in Section 6.9. However, even in the setting of optimizing quadratic functions, gradient descent can take an excessively long time to converge to an accurate approximation to the minimizer. It turns out that by cleverly modifying the direction used in the descent step, we can dramatically accelerate the convergence rate in certain settings. The resulting method is known as the conjugate gradient method; we introduce the main ideas in this section, but postpone a convergence analysis to Section 11.3.

The basic ideas can be explained in the context of the usual quadratic objective function (6.44), where we assume that the $n \times n$ coefficient matrix H is symmetric, positive definite, and hence there is a unique minimizer, namely the solution $\mathbf{x}^\star$ to the linear system $H\mathbf{x}^\star = \mathbf{b}$. As noted above, if H is ill conditioned, the gradient descent algorithm (6.46) will converge too slowly to be of practical use. One modification discussed above is to precondition the algorithm by employing a different inner product to compute the gradient. Moreover, it was noted that the optimal inner product is the one based on the coefficient matrix H itself:

$$\langle \mathbf{x}, \mathbf{y}\rangle_H = \mathbf{x}^T H \mathbf{y}. \tag{6.63}$$

However, the resulting gradient descent algorithm is unusable since it assumes we can already solve the linear system.

The *conjugate gradient method*, which was first developed in 1952 by Hestenes and Stiefel, [103], uses a different tactic. It retains the inner product (6.63) defined by the coefficient matrix. Two vectors $\mathbf{x}, \mathbf{y} \in \mathbb{R}^n$ that are orthogonal under this inner product, i.e., $\langle\, \mathbf{x}, \mathbf{y} \,\rangle_H = 0$ are said to be *conjugate*, whence the name of the algorithm. In outline, the method successively generates a sequence of mutually conjugate vectors $\mathbf{v}_1, \ldots, \mathbf{v}_n$ that form an H orthogonal basis of $\mathbb{R}^n$. The solution vector $\mathbf{x}^\star$ that minimizes $P(\mathbf{x})$ defined in (6.44) or, equivalently, solves the linear system $H\mathbf{x}^\star = \mathbf{b}$ is written in terms of the conjugate vectors

$$\mathbf{x}^\star = \mathbf{x}_0 + t_1 \mathbf{v}_1 + \cdots + t_n \mathbf{v}_n, \tag{6.64}$$

where $\mathbf{x}_0$ is some initial approximation to the solution. In view of the orthogonality condition, the coordinates of the solution vector are

$$t_k = \frac{\langle\, \mathbf{x}^\star - \mathbf{x}_0, \mathbf{v}_k \,\rangle_H}{\|\,\mathbf{v}_k\,\|_H^2}. \tag{6.65}$$

The conjugate gradient algorithm, to be derived below, computes the t_k and $\mathbf{v}_k$ iteratively, so that the k-th approximation to the solution is

$$\mathbf{x}_k = \mathbf{x}_0 + t_1 \mathbf{v}_1 + \cdots + t_k \mathbf{v}_k, \qquad \text{or, equivalently} \qquad \mathbf{x}_k = \mathbf{x}_{k-1} + t_k \mathbf{v}_k.$$

The vector $\mathbf{x}_k$ is obtained from $\mathbf{x}_{k-1}$ by minimizing the *residual vector*

$$\mathbf{r}_k = \mathbf{b} - H\mathbf{x}_k,$$

which serves as an estimate of the error in the k-th approximation. The secret is not to try to specify the conjugate basis vectors in advance, but rather to successively construct them during the course of the algorithm.

We begin with an initial guess $\mathbf{x}_0$ — for example, $\mathbf{x}_0 = \mathbf{0}$. According to (6.45) the residual vector $\mathbf{r}_0 = \mathbf{b} - H\mathbf{x}_0$ is the negative of the Euclidean gradient of P at the point $\mathbf{x}_0$, and hence indicates the direction of steepest decrease. We begin by updating our original guess by moving in this direction, taking $\mathbf{v}_1 = \mathbf{r}_0$ as our first conjugate direction. The next iterate is $\mathbf{x}_1 = \mathbf{x}_0 + t_1 \mathbf{v}_1$, and we choose the parameter t_1 so that the corresponding residual vector

$$\mathbf{r}_1 = \mathbf{b} - H\mathbf{x}_1 = \mathbf{b} - H\mathbf{x}_0 - t_1 H\mathbf{v}_1 = \mathbf{r}_0 - t_1 H\mathbf{v}_1 \tag{6.66}$$

is as close to $\mathbf{0}$ (in the Euclidean norm) as possible. This occurs when $\mathbf{r}_1$ is orthogonal to $\mathbf{r}_0$ (why?), and so we require

$$0 = \mathbf{r}_0 \cdot \mathbf{r}_1 = \mathbf{r}_0^T \mathbf{r}_1 = \mathbf{r}_0^T \mathbf{r}_0 - t_1 \mathbf{r}_0^T H\mathbf{v}_1 = \|\,\mathbf{r}_0\,\|_2^2 - t_1 \mathbf{v}_1^T H\mathbf{v}_1 = \|\,\mathbf{r}_0\,\|_2^2 - t_1 \|\,\mathbf{v}_1\,\|_H^2. \tag{6.67}$$

Therefore, we set

$$t_1 = \frac{\|\,\mathbf{r}_0\,\|_2^2}{\|\,\mathbf{v}_1\,\|_H^2}. \tag{6.68}$$

We can assume that $t_1 \neq 0$, since otherwise the residual $\mathbf{r}_0 = \mathbf{0}$, which would imply $\mathbf{x}_0 = \mathbf{x}^\star$ is the exact solution of the linear system, and there would be no reason to continue the procedure.

The gradient descent algorithm would tell us to update $\mathbf{x}_1$ by moving in the residual direction $\mathbf{r}_1$. In the conjugate gradient algorithm, we instead choose a direction $\mathbf{v}_2$ which is conjugate, meaning H–orthogonal, to the first direction $\mathbf{v}_1 = \mathbf{r}_0$. Thus, as in the Gram–Schmidt process, we modify the residual direction by setting $\mathbf{v}_2 = \mathbf{r}_1 + s_1 \mathbf{v}_1$, where the scalar factor s_1 is determined by the imposed orthogonality requirement:

$$0 = \langle\, \mathbf{v}_1, \mathbf{v}_2 \,\rangle_H = \langle\, \mathbf{v}_1, \mathbf{r}_1 + s_1 \mathbf{v}_1 \,\rangle_H = \langle\, \mathbf{v}_1, \mathbf{r}_1 \,\rangle_H + s_1 \langle\, \mathbf{v}_1, \mathbf{v}_1 \,\rangle_H = \langle\, \mathbf{r}_1, \mathbf{v}_1 \,\rangle_H + s_1 \|\,\mathbf{v}_1\,\|_H^2,$$

and hence we fix

$$s_1 = -\frac{\langle \mathbf{r}_1, \mathbf{v}_1 \rangle_H}{\|\mathbf{v}_1\|_H^2}.$$

Now, in view of (6.66) and the orthogonality of $\mathbf{r}_0$ and $\mathbf{r}_1$,

$$\langle \mathbf{r}_1, \mathbf{v}_1 \rangle_H = \mathbf{r}_1^T H \mathbf{v}_1 = \mathbf{r}_1^T \left(\frac{\mathbf{r}_0 - \mathbf{r}_1}{t_1} \right) = -\frac{1}{t_1}\|\mathbf{r}_1\|_2^2,$$

while, by (6.68),

$$\|\mathbf{v}_1\|_H^2 = \frac{1}{t_1}\|\mathbf{r}_0\|_2^2.$$

Therefore, the second conjugate direction is given by

$$\mathbf{v}_2 = \mathbf{r}_1 + s_1 \mathbf{v}_1, \qquad \text{where} \qquad s_1 = \frac{\|\mathbf{r}_1\|_2^2}{\|\mathbf{r}_0\|_2^2}. \tag{6.69}$$

We then update

$$\mathbf{x}_2 = \mathbf{x}_1 + t_2 \mathbf{v}_2$$

so as to make the corresponding residual vector

$$\mathbf{r}_2 = \mathbf{b} - H\mathbf{x}_2 = \mathbf{b} - H\mathbf{x}_1 - t_2 H\mathbf{v}_2 = \mathbf{r}_1 - t_2 H\mathbf{v}_2$$

as small as possible in the Euclidean norm, which is accomplished by requiring it to be orthogonal to $\mathbf{r}_1$. Thus, using (6.69) and the H–orthogonality of $\mathbf{v}_1$ and $\mathbf{v}_2$, we have

$$0 = \mathbf{r}_1^T \mathbf{r}_2 = \|\mathbf{r}_1\|_2^2 - t_2 \mathbf{r}_1^T H \mathbf{v}_2 = \|\mathbf{r}_1\|_2^2 - t_2 \langle \mathbf{r}_1, \mathbf{v}_2 \rangle_H$$
$$= \|\mathbf{r}_1\|_2^2 - t_2 \langle \mathbf{v}_2 - s_1 \mathbf{v}_1, \mathbf{v}_2 \rangle_H = \|\mathbf{r}_1\|_2^2 - t_2 \|\mathbf{v}_2\|_H^2,$$

and so

$$t_2 = \frac{\|\mathbf{r}_1\|_2^2}{\|\mathbf{v}_2\|_H^2}.$$

Again, we can assume that $t_2 \neq 0$, as otherwise $\mathbf{r}_1 = \mathbf{0}$ and $\mathbf{x}_1$ would be the exact solution, so the algorithm should be terminated.

Continuing in this manner, at the k-th stage, we have already constructed the conjugate vectors $\mathbf{v}_1, \ldots, \mathbf{v}_k$, and the solution approximation $\mathbf{x}_k$ as a suitable linear combination of them. The next conjugate direction is given by

$$\mathbf{v}_{k+1} = \mathbf{r}_k + s_k \mathbf{v}_k, \qquad \text{where} \qquad s_k = \frac{\|\mathbf{r}_k\|_2^2}{\|\mathbf{r}_{k-1}\|_2^2} \tag{6.70}$$

results from the H–orthogonality requirement: $\langle \mathbf{v}_i, \mathbf{v}_k \rangle_H = 0$ for $i < k$. The updated solution approximation

$$\mathbf{x}_{k+1} = \mathbf{x}_k + t_{k+1} \mathbf{v}_{k+1}, \qquad \text{where} \qquad t_{k+1} = \frac{\|\mathbf{r}_k\|_2^2}{\|\mathbf{v}_{k+1}\|_H^2} \tag{6.71}$$

is then specified so as to make the corresponding residual

$$\mathbf{r}_{k+1} = \mathbf{b} - H\mathbf{x}_{k+1} = \mathbf{r}_k - t_{k+1} H\mathbf{v}_{k+1} \tag{6.72}$$

as small as possible, by requiring that it be orthogonal to $\mathbf{r}_k$.

Starting with an initial guess $\mathbf{x}_0$, the iterative equations (6.70), (6.71) implement the *conjugate gradient method*. Observe that the algorithm does not require solving any linear systems: apart from multiplication of a matrix times a vector to evaluate $H\mathbf{v}_k$, all other operations are rapidly evaluated Euclidean dot products. The method produces a sequence of successive approximations $\mathbf{x}_1, \mathbf{x}_2, \ldots$ to the solution $\mathbf{x}^\star$, and so the iteration can be stopped as soon as a desired solution accuracy is reached — which can be assessed by comparing how close the successive iterates are to each other. Moreover, *the conjugate gradient method does eventually terminate at the exact solution*[6] because, as remarked at the outset, there are at most n conjugate directions, forming a orthogonal basis of $\mathbb{R}^n$ for the inner product induced by H. Therefore,

$$\mathbf{x}_n = \mathbf{x}_0 + t_1 \mathbf{v}_1 + \cdots + t_n \mathbf{v}_n = \mathbf{x}^\star$$

must be the solution since its residual $\mathbf{r}_n = \mathbf{b} - H\mathbf{x}_n$ is orthogonal to all the conjugate basis vectors $\mathbf{v}_1, \ldots, \mathbf{v}_n$, and hence must be $\mathbf{0}$.

Example 6.27. Consider the linear system $H\mathbf{x} = \mathbf{b}$ with

$$H = \begin{pmatrix} 3 & -1 & 0 \\ -1 & 2 & 1 \\ 0 & 1 & 1 \end{pmatrix}, \qquad \mathbf{b} = \begin{pmatrix} 1 \\ 2 \\ -1 \end{pmatrix}.$$

The exact solution is $\mathbf{x}^\star = (2, 5, -6)^T$. Let us implement the method of conjugate gradients, starting with the initial guess $\mathbf{x}_0 = (0,0,0)^T$. The corresponding residual vector is merely $\mathbf{r}_0 = \mathbf{b} - H\mathbf{x}_0 = \mathbf{b} = (1, 2, -1)^T$. The first conjugate direction is $\mathbf{v}_1 = \mathbf{r}_0 = (1, 2, -1)^T$, and we use formula (6.68) to obtain the updated approximation to the solution

$$\mathbf{x}_1 = \mathbf{x}_0 + \frac{\|\mathbf{r}_0\|_2^2}{\|\mathbf{v}_1\|_H^2}\, \mathbf{v}_1 = \frac{6}{4} \begin{pmatrix} 1 \\ 2 \\ -1 \end{pmatrix} = \begin{pmatrix} \frac{3}{2} \\ 3 \\ -\frac{3}{2} \end{pmatrix}.$$

In the next stage of the algorithm, we compute the corresponding residual $\mathbf{r}_1 = \mathbf{b} - H\mathbf{x}_1 = \left(-\frac{1}{2}, -1, -\frac{5}{2}\right)^T$. The conjugate direction is

$$\mathbf{v}_2 = \mathbf{r}_1 + \frac{\|\mathbf{r}_1\|_2^2}{\|\mathbf{r}_0\|_2^2}\, \mathbf{v}_1 = \begin{pmatrix} -\frac{1}{2} \\ -1 \\ -\frac{5}{2} \end{pmatrix} + \frac{\frac{15}{2}}{6} \begin{pmatrix} 1 \\ 2 \\ -1 \end{pmatrix} = \begin{pmatrix} \frac{3}{4} \\ \frac{3}{2} \\ -\frac{15}{4} \end{pmatrix},$$

which, as designed, satisfies the conjugacy condition $\langle \mathbf{v}_1, \mathbf{v}_2 \rangle_H = \mathbf{v}_1^T H \mathbf{v}_2 = 0$. Each entry of the ensuing approximation

$$\mathbf{x}_2 = \mathbf{x}_1 + \frac{\|\mathbf{r}_1\|_2^2}{\|\mathbf{v}_2\|_H^2}\, \mathbf{v}_2 = \begin{pmatrix} \frac{3}{2} \\ 3 \\ -\frac{3}{2} \end{pmatrix} + \frac{\frac{15}{2}}{\frac{27}{4}} \begin{pmatrix} \frac{3}{4} \\ \frac{3}{2} \\ -\frac{15}{4} \end{pmatrix} = \begin{pmatrix} \frac{7}{3} \\ \frac{14}{3} \\ -\frac{17}{3} \end{pmatrix} \approx \begin{pmatrix} 2.3333 \\ 4.6667 \\ -5.6667 \end{pmatrix}$$

---

[6]This discussion assumes exact, or very high precision, arithmetic. In floating point precision, the computed directions $\mathbf{v}_1, \mathbf{v}_2, \ldots, \mathbf{v}_k$ may not exactly satisfy the conjugacy condition, due to floating point roundoff errors, and this can affect the convergence of the conjugate gradient method, though an analysis is outside the scope of this book; see, for example, [223]. There are many techniques for addressing this in practice, such as reorthogonalization [129].

is now within $\frac{1}{3}$ of the exact solution $\mathbf{x}^\star$.

Since we are dealing with a 3×3 system, we will recover the exact solution by one more iteration of the algorithm. The new residual is $\mathbf{r}_2 = \mathbf{b} - H\mathbf{x}_2 = \left(-\frac{4}{3}, \frac{2}{3}, 0\right)^T$. The final conjugate direction is

$$\mathbf{v}_3 = \mathbf{r}_2 + \frac{\|\mathbf{r}_2\|_2^2}{\|\mathbf{r}_1\|_2^2}\,\mathbf{v}_2 = \begin{pmatrix} -\frac{4}{3} \\ \frac{2}{3} \\ 0 \end{pmatrix} + \frac{20}{9}\begin{pmatrix} \frac{3}{4} \\ \frac{3}{2} \\ -\frac{15}{4} \end{pmatrix} = \begin{pmatrix} -\frac{10}{9} \\ \frac{10}{9} \\ -\frac{10}{9} \end{pmatrix},$$

which, as you can check, is conjugate to both $\mathbf{v}_1$ and $\mathbf{v}_2$. The solution is obtained from

$$\mathbf{x}_3 = \mathbf{x}_2 + \frac{\|\mathbf{r}_2\|_2^2}{\|\mathbf{v}_3\|_H^2}\,\mathbf{v}_3 = \begin{pmatrix} \frac{7}{3} \\ \frac{14}{3} \\ -\frac{17}{3} \end{pmatrix} + \frac{\frac{20}{9}}{\frac{200}{27}}\begin{pmatrix} -\frac{10}{9} \\ \frac{10}{9} \\ -\frac{10}{9} \end{pmatrix} = \begin{pmatrix} 2 \\ 5 \\ -6 \end{pmatrix}. \qquad \blacktriangle$$

In larger examples, one would not carry through the algorithm to the bitter end since a decent approximation to the solution is typically obtained with only a few iterations. The result can be a substantial saving in computational time and effort required to produce an approximation to the solution. We study the conjugate gradient method further in Section 11.3, where we prove a convergence rate and show that it is significantly faster than gradient descent, especially for ill-conditioned matrices. For further developments and applications, see [56, 230, 245]. We also mention that there are various generalizations of the conjugate gradient method to the fully nonlinear setting, meaning that the objective function is not quadratic and so its gradient is not linear [96].

Exercises

5.1. Solve the following linear systems by the conjugate gradient method, keeping track of the residual vectors and solution approximations as you iterate.

$(a)\,\heartsuit\ \begin{pmatrix} 3 & -1 \\ -1 & 5 \end{pmatrix}\mathbf{x} = \begin{pmatrix} 2 \\ 1 \end{pmatrix}, \quad (b)\ \begin{pmatrix} 2 & 1 \\ 1 & 1 \end{pmatrix}\mathbf{x} = \begin{pmatrix} -3 \\ 1 \end{pmatrix}, \quad (c)\,\heartsuit\ \begin{pmatrix} 6 & 2 & 1 \\ 2 & 3 & -1 \\ 1 & -1 & 2 \end{pmatrix}\mathbf{x} = \begin{pmatrix} 1 \\ 0 \\ -2 \end{pmatrix},$

$(d)\,\diamondsuit\ \begin{pmatrix} 6 & -1 & -1 & 5 \\ -1 & 7 & 1 & -1 \\ -1 & 1 & 3 & -3 \\ 5 & -1 & -3 & 6 \end{pmatrix}\mathbf{x} = \begin{pmatrix} 1 \\ 2 \\ 0 \\ -1 \end{pmatrix}, \quad (e)\ \begin{pmatrix} 5 & 1 & 1 & 1 \\ 1 & 5 & 1 & 1 \\ 1 & 1 & 5 & 1 \\ 1 & 1 & 1 & 5 \end{pmatrix}\mathbf{x} = \begin{pmatrix} 4 \\ 0 \\ 0 \\ 0 \end{pmatrix}.$

5.2. According to [181], the $n \times n$ Hilbert matrix H_n, whose (i, j) entry is $1/(i + j - 1)$ — see also (4.74) — is positive definite, and hence we can apply the conjugate gradient method to solve the linear system $H_n\mathbf{x} = \mathbf{b}$. For the values $n = 5, 10, 30$, let $\mathbf{x}^\star \in \mathbb{R}^n$ be the vector with all entries equal to 1. (a) Compute $\mathbf{b} = H_n\mathbf{x}^\star$. (b) Use QR to solve $H_n\mathbf{x} = \mathbf{b}$. How close is your solution to $\mathbf{x}^\star$? (c) Does the conjugate gradient algorithm do any better?

5.3. Try applying the conjugate gradient method to the linear system $-x + 2y + z = -2$, $y + 2z = 1$, $3x + y - z = 1$. Do you obtain the solution? Why or why not?

5.4. $\diamondsuit$ *True or false*: If the residual vector satisfies $\|\mathbf{r}\|_2 < .01$, then $\mathbf{x}$ approximates the solution to within two decimal places.

5.5. ♡ Use the conjugate gradient method to solve the system $A\mathbf{u} = \mathbf{e}_5$ with coefficient matrix

$$
A = \begin{pmatrix}
4 & -1 & 0 & -1 & 0 & 0 & 0 & 0 & 0 \\
-1 & 4 & -1 & 0 & -1 & 0 & 0 & 0 & 0 \\
0 & -1 & 4 & 0 & 0 & -1 & 0 & 0 & 0 \\
-1 & 0 & 0 & 4 & -1 & 0 & -1 & 0 & 0 \\
0 & -1 & 0 & -1 & 4 & -1 & 0 & -1 & 0 \\
0 & 0 & -1 & 0 & -1 & 4 & 0 & 0 & -1 \\
0 & 0 & 0 & -1 & 0 & 0 & 4 & -1 & 0 \\
0 & 0 & 0 & 0 & -1 & 0 & -1 & 4 & -1 \\
0 & 0 & 0 & 0 & 0 & -1 & 0 & -1 & 4
\end{pmatrix}.
$$

How many iterations do you need to obtain the solution that is accurate to 2 decimal places? *Remark*: This matrix arises in the numerical discretization of the two-dimensional Laplace partial differential equation, of great importance in many applications, [180].

6.6   The Second Derivative Test

As in the scalar case, the status of a critical point — minimizer, maximizer, or neither — can often be resolved by analyzing the second order derivatives of the objective function at the point. This is one place where we need to tighten our underlying smoothness assumptions.

Definition 6.28. A function $F(\mathbf{x}) = F(x_1, \ldots, x_n)$ is said to be order n *continuously differentiable*, written $F \in \mathrm{C}^n$, if F and all its partial derivatives up to order n are continuous.

Thus, "continuously differentiable" in the previous section is equivalent to $F \in \mathrm{C}^1$. If $F \in \mathrm{C}^n$, then it satisfies the condition of "equality of mixed partials" meaning it does not matter in which order the partial derivatives of order $\leq n$ are taken, cf. [4, 158].

In multivariable calculus, the "second derivative" of a scalar-valued function $F(\mathbf{x}) = F(x_1, \ldots, x_n)$ is represented by its $n \times n$ *Hessian matrix*[7], whose entries are all its second order partial derivatives:

$$
\nabla^2 F(\mathbf{x}) = \begin{pmatrix}
\dfrac{\partial^2 F}{\partial x_1^2} & \dfrac{\partial^2 F}{\partial x_2\, \partial x_1} & \cdots & \dfrac{\partial^2 F}{\partial x_n\, \partial x_1} \\[2ex]
\dfrac{\partial^2 F}{\partial x_1\, \partial x_2} & \dfrac{\partial^2 F}{\partial x_2^2} & \cdots & \dfrac{\partial^2 F}{\partial x_n\, \partial x_2} \\[2ex]
\vdots & \vdots & \ddots & \vdots \\[2ex]
\dfrac{\partial^2 F}{\partial x_1\, \partial x_n} & \dfrac{\partial^2 F}{\partial x_2\, \partial x_n} & \cdots & \dfrac{\partial^2 F}{\partial x_n^2}
\end{pmatrix},
\tag{6.73}
$$

where the partial derivatives are all evaluated at $\mathbf{x}$. When $F \in \mathrm{C}^2$ has continuous second order partial derivatives, its mixed partial derivatives are equal, $\partial^2 F/\partial x_i\, \partial x_j = \partial^2 F/\partial x_j\, \partial x_i$, and hence its Hessian matrix is symmetric: $\nabla^2 F(\mathbf{x}) = \nabla^2 F(\mathbf{x})^T$.

---

[7] Named after the nineteenth century German mathematician Ludwig Otto Hesse. Interestingly, the paper where he introduced the Hessian matrix was devoted to the "proof" of a false theorem; see [179].

Remark 6.29. We can view the Hessian as a matrix/vector version of the second derivative of F in the following way. For a vector valued function $G \colon \mathbb{R}^n \to \mathbb{R}^m$, with components $G_i \colon \mathbb{R}^n \to \mathbb{R}$ for $i = 1, \ldots, m$, we define its *Jacobian matrix* at $\mathbf{x} \in \mathbb{R}^n$ to be the $m \times n$ matrix

$$
\mathbf{D}G(\mathbf{x}) =
\begin{pmatrix}
\dfrac{\partial G_1}{\partial x_1} & \dfrac{\partial G_1}{\partial x_2} & \cdots & \dfrac{\partial G_1}{\partial x_n} \\[2ex]
\dfrac{\partial G_2}{\partial x_1} & \dfrac{\partial G_2}{\partial x_2} & \cdots & \dfrac{\partial G_2}{\partial x_n} \\[2ex]
\vdots & \vdots & \ddots & \vdots \\[2ex]
\dfrac{\partial G_m}{\partial x_1} & \dfrac{\partial G_m}{\partial x_2} & \cdots & \dfrac{\partial G_m}{\partial x_n}
\end{pmatrix},
\tag{6.74}
$$

where again all the partial derivatives are evaluated at $\mathbf{x}$. We use the bold notation $\mathbf{D}$ for the Jacobian to distinguish it from the gradient ∇ and matrices denoted D. When $F \colon \mathbb{R}^n \to \mathbb{R}$ is a scalar function, its Jacobian is a row vector, namely, the transpose of the standard gradient vector (6.27), so $\mathbf{D}F = \nabla F^T$. By letting

$$
G(\mathbf{x}) = \nabla F(\mathbf{x}) = \left(\frac{\partial F}{\partial x_1}, \frac{\partial F}{\partial x_2}, \; \cdots \; , \frac{\partial F}{\partial x_n} \right)^T
$$

be the standard gradient, which is a column vector, we can write its Hessian as the Jacobian of its gradient:

$$
\nabla^2 F(\mathbf{x}) = \mathbf{D}(\nabla F)(\mathbf{x}),
\tag{6.75}
$$

an expression we will use in some computations later in this chapter.

For later use, we also record here the multivariable version of the *chain rule* that involves Jacobian matrices. Given $F \colon \mathbb{R}^m \to \mathbb{R}^k$ and $G \colon \mathbb{R}^n \to \mathbb{R}^m$ then the Jacobian of their composition $F \circ G \colon \mathbb{R}^m \to \mathbb{R}^k$ equals the product of their individual Jacobians, evaluated at the appropriate points:

$$
\mathbf{D}(F \circ G)(\mathbf{x}) = \mathbf{D}F(G(\mathbf{x})) \, \mathbf{D}G(\mathbf{x}).
\tag{6.76}
$$

Since the gradient of a scalar-valued function $F \colon \mathbb{R}^m \to \mathbb{R}$ is the transpose of its Jacobian, the chain rule in this case has the alternative form:

$$
\nabla(F \circ G)(\mathbf{x}) = \mathbf{D}G(\mathbf{x})^T \, \nabla F(G(\mathbf{x})),
\tag{6.77}
$$

which is obtained by taking the transpose of both sides of (6.76). ▲

According to Proposition 6.4, a local minimum of a scalar function requires positivity of its second derivative. For a function of several variables, the corresponding condition is that the Hessian matrix be positive definite, as per Definition 4.1. More specifically, the multi-dimensional version of the second derivative test for a local minimizer is stated as follows. As in the scalar case, the proof is based on a second order Taylor expansion, and appears at the end of this section.

> **Theorem 6.30.** *Let $F(\mathbf{x}) = F(x_1, \ldots, x_n)$ be a real-valued, twice continuously differentiable function. If $\mathbf{x}^\star$ is a local minimizer for F, then it is necessarily a critical point, so $\nabla F(\mathbf{x}^\star) = \mathbf{0}$. Moreover, the Hessian matrix (6.73) must be positive semidefinite at the minimizer, so $\nabla^2 F(\mathbf{x}^\star) \geq 0$. Conversely, if $\mathbf{x}^\star$ is a critical point with positive definite Hessian matrix $\nabla^2 F(\mathbf{x}^\star) > 0$, then $\mathbf{x}^\star$ is a strict local minimizer.*

For example, at every $\mathbf{x} \in \mathbb{R}^n$, the quadratic polynomial (6.10) has constant Hessian, which equals the coefficient matrix, $H = \nabla^2 F(\mathbf{x})$. In general, a maximum requires a negative semidefinite Hessian matrix. If, moreover, the Hessian at the critical point is negative definite, then the critical point is a strict local maximizer. If the Hessian matrix is indefinite, then the critical point is a saddle point — neither minimizer nor maximizer. In general, a critical point is called *nondegenerate* if the Hessian matrix is nonsingular. In the borderline case, when the Hessian is only positive or negative semidefinite at the critical point, the second derivative test is inconclusive, and resolving the nature of the critical point requires more detailed knowledge of the objective function, e.g., its higher order derivatives (when they exist).

Example 6.31. The function

$$F(x,y) = x^2 + y^2 - y^3 \qquad \text{has gradient} \qquad \nabla F(x,y) = \begin{pmatrix} 2x \\ 2y - 3y^2 \end{pmatrix}.$$

The critical point equation $\nabla F = \mathbf{0}$ has two solutions: $\mathbf{x}_1^\star = (0,0)^T$ and $\mathbf{x}_2^\star = (0,\tfrac{2}{3})^T$. The Hessian matrix of the objective function is

$$\nabla^2 F(x,y) = \begin{pmatrix} 2 & 0 \\ 0 & 2 - 6y \end{pmatrix}.$$

At the first critical point, the Hessian $\nabla^2 F(0,0) = \begin{pmatrix} 2 & 0 \\ 0 & 2 \end{pmatrix}$ is positive definite. Therefore, the origin is at a strict local minimum. On the other hand, $\nabla^2 F\left(0,\tfrac{2}{3}\right) = \begin{pmatrix} 2 & 0 \\ 0 & -2 \end{pmatrix}$ is indefinite, and hence $\mathbf{x}_2^\star = \left(0,\tfrac{2}{3}\right)^T$ a nondegenerate saddle point. The origin is, in fact, only a local minimum, since $F(0,0) = 0$, whereas $F(0,y) < 0$ for all $y > 1$. Thus, this particular function has no global minimum or maximum on $\mathbb{R}^2$.

Next, consider the function

$$F(x,y) = x^2 + y^4, \qquad \text{with gradient} \qquad \nabla F(x,y) = \begin{pmatrix} 2x \\ 4y^3 \end{pmatrix}.$$

The only critical point is the origin $x = y = 0$, which is a strict global minimizer because $F(x,y) > 0 = F(0,0)$ for all $(x,y) \neq (0,0)^T$. However, its Hessian matrix

$$\nabla^2 F(x,y) = \begin{pmatrix} 2 & 0 \\ 0 & 12y^2 \end{pmatrix}$$

is only positive semidefinite at the origin, since $\nabla^2 F(0,0) = \begin{pmatrix} 2 & 0 \\ 0 & 0 \end{pmatrix}$, and hence the origin is a degenerate critical point.

On the other hand, the origin is also the only critical point for the function

$$F(x,y) = x^2 + y^3 \qquad \text{with} \qquad \nabla F(x,y) = \begin{pmatrix} 2x \\ 3y^2 \end{pmatrix}.$$

The Hessian matrix is

$$\nabla^2 F(x,y) = \begin{pmatrix} 2 & 0 \\ 0 & 6y \end{pmatrix}, \qquad \text{and so} \qquad \nabla^2 F(0,0) = \begin{pmatrix} 2 & 0 \\ 0 & 0 \end{pmatrix}$$

is the same positive semidefinite matrix at the critical point. However, in this case $(0,0)$ is not a local minimizer; indeed $F(0,y) < 0 = F(0,0)$ whenever $y < 0$, and so there exist points arbitrarily close to the origin where F takes on smaller values. The origin is, in fact, a degenerate saddle point.

Finally, the quadratic function

$$F(x,y) = x^2 - 2xy + y^2 \qquad \text{has gradient} \qquad \nabla F(x,y) = \begin{pmatrix} 2x - 2y \\ -2x + 2y \end{pmatrix},$$

and so every point on the line $x = y$ is a critical point. The Hessian matrix

$$\nabla^2 F(x,y) = \begin{pmatrix} F_{xx} & F_{xy} \\ F_{xy} & F_{yy} \end{pmatrix} = \begin{pmatrix} 2 & -2 \\ -2 & 2 \end{pmatrix}$$

is positive semidefinite everywhere. Since $F(x,x) = 0$, while $F(x,y) = (x-y)^2 > 0$ when $x \neq y$, each of these critical points is a non-isolated, and hence non-strict, degenerate local minimizer. Thus, comparing the preceding examples, we deduce that a semidefinite Hessian matrix is unable to distinguish between different types of degenerate critical points. $\blacktriangle$

Finally, the reader should always keep in mind that first and second derivative tests only determine the local behavior of the function near the critical point. They cannot be used to determine whether or not we are at a global minimum, which requires additional analysis, and, often, a fair amount of ingenuity.

Proof of Theorem 6.30: Given $\mathbf{x}, \mathbf{y} \in \mathbb{R}^n$, consider the scalar function

$$g(t) = F(\mathbf{z}) \quad \text{where} \quad \mathbf{z} = (1-t)\mathbf{x} + t\mathbf{y}, \qquad \text{so that} \qquad g(0) = F(\mathbf{x}), \quad g(1) = F(\mathbf{y}).$$

We apply the Taylor formula (6.4): with $(a,s,t) \longmapsto (0,t,1)$, which reduces to

$$g(1) = g(0) + g'(0) + \tfrac{1}{2}\, g''(t) \qquad \text{for some} \qquad 0 \leq t \leq 1. \tag{6.78}$$

Noting that $d\mathbf{z}/dt = \mathbf{y} - \mathbf{x}$, we use the chain rule to compute the derivatives:

$$g'(t) = \sum_{i=1}^{n} \frac{\partial F}{\partial x_i}(\mathbf{z})\,(y_i - x_i) = \nabla F(\mathbf{z}) \cdot (\mathbf{y} - \mathbf{x}),$$

$$g''(t) = \sum_{i,j=1}^{n} \frac{\partial^2 F}{\partial x_i \partial x_j}(\mathbf{z})\,(y_i - x_i)(y_j - x_j) = (\mathbf{y} - \mathbf{x})^T \nabla^2 F(\mathbf{z})\,(\mathbf{y} - \mathbf{x}).$$

Substituting into (6.78) produces the *first order Taylor formula* for functions of several variables:

$$F(\mathbf{y}) = F(\mathbf{x}) + \nabla F(\mathbf{x}) \cdot (\mathbf{y} - \mathbf{x}) + \tfrac{1}{2}(\mathbf{y} - \mathbf{x})^T \nabla^2 F(\mathbf{z})\,(\mathbf{y} - \mathbf{x}), \tag{6.79}$$

for some $\mathbf{z} = (1-t)\mathbf{x} + t\mathbf{y}$, with $0 \leq t \leq 1$, lying on the line segment connecting $\mathbf{x}$ and $\mathbf{y}$.

In particular, if $\mathbf{x} = \mathbf{x}^\star$ is a local minimizer of F, then the function $g(t) = F(\mathbf{x}^\star + t\mathbf{y})$ has a local minimum at $t = 0$, and hence must satisfy

$$g'(0) = \nabla F(\mathbf{x}^\star) \cdot (\mathbf{y} - \mathbf{x}^\star) = 0, \qquad g''(0) = (\mathbf{y} - \mathbf{x}^\star)^T \nabla^2 F(\mathbf{x}^\star)\,(\mathbf{y} - \mathbf{x}^\star) \geq 0. \tag{6.80}$$

Since this holds for any $\mathbf{y} \in \mathbb{R}^n$, the first condition leads to the critical point equation $\nabla F(\mathbf{x}^\star) = \mathbf{0}$, while the second condition requires that $\nabla^2 F(\mathbf{x}^\star)$ be positive semidefinite, proving the first part of the theorem.

Conversely, if $\mathbf{x} = \mathbf{x}^\star$ is a critical point of F, the gradient term in the Taylor formula (6.79) vanishes, and hence

$$F(\mathbf{y}) = F(\mathbf{x}^\star) + \tfrac{1}{2}(\mathbf{y} - \mathbf{x}^\star)^T \nabla^2 F(\mathbf{z})\,(\mathbf{y} - \mathbf{x}^\star), \tag{6.81}$$

for some $\mathbf{z}$ lying on the line segment connecting $\mathbf{x}^\star$ and $\mathbf{y}$. Now, if $\nabla^2 F(\mathbf{x}^\star)$ is positive definite, then, by continuity — see Exercise 1.15 in Chapter 4 — $\nabla^2 F(\mathbf{z})$ is also positive definite for $\mathbf{z}$ sufficiently close to $\mathbf{x}^\star$. Thus, (6.81) implies $F(\mathbf{y}) > F(\mathbf{x}^\star)$ whenever $\mathbf{y} \neq \mathbf{x}^\star$, and hence also $\mathbf{z}$, lie sufficiently close to $\mathbf{x}^\star$. We conclude that $\mathbf{x}^\star$ is a strict local minimizer.   ∎

Setting $\mathbf{v} = \mathbf{y} - \mathbf{x}$ in the proof of Theorem 6.30, we find

$$\left. \frac{d^2}{dt^2} \right|_{t=0} F(\mathbf{x} + t\mathbf{v}) = \left(\nabla^2 F(\mathbf{x}) \mathbf{v} \right) \cdot \mathbf{v}. \tag{6.82}$$

As with our generalization of the gradient in Section 6.3, we can use this observation to define the Hessian with respect to a general inner product.

Definition 6.32. The *Hessian* of a real-valued function $F \colon \mathbb{R}^n \to \mathbb{R}$ at the point $\mathbf{x} \in \mathbb{R}^n$ with respect to the inner product $\langle \cdot, \cdot \rangle$, denoted again by $\nabla^2 F(\mathbf{x})$, is the $n \times n$ matrix defined by the equality

$$\left. \frac{d^2}{dt^2} \right|_{t=0} F(\mathbf{x} + t\mathbf{v}) = \langle \nabla^2 F(\mathbf{x}) \mathbf{v}, \mathbf{v} \rangle \qquad \text{for all} \qquad \mathbf{v} \in \mathbb{R}^n. \tag{6.83}$$

From now on, we will use $\nabla^2 F(\mathbf{x})$ to denote the Hessian with respect to an inner product, when necessary using $\nabla_C^2 F(\mathbf{x})$ to indicate the inner product $\langle \mathbf{x}, \mathbf{y} \rangle_C = \mathbf{x}^T C \mathbf{y}$. As with gradients, the standard Hessian is taken with respect to the dot product, so $C = I$, and denoted $\nabla_2^2 F$. To derive an expression for the general Hessian $\nabla_C^2 F$, we equate (6.82), (6.83):

$$\left(\nabla^2 F(\mathbf{x}) \mathbf{v} \right) \cdot \mathbf{v} = \langle \nabla_C^2 F(\mathbf{x}) \mathbf{v}, \mathbf{v} \rangle_C = \mathbf{v}^T \nabla_C^2 F(\mathbf{x})^T C \mathbf{v} = \left(C \nabla_C^2 F(\mathbf{x}) \mathbf{v} \right) \cdot \mathbf{v}.$$

Since this holds for all $\mathbf{v} \in \mathbb{R}^n$, we conclude that

$$\nabla_C^2 F(\mathbf{x}) = C^{-1} \nabla_2^2 F(\mathbf{x}), \tag{6.84}$$

which is in direct analogy with the corresponding formula for the gradient $\nabla_C F$ given in (6.36). It is worth noting that, in view of (6.75), the general Hessian can also be expressed in Jacobian form:

$$\nabla_C^2 F(\mathbf{x}) = C^{-1} \mathbf{D}(\nabla_2 F)(\mathbf{x}) = \mathbf{D}(C^{-1} \nabla_2 F)(\mathbf{x}) = \mathbf{D}(\nabla_C F)(\mathbf{x}), \tag{6.85}$$

since C is a constant matrix. Keep in mind that, unless $C = I$, the Hessian matrix $\nabla_C^2 F$ is *not* symmetric, but, as in Proposition 4.19, it is *self-adjoint* with respect to the inner product determined by C. It follows that $\nabla_C^2 F(\mathbf{x})$ is positive (semi)definite if and only if $\nabla_2^2 F$ is positive (semi)definite. Thus, Theorem 6.30 continues to hold when $\nabla_2^2 F$ is replaced by $\nabla_C^2 F$.

Example 6.33. As an example, we compute the Hessian of the general quadratic function

$$F(\mathbf{x}) = \tfrac{1}{2} \langle H\mathbf{x}, \mathbf{x} \rangle_C - \langle \mathbf{b}, \mathbf{x} \rangle_C + c = \tfrac{1}{2} \mathbf{x}^T H^T C \mathbf{x} - \mathbf{b}^T C \mathbf{x} + c,$$

where the matrix H is self-adjoint for the C inner product. The standard Hessian matrix is given by $\nabla_2^2 F(\mathbf{x}) = H^T C = CH$, which is symmetric as a consequence of the self-adjointness of H. Moreover, by (6.84), $\nabla_C^2 F(\mathbf{x}) = H$.   ▲

Exercises

6.1. When possible, use Theorem 6.30 to determine the status of the critical points you found in Exercises 3.3 and 3.4.

6.2. Let $f(x) \in C^4$ be a scalar function. (a) ♡ Suppose that $f'(x^\star) = f''(x^\star) = 0$, but $f'''(x^\star) \neq 0$. Prove that $x^\star$ cannot be a local minimizer or maximizer of $f(x)$.
(b) ◇ Suppose that $f'(x^\star) = f''(x^\star) = f'''(x^\star) = 0$, while $f''''(x^\star) > 0$. Is $x^\star$ necessarily a local (*i*) maximizer, (*ii*) minimizer, (*iii*) neither, or (*iv*) cannot tell with this information alone?

6.3. Let $f(x) \in C^1$ be restricted to a bounded closed interval $I = [a, b]$. (a) Show that if the boundary point a is a local minimizer of f on I then $f'(a) \geq 0$. Furthermore, if $f'(a) > 0$, then a is a strict local minimizer of f on I. (b) Formulate similar conditions for the right hand endpoint b.

6.4. ♡ Give an example of a quadratic function $Q(x, y)$ of two variables that has no critical points. If your answer is an affine function, try harder. What can you say about the graph of $Q(x, y)$?

6.5. ◇ Prove that a critical point with indefinite Hessian matrix (either nonsingular or singular) cannot be a local minimizer or local maximizer for the objective function.

6.6. Can a critical point with a (not identically zero) positive semidefinite Hessian be a local maximizer?

6.7. ◇ Let $f(x)$ be a C^2 scalar function, and define $F(x) = [f(x)]^2$. (a) Explain why every solution x^* to the equation $f(x) = 0$ is a global minimizer of $F(x)$. (b) Under what conditions is a solution x^* a nondegenerate minimizer? (c) Find all critical points of $F(x)$. Which are local minimizers?

6.7 Convex Functions

Determining the minima of complicated functions, especially those defined on high dimensional spaces, can be quite difficult. The innate challenges of optimization can be substantially mitigated when the objective function satisfies a convexity condition that we now introduce and develop. Such functions play a important role in our applications.

We begin by introducing the basic geometric concept of a convex subset of Euclidean space.

> **Definition 6.34.** A set $\Omega \subset \mathbb{R}^n$ is *convex* if the line segment connecting two points in the set is also contained therein:
>
> $$\{\, t\mathbf{x} + (1-t)\mathbf{y} \mid 0 \leq t \leq 1 \,\} \subset \Omega \qquad \text{for all} \qquad \mathbf{x} \neq \mathbf{y} \in \Omega. \tag{6.86}$$
>
> The set is called *strictly convex* if the interior of the segment, meaning all except its endpoints, lies in the interior of Ω.

For example, $\mathbb{R}^n$ itself is strictly convex, as is any open ball $\{\, \|\mathbf{x} - \mathbf{a}\| < r \,\}$ for $r > 0$ and $\mathbf{a} \in \mathbb{R}^n$, where $\|\cdot\|$ can be any norm. On the other hand, the closed ball $\{\, \|\mathbf{x} - \mathbf{a}\| \leq r \,\}$ is convex, but not necessarily strictly convex; it is strictly convex for the Euclidean norm, but

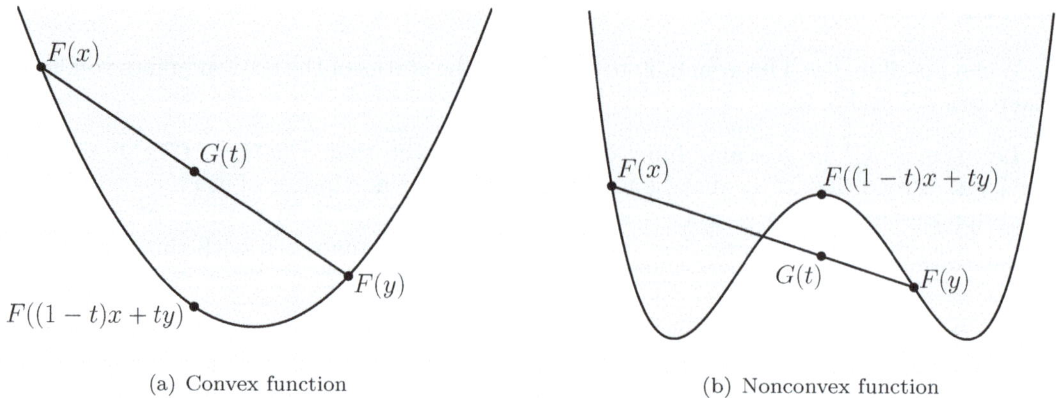

(a) Convex function   (b) Nonconvex function

Figure 6.4: An illustration of the definition of a convex function, Definition 6.35. In (a) we show a strictly convex function, and in (b) we show a nonconvex function where (6.87) does not hold. In the figures, we write $G(t) = (1 - t)\,F(\mathbf{x}) + t\,F(\mathbf{y})$. Notice that the secant line must stay within the shaded region above the graph for the function to be convex.

not for the 1 or ∞ norms since the line segment connecting points on the same flat part of the boundary will not lie in its interior, i.e., the corresponding open ball.

Definition 6.35. A real-valued function $F\colon \mathbb{R}^n \to \mathbb{R}$ is *convex* if the domain lying above its graph (known as its *epigraph*) is a convex subset of $\mathbb{R}^{n+1}$. In other words, given any $\mathbf{x} \neq \mathbf{y}$ in the domain of F, convexity requires

$$F\big((1 - t)\mathbf{x} + t\mathbf{y}\big) \le (1 - t)\,F(\mathbf{x}) + t\,F(\mathbf{y}) \qquad \text{for all} \qquad 0 \le t \le 1. \tag{6.87}$$

Strict convexity of F is defined by imposing the strict inequality in (6.87) whenever $0 < t < 1$.

Remark. A function $F(\mathbf{x})$ is called *concave* if its negative $-F(\mathbf{x})$ is convex.   ▲

The left hand side of the convexity inequality (6.87) corresponds to the value of F on the line segment connecting $\mathbf{x}$ to $\mathbf{y}$, while the right hand side parametrizes the secant line segment connecting the points $(\mathbf{x}, F(\mathbf{x}))$ and $(\mathbf{y}, F(\mathbf{y}))$ on the graph of F. Thus, convexity requires that, on each line segment in the domain of F, its graph lies on or below the corresponding secant line; strict convexity requires it lies strictly below except at the endpoints $\mathbf{x}, \mathbf{y}$. Figure 6.4 gives an illustration of convex and nonconvex functions, plotting the values of $F\big((1 - t)\mathbf{x} + t\mathbf{y}\big)$ and the secant line $G(t) = (1 - t)\,F(\mathbf{x}) + t\,F(\mathbf{y})$ used in Definition 6.35. It is also important to note that if we set $\mathbf{x} = \mathbf{0}$ in (6.87) and then replace $\mathbf{y}$ by $\mathbf{x}$, we deduce

$$F(t\mathbf{x}) \le t\,F(\mathbf{x}) \quad \text{for all} \;\; 0 \le t \le 1, \quad \text{provided} \;\; F(\mathbf{0}) = 0. \tag{6.88}$$

Example 6.36. Consider the quadratic function

$$F(\mathbf{x}) = \tfrac{1}{2}\mathbf{x}^T H \mathbf{x} - \mathbf{x}^T \mathbf{b} + c, \tag{6.89}$$

where H is a symmetric matrix. A short calculation shows that

$$F\big((1 - t)\mathbf{x} + t\mathbf{y}\big) = (1 - t)\,F(\mathbf{x}) + t\,F(\mathbf{y}) - \tfrac{1}{2}t(1 - t)\,(\mathbf{y} - \mathbf{x})^T H\,(\mathbf{y} - \mathbf{x}).$$

Thus F is convex if and only if the last term (including the minus sign) is ≤ 0 for all $\mathbf{x}, \mathbf{y}$ and all $0 \leq t \leq 1$, which is equivalent to the condition that H be positive semidefinite. In particular, setting $H = O$, we conclude that any affine function is convex. By the same reasoning, F is strictly convex if and only if H is positive definite. So affine functions are convex, but not strictly so. On the other hand, setting $\mathbf{b} = \mathbf{0}$ and $c = 0$, and letting H be positive definite, the corresponding squared norm function $2F(\mathbf{x}) = \|\mathbf{x}\|_H^2 = \mathbf{x}^T H \mathbf{x}$ is strictly convex. ▲

The reader is asked to prove the following result in Exercise 7.10.

Lemma 6.37. *If $F, G \colon \mathbb{R}^n \to \mathbb{R}$ are both convex, and $0 \leq a, b \in \mathbb{R}$, then the linear combination $a\,F + b\,G$ is also convex.*

Another important result that follows from the definition of convexity is *Jensen's inequality*, which the reader is asked to prove in Exercise 7.16.

Theorem 6.38 (Jensen's Inequality)**.** *Let $F \colon \mathbb{R}^n \to \mathbb{R}$ be convex. Let $t_1, \ldots, t_m \geq 0$ with $t_1 + \cdots + t_m = 1$. Then, for any $\mathbf{x}_1, \ldots, \mathbf{x}_m \in \mathbb{R}^n$,*

$$F\left(\sum_{i=1}^{m} t_i \mathbf{x}_i \right) \leq \sum_{i=1}^{m} t_i F(\mathbf{x}_i). \tag{6.90}$$

Notice that Jensen's inequality with $m = 2$ is exactly the definition of convexity given in Definition 6.35 — indeed, set $t = t_1$ and then note that $t_2 = 1 - t_1 = 1 - t$, and so (6.90) is equivalent to (6.87). Thus, Jensen's inequality extends the definition of convexity to more than 2 points.

For the rest of this section, we fix an inner product $\langle \cdot, \cdot \rangle$ and induced norm $\|\cdot\|$ on $\mathbb{R}^n$, which need not be Euclidean. If the function F is continuously differentiable, convexity can be alternatively characterized by the statement that its graph lies above its tangent space at each point; see Figure 6.5.

Theorem 6.39. *Let $F \colon \mathbb{R}^n \to \mathbb{R}$ be continuously differentiable. Then F is convex if and only if*

$$F(\mathbf{y}) \geq F(\mathbf{x}) + \langle \nabla F(\mathbf{x}), \mathbf{y} - \mathbf{x} \rangle \qquad \text{for all} \qquad \mathbf{x}, \mathbf{y} \in \mathbb{R}^n. \tag{6.91}$$

Remark. Fixing $\mathbf{x}$, the right hand side of (6.91), as a function of $\mathbf{y}$, defines the *tangent space* to the graph of F at $\mathbf{x}$; in particular, when $n = 1$ it defines the tangent line. ▲

Proof. First, if $0 < t \leq 1$, then dividing both sides of the convexity inequality (6.87) by t yields

$$F(\mathbf{y}) \geq F(\mathbf{x}) + \frac{F\big((1-t)\mathbf{x} + t\mathbf{y}\big) - F(\mathbf{x})}{t} = F(\mathbf{x}) + \frac{F\big(\mathbf{x} + t(\mathbf{y} - \mathbf{x})\big) - F(\mathbf{x})}{t}.$$

In the limit as $t \to 0^+$, the second term converges to

$$\lim_{t^+ \to 0} \frac{F\big(\mathbf{x} + t(\mathbf{y} - \mathbf{x})\big) - F(\mathbf{x})}{t} = \frac{d}{dt} F\big(\mathbf{x} + t(\mathbf{y} - \mathbf{x})\big) \bigg|_{t=0} = \langle \nabla F(\mathbf{x}), \mathbf{y} - \mathbf{x} \rangle,$$

by the definition (6.33) of the gradient, thus establishing (6.91).

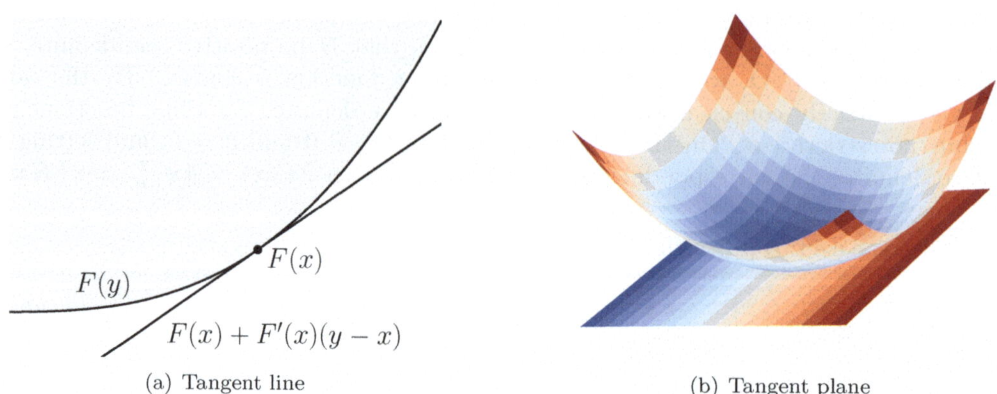

(a) Tangent line   (b) Tangent plane

Figure 6.5: An illustration of Theorem 6.39, which states that the graph of a convex function $F(y)$ lies above (a) its tangent line $G(y) = F(x) + F'(x)(y - x)$ centered at any point $x \in \mathbb{R}$ when F depends on one variable, and (b) its tangent plane when F depends on two variables.

To prove the converse, set

$$\mathbf{z} = (1 - t)\mathbf{x} + t\mathbf{y}.$$

Then (6.91) implies

$$F(\mathbf{x}) \geq F(\mathbf{z}) + \langle \nabla F(\mathbf{z}), \mathbf{x} - \mathbf{z} \rangle, \quad \text{and} \quad F(\mathbf{y}) \geq F(\mathbf{z}) + \langle \nabla F(\mathbf{z}), \mathbf{y} - \mathbf{z} \rangle.$$

Let us multiply the first inequality by $1 - t$ and the second by t, noting that $0 \leq t \leq 1$. When we add the resulting inequalities, the terms involving $\nabla F(\mathbf{z})$ cancel out and we are left with

$$(1 - t)F(\mathbf{x}) + tF(\mathbf{y}) \geq F(\mathbf{z}),$$

which, in view of the formula for $\mathbf{z}$, is the convexity condition (6.87).   ■

In our applications to optimization, the convexity inequality (6.91) is used to compare the value of $F(\mathbf{x})$ to the optimal value $F(\mathbf{x}^\star)$. In particular, if we set $\mathbf{y} = \mathbf{x}^\star$, and rearrange (6.91), we obtain

$$F(\mathbf{x}) - F(\mathbf{x}^\star) \leq \langle \nabla F(\mathbf{x}), \mathbf{x} - \mathbf{x}^\star \rangle. \tag{6.92}$$

As a consequence of (6.91), we deduce the following important result concerning minima of convex functions.

Proposition 6.40. *If $\mathbf{x}^\star$ is a critical point of a convex function, then it is a global minimizer.*

Proof. Indeed, if $\nabla F(\mathbf{x}^\star) = 0$, then (6.91) implies $F(\mathbf{y}) \geq F(\mathbf{x}^\star)$ for any $\mathbf{y} \in \mathbb{R}^n$, and hence $\mathbf{x}^\star$ is a global minimizer.   ■

In general, a convex function can have more than one global minimizer (e.g., every point is a minimizer of a constant function), or it can fail to have a global minimizer. When a global minimizer exists, its uniqueness requires an additional condition on F.

Theorem 6.41. *If $F \colon \mathbb{R}^n \to \mathbb{R}$ is strictly convex, then it has at most one critical point, which, when it exists, is its global minimizer.*

Proof. Suppose $\mathbf{y}^* \neq \mathbf{x}^*$ is another critical point. Proposition 6.40 implies they are both global minimizers, so $F(\mathbf{y}^*) = F(\mathbf{x}^*) \leq F(\mathbf{x})$ for all $\mathbf{x} \in \mathbb{R}^n$. Now, the strict version of (6.87) implies

$$F\big((1-t)\mathbf{x}^* + t\mathbf{y}^*\big) < (1-t)F(\mathbf{x}^*) + tF(\mathbf{y}^*) = F(\mathbf{x}^*) \qquad \text{whenever} \qquad 0 < t < 1,$$

which contradicts our assumption that $\mathbf{x}^*$ is a global minimizer. $\blacksquare$

Warning: Not every convex function, or even every strictly convex function has a global minimum. An example of the latter is the scalar function $F(x) = e^x$.

In general, it can often be difficult to test whether a function is convex using Definition 6.35 or Theorem 6.39. When the function F is twice continuously continuously differentiable, we can use the second derivative, i.e., its Hessian, to test for convexity, which is often simpler to check in practice.

> **Theorem 6.42.** *Let $F: \mathbb{R}^n \to \mathbb{R}$ be twice continuously differentiable. Then F is convex if and only if its Hessian matrix is positive semidefinite at each point:* $\nabla^2 F(\mathbf{x}) \geq 0$.

For example, the quadratic function (6.89) has Hessian matrix equal to H, and is thus convex if and only if H is positive semidefinite.

Proof. Fixing $\mathbf{x}$, consider the function

$$G(\mathbf{y}) = F(\mathbf{y}) + \langle\, \nabla F(\mathbf{x}), \mathbf{x} - \mathbf{y}\, \rangle.$$

Note that $G(\mathbf{y})$ is the sum of $F(\mathbf{y})$ and an affine function of $\mathbf{y}$, and hence by Lemma 6.37, G is also convex. Moreover, taking the gradient with respect to $\mathbf{y}$ and keeping $\mathbf{x}$ fixed,

$$\nabla G(\mathbf{y}) = \nabla F(\mathbf{y}) - \nabla F(\mathbf{x}), \qquad \text{and hence} \qquad \nabla G(\mathbf{x}) = \mathbf{0},$$

which means that $\mathbf{x}$ is a critical point of G. Thus, Proposition 6.40 implies that $\mathbf{x}$ is a global minimizer. Theorem 6.30 implies that the Hessian matrix of G at $\mathbf{x}$ must be positive semidefinite. But $\nabla^2 G(\mathbf{y}) = \nabla^2 F(\mathbf{y})$ for all $\mathbf{y}$, and hence $\nabla^2 F(\mathbf{x}) = \nabla^2 G(\mathbf{x}) \geq 0$. As for the converse, positive semidefiniteness of $\nabla^2 F$ implies that the last term in the first order Taylor formula (6.79) is ≥ 0, which immediately yields the convexity inequality (6.91). $\blacksquare$

Remark. The second part of the proof shows that if the Hessian matrix is everywhere positive definite, then the function is strictly convex. However, it is not true that strict convexity implies positive definiteness of the Hessian. For example, the scalar function $f(x) = x^4$ is strictly convex, but has vanishing second derivative at the origin. $\blacktriangle$

Example 6.43. The negative logarithm $f(x) = -\log x = \log(1/x)$ has positive second derivative, $f''(x) = 1/x^2 > 0$, and hence, in accordance with the shape of its graph, is strictly convex on its domain $\{x > 0\}$; see Figure 6.6(a).

On the domain $\Omega = \{x, y > 0\}$, the *relative entropy* function

$$F(x,y) = -x\log(y/x) = x\log(x/y) = x\log x - x\log y \tag{6.93}$$

has positive semidefinite Hessian (see Exercise 7.15)

$$\nabla_2^2 F(x,y) = \begin{pmatrix} 1/x & -1/y \\ -1/y & x/y^2 \end{pmatrix} \tag{6.94}$$

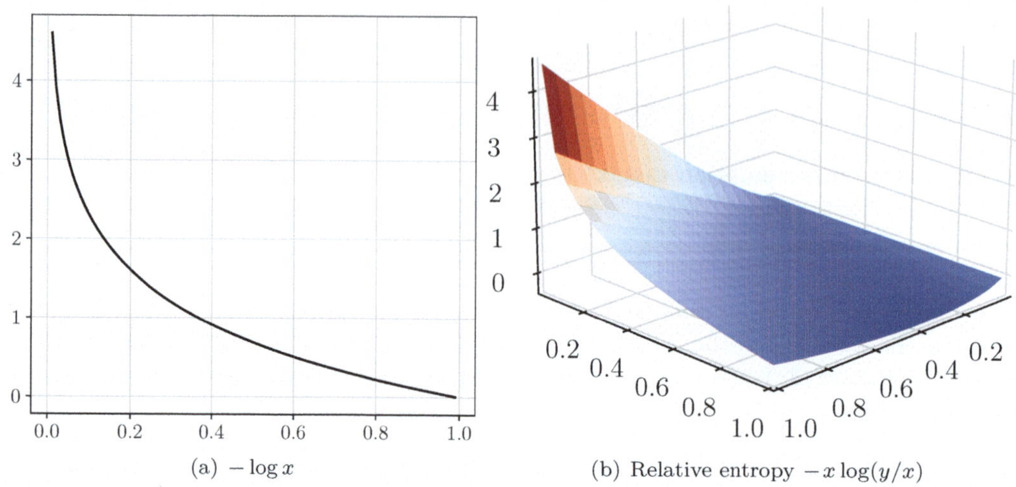

Figure 6.6: Plots of the convex functions (a) $f(x) = -\log x$ (b) $F(x,y) = -x\log(y/x)$.

at each point, and hence is convex. On the other hand, (6.93) is not strictly convex since it depends linearly on x along the rays $y = \lambda x$; see Figure 6.6(b).

In higher dimensions, the *relative entropy*, also known as the *Kullback-Leibler divergence*, between two probability vectors with all positive entries is obtained by summing the relative entropies of their respective components:

$$F(\mathbf{x},\mathbf{y}) = \sum_{i=1}^{n} x_i \log \frac{x_i}{y_i} = \sum_{i=1}^{n} (x_i \log x_i - x_i \log y_i), \qquad \mathbf{x},\mathbf{y} \in \Omega,$$

$$\text{where} \qquad \Omega = \left\{ \mathbf{x} \in \mathbb{R}^n \mid x_i > 0,\ i = 1,\ldots,n,\ x_1 + \cdots + x_n = 1 \right\}.$$

(6.95)

Since it is the sum of convex functions, Lemma 6.37 implies that it is also convex. According to Exercise 7.7, the Kullback-Liebler divergence is also always nonnegative, that is $F(\mathbf{x},\mathbf{y}) \geq 0$ for all $\mathbf{x},\mathbf{y} \in \Omega$, and $F(\mathbf{x},\mathbf{y}) = 0$ when $\mathbf{x} = \mathbf{y}$, making it a reasonable candidate for a notion of distance between such vectors.   ▲

6.7.1   Some Inequalities

We take a moment here to discuss some applications of the theory of convex functions we have developed so far, by establishing some basic inequalities that are used throughout mathematical analysis. We begin with *Young's inequality*.

Theorem 6.44 (Young's inequality). *Let $p,q > 1$ be any two numbers satisfying*

$$\frac{1}{p} + \frac{1}{q} = 1.$$

(6.96)

Then

$$ab \leq \frac{a^p}{p} + \frac{b^q}{q} \qquad \text{for any} \qquad a,b \geq 0.$$

(6.97)

Proof. According to Exercise 7.1, the exponential function $\exp(x) = e^x$ is strictly convex. Thus, if $a, b > 0$,

$$ab = \exp\big(\log(ab)\big) = \exp\big(\log a + \log b\big) = \exp\left(\frac{1}{p}\log a^p + \frac{1}{q}\log b^q\right)$$

$$\leq \frac{1}{p}\exp\big(\log(a^p)\big) + \frac{1}{q}\exp\big(\log(b^q)\big) = \frac{a^p}{p} + \frac{b^q}{q},$$

where we used the defining equation, (6.91), of convexity with $t = 1/q$ and $1 - t = 1/p$, based on (6.96). The case where $a = 0$ or $b = 0$ is trivially true. $\blacksquare$

Taking $p = q = 2$ in Young's inequality (6.97) yields *Cauchy's inequality*

$$ab \leq \tfrac{1}{2}a^2 + \tfrac{1}{2}b^2, \tag{6.98}$$

which is valid for all $a, b \in \mathbb{R}$. Cauchy's inequality can alternatively be proved by expanding and rearranging the inequality $(a - b)^2 \geq 0$.

Young's inequality (6.97) has several important applications, the first of which is *Hölder's inequality*.

Theorem 6.45 (Hölder's inequality). *Let $1 \leq p, q \leq \infty$ satisfy (6.96), where, by convention, when $p = 1$ we set $q = \infty$, and vice versa. Then*

$$\mathbf{x} \cdot \mathbf{y} \leq \|\mathbf{x}\|_p \|\mathbf{y}\|_q \qquad \textit{for all} \qquad \mathbf{x}, \mathbf{y} \in \mathbb{R}^n. \tag{6.99}$$

Proof. We assume $1 < p, q < \infty$, leaving the remaining case to the reader as Exercise 7.17. Since the inequality is trivial whenever $\mathbf{x}$ or $\mathbf{y}$ is the zero vector, we further assume $\mathbf{x}, \mathbf{y} \neq \mathbf{0}$. Dividing both sides of (6.99) by $\|\mathbf{x}\|_p \|\mathbf{y}\|_q$ and setting $\mathbf{u} = \mathbf{x}/\|\mathbf{x}\|_p$, $\mathbf{v} = \mathbf{y}/\|\mathbf{y}\|_q$, so that $\|\mathbf{u}\|_p = \|\mathbf{v}\|_q = 1$, it suffices to prove $\mathbf{u} \cdot \mathbf{v} \leq 1$ under these conditions. We write out the dot product and use Young's inequality on each summand:

$$\mathbf{u} \cdot \mathbf{v} = \sum_{i=1}^{n} u_i v_i \leq \sum_{i=1}^{n} \left(\frac{u_i^p}{p} + \frac{v_i^q}{q}\right) = \frac{\|\mathbf{u}\|_p^p}{p} + \frac{\|\mathbf{v}\|_q^q}{q} = \frac{1}{p} + \frac{1}{q} = 1. \qquad \blacksquare$$

Note that when $p = q = 2$, Hölder's inequality reduces to the Cauchy–Schwarz inequality (2.27), and hence can be regarded as its generalization to the p and q norms when subject to the algebraic relation (6.96).

Equipped with Holder's inequality, we can prove *Minkowski's inequality*, which establishes the triangle inequality for the p norms, thus proving that they do define norms on $\mathbb{R}^n$.

Theorem 6.46 (Minkowski's inequality). *Let $1 \leq p \leq \infty$. Then*

$$\|\mathbf{x} + \mathbf{y}\|_p \leq \|\mathbf{x}\|_p + \|\mathbf{y}\|_p \qquad \textit{for all} \qquad \mathbf{x}, \mathbf{y} \in \mathbb{R}^n. \tag{6.100}$$

Proof. We have already established the triangle inequality for $p = 1, 2, \infty$, so we may assume $1 < p < \infty$; observe that the corresponding $1 < q < \infty$ satisfying (6.96) is $q = p/(p-1)$. Now, note that

$$\|\mathbf{x} + \mathbf{y}\|_p^p = \sum_{i=1}^{n} |x_i + y_i|\,|x_i + y_i|^{p-1} \leq \sum_{i=1}^{n} |x_i|\,|x_i + y_i|^{p-1} + \sum_{i=1}^{n} |y_i|\,|x_i + y_i|^{p-1}. \tag{6.101}$$

Let $\widetilde{\mathbf{x}}, \mathbf{z} \in \mathbb{R}^n$ have respective components $\widetilde{x}_i = |x_i|$, $z_i = |x_i + y_i|^{p-1}$, so that

$$\|\widetilde{\mathbf{x}}\|_p = \|\mathbf{x}\|_p, \quad \|\mathbf{z}\|_q = \left(\sum_{i=1}^n |x_i + y_i|^{(p-1)q} \right)^{1/q} = \left(\sum_{i=1}^n |x_i + y_i|^p \right)^{(p-1)/p} = \|\mathbf{x} + \mathbf{y}\|_p^{p-1}.$$

Thus, by Hölder's inequality (6.99),

$$\sum_{i=1}^n |x_i| \, |x_i + y_i|^{p-1} = \widetilde{\mathbf{x}} \cdot \mathbf{z} \le \|\widetilde{\mathbf{x}}\|_p \, \|\mathbf{z}\|_q = \|\mathbf{x} + \mathbf{y}\|_p^{p-1} \, \|\mathbf{x}\|_p.$$

Interchanging $\mathbf{x}$ and $\mathbf{y}$ shows that

$$\sum_{i=1}^n |y_i| \, |x_i + y_i|^{p-1} \le \|\mathbf{x} + \mathbf{y}\|_p^{p-1} \, \|\mathbf{y}\|_p.$$

Inserting these estimates into (6.101) produces

$$\|\mathbf{x} + \mathbf{y}\|_p^p \le \|\mathbf{x} + \mathbf{y}\|_p^{p-1} \left(\|\mathbf{x}\|_p + \|\mathbf{y}\|_p \right). \tag{6.102}$$

Minkowski's inequality (6.100) is trivially true when $\mathbf{x} + \mathbf{y} = \mathbf{0}$. Otherwise, we can divide both sides of (6.102) by $\|\mathbf{x} + \mathbf{y}\|_p^{p-1}$ to complete the proof. ∎

6.7.2   Strong Convexity

Finally, we introduce a more restrictive notion of convexity that is important in the convergence analysis of gradient descent.

> **Definition 6.47** (Strong convexity)**.** Let $\mu > 0$. A real-valued function $F \colon \mathbb{R}^n \to \mathbb{R}$ is said to be *μ-strongly convex* if the function
>
> $$G(\mathbf{x}) = F(\mathbf{x}) - \tfrac{1}{2}\mu \|\mathbf{x}\|^2 \tag{6.103}$$
>
> is convex. We will say F is *strongly convex* if it is μ-strongly convex for some $\mu > 0$.

Remark. By the equivalence of norms on $\mathbb{R}^n$, as formulated in Theorem 2.36, the property of being strongly convex is independent of the choice of norm., although the value of the strong convexity constant μ is norm dependent. See Example 6.49 below. ▲

If F is μ-strongly convex then $F(\mathbf{x}) = G(\mathbf{x}) + \tfrac{1}{2}\mu\|\mathbf{x}\|^2$ is the sum of a convex and a strictly convex function — see Example 6.36 — and hence is also strictly convex. An important property of a strongly convex function is that it is not "too flat" anywhere, as illustrated in the following examples.

Example 6.48. The zero function $F(\mathbf{x}) = 0$ is convex, but not strongly convex for any $\mu > 0$ since $F(\mathbf{x}) - \tfrac{1}{2}\mu\|\mathbf{x}\|^2 = -\tfrac{1}{2}\mu\|\mathbf{x}\|^2$ is not convex when $\mu > 0$; indeed, it is concave, i.e., the negative of a convex function. Likewise, the quadratic function $F(\mathbf{x}) = \tfrac{1}{2}a\|\mathbf{x}\|^2$ is convex, but is only μ-strongly convex when $a \ge \mu$. ▲

Example 6.49. Consider the general quadratic function

$$F(\mathbf{x}) = \tfrac{1}{2}\langle H\mathbf{x}, \mathbf{x} \rangle + \langle \mathbf{b}, \mathbf{x} \rangle + c,$$

where H is self-adjoint in the inner product $\langle\,\cdot\,,\cdot\,\rangle$. Then by Example 6.33 the Hessian of $G(\mathbf{x}) = F(\mathbf{x}) - \frac{1}{2}\mu\|\mathbf{x}\|^2$ is given by $\nabla^2 G(\mathbf{x}) = H - \mu\mathrm{I}$. Thus, by Theorem 6.42, F is μ-strongly convex whenever $H - \mu\mathrm{I}$ is positive semidefinite, which requires that H be positive definite. According to Theorem 5.32, this is equivalent to the condition that $0 < \mu \leq \lambda_{min}(H)$, and hence F is $\lambda_{min}(H)$-strongly convex, where $\lambda_{min}(H)$ is the smallest eigenvalue of H.

This observation has a natural extension to general functions. By Theorem 6.42, a function $F\colon \mathbb{R}^n \to \mathbb{R}$ is strongly convex if and only if the matrix $\nabla^2 F(\mathbf{x}) - \mu\mathrm{I}$, which is the Hessian of (6.103), is positive semidefinite for all $\mathbf{x} \in \mathbb{R}^n$, which implies that $0 < \mu \leq \lambda_{min}(\nabla^2 F(\mathbf{x}))$. In particular, F is μ-strongly convex provided

$$0 < \mu = \min_{\mathbf{x}} \lambda_{min}(\nabla^2 F(\mathbf{x})). \qquad (6.104)$$

If we fix the inner product to be $\langle \mathbf{x}, \mathbf{y} \rangle_C = \mathbf{x}^T C \mathbf{y}$, then (6.104) says that F is μ-strongly convex with

$$\mu = \min_{\mathbf{x}} \lambda_{min}(\nabla^2_C F(\mathbf{x})) = \min_{\mathbf{x}} \lambda_{min}(C^{-1}\nabla^2_2 F(\mathbf{x})).$$

At this point, it would seem that we can make the intriguing observation that if we choose $C = \nabla^2 F(\mathbf{x})$ then F would be strongly convex with constant $\mu = 1$! However, this is not exactly true since we are not allowed to take C to be a variable function of $\mathbf{x}$ when setting up the inner product. It is nonetheless common in optimization to take the preconditioner to be $C = \nabla^2 F(\mathbf{x}_k)$, or some approximation thereof, where $\mathbf{x}_k$ is the current iterate for an optimization algorithm, e.g., gradient descent.　▲

We now state an alternative characterization of strong convexity.

Theorem 6.50. *A continuously differentiable function $F\colon \mathbb{R}^n \to \mathbb{R}$ is μ-strongly convex if and only if*

$$F(\mathbf{y}) \geq F(\mathbf{x}) + \langle \nabla F(\mathbf{x}), \mathbf{y} - \mathbf{x} \rangle + \tfrac{1}{2}\mu\|\mathbf{x} - \mathbf{y}\|^2 \quad \textit{for all} \quad \mathbf{x}, \mathbf{y} \in \mathbb{R}^n. \qquad (6.105)$$

Proof. According to Theorem 6.39, the function (6.103) is convex if and only if

$$G(\mathbf{y}) \geq G(\mathbf{x}) + \langle \nabla G(\mathbf{x}), \mathbf{y} - \mathbf{x} \rangle \quad \text{for all} \quad \mathbf{x}, \mathbf{y} \in \mathbb{R}^n.$$

Since $\nabla G(\mathbf{x}) = \nabla F(\mathbf{x}) - \mu\mathbf{x}$, this holds in turn if and only if

$$F(\mathbf{y}) - \tfrac{1}{2}\mu\|\mathbf{y}\|^2 \geq F(\mathbf{x}) - \tfrac{1}{2}\mu\|\mathbf{x}\|^2 + \langle \nabla F(\mathbf{x}), \mathbf{y} - \mathbf{x} \rangle - \mu\langle \mathbf{x}, \mathbf{y} - \mathbf{x} \rangle$$

for all $\mathbf{x}, \mathbf{y} \in \mathbb{R}^n$. Rearranging terms, we find that this is equivalent to (6.105).　∎

An important consequence of Theorems 6.50 and 6.41 is that a strongly convex function always admits a minimum.

Theorem 6.51. *If $F\colon \mathbb{R}^n \to \mathbb{R}$ is strongly convex, then F has a unique global minimizer $\mathbf{x}^\star \in \mathbb{R}^n$.*

Proof. By replacing $F(\mathbf{x})$ with $F(\mathbf{x}) - F(\mathbf{0})$, we can assume that $F(\mathbf{0}) = 0$. Let $\mathbf{x}_0$ denote a minimizer of F over the unit ball $B_1 = \{\|\mathbf{x}\| \leq 1\}$, so that $F(\mathbf{x}_0) \leq F(\mathbf{x})$ for all $\mathbf{x} \in B_1$. Existence of a minimizer follows from the compactness of the unit ball; see Theorem 2.35.

By definition of strong convexity, the function $G(\mathbf{x}) = F(\mathbf{x}) - \frac{1}{2}\mu\|\mathbf{x}\|^2$ is convex for some $\mu > 0$. Moreover, $G(\mathbf{0}) = F(\mathbf{0}) - \frac{1}{2}\mu\|\mathbf{0}\|^2 = 0$, and thus (6.88) implies

$$F(t\mathbf{x}) - \tfrac{1}{2}\mu\|t\mathbf{x}\|^2 = G(t\mathbf{x}) \le t\,G(\mathbf{x}) = t\,F(\mathbf{x}) - \tfrac{1}{2}t\mu\|\mathbf{x}\|^2, \qquad \mathbf{x} \in \mathbb{R}^n, \qquad 0 \le t \le 1,$$

which, provided $t \ne 0$, can be rearranged to read

$$F(\mathbf{x}) \ge \frac{1}{t}F(t\mathbf{x}) + \frac{\mu}{2}(1-t)\|\mathbf{x}\|^2, \qquad 0 < t \le 1.$$

Now, suppose $\|\mathbf{x}\| \ge 1$. Set $t = 1/\|\mathbf{x}\|$ so that $0 < t \le 1$ and $t\mathbf{x} \in B_1$, and hence $F(t\mathbf{x}) \ge F(\mathbf{x}_0)$. Thus, by the preceding inequality,

$$F(\mathbf{x}) \ge F(\mathbf{x}_0)\|\mathbf{x}\| + \frac{\mu}{2}\left(1 - \frac{1}{\|\mathbf{x}\|}\right)\|\mathbf{x}\|^2 = \frac{\mu}{2}\|\mathbf{x}\|\left[\|\mathbf{x}\| - \left(1 - \frac{2F(\mathbf{x}_0)}{\mu}\right)\right].$$

Thus, since $F(\mathbf{x}_0) \le F(\mathbf{0}) = 0$, if

$$\|\mathbf{x}\| \ge r := 1 - \frac{2F(\mathbf{x}_0)}{\mu} \ge 1, \qquad \text{then} \qquad F(\mathbf{x}) \ge 0.$$

Finally, given r, let $\mathbf{x}^\star$ denote a minimizer of F over the ball $B_r = \{\|\mathbf{x}\| \le r\}$, so that $F(\mathbf{x}^\star) \le F(\mathbf{x})$ for all $\mathbf{x} \in B_r$ and, in particular, $F(\mathbf{x}^\star) \le F(\mathbf{0}) = 0$. The preceding argument tells us that $F(\mathbf{x}^\star) \le 0 \le F(\mathbf{x})$ for all $\mathbf{x} \in \mathbb{R}^n \setminus B_r$ also, and hence $\mathbf{x}^\star$ is a global minimizer of F, whose uniqueness is assured by Theorem 6.41. $\blacksquare$

We next establish the *Polyak-Lojasiewicz* (PL) *inequality* [153, 186], which will be a key ingredient in the proof of the linear convergence rate for gradient descent on strongly convex functions.

Theorem 6.52. *Let F be continuously differentiable and μ-strongly convex, and let $\mathbf{x}^\star$ be its global minimizer. Then,*

$$F(\mathbf{x}) - F(\mathbf{x}^\star) \le \frac{1}{2\mu}\|\nabla F(\mathbf{x})\|^2 \qquad \text{for all} \qquad \mathbf{x} \in \mathbb{R}^n. \tag{6.106}$$

Proof. Fix $\mathbf{x} \in \mathbb{R}^n$. We minimize both sides of the strong convexity inequality (6.105) over $\mathbf{y} \in \mathbb{R}^n$ to find that

$$F(\mathbf{x}^\star) = \min_{\mathbf{y}} F(\mathbf{y}) \ge F(\mathbf{x}) + \min_{\mathbf{y}}\left[\langle \nabla F(\mathbf{x}), \mathbf{y} - \mathbf{x}\rangle + \tfrac{1}{2}\mu\|\mathbf{x} - \mathbf{y}\|^2\right]. \tag{6.107}$$

The expression to be minimized is a quadratic polynomial in $\mathbf{y}$ whose quadratic term, namely $\frac{1}{2}\mu\|\mathbf{y}\|^2$ is positive definite. Using the methods in Section 6.2, we find that $\mathbf{y} = \mathbf{x} - \mu^{-1}\nabla F(\mathbf{x})$ is its global minimizer. Substituting the value of $\mathbf{y}$ into (6.107) yields

$$F(\mathbf{x}^\star) \ge F(\mathbf{x}) - \frac{1}{\mu}\|\nabla F(\mathbf{x})\|^2 + \frac{1}{2\mu}\|\nabla F(\mathbf{x})\|^2 = F(\mathbf{x}) - \frac{1}{2\mu}\|\nabla F(\mathbf{x})\|^2. \qquad \blacksquare$$

Remark 6.53. If F is continuously differentiable and μ-strongly convex, its unique minimizer $\mathbf{x}^\star$ satisfies $\nabla F(\mathbf{x}^\star) = \mathbf{0}$. Replacing $\mathbf{x}$ and $\mathbf{y}$ by $\mathbf{x}^\star$ and $\mathbf{x}$, respectively, in (6.105) produces

$$\frac{\mu}{2}\|\mathbf{x} - \mathbf{x}^\star\|^2 \le F(\mathbf{x}) - F(\mathbf{x}^\star). \tag{6.108}$$

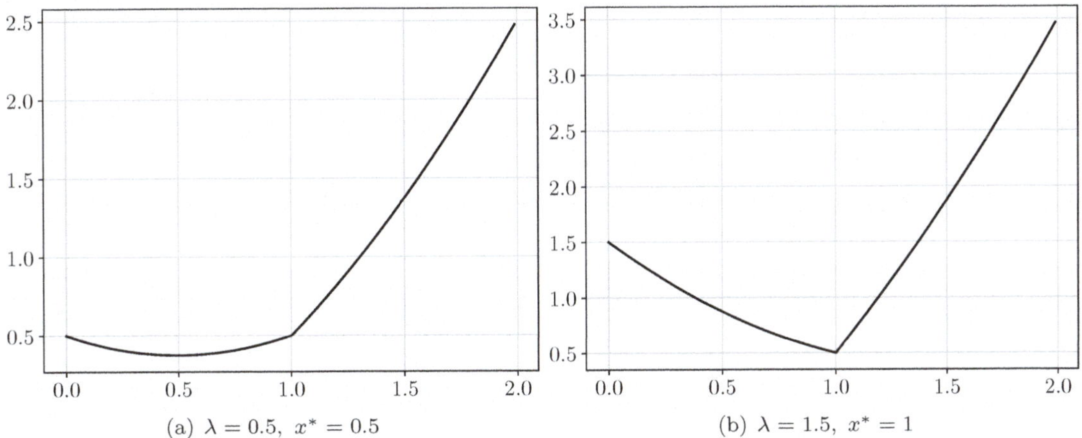

(a) $\lambda = 0.5,\ x^* = 0.5$

(b) $\lambda = 1.5,\ x^* = 1$

Figure 6.7: Plots of the function $F(x) = \frac{1}{2}x^2 + \lambda\,|x-1|$

This inequality shows that, for a strongly convex function, the distance to the minimizer $\|\mathbf{x} - \mathbf{x}^\star\|$ is controlled by the difference in the objective values $F(\mathbf{x}) - F(\mathbf{x}^\star)$. Note that (6.108) gives another proof of uniqueness of minimizers for strongly convex functions when F is continuously differentiable, as $F(\mathbf{x}) = F(\mathbf{x}^\star)$ implies $\mathbf{x} = \mathbf{x}^\star$. We also mention that combining (6.106) with (6.108) produces the useful inequality

$$\mu^2 \|\mathbf{x} - \mathbf{x}^\star\| \leq \|\nabla F(\mathbf{x})\|. \tag{6.109}$$

Hence, we can also control the distance from a point $\mathbf{x}$ to the minimizer $\mathbf{x}^\star$ by the norm of the gradient, which shows that if gradient descent finds a point $\mathbf{x}$ with $\nabla F(\mathbf{x}) \approx \mathbf{0}$, then $\mathbf{x}$ is close to the global minimizer $\mathbf{x}^\star$. ▲

Example 6.54. Let us consider the problem of minimizing the scalar function

$$F(x) = \tfrac{1}{2}x^2 + \lambda\,|x-1|, \tag{6.110}$$

where $\lambda > 0$ is a parameter. The function F is strongly convex, but is not differentiable at $x = 1$. Theorem 6.51 guarantees that F has a unique global minimizer $x^* \in \mathbb{R}$, but since F is not differentiable everywhere, we cannot simply set $F'(x^*) = 0$ to find x^*. When the derivative exists it is equal to

$$F'(x) = \begin{cases} x + \lambda, & x > 1, \\ x - \lambda, & x < 1. \end{cases}$$

We know that F can have at most one critical point, and the global minimum is located at the critical point when one exists. Thus, if F has no critical points then the global minimizer must be $x^* = 1$, the point of non-differentiability.

To determine whether or not F has a critical point, first note that since $F'(x) > 0$ when $x > 1$, there are no critical points larger than 1. As for $x < 1$, a critical point must satisfy $1 > x = \lambda$. We conclude that the minimizer is either the critical point $x^* = \lambda$ when $0 < \lambda < 1$, or is $x^* = 1$ when $\lambda \geq 1$. In other words, $x^* = \min\{\lambda, 1\}$. We show plots of the function F for $\lambda = 0.5$ and $\lambda = 1.5$ in Figure 6.7, illustrating how the minimizer can be at a point of non-differentiability. ▲

The field of convex optimization is vast, and further results can be found, for instance, in [31].

Exercises

7.1. Prove that e^x is strictly convex, but not strongly convex.

7.2. ♡ Show that $-\log x$ is strictly convex when $x > 0$. Use this to prove $\log x \leq x - 1$ for all $x > 0$, with equality if and only if $x = 1$.

7.3. ◇ For $\beta > 0$ define the *softplus function* $f_\beta \colon \mathbb{R} \to \mathbb{R}$ by $f_\beta(x) = \dfrac{1}{\beta} \log(1 + e^{\beta x})$. Show that f_β is strictly convex and, moreover, $\lim_{\beta \to \infty} f_\beta(x) = x_+ := \max\{x, 0\}$. Thus, f_β is a smooth convex approximation to the *plus function* (also known as the *ReLU function*) $g(x) = x_+$, hence the name "softplus".

7.4. Determine whether the following functions are (i) convex; (ii) strictly convex:
$$(a)\,♡\; x, \quad (b)\,♡\; x^2, \quad (c)\,♡\; x^3, \quad (d)\; |x|, \quad (e)\,◇\; |x|^3, \quad (f)\; 1/(1 + x^2).$$

7.5. ◇ Let $\alpha \in \mathbb{R}$. For which values of α is the scalar function $F(x) = x^\alpha$ on the domain $\Omega = \{x > 0\}$ (a) convex? (b) strictly convex? (c) strongly convex?

7.6. ♡ Let $\|\cdot\|$ be any norm on $\mathbb{R}^n$. Show that the norm function $F(\mathbf{x}) = \|\mathbf{x}\|$ is convex.

7.7. Let $F(\mathbf{x}, \mathbf{y})$ be the Kullback–Leibler divergence between probability vectors $\mathbf{x}, \mathbf{y} \in \Omega$, as defined in (6.95). Show that $F(\mathbf{x}, \mathbf{y}) \geq 0$ with equality if and only if $\mathbf{x} = \mathbf{y}$. *Hint:* Use Exercise 7.2.

7.8. (a) ♡ Prove that if $F \colon \mathbb{R}^n \to \mathbb{R}$ is continuously differentiable, and both F and $-F$ are convex, then F is an affine function. (b) ◇ Is the result also valid for general functions F? If so, prove it. If not, find an explicit counterexample.

7.9. Let $F \colon \mathbb{R}^n \to \mathbb{R}$ be convex and let $\mathbf{x}, \mathbf{y} \in \mathbb{R}^n$. Assume there exists $0 < t^* < 1$ such that $F\big((1 - t^*)\mathbf{x} + t^*\mathbf{y}\big) = (1 - t^*)\,F(\mathbf{x}) + t^*F(\mathbf{y})$. Show that $F\big((1 - t)\mathbf{x} + t\mathbf{y}\big) = (1 - t)\,F(\mathbf{x}) + t\,F(\mathbf{y})$ for all $0 \leq t \leq 1$. Thus, the only way that F can fail to be strictly convex between two points $\mathbf{x}$ and $\mathbf{y}$ is when F is an affine function between those points.

7.10. ♡ Prove Lemma 6.37.

7.11. ◇ Prove that if $F, G \colon \mathbb{R}^n \to \mathbb{R}$ are convex, then so is $H(\mathbf{x}) = \max\{\,F(\mathbf{x}), G(\mathbf{x})\,\}$.

7.12. ◇ Let $F \colon \mathbb{R}^n \to \mathbb{R}$ be a convex function. Suppose $G \colon \mathbb{R} \to \mathbb{R}$ is a nondecreasing scalar convex function, so $G(x) \leq G(y)$ whenever $x \leq y$. Show that the composition $H = G \circ F$ is a convex function. Is this true when G is allowed to be decreasing?

7.13. ♡ Let $F \colon [0, \infty) \to \mathbb{R}$ be a convex function satisfying $F(0) = 0$. Show that F is *superadditive*, which means that $F(x) + F(y) \leq F(x + y)$ for all $x, y \geq 0$. *Hint:* Use (6.88) to show that $F(x) \leq \dfrac{x}{x + y}\, F(x + y)$ and $F(y) \leq \dfrac{y}{x + y}\, F(x + y)$.

7.14. *True or false:* Every strictly convex function is differentiable.

7.15. ◇ Show that when $a > 0$, the matrix $A = \begin{pmatrix} a & -1 \\ -1 & a^{-1} \end{pmatrix}$ is positive semidefinite by directly showing that $\mathbf{x}^T A \mathbf{x} \geq 0$ for all $\mathbf{x}$. Use this to show that the Hessian of the relative entropy function (6.94) is positive semidefinite, so the relative entropy function is convex.

7.16. Use induction on m to prove Jensen's inequality (6.90).

7.17. ♡ Prove Hölder's inequality (6.99) when $p = \infty$ and $q = 1$.

7.18. ◊ Let $1 \le s \le r \le t \le \infty$ and $0 \le \theta \le 1$, and assume that $\dfrac{1}{r} = \dfrac{\theta}{s} + \dfrac{1-\theta}{t}$, where we interpret $1/\infty = 0$. Use Hölder's inequality (6.99) to prove *Littlewood's interpolation inequality*

$$\|\mathbf{x}\|_r \le \|\mathbf{x}\|_s^{\theta} \|\mathbf{x}\|_t^{1-\theta} \qquad \text{for} \qquad \mathbf{x} \in \mathbb{R}^n. \tag{6.111}$$

7.19. Show that for $\mathbf{x} \in \mathbb{R}^n$ and $1 \le p \le \infty$, $\|\mathbf{x}\|_p \le \|\mathbf{x}\|_1 \le n^{1-1/p}\|\mathbf{x}\|_p$, where for $p = \infty$ we set $1 - 1/p = 1$. *Hint:* For the first inequality, use the superadditivity, as in Exercise 7.13, of the convex function $g(t) = t^p$, and use Hölder's inequality (6.99) for the other.

7.20. ♡ Show that $F \colon \mathbb{R}^n \to \mathbb{R}$ is μ-strongly convex if and only if

$$F\big((1-t)\mathbf{x} + t\mathbf{y}\big) + \tfrac{1}{2}\mu t(1-t)\|\mathbf{x} - \mathbf{y}\|^2 \le (1-t)F(\mathbf{x}) + tF(\mathbf{y}) \tag{6.112}$$

holds for all $\mathbf{x}, \mathbf{y} \in \mathbb{R}^n$ and $0 \le t \le 1$.

7.21. Let $1 \le p, q \le \infty$ satisfy (6.96). Show that $\|\mathbf{x}\|_p = \max\{\mathbf{x} \cdot \mathbf{y} \mid \|\mathbf{y}\|_q = 1\}$.

7.22. Let $1 \le p, q \le \infty$ satisfy (6.96). Use Exercise 7.21 to show that $\|A\|_p = \|A^T\|_q$ for any matrix A. *Hint:* At one point you will have to exchange two max operations, which you may do without justification.

6.8 Lipschitz Continuity

In many applications of analysis, mere continuity of functions is too weak a hypothesis for proving significant results. On the other hand, differentiability can be overly restrictive, and, for a variety of reasons, one would like to weaken it in order to extend the range of usable functions. An intermediate condition that arises in many fields is known as Lipschitz continuity, named after the nineteenth century German analyst Rudolf Lipschitz. In this section, we define the concept and present some useful properties of such functions. The basic definition relies on a choice of norm $\|\mathbf{x}\|$ on $\mathbb{R}^n$. This could be, for example, a p norm, with the most common cases being $p = 1, 2$, or ∞, or a norm based on weighted inner products, which we use for preconditioned gradient descent.

> **Definition 6.55.** Let $\Omega \subset \mathbb{R}^n$. A real-valued function $F \colon \Omega \to \mathbb{R}$ is called *Lipschitz continuous* on Ω if there exists a nonnegative real number $\lambda \ge 0$ such that
>
> $$|F(\mathbf{x}) - F(\mathbf{y})| \le \lambda \|\mathbf{x} - \mathbf{y}\| \qquad \text{for all} \qquad \mathbf{x}, \mathbf{y} \in \Omega. \tag{6.113}$$
>
> The smallest such constant λ is called the *Lipschitz constant* of F, and denoted[8]
>
> $$\mathrm{Lip}(F) = \mathrm{Lip}_{\Omega}(F) = \max\left\{ \frac{|F(\mathbf{x}) - F(\mathbf{y})|}{\|\mathbf{x} - \mathbf{y}\|} \;\middle|\; \mathbf{x} \ne \mathbf{y} \in \Omega \right\}. \tag{6.114}$$
>
> We drop the Ω subscript when the choice of domain is clear, and usually this is $\Omega = \mathbb{R}^n$.

Remark. While the magnitude of the Lipschitz constant $\mathrm{Lip}(F)$ will depend on the choice of norm, the equivalence of norms on $\mathbb{R}^n$, cf. Theorem 2.36, implies that the property of being Lipschitz continuous is independent of the underlying norm. ▲

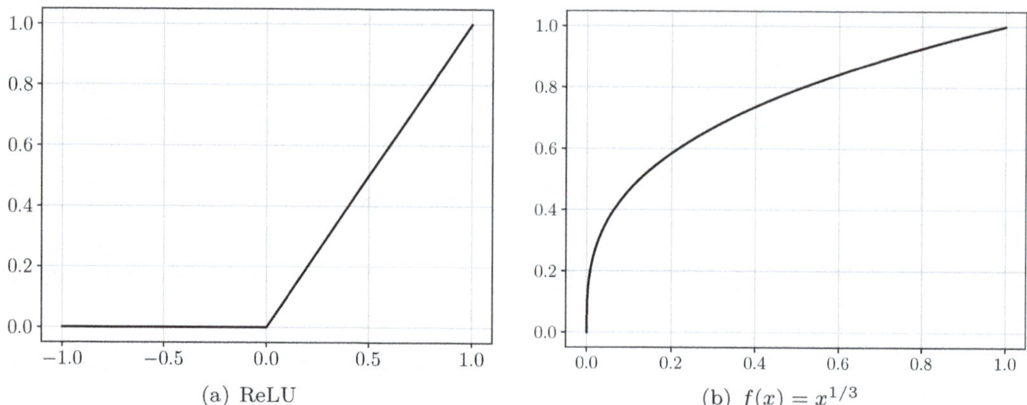

(a) ReLU   (b) $f(x) = x^{1/3}$

Figure 6.8: Illustration of (a) the Lipschitz function ReLU, $f(x) = \max\{x, 0\}$, and (b) the function $f(x) = x^{1/3}$, which is not Lipschitz on any interval containing the origin.

Example 6.56. The *rectified linear unit*, or *ReLU* for short, is the simple piecewise linear scalar function $f\colon \mathbb{R} \to \mathbb{R}$ given by

$$f(x) = \max\{x, 0\} = x_+. \tag{6.115}$$

(It is also known as the *plus function*; see Exercise 7.3.) This elementary function plays an essential role throughout modern machine learning, and is depicted in Figure 6.8(a). Using $|x|$ as the norm on $\mathbb{R}$, it is not hard to see that

$$|f(x) - f(y)| \leq |x - y| \qquad \text{for all} \qquad x, y \in \mathbb{R}.$$

Thus, the ReLU function is Lipschitz continuous with Lipschitz constant $\mathrm{Lip}(f) = 1$. Observe that it is not differentiable at the origin, underscoring the fact that Lipschitz continuity is more general than differentiability. We also note that the ReLU function is convex, but not strictly convex. See Exercise 7.3 for a smooth convex approximation known as the softmax function.

On the other hand, the cube root function $f(x) = x^{1/3}$ is continuous, but *not* Lipschitz continuous on any interval containing the origin. Indeed,

$$|f(x) - f(0)| = |x^{1/3}| \not\leq \lambda |x|,$$

for any constant λ when x is small. In effect, this is a consequence of the fact that its derivative $|f'(x)| = x^{-2/3} \to \infty$ is unbounded as $x \to 0$, as illustrated in Figure 6.8(b).

Finally the quadratic function $f(x) = x^2$ is Lipschitz continuous on any bounded interval $I \subset \mathbb{R}$. Indeed, we can write

$$|f(x) - f(y)| = |x^2 - y^2| = |(x - y)(x + y)| \leq |x + y|\,|x - y|.$$

On any bounded interval I, there exists $\lambda \geq 0$ such that $|x + y| \leq \lambda$ for all $x, y \in I$, and so f is Lipschitz continuous on I with Lipschitz constant $\mathrm{Lip}(f) \leq \lambda$. However, it is not Lipschitz continuous on all of $\mathbb{R}$, or, for that matter, on any unbounded interval.   ▲

---

[8]The max in (6.114) is actually the supremum. See the Preface for our conventions in this regard.

It turns out that every continuously differentiable function with bounded gradient is Lipschitz continuous.

Lemma 6.57. *Let $\Omega \subset \mathbb{R}^n$ be an open convex domain, and suppose that $F \in C^1(\Omega)$ is continuously differentiable on Ω with bounded partial derivatives. Then F is Lipschitz continuous with Lipschitz constant*[9]

$$\mathrm{Lip}(F) = \max\left\{\, \|\nabla F(\mathbf{x})\| \mid \mathbf{x} \in \Omega \,\right\}. \tag{6.116}$$

Remark. If Ω is closed and bounded, i.e., compact, and F is the restriction to Ω of a continuously differentiable function on an open set that contains Ω, then the bounded derivatives condition holds. ▲

Proof. Let

$$\lambda = \max\left\{\, \|\nabla F(\mathbf{x})\| \mid \mathbf{x} \in \Omega \,\right\},$$

and let $\mathbf{x}, \mathbf{y} \in \Omega$. Since Ω is convex, the line segment between $\mathbf{x}$ and $\mathbf{y}$ belongs to Ω. Thus, by the definition of the gradient and the Cauchy–Schwarz inequality (2.27),

$$F(\mathbf{y}) - F(\mathbf{x}) = \int_0^1 \frac{d}{dt} F(\mathbf{x} + t(\mathbf{y} - \mathbf{x}))\, dt = \int_0^1 \langle \nabla F(\mathbf{x} + t(\mathbf{y} - \mathbf{x})), \mathbf{y} - \mathbf{x} \rangle\, dt$$

$$\leq \int_0^1 \|\nabla F(\mathbf{x} + t(\mathbf{y} - \mathbf{x}))\|\, \|\mathbf{y} - \mathbf{x}\|\, dt \leq \lambda \|\mathbf{x} - \mathbf{y}\|.$$

Thus, $\mathrm{Lip}(F) \leq \lambda$. As for the opposite inequality, note that for any $\mathbf{z} \in \mathbb{R}^n$,

$$\langle \nabla F(\mathbf{x}), \mathbf{z} \rangle = \frac{d}{dt} F(\mathbf{x} + t\mathbf{z})\bigg|_{t=0} = \lim_{t \to 0} \frac{F(\mathbf{x} + t\mathbf{z}) - F(\mathbf{x})}{t} \leq \mathrm{Lip}(F)\, \|\mathbf{z}\|.$$

If $\mathbf{z} = \nabla F(\mathbf{x}) \neq \mathbf{0}$, this yields $\|\nabla F(\mathbf{x})\| \leq \mathrm{Lip}(F)$, which also trivially holds when $\nabla F(\mathbf{x}) = \mathbf{0}$. The proof is completed by maximizing over $\mathbf{x} \in \Omega$. ∎

Remark 6.58. If the domain Ω is not convex, then Lemma 6.57 may not hold. As an example, consider the domain $\Omega = (-1, 0) \cup (0, 1)$, which is the interval $(-1, 1)$ with the point $\{0\}$ removed. The function f defined by $f(x) = 0$ for $x < 0$ and $f(x) = 1$ for $x > 0$ has $f'(x) = 0$ for all $x \in \Omega$, since the point $x = 0$ where the derivative is undefined does not belong to Ω. However, f is not Lipschitz on Ω. Indeed, for any $x > 0$ we have

$$\frac{f(x) - f(-x)}{2x} = \frac{1 - 0}{2x} = \frac{1}{2x},$$

which is unbounded as $x \to 0^+$.

On the other hand, Lemma 6.57 has a natural extension to sufficiently regular *connected*[10] domains Ω. In this case, the same result is true except that the Lipschitz constant satisfies

$$\mathrm{Lip}(F) \leq \beta \max\left\{\, \|\nabla F(\mathbf{x})\| \mid \mathbf{x} \in \Omega \,\right\}, \tag{6.117}$$

---

[9] As in Definition 6.55, the max in (6.116) is actually the supremum, since Ω is an open set; see the discussion of our convention in the Preface.

[10] Here, connected (sometimes called pathwise connected) means that every $\mathbf{x}, \mathbf{y} \in \Omega$ can be connected by a continuous path belonging to Ω.

where $\beta \geq 1$ is a constant that depends solely on the shape of the domain Ω, and not on F. For a precise proof, we refer to [71, Chapter 5]. In particular, connectivity of the domain does not suffice, by itself, for (6.117) to hold, see Exercise 8.14.[11]   ▲

In the scalar case, there is a useful extension of this result. A function $f\colon \mathbb{R} \to \mathbb{R}$ is called *piecewise continuously differentiable* if it is continuous, and its derivative $f'(x)$ is *piecewise continuous*, meaning that f' is continuous except at finitely many points[12] $a_1, \ldots, a_n \in \mathbb{R}$, where it has right and left handed limits $f'(a_i^+) \neq f'(a_i^-)$. Examples of such functions include the plus function x_+, as in (6.115), and the absolute value function $|x|$. The proof of the following proposition is left as Exercise 8.6.

Proposition 6.59. *Suppose that $f\colon \mathbb{R} \to \mathbb{R}$ is piecewise continuously differentiable and its derivative is bounded: $|f'(x)| \leq \lambda$. Then f is Lipschitz continuous.*

More generally, a vector-valued function $F\colon \Omega \subset \mathbb{R}^n \to \mathbb{R}^m$ is called *Lipschitz continuous* if

$$\| F(\mathbf{x}) - F(\mathbf{y}) \| \leq \lambda \| \mathbf{x} - \mathbf{y} \| \qquad \text{for all} \qquad \mathbf{x}, \mathbf{y} \in \Omega \tag{6.118}$$

for some $\lambda \geq 0$. Again, the Lipschitz constant $\mathrm{Lip}(F)$ is the smallest such λ. We are allowed to choose different norms on $\mathbb{R}^n$ and $\mathbb{R}^m$; as in the scalar case, the property of being Lipschitz continuous is independent of the choice of norm, although the values of the Lipschitz constant will be norm dependent. When F is continuously differentiable and Ω is convex, an argument similar to Lemma 6.57 shows that its Lipschitz constant can be identified with the maximum of the induced matrix norm of its Jacobian matrix, that is

$$\mathrm{Lip}(F) = \max \left\{ \| \mathbf{D}F(\mathbf{x}) \| \mid \mathbf{x} \in \Omega \right\}. \tag{6.119}$$

Remark 6.60. In optimization, we are particularly interested in the situation where the function $F\colon \Omega \subset \mathbb{R}^n \to \mathbb{R}$ is continuously differentiable, real-valued, and has Lipschitz continuous gradient $\nabla F\colon \Omega \to \mathbb{R}^n$, whereby

$$\| \nabla F(\mathbf{x}) - \nabla F(\mathbf{y}) \| \leq \lambda \| \mathbf{x} - \mathbf{y} \|, \qquad \mathbf{x}, \mathbf{y} \in \Omega, \tag{6.120}$$

for some $\lambda \geq 0$. Typically, we work with the smallest such constant: $\lambda = \mathrm{Lip}(\nabla F)$.

Suppose now that F is twice continuously differentiable, so that ∇F is continuously differentiable. Then, since the Jacobian of the gradient is the Hessian, cf. (6.75), it follows from (6.119) that

$$\mathrm{Lip}(\nabla F) = \max \left\{ \| \nabla^2 F(\mathbf{x}) \| \mid \mathbf{x} \in \Omega \right\},$$

provided Ω is convex. In particular, if F is a convex function, its Hessian is self-adjoint and positive semidefinite. Theorem 5.45 implies that[13] $\| \nabla^2 F(\mathbf{x}) \| = \lambda_{max}(\nabla^2 F(\mathbf{x}))$. If we fix the inner product $\langle \mathbf{x}, \mathbf{y} \rangle_C = \mathbf{x}^T C \mathbf{y}$, where C is symmetric, positive definite, then by (6.84),

$$\mathrm{Lip}(\nabla_C F) = \max \left\{ \lambda_{max}(C^{-1} \nabla_2^2 F(\mathbf{x})) \mid \mathbf{x} \in \Omega \right\}. \tag{6.121}$$

Inspecting this expression, it is quite natural to attempt to take $C = \nabla_2^2 F(\mathbf{x})$, for which it appears that $\mathrm{Lip}(\nabla_C F) = 1$. However, as we pointed out earlier in Example 6.49, the matrix

---

[11] In essence, there needs to exist a path between any $\mathbf{x}, \mathbf{y} \in \Omega$ that lives inside Ω and has length at most $\beta \| \mathbf{x} - \mathbf{y} \|$, which can be guaranteed by a variety of regularity assumptions on Ω; see [71].

[12] This assumption can be relaxed to allow infinitely many points as long as there is no accumulation point.

[13] When F is not convex, the same statement holds with σ_{max} replacing λ_{max}, provided we interpret σ_{max} with respect to the same inner product used to define the Hessian, or by replacing λ_{max} with the spectral norm.

C defining the inner product is not allowed to depend on $\mathbf{x}$. On the other hand, this again suggests that, during iteration, a good choice is $C = \nabla^2 F(\mathbf{x}_k)$, where $\mathbf{x}_k$ is the current iterate of our optimization algorithm. ▲

We next note that the Lipschitz condition on the gradient of a function ensures that there is a first order Taylor expansion.

Proposition 6.61. *Let $\Omega \subset \mathbb{R}^n$ be convex. Given $F \colon \Omega \to \mathbb{R}$ such that ∇F is Lipschitz continuous, then*

$$\big| F(\mathbf{y}) - F(\mathbf{x}) - \langle \nabla F(\mathbf{x}), \mathbf{y} - \mathbf{x} \rangle \big| \leq \tfrac{1}{2} \mathrm{Lip}(\nabla F) \, \|\mathbf{y} - \mathbf{x}\|^2 \tag{6.122}$$

for all $\mathbf{x}, \mathbf{y} \in \Omega$.

Proof. Let $\mathbf{x}, \mathbf{y} \in \Omega$. Since Ω is convex, the line segment

$$[\mathbf{x}, \mathbf{y}] = \{ \mathbf{x} + t\,(\mathbf{y} - \mathbf{x}) \mid 0 \leq t \leq 1 \}$$

connecting $\mathbf{x}$ and $\mathbf{y}$ belongs to Ω. Using the fundamental theorem of calculus and the multivariable chain rule, we can write

$$\begin{aligned}
F(\mathbf{y}) - F(\mathbf{x}) &= \int_0^1 \frac{d}{dt} F\big(\mathbf{x} + t\,(\mathbf{y} - \mathbf{x})\big)\, dt \\
&= \int_0^1 \big\langle \nabla F\big(\mathbf{x} + t\,(\mathbf{y} - \mathbf{x})\big), \mathbf{y} - \mathbf{x} \big\rangle\, dt = \langle \nabla F(\mathbf{x}), \mathbf{y} - \mathbf{x} \rangle + R(\mathbf{x}, \mathbf{y}),
\end{aligned}$$

where

$$R(\mathbf{x}, \mathbf{y}) = \int_0^1 \big\langle \nabla F\big(\mathbf{x} + t\,(\mathbf{y} - \mathbf{x})\big) - \nabla F(\mathbf{x}), \mathbf{y} - \mathbf{x} \big\rangle\, dt.$$

Applying the Cauchy-Schwarz inequality (2.27) and then invoking the Lipschitz continuity of ∇F produces

$$\begin{aligned}
|R(\mathbf{x}, \mathbf{y})| &\leq \int_0^1 \big\| \nabla F\big(\mathbf{x} + t\,(\mathbf{y} - \mathbf{x})\big) - \nabla F(\mathbf{x}) \big\| \, \|\mathbf{y} - \mathbf{x}\| \, dt \\
&\leq \int_0^1 \mathrm{Lip}(\nabla F) \, \|\mathbf{x} + t\,(\mathbf{y} - \mathbf{x}) - \mathbf{x}\| \, \|\mathbf{y} - \mathbf{x}\| \, dt \\
&= \mathrm{Lip}(\nabla F) \, \|\mathbf{y} - \mathbf{x}\|^2 \int_0^1 t \, dt = \tfrac{1}{2} \mathrm{Lip}(\nabla F) \, \|\mathbf{y} - \mathbf{x}\|^2. \qquad \blacksquare
\end{aligned}$$

Remark 6.62. A careful examination of the proof of Proposition 6.61 reveals that we have in fact proved the stronger result

$$\big| F(\mathbf{y}) - F(\mathbf{x}) - \langle \nabla F(\mathbf{x}), \mathbf{y} - \mathbf{x} \rangle \big| \leq \tfrac{1}{2} \mathrm{Lip}_{[\mathbf{x}, \mathbf{y}]}(\nabla F) \, \|\mathbf{y} - \mathbf{x}\|^2, \tag{6.123}$$

i.e., we only need the Lipschitz constant of F on the line segment between $\mathbf{x}$ and $\mathbf{y}$. Using big O notation[14], we can write the latter inequality as

$$F(\mathbf{y}) = F(\mathbf{x}) + \nabla F(\mathbf{x}) \cdot (\mathbf{y} - \mathbf{x}) + \mathrm{O}\big(\mathrm{Lip}_{[\mathbf{x}, \mathbf{y}]}(\nabla F) \, \|\mathbf{y} - \mathbf{x}\|^2 \big), \tag{6.124}$$

which can be regarded as a first order Taylor expansion for the function F. ▲

---

[14]See the Preface for details.

Remark 6.63. It is also important to point out that Proposition 6.61 has a natural extension to vector-valued functions $G\colon \mathbb{R}^n \to \mathbb{R}^n$. In this case, if the Jacobian $\mathbf{D}G$ is Lipschitz continuous, as in (6.118), then, by a similar argument as used in the proof of Proposition 6.61,

$$\|G(\mathbf{y}) - G(\mathbf{x}) - \mathbf{D}G(\mathbf{x})(\mathbf{y} - \mathbf{x})\| \le \tfrac{1}{2}\,\mathrm{Lip}(\mathbf{D}G)\,\|\mathbf{y} - \mathbf{x}\|^2, \qquad \mathbf{x}, \mathbf{y} \in \mathbb{R}^n. \tag{6.125}$$

An important consequence involves taking $G(\mathbf{x}) = \nabla F(\mathbf{x})$ for a function $F\colon \mathbb{R}^n \to \mathbb{R}$. Then, since $\mathbf{D}G(\mathbf{x}) = \nabla^2 F(\mathbf{x})$, (6.125) becomes

$$\|\nabla F(\mathbf{y}) - \nabla F(\mathbf{x}) - \nabla^2 F(\mathbf{x})(\mathbf{y} - \mathbf{x})\| \le \tfrac{1}{2}\,\mathrm{Lip}(\nabla^2 F)\,\|\mathbf{y} - \mathbf{x}\|^2, \qquad \mathbf{x}, \mathbf{y} \in \mathbb{R}^n. \tag{6.126}$$

▲

Exercises

8.1. Determine whether or not the following scalar functions are Lipschitz continuous on $\mathbb{R}$. If so, find their Lipschitz constant.
  $(a)\,\heartsuit\ |x| + |x - 1|$, $(b)\ x^{2/3}$, $(c)\,\heartsuit\ \mathrm{sign}\,x$, $(d)\,\heartsuit\ e^x$, $(e)\,\diamondsuit\ e^{-x^2}$, $(f)\ \tanh x$.

8.2. Do your answers to Exercise 8.1 change if the domain is restricted to $[-1, 1]$?

8.3. $\diamondsuit$ For what values of α is the function $F(x) = |x|^\alpha$ continuous? Lipschitz continuous? Differentiable?

8.4. Let $\|\cdot\|$ be a norm on $\mathbb{R}^n$. Prove that $F(\mathbf{x}) = \|\mathbf{x}\|$ is Lipschitz continuous with Lipschitz constant $\lambda = 1$.

8.5. Determine whether or not the following functions on $\mathbb{R}^2$ are Lipschitz continuous.
  $(a)\,\heartsuit\ |x - y|$, $(b)\ \max\{|x|, |y|\}$, $(c)\,\heartsuit\ x^2 - y^2$, $(d)\,\diamondsuit\ \exp(-x^2 - y^2)$.

8.6. Prove Proposition 6.59.

8.7. Prove that the vector-valued function $F(x, y) = \begin{pmatrix} |x - y| \\ \max\{|x|, |y|\} \end{pmatrix}$ is Lipschitz continuous. Find its Lipschitz constant with respect to the 1 norm on $\mathbb{R}^2$.

8.8. $\heartsuit$ Suppose the scalar-valued functions $F_1, F_2\colon \mathbb{R}^2 \to \mathbb{R}$ are both Lipschitz continuous.
  (a) Prove that the vector-valued function $F(x, y) = (F_1(x, y), F_2(x, y))^T$ is Lipschitz continuous. (b) Is the converse to part (a) valid?

8.9. $\diamondsuit$ Let F be μ-strongly convex, and let $\lambda = \mathrm{Lip}(\nabla F)$ be the Lipschitz constant of ∇F. (a) Show that $\mu \le \lambda$. (b) Show that $\mu = \lambda$ if and only if $F(\mathbf{x}) = F(\mathbf{x}^\star) + \tfrac{1}{2}\mu\|\mathbf{x} - \mathbf{x}^\star\|^2$, where $\mathbf{x}^\star$ is the unique global minimizer of F.

8.10. $\heartsuit$ Prove that the property of Lipschitz continuity does not depend on the underlying norm on $\mathbb{R}^n$.

8.11. A function $F\colon \mathbb{R}^n \to \mathbb{R}^n$ defines a *contraction* if it has Lipschitz constant $\mathrm{Lip}(F) < 1$. Prove that a contraction can only have one fixed point, meaning a point $\mathbf{x}^\star \in \mathbb{R}^n$ such that $F(\mathbf{x}^\star) = \mathbf{x}^\star$.

8.12. *True or false:* $(a)\,\heartsuit$ A convex scalar function is Lipschitz continuous.
  $(b)\,\diamondsuit$ A strictly convex scalar function is Lipschitz continuous.
  (c) A Lipschitz continuous scalar function is convex.

8.13. Prove the inequalities (6.125) and (6.126).

8.14. ♡ Give an example to show that (6.117) does not hold in general on connected domains. *Hint*: Take the domain to be a disk in $\mathbb{R}^2$ centered at the origin with the negative x axis removed.

8.15. ◇ We say a differentiable function $F\colon \mathbb{R}^n \to \mathbb{R}$ has a *Hölder continuous* gradient ∇F with exponent $0 < \gamma < 1$ if there exists $\lambda > 0$ such that

$$\|\nabla F(\mathbf{x}) - \nabla F(\mathbf{y})\| \le \lambda \|\mathbf{x} - \mathbf{y}\|^\gamma \qquad \text{for all} \qquad \mathbf{x}, \mathbf{y} \in \mathbb{R}^n. \tag{6.127}$$

The smallest such $\lambda > 0$ is called the *Hölder seminorm* of F, and denoted $H_\gamma(\nabla F)$. Show that $F(\mathbf{x}) = \|\mathbf{x}\|^{1+\gamma}$ has a Hölder continuous gradient with exponent γ for $0 < \gamma < 1$, and that $H_\gamma(\nabla F) \le 2\,(1 + \gamma^{-1})$.
Warning: This problem is rather difficult. You may want to prove it just for $n = 1$, for which Exercise 7.13 is helpful, and $H_\gamma(F') \le 1 + \gamma$.

8.16. Suppose that $F\colon \mathbb{R}^n \to \mathbb{R}$ has a Hölder continuous gradient with exponent $0 < \gamma < 1$, as in Exercise 8.15. Show that the Taylor expansion

$$|\,F(\mathbf{y}) - F(\mathbf{x}) - \langle \nabla F(\mathbf{x}), \mathbf{y} - \mathbf{x} \rangle\,| \le \frac{1}{1+\gamma}\, H_\gamma(\nabla F)\, \|\mathbf{x} - \mathbf{y}\|^{1+\gamma} \tag{6.128}$$

holds for all $\mathbf{x}, \mathbf{y} \in \mathbb{R}^n$.

6.9 Basic Convergence Results

In this section, we begin our study of the convergence of the gradient descent algorithm to a (local) minimizer of the objective function. We already saw a simple convergence result for quadratic functions in Example 6.25, where gradient descent converges at a linear rate, which depends on the condition number of the coefficient matrix. The goal of this section is to extend these basic ideas to more general functions subject to some basic convexity and smoothness assumptions. As before, the results are stated when the domain of the function is all of $\mathbb{R}^n$, but remain valid locally on open subsets thereof. We also remind the reader that in this chapter we work with a general inner product $\langle \cdot, \cdot \rangle$ on $\mathbb{R}^n$ along with the induced norm $\|\cdot\|$ and gradient ∇F. Additional convergence results can be found in Chapter 11.

We first prove a preliminary lemma, which shows that, for a sufficiently small time step, a suitably smooth objective function must decrease with each step of gradient descent.

Lemma 6.64. *Let ∇F be Lipschitz continuous. Then for $0 < \alpha \le \mathrm{Lip}(\nabla F)^{-1}$,*

$$F\big(\mathbf{x} - \alpha \nabla F(\mathbf{x})\big) \le F(\mathbf{x}) - \frac{\alpha}{2}\, \|\nabla F(\mathbf{x})\|^2. \tag{6.129}$$

Proof. We set $\mathbf{y} = \mathbf{x} - \alpha\, \nabla F(\mathbf{x})$ and use Proposition 6.61 to obtain

$$\begin{aligned}
F(\mathbf{y}) &\le F(\mathbf{x}) + \langle \nabla F(\mathbf{x}), \mathbf{y} - \mathbf{x} \rangle + \tfrac{1}{2}\, \mathrm{Lip}(\nabla F)\, \|\mathbf{x} - \mathbf{y}\|^2 \\
&= F(\mathbf{x}) - \alpha \langle \nabla F(\mathbf{x}), \nabla F(\mathbf{x}) \rangle + \tfrac{1}{2}\, \mathrm{Lip}(\nabla F)\, \|-\alpha \nabla F(\mathbf{x})\|^2 \\
&= F(\mathbf{x}) + \big[\tfrac{1}{2}\, \mathrm{Lip}(\nabla F)\, \alpha^2 - \alpha\big]\, \|\nabla F(\mathbf{x})\|^2.
\end{aligned} \tag{6.130}$$

Since we assumed that $\alpha \le \mathrm{Lip}(\nabla F)^{-1}$, we have $\mathrm{Lip}(\nabla F)\alpha^2 \le \alpha$, and hence the inequality (6.130) implies (6.129). ∎

Lemma 6.64 guarantees that gradient descent will strictly decrease the objective function F provided the time step is sufficiently small, namely, $\alpha \leq \mathrm{Lip}(\nabla F)^{-1}$. The amount of decrease — the second term in (6.129) — depends on the size of the time step α and the squared norm of the gradient. In particular, if $\nabla F(\mathbf{x}) = \mathbf{0}$, then we are at a critical point, which need not be a minimizer, and gradient descent will not decrease F.

We are now equipped to prove our first convergence result.

Theorem 6.65 (Local sublinear convergence). *Assume that ∇F is Lipschitz continuous and let $0 < \alpha \leq \mathrm{Lip}(\nabla F)^{-1}$. Let $\mathbf{x}_k$, for $k \geq 0$, be the iterations of the gradient descent algorithm (6.39) with fixed time step $\alpha_k = \alpha$. Then, for any integer $k \geq 1$,*

$$\min_{0 \leq j \leq k-1} \|\nabla F(\mathbf{x}_j)\|^2 \leq \frac{2}{\alpha k} \left[F(\mathbf{x}_0) - F(\mathbf{x}_k) \right]. \tag{6.131}$$

Proof. By Lemma 6.64 we have

$$\frac{\alpha}{2} \|\nabla F(\mathbf{x}_j)\|^2 \leq F(\mathbf{x}_j) - F(\mathbf{x}_{j+1}). \tag{6.132}$$

Summing from $j = 0$ to $k - 1$ yields

$$\frac{\alpha}{2} \sum_{j=0}^{k-1} \|\nabla F(\mathbf{x}_j)\|^2 \leq \sum_{j=0}^{k-1} \left[F(\mathbf{x}_j) - F(\mathbf{x}_{j+1}) \right] = F(\mathbf{x}_0) - F(\mathbf{x}_j),$$

since the sum telescopes. The inequality (6.131) follows from the evident lower bound

$$\sum_{j=0}^{k-1} \|\nabla F(\mathbf{x}_j)\|^2 \geq k \min_{0 \leq j \leq k-1} \|\nabla F(\mathbf{x}_j)\|^2. \qquad \blacksquare$$

Theorem 6.65 shows that after k steps of gradient descent, we are guaranteed to find a point $\mathbf{x}_j$, for some $0 \leq j \leq k - 1$, for which $\|\nabla F(\mathbf{x}_j)\|^2 = \mathrm{O}\left(k^{-1}\right)$. It is important to point out that j may not be equal to $k - 1$, i.e., $\mathbf{x}_j$ may not be the most recent gradient descent iterate. Also, the convergence rate $\mathrm{O}\left(k^{-1}\right)$ is rather slow, and the estimate (6.131) is referred to as *sublinear convergence*. Nevertheless, under the stated rather mild assumptions on F, Theorem 6.65 demonstrates that gradient descent converges to a critical point of F in the sense that

$$\lim_{k \to \infty} \min_{0 \leq j \leq k-1} \|\nabla F(\mathbf{x}_j)\|^2 = 0.$$

Since we made no assumptions about F, aside from Lipschitz continuity of its gradient, there may be critical points that are not global minimizers of F. In particular, Theorem 6.65 does not guarantee that gradient descent converges to a (local or global) minimum of F.

When the objective function F is convex, Proposition 6.40 tells us that any critical point is, in fact, a global minimizer. In this case, Theorem 6.65 can be improved to show that gradient descent converges to a global minimizer of F at the same sublinear rate.

Theorem 6.66 (Global sublinear convergence). *Assume F is convex with a global minimizer $\mathbf{x}^\star$, and that ∇F is Lipschitz continuous. Let $0 < \alpha \leq \mathrm{Lip}(\nabla F)^{-1}$. Then, for any integer $k \geq 1$, we have*

$$F(\mathbf{x}_k) - F(\mathbf{x}^\star) \leq \frac{\|\mathbf{x}_0 - \mathbf{x}^\star\|^2}{2 \alpha k}. \tag{6.133}$$

Proof. We start by rearranging (6.132) and subtracting $F(\mathbf{x}^\star)$ from both sides to obtain

$$
\begin{aligned}
F(\mathbf{x}_{j+1}) - F(\mathbf{x}^\star) &\leq F(\mathbf{x}_j) - F(\mathbf{x}^\star) - \frac{\alpha}{2} \|\nabla F(\mathbf{x}_j)\|^2 \\
&\leq \langle \nabla F(\mathbf{x}_j), \mathbf{x}_j - \mathbf{x}^\star \rangle - \frac{\alpha}{2} \|\nabla F(\mathbf{x}_j)\|^2,
\end{aligned}
\tag{6.134}
$$

where the second inequality follows from the convexity inequality (6.92). We then use the definition of gradient descent to replace $\nabla F(\mathbf{x}_j) = (\mathbf{x}_j - \mathbf{x}_{j+1})/\alpha$, and complete the square, producing

$$
F(\mathbf{x}_{j+1}) - F(\mathbf{x}^\star) \leq \frac{\langle \mathbf{x}_j - \mathbf{x}_{j+1}, \mathbf{x}_j - \mathbf{x}^\star \rangle}{\alpha} - \frac{\|\mathbf{x}_j - \mathbf{x}_{j+1}\|^2}{2\alpha} = \frac{\|\mathbf{x}_j - \mathbf{x}^\star\|^2 - \|\mathbf{x}_{j+1} - \mathbf{x}^\star\|^2}{2\alpha}.
$$

We now sum both sides of this inequality from $j = 0$ to $k - 1$, and use the fact that the sum on the right hand side telescopes:

$$
\begin{aligned}
\sum_{j=0}^{k-1} \left[F(\mathbf{x}_{j+1}) - F(\mathbf{x}^\star) \right] &\leq \frac{1}{2\alpha} \sum_{j=0}^{k-1} \left[\|\mathbf{x}_j - \mathbf{x}^\star\|^2 - \|\mathbf{x}_{j+1} - \mathbf{x}^\star\|^2 \right] \\
&= \frac{\|\mathbf{x}_0 - \mathbf{x}^\star\|^2 - \|\mathbf{x}_k - \mathbf{x}^\star\|^2}{2\alpha} \leq \frac{\|\mathbf{x}_0 - \mathbf{x}^\star\|^2}{2\alpha}.
\end{aligned}
\tag{6.135}
$$

By (6.134), we know that $F(\mathbf{x}_j) - F(\mathbf{x}^\star)$ is decreasing as j increases, and so

$$
F(\mathbf{x}_k) - F(\mathbf{x}^\star) \leq F(\mathbf{x}_j) - F(\mathbf{x}^\star) \qquad \text{for all} \qquad 0 \leq j \leq k,
$$

and hence

$$
k \left[F(\mathbf{x}_k) - F(\mathbf{x}^\star) \right] \leq \sum_{j=0}^{k-1} \left[F(\mathbf{x}_{j+1}) - F(\mathbf{x}^\star) \right].
$$

Using this lower bound for the left hand side of (6.135) completes the proof. ∎

Remark. The optimal convergence rate in Theorem 6.66 can be obtained by choosing the time step $\alpha = \mathrm{Lip}(\nabla F)^{-1}$, by which we obtain

$$
F(\mathbf{x}_k) - F(\mathbf{x}^\star) \leq \frac{\mathrm{Lip}(\nabla F) \|\mathbf{x}_0 - \mathbf{x}^\star\|^2}{2k}.
\tag{6.136}
$$

Thus, the convergence rate of gradient descent is governed by the Lipschitz constant of the gradient of the objective function, and the rate of convergence is $\mathrm{O}\left(k^{-1}\right)$ after k iterations. However, as noted above, the resulting sublinear convergence rate is very slow. Minimization of F to within $\varepsilon > 0$ of the optimal value $F(\mathbf{x}^\star)$ requires $k = \mathrm{O}(\varepsilon^{-1})$ steps. For example, to minimize F to within $\varepsilon = 10^{-6}$ accuracy requires roughly 10^6, or 1 million, iterations. The reason for this slow convergence is that a general convex function F may be very flat near a minimizer $\mathbf{x}^\star$ — think of the function $F(x) = x^4$ — and hence its gradient ∇F becomes extremely small as the iterates converge. The conclusion is that, absent further assumptions on the objective function, gradient descent may proceed arbitrarily slowly. In fact, the preceding convergence rate is optimal for general convex functions; see Example 6.67.

We contrast this with Example 6.25, where application of gradient descent to a quadratic function with positive definite coefficient matrix results in a linear convergence rate $\mathrm{O}(\beta^k)$ for some $0 < \beta < 1$. In this case, in order to ensure an $\mathrm{O}(\varepsilon)$ error, we need to set $\beta^k = \varepsilon$, which requires $k = \log \varepsilon / \log \beta$ iterations. As a concrete example, suppose that $\beta = \frac{1}{2}$ and we wish to minimize F up to the same $\varepsilon = 10^{-6}$ accuracy as above. Then we would require on the order of $k = \log(10^{-6}) / \log(\frac{1}{2}) = 6 \log_2(10) \leq 24$ iterations.

Example 6.67. Consider the function

$$F(x) = \frac{|x|^p}{p} \tag{6.137}$$

for $p > 2$, which is convex but not strongly convex. Let us start gradient descent at some $x_0 \in (0,1)$. The gradient descent iteration with a fixed time step α is given by

$$x_{k+1} = x_k - \alpha x_k^{p-1}, \tag{6.138}$$

provided $x_k \geq 0$. In fact, we can ensure $x_k > 0$ for all k by restricting the time step $\alpha \leq 1$, since in this case, whenever $x_k \in (0,1)$ we have $0 < \alpha x_k^{p-1} < x_k$, which implies $x_{k+1} \in (0,1)$. Furthermore, we have $1 > x_0 > x_1 > x_2 > \cdots > x_k > 0$. As we show below, gradient descent converges as $k \to \infty$ to the minimizer, that is $x_k \to 0^+$, but the convergence rate is very slow, especially for large p.

To explicitly establish a convergence rate, we rearrange (6.138) to read

$$\alpha = x_j^{1-p}\,(x_j - x_{j+1}),$$

and, for $k \geq 1$, sum both sides from $j = 0$ to $j = k - 1$, to obtain

$$\alpha k = \sum_{j=0}^{k-1} x_j^{1-p}\,(x_j - x_{j+1}). \tag{6.139}$$

The right hand side is a right-point rule Riemann sum for the integral of x^{1-p} from $x = x_k$ to $x = x_0$ using the intervals $[x_k, x_{k-1}], \ldots, [x_1, x_0]$, which are of varying size. Since x^{1-p} is a decreasing function, the right-point rule is an underestimate of the integral and so

$$\alpha k \leq \int_{x_k}^{x_0} x^{1-p}\, dx = \frac{x_k^{2-p} - x_0^{2-p}}{p-2}.$$

It follows that

$$x_k \leq \left(\frac{1}{(p-2)\,\alpha k + x_0^{2-p}} \right)^{1/(p-2)}. \tag{6.140}$$

This holds for $k = 0$ as well, trivially, since the right hand side is x_0 in this case.

The argument above shows that x_k converges to the minimizer $x^\star = 0$ at a rate at least as fast as $O\big((1/k)^{1/(p-2)}\big)$. We can also obtain a similar lower bound, which shows that this rate is sharp and correct. For this, suppose that $x_0^{p-2}\alpha \leq \frac{1}{2}$, which can be satisfied if, for example, $\alpha \leq \frac{1}{2}$ or $x_0^{p-2} \leq \frac{1}{2}$ and $\alpha \leq 1$. Then since $x_j \leq x_0$ we have

$$x_{j+1} = x_j - \alpha x_j^{p-1} = (1 - \alpha x_j^{p-2})x_j \geq (1 - \alpha x_0^{p-2})x_j \geq \tfrac{1}{2} x_j,$$

and so $x_j \leq 2 x_{j+1}$. Plugging this into (6.139) yields

$$\alpha k \geq 2^{1-p} \sum_{j=0}^{k-1} x_{j+1}^{1-p}\,(x_j - x_{j+1}). \tag{6.141}$$

This is now a left-point rule for a Riemann sum, which is an overestimate, and so

$$\alpha k \geq 2^{1-p} \int_{x_k}^{x_0} x^{1-p}\, dx = \frac{x_k^{2-p} - x_0^{2-p}}{2^{p-1}\,(p-2)}.$$

Therefore

$$x_k \geq \left(\frac{1}{2^{p-1}\,(p-2)\,\alpha k + x_0^{2-p}} \right)^{1/(p-2)}. \tag{6.142}$$

This lower bound is analogous to the upper bound in (6.140), except that the presence of the possibly large constant 2^{p-1} makes it potentially significantly smaller, even though both bounds have the same $\mathrm{O}\big((1/k)^{1/(p-2)} \big)$ scaling.

The preceding rate of convergence is very slow, especially as $p \to \infty$, where we cannot expect to obtain an algebraic convergence rate of the form $\mathrm{O}(1/k^\beta)$ for *any* $\beta > 0$ for the *iterates* x_k of gradient descent among the class of convex functions. The reason for this here is that the function F may become flat near the minimizer at $x = 0$, in which case its derivative F' becomes extremely small and gradient descent proceeds very slowly. Note, however, that by (6.140) we have

$$F(x_k) = \frac{x_k^p}{p} \leq \frac{1}{p} \left(\frac{1}{(p-2)\,\alpha k + x_0^{2-p}} \right)^{p/(p-2)},$$

which is slightly better than the $\mathrm{O}(1/k)$ rate from Theorem 6.66, but matches it as $p \to \infty$. The reason we get a better rate for $F(x_k) \to 0$, compared to $x_k \to 0$, is that while the flatness of F slows down convergence of the iterates x_k to the minimizer $x^\star = 0$, it speeds up convergence of the *values* $F(x_k)$. This tradeoff is what allows us to prove the $\mathrm{O}(1/k)$ rate in such a general setting, and this example indicates that Theorem 6.66 cannot be improved over the class of convex functions. ▲

We now turn to our final result in this section, which shows that, since they cannot be too *flat*, the faster linear convergence rate for quadratic functions found in Example 6.25 can be extended to strongly convex functions.

Theorem 6.68. *Assume that F is μ-strongly convex, ∇F is Lipschitz continuous, and let $0 < \alpha \leq \mathrm{Lip}(\nabla F)^{-1}$. Then for any integer $k \geq 0$ we have*

$$F(\mathbf{x}_k) - F(\mathbf{x}^\star) \leq (1 - \alpha\mu)^k \left[F(\mathbf{x}_0) - F(\mathbf{x}^\star) \right]. \tag{6.143}$$

Proof. Observe first that $\mu \leq \mathrm{Lip}(\nabla F)$ — see Exercise 8.9 — and so $1 - \alpha\mu \geq 0$. Rearranging the inequality (6.106) in Corollary 6.52 yields

$$-\tfrac{1}{2} \| \nabla F(\mathbf{x}_k) \|^2 \leq -\mu \left[F(\mathbf{x}_k) - F(\mathbf{x}^\star) \right].$$

We now insert this into the inequality established in Lemma 6.64 to obtain

$$F(\mathbf{x}_{k+1}) \leq F(\mathbf{x}_k) - \frac{\alpha}{2} \| \nabla F(\mathbf{x}_k) \|^2 \leq F(\mathbf{x}_k) - \alpha\mu \left[F(\mathbf{x}_k) - F(\mathbf{x}^\star) \right].$$

Subtracting $F(\mathbf{x}^\star)$ from both sides and rearranging terms yields

$$F(\mathbf{x}_{k+1}) - F(\mathbf{x}^\star) \leq (1 - \alpha\mu) \left[F(\mathbf{x}_k) - F(\mathbf{x}^\star) \right]. \tag{6.144}$$

We then iterate the inequality (6.144) to obtain (6.143). ∎

Remark 6.69. It is natural to ask how fast the iterates $\mathbf{x}_k$ are converging to the minimizer $\mathbf{x}^\star$. Since F is μ-strongly convex, we can combine Theorem 6.68 with (6.108) to obtain

$$\frac{\mu}{2} \| \mathbf{x}_k - \mathbf{x}^\star \|^2 \leq F(\mathbf{x}_k) - F(\mathbf{x}^\star) \leq (1 - \alpha\mu)^k \left[F(\mathbf{x}_0) - F(\mathbf{x}^\star) \right]. \tag{6.145}$$

We conclude that $\mathbf{x}_k$ converges to $\mathbf{x}^\star$ at the same linear convergence rate. ▲

Remark 6.70. Taking the largest possible value $\alpha = \mathrm{Lip}(\nabla F)^{-1}$ in (6.143) yields the linear convergence rate $1 - \tau$, where $\tau = \mu\,\mathrm{Lip}(\nabla F)^{-1}$ for gradient descent on strongly convex functions. Recalling the discussions in Example 6.49 and Remark 6.60,

$$\tau = \frac{\mu}{\mathrm{Lip}(\nabla F)} = \frac{\min_{\mathbf{x}} \lambda_{min}(C^{-1}\nabla_2^2 F(\mathbf{x}))}{\max_{\mathbf{x}} \lambda_{max}(C^{-1}\nabla_2^2 F(\mathbf{x}))}. \tag{6.146}$$

When $F(\mathbf{x}) = \frac{1}{2}\mathbf{x}^T H \mathbf{x} - \mathbf{f}^T \mathbf{x} + c$ is a quadratic function, this matches the discussion in Example 6.25, and the optimal choice for the preconditioner is $C = H$, which makes $\tau = 1$ and convergence immediate after one iteration. When F is strongly convex, but not necessarily quadratic, then (6.146) again suggests that a good choice for the preconditioner at the k-th step of gradient descent is $C_k = \nabla_2^2 F(\mathbf{x}_k)$. Then if one restricts the definition of τ to points $\mathbf{x}$ that are nearby $\mathbf{x}_k$, one would hope to obtain $\tau \approx 1$, and much faster convergence. The ensuing convergence analysis is the subject of Newton's method, which will be undertaken in the following section.   ▲

Exercises

9.1. ♡ For F_1 and F_2 from Exercise 4.1, compute the Lipschitz constant of the gradient and determine the rate of convergence of gradient descent, according to Theorem 6.66, with the optimal choice of time step. Use the Euclidean norm and dot product. Do the theoretical convergence rates match up with the experimental rates determined in Exercise 4.1?

9.2. ♡ Repeat Exercise 9.1, except this time compare the linear convergence rates provided by Theorem 6.68. Use the Euclidean norm and dot product.

9.3. ♡ Find a preconditioning matrix C so that preconditioned gradient descent on F_2 from Exercise 4.1 is equivalent to ordinary gradient descent on F_1 (from the same exercise), and thus admits the same convergence rate.

9.4. ◇ Modify the proof of Theorem 6.68 to show that if $0 < \alpha \leq 2/\mathrm{Lip}(\nabla F)$ then

$$F(\mathbf{x}_k) - F(\mathbf{x}^\star) \leq \left[1 - 2\alpha\mu\left(1 - \frac{\mathrm{Lip}(\nabla F)\alpha}{2}\right)\right]^k \left[F(\mathbf{x}_0) - F(\mathbf{x}^\star)\right].$$

9.5. ◇ Suppose that $F\colon \mathbb{R}^n \to \mathbb{R}$ has a Hölder continuous gradient ∇F with exponent $0 < \gamma < 1$, as defined in Exercise 8.15. (a) Use Exercise 8.16 to prove the descent inequality

$$F(\mathbf{x} - \alpha\,\nabla F(\mathbf{x})) \leq F(\mathbf{x}) - \alpha\,\|\nabla F(\mathbf{x})\|^2 + \frac{\alpha^{1+\gamma}}{1+\gamma}\,\|\nabla F(\mathbf{x})\|^{1+\gamma}. \tag{6.147}$$

(b) Let $\mathbf{x}_{k+1} = \mathbf{x}_k - \alpha_k \nabla F(\mathbf{x}_k)$ be the iterations of gradient descent. Show that if

$$\alpha_k^\gamma \leq \frac{1+\gamma}{H_\gamma(\nabla F)}\,\|\nabla F(\mathbf{x}_k)\|^{1-\gamma}, \tag{6.148}$$

then $F(\mathbf{x}_{k+1}) \leq F(\mathbf{x}_k)$.

9.6. Suppose that $F \colon \mathbb{R}^n \to \mathbb{R}$ is a convex function that admits a global minimizer $\mathbf{x}^\star$ and whose gradient ∇F is Hölder continuous with exponent $0 < \gamma < 1$, as defined in Exercise 8.15. Assume a *nonincreasing* sequence of time steps $\alpha_0 \geq \alpha_1 \geq \alpha_2 \geq \cdots$ is chosen in gradient descent so that

$$\alpha_k^\gamma \leq \frac{1+\gamma}{2 H_\gamma(\nabla F)} \, \|\nabla F(\mathbf{x}_k)\|^{1-\gamma}. \tag{6.149}$$

Follow the proof of Theorem 6.66 to show that

$$F(\mathbf{x}_k) - F(\mathbf{x}^\star) \leq \frac{\|\mathbf{x}_0 - \mathbf{x}^\star\|^2}{2\,k\,\alpha_k}. \tag{6.150}$$

9.7. Let $F(\mathbf{x}) = \|\mathbf{x}\|^{1+\gamma}$ where $0 < \gamma < 1$. Recall from Exercise 8.15 that ∇F is Hölder continuous with exponent γ. Show that it is possible to choose a decreasing sequence of time steps $\alpha_0 \geq \alpha_1 \geq \alpha_2 \geq \cdots$ so that the iterates $\mathbf{x}_k$ of gradient descent on F satisfy

$$F(\mathbf{x}_k) - F(\mathbf{x}^*) \leq \left(\frac{\|\mathbf{x}_0 - \mathbf{x}^*\|^2}{2\,c\,k} \right)^{1+\gamma/2}. \tag{6.151}$$

where $c > 0$ is a constant depending only on γ. Notice that this rate nicely interpolates between the case of a Lipschitz gradient where $\gamma = 1$ and the rate is $O(1/k)$, and Hölder gradients. It also suggests that for nonsmooth optimization, where $\gamma = 0$ and the gradient is not Hölder continuous, we may expect to obtain a convergence rate of the form $O(1/\sqrt{k})$. *Hint*: Do Exercises 8.15 and 9.6 first; then choose α_k to saturate the inequality (6.149).

6.10   Newton's Method

Python Notebook: Newton's Method (.ipynb)

In this section, we turn to one of the oldest iterative methods for approximating extremizers and solutions to systems of equations, originally proposed by Isaac Newton and his contemporary Joseph Raphson.

Recall the discussion in Remark 6.70, which suggests that the optimal preconditioner for gradient descent is the Hessian matrix $\nabla_2^2 F(\mathbf{x}_k)$. This would lead to *Hessian preconditioned gradient descent*, given by

$$\mathbf{x}_{k+1} = \mathbf{x}_k - \alpha_k \left[\nabla_2^2 F(\mathbf{x}_k) \right]^{-1} \nabla_2 F(\mathbf{x}_k). \tag{6.152}$$

In this section, we will assume that F is strongly convex, so that $\nabla_2^2 F$ is positive definite and hence invertible. We may expect the iteration (6.152) to exhibit a faster convergence rate than gradient descent, since the preconditioning is in some sense optimal. However, the analysis from Section 6.9 does not hold in the setting where the preconditioner is changing at each step.

It turns out that with a uniform choice of time step $\alpha_k = 1$, (6.152) becomes the classical *Newton's method*, also known as the *Newton-Raphson method*. Newton's method is a general iterative numerical root finding method that can be used to solve systems of nonlinear equations $G(\mathbf{x}) = \mathbf{0}$. In the context of optimization it solves the equation $\nabla F(\mathbf{x}) = \mathbf{0}$ satisfied by the critical points of F. In general, Newton's method takes the form

$$\mathbf{x}_{k+1} = \mathbf{x}_k - \left[\mathbf{D}G(\mathbf{x}_k) \right]^{-1} G(\mathbf{x}_k), \tag{6.153}$$

and, in favorable circumstances, the iterates $\mathbf{x}_k$ will converge to a solution to the system $G(\mathbf{x}) = \mathbf{0}$, cf. [105, 230].

Newton's method can be derived from an alternative perspective, which is also useful for understanding why the method converges faster than gradient descent. By Exercise 4.6, we can rewrite gradient descent (6.39) in the form

$$\mathbf{x}_{k+1} = \operatorname*{argmin}_{\mathbf{x}} \left\{ F(\mathbf{x}_k) + \langle \nabla F(\mathbf{x}_k), \mathbf{x} - \mathbf{x}_k \rangle + \frac{1}{2\alpha_k} \| \mathbf{x} - \mathbf{x}_k \|^2 \right\}. \tag{6.154}$$

The function on the right hand side of (6.154) is the tangent space linear approximation of F at $\mathbf{x}_k$, plus an additional quadratic term that prevents $\mathbf{x}_{k+1}$ from deviating too far from $\mathbf{x}_k$. Without this quadratic term, the objective would be a linear (or rather, affine) function that has no minimal value. Thus, each step of gradient descent can be viewed as minimizing the *linear* approximation of F, with a constraint on the distance moved.

The role of the quadratic term $\| \mathbf{x} - \mathbf{x}_k \|^2$ in (6.154) is only to restrict movement of the iterates, so that the linear approximation remains valid, and it does not help to approximate F in any way. This can help to explain why gradient descent can be slow to converge, and a straightforward way to improve this would be to replace the right hand side of (6.154) by a second order Taylor expansion, obtained by setting $\mathbf{z} = \mathbf{x}$ in (6.79) to obtain the second order Taylor terms, that better approximates the underlying objective function F. Fixing the dot product as our inner product, we arrive at the iterative scheme

$$\mathbf{x}_{k+1} = \operatorname*{argmin}_{\mathbf{x} \in \mathbb{R}^n} \left\{ F(\mathbf{x}_k) + \nabla_2 F(\mathbf{x}_k) \cdot (\mathbf{x} - \mathbf{x}_k) + \tfrac{1}{2} (\mathbf{x} - \mathbf{x}_k)^T \nabla_2^2 F(\mathbf{x}_k) (\mathbf{x} - \mathbf{x}_k) \right\}, \tag{6.155}$$

which amounts to minimizing the second order Taylor expansion of F at each iteration. When F is strongly convex, the quadratic term in (6.155) is sufficient to constrain the optimization problem for $\mathbf{x}_{k+1}$, and an additional penalty term is not needed — although see [174] for cubically constrained Newton methods. Since the expression on the right hand side of (6.155) is a quadratic function of $\mathbf{x}$, we can easily minimize it — see Exercise 10.5 — and we find that it prescribes

$$\mathbf{x}_{k+1} = \mathbf{x}_k - [\nabla_2^2 F(\mathbf{x}_k)]^{-1} \nabla_2 F(\mathbf{x}_k), \tag{6.156}$$

which is exactly the Hessian preconditioned gradient descent algorithm (6.152) with uniform time step $\alpha_k = 1$.

Example 6.71. The one-dimensional version of Newton's method is particularly simple. For a scalar function $g \colon \mathbb{R} \to \mathbb{R}$, (6.153) becomes

$$x_{k+1} = x_k - \frac{g(x_k)}{g'(x_k)}. \tag{6.157}$$

A famous example is the use of Newton's method to compute the square root $\sqrt{a}$ of a positive real number $a > 0$. Using[15] $g(x) = x^2 - a$, whose roots are $x = \pm\sqrt{a}$, (6.156) is

$$x_{k+1} = x_k - \frac{g(x_k)}{g'(x_k)} = x_k - \frac{x_k^2 - a}{2x_k} = \frac{2x_k^2 - x_k^2 + a}{2x_k} = \frac{1}{2}\left(x_k + \frac{a}{x_k}\right). \tag{6.158}$$

According to Exercise 10.4, provided x_0 is reasonably chosen (any positive number will do), the iterates x_k converge *very quickly* to the square root of a, meaning that the convergence is quadratic, as discussed below. This iterative scheme has often been called the *Babylonian*

---

[15]Here, we apply the root-finding version of Newton's method (6.153). We can equivalently use the optimization version (6.156) for the function $f(x) = \tfrac{1}{3}x^3 - ax$, chosen so that $f'(x) = g(x)$.

method, though there is little evidence it was known to the Babylonians. The first century Greek mathematician Hero (or Heron) of Alexandria described it in his AD 60 work *Metrica* and so it is also known as *Heron's method*, [101].

Heron's method for finding square roots can also be used to compute the matrix square root of a positive definite matrix H, as formulated in Definition 5.37. We start with an initial guess X_0, say $X_0 = H$ or $X_0 = I$, and then iterate

$$X_{k+1} = \tfrac{1}{2}(X_k + X_k^{-1}H). \tag{6.159}$$

Under some conditions on the eigenvalues of H, the matrix version also converges quadratically to the matrix square root of H. However, the method has poor numerical stability and other iterative methods are preferred. We refer the reader to [104] for details. ▲

Example 6.72. Let us apply Newton's method to the quadratic function

$$F(\mathbf{x}) = \tfrac{1}{2}\mathbf{x}^T H \mathbf{x} - \mathbf{f}^T \mathbf{x} + c,$$

where H is positive definite. Since $\nabla_2 F(\mathbf{x}) = H\mathbf{x} - \mathbf{f}$ and $\nabla_2^2 F(\mathbf{x}) = H$, from any initial guess $\mathbf{x}_0$ the first iteration of Newton's method (6.156) produces

$$\mathbf{x}_1 = \mathbf{x}_0 - H^{-1}(H\mathbf{x}_0 - \mathbf{f}) = \mathbf{x}_0 - \mathbf{x}_0 + H^{-1}\mathbf{f} = H^{-1}\mathbf{f}.$$

Hence, Newton's method converges in a single iteration. On the other hand, this requires us to compute $H^{-1}\mathbf{f}$, or rather, solve the linear system $H\mathbf{x} = \mathbf{f}$. In other words, Newton's method for a quadratic function reduces to the original problem of finding the solution to the linear system that characterizes the minimizer, and is thus of no help. ▲

We now formulate a convergence result for Newton's method. For simplicity, we work only with the dot product and Euclidean norm when defining strongly convex and Lipschitz functions.

Theorem 6.73. *Let* $F \colon \mathbb{R}^n \to \mathbb{R}$ *be* μ-*strongly convex, with Lipschitz continuous Hessian* $\nabla_2^2 F$. *Assume that*

$$\beta := \frac{\mathrm{Lip}(\nabla_2^2 F)}{2\mu^2}\|\nabla_2 F(\mathbf{x}_0)\|_2 < 1. \tag{6.160}$$

Then, as $k \to \infty$, *the iterates* $\mathbf{x}_k$ *in Newton's method (6.156) converge to the minimizer* $\mathbf{x}^\star$ *of* F. *Furthermore for any* $k \geq 0$, *we have the estimate*

$$\|\mathbf{x}_k - \mathbf{x}^\star\|_2 \leq \frac{2\mu\beta^{2^k}}{\mathrm{Lip}(\nabla_2^2 F)}. \tag{6.161}$$

Proof. Define the k-th *error*

$$\varepsilon_k = \frac{\mathrm{Lip}(\nabla_2^2 F)}{2\mu^2}\|\nabla_2 F(\mathbf{x}_k)\|_2, \tag{6.162}$$

so that $\beta = \varepsilon_0$. The overall strategy of the proof will be to show that

$$\varepsilon_{k+1} \leq \varepsilon_k^2 \quad \text{for all} \quad k \geq 0. \tag{6.163}$$

Indeed, iterating (6.163), we obtain $\varepsilon_k \le \varepsilon_0^{2^k} = \beta^{2^k}$ for all $k \ge 0$. Substituting the definition (6.162) of ε_k yields

$$\|\nabla_2 F(\mathbf{x}_k)\|_2 \le \frac{2\,\mu^2 \beta^{2^k}}{\mathrm{Lip}(\nabla^2 F)}.$$

The proof of (6.161) is completed by combining this with the estimate in (6.109). Assuming $\beta < 1$, we deduce that $\|\mathbf{x}_k - \mathbf{x}^\star\|_2 \to 0$ as $k \to \infty$, and so Newton's method converges to the minimizer of F.

To prove (6.163), we first note that the Newton iteration (6.156) satisfies

$$\nabla_2 F(\mathbf{x}_k) + \nabla_2^2 F(\mathbf{x}_k)(\mathbf{x}_{k+1} - \mathbf{x}_k) = \mathbf{0}.$$

Therefore, by (6.126),

$$
\begin{aligned}
\varepsilon_{k+1} &= \frac{\mathrm{Lip}(\nabla_2^2 F)}{2\,\mu^2}\,\|\nabla_2 F(\mathbf{x}_{k+1})\|_2 \\[4pt]
&= \frac{\mathrm{Lip}(\nabla_2^2 F)}{2\,\mu^2}\,\|\nabla_2 F(\mathbf{x}_{k+1}) - \nabla_2 F(\mathbf{x}_k) - \nabla_2^2 F(\mathbf{x}_k)(\mathbf{x}_{k+1} - \mathbf{x}_k)\|_2 \\[4pt]
&\le \frac{\mathrm{Lip}(\nabla_2^2 F)}{2\mu^2}\left(\frac{\mathrm{Lip}(\nabla_2^2 F)}{2}\,\|\mathbf{x}_{k+1} - \mathbf{x}_k\|_2^2\right) = \frac{\mathrm{Lip}(\nabla_2^2 F)^2}{4\,\mu^2}\,\|\,[\nabla_2^2 F(\mathbf{x}_k)]^{-1}\nabla_2 F(\mathbf{x}_k)\|_2^2 \\[4pt]
&\le \frac{\mathrm{Lip}(\nabla_2^2 F)^2}{4\,\mu^2}\,\|\,[\nabla_2^2 F(\mathbf{x}_k)]^{-1}\|_2^2\,\|\nabla_2 F(\mathbf{x}_k)\|_2^2 \le \frac{\mathrm{Lip}(\nabla_2^2 F)^2}{4\,\mu^4}\,\|\nabla_2 F(\mathbf{x}_k)\|_2^2 = \varepsilon_k^2,
\end{aligned}
$$

where in the last line, we used (6.104) to replace

$$\|\,[\nabla_2^2 F(\mathbf{x}_k)]^{-1}\|_2 = \lambda_{max}\big(\nabla_2^2 F(\mathbf{x}_k)^{-1}\big) = \lambda_{min}\big(\nabla_2^2 F(\mathbf{x}_k)\big)^{-1} \le \mu^{-1}.$$

This establishes (6.163), and hence completes the proof.   ∎

Remark 6.74. The convergence rate established in Theorem 6.73 is called *quadratic convergence*, since, by (6.163), the error at each iteration is less than a multiple — here the multiple is 1 — of the square of the previous error. Quadratic convergence is extremely fast and typically only takes a handful of iterations to converge. Roughly speaking, each iterate of a quadratically convergent scheme doubles the number of accurate decimal digits in the approximation to the minimizer. However, in order to achieve quadratic convergence of Newton's method, the condition $\beta < 1$ that guarantees convergence as a consequence of Theorem 6.73 must be satisfied, which essentially states that we must initialize Newton's method sufficiently close to the minimizer, in order that $\|\nabla_2 F(\mathbf{x}_0)\|_2$ be sufficiently small. If the initial guess $\mathbf{x}_0$ is too far away, Newton's method may not converge; see [185] for a discussion and plots of the striking fractal behavior of Newton's method, for scalar complex functions, outside the regions of convergence. We also mention that if F is not strongly convex, the convergence rate may not be quadratic, though it often still represents an improvement over gradient descent; see Example 6.76.

In practice, Newton's method is often modified with the inclusion of a time step, in the form originally introduced in (6.152) at the start of this section. With a good adaptive selection of the time step α_k Newton's method is provably convergent from any initial guess $\mathbf{x}_0$, [31], except that the method may take many steps before it enters the quadratic convergence regime where $\beta < 1$; see Exercise 10.4. There are other ways to guarantee global convergence of Newton's method, such as adding cubic constraints; see [174].

It is also important to point out that Newton's method requires inverting the Hessian matrix $\nabla_2^2 F$, or at least solving the linear system

$$\nabla_2^2 F(\mathbf{x}_k)\,(\mathbf{x}_{k+1} - \mathbf{x}_k) = -\nabla_2 F(\mathbf{x}_k),$$

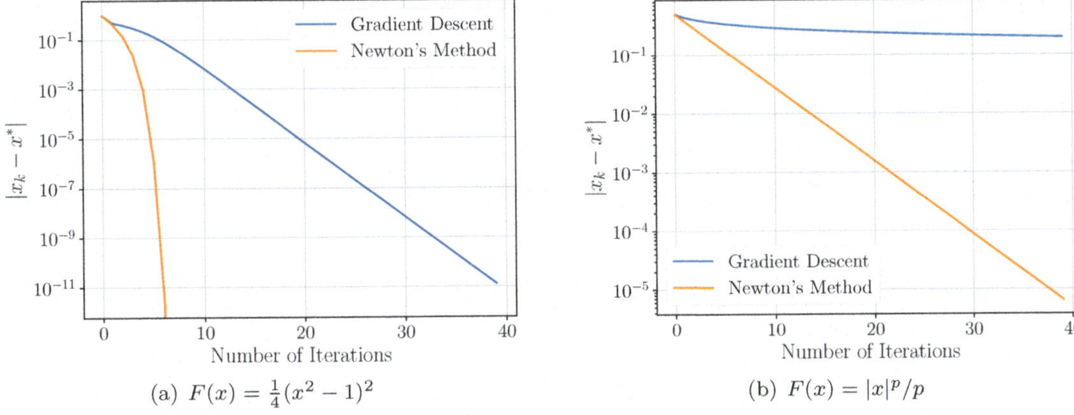

Figure 6.9: Comparison of Newton's method and gradient descent on two functions $F\colon \mathbb{R} \to \mathbb{R}$. In (a) we use the double well potential $F(x) = \frac{1}{4}(x^2 - 1)^2$, which is strongly convex in a region around each of its minimizers $x = \pm 1$, and we correspondingly see linear convergence for gradient descent and quadratic convergence for Newton's method. In (b) we use the non-strongly convex function $F(x) = |x|^p/p$ with $p = 5$ that is discussed in Example 6.76. In this case, we see sublinear convergence for gradient descent, and linear convergence for Newton's method.

at each time step. Thus, while Newton's method may require far fewer iterations to converge, compared to gradient descent, the computational complexity of each iteration is much greater. In some very high dimensional applications, including training deep neural networks, it is computationally intractable to even construct the Hessian, much less its inverse! Choosing the appropriate algorithm for given optimization problem is a challenging task and requires a deep understanding of many different techniques. We defer more advanced methods to Chapter 11. ▲

To compare Newton's method with gradient descent, we consider two toy examples.

Example 6.75. First, we consider the double-well potential

$$F(x) = \tfrac{1}{4}(x^2 - 1)^2,\tag{6.164}$$

whose graph looks very similar to that of Figure 6.4(b). Double-well potentials arise in quantum mechanics [99] and various areas of computational mathematics, in particular the computation of the shapes of soap bubbles [33]. The function F has global minimizers at $x = \pm 1$. We initialize gradient descent and Newton's method at $x_0 = 2$, and expect them to converge to the minimizer at $x = 1$. We use a time step of $\alpha = 0.25$ for gradient descent and run both methods for $k = 40$ steps. The results are in Figure 6.9(a), and we see that gradient descent gives the expected linear convergence rate, while Newton's method exhibits much faster quadratic convergence. ▲

Example 6.76. As a second example, we consider the function $F(x) = |x|^p/p$ for some $p > 2$, so it is convex but not strongly convex. We showed in Example 6.67 that gradient descent converges at a rate of $F(x_k) = O(1/k^{p/(p-2)})$, which, for large p, is close to the $O(1/k)$ rate guaranteed in Theorem 6.66, while the rate of convergence of the iterates $x_k = O(1/k^{1/(p-2)})$ completely degenerates as $p \to \infty$.

Newton's method for minimizing F corresponds to

$$x_{k+1} = x_k - \frac{F'(x_k)}{F''(x_k)} = x_k - \frac{x_k^{p-1}}{(p-1)x_k^{p-2}} = \frac{p-2}{p-1}x_k,$$

provided $x_k > 0$. Therefore,

$$x_k = \beta^k x_0, \qquad \text{where} \qquad \beta = \frac{p-2}{p-1} < 1,$$

and hence Newton's method converges at the linear rate β; in particular, the rate is not quadratic. While the rate β degenerates to $\beta = 1$ as $p \to \infty$, we note that the rate for $F(x_k)$ is

$$F(x_k) = \frac{x_k^p}{p} = \frac{1}{p}\left(1 - \frac{1}{p-1}\right)^{pk} x_0^p \le \frac{x_0^p}{p} e^{-k},$$

where the final inequality uses the estimate

$$1 - x \le e^{-x}, \qquad \text{which implies} \qquad \left(1 - \frac{1}{p-1}\right)^{pk} \le \left(1 - \frac{1}{p}\right)^{pk} \le e^{-k}.$$

Hence, the convergence for $F(x_k)$ is linear with a rate of $\beta = e^{-1}$ that is independent of p.

We illustrate the sublinear convergence of gradient descent and linear convergence of Newton's method for this example by taking $p = 5$ and $\alpha = 1$ for gradient descent in Figure 6.9(b). Thus, for functions F that are not strongly convex, Newton's method may no longer converge quadratically, but in some cases still offers a substantial improvement over the very slow sublinear convergence rate offered by gradient descent. ▲

Exercises

10.1. Implement Newton's method in Python for the following scalar functions and investigate rates of convergence. (a) $f(x) = x^2 + (x-1)^4$, (b) $f(x) = e^{x^2}$ (c) $f(x) = \sin(x)^2$.

10.2. Use Newton's Method to find all points of intersection of the following pairs of plane curves: (a) ♡ $x^2 + y^2 = 1$, $xy = \frac{1}{2}$, (b) ♡ $x^3 + y^3 = 1$, $x^2 - y^2 = 1$,
  (c) ◇ $x^2 + \frac{1}{3}y^2 = 1$, $x^2 + \frac{1}{4}x + 2y^2 - \frac{1}{4}y = 5$, (d) $y = x^2 - 3x - 5$, $x = -2y^2 + 6y$.
Hint: Sketching the curves will help you decide where to start the iterations.

10.3. Use Newton's Method to find all common points of intersection of the following three surfaces: $x^2 + y^2 + z^2 = 1$, $z = x^2 + y^2$, $x + y + z = 1$.

10.4. ◇ In this exercise, you will prove that the Babylonian method, which was introduced in Example 6.71, converges quadratically to the square root of a. To do this, show that the error $\varepsilon_n = x_n/\sqrt{a} - 1$ satisfies

$$\varepsilon_{n+1} = \frac{\varepsilon_n^2}{2(\varepsilon_n + 1)}. \tag{6.165}$$

Use this to show that $\varepsilon_n \ge 0$ for $n \ge 1$. Then show that both

$$\varepsilon_{n+1} \le \tfrac{1}{2}\varepsilon_n \quad \text{and} \quad \varepsilon_{n+1} \le \tfrac{1}{2}\varepsilon_n^2 \quad \text{hold for all} \quad n \ge 1.$$

Use the first inequality to show that $\varepsilon_n \le 2^{-(n-1)}\varepsilon_1$, which shows that $\varepsilon_n \to 0$ as $n \to \infty$. Given this, the second inequality implies quadratic convergence.

10.5. ♡ Prove that (6.155) is equivalent to (6.156).

10.6. Consider the nonlinear system $x^3 - 9xy^2 = 1$, $x^2 - y^2 = 0$. (a) Find all the solutions by hand. For the remaining parts of the problem choose one of your solutions to work with. (b) Use gradient descent, with a suitable initial value, to approximate your chosen solution. How many iterations are needed to obtain 10 decimal place accuracy? (c) Write down the equations for Newton's Method applied to this system. (d) Suppose you start the Newton iterations with the same initial guess. Approximately how many iterations would you anticipate needing in order to get 10 decimal place accuracy in your solution? Check your prediction by running the algorithm.

10.7.$\diamond$ (a) Show that $F(x, y) = x^2 + y^6$, has a unique global minimizer at $(x^*, y^*)^T = (0, 0)^T$.

(b) Write Python code to minimize F by gradient descent starting at $(x_0, y_0)^T = (1, 1)^T$. By trial and error, find the largest time step α for which gradient descent is stable and convergent.

(c) Write Python code for preconditioned gradient descent using the Hessian matrix $\nabla^2 F$ as the preconditioner C (i.e., Newton's method). By trial and error, find the largest time step α for which preconditioned gradient descent is stable and convergent.

(d) Plot $x_k^2 + y_k^2$ versus the number of iterations of gradient descent for both methods on the same plot. Which one converges faster?

10.8. $\heartsuit$ Given a descent direction $\mathbf{v}$ for an optimization method — for gradient descent $\mathbf{v} = -\nabla F(\mathbf{x})$, while for Newton's method $\mathbf{v} = -\nabla_2^2 F(\mathbf{x})^{-1} \nabla_2 F(\mathbf{x})$ — a *backtracking line search* aims to choose the best time step α to minimize the function F along the descent direction, that is, to minimize $F(\mathbf{x} + \alpha \mathbf{v})$ over α. The backtracking line search has two parameters $0 < \gamma \leq \frac{1}{2}$ and $0 < \beta < 1$, and chooses $\alpha = \beta^k$, where $k \geq 0$ is the smallest nonnegative integer such that

$$F(\mathbf{x} + \beta^k \mathbf{v}) \leq F(\mathbf{x}) + \gamma \beta^k \langle \nabla F(\mathbf{x}), \mathbf{v} \rangle. \tag{6.166}$$

In practice, one starts with $k = 0$, and then iteratively increases $k = 1, 2, \ldots$ until the inequality (6.166) holds.

(a) Assume ∇F is Lipschitz continuous, and the descent direction is $\mathbf{v} = -\nabla F(\mathbf{x})$. Show that there exists an integer $k \geq 0$ such that (6.166) holds. That is, the backtracking line search will eventually terminate. *Hint*: Use Lemma 6.64.

(b) Implement the backtracking line search in Python when $F(\mathbf{x}) = x_1^3 + 10 x_2^2$. Try gradient descent, where $\mathbf{v} = -\nabla F(\mathbf{x}_k)$, and Newton's method where $\mathbf{v} = -\nabla_2^2 F(\mathbf{x}_k)^{-1} \nabla_2 F(\mathbf{x}_k)$. In both cases, after conducting the backtracking line search, the update is $\mathbf{x}_{k+1} = \mathbf{x}_k + \beta^k \mathbf{v}$. Starting from $\mathbf{x}_0 = (1, 1)^T$, you should observe faster convergence with the backtracking line search with good choices of parameters: $\gamma = 0.5$ and $\beta = 0.9$ are reasonable.

Chapter 7

Introduction to Machine Learning and Data

The primary goal of this text is to understand and apply the mathematics of linear algebra and optimization to develop machine learning and data analysis, which will form the focus of the second half of the text. Machine learning refers to a class of algorithms that learn to complete tasks, such as image classification, face recognition, text generation, etc., from examples or experience, and are not explicitly programmed with a list of instructions to follow. For example, to perform handwritten digit recognition with a machine learning algorithm, one would provide many examples (sometimes hundreds or thousands) of images of handwritten digits and their known labels, and the algorithm will attempt to learn a general rule that is able to to correctly label new instances.

In this chapter, we describe the field in some detail and introduce several basic and important methods that are used in machine learning and data analysis. Their performance is examined by applying them to some publicly available data sets. The goal here is to be both introductory and illustrative, and thereby provide the foundation and motivation for the more advanced methods to be presented later, including the graph-based learning methods developed in Chapter 9, and deep neural networks developed in Chapter 10.

Note: Throughout this chapter, unless otherwise specified, we will use $\|\cdot\|$ to refer to the standard Euclidean norm,

7.1 Basics of Machine Learning and Data

The primary object of study in machine learning, statistics, and many other fields of science, engineering, finance, social sciences, and beyond, is data. Thus, the first order of business is to specify precisely what we mean by "data". Each object under investigation is characterized by one of more measurements of its properties, which are often referred to as *features*. For us, the measurements will always be real-valued scalars. If a measurement is a vector-valued quantity, for example the position of a body in three-dimensional space, each component is viewed as an individual measurement. If the measurement is discrete, for example some physical trait is either present or absent, it will still be represented by a real-valued quantity that is restricted to discrete values, e.g., 0 or 1, representing the different possibilities. Thus, an object's measurements form a vector $\mathbf{x} = (x_1, \ldots, x_n)^T \in \mathbb{R}^n$ whose components x_i are the individual measurements. Keep in mind that all measurements are, to some degree, approximate, and can be corrupted by experimental error, noise, numerical approximation,

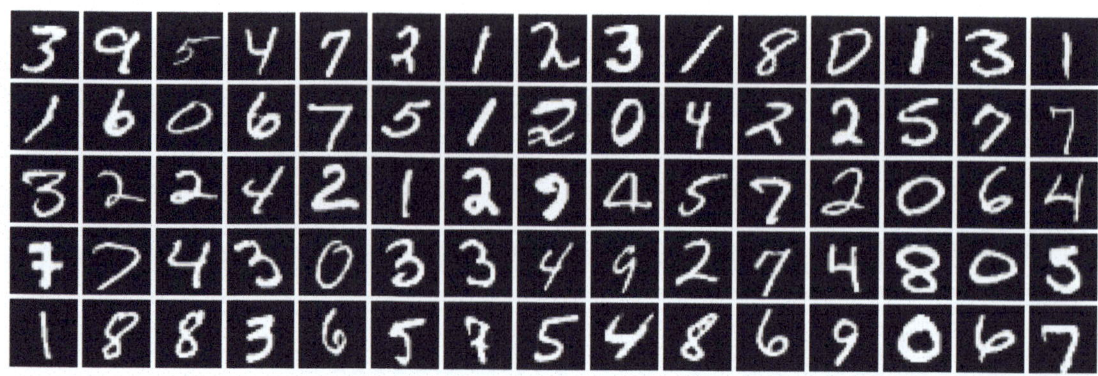

Figure 7.1: Examples of some of the MNIST digits. Each is a 28×28 pixel gray scale image that contains a single handwritten digit.

and so on. An important example is when the object is a digital image. For a two-dimensional black and white image, each measurement represents an individual pixel, with $0 \leq x_i \leq 1$ indicating its gray scale value, where 0 represents black and 1 represents white. For a color image, the pixel measurements have 3 or 4 components, also known as channels, representing color saturation and darkness in a color space, such as RGB or CMYK, while hyperspectral satellite images can have hundreds of channels. Of course, a two-dimensional (rectangular) image is most naturally represented by a matrix of pixel values, but this can be converted into a vector by our usual identification of matrices and vectors.[1] Similar remarks apply to three-dimensional images, videos, three-dimensional videos, and so on. Observe that the number of measurements (pixels and perhaps their colors) can be gigantic in the latter instances.

A simple example of an image data set is MNIST[2], which we will often use for illustrating machine learning throughout the book. The MNIST data set contains 70,000 grayscale images of handwritten digits 0 through 9. Figure 7.1 shows an example of some images from the MNIST data set. Each image is quite small, containing only $28 \times 28 = 784$ pixels, and the data representing each image is the vector $\mathbf{x} \in \mathbb{R}^{784}$ containing the grayscale pixel values. Each image in the MNIST data set also comes with a prescribed label[3] $y \in \{0, 1, \ldots, 9\}$ indicating which digit is depicted in the image. The goal of a machine learning classifier trained on MNIST is to predict the label of each digit image — that is, to perform *optical character recognition*, which is very commonly employed for many tasks, including archiving old newspapers or books, and teaching self-driving cars to read street signs and house numbers.

Another example is the diabetes data set originally presented in [69], and available through the Python package `scikit-learn` as well as other sources.[4] This data set is used as a prototype for studying medical data analysis, where the goal is to make useful predictions about patients. In this context, the feature vector $\mathbf{x}$ for each patient may include data such

---

[1] However, it may not be desirable to convert images into vectors, since one loses the spatial structure of the image. In Chapters 9 and 10 we develop machine learning methods for computer vision that do not treat images as vectors.

[2] The MNIST data set is available online `http://yann.lecun.com/exdb/mnist/`. It was created in 1994 using a mixture of several data sets curated by the National Institute of Standards and Technology (NIST). It has become somewhat of a benchmark within machine learning for offering a way to quickly evaluate new algorithms. Classification of MNIST digits is now an "easy" problem in machine learning, with the best modern deep learning methods attaining 99.87% accuracy [35].

[3] See the following section for details on how labels are prescribed and handled in machine learning.

[4] See `https://www4.stat.ncsu.edu/~boos/var.select/diabetes.html`

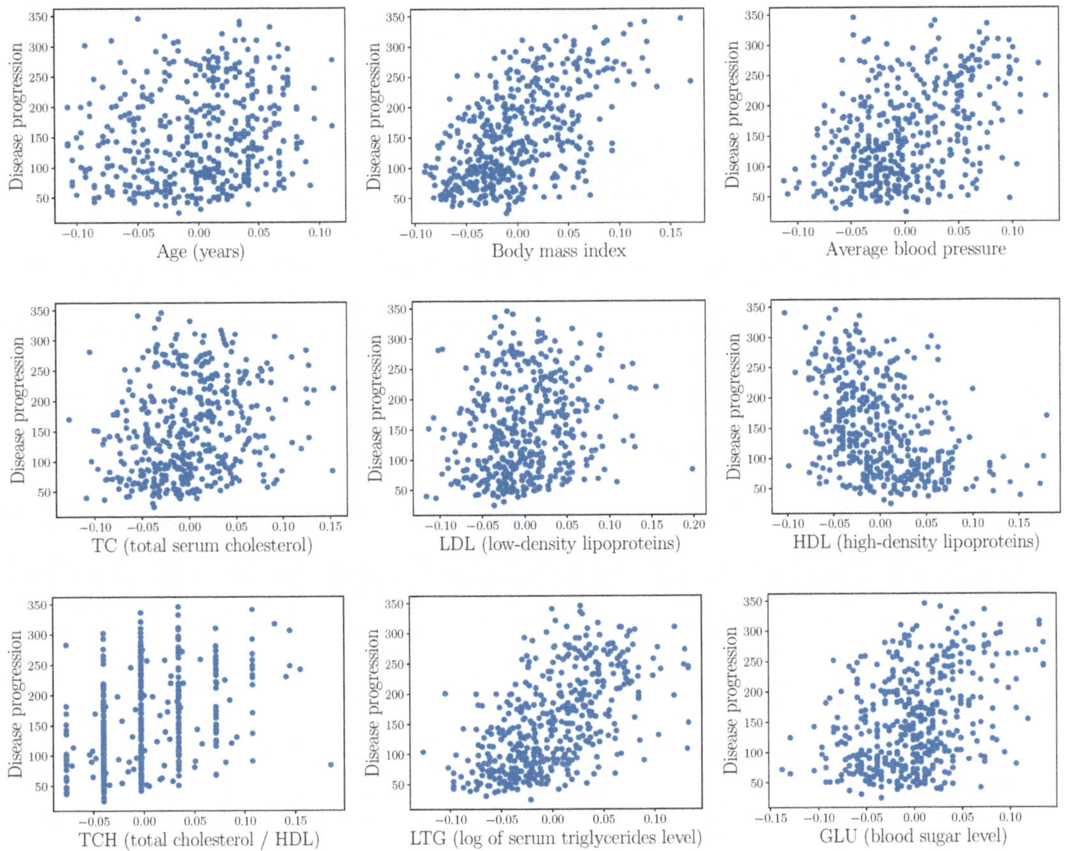

Figure 7.2: Plots of the diabetes disease progression versus the various feature variables in the regression problem. None of the variables themselves offer a particular strong prediction of disease progression. Note that all of the features (on the x-axis) have been normalized to have mean zero and variance $1/442$.

as age, body mass index, and blood serum measurements. In the diabetes data set, there are 442 patients, and each patient has 9 measurements, so we have a patient feature vector $\mathbf{x} \in \mathbb{R}^9$. For each patient there is an additional positive number $0 < y \in \mathbb{R}$ that measures disease progression, with larger numbers indicating a more serious onset, and the goal is to predict the disease progression from the patient data. Figure 7.2 shows plots of the disease progression versus each feature in the diabetes data set. From these plots, it is difficult to see whether any of these individual features are useful, on their own for prediction. In Figure 7.3 we show plots of some pairs of features, where the color of each data point indicates disease progression. Examining the plots indicates that there may be some potential for using multiple features as predictors.

In general, we are given m objects of a similar nature, each represented by a measurement vector, and so by *data* we specifically mean a collection of vectors $\mathbf{x}_1, \ldots, \mathbf{x}_m \in \mathbb{R}^n$, also known as *data points* or *data vectors*. This requires that *all* the objects under study have the same set of measurements, and that we know *all* their values. Ensuring that the data set is complete in this manner may require some preprocessing. For example, images are often of different sizes and/or involving differing numbers of pixels. In such cases, cropping,

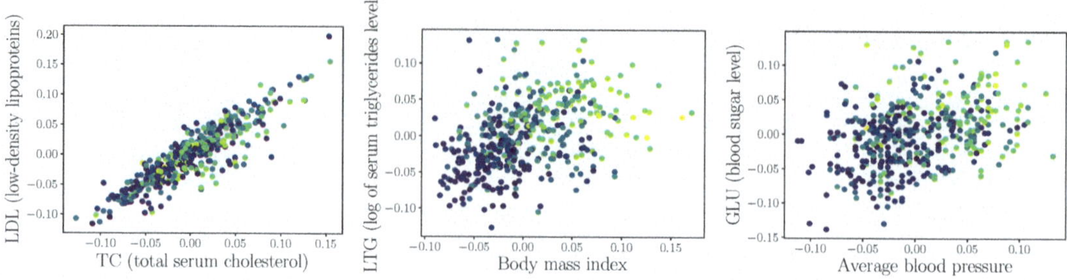

Figure 7.3: Three plots of pairs of variables with the diabetes disease progression shown as the color of each data point. The right two plots show some promise that pairs of variables can be used in combination to predict disease progression. The leftmost plot shows a strong correlation between two variables for measuring cholesterol.

enlarging, or subsampling the images may be required in order to ensure that they all have the same pixel configuration. Extending the analysis to more general missing or unavailable or mismatched data is a very active area of contemporary research, which we unfortunately do not have space to examine here. We refer the interested reader to [70, 77] and the references therein.

It will be convenient to assemble the data vectors $\mathbf{x}_1, \ldots, \mathbf{x}_m$ into a matrix, known as the *data matrix*, and denoted by X. For various reasons, it is more convenient to let the individual data points be the *rows* of the data matrix. Thus, we define

$$X = \begin{pmatrix} \mathbf{x}_1^T \\ \mathbf{x}_2^T \\ \vdots \\ \mathbf{x}_m^T \end{pmatrix} = \begin{pmatrix} x_{11} & x_{12} & \cdots & x_{1n} \\ x_{21} & x_{22} & \cdots & x_{2n} \\ \vdots & \vdots & \ddots & \vdots \\ x_{m1} & x_{m2} & \cdots & x_{mn} \end{pmatrix}. \tag{7.1}$$

Observe that X has size $m \times n$, where we make the blanket convention that m represents the number of data points and n the number of measurements. Thus, the entry x_{ij} indicates the j-th measurement of the i-th object in our data set. The columns of the data matrix X are the *measurement vectors*; thus the j-th column, denoted $\mathbf{v}_j = (x_{1j}, \ldots, x_{mj})^T$, contains all the measurements of the j-th quantity. The data matrix can also be written in the alternative forms

$$X = \sum_{i=1}^{m} \mathbf{e}_i \mathbf{x}_i^T = \sum_{j=1}^{n} \mathbf{v}_j \mathbf{e}_j^T, \tag{7.2}$$

where in the formula above, $\mathbf{e}_i \in \mathbb{R}^m$ and $\mathbf{e}_j \in \mathbb{R}^n$.

7.1.1   Mean, Variance, and Covariance

Let us now give a brief description of basic statistical concepts associated with data. Suppose $\mathbf{v} = (v_1 \, v_2 \ldots v_m)^T \in \mathbb{R}^m$ is one of the columns of our data matrix, representing a collection of m measurements of a single physical quantity, e.g., the distance to a star as measured by various physical apparatuses, the speed of a car at a given instant measured by a collection of instruments, a person's blood pressure or IQ as measured by a series of tests, etc. Experimental error, statistical fluctuations, quantum mechanical effects, numerical approximations, and the like imply that the individual measurements will almost certainly not precisely agree. Nevertheless, one wants to know the most likely value of the measured quantity and the degree of confidence that one has in the proposed value. A variety of statistical tests have been devised to resolve these issues, and we refer the interested reader to, for example, [110, 207, 241].

The most basic collective quantity of such a set of measurements is its *mean*, which is the average of its entries:

$$\overline{v} = \frac{v_1 + \cdots + v_m}{m} = \frac{1}{m} \mathbf{1}^T \mathbf{v}. \tag{7.3}$$

Here $\mathbf{1} = (1, \ldots, 1)^T \in \mathbb{R}^m$ is the column vector containing all 1's, so $\mathbf{1}^T$ is the corresponding row vector. Barring some inherent statistical or experimental bias, the mean can be viewed as the most likely value, known as the *expected value*, of the quantity being measured, and thus the best bet for its actual value. Once the mean has been computed, it will be helpful to *center* the measurements to have *mean zero*, which is done by subtracting off the mean from each entry. The resulting *centered measurement vector* will be denoted by an underbar:

$$\underline{\mathbf{v}} = (\underline{v}_1 \ \underline{v}_2 \ \cdots \ \underline{v}_m)^T = (v_1 - \overline{v}, \ldots, v_m - \overline{v})^T = \mathbf{v} - \overline{v}\,\mathbf{1} = J\mathbf{v}, \tag{7.4}$$

where, in view of (7.3), the $m \times m$ matrix

$$J = I - \frac{1}{m} \mathbf{1}\mathbf{1}^T = \begin{pmatrix} (m-1)/m & -1/m & -1/m & \cdots & -1/m \\ -1/m & (m-1)/m & -1/m & \cdots & -1/m \\ -1/m & -1/m & (m-1)/m & \cdots & -1/m \\ \vdots & \vdots & \vdots & \ddots & \vdots \\ -1/m & -1/m & -1/m & \cdots & (m-1)/m \end{pmatrix}, \tag{7.5}$$

is known as the *centering matrix*. We note that $J\mathbf{1} = \mathbf{0}$; in fact $\ker J$ is one-dimensional, spanned by the ones vector $\mathbf{1}$, and hence $\operatorname{rank} J = m - 1$. Moreover, this implies that J is an idempotent matrix, meaning that $J^2 = J$, and, in fact, represents orthogonal projection onto the subspace $V_0 = \{v_1 + \cdots + v_m = 0\} \subset \mathbb{R}^m$ consisting of all mean zero measurement vectors.

Given a data matrix X, the (row) vector containing the various measurement means is

$$\overline{\mathbf{v}}^T = (\overline{v}_1, \ldots, \overline{v}_n) = \frac{1}{m} \mathbf{1}^T X, \tag{7.6}$$

where $\overline{v}_j$ is the mean of the j-th measurement vector, i.e., the j-th column of X. Centering each of the columns of X by subtracting its mean is equivalent to multiplying X on the left by the centering matrix (7.5); the result is the *centered data matrix*

$$\underline{X} = X - \mathbf{1}\overline{\mathbf{v}}^T = JX, \tag{7.7}$$

each of whose columns has mean zero. In terms of the data points, that is the rows $\mathbf{x}_i^T$ of X, the column vector containing the means corresponding to (7.6) is also given by

$$\overline{\mathbf{v}} = \frac{1}{m} X^T \mathbf{1} = \frac{1}{m} \sum_{i=1}^{m} \mathbf{x}_i =: \overline{\mathbf{x}}, \tag{7.8}$$

and the rows of

$$\underline{X} = JX = \begin{pmatrix} \underline{\mathbf{x}}_1^T \\ \underline{\mathbf{x}}_2^T \\ \vdots \\ \underline{\mathbf{x}}_m^T \end{pmatrix} = \begin{pmatrix} \mathbf{x}_1^T - \overline{\mathbf{x}}^T \\ \mathbf{x}_2^T - \overline{\mathbf{x}}^T \\ \vdots \\ \mathbf{x}_m^T - \overline{\mathbf{x}}^T \end{pmatrix} \tag{7.9}$$

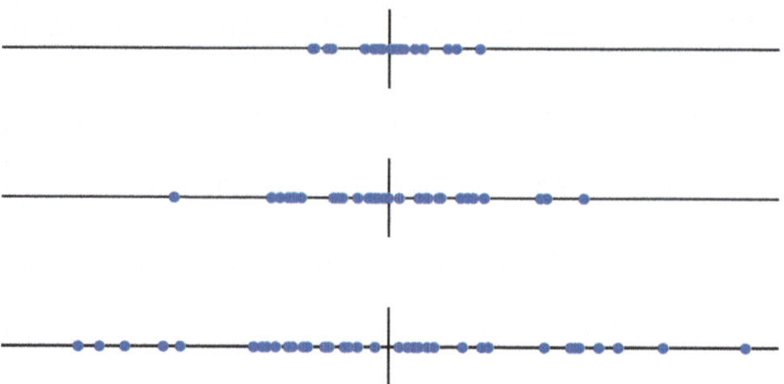

Figure 7.4: One-dimensional Scatter Plots

are the corresponding *centered data points.*

The *variance* of a set of measurements $\mathbf{v} = (\, v_1 \; v_2 \; \ldots \; v_m \,)^T$ tells us how widely they are "scattered" about their mean $\overline{v}$. This is quantified by summing the squares of their deviations from the mean, and denoted

$$\sigma_v^2 = \nu \left[\, (v_1 - \overline{v})^2 + \; \cdots \; + (v_m - \overline{v})^2 \,\right] = \nu \, \|\mathbf{v} - \overline{v}\,\mathbf{1}\|^2 = \nu \, \|\, J\mathbf{v}\,\|^2, \tag{7.10}$$

where $\nu > 0$ is a certain specified prefactor, which can assume different values depending upon one's statistical objectives; common examples are

(a) $\nu = 1$ for the unnormalized variance, or sum of squares;

(b) $\nu = 1/m$ for the "naïve" variance;

(c) $\nu = 1/(m-1)$ (assuming $m > 1$, i.e., there are at least 2 measurements) for an unbiased version;

(d) $\nu = 1/(m+1)$ for the minimal mean squared estimation of variance; and

(e) more exotic choices, e.g., if one desires an unbiased estimation of standard deviation instead of variance, cf. [110, p. 349].

Fortunately, apart from the resulting numerical values, much of the underlying analysis is independent of the prefactor.

The square root of the variance is known as the *standard deviation*, and denoted by

$$\sigma = \sigma_v = \sqrt{\nu} \, \|\, J\mathbf{v}\,\|. \tag{7.11}$$

The variance and standard deviation measure how far, on average, the values $v_1, \ldots, v_m$ deviate from their mean $\overline{v}$. When the variance and standard deviation are small, the measurements are tightly clustered around the mean value, while when they are large, some (or many) measurements lie far away from the mean. Figure 7.4 contains several *scatter plots*, in which each real-valued measurement is indicated by a dot and their mean is represented by a small vertical bar. The top plot shows data with relatively small variance, since the measurements are closely clustered about their mean, whereas on the bottom plot, the variance is large because the data is fairly spread out.

It is often useful in machine learning and data analysis tasks to *normalize* measurement vectors to have mean zero and unit variance, so $\overline{v} = 0$, $\sigma_v = 1$. This accomplished by subtracting the mean from each of the entries, and dividing by the standard deviation, which amounts to defining a new measurement vector

$$\widehat{\mathbf{v}} = \frac{\mathbf{v}}{\sigma_v} = \frac{\mathbf{v} - \overline{v}\,\mathbf{1}}{\sigma_v}, \qquad \text{so that} \qquad \widehat{v}_i = \frac{v_i - \overline{v}}{\sigma_v}. \tag{7.12}$$

The reader is encouraged to check that the measurement vector (7.12) has mean zero and standard deviation of one, that is

$$\mathbf{1}^T\,\widehat{\mathbf{v}} = 0 \qquad \text{and} \qquad \nu\,\widehat{\mathbf{v}}^T\,\widehat{\mathbf{v}} = 1.$$

Let $\sigma_1, \ldots, \sigma_n$ denote the standard deviations of the measurement vectors $\mathbf{v}_1, \ldots, \mathbf{v}_n$, and define the diagonal *standard deviation matrix*[5] $\Sigma = \operatorname{diag}(\sigma_1, \ldots, \sigma_n)$. The normalized (mean zero and variance one) measurement vectors form the columns of the *normalized data matrix*

$$\widehat{X} = (\widehat{\mathbf{v}}_1, \ldots, \widehat{\mathbf{v}}_n) = \underline{X}\Sigma^{-1} = JX\Sigma^{-1}. \tag{7.13}$$

The fact that the columns of $\widehat{X}$ all have mean zero is equivalent to the statement that $\mathbf{1} \in \operatorname{coker} \widehat{X}$, i.e., $\mathbf{1}^T\widehat{X} = \mathbf{0}$. We will call column vectors $\widehat{\mathbf{x}}_i = (\widehat{x}_{i1}, \ldots, \widehat{x}_{in})$ corresponding to the rows of $\widehat{X}$ the *normalized data points*, so that their entries are given by

$$\widehat{x}_{ij} = \widehat{v}_{ij} = \frac{x_{ij} - \overline{v}_j}{\sigma_j}. \tag{7.14}$$

The normalized data points are depicted for the diabetes data set in Figure 7.2, where the authors of [69] used the unnormalized variance by setting $\nu = 1$. In practice, the importance of normalization is to ensure that all of the measurements are on the same scale, so that a machine learning algorithm does not pay attention to one measurement over another simply because its values are larger.

Now suppose we make measurements of several different physical quantities. The individual variances themselves may fail to capture many important features of the resulting data set. For example, Figure 7.5 shows the scatter plots of data sets each representing simultaneous measurements of two quantities, as specified by their horizontal and vertical coordinates. All have the same variances, both individual and cumulative, but clearly represent different interrelationships between the two measured quantities. In the central plot, they are completely uncorrelated, while on either side they are progressively more correlated (or anti-correlated), meaning that the value of the first measurement is a strong indicator of the value of the second.

This motivates introducing what is known as the *covariance* σ_{vw} between a pair of measurement vectors $\mathbf{v} = (v_1, v_2, \ldots, v_m)^T$ and $\mathbf{w} = (w_1, w_2, \ldots, w_m)^T$ to be the expected value of the product of the deviations from their respective means $\overline{v}, \overline{w}$. In other words, their covariance

$$\sigma_{vw} = \nu \sum_{k=1}^{m} (v_k - \overline{v})(w_k - \overline{w}) = \nu\,(\mathbf{v} - \overline{v}\,\mathbf{1})\cdot(\mathbf{w} - \overline{w}\,\mathbf{1}) = \nu\,(J\mathbf{v})\cdot(J\mathbf{w}) = \nu\,\mathbf{v}^T J\mathbf{w}, \tag{7.15}$$

is, up to a factor, the dot product of their centered counterparts. In the final formula, we use the fact that the centering matrix (7.5) is symmetric and idempotent. Note that, in view of (7.10), the covariance of a set of measurements with itself is its variance: $\sigma_{vv} = \sigma_v^2$. The *correlation* between the two measurement sets is then defined as

$$\rho_{vw} = \frac{\sigma_{vw}}{\sigma_v\,\sigma_w}, \tag{7.16}$$

---

[5] *Warning*: In this section, Σ is *not* a singular value matrix.

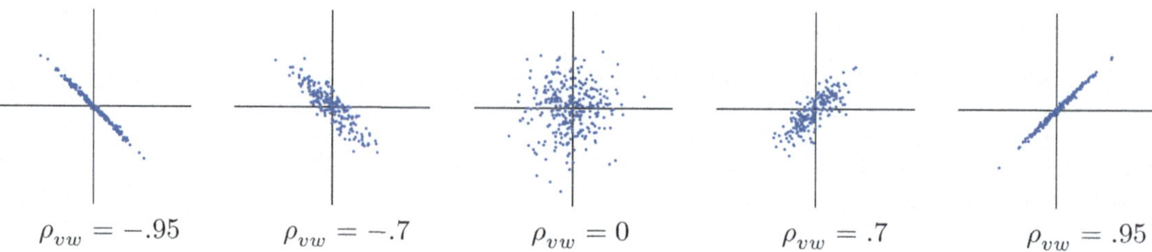

Figure 7.5: Correlations of Data Points in the Plane

and is independent of the prefactor ν. There is an overall bound on the correlation, since the Cauchy–Schwarz inequality (2.27) implies that

$$|\sigma_{vw}| \leq \sigma_v \sigma_w \qquad \text{and hence} \qquad -1 \leq \rho_{vw} \leq 1. \tag{7.17}$$

The closer ρ_{vw} is to $+1$, the more the measurements are correlated; the closer to -1, the more they are anti-correlated, while $\rho_{vw} = 0$ when the measurements are uncorrelated. In Figure 7.5, each scatter plot is labeled by its correlation. Statistically independent measurements are automatically uncorrelated, but the converse is not necessarily true, since correlation only indicates linear dependencies, and it is possible for measurements to be nonlinearly related but nevertheless have zero correlation.

More generally, suppose we have an $m \times n$ data matrix X. Let $\underline{X} = JX$ denote the corresponding centered data matrix, as in (7.7). We define the $n \times n$ *covariance matrix*

$$S_X = \nu \underline{X}^T \underline{X} = \nu X^T J X. \tag{7.18}$$

The entries of the covariance matrix are exactly the pairwise covariances of the individual measurements, i.e., the columns of X:

$$s_{ij} = \sigma_{v_i v_j} = \nu \sum_{k=1}^{m} (v_{ki} - \overline{v}_i)(v_{kj} - \overline{v}_j), \qquad \text{for} \qquad i, j = 1, \ldots, n. \tag{7.19}$$

Its diagonal entries are the individual variances: $s_{ii} = \sigma_{v_i v_i} = \sigma_{v_i}^2$. In particular, the trace of the covariance matrix

$$\operatorname{tr} S_X = \sum_{i=1}^{n} \sigma_{v_i}^2, \tag{7.20}$$

is a measure of the *total variance* of the data. When the covariance matrix is diagonal, so $S_X = \operatorname{diag}(\sigma_{v_1}^2, \ldots, \sigma_{v_n}^2)$, then all the measurements are *uncorrelated*.

The covariance matrix (7.18) is clearly symmetric, $S_X = S_X^T$. It is also a Gram matrix, so Theorem 4.12 tells us that the covariance matrix is always positive semi-definite: $S_X \geq 0$; however, it need not be positive definite. Indeed, since the rows of the centered data matrix $\underline{X}$ sum to zero, the rank of $\underline{X}$ is at most $m - 1$, and therefore

$$\operatorname{rank} S_X = \operatorname{rank} \underline{X} \leq m - 1. \tag{7.21}$$

In particular, in the case where we have fewer measurements than quantities to measure, i.e., $m \leq n$, the covariance matrix has rank at most $m - 1 \leq n - 1$ and is thus a singular $n \times n$ matrix. This is precisely the setting of high dimensional data, where the dimension n exceeds the number of data points m. Even when $m \geq n$, an underlying low dimensional structure in the data can render the covariance matrix singular. It is also important to point out that the covariance matrix can be expressed as a sum of rank one matrices of the form

$$S_X = \nu \sum_{i=1}^{m} \underline{\mathbf{x}}_i \, \underline{\mathbf{x}}_i^T = \nu \sum_{i=1}^{m} \left(\mathbf{x}_i - \overline{\mathbf{x}} \right) \left(\mathbf{x}_i - \overline{\mathbf{x}} \right)^T. \tag{7.22}$$

The expression for the covariance matrix in (7.22) allows us to see the contribution of each data point $\mathbf{x}_i$. The proof is left to Exercise 1.3.

7.1.2 Labels and Learning from Data

As noted above, in addition to the data that has been assembled, some or all of the objects or, equivalently, data points, come with a known label. For example, in image classification, if $\mathbf{x}_i$ represents the pixel values in a particular image, the label $\mathbf{y}_i$ could indicate what is in the image, e.g., a dog, a cat, an automobile, etc. For other problems, such as automatic image annotation, the label $\mathbf{y}_i$ encodes a caption for the image $\mathbf{x}_i$. In medical data analysis (e.g., the diabetes data set), the label $\mathbf{y}_i$ may record the amount of disease progression. There is not much loss of generality in assuming our data points and labels live in Euclidean space, so $\mathbf{x}_i \in \mathbb{R}^n$ and $\mathbf{y}_i \in \mathbb{R}^c$, respectively, since more abstract data is normally embedded in Euclidean space before applying machine learning algorithms. For captions or other types of label text, one uses any convenient word to vector encoding [43].

Machine learning prediction tasks can be either *classification* or *regression*. In a *classification problem*, the goal is to predict a *discrete* quantity, such as the class that an image belongs to, e.g., the digit appearing in an MNIST image. As such, in classification problems, the labels $\mathbf{y}_i$ are chosen from a discrete set, which is usually the set of one-hot vectors $\mathbf{e}_1, \dots, \mathbf{e}_c$, which are just the standard basis vectors in $\mathbb{R}^c$, as in (1.3). The vector $\mathbf{e}_j$ represents the j-th class, out of a total of c different classes. For example, in machine learning analysis, the labels used in the MNIST image data set illustrated in Figure 7.1 are taken to be the one-hot vectors $\mathbf{e}_1, \dots, \mathbf{e}_{10} \in \mathbb{R}^{10}$ and not the digits $0, \dots, 9$.

In a *regression problem*, the goal is to predict a *continuous* quantity, such as the amount of disease progression in the diabetes data set. In this case, the labels can assume a range of values in $\mathbb{R}^c$. For another example, the data points $\mathbf{x}_i$ could represent weather data, such as temperature, humidity, and/or precipitation, measured each day or hour over a period of time, and the labels $\mathbf{y}_i$ could represent crime rates, with the goal of understanding how they are affected by weather and time of day.

The key goal of a machine learning algorithm is to learn patterns and relationships between the data and labels, so that accurate and informative predictions can be made. In general, not all the data points may have known labels, and so there are three sub-fields within machine learning, depending on how much labeled data is available.

- *Fully supervised learning* refers to when all the data points are labeled.

- *Semi-supervised learning* refers to when some, but not all of the data points are labeled.

- *Unsupervised learning* refers to when none of the data points are labeled.

In fully supervised learning, we typically expect to be in possession of a large amount of labeled data, and the goal is to learn how to predict the labels of the data points. We discuss more about how this is done in Section 7.1.3 below. The semi-supervised setting is typically used when relatively few labeled data points are available. This is common in practical situations, since labeling data can often be costly, as it usually requires human expertise, whereas unlabeled data tends to be abundant and virtually free. Semi-supervised learning uses both the labeled and unlabeled data to make better predictions than would be

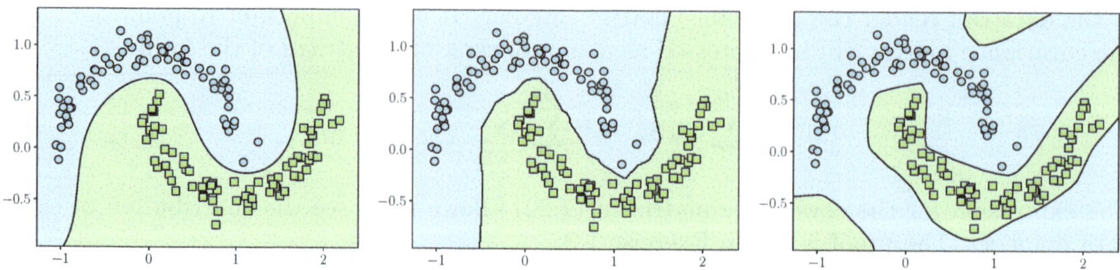

Figure 7.6: Illustration of different learned functions for the same classification data set. The colored regions in the plots indicate the regions where the classification function F predicts one class or the other. Clearly the classification function that correctly classifies a given data set is far from unique.

made with fully supervised learning applied to the labeled data alone. Unsupervised learning methods attempt to uncover structure in the data itself, e.g., clustering similar data points, and will be the topic of Section 7.5. We will discuss the semi-supervised and unsupervised settings in a bit more depth in Section 7.1.6.

7.1.3 Fully Supervised Learning

A fully supervised machine learning algorithm uses a completely labeled *training data set* $(\mathbf{x}_1, \mathbf{y}_1), \ldots, (\mathbf{x}_m, \mathbf{y}_m) \in \mathbb{R}^n \times \mathbb{R}^c$, consisting of data points with known labels, in order to learn a function

$$F \colon \mathbb{R}^n \longrightarrow \mathbb{R}^c \tag{7.23}$$

that maps data points to their labels. In other words, the goal is to find a "good" map (7.23) that attempts to generalize the rule $F(\mathbf{x}_i) = \mathbf{y}_i$ for $i = 1, \ldots, m$, so that if $\mathbf{x} \in \mathbb{R}^n$, then $\mathbf{y} = F(\mathbf{x})$ predicts its label. Clearly there are many choices for the function F, so the learned function is far from unique. For example, in Figure 7.6 we show three classification functions that correctly classify a given data set consisting of data points belonging to one of two classes. The colors indicate the regions in the plane that are predicted to be in one class or the other. We also remark that the value of the function at a prescribed data point, $F(\mathbf{x}_i)$, need not necessarily agree with its label, $\mathbf{y}_i$ — many machine learning algorithms can deal with mislabeled data and noise.

In practice, F is normally chosen from a specified class of parameterized functions $F(\mathbf{x}; \mathbf{w})$, where $\mathbf{w} = (w_1, w_2, \ldots, w_N)^T \in \mathbb{R}^N$ are the parameters. For example, it could be a linear function $F(\mathbf{x}; W) = W\mathbf{x}$, where the parameters are the $N = nc$ entries of a $c \times n$ matrix $W \in \mathcal{M}_{c \times n} \simeq \mathbb{R}^{cn}$, or it could be the output of a neural network, where $\mathbf{w}$ contains the weights and biases of all the neurons; see Chapter 10.

The goal of learning is to find parameters $\mathbf{w}$ that fit the data as well as possible. Typically, this is achieved by minimizing a real-valued *total loss function* of the form

$$\mathcal{L}(\mathbf{w}) = \frac{1}{m} \sum_{i=1}^{m} \ell\big(F(\mathbf{x}_i; \mathbf{w}), \mathbf{y}_i\big), \tag{7.24}$$

where $\ell \colon \mathbb{R}^c \times \mathbb{R}^c \to \mathbb{R}$ is a prescribed *loss function* that measures how close the predicted value $F(\mathbf{x}_i; \mathbf{w})$ is to the label $\mathbf{y}_i$. By minimizing the total loss function (7.24), we are attempting to tune the weights $\mathbf{w}$ so that, in the ideal case $F(\mathbf{x}_i; \mathbf{w}) = \mathbf{y}_i$ for all i, or, more generally, make their values as close as possible. The process of minimizing the total loss $\mathcal{L}$ is called *training*.

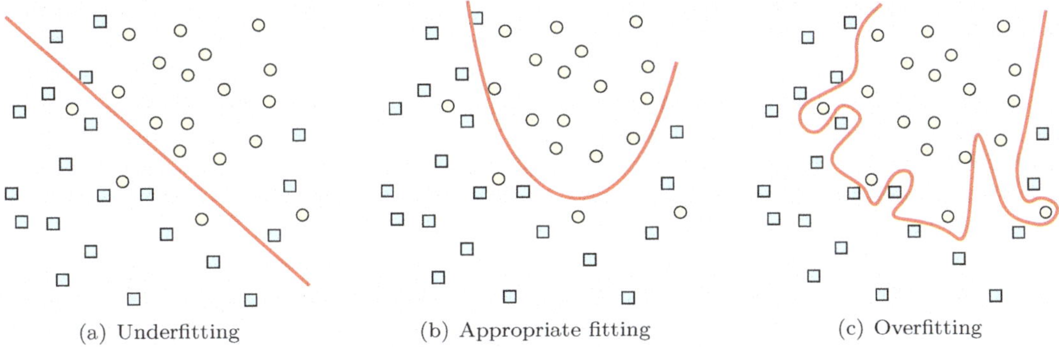

(a) Underfitting   (b) Appropriate fitting   (c) Overfitting

Figure 7.7: Example of underfitting, appropriate fitting, and overfitting. The decision about what type of fit is correct, and what constitutes an overfit or underfit, is context dependent. In a setting where some of the data points are expected to be noisy, the fit in (b) may in fact be an overfit, and (a) may be preferable.

Possible choices for the loss function ℓ in regression problems include the 2 *loss*, based on the squared distance between points in the Euclidean norm (2.2),

$$\ell(\mathbf{z}, \mathbf{y}) = \|\mathbf{z} - \mathbf{y}\|^2 = \sum_{i=1}^{c} (z_i - y_i)^2, \tag{7.25}$$

and the 1 *loss*, which is the distance between points measured in the 1 norm (2.59),

$$\ell(\mathbf{z}, \mathbf{y}) = \|\mathbf{z} - \mathbf{y}\|_1 = \sum_{i=1}^{c} |z_i - y_i|. \tag{7.26}$$

Other notions of distance can be profitably employed. In classification problems, it is often the case that the output $\mathbf{z} = F(\mathbf{x}; \mathbf{w})$ is interpreted as a probability vector — see Definition 5.60 — where $0 \leq z_i \leq 1$ is the probability that $\mathbf{x}$ belongs to the i-th class, and $\mathbf{1}^T \mathbf{z} = z_1 + \cdots + z_c = 1$ because $\mathbf{z}$ must belong to one of the classes. In this case, it is common to use the *negative log-likelihood loss*, also called the *cross-entropy loss*, which is given by

$$\ell(\mathbf{z}, \mathbf{y}) = - \sum_{i=1}^{c} y_i \log z_i, \tag{7.27}$$

where we interpret $0 \log 0 = 0$. Note that in classification, $\mathbf{y}$ is usually a one-hot vector, say $\mathbf{y} = \mathbf{e}_j$, and so we can rewrite the negative log-likelihood loss as

$$\ell(\mathbf{z}, \mathbf{e}_j) = - \log z_j = \log(1/z_j).$$

Hence, minimizing the negative log-likelihood loss encourages the output probability z_j corresponding to the correct label to be as close to 1 as possible. The negative log likelihood loss has an information theoretic interpretation as measuring the amount of common information between two probability distributions; see [49] for details. Additional motivation for the negative log likelihood loss is given in Chapter 10.

7.1.4 Overfitting and Generalization

While the immediate goal of training is to minimize the total loss (7.24), the real objective is to learn a function $F(\mathbf{x}; \mathbf{w})$ — that is, prescribe its parameters $\mathbf{w}$ — that correctly classifies new data points that have not been seen and are not included in the training data. *Generalization error* refers to the difference between an algorithm's performance on the data it was trained on, and its performance on new, previously unseen data. A model with small generalization error is said to *generalize well*. Generally speaking, if the parametrization of F has too few degrees of freedom, i.e., too few parameters, then F may not fit the training data well, which is called *underfitting*, and thus will probably not perform well on new data points.[6] If there are too many degrees of freedom then F may be *overfitting* the training data, meaning that it performs well on the training data, but has large generalization error (i.e., it performs poorly on new, unseen data points). The goal is to find a function F that *correctly fits* the training data, in the sense that it gives the simplest explanation for the observed trends, and is most likely to generalize to new data. Figure 7.7 sketches examples of underfitting, overfitting, and a correct fit for some training data (the orange points). It is often the case that what constitutes a correct fit is context dependent, and in the setting of high noise levels, the underfitted example in Figure 7.7(a) could be interpreted as a correct (or close to correct) fit.

In order to prevent overfitting it is common to augment the loss with a *regularizer*, which is a scalar-valued function $\mathcal{R}\colon \mathbb{R}^N \to \mathbb{R}$ that depends on the weights. The *augmented* or *regularized loss function* takes the form

$$\mathcal{L}_\lambda(\mathbf{w}) = \mathcal{L}(\mathbf{w}) + \lambda \mathcal{R}(\mathbf{w}), \tag{7.28}$$

where $\lambda > 0$ is a *hyperparameter*, meaning that it is *not* optimized during training and is either fixed in advance, or is otherwise tuned by the user — see the discussion in the following subsection. The role of the regularizer $\mathcal{R}(\mathbf{w})$ is to bias the solution towards selecting weights that are less likely overfit the training data, and thus generalize well. One way to do this is to ask that as many of the components of $\mathbf{w}$ as possible vanish, so that the resulting function has very few degrees of freedom. However, this objective is often hard to work with, and a reasonable proxy is to choose $\mathcal{R}(\mathbf{w})$ so as to penalize the size of $\mathbf{w}$ in some way. The specific form of the regularizer depends on the machine learning model, but a common choice is a norm $\mathcal{R}(\mathbf{w}) = \|\mathbf{w}\|$, or a squared norm $\mathcal{R}(\mathbf{w}) = \|\mathbf{w}\|^2$, both of which will appear in Section 7.2 in the context of linear regression, and in Section 7.3 in the context of support vector machines.

Since machine learning models learn from examples, the means by which the models arrive at their predictions can be hard to interpret, and, consequently, models can overfit in many different ways, which can often be difficult to understand. However, a general principle is that overfitting amounts to using spurious unimportant details to make predictions, instead of learning more general patterns that are likely to generalize well. Indeed, data sets may contain extraneous information that correlates with the desired predictions, but is not useful for generalizing to new data. For example, suppose our goal is to predict whether an image contains a cat or a dog, and further suppose that the dog images were all captured outside on sunny days, while the cat images were captured inside on rainy days. One may then classify the images by measuring their overall brightness, thereby detecting whether they were captured indoors (and hence a cat) or outdoors (and hence a dog). These are certainly not good ways of distinguishing dogs from cats, and will clearly not generalize to other settings. This suggests another way to combat overfitting is to augment the training set by including additional copies of the training images where certain features, such as image brightness, orientation, scale, etc., are adjusted at random, in order to enable the machine learning model to ignore them during classification. This technique is called *data augmentation* and is discussed, along with other regularization techniques, in Chapter 10.

---

[6]Technically speaking, an underfitting model can still have small generalization error provided it performs similarly on new data as it did on the training set, even if this performance is poor.

7.1.5 The Train–Test Split and Hyperparameters

In practical applications of machine learning, the generalization error is measured by splitting the data set into two subsets. Usually the split is done at random. The first subset is the *training data*, and is used to "train" the machine learning model through minimizing the total loss by suitably adjusting its parameters. The second subset is called the *testing data*, and is used to evaluate the performance of the trained model on data that was unseen during training. Usually the training set is much larger than the testing set, e.g., we reserve 25% of the data for the testing set and use the other 75% for training. If the trained model performs well on the testing data, or gives similar performance as it did on the training data, then the model can reasonably be expected to generalize to new data. If the testing accuracy is much lower than the training accuracy, then this is an indication that the model is overfitting and will not generalize well.

Using only a single randomized train-test split of the data set subjects the evaluation of generalization error to random chance. The chosen train-test split could be relatively fortunate (or unfortunate) for the performance of the algorithm. To get a more accurate evaluation of algorithm performance, it is common to use many train-test splits and average the performance over all of them. This can also be done by random selection. Alternatively, one can use a k-fold cross validation, which randomly splits the data into k equally sized subsets, called *folds*, and then forms k train-test splits by taking each fold to be a testing set, and the rest of the data as the training set. A k-fold cross-validation ensures that all data points appear in the testing set exactly once.

Many algorithms include one or more *hyperparameters*, which affect the behavior of the model but are specified directly by the user, and not set during training. We will see many examples of hyperparameters in this chapter. One example is the parameter λ in an augmented loss function (7.28); another is the number k of neighbors in the k-nearest neighbor classifier in Section 7.4. Hyperparameters also arise in *ensemble learning*, which refers to the technique of training multiple machine learning models for the same task, and combining their results to obtain an improved model. Here, the choice of which models to use and how to combine their predictions will involve several hyperparameters. Hyperparameter tuning can greatly improve the performance of machine learning algorithms. However, it is very important to use only the training set to tune them, so that the testing set remains an unbiased evaluation of model performance. Another common means of optimizing hyperparameters is to hold out another set of data, called the *validation data set*, which is used during training to select the best hyperparameters. Thus, it is also common to split the data set into three subsets: training, validation, and testing.

However the data is split, it is extremely important to ensure there is no contamination of data between the training and testing data sets. For example, if the training and testing sets are identical, the model will automatically perform well on the latter, but this tells us nothing about its performance when confronted with genuinely new data. There are subtleties in the train-test split that can lead to inadvertent contamination. Some data sets may contain duplicated data points, or data points where a large part of the feature vectors are the same. There are further potentially less immediately evident issues: for instance, in our work on classification of broken bones [258], we have observed that placing data from different fragments that come from the same bone in the training and testing sets can lead to contamination. Unfortunately, misuse of machine learning algorithms, through train-test contamination or using the testing set to tune hyperparameters, can be found in a number of papers in the applied literature [36], in which the claimed results and accuracies cannot be trusted.

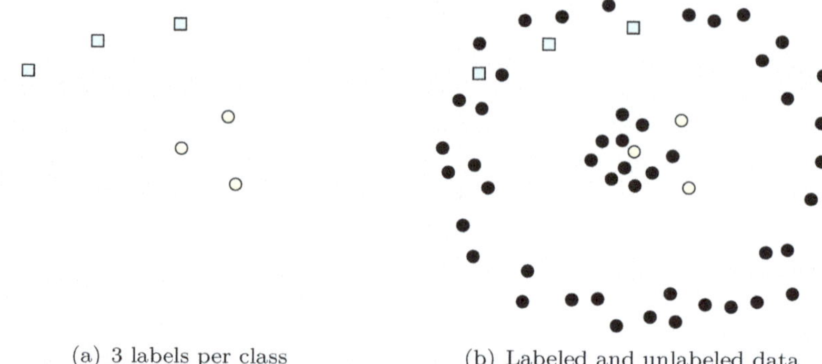

(a) 3 labels per class   (b) Labeled and unlabeled data

Figure 7.8: Example showing how the unlabeled data (the black points) can be useful for training a classifier. Without the unlabeled data, one cannot see the natural geometry and cluster structure in the data set.

7.1.6   Semi-supervised and Unsupervised Learning

Fully supervised learning typically requires an abundance of labeled training data. In many applications, such as medical images or other data that requires human input to label, labeled training examples are costly to obtain, and it is desirable to have algorithms that can achieve good performance with far fewer labeled examples than are required in fully-supervised learning. *Semi-supervised learning* uses both labeled and unlabeled data to obtain higher performance at lower labeling rates. In this setting, we still have a set of labeled training data $(\mathbf{x}_1, \mathbf{y}_1), (\mathbf{x}_2, \mathbf{y}_2), \ldots, (\mathbf{x}_l, \mathbf{y}_l)$, but the number l of labeled training points may be small. Additionally, we assume we have access to a large amount of unlabeled data $\mathbf{x}_{l+1}, \mathbf{x}_{l+2}, \ldots, \mathbf{x}_m$, where $m \gg l$. The goal is to use the additional unlabeled data to train a better classifier than one would obtain through fully supervised learning based on only the limited labeled data. In many applications, like image classification, speech recognition, or text generation, unlabeled data is abundant and essentially free, so it is natural to attempt to make use of this additional information in some way.

To see why unlabeled data may be useful in classification, consider the data points in Figure 7.8(a), which constitute six data points in $\mathbb{R}^2$ split into two classes (blue square and yellow circle). If we only use these six data points to train a classifier in the fully supervised setting, then we have very little information and the trained classifier is unlikely to generalize well. If, on the other hand, we have access to unlabeled data, which are shown as the black points in Figure 7.8(b), then we can use this to inform our classifier, which in this case would split the inner circle of data from the outer one. In some sense, the unlabeled data gives additional information about the underlying structure of the data set that will be seen when the algorithm is evaluated on new data.

Semi-supervised learning comes in two variations. The first is the *inductive* setting, where one still learns a general rule $F\colon \mathbb{R}^n \to \mathbb{R}^c$ that aims to generalize the training data, while using properties of the unlabeled data. The second is the *transductive* setting, where we only learn labels for the additional unlabeled data points $\mathbf{x}_{l+1}, \ldots, \mathbf{x}_m$, i.e., the black points in Figure 7.8(b). The transductive setting does not learn a general rule, and the classifier cannot be immediately applied to new data without retraining, or by adopting some simple heuristic, like choosing the label of the closest data point for which a label prediction exists.

In contrast, *unsupervised learning* algorithms use only a set of unlabeled data points $\mathbf{x}_1, \mathbf{x}_2, \ldots, \mathbf{x}_m$ for learning. Common tasks include clustering, dimension reduction, and data visualization, which arise in nearly all applications of data science and machine learning, an example being the visualization of RNA data [147].

Exercises

1.1. Find the mean, the variance, and the standard deviation of the following data sets. You can set $\nu = 1$ when computing the latter.
$(a)\,\heartsuit\ 1.1, 1.3, 1.5, 1.55, 1.6, 1.9, 2, 2.1; \quad (b)\ 2., .9, .7, 1.5, 2.6, .3, .8, 1.4; \quad (c)\,\heartsuit\ -2.9, -.5, .1, -1.5,$
$-3.6, 1.3, .4, -.7; \quad (d)\,\diamond\ 1.1, .2, .1, .6, 1.3, -.4, -.1, .4; \quad (e)\ .9, -.4, -.8, .2, 1., -1.6, -1.2, -.7.$

1.2. Show that the centering matrix J is $(a)\,\heartsuit$ positive semi-definite, $(b)\,\heartsuit$ idempotent, so $J^2 = J$, $(c)\,\diamond$ has one-dimensional kernel spanned by $\mathbf{1}$, and hence is not positive definite, and (d) has rank $m - 1$.

1.3. $\diamond$ Prove formula (7.22).

1.4. Suppose we define the covariance with respect to an inner product, that is

$$\sigma_{vw} = \nu \langle J\mathbf{v}, J\mathbf{w} \rangle = \nu \mathbf{v}^T J C J \mathbf{w},$$

where C is the positive definite matrix defining the inner product. Show that associated covariance matrix is $\nu \underline{X}^T C \underline{X}$, whose (i, j) entry is the inner product covariance of the i-th and j-th columns of $\underline{X}$.

1.5. $\diamond$ Find a formula like (7.22) for the inner product covariance matrix $\underline{X}^T C \underline{X}$ from Exercise 1.4.

1.6. $\heartsuit$ Suppose we have a collection of data points $\mathbf{x}_1, \ldots, \mathbf{x}_m$ lying along a line spanned by the unit vector $\mathbf{u}$, that is each $\mathbf{x}_i = s_i \mathbf{u}$ for some $s_i \in \mathbb{R}$. Show that the covariance matrix of this data is $S_X = \sigma_s^2 \mathbf{u}\mathbf{u}^T$, where σ_s^2 is the variance of the weights $\mathbf{s} = (s_1, \ldots, s_m)$.

1.7. What happens in Exercise 1.6 if there are two linearly independent unit vector directions, $\mathbf{u}$ and $\mathbf{v}$, such that each data point lies along a line in one direction or the other — that is, there are weights $s_i \in \mathbb{R}$ such that for each i we have either $\mathbf{x}_i = s_i \mathbf{u}$ or $\mathbf{x}_i = s_i \mathbf{v}$? Can you write a simple formula for the covariance matrix involving $\mathbf{u}\mathbf{u}^T$ and $\mathbf{v}\mathbf{v}^T$?

7.2 Linear Regression

Python Notebook: Linear Regression (.ipynb)

In many respects, the simplest class of functions to use in machine learning algorithms are linear functions $F\colon \mathbb{R}^n \to \mathbb{R}$, which, by Theorem 3.33, have the form

$$F(\mathbf{x}; \mathbf{w}) = \mathbf{x} \cdot \mathbf{w} = \sum_{i=1}^{n} x_i w_i, \tag{7.29}$$

where the vector $\mathbf{w} \in \mathbb{R}^n$ contains the *parameters* in F. The linear function F is uniquely characterized by the parameter vector $\mathbf{w} \in \mathbb{R}^n$. *Linear regression* seeks a linear function that

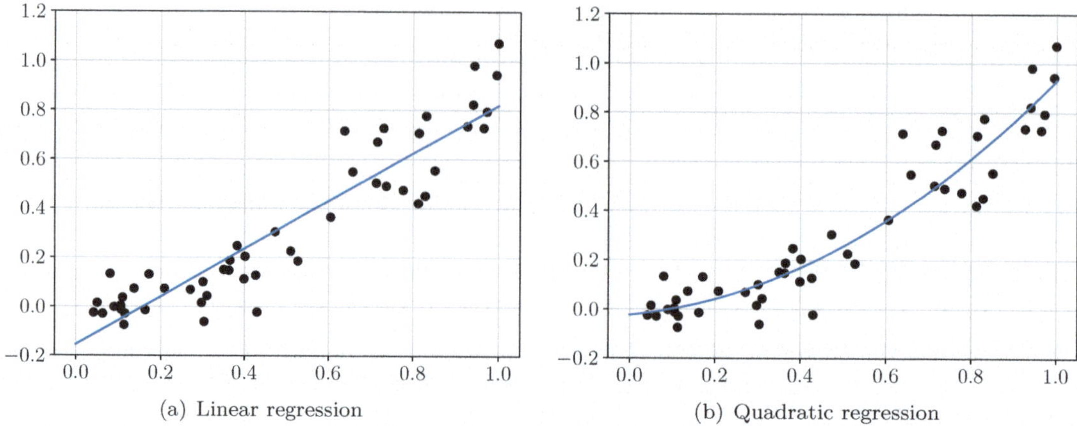

(a) Linear regression   (b) Quadratic regression

Figure 7.9: Example of (a) linear regression on a toy data set, which amounts to finding the line of best fit, and (b) quadratic polynomial regression, which is described at the end of this section.

best approximates a data set consisting of data points $\mathbf{x}_1, \ldots, \mathbf{x}_m \in \mathbb{R}^n$ and associated *scalar* labels[7] $y_1, \ldots, y_m \in \mathbb{R}$, also known as *outputs*. That is, we aim to find $\mathbf{w} \in \mathbb{R}^n$ such that $F(\mathbf{x}_i; \mathbf{w}) \approx y_i$ for all i. The goal of linear regression is to uncover underlying trends in the data in the setting where the measurements y_i or the data $\mathbf{x}_i$ may be noisy or corrupted, so we don't expect to exactly fit the data. Indeed, linear functions have relatively few degrees of freedom, and can thus be expected to ignore the noise in favor of learning broad trends, and hence tend not to overfit.

Remark. A more general affine function $F(\mathbf{x}; \mathbf{w}, b) = \mathbf{x} \cdot \mathbf{w} + b$ can be handled by extending $\mathbf{x}$ and $\mathbf{w}$ to vectors in $\mathbb{R}^{n+1}$ by setting $x_{n+1} = 1$ and $w_{n+1} = b$, and therefore, there is no loss of generality in restricting our attention to linear functions here.   ▲

One of the simplest ways to find the best linear function that fits the data is to choose the weight vector $\mathbf{w}$ to minimize the mean squared error loss function

$$\mathcal{L}(\mathbf{w}) = \sum_{i=1}^{m} (\mathbf{x}_i \cdot \mathbf{w} - y_i)^2 = \| X\mathbf{w} - \mathbf{y} \|^2, \tag{7.30}$$

over the possible choices of $\mathbf{w} \in \mathbb{R}^n$. Here X denotes the data matrix (7.1), whose rows are the transposed data vectors $\mathbf{x}_i^T$, and $\mathbf{y} = (y_1, \ldots, y_m)^T$ denotes the *target vector*. Thus, linear regression with the mean squared error loss function requires solving the least squares problem

$$\min_{\mathbf{w}} \| X\mathbf{w} - \mathbf{y} \|^2, \tag{7.31}$$

which we studied in Section 6.2. The parameters can be found, for example, by either solving the associated normal equations (6.17), or by using the general QR algorithm in Theorem 4.47.

In Figure 7.9(a) we show a simple example of linear regression applied to a toy data set, in which case we are simply finding the line of best fit through a collection of data points. In Figure 7.9(b) we show the results of *quadratic regression* on the same toy data, which may fit the general trends better. Quadratic regression can be viewed as an instance of linear

---

[7]Recall that in the regression setting, the labels are not restricted to discrete values associated with classes, as in classification, but are instead allowed to take on any real values.

Feature	Weight
Age (years)	-51.59
Body mass index	562.31
Average blood pressure	307.01
TC (total serum cholesterol)	-295.60
LDL (low-density lipoproteins)	71.97
TCH (total cholesterol / HDL)	24.94
LTG (log of serum triglycerides level)	625.95
GLU (blood sugar level)	102.48
Offset b	151.14

Table 7.10: Features and weights from linear regression performed on the diabetes data set.

regression where the data is augmented with the squares x_i^2 of each data point; we describe this in more detail at the end of this section.

We now turn to an example of linear regression on the diabetes data set, introduced earlier. We randomly split the data into a training set consisting of 2/3 of the data and placed the remaining data in the testing set. The square roots of the training loss and testing loss were 50.8 and 59.24, respectively. Since the scale of the disease progression variable (see Figure 7.2) ranges from 50 up to 350, this is a reasonably good result, indicating that we have some ability to predict disease progression, but cannot do so with high accuracy. Table 7.10 shows the components of the optimal weight vector $\mathbf{w}$ corresponding to each feature, which allows us to gain some insights into which are more important for prediction. In this case, by comparing their absolute values, we deduce that the two most important features are LTG and body mass index, both of which are positively correlated with disease progression.

7.2.1 Ridge Regression

In practice, it is common to *regularize* the least squares loss (7.30) following the general procedure in (7.28). Regularization can help combat the effects of noise, and, in addition, produce a unique solution even when the solution to the original least squares problem is not unique.

The simplest such regularizer is the squared Euclidean norm of $\mathbf{w}$, yielding what is known as the *ridge regression*[8] problem

$$\min_{\mathbf{w}} \left\{ \| X\mathbf{w} - \mathbf{y} \|^2 + \lambda \| \mathbf{w} \|^2 \right\}, \tag{7.32}$$

where $\lambda > 0$ is a hyperparameter that controls the strength of the regularization. Taking a larger value for λ will bias the regression to select weights $\mathbf{w}$ with smaller norms. Ridge regression is also called *Tikhonov regularization*, named after the twentieth century Russian mathematician Andrey Tikhonov.

To minimize the ridge regression loss, we expand the squared norms to obtain the equivalent problem

$$\min_{\mathbf{w}} \left\{ \mathbf{w}^T (X^T X + \lambda \, \mathrm{I}) \mathbf{w} - 2\mathbf{w}^T X^T \mathbf{y} + \| \mathbf{y} \|^2 \right\}.$$

This is a quadratic minimization problem of the form analyzed in Theorem 6.7, with $H = X^T X + \lambda \mathrm{I}$ and $\mathbf{f} = X^T \mathbf{y}$. In this case, since $\lambda > 0$, the matrix H is always positive definite

---

[8]For an explanation of origins of the term *ridge*, we refer to [108].

— see Exercise 1.7 — and so the ridge regression problem has a unique solution

$$\mathbf{w}_\lambda = (X^T X + \lambda\, \mathrm{I})^{-1} X^T \mathbf{y}. \tag{7.33}$$

When $\lambda = 0$ and $X^T X$ is not invertible, we define $\mathbf{w}_0$ to be the unique least squares solution of $X\mathbf{w} = \mathbf{y}$ with minimal Euclidean norm; see Theorem 6.11 for details.

It turns out we can express the solution $\mathbf{w}_\lambda$ of the ridge regression problem in terms of the singular value decomposition of the data matrix X, as presented in Theorem 5.75. Here $C = \mathrm{I}$, since we are using the Euclidean norm and dot product.

> **Theorem 7.1.** Let $X = P\Sigma Q^T$ be the (dot product) singular value decomposition of X, and let $\lambda \geq 0$. Then the solution $\mathbf{w}_\lambda$ of the ridge regression problem given by (7.33) can be written as
>
> $$\mathbf{w}_\lambda = R\,\mathbf{y}, \quad \text{where} \quad R = QDP^T, \quad D = (\Sigma^2 + \lambda\,\mathrm{I})^{-1}\Sigma = (\Sigma + \lambda\Sigma^{-1})^{-1}. \tag{7.34}$$

Proof. We note that, by (7.33), $\mathbf{w}_\lambda$ satisfies

$$X^T X\mathbf{w}_\lambda + \lambda\mathbf{w}_\lambda = X^T\mathbf{y}, \qquad \text{hence} \qquad \lambda\mathbf{w}_\lambda = X^T(\mathbf{y} - X\mathbf{w}_\lambda). \tag{7.35}$$

This implies that $\mathbf{w}_\lambda \in \mathrm{img}\, X^T$ when $\lambda > 0$, which also holds when $\lambda = 0$, because, according to Theorem 6.11, the least squares solution of $X\mathbf{w} = \mathbf{y}$ with minimal Euclidean norm also satisfies $\mathbf{w}_0 \in \mathrm{img}\, X^T = \mathrm{coimg}\, X$. Since $X^T = Q\Sigma P^T$, and $Q^T Q = \mathrm{I}$, because the columns of Q are orthonormal, this means that, for some $\mathbf{z} \in \mathbb{R}^r$,

$$\mathbf{w}_\lambda = Q\mathbf{z} = QQ^T Q\mathbf{z} = QQ^T\mathbf{w}_\lambda. \tag{7.36}$$

Since $P^T P = \mathrm{I}$, we also have $X^T X = Q\Sigma^2 Q^T$, and so we can write (7.35) as

$$Q\Sigma^2 Q^T\mathbf{w}_\lambda + \lambda QQ^T\mathbf{w}_\lambda = X^T\mathbf{y}, \quad \text{which is equivalent to} \quad Q(\Sigma^2 + \lambda\,\mathrm{I})Q^T\mathbf{w}_\lambda = X^T\mathbf{y}.$$

Multiplying the latter equation by $(\Sigma^2 + \lambda\,\mathrm{I})^{-1}Q^T$ on the left, we obtain

$$Q^T\mathbf{w}_\lambda = (\Sigma^2 + \lambda\,\mathrm{I})^{-1}Q^T X^T\mathbf{y} = (\Sigma^2 + \lambda\,\mathrm{I})^{-1}\Sigma P^T\mathbf{y}.$$

Thus, by (7.36),

$$\mathbf{w}_\lambda = QQ^T\mathbf{w}_\lambda = Q(\Sigma^2 + \lambda\,\mathrm{I})^{-1}\Sigma P^T\mathbf{y} = QDP^T\mathbf{y} = R\,\mathbf{y}. \qquad\blacksquare$$

Remark 7.2. Let $\sigma_1, \ldots, \sigma_r$ denote the singular values of X, ordered from largest to smallest, which are the diagonal entries of Σ. Then $D = (\Sigma^2 + \lambda\,\mathrm{I})^{-1}\Sigma$ is a diagonal matrix with entries

$$d_{ii} = \frac{\sigma_i}{\sigma_i^2 + \lambda} = \frac{1}{\sigma_i + \lambda\sigma_i^{-1}}. \tag{7.37}$$

Thus, by Theorem 7.1 the solution of the ridge regression problem (7.34) can also be written as

$$\mathbf{w}_\lambda = \sum_{k=1}^{r} \frac{\sigma_k}{\sigma_k^2 + \lambda}\,(\mathbf{p}_k \cdot \mathbf{y})\,\mathbf{q}_k, \qquad \lambda \geq 0, \tag{7.38}$$

where $\mathbf{p}_1, \ldots, \mathbf{p}_r \in \mathbb{R}^m$ are the columns of P, and $\mathbf{q}_1, \ldots, \mathbf{q}_r \in \mathbb{R}^n$ the columns of Q.

We also note that by equating (7.33) with (7.34), we obtain

$$(X^T X + \lambda\, I)^{-1} X^T = Q\, D\, P^T = R,$$

which is exactly the singular value decomposition of the ridge regression solution matrix, and the entries (7.37) are its singular values. Note that the columns of P are its singular vectors, whereas the columns of Q are the singular vectors of the data matrix X. From this, we can see that the condition number of the solution matrix is improved for large values of the regularization parameter λ. Indeed, for $\lambda \geq \sigma_1^2 > 0$, the singular values d_{ii} of the solution matrix are ordered the same way[9] as the singular values of X, that is $d_{11} \geq \cdots \geq d_{nn}$. Therefore, the condition number of the solution matrix is

$$\kappa(R) = \frac{d_{11}}{d_{nn}} = \frac{\sigma_1}{\sigma_n}\left(\frac{\sigma_n^2 + \lambda}{\sigma_1^2 + \lambda}\right) \leq \frac{\sigma_1}{\sigma_n} = \kappa(X), \tag{7.39}$$

with equality if and only if $\sigma_1 = \sigma_n$, the latter meaning that all singular values are the same, and so Σ is a multiple of the identity matrix.[10] On the other hand, when $\lambda < \sigma_1^2$ it is not immediately clear whether the condition number is improved. $\blacktriangle$

As a simple consequence of Theorem 7.1, we can show that ridge regression approximates the minimal Euclidean norm least squares solution when $\lambda > 0$ is small.

Corollary 7.3. *For $\lambda \geq 0$, the solution (7.34) of the ridge regression problem (7.32) satisfies*

$$\|\mathbf{w}_\lambda\| \leq \|\mathbf{w}_0\| \qquad and \qquad \|\mathbf{w}_\lambda - \mathbf{w}_0\| \leq \frac{\lambda}{\sigma^2}\|\mathbf{w}_0\|, \tag{7.40}$$

where $\sigma = \sigma_{min}(X) = \sigma_r > 0$ is the smallest singular value of X.

Proof. By (7.38) and the orthonormality of the $\mathbf{q}_k$, we can write

$$\|\mathbf{w}_\lambda\|^2 = \sum_{k=1}^{r} \frac{\sigma_k^2}{(\sigma_k^2 + \lambda)^2}(\mathbf{p}_k \cdot \mathbf{y})^2 \leq \sum_{k=1}^{r} \frac{\sigma_k^2}{\sigma_k^4}(\mathbf{p}_k \cdot \mathbf{y})^2 = \|\mathbf{w}_0\|^2,$$

proving the first inequality. Furthermore, again using (7.38) we have

$$\mathbf{w}_\lambda - \mathbf{w}_0 = \sum_{k=1}^{r}\left(\frac{\sigma_k}{\sigma_k^2 + \lambda} - \frac{1}{\sigma_k}\right)(\mathbf{p}_k \cdot \mathbf{y})\,\mathbf{q}_k = -\sum_{k=1}^{r}\frac{\lambda}{\sigma_k}\left(\frac{1}{\sigma_k^2 + \lambda}\right)(\mathbf{p}_k \cdot \mathbf{y})\,\mathbf{q}_k.$$

Therefore, since $\lambda \geq 0$,

$$\|\mathbf{w}_\lambda - \mathbf{w}_0\|^2 = \sum_{k=1}^{r}\frac{\lambda^2}{\sigma_k^2}\left(\frac{1}{\sigma_k^2 + \lambda}\right)^2(\mathbf{p}_k \cdot \mathbf{y})^2 \leq \frac{\lambda^2}{\sigma_r^4}\sum_{k=1}^{r}\frac{1}{\sigma_k^2}(\mathbf{p}_k \cdot \mathbf{y})^2 = \frac{\lambda^2}{\sigma_r^4}\|\mathbf{w}_0\|^2. \quad \blacksquare$$

There are many other forms of regularization that can be used in linear regression. In the ridge regression formulation (7.32), we can use $\|B\mathbf{w}\|^2$, where B is an $k \times n$ matrix, instead of $\|\mathbf{w}\|^2$ in the regularization term. In this case, according to Exercise 2.3, the solution is given by

$$\mathbf{w}_\lambda = (X^T X + \lambda\, B^T B)^{-1} X^T \mathbf{y}, \tag{7.41}$$

---

[9]This is because the function $f(x) = x/(x^2 + \lambda)$ is increasing for $x^2 < \lambda$.

[10]In this case, in view of Example 5.78, $X^T X$ is an idempotent matrix that orthogonality projects onto the subspace $\operatorname{coimg} X \subset \mathbb{R}^n$, i.e., the subspace spanned by the data vectors.

provided that $X^T X + \lambda B^T B$ is nonsingular. For example, if nullity $B = 0$, then the Gram matrix $B^T B$ is positive definite and hence nonsingular, and thus so is the matrix $X^T X + \lambda B^T B$ when $\lambda > 0$. An important example arises when $X = I$, so the ridge regression problem becomes

$$\min_{\mathbf{w}} \left\{ \| \mathbf{w} - \mathbf{y} \|^2 + \lambda \| B \mathbf{w} \|^2 \right\}, \qquad \text{with solution} \qquad \mathbf{w}_\lambda = (I + \lambda B^T B)^{-1} \mathbf{y}. \qquad (7.42)$$

This type of linear regression problem is used, for example, to remove noise from signals and images. In this case, we take $\mathbf{w}$ to be the vector containing all of the pixel values in an image, which are the main parameters controlling how an image appears visually. The noisy signal or image is given in the vector $\mathbf{y}$, and the solution $\mathbf{w}$ is the denoised (i.e., improved) image. The choice of B is made so that $\| B \mathbf{w} \|^2$ measures the amount of noise in the image $\mathbf{w}$. These ideas are explored in more detail in Chapter 9.

7.2.2   Lasso Regression

One can also replace the Euclidean norm in the ridge regression regularization term by other types of norms. If the norm comes from an inner product, so that $\| \mathbf{w} \|^2 = \mathbf{w}^T C \mathbf{w}$ where C is symmetric, positive definite, this reduces to the previous regularization (7.41) with $B = C^{1/2}$. In this section we investigate what happens when using the 1 norm instead, denoted as $\| \cdot \|_1$. We recall that throughout this chapter we use $\| \cdot \|$ to denote the Euclidean norm (or 2 norm).

By way of motivation, we observe that ridge regression can often produce solutions where many or all of the weights in the solution vector $\mathbf{w} \in \mathbb{R}^n$ are moderate or large in size, especially in settings where there are a large number $n \gg 0$ of different types of measurements that are highly correlated; see Exercises 2.2, 2.5. It is often desirable in practice to obtain a regression that uses as few of the measurements as possible, so as to give a simpler explanation for relationships between variables. In other words, we seek a weight vector $\mathbf{w}$ that has many *zero* entries that still fits the data well.

Lasso, which is an acronym for *least absolute shrinkage and selection operator*, addresses this issue by using the 1 norm

$$\| \mathbf{w} \|_1 = | w_1 | + \cdots + | w_n |$$

for regularization. The lasso regression problem thus corresponds to

$$\min_{\mathbf{w}} \left\{ \| X \mathbf{w} - \mathbf{y} \|^2 + \lambda \| \mathbf{w} \|_1 \right\}, \qquad (7.43)$$

where the first term is the squared Euclidean norm. This regularization produces solutions that tend to be sparser, meaning that, in comparison with ridge regression, they place large weights w_i on fewer features x_i, and set many weights to be *exactly* zero. This can be useful when there is noise or a large number of features, many of which are highly correlated.

Example 7.4. As an example, we consider a toy regression problem with $m = 64$ data points $\mathbf{x}_1, \ldots, \mathbf{x}_{64} \in \mathbb{R}^{32}$ whose coordinates (i.e., the measurements) are, in the first case, drawn independently at random, as Gaussian random variables with mean zero and unit variance. The target labels are given by the average of all measurements, so $y_i = \mathbf{1} \cdot \mathbf{x}_i / m$. In this setting, the measurements have no correlation, and both ridge and lasso regression produce regression vectors very close to $\mathbf{w} = \mathbf{1}/m$ — the coefficients are depicted in Figure 7.11(a). We used $\lambda = 0.01$ for ridge regression and $\lambda = 0.001$ for lasso.

In the second setting, we introduce a high degree of correlation among the measurements, by drawing 4 of the measurements at random, then duplicating these measurements 8 times

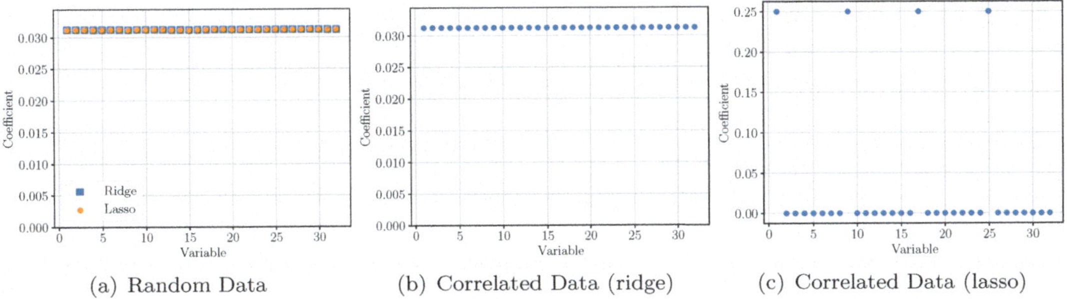

(a) Random Data (b) Correlated Data (ridge) (c) Correlated Data (lasso)

Figure 7.11: A comparison of ridge and lasso regression on uncorrelated and correlated data.

each, to get a measurement vector in $\mathbb{R}^{32}$. We arrange the measurements into 4 blocks of 8 identical measurements, so each $\mathbf{x}_i$ has the form

$$\mathbf{x}_i = (a_i, \ldots, a_i, b_i, \ldots, b_i, c_i, \ldots, c_i, d_i, \ldots, d_i)^T \in \mathbb{R}^{32},$$

where a_i, b_i, c_i, d_i are repeated 8 times each. In this case, each measurement is perfectly correlated with 7 other measurements. We show the result of ridge regression in Figure 7.11(b) and lasso in Figure 7.11(c). We see that ridge regression again produces the uniform weight vector $\mathbf{w} = 1/m$, placing equal weights on all of the measurements. On the other hand, lasso recognizes the correlations, and thus is able to produce a sparse weight vector $\mathbf{w} = \frac{1}{4}(\mathbf{e}_1 + \mathbf{e}_9 + \mathbf{e}_{17} + \mathbf{e}_{25})$ that does not utilize any repeated measurement twice. Note, however, that in this case the solution to the lasso regression problem (7.43) is *not* unique — indeed, we could have selected *any* of the measurements in each block of 8, or taken any convex combination of them; see Exercise 2.5. In this case, the minimizer found depends on the optimization method that is used; we used `sklearn.linear_model.lasso`, which uses the coordinate descent shrinkage method to be discussed in Section 7.2.3. ▲

We now aim to study how lasso regression can produce sparse regression vectors $\mathbf{w}$, which is characteristically different than ridge regression. To understand this, we define the *shrinkage function* or, as is often designated, *operator*.

Definition 7.5. Given a positive real number $\lambda > 0$, the *shrinkage operator* is the scalar function $\mathrm{Shrink}_\lambda \colon \mathbb{R} \to \mathbb{R}$ defined by

$$\mathrm{Shrink}_\lambda(x) = (\mathrm{sign}\, x) \max\left\{0, |x| - \tfrac{1}{2}\lambda\right\}. \tag{7.44}$$

The shrinkage operator acts coordinatewise on vectors $\mathbf{x} \in \mathbb{R}^n$, so that the coordinates of $\mathbf{z} = \mathrm{Shrink}_\lambda(\mathbf{x})$ are exactly $z_i = \mathrm{Shrink}_\lambda(x_i)$ for $i = 1, \ldots, n$.

The shrinkage operator with $\lambda = 1$ is depicted in Figure 7.12(a). Essentially, when applied to an input x, it decreases (shrinks) the absolute value $|x|$ by $\frac{1}{2}\lambda$, but then gives an output of 0 if the resulting number is negative, which occurs when $|x| < \frac{1}{2}\lambda$. The shrinkage operator arises because it solves the following relatively simple scalar minimization problem, which is in essence the one-dimensional version of lasso minimization.

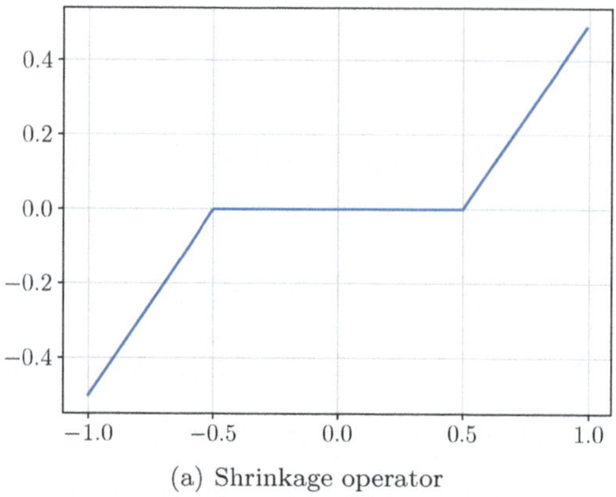

(a) Shrinkage operator

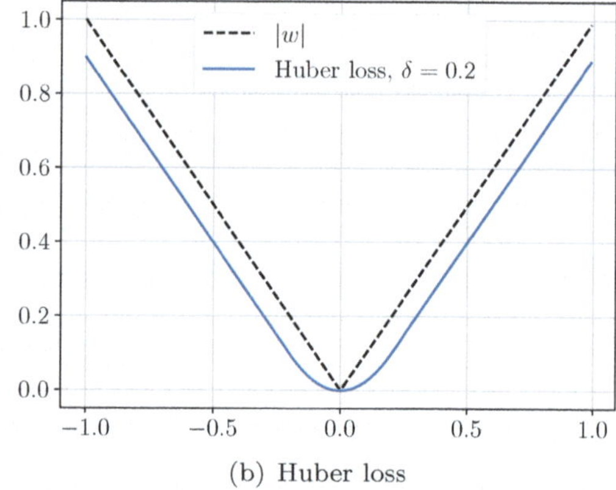

(b) Huber loss

Figure 7.12: Illustration of (a) the shrinkage operator Shrink_λ with $\lambda = 1$, and (b) the Huber loss $g_H(w)$ with $\delta = 0.2$.

Lemma 7.6. *Given $x \in \mathbb{R}$ and $\lambda \geq 0$, the solution of the minimization problem*

$$\min_w \left\{ (w - x)^2 + \lambda \, | \, w \, | \right\} \qquad \text{is given by} \qquad w^\star = \text{Shrink}_\lambda(x). \tag{7.45}$$

We defer the proof of Lemma 7.6 until the end of this section, although the motivated reader may like to try their hand at proving it before reading our version.

We can now illuminate the difference between lasso and ridge regression for a special class of data matrices X, namely those with orthonormal columns, so that $X^T X = I$. (These are matrices whose singular values are all the same, namely $\sigma_1 = \cdots = \sigma_r = 1$, where $r = \text{rank}\, X$; see Example 5.78 for more details.)

Theorem 7.7. *Assume that the columns of X are orthonormal. Let $\mathbf{w}_0 = X^T \mathbf{y}$ be the least squares solution of $X\mathbf{w} = \mathbf{y}$. Then the solution of the ridge regression problem (7.32) is given by*

$$\mathbf{w}_\lambda = \frac{\mathbf{w}_0}{1 + \lambda}, \tag{7.46}$$

while the solution of the lasso regression problem (7.43) is given by

$$\mathbf{w}_\lambda = \text{Shrink}_\lambda(\mathbf{w}_0). \tag{7.47}$$

Theorem 7.7 shows that ridge regression for such data simply decreases all the weights in $\mathbf{w}_0$ by the same scaling factor $1/(1 + \lambda)$. In particular, ridge regression will not set any weights to zero, unless they were already zero in the least squares solution. On the other hand, lasso regression uses the shrinkage operator Shrink_λ, which decreases all the weights by the same amount $\frac{1}{2}\lambda$, and sets any weights whose absolute value is less than $\frac{1}{2}\lambda$ to zero.

Proof of Theorem 7.7. To prove (7.46), we note that (7.33) with $X^T X = I$ amounts to

$$\mathbf{w}_\lambda = (I + \lambda I)^{-1} X^T \mathbf{y} = \frac{\mathbf{w}_0}{1 + \lambda}.$$

To prove (7.47), we note that when $X^T X = I$, we have

$$\| X\mathbf{w} - \mathbf{y} \|^2 = \mathbf{w}^T X^T X \mathbf{w} - 2\mathbf{w}^T X^T \mathbf{y} + \| \mathbf{y} \|^2$$
$$= \| \mathbf{w} \|^2 - 2\mathbf{w}^T \mathbf{w}_0 + \| \mathbf{y} \|^2 = \| \mathbf{w} - \mathbf{w}_0 \|^2 - \| \mathbf{w}_0 \|^2 + \| \mathbf{y} \|^2.$$

Therefore, the lasso regression objective in (7.43) can be written as

$$\| X\mathbf{w} - \mathbf{y} \|^2 + \lambda \| \mathbf{w} \|_1 = \sum_{i=1}^{n} \big[(w_i - w_{0,i})^2 + \lambda\, | w_i | \big] - \| \mathbf{w}_0 \|^2 + \| \mathbf{y} \|^2.$$

We can thus minimize the individual summands, and, courtesy of Lemma 7.6, the minimizers are $w_i = \mathrm{Shrink}_\lambda(w_{0,i})$. ■

It is also important to point out that there exist more exotic methods that combine two or more different norms. An example is *elastic net regression* which solves the problem

$$\min_{\mathbf{w}} \big\{ \| X\mathbf{w} - \mathbf{y} \|^2 + \lambda_1 \| \mathbf{w} \|_1 + \lambda_2 \| \mathbf{w} \|^2 \big\}, \tag{7.48}$$

with $\lambda_1, \lambda_2 > 0$. In elastic net regression, the additional squared Euclidean norm makes the objective function strongly convex, and so (unlike lasso; see below) it has a unique minimizer $\mathbf{w}^\star$, and gradient descent converges quickly to the unique minimizer; see Section 6.9. This makes elastic net regression an effective compromise between the sparse and interpretable (meaning fewer features) lasso regression, and the computationally efficient ridge regression.

Proof of Lemma 7.6. Let
$$f_x(w) = (w - x)^2 + \lambda\, | w |$$

denote the objective function in (7.45). For fixed x, it is easily seen to be a strongly convex function of w, and hence there exists a unique minimizer $w^\star$; see Example 6.54 for a closely related minimization problem. Let us first assume $x \geq 0$. Then $f_x(w) > f_x(0) = x^2$ when $w < 0$, and hence $w^\star \geq 0$. Now, the minimizer is either $w^\star = 0$, where the objective is not differentiable, or is at a critical point $w^\star > 0$, where

$$f_x'(w^\star) = 2\,(w^\star - x) + \lambda = 0, \quad \text{or, equivalently,} \quad w^\star = x - \tfrac{1}{2}\lambda. \tag{7.49}$$

If $x \leq \tfrac{1}{2}\lambda$, then there is no critical point with $w^\star > 0$, and so the minimizer must be $w^\star = 0$. On the other hand, if $x > \tfrac{1}{2}\lambda$, then (7.49) defines the minimizer $w^\star$. In other words,

$$w^\star = \max\big\{ 0,\, x - \tfrac{1}{2}\lambda \big\} = \mathrm{Shrink}_\lambda(x) \quad \text{when} \quad x \geq 0.$$

To handle the case $x \leq 0$, we merely note that $f_x(w) = f_{-x}(-w)$, and hence, by the preceding argument,

$$w^\star = -\max\big\{ 0,\, -x - \tfrac{1}{2}\lambda \big\} = \mathrm{Shrink}_\lambda(x) \quad \text{when} \quad x \leq 0 \text{ also.} \quad ■$$

7.2.3 Optimization Aspects

We now turn to a brief discussion of how to solve the lasso regression problem computationally. The objective function (7.43) is convex, but is not quadratic due to the presence of the term involving the 1 norm. Therefore, we cannot compute the minimizer with an explicit formula as we did for ridge regression. The objective function is also not, in general, strongly or strictly convex, so, unlike ridge or elastic net regression, minimizers are not necessarily unique.

Thus, we must resort to an iterative optimization algorithm to find suitable weights $\mathbf{w}$ that approximately minimize the lasso objective function.

A natural approach is to use gradient descent, which was developed in Chapter 6. However, one issue is that the 1 norm $\|\mathbf{w}\|_1$ is not differentiable when $w_i = 0$ for any i, and does not have a Lipschitz continuous gradient. Consequently, the convergence results in Section 6.9 do not apply.[11] One way to resolve this is to replace the 1 norm with a smooth approximation, and solve the problem

$$\min_{\mathbf{w}} \left\{ \|X\mathbf{w} - \mathbf{y}\|_2^2 + \lambda \sum_{i=1}^{n} g(w_i) \right\}, \tag{7.50}$$

where $g(w)$ is some smooth and convex approximation of the absolute value function $|w|$. Two examples are

$$g_S(w) = \sqrt{w^2 + \delta^2}, \quad \text{and} \quad g_H(w) = \begin{cases} \frac{1}{2}\delta^{-1}w^2, & |w| \le \delta, \\ |w| - \frac{1}{2}\delta, & |w| > \delta, \end{cases} \tag{7.51}$$

where, in both cases, $\delta > 0$ is a small hyperparameter. The second approximation is called the *Huber loss*, and simply modifies the absolute value function near the origin to make it quadratic there; see Figure 7.12(b) for an illustration. Fixing a small value of δ and choosing either approximation yields an optimization problem (7.50) for which the objective function is convex with a Lipschitz continuous gradient — see Exercise 2.1 — and hence the gradient descent algorithms and convergence results from Chapter 6 can be directly applied. However, one has to use a small time step α proportional to δ, which can lead to slow convergence, and this approximation is changing the lasso problem, which may not be desirable.

An alternative approach that is highly effective is to apply the proximal gradient descent method discussed in Section 6.4.1. Taking $G(\mathbf{w}) = \|X\mathbf{w} - \mathbf{y}\|_2^2$ and $H(\mathbf{w}) = \|\mathbf{w}\|_1$, the proximal gradient descent iteration (6.61) amounts to

$$\mathbf{w}_{k+1} = \operatorname*{argmin}_{\mathbf{w}} \left[\frac{1}{2\alpha_k \lambda} \|\mathbf{w} - \mathbf{z}_k\|^2 + \|\mathbf{w}\|_1 \right], \quad \text{where} \quad \mathbf{z}_k = \mathbf{w}_k - 2\alpha_k X^T(X\mathbf{w}_k - \mathbf{y}),$$

and we used the fact that $\nabla G(\mathbf{w}) = 2X^T(X\mathbf{w} - \mathbf{y})$. As in Theorem 7.7, the solution $\mathbf{w}_{k+1}$ is obtained by applying the shrinkage operator with parameter $2\alpha_k\lambda$ to the vector $\mathbf{z}_k$; that is

$$\mathbf{w}_{k+1} = \operatorname{Shrink}_{2\alpha_k\lambda}(\mathbf{z}_k). \tag{7.52}$$

The resulting algorithm is known as the *iterative shrinkage–thresholding algorithm* (ISTA), and is a highly efficient method for optimizing the lasso objective function, as well as a wide range of related problems [17].

In fact, an even more efficient way to use shrinkage when optimizing the lasso objective is via *coordinate descent*, whereby we iteratively minimize over each regression coefficient w_i, while keeping the others, i.e., w_j for $j \ne i$, fixed. To see how this is done, let $\mathbf{v}_1, \ldots, \mathbf{v}_n \in \mathbb{R}^m$ denote the columns of the data matrix X, and note that the lasso objective (7.43) can be expressed as (see Exercise 2.6)

$$\|X\mathbf{w} - \mathbf{y}\|^2 + \lambda\|\mathbf{w}\|_1 = \|\mathbf{v}_i\|^2 w_i^2 - 2w_i b_i + \lambda|w_i| + \|\mathbf{y}\|^2, \tag{7.53}$$

for any $i = 1, \ldots, n$, where

$$b_i = \mathbf{v}_i \cdot \left(\mathbf{y} - \sum_{j \ne i} w_j \mathbf{v}_j \right).$$

---

[11]Gradient descent can be applied to certain nondifferentiable functions, with the notion of gradient replaced by *subgradient*; see [10].

Minimizing the expression on the right hand side of (7.53) is exactly the shrinkage problem solved in Lemma 7.6, whose solution is given in closed form by

$$w_i = \mathrm{Shrink}_{\lambda/\|\mathbf{v}_i\|^2}\big(b_i / \|\mathbf{v}_i\|^2\big). \tag{7.54}$$

One pass of coordinate descent for optimizing the lasso objective iteratively applies the shrinkage formula (7.54) for $i = 1, \ldots, n$. This is repeated, as desired, until a convergence criterion is met. The reader may notice there are similarities between coordinate descent and stochastic gradient descent, which is introduced and studied rigorously in Chapter 11.

We explore the computational aspects of the ISTA and coordinate descent algorithms in Exercises 2.9, 2.10.

7.2.4 Kernel Regression

All of the linear regression techniques discussed above can also be applied to any *function* of the data matrix X. That is, given a function $\phi \colon \mathbb{R}^n \to \mathbb{R}^d$, often called a *feature map*, that transforms our data points $\mathbf{x}_i$ into $\mathbf{z}_i = \phi(\mathbf{x}_i)$, we can define the *transformed $m \times d$ data matrix Z* by

$$Z = \begin{pmatrix} \mathbf{z}_1^T \\ \mathbf{z}_2^T \\ \vdots \\ \mathbf{z}_m^T \end{pmatrix} = \begin{pmatrix} z_{11} & z_{12} & \cdots & z_{1d} \\ z_{21} & z_{22} & \cdots & z_{2d} \\ \vdots & \vdots & \ddots & \vdots \\ z_{m1} & z_{m2} & \cdots & z_{md} \end{pmatrix}, \tag{7.55}$$

and can then apply any of the above linear regression techniques with Z in place of X. This amounts to a regression function $F : \mathbb{R}^n \to \mathbb{R}$ of the form

$$F(\mathbf{x};\mathbf{w}) = \phi(\mathbf{x}) \cdot \mathbf{w} = \sum_{i=1}^{d} w_i \phi_i(\mathbf{x}),$$

where $\phi(\mathbf{x}) = \big(\phi_1(\mathbf{x}), \ldots, \phi_d(\mathbf{x})\big)$. This allows us to produce various types of *nonlinear regression* and forms a special case of *kernel regression*,[12] which will be discussed further in Section 7.6.

A classic example is *polynomial regression*, where our data points $x_i \in \mathbb{R}$ are scalars, and we take $\phi(x) = (1, x, x^2, \ldots, x^k) \in \mathbb{R}^d$ where $d = k + 1$, which is designed in order to contain all the terms in a k-th degree polynomial. Then the transformed data matrix

$$Z = \begin{pmatrix} 1 & x_1 & \cdots & x_1^k \\ 1 & x_2 & \cdots & x_2^k \\ \vdots & \vdots & \ddots & \vdots \\ 1 & x_m & \cdots & x_m^k \end{pmatrix}, \tag{7.56}$$

is known as a *Vandermonde matrix*. Such matrices are fundamental in the development of classical methods of polynomial interpolation and approximation, cf. [181]. The corresponding regression function has the form of a k-th degree polynomial:

$$F(x;\mathbf{w}) = w_0 + w_1 x + w_2 x^2 + \cdots + w_k x^k, \qquad \text{where} \qquad \mathbf{w} = (w_0, \ldots, w_k) \in \mathbb{R}^{k+1}.$$

---

[12] Here the term "kernel" refers to a certain type of function and not to the kernel of a matrix; see the remark after Definition 7.16.

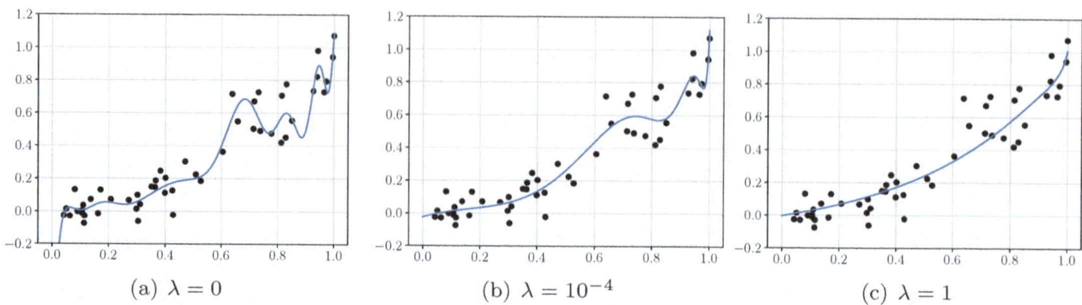

(a) $\lambda = 0$ (b) $\lambda = 10^{-4}$ (c) $\lambda = 1$

Figure 7.13: Example of polynomial ridge regression with a degree 50 polynomial on the toy data set from Figure 7.9. We can see that when there is no regularization, the polynomial fit exhibits more degrees of freedom, while as we increase λ the complexity of the regression decreases. By increasing λ to $\lambda = 1$, we see a result very similar to the quadratic regression in Figure 7.9.

In Figure 7.13 we show an example of polynomial ridge regression with a degree 50 polynomial on the toy data set with $m = 50$ data points from Figure 7.9. We can see that the predicted trend becomes *simpler* as we increase the strength λ of the regularization.

Remark 7.8. In Figure 7.13(a) there are 50 data points. Thus, in theory, [181], there exists a degree 49 degree polynomial that exactly fits all the data points. However, the result in Figure 7.13(a), which utilizes a degree 50 polynomial, does not exactly fit the data; in fact, it fits almost none of the data points. This is due to the fact that the Vandermonde matrix with $k = 50$ is extremely ill-conditioned, so the normal equations are difficult/impossible to solve with precision. For this reason, it is common in polynomial interpolation to use other types polynomial basis functions for which the corresponding data matrix is better conditioned. One such example is the Chebyshev polynomials defined in (11.28).   ▲

Exercises

2.1. Show that the Huber loss $g_H(w)$ in (7.51) has a Lipschitz continuous derivative.

2.2. ♡ This exercise compares the 1 norm and 2 norm in regression.

(a) Show that the solution of the optimization problem $\min\{\, \|\mathbf{w}\|_2 \mid \mathbf{1} \cdot \mathbf{w} = 1 \,\}$ is given by $\mathbf{w} = 1/n$. This shows that under a constraint on the total *mass* of the weights, i.e., $\mathbf{w} \cdot \mathbf{1} = 1$, the 2 norm prefers to assign weights equally across all features. *Hint:* Write $\mathbf{z} = \mathbf{w} - 1/n$ and convert it into the equivalent optimization problem $\min\{\, \|\mathbf{z} + 1/n\|_2 \mid \mathbf{1} \cdot \mathbf{z} = 0 \,\}$ whose optimal solution is $\mathbf{z} = \mathbf{0}$.

(b) Show that the solution of the optimization problem $\min\{\, \|\mathbf{w}\|_1 \mid \mathbf{1} \cdot \mathbf{w} = 1 \,\}$ is any vector $\mathbf{w}$ satisfying the constraint $\mathbf{1} \cdot \mathbf{w} = 1$ that has nonnegative entries, i.e., $w_i \geq 0$ for all i. Hence, the 1 norm does not place any preference on how the mass is distributed among features, and the sparse solution $\mathbf{w} = \mathbf{e}_1$ is as equally preferred as the nonsparse solution $\mathbf{w} = 1/n$. This fact allows lasso and elastic net to find sparser solutions when they exist.

2.3. ♡ Consider the general ridge regression problem $\min_{\mathbf{w}}\{\, \|X\mathbf{w} - \mathbf{y}\|^2 + \lambda \|B\mathbf{w}\|^2 \,\}$. Show that the solution is unique and is given by (7.41) when $X^T X + \lambda B^T B$ is nonsingular. Explain how this solves the ridge regression problem (7.42).

2.4. Write Python code to solve the ridge regression problem using SVD, as established in Theorem 7.1. Compare your solution to the one using `numpy.linalg.solve` from the Python notebook in this section.

2.5. $\diamondsuit$ Suppose we have a linear regression problem with data matrix X and target vector $\mathbf{y}$ in which the first k columns of X are identical — that is, the first k measurements are all duplicates of each other. (a) Show that the solution $\mathbf{w}$ of the ridge regression problem sets $w_1 = \cdots = w_k$. (b) Show that there exists a solution $\mathbf{w}$ of the lasso regression problem where each of $w_1, \ldots, w_k$ has the same sign. (c) Assume that a solution $\mathbf{w}$ of lasso regression satisfies $w_1, \ldots, w_k > 0$. Show that *any* other coefficient vector $\widetilde{\mathbf{w}}$ for which $\widetilde{w}_1, \ldots, \widetilde{w}_k > 0$, $\widetilde{w}_1 + \cdots + \widetilde{w}_k = w_1 + \cdots + w_k$, and $\widetilde{w}_i = w_i$ for $i = k+1, \ldots, n$ is also a solution of the lasso regression problem.

2.6. $\heartsuit$ Prove that (7.53) holds.

2.7. Use `sklearn` to apply lasso regression to the diabetes data set and compare against the results in Table 7.10.

2.8. Write Python code that solves the lasso regression problem with the Huber loss $g_H(w)$ in (7.51) with $\delta > 0$ using gradient descent. Test your code on the diabetes data set.

2.9. $\heartsuit$ Write Python code that solves the lasso regression problem using the iterative shrinkage–thresholding algorithm (ISTA). Test your code on the diabetes data set.

2.10. Repeat Exercise 2.9 using coordinate descent and compare against ISTA.

2.11. Let $x_1, \ldots, x_{k+1} \in \mathbb{R}$ be distinct real numbers, i.e., $x_i \neq x_j$ when $i \neq j$. (a) $\heartsuit$ Prove that the corresponding $(k+1) \times (k+1)$ Vandermonde matrix (7.56) is nonsingular. *Hint:* Prove that $\ker Z = \{\mathbf{0}\}$, by using the fact that a nonzero polynomial of degree k can have at most k roots. (b) $\diamondsuit$ Given data points $y_1, \ldots, y_{k+1} \in \mathbb{R}$, an *interpolating polynomial $p(x)$* satisfies $p(x_i) = y_i$ for all $i = 1, \ldots, k+1$. Prove that there exists a unique interpolating polynomial of degree $\leq k$ for any collection of data points. *Hint:* Write the interpolation conditions in vectorial form using the Vandermonde matrix.

2.12. Write Python code for polynomial ridge regression on scalar variables, as described at the end of the section. Pick one of the features from the diabetes data set and run polynomial regression with different degree polynomials. Can you fit the data better than with linear regression? Are your results interpretable?

2.13. Apply linear regression to another regression data set in `sklearn.datasets`.

7.3   Support Vector Machines (SVM)

Python Notebook: Support Vector Machines (.ipynb)

Just as it was natural to use linear functions for regression in Section 7.2, we can also use them for classification. We will focus at first on binary classification, meaning that there are only two classes, and will address the case of three or more classes at the end of the section.

Indeed, there are standard machine learning approaches for constructing multiclass classifiers out of binary classifiers. For binary classification, we take the labels to be $y_i \in \{-1, 1\}$, where $y_i = 1$ indicates one class and $y_i = -1$ indicates the other.[13] As before, each data point also has an associated feature vector $\mathbf{x}_i \in \mathbb{R}^n$, and we are in the fully supervised context where all data points are labeled.

In this section, we will use a linear, or, rather, affine *classification function* of the form

$$F(\mathbf{x}; \mathbf{w}, b) = \mathbf{x} \cdot \mathbf{w} - b, \qquad (7.57)$$

which has *weights* $\mathbf{0} \neq \mathbf{w} \in \mathbb{R}^n$ and a *bias* $b \in \mathbb{R}$. Now, we could certainly treat the classification problem as a linear regression problem, as in Section 7.2, since the labels $y_i \in \mathbb{R}$ are real-valued, and then we could minimize the mean squared error. However, since we know the labels are discrete and only take on values -1 and 1, it is not necessary that the learned function F actually fit the data that well in the sense that $F(\mathbf{x}_i; \mathbf{w}, b) \approx y_i$. In fact, it may be prohibitively hard to do this with an affine function.

Ultimately, we would like to make a discrete prediction of $+1$ or -1, so it is natural to use the *sign* of $F(\mathbf{x}_i; \mathbf{w}, b)$ as the class prediction, instead of its precise value,. This provides more flexibility as to how the linear function fits the data. That is, $F(\mathbf{x}; \mathbf{w}, b) > 0$ indicates that $\mathbf{x}$ belongs to class 1, while $F(\mathbf{x}; \mathbf{w}, b) < 0$ indicates class -1. Thus, a linear classifier that properly classifies the data points $(\mathbf{x}_1, y_1), \ldots, (\mathbf{x}_m, y_m)$ must satisfy

$$y_i(\mathbf{x}_i \cdot \mathbf{w} - b) > 0 \qquad \text{for all} \qquad i = 1, \ldots, m. \qquad (7.58)$$

The *decision boundary* separating the two classes is

$$\mathcal{D} = \{\mathbf{x} \in \mathbb{R}^n \mid \mathbf{x} \cdot \mathbf{w} = b\} \subset \mathbb{R}^n, \qquad (7.59)$$

which is an $(n - 1)$-dimensional affine subspace of $\mathbb{R}^n$, also called an (affine) *hyperplane*. Keep in mind that the weights and bias are not uniquely determined by the hyperplane; see Exercise 2.7. Thus, in order to find a linear classifier that correctly classifies all of the data points, there would need to exist a dividing hyperplane the splits the data in the sense that the first class lies on one side of the hyperplane and the second class on the other. When such a dividing hyperplane exists, the data is called *linearly separable*. Linear separability is a rather simple geometrical configuration that arises in some applications, but is certainly not a property universally enjoyed by all data sets, and thus it may well not be possible to find a linear classifier that separates the data. Strategies for dealing with such more general situations will be discussed below.

Figure 7.14 shows an example. When the data vectors are linearly separable, there are, in general, many different hyperplane decision boundaries that can separate the two classes; in fact, there are typically infinitely many. Figure 7.14(b) shows three different examples of linear decision boundaries. Since the testing data may vary somewhat from tthe raining data, in order to ensure our model generalizes, it is preferable to choose a decision boundary that lies as far away as possible from the training data vectors, so that a small change in their positions does not alter the labels.

The margin of the classifier measures exactly how far the decision boundary can be moved before encountering a data point, and is a measure of how robust the classifier is to perturbations. To formulate this mathematically, we encode a margin in the output of our classifier, and instead of (7.58), by suitably rescaling $\mathbf{w}$ and adjusting b, we ask that

$$y_i(\mathbf{x}_i \cdot \mathbf{w} - b) \geq 1 \qquad \text{for all} \qquad i = 1, \ldots, m. \qquad (7.60)$$

---

[13]There is no advantage in binary classification to using one-hot vectors to represent the classes; see Exercise 3.1.

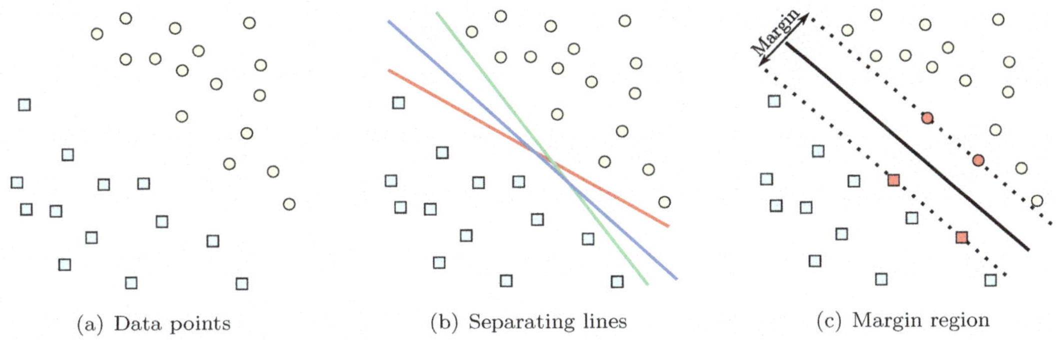

<table>
<tr><td>(a) Data points</td><td>(b) Separating lines</td><td>(c) Margin region</td></tr>
</table>

Figure 7.14: Example of (a) a linearly separable data set, (b) many different separating lines, and (c) the maximal margin classifier selected by SVM with the support vectors colored in red.

The *margin region* is the set

$$\mathcal{M} = \{\mathbf{x} \in \mathbb{R}^m \mid |\mathbf{x} \cdot \mathbf{w} - b| < 1\}, \tag{7.61}$$

and is depicted in Figure 7.14(c). The margin region should not contain any data points, provided the data is linearly separable. We define the *margin* to be the width of the margin region $\mathcal{M}$, and the goal is to find a classifier with the *largest margin*. We can compute the margin by taking a point a in the decision boundary, so $\mathbf{a} \in \mathcal{D}$ or, equivalently, $\mathbf{a} \cdot \mathbf{w} = b$, and look for δ such that

$$\left(\mathbf{a} + \delta \frac{\mathbf{w}}{\|\mathbf{w}\|}\right) \cdot \mathbf{w} - b = \pm 1.$$

The left hand side simplifies to $\delta \|\mathbf{w}\|$, and so $\delta = \pm 1/\|\mathbf{w}\|$. Thus, the resulting margin is

$$2\delta = \frac{2}{\|\mathbf{w}\|}.$$

The *support vector machine* (SVM) classifier for linearly separable data seeks the linear classifier with the largest margin, i.e., the smallest $\|\mathbf{w}\|$. This leads to the optimization problem

$$\min_{\mathbf{w} \in \mathbb{R}^n} \left\{ \|\mathbf{w}\|^2 \mid b \in \mathbb{R}, \quad y_i(\mathbf{x}_i \cdot \mathbf{w} - b) \geq 1, \quad i = 1, \ldots, m \right\}. \tag{7.62}$$

The *support vectors* are any data points $\mathbf{x}_i$ that lie on the boundary of the margin region, that is, those that satisfy $\mathbf{x}_i \cdot \mathbf{w} - b = \pm 1$; they are colored red in Figure 7.14(c). The formulation (7.62) is a constrained optimization problem that can be solved with a variety of methods, some of which, e.g., gradient descent, are covered in this book. We will postpone discussion of the optimization aspects of SVM until the soft-margin version is introduced below.

Example 7.9. Consider a simple data set with two points $\mathbf{x}_1 = \mathbf{z}$ and $\mathbf{x}_2 = -\mathbf{z}$ with labels $y_1 = 1$ and $y_2 = -1$. Then the SVM problem (7.62) becomes

$$\min_{\mathbf{w} \in \mathbb{R}^n} \left\{ \|\mathbf{w}\|^2 \mid b \in \mathbb{R}, \quad \mathbf{z} \cdot \mathbf{w} - b \geq 1, \quad \mathbf{z} \cdot \mathbf{w} + b \geq 1 \right\}. \tag{7.63}$$

We claim that the solution is $\mathbf{w} = \mathbf{z}/\|\mathbf{z}\|^2$ and $b = 0$, so the SVM classification decision reduces to checking the sign of $\mathbf{x} \cdot \mathbf{z}$. To see this, we average the two constraints in (7.63) to

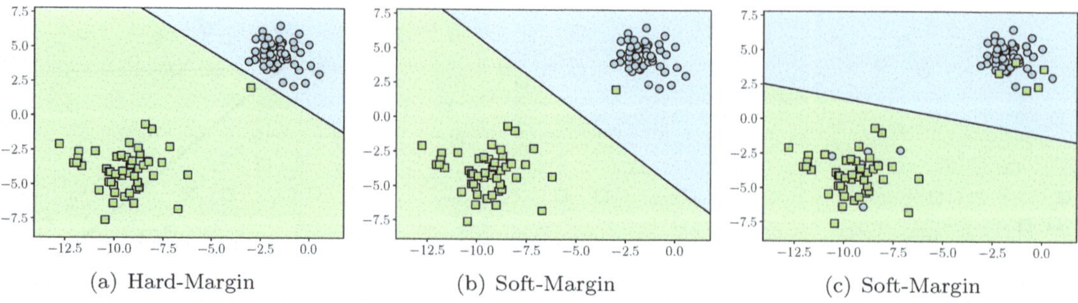

(a) Hard-Margin   (b) Soft-Margin   (c) Soft-Margin

Figure 7.15: Examples of (a) hard-margin and (b), (c) soft-margin linear SVM for different data sets with some corruption in the labels. In (b), the soft-margin SVM classifier ignores the green square that is close to the other cluster, while in (c) we see how soft-margin SVM can classify data that is not linearly separable.

find that $\mathbf{z} \cdot \mathbf{w} \geq 1$. We can write the optimal $\mathbf{w}$ in the form $\mathbf{w} = \lambda \mathbf{z} + \mathbf{v}$, where $\mathbf{v} \in \mathbf{z}^{\perp}$, i.e., $\mathbf{v} \cdot \mathbf{z} = 0$. Then

$$\|\mathbf{w}\|^2 = \lambda^2 \|\mathbf{z}\|^2 + \|\mathbf{v}\|^2 \geq \lambda^2 \|\mathbf{z}\|^2, \qquad \mathbf{w} \cdot \mathbf{z} = \lambda \|\mathbf{z}\|^2.$$

Therefore the constraint $\mathbf{z} \cdot \mathbf{w} \geq 1$ amounts to $\lambda \geq 1/\|\mathbf{z}\|^2$ and any feasible $\mathbf{w}$ must satisfy $\|\mathbf{w}\|^2 \geq 1/\|\mathbf{z}\|^2$. Setting $\mathbf{w} = \mathbf{z}/\|\mathbf{z}\|^2$ achieves this lower bound, and is compatible with the constraints in (7.63) once we set $b = 0$. ▲

Oftentimes data is not linearly separable, such as that displayed in Figure 7.7. In this case, the optimization problem (7.62) for SVM has no feasible weights $\mathbf{w}$ and thus has no solution. In this case, we reformulate SVM with a *soft margin* in the form

$$\min_{\mathbf{w},b} \left\{ \lambda \|\mathbf{w}\|^2 + \frac{1}{m} \sum_{i=1}^{m} \left(1 - y_i(\mathbf{x}_i \cdot \mathbf{w} - b) \right)_+ \right\}, \tag{7.64}$$

where, as above, $a_+ = \max\{a, 0\}$, while $\lambda > 0$ is a hyperparameter. When the constraint $y_i(\mathbf{x}_i \cdot \mathbf{w} - b) \geq 1$ is satisfied, the additional term $(1 - y_i(\mathbf{x}_i \cdot \mathbf{w} - b))_+$ is zero. When it is not satisfied, $(1 - y_i(\mathbf{x}_i \cdot \mathbf{w} - b))_+$ represents the Euclidean distance $\mathbf{x}_i$ would need to be moved in order to satisfy the constraint, so it is a natural quantity to minimize. The hyperparameter $\lambda > 0$ allows us to trade off between enforcing separability and maximizing the margin. Figure 7.15 gives an illustration of SVM decision boundaries for different data sets. In Figure 7.15(a) we show the result of hard-margin SVM, which is required to separate the data linearly, and is thus highly sensitive to the one outlying green square that is near the blue cluster. In Figure 7.15(b) we show soft-margin SVM applied to the same data set, which is able to ignore the outlying data point, since this drastically improves the margin while minimally affecting the soft separability criterion. In Figure 7.15(c) we show a data set that is not linearly separable due to a few data points having incorrect labels in each cluster. In this case, hard-margin SVM is not applicable, since there is no linear decision boundary that separates the data perfectly, while soft-margin SVM is able to train a reasonable classifier.

The soft-margin formulation of SVM given in (7.64) is one of the most common versions used in applications. It is worth noting that it uses the same regularizer as ridge regression (7.32), but the loss function is different. We also mention that there are many other varieties of SVM based on how one chooses to regularize — in fact, there is a 1 norm version of SVM that replaces the regularizer $\lambda \|\mathbf{w}\|^2$ with $\lambda \|\mathbf{w}\|_1$, similar in spirit to lasso regularization; see [267] for further developments.

7.3.1 Optimization Aspects

We briefly discuss here the issue of computing a solution of soft-margin SVM, which is a convex optimization problem. One common approach is the dual optimization method, which is discussed in Section 7.6. A more straightforward approach is to solve the optimization problem with gradient descent. However, as was the case with lasso in Section 7.2, the presence of the term a_+ in the SVM objective function renders the gradient non-Lipschitz, and we cannot guarantee convergence using the results in Chapter 6. To handle this, we replace the soft-margin SVM problem by the smoothed approximate problem $\min_{\mathbf{w},b} \mathcal{L}(\mathbf{w}, b)$, where

$$\mathcal{L}(\mathbf{w}, b) = \lambda \|\mathbf{w}\|^2 + \frac{1}{m} \sum_{i=1}^{m} \frac{1}{\beta} \log\left(1 + \exp\left[\beta\left(1 - y_i(\mathbf{x}_i \cdot \mathbf{w} - b)\right)\right]\right), \tag{7.65}$$

which involves the *softplus function*

$$f_\beta(x) = \frac{1}{\beta} \log(1 + e^{\beta x}) \tag{7.66}$$

from Exercise 7.3 in Chapter 6. Here $\beta > 0$ is a hyperparameter, and the softplus function is a smooth convex approximation to the plus function x_+ in the sense that $\lim_{\beta \to \infty} f_\beta(x) = x_+$. Thus, larger values of β yield better approximations of the true soft-margin SVM problem. The objective function in (7.65) is convex with Lipschitz continuous gradient and is thus amenable to solution via gradient descent; see Exercise 3.9. However, we warn the reader that, while a minimizer exists, it may not be unique, and there are even degenerate cases where $\mathbf{w} = \mathbf{0}$ is optimal, and SVM predicts the same class for all data points, cf. [194]!

7.3.2 Multiclass Support Vector Machines

So far we have considered only binary classification, where we just have two classes. In general, we often have a *multiclass* classification problem, where there are three or more classes. For example, the MNIST classification problem has 10 classes. Fortunately, there are simple and general techniques for constructing a multiclass classifier by combining a number of binary classifiers. The two common techniques are called *one-vs-one*, and *one-vs-rest*, the latter sometimes called *one-vs-all*. In the one-vs-one framework with c classes, we train[14]
$M = \binom{c}{2} = \frac{c(c-1)}{2}$ binary classifiers on each pair of classes. This gives M class predictions
for every data point, and we can take a majority vote to determine the final class prediction. In cases, like SVM, where the classifier is a scoring function that is thresholded to obtain classifications, it is also common to average the raw outputs of the scoring function, which is akin to a weighted majority vote, where the weights are influenced by the confidence of each classifier. The one-vs-one framework can be applied to *any* binary classifier.

In contrast, the *one-vs-rest* framework only works with classifiers, like SVM, that make an initial continuous prediction score, which is then thresholded to a label prediction, with the idea being that a larger score indicates a more confident prediction of the class +1, and a large negative score is a more confident prediction of the class −1. In the one-vs-rest framework, we train c binary classifiers, with the i-th classifier trained to predict whether each data point is in class i, or not. That is, for the i-th classifier we assign labels of +1 to data points in the i-th class, and labels of −1 to all other data points. Let us denote the classification function for the i-th class by $F_i(\mathbf{x})$, which for SVM is the linear function $F_i(\mathbf{x}) = \mathbf{x} \cdot \mathbf{w}_i - b_i$ *before* thresholding to a label prediction. Larger values of $F_i(\mathbf{x})$ indicate higher confidence

---

[14]Recall from the preface that $\binom{j}{i} = \frac{j!}{i!\,(n-i)!}$, where $0 \le i \le j$, is the usual *binomial coefficient*.

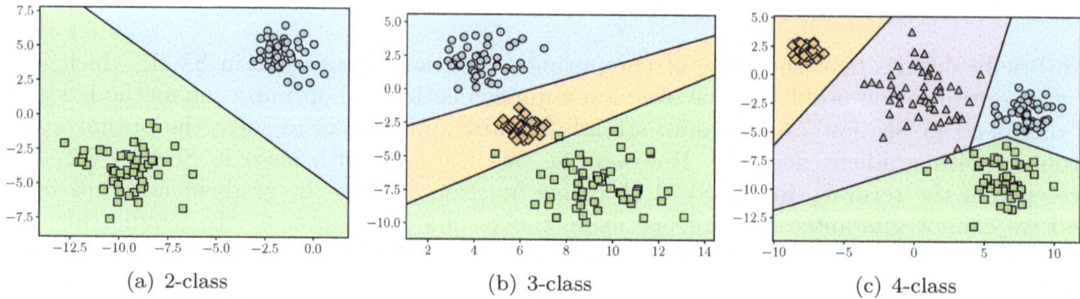

(a) 2-class　　　　　　(b) 3-class　　　　　　(c) 4-class

Figure 7.16: Examples of linear SVM for (a) binary classification, (b) 3-way classification, and (c) 4-way classification.

Digit	1	2	3	4	5	6	7	8	9
0	99.8	97.6	98.4	99.3	97.1	97.8	98.8	98.0	98.2
1		97.2	98.4	98.7	98.0	99.0	98.3	96.8	99.2
2			95.6	96.3	95.1	95.6	96.7	95.3	95.6
3				98.7	91.6	98.2	97.6	90.9	96.8
4					97.1	98.5	97.9	97.8	89.6
5						95.7	98.0	92.4	96.2
6							99.3	96.5	99.4
7								96.7	91.7
8									94.0

Table 7.17: Accuracy for binary SVM with linear kernel classifiers trained on pairs of MNIST digits. We trained the classifier using 1% of the data as training (about 70 images per digit) with the other 99% used as testing data.

that the data point $\mathbf{x}$ belongs to the i-th class. The one-vs-rest approach makes predictions by choosing the index i for which $F_i(\mathbf{x})$ is largest, that is, the label prediction $\widehat{y}$ for a data point $\mathbf{x}$ is

$$\widehat{y} = \operatorname*{argmax}_{1 \leq i \leq c} F_i(\mathbf{x}). \tag{7.67}$$

Ties can be broken by any consistent method. Clearly the one-vs-rest approach will not work for classifiers that only give binary predictions $F_i(\mathbf{x}) \in \{-1, 1\}$, since we will almost surely have ties that cannot be meaningfully broken. The one-vs-rest framework requires training fewer classifiers than one-vs-one, and can often be preferable for this reason. In common implementations of SVM, the one-vs-rest approach is normally used. Figures 7.16(a), 7.16(b) and 7.16(c) show toy examples of 2-class, 3-class and 4-class SVM. Notice how the decision regions are piecewise linear, which is due to the label decision (7.67) being a maximum of linear functions.

Classification of MNIST Digits

We now turn to some experiments with classification of MNIST digits. We start with binary classification of pairs of digits. Table 7.17 shows the accuracy for each digit-pair using a soft-margin SVM with $\lambda = 1$. We use only 1% of data for training the SVM and reserve 99% for testing. The accuracy results are very good for a majority of pairs of digits, indicating that most pairs are approximately linearly separable.

Training data size	0.1%	1%	10%	20%	40%	85.7%
Training accuracy (%)	100	100	99.89	99.33	98.41	97.12
Testing accuracy (%)	64.05	89.24	93.01	93.81	95.78	96.83

Table 7.18: Training and testing accuracy of linear kernel SVM on MNIST using different amounts of training data, given as percentages of the full 70,000 MNIST images. We always use the same 10,000 testing images. We can see that SVM overfits when provided with too few training examples.

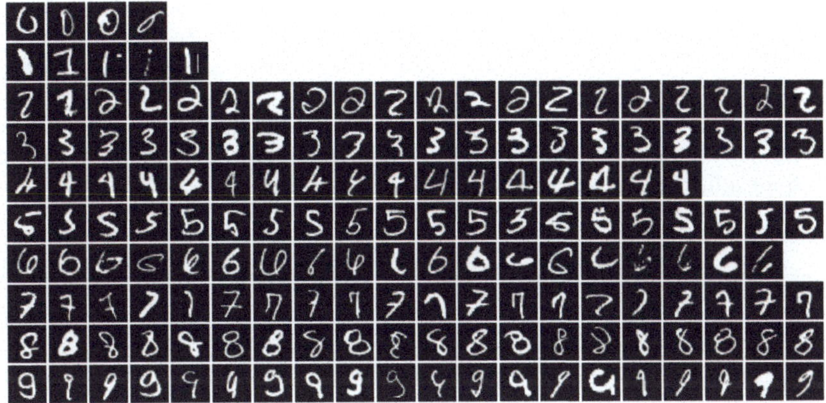

Figure 7.19: Some examples of digits from each class that are misclassified by linear SVM. We show up to 20 misclassified images in each case, depicting fewer if less than 20 were incorrect.

When applied to the full MNIST data set with 10 classes in the standard setting of $60,000$ training images and $10,000$ testing images, SVM yields training accuracy of 97.12% and testing accuracy of 96.83%, indicating a good result with a very small amount of overfitting. We show some of the misclassified images for each digit in Figure 7.19. In Table 7.18 we show the result of training with less training data, still using the same $10,000$ test images. Here, the training data percentage refers to a percentage of the total $70,000$ images, so 0.1% corresponds to 70 images, while 85.7% refers to the usual $60,000$ image training set. We can see that there is a substantial amount of overfitting when the training set is small, and the overfitting reduces consistently as the size of the training set increases. The results in Table 7.18 were generated with a single random train/test split for each experiment, and are likely to exhibit some variation with each run. The exact accuracy numbers are not as important as the trends, and more precise measurements of model performance using cross validation are left to the exercises.

The Confusion Matrix

When performing multiclass classification, the overall accuracy is not the only score worth reporting. It is also important to understand which classes are being classified better (or worse), and when mistakes are made, which classes are most often mistaken for each other. All of this information is conveyed in the *confusion matrix* of a classifier. When there are c classes, the confusion matrix C is a $c \times c$ matrix whose (i, j) entry c_{ij} is the number of testing data points from class i that were predicted to be in class j. A perfect classification occurs when C is a diagonal matrix, as the off-diagonal elements correspond to incorrect classifications. We show in Table 7.20 the confusion matrix for the preceding classification of MNIST digits.

DIGIT	0	1	2	3	4	5	6	7	8	9
0	976	0	0	0	0	2	1	0	1	0
1	0	1130	0	2	0	1	0	0	2	0
2	4	1	998	6	3	1	3	4	12	0
3	0	0	8	974	0	12	0	3	11	2
4	0	0	5	0	965	1	1	0	1	9
5	4	1	0	27	3	834	9	0	13	1
6	5	2	4	0	2	6	939	0	0	0
7	2	4	16	1	0	0	0	989	0	16
8	1	3	6	16	3	14	3	2	924	2
9	2	2	1	6	25	3	0	11	5	954

Table 7.20: Confusion matrix for classification of MNIST digits using linear kernel SVM with 60,000 training examples. We see many common errors are understandable, such as mistaking a 3 for an 8, 4 for a 9, or an 8 for a 5.

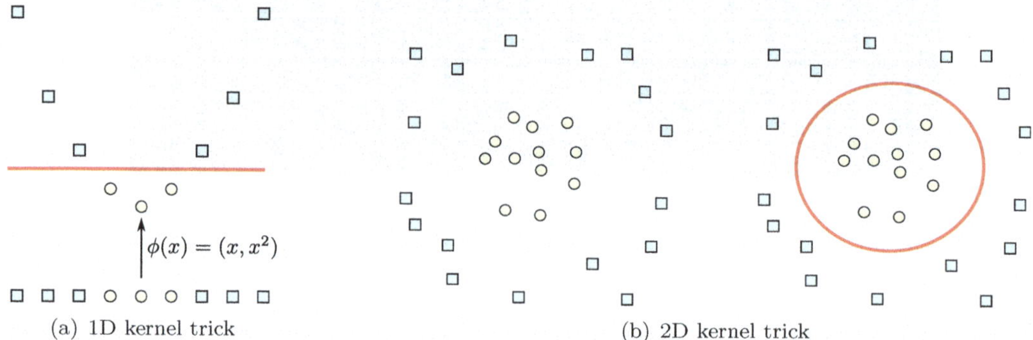

(a) 1D kernel trick

(b) 2D kernel trick

Figure 7.21: Example of (a) a one-dimensional data set that is linearly separable under the feature mapping $\phi(x) = (x, x^2)$, and (b) a two-dimensional data set that is linearly separable under the embedding $\phi(x) = (x_1, x_2, x_1^2 + x_2^2)$. In both cases we show a possible SVM decision boundary in red. Neither data set is linearly separable without a feature embedding.

7.3.3 Kernel Support Vector Machines

We have so far restricted our attention to data that is linearly separable, or approximately so, and the version of SVM we have described is called *linear SVM* or *SVM with linear kernel*. (As before, the term "kernel" as used here does not refer to the kernel of a matrix.) Many data sets are not linearly separable, even approximately; as a simple example in one dimension, see the data in Figure 7.21(a).

As was the case with linear regression in Section 7.2, we do not need to apply SVM directly to the data points $\mathbf{x}_i$, but can apply SVM to *any* transformation of the data points using a feature map $\phi \colon \mathbb{R}^n \to \mathbb{R}^d$. The goal here is to choose the feature map so that the transformed data is linearly separable in the (usually higher dimensional) feature space. Linear soft-margin SVM is then applied to the feature data $\mathbf{z}_i = \phi(\mathbf{x}_i)$, which leads to a *nonlinear* classification function of the form

$$F(\mathbf{x}; \mathbf{w}, b) = \phi(\mathbf{x}) \cdot \mathbf{w} - b. \tag{7.68}$$

This is a special case of what is called *kernel SVM*, and it allows SVM to handle more complicated nonlinear decision boundaries.

We postpone a detailed analysis and discussion of kernel methods until Section 7.6, and for now we will focus on a simple example. Consider the one-dimensional data in Figure 7.21(a), which is not linearly separable. If the origin is located at the center yellow circle, then we can use the quadratic feature map $\mathbf{z} = \phi(x) = (x, x^2)$, for which $\phi \colon \mathbb{R} \to \mathbb{R}^2$, to lift the data onto a parabola in $\mathbb{R}^2$. In the feature space $\mathbb{R}^2$, the data is linearly separable by a line of the form $x_2 = b$. In Figure 7.21(b) we show an example of data that is not linearly separable in two dimensions. Here, we can use the feature map $\mathbf{z} = \phi(\mathbf{x}) = (x_1, x_2, x_1^2 + x_2^2)$ to lift the data to a paraboloid in $\mathbb{R}^3$, provided, again, that the origin is near the center of the yellow circles. Then a linear decision boundary of the form $z_3 = b$ can be used to separate the two classes. Note that the projection of the intersection of this linear decision boundary with the image paraboloid in $\mathbb{R}^3$ back to the original data space, namely $\phi^{-1}\{z_3 = c\} \subset \mathbb{R}^2$ is the red circle $x_1^2 + x_2^2 = b$ depicted in Figure 7.21(b). Thus, a linear decision boundary in the higher dimensional feature space can represent a nonlinear decision boundary in the original space.

In the examples given above, the feature map was constructed by hand using knowledge of the data set. In general, it is very hard to find a suitable feature map ϕ that separates the data linearly. The *kernel trick* provides a way to apply the ideas described above, *without* the burden of constructing the required feature map ϕ. We discuss this further in Section 7.6.

Exercises

3.1. ♡ Let $F \colon \mathbb{R}^n \to \mathbb{R}^2$, so $F(\mathbf{x}) = (F_1(\mathbf{x}), F_2(\mathbf{x}))^T$, be the output of a binary classifier, which we assume to be a probability vector. Explain why we only need to learn the scalar-valued function $F_1(\mathbf{x})$ and threshold at $F_1(\mathbf{x}) = 0.5$ to perform binary classification.

3.2. ♡ Given the data matrix $X = \begin{pmatrix} 0 & 0 \\ 1 & 0 \\ 0 & 1 \end{pmatrix}$ and labels $\mathbf{y} = \begin{pmatrix} -1 \\ 1 \\ 1 \end{pmatrix}$ find three separating hyperplanes, and find the maximal margin SVM classifier.

3.3. ♡ Use `sklearn` in Python to train a soft-margin SVM to be applied to the data matrix

$$X = \begin{pmatrix} 0 & 0 & 0 \\ 1 & -1 & 1 \\ 0 & 1 & 1 \\ 2 & -2 & 3 \end{pmatrix} \quad \text{and labels } \mathbf{y} = \begin{pmatrix} -1 \\ 1 \\ -1 \\ 1 \end{pmatrix}.$$ What are the values for $\mathbf{w}$ and b? The Python

notebook for this section will be helpful, but keep in mind that, in Python, the labels must be nonnegative, so use $\mathbf{y} = (0, 1, 0, 1)^T$ and adapt your result.

3.4. Use `sklearn` in Python to train a soft-margin SVM on a random data matrix consisting of $m = 100$ data points in dimension $n = 2$, with random binary labels. Use the `numpy.random` package to generate random data and labels. How well does the SVM fit the data, i.e., is the data linearly separable? Try this again with a much higher dimension $n \gg 2$. How high do you need to make the dimension before the randomized data is linearly separable?

3.5. Apply SVM to a classification data set contained in `sklearn.datasets`.

3.6. ◇ Consider the two-point data set from Example 7.9 in the context of the soft-margin SVM problem (7.64). Show that a solution is given by

$$\mathbf{w} = \begin{cases} \mathbf{z}/(2\lambda), & \lambda \geq \tfrac{1}{2}\|\mathbf{z}\|^2, \\ \mathbf{z}/\|\mathbf{z}\|^2, & 0 < \lambda \leq \tfrac{1}{2}\|\mathbf{z}\|^2, \end{cases} \quad \text{and} \quad b = 0.$$

Thus, at least in this example, for sufficiently small $\lambda > 0$ the soft-margin SVM problem gives the same solution as the hard-margin problem (7.62).

3.7. Produce a table like Table 7.18 using k-fold cross validation to produce more reliable measures of model performance.

3.8. ♡ Consider a linearly separable data set, where there exists a solution $(\mathbf{w}_0, b_0)$ of the hard-margin SVM problem (7.62). Let $(\mathbf{w}_\lambda, b_\lambda)$ be a solution of the soft-margin problem (7.64) for $\lambda > 0$. (a) Show that $\|\mathbf{w}_\lambda\| \leq \|\mathbf{w}_0\|$. (b) Show that $y_i(\mathbf{x}_i \cdot \mathbf{w}_\lambda - b_\lambda) + \lambda m \|\mathbf{w}_0\|^2 \geq 1$. Therefore, in the linearly separable case, the soft-margin SVM problem with small λ provides a good approximation to the solution of the hard-margin problem.

3.9. ◇ Let $\mathcal{L}(\mathbf{w}, b)$ be the objective function of the soft-margin SVM with softplus loss (7.65). (a) Show that

$$\nabla_{\mathbf{w}} \mathcal{L}(\mathbf{w}, b) = 2\lambda \mathbf{w} - \frac{1}{m} \sum_{i=1}^{m} \frac{y_i \mathbf{x}_i}{1 + e^{-\beta(1 - y_i(\mathbf{x}_i \cdot \mathbf{w} - b))}},$$

$$\nabla_{b} \mathcal{L}(\mathbf{w}, b) = \frac{1}{m} \sum_{i=1}^{m} \frac{y_i}{1 + e^{-\beta(1 - y_i(\mathbf{x}_i \cdot \mathbf{w} - b))}}.$$

(b) Write Python code to solve the soft-margin SVM (7.65) using gradient descent. Test your program at first on some synthetic data, like the two-point problem given in Example 7.9. Then test your algorithm on pairs of MNIST digits. Try different pairs; which are easiest to separate? You can use the notebook

Python Notebook: Support Vector Machine Homework (.ipynb)

Challenge: Use stochastic gradient descent, covered in Section 11.5.

3.10. Reproduce the results in Figure 7.21(b) using linear kernel SVM from `sklearn` in Python with the feature map $\phi(\mathbf{x}) = \left(x_1, x_2, x_1^2 + x_2^2 \right)^T$.

7.4 k–Nearest Neighbor Classification

Python Notebook: *k-nearest neighbor classification* (.ipynb)

One of the simplest approaches to fully supervised classification is *nearest neighbors*, which classifies data points according to the label of the closest point in the training set. More precisely, given data points $\mathbf{x}_1, \ldots, \mathbf{x}_m \in \mathbb{R}^n$ and corresponding labels $\mathbf{y}_1, \ldots, \mathbf{y}_m \in \mathbb{R}^c$, a nearest neighbor classifier predicts the class label $\mathbf{y}$ corresponding to a new data point $\mathbf{x} \in \mathbb{R}^n$ by taking the label vector $\mathbf{y}_i$ of the closest data point $\mathbf{x}_i$. The measure of closeness is often based on the Euclidean distance $\|\mathbf{x} - \mathbf{x}_i\|$, but, as we discuss below, other notions of distance can be utilized. Figure 7.22 shows an example of how a nearest neighbor classifier works. To classify the black point at the center of the figure, we look for its nearest neighbor, which in this case is a yellow circle, and so we predict that the black point should be a yellow circle.

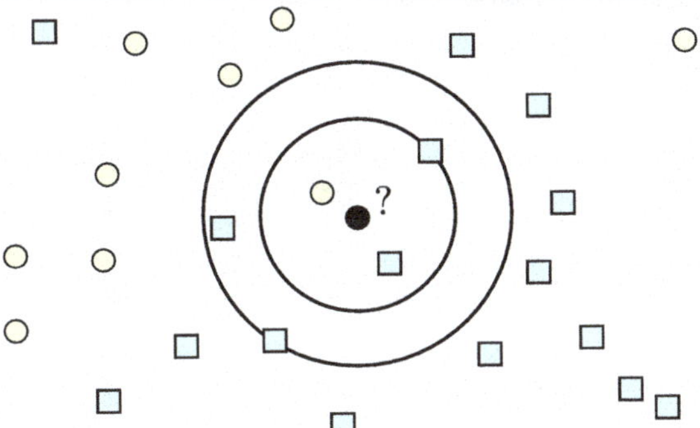

Figure 7.22: Illustration of how a k-nearest neighbor classifier works. To decide on a class prediction for the black point at the center, a 1-nearest neighbor classifier would assign the label of the closest neighbor in the Euclidean distance, that is, it would predict the yellow circle. On the other hand, a majority votes k-nearest neighbor classifier with $k = 3, 4$, or 5 would predict a blue square. The black circles in the figure show the 3-nearest neighbor and 5-nearest neighbor balls, respectively.

However, upon examining Figure 7.22, it would seem more natural to predict the black point to be a blue square, since it lies close to many other blue squares, and perhaps the closest yellow circle is just a noisy label. This illustrates one drawback of a nearest neighbor classifier, namely, it is highly sensitive to noise in the data set. To obtain a more robust classifier, we can use a k-nearest neighbor classifier, which incorporates label information from each of the k nearest data points for some specified $k \geq 1$. A very simple way to do this is with a majority vote, which is simplest when k is odd, as otherwise we may have to break ties. In Figure 7.22, the circles depict the disks containing the $k = 3$ and $k = 5$ nearest neighbors. In each case, the blue squares would win the majority vote and the point would be classified as a blue square.

In Figure 7.23 we show an example of a k-nearest neighbor classifier on a toy data set with some corruption in the labels. The data set contains two classes, one is a cluster near the origin and the other is a ring. The data set has 100 data points and we randomly corrupted 15 of them by flipping them to the opposite class. We can see that the nearest neighbor classifier, $k = 1$, is highly sensitive to such corruptions, while the $k = 10$ nearest neighbor classifier ignores the corrupted data points, since they do not form the majority in the voting algorithm.

A majority vote algorithm certainly has disadvantages, especially when k is large, since the closest points are given the same voting weight as those lying further away. A common way to address this is to assign nonnegative weights $w_1, \ldots, w_k$ to the k nearest neighbors $\mathbf{x}_1, \ldots, \mathbf{x}_k$ of $\mathbf{x}$, and then calculate a weighted average of the neighboring label vectors:

$$\mathbf{y} = \frac{w_1 \mathbf{y}_1 + \cdots + w_k \mathbf{y}_k}{w_1 + \cdots + w_k}. \tag{7.69}$$

The average label vector (7.69) will generally not be a one-hot vector. Thus, to specify the label, we can simply project to the closest one-hot vector by choosing the class corresponding to the largest component of $\mathbf{y}$. This is equivalent to a weighted majority vote algorithm, weighted by w_i. Common choices for the weights includes the *Gaussian weights*

$$w_i = \exp\left(-\frac{\|\mathbf{x} - \mathbf{x}_i\|^2}{2\sigma^2}\right), \tag{7.70}$$

where the parameter σ is usually set as some multiple of the distance to the k-th nearest

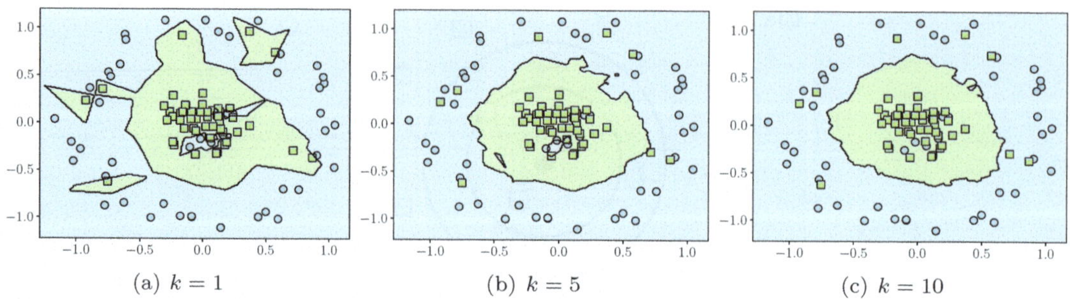

(a) $k = 1$ (b) $k = 5$ (c) $k = 10$

Figure 7.23: Illustration of how a nearest neighbor classifier ($k = 1$) is highly sensitive to noise, and that increasing the number of neighbors k renders the classifier more robust. The data set was generated with two classes, one at the center (squares) and the other a surrounding ring (circles); then 15 points chosen at random were flipped to the opposite class.

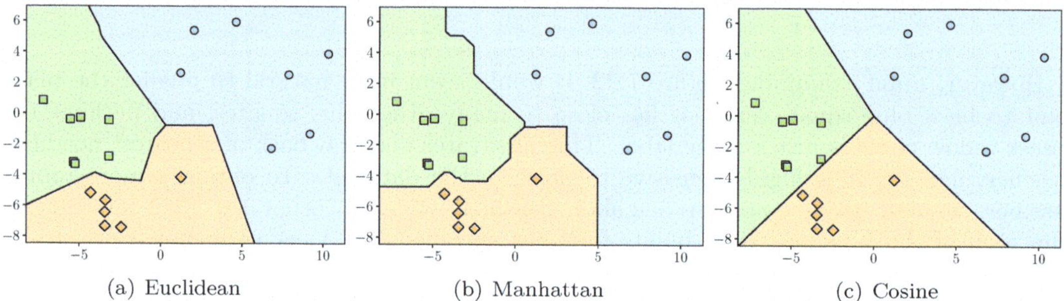

(a) Euclidean   (b) Manhattan   (c) Cosine

Figure 7.24: Illustration of using different notions of distance with a nearest neighbor classifier. In (a) we show the Euclidean distance, in (b) we show the 1 norm, or Manhattan, distance, and in (c) we show the cosine distance. Notice in (c) that the decision boundaries are rays emanating from the origin.

neighbor, and the *inverse distance weight*

$$w_i = \frac{1}{\|\mathbf{x} - \mathbf{x}_i\|^p},$$

(7.71)

where the exponent $p > 0$ is a hyperparameter. Introducing a weighting allows the k-nearest neighbor classifier to be used for regression tasks as well, simply by omitting the final step where we project to the nearest label vector. The resulting *k-nearest neighbor regression algorithm* is one of the simplest methods for performing non-parametric regression.[15]

As we hinted at above, the k-nearest neighbor classifier is not restricted to using the Euclidean distance between points. Indeed, we may use *any* notion of distance on $\mathbb{R}^n$, or even quantities that do not strictly represent distances. One common alternative to the Euclidean norm is the 1 norm producing the *Manhattan* or *city block distance*; another is the *cosine distance* (2.76) that we introduced in Section 2.7, which depends only on the angle between the vectors. Figure 7.24 shows an example of using nearest neighbor classifiers with different notions of distance to classify a toy data set. For the cosine distance, we see that the decision boundaries are all rays that start at the origin.

---

[15] *Non-parametric* refers to methods that do not learn a general function involving parameters.

Training data size	0.1%	1%	10%	20%	40%	85.7%
Training accuracy: Euclidean (%)	61.43	87.57	94.91	96.21	96.86	97.59
Testing accuracy: Euclidean (%)	44.28	84.91	93.81	95.51	96.40	97.37
Training accuracy: Cosine (%)	72.86	89.71	95.63	96.79	97.32	97.95
Testing accuracy: Cosine (%)	52.45	87.71	95.27	96.43	97.06	97.75

Table 7.25: Training and testing accuracy of a k-nearest neighbor classifier with $k = 9$ on MNIST using different amounts of training data, given as percentages of the full 70,000 MNIST images. We always use the same 10,000 testing images. We notice there is very little overfitting once the training set size is at least 1%, and that the cosine similarity generally outperforms the Euclidean distance by a small amount.

Digit	0	1	2	3	4	5	6	7	8	9
0	975	1	0	0	0	0	3	1	0	0
1	0	1132	2	1	0	0	0	0	0	0
2	11	1	1004	1	1	0	0	11	3	0
3	2	0	3	987	1	2	0	6	5	4
4	2	4	0	0	945	0	6	0	1	24
5	4	1	0	7	1	864	7	1	4	3
6	4	3	0	0	1	0	950	0	0	0
7	1	12	5	0	0	0	0	996	0	14
8	5	2	2	8	1	2	2	3	946	3
9	4	5	1	5	4	1	1	7	5	976

Table 7.26: Confusion matrix for classification of MNIST digits using a k-nearest neighbor classifier ($k = 9$) with cosine distance with $60,000$ training examples. Some commonly confused pairs of digits include $(4, 9)$, $(7, 9)$, $(7, 1)$, and $(7, 2)$.

We now investigate using k-nearest neighbors classification on the MNIST data set. Table 7.25 shows the results of a k-nearest neighbor classifier with $k = 9$ with various amounts of labeled training data using both Euclidean and cosine distances. We see that the accuracy is worse at low label rates, but gives good performance when provided access to an abundance of labeled data, in fact, performing slightly better than the SVM results in Table 7.18. We see very little overfitting, especially at lower label rates, and slightly better performance when using the cosine distance. In Table 7.26, we display the confusion matrix for the cosine distance with $60,000$ training data points.

Finally, we remark that, in contrast to SVM and linear regression, a k-nearest neighbor classifier does not require kernel methods to learn nonlinear decision boundaries, provided there is a sufficient amount of training data. In Figure 7.27 we show a toy example on the two moons data set with different amounts of training data. The training data points are given shown in red in the figure. When the number of training data points is small, they are not sufficiently representative of the testing data and the classifier is unable to identify the underlying nonlinear geometry. As the number of training data points increases, the situation improves, and the decision boundary is better resolved, although requiring a relatively large amount of training data to do so.

Of course, the k-nearest neighbor classifier is predicated on the notion that there is a measure of distance, be it Euclidean, Manhattan, cosine, or some other, that correlates with class membership; in the absence of any such notion, more sophisticated techniques are required, such as the kernel methods discussed in Section 7.6.

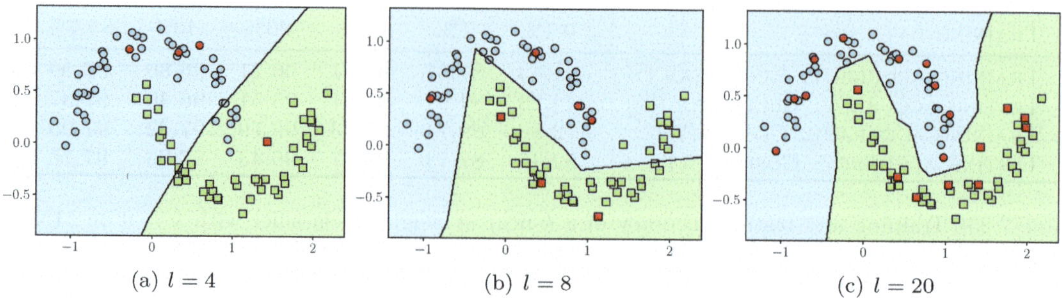

(a) $l = 4$ (b) $l = 8$ (c) $l = 20$

Figure 7.27: Example of a nearest neighbor classifier on the toy two moons data set with $l = 4, 8$, and 20 training data points, out of a total of $m = 100$ data points. The training data points are depicted in red, and the remaining points are considered testing data points.

7.4.1 Computational Aspects

The previous examples of linear regression and SVM in this chapter work by learning a parametrized function F that can be applied to new data points. Machine learning algorithms that work this way are called *parametric methods*. In contrast, the k-nearest neighbor classifier is *non-parametric*, since it learns no such parameterized function, and must retain the entire training data for use in testing. In particular, it does not do much work at training time, and leaves most of the computations for the testing phase, where the model is used to predict labels at previously unseen data points.[16] For this reason, it is sometimes called a *lazy classifier*.

Training a k-nearest neighbor classifier does not require optimization, as was the case for linear regression and SVM. The main computational costs occur in the testing phase, when we need to find the k nearest neighbors of a given data point $\mathbf{x}$. If we have m training data points $\mathbf{x}_1, \ldots, \mathbf{x}_m$, then a brute force search would involve computing all m distances $\| \mathbf{x} - \mathbf{x}_i \|$, which will take on the order of $O(m)$ operations for vectors in $\mathbb{R}^n$ (not counting operations for keeping track of the k smallest distances). This can be computationally intractable when the size m of the training set is large. One approach to accelerate computations is to subsample the training set to produce a much smaller set that is still representative of the training data, e.g., using k-means, which will be introduced in the following section.

In some situations there exist faster algorithms for querying nearest neighbors that are based on more sophisticated data structures. For data in a relatively low dimension, the k-d tree [80] data structure is widely used for efficiently querying nearest neighbors, and can return the nearest neighbor of a given data point $\mathbf{x}$ in $O(\log m)$ computational time, on average, which is a substantial improvement over $O(m)$. However, when dealing with high dimensional data, a k-d tree becomes rather inefficient because the constant inside the $O(\log m)$ complexity hides the dependence on dimension, and the resulting algorithm is no better than a brute force search [151]. In the high dimensional setting, there are various fast approximate nearest neighbor algorithms, which may not necessarily find the closest neighboring point, but include certain guarantees on how large its mistakes can be. We refer the interested reader to [7, 151] for more details.

---

[16] A k-nearest neighbor classifier may do some pre-computation in the training phase, such as the construction of certain data structures over the training set in order to make nearest neighbor queries efficient at test time.

Exercises

4.1. ♡ Let $x_1, \ldots, x_m \in \mathbb{R}$ be a collection of one dimensional data points. Show that we can compute the nearest neighbor of every point x_i in the data set in $O(m \log m)$ operations, compared to the $O(m^2)$ operations it would take to compare distances between every pair of data points $|x_i - x_j|$.

Hint: Recall that the computational complexity of sorting m numbers is $O(m \log m)$ [47].

4.2. ◇ Let $\mathbf{x}_1, \ldots, \mathbf{x}_m \in \mathbb{R}^2$. Use the ideas in Exercise 4.1 to construct an algorithm for finding the nearest neighbor of each $\mathbf{x}_i$ in the cosine distance in $O(m \log m)$ computational time.

4.3. Using Python and `sklearn`, apply the k-nearest neighbor classifier to one of the classification data sets in `sklearn.datasets`. Investigate what happens when you use a variety of distances and numbers of neighbors.

4.4. Repeat Exercise 4.3 for k-nearest neighbor regression, using a regression data set, such as the diabetes data set.

4.5. ♡ Let $\mathbf{x}_1, \ldots, \mathbf{x}_m \in \mathbb{R}^n$ and $\mathbf{y}_1, \ldots, \mathbf{y}_m \in \mathbb{R}^c$ denote the training data for a k-nearest neighbor classifier. Define

$$G(\mathbf{x}) = \begin{cases} 1, & \|\mathbf{x}\| \leq 1, \\ 0, & \text{otherwise.} \end{cases}$$

Show that there exists a function $H : \mathbb{R}^n \to \mathbb{R}$, which may depend on the training data, such that the classification decision of a uniformly weighted k-nearest neighbor classifier using the norm $\|\cdot\|$ can be deduced from the function

$$F(\mathbf{x}) = \sum_{i=1}^{m} G\left(\frac{\mathbf{x} - \mathbf{x}_i}{H(\mathbf{x})}\right) \mathbf{y}_i. \tag{7.72}$$

How does the classification decision relate to $F(\mathbf{x})$? You can assume the k-th nearest neighbor of $\mathbf{x}$ is unique, so no ties have to be broken.

4.6. ◇ Let $\Omega \subset \mathbb{R}^n$ and let $F : \Omega \to \mathbb{R}$ be Lipschitz continuous with Lipschitz contant $\mathrm{Lip}(F)$, and assume F is the underlying ground truth for a regression problem. Let $\mathbf{x}_1, \ldots, \mathbf{x}_m \in \Omega$ be training examples and let

$$\varepsilon = \max_{x \in \Omega} \min_{1 \leq i \leq m} \|\mathbf{x} - \mathbf{x}_i\|.$$

The value $\varepsilon > 0$ measures how well the training set covers Ω, with smaller values providing better coverage. Show that a nearest neighbor regression algorithm will make errors no larger than $\mathrm{Lip}(F)\,\varepsilon$.

7.5 k–Means Clustering

Python Notebook: *k*-means clustering (.ipynb)

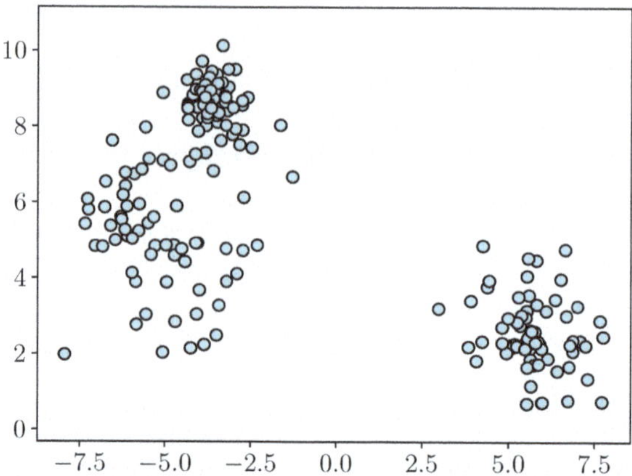

Figure 7.28: An example of a point cloud that has three clusters, one of which is substantially separated from the other two.

We now turn to the problem of clustering, or grouping, unlabeled data, which is one of the most important unsupervised machine learning problems. Figure 7.28 shows a sample data set consisting of 200 points that appear to belong to three distinct clusters. Two of the clusters are fairly close to each other, whereas the third is relatively isolated. The goal of clustering is to separate the points in Figure 7.28 into these three natural clusters. In general, when working with real data and not synthetic examples, it is difficult to visualize the "natural clusters", and it can be difficult to define what constitutes a good or bad clustering, since there are various natural ways to group data, or even to decide how many clusters are present.

In this section we will introduce and study a relatively simple and widely used algorithm known as *k-means clustering*, which works well in certain contexts, and is the foundation for more sophisticated clustering techniques, like spectral clustering, which is introduced in Chapter 9. The k-means clustering algorithm aims to find a single good representative point from each of k clusters. The data set is then clustered into k groups by assigning each data point to the cluster corresponding to the closest such representative point, as measured, usually, by the Euclidean distance. The hyperparameter k is specified in advance, although one can subsequently compare clusterings using several different values of k.

To describe the setting mathematically, let $\mathbf{x}_1, \mathbf{x}_2, \ldots, \mathbf{x}_m$ be a data set consisting of m points in $\mathbb{R}^n$. Let $\mathbf{c}_1, \mathbf{c}_2, \ldots, \mathbf{c}_k \in \mathbb{R}^n$ be the cluster centers, also called the cluster "means", hence the name of the algorithm, which are to be determined. Once the means are specified, the individual clusters consist of the data points that lie closer to a given mean than to any other. In other words, the j-th cluster is

$$C_j = \left\{ \mathbf{x}_i \,\middle|\, \|\mathbf{c}_j - \mathbf{x}_i\| = \min_{1 \le \ell \le k} \|\mathbf{c}_\ell - \mathbf{x}_i\| \right\}, \qquad j = 1, \ldots, k, \qquad (7.73)$$

where $\|\mathbf{c}_j - \mathbf{x}_i\|$ denotes the Euclidean distance between the cluster center $\mathbf{c}_j$ and the data point $\mathbf{x}_i$. We note that a point $\mathbf{x}_i$ may be equally close to more than one cluster center, and in this case we can make any reasonable choice of which cluster to assign it to. In this section, for specificity, we will choose the cluster whose index j is smallest to break ties. We note that it is certainly possible that one or more of the clusters C_j are empty, depending on how the means $\mathbf{c}_1, \ldots, \mathbf{c}_k$ are chosen.

In k-means clustering, the choice of the cluster centers $\mathbf{c}_1, \ldots, \mathbf{c}_k$ is guided by the task of minimizing the *k-means clustering energy*

$$E_{\mathrm{km}}(\mathbf{c}_1, \mathbf{c}_2, \ldots, \mathbf{c}_k) = \sum_{i=1}^{m} \min_{1 \le j \le k} \| \mathbf{c}_j - \mathbf{x}_i \|^2. \tag{7.74}$$

The k-means clustering energy measures how well the cluster centers $\mathbf{c}_1, \ldots, \mathbf{c}_k$ represent the data set in the squared Euclidean distance. Using our definition of clusters given in (7.73), we note that the k-means clustering energy can also be written as

$$E_{\mathrm{km}}(\mathbf{c}_1, \mathbf{c}_2, \ldots, \mathbf{c}_k) = \sum_{j=1}^{k} \sum_{\mathbf{x} \in C_j} \| \mathbf{c}_j - \mathbf{x} \|^2, \tag{7.75}$$

so we are merely summing the squared Euclidean distances from each point to its assigned cluster representative.

It turns out that minimizing the k-means clustering energy E_{km} over the choice of cluster centers is a very hard computational problem; indeed it has been shown to be NP-hard [3]. Furthermore it is not straightforward to apply gradient descent, since the min operation in the k-means clustering energy (7.74) is not differentiable, and not easily regularized as we did for lasso in Section 7.2 and soft-margin SVM in Section 7.3. However, there is a simple algorithm that monotonically decreases the clustering energy, is provably convergent, and often gives good clustering results even though it may not minimize E_{km}. The algorithm is usually called the *k-means algorithm*; it is also often called *Lloyd's algorithm*, named after Stuart P. Lloyd, who invented the algorithm in 1957, although he did not publish it until 1982, [152]. The steps of the k-means algorithm are outlined below.

The k-means Algorithm:

Choose, at random, *distinct* initial values for the cluster centers $\mathbf{c}_1^0, \mathbf{c}_2^0, \ldots, \mathbf{c}_k^0$, that are selected from the data points $\mathbf{x}_1, \ldots, \mathbf{x}_m$. Then iterate the steps below, for $t = 0, 1, 2, 3, \ldots$, until convergence.

1. Update the clusters:

$$C_j^t = \left\{ \mathbf{x}_i \ \middle| \ \| \mathbf{c}_j^t - \mathbf{x}_i \| = \min_{1 \le \ell \le k} \| \mathbf{c}_\ell^t - \mathbf{x}_i \| \right\}, \qquad j = 1, \ldots, k. \tag{7.76}$$

2. Update the cluster centers:

$$\mathbf{c}_j^{t+1} = \frac{1}{\# C_j^t} \sum_{\mathbf{x} \in C_j^t} \mathbf{x}, \qquad j = 1, \ldots, k. \tag{7.77}$$

Here, $\# C_j^t$ denotes the number of points in the jth cluster C_j^t at step t of the algorithm, and hence $\mathbf{c}_j^{t+1}$ is exactly the mean of the jth cluster. By definition, the k-means algorithm *converges* when the clusters (and hence the cluster means) do not change from one iteration to the next, that is $C_j^t = C_j^{t+1}$ for all $j = 1, \ldots, k$.

The k-means algorithm generates a sequence of clusterings $C_j^0, C_j^1, C_j^2, \ldots$ and cluster means $\mathbf{c}_j^0, \mathbf{c}_j^1, \mathbf{c}_j^2, \ldots$, for each $j = 1, \ldots, k$, that get progressively better in the sense that the k-means clustering energy (7.74) is decreasing. The two steps of the k-means algorithm view clustering from different perspectives, with the first step considering the cluster membership of each point, and the second step considering the locations of the cluster centers. In fact, as we will show below, each step of the k-means algorithm is optimal from its own perspective, in terms of minimizing E_{km}. That is, if the cluster centers are fixed, then Step 1 chooses the

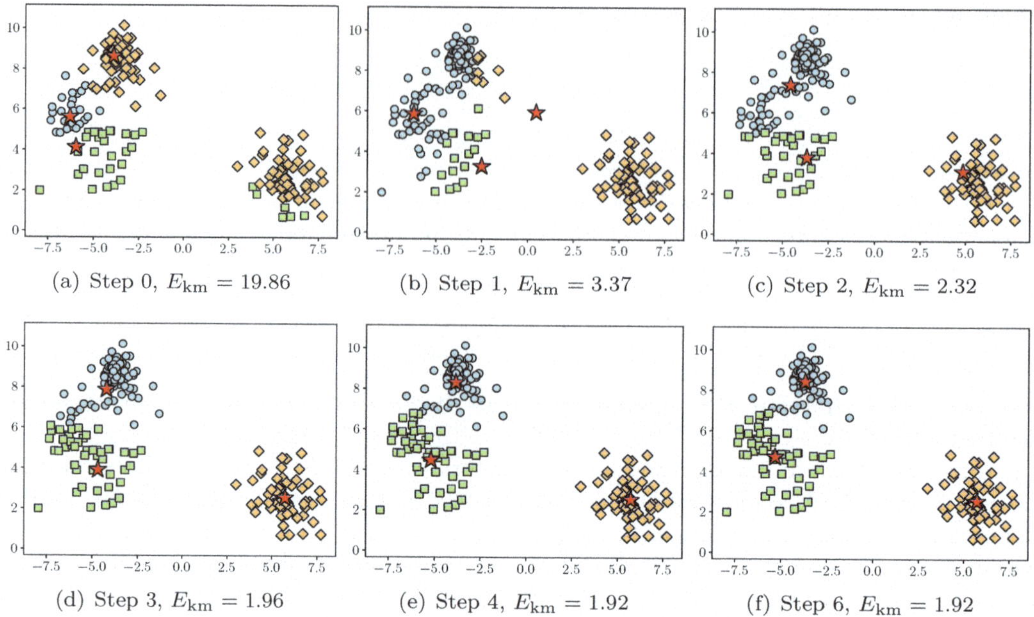

(a) Step 0, $E_{\mathrm{km}} = 19.86$ (b) Step 1, $E_{\mathrm{km}} = 3.37$ (c) Step 2, $E_{\mathrm{km}} = 2.32$

(d) Step 3, $E_{\mathrm{km}} = 1.96$ (e) Step 4, $E_{\mathrm{km}} = 1.92$ (f) Step 6, $E_{\mathrm{km}} = 1.92$

Figure 7.29: An illustration of the intermediate steps in the k-means clustering algorithm and the corresponding values of the k-means energy E_{km}. The red stars are the cluster centers. The algorithm converged in 6 steps, but steps 3–6 show very little change in the clustering.

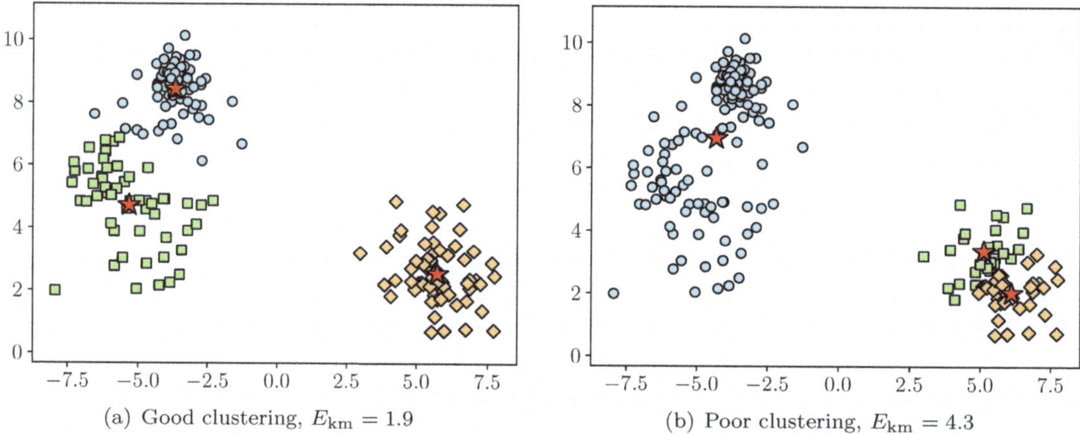

(a) Good clustering, $E_{\mathrm{km}} = 1.9$ (b) Poor clustering, $E_{\mathrm{km}} = 4.3$

Figure 7.30: Examples of good and poor clustering results obtained by the k-means algorithm using different initial conditions.

optimal cluster assignment, while if the clusters are fixed, then Step 2 chooses the optimal cluster centers. As we will prove in this section, before termination (convergence), the k-means algorithm monotonically decreases the k-means energy, and is guaranteed to eventually converge in a finite number of steps. In practice, it usually converges quite rapidly, although a mathematical analysis of the speed of convergence is beyond the scope of this book; we refer the interested reader to [29] for details. We also mention that in dimension $n = 1$, there is an efficient and globally optimal algorithm for 2-means clustering; see Exercise 5.10.

We show in Figure 7.29 an illustration of some of the intermediate steps in applying k-means clustering with $k = 3$, i.e., the 3-means clustering algorithm, to the point cloud from Figure 7.28. In the figure captions we also show the values of the k-means clustering energy E_{km}, which decrease with each iteration, with the most substantial decreases occurring early on. The algorithm converged in 6 steps to a good clustering, although this result depends on the randomized initial condition. For some initializations the algorithm converged in fewer iterations, sometimes as few as three, while for other initializations it took longer. The final clustering can also depend on the choice of initial condition. We show in Figure 7.30 an example of good and poor clusterings obtained by 3-means clustering of the same point cloud. In practice, one can run the k-means algorithm many times from different random initializations, and choose the clustering that results in the smallest value for the k-means clustering energy. In Figure 7.30, the energy of the poor clustering is more than double the energy of the good one. Alternatively, there are ways to initialize the clusters centers in the k-means algorithm that can provably give better results than random initialization. One such technique that is widely used is called k-means++ [6].

The careful reader may have noticed that we haven't fully specified the k-means algorithm, since we have not decided what to do if any one of the clusters becomes empty after an iteration. This rarely happens, especially when there are far fewer clusters than data points, i.e., $k \ll m$, but it remains a possibility.

Example 7.10. Consider the one dimensional data set consisting of the six data points $x_1 = 0$, $x_2 = 2$, $x_3 = 3$, $x_4 = 16$, $x_5 = 18$, $x_6 = 30$, and assume the initial cluster centers for the 3-means clustering algorithm are $c_1^0 = x_1$, $c_2^0 = x_3$, $c_3^0 = x_6$. The initial clusters are then

$$C_1^0 = \{x_1\} = \{0\}, \qquad C_2^0 = \{x_2, x_3, x_4\} = \{2, 3, 16\}, \qquad C_3^0 = \{x_5, x_6\} = \{18, 30\}.$$

The cluster centers after one iteration are

$$c_1^1 = 0, \qquad c_2^1 = \tfrac{1}{3}(2 + 3 + 16) = 7 \qquad c_3^1 = \tfrac{1}{2}(18 + 30) = 24,$$

and the updated clusters are given by

$$C_1^1 = \{x_1, x_2, x_3\} = \{0, 2, 3\}, \qquad C_2^1 = \varnothing, \qquad C_3^1 = \{x_4, x_5, x_6\} = \{16, 18, 30\}. \qquad \blacktriangle$$

When a cluster becomes empty during the k-means algorithm, the update formula (7.77) is invalid and we have to decide how to proceed. Before addressing this, we show that empty clusters can only occur when $k \geq 3$.

Lemma 7.11. *In the $k = 2$-means algorithm, for all $t \geq 0$ the cluster centers $\mathbf{c}_1^t$ and $\mathbf{c}_2^t$ are distinct, that is $\mathbf{c}_1^t \neq \mathbf{c}_2^t$, and the clusters C_1^t and C_2^t are nonempty.*

Proof. We will prove the lemma by induction, the initial case when $t = 0$ being automatically true by our initialization of $\mathbf{c}_1^0$ and $\mathbf{c}_2^0$ being chosen as two different points from our data set $\mathbf{x}_1, \ldots, \mathbf{x}_m$, which in particular, means that neither cluster C_1^0 nor C_2^0 can be empty, since they contain $\mathbf{c}_1^0$ and $\mathbf{c}_2^0$, respectively.

The proof is based on the fact that in the 2-means algorithm, whenever $\mathbf{c}_1^t$ and $\mathbf{c}_2^t$ are distinct, the set of points in $\mathbb{R}^n$ that are equidistant from $\mathbf{c}_1^t$ and $\mathbf{c}_2^t$ forms a hyperplane, i.e., an affine subspace of dimension $n - 1$; see Exercise 5.3. Thus, by setting

$$\mathbf{w}^t = \mathbf{c}_2^t - \mathbf{c}_1^t, \qquad b^t = \frac{\|\mathbf{c}_2^t\|^2 - \|\mathbf{c}_1^t\|^2}{2}, \qquad \text{we have} \qquad \begin{aligned} C_1^t &= \{\, i \mid \mathbf{x}_i \cdot \mathbf{w}^t \leq b^t \,\}, \\ C_2^t &= \{\, i \mid \mathbf{x}_i \cdot \mathbf{w}^t > b^t \,\}. \end{aligned}$$

In other words, the 2-means clustering at each step can be viewed as the simple two-point SVM problem, as we discussed in Example 7.9.

Now, assume, by way of induction, that $\mathbf{c}_1^t \neq \mathbf{c}_2^t$ and that both C_1^t and C_2^t are nonempty. By the definition of $\mathbf{c}_1^{t+1}$ and $\mathbf{c}_2^{t+1}$ in the k-means algorithm we have

$$\mathbf{c}_1^{t+1} \cdot \mathbf{w}^t = \frac{1}{\#C_1^t} \sum_{\mathbf{x} \in C_1^t} \mathbf{x} \cdot \mathbf{w}^t \leq \frac{1}{\#C_1^t} \sum_{\mathbf{x} \in C_1^t} b^t \leq b^t. \tag{7.78}$$

A similar computation shows that $\mathbf{c}_2^{t+1} \cdot \mathbf{w}^t > b^t$, and hence $\mathbf{c}_1^{t+1} \neq \mathbf{c}_2^{t+1}$.

To complete the proof we just need to show that C_1^{t+1} and C_2^{t+1} are nonempty. Assume, by way of contradiction, that C_2^{t+1} is empty, which means that $\mathbf{x}_i \cdot \mathbf{w}^{t+1} \leq b^{t+1}$ for all $i = 1, \ldots, m$. But then, by a computation similar to (7.78), we would have that $\mathbf{c}_2^{t+1} \cdot \mathbf{w}^{t+1} \leq b^{t+1}$. This contradicts the fact that $\mathbf{c}_2^{t+1} \cdot \mathbf{w}^{t+1} > b^{t+1}$, which holds by definition of $\mathbf{w}^{t+1}$ and b^{t+1}. Hence, C_2^{t+1} is nonempty. A similar argument can be applied to C_1^{t+1}; see Exercise 5.6. ∎

We now return to the issue of what to do in the k-means algorithm when a cluster becomes empty. One option is to remove the cluster and proceed with the $(k-1)$-means algorithm. By Lemma 7.11 we know that this process will never reduce to the trivial case of $k = 1$, so this is a reasonable approach. However, it is possible that a cluster that becomes empty at one iteration can become nonempty again in a future iteration; see Exercise 5.7. In other words, a cluster becoming empty during k-means does not necessarily mean that the data set has fewer clusters and that k should be reduced. Thus, another natural approach to handle empty clusters is to set $\mathbf{c}_j^{t+1} = \mathbf{c}_j^t$ and continue the k-means iteration with an empty cluster, to allow for the possibility that it will become nonempty in the future. This is the approach we will take in our analysis. If the cluster remains empty at convergence of the k-means algorithm, then it can be removed at that point. Other approaches are commonly taken in practice, such as re-initializing any empty cluster centers $\mathbf{c}_j^t$ either at random, or by choosing a data point far from the largest clusters, but analyzing these approaches adds additional technicalities that lie outside the scope of this book.

While the k–means algorithm often gives good results, it can converge to a local minimizer of the k-means clustering energy (7.74), and thus does not necessarily find a global minimizer. Nevertheless, we can prove that the k-means algorithm always reduces the energy E_{km} and converges in a finite number of iterations. The proof requires a preliminary lemma, which shows that the mean vector minimizes the sum of squared distances to the cluster center.

Lemma 7.12. *Let $\mathbf{x}_1, \ldots, \mathbf{x}_m \in \mathbb{R}^n$. The unique global minimizer of the quadratic function*

$$Q(\mathbf{y}) = \sum_{i=1}^{m} \|\mathbf{y} - \mathbf{x}_i\|^2 \quad \text{is the mean vector} \quad \mathbf{y} = \mathbf{c} = \frac{1}{m} \sum_{i=1}^{m} \mathbf{x}_i. \tag{7.79}$$

Proof. The result follows from our general minimization Theorem 6.7 for quadratic functions, but for completeness we give an easy direct proof here. It suffices to write

$$Q(\mathbf{y}) = \sum_{i=1}^{m} \left(\|\mathbf{y}\|^2 - 2\mathbf{y}^T \mathbf{x}_i + \|\mathbf{x}_i\|^2 \right)$$

$$= m\left(\|\mathbf{y}\|^2 - 2\mathbf{y}^T \mathbf{c} \right) + \sum_{i=1}^{m} \|\mathbf{x}_i\|^2 = m\|\mathbf{y} - \mathbf{c}\|^2 + \left(\sum_{i=1}^{m} \|\mathbf{x}_i\|^2 - m\|\mathbf{c}\|^2 \right).$$

Note that the final term in parentheses is independent of $\mathbf{y}$, and so $Q(\mathbf{y})$ is clearly minimized when the initial term vanishes, whence $\mathbf{y} = \mathbf{c}$. ∎

We can now prove convergence of the k-means algorithm.

Theorem 7.13. *The k-means algorithm descends on the energy (7.74); that is*

$$E_{\mathrm{km}}(\mathbf{c}_1^{t+1}, \mathbf{c}_2^{t+1}, \ldots, \mathbf{c}_k^{t+1}) \leq E_{\mathrm{km}}(\mathbf{c}_1^t, \mathbf{c}_2^t, \ldots, \mathbf{c}_k^t). \tag{7.80}$$

Furthermore, we have equality in (7.80) if and only if $\mathbf{c}_j^{t+1} = \mathbf{c}_j^t$ for all $j = 1, \ldots, k$. We conclude that the k-means algorithm converges in a finite number of iterations, meaning that the cluster centers no longer change.

Proof. The proof is based on re-writing the k-means energy as a sum over the disjoint clusters, as we did in (7.75):

$$E_{\mathrm{km}}(\mathbf{c}_1^t, \ldots, \mathbf{c}_k^t) = \sum_{j=1}^{k} \sum_{\mathbf{x} \in C_j^t} \| \mathbf{c}_j^t - \mathbf{x} \|^2,$$

where some of the clusters C_j^t may be empty, so the corresponding sums have no terms. If the cluster C_j^t is not empty, $\mathbf{c}_j^{t+1}$ is its mean, and so Lemma 7.12 implies that

$$\sum_{\mathbf{x} \in C_j^t} \| \mathbf{c}_j^{t+1} - \mathbf{x} \|^2 \leq \sum_{\mathbf{x} \in C_j^t} \| \mathbf{c}_j^t - \mathbf{x} \|^2$$

with equality if and only if $\mathbf{c}_j^{t+1} = \mathbf{c}_j^t$. Empty clusters do not contribute to the sum, and we recall our rule that $\mathbf{c}_j^{t+1} = \mathbf{c}_j^t$ if C_j^t is empty. Therefore,

$$E_{\mathrm{km}}(\mathbf{c}_1^t, \ldots, \mathbf{c}_k^t) = \sum_{j=1}^{k} \sum_{\mathbf{x} \in C_j^t} \| \mathbf{c}_j^t - \mathbf{x} \|^2 \geq \sum_{j=1}^{k} \sum_{\mathbf{x} \in C_j^t} \| \mathbf{c}_j^{t+1} - \mathbf{x} \|^2$$

with equality if and only if $\mathbf{c}_j^{t+1} = \mathbf{c}_j^t$ for $j = 1, \ldots, k$. Finally, we note that

$$\sum_{j=1}^{k} \sum_{\mathbf{x} \in C_j^t} \| \mathbf{c}_j^{t+1} - \mathbf{x} \|^2 \geq \sum_{j=1}^{k} \sum_{\mathbf{x} \in C_j^t} \min_{1 \leq \ell \leq k} \| \mathbf{c}_\ell^{t+1} - \mathbf{x} \|^2 = E_{\mathrm{km}}(\mathbf{c}_1^{t+1}, \ldots, \mathbf{c}_k^{t+1}),$$

with equality if $\mathbf{c}_j^{t+1} = \mathbf{c}_j^t$ for $j = 1, \ldots, k$, which establishes (7.80).

We now show that this implies convergence of the k-means algorithm, meaning that eventually, after a finite number of steps, the cluster centers do not change. Note that if $\mathbf{c}_j^{t+1} \neq \mathbf{c}_j^t$ for some j, and so the algorithm has not converged, then, as we proved above, the energy is strictly decreasing:

$$E_{\mathrm{km}}(\mathbf{c}_1^{t+1}, \ldots, \mathbf{c}_k^{t+1}) < E_{\mathrm{km}}(\mathbf{c}_1^t, \ldots, \mathbf{c}_k^t).$$

This implies that, prior to convergence, we can never revisit the same clustering $C_1^t, C_2^t, \ldots, C_k^t$ at any step, because the k-means energy associated with any subsequent clustering must be strictly less than the current energy. Since there are only a finite number of possible ways to cluster the data set into k groups, and the k-means algorithm cannot revisit any given clustering, it must eventually converge in a finite number of iterations. ∎

Remark 7.14. The convergence proof in Theorem 7.13 is non-quantitative, meaning it does not say anything about how many iterations the k-means algorithm may take to converge. In practice, the algorithm tends to converge very quickly, in only a handful of iterations, and

DIGIT	1	2	3	4	5	6	7	8	9
0	99.1	95.0	95.0	96.5	86.8	93.0	96.9	94.5	95.6
1		92.7	96.1	97.9	91.1	96.4	95.8	94.3	97.2
2			89.7	95.7	94.6	93.8	95.6	91.3	95.7
3				97.3	66.3	97.6	96.4	80.0	94.0
4					88.2	95.6	95.2	95.7	52.7
5						91.4	87.2	52.3	55.7
6							99.1	96.6	98.9
7								95.9	60.5
8									92.7

Table 7.31: Accuracy for binary (2-means) clustering of pairs of MNIST digits. We see most pairs of digits are easy to separate, while a few pairs, such as (4,9), (5,9), (7,9), and (5,8) are more difficult.

there is some theoretical work explaining this phenomenon [29], but it is possible for it to take substantially longer. Indeed, in the worst case, our proof of Theorem 7.13 indicates the algorithm may visit every possible clustering before converging. Even for the 2-means problem with m points, there are $2^{m-1} - 1$ possible ways to cluster the data (not allowing empty clusters), so the search space and consequential potential convergence time are exponentially large.

We also recall that convergence of the k-means algorithm simply means that the cluster centers stop changing from one iteration to the next. This does not mean the algorithm has converged to a bona fide minimizer of the k-means energy (7.74), and in general the algorithm does not find the global minimizer. As we saw in Figure 7.30, due to the random choice of initialization, the algorithm may converge to different clusterings each time it is executed. There are good sophisticated randomized initialization strategies that can be used to guarantee, with high probability, that poor results will be avoided; see [6] for details. ▲

Experiments With MNIST

We now consider a brief application of k-means clustering to real data. Again, we use the MNIST data set of handwritten digits, and we evaluate the 2-means algorithm for clustering pairs of MNIST digits. We consider all pairs of MNIST digits, which is 45 binary clustering problems, each with around 14000 data points in dimension $\mathbb{R}^{784} = \mathbb{R}^{28 \times 28}$. The k-means algorithm converged very quickly, in around 15 iterations taking around 1 second per clustering problem. Table 7.31 shows the clustering accuracy obtained by the 2-means algorithm for each pair of MNIST digits. The numbers can vary depending on the choice of initial condition. We can see that many pairs of digits are very easy to cluster into the correct classes with the 2-means algorithm, while a handful of pairs of digits, such as (4,9), (5,8), (5,9), (7,9), are more challenging.

It is also natural to run the 10-means algorithm on the whole MNIST data set. When evaluating clustering performance with more than two clusters, accuracy is not a useful metric to use, since clustering and classification are generally different tasks. Each class in a data set, i.e., each digit in MNIST, may in fact consist of several different clusters, and k-means may split a digit into two or more clusters (e.g., based on differences in how the digit is written) while grouping two similar digits into one cluster. A more useful notion of clustering performance is clustering purity. The *purity* of each found cluster is the largest number of data points in that cluster belonging to the same class. The *clustering purity* is the average

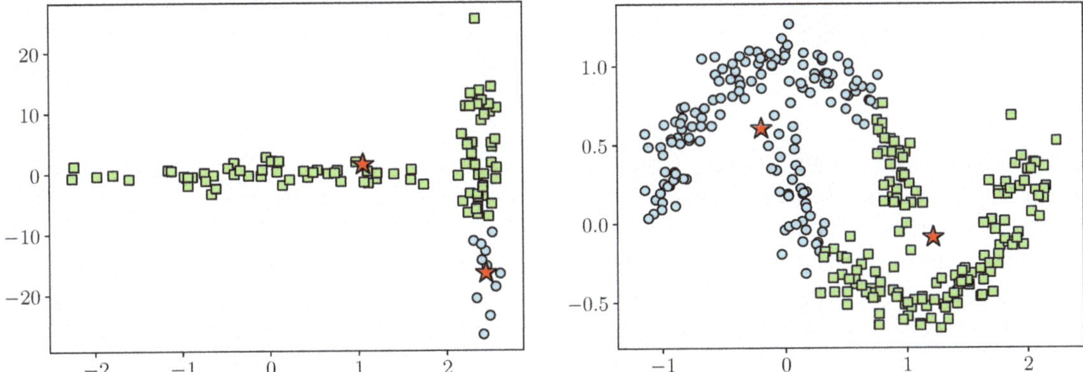

Figure 7.32: Two examples of data sets that are difficult to cluster properly with k–means. On the left, the clusters are on different scales and on the right the clusters have nonconvex shapes so that a single data point is not a good Euclidean representative of the cluster.

purity of all clusters. Mathematically, suppose we have a clustering of a data set into k clusters $C_1, \ldots, C_k$ and suppose there are C classes $S_1, \ldots, S_C$. Then

$$\text{Clustering Purity} = \frac{1}{m} \sum_{i=1}^{k} \max_{1 \leq j \leq C} \#(C_i \cap S_j). \tag{7.81}$$

The 10-means clustering algorithm applied to MNIST results in a clustering purity of 58.5%. Since we may expect there are more than 10 clusters in MNIST, we also ran the 14-means algorithm, which produced a clustering with purity 65.7%. In general, as the number of clusters increases, the clustering purity may increase as well, to the point where the clustering purity will be 100% when there are $k = m$ clusters, with one point in each cluster. There are other measures of clustering performance that are also used, including Rand index, F-measure, and homogeneity score; for these, we refer to [81].

Failures and Extensions

The k–means clustering algorithm may perform poorly when dealing with certain, more complicated, cluster geometries. In Figure 7.32, we show two such examples. In the first case, the clusters have different scales, with one on the order of 10 times larger than the other. In the second case, the famous *two-moons* data set, the two clusters have nonconvex geometries that are not well-represented by a single cluster center. We will introduce techniques, in particular, spectral clustering, that can handle more complicated cluster structures in Chapter 9. The success of k-means on pairs of MNIST digits indicates that the cluster structure of MNIST is, to a large degree, particularly simple.

Finally we comment on some extensions of the k-means algorithm, leaving much of the work to the exercises. First, it is common to use other norms than the Euclidean norm in the k-means energy (7.74). Second, the k-means clustering algorithm can be sensitive to outliers, since the squared distance pays more attention to large deviations from the cluster than to small ones. To address this, we can drop the square, and replace the k-means energy (7.74) by the *robust k-means energy*

$$E_{\text{robust}}(\mathbf{c}_1, \mathbf{c}_2, \ldots, \mathbf{c}_k) = \sum_{i=1}^{m} \min_{1 \leq j \leq k} \| \mathbf{x}_i - \mathbf{c}_j \|. \tag{7.82}$$

Again, we must replace the cluster center update step with the minimization of

$$\sum_{\mathbf{x}\in C_j^t} \|\mathbf{x} - \mathbf{c}\|, \tag{7.83}$$

over the choice of $\mathbf{c}$. Any minimizer of (7.83) is called a *geometric median* of the cluster C_j^t. When the 1 norm is used in the robust k-means energy (7.82), the corresponding algorithm is called *k-medians clustering*; see Exercise 5.9.

Finally, another variation of the k-means algorithm that is more robust to outliers is the *k-medoids algorithm*, which restricts the choice of the cluster center to be a data point. That is, one replaces the second step of the k-means algorithm with choosing $\mathbf{c}_j^{t+1} = \mathbf{x}_i$, where $\mathbf{x}_i$ solves

$$\min_{1\le i\le m} \sum_{\mathbf{x}\in C_j^t} \|\mathbf{x} - \mathbf{x}_i\|^2.$$

One advantage is that this choice avoids empty clusters. It is also possible to apply the k-medoids algorithm in a situation where we only have a notion of distance between each pair of data points. That is, we only have access to an $m \times m$ *distance matrix D* with entries $d_{i\ell} = d(\mathbf{x}_i, \mathbf{x}_\ell)$, where d is a distance function (not necessarily based on a norm). In this case, the cluster center step becomes

$$\min_{1\le i\le m} \sum_{\mathbf{x}_\ell\in C_j^t} d(\mathbf{x}_\ell, \mathbf{x}_i)^2,$$

As before, we can also consider robust variants of k-medoids where we minimize the sum of distances, instead of squared distances; see Exercise 5.8.

Exercises

5.1. ♡ Given a data set with m points, prove that there are $2^{m-1} - 1$ possible ways to cluster the data into 2 nonempty clusters. *Remark*: The generalization of this result to k clusters is provided by the Stirling numbers of the second kind, cf. [92].

5.2. Test the k means algorithm on another data set available in the Python packages `sklearn` or `graphlearning`. For example, try the FashionMNIST data set in `graphlearning`, or the Olivetti faces data set from `sklearn.datasets`.

5.3. ◇ Let $\mathbf{c}_1 \ne \mathbf{c}_2 \in \mathbb{R}^n$, and set

$$C_1 = \{\,\mathbf{x}\mid \|\mathbf{x} - \mathbf{c}_1\| \le \|\mathbf{x} - \mathbf{c}_2\|\,\}, \qquad C_2 = \{\,\mathbf{x}\mid \|\mathbf{x} - \mathbf{c}_2\| < \|\mathbf{x} - \mathbf{c}_1\|\,\}.$$

Show that

$$C_1 = \{\,\mathbf{x}\mid \mathbf{x}\cdot\mathbf{w} \le b\,\}, \qquad C_2 = \{\,\mathbf{x}\mid \mathbf{x}\cdot\mathbf{w} > b\,\},$$

where $\mathbf{w} = \mathbf{c}_2 - \mathbf{c}_1$ and $b = \frac{1}{2}\big(\|\mathbf{c}_2\|^2 - \|\mathbf{c}_1\|^2\big)$.

5.4. ♡ Consider Exercise 5.3 in dimension $n = 1$ and assume $c_1 < c_2$. Show that

$$C_1 = \{\,x\mid x \le \tfrac{1}{2}(c_1 + c_2)\,\}, \qquad C_2 = \{\,x\mid x > \tfrac{1}{2}(c_1 + c_2)\,\}.$$

5.5. Formulate Lloyd's algorithm for the weighted k means objective

$$E_{\mathrm{km}}(\mathbf{c}_1, \mathbf{c}_2, \ldots, \mathbf{c}_k) = \sum_{i=1}^{m} w_i \min_{1 \le j \le k} \| \mathbf{c}_j - \mathbf{x}_i \|^2, \tag{7.84}$$

where $w_1, \ldots, w_m > 0$ are positive weights.

5.6. ♡ Complete the proof of Lemma 7.11 by showing that C_1^{t+1} is nonempty.

5.7. Give an example of the 4-means algorithm in $n = 1$ dimension, where a cluster becomes empty on the first iteration, and then nonempty again on the second iteration. *Hint*: Modify Example 7.10 by adding additional data points with very large values, and an additional cluster center.

5.8. Formulate the k-medoids algorithm for a general distance function. In particular, define the k-medoids energy and show that the k-medoids algorithm decreases the energy at each iteration. Are you able to prove convergence as we did for k-means in Theorem 7.13?

5.9. ♡ (Robust k-means clustering) The exercise is focused on the robust k-means algorithm, which is guided by minimizing (7.82). We start with distinct randomized initial values for the means $\mathbf{c}_1^0, \mathbf{c}_2^0, \ldots, \mathbf{c}_k^0$ chosen from the data set, and iterate the steps below until convergence.

(i) Update the clusters as in (7.76).

(ii) Update the cluster centers

$$\mathbf{c}_j^{t+1} \in \operatorname*{argmin}_{\mathbf{c}} \sum_{\mathbf{x} \in C_j^t} \| \mathbf{x} - \mathbf{c} \|. \tag{7.85}$$

(a) Show that the robust k-means algorithm descends on the energy E_{robust}.

(b) The cluster center $\mathbf{c}_j^{t+1}$ does not admit a closed form expression and is sometimes inconvenient to work with in practice. Consider changing the Euclidean norm in (7.82) to the 1 *norm* and redefine E_{robust} as

$$E_{\mathrm{robust}}(\mathbf{c}_1, \mathbf{c}_2, \ldots, \mathbf{c}_k) = \sum_{i=1}^{m} \min_{1 \le j \le k} \| \mathbf{c}_j - \mathbf{x}_i \|_1. \tag{7.86}$$

This is called k-*medians clustering*. Formulate both steps of the k-medians algorithm so that it descends on the k medians clustering energy (7.86). In particular, show that the cluster centers $\mathbf{c}_j^{t+1}$ are the coordinatewise medians of the points $\mathbf{x} \in C_j^t$, which are simple to compute.

(c) Can you think of any reasons why the Euclidean norm would be preferred over the 1 norm in the k-means energy?

(d) *Challenge*: Implement the robust k-medians algorithm in Python.

5.10. ◇ (Optimal clustering in 1D) We consider here the 2-means clustering algorithm in dimension $n = 1$. Let $x_1, x_2, \ldots, x_m \in \mathbb{R}$ and recall the 2-means energy is

$$E(c_1, c_2) = \sum_{i=1}^{m} \min\{ (x_i - c_1)^2, (x_i - c_2)^2 \}.$$

Throughout the question we assume that the x_i are ordered so that $x_1 \le x_2 \le \cdots \le x_m$. For $1 \le j \le m - 1$, we define

$$\mu_j^- = \frac{1}{j} \sum_{i=1}^{j} x_i, \qquad \mu_j^+ = \frac{1}{m-j} \sum_{i=j+1}^{m} x_i, \qquad F_j = \sum_{i=1}^{j} (x_i - \mu_j^-)^2 + \sum_{i=j+1}^{m} (x_i - \mu_j^+)^2.$$

(a) Suppose that the 2-means algorithm converges to cluster centers (c_1, c_2). Show that there exists $1 \leq j \leq m - 1$ such that $E(c_1, c_2) = E(\mu^-(j), \mu^+(j)) = F_j$. Thus, minimizing F_j over $j = 1, \ldots, m - 1$, and setting $c_1 = \mu_{j_*}^-$ and $c_2 = \mu_{j_*}^+$, where j_* is a minimizer of F_j, will give a solution at least as good as the 2-means algorithm. The rest of the exercise will focus on minimizing F_j.

(b) By part (a) we can replace the 2-means problem with minimizing F_j. We will now show how to do this efficiently. In this part, show that

$$F_j = \sum_{i=1}^{m} x_i^2 - j\,(\mu_j^-)^2 - (m - j)\,(\mu_j^+)^2.$$

Thus, minimizing F_j is equivalent to maximizing $G_j = j\,(\mu_j^-)^2 + (m - j)\,(\mu_j^+)^2$.

(c) Show that we can maximize G (i.e., find j_* with $G_j \leq G_{j_*}$ for all j) in $\mathrm{O}(m \log m)$ computations. *Hint*: First show that

$$\mu_{j+1}^- = \frac{j}{j+1}\,\mu_j^- + \frac{x_{j+1}}{j+1}, \qquad \mu_{j+1}^+ = \frac{m - j}{m - j - 1}\,\mu_j^+ - \frac{x_{j+1}}{m - j - 1}.$$

(d) Explain how these formulas allow you to compute $G(1), G(2), \ldots, G(m-1)$ recursively in $\mathrm{O}(m \log m)$ operations, at which point the maximum is found by brute force. *Hint*: Most computations are $\mathrm{O}(m)$, the only thing that takes $\mathrm{O}(m \log m)$ is the initial step of sorting the points in order from smallest to largest.

(e) Implement the method described in the previous four parts in Python, and test it on some synthetic 1D data. For example, try a mixture of two Gaussians with different means.

5.11. The optimal 2-means one dimensional clustering from Exercise 5.10 can be applied to higher dimensional data by projecting the data (randomly) to one dimension. That is, if we have data points $\mathbf{x}_1, \ldots, \mathbf{x}_m \in \mathbb{R}^n$, we pick a random unit vector $\mathbf{v} \in \mathbb{R}^n$, and define the projected data points $y_i = \mathbf{x}_i \cdot \mathbf{v}$ for $i = 1, \ldots, m$. We then apply the optimal algorithm from Exercise 5.10(d) to the projected data points $y_1, \ldots, y_m$. Implement this method in Python and experiment on clustering pairs of MNIST digits. You may have to try several random projections to get a good clustering. How can you measure the quality of the clustering in order to compare over each random projection? This clustering method is called *random projection clustering*; see [98] for more details.

7.6   Kernel Methods

In this section we study *kernel methods* in machine learning, which we earlier encountered in a simplified setting. These provide a robust framework for casting the feature map idea introduced in Sections 7.2.4 and 7.3.3 into a practical machine learning method. Recall that, in the context of regression and SVM, we showed how it can be useful to use a feature map $\phi : \mathbb{R}^n \to \mathbb{R}^d$ to transform the given data points $\mathbf{x}_1, \ldots, \mathbf{x}_m \in \mathbb{R}^n$ into the features vectors $\mathbf{z}_i = \phi(\mathbf{x}_i) \in \mathbb{R}^d$ before applying machine learning algorithms. This allowed us to easily cast polynomial regression in the same mathematical language as linear regression in Section 7.2.4. In Section 7.3.3 we showed that carefully chosen features maps ϕ could linearly separate some toy data sets so that linear SVM could be applied successfully, even when the original data $\mathbf{x}_1, \ldots, \mathbf{x}_m$ were not linearly separable. These toy applications of the feature map idea involved highly specialized choices of ϕ, and do not generalize easily to real data sets.

The main issue with this approach is that it is difficult to produce a good feature map by hand.[17] Furthermore, a good feature space $\mathbb{R}^d$ — one that, say, linearly separates the data — may be very high dimensional, and the resulting computations are often not tractable. The key insight behind kernel methods is to dispense with the feature map altogether, and instead to work with a *kernel function*. Kernel functions originally arose in functional analysis and, in particular, integral equations, which was also the source of the Fredholm alternative, and kernel methods in machine learning rely on this existing functional analysis framework. We begin here with the basic definitions and associated constructions, and then go on to give applications to kernel SVM and kernel regression. In later chapters we will study other kernel methods, such as kernel PCA in Chapter 8, and draw connections between deep learning and kernel methods in Chapters 10 and 11.

The definition of a general kernel function is elementary.

Definition 7.15. A *kernel function* is a symmetric function $\mathcal{K}\colon \mathbb{R}^n \times \mathbb{R}^n \to \mathbb{R}$.

Thus, $\mathcal{K}(\mathbf{x}, \mathbf{y})$ assigns a real number to each pair of vectors $\mathbf{x}, \mathbf{y} \in \mathbb{R}^n$. Symmetry is the requirement that

$$\mathcal{K}(\mathbf{x}, \mathbf{y}) = \mathcal{K}(\mathbf{y}, \mathbf{x}) \qquad \text{for all} \qquad \mathbf{x}, \mathbf{y} \in \mathbb{R}^n. \tag{7.87}$$

In this book, we will always assume that the kernel function is continuous, although many of our constructions can be straightforwardly extended to mildly discontinuous kernel functions.

Generally speaking, a kernel function encodes a notion of *similarity* between pairs of data points, where the notion of similarity may vary from task to task. A simple example is the *distance kernel function*

$$\mathcal{K}(\mathbf{x}, \mathbf{y}) = \|\mathbf{x} - \mathbf{y}\|^p, \tag{7.88}$$

where $\|\cdot\|$ can be any norm on $\mathbb{R}^n$, e.g., Euclidean norm, 1 norm, etc., and $p \in \mathbb{R}$ can be any real-valued exponent[18] Another example is the dot product kernel function

$$\mathcal{K}(\mathbf{x}, \mathbf{y}) = \mathbf{x} \cdot \mathbf{y} = \mathbf{x}^T \mathbf{y}, \tag{7.89}$$

which is called the *linear kernel function*. The kernel function $\mathcal{K}_\phi$ associated to the feature map $\phi\colon \mathbb{R}^n \to \mathbb{R}^d$, called the *feature map kernel function*, is given by

$$\mathcal{K}_\phi(\mathbf{x}, \mathbf{y}) = \phi(\mathbf{x}) \cdot \phi(\mathbf{y}) = \sum_{i=1}^d \phi_i(\mathbf{x})\, \phi_i(\mathbf{y}), \tag{7.90}$$

and plays a particularly important role. Here $\phi(x) = \big(\phi_1(\mathbf{x}), \ldots, \phi_d(\mathbf{x})\big)$, so that each $\phi_i\colon \mathbb{R}^n \to \mathbb{R}$. One could, of course, replace the dot product in (7.89) or (7.90) by a more general inner product, but it turns out this does not extend the class of feature map kernel functions, since one can suitably modify the feature map ϕ to reduce back to (7.90); see Exercise 6.1. We will see more examples of kernel functions later on.

Given a kernel function $\mathcal{K}$, let us generalize the Gram matrix construction (4.11) by replacing the inner products by the kernel function.

Definition 7.16. Let $\mathcal{K}\colon \mathbb{R}^n \times \mathbb{R}^n \to \mathbb{R}$ be a kernel function. The *kernel matrix* associated with data points $\mathbf{x}_1, \ldots, \mathbf{x}_m \in \mathbb{R}^n$ is the $m \times m$ matrix $K = K(\mathbf{x}_1, \ldots, \mathbf{x}_m)$ whose (i, j) entry is $\mathcal{K}(\mathbf{x}_i, \mathbf{x}_j)$.

---

[17] Good feature maps can be *learned* from the data, and this is one way to view deep neural networks, which are the topic of Chapter 10.

[18] Although when $p < 0$, the distance kernel has a singularity when $\mathbf{x} = \mathbf{y}$.

Warning: A kernel matrix has nothing to do with the kernel of a matrix. One can even talk of the kernel of a kernel matrix, $\ker K$, which consists of all the vectors $\mathbf{v} \in \mathbb{R}^m$ such that $K\mathbf{v} = \mathbf{0}$. The clash in standard terminology is unfortunate, and hopefully will not cause undue confusion.

Explicitly, the kernel matrix takes the form

$$
K = \begin{pmatrix}
\mathcal{K}(\mathbf{x}_1, \mathbf{x}_1) & \mathcal{K}(\mathbf{x}_1, \mathbf{x}_2) & \cdots & \mathcal{K}(\mathbf{x}_1, \mathbf{x}_m) \\
\mathcal{K}(\mathbf{x}_2, \mathbf{x}_1) & \mathcal{K}(\mathbf{x}_2, \mathbf{x}_2) & \cdots & \mathcal{K}(\mathbf{x}_2, \mathbf{x}_m) \\
\vdots & \vdots & \ddots & \vdots \\
\mathcal{K}(\mathbf{x}_m, \mathbf{x}_1) & \mathcal{K}(\mathbf{x}_m, \mathbf{x}_2) & \cdots & \mathcal{K}(\mathbf{x}_m, \mathbf{x}_m)
\end{pmatrix}. \tag{7.91}
$$

Symmetry of the kernel function implies symmetry of its kernel matrices: $K^T = K$. If $\mathcal{K}(\mathbf{x}, \mathbf{y}) = \mathbf{x} \cdot \mathbf{y}$ is the linear dot product kernel function, then the associated kernel matrix (7.91) coincides with the (dot product) Gram matrix constructed from the points $\mathbf{x}_1, \ldots, \mathbf{x}_m$, namely $K = XX^T$, where X is the data matrix (7.1). (Keep in mind that the data points $\mathbf{x}_i$ are the *rows* of X.) Generalizing the inner product produces a generalized Gram matrix, cf. (4.17). The same is true if we consider the data matrix consisting of feature vectors

$$
Z = \begin{pmatrix}
\phi(\mathbf{x}_1)^T \\
\phi(\mathbf{x}_2)^T \\
\vdots \\
\phi(\mathbf{x}_m)^T
\end{pmatrix} = \begin{pmatrix}
\mathbf{z}_1^T \\
\mathbf{z}_2^T \\
\vdots \\
\mathbf{z}_m^T
\end{pmatrix}, \tag{7.92}
$$

Then the kernel matrix K associated with the feature map kernel function (7.90) is the Gram matrix $K = ZZ^T$, and is thus positive semi-definite. While a kernel function encodes a notion of similarity, the kernel matrix is the explicit realization of the kernel-based similarity on a data set. For this reason, a kernel matrix is closely related to a *similarity matrix*, which appears in graph-based learning, the subject of Chapter 9.

Now, the key idea behind kernel methods is to replace the problem of finding a feature map ϕ with that of choosing a kernel function $\mathcal{K}$. To see how this can be done, let $\mathbf{x}_1, \ldots, \mathbf{x}_m \in \mathbb{R}^n$ be given data points and let $\phi \colon \mathbb{R}^n \to \mathbb{R}^d$ be a feature map producing feature vectors $\mathbf{z}_i = \phi(\mathbf{x}_i)$. We consider a general loss function applied to the feature representations of the data, of the form

$$
\mathcal{L}(\mathbf{w}) = L\big(\phi(\mathbf{x}_1) \cdot \mathbf{w}, \ldots, \phi(\mathbf{x}_m) \cdot \mathbf{w}\big) + R\big(\|\mathbf{w}\|^2\big), \tag{7.93}
$$

where $L \colon \mathbb{R}^m \to \mathbb{R}$ and $R \colon \mathbb{R} \to \mathbb{R}$ are given functions with R *nondecreasing*. Many machine learning problems can be cast as minimizing a loss of the form (7.93), such as ridge regression (7.32), as well as soft-margin SVM (7.64). A key result is the following.

Theorem 7.17 (Representer Theorem). *If R is nondecreasing and the loss $\mathcal{L}$ in (7.93) admits a minimizer, then there exists a minimizer $\mathbf{w} \in \mathbb{R}^d$ of the form*

$$
\mathbf{w} = \sum_{i=1}^m c_i \phi(\mathbf{x}_i), \tag{7.94}
$$

for some coefficients $c_1, \ldots, c_m \in \mathbb{R}$.

Proof. Let $\mathbf{w} \in \mathbb{R}^d$ be a minimizer of (7.93) and let

$$V = \operatorname{span}\{\phi(\mathbf{x}_1), \ldots, \phi(\mathbf{x}_m)\} \subset \mathbb{R}^d.$$

Then we can decompose $\mathbf{w} = \mathbf{v} + \mathbf{u}$ where $\mathbf{v} \in V$, $\mathbf{u} \in V^\perp$. Since $\mathbf{u} \cdot \phi(\mathbf{x}_i) = 0$ for all i we have

$$L\big(\phi(\mathbf{x}_1) \cdot \mathbf{w}, \ldots, \phi(\mathbf{x}_m) \cdot \mathbf{w}\big) = L\big(\phi(\mathbf{x}_1) \cdot \mathbf{v}, \ldots, \phi(\mathbf{x}_m) \cdot \mathbf{v}\big).$$

Since $\|\mathbf{w}\|^2 = \|\mathbf{v}\|^2 + \|\mathbf{u}\|^2$ and R is nondecreasing we have $R(\|\mathbf{v}\|^2) \leq R(\|\mathbf{w}\|^2)$. Therefore $\mathcal{L}(\mathbf{v}) \leq \mathcal{L}(\mathbf{w})$ and so $\mathbf{v}$ is also a minimizer of $\mathcal{L}$. Since $\mathbf{v} \in V$ we can by definition write $\mathbf{v}$ in the form (7.94). $\blacksquare$

The Representer Theorem 7.17 allows us to rewrite the loss $\mathcal{L}(\mathbf{w})$, for the purpose of optimizing it, in terms of the kernel matrix $K \in \mathcal{M}_{m \times m}$ associated with the feature map kernel function $\mathcal{K}_\phi$, as per Definition 7.16. Indeed, since the minimizer $\mathbf{w}$ has the form (7.94), for any $\mathbf{x} \in \mathbb{R}^n$ we can write

$$\phi(\mathbf{x}) \cdot \mathbf{w} = \phi(\mathbf{x}) \cdot \left(\sum_{j=1}^{m} c_j \phi(\mathbf{x}_j)\right) = \sum_{j=1}^{m} c_j\, \phi(\mathbf{x}) \cdot \phi(\mathbf{x}_j) = \sum_{j=1}^{m} c_j\, \mathcal{K}_\phi(\mathbf{x}, \mathbf{x}_j). \tag{7.95}$$

Thus, letting $\mathbf{x} = \mathbf{x}_i$ be one the data points,

$$\phi(\mathbf{x}_i) \cdot \mathbf{w} = \sum_{j=1}^{m} c_j\, \mathcal{K}_\phi(\mathbf{x}, \mathbf{x}_j) = \sum_{j=1}^{m} k_{ij}\, c_j.$$

Letting $\mathbf{c} = (c_1, \ldots, c_m) \in \mathbb{R}^m$ we have $\phi(\mathbf{x}_i) \cdot \mathbf{w} = (K\mathbf{c})_i$. Thus, taking dot products of the feature vectors $\phi(\mathbf{x}_i)$ with $\mathbf{w}$ amounts to matrix multiplication with the kernel matrix. This allows us to write the first term in the loss $\mathcal{L}(\mathbf{w})$ as

$$L\big(\phi(\mathbf{x}_1) \cdot \mathbf{w}, \ldots, \phi(\mathbf{x}_m) \cdot \mathbf{w}\big) = L(K\mathbf{c}),$$

provided $\mathbf{w}$ has the form (7.94). The second term in the loss, which is the regularizer, can also be written in terms of the kernel matrix K, since

$$\|\mathbf{w}\|^2 = \mathbf{w} \cdot \mathbf{w} = \left(\sum_{i=1}^{m} c_i\, \phi(\mathbf{x}_i)\right) \cdot \left(\sum_{j=1}^{m} c_j\, \phi(\mathbf{x}_j)\right)$$

$$= \sum_{i,j=1}^{m} c_i\, c_j\, \phi(\mathbf{x}_i)\, \phi(\mathbf{x}_j) = \sum_{i,j=1}^{m} c_i\, c_j\, k_{ij} = \mathbf{c}^T K\, \mathbf{c}.$$

Therefore, when $\mathbf{w}$ has the form (7.94), we can write the loss $\mathcal{L}(\mathbf{w})$ as

$$\mathcal{L}(\mathbf{w}) = L(K\mathbf{c}) + R(\mathbf{c}^T K\, \mathbf{c}). \tag{7.96}$$

By the Representer Theorem 7.17, there always exists a minimizer of this form, and so instead of minimizing $\mathcal{L}(\mathbf{w})$ over $\mathbf{w} \in \mathbb{R}^n$, we can minimize (7.96) over the choice of $\mathbf{c} \in \mathbb{R}^m$. That is, we converted our machine learning problem into the *kernel minimization problem*

$$\min_{\mathbf{c}}\{L(K\mathbf{c}) + R(\mathbf{c}^T K\mathbf{c})\}. \tag{7.97}$$

Notice that while $\mathbf{w} \in \mathbb{R}^d$ may lie in an extremely high dimensional space, the vector $\mathbf{c}$ that we optimize over in the kernel problem (7.97) lies in $\mathbb{R}^m$, where m is the number of data points. Thus, by switching perspectives to the kernel minimization problem (7.97), we do not need to work in the high dimensional feature space.

It is important to note that for ridge regression and soft-margin SVM, the classification or regression function is the linear function $F(\mathbf{x};\mathbf{w}) = \phi(\mathbf{x})\cdot\mathbf{w}$ of the features $\phi(\mathbf{x})$, provided we absorb the bias b into the weights as we did for regression in Section 7.2. When $\mathbf{w}$ has the form (7.94), we can use (7.95) to write this as

$$F(\mathbf{x};\mathbf{w}) = \phi(\mathbf{x})\cdot\mathbf{w} = \sum_{j=1}^{m} c_j\, \mathcal{K}_\phi(\mathbf{x},\mathbf{x}_j). \tag{7.98}$$

Thus, from the kernel perspective, the learned function F is a linear combination of kernel functions associated with the feature map ϕ centered at the data points, where the coefficients c_j in the combination are the parameters that are to be learned.

Now, the kernel minimization problem (7.97) depends only on the kernel matrix, which need not be constructed explicitly with a feature map. In fact, we do exactly the opposite. In kernel methods in machine learning, the kernel matrix K in (7.97) is constructed by choosing a kernel function and applying it to the data points $\mathbf{x}_1,\ldots,\mathbf{x}_m$, instead of choosing a feature map. This is colloquially known as the *kernel trick*. Some important kernel functions that are often used in practice are given below.

$$\text{Polynomial Kernel Function:}\quad \mathcal{K}(\mathbf{x},\mathbf{y}) = (\mathbf{x}\cdot\mathbf{y} + b)^k,$$

$$\text{Radial Basis Kernel Function:}\quad \mathcal{K}(\mathbf{x},\mathbf{y}) = \exp\!\big(-\gamma\,\|\mathbf{x}-\mathbf{y}\|^2\big),$$

$$\text{Sigmoid Kernel Function:}\quad \mathcal{K}(\mathbf{x},\mathbf{y}) = \tanh\!\big(\kappa\,\mathbf{x}\cdot\mathbf{y} + c\big).$$

Here, the quantities $b,\kappa,c \in \mathbb{R}$, $k \in \mathbb{N}$, and $\gamma > 0$, are kernel parameters, which are often considered hyperparameters in machine learning tasks.

Each of the kernel functions above is clearly symmetric, so they satisfy Definition 7.15. However, it is unclear from how they are defined whether any of these kernel functions are feature map kernel functions themselves; that is, does there exist a feature map ϕ so that $\mathcal{K} = \mathcal{K}_\phi$? In other words, is the problem of choosing a feature map equivalent to the problem of choosing a kernel function? In practice, this is an important question, since the kernel matrix K associated to a feature map kernel function is a Gram matrix and is thus positive semi-definite, which is a requirement for the kernel problem (7.97) to be a *convex* optimization problem with a well-defined solution, where we recall that the term $\mathbf{c}^T K \mathbf{c}$ is convex if and only if K is positive semi-definite.

The preceding discussion motivates the following definition.

Mercer kernels are named after the early twentieth century English mathematician James Mercer. Explicitly, a Mercer kernel must satisfy

$$\mathbf{c}^T K \mathbf{c} = \sum_{i,j=1}^{m} c_i c_j\, \mathcal{K}(\mathbf{x}_i,\mathbf{x}_j) \geq 0, \tag{7.99}$$

for all $\mathbf{x}_1,\ldots,\mathbf{x}_m \in \mathbb{R}^n$, all $\mathbf{c} = (c_1,\ldots,c_m) \in \mathbb{R}^m$, and any $m = 1,2,3,\ldots$. In particular, taking $m = 1$, a Mercer kernel must satisfy $\mathcal{K}(\mathbf{x},\mathbf{x}) \geq 0$. This restricts $\kappa > 0$ and $c \geq 0$ in the sigmoid kernel; see Exercise 6.4.

Since the kernel matrices associated with the linear kernel function and with the feature map kernel function are Gram matrices, they are automatically positive semi-definite, and hence both are Mercer kernels. It turns out that these are essentially all of the Mercer kernels. That is, at an informal level, a kernel function $\mathcal{K}$ is a Mercer kernel if and only if it is a feature map kernel function $\mathcal{K} = \mathcal{K}_\phi$. To be more precise, we state, but do not prove, the following theorem, originally due to Mercer [164].

Theorem 7.19. *If $\mathcal{K}$ is a Mercer kernel, then there exist continuous scalar-valued functions $\phi_1, \phi_2, \phi_3, \ldots$, so $\phi_i \colon \mathbb{R}^n \to \mathbb{R}$, such that*

$$\mathcal{K}(\mathbf{x}, \mathbf{y}) = \sum_{i=1}^{\infty} \phi_i(\mathbf{x})\, \phi_i(\mathbf{y}) \qquad \text{for all} \qquad \mathbf{x}, \mathbf{y} \in \mathbb{R}^n. \tag{7.100}$$

Note that the right hand side of (7.100) can be interpreted as the dot product between the *infinite-dimensional* feature map $\phi(\mathbf{x}) = (\phi_1(\mathbf{x}), \phi_2(\mathbf{x}), \ldots)$ evaluated at $\mathbf{x}$ and $\mathbf{y}$. Thus, Theorem 7.19 shows that any Mercer kernel can be expressed as a feature map kernel function, provided we allow infinite-dimensional feature maps.

The proof of Mercer's theorem would take us too far afield, in view of our avoidance of infinite-dimensional vector spaces. The main idea is that (7.100) is the infinite-dimensional version of the spectral decomposition of a symmetric positive semi-definite matrix K, which can be written in the form

$$K = \sum_{i=1}^{n} \lambda_i\, \mathbf{q}_i\, \mathbf{q}_i^T = \sum_{i=1}^{n} \mathbf{p}_i\, \mathbf{p}_i^T,$$

where $\mathbf{p}_i = \sqrt{\lambda_i}\, \mathbf{q}_i$. The last step of absorbing the eigenvalues into $\mathbf{p}_i$ by taking their square root can only be done when $\lambda_i \geq 0$, i.e., K is positive semi-definite. The generalization of this result to Mercer kernels is the subject of functional analysis; in particular, the spectral theory of compact self-adjoint operators in infinite-dimensional Hilbert space. The original proofs for Mercer kernels can be found in [164], and a modern treatment is available in most functional analysis textbooks, e.g., [124].

Given Theorem 7.19, Mercer kernels play a very important role in kernel methods, and it is generally preferable to use Mercer kernels whenever possible. Hence, it is useful to have a set of tools for determining when a given kernel function is a Mercer kernel. The following proposition collects several useful operations that preserve the property of being a Mercer kernel.

Proposition 7.20. *Let $\mathcal{K}$, $\mathcal{K}_1$, and $\mathcal{K}_2$ be Mercer kernels. Then the following are also Mercer kernels.*

    (i)   The constant multiple $a\mathcal{K}$ for any $a > 0$, .

    (ii) The sum $\mathcal{K}_1 + \mathcal{K}_2$.

    (iii) The product kernel $\mathcal{K}_1(\mathbf{x}, \mathbf{y})\, \mathcal{K}_2(\mathbf{x}, \mathbf{y})$.

    (iv) The power kernel $\mathcal{K}(\mathbf{x}, \mathbf{y})^d$ for a positive integer d.

    (v) The exponential kernel $\exp\big(\mathcal{K}(\mathbf{x}, \mathbf{y})\big)$.

    (vi) The rescaled kernel $F(\mathbf{x})\, \mathcal{K}(\mathbf{x}, \mathbf{y})\, F(\mathbf{y})$ for any function $F \colon \mathbb{R}^n \to \mathbb{R}$.

Proof. The proofs of (*i*) and (*ii*) are immediate, and (*iii*) is a consequence of the Schur Product Theorem 5.38, since the product kernel is exactly the Hadamard matrix product. (*iv*) follows directly from (iii). To prove (v) we use a Taylor expansion to write

$$\exp\big(\mathcal{K}(\mathbf{x},\mathbf{y})\big) = \lim_{k\to\infty} \sum_{m=0}^{k} \frac{\mathcal{K}(\mathbf{x},\mathbf{y})^m}{m!}.$$

By properties (i), (ii), and (iv), the partial sums for any $k \geq 1$ are nonnegative kernel functions. The reader can easily verify that the limit of a sequence of nonnegative kernel functions is again nonnegative, and hence a Mercer kernel. Finally, to prove (vi), we simply write

$$\sum_{i,j=1}^{m} a_i\, a_j\, F(\mathbf{x}_i)\, \mathcal{K}(\mathbf{x}_i,\mathbf{x}_j)\, F(\mathbf{x}_j) = \sum_{i,j=1}^{m} b_i\, b_j\, \mathcal{K}(\mathbf{x}_i,\mathbf{x}_j),$$

where $b_i = a_i F(\mathbf{x}_i)$, and use the fact that $\mathcal{K}$ is a Mercer kernel. ∎

We can use Proposition 7.20 to prove that both the polynomial kernel function (with $b = 0$) and radial basis function kernel are Mercer kernels; see Exercises 6.2 and 6.3. On the other hand, the sigmoid kernel function is not a Mercer kernel; see Exercise 6.4.

The next two subsections explain how kernel methods can be applied to ridge regression and SVM. In Chapter 8, we explain how to use kernel methods in the context of principal component analysis (PCA). Kernel methods are also connected to deep learning, which is discussed in Chapter 10. Kernel methods can also be applied to many other machine learning methods; see Exercise 6.8 for an application of kernel methods to k-means clustering.

7.6.1 Kernel Regression

The ridge regression problem (7.32) corresponds to the loss function $\mathcal{L}(\mathbf{w})$ in (7.96), with $L(\mathbf{z}) = \|\mathbf{z} - \mathbf{y}\|^2$ and $R(t) = \lambda t$. Thus, the kernel version of ridge regression, that is (7.97) with these choices for L and R, is given by

$$\min_{\mathbf{c}} \big\{\, \|K\mathbf{c} - \mathbf{y}\|^2 + \lambda\, \mathbf{c}^T K \mathbf{c} \,\big\}, \tag{7.101}$$

where K is the kernel matrix associated to the data points $\mathbf{x}_1,\ldots,\mathbf{x}_m$ and with a chosen kernel function $\mathcal{K}(\mathbf{x},\mathbf{y})$; that is $k_{ij} = \mathcal{K}(\mathbf{x}_i,\mathbf{x}_j)$. This is of the same form as a general ridge regression problem, but it can be simplified. Let $\mathbf{b} = K^{\frac{1}{2}}\mathbf{c}$ so that we can rewrite (7.101) as

$$\min_{\mathbf{b}} \big\{\, \|K^{1/2}\mathbf{b} - \mathbf{y}\|^2 + \lambda\, \|\mathbf{b}\|^2 \,\big\}.$$

If $\lambda > 0$, then by (7.33) the unique solution of this problem is

$$\mathbf{b} = (K + \lambda\,\mathrm{I})^{-1} K^{1/2} \mathbf{y} = K^{1/2}(K + \lambda\,\mathrm{I})^{-1}\mathbf{y},$$

where we leave the second equality to the reader to verify in Exercise 6.5. Equating this with $\mathbf{b} = K^{1/2}\mathbf{c}$ we see a natural choice for $\mathbf{c}$ is

$$\mathbf{c} = (K + \lambda\,\mathrm{I})^{-1}\mathbf{y}. \tag{7.102}$$

If K is invertible, then this solution is unique. However, if K is singular, then any vector $\mathbf{c}+\mathbf{v}$ where $\mathbf{v} \in \ker K$ also solves (7.101), since it results in the same vector $\mathbf{b}$ upon multiplying by $K^{1/2}$, since $\ker K^{1/2} = \ker K$.

Now, we can also minimize (7.101) directly, which by (7.41) gives the solution

$$\mathbf{c} = (K^2 + \lambda K)^{-1} K \mathbf{y} = (K + \lambda\,\mathrm{I})^{-1}\mathbf{y}, \tag{7.103}$$

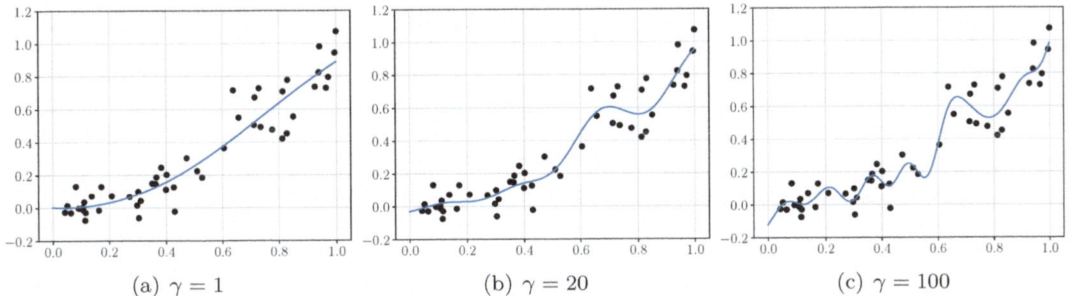

(a) $\gamma = 1$ (b) $\gamma = 20$ (c) $\gamma = 100$

Figure 7.33: An example of kernel ridge regression with the radial basis function kernel and $\lambda = 0.01$. The data is the same as in Figure 7.9.

provided that $K^2 + \lambda K$ is invertible, although this is not needed for the ultimate formula (7.102); it is only used in the intermediate step. Invertibility of $K^2 + \lambda K$ is equivalent to invertibility of K, which may not hold. In the case that K is not invertible, it follows from Theorem 6.7 that every minimizer of (7.101) is a solution of

$$(K^2 + \lambda K)\mathbf{c} = K\mathbf{y}. \tag{7.104}$$

In this case we select the solution given by (7.102), though it may not be the minimal norm solution; see Exercise 6.7.

We give a brief experiment with kernel regression with the radial basis function kernel. By (7.98), this means we are using functions of the form

$$F(\mathbf{x}) = \sum_{j=1}^{m} c_j\, e^{-\gamma \| \mathbf{x} - \mathbf{x}_j \|^2},$$

for regression, where the learned coefficients $\mathbf{c} = (c_1, \ldots, c_m)$ are given by (7.102). In Figure 7.33 we show an example of kernel ridge regression with the radial basis function kernel on the simple toy data set from Figure 7.9. We set $\lambda = 0.01$, and show three different values of γ, illustrating how the parameter influences the smoothness of the regression.

7.6.2 Kernel Support Vector Machines

We now turn to kernel SVM. While it can be formulated using the representer theorem 7.17, there is a sharper version of the result for soft-margin SVM. Let $\phi \colon \mathbb{R}^n \to \mathbb{R}^d$ be a feature map and consider the soft-margin SVM problem[19] applied to the features:

$$\min_{\mathbf{w},b} \left\{ \lambda \| \mathbf{w} \|^2 + \frac{1}{m} \sum_{i=1}^{m} \left[1 - y_i(\phi(\mathbf{x}_i) \cdot \mathbf{w} - b) \right]_+ \right\}. \tag{7.105}$$

The following result reformulates the representer theorem for SVM.

---

[19] As mentioned in the introduction, we will sometimes be a bit sloppy, and use minimum and maximum even when, to be technically correct, one should really use infimum and supremum. Readers who are familiar with the latter can easily make the required substitutions.

Theorem 7.21 (SVM Representer Theorem). *There exists a minimizer of the soft-margin kernel SVM problem* (7.105) *of the form*

$$\mathbf{w}_* = \frac{1}{2\lambda m} \sum_{i=1}^{m} c_i\, y_i\, \phi(\mathbf{x}_i), \tag{7.106}$$

for coefficients $c_1, \ldots, c_m$ *that satisfy* $0 \le c_i \le 1$ *and* $\mathbf{c} \cdot \mathbf{y} = c_1 y_1 + \cdots + c_m y_m = 0.$

Proof. We start with the simple observation that for any $a \in \mathbb{R}$,

$$a_+ = \max\{0, a\} = \max_{0 \le c \le 1} ca.$$

Indeed, when $a > 0$, then the maximum occurs at $c = 1$ and $a_+ = a$. When $a < 0$ the maximum occurs at $c = 0$ and $a_+ = 0$, as desired. When $a = 0$, any c will do. This allows us to write the SVM minimization problem (7.105) as

$$\min_{\mathbf{w},b} \max_{\mathbf{c}} \left\{ \lambda \|\mathbf{w}\|^2 + \frac{1}{m} \sum_{i=1}^{m} c_i \big[1 - y_i \left(\phi(\mathbf{x}_i) \cdot \mathbf{w} - b \right) \big] \right\}, \tag{7.107}$$

where $\mathbf{c} = (c_1, \ldots, c_m)$ is subject to $0 \le c_i \le 1$.

We now appeal to the following min-max theorem[20], due to Fan, [72], to swap the min and the max, and solve instead the problem

$$\max_{\mathbf{c}} \min_{\mathbf{w},b} \left\{ \lambda \|\mathbf{w}\|^2 + \frac{1}{m} \sum_{i=1}^{m} c_i \left(1 - y_i \phi(\mathbf{x}_i) \cdot \mathbf{w} \right) + \frac{b}{m} \sum_{i=1}^{m} c_i y_i \right\}. \tag{7.108}$$

Theorem 7.22. *Let $D \subset \mathbb{R}^m$ be compact, i.e., closed and bounded. Let $F: D \times \mathbb{R}^n \to \mathbb{R}$ be continuous, and suppose that $F(\mathbf{x}, \mathbf{y})$ is convex as a function of $\mathbf{x}$ for each fixed $\mathbf{y} \in \mathbb{R}^m$, while $F(\mathbf{x}, \mathbf{y})$ is concave, i.e., $-F(\mathbf{x}, \mathbf{y})$ is convex, as a function of $\mathbf{y}$ for each fixed $\mathbf{x} \in \mathbb{R}^n$. Then,*

$$\min_{\mathbf{y} \in \mathbb{R}^n} \max_{\mathbf{x} \in D} F(\mathbf{x}, \mathbf{y}) = \max_{\mathbf{x} \in D} \min_{\mathbf{y} \in \mathbb{R}^n} F(\mathbf{x}, \mathbf{y}). \tag{7.109}$$

In light of (7.108), we now see that any optimal $\mathbf{c}$ must satisfy $c_1 y_1 + \cdots + c_m y_m = 0$, as otherwise the objective function will not be bounded from below since we can choose b to make it as large negative as desired. Thus, we may consider the equivalent problem

$$\max_{\mathbf{c}} \min_{\mathbf{w}} \left\{ \lambda \|\mathbf{w}\|^2 + \frac{1}{m} \sum_{i=1}^{m} c_i \left(1 - y_i \phi(\mathbf{x}_i) \cdot \mathbf{w} \right) \right\}. \tag{7.110}$$

subject to the constraints $0 \le c_i \le 1$ and $\mathbf{c} \cdot \mathbf{y} = 0$. As a function of the $\mathbf{w}$ the objective is quadratic, and, by Theorem 6.7, the optimal $\mathbf{w}$ is given by (7.106).   ∎

---

[20] Fan's theorem is a bit more general than the stated result. An even more general min-max theorem can be found in [216]. The assumed continuity of the function is important. Although it can be slightly weakened, in general one can construct discontinuous counterexamples to the min-max equation (7.109).

If we continue the reasoning in the proof of Theorem 7.21, and plug our formula (7.106) for $\mathbf{w}$ back into (7.110), we obtain what is called the *dual optimization problem*

$$\max_{\mathbf{c}} \left\{ \frac{1}{m} \sum_{i=1}^{m} c_i - \frac{1}{4\lambda m^2} \sum_{i,j=1}^{m} c_i c_j y_i y_j \, \mathcal{K}_\phi(\mathbf{x}_i, \mathbf{x}_j) \right\}, \tag{7.111}$$

subject to $0 \le c_i \le 1$ and $\mathbf{c} \cdot \mathbf{y} = 0$, where $\mathcal{K}_\phi(\mathbf{x}, \mathbf{y}) = \phi(\mathbf{x}) \cdot \phi(\mathbf{y})$ is the feature map kernel function. Now, notice that b does not appear in the dual problem (7.111). To determine b, we return to (7.107), from which we can see that whenever $0 < c_i < 1$ we must have $1 - y_i(\phi(\mathbf{x}_i) \cdot \mathbf{w} - b) = 0$. Hence, to find b, we simply have to find a c_i that is strictly between 0 and 1 — that is, find a support vector[21] — and then solve for b in the equation

$$y_i \big(\phi(\mathbf{x}_i) \cdot \mathbf{w} - b \big) = 1.$$

Rearranging and using (7.106) we have

$$b = \phi(\mathbf{x}_i) \cdot \mathbf{w} - \frac{1}{y_i} = \frac{1}{2\lambda m} \sum_{j=1}^{m} c_j y_j \, \mathcal{K}_\phi(\mathbf{x}_i, \mathbf{x}_j) - y_i,$$

as $y_i \in \{-1, 1\}$ so $1/y_i = y_i$. Hence, we can also compute b using only the kernel function $\mathcal{K}_\phi$. After solving the kernel SVM problem (7.111), the classification function becomes

$$F(\mathbf{x}) = \mathbf{w} \cdot \phi(\mathbf{x}) - b = \frac{1}{2\lambda m} \sum_{i=1}^{n} c_i y_i \, \mathcal{K}_\phi(\mathbf{x}, \mathbf{x}_i) - b. \tag{7.112}$$

Remark 7.23. We note that any data points $\mathbf{x}_i$ for which the optimal coefficient is $c_i = 0$ can essentially be omitted from the dual optimization problem (7.111) without affecting the optimal values of the other coefficients. Thus, the solution of the soft-margin SVM problem depends only on those data points for which $c_i > 0$, which are exactly those data points that saturate or violate the margin constraint. ▲

Now we apply the kernel trick of choosing the kernel function $\mathcal{K}$ instead of the feature map. For any chosen $\mathcal{K}$, such as the radial basis function kernel, we compute the kernel matrix K with entries $k_{ij} = \mathcal{K}(\mathbf{x}_i, \mathbf{x}_j)$ and solve the kernel SVM problem

$$\max_{\mathbf{c}} \left\{ \frac{1}{m} \sum_{i=1}^{m} c_i - \frac{1}{4\lambda m^2} \sum_{i,j=1}^{m} k_{ij} c_i c_j y_i y_j \right\}. \tag{7.113}$$

We then find a $c_i \in (0, 1)$ and compute

$$b = \frac{1}{2\lambda m} \sum_{j=1}^{m} c_j y_j k_{ij} - y_i. \tag{7.114}$$

The classification function $F(\mathbf{x})$ for the general kernel function has the form (7.112) where $\mathcal{K}$ replaces $\mathcal{K}_\phi$.

---

[21]The existence of support vectors, i.e., a value of i such that $0 < c_i < 1$, is a more subtle question, and is studied in the literature on primal-dual optimization; see [57].

Digit	0	1	2	3	4	5	6	7	8	9
0	973	0	1	0	0	2	1	1	2	0
1	0	1126	3	1	0	1	1	1	2	0
2	6	1	1006	2	1	0	2	7	6	1
3	0	0	2	995	0	2	0	5	5	1
4	0	0	5	0	961	0	3	0	2	11
5	2	0	0	9	0	871	4	1	4	1
6	6	2	0	0	2	3	944	0	1	0
7	0	6	11	1	1	0	0	996	2	11
8	3	0	2	6	3	2	2	3	950	3
9	3	4	1	7	10	2	1	7	4	970

Table 7.34: Confusion matrix for the classification performance of SVM with radial basis function kernel on the full MNIST data set. The overall accuracy is 97.92%. The entry C_{ij} of the confusion matrix records how many testing images from class i were predicted in class j. For example, from row 0 we see that 973 zeros were classified correctly, one zero was incorrectly classified as a 2, two zeros were incorrectly classified as 5s, one zero was incorrectly classified as a 6, another was incorrectly classified as an 7, and 2 were misclassified as 8s.

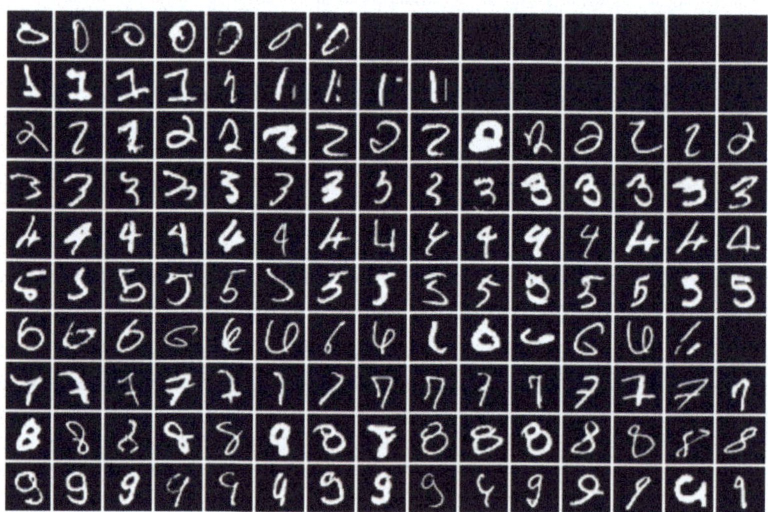

Figure 7.35: An example of some images from each MNIST class that were classified incorrectly by SVM with radial basis function kernel. We show 15 images from each class, with the exception of classes 0, 1, and 6, where less than 15 were misclassified overall.

In practice, SVM is usually trained by solving the optimization problem (7.113). This is a constrained quadratic optimization problem. The constraints $0 \leq c_i \leq 1$ form a *convex* set, and so projected gradient ascent, where we alternate steps of gradient ascent with clipping the values of c_i to the interval $[0, 1]$ is a reasonable approach. However, since the objective function is quadratic, more efficient optimization methods known as *quadratic programming* [177] can be used.

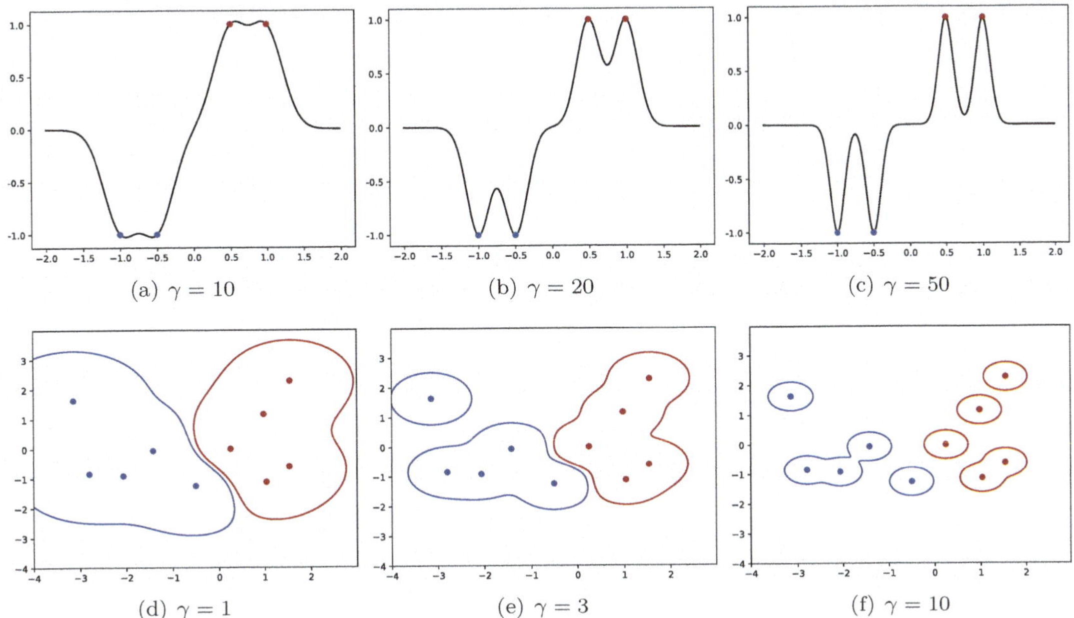

Figure 7.36: Radial basis kernel SVM decision functions F for different values of γ. As we increase γ, the decision function becomes more localized around labeled data points. In the bottom row with two dimensional data, we plot the contours $F = \pm 0.1$ of the decision function.

We trained SVM with a radial basis function kernel on the MNIST data set for handwritten digit classification, using the standard 10,000 image testing set and 60,000 image training set. The method achieved 97.92% testing accuracy, which is slightly better than the 96.83% we obtained with linear kernel SVM. Table 7.34 shows the *confusion matrix* for the classification performance on the testing dataset. We see very few misclassified images; common mistakes occur with the pairs $(4, 9)$, $(7, 2)$, $(7, 9)$ and $(5, 3)$. We also show some of the misclassified images for each class in Figure 7.35.

To gain some more geometric intuition about kernel SVM, consider, for example, the form of the classification function (7.112) for the radial basis function kernel, which has the form

$$F(\mathbf{x}) = \frac{1}{2\lambda m} \sum_{i=1}^{n} c_i y_i \exp\left(-\gamma \|\mathbf{x}_i - \mathbf{x}\|^2\right) - b.$$

Since the coefficients c_i are nonnegative, and $y_i \in \{-1, 1\}$ are binary labels, the value of $F(\mathbf{x})$ can be interpreted as a weighted average of nearby labels. Alternatively, we can view F as a linear superposition of Gaussian kernel functions attached to each training point scaled by c_i and carrying along the sign y_i corresponding to which class the training point belongs to. Notice again that only coefficients with $c_i > 0$ influence the classification function F, since otherwise $c_i = 0$. We illustrate the decision function F for radial basis kernel SVM applied to toy one and two dimensional data in Figure 7.36. We can see in the figure that as we increase γ, the decision function becomes more localized around the labeled data points, limiting the propagation of label information, while when γ is smaller, the kernel functions are "wider", and propagate labels further.

Exercises

6.1. ♡ Consider the inner product feature map kernel $\mathcal{K}_\phi(\mathbf{x}, \mathbf{y}) = \langle \phi(\mathbf{x}), \phi(\mathbf{y}) \rangle$. Show that there exists a feature map ψ such that $\mathcal{K}_\phi(\mathbf{x}, \mathbf{y}) = \psi(\mathbf{x}) \cdot \psi(\mathbf{y})$, so that there is no loss of generality in using the dot product in the definition of feature map kernels.

6.2. Prove that the polynomial kernel function is a Mercer kernel when $b = 0$.

6.3. ◇ Assume we are in dimension $n = 1$.
(a) Show that the radial basis function kernel $\mathcal{K}(x, y) = e^{-\gamma(x-y)^2}$ can be expressed as

$$\mathcal{K}(x, y) = \sum_{i=0}^{\infty} \phi_i(x) \cdot \phi_i(y), \qquad \text{where} \qquad \phi_i(x) = e^{-\gamma x^2} x^i \sqrt{\frac{(2\gamma)^i}{i!}}.$$

Hint: Write $\mathcal{K}(x, y) = e^{-\gamma x^2} e^{-\gamma y^2} e^{2\gamma x y}$, and use a Taylor expansion on the last term.
(b) Can you find a similar formula in dimension $n = 2$ or $n \geq 3$?

6.4. ♡ Let $\mathcal{K}$ be the sigmoid kernel function. (a) Find $\mathbf{x}$ so that $\mathcal{K}(\mathbf{x}, \mathbf{x}) < 0$ when $\kappa < 0$ or $c < 0$. (b) Show that $\mathcal{K}$ is not a Mercer kernel even when $c > 0$ and $\kappa > 0$. *Hint:* Look for an $m = 2$ counterexample.

6.5. Suppose K is symmetric positive semi-definite and let $\lambda, s > 0$. Show that

$$K^s (K + \lambda\,\mathrm{I})^{-1} = (K + \lambda\,\mathrm{I})^{-1} K^s.$$

6.6. Assume that R is strictly increasing in Theorem 7.17. Show that every minimizer of $\mathcal{L}$ has the form (7.94).

6.7. ♡ Show that $\mathbf{c}$ defined by (7.102) is the minimal Euclidean norm solution of (7.104) when $\mathbf{y} \in \mathrm{img}\, K$. What happen when $\mathbf{y} \notin \mathrm{img}\, K$?

6.8. ◇ This exercise will develop the kernel k-means algorithm. Let $\mathbf{x}_1, \ldots, \mathbf{x}_m \in \mathbb{R}^n$ be data points, let $\phi \colon \mathbb{R}^n \to \mathbb{R}^d$ be a feature map, and let $\mathbf{z}_i = \phi(\mathbf{x}_i)$ be the feature vectors associated with the data points. Let $K \in \mathcal{M}_{m \times m}$ be the associated kernel matrix, with entries $k_{ij} = \phi(\mathbf{x}_i) \cdot \phi(\mathbf{x}_j) = \mathbf{z}_i \cdot \mathbf{z}_j$. The kernel k-means algorithm applies the standard k-means algorithm to the feature vectors $\mathbf{z}_1, \ldots, \mathbf{z}_m$, and in order to formulate this for general kernel functions we need to show that each step can be expressed in a way that depends only on the kernel matrix.
To do this, let $C \subset \{1, \ldots, m\}$ be any cluster, with mean vector $\mathbf{c} = (1/\#C) \sum_{i \in C} \mathbf{z}_i$. Let $\mathbf{1}_C = (1/\#C) \sum_{i \in C} \mathbf{e}_i \in \mathbb{R}^m$ denote its *normalized indicator vector*. Define the kernel matrix inner product $\langle \cdot, \cdot \rangle_K$ on $\mathbb{R}^m$ by $\langle \mathbf{x}, \mathbf{y} \rangle_K = \mathbf{x}^T K \mathbf{y}$, with corresponding norm $\|\mathbf{x}\|_K^2 = \mathbf{x}^T K \mathbf{x}$.
(a) Show that $\|\mathbf{z}_i\| = \|\mathbf{e}_i\|_K$, $\mathbf{z}_i \cdot \mathbf{c} = \langle \mathbf{e}_i, \mathbf{1}_C \rangle_K$, and $\|\mathbf{c}\| = \|\mathbf{1}_C\|_K$.
(b) Show that the Euclidean distance from any feature vector $\mathbf{z}_i$ to the mean vector $\mathbf{c}$ can be written as $\|\mathbf{z}_i - \mathbf{c}\| = \|\mathbf{e}_i - \mathbf{1}_C\|_K$. *Hint:* Expand $\|\mathbf{z}_i - \mathbf{c}\|^2$ and use part (a).
(c) Explain how to use part (b) to formulate a kernel k-means algorithm that works for a general kernel function, such as the radial basis function kernel.

6.9. ◇ Implement the kernel k-means algorithm from Exercise 6.8 in Python and apply it to some toy data sets for which k-means performs poorly, such as the two moons data set.

Chapter 8

Principal Component Analysis

Singular values and vectors underlie contemporary statistical data analysis. In particular, the method of *principal component analysis* (PCA) has assumed an ever increasing role in a wide range of applications, including machine learning, image processing, speech recognition, face recognition, data mining, semantics, and health informatics; see [94, 121, 122] and the references therein. The earliest descriptions of the method are to be found in the first half of the twentieth century in the work of the statisticians Karl Pearson, [184], and Harold Hotelling, [114].

PCA is used to simplify data by looking for linear, or rather affine, relationships between the measurements of different data points. Mathematically, PCA amounts to projection onto the top singular vectors of a centered version of the data matrix X associated to a data set, which are called the *principal components* of the data. The key idea behind PCA is that the singular vectors associated with larger singular values represent important correlations in the data, while those with smaller singular values indicate relatively unimportant features or noise. Projecting the data onto the principal components yields an effective dimensionality reduction algorithm, which is widely employed in data analysis tasks and other applications, such as visualization of high dimensional data sets and image compression.

In this chapter, we introduce the basics of PCA, and then show how it can be interpreted as finding the best affine subspace approximating a collection of data points. We also explore robust versions and applications to image compression, along with several related methods, including kernel PCA, linear discriminant analysis (LDA) and multidimensional scaling (MDS).

Note: Throughout this chapter, $\|\cdot\|$ denotes the Euclidean norm.

8.1 The Principal Components

Python Notebook: Intro to PCA (.ipynb)

Let us begin by recalling how we handle data, as described in detail in Section 7.1. The data is assembled into an $m \times n$ data matrix X — see (7.1) and the ensuing discussion for details — whose rows $\mathbf{x}_1^T, \ldots, \mathbf{x}_m^T$ are the data points, and whose columns $\mathbf{v}_1, \ldots, \mathbf{v}_n$ are the measurement vectors. As in (7.7), we center the measurement vectors by subtracting their mean. The resulting centered data matrix $\underline{X} = JX$ is obtained by premultiplying the data matrix by the centering matrix (7.5); its columns, the centered measurement vectors, are all of mean zero. In this chapter, we always take $\nu = 1$ in the definition of variance given in (7.10), noting that other choices do not significantly affect any of the results.

High dimensional data sets, where $n \geq 4$, are prevalent in applications, but are generally difficult to visualize and work with directly. Standard or random projections of high-dimensional data onto two- or three-dimensional subspaces give some limited insight, but the results are highly dependent on the direction of projection and tend to obscure any underlying structure. For example, projecting the data sets in Figure 7.5 onto the x- and y-axes produces more or less the same results, thereby hiding the variety of two-dimensional correlations. In this section, we outline a more systematic approach that projects the data along *important* directions, called the *principal components* of the data, which is useful for simplifying and visualizing high dimensional data.

The basic idea behind *principal component analysis*, often abbreviated PCA, is to look for directions in which the variance of the data is largest. To see how to do this, we recall the definition of the $n \times n$ *covariance matrix* from (7.18) corresponding to the data matrix X, which (since $\nu = 1$) is the $n \times n$ Gram matrix associated with the centered data matrix:

$$S_X = \underline{X}^T \underline{X}.$$

Its entries are the pairwise covariances of the individual measurements, i.e., the columns of X. In particular, the diagonal entries are the individual variances. The covariance matrix of a data set encodes the information concerning the possible linear dependencies and interrelationships among the data points.

Given an $m \times n$ data matrix X, we define the first principal direction as that in which the data experiences the most variance. By "direction", we mean a line through the origin in $\mathbb{R}^n$, and the variance is computed from the orthogonal projection of the data measurements onto the line. Each line is spanned by a unit vector[1] $\mathbf{u} = (u_1, u_2, \ldots, u_n)^T$ with $\|\mathbf{u}\| = 1$. The coordinates[2] of the orthogonal projection — see formula (2.39) — of the data matrix X in the direction $\mathbf{u}$ are given by the vector

$$X\mathbf{u} = (\mathbf{x}_1 \cdot \mathbf{u}, \mathbf{x}_2 \cdot \mathbf{u}, \ldots, \mathbf{x}_m \cdot \mathbf{u})^T.$$

Our goal is to find a direction $\mathbf{u}$ that maximizes the *variance* of the projected measurements $X\mathbf{u}$. By (7.10) and (7.7), the variance of the projected data $X\mathbf{u}$ is

$$\sigma_{\mathbf{u}}^2 = \|JX\mathbf{u}\|^2 = \|\underline{X}\mathbf{u}\|^2 = \mathbf{u}^T \underline{X}^T \underline{X} \mathbf{u} = \mathbf{u}^T S_X \mathbf{u}. \tag{8.1}$$

Thus, our aim is to maximize the projected variance (8.1) over all possible choices of directions $\mathbf{u} \in \mathbb{R}^n$ which satisfy $\|\mathbf{u}\| = 1$. This is precisely the maximization problem that was solved by Theorem 5.43. We thus immediately deduce that the first principal direction is given by the dominant unit eigenvector $\mathbf{u} = \mathbf{q}_1$ of the covariance matrix S_X, and the maximum variance is the dominant, or largest, eigenvalue of the covariance matrix, namely, $\max_{\mathbf{u}} \sigma_{\mathbf{u}}^2 = \lambda_{max}(S_X)$. Since $S_X = \underline{X}^T \underline{X}$, we can also interpret — see Definition 5.72 — the maximum variance as the square of the maximal singular value of $\underline{X}$ (using the dot product), that is

$$\max_{\mathbf{u}} \sigma_{\mathbf{u}}^2 = \lambda_{max}(S_X) = \sigma_{max}(\underline{X})^2. \tag{8.2}$$

---

[1] Actually, there are two unit vectors, namely $\pm\mathbf{u}$, in each line, but it doesn't matter which one we choose.

[2] The orthogonal projection of $\mathbf{x}_i$ is, in fact, the vector $\mathbf{u}\mathbf{u}^T\mathbf{x}_i = (\mathbf{x}_i \cdot \mathbf{u})\mathbf{u} \in \mathbb{R}^n$, but we instead use its coordinate $\mathbf{x}_i \cdot \mathbf{u}$ to facilitate the identification with a one-dimensional data set.

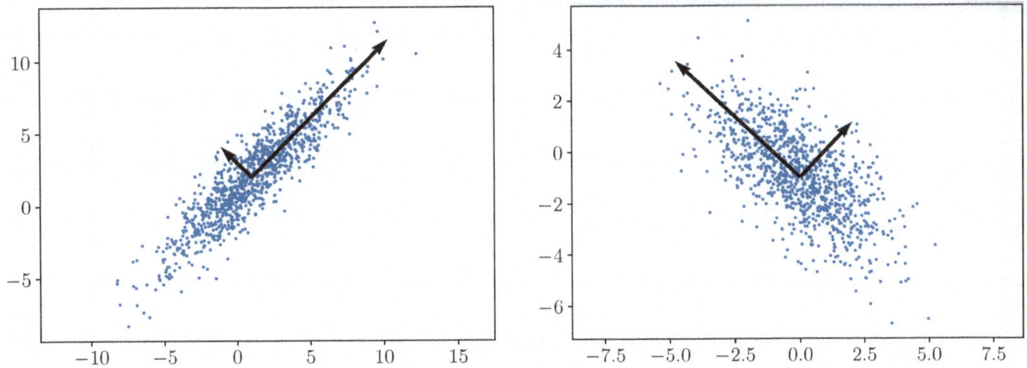

Figure 8.1: Illustration of principal components for two different data sets. We plot the two principal components with length proportional to the associated principal standard deviation σ_i, i.e., the associated singular value, for visualization purposes.

The first principal direction $\mathbf{q}_1$ is also the dominant unit singular vector of the centered data matrix $\underline{X}$.

The second principal direction is assumed to be orthogonal to the first, so as to avoid contaminating it with the already noted direction of maximal variance, and is to be chosen so that the variance of its projected measurements is maximized among all such orthogonal directions. Thus, the second principal direction will maximize $\sigma_{\mathbf{u}}^2$, as given by (8.1), over all unit vectors $\mathbf{u}$ satisfying $\mathbf{u} \cdot \mathbf{q}_1 = 0$. More generally, given the first $j - 1$ principal directions $\mathbf{q}_1, \ldots, \mathbf{q}_{j-1}$, the j-th principal component is in the direction $\mathbf{u} = \mathbf{q}_j$ that maximizes the variance

$$\sigma_{\mathbf{u}}^2 = \mathbf{u}^T S_X \mathbf{u} \quad \text{over all vectors } \mathbf{u} \text{ satisfying} \quad \|\mathbf{u}\| = 1, \quad \mathbf{u} \cdot \mathbf{q}_1 = \cdots = \mathbf{u} \cdot \mathbf{q}_{j-1} = 0.$$

Theorem 5.47 immediately implies that $\mathbf{q}_j$ is a unit eigenvector of S_X associated with its j-th largest eigenvalue $\lambda_j(S_X)$, or, equivalently, the j-th singular vector of the centered data matrix $\underline{X}$. As in (8.2), the variance in the j-th principal component direction is given by $\lambda_j(S_X) = \sigma_j(\underline{X})^2$. We summarize this discussion in the following theorem.

Theorem 8.1. *The j-th principal direction of a data matrix X is the j-th unit singular vector $\mathbf{q}_j$ of the centered data matrix $\underline{X}$. The corresponding principal standard deviation σ_j is the j-th singular value $\sigma_j = \sigma_j(\underline{X})$.*

Figure 8.1 shows the principal components for some toy data sets in $n = 2$ dimensions. In the plots, the lengths of the principal component vectors are proportional to the corresponding singular value σ_i of the centered data matrix $\underline{X}$, so that they indicate the amount of variance of the data in that direction. For a concrete application to real data, we consider performing PCA on each digit from the MNIST dataset. Figure 8.2 shows the mean images from each class, along with the top $k = 14$ principal components from each digit. These principal components describe the main differences in the ways that people write each digit, and a majority of the MNIST digits can be expressed as linear combinations of the mean digit and the top principal components. We will study this approximation power in the next subsections.

In applications, one designates a certain number, say $k \le r = \operatorname{rank} \underline{X}$, of the dominant (largest) variances $\sigma_1^2 \ge \sigma_2^2 \ge \cdots \ge \sigma_k^2$, as "principal" and the corresponding unit singular

Figure 8.2: The mean image and first 14 principal components from the MNIST data set. These are called *eigendigits*.

vectors $\mathbf{q}_1, \ldots, \mathbf{q}_k$ as the *principal directions*. The value of k depends on the user and on the application. For example, in visualization, we choose $k = 2$ or 3 in order to plot these components of the data in the plane or in space. More generally, one could specify k based on some overall size threshold, or where there is a perceived gap in the magnitudes of the variances. Another choice is to designate the principal variances as those that make up some large fraction $\mu \in [0, 1]$ (e.g., $\mu = 0.95$), of the total variance. That is, if $\underline{X}$ has rank r, then we choose the smallest $k \leq r$ so that

$$\sum_{i=1}^{k} \sigma_i^2 \geq \mu \left(\sum_{i=1}^{r} \sigma_i^2 \right) = \mu \operatorname{tr}\left(\underline{X}^T \underline{X} \right), \tag{8.3}$$

or, equivalently,[3]

$$\frac{1}{\operatorname{tr}\left(\underline{X}^T \underline{X} \right)} \sum_{i=1}^{k} \sigma_i^2 \geq \mu. \tag{8.4}$$

The selected value of k will give an approximate rank of the covariance matrix S_X and hence the centered data matrix $\underline{X}$, thereby indicating that the centered data points all lie (approximately) on a k-dimensional subspace or, equivalently, the original data points all approximately lie on a k-dimensional affine subspace. Further, the variance in any direction orthogonal to principal directions is relatively small, and hence relatively unimportant. As

---

[3]Note that $\operatorname{tr}\left(\underline{X}^T \underline{X} \right) = 0$ if and only if $\underline{X} = O$ is the zero matrix, i.e., all measurements of each quantity coincide. We ignore this trivial and unrealistic case in what follows.

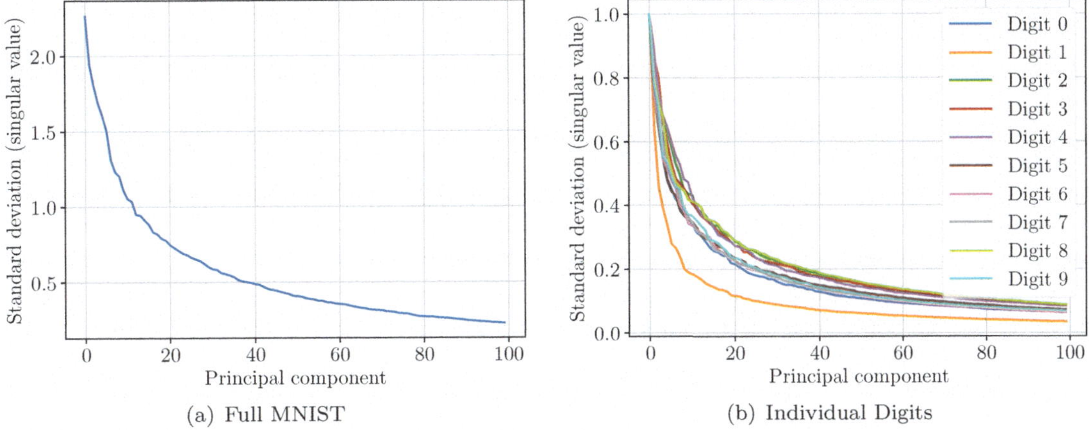

(a) Full MNIST (b) Individual Digits

Figure 8.3: The first 100 singular values, or principal standard deviations, of (a) the MNIST data set and (b) each digit from the MNIST data set. We see the singular values decay very quickly, indicating the MNIST data matrix can be well-approximated by a low rank matrix. The singular values for the digit 1 decay the most quickly. In (b), we normalized the singular values so that $\sigma_1 = 1$ in order to compare the rate of decay between digits.

DIGITS	0	1	2	3	4	5	6	7	8	9	ALL
80%	26	13	39	38	34	35	28	28	41	29	42
90%	61	35	80	79	75	74	61	65	81	62	86
95%	115	70	142	137	134	129	112	121	136	111	153
99%	270	176	304	288	287	282	257	269	276	243	330

Table 8.4: Number of principal components required to achieve 80%, 90%, 95%, and 99% of the variability of each MNIST digit separately, and the entire data set as a whole. We see some variability in the number of required principal components, with the digit 1 requiring the fewest, while the digit 8 generally requires the most.

a consequence, dimensional reduction by orthogonally projecting the data vectors onto the k-dimensional subspace spanned by the principal directions $\mathbf{q}_1, \ldots, \mathbf{q}_k$, serves to eliminate significant redundancies.

In Figure 8.3 we show the first 100 singular values for the MNIST data set, depicting the singular values for each digit, along with the singular values of the whole data set. We see the singular values decay very quickly, indicating a relatively low rank structure in this data set. In Table 8.4 we report the number of principal components necessary to capture 80%, 90%, 95%, and 99% of the variability in the MNIST data set, on a per-digit basis, and for the whole data set. We clearly see some variability across the digits, with the digit 1 requiring the fewest principal components. For example, we can capture 95% of the variability of the digit 1 using 70 principal components, which is significantly smaller than the $n = 784$ dimensions present in the original data set.

Remark 8.2. In practice, we can compute the top singular vectors with an iterative indirect method, such as the power method (for the top singular vector) or the orthogonal iteration, as described in Section 5.6 (for the top k singular vectors) applied to the covariance matrix.[4]

---

[4]We can also use more advanced iterative techniques, such as the Arnoldi method, cf. [181, 205], which are not discussed in this book.

In particular, we do not need to compute all of the singular vectors of $\underline{X}$, which may be computationally intractable in high dimensional settings. We simply need to continue computing singular vectors iteratively until (8.4) holds. In addition, in the setting where $n \gg m$ — the dimension is much larger than the number of data points — we may appeal to Proposition 5.76 (in the dot product setting), which says that $\underline{X}$ and $\underline{X}^T$ have the same singular values, and hence we can instead compute the eigenvectors and eigenvalues of the smaller $m \times m$ matrix $\underline{X}\,\underline{X}^T$, in which we have simply multiplied $\underline{X}$ and $\underline{X}^T$ in the opposite order. Moreover, their corresponding eigenvectors are easily related; see equation (5.104). This idea will be useful later in kernel PCA, to be discussed in Section 8.1.1.   ▲

In applications of PCA to dimension reduction, the data is projected onto the subspace spanned by the top k principal components, retaining only the *coordinates* of this projection. Defining the $n \times k$ matrix $Q_k = (\mathbf{q}_1, \ldots, \mathbf{q}_k)$, whose columns are the top k principal components, the *PCA coordinates* are the entries of the $m \times k$ projected data matrix

$$Y_k = X\,Q_k. \tag{8.5}$$

Note that while the original data points lie in $\mathbb{R}^n$, the PCA coordinate data points — that is, the rows of Y_k — lie in $\mathbb{R}^k$. In particular, this allows for PCA to be used for data visualization when $k = 2$ or $k = 3$. Figure 8.5 shows the results of dimension reduction to $k = 2$ dimensions for a subset of 5000 images from the MNIST data set, which originally has $n = 784$ dimensions and 70000 images. In the plots, we show the first two PCA coordinates, where each point corresponds to an MNIST image and the color indicates the class (i.e., digit). We observe that PCA dimension reduction separates the digits well up to about four digits, after which we see substantial overlap in the visualization.

To further understand the PCA coordinates, we provide a basic result that characterizes how the covariance matrix changes under a linear transformation of the data. The proof of the following formula is left as Exercise 1.5.

> **Proposition 8.3.** *Let X be an $m \times n$ data matrix and let W be an $n \times p$ matrix. Then the covariance matrices of X and $Y = XW$ are related by*
>
> $$S_Y = W^T S_X W. \tag{8.6}$$

We will use Proposition 8.3 to see how the covariance matrix transforms when projecting the data along the principal component directions via (8.5). Let

$$S_X = Q \Lambda Q^T$$

be the spectral decomposition (5.31) of the covariance matrix, where the columns of the orthogonal matrix $Q = (\,\mathbf{q}_1 \ \cdots \ \mathbf{q}_n\,)$ are the orthonormal eigenvector basis, while

$$\Lambda = \mathrm{diag}\,(\lambda_1, \ldots, \lambda_n) = \mathrm{diag}\,(\sigma_1^2, \ldots, \sigma_r^2, 0, \ldots, 0), \qquad \text{with} \qquad r = \mathrm{rank}\,\underline{X} = \mathrm{rank}\,S_X,$$

is the diagonal eigenvalue matrix whose nonzero entries are the squared singular values of the centered data matrix. Assuming $k \le r$, according to Proposition 8.3, the covariance matrix of the projected PCA coordinate data Y_k is given by

$$S_{Y_k} = Q_k^T S_X Q_k = Q_k^T Q \Lambda Q^T Q_k = (Q^T Q_k)^T \Lambda Q^T Q_k = \Sigma_k^2, \tag{8.7}$$

where $\Sigma_k = \mathrm{diag}\,(\sigma_1, \ldots, \sigma_k)$. In the computation above, we used the fact that, owing to the orthogonality of the eigenvectors, the submatrix formed by the first k rows of the $n \times k$

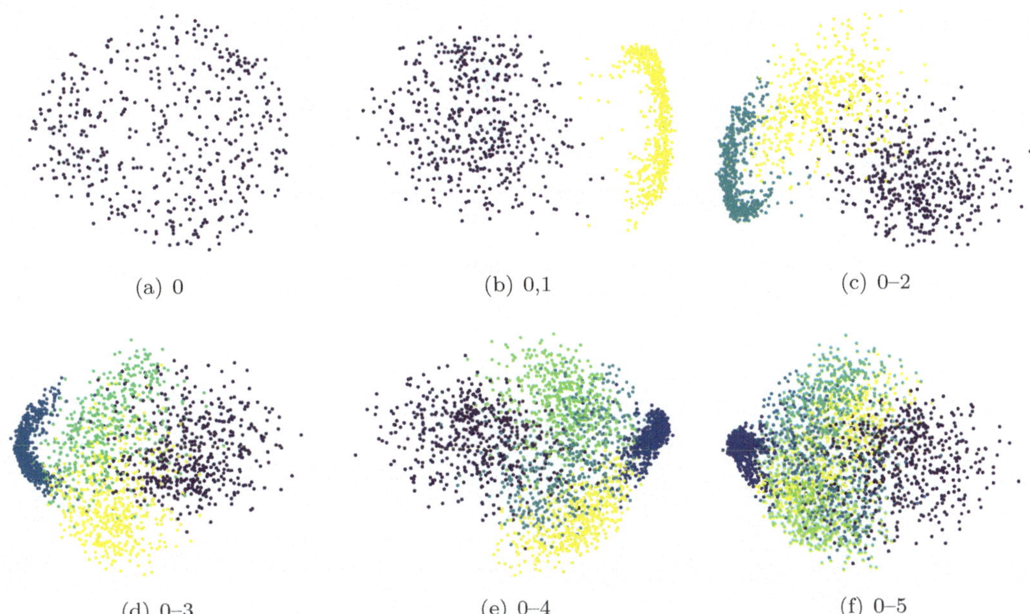

(a) 0   (b) 0,1   (c) 0–2

(d) 0–3   (e) 0–4   (f) 0–5

Figure 8.5: Plots of $k = 2$ PCA coordinates of the MNIST dataset, which allow for dimension reduction and visualization in two dimensions. The plots are colored by the underlying digit label. We start with just the zeros, and incrementally add digits up to 5. We note that PCA is able to well-separate the digits from 0 to 3, but, when we add digits 4 and 5, there is a significant amount of overlap between clusters.

matrix $Q_k^T Q_k$ is the $k \times k$ identity matrix, while the last $n - k$ rows are identically zero, and so only the top k squared singular values are selected in the last equality. Thus, in the PCA coordinates, the covariance matrix S_{Y_k} is diagonal, and hence the principal components are uncorrelated!

A linear transformation that removes the correlations in such data is called a *decorrelation transformation*. In geometric terms, the original data tends to form an (approximate) ellipsoid in the high-dimensional data space, and the principal directions are aligned with its principal semi-axis, thereby conforming to and exposing the intrinsic geometry of the data set. The reader can observe the decorrelation in the MNIST PCA coordinate plots in Figure 8.5, which indeed appear to be uncorrelated as in the middle plot in Figure 7.5.

Inspecting (8.7), we can, in fact, go further and define *normalized PCA coordinates* for which the covariance matrix is the identity by setting

$$Z_k = X Q_k \Sigma_k^{-1} = Y_k \Sigma_k^{-1}. \tag{8.8}$$

In this case, we can again use Proposition 8.3 to obtain that the covariance matrix of the data matrix Z_k is

$$S_{Z_k} = \Sigma_k^{-1} S_{Y_k} \Sigma_k^{-1} = \Sigma_k^{-1} \Sigma_k^2 \Sigma_k^{-1} = I. \tag{8.9}$$

A decorrelation transformation that sets all variances to be equal to one is called a *whitening transformation* or a *sphering transformation*. The data in the matrix Z_k should in general look like the plot in the middle of Figure 7.5 or Figure 8.5(a).

PCA is also used in machine learning as a preprocessing dimension reduction step before applying a machine learning algorithm, such as classification or regression. Reducing the

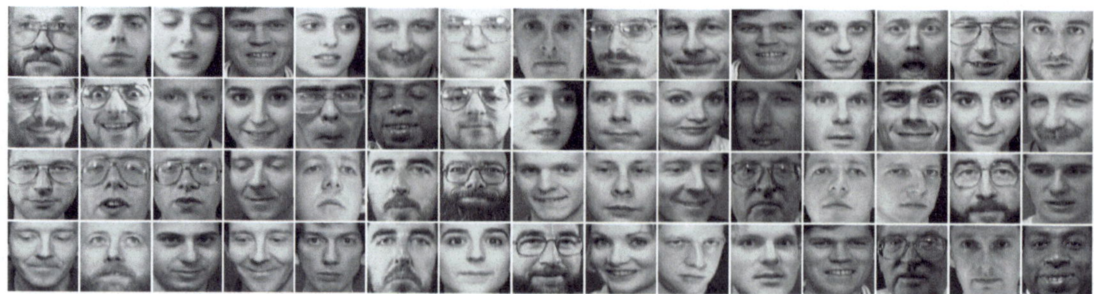

Figure 8.6: Some example of images from the Olivetti face data set.

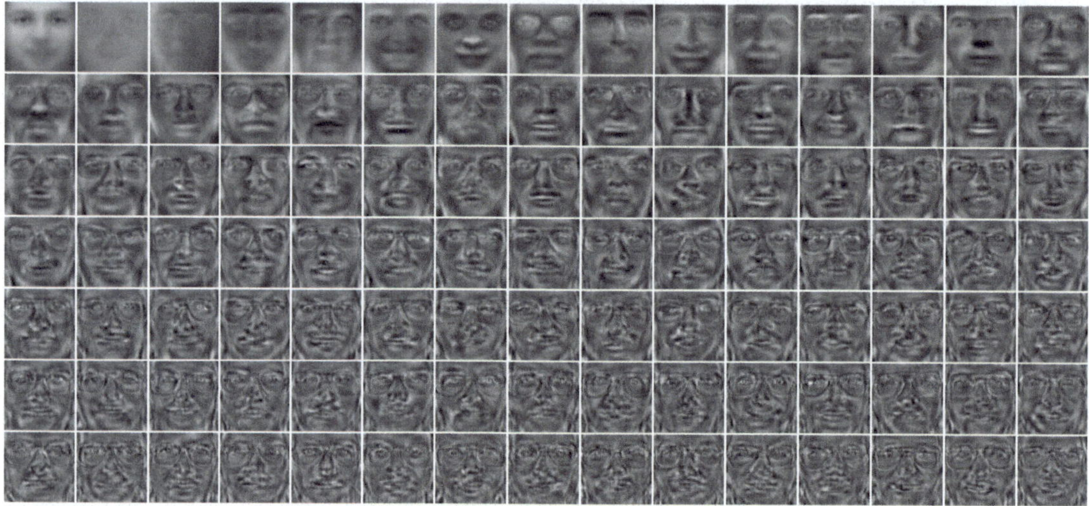

Figure 8.7: The average face (top left) and top 104 principal components of the face dataset, which are often called *eigenfaces*.

dimension allows the downstream machine learning task to focus on the most important features in the data and ignore spurious details such as noise, which can thereby speed up computations, while helping to prevent overfitting. A famous application of this technique is to face recognition [217, 232], which is the task of identifying when two images of a face are the same, when they could be taken from different angles, have different expressions, or be subject to different lighting conditions. We show in Figure 8.6 a random sample of some of the faces images in the Olivetti face data set, which is available from `sklearn` in Python.[5] The data set has 400 grayscale 64×64 pixel images of faces, from 40 different subjects. The 10 images per subject are taken from slightly different angles, with different expressions, and sometimes with or without glasses. The top left plot in Figure 8.7 shows the mean face image, which is followed by the top $k = 104$ principal component images, referred to as *eigenfaces*. The eigenfaces represent the main modes of variation in facial features, at least for this data set. We refer the reader to Exercise 1.7 for the application of eigenfaces to facial recognition.

---

[5]The images were taken between April 1992 and April 1994 at the AT&T Cambridge Laboratory and are described here: `https://cam-orl.co.uk/facedatabase.html`.

8.1.1 Kernel Principal Component Analysis

PCA performs well at uncovering linear structure in data sets, that is, when clusters are well-separated by linear decision boundaries the embedding learned by PCA preserves much of this structure. We refer to, for example, Figure 8.5. On the other hand, PCA can perform poorly on data that has nonlinear geometry and structure — we give examples of this below in Figure 8.8. As was the case with linear SVM and ridge regression in Chapter 7, we can use kernel methods to improve PCA in such settings.

Let $\mathbf{x}_1, \ldots, \mathbf{x}_m \in \mathbb{R}^n$ be a collection of data points. Let $\phi \colon \mathbb{R}^n \to \mathbb{R}^d$ be a feature map, which usually embeds the data into a much higher dimensional space, so $d \gg n$. To derive *kernel PCA*, we consider applying PCA to the feature vectors $\mathbf{z}_i = \phi(\mathbf{x}_i)$ for $i = 1, \ldots, m$. Let Z be the corresponding data matrix defined in (7.55). Then the covariance matrix of Z, see (7.18), is given by

$$S_Z = \underline{Z}^T \underline{Z} = Z^T J Z,$$

since the centering matrix is idempotent: $J = J^2$. This does not seem so helpful, since we have not produced the kernel matrix $K = ZZ^T$, as described in Section 7.6. Even without the centering step, we would have the matrix $Z^T Z$, which is still not the kernel matrix. The key idea in kernel PCA is to use the observation in Remark 8.2 that we can compute the eigenvectors and eigenvalues of the covariance matrix by multiplying $\underline{Z}$ and $\underline{Z}^T$ in the *opposite* order. That is, we can compute the eigenvectors of the matrix

$$\underline{Z}\,\underline{Z}^T = JZ(JZ)^T = JZZ^T J = JKJ, \qquad \text{where} \qquad K = ZZ^T, \tag{8.10}$$

and deduce the eigenvectors of S_Z from these — below, we show how to do this in more detail than what was described in Remark 8.2. Hence, in kernel PCA, the centered kernel matrix JKJ replaces the covariance matrix.

Since J is a symmetric matrix and the kernel matrix K is positive semidefinite, the centered matrix JKJ is also symmetric positive semidefinite. Let $\mathbf{p}_1, \ldots, \mathbf{p}_m \in \mathbb{R}^m$ be its orthonormal eigenvectors with corresponding eigenvalues $\lambda_1 \geq \lambda_2 \geq \cdots \geq \lambda_m \geq 0$. Since $\underline{Z}\,\underline{Z}^T \mathbf{p}_i = \lambda_i\, \mathbf{p}_i$, setting $\mathbf{v}_i = \underline{Z}^T \mathbf{p}_i$ we find that

$$S_Z \mathbf{v}_i = \underline{Z}^T \underline{Z} \mathbf{v}_i = \underline{Z}^T \underline{Z}\,\underline{Z}^T \mathbf{p}_i = \lambda_i \underline{Z}^T \mathbf{p}_i = \lambda_i \mathbf{v}_i.$$

Thus, $\mathbf{v}_1, \ldots, \mathbf{v}_m$ are the top m eigenvectors of the covariance matrix S_Z, and they have the same eigenvalues $\lambda_1, \ldots, \lambda_m$. Since the rank of S_Z is at most m, when $d > m$, the remaining eigenvalues are zero. Since

$$\|\mathbf{v}_i\|^2 = \|\underline{Z}^T \mathbf{p}_i\|^2 = \mathbf{p}_i^T \underline{Z}\,\underline{Z}^T \mathbf{p}_i = \lambda_i \|\mathbf{p}_i\|^2 = \lambda_i,$$

to obtain unit principal component vectors, we set

$$\mathbf{q}_i - \frac{\mathbf{v}_i}{\|\mathbf{v}_i\|} = \lambda_i^{-1/2} \underline{Z}^T \mathbf{p}_i \qquad \text{so that} \qquad \|\mathbf{q}_i\| = 1.$$

In other words, $\mathbf{q}_1, \ldots, \mathbf{q}_m$ are the top m principal components of the feature vector data $\mathbf{z}_1, \ldots, \mathbf{z}_m$.

One further simplification can be made. Since $J\mathbf{1} = 0$, the ones vector $\mathbf{1}$ is a null eigenvector of the centered kernel matrix JKJ. Any eigenvector with positive eigenvalue is necessarily orthogonal to $\mathbf{1}$, so if $\lambda_i > 0$ then $\mathbf{p}_i \cdot \mathbf{1} = 0$, and so $J\mathbf{p}_i = \mathbf{p}_i$. This implies that

$$\mathbf{q}_i = \lambda_i^{-1/2} \underline{Z}^T \mathbf{p}_i = \lambda_i^{-1/2} Z^T J \mathbf{p}_i = \lambda_i^{-1/2} Z^T \mathbf{p}_i.$$

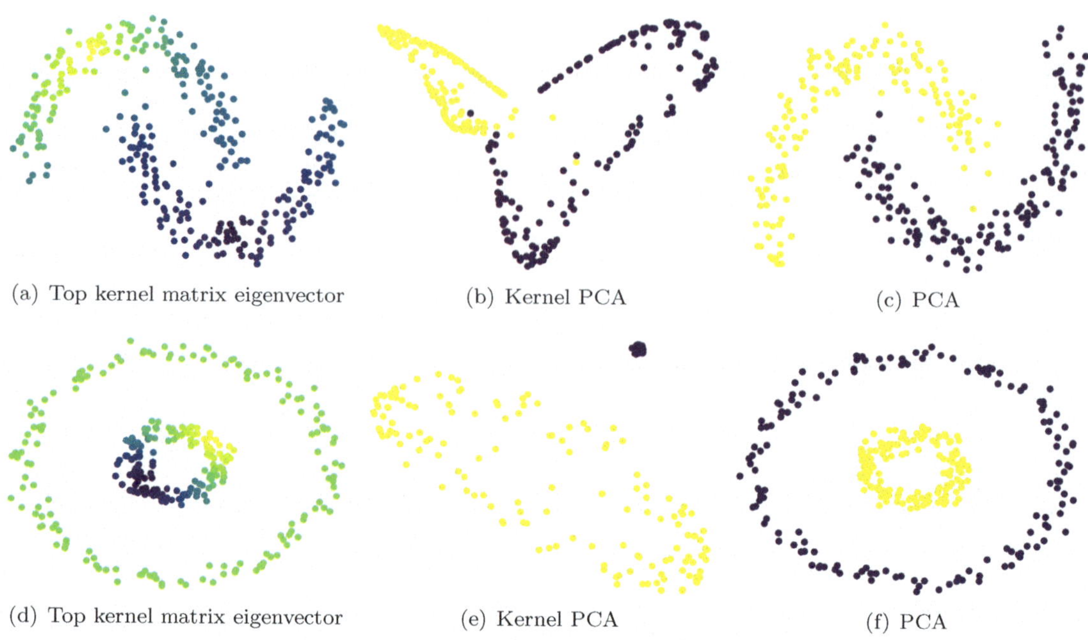

(a) Top kernel matrix eigenvector   (b) Kernel PCA   (c) PCA

(d) Top kernel matrix eigenvector   (e) Kernel PCA   (f) PCA

Figure 8.8: Comparison of kernel PCA with radial basis function kernel and PCA on the two moons and circles data sets.

Thus, we can omit the centering of the data matrix Z when defining the principal components $\mathbf{q}_i$, provided the corresponding eigenvalues are positive, which are the only principal components of interest anyway. We summarize these results in the following theorem.

Theorem 8.4 (Kernel PCA). *Let Z be the transformed data matrix (7.55) associated with the feature vectors $\phi(\mathbf{x}_1), \ldots, \phi(\mathbf{x}_m)$. Let $\mathbf{p}_1, \ldots, \mathbf{p}_m \in \mathbb{R}^m$ be the orthonormal eigenvectors, with corresponding eigenvalues $\lambda_1 \geq \lambda_2 \geq \cdots \geq \lambda_m \geq 0$, of the centered kernel matrix JKJ, where $K = ZZ^T$. If $\lambda_k > 0$, then the top k principal components of Z are $\mathbf{q}_i = \lambda_i^{-1/2} JZZ^T \mathbf{p}_i$ with corresponding singular value $\sigma_i^2 = \lambda_i$ for $i = 1, \ldots, k$.*

We now use the *kernel trick* of choosing the kernel function $\mathcal{K}$ instead of the feature map ϕ, as we did in Section 7.6. Given a choice of kernel function $\mathcal{K}$, we compute the entries $k_{ij} = \mathcal{K}(\mathbf{x}_i, \mathbf{x}_j)$ of the kernel matrix $K \in \mathcal{M}_{m \times m}$, and then compute the orthonormal eigenvectors $\mathbf{p}_1, \ldots, \mathbf{p}_m$ and eigenvalues $\lambda_1 \geq \lambda_2 \geq \cdots \geq \lambda_m \geq 0$ of the centered kernel matrix JKJ. However, at this point we cannot go further and compute the principal components $\mathbf{q}_i$ from Theorem 8.4, since we do not have a feature map ϕ or feature vectors $\phi(\mathbf{x}_i)$ from which to build the matrix Z — indeed, this is the whole point of the kernel trick! Even if we could find the feature map ϕ corresponding to the chosen kernel function, as we pointed out in Section 7.6, it can map to a *very* high dimensional feature space — indeed, in general it will be infinite-dimensional.

Instead, we note that the entries of the vectors $\mathbf{p}_1, \ldots, \mathbf{p}_m \in \mathbb{R}^m$ are associated with the data points $\mathbf{x}_1, \ldots, \mathbf{x}_m$, so we can visualize them on the data points themselves. This is because the entries of the kernel matrix $k_{ij} = \mathcal{K}(\mathbf{x}_i, \mathbf{x}_j)$ are associated with pairs of data points, and each $\mathbf{p}_i$ is an eigenvector of K. To see an illustrative example, we ran kernel

PCA with the radial basis function kernel with $\gamma = 10$ on the toy two moons and circles data sets, both of which exhibit nonlinearly separable cluster structure that cannot be uncovered with PCA. Figure 8.8 shows a visualization of the top eigenvector $\mathbf{p}_1$ on both data sets. The colors of the data points in the figures on the left correspond to the values of the coordinates of $\mathbf{p}_1 \in \mathbb{R}^m$ on each of the $m = 300$ data points. We see that the values are large in one cluster and small in the other, indicating that the top kernel matrix eigenvector can separate the clusters well.

To project the data to kernel PCA coordinates, we can simply take the top k eigenvectors $\mathbf{p}_1, \ldots, \mathbf{p}_k$ and assemble them into the columns of a matrix $P_k = (\mathbf{p}_1, \ldots, \mathbf{p}_k) \in \mathcal{M}_{m \times k}$. Then the *rows* of the matrix P_k are the embeddings of each data point into $\mathbb{R}^k$. In Figure 8.8 we show the embedding into $k = 2$ dimensions in the middle figures. The x-coordinate in the figure is exactly the top principal component $\mathbf{p}_1$ illustrated by the colors of data points in the figures on the left. The y-coordinate corresponds to $\mathbf{p}_2$. We see that kernel PCA is able to linearly separate the moons and circles, while ordinary PCA is not able to do this, and leaves the data sets largely unchanged.

Kernel PCA is closely related to spectral methods in graph-based learning, such as spectral clustering and spectral embeddings, which will be discussed in detail in Chapter 9.

Exercises

Note: Recall that the variance prefactor is set to $\nu = 1$.

1.1. ♡ Construct the 5×5 covariance matrix for the data set from Exercise 1.1 in Chapter 7, and find its principal variances, principal standard deviations, and principal directions. What do you think is the dimension of the subspace the data lies in?

1.2. ♡ Using the Euclidean norm, compute a fairly dense sample of points on the unit sphere $S = \{\mathbf{x} \in \mathbb{R}^3 \mid \|\mathbf{x}\| = 1\}$. (a) Set $\mu = .95$ in (8.3), and then find the principal components of your data set. Do they indicate the two-dimensional nature of the sphere? If not, why not? (b) Now look at the subset of your data that is within a distance $r > 0$ of the north pole, i.e., $\|\mathbf{x} - (0, 0, 1)^T\| \leq r$, and compute its principal components. How small does r need to be to reveal the actual dimension of S? Interpret your calculations.

1.3. For each of the following subsets $S \subset \mathbb{R}^3$: (*i*) Compute a fairly dense sample of data points $\mathbf{z}_i \in S$; (*ii*) find the principal components of your data set, using $\mu = .95$ in the criterion in (8.3); (*iii*) using your principal components, estimate the dimension of the set S. Does your estimate coincide with the actual dimension? If not, explain any discrepancies.
    (a) The line segment $S = \{(t + 1, 3t - 1, -2t)^T \mid -1 \leq t \leq 1\}$;
    (b) the set of points $\mathbf{z}$ on the three coordinate axes with Euclidean norm $\|\mathbf{z}\| \leq 1$;
    (c) the set of "probability vectors" $S = \{(x, y, z)^T \mid 0 \leq x, y, z \leq 1,\ x + y + z = 1\}$;
    (d) the unit ball $S - \{\|\mathbf{z}\| \leq 1\}$ for the Euclidean norm;
    (e) the unit sphere $S = \{\|\mathbf{z}\| = 1\}$ for the Euclidean norm;
    (f) the unit ball $S = \{\|\mathbf{z}\|_\infty \leq 1\}$ for the ∞ norm;
    (g) the unit sphere $S = \{\|\mathbf{z}\|_\infty = 1\}$ for the ∞ norm.

1.4. ♡ Show that the first principal direction $\mathbf{q}_1$ can be characterized as the direction of the line that minimizes the sums of the squares of its distances to the data points. *Hint*: Use Theorem 5.43.

1.5. ♡ Prove Proposition 8.3.

1.6. Write Python code to apply PCA as a preprocessing step on the *training* set in a classification problem using support vector machines from Section 7.3 or *k*-nearest neighbors from Section 7.4. Pick one of the classification data sets from `sklearn.datasets`, or use MNIST. How does the accuracy change with the number of principal components used?

1.7. Write Python code to use the eigenfaces approach for facial recognition on the Olivetti faces data set. After using PCA for dimension reduction, use a nearest neighbor classifier in the PCA coordinates. Make sure to apply PCA to the *training set* after a train-test split. Try both decorrelation and whitening transformations. Which works best? Use the Python notebook from this section to get started.

1.8. $\diamond$ Write Python code to compute the top principal component for MNIST and the Olivetti face data set using the power method described in Section 5.6. You can start with the notebook below.

Python Notebook: Numerical Computation of Eigenvalues (.ipynb)

1.9. $\diamond$ Write Python code to compute the top k principal components for MNIST and the Olivetti face data set using the orthogonal iteration method described in Section 5.6. Start with the notebook from Exercise 1.8.

1.10. Write Python code to implement kernel PCA and test it on the two moons and circles data sets, which are available through the `sklearn` Python package.

8.2   The Best Approximating Subspace

In the previous section, we saw that PCA can be defined by looking for directions of maximum variability in the data matrix. In this section, we perform a mathematical study of PCA from an optimization perspective. In particular, we will interpret the *affine subspace* determined by top k principal components as forming the best k dimensional *linear approximation* of the data set. Optimality is measured by minimization of the sum of squared distances between each data point and its projection to the affine subspace. We will also show, in a similar vein, that PCA can be interpreted as finding the bast rank k approximation of the data matrix in the Frobenius matrix norm.

To linearly approximate a data set, we cannot, in general, work with linear subspaces of $\mathbb{R}^n$, since a data set may not be located anywhere near the origin. Instead, we work with affine subspaces, which include lines, planes, etc., that do not pass through the origin. Recalling Exercise 2.7:

Definition 8.5. An *affine subspace* $W \subset \mathbb{R}^n$ is a set of the form

$$W = \mathbf{a} + V = \{ \mathbf{a} + \mathbf{v} \mid \mathbf{v} \in V \}, \tag{8.11}$$

where $\mathbf{a} \in \mathbb{R}^n$ and $V \subset \mathbb{R}^n$ is a (linear) subspace. We refer to $\mathbf{a}$ as the *offset* of the affine subspace. The *dimension* of W is equal to the dimension of V.

Keep in mind that an affine subspace is not generally a subspace in the ordinary sense since, unless it contains the origin, it is not closed under vector addition and scalar multiplication.

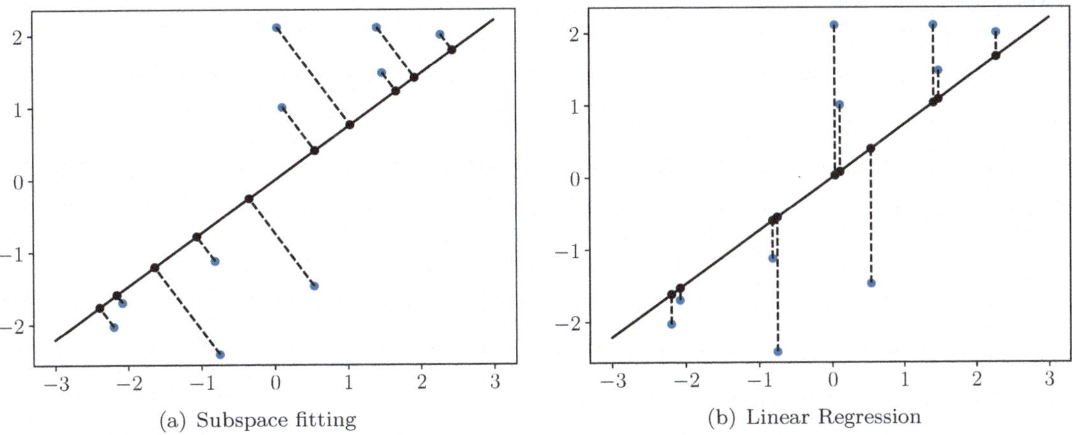

(a) Subspace fitting
(b) Linear Regression

Figure 8.9: Comparison of subspace fitting and linear regression for finding a line of best fit. Subspace fitting (i.e., minimizing (8.13) measures errors orthogonally to the subspace, while linear regression identifies a target variable (in this case y) and measures errors in this variable (in this case vertically).

Indeed, W is a subspace if and only if $\mathbf{a} \in V$, in which case $W = V$. More generally, $W = \mathbf{a} + V = \mathbf{b} + V$ if and only if $\mathbf{a} - \mathbf{b} \in V$. Thus the offset $\mathbf{a}$ of an affine subspace is not unique, and can be taken to be any point $\mathbf{a} \in W$.

Example 8.6. The line $L = \{y = 2x + 1\} \subset \mathbb{R}^2$ is an affine subspace, given by $L = \mathbf{a} + V$, where $V = \{y = 2x\}$ and $\mathbf{a} = (0, 1)^T$. More generally, we can take $\mathbf{a}$ to be any point lying in L. Similarly, the plane $P = \{2x - y - z = 3\} \subset \mathbb{R}^3$ is the affine subspace $P = \mathbf{a} + V$, where $V = \{2x - y - z = 0\}$ and, for example, $\mathbf{a} = (0, 0, -3)^T$. $\blacktriangle$

The closeness of a set of data points and an affine subspace can be measured in a variety of ways. The easiest one to deal with, and hence the choice in many (but not all) applications, is based on the squared Euclidean distance. First, as with the distance between a point and an ordinary subspace, we set the distance between a point and an affine subspace to be the minimum distance from the point to any point therein:

$$\operatorname{dist}(\mathbf{x}, W) = \min \{ \, \|\mathbf{x} - \mathbf{y}\| \mid \mathbf{y} \in W \, \}. \tag{8.12}$$

Soon we will see how to use our orthogonal projection formulas to easily compute this distance. Given points $\mathbf{x}_1, \ldots, \mathbf{x}_m \in \mathbb{R}^n$ and an affine subspace $W \subset \mathbb{R}^n$ we define the *squared distance energy function* to be

$$E(W; \mathbf{x}_1, \ldots, \mathbf{x}_m) = \sum_{i=1}^{m} \operatorname{dist}(\mathbf{x}_i, W)^2. \tag{8.13}$$

Our goal is to find the affine subspace $W = \mathbf{a} + V$ that minimizes the energy (8.13). As noted above, the use of squared distance makes the analysis much easier than using, say, just the sum of the distances, although, as we explore in Section 8.2.1, the sum of distances is more robust. On the other hand, one can replace the Euclidean distance by any other distance based on an inner product on $\mathbb{R}^n$ without appreciable complications; the required analysis is delegated to the exercises. While this problem shares some similarities with linear regression, studied in Section 7.2, they are fundamentally different in their objectives and how they measure error. Figure 8.9 shows a simple example of this difference for finding a line of best fit.

We now proceed to study the minimization of the squared distance energy (8.13). First let us recall, (4.54), the orthogonal projection matrix $P = UU^T$, where the columns of $U = (\mathbf{u}_1 \ \ldots \ \mathbf{u}_k)$ form an orthonormal basis of V, so that $P\mathbf{x}$ is the orthogonal projection of $\mathbf{x}$ onto V. We also recall the definition of $R = I - P$ from (4.58), often called the *residual matrix*, which has the property that $R\mathbf{x} = \mathbf{x} - P\mathbf{x}$ is the difference between $\mathbf{x}$ and its orthogonal projection onto V, and can be identified with the orthogonal projection of $\mathbf{x}$ onto the orthogonal complement $V^\perp$; for details see (4.55), (4.59). These definitions allow us to easily compute the distance from a point $\mathbf{x}$ to an affine subspace W.

Lemma 8.7. *Let $\mathbf{x} \in \mathbb{R}^n$, and let $W = \mathbf{a} + V$ be an affine subspace. Then*

$$\operatorname{dist}(\mathbf{x}, W) = \| R(\mathbf{x} - \mathbf{a}) \|. \tag{8.14}$$

Proof. It suffices to note that $\operatorname{dist}(\mathbf{x}, \mathbf{a} + V) = \operatorname{dist}(\mathbf{x} - \mathbf{a}, V)$ by the translational invariance of distance. Thus, the result follows from the closest point Theorem 2.25. ∎

Now, the first step in our analysis of the optimality of PCA is to determine the best value of the offset $\mathbf{a}$.

Lemma 8.8. *Let $\mathbf{x}_1, \ldots, \mathbf{x}_m \in \mathbb{R}^n$, and let $V \subset \mathbb{R}^n$ be a fixed subspace. Then an offset $\mathbf{a} \in \mathbb{R}^n$ that minimizes the squared energy (8.13) over all affine subspaces of the form $\mathbf{a} + V$ is the mean of the data points:*

$$\mathbf{a} = \overline{\mathbf{x}} = \frac{1}{m} \sum_{i=1}^{m} \mathbf{x}_i. \tag{8.15}$$

Proof. We compute, using (8.14),

$$\sum_{i=1}^{m} \operatorname{dist}(\mathbf{x}_i, W)^2 = \sum_{i=1}^{m} \| R(\mathbf{x}_i - \mathbf{a}) \|^2 = \sum_{i=1}^{m} \left(\| R\mathbf{x}_i \|^2 - 2 (R\mathbf{x}_i) \cdot (R\mathbf{a}) + \| R\mathbf{a} \|^2 \right)$$

$$= \sum_{i=1}^{m} \| R\mathbf{x}_i \|^2 - 2m (R\overline{\mathbf{x}}) \cdot (R\mathbf{a}) + m \| R\mathbf{a} \|^2$$

$$= \left(\sum_{i=1}^{m} \| R\mathbf{x}_i \|^2 - m \| R\overline{\mathbf{x}} \|^2 \right) + m \| R(\mathbf{a} - \overline{\mathbf{x}}) \|^2.$$

The initial terms in parentheses are independent of $\mathbf{a}$, and hence the energy is minimized when the last term vanishes, which requires

$$R(\mathbf{a} - \overline{\mathbf{x}}) = \mathbf{0}, \qquad \text{and hence} \qquad \mathbf{a} = \overline{\mathbf{x}} + \mathbf{v}, \qquad \text{where} \quad \mathbf{v} \in \ker R = V.$$

In order to set $\mathbf{a}$ independently of the subspace V, we choose $\mathbf{v} = \mathbf{0}$ above, since regardless of the choice of V, we know that $\mathbf{0} \in V$, since V is a subspace. This yields $\mathbf{a} = \overline{\mathbf{x}}$. ∎

Thus, we can center our data by subtracting the mean from the data points — exactly as is done in PCA — by setting $\mathbf{y}_i = \mathbf{x}_i - \overline{\mathbf{x}}$, $i = 1, \ldots, m$, and noting that $\overline{\mathbf{y}} = \mathbf{0}$. By translation invariance, the energy (8.13) becomes

$$E(V; \mathbf{y}_1, \ldots, \mathbf{y}_m) = \sum_{i=1}^{m} \operatorname{dist}(\mathbf{y}_i, V)^2, \tag{8.16}$$

and our goal now is to minimize over all k-dimensional subspaces $V \subset \mathbb{R}^n$. The key result — Theorem 8.9 below — is that the minimizing subspace is the one spanned by the first k singular vectors of the data matrix $Y = (\mathbf{y}_1 \ldots \mathbf{y}_m)$ or, equivalently, the first k principal components of the data (since $Y = \underline{X}$ is exactly the centered data matrix introduced earlier).

> **Theorem 8.9.** *Let $k \le r = \operatorname{rank} Y$. Then the k-dimensional subspace $V_k \subset \mathbb{R}^n$ that minimizes the squared distance energy (8.16) is the one spanned by the top k singular vectors $\mathbf{q}_1, \ldots, \mathbf{q}_k$ of the centered data matrix Y, or equivalently, the top k eigenvectors of the Gram matrix $S = Y^T Y$. Furthermore, the minimal energy is given by*
>
> $$E(V_k; \mathbf{y}_1, \ldots, \mathbf{y}_m) = \sum_{i=1}^{m} \operatorname{dist}(\mathbf{y}_i, V_k)^2 = \operatorname{tr} S - \sum_{i=1}^{k} \lambda_i = \sum_{i=k+1}^{n} \lambda_i = \sum_{i=k+1}^{r} \sigma_i^2, \quad (8.17)$$
>
> *where $\lambda_1, \ldots, \lambda_n$ are the eigenvalues of S and $\sigma_1, \ldots, \sigma_r$ are the singular values of Y, both arranged in decreasing order.*

Remark. We note that when Y is a centered data matrix, the matrix $S = Y^T Y$ in Theorem 8.9 is the covariance matrix of the data. However, the theorem does not require this, and holds even when Y is not centered. ▲

Example 8.10. As an example of Theorem 8.9, suppose the data points all lie along the line spanned by a unit vector $\mathbf{u} \in \mathbb{R}^n$, so $\mathbf{y}_i = c_i \mathbf{u}$ for some scalars $c_1, \ldots, c_m$. Let us write $\mathbf{c} = (c_1, \ldots, c_m)^T \in \mathbb{R}^m$. Then the data matrix is given by

$$Y = (\mathbf{y}_1 \ldots \mathbf{y}_m)^T = (c_1 \mathbf{u} \ldots c_m \mathbf{u})^T = \mathbf{c}\, \mathbf{u}^T.$$

Thus, Y is a rank one matrix with singular value decomposition

$$Y = \sigma \mathbf{p}\mathbf{q}^T, \qquad \text{where} \qquad \sigma = \|\mathbf{c}\|, \qquad \mathbf{p} = \frac{\mathbf{c}}{\|\mathbf{c}\|}, \qquad \mathbf{q} = \mathbf{u},$$

since $\mathbf{u}$ was assumed to be a unit vector. The top (in fact, the only) singular vector is $\mathbf{q} = \mathbf{u}$, which spans the line containing all data points, and so the minimal energy is identically zero. The covariance matrix is

$$S = Y^T Y = \mathbf{u}\mathbf{c}^T \mathbf{c}\mathbf{u}^T = \|\mathbf{c}\|^2 \mathbf{u}\mathbf{u}^T,$$

whose eigenvalues are $\lambda_1 = \|\mathbf{c}\|^2$, with eigenvector $\mathbf{q}_1 = \mathbf{u}$, and $\lambda_i = 0$ for $i \ge 2$. Keep in mind that the null eigenvectors are not uniquely determined; any orthonormal basis of $\ker S = \ker Y = \mathbf{u}^\perp$ will do. ▲

Proof of Theorem 8.9. Let $\mathbf{u}_1, \ldots, \mathbf{u}_k$ be an orthonormal basis for V. Let $U = (\mathbf{u}_1 \ldots \mathbf{u}_k)$ and let $P = U U^T$ be the corresponding orthogonal projection matrix, cf. (2.39). Then the squared distance energy is given by

$$\sum_{i=1}^{m} \operatorname{dist}(\mathbf{y}_i, V)^2 = \sum_{i=1}^{m} \|R\mathbf{y}_i\|^2 = \sum_{i=1}^{m} \|\mathbf{y}_i - U U^T \mathbf{y}_i\|^2$$

$$= \sum_{i=1}^{m} \left(\|\mathbf{y}_i\|^2 - 2\mathbf{y}_i^T U U^T \mathbf{y}_i + \mathbf{y}_i^T U U^T U U^T \mathbf{y}_i \right)$$

$$= \sum_{i=1}^{m} \|\mathbf{y}_i\|^2 - \sum_{i=1}^{m} \mathbf{y}_i^T U U^T \mathbf{y}_i = \operatorname{tr} S - \sum_{i=1}^{m} \|U^T \mathbf{y}_i\|^2,$$

where we used (4.57) to replace $UU^TUU^T = UU^T$. The first summation is independent of the subspace V, and hence we need only minimize the second summation, or, equivalently, maximize

$$
\sum_{i=1}^{m} \| U^T \mathbf{y}_i \|^2 = \sum_{i=1}^{m} \sum_{j=1}^{k} (\mathbf{u}_j^T \mathbf{y}_i)^2 = \sum_{j=1}^{k} \sum_{i=1}^{m} (\mathbf{y}_i^T \mathbf{u}_j)^2 = \sum_{j=1}^{k} \| Y \mathbf{u}_j \|^2 = \sum_{j=1}^{k} \mathbf{u}_j^T Y^T Y \mathbf{u}_j.
$$

We now insert the spectral factorization

$$
S = Y^T Y = Q \Lambda Q^T, \qquad \text{where} \qquad Q = (\mathbf{q}_1 \ \cdots \ \mathbf{q}_n), \qquad \Lambda = \operatorname{diag}(\lambda_1, \ldots, \lambda_n),
$$

into the preceding formula:

$$
\sum_{j=1}^{k} \mathbf{u}_j^T Y^T Y \mathbf{u}_j = \sum_{j=1}^{k} \mathbf{u}_j^T Q \Lambda Q^T \mathbf{u}_j = \sum_{i=1}^{n} \lambda_i \sum_{j=1}^{k} (\mathbf{q}_i^T \mathbf{u}_j)^2 = \sum_{i=1}^{n} \lambda_i \| P \mathbf{q}_i \|^2, \tag{8.18}
$$

where the last equality follows from (2.41). Since the eigenvalues are arranged from largest to smallest, we claim that (8.18) is maximized by setting $\mathbf{u}_i = \mathbf{q}_i$ for $i = 1, \ldots, k$, which produces $P\mathbf{q}_i = \mathbf{q}_i$ for $i = 1, \ldots, k$ and $P\mathbf{q}_i = \mathbf{0}$ for $i \geq k + 1$, and so

$$
\sum_{i=1}^{m} \| U^T \mathbf{y}_i \|^2 = \sum_{i=1}^{m} \lambda_i \| \mathbf{q}_i \|^2 = \sum_{i=1}^{k} \lambda_i,
$$

which gives the optimal energy in (8.17). That this choice is optimal is an application of Lemma 8.11 below with $\alpha_i := \| P \mathbf{q}_i \|^2$. Indeed, since P is a projection, we have

$$
0 \leq \alpha_i = \| P \mathbf{q}_i \|^2 \leq \| \mathbf{q}_i \|^2 \leq 1, \qquad i = 1, \ldots, n,
$$

and we easily check that

$$
\sum_{i=1}^{n} \alpha_i = \sum_{i=1}^{n} \| P \mathbf{q}_i \|^2 = \sum_{j=1}^{k} \sum_{i=1}^{n} (\mathbf{q}_i \cdot \mathbf{u}_j)^2 = \sum_{j=1}^{k} \| \mathbf{u}_j \|^2 = k,
$$

since $\mathbf{q}_1, \ldots, \mathbf{q}_n$ is an orthonormal basis of $\mathbb{R}^n$. Thus, the values of α_i satisfy the assumptions in Lemma 8.11, which completes the proof. $\blacksquare$

Lemma 8.11. *Let $\lambda_1 \geq \lambda_2 \geq \cdots \geq \lambda_n \geq 0$. Suppose $0 \leq \alpha_i \leq 1$ and $\alpha_1 + \cdots + \alpha_n = k$. Then the weighted sum $s = \alpha_1 \lambda_1 + \cdots + \alpha_n \lambda_n$ is maximized over all possible choices of $\alpha_1, \ldots, \alpha_n$ by setting $\alpha_1 = \cdots = \alpha_k = 1$ and $\alpha_{k+1} = \cdots = \alpha_n = 0$, with maximal value $s_{\max} = \lambda_1 + \cdots + \lambda_k$.*

Proof. Since the λ_i are in decreasing order,

$$
s = \sum_{i=1}^{n} \alpha_i \lambda_i = \sum_{i=1}^{k} \lambda_i - \sum_{i=1}^{k} (1 - \alpha_i) \lambda_i + \sum_{i=k+1}^{n} \alpha_i \lambda_i
$$

$$
\leq \sum_{i=1}^{k} \lambda_i - \lambda_k \sum_{i=1}^{k} (1 - \alpha_i) + \lambda_k \sum_{i=k+1}^{n} \alpha_i = \sum_{i=1}^{k} \lambda_i + \lambda_k \left(\sum_{i=1}^{n} \alpha_i - k \right) = \sum_{i=1}^{k} \lambda_i,
$$

with equality clearly achieved by the indicated choices of α_i. $\blacksquare$

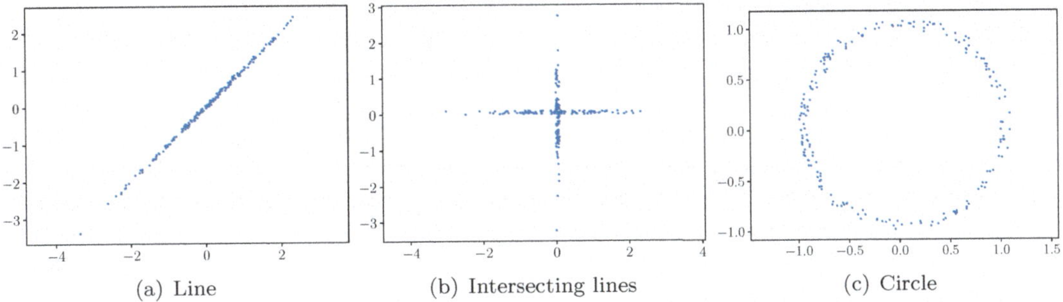

(a) Line   (b) Intersecting lines   (c) Circle

Figure 8.10: Examples of data sets with low dimensional structure.

Theorem 8.9 shows that if a data set is well-represented by a low dimensional linear subspace, then that subspace can be discovered through PCA, or equivalently, singular value decomposition. It is interesting to observe, however, that data sets can have low dimensional structures that are not linear and this cannot be uncovered through PCA. As an example, we show plots of three data sets in Figure 8.10. All three exhibit some form of low dimensional structure, but only the first plot, of data along a line, will be detected through PCA, while it does not reveal the low dimensional structure of data sets that lie on intersections of lines, or on nonlinear curves and/or surfaces. We saw briefly in Section 8.1.1 that kernel methods can be utilized with PCA to uncover nonlinear geometry and structure. We will see further examples of this in Section 8.5 and Chapter 9. We could also apply PCA to subsets of the data, i.e., to a data point and its nearest neighbors, in order to try and discover nonlinear low dimensional structures. Each of the data sets in Figure 8.10 resembles a line at small scales (with the exception of a neighborhood of the crossing point in (b)). This localized application of PCA can uncover nonlinear structures in data, and is related to graph-based learning techniques discussed in Chapter 9.

Remark 8.12. Inspecting the proof of Theorem 8.9, we see there are several alternative ways to interpret the top k singular vectors $Q_k = (\, \mathbf{q}_1 \, \dots \, \mathbf{q}_k \,)$ of a matrix Y. First, noting that $P\mathbf{y}_i = UU^T \mathbf{y}_i$ is the transpose of the i-th row of $YP^T = YP$, the first full line equation in the proof can be written as

$$\sum_{i=1}^{m} \mathrm{dist}(\mathbf{y}_i, V)^2 = \sum_{i=1}^{m} \| \mathbf{y}_i - P\mathbf{y}_i \|^2 = \| Y - YP \|_F^2 = \| Y - YUU^T \|_F^2, \qquad (8.19)$$

where $\| \cdot \|_F$ denotes the Frobenius norm; see (4.87). Thus, by Theorem 8.9, the solution of the optimization problem

$$\min \left\{ \, \| Y - YUU^T \|_F^2 \mid U^TU = I \, \right\}, \qquad (8.20)$$

over $n \times k$ matrices U is exactly $U = Q_k$.

If we look further into the proof, we notice the equivalent problem of maximizing the left hand side of (8.18), which, according to Exercise 2.1, can be written as

$$\sum_{j=1}^{k} \mathbf{u}_j^T Y^T Y \mathbf{u}_j = \mathrm{tr}\,(U^T Y^T YU) = \mathrm{tr}\,(U^T SU) \qquad (8.21)$$

where we recall $S = Y^T Y$. Thus, the solution of the optimization problem

$$\max \left\{ \, \mathrm{tr}\,(U^T SU) \mid U^TU = I \, \right\} \qquad (8.22)$$

over $n \times k$ matrices U with orthonormal columns is again exactly $U = Q_k$.

When $Y = \underline{X}$ is a centered data matrix, $S = S_X$ is the covariance matrix of the data, and by Proposition 8.3, $U^T S_X U$ is the covariance matrix of the transformed data $Z = XU$, i.e., $S_Z = U^T S_X U$. The quantity being optimized in PCA is thus $\operatorname{tr}(U^T S_X U) = \operatorname{tr} S_Z$, which is the *total variance* of the transformed data $Z = XU$; see (7.20). Thus, PCA is finding the linear transformation that maximizes the total variance in PCA coordinates. Recalling that PCA was initially defined in a greedy way, by sequentially choosing directions that maximize variance, it is remarkable that the chosen directions are also optimal for maximizing the *total variance* after k directions have been chosen. ▲

Underlying Theorem 8.9 and Remark 8.12 is a more fundamental result, known as the *Schmidt–Eckart–Young–Mirsky Theorem*, that is worth stating on its own.

Theorem 8.13. *Let $X \in \mathcal{M}_{m \times n}$ be a rank r matrix, and denote its singular value decomposition by*

$$X = \sigma_1 \mathbf{p}_1 \mathbf{q}_1^T + \cdots + \sigma_r \mathbf{p}_r \mathbf{q}_r^T. \tag{8.23}$$

For any $k \leq r$, the best approximating matrix of X in the Frobenius norm with rank at most k is the truncated singular value decomposition matrix

$$X_k = \sigma_1 \mathbf{p}_1 \mathbf{q}_1^T + \cdots + \sigma_k \mathbf{p}_k \mathbf{q}_k^T. \tag{8.24}$$

That is,

$$\|X - X_k\|_F \leq \|X - A\|_F \tag{8.25}$$

for all $m \times n$ matrices A with $\operatorname{rank} A \leq k$. Furthermore, the error is given by

$$\|X - X_k\|_F^2 = \sum_{i=k+1}^{r} \sigma_i^2. \tag{8.26}$$

Remark 8.14. Theorem 8.13 was originally proved by Erhard Schmidt in 1907, [208], in the setting of infinite-dimensional vector spaces. The result was independently rediscovered in 1936 in the finite dimensional setting by Carl Eckart and Gale Young, [68]. Later, Leon Mirsky, [168], found a generalization. It is often misattributed to only the last three authors. ▲

Proof. Let $A = (\mathbf{a}_1 \ldots \mathbf{a}_m)^T$ and $X = (\mathbf{x}_1 \ldots \mathbf{x}_m)^T$, so $\mathbf{a}_i^T$ and $\mathbf{x}_i^T$ denote the i-th rows of A and X, respectively. Then, using (4.87), the squared Frobenius norm of $X - A$ is given by

$$\|X - A\|_F^2 = \sum_{i=1}^{m} \|\mathbf{x}_i - \mathbf{a}_i\|^2.$$

For the moment we assume that $\operatorname{rank} A = k$. Let $V = \operatorname{coimg} A \subset \mathbb{R}^n$ be the k-dimensional subspace spanned by $\mathbf{a}_1, \ldots, \mathbf{a}_m$, and let $P = UU^T$ be the orthogonal projection matrix projecting $\mathbb{R}^n$ onto V, so U is the $n \times k$ matrix whose columns form an orthonormal basis for V. We can replace the rows of A by the projections of the rows of X onto V. Indeed, since the orthogonal projection $P\mathbf{x}_i$ is the closest point in V to $\mathbf{x}_i$, we have

$$\|X - A\|_F^2 = \sum_{i=1}^{m} \|\mathbf{x}_i - \mathbf{a}_i\|^2 \geq \sum_{i=1}^{m} \|\mathbf{x}_i - P\mathbf{x}_i\|^2 = \|X - XP\|_F^2 = \|X - XUU^T\|_F^2.$$

According to Remark 8.12, the minimizer of the right hand side over all $U \in \mathcal{M}_{n \times k}$ with $U^T U = I$ is $U = Q_k = (\mathbf{q}_1 \; \cdots \; \mathbf{q}_k)$. We conclude that the best approximating matrix is

$$A = X Q_k Q_k^T = \sum_{i=1}^{r} \sigma_i \mathbf{p}_i \mathbf{q}_i^T Q_k Q_k^T = \sum_{i=1}^{k} \sigma_i \mathbf{p}_i \mathbf{e}_i Q_k^T = \sum_{i=1}^{k} \sigma_i \mathbf{p}_i \mathbf{q}_i^T = X_k,$$

where the sum reduced to $i = 1, \ldots, k$ in the third equality since, by orthogonality, $\mathbf{q}_i^T Q_k = \mathbf{0}$ for $i > k$. Theorem 8.9 implies that the error is given by

$$\| X - X_k \|_F^2 = \sum_{i=k+1}^{r} \sigma_i^2 = \operatorname{tr}(X^T X) - \sum_{i=1}^{k} \sigma_i^2. \tag{8.27}$$

Finally, since the error on the right hand side of (8.27) decreases as the rank k increases, we may relax the condition $\operatorname{rank} A = k$ to $\operatorname{rank} A \leq k$, since we will be assured that the best approximating matrix will have rank k. ∎

It turns out the same result is true in the matrix 2 norm, i.e., the spectral norm, except that the formula for the error is different.

Theorem 8.15. *In the context of Theorem 8.13, the best approximating matrix in the matrix 2 norm with rank at most k is the truncated singular value decomposition matrix X_k. In particular, we have*

$$\sigma_{k+1} = \| X - X_k \| \leq \| X - A \| \tag{8.28}$$

for all $m \times n$ matrices A with $\operatorname{rank} A \leq k$.

Proof. Let $\widetilde{\Sigma}_k$ denote the $r \times r$ diagonal matrix whose first k diagonal entries are $\sigma_1, \ldots, \sigma_k$ and whose last $r - k$ diagonal entries are all 0. Clearly $X_k = P \widetilde{\Sigma}_k Q^T$ since the additional zero entries have no effect on the product. Moreover, $\Sigma - \widetilde{\Sigma}_k$ is a diagonal matrix whose first k diagonal entries are all 0 and whose last $r - k$ diagonal entries are $\sigma_{k+1}, \ldots, \sigma_r$. Thus, the difference $X - X_k = P(\Sigma - \widetilde{\Sigma}_k) Q^T$ has singular values $\sigma_{k+1}, \ldots, \sigma_r$. Since σ_{k+1} is the largest of these, Theorem 5.79 implies that $\| X - X_k \|_2 = \sigma_{k+1}$.

We now prove that this is the smallest possible among all $m \times n$ matrices A of rank k. For such a matrix, according to the Fundamental Theorem 4.24, $\dim \ker A = n - k$. Let $V_{k+1} \subset \mathbb{R}^n$ denote the $(k+1)$-dimensional subspace spanned by the first $k + 1$ singular vectors $\mathbf{q}_1, \ldots, \mathbf{q}_{k+1}$ of X. Since the dimensions of the subspaces V_{k+1} and $\ker A$ sum up to $k + 1 + n - k = n + 1 > n$, their intersection is a nontrivial subspace, and hence we can find a nonzero unit vector

$$\mathbf{u}^\star = c_1 \mathbf{q}_1 + \cdots + c_{k+1} \mathbf{q}_{k+1} \in V_{k+1} \cap \ker A.$$

Thus, since $\mathbf{q}_1, \ldots, \mathbf{q}_{k+1}$ are orthonormal,

$$\| \mathbf{u}^\star \|_2 = \| \mathbf{u}^\star \| = c_1^2 + \cdots + c_{k+1}^2 = 1, \quad \text{and, moreover,} \quad A \mathbf{u}^\star = \mathbf{0}.$$

Therefore,

$$(X - A) \mathbf{u}^\star = X \mathbf{u}^\star = c_1 X \mathbf{q}_1 + \cdots + c_{k+1} X \mathbf{q}_{k+1} = c_1 \sigma_1 \mathbf{p}_1 + \cdots + c_{k+1} \sigma_{k+1} \mathbf{p}_{k+1}.$$

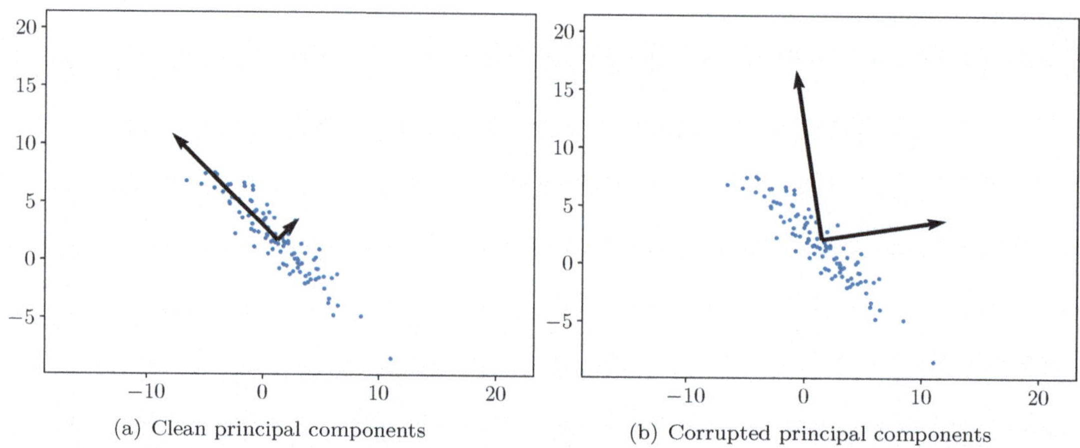

(a) Clean principal components   (b) Corrupted principal components

Figure 8.11: An illustration of how PCA is sensitive to outliers. In (a) we show the principal components on the clean dataset and in (b) we show the corrupted principal components where the data is corrupted by a single outlying point at location $(20, 40)$ (not depicted).

Since $\mathbf{p}_1, \ldots, \mathbf{p}_{k+1}$ are also orthonormal,

$$\| (X - A)\mathbf{u}^\star \|_2^2 = c_1^2 \sigma_1^2 + \cdots + c_{k+1}^2 \sigma_{k+1}^2 \geq (c_1^2 + \cdots + c_{k+1}^2) \sigma_{k+1}^2 = \sigma_{k+1}^2.$$

Thus, using the definition (4.75) of the Euclidean matrix norm

$$\| X - A \|_2 = \max \{ \, \| (X - A)\mathbf{u} \|_2 \mid \| \mathbf{u} \|_2 = 1 \, \} \geq \| (X - A)\mathbf{u}^\star \|_2 \geq \sigma_{k+1}.$$

This proves that σ_{k+1} minimizes $\| X - A \|_2$ among all rank k matrices A. Finally, as before, because the error decreases as the rank k increases, one cannot do any better with a matrix of lower rank.   ∎

8.2.1   Robust Subspace Recovery

PCA can be sensitive to outliers in the data set, due to its use of the mean squared error in (8.13), which strongly penalizes outliers. Figure 8.11 shows an example of how a single outlier can skew the result of PCA so that the principal directions fit the majority of the data very poorly. In Figure 8.11(b) there is an outlying data point at the location $(20, 40)$ that is not depicted in the image. The same kind of corruption happens with real data sets; in Figure 8.12(a) we show the principal component images for the Olivetti data set corrupted with a single "5" digit from the MNIST data set. Two of the top 10 principal components are clearly corrupted by this single outlying image.

Data sets are ordinarily expected to include noise and, possibly, errors, so the sensitivity of PCA to such outliers is an important consideration in practical applications. One possible approach is to attempt to remove outliers before applying PCA. This can, however, be difficult, and an alternative is to develop more robust versions of PCA that are not as sensitive to outliers, so that their removal is unnecessary. Many variants of robust PCA have been proposed in the literature [141, 142, 238]. The approach proposed in [141] is based on minimizing the p norm of the vector containing the distance energies. Raising to the power p produces

$$E_p(W; \mathbf{x}_1, \ldots, \mathbf{x}_m) = \sum_{i=1}^{m} \operatorname{dist}(\mathbf{x}_i, W)^p, \tag{8.29}$$

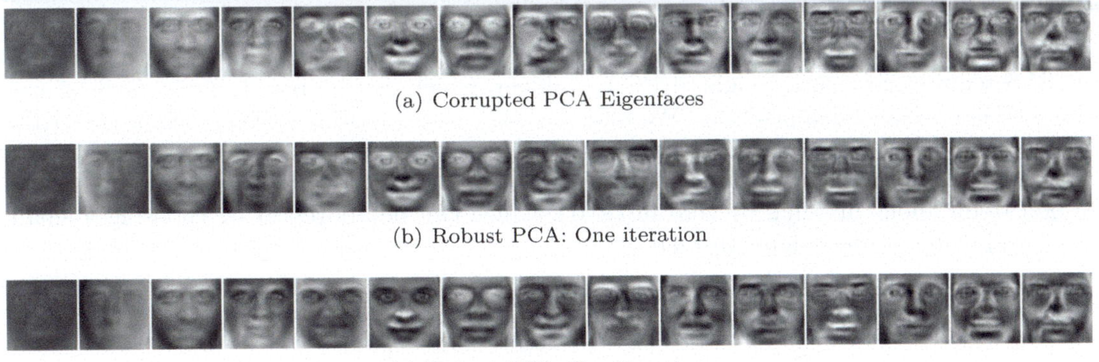

(a) Corrupted PCA Eigenfaces

(b) Robust PCA: One iteration

(c) Robust PCA: Two iterations

Figure 8.12: Comparison of PCA and robust subspace recovery on the Olivetti faces dataset that is corrupted by adding a single digit "5" from the MNIST dataset.

where dist is defined in (8.12), which is, as before, to be minimized over the affine space $W \subset \mathbb{R}^n$, . By decreasing the value of p — it is common to take $p = 1$ — we can place a lower penalty on severe outliers, and thereby achieve a more robust performance. However, it is more computationally challenging to minimize E_p, compared to the mean-squared error E_2 from (8.13), since there is no longer a simple relationship with the eigenvectors of the covariance matrix.

There are various strategies one can adopt to minimize the robust subspace energy (8.29). For example, it is possible to use gradient descent; however, we are minimizing over the set of matrices whose columns must satisfy the orthogonality constraint $U^T U = I$. This is a nonlinear constraint that is not straightforward to address, and its implementation lies beyond the scope of this book. Instead, we will take a simpler approach called *iteratively reweighted least squares* (IRLS), which involves solving a weighted PCA problem at every iteration. To describe the IRLS method, we will work in the more general setting of minimizing the energy

$$E_\varphi(W; \mathbf{x}_1, \ldots, \mathbf{x}_m) = \sum_{i=1}^m \varphi\big(\operatorname{dist}(\mathbf{x}_i, W)\big), \qquad (8.30)$$

where $\varphi \colon [0, \infty) \to [0, \infty)$ is a nondecreasing scalar function and $W \subset \mathbb{R}^n$ is an affine subspace. Choosing $\varphi(s) = s^2$ yields the PCA energy (8.13), while $\varphi(s) = s^p$ yields the energy function E_p given in (8.29).

For many choices of φ, the energy (8.30) can be minimized with an IRLS approach, which iteratively solves a sequence of weighted least squares problems. Let W_0 be the affine space found by ordinary PCA on the data, that is, the minimizer of E_2. Then for each $k \geq 0$, given W_k, we define the weight vector $\mathbf{c}_k = \big(c_{k,1}, \ldots, c_{k,m}\big)^T \in \mathbb{R}^m$ with entries

$$c_{k,i} = \frac{\varphi\big(\operatorname{dist}(\mathbf{x}_i, A)\big)}{\operatorname{dist}(\mathbf{x}_i, W_k)^2}. \qquad (8.31)$$

We then define $W_{k+1} \subset \mathbb{R}^n$ to be the solution of the weighted PCA problem

$$W_{k+1} = \operatorname*{argmin}_W \sum_{i=1}^m c_{k,i} \operatorname{dist}(\mathbf{x}_i, W)^2. \qquad (8.32)$$

Solving the weighted PCA problem (8.32) simply requires weighting the covariance matrix; we refer to Exercise 2.5 for details. The main idea behind IRLS is that if the iterations converge, so that if $W := W_k = W_{k+1}$ in (8.32), then the definition of the weights reduces the energy to exactly E_φ, which is the one we intended to minimize. In addition, each iteration of IRLS requires solving a weighted PCA problem, which can be readily performed by use of a singular value decomposition. We refer to [141] for a proof of convergence of the IRLS iterations, which depends on taking φ to be a suitable regularization of $\varphi(s) = s^p$ when $p = 1$.

In practice the method often gives good results after only a few iterations. We often need to be careful about dividing by zero in (8.31), which can be addressed by choosing a small $\varepsilon > 0$ and defining the weights instead by

$$w_{k,i} = \frac{\varphi\big(\,\mathrm{dist}(\mathbf{x}_i, W)\,\big)}{\max\big\{\,\mathrm{dist}(\mathbf{x}_i, W_k)^2, \varepsilon\,\big\}}.$$

Figure 8.12(c) shows the principal components obtained by the first two iterations of robust subspace recovery via IRLS with $\varepsilon = 10^{-8}$. The corrupting MNIST digit appears in only 1 principal component after the first iteration, and is completely removed from the top 15 principal components after the second iteration.

Exercises

2.1. ♡   Verify equation (8.21).

2.2. Implement the IRLS method for robust subspace recovery in Python and test it when outliers are added to (a) some some synthetic data sets; (b) the Olivetti face data set.

2.3. ♡   This exercise considers the problem of fitting the best subspace in a general inner product norm $\|\mathbf{x}\|_C = \sqrt{\langle \mathbf{x}, \mathbf{x} \rangle_C} = \sqrt{\mathbf{x}^T C \mathbf{y}}$, where C is symmetric, positive definite. Given points $\mathbf{x}_1, \ldots, \mathbf{x}_m \in \mathbb{R}^n$, let $X = (\mathbf{x}_1, \ldots, \mathbf{x}_m)^T$ be the corresponding data matrix. Then, given a subspace $V \subset \mathbb{R}^n$, define the distance and squared energy

$$\mathrm{dist}_C(\mathbf{x}, V) = \min\{\|\mathbf{x} - \mathbf{y}\|_C \mid \mathbf{y} \in V\}, \qquad E_C(V; \mathbf{x}_1, \ldots, \mathbf{x}_m) = \sum_{i=1}^{m} \mathrm{dist}_C(\mathbf{x}_i, V)^2.$$

    (a) Show that the k-dimensional subspace minimizing E_C is the one spanned by the top k eigenvectors $\mathbf{q}_1, \ldots, \mathbf{q}_k$ of the matrix $S = X^T X C$.
    (b) What happens if we minimize over affine subspaces $W = \mathbf{a} + V$? What choice of $\mathbf{a}$ is optimal?
    (c) Formulate equivalent optimization principles as was done in Remark 8.12.

2.4. A matrix norm $\|\cdot\|$ on $\mathcal{M}_{m \times n}$ is called *orthogonally invariant* if $\|PAQ\| = \|A\|$ for all orthogonal matrices $P \in \mathcal{M}_{m \times m}$, $Q \in \mathcal{M}_{n \times n}$, and all $A \in \mathcal{M}_{m \times n}$. Mirsky [168] showed that truncated SVD provides the best low rank approximation to a matrix in *any* orthogonally invariant norm, thus generalizing Theorems 8.13 and 8.15. (a) Show that the spectral matrix norm is orthogonally invariant. (b) Show that the Frobenius matrix norm is orthogonally invariant. (c) Can you construct other matrix norms that are orthogonally invariant? *Hint:* Consider norms that are defined directly as functions of the singular values.

2.5. ◊   Consider the weighted PCA energy

$$E_{\mathbf{c}}(W; \mathbf{c}_1, \ldots, \mathbf{c}_m) = \sum_{i=1}^{m} c_i \, \mathrm{dist}(\mathbf{x}_i, W)^2,$$

where $\mathbf{c} = (c_1, c_2, \ldots, c_m)$ are nonnegative numbers (weights), and W is an affine subspace.
(a) Show that $E_{\mathbf{c}}$ is minimized over the offset $\mathbf{a}$ of the affine subspace $W = \mathbf{a} + V$ by setting

$$\mathbf{a} = \frac{c_1 \mathbf{x}_1 + \cdots + c_m \mathbf{x}_m}{c_1 + \cdots + c_m}.$$

(b) By part (a), we can reduce the problem to minimizing $E_{\mathbf{c}}$ over linear subspaces V.
Show that the weighted energy $E_{\mathbf{c}}$ is minimized over all k-dimensional subspaces $V \subset \mathbb{R}^n$ by
setting $V = \mathrm{span}\{\mathbf{q}_1, \mathbf{q}_2, \ldots, \mathbf{q}_k\}$, where $\mathbf{q}_1, \mathbf{q}_2, \ldots, \mathbf{q}_n$ are the orthonormal eigenvectors of
the weighted covariance matrix

$$S_{\mathbf{c}} = c_1 \mathbf{x}_1 \mathbf{x}_1^T + \cdots + c_m \mathbf{x}_m \mathbf{x}_m^T = X^T C X$$

with eigenvalues $\lambda_1 \geq \lambda_2 \geq \cdots \geq \lambda_n$, where $C = \mathrm{diag}(c_1, \ldots, c_m)$, and that the minimal
energy is $E_{\mathbf{c}} = \lambda_{k+1} + \cdots + \lambda_n$.

8.3 PCA-based Compression

Python Notebook: PCA-based compression (.ipynb)

Given a data matrix $X \in \mathcal{M}_{m \times n}$, consisting of m data points in $\mathbb{R}^n$, let us write its
singular value decomposition as

$$X = P \Sigma Q = \sum_{i=1}^{r} \sigma_i \mathbf{p}_i \mathbf{q}_i^T, \tag{8.33}$$

where $r = \mathrm{rank}\, X$. The singular values are ordered from largest, the most important, to
smallest. Indeed, by the Schmidt–Eckart–Young–Mirsky Theorem 8.13, the truncated singular value decomposition

$$X_k = P_k \Sigma_k Q_k = \sum_{i=1}^{k} \sigma_i \mathbf{p}_i \mathbf{q}_i^T, \tag{8.34}$$

where

$$P_k = (\mathbf{p}_1 \, \cdots \, \mathbf{p}_k), \qquad \Sigma_k = \mathrm{diag}(\sigma_1, \ldots, \sigma_k), \qquad Q_k = (\mathbf{q}_1 \, \cdots \, \mathbf{q}_k),$$

is the best rank k approximation of X in the Frobenius norm, which is simply the sum of
squared errors; see Definition 4.58. (According to Theorem 8.15, it is also the best rank k
approximation in the Euclidean norm.) Furthermore, we can store the matrix X_k by simply
recording the matrices P_k, Σ_k, Q_k, which, when k is small, are smaller in size than the matrix
X. Thus, it is natural to use SVD for the task of *data compression*, where the goal is to store
the matrix X in a compressed form that has smaller storage requirements than keeping all
$m \times n$ entries, while discarding as little information as possible along the way. Using SVD
for data compression discards the smaller singular values and vectors, which often represent
noise or spurious and unimportant details.

To compress a data matrix X, we perform an SVD, or, equivalently, an eigendecomposition
of $X^T X$, to find the top k singular vectors and then form the matrix Q_k. The compression
and decompression steps are then given by

$$\text{Compression: } C_k = X Q_k, \qquad \text{Decompression: } X_k = C_k Q_k^T, \tag{8.35}$$

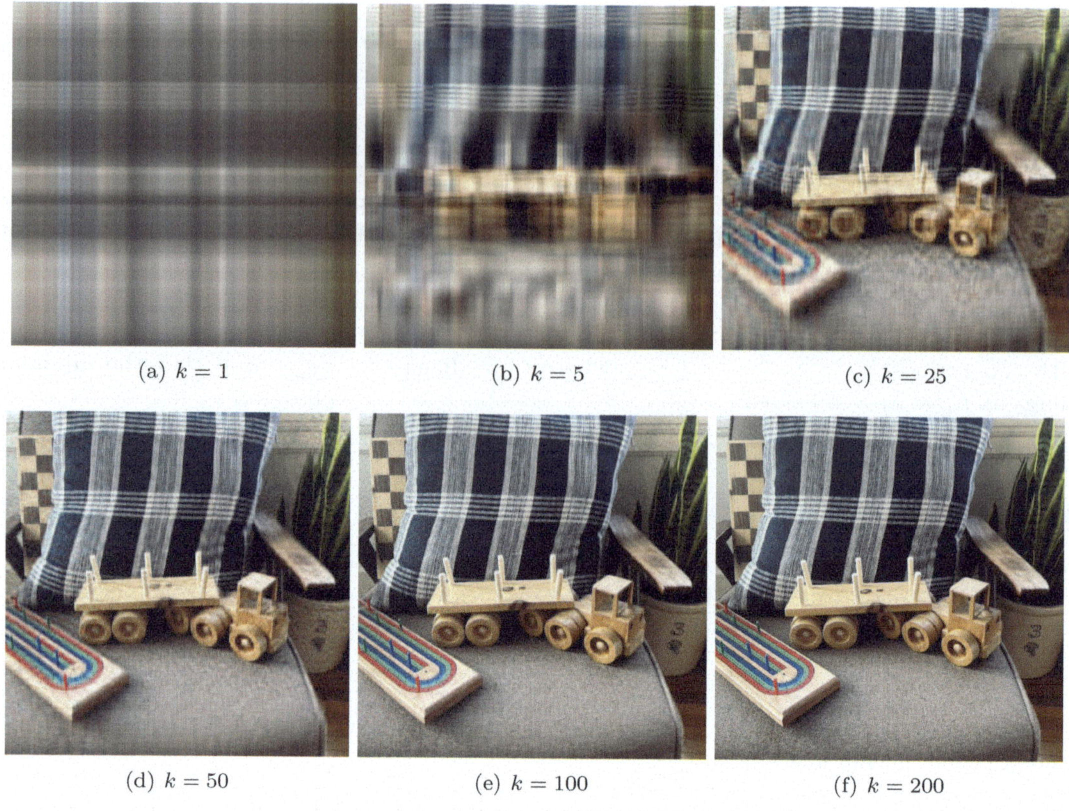

(a) $k = 1$ (b) $k = 5$ (c) $k = 25$

(d) $k = 50$ (e) $k = 100$ (f) $k = 200$

Figure 8.13: Truncated singular value decompositions of an image matrix used for compression. In each figure we keep the top k singular vectors.

where C_k represents the compressed data. Note that, as a consequence of the SVD formula (8.33), $C_k = P_k \Sigma_k$, and hence X_k is exactly the truncated SVD matrix (8.34). The compressed data matrix C_k is of size $m \times k$, while the singular vector matrix Q_k has size $n \times k$. Therefore, to store the compressed data we need to store $k(m + n)$ numbers. On the other hand, to store the original data matrix requires storing mn numbers, and hence the *compression ratio* — the ratio of the size of the compressed data to the original data — is given by

$$\text{Compression Ratio} = \frac{k(m + n)}{mn} = k \left(\frac{1}{m} + \frac{1}{n} \right). \tag{8.36}$$

In other words, the amount of compression depends linearly on the choice of k. By the Schmidt–Eckart–Young–Mirsky Theorem 8.13, the error between the compressed and original data in the Frobenius norm is given by (8.26), and thus, our ability to compress the data without significant error is controlled by how quickly the singular values decay, or equivalently, how close the data matrix X is to a low rank matrix — in this case, one of rank at most k.

A common application is image compression, where X is an array of pixel values in a digital image. Figure 8.13 shows the result of approximating an image matrix X by its truncated SVD, which shows good image reconstructions provided we do not use too few singular vectors. If the image is square, so $X \in \mathcal{M}_{n \times n}$, then $m = n$ and the compression ratio is k/n, or n to k. (For color images we can compress each color channel separately, or

Figure 8.14: A 512×512 color image and its decomposition into blocks of size 32×32.

treat the color channels as additional pixels.) We note that this naïve application of SVD compression is essentially compressing the *rows* of the image, and does not take into account vertical correlations between pixels. In addition, the rows span the entire image and there is little reason to expect they have a great deal of low dimensional structure that would be useful when compressing.

A better way to split up an image for compression is to use blocks that are localized in space. Away from edges and texture, the pixel intensities tend to not vary much in local sections of an image, and so small blocks can often be well-approximated by a low dimensional subspace. As an application, we work with the 512×512 color image shown in Figure 8.14. We use 8×8 pixel blocks in a regular grid, so the image contains $64 \times 64 = 4096$ blocks, each containing $8 \times 8 = 64$ pixels, each with red, green and blue values. Figure 8.14 shows the image broken down into 8×8 blocks. Working with blocks instead of rows requires a small amount of preprocessing. Splitting it into blocks produces a matrix X of size 4096×192, since there are three color channels and $192 = 64 \times 3$. We apply SVD compression to this matrix, instead of to the image itself. After this preprocessing, the compression proceeds exactly the same as before, and the decompressed image then needs to be reconstructed from its blocks.

The reconstruction error in image compression is measured with the *peak signal to noise ratio* (PSNR). PSNR is based on the mean squared error, which is the rescaled Frobenius norm of the difference:

$$\text{MSE} = \frac{1}{mn} \| X - X_k \|_F^2, \tag{8.37}$$

where X is the original image and X_k is the reconstructed image after compression. Both images have size $m \times n$. PSNR also requires the peak signal value, S_{peak}, which is the largest possible value of the pixel intensity. We work with images scaled to have pixel intensities in the unit interval, so $S_{\text{peak}} = 1$. The PSNR is then given by

$$\text{PSNR} = 10 \log_{10} \left(\frac{S_{\text{peak}}^2}{MSE} \right), \tag{8.38}$$

and is measured in decibels (dB). PSNR values of 20 dB to 30 dB are very low quality images, while 30 dB to 50 dB are respectable, and above 50 dB represent very good quality

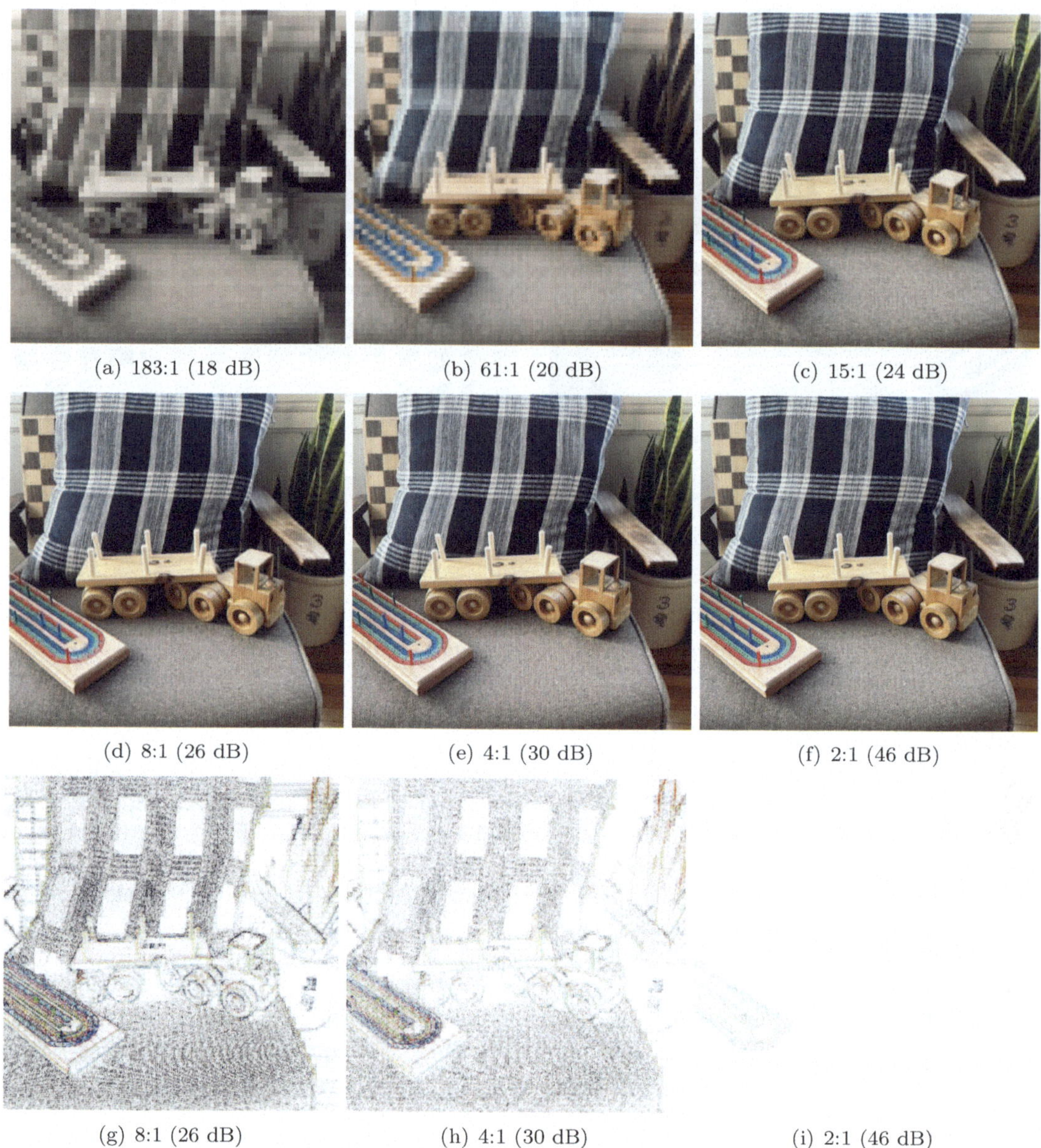

(a) 183:1 (18 dB)   (b) 61:1 (20 dB)   (c) 15:1 (24 dB)

(d) 8:1 (26 dB)   (e) 4:1 (30 dB)   (f) 2:1 (46 dB)

(g) 8:1 (26 dB)   (h) 4:1 (30 dB)   (i) 2:1 (46 dB)

Figure 8.15: Examples of PCA-based image compression at different compression ratios. The last three images show the difference images between the original and compressed for the highest PSNR and lowest compression ratios, where white indicates zero error, and larger errors are indicated by darker colors (the errors are magnified by a factor of 5 for visibility).

compressions. For color images, the MSE is averaged over color channels before the PSNR computation.

Figure 8.15 shows the compressed and difference images at three different compression ratios, with PSNR ranging from 34 dB up to 51 dB. The reader should note the blocking-type artifacts at higher compression ratios. These are caused by the decomposition of the image into blocks, which allows for neighboring reconstructed blocks to differ along the block

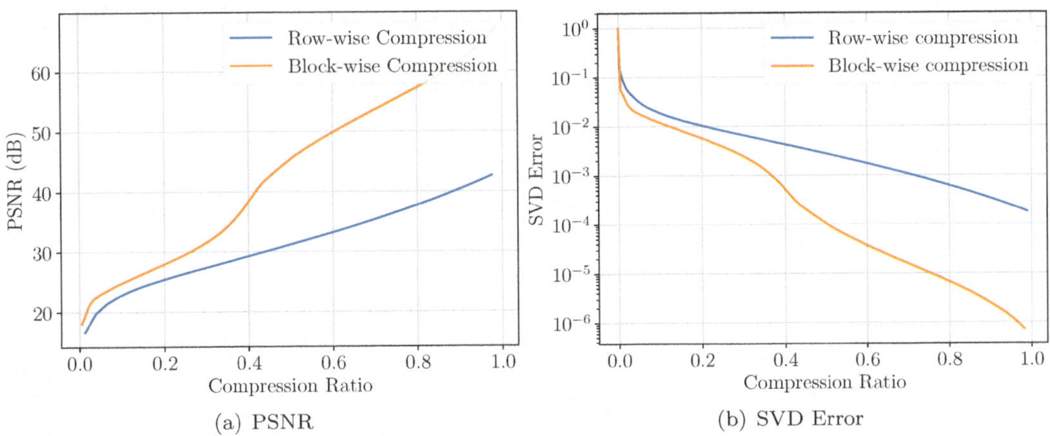

(a) PSNR

(b) SVD Error

Figure 8.16: In (a) we show PSNR vs Compression Ratio for block-wise and row-wise compression, and in (b) we plot the SVD error (8.26) guaranteed by Theorem 8.13, normalized by the trace of the covariance matrix, against the compression ratio. Both plots show the same data, just expressed in different ways.

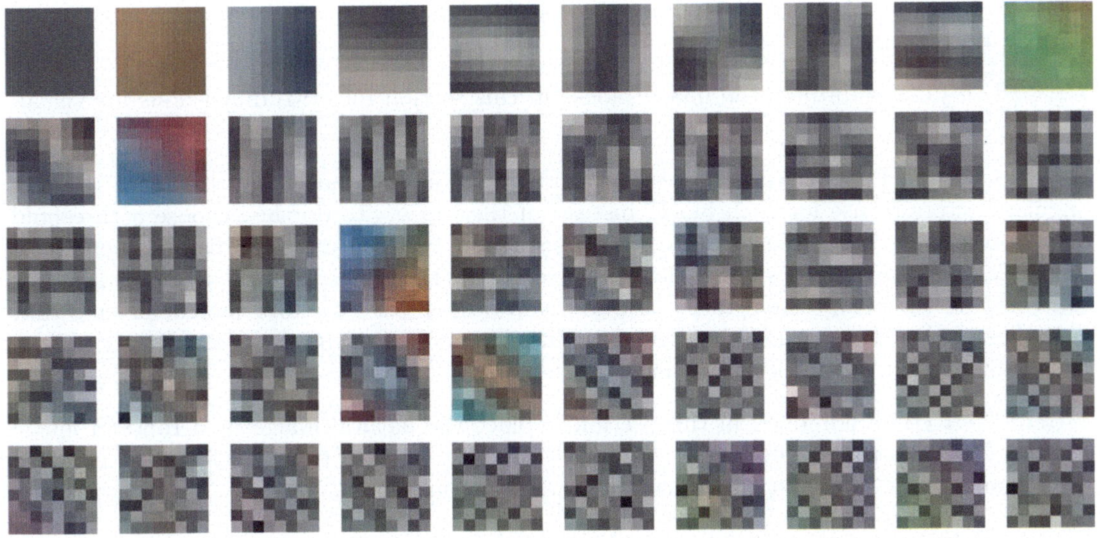

Figure 8.17: The first 50 principal components extracted via PCA on 8×8 pixel blocks.

boundary, causing a discontinuity in the image. In the worst case of Figure 8.15(a), we are keeping only the first ($k = 1$) principal component, so the compression algorithm approximates the image by a constant value in each block, and ends up discarding the color information, thereby producing an approximately grayscale image. In Figure 8.16(a) we plot the PSNR versus compression ratio for block-based image compression and row-wise image compression, showing the advantage of working with blocks instead of rows. Equivalently, in Figure 8.16(b) we show the SVD error computed by (8.26), normalized by the trace of the covariance matrix. Both plots show the same data, just computed and presented differently.

One of the most common image compression algorithms, JPEG (short for Joint Photographic Experts Group), uses an approach that shares some similarities with the block-based compression described in this section. The original JPEG algorithm also breaks the image into blocks, and exhibits the same blocking artifacts[6] that we saw in Figure 8.15. The main difference is that the JPEG algorithm uses the discrete cosine transform (DCT), a version of the discrete Fourier transform (DFT) discussed in Chapter 9, which decomposes a signal or image into pure frequencies. JPEG also uses clever methods for deciding which components to keep and which to discard on a block-by-block basis. The DCT is a hand designed transformation, i.e., it is not learned from the data as PCA is. While it has similar properties as SVD, it can be computed far more efficiently than PCA using the fast Fourier transform (FFT); see Section 9.10.6. For image and audio data, the majority of the information is contained in the first few DCT coefficients, which encode the low frequency content locally in space or time. PCA has the same characteristics; for example, we show in Figure 8.17 the first 50 principal components obtained by applying PCA to image blocks. The principal components start off as low frequency, smooth, features, while the later components describe more high frequency content, like texture.

Exercises

3.1. ♡ Generate a plot of the singular values for the rows versus blocks of the image in this section. Which ones decay faster?

3.2. ◊ Use PCA to project the image blocks in this section into two dimensions for visualization.

3.3. Modify the PCA-based compression algorithm from this notebook to choose the *best* singular vectors to use for each block, instead of the top k. To do this, choose a threshold $\mu > 0$, project the image blocks onto all of the singular vectors, and then discard (i.e., set to zero) any coefficient that is smaller than μ. Reconstruct the image from the truncated blocks, and compute the compression ratio assuming you do not have to store the coefficients that were thresholded to zero, and that you don't need to store the singular vectors.[7] How does this compare with the block-based method?

3.4. In the Python notebook in this section, replace the natural image by a random image generated with `numpy.random.rand`. Plot the singular values for the rows and blocks. Do you see any decay? Are you able to compress the random image?

3.5. ♡ In this exercise, you will extend the PCA-based compression algorithm from this section to audio compression. Complete the parts (a) through (c); the notebook below will help you get started.

Python Notebook: Audio Compression (.ipynb)

   (a) Use the block-based image compression algorithm described in this section for audio compression. You can use any audio file you like; thePython notebook linked above

---

[6]More recent versions of JPEG make use of wavelets, and do not exhibit blocking artifacts [233].

[7]The setting is that you learn good singular vectors, and then share them between the encoder and decoder, so only the coefficients must be transmitted/stored.

downloads a classical music sample from the textbook GitHub website. A stereo audio signal is an array of size $n \times 2$, where n is the number of samples. Use blocks of size $N \times 2$ for compression.

(b) Plot the top $k = 10$ or so principal components. They should look suspiciously like sinusoids.

(c) When you play back the compressed audio file, you will likely hear some static noise artifacts, even at very low compression rates. These are caused by blocking artifacts, where the signals do not match up on the edges of the blocks used for compression, which introduces discontinuities into the signal. This is similar to the blocking artifacts we observed in image compression in this section, however, the artifacts are more noticeable in audio than in images.

To fix this, audio compression algorithms use overlapping blocks, and apply a windowing function in order to smoothly patch together the audio in each block. The blocks are structured so that half of the first block overlaps with half of the second block, and so on. To implement this in Python, just shift the signal by half of the block width, and apply the `image_to_patches` function on the original and shifted signals. Then compress and decompress both signals. After decompressing, and before converting back from the block format to the audio signal, you'll need to multiply by a windowing function to smooth the transition between blocks. If the block size is $N \times 2$, then each channel should be multiplied by a window function w_i, $i = 0, 1, \ldots, N - 1$. A common window function that is used, for example, in mp3 compression, is

$$ w_i = \sin^2 \left[\left(i + \frac{1}{2} \right) \frac{\pi}{N} \right]. $$

After you decompress and apply the window, undo the shift and add the signals together to get the decompressed audio. Does this improve the audio quality? As a note, in order to make sure the shifted signals add up correctly, we need that

$$ w_i + w_{i+N/2} = 1. $$

As an exercise, the reader should check that the window function above satisfies this condition, which is called the *Princen-Bradley condition*.

8.4 Linear Discriminant Analysis

Python Notebook: LDA (.ipynb)

The principal directions found in PCA are not necessarily good directions for separating, or discriminating, between the different classes or clusters in a data set. In fact, choosing directions that maximize the variance in the data may be completely at odds with choosing directions that discriminate well between classes. Consider the toy data sets in Figure 8.18, which consist of two clusters and their projections onto the x and y axes. In the figure on the left, the two clusters are isotropic and the largest direction of variation in the data, i.e., the

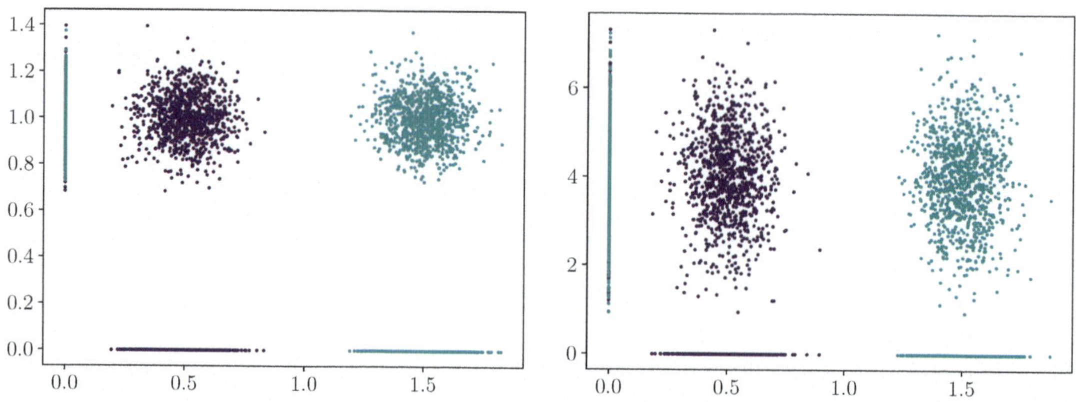

Figure 8.18: Illustration of projecting two data sets to the x and y axes. On the left, the top principal component of the data lies very close to the direction $\mathbf{e}_1$, which perfectly separates the two clusters, while on the right, the top principal component is very close to the direction of $\mathbf{e}_2$, which does not separate the clusters.

top principal component, is between the two clusters. Indeed, in this case the top principal component is the vector $\mathbf{q}_1 \approx (0.9999, -0.0047)^T$ to 4 decimal places, which points along the x-axis and projection onto $\mathbf{q}_1$, essentially the x-axis, perfectly separates the data. On the other hand, in the figure on the right, the tall clusters are not isotropic, and the top principal component of the data set points roughly in the y direction; more precisely, it is $\mathbf{q}_1 \approx (-0.0204, 0.9998)^T$. Now projecting onto $\mathbf{q}_1$ is roughly the same as projecting onto the y axis, which mixes the two classes together and does not allow us to discriminate between them in the PCA coordinates. If our downstream task after dimension reduction is clustering or classification, it is clearly preferable to produce an embedding that separates, or discriminates, between classes as much as is possible.

In the example data set on the right side in Figure 8.18, it is in fact the direction of *smallest variation* that separates the clusters, but this is merely a coincidence, and is not a rule of thumb we can rely on in general situations. Indeed, the directions of smallest variation often correspond to noise. In order to find directions that discriminate well between classes, we must assume that we are in possession of some information about the classes in the data. Here, we assume we are in the fully supervised classification setting, where we have a training set consisting of data points $\mathbf{x}_1, \ldots, \mathbf{x}_m \in \mathbb{R}^n$, which, as usual, are assembled into an $m \times n$ data matrix X, along with class labels $y_1, \ldots, y_m$, which, for this purpose, are integers between 1 and c denoting which class, out of c classes in total, each data point belongs to. Let

$$C_i = \left\{ j \mid y_j = i \right\},$$

denote the indices of the data points in class i, and let $m_i = \#C_i$ be the number of data points in class i. We also let $X_1, \ldots, X_c$ denote the submatrices of X corresponding to the data points in each class, so X_i is an $m_i \times n$ matrix whose rows contain the data points in class i.

Before going further, we need to introduce various class-based covariance matrices. Let

$$\mathbf{c}_i = \frac{1}{m_i} \sum_{j \in C_i} \mathbf{x}_j$$

be the mean of the i-th class. Then, according to (7.22), the corresponding *class covariance matrix* is given by

$$S_{X_i} = \sum_{j \in C_i} (\mathbf{x}_j - \mathbf{c}_i)(\mathbf{x}_j - \mathbf{c}_i)^T,$$

where we recall we have taken the prefactor $\nu = 1$ in this section. The *within class covariance matrix* is the sum of the class covariance matrices:

$$S_w = S_{X_1} + \cdots + S_{X_c}. \tag{8.39}$$

We define the *between class covariance matrix* by the identity

$$S_b = S_X - S_w, \qquad \text{or, equivalently,} \qquad S_X = S_w + S_b, \tag{8.40}$$

where

$$S_X = \underline{X}^T \underline{X} = \sum_{i=1}^m (\mathbf{x}_i - \overline{\mathbf{x}})(\mathbf{x}_i - \overline{\mathbf{x}})^T, \qquad \overline{\mathbf{x}} = \frac{1}{m}\sum_{i=1}^m \mathbf{x}_i$$

is the usual total covariance matrix of the data. It turns out that the between class covariance has a more convenient form.

Proposition 8.16. *The between class covariance matrix is given by*

$$S_b = \sum_{i=1}^c m_i (\mathbf{c}_i - \overline{\mathbf{x}})(\mathbf{c}_i - \overline{\mathbf{x}})^T. \tag{8.41}$$

Thus, S_b is simply a weighted covariance matrix of the class means.

Proof. Let us note first that the mean of all the data points can be re-expressed as

$$\overline{\mathbf{x}} = \frac{1}{m}\sum_{i=1}^m \mathbf{x}_i = \frac{1}{m}\sum_{i=1}^c \sum_{j \in C_i} \mathbf{x}_i = \frac{1}{m}\sum_{i=1}^c m_i \mathbf{c}_i.$$

Using this and the definitions of S_w and S_X we compute

$$S_b = S_X - S_w = \sum_{i=1}^m (\mathbf{x}_i - \overline{\mathbf{x}})(\mathbf{x}_i - \overline{\mathbf{x}})^T - \sum_{i=1}^c S_{X_i}$$

$$= \sum_{i=1}^c \sum_{j \in C_i} \left[(\mathbf{x}_j - \overline{\mathbf{x}})(\mathbf{x}_j - \overline{\mathbf{x}})^T - (\mathbf{x}_j - \mathbf{c}_i)(\mathbf{x}_j - \mathbf{c}_i)^T \right]$$

$$= \sum_{i=1}^c \sum_{j \in C_i} \left(\mathbf{x}_j \mathbf{x}_j^T - \overline{\mathbf{x}}\mathbf{x}_j^T - \mathbf{x}_j \overline{\mathbf{x}}^T + \overline{\mathbf{x}}\,\overline{\mathbf{x}}^T - \mathbf{x}_j \mathbf{x}_j^T + \mathbf{c}_i \mathbf{x}_j^T + \mathbf{x}_j \mathbf{c}_i^T - \mathbf{c}_i \mathbf{c}_i^T \right)$$

$$= \sum_{i=1}^c m_i \left(-\overline{\mathbf{x}}\mathbf{c}_i^T - \mathbf{c}_i \overline{\mathbf{x}}^T + \overline{\mathbf{x}}\,\overline{\mathbf{x}}^T + \mathbf{c}_i \mathbf{c}_i^T + \mathbf{c}_i \mathbf{c}_i^T - \mathbf{c}_i \mathbf{c}_i^T \right) = \sum_{i=1}^c m_i (\mathbf{c}_i - \overline{\mathbf{x}})(\mathbf{c}_i - \overline{\mathbf{x}})^T,$$

as required. $\blacksquare$

In order to find a discriminating direction, we will look for a unit vector $\mathbf{u} \in \mathbb{R}^n$ such that the within class variance $\mathbf{u}^T S_w \mathbf{u}$ is small and the between class variance $\mathbf{u}^T S_b \mathbf{u}$ is large, recalling from (8.1) that $\mathbf{u}^T S_X \mathbf{u}$ represents the variance of the data matrix X in the direction

u. This will ensure maximal separation between classes, while bringing each class more tightly together by making the within class variance small. A natural quantity to *maximize* is thus the ratio of these two quantities, namely

$$\frac{\mathbf{u}^T S_b \mathbf{u}}{\mathbf{u}^T S_w \mathbf{u}}, \tag{8.42}$$

which is often called the *class separation*. If we assume that S_w is positive definite, then the class separation ratio coincides with the generalized Rayleigh quotient appearing in Theorem 5.50. Using the case $k = 1$ in that result, this implies that the maximizing direction $\mathbf{u}$ is a generalized eigenvector of the matrix pair S_b, S_w, meaning that it satisfies the generalized eigenvalue problem

$$S_b \mathbf{u} = \lambda S_w \mathbf{u}, \tag{8.43}$$

with $\lambda = \lambda_{max}(S_b, S_w)$ the largest generalized eigenvalue, which is the maximal value of the class separation quotient, and $\mathbf{u} = \mathbf{q}_1$ the corresponding generalized eigenvector.

Returning to the example on the right hand side in Figure 8.18, the top discriminating direction is $\mathbf{u} \approx (0.999997, 0.002434)^T$, which nearly points in the x direction, as we expected, and is able to perfectly separate the two clusters. The top discriminating direction for the left figure also points along the x-axis.

Now, it is possible that the within class covariance matrix S_w is singular, and then the preceding discussion is no longer valid. In this case the kernel of the matrix S_w is nontrivial, so there are directions $\mathbf{0} \neq \mathbf{u} \in \ker S_w$ for which the class separation (8.42) is infinity! Thus, in this case, maximizing the class separation is not a well-posed problem. More generally, S_w could be close to singular, that is, it could have a large condition number, and optimizing the class separation could be challenging numerically. There are several ways to address this. We can regularize the covariance matrix S_w by replacing it with

$$S_{w,\lambda} = S_w + \lambda I,$$

where $\lambda > 0$ is a parameter. This simply adds a small amount of noise (i.e., variance) in all directions to make $S_{w,\lambda}$ positive definite. This is a special case of *covariance shrinkage*, and is very simple to implement in practice. Indeed, very small values of λ are usually sufficient, so we do not need to over-regularize. Another option is to preprocess the data by applying PCA in order to reduce the dimension sufficiently so that all directions have positive variance in at least one class, and so S_w is positive definite in the reduced dimensional space. The amount of dimension reduction required is data dependent and requires some user input. It is explored further in Exercise 4.2.

As we did with PCA, after finding the top discriminating direction $\mathbf{q}_1$, we look for additional discriminating directions by maximizing the class separation over directions orthogonal to those already found, where we use the inner product $\langle \mathbf{x}, \mathbf{y} \rangle_{S_w} = \mathbf{x}^T S_w \mathbf{y}$ based on S_w, which is assumed to be positive definite. (Otherwise we can replace S_w by $S_{w,\lambda}$.) That is, once we have found the first $j - 1$ discriminating directions $\mathbf{q}_1, \ldots, \mathbf{q}_{j-1}$, the j-th direction is the solution of

$$\max \left\{ \mathbf{u}^T S_b \mathbf{u} \mid \|\mathbf{u}\|_{S_w} = 1, \ \langle \mathbf{u}, \mathbf{q}_1 \rangle_{S_w} = \cdots = \langle \mathbf{u}, \mathbf{q}_{j-1} \rangle_{S_w} = 0 \right\}. \tag{8.44}$$

By Theorem 5.47, the j-th *discriminating direction* is exactly the j-th unit generalized eigenvector, satisfying (8.43). Unlike PCA, we cannot continue this process indefinitely. The between class covariance matrix S_b is a covariance matrix over c points, namely the means of the classes, and so by (7.21) it has rank at most $c - 1$. Thus, there are at most $c - 1$ discriminating directions that contain any useful information, since beyond this, all vectors

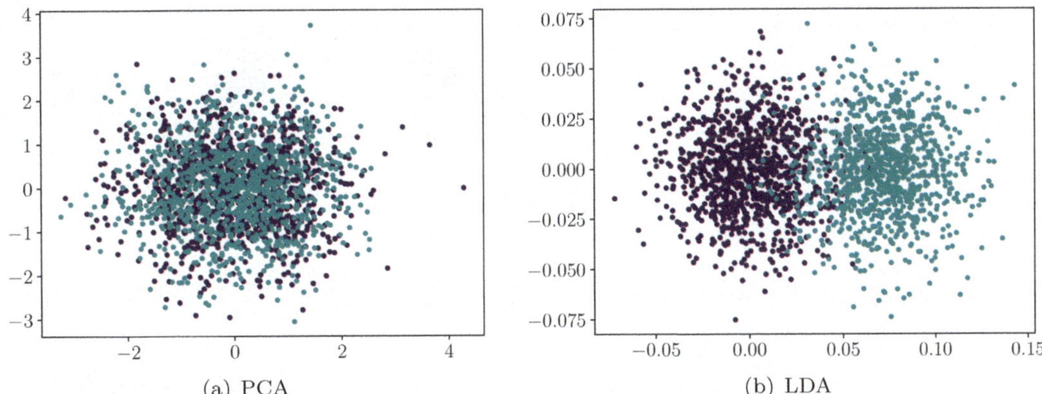

Figure 8.19: Example of PCA vs LDA for embedding a three dimensional data set with two clusters into the plane for visualization.

orthogonal to the first $c - 1$ directions solve the optimization problem (8.44) with class separation of zero, which is no longer informative. In fact, if $\operatorname{rank} S_b = r < c - 1$, then the null directions $\mathbf{q}_{r+1}, \ldots, \mathbf{q}_{c-1}$ are not uniquely determined.

Let us summarize the discussion above in the following theorem.

Theorem 8.17. *Assuming S_w is positive definite, the $c - 1$ discriminating directions $\mathbf{q}_1, \ldots, \mathbf{q}_{c-1}$ are the top $c - 1$ unit eigenvectors of the generalized eigenvalue problem $S_b \mathbf{q}_i = \lambda_i S_w \mathbf{q}_i.$*

As we did with PCA, we form the matrix $Q_k = (\mathbf{q}_1 \ \ldots \ \mathbf{q}_k)$, for any $k \leq c-1$, and project the data matrix X onto the top k discriminating directions by computing

$$Y = X Q_k \in \mathcal{M}_{m \times k}.$$

For two-dimensional visualizations of data, we just take $k = 2$, i.e., the top two discriminating directions. The process of projecting the data onto the discriminating directions is usually referred to as *linear discriminant analysis* (LDA), though the term LDA is also often used to refer to some specific classification techniques that use LDA as a preprocessing step, which we discuss further below.

As an example of LDA, and how it differs from PCA, we first consider a simple toy example where we have a data set with $n = 2000$ points in $\mathbb{R}^3$ with two classes, each of which follows a Gaussian distribution with unit variance in x_2, x_3, and variance of 0.1 in x_1. The means of the two clusters differ by 1 in the x_1 direction. This is a higher dimensional version of the initial example in this section appearing in Figure 8.18. Figure 8.19 shows the results of using PCA and LDA on this data set to visualize it in $\mathbb{R}^2$. The direction between the two clusters has less variance than orthogonal directions, thus the two classes are completely mixed along the first two principal components in Figure 8.19(a), whereas LDA in 8.19(b) does a noticeably better job separating the classes. We should note that there are only two classes in this example, so there is technically only one discriminating direction, and the second one is not uniquely determined. Here, the choice of the second direction is irrelevant, since we can see in Figure 8.19(b) that the class separation is completely determined by the first direction, i.e., the separation is along the x axis.

We proceed to an example with real data. In Figure 8.20 we show the projection of various subsets of the MNIST digits onto the top two discriminating directions. We use covariance

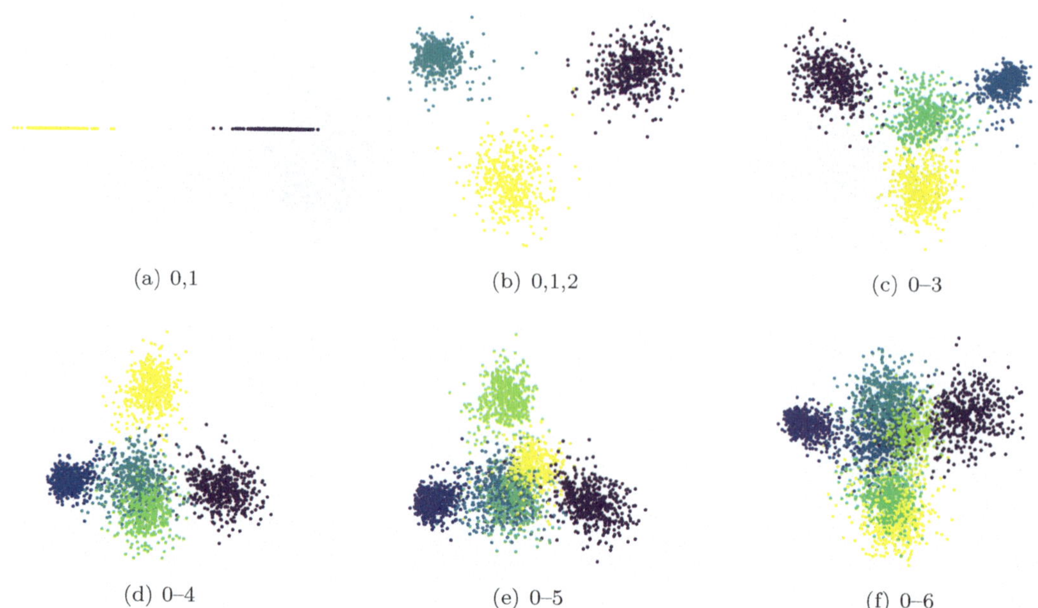

(a) 0,1   (b) 0,1,2   (c) 0–3

(d) 0–4   (e) 0–5   (f) 0–6

Figure 8.20: Plots of subsets of the MNIST dataset reduced to 2 dimensions through LDA. The reader may wish to compare to the PCA plots in Figure 8.5.

shrinkage with $\lambda = 10^{-10}$ to ensure that the within class covariance matrix S_w is nonsingular. Comparing to the PCA embedding in Figure 8.5, we see the classes are separated better, although some classes start mixing after the digit 4 is introduced. The digits are colored from dark purple (digit 0) through yellow (highest digit). It is interesting to note that for the digits $0 - 4$ and $0 - 5$, those with the most overlap are 2 and 3, which are perhaps the most similar digits in those ranges. The reader should keep in mind, when comparing to the PCA plots in Figure 8.5, that LDA is using knowledge of the class labels of each data point, thus LDA is *supervised*, while PCA is an *unsupervised* algorithm that does not use any label information.

As we mentioned briefly above, LDA is normally used as a preprocessing dimension reduction step before applying a fully supervised classification algorithm. Separating the classes in the reduced space can make it easier to fit the data with a given classification algorithm. The standard algorithm used with LDA is called the *Bayes classifier*, which is the optimal classifier in a statistical sense, in that it minimizes the probability of misclassification under the assumption that each cluster follows a Gaussian distribution with the same covariance matrix. We do not cover the Bayes classifier in this book, and refer the reader to [26], while noting that it is not necessary to use the Bayes classifier, and we can combine any classification method with LDA.

When using LDA for classification, it is import to apply LDA to the training data *after* performing a train/test split, since we cannot allow the testing data and its labels to be used in any way during training. We can then save the LDA components and project the testing data in the same way during the testing phase. In Table 8.21 we show the training and testing accuracy for LDA applied to the classification of MNIST digits. Since there are 10 classes, LDA dimension reduction reduces the data to $\mathbb{R}^9$. We show the results of both the standard LDA classification algorithm, which uses the optimal Bayes classifier along with the combination of LDA for dimension reduction and linear SVM for classification. Both methods perform similarly. It's interesting to note that the testing accuracy does not change substantially once the training set contains 10% of the data. Also, comparing to the results of

Training data size	0.1%	1%	10%	20%	40%	85.7%
Training acc: LDA-Bayes (%)	97.14	100	89.69	88.30	87.64	87.36
Testing acc: LDA-Bayes (%)	54.95	53.78	86.03	86.81	87.60	87.86
Training acc: LDA-SVM (%)	100	100	89.16	87.79	87.13	86.76
Testing acc: LDA-SVM (%)	59.69	53.62	85.06	86.46	87.00	87.26
Training acc: PCA-SVM (%)	100	88.57	82.93	83.04	82.98	82.74
Testing acc: PCA-SVM (%)	58.24	80.43	82.75	83.26	83.34	83.41

Table 8.21: Training and testing accuracy of LDA for classification (using the optimal Bayes classifier) and the combination of LDA or PCA dimension reduction followed by SVM classification on the MNIST data set. The testing accuracies are comparable between the two methods that use LDA. With PCA, the performance is better with small training sets, but worse when using larger sets.

SVM in Table 7.18, we see that the LDA results are uniformly worse than SVM applied to the original data set, indicating some information is lost through the LDA dimension reduction. This should not be surprising, since the data is reduced to a 9 dimensional space, down from an original 784 dimensions! In Table 8.21 we also compare with using PCA to reduce the data to $\mathbb{R}^9$ before applying SVM. For larger training set sizes, we can see the advantage of the class separation objective in LDA, while with smaller training sets, PCA works better.

Remark 8.18. As we did with PCA in Remark 8.12, we can formulate versions of LDA that optimize over all the discriminant directions simultaneously, instead of one at a time. Recall from Remark 8.12 that $\mathrm{tr}\,(U^T S_b U)$ and $\mathrm{tr}\,(U^T S_w U)$ represent the total variance of the means and classes under the transformation $Y = XU$. Thus, to find $k \leq c - 1$ discriminating directions, it is natural to replace the problem of maximizing the class separation ratio (8.42) over directions $\mathbf{u}$, with maximizing the trace version [146]

$$\frac{\mathrm{tr}\,\left(U^T S_b U \right)}{\mathrm{tr}\,\left(U^T S_w U \right)} \tag{8.45}$$

over $n \times k$ matrices U (subject to an orthogonality constraint we'll discuss below). Another objective function that is often used in LDA [119] is

$$\mathrm{tr}\left[\left(U^T S_w U \right)^{-1} \left(U^T S_b U \right) \right]. \tag{8.46}$$

We clearly have to impose some orthogonality constraint on U to make sense of either problem. If we maximize either (8.45) or (8.46) subject to the dot product orthogonality constraint $U^T U = I$, as in [119], then there is no simple solution formula, and, in particular, the solution is not the simple LDA method described in this section unless $S_w = I$. Indeed, the LDA components $\mathbf{q}_1, \ldots, \mathbf{q}_k$ are generalized eigenvectors, and hence orthogonal with respect to the inner product $\langle \mathbf{x}, \mathbf{y} \rangle_{S_w} = \mathbf{x}^T S_w \mathbf{y}$, so the matrix $Q_k = (\mathbf{q}_1 \cdots \mathbf{q}_k)$ satisfies $Q_k^T S_w Q_k = I$ and not $Q_k^T Q_k = I$ — again, unless $S_w = I$. On the other hand, the constraint $U^T S_w U = I$ is more natural, as it agrees with our definition of LDA in this section, and in this case maximizing either (8.45) or (8.46) leads trivially to the same optimization problem

$$\max \left\{ \mathrm{tr}\,\left(U^T S_b U \right) \mid U^T S_w U = I \right\}, \tag{8.47}$$

whose solution is exactly the matrix $U = Q_k$ containing the top k discriminating directions; see Exercise 4.5. ▲

Exercises

4.1. Write Python code to apply LDA as a preprocessing step on the *training* set in a classification problem using support vector machines (SVM) from Section 7.3, or k-nearest neighbors from Section 7.4. Pick one of the classification data sets from `sklearn.datasets`, or use MNIST. How does the accuracy change with the number of principal components used? Compare against PCA from Exercise 1.6.

4.2. ◇ Another approach to handle a singular within class covariance matrix S_w is to consider the constrained optimization problem

$$\max\left\{ \mathbf{u}^T S_b \mathbf{u} \mid \|\mathbf{u}\|_{S_w} = 1, \quad \mathbf{u} \in \mathrm{img}\, S_w \right\} \tag{8.48}$$

for finding the best discriminating direction. The difference with (8.44) with $j = 1$ is that we have added the additional restriction that $\mathbf{u} \in \mathrm{img}\, S_w$.

(a) Show that the maximal value of (8.48) is finite, even when S_w is singular. *Hint*: Recall that $\mathrm{img}\, S_w$ is the orthogonal complement of $\ker S_w$.

(b) Let $r = \mathrm{rank}\, S_w$ and let $V = (\,\mathbf{p}_1 \cdots \mathbf{p}_r\,)$, where $\mathbf{p}_1, \ldots, \mathbf{p}_r$ are the eigenvectors of S_w with nonzero eigenvalues. Show that the solution $\mathbf{u}$ of (8.48) has the form $\mathbf{u} = V\mathbf{c}$, where $\mathbf{c}$ is the top eigenvector of the generalized eigenvalue problem $V^T S_b V \mathbf{c} = \lambda V^T S_w V \mathbf{c}$, which is the projection of (8.43) onto the subspace $\mathrm{img}\, S_w$.

4.3. Write Python code to implement the version of LDA where a singular within class covariance matrix is handled by first performing PCA to sufficiently reduce the dimensionality of the data so that the within class covariance is nonsingular. Compare against covariance shrinkage on MNIST data.

4.4. ♡ Write Python code to implement the version of LDA where a singular within class covariance is handled according to Exercise 4.2, and compare to covariance shrinkage on MNIST data. *Hint*: Instead of trying to figure out exactly which singular values are zero, a more numerically stable approach is to truncate all singular values less than a threshold $\varepsilon > 0$ to zero. In this problem, it works well to, for example, just take the top 100 eigenvectors of S_w on MNIST.

4.5. ♡ Show that the solution of (8.47) is the matrix $U = Q_k$ whose columns are the top k discriminating directions. *Hint*: See Remark 8.12.

8.5   Multidimensional Scaling (MDS)

Python Notebook: MDS (.ipynb)

We have seen that PCA and LDA can be used for embedding data sets into a lower dimensional space, including visualization of data in two and three dimensions. The objective of PCA is to maximize the amount of variability captured in the embedding, while the objective of LDA is to maximize the separation between classes. In many real-world problems, it is important to correctly capture the *pairwise distances* between data points in the embedding or visualization. For example, the data points may be locations (e.g., cities or countries), and we aim to ensure the visualized distances between locations matches the true distance as closely as possible. We may also have data for which we only have access to some notion of distance between pairs of data points, and not the data points themselves, and wish to embed the data in such a way that these distances are realized.

The problem studied in this section is the following: Given data points $\mathbf{x}_1, \ldots, \mathbf{x}_m \in \mathbb{R}^n$, can we find corresponding points $\mathbf{z}_1, \ldots, \mathbf{z}_m \in \mathbb{R}^k$ where $k \ll n$ such that the pairwise distances are the same, so

$$\|\mathbf{x}_i - \mathbf{x}_j\| = \|\mathbf{z}_i - \mathbf{z}_j\| \qquad \text{for all} \qquad i, j = 1, \ldots, m?$$

The points $\mathbf{z}_1, \ldots, \mathbf{z}_m$ will then serve as a low dimensional embedding of the data $\mathbf{x}_1, \ldots, \mathbf{x}_m$ that preserves pairwise distances, and is referred to as an *isometric embedding* of the data set. The main problem we will study in this section is how to determine when there exists an isometric embedding of a data set into $\mathbb{R}^k$ and how do we compute the embedded points. When embeddings do not exist, we will look for approximate isometric embeddings that allow some small amounts of distortion. For simplicity, we work exclusively with the Euclidean distance in this section, although the constructions can be straightforwardly extended to any distance based on an inner product norm.

First, it's important to point out that not all data sets can be isometrically embedded in a lower dimensional space.

Example 8.19. Consider three points $\mathbf{x}_1, \mathbf{x}_2, \mathbf{x}_3 \in \mathbb{R}^2$ that are the vertices of an equilateral triangle with unit side length. The three points are all equidistant from each other, so $\|\mathbf{x}_i - \mathbf{x}_j\| = 1$ for all $i \neq j$. On the other hand, as you are asked to prove in Exercise 5.5, there do not exist three points $z_1, z_2, z_3 \in \mathbb{R}$ with $|z_i - z_j| = 1$ for $i \neq j$.

In general, in $\mathbb{R}^n$ we can construct $n + 1$ points that are equidistant from each other, but no more than this. We give a construction here, and postpone the proof that $n + 1$ points is maximal to Example 8.25 after Corollary 8.23. We start with the standard basis vectors $\mathbf{e}_1, \ldots, \mathbf{e}_n$ which are equidistant, with $\|\mathbf{e}_i - \mathbf{e}_j\| = \sqrt{2}$ for all $i \neq j$. To construct an additional equidistant point $\mathbf{x}$, we need that

$$\|\mathbf{x} - \mathbf{e}_i\|^2 = \|\mathbf{x} - \mathbf{e}_j\|^2 \qquad \text{for all} \qquad i, j.$$

Expanding both sides and simplifying, this reduces to $x_i = \mathbf{x} \cdot \mathbf{e}_i = \mathbf{x} \cdot \mathbf{e}_j = x_j$, and hence $\mathbf{x} = (\lambda, \ldots, \lambda)^T = \lambda \mathbf{1}$ for some scalar λ. The value of λ is fixed by solving the quadratic equation

$$\|\mathbf{x} - \mathbf{e}_i\|^2 = (\lambda - 1)^2 + (n - 1)\lambda^2 = \|\mathbf{e}_i - \mathbf{e}_j\|^2 \qquad \text{for} \qquad \lambda = \lambda_\pm := \frac{1 \pm \sqrt{n+1}}{n}. \qquad (8.49)$$

There are thus precisely two choices for the $(n + 1)$-st point, namely $\mathbf{x}_\pm = \lambda_\pm \mathbf{1}$, and either one will do. The reason there are two choices for λ is that $\mathbf{e}_1, \ldots, \mathbf{e}_n$ all lie on the hyperplane $x_1 + \cdots + x_n = 1$, whose normal vector is given by $\mathbf{1}$. The additional point $\mathbf{x}_+ = \lambda_+ \mathbf{1}$ is chosen by moving orthogonally to the plane, in the normal direction, a suitable distance so that the equidistant property holds, and we can do this by moving in either direction.

Moreover, there cannot be an $(n + 2)$-nd equidistant point, since by the above construction, the only option is to take both $\mathbf{x}_+$ and $\mathbf{x}_-$, but $\|\mathbf{x}_+ - \mathbf{x}_-\| \neq \sqrt{2}$. The equidistant configurations $\mathbf{e}_1, \ldots, \mathbf{e}_n, \mathbf{x}_\pm$ form the vertices of a regular n-dimensional tetrahedron. In fact, any other equidistant collection of $n + 1$ points in $\mathbb{R}^n$, including $\mathbf{e}_1, \ldots, \mathbf{e}_n, \mathbf{x}_-$, can be obtained by applying a rigid motion (rotation and translation) and uniform scaling to the basic tetrahedron $\mathbf{e}_1, \ldots, \mathbf{e}_n, \mathbf{x}_+$; a justification can be found in Exercise 5.6. $\blacktriangle$

In order to mathematically study the isometric embedding problem through linear algebra, we define the squared distance matrix corresponding to a data set.

Definition 8.20. Given a data matrix $X = (\mathbf{x}_1, \ldots, \mathbf{x}_m)^T \in \mathcal{M}_{m \times n}$ we define the *squared Euclidean distance matrix* $D_X \in \mathcal{M}_{m \times m}$ with entries

$$d_{ij} = \|\mathbf{x}_i - \mathbf{x}_j\|^2. \tag{8.50}$$

We will often drop the qualifier *squared* and simply refer to D_X as a Euclidean distance matrix. We can define a distance matrix with respect to any norm, but the results in this section will generalize only to norms that are induced by inner products. In terms of distance matrices, the isometric embedding problem posed in this section is whether we can find another data matrix $Y \in \mathcal{M}_{m \times k}$ with $k \ll n$ for which $D_X = D_Y$, or, more generally, $D_X \approx D_Y$ if we allow for some distortion.

Since it does not take any additional effort, we will shift to studying a slightly more general problem. Namely, when is a given matrix $D \in \mathcal{M}_{m \times m}$ a Euclidean distance matrix, i.e., does there exist $n \geq 1$ and $X \in \mathcal{M}_{m \times n}$ such that $D = D_X$? That is, instead of working with a matrix D_X that is already a Euclidean distance matrix for high dimensional data that we wish to embed or visualize in low dimensions, we will work with a general matrix D that could arise by other means, such as a distance matrix in another norm, or a matrix that encodes some non-metric type of distance or similarity between points, e.g., the cosine distance.

The key idea is to first *center* the distance matrix D by applying the centering matrix $J = I - (1/m)\mathbf{1}\mathbf{1}^T$ that appeared in (7.5) on both sides; that is, the matrix JDJ will be a key object of study. We remind the reader that the same centering appeared earlier in the context of kernel PCA in Section 8.1.1.

Proposition 8.21. *Let $X \in \mathcal{M}_{m \times n}$ be a data matrix, and $\underline{X} = JX$ the corresponding centered data matrix. Then,*

$$\underline{X}\,\underline{X}^T = -\tfrac{1}{2} J D_X J. \tag{8.51}$$

Moreover, given any symmetric matrix $D \in \mathcal{M}_{m \times m}$ with zero diagonal,

$$JDJ = JD_X J \qquad \text{if and only if} \qquad D = D_X. \tag{8.52}$$

Proof. To prove (8.51), note that $d_{ij} = \|\mathbf{x}_i - \mathbf{x}_j\|^2 = \|\mathbf{x}_i\|^2 - 2\,\mathbf{x}_i \cdot \mathbf{x}_j + \|\mathbf{x}_j\|^2$, and hence

$$D_X = \mathbf{v}\mathbf{1}^T - 2XX^T + \mathbf{1}\mathbf{v}^T, \qquad \text{where} \qquad \mathbf{v} = \left(\|\mathbf{x}_1\|^2, \ldots, \|\mathbf{x}_m\|^2\right)^T \in \mathbb{R}^m.$$

Since $J = J^T$ is symmetric and $J\mathbf{1} = \mathbf{0}$,

$$J D_X J = J\mathbf{v}\mathbf{1}^T J - 2JXX^T J + J\mathbf{1}\mathbf{v}^T J = -2\,JX\,(JX)^T = -2\underline{X}\,\underline{X}^T.$$

To prove the second part, suppose that

$$J(D - D_X)J = O.$$

Since the kernel of J is one-dimensional, spanned by $\mathbf{1}$, this implies that there exists $\mathbf{v} \in \mathbb{R}^m$ so that

$$(D - D_X)J = \mathbf{1}\mathbf{v}^T.$$

Multiplying by $\mathbf{1}$ on the right, we obtain $\mathbf{v}^T \mathbf{1} = \mathbf{0}$, and so $J\mathbf{v} = \mathbf{v}$. Therefore

$$(D - D_X - \mathbf{1}\mathbf{v}^T)J = O.$$

By a similar argument, there exists $\mathbf{w} \in \mathbb{R}^m$ such that

$$D - D_X = \mathbf{1}\mathbf{v}^T + \mathbf{w}\mathbf{1}^T.$$

Since the diagonal entries of D and D_X are zero, the diagonal entries of $\mathbf{1}\mathbf{v}^T + \mathbf{w}\mathbf{1}^T$, which are the entries of $\mathbf{v} + \mathbf{w}$, are also zero, and so $\mathbf{v} = -\mathbf{w}$, which yields

$$D - D_X = \mathbf{1}\mathbf{v}^T - \mathbf{v}\mathbf{1}^T.$$

Since $D - D_X$ is symmetric, we must have $\mathbf{v} = \mathbf{0}$, which completes the proof.   ∎

Proposition 8.21 establishes that the problem of determining whether a given matrix D is a Euclidean distance matrix is equivalent to finding a data matrix X for which

$$-\tfrac{1}{2} JDJ = \underline{X}\,\underline{X}^T. \tag{8.53}$$

Then (8.52) implies $D = D_X$, provided D is symmetric and has zero diagonal, which is clearly a prerequisite for a distance matrix.

Let us remark that $\underline{X}\,\underline{X}^T$ is a Gram matrix whose entries

$$(\mathbf{x}_i - \overline{\mathbf{x}}) \cdot (\mathbf{x}_j - \overline{\mathbf{x}}), \qquad \text{where} \qquad \overline{\mathbf{x}} = \frac{1}{m} \sum_{i=1}^{m} \mathbf{x}_i$$

are the dot products of the centered data points. It is important to point out that $\underline{X}\,\underline{X}^T$ is *not* the covariance matrix of X, since the centered data matrices are multiplied in the wrong order (recall $S_X = \underline{X}^T\underline{X}$). It is also not the covariance matrix of X^T, since this would involve centering the *columns* of X, and not the rows. Nevertheless, according to Proposition 5.76, its nonzero eigenvalues — the squares of the singular values of $\underline{X}^T$ — are the same as those of the covariance matrix $S_X = \underline{X}^T\underline{X}$ — the squares of the singular values of $\underline{X}$. We also mention that the matrix $\underline{X}\,\underline{X}^T$ is the same matrix that made an appearance in Remark 8.2, as well as in kernel PCA in Section 8.1.1.

Given D, the existence of the matrix $\underline{X}$ in (8.53) is essentially a question of matrix factorization. This leads us to our main result in this section, which gives a complete characterization of when a matrix is a Euclidean distance matrix. The result was originally proved in 1935 by Isaac Schoenberg [209], and, slightly later, independently discovered by Gale Young and Alston Householder [259].

> **Theorem 8.22.** *A matrix $D \in \mathcal{M}_{m \times m}$ is a Euclidean distance matrix if and only if $D = D^T$ is symmetric, has zeros on the diagonal, and the matrix product JDJ is negative semidefinite.*

Proof. Every Euclidean distance matrix D_X is symmetric, has zeros along the diagonal, and, by (8.51), $JD_X J = -2\,\underline{X}\,\underline{X}^T$, which is negative semidefinite, being the negative of the positive semidefinite Gram matrix $\underline{X}\,\underline{X}^T$.

To establish the other direction, let us write $H = -\tfrac{1}{2} JDJ$, which is, by assumption, positive semidefinite. By (5.32) we have the (reduced) spectral decomposition

$$H = Q_k \Lambda_k Q_k^T, \tag{8.54}$$

where $k = \operatorname{rank} H$, while the diagonal matrix $\Lambda_k = \operatorname{diag}(\lambda_1, \ldots, \lambda_k)$ contains its nonzero eigenvalues $\lambda_1 \geq \cdots \geq \lambda_k > 0$, while the columns of $Q_k = (\mathbf{q}_1 \cdots \mathbf{q}_k) \in \mathcal{M}_{n \times k}$ are the corresponding unit eigenvectors. Define

$$X = Q_k \Lambda_k^{1/2} \in \mathcal{M}_{n \times k}, \qquad \text{whereby} \qquad H = XX^T. \tag{8.55}$$

Since $\mathbf{1} \in \ker J$, we have $H\mathbf{1} = \mathbf{0}$, and so

$$\lambda_i \mathbf{q}_i^T \mathbf{1} = (H\mathbf{q}_i)^T \mathbf{1} = \mathbf{q}_i^T (H\mathbf{1}) = \mathbf{0}, \qquad i = 1, \ldots, k,$$

which, since $\lambda_i > 0$, implies $\mathbf{q}_i^T \mathbf{1} = \mathbf{0}$. Thus, $JQ_k = Q_k$, and hence $\underline{X} = JX = X$, i.e., X is already a centered data matrix. Further, by (8.51),

$$-\tfrac{1}{2} JDJ = H = XX^T = \underline{X}\,\underline{X}^T = -\tfrac{1}{2} JD_X J, \quad \text{and so} \quad JDJ = JD_X J.$$

Finally, (8.52) yields $D = D_X$. $\blacksquare$

Inspecting the proof of Theorem 8.22, we immediately deduce a result that tells us how to construct the optimal embedding.

Corollary 8.23. *Let $D \in \mathcal{M}_{m \times m}$ be a Euclidean distance matrix, and let $k = \operatorname{rank}(JDJ)$. Then $D = D_X$ where $X \in \mathcal{M}_{m \times k}$ is given by (8.55), based on the spectral decomposition (8.54) of $H = -\tfrac{1}{2} JDJ$. Furthermore, if $X \in \mathcal{M}_{m \times n}$ has Euclidean distance matrix $D = D_X$, then necessarily $n \geq k$.*

Proof. The first statement follows directly from Theorem 8.22 and its proof. As for the second statement, if $X \in \mathcal{M}_{m \times n}$, so is $\underline{X} \in \mathcal{M}_{m \times n}$. Thus,

$$n \geq \operatorname{rank} \underline{X} = \operatorname{rank}(\underline{X}\,\underline{X}^T) = \operatorname{rank}(JDJ) \geq k. \qquad \blacksquare$$

We further note that, in fact, the ranks of the Euclidean distance matrix and its centered version cannot be too far apart.

Proposition 8.24. *Suppose $D \in \mathcal{M}_{m \times m}$ and $\operatorname{rank} D = r$. Then*

$$\max\{r - 2, 0\} \leq \operatorname{rank}(JDJ) \leq \min\{r, m - 1\}. \tag{8.56}$$

Proof. Since $\operatorname{rank} J = m - 1$, by the Sylvester inequalities (4.42),

$$\max\{r - 1, 0\} \leq \operatorname{rank}(DJ) \leq \min\{r, m - 1\},$$

and hence

$$\max\{r - 2, 0\} \leq \max\{\operatorname{rank}(DJ) - 1, 0\} \leq \operatorname{rank}(JDJ)$$
$$\leq \min\{m - 1, \operatorname{rank}(DJ)\} \leq \min\{r, m - 1\}. \qquad \blacksquare$$

Example 8.25. We return to the problem considered in Example 8.19 of embedding equidistant points in Euclidean space. Suppose we have m data points that are equidistant from each other, which we can, by rescaling, take to be unit distance without loss of generality. The Euclidean distance matrix for such a data set contains has all its off diagonal entries equal to 1, and its diagonal entries equal to 0; thus $D = \mathbf{1}\mathbf{1}^T - \mathrm{I}$, where $\mathbf{1} \in \mathbb{R}^m$ is the ones vector. Since $J\mathbf{1} = 0$ and $J^2 = J$, we have

$$JDJ = J(\mathbf{1}\mathbf{1}^T - \mathrm{I})J = -J^2 = -J.$$

The centering matrix J is positive semidefinite with rank $J = m - 1$; see Exercise 1.2 in Chapter 7. Therefore, by Corollary 8.23, we can isometrically embed m equidistant points in $\mathbb{R}^{m-1}$, but not in any lower dimensional space. Stated differently, by setting $n = m - 1$, there can be at most $n + 1$ equidistant points in $\mathbb{R}^n$. ▲

Corollary 8.23 gives us a recipe for constructing an isometric embedding for a Euclidean distance matrix D. We compute the spectral decomposition of the matrix $H = -\frac{1}{2}JDJ$, discard the zero eigenvalues and eigenvectors, and define $X = Q_k \Lambda_k^{1/2} \in \mathcal{M}_{m \times k}$. Furthermore, the lowest dimensional space in which we can isometrically embed the data is $\mathbb{R}^k$, where $k = \mathrm{rank}(JDJ)$. In practice k may be quite large, and it may be desirable to obtain a low dimensional embedding that only approximately preserves distances, meaning that it has the least distortion possible. In this case, we can look for $X \in \mathcal{M}_{m \times k}$ that minimizes

$$\left\| \tfrac{1}{2}JDJ + \underline{X}\,\underline{X}^T \right\|_F^2 = \left\| \tfrac{1}{2}J(D - D_X)J \right\|_F^2.$$

That is, we look for embedded points X whose distance matrix D_X is as close to D as possible. Since $\mathrm{rank}(\underline{X}\,\underline{X}^T) \le k$, the Schmidt–Eckart–Young–Mirsky Theorem 8.13 guarantees that the beset choice is the truncated SVD[8] of $H = -\frac{1}{2}JDJ$, that is

$$\underline{X}\,\underline{X}^T = P_k \Lambda_k P_k^T,$$

where the columns of P_k are the top k eigenvectors of H, and Λ_k is the diagonal matrix containing the corresponding eigenvalues. This is achieved by setting

$$X = P_k \Lambda_k^{1/2} \in \mathcal{M}_{m \times k}, \tag{8.57}$$

which, as before, is a centered data matrix, so $X = \underline{X}$.

Remark 8.26. There are two close connections between MDS and PCA. First, when $D = D_X$ is itself a Euclidean distance matrix for $X \in \mathcal{M}_{m \times n}$, and our goal is to find a lower dimensional data matrix $Y \in \mathcal{M}_{m \times k}$ for which $D_Y = D_X$, MDS is equivalent to PCA. This holds even in the setting where we allow distortion, so $D_Y \approx D_X$. To see this, let $r = \mathrm{rank}\,\underline{X}$ and take the singular value decomposition $\underline{X} = P\Sigma Q^T$, where $P \in \mathcal{M}_{m \times r}$, $Q \in \mathcal{M}_{n \times r}$, and $\Sigma \in \mathcal{M}_{r \times r}$. Then, by (8.51),

$$-\tfrac{1}{2}JD_XJ = \underline{X}\,\underline{X}^T = P\Sigma Q^T Q\Sigma P^T = P\Sigma^2 P^T,$$

which is exactly the spectral decomposition used in Corollary 8.23 with $\Sigma^2 = \Lambda$, and, in

---

[8]Since the matrix is positive semidefinite, the truncated SVD and eigendecompositions are the same.

particular, $r = \operatorname{rank}(JD_X J)$. Thus, Corollary 8.23 guarantees we can isometrically embed X into $\mathbb{R}^r$, where $r = \operatorname{rank}\underline{X}$, by setting $Y = P\Sigma = \underline{X}Q$ to be the projection of the centered data matrix $\underline{X}$ onto the top r principal components, which in this case correspond to all of its singular vectors. When we allow for distortion and map X into a lower dimensional space $\mathbb{R}^k$ with $k < r$ via (8.57), setting $Y = P_k\Sigma_k$, we are simply projecting onto the top k principal components. Thus, PCA exactly solves the isometric embedding problem for a data matrix X, in both the clean and distorted settings. In particular, we can only find an isometric embedding into $\mathbb{R}^k$ when the high dimensional data is contained in a k-dimensional subspace of $\mathbb{R}^n$.

The second connection is to kernel PCA, introduced in Section 8.1.1. If the matrix D is negative semidefinite, then $K = -D$ is positive semidefinite and we can view it as a kernel matrix, even though it may not have been constructed in this way. This is reasonable intuitively, since kernel matrices should measure similarity between data points, while the distance matrix D is inversely proportional to similarity. In this case, JKJ is positive semidefinite, and both MDS and kernel PCA are identical — both work by projecting onto the top eigenvectors of JKJ. On the other hand, if D is not negative semidefinite, then there is no direct connection to kernel PCA.   ▲

One of the most important applications of isometric embeddings occurs when the given matrix $D \in \mathcal{M}_{m \times m}$ was *not* initially constructed as a Euclidean distance matrix, but was instead obtained by measuring some notion of distance between data points that may come from another norm, or be non-metric like the cosine distance. We will also see applications in Chapter 9 where the distance matrix corresponds to pairwise shortest path distances on graphs. In this case, D may not be a Euclidean distance matrix, i.e., JDJ may not be negative semidefinite, so the previous results do not apply. Nevertheless, we would like to obtain an embedding into Euclidean space that preserves the distances in D as much as is possible.

To proceed in this setting, let us work directly with the centered matrix, which we denote by

$$H = -\tfrac{1}{2}JDJ \in \mathcal{M}_{m \times m}. \tag{8.58}$$

Recall that when $D = D_X$ is a Euclidean distance matrix, we have by Proposition 8.21 that $H = \underline{X}\,\underline{X}^T$, so the entries h_{ij} of H are the inner products between centered data points. This is a notion of similarity — in fact, it is related to the cosine similarity introduced in Section 7.4, since, in contrast to a distance, the value of h_{ij} is larger when the data points i and j are more similar, and smaller when they are less similar. Thus, in the general setting when $D \neq D_X$, we refer to H as a *similarity matrix*.

We will proceed by assuming we are given a symmetric similarity matrix H, which may or may not be produced by centering a distance matrix. In practice, we may construct H by measuring any reasonable notion of similarity between data points. Motivated by the results and discussions in this section, to embed the similarity matrix into $\mathbb{R}^k$ in a way that preserves the similarities, we seek a data matrix X that solves the minimization problem

$$\min\left\{ \, \| H - XX^T \|_F^2 \; \big| \; X \in \mathcal{M}_{m \times k} \right\}. \tag{8.59}$$

While this problem looks like the one solved by Theorem 8.13, this is not the case, since XX^T is positive semidefinite, being a Gram matrix, but H may not be, so we cannot take XX^T to be its truncated SVD or eigendecomposition. Essentially, the problem (8.59) is a constrained version of Theorem 8.13, where the minimization is taken over positive semidefinite matrices of rank k. It turns out we can extend Theorem 8.13 to this setting.

Theorem 8.27. *Let $H \in \mathcal{M}_{m \times m}$ be symmetric, and let*

$$H = \sum_{i=1}^{m} \lambda_i \mathbf{p}_i \mathbf{p}_i^T$$

be its spectral decomposition, where its eigenvalues are ordered from largest to smallest, $\lambda_1 \geq \lambda_2 \geq \cdots \geq \lambda_m$. Then the best positive semidefinite matrix approximating H in the Frobenius norm with rank at most k is the matrix

$$H_k = \sum_{i=1}^{k} \lambda_i^+ \mathbf{p}_i \mathbf{p}_i^T,$$

where $\lambda_i^+ = \max\{\lambda_i, 0\}$. That is,

$$\|H - H_k\|_F \leq \|H - K\|_F$$

for all positive semidefinite K with $\operatorname{rank} K \leq k$.

Proof. Let $H = P \Lambda_H P^T$ be the spectral decomposition of H. Let K be positive semidefinite with spectral decomposition $K = Q \Lambda_K Q^T$. Then, by von Neumann's trace inequality (5.65),

$$\|H - K\|_F^2 = \operatorname{tr}(H - K)^2 = \operatorname{tr}(H^2) - 2\operatorname{tr}(HK) + \operatorname{tr}(K^2)$$

$$\geq \sum_{i=1}^{m} \lambda_i(H)^2 - 2\sum_{i=1}^{m} \lambda_i(H)\lambda_i(K) + \sum_{i=1}^{m} \lambda_i(K)^2 = \sum_{i=1}^{m} \left[\lambda_i(H) - \lambda_i(K)\right]^2$$

$$= \|\Lambda_H - \Lambda_K\|_F^2 = \|P\Lambda_H P^T - P\Lambda_K P^T\|_F^2 = \|H - P\Lambda_K P^T\|_F^2,$$

where, in the last line, we used the fact that $\|A\|_F = \|PAP^T\|_F$ since P is an orthogonal matrix; see Exercise 8.6 in Chapter 4. Therefore, we can replace K with $L = P\Lambda_K P^T$ and achieve a smaller or equal Frobenius norm. Furthermore, L is also positive semidefinite, and has the same rank as K. In fact, L has all the same eigenvalues as K, and we have just exchanged its eigenvectors for those of H.

Therefore, we may assume K has the form $P\Lambda_K P^T$ and we can minimize

$$\|H - K\|_F^2 = \|\Lambda_H - \Lambda_K\|_F^2 = \sum_{i=1}^{m} \left[\lambda_i(H) - \lambda_i(K)\right]^2,$$

over the choices of the eigenvalues $\lambda_1(K) \geq \cdots \geq \lambda_n(K) \geq 0$. Since K is restricted to have rank at most k, it has at most k nonzero eigenvalues. As the eigenvalues of H are also ordered from largest to smallest, the best choice is to set $\lambda_i(K) = \max\{\lambda_i(H), 0\} = \lambda_i^+(H)$ for $i = 1, \ldots, k$, and $\lambda_i(K) = 0$ for $i = k+1, \ldots, m$. $\blacksquare$

By Theorem 8.27, the solution to (8.59) satisfies

$$XX^T = \sum_{i=1}^{k} \lambda_i^+ \mathbf{p}_i \mathbf{p}_i^T = P_k \Lambda_k^+ P_k,$$

where $P_k = (\mathbf{p}_1 \ \cdots \ \mathbf{p}_k)$ contains the top k eigenvectors of H, and $\Lambda_k^+ = \operatorname{diag}(\lambda_1^+, \ldots, \lambda_k^+)$ are the positive parts of the top k eigenvalues, which can be obtained by taking the embedded points to be the rows of

$$X = P_k (\Lambda_k^+)^{1/2} \in \mathcal{M}_{m \times k}. \tag{8.60}$$

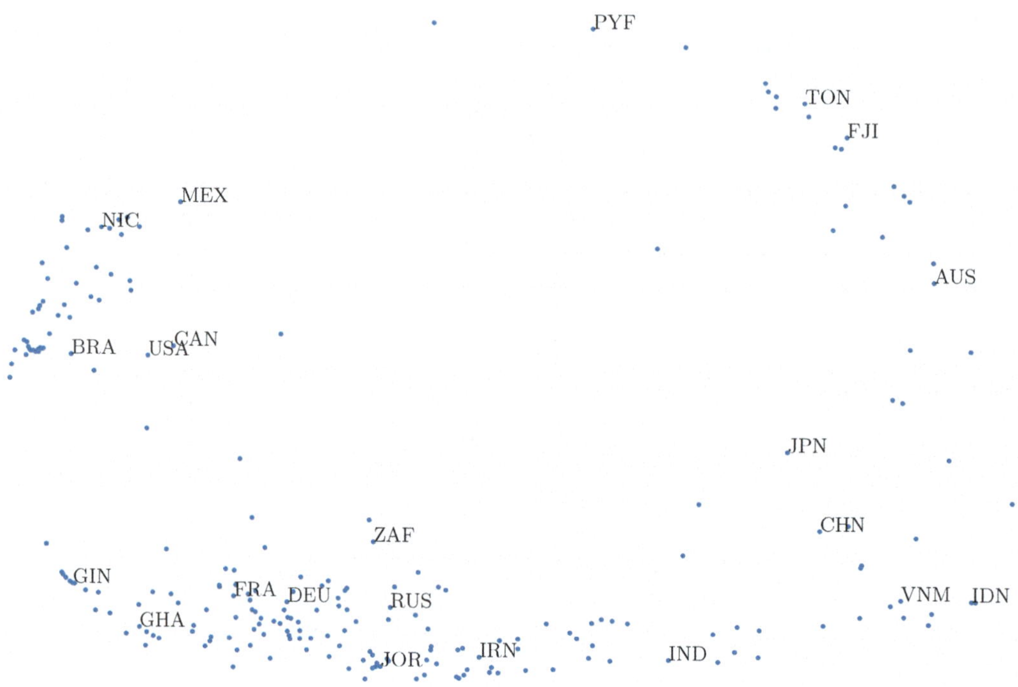

Figure 8.22: Using MDS to display all countries on Earth in a two-dimensional plot. Some of the country codes are displayed.

The embedding X in (8.60) is referred to as *classical multidimensional scaling* or *classical MDS*. Classical MDS essentially uses the embedding from Corollary 8.23, except that we take the positive parts of the eigenvalues first, which allows us to take the square root. There are many other versions of MDS based on minimizing quantities similar to (8.59), such as $\| D - D_X \|_F$, some of which require more sophisticated optimization procedures to solve; we refer the reader to [28] for more details.

We now turn to some examples with real data. In Figure 8.22 we show a visualization of all the countries on Earth using MDS to preserve their pairwise distances. The pairwise distances in this case are the geodesic (great circle) distances on the globe between the most populous cities in each country. In particular, because we are using geodesic distances, the distance matrix is *not* a Euclidean distance matrix. We show some of the country names in the figure, and we can see, as expected, that the embedding does a reasonably good job of keeping neighboring and nearby countries close together in the visualization. Here, we are working with a distance matrix, so we performed the centering step (8.58).

For our second example we return to visualization of the MNIST data set, to which we applied PCA and LDA previously; see Figures 8.5 and 8.20. As discussed in Remark 8.26, if we use MDS on the MNIST data set with pairwise Euclidean distances, we will simply recover the same result as with PCA. Instead, here we consider two non-metric similarity matrices. First, consider the cosine similarity

$$h_{\cos}(\mathbf{x}, \mathbf{y}) = \frac{\mathbf{x} \cdot \mathbf{y}}{\|\mathbf{x}\| \, \|\mathbf{y}\|} = 1 - d_{\cos}(\mathbf{x}, \mathbf{y}),$$

where $d_{\cos}$ is the cosine distance defined in (2.76). We plot the corresponding MNIST visualizations of the first few digits in Figure 8.23(a). We also consider similarity based on a

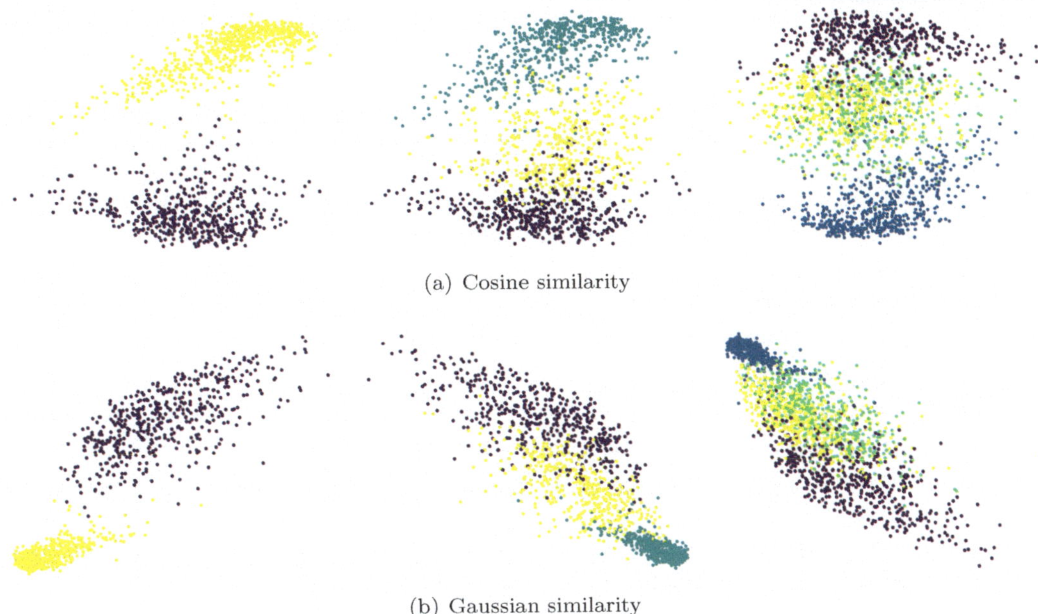

(a) Cosine similarity

(b) Gaussian similarity

Figure 8.23: Multidimensional scaling for visualizing MNIST digits using cosine similarity and the Gaussian similarity.

Gaussian kernel function, given by

$$h(\mathbf{x}, \mathbf{y}) = \exp\left(-\gamma \|\mathbf{x} - \mathbf{y}\|^2\right), \qquad \text{where} \qquad \gamma = \frac{1}{n}. \tag{8.61}$$

The Gaussian MDS embedding is shown in Figure 8.23(b). In both cases, we work directly with the similarity matrix H, and do not perform the centering step as in classical MDS.

Multidimensional scaling has been used for many other problems. A famous example is in the visualization of members of the US Congress based on their voting records [188]. We refer the interested reader to [28] for more details.

Exercises

5.1. Choose a data set from `sklearn` and apply MDS for visualization.

5.2. Apply MDS for visualization of the FashionMNIST data set in `graphlearning`.

5.3. ♡ Let C be a positive definite symmetric matrix and define the distance matrix D_X^C to be the distance matrix in the norm $\|\mathbf{x}\|_C = \sqrt{\mathbf{x}^T C \mathbf{x}}$, with entries $d_{ij} = \|\mathbf{x}_i - \mathbf{x}_j\|_C^2$. Generalize Proposition 8.21 and Theorem 8.22 to this setting. In particular, how do you construct the optimal isometric embedding in this case?

5.4. Let C be a positive definite symmetric matrix. Show that the solution of

$$\min\left\{ \|H - XCX^T\|_F^2 \mid X \in \mathcal{M}_{m \times k} \right\} \tag{8.62}$$

is given by $X = P_k (\Lambda_k^+)^{1/2} C^{-1/2}$, where $P_k = (\mathbf{p}_1 \cdots \mathbf{p}_k)$ contains the top k eigenvectors of H, and $\Lambda_k^+ = \mathrm{diag}\left(\lambda_1^+, \ldots, \lambda_k^+\right)$ are the positive parts of the top k eigenvalues.

5.5. ♡   Show that there do not exist three points $z_1, z_2, z_3 \in \mathbb{R}$ that satisfy
$$|z_1 - z_2| = |z_1 - z_3| = |z_2 - z_3| = 1.$$

5.6. ◇   In this exercise, we prove that any set of $n+1$ equidistant points in $\mathbb{R}^n$ can be mapped, by a combination of scaling and rigid motion, to the standard set $\mathbf{e}_1, \ldots, \mathbf{e}_n, \lambda_+ \mathbf{1}$ where λ_+ is given by (8.49).

Thus, suppose $\mathbf{x}_1, \ldots, \mathbf{x}_{n+1} \in \mathbb{R}^n$ satisfy $\|\mathbf{x}_j - \mathbf{x}_k\| = D$ for some $D > 0$ and all $j \neq k$. Complete the following steps to justify the preceding claim.

(a) First, explain how to use a uniform scaling $\widetilde{\mathbf{x}}_k = \mu \mathbf{x}_k$, where $\mu > 0$, to make $\|\widetilde{\mathbf{x}}_j - \widetilde{\mathbf{x}}_k\| = \sqrt{2}$ for all $j \neq k$.

(b) Next apply the translation $\mathbf{y}_j = \widetilde{\mathbf{x}}_j + \mathbf{a}$ where $\mathbf{a} = -\widetilde{\mathbf{x}}_1$. Prove that $\mathbf{y}_1 = \mathbf{0}$, while $\|\mathbf{y}_j\| = \sqrt{2}$ and $\mathbf{y}_j \cdot \mathbf{y}_k = 1$ for all $2 \leq j \neq k \leq n$.

(c) Let $Y = (\,\mathbf{y}_2 \; \cdots \; \mathbf{y}_n\,) \in \mathcal{M}_{n \times (n-1)}$. Prove that $\mathbf{w} = -(\mathbf{y}_2 + \cdots + \mathbf{y}_n)/n$ is the minimal norm solution to the linear system $Y^T \mathbf{w} = -\mathbf{1} \in \mathbb{R}^{n-1}$, with $\|\mathbf{w}\| = \sqrt{(n-1)/n} < 1$.

(d) Use part (c) to explain why one can find $\mathbf{q}_1 \in \mathbb{R}^n$ such that $\mathbf{y}_j \cdot \mathbf{q}_1 = -1$ for all $j = 2, \ldots, n$ and $\|\mathbf{q}_1\| = 1$.

(e) Set $\mathbf{q}_k = \mathbf{y}_k + \mathbf{q}_1$ for $k = 2, \ldots, n$. Prove that $Q = (\,\mathbf{q}_1 \; \mathbf{q}_2 \; \cdots \; \mathbf{q}_n\,)$ is an orthogonal matrix that maps $Q \mathbf{e}_k = \mathbf{q}_k$ and hence $Q^T \mathbf{y}_k + \mathbf{e}_1 = \mathbf{e}_k$ for $k = 1, \ldots, n$.

(f) Explain why $Q^T \mathbf{y}_{n+1} + \mathbf{e}_1 = \lambda_\pm \mathbf{1}$ to conclude that the affine map $F[\mathbf{y}] = Q^T \mathbf{y} + \mathbf{e}_1$ takes $\mathbf{y}_1 = 0, \mathbf{y}_2, \ldots, \mathbf{y}_n, \mathbf{y}_{n+1}$ to either $\mathbf{e}_1, \ldots, \mathbf{e}_n, \lambda_+ \mathbf{1}$ or $\mathbf{e}_1, \ldots, \mathbf{e}_n, \lambda_- \mathbf{1}$

(g) Finally, use a reflection through the hyperplane $x_1 + \cdots + x_n = 1$ to map the other configuration $\mathbf{e}_1, \ldots, \mathbf{e}_n, \lambda_- \mathbf{1}$ to the standard one $\mathbf{e}_1, \ldots, \mathbf{e}_n, \lambda_+ \mathbf{1}$.

(h) Deduce that the combination of affine maps in the previous parts produces an affine map of the form $F(\mathbf{x}) = \mu \widetilde{Q} \mathbf{x} + \mathbf{a}$, where $\mu > 0$ is a scaling, $\widetilde{Q}$ is an orthogonal matrix, and $\mathbf{a}$ represents a translation, which maps $\mathbf{x}_1, \ldots, \mathbf{x}_{n+1}$ to $\mathbf{e}_1, \ldots, \mathbf{e}_n, \lambda_+ \mathbf{1}$. If $\widetilde{Q}$ is a reflection and one desires a rigid motion instead, explain how to construct a reflection that preserves all the points $\mathbf{e}_1, \ldots, \mathbf{e}_n, \lambda_+ \mathbf{1}$ and, by composition, converts $\widetilde{Q}$ into a rotation matrix.

Chapter 9

Graph Theory and Graph-based Learning

In this chapter, we cover the basics of graph theory followed by some of the graph-based machine learning algorithms arising in applications. By a "graph"[1], we mean a combinatorial object consisting of a finite number of points, known as nodes or vertices, and a finite number of edges, each of which connects two of the nodes. In addition to its many roles in a broad range of mathematics, graph theory finds applications to a wide variety of applied problems, including the analysis of network data, such as communication, social, biological, or academic networks, or more broadly, the internet, molecular property prediction and drug discovery, Markov processes, image processing and computer vision, and geometric structures. The applications of very large graphs, e.g., with millions or billions of nodes, or of very large data sets consisting of millions of graphical objects, are playing an increasingly important role in modern data analysis, machine learning, and computer science.

A main focus of this chapter is spectral graph theory, [42,219], which refers to the study of the properties of graphs that are captured by their spectrum, meaning the set of eigenvalues of certain naturally associated matrices, in particular the graph Laplacian matrix. Applications include spectral embedding, spectral clustering, diffusion on graphs, the PageRank algorithm, graph-based semi-supervised learning, and various graph-based visualization and dimension reduction techniques. We will also see how the discrete Fourier transform can be viewed through the lens of the graph Laplacian spectrum. Later, in Chapter 10, we introduce graph neural networks, which build upon the theory developed in this chapter.

Throughout this chapter, we use $\| \mathbf{x} \| = \sqrt{\mathbf{x} \cdot \mathbf{x}}$ to denote the Euclidean norm of $\mathbf{x}$, also known as the 2 norm. When we, on occasion, use other inner products and induced norms, we will denote them by $\langle \mathbf{x}, \mathbf{y} \rangle_C = \mathbf{x}^T C \mathbf{y}$ and $\| \mathbf{x} \|_C = \sqrt{\langle \mathbf{x}, \mathbf{y} \rangle_C}$, as usual, for some symmetric positive definite matrix C. As a warning to the reader, while we usually enumerate the eigenvalues of matrices in order of largest to smallest, in this chapter, solely when dealing with graph Laplacian matrices, we will order the eigenvalues from smallest to largest, i.e., $\lambda_1 \leq \cdots \leq \lambda_m$. This is done both to match the conventions in the literature, and because, as we shall see later on in this chapter, the eigenvectors with smallest eigenvalues are, generally speaking, the most significant in applications.

---

[1] Not to be confused with the graph of a function, which is a completely different concept.

9.1 Graphs and Digraphs

Python Notebook: Intro to Graphs (.ipynb)

We begin with the basic definitions. A *graph* consists of a finite number of points, called *nodes* or *vertices*, along with finitely many lines or curves connecting them, called *edges*. Each edge connects exactly two nodes, which are its endpoints. Two nodes are *adjacent* or *neighbors* if there is an edge connecting them. A graph can be viewed as an electrical network: the edges represent the wires, whose ends are joined together at the nodes. The graph encodes the topology — meaning interconnectedness — of the network, but not its geometry or physics: lengths or shapes of the wires, their resistances, etc.

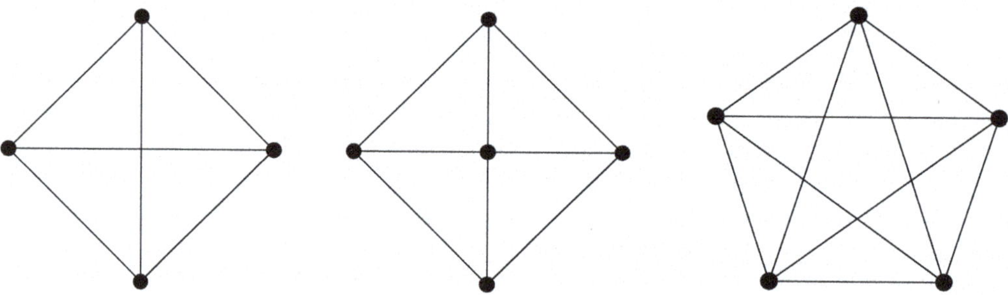

Figure 9.1: Three Different Graphs

Some examples of graphs can be seen in Figure 9.1; the nodes are the black dots and the edges are the lines connecting them. In a planar representation of a graph, the edges are allowed to cross over each other at non-nodal points without meeting — think of a network where the insulated wires lie on top of each other, but do not interconnect. Thus, the first graph has 5 nodes and 8 edges; the second has 4 nodes and 6 edges — the two central edges do not meet; the final graph has 5 nodes and 10 edges, and the edges forming the diagonals do not intersect.

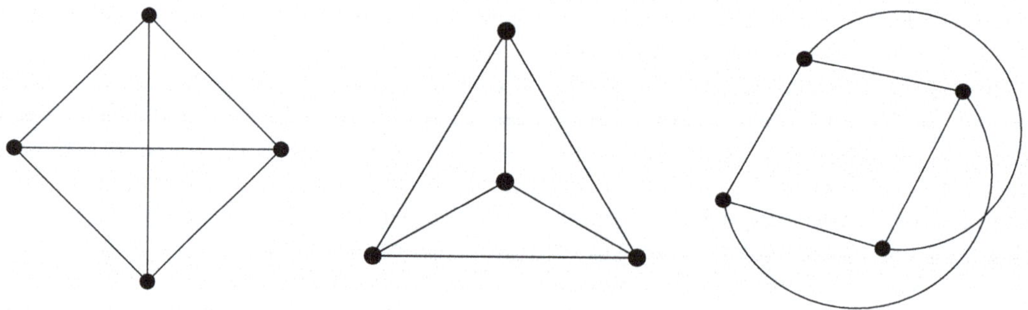

Figure 9.2: Three Versions of the Same Graph

Two graphs are considered to be the same if there is a one-to-one correspondence between their edges and their nodes, so that matched edges connect matched nodes. In an electrical network, moving the nodes and wires around without cutting or rejoining will have no effect on the underlying graph. Consequently, there are many ways to draw a given graph; three representations of one and the same graph appear in Figure 9.2. The reader should be able to figure out how each one can be identified with the others.

We will use calligraphic letters, e.g., $\mathcal{G}$, to denote graphs, and write $\mathcal{G} = (\mathcal{N}, \mathcal{E})$ where $\mathcal{N}$ denotes the nodes and $\mathcal{E}$ the edges. We let $m = \#\mathcal{N}$ denote the number of nodes; often we will label the nodes by integers, and so can identify $\mathcal{N} \simeq \{1, \ldots, m\}$. Similarly, we let $e = \#\mathcal{E}$ denote the number of edges, which can also be labeled by integers, so $\mathcal{E} \simeq \{1, \ldots, e\}$. Alternatively, an edge that connects nodes i and j can be denoted by $\varepsilon = (i, j)$, the order of the two nodes not mattering. A graph is called *simple* if every edge connects two *distinct* nodes, i.e., $i \neq j$ in the preceding notation, so no edge forms a *loop* that connects a node to itself, and, moreover, two distinct nodes are connected by at most one edge. All the graphs in Figures 9.1 and 9.2 are simple.

Example 9.1. An important example is the *complete graph* $\mathcal{G}_m$ on m nodes. It has one edge joining every distinct pair of nodes, and hence a total of $n = \begin{pmatrix} m \\ 2 \end{pmatrix} = \dfrac{m(m-1)}{2}$ edges.

For example, the second and third graphs in Figure 9.1 represent the complete graph on, respectively, 4 and 5 nodes. ▲

It is often convenient to assign a direction to each edge in a graph. The direction or orientation will be fixed by identifying the node the edge "starts" at, known as its *tail*, and the node it "ends" at, known as its *head*. Sometimes the direction is specified by what the graph and its edges represent, while in other contexts, the direction can be assigned arbitrarily. For example, in an electrical circuit, there is no a priori assignment of direction to a wire represented by an edge. But once we assign a direction, a current along that wire will be positive if it moves in the same direction, i.e., goes from the tail node to the head node, and negative if it moves in the opposite direction. The direction of the edge does *not* dictate the direction of the current — it just fixes which direction positive and negative values represent.

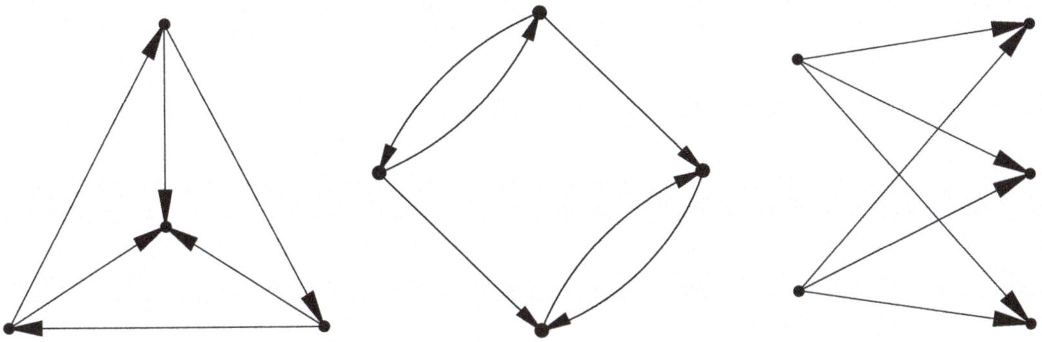

Figure 9.3: Some Digraphs

A graph with directed edges is known as a *directed graph*, or *digraph* for short. Examples of digraphs can be seen in Figure 9.3; the edge directions are represented by arrows. For a digraph, an edge $\varepsilon = (i, j)$ indicates that it is directed *from* node i to node j, while for an undirected graph it merely indicates there is an edge between i and j. Every digraph $\widehat{\mathcal{G}} = (\mathcal{N}, \widehat{\mathcal{E}})$ has an underlying undirected graph $\mathcal{G} = (\mathcal{N}, \mathcal{E})$, where the edges in $\mathcal{E}$ are given by forgetting the directions of the edges in $\widehat{\mathcal{E}}$. Conversely, given an undirected graph $\mathcal{G}$ one can construct a directed graph $\widehat{\mathcal{G}}$ with the same vertices and edges by assigning a direction to each of the edges. In some applications, the assigned directions are determined by the nature of the data the digraph represents. In other situations, one assigns directions in any convenient manner. Indeed, many of our results do not depend on the choice of directions, and so the assignation, when required, is unimportant.

A digraph without loops and with at most one directed edge from a node i to a node $j \neq i$ is called a *simple digraph*. Our definition of a simple digraph does allow two directed edges between two distinct nodes i and j provided they go in the opposite directions. All the digraphs in Figure 9.3 are simple. Two nodes in a digraph are *adjacent* if there is a directed edge going from the first to the second; thus, unlike graphs, node i might be adjacent to node j without node j being adjacent to node i. If the digraph is allowed to have multiple edges connected the same two nodes it is called a *multidigraph* or *quiver*. For example, at any instant in time, the internet can be viewed as a gigantic quiver, in which each node represents a webpage, and each edge represents an existing link from one page to another; in this case its direction is clear, and one may well have multiple edges representing links in both directions.

In this text, we will almost always work with simple graphs and simple digraphs, and so drop the *simple* qualifier from now on. See below for methods for converting a (di)graph with multiple edges to an essentially equivalent (di)graph.

The structure of a graph or digraph can be entirely encoded in an associated square matrix.

> **Definition 9.2.** Given a simple digraph on m nodes, the *adjacency matrix* is the $m \times m$ matrix A whose off-diagonal entry a_{ij}, for $i \neq j$, is equal to 1 if there is an edge from node i to node j, and 0 otherwise. The diagonal entries of the adjacency matrix are all zero: $a_{ii} = 0$. The adjacency matrix of an undirected graph is symmetric, $A = A^T$, with $a_{ij} = a_{ji} = 1$ if and only if nodes i and j are connected by an edge, while $a_{ij} = a_{ji} = 0$ when they are not connected.

For example, the adjacency matrix of the graph in Figure 9.2, which is the complete graph on four nodes, is given by

$$A = \begin{pmatrix} 0 & 1 & 1 & 1 \\ 1 & 0 & 1 & 1 \\ 1 & 1 & 0 & 1 \\ 1 & 1 & 1 & 0 \end{pmatrix}.$$

Because, for this graph, all nodes have the same connectivity, it does not matter how we label them. Similarly, labeling the nodes in order from top to bottom and, when at the same height, from left to right, the adjacency matrices of the digraphs in Figure 9.3 are, respectively,

$$\begin{pmatrix} 0 & 1 & 0 & 1 \\ 0 & 0 & 0 & 0 \\ 1 & 1 & 0 & 0 \\ 0 & 1 & 1 & 0 \end{pmatrix}, \quad \begin{pmatrix} 0 & 1 & 1 & 0 \\ 1 & 0 & 0 & 1 \\ 0 & 0 & 0 & 1 \\ 0 & 0 & 1 & 0 \end{pmatrix}, \quad \begin{pmatrix} 0 & 0 & 0 & 0 & 0 \\ 1 & 0 & 1 & 0 & 1 \\ 0 & 0 & 0 & 0 & 0 \\ 1 & 0 & 1 & 0 & 1 \\ 0 & 0 & 0 & 0 & 0 \end{pmatrix}.$$

Observe that the positions of the ones in an adjacency matrix completely describes the structure of the graph or digraph. However, there are many applications, described in Section 9.1.1 below, where some edges in the graph have more importance than others. In this case, we can assign a *positive weight* to each edge, where larger weights mean the edges are more important, and smaller weights indicate less importance. When an edge has weight zero that indicates its absence in the (di)graph.

Definition 9.3. Given a simple digraph on m nodes, an associated *weight matrix* is an $m \times m$ matrix whose off-diagonal entry w_{ij}, for $i \neq j$, is zero if there is no edge from node i to node j, and a positive number $w_{ij} > 0$ when there is an edge from node i to node j, with the value w_{ij} prescribing the *weight* of the edge. The diagonal entries of the weight matrix are all zero: $w_{ii} = 0$.

A graph with an accompanying weight matrix W is called a *weighted graph*, or *weighted digraph* if the graph is directed. Note that the weight matrix of a weighted graph is necessarily symmetric: $W = W^T$, with $w_{ij} = w_{ji} > 0$ if and only if nodes i and j are connected by an edge. On the other hand, for a weighted digraph, we may have edges from node i to j and from node j to i with different edge weights, in which case $w_{ij} \neq w_{ji}$ and so the weight matrix W is not symmetric. An unweighted graph or digraph can be thought of as weighted with all edge weights equal to 1, and thus its adjacency matrix is its weight matrix: $W = A$. As with the adjacency matrix, the weight matrix for a weighted graph completely describes the graph structure. We will generally always work with weighted graphs or weighted digraphs, since this contains as a special case unweighted (di)graphs. Some results and algorithms in this chapter work on weighted digraphs, and some hold only on weighted graphs, and we will clearly specify with which we are working.

Given a weighted digraph, there is an underlying weighted graph that is obtained by *forgetting* the directions of the edges. When two directed edges connect the same pair of nodes, they are replaced by a single edge whose weight is the sum of both directed edge weights.

Definition 9.4. Given a weighted digraph $\widehat{\mathcal{G}}$ with weight matrix $\widehat{W}$, the *underlying weighted graph* $\mathcal{G}$ has the symmetric weight matrix $W = \widehat{W} + \widehat{W}^T$.

Example 9.5. Consider the weighted digraph with $m = 4$ nodes with weight matrix

$$\widehat{W} = \begin{pmatrix} 0 & 1 & 2 & 0 \\ 3 & 0 & 0 & 2 \\ 0 & 0 & 0 & 4 \\ 0 & 0 & 1 & 0 \end{pmatrix}.$$

The directed graph structure is the same as the digraph in the middle of Figure 9.3, except with weights attached to each edge — as an exercise the reader may wish to label the edges with the corresponding weights from W. The underlying weighted graph, which is the square graph with 4 edges connecting 4 nodes, has weight matrix

$$W = \begin{pmatrix} 0 & 4 & 2 & 0 \\ 4 & 0 & 0 & 2 \\ 2 & 0 & 0 & 5 \\ 0 & 2 & 5 & 0 \end{pmatrix}. \qquad \blacktriangle$$

Remark 9.6. More generally, given a non-simple (un)weighted graph or digraph without loops, we can construct an equivalent weighted graph or digraph by combining multiple (directed) edges that connect the same two nodes into a single edge by simply summing the associated weights. $\qquad \blacktriangle$

The *degree* of a node is an important measure of the graph's local connectivity there. Recall that $\mathbf{1} \in \mathbb{R}^m$ denotes the ones vector all of whose entries equal 1.

Definition 9.7. Given a weighted graph or digraph $\mathcal{G}$ with weight matrix W, the *weighted degree* of node i is the sum of the weights of all edges originating at node i:

$$d_i = \sum_{j=1}^{m} w_{ij}. \tag{9.1}$$

The *weighted degree vector* is the vector

$$\mathbf{d} = (d_1, d_2, \ldots, d_m)^T = W\mathbf{1} \in \mathbb{R}^m \tag{9.2}$$

containing the degrees as entries. The *weighted degree matrix* is the $m \times m$ diagonal matrix $D = \mathrm{diag}(d_1, \ldots, d_m)$ containing the degrees of the nodes.

We are using the convention that the degree measures the *outgoing edges* from node i. Clearly an isolated node containing no outgoing edges has degree 0, though it may have incoming edges. Consequently, the degree matrix D is invertible if and only if the digraph contains no isolated nodes. In the case of an unweighted simple graph or digraph, all nonzero weights are $w_{ij} = 1$, so the weight matrix coincides with the adjacency matrix, $W = A$, and the degree of a node is exactly the number of its neighbors, i.e., adjacent nodes. Henceforth, we will usually drop the adjective "weighted" in the above definitions, and speak of the degrees of the nodes and the degree matrix.

For a digraph, there is an equivalent definition of *incoming* degree $\widetilde{d}_i$ that measures the edges terminating at node i, which is obtained by replacing w_{ij} by w_{ji} in (9.1). Thus, in analogy with (9.2), the *incoming degree vector* is $\widetilde{\mathbf{d}} = W^T\mathbf{1}$. In the case of a graph, so $W = W^T$ is symmetric, the degrees are the same: $\widetilde{\mathbf{d}} = \mathbf{d}$. More generally, we will call a digraph *balanced*[2] if the incoming and outgoing degrees are equal at all nodes, so $\widetilde{\mathbf{d}} = \mathbf{d}$. The weight matrix of a balanced digraph is not necessarily symmetric, but does have the property that its row sums equal its corresponding column sums.

In practical applications of graphs, we may also possess some additional information associated with each node in the graph, which are known as *node features*. We will write $\mathbf{x}_1, \ldots, \mathbf{x}_m \in \mathbb{R}^n$ for the node features. If the nodes correspond to images, the features may be the pixel values in the images, or some information extracted from the image, such as the image classification or an image annotation. If the nodes correspond to websites, the features may encode the type of website, or provide some summary statistics about the content of the website. The next section gives further applications and examples of graphs in the real world.

A *walk* in a weighted digraph is an ordered list of edges $\varepsilon_1, \varepsilon_2, \ldots, \varepsilon_k$ connecting adjacent nodes $m_1, m_2, \ldots, m_{k+1}$ so that edge $\varepsilon_i = (m_i, m_{i+1})$ connects node m_i to node m_{i+1} with $w_{m_i, m_{i+1}} > 0$. That is, a walk must consistently follow directed edges in the prescribed directions. The same holds for a weighted graph, except now one does not need to pay attention to edge directions. A *trail* is a walk in which all the edges are distinct, so $\varepsilon_i \neq \varepsilon_j$ for $i \neq j$. A *path* is a trail for which the nodes are also distinct, so $m_i \neq m_j$ for $i \neq j$. While an edge cannot be repeated in a trail, a node may be — whereas in a path, no edge or node can be repeated. For instance, in the graph in Figure 9.4(b), with the nodes and edges labeled as indicated, one walk starts at node 1, then goes in order along the edges labeled as $1, 4, 3, 2$, successively passing through the nodes $1, 2, 4, 1, 3$. The edges are distinct, while the nodes are not, so this walk is also a trail, but not a path. The walk starting at node 1 and following

---

[2]In [111], it is shown that an unweighted digraph can be balanced by a suitable choice of weights if and only if every edge is contained in a circuit, as defined below.

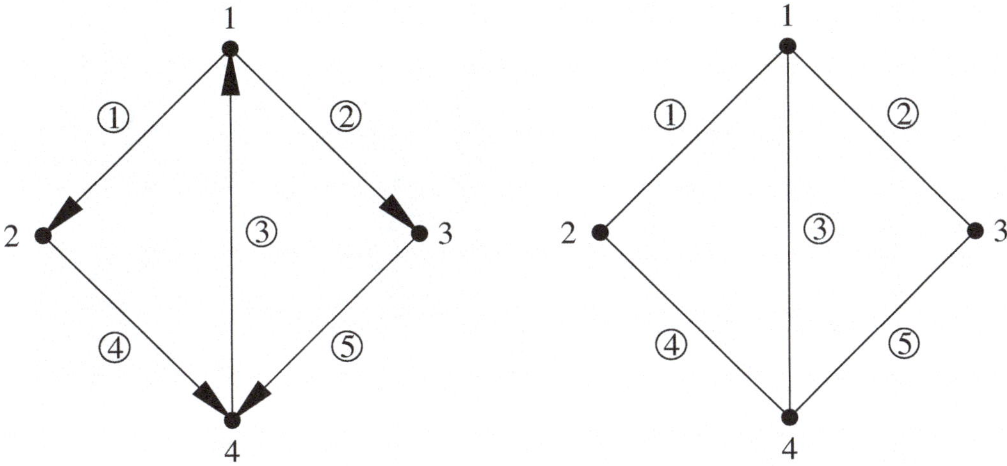

Figure 9.4: A Digraph and its Underlying Graph.

the edges $2, 5$ is a path passing through nodes $1, 3, 4$. For the digraph, there is no path that visits all the nodes; on the other hand, the underlying graph has such a path, namely one that goes along edges $1, 4, 5$ which is allowed since now one does not need to pay attention to the orientation of the edges.

A *circuit* is a trail, connecting adjacent nodes $m_1, m_2, \ldots, m_{k+1}$ by edges, that ends up where it began, i.e., $m_{k+1} = m_1$. Again, while each edge in the circuit is only traversed once, the circuit can visit a node multiple times. For example, the circuit in Figure 9.4(b) consisting of edges $1, 4, 3$ starts at node 1, then goes to nodes $2, 4$ in order, and finally returns to node 1. In a circuit, the choice of starting node is not important, and we identify circuits that go around the edges in the same order. Thus, for example, the edges $4, 3, 1$ represent the same circuit as above. Observe that the edges $1, 4, 5, 2$ form a circuit in the underlying graph, but not in the digraph since their directions are not consistent. In the case of a graph, the direction the circuit is traversed is also not important, so $5, 4, 1, 2$ represents the same circuit.

A graph or digraph is *connected*[3] if one can get from any node to any other node by a path. Any graph containing an *isolated node*, meaning one that has degree 0 and hence does not have any outgoing edges, is automatically disconnected. We note that every graph can be decomposed into the disjoint union of a finite number of connected subgraphs, known as the *connected components* of $\mathcal{G}$, each disconnected from the others, i.e., they have no nodes in common and there is no path from a node in one component to a node in a different component. A connected graph $\mathcal{G}$ has only one connected component. At the other extreme, a graph is *totally disconnected* if it has no edges, and hence m connected components, namely its nodes, all of which are isolated. Such a graph has a zero weight matrix: $W = O$.

Let $\mathcal{G} = (\mathcal{N}, \mathcal{E})$ be a graph or digraph with m nodes. Given a subset $\mathcal{S} \subset \mathcal{N}$ of the nodes, the *indicator vector* associated with $\mathcal{S}$ is the vector $\mathbf{1}_{\mathcal{S}} \in \mathbb{R}^m$ whose i-th entry equals to 1 if node i belongs to $\mathcal{S}$ and equals 0 otherwise; in other words,

$$\mathbf{1}_{\mathcal{S}} = \sum_{i \in \mathcal{S}} \mathbf{e}_i, \tag{9.3}$$

---

[3]In the literature, this is some times called *strongly connected*. A *weakly connected* digraph is one whose underlying graph is connected, which does not necessarily imply that the digraph is strongly connected.

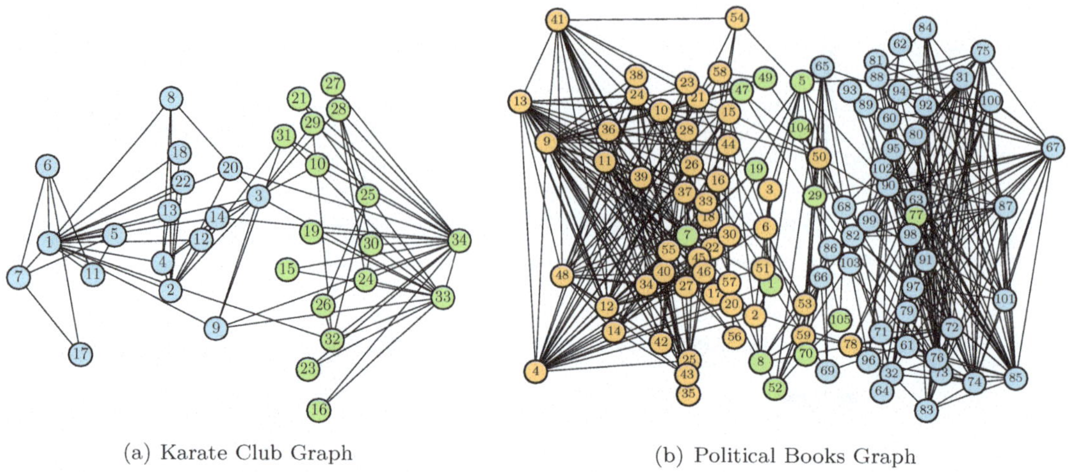

(a) Karate Club Graph   (b) Political Books Graph

Figure 9.5: Visualizations of (a) Zachary's karate club graph and (b) the political books graph. In (a) nodes are colored by group membership after the club split in two and in (b) the nodes (i.e., books) are colored by political leaning; blue is liberal, green is neutral and orange is conservative.

where $\mathbf{e}_1, \ldots, \mathbf{e}_m \in \mathbb{R}^m$ are the standard basis or one-hot vectors (1.3). In particular, an individual one-hot vector $\mathbf{e}_i$ is the indicator of its respective node, while the indicator vector of the entire graph is the ones vector: $\mathbf{1}_{\mathcal{N}} = \mathbf{1} = \mathbf{e}_1 + \cdots + \mathbf{e}_m$. We will sometimes write $\mathbf{1}_{\mathcal{G}} = \mathbf{1}_{\mathcal{N}}$ and similarly for subgraphs $\mathcal{H} = (\mathcal{S}, \mathcal{F}) \subset \mathcal{G} = (\mathcal{N}, \mathcal{E})$, meaning that its nodes and edges are subsets of those of $\mathcal{G}$, so $\mathcal{S} \subset \mathcal{N}$ and $\mathcal{F} \subset \mathcal{E}$. The weights on the edges of a subgraph are inherited from those on the edges of the graph, so $W_{\mathcal{H}}$ is the submatrix obtained by deleting the rows and columns of $W_{\mathcal{G}}$ corresponding to nodes that are not in $\mathcal{H}$.

Let $\mathcal{S}_1, \ldots, \mathcal{S}_k \subset \mathcal{N}$ be a collection of mutually disjoint nonempty subsets of the nodes — for example, the different connected components of the graph — so that $\mathcal{S}_i \cap \mathcal{S}_j = \varnothing$ for all $i \neq j$. The corresponding indicator vectors are easily seen to be mutually orthogonal under the dot product: $\mathbf{1}_{\mathcal{S}_i} \cdot \mathbf{1}_{\mathcal{S}_j} = 0$ for $i \neq j$, and hence, according to Theorem 2.18, are linearly independent. We will sometimes find it more convenient to use the *normalized indicator vectors*

$$\mathbf{u}_{\mathcal{S}} = \frac{\mathbf{1}_{\mathcal{S}}}{\sqrt{\#\mathcal{S}}}, \quad \text{where} \quad \|\mathbf{1}_{\mathcal{S}}\|^2 = \#\mathcal{S} \text{ is the number of nodes in } \mathcal{S}. \tag{9.4}$$

denotes the number of nodes in $\mathcal{S}$. These are defined so that they are unit vectors under the Euclidean norm: $\|\mathbf{u}_{\mathcal{S}}\| = 1$. In particular, the normalized indicator vector of the entire graph is $\mathbf{u}_{\mathcal{G}} = \mathbf{u}_{\mathcal{N}} = 1/\sqrt{m}$, where $m = \#\mathcal{N}$.

9.1.1   Graphs in Applications

We now introduce some examples of graphs that arise in real-world applications. These examples will be used to illustrate techniques and algorithms throughout this and the following chapters.

Our first example is Zachary's karate club graph [261], which consists of $m = 34$ nodes and $e = 78$ edges. The nodes of the graph represent the members of a university karate club, and there is an edge between two members of the club if they interacted socially outside of the club. The graph was first introduced in an anthropological study [261] of conflict and fission in small groups. The karate club was observed by researchers during a period of rising tensions between the club president and instructor over the price of lessons. The entire club became divided over the issue and eventually split into two clubs of equal size. The label for each member i in the graph corresponds to the club that the member joined after the fracture. One common task with the karate graph is to use the graph structure to predict how the club fractured, under the assumption that members with social interactions outside the club may be likely to split the same way. This is a special case of the *graph clustering* problem, which in this case is also known as *community detection*, since the edges correspond to social interactions [76]. We show a visualization of the karate club graph in Figure 9.5(a), where the nodes are colored by their labels, and the positions of the nodes are based on a spectral embedding of the graph, which we discuss in Section 9.7.[4] We will use Zachary's karate club as a toy real-world data set throughout this section.

Another example of a small real-world graph is Krebs' political books graph, which was first introduced in [175]. The nodes of the graph correspond to 105 books on American politics that were sold on Amazon. There are edges between pairs of books that are frequently purchased by the same customer, and each book has a label of "liberal", "neutral", or "conservative", based on its political leaning. There are 43 liberal books, 49 conservative books, and 13 neutral or bipartisan books. A natural question is whether the political identification of the books is reflected in the graph structure, or rather, can we predict which books are similar politically based simply on how often the books are purchased together by the same customer. Figure 9.5(b) gives a simple illustration of the political books graph colored by political ideology.

Graphs also find applications in network data. One example noted above is the internet, where each website is a node in the graph, and edges correspond to hyperlinks between websites. A similar application is found in data bases of academic journal articles, where each node corresponds to an academic paper, and the edges correspond to citations between papers. One widely used data set is PubMed [255], which is an online data base of medical research papers. The PubMed citation graph has 19,717 nodes, representing academic papers, and 44,338 edges, representing citations between papers. The version of PubMed used most often in practice is an unweighted undirected graph — if paper i cites paper j, or j cites i, then there is an edge between nodes i and j — and so the (symmetric) weight and adjacency matrices are the same: $W = A$. (One could, of course, introduce the directed version, where the edges are directed according to who cites whom.) Each paper i in PubMed has a label y_i selected from one of three classes that corresponds to its subject classification, as well as an associated feature vector $\mathbf{x}_i \in \mathbb{R}^n$ that records the frequency with which particular key-words appear in the paper and abstract. One task to be implemented on PubMed is *node classification*; that is, given the subject classification for some of the papers, predict the subject classification for the others using the underlying graph structure and node features. The underlying assumption is that papers in the same subject area should be more likely to cite each other than papers in different areas. Likewise, papers in the same class may be likely to use the same key-words, making the feature vectors useful for classification. Problems of this nature fall under the umbrella of *graph-based semi-supervised learning* to be discussed in Section 9.9.

Another recent application of graph-based learning is in molecular property prediction and drug discovery [27, 118]. Every molecule can be described by a graph, whose nodes are the constituent atoms and whose edges correspond to bonds between them. In Figure 9.6 we show the simple graphical structures for carbon dioxide, ethanol, and benzene, where the atomic labels are C for carbon, H for hydrogen, and O for oxygen. A currently active area of research is molecule property prediction, which refers to the task of predicting how a molecule

---

[4]In particular, the x-coordinate in the embedding uses projection onto the line spanned by the Fiedler vector, cf. Definition 9.23, which is used for clustering nodes, while the y-coordinate was chosen at random in order spread out the images of the nodes.

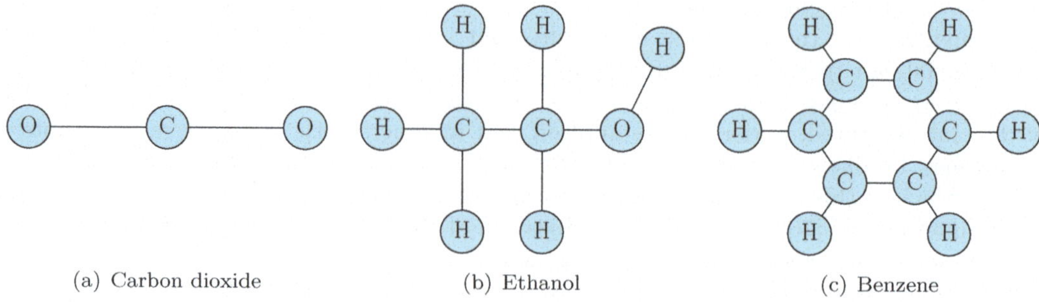

Figure 9.6: Illustration of how molecules can be represented as graphs.

Figure 9.7: Examples of MNIST digits viewed as graphs.

will interact with its environment based on its molecular structure, along with drug discovery, which is the problem of generating new molecules that have desired molecular properties. In machine learning terminology, we are interested in *classification* of entire molecules, which is different from the problem of predicting node labels introduced above. Here, the input to the machine learning algorithm is a graph, representing the molecule, and the output is a classification thereof.

Graphs also appear naturally in image processing and computer vision. Any digital image can be endowed with a graph structure by associating each pixel with a node in the graph, and assigning edges between neighboring pixels, or sometimes between *similar pixels*, the latter being a special case of *similarity graphs* to be discussed below. Figure 9.7 shows examples of three MNIST digits viewed as graphs. In this case, there are $m = 784 = 28^2$ nodes, where each node represents a pixel in the image and is connected to its 4 immediately adjacent neighboring pixels, although the edge and corner pixels obviously connect to fewer neighbors. Each node i can be assigned a feature vector $\mathbf{x}_i$ that includes all of the color information for that pixel, which in the case of MNIST is simply a number $0 \leq x_i \leq 1$ indicating pixel intensity. In Figure 9.7, we plot the pixel intensities on a color scale where purple is darkest and yellow is brightest. We may also choose to connect pixels that are further away than adjacent neighbors, and may assign weights to the graph based on similarity in pixel values; we refer to Section 9.7.2 for more details.

A closely related situation is the graph representing a triangulated surface $S \subset \mathbb{R}^3$, in which the surface is approximated by a large number of small triangles, whose vertices are the nodes in the underlying graph, while the graph's edges are those of the triangles; an

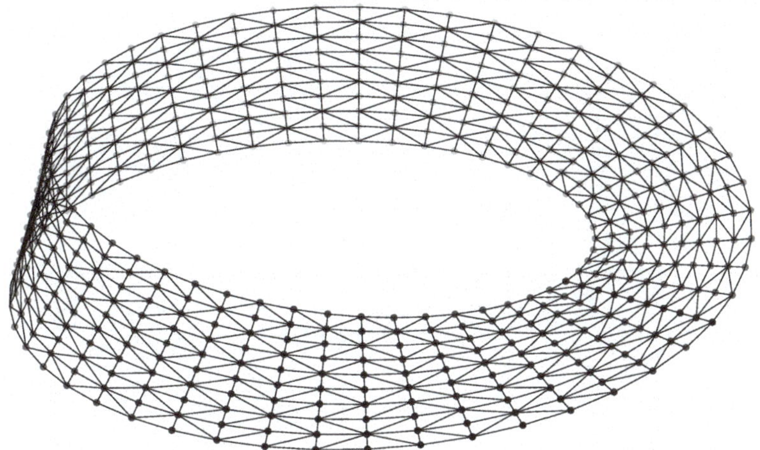

Figure 9.8: An example of a triangulated surface, which can be viewed as a graph, where two vertices are adjacent if they belong to the same triangle.

example of a triangulated Möbius strip is shown in Figure 9.8.

9.1.2   Similarity Graphs

Another common way for graphs to appear in applications is a though the construction of a *similarity graph* over a given data set of m data points $\mathbf{x}_1, \ldots, \mathbf{x}_m$. In a similarity graph, we identify each $\mathbf{x}_i$ with a node in the graph. Two nodes that correspond to a pair of data points $\mathbf{x}_i, \mathbf{x}_j$ that are sufficiently similar are connected by an edge, and in this manner we construct a graph. In this initial construction, the connected components are the data *clusters* that contain all data points that are similar to each other — even though they may not be directly connected by an edge, they are connected by a path of pairwise similar data points. For example, one cluster might contain all the cat images and another all the dog images in our image data set, in which only very similar cat or dog images are directly connected, and no cat image is connected to a dog image.

Of course, the preceding bipartite assignment of edges — either the two data points are similar, and there is an edge connecting their nodes, or they are not, and there is no edge — is overly simplistic for most real world data sets. Data points and images can be more similar or less similar, and so their degree of similarity should be measured on a variable scale. This is done by assigning a weight $w_{ij} = w_{ji}$ to the edge that connects nodes i and j, which measures the similarity of the corresponding data points at its two nodes, leading to a *weighted graph*. As above, we will always assume that the weights are nonnegative, so $w_{ij} \geq 0$, and identify a zero weight to mean that there is no edge connecting nodes i and j. For example, if nodes i and j correspond to very similar images, say two dogs, then the weight is large, while if they are very dissimilar, say a dog and a house, then the assigned weight is small or even zero. In particular, $w_{ii} = 0$, since we assume the graph contains no loops.[5]

When dealing with a small data set, one could envision assigning the weights manually by inspection of the data. But in the large data sets required for machine learning and other real world applications, this is impractical, and one needs to be able to assign weights automatically using an algorithm that measures the similarity of the data points. There is

---

[5] Of course a data point is extremely similar to itself, but encoding this in our assignment of weights does not aid in the subsequent analysis.

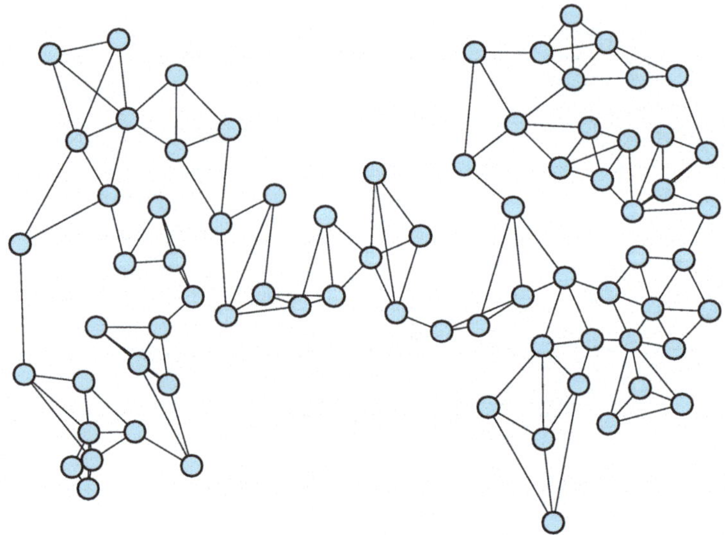

Figure 9.9: An example of a k-nearest neighbor graph with $k = 3$ constructed over a data set $\mathbf{x}_1, \ldots, \mathbf{x}_m \in \mathbb{R}^2$, $m = 75$.

a range of methods that have been employed for computing the edge weights, whose use depends on the nature of the application. One can even experiment with a variety of weight assignment algorithms so as to optimize performance of the chosen graph-based learning task. Let us present some of the possibilities.

Algorithms for assigning weights are almost always based on a choice of norm on the Euclidean space that contains all the data. In other words, the data points are assumed to all lie in the same n-dimensional space, $\mathbf{x}_i \in \mathbb{R}^n$, where n may be very large. For example, if each data point $\mathbf{x}_i$ represents a two-dimensional image, then the dimension n might equal the number of pixels in the case of gray scale images, or 3 or 4 times this number in the case of color images. The data space $\mathbb{R}^n$ will be endowed with a measure of distance, usually coming from an underlying norm $\|\cdot\|$, in order to provide a mechanism for comparing the data points, that is, determining how close they are to each other. Thus the distance between data points $\mathbf{x}_i, \mathbf{x}_j \in \mathbb{R}^n$ is given by $d(\mathbf{x}_i, \mathbf{x}_j) = \|\mathbf{x}_i - \mathbf{x}_j\|$. The smaller the distance, the closer the data points, and hence the larger their assigned edge weight w_{ij} should be, while those that are far apart will have very small or even zero weight.

A simple choice is to connect a pair of nodes when their data points lie sufficiently close to each other, so we set

$$w_{ij} = \begin{cases} 1, & 0 < \|\mathbf{x}_i - \mathbf{x}_j\| < r, \\ 0, & \text{otherwise}, \end{cases} \tag{9.5}$$

where $r > 0$ is a fixed constant. In this case, the weights represent an unweighted graph, with edges only connecting nearby data points. Beyond that, there are several common choices of variable weighting of edges. One is to simply use the inverse distances to some power:

$$w_{ij} = \|\mathbf{x}_i - \mathbf{x}_j\|^{-\alpha} \qquad \text{or, perhaps better,} \qquad w_{ij} = \frac{1}{1 + \beta \|\mathbf{x}_i - \mathbf{x}_j\|^{\alpha}}, \tag{9.6}$$

for some $\alpha > 0$ and $\beta > 0$, the latter version avoiding blow-up of the denominator as $\mathbf{x}_i \to \mathbf{x}_j$.

Another common choice is to use *Gaussian weights*

$$w_{ij} = \begin{cases} \exp\left(-\dfrac{\|\mathbf{x}_i - \mathbf{x}_j\|^2}{2\varepsilon^2}\right) & i \neq j, \\[2mm] 0, & i = j, \end{cases} \tag{9.7}$$

based on the normal distribution of their distances. The parameter ε, called the *connectivity scale* in this context, serves to control how close the data points must be in order that their weight be relatively large, meaning, in this case, near 1. One can further replace the exponential function by other functions of the interpoint distances that are close or equal to zero when the points are far apart. Notice that if we did not zero out the diagonal weights $w_{ii} = 0$ in (9.7), then (9.7) would be exactly the radial basis function kernel matrix associated with the data points $\mathbf{x}_1, \ldots, \mathbf{x}_m$ discussed in Section 7.6.

For real world data sets, using the same connectivity scale ε for all pairs of data points $\mathbf{x}_i, \mathbf{x}_j$ leads to a graph with a very large number of edges in areas of high density, and too few in sparse locations. In practice, we will allow the connectivity length scale $\varepsilon = \varepsilon_{ij}$ to vary with the data points involved. One particular application of this is a k-nearest neighbors, or k-nn, graph. Given a data point $\mathbf{x}_i$, we arrange the other data points in increasing order of their distances from it, so[6]

$$0 = d(\mathbf{x}_i, \mathbf{x}_i) < d(\mathbf{x}_{j_1}, \mathbf{x}_i) \leq d(\mathbf{x}_{j_2}, \mathbf{x}_i) \leq d(\mathbf{x}_{j_3}, \mathbf{x}_i) \leq \cdots .$$

The *k-nearest neighbors* of the point $\mathbf{x}_i$ are the first k data points in this list, excluding $\mathbf{x}_i$ itself, namely, $\mathbf{x}_{j_1}, \mathbf{x}_{j_2}, \ldots, \mathbf{x}_{j_k}$. Fixing k, we then assign the weight $w_{ij} = 1$ if $\mathbf{x}_j$ is one of these k nearest neighbors of $\mathbf{x}_i$ and 0 if not; in particular $w_{ii} = 0$ since $\mathbf{x}_i$ is not viewed as a neighbor of itself. However, the resulting weights need not be symmetric, so possibly $w_{ij} \neq w_{ji}$, since $\mathbf{x}_j$ might be one of the k nearest neighbors of $\mathbf{x}_i$ while $\mathbf{x}_i$ does not belong to the set of k nearest neighbors of $\mathbf{x}_j$. The k-nn relation can be symmetrized in many ways; for example, we can set $w_{ij} = 1$ if either $\mathbf{x}_j$ is a k-nearest neighbor of $\mathbf{x}_i$, or vice versa. Or we can be more strict and set $w_{ij} = 1$ if both $\mathbf{x}_i$ and $\mathbf{x}_j$ are among the k nearest neighbors of each other. We can also easily define a symmetric weighted k-nearest neighbor graph by, for example, the definition

$$w_{ij} = \exp\left(-\frac{\|\mathbf{x}_i - \mathbf{x}_j\|^2}{2\varepsilon_i \varepsilon_j}\right) \tag{9.8}$$

for $i \neq j$, where ε_i denotes the distance from $\mathbf{x}_i$ to its kth nearest neighbor, or some scalar multiple of this distance. Of course, there are other ways to define a symmetric k-nn graph, for example, we may use

$$w_{ij} = \exp\left(-\frac{\|\mathbf{x}_i - \mathbf{x}_j\|^2}{2\varepsilon_i^2}\right) + \exp\left(-\frac{\|\mathbf{x}_i - \mathbf{x}_j\|^2}{2\varepsilon_j^2}\right). \tag{9.9}$$

Regardless of the choice of weights in a k-nn graph, the key idea is that the bandwidth of the graph adjusts locally to the density of the point cloud.

The Gaussian weights (9.7) and (9.8) technically produce a complete graph where all pairs of nodes are connected by edges, although some weights may be vanishingly small if their connecting nodes are far apart. It is common in this case to decide on some threshold

---

[6]When points are at equal distance, one can employ any convenient strategy, such as random choice, to assign the order in which to place them.

$\theta > 0$ and set $w_{ij} = 0$ whenever

$$\frac{\|\mathbf{x}_i - \mathbf{x}_j\|^2}{2\varepsilon_i\varepsilon_j} > \theta.$$

A good choice of θ can produce a *sparse graph*, where many entries in the weight matrix are zero, meaning that the graph has few edges, which is easier to work with computationally, since the zero entries do not need to be stored in memory, or computed with. Furthermore, we can make use of fast approximate nearest neighbor searches to construct a sparse k-nn graph in far less time than the $\mathrm{O}(n^2)$ computation time required to compare all pairs of points [170]. Throughout this section, we will demonstrate many algorithms on the MNIST data set by constructing a sparse k-nearest neighbor graph using Euclidean distances between pixel values, as described above. We will also see a variant of the k-nearest neighbor graph in Section 9.8 that is constructed using a notion of *perplexity*.

There are also examples of graph constructions where the edge weights w_{ij} are *learned* from the data. A common example is in the transformer neural network architecture, which is the foundational deep learning model powering large language models that have experienced tremendous success in natural language processing, among other problems. As part of the transformer architecture, a complete graph is constructed with weights given by

$$w_{ij} = \exp\left(\beta \mathbf{x}_i^T V \mathbf{x}_j\right),$$

where β is a parameter and $V \in \mathcal{M}_{m\times m}$ is a matrix whose entries are tunable parameters, meaning that they are learned from training data for a particular task. We will discuss transformers in more detail in Section 10.5.

Exercises

1.1. Sketch the graphs corresponding to the following adjacency matrices.

$$(a)\;\heartsuit\;\begin{pmatrix} 0 & 1 & 0 \\ 1 & 0 & 1 \\ 0 & 1 & 0 \end{pmatrix}; \quad (b)\;\begin{pmatrix} 0 & 1 & 1 & 1 & 1 \\ 1 & 0 & 1 & 1 & 1 \\ 1 & 1 & 0 & 1 & 1 \\ 1 & 1 & 1 & 0 & 1 \\ 1 & 1 & 1 & 1 & 0 \end{pmatrix}; \quad (c)\;\heartsuit\;\begin{pmatrix} 0 & 1 & 0 & 1 & 0 & 0 \\ 1 & 0 & 0 & 0 & 1 & 0 \\ 0 & 0 & 0 & 1 & 0 & 0 \\ 1 & 0 & 1 & 0 & 1 & 0 \\ 0 & 1 & 0 & 1 & 0 & 1 \\ 0 & 0 & 0 & 0 & 1 & 0 \end{pmatrix}.$$

1.2. Sketch the digraphs corresponding to the following adjacency matrices.

$$(a)\;\heartsuit\;\begin{pmatrix} 0 & 1 & 0 \\ 0 & 0 & 1 \\ 1 & 0 & 0 \end{pmatrix}; \quad (b)\;\diamondsuit\;\begin{pmatrix} 0 & 1 & 0 \\ 1 & 0 & 1 \\ 0 & 1 & 0 \end{pmatrix}; \quad (c)\;\heartsuit\;\begin{pmatrix} 0 & 1 & 0 & 1 \\ 0 & 0 & 1 & 0 \\ 1 & 0 & 0 & 0 \\ 0 & 1 & 1 & 0 \end{pmatrix}; \quad (d)\;\begin{pmatrix} 0 & 0 & 1 & 0 & 0 \\ 0 & 0 & 1 & 1 & 0 \\ 1 & 0 & 0 & 0 & 1 \\ 0 & 0 & 0 & 0 & 0 \\ 1 & 0 & 1 & 0 & 0 \end{pmatrix}.$$

1.3. Write out an adjacency matrix for the following digraphs.

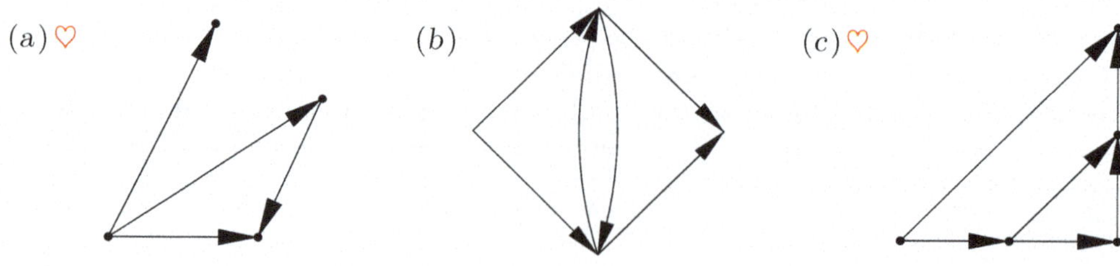

$(a)\;\heartsuit$ (b) $(c)\;\heartsuit$

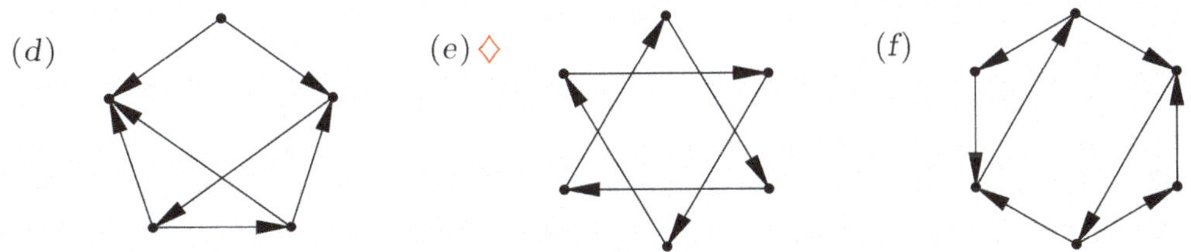

1.4. Write out an adjacency matrix for graphs given by the edges of the Platonic solids: $(a)\,\heartsuit$ tetrahedron, $(b)\,\heartsuit$ cube, $(c)\,\diamondsuit$ octahedron, (d) dodecahedron, and (e) icosahedron.

1.5. An image consists of a rectangular array of pixels. Construct a graph based on such an image, whose edges connect neighboring pixels. Write down the adjacency matrix for a couple of small examples, e.g., 3×3 or 3×4 arrays.

1.6. $\heartsuit$ *True or false*: Let A be the adjacency matrix for an unweighted digraph. Then the underlying unweighted graph has adjacency matrix $A = \widehat{A} + \widehat{A}^T$.

1.7. Let A, B be the adjacency matrices for graphs $\mathcal{G}, \mathcal{H}$. *True or false*: If $A \neq B$ then $\mathcal{G} \neq \mathcal{H}$.

1.8. Find edge weights that make the digraph in Figure 9.4(a) balanced.

1.9. $\diamondsuit$ (a) Explain why a digraph which has a node that only has outgoing edges or only has incoming edges cannot be balanced by any choice of positive edge weights. (b) Find an example of an unweighted digraph which has one or more incoming and one or more outgoing edges at each node that cannot be balanced by any assignment of nonzero edge weights.

1.10. $\diamondsuit$ Construct a weighted graph in which, for some $k \geq 1$, node j is one of the k nearest neighbors of node i whereas node i is *not* one of the k nearest neighbors of node j. Can you find an unweighted example?

1.11. A connected graph is called a *tree* if it has no circuits. (a) Find an adjacency matrix for each of the following trees:

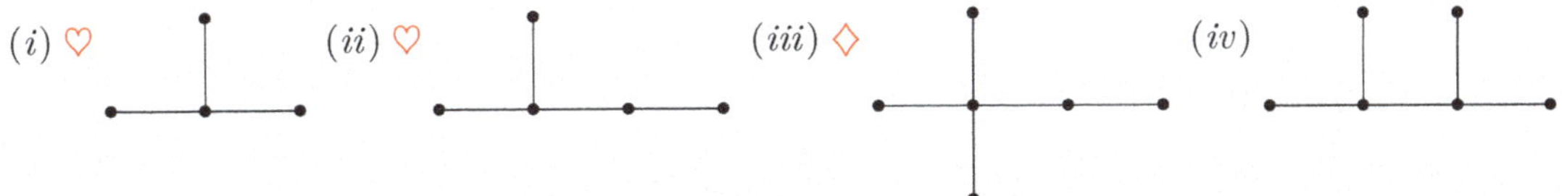

(b) $\diamondsuit$ Draw all distinct trees with 5 nodes, and write down the corresponding adjacency matrices. (c) Prove that any two nodes in a tree are connected by one and only one path.

1.12. Let $\mathcal{G} = (\mathcal{N}, \mathcal{E})$ be a connected graph with m nodes. A *spanning tree* is a subgraph $\mathcal{T} = (\mathcal{N}, \mathcal{D})$ that contains all the vertices of $\mathcal{G}$, and a subset of the edges $\mathcal{D} \subset \mathcal{E}$ with the property that it has no circuits, and so, as in Exercise 1.11 forms a tree. Prove that every connected graph has at least one spanning tree. *Hint*: To construct a spanning tree, use the following inductive procedure. Start with a single node and no edges, so $\mathcal{T}_1 = (\{i\}, \varnothing)$ for any convenient $1 \leq i \leq m$. Then, for $k = 1, \ldots, m - 1$, let $\mathcal{T}_{k+1} = (\mathcal{N}_{k+1}, \mathcal{D}_{k+1})$ be obtained from $\mathcal{T}_k = (\mathcal{N}_k, \mathcal{D}_k)$ by appending an edge $\varepsilon \notin \mathcal{D}_k$ that connects a node in $\mathcal{T}_k$ to a node $j \notin \mathcal{T}_k$, so that $\mathcal{N}_{k+1} = \{j\} \cup \mathcal{N}_k$, $\mathcal{D}_{k+1} = \{\varepsilon\} \cup \mathcal{D}_{k+1}$. Show that (i) Such an edge exists, and (ii) $\mathcal{T}_{k+1}$ has no circuits. Conclude that $\mathcal{T}_m$ is a spanning tree.

9.2　The Incidence Matrix

In this section, we discuss another important way to represent the structure of a digraph. Consider a digraph consisting of m nodes connected by e edges. The associated *incidence matrix* is an $e \times m$ matrix N whose rows are indexed by the edges and whose columns are indexed by the nodes. If edge k starts at node i and ends at node j, then row k of the incidence matrix will have $+1$ in its (k, i) entry and -1 in its (k, j) entry; all other entries in the row are zero. Our convention is that $+1$ represents its tail node and -1 its head node.

Example 9.8. A simple example is the digraph in Figure 9.4(a), which consists of five edges joined at four different nodes. Its 5×4 incidence matrix is

$$N = \begin{pmatrix} 1 & -1 & 0 & 0 \\ 1 & 0 & -1 & 0 \\ -1 & 0 & 0 & 1 \\ 0 & 1 & 0 & -1 \\ 0 & 0 & 1 & -1 \end{pmatrix}. \tag{9.10}$$

Thus the first row of N tells us that the first edge starts at node 1 and ends at node 2. Similarly, row 2 says that the second edge goes from node 1 to node 3, and so on. Clearly, one can completely reconstruct any digraph from its incidence matrix. ▲

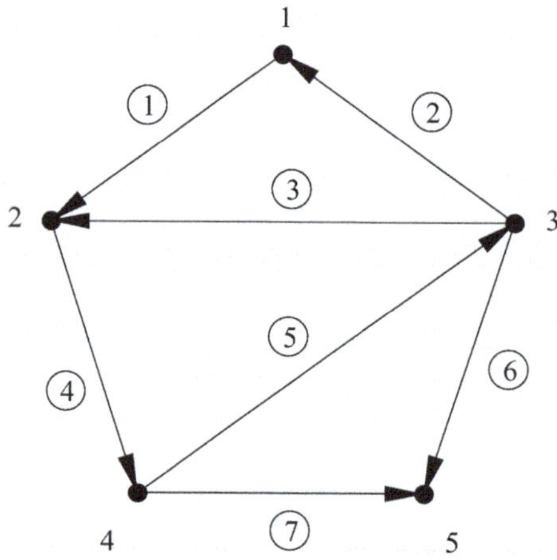

Figure 9.10: A Labeled Digraph

Example 9.9. The matrix

$$N = \begin{pmatrix} 1 & -1 & 0 & 0 & 0 \\ -1 & 0 & 1 & 0 & 0 \\ 0 & -1 & 1 & 0 & 0 \\ 0 & 1 & 0 & -1 & 0 \\ 0 & 0 & -1 & 1 & 0 \\ 0 & 0 & 1 & 0 & -1 \\ 0 & 0 & 0 & 1 & -1 \end{pmatrix}. \tag{9.11}$$

qualifies as an incidence matrix of a simple graph because each row contains a single $+1$, a single -1, and the other entries are 0. Let us construct the digraph corresponding to N. Since N has five columns, there are five nodes in the digraph, which we label by the numbers

$1, 2, 3, 4, 5$. Since it has seven rows, there are 7 edges. The first row has its $+1$ in column 1 and its -1 in column 2, and so the first edge goes from node 1 to node 2. Similarly, the second edge corresponds to the second row of N and so goes from node 3 to node 1. The third row of N indicates an edge from node 3 to node 2; and so on. In this manner, we construct the digraph drawn in Figure 9.10. ▲

The incidence matrix serves to encode important geometric information about the digraph it represents. In particular, its kernel and cokernel have topological significance. For example, the kernel of the incidence matrix (9.10) is one-dimensional, spanned by the ones vector $\mathbf{1} = (1, 1, 1, 1)^T$, and represents the fact that the sum of the entries in any given row of N is zero. More generally, the kernel of an incidence matrix is spanned by the indicator vectors associated with the connected components of the underlying graph, and hence its dimension counts the number of connected components.

Theorem 9.10. *The kernel of the incidence matrix N of a digraph $\widehat{G}$ has a basis consisting of the indicator vectors $\mathbf{1}_{\mathcal{H}}$, cf. (9.3), of the connected components $\mathcal{H}$ of the underlying graph G. Therefore,* nullity N *equals the number of connected components of G. In particular, if G is connected, then* nullity $N = 1$ *and* $\ker N$ *has a single basis element, namely the indicator vector* $\mathbf{1}_G = \mathbf{1} = (1, 1, \dots, 1)^T$.

Proof. If edge k connects node i to node j, then the k-th equation in $N\mathbf{z} = \mathbf{0}$ is $z_i - z_j = 0$, or, equivalently, $z_i = z_j$. The same equality holds, by a simple induction, if the nodes i and j are connected by a path in the underlying graph G. Therefore, the entries of $\mathbf{z}$ must be equal at all nodes belonging to each connected component $\mathcal{H} \subset G$, and hence

$$\mathbf{z} = \sum_S c_{\mathcal{H}} \mathbf{1}_{\mathcal{H}}, \qquad \text{for} \qquad c_{\mathcal{H}} \in \mathbb{R},$$

is a linear combination of the indicator vectors of connected components, which hence span $\ker N$. We already noted their linear independence, and thus they form a basis for the kernel. ∎

Applying the rank/nullity Theorem 3.9, we immediately deduce the following:

Corollary 9.11. *Let N be the incidence matrix for a digraph $\widehat{G}$ with m nodes. If the underlying graph G has s connected components then* rank $N = m - s$.

Next, let us look at the cokernel[7] of the incidence matrix, which is the kernel of its transpose. Consider the particular example (9.10) corresponding to the digraph in Figure 9.4(a). Let us compute the kernel of the transposed incidence matrix

$$N^T = \begin{pmatrix} 1 & 1 & -1 & 0 & 0 \\ -1 & 0 & 0 & 1 & 0 \\ 0 & -1 & 0 & 0 & 1 \\ 0 & 0 & 1 & -1 & -1 \end{pmatrix}.$$

Solving the homogeneous system $N^T \mathbf{y} = \mathbf{0}$ (either by hand or using the QR solution method), we find that $\operatorname{coker} N = \ker N^T$ is spanned by the two vectors

$$\mathbf{y}_1 = (1, \ 0, \ 1, \ 1, \ 0)^T, \qquad \mathbf{y}_2 = (0, \ 1, \ 1, \ 0, \ 1)^T.$$

---

[7] As above, we use the dot product as our underlying inner product.

Each of these vectors represents a *circuit* in the underlying graph $\mathcal{G}$. Keep in mind that their entries are indexed by the edges, so a nonzero entry indicates the direction to traverse the corresponding edge. For example, $\mathbf{y}_1$ corresponds to the circuit that starts out along edge 1, then goes along edge 4 and finishes by going along edge 3 in the reverse direction, which is indicated by the minus sign in its third entry. Similarly, $\mathbf{y}_2$ represents the circuit consisting of edge 2, followed by edge 5, and then edge 3. The fact that $\mathbf{y}_1$ and $\mathbf{y}_2$ are linearly independent vectors says that the two circuits are "independent".

The general element of coker N is a linear combination $c_1\mathbf{y}_1 + c_2\mathbf{y}_2$. Certain values of the constants lead to other types of circuits; for example, $-\mathbf{y}_1$ represents the same circuit as $\mathbf{y}_1$, but traversed in the opposite direction; it is a circuit for the underlying graph, but not the digraph. Another example is

$$\mathbf{y}_1 - \mathbf{y}_2 = (1, \;\; -1, \;\; 0, \;\; 1, \;\; -1)^T,$$

which represents the square circuit going around the outside of the digraph along edges $1, 4, 5, 2$, the fifth and second edges taken in the reverse direction. We can view this circuit as a combination of the two triangular circuits; when we add them together, the middle edge 3 is traversed once in each direction, which effectively "cancels" its contribution. (A similar cancellation occurs in the calculus of line integrals, [1].) Other combinations represent "virtual" circuits; for instance, one can "interpret" $2\mathbf{y}_1 - \frac{1}{2}\mathbf{y}_2$ as two times around the first triangular circuit plus one-half of the other triangular circuit, taken in the reverse direction — whatever that might mean.

In general, given a directed graph $\widehat{\mathcal{G}} = (\mathcal{N}, \widehat{\mathcal{E}})$ with at most one directed edge between each pair of vertices, let $\mathcal{G} = (\mathcal{N}, \mathcal{E})$ denote the underlying undirected simple graph. The vectors $\mathbf{v} \in \mathbb{R}^e$ belonging to the "edge space" have their entries indexed by the edges $\varepsilon \in \mathcal{E}$. Given a circuit $\mathcal{C} \subset \mathcal{E}$ of the graph $\mathcal{G}$, define the corresponding *circuit vector* $\mathbf{v} = \mathbf{v}_\mathcal{C} \in \mathbb{R}^e$ to have entries v_ε equal to $+1$ if the edge ε belong to the circuit and is traversed in the same direction as its orientation in $\widehat{\mathcal{G}}$, or -1 if it is traversed in the opposite direction, or 0 if ε does not belong to the circuit. Thus, a circuit vector corresponds to a circuit in the digraph $\widehat{\mathcal{G}}$ if and only if all its entries are $+1$.

Proof. By the construction of the incidence matrix $N = (\mathbf{c}_1 \; \ldots \; \mathbf{c}_m)$, for each node $i = 1, \ldots, m$, the nonzero entries in the corresponding column $\mathbf{c}_i$ are in the positions indexed by the edges containing i, and equal $+1$ if the edge starts at i or -1 if it ends there. We claim that $\mathbf{v}_\mathcal{C} \cdot \mathbf{c}_i = \mathbf{v}_\mathcal{C}^T \mathbf{c}_i = 0$ for all $i = 1, \ldots, m$, which serves to prove $\mathbf{v}_\mathcal{C}^T N = \mathbf{0}$ and hence $\mathbf{v}_\mathcal{C} \in \operatorname{coker} N$.

Proving the claim requires checking several cases. (To understand the argument, it is recommended that the reader look at the preceding example.) First, if node i does not belong to $\mathcal{C}$, then $v_\varepsilon = 0$ whenever ε contains node i; thus every summand in the dot product $\mathbf{v}_\mathcal{C} \cdot \mathbf{c}_i$ is 0, and hence the claim follows immediately. On the other hand, if node i belongs to $\mathcal{C}$, then there are precisely two nonzero terms in the dot product $\mathbf{v}_\mathcal{C} \cdot \mathbf{n}_i$, namely those corresponding to the two edges in $\mathcal{C}$ containing node i. If these two edges are in the same direction as the edges in $\widehat{\mathcal{G}}$, then both entries in $\mathbf{v}_\mathcal{C}$ are $+1$, while the corresponding two entries of $\mathbf{n}_i$ are $+1$ and -1 since one edge at node i must be incoming and one must be outgoing. Thus, $\mathbf{v}_\mathcal{C} \cdot \mathbf{n}_i = 1 \cdot 1 + 1 \cdot (-1) = 0$. There are three other possibilities, depending on the relative orientations of the two edges under consideration, and, as the reader can check, these similarly give zero for the dot product. This completes the proof of the claim and hence the lemma. ∎

> **Theorem 9.13.** *Let $\mathcal{G}$ be a connected simple graph with m nodes and e edges. Let $\widehat{\mathcal{G}}$ be a digraph obtained from $\mathcal{G}$ by choosing orientations for the edges, and let N be its incidence matrix. Then $\operatorname{coker} N$ has a basis consisting of $e - m + 1$ independent circuit vectors. Moreover, any other circuit vector is a linear combination of the basis circuit vectors.*

Proof. Given $\mathcal{G} = (\mathcal{N}, \mathcal{E})$, let $\mathcal{T} = (\mathcal{N}, \mathcal{D})$ be a spanning tree, as defined and constructed in Exercise 1.12, whereby $\mathcal{T}$ contains all the nodes in $\mathcal{G}$, while its edges $\mathcal{D} \subset \mathcal{E}$ and, furthermore, it has no circuits. Given an edge $\varepsilon = (i, j) \notin \mathcal{D}$ that does not belong to the spanning tree, according to Exercise 1.11(c), there is a unique path $\mathcal{P} = \{j = \varepsilon_1, \varepsilon_2, \ldots, \varepsilon_{k-1}, \varepsilon_k = i\} \subset \mathcal{D}$ contained in the tree that connects node j to node i. Clearly $\mathcal{C}_\varepsilon = \{\varepsilon, \varepsilon_1, \varepsilon_2, \ldots, \varepsilon_k\} \subset \mathcal{E}$ forms a circuit in $\mathcal{G}$. Let $\mathbf{v}_\varepsilon = \mathbf{v}_{\mathcal{C}_\varepsilon} \in \mathbb{R}^e$ denote the corresponding circuit vector. We claim that the collection of all such circuit vectors $\mathbf{v}_\varepsilon$ corresponding to all edges $\varepsilon \in \mathcal{E} \setminus \mathcal{D}$ that do not belong to the spanning tree forms a basis for $\operatorname{coker} N$.

First, Lemma 9.12 assures us that $\mathbf{v}_\varepsilon \in \operatorname{coker} N$. Next, to prove linear independence, consider the linear combination

$$\mathbf{v} = \sum_{\varepsilon \in \mathcal{E} \setminus \mathcal{T}} c_\varepsilon \mathbf{v}_\varepsilon \tag{9.12}$$

for scalars $c_\varepsilon \in \mathbb{R}$. By construction, each edge $\varepsilon \in \mathcal{E} \setminus \mathcal{D}$ only appears in one such circuit $\mathcal{C}_\varepsilon$, and hence the entry of $\mathbf{v}$ corresponding to that edge is $\pm c_\varepsilon$. Thus the only way the linear combination (9.12) can equal $\mathbf{0}$ is if all $c_\varepsilon = 0$. Finally, by using Exercise 2.8 or referring to the construction of the spanning tree in Exercise 1.12, $\mathcal{T}$ has exactly $m - 1$ edges. Thus, the number of vectors $\mathbf{v}_\varepsilon$ corresponding to the edges not in $\mathcal{T}$ is $e - (m - 1)$. On the other hand, according to Theorem 4.24 and Corollary 9.11, and because we are assuming $\mathcal{G}$ is connected,

$$\dim \operatorname{coker} N = e - \operatorname{rank} N = e - (m - 1) = e - m + 1$$

Thus, because the $\mathbf{v}_\varepsilon$ form a linear independent set of $e - m + 1$ vectors in the space, according to Theorem 1.19, they necessarily form a basis. In other words, the circuits $\mathcal{C}_\varepsilon$ corresponding to the edges that do not belong to the spanning tree $\mathcal{T}$ form a complete set of independent circuits in $\mathcal{G}$.

Moreover, given any other circuit $\mathcal{C}$ in $\mathcal{G}$, Lemma 9.12 implies that its circuit vector $\mathbf{v}_{\mathcal{C}} \in \operatorname{coker} N$, and hence must be a linear combination of the basic circuit vectors. In fact, it is not hard to see that

$$\mathbf{v}_{\mathcal{C}} = \sum_{\varepsilon \in \mathcal{C} \setminus \mathcal{D}} \pm \mathbf{v}_\varepsilon, \tag{9.13}$$

the sign depending upon whether the edge $\varepsilon \in \mathcal{C}_\varepsilon \setminus \mathcal{D}$ not in the spanning tree is traversed in the same or the opposite direction in the circuit $\mathcal{C}$. $\blacksquare$

A direct consequence of this theorem is the following remarkable result, first discovered by the extraordinarily prolific eighteenth-century Swiss mathematician Leonhard Euler.[8] For any graph $\mathcal{G}$,

$$\# \text{ nodes } + \# \text{ independent circuits } = \# \text{ edges } + \# \text{ connected components.} \tag{9.14}$$

The case when $\mathcal{G}$ is connected, in which case the last term is 1, follows immediately from Theorem 9.13 since independent circuits $= \dim \operatorname{coker} N = e - m + 1$. The general case is established by summing the formulas for each connected component.

---

[8]Pronounced "Oiler". Euler spent most of his career in Germany and Russia.

Remark. If the graph is *planar*, meaning that it can be drawn in the plane without any edges crossing over each other, then the number of independent circuits is equal to the number of "holes", i.e., the number of distinct regions bounded by the edges of the graph. For example, the pentagonal digraph in Figure 9.10 bounds three triangles, and so has three independent circuits. ▲

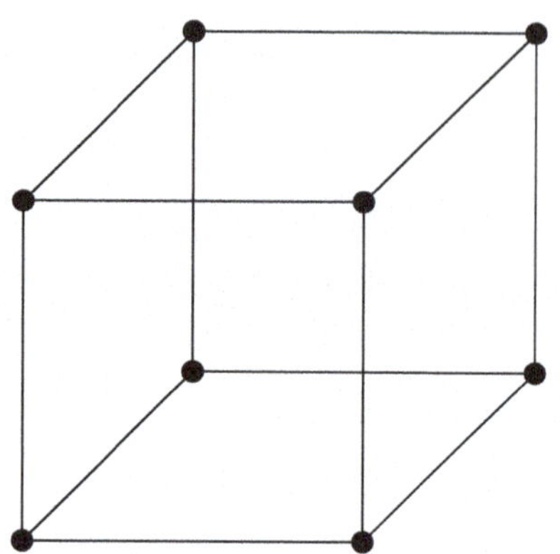 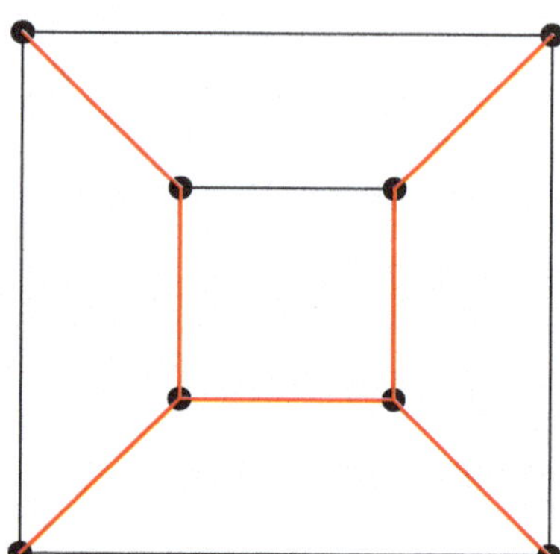

Figure 9.11: A Cubical Graph

Example 9.14. Consider the graph corresponding to the edges of a cube, as illustrated in Figure 9.11, where the second figure represents the same graph squashed down onto a plane. The graph has 8 nodes and 12 edges. Since it is connected, Euler's formula (9.14) tells us that there are $5 = 12 - 8 + 1$ independent circuits. These correspond to the interior square and the four trapezoids in the planar version of the digraph, and hence to circuits around 5 of the 6 faces of the cube. The "missing" face does indeed define a circuit, but it can be represented as a linear combination of the other five circuits, and so is not independent; see Exercise 2.6.

In the second figure, a spanning tree is indicated in red. Labeling the nodes in order from left to right and from top to bottom, the 5 independent circuits resulting from the construction used to prove Theorem 9.13 contain the following sets of nodes: 124653, 1753, 2864, 3465, 7865, where the first edge in each circuit is the one that does not belong to the spanning tree. The reader may enjoy trying to write each of the circuits from the preceding construction in terms of the basic circuits arising from the spanning tree; see Exercise 2.6. Other choices of spanning tree lead to other collections of 5 circuits that also form bases for the cokernel of the incidence matrix. ▲

Exercises

2.1. ◇ (a) Draw the digraph corresponding to the 6×7 incidence matrix whose nonzero (i, j) entries equal 1 if $j = i$ and -1 if $j = i + 1$, for $i = 1$ to 6. (b) Find a basis for its kernel and cokernel. (c) How many independent circuits are in the digraph? Can you identify them?

2.2. Verify Euler's formula for each of the Platonic solids of Exercise 1.4.

2.3. Draw the digraph represented by the following incidence matrices:

$$(a)\;\heartsuit\;\begin{pmatrix} -1 & 0 & 1 & 0 \\ 1 & 0 & 0 & -1 \\ 0 & -1 & 1 & 0 \\ 0 & 1 & 0 & -1 \end{pmatrix},\quad (b)\;\begin{pmatrix} 1 & 0 & -1 & 0 \\ 0 & 1 & 0 & -1 \\ -1 & 1 & 0 & 0 \\ 0 & 0 & 1 & -1 \end{pmatrix},\quad (c)\;\heartsuit\;\begin{pmatrix} 0 & 1 & 0 & 0 & -1 \\ -1 & 0 & 1 & 0 & 0 \\ 0 & 0 & 0 & -1 & 1 \\ 0 & -1 & 1 & 0 & 0 \end{pmatrix},$$

$$(d)\;\diamondsuit\;\begin{pmatrix} -1 & 0 & 1 & 0 & 0 \\ 0 & -1 & 0 & 1 & 0 \\ 1 & -1 & 0 & 0 & 0 \\ 0 & 0 & 0 & -1 & 1 \\ 0 & 0 & -1 & 0 & 1 \end{pmatrix},\quad (e)\;\begin{pmatrix} 0 & 1 & -1 & 0 & 0 & 0 & 0 \\ -1 & 0 & 0 & 1 & 0 & 0 & 0 \\ 0 & 0 & 0 & -1 & 1 & 0 & 0 \\ 0 & -1 & 0 & 0 & 0 & 0 & 1 \\ 0 & 0 & -1 & 0 & 0 & 1 & 0 \\ 0 & 0 & 0 & 0 & 0 & 1 & -1 \end{pmatrix}.$$

2.4. For each of the digraphs in Exercise 1.3, see if you can determine a collection of independent circuits of the underlying graph. Verify your answer by writing out the incidence matrix and constructing a suitable basis of its cokernel.

2.5.$\heartsuit$ A *complete graph* $\mathcal{G}_m$ on m nodes has one edge joining every distinct pair of nodes. (a) Draw $\mathcal{G}_3$, $\mathcal{G}_4$ and $\mathcal{G}_5$. (b) Choose an orientation for each edge and write out the resulting incidence matrix of each digraph. (c) How many edges does $\mathcal{G}_n$ have? (d) How many independent circuits? (e) Find a spanning tree and the corresponding basic circuits.

2.6. (a) Choose orientations for the edges of the cubical digraph in Figure 9.11, and then write down the corresponding incidence matrix. (b) Write down the cokernel basis vectors associated with the 5 circuits corresponding to the interior square and the four trapezoids. Then write the circuit vector corresponding to the exterior square as a linear combination of your basis circuit vectors, and interpret your result geometrically. (c) Write down the cokernel basis vectors associated with the circuits determined by the spanning tree used in Example 9.14. Then write down the circuit vectors found in part (b) as a linear combinations of the 5 basis spanning tree circuit vectors, and interpret your result geometrically.   (d) Construct a different spanning tree. Then determine the corresponding basis circuit vectors, and repeat part (c) for this basis of the cokernel of the incidence matrix.

2.7.$\diamondsuit$ Prove that a graph with m nodes and m edges must have at least one circuit.

2.8. Prove that a connected graph that has m nodes is a tree (see Exercise 1.11) if and only if it has precisely $m - 1$ edges.

2.9.$\diamondsuit$ A digraph is called *acyclic* if it has no circuits. Find an acyclic digraph whose underlying graph is not a tree, as defined in Exercise 1.11. In other words, the digraph has no circuits, but its underlying graph does.

2.10. Give an example of a simple digraph whose underlying graph is not simple.

2.11.$\heartsuit$ *True or false*: If N and $\widetilde{N}$ are incidence matrices of the same size and

$$\operatorname{coker} N = \operatorname{coker} \widetilde{N},$$ then the corresponding digraphs are equivalent.

2.12. If the labeling of the nodes, and edges in a digraph is changed, how does this affect the incidence matrix N? What about $\ker N$ and $\operatorname{img} N$?

2.13. (a) Sketch (by hand) a digraph with 5 nodes and 2 connected components. (b) Define the incidence matrix N as a **numpy** array in Python. (c) Use **numpy.linalg.svd** to find vectors that span the kernel of N, which have singular value zero (according to the **numpy** convention). Are they indicator vectors of the connected components? (d) Try the same thing with a digraph that has 3 or more connected components. (You'll need a few more nodes.)

9.3   The Graph Laplacian

In this section, we introduce a fundamental matrix associated with a (weighted) graph or digraph, known as the graph Laplacian. The properties of the graph Laplacian matrix, particularly its eigenvalues, play a foundational role in graph-based machine learning algorithms.

Let $\mathcal{G}$ be a weighted digraph with m nodes. Let W, D be its $m \times m$ weight matrix and diagonal degree matrix, as introduced in Definitions 9.3 and 9.7, respectively.

Definition 9.15. The *graph Laplacian matrix* of a weighted digraph is the $m \times m$ matrix

$$L = D - W. \tag{9.15}$$

Since the diagonal entries of W vanish, $w_{ii} = 0$, the entries of the graph Laplacian are as follows:

$$l_{ij} = \begin{cases} d_i = \sum_{k=1}^{m} w_{ik}, & i = j, \\ -w_{ij}, & i \neq j. \end{cases} \tag{9.16}$$

The diagonal entries of the graph Laplacian are exactly the weighted degrees of the nodes, while its off-diagonal entries are minus the edge weights. If the digraph is actually a graph, so $W = W^T$ is symmetric, then the graph Laplacian matrix $L = L^T$ is also symmetric. The matrix (9.15) is often referred to as the *combinatorial* or *unnormalized* graph Laplacian, to distinguish it from certain normalized versions introduced later in Section 9.6.

Remark. The name "graph Laplacian" comes from the fact that it represents a discrete analogue of the *Laplace differential operator* or *Laplacian*, denoted by Δ, and named in honor of the influential eighteenth-century French mathematician and cosmologist Pierre–Simon Laplace. The Laplace operator and the associated Laplace partial differential equation play an absolutely fundamental role throughout mathematics and its manifold applications, [180]. In particular, if the graph represents a rectangular planar (or higher dimensional) grid, the associated graph Laplacian matrix can be identified (modulo rescaling) with the standard finite difference numerical discretization of the Laplace operator; see Exercise 3.7. However, it is important to point out that the "sign" of the graph Laplacian is the opposite of the Laplace differential operator (by standard conventions), in the sense that, as we will see below, L is positive semidefinite, while Δ turns out to be negative semidefinite.   ▲

For the rest of this section, let focus on the case when $\mathcal{G}$ is an undirected graph, so that both its weight and graph Laplacian matrices are symmetric: $W = W^T$, $L = L^T$. One of the most important, and sometimes defining, properties of the graph Laplacian is the following formula.

Proposition 9.16. *Let L be the graph Laplacian matrix for a weighted graph with m nodes. Then*

$$\mathbf{x}^T L \mathbf{x} = \frac{1}{2} \sum_{i,j=1}^{m} w_{ij} (x_i - x_j)^2 \qquad \text{for any} \qquad \mathbf{x} \in \mathbb{R}^m. \tag{9.17}$$

Proof. Since $L = D - W$,

$$\mathbf{x}^T L \mathbf{x} = \mathbf{x}^T D \mathbf{x} - \mathbf{x}^T W \mathbf{x} = \sum_{i=1}^{m} d_i x_i^2 - \sum_{i,j=1}^{m} w_{ij} x_i x_j = \sum_{i,j=1}^{m} w_{ij} x_i^2 - \sum_{i,j=1}^{m} w_{ij} x_i x_j$$

$$= \frac{1}{2} \sum_{i,j=1}^{m} w_{ij} x_i^2 - \sum_{i,j=1}^{m} w_{ij} x_i x_j + \frac{1}{2} \sum_{i,j=1}^{m} w_{ij} x_j^2 = \frac{1}{2} \sum_{i,j=1}^{m} w_{ij} (x_i - x_j)^2,$$

where we used the symmetry of the weight matrix, so $w_{ij} = w_{ji}$, to split the first sum into two equal pieces. ∎

Since the right hand side of (9.17) is clearly ≥ 0, we immediately deduce:

Corollary 9.17. *The graph Laplacian matrix of a weighted graph is symmetric and positive semidefinite.*

Note that the graph Laplacian can never be positive definite since, for example, the right hand side of (9.17) vanishes when all the x_i are equal. The quantity in (9.17) is of fundamental importance in graph theory and applications of graph-based learning.

Definition 9.18. Given a weighted graph $\mathcal{G}$, the quadratic form

$$E(\mathbf{x}) = \frac{1}{2} \mathbf{x}^T L \mathbf{x} = \frac{1}{4} \sum_{i,j=1}^{m} w_{ij} (x_i - x_j)^2, \qquad \mathbf{x} \in \mathbb{R}^m, \tag{9.18}$$

associated with its graph Laplacian matrix L is known as the *Dirichlet energy.*

The Dirichlet energy (9.18) is the graph-theoretic analog of the physical energy principle for the Laplace partial differential equation, and is named after the nineteenth-century German analyst Johann Peter Gustav Lejeune Dirichlet. The minimum value of the Dirichlet energy (9.18) is 0, since we can set all x_i to be equal; this corresponds to a scalar multiple of the null vector $\mathbf{1} = \mathbf{1}_{\mathcal{G}} \in \ker L$, which belongs to the kernel whether or not the graph is connected. Later, to accommodate data points at the nodes, we will introduce a vectorized version of the Dirichlet energy; see (9.126).

Remark 9.19. If $\mathcal{G}$ is a digraph with non-symmetric weight matrix $W \neq W^T$, outgoing degree vector $\mathbf{d} = W \mathbf{1}$ and degree matrix $D = \operatorname{diag} \mathbf{d}$. The associated graph Laplacian matrix $L = D - W$ is not symmetric; furthermore, its symmetrization

$$L_s = L^T + L = 2D - W - W^T \tag{9.19}$$

is not necessarily positive semidefinite. In fact, as the following result shows, this is the case if and only if the digraph is *balanced,* meaning that at each node the outgoing degree equals the incoming degree: $\mathbf{d} = \widetilde{\mathbf{d}}$, where $\widetilde{\mathbf{d}} = W^T \mathbf{1}$ is the outgoing degree vector, with $\widetilde{D} = \operatorname{diag} \widetilde{\mathbf{d}}$ the corresponding diagonal matrix.

Proposition 9.20. *If $\mathcal{G}$ is balanced, then its symmetrized graph Laplacian $L_s = L^T + L$ is positive semidefinite. Conversely, if $\mathcal{G}$ is not balanced, then L_s is indefinite.*

Proof. To prove the first statement, note that $W + W^T$ is the symmetric weight matrix for the undirected graph $\widehat{\mathcal{G}}$ whose edges have weights $w_{ij} + w_{ji}$ and hence has degree matrix

$D + \widetilde{D}$ and Laplacian $\widehat{L} = D + \widetilde{D} - W - W^T$. Thus, if $\mathcal{G}$ is balanced, $D = \widetilde{D}$, and so $L_s = \widehat{L}$ is the graph Laplacian for $\widehat{\mathcal{G}}$, and hence, by Corollary 9.17, positive semidefinite.

As for the converse, note that

$$L_s \mathbf{1} = 2 D \mathbf{1} - W \mathbf{1} - W^T \mathbf{1} = \mathbf{d} - \widetilde{\mathbf{d}}, \qquad \text{and hence} \qquad \mathbf{1}^T L_s \mathbf{1} = \mathbf{1} \cdot \mathbf{d} - \mathbf{1} \cdot \widetilde{\mathbf{d}} = 0.$$

Thus, given $t \in \mathbb{R}$, and $1 \leq i \leq m$,

$$(\mathbf{1} + t\,\mathbf{e}_i)^T L_s (\mathbf{1} + t\,\mathbf{e}_i) = 2t\,\mathbf{e}_i^T L_s \mathbf{1} + t^2\,\mathbf{e}_i^T L_s\,\mathbf{e}_i = 2t\,(d_i - \widetilde{d}_i) + 2t^2\,d_i. \tag{9.20}$$

If $d_i \neq \widetilde{d}_i$, then the quadratic function on the right hand side of (9.20) is both positive and negative as t ranges over $\mathbb{R}$, and hence L_s is indefinite. ▲

We now give an alternative characterization of the graph Laplacian associated with an undirected graph $\mathcal{G}$. Let $c_k > 0$ be the weight associated with the k-th edge so that if it connects nodes i_k and j_k, then $c_k = w_{i_k,j_k}$. We use the edge weights to construct a weighted inner product on the edge space $\mathbb{R}^e$, where, as above, e denotes the number of edges. Namely, let $C = \operatorname{diag}(c_1, \ldots, c_e)$ denote the $e \times e$ diagonal *edge weight matrix*. The corresponding weighted inner product is given by

$$\langle \mathbf{v}, \mathbf{w} \rangle_C = \mathbf{v}^T C \mathbf{w} = \sum_{k=1}^{e} c_k v_k w_k, \qquad \mathbf{v}, \mathbf{w} \in \mathbb{R}^e. \tag{9.21}$$

Let us fix a direction to every edge in the graph $\mathcal{G}$ and let N be the associated incidence matrix. This gives a digraph whose underlying graph is $\mathcal{G}$, and which has at most one directed edge between every pair of nodes. Independent of the choices we make for the directions of the edges, we have the following result connecting the graph Laplacian to the incidence matrix.

Proposition 9.21. *Let $\mathcal{G}$ be a weighted graph and let N be the incidence matrix constructed by selecting directions for each edge. Let $C = \operatorname{diag}(c_1, \ldots, c_e)$ denote the $e \times e$ diagonal edge weight matrix. Then its graph Laplacian is equal to*

$$L = N^T C N. \tag{9.22}$$

Proof. Suppose the k-th edge goes from i_k to j_k. Fix $\mathbf{x} \in \mathbb{R}^m$, and let $\mathbf{y} = N\mathbf{x}$, so that $y_k = x_{j_k} - x_{i_k}$ and $c_k = w_{i_k,j_k}$. Therefore, by Proposition 9.16,

$$\mathbf{x}^T N^T C N \mathbf{x} = \mathbf{y}^T C \mathbf{y} = \sum_{k=1}^{e} w_{i_k j_k}(x_{i_k} - x_{j_k})^2 = \frac{1}{2} \sum_{i,j=1}^{m} w_{ij}(x_i - x_j)^2 = \mathbf{x}^T L \mathbf{x}.$$

The prefactor $\frac{1}{2}$ appears because the second sum counts every edge twice. Since this holds for all $\mathbf{x} \in \mathbb{R}^m$, and both L and $N^T C N$ are symmetric, we conclude that $L = N^T C N$; see Exercise 1.14 in Chapter 4 for justification. ■

Recalling formula (4.17), we deduce that the graph Laplacian L is a Gram matrix whose entries are the edge inner products (9.21) between the columns of the incidence matrix. Note also that, the graph Laplacian depends only on the underlying edge weights, and has the same formula (9.22) no matter which orientations are assigned to the edges.

Remark. The decomposition (9.22) is the discrete analogue of the fact that the Laplacian differential operator can be written as the composition of the divergence and gradient operators: $\Delta = \operatorname{div} \circ \operatorname{grad}$; see [181] for further details. ▲

The following result is an immediate consequences of Theorems 4.12 and 9.10.

Theorem 9.22. *Let $\mathcal{G}$ be a weighted graph and let L be its graph Laplacian matrix. Then $\ker L$ has a basis consisting of the indicator vectors (9.3) of its connected components, and so* nullity L *equals number of connected components. In particular, if $\mathcal{G}$ is connected, then $\ker L$ is one-dimensional, spanned by the indicator vector* $\mathbf{1}_{\mathcal{G}} = (1, 1, \ldots, 1)^T.$

The eigenvectors of the graph Laplacian play a crucial role in applications of graph-based learning. Let $\mathcal{G}$ be a graph with m nodes. The corresponding graph Laplacian matrix L is symmetric positive semidefinite, and so, by Theorem 5.29, is diagonalizable, and its eigenvectors form an orthonormal basis for $\mathbb{R}^m$. We order the eigenvalues from smallest to largest,

$$0 = \lambda_1 \le \lambda_2 \le \lambda_3 \le \cdots \le \lambda_m.$$

Let $\mathbf{u}_1, \ldots, \mathbf{u}_m \in \mathbb{R}^m$ be the corresponding orthonormal eigenvectors. Theorem 9.22 tells us that the multiplicity of the zero eigenvalue, say $1 \le k \le m$, equals the number of connected components in $\mathcal{G}$. When the graph is connected, the zero eigenvalue λ_1 is simple, and the corresponding normalized eigenvector is[9] $\mathbf{u}_1 = \mathbf{u}_{\mathcal{G}} = 1/\sqrt{m}$. When λ_1 is not simple, of multiplicity $k \ge 2$, so $0 = \lambda_1 = \cdots = \lambda_k < \lambda_{k+1}$, then the first k eigenvectors $\mathbf{u}_1, \mathbf{u}_2, \ldots, \mathbf{u}_k$ can be chosen to be *any* orthonormal basis for the kernel of L; that is, there is no unique way to define $\mathbf{u}_1, \ldots, \mathbf{u}_k$. In this book, we will *always* choose $\mathbf{u}_1 = 1/\sqrt{m}$, and allow $\mathbf{u}_2, \ldots, \mathbf{u}_k \in \ker L$ to be any collection of orthonormal vectors that are orthogonal to $\mathbf{u}_1$. Furthermore, according to Theorem 5.75 and Proposition 9.21, we can interpret the eigenvectors $\mathbf{u}_{k+1}, \ldots, \mathbf{u}_m$ corresponding to the *nonzero* eigenvalues as the singular vectors of the incidence matrix N under the edge inner product (9.21).

Since the first eigenvector $\mathbf{u}_1$ is always trivial, the next smallest or "subminimal" eigenvalue and corresponding eigenvector play a particularly important role.

Definition 9.23. The subminimal eigenvalue of the graph Laplacian is called the *Fiedler eigenvalue*, or sometimes the *Fiedler value*, and denoted $\lambda_F = \lambda_2$. If $\lambda_F > 0$, then the eigenspace

$$V_F = \ker (L - \lambda_F I)$$

is known as the *Fiedler subspace*. If $\lambda_F = 0 = \lambda_1$, then the *Fiedler subspace* is defined to be the orthogonal complement to $\mathbf{u}_1 = 1/\sqrt{m}$ in $\ker L$; in other words,

$$V_F = \{\, \mathbf{v} \in \ker L \mid \mathbf{v} \cdot \mathbf{u}_1 = 0 \,\} = \{\, \mathbf{v} \in \ker L \mid v_1 + \cdots + v_m = 0 \,\}.$$

(When $\lambda_F > 0$, the vectors in V_F are automatically orthogonal to $\mathbf{u}_1$.) Every unit vector in the Fiedler subspace, so $\mathbf{u} \in V_F$ and $\|\mathbf{u}\| = 1$, is called a *Fiedler vector*.

The Fiedler vectors and subspace are named after Czech mathematician Miroslav Fiedler, who made many fundamental contributions to linear algebra and graph theory.

Example 9.24. Consider the graph defined by the weight matrix

$$W = \begin{pmatrix} 0 & 1 & 0 & 0 \\ 1 & 0 & 0 & 0 \\ 0 & 0 & 0 & 1 \\ 0 & 0 & 1 & 0 \end{pmatrix},$$

---

[9] We could also choose $\mathbf{u}_1 = -1/\sqrt{m}$, but for specificity we will use the plus sign throughout.

which corresponds to a graph with 4 nodes and two connected components $\{1,2\}$ and $\{3,4\}$. In this case, the graph Laplacian matrix is

$$L = \begin{pmatrix} 1 & -1 & 0 & 0 \\ -1 & 1 & 0 & 0 \\ 0 & 0 & 1 & -1 \\ 0 & 0 & -1 & 1 \end{pmatrix},$$

and the kernel of L is two-dimensional:

$$\ker L = \operatorname{span}\{\mathbf{u}_1, \mathbf{u}_2\}, \quad \text{where} \quad \mathbf{u}_1 = \tfrac{1}{2}\mathbf{1} = \tfrac{1}{2}(1,1,1,1)^T, \quad \mathbf{u}_2 = \tfrac{1}{2}(1,1,-1,-1)^T.$$

In this case, the first two eigenvalues of L are $\lambda_1 = \lambda_2 = 0$ and so the Fiedler value is $\lambda_F = 0$. The Fiedler subspace is spanned by $\mathbf{u}_2$, and the Fiedler vectors are $\mathbf{u} = \pm\mathbf{u}_2$. Notice that the sign of the entries of the Fiedler vector indicates which of the two connected components in the graph the corresponding node belongs to.

In this case, we can easily compute the remaining eigenvectors

$$\mathbf{u}_3 = \tfrac{1}{\sqrt{2}}(1,-1,0,0)^T, \qquad \mathbf{u}_4 = \tfrac{1}{\sqrt{2}}(0,0,1,-1)^T,$$

each of which have eigenvalues $\lambda_3 = \lambda_4 = 2$. Notice here that the signs of the nonzero components of $\mathbf{u}_3$ and $\mathbf{u}_4$ further *split* the connected components of the graph. ▲

The Fiedler value and associated Fiedler vectors can be characterized by the general minimization principle provided in Theorem 5.47. Applications will appear below.

Theorem 9.25. *The Fiedler value of a graph is characterized by*

$$\lambda_F = \min\left\{ \mathbf{x}^T L \mathbf{x} \mid \|\mathbf{x}\| = 1, \ \mathbf{1}\cdot\mathbf{x} = 0 \right\}. \tag{9.23}$$

Every vector achieving the minimum is a Fiedler vector.

Since the Fiedler value vanishes when the graph is disconnected, its magnitude can be interpreted as a measure of how weakly or strongly the graph is connected. The smaller $\lambda_F \geq 0$ is, the closer $\mathcal{G}$ is, in some vague sense, to a disconnected graph. This is borne out by numerical experiments, which demonstrate that a connected graph with a small Fiedler value can be disconnected by removing a relatively small number of its edges. This observation forms the basis for *spectral clustering*, to be developed in Sections 9.4 and 9.7.2.

Example 9.26. Consider the unweighted graph sketched in Figure 9.12. Using the indicated node labels, the corresponding weight and degree matrices are

$$W = \begin{pmatrix} 0 & 1 & 1 & 1 & 0 & 0 & 0 & 0 \\ 1 & 0 & 1 & 1 & 0 & 0 & 0 & 0 \\ 1 & 1 & 0 & 1 & 0 & 0 & 0 & 0 \\ 1 & 1 & 1 & 0 & 1 & 0 & 0 & 0 \\ 0 & 0 & 0 & 1 & 0 & 1 & 1 & 0 \\ 0 & 0 & 0 & 0 & 1 & 0 & 1 & 1 \\ 0 & 0 & 0 & 0 & 1 & 1 & 0 & 1 \\ 0 & 0 & 0 & 0 & 0 & 1 & 1 & 0 \end{pmatrix}, \qquad D = \begin{pmatrix} 3 & 0 & 0 & 0 & 0 & 0 & 0 & 0 \\ 0 & 3 & 0 & 0 & 0 & 0 & 0 & 0 \\ 0 & 0 & 3 & 0 & 0 & 0 & 0 & 0 \\ 0 & 0 & 0 & 4 & 0 & 0 & 0 & 0 \\ 0 & 0 & 0 & 0 & 3 & 0 & 0 & 0 \\ 0 & 0 & 0 & 0 & 0 & 3 & 0 & 0 \\ 0 & 0 & 0 & 0 & 0 & 0 & 3 & 0 \\ 0 & 0 & 0 & 0 & 0 & 0 & 0 & 2 \end{pmatrix}.$$

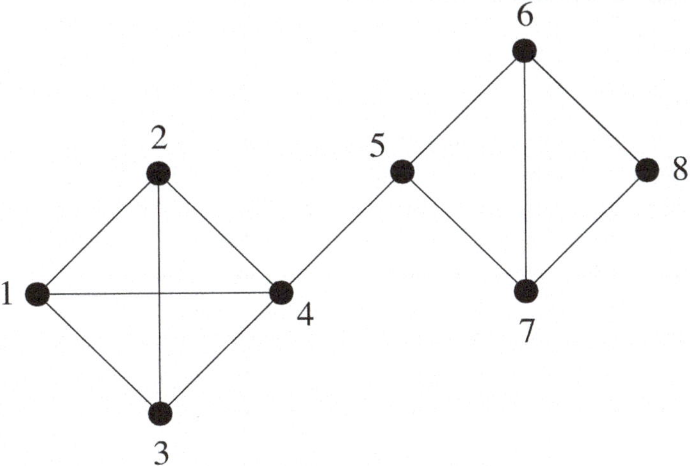

Figure 9.12: An Almost Disconnected Graph

Hence its graph Laplacian is

$$
L = D - W =
\begin{pmatrix}
3 & -1 & -1 & -1 & 0 & 0 & 0 & 0 \\
-1 & 3 & -1 & -1 & 0 & 0 & 0 & 0 \\
-1 & -1 & 3 & -1 & 0 & 0 & 0 & 0 \\
-1 & -1 & -1 & 4 & -1 & 0 & 0 & 0 \\
0 & 0 & 0 & -1 & 3 & -1 & -1 & 0 \\
0 & 0 & 0 & 0 & -1 & 3 & -1 & -1 \\
0 & 0 & 0 & 0 & -1 & -1 & 3 & -1 \\
0 & 0 & 0 & 0 & 0 & -1 & -1 & 2
\end{pmatrix}.
\tag{9.24}
$$

To four decimal places, the eigenvalues are $0., 0.3187, 2.3579, 4., 4., 4., 4., 5.3234$. The relatively small value of $\lambda_2 = .3187$ indicates the graph is not especially well connected. Indeed, we can disconnect it by removing just the one edge connecting nodes 4 and 5. The resulting disconnected graph Laplacian is the block diagonal matrix

$$
\widetilde{L} =
\begin{pmatrix}
3 & -1 & -1 & -1 & 0 & 0 & 0 & 0 \\
-1 & 3 & -1 & -1 & 0 & 0 & 0 & 0 \\
-1 & -1 & 3 & -1 & 0 & 0 & 0 & 0 \\
-1 & -1 & -1 & 3 & 0 & 0 & 0 & 0 \\
0 & 0 & 0 & 0 & 2 & -1 & -1 & 0 \\
0 & 0 & 0 & 0 & -1 & 3 & -1 & -1 \\
0 & 0 & 0 & 0 & -1 & -1 & 3 & -1 \\
0 & 0 & 0 & 0 & 0 & -1 & -1 & 2
\end{pmatrix}
$$

whose spectrum is the union of the spectra of the two constituent connected components: the upper left 4×4 block has a triple eigenvalue of 4 and a zero eigenvalue, while the lower right 4×4 block has eigenvalues $4, 2, 2, 0$. Note that these values are fairly close to those of the original connected graph. Such observations are even more striking when one is dealing with much larger graphs; see Section 9.7.2 for further discussion. ▲

Example 9.27. Consider the complete graph $\mathcal{G}_m$ on m nodes. It has one edge joining every distinct pair of nodes, and hence is *the* most connected simple graph. Its unweighted graph Laplacian is easily constructed, and is the $m \times m$ matrix $L_m = m\,\mathbf{I} - E$, where $E = \mathbf{1}\,\mathbf{1}^T$

is the $m \times m$ matrix with every entry equal to 1. Since $\dim \ker E = m - 1$ (why?), we see that L_m has only one nonzero eigenvalue, namely $\lambda_2 = \cdots = \lambda_m = m$, of multiplicity $m - 1$ along with its zero eigenvalue $\lambda_1 = 0$.

Motivated by this observation, graphs whose nonzero eigenvalues are close together are, in a certain sense, very highly connected, and are known as *expander graphs*. Expander graphs have many remarkable properties, which underlie their applications in communication networks, error-correcting codes, fault-tolerant circuits, pseudo-random number generators, Markov processes, statistical physics, as well as more theoretical disciplines such as group theory and geometry, [112]. ▲

To compute the Fiedler vector for the large graphs arising in practical applications, we can adapt the power method, introduced in Section 5.6. The power method applied directly to L would compute the largest eigenvalue $\lambda_m = \lambda_{max}(L)$ and corresponding eigenvector. To compute a Fiedler vector, we perform a spectral shift of the graph Laplacian by defining the matrix

$$K = \lambda I - L,$$

where $\lambda > 0$ is any positive number satisfying $\lambda \geq \lambda_m$. If $\mathbf{u}_1, \ldots, \mathbf{u}_m$ are the eigenvectors of L with eigenvalues $0 = \lambda_1 \leq \lambda_2 \leq \cdots \leq \lambda_m$, then the vectors $\mathbf{u}_i$ are also the eigenvectors of K — since L and K differ by a multiple of the identity matrix — but the eigenvalues of K are $\mu_i = \lambda - \lambda_i \geq 0$. Thus, the top unit eigenvector of K corresponding to the largest eigenvalue (in absolute value) $\lambda - \lambda_1 = \lambda$ is (up to sign) the normalized ones vector $\mathbf{u}_1 = 1/\sqrt{m}$. The second eigenvector corresponding to $\lambda - \lambda_2$ is a Fiedler vector $\mathbf{u}_2$. This allows us to compute the Fiedler vector with a version of the *renormalized power method* (5.86), namely

$$\mathbf{x}_{k+1} = \frac{K\,\mathbf{x}_k}{\|\,K\,\mathbf{x}_k\,\|} = \frac{\lambda\,\mathbf{x}_k - L\,\mathbf{x}_k}{\|\,\lambda\,\mathbf{x}_k - L\,\mathbf{x}_k\,\|}, \tag{9.25}$$

starting from a vector $\mathbf{x}_0 \in \mathbb{R}^m$ that is orthogonal to $\mathbf{u}_1$, i.e., with $\mathbf{x}_0 \cdot \mathbf{1} = 0$, and for which $\mathbf{x}_0 \cdot \mathbf{u}_2 \neq 0$. Such an initial vector $\mathbf{x}_0$ can typically be obtained by selecting $\mathbf{y} \in \mathbb{R}^m$ at random and setting $\mathbf{x}_0 = \mathbf{y} - (\mathbf{y} \cdot \mathbf{1})\mathbf{1}/m$ — that is, setting $\mathbf{x}_0$ to have mean zero. In theory we will have $\mathbf{x}_k \cdot \mathbf{1} = 0$ for all k, but if the power method iterations proceed for a sufficiently long time, roundoff errors may contaminate this condition, so it may be necessary to center $\mathbf{x}_k$ every so often, by subtracting off the mean $(\mathbf{x}_k \cdot \mathbf{1})\mathbf{1}/m$ before proceeding to the next iteration.

In order to use the iterative scheme (9.25) to compute a Fiedler vector, we need to be able to select $\lambda \geq \lambda_{max}(L)$. The following lemma provides a simple upper bound for $\lambda_{max}(L)$ that can be used for this purpose.

Lemma 9.28. *Let $\mathcal{G}$ be a graph with m nodes. Then the largest eigenvalue $\lambda_{max}(L)$ of the graph Laplacian L is bounded by twice the maximal node degree:*

$$\lambda_{max}(L) \leq 2 \max\{d_1, \ldots, d_m\}. \tag{9.26}$$

Proof. By Theorem 5.43 and Proposition 9.16,

$$\lambda_{max}(L) = \max_{\|\mathbf{x}\|=1} \mathbf{x}^T L \mathbf{x} = \max_{\|\mathbf{x}\|=1} \frac{1}{2} \sum_{i,j=1}^{m} w_{ij}(x_i - x_j)^2. \tag{9.27}$$

By Cauchy's inequality (6.98),

$$(x_i - x_j)^2 = x_i^2 - 2x_i x_j + x_j^2 \leq x_i^2 + x_i^2 + x_j^2 + x_j^2 = 2(x_i^2 + x_j^2), \tag{9.28}$$

and therefore

$$\frac{1}{2}\sum_{i,j=1}^{m} w_{ij}\,(x_i - x_j)^2 \le \sum_{i,j=1}^{m} w_{ij}\,(x_i^2 + x_j^2) = \left(\sum_{i=1}^{m} x_i^2\right)\left(\sum_{j=1}^{m} w_{ij}\right) + \left(\sum_{j=1}^{m} x_j^2\right)\left(\sum_{i=1}^{m} w_{ij}\right)$$

$$= \sum_{i=1}^{m} d_i\,x_i^2 + \sum_{j=1}^{m} d_j\,x_j^2 = 2\sum_{i=1}^{m} d_i\,x_i^2 \le 2\left(\max_{1\le i\le m} d_i\right)\|\mathbf{x}\|^2.$$

Substituting this inequality into (9.27) completes the proof. ∎

In general, the bound in Lemma 9.28 may well not be tight, so the inequality in (9.26) may be strict. For example, the bound is tight in Example 9.24, where the maximum degree is 1 and the maximum eigenvalue is $\lambda_4 = 2$, but not in Example 9.26, where the maximum degree is 4 and the maximum eigenvalue is $\lambda_8 = 5.3234$.

Remark 9.29. The proof of Lemma 9.28 gives some insight into the structure of the top eigenvectors of the graph Laplacian. To explain this in a simple setting, suppose that the degree $d_i = d$ is constant over the graph. Then the only place an inequality arises in the proof of Lemma 9.28 is the estimate in (9.28). Suppose for a moment that all of the entries of $\mathbf{x}$ have absolute value 1, so $x_i = \pm 1$ for all i. Then we have equality in (9.28) if and only if $x_i = 1$ and $x_j = -1$, or vice versa, in which case both sides of the inequality are equal to 4. Since this estimate is only used across edges in the graph, as it is multiplied by w_{ij} in the next step of the proof, this indicates that the highest eigenvectors of the graph Laplacian are vectors whose entries oscillate very rapidly over the graph, in sense that they change sign across as many edges as possible. We will see a more concrete illustration of this phenomenon when the discrete Fourier transform is introduced in Section 9.10. ▲

Exercises

3.1. Choose a direction for each of the edges and write down the incidence matrix N for the graph sketched in Figure 9.12. Verify that its graph Laplacian (9.24) equals $L = N^T N$.

3.2. Determine the graph Laplacian and its spectrum for the graphs with adjacency matrices listed in Exercise 1.1.

3.3. ◇ Suppose we allow loops in a graph, by allowing $w_{ii} > 0$. Show that the graph Laplacian matrix $L = D - W$ does not depend on the diagonal values w_{ii} of the weight matrix W, and so the graph Laplacian does not *see* loops.

3.4. Suppose that $\widehat{\mathcal{G}}$ is a weighted digraph with weight matrix $\widehat{W}$. Let L be the symmetric graph Laplacian matrix for the underlying weighted graph $\mathcal{G}$. Show that

$$\mathbf{x}^T L \mathbf{x} = \sum_{i,j=1}^{m} \widehat{w}_{ij}\,(x_i - x_j)^2.$$

3.5. ♡ In Proposition 9.21, assume that N is the incidence matrix for a weighted digraph $\widehat{\mathcal{G}}$, without the restriction that each pair of nodes (i, j) has at most one directed edge between them. Show that $L = N^T C N$ is the graph Laplacian for the underlying weighted graph $\mathcal{G}$.

3.6. ♡ Let $\mathcal{G}$ be a connected graph with m nodes and with graph Laplacian matrix L. Let $P = (\mathrm{I} \ -\mathbf{1})$ be the $(m-1) \times m$ matrix whose first $m-1$ columns form the $(m-1) \times (m-1)$ identity matrix and whose last column has all -1 entries.

    (a) Show that the $(m-1) \times (m-1)$ matrix PLP^T is positive definite.

    (b) Let $\mathbf{b} \in \mathbb{R}^m$ satisfy $\mathbf{b} \cdot \mathbf{1} = 0$, and let $\mathbf{y} \in \mathbb{R}^m$ be the unique solution of $PLP^T\mathbf{y} = P\mathbf{b}$. Show that $\mathbf{x} = P^T\mathbf{y}$ solves $L\mathbf{x} = \mathbf{b}$ and $\mathbf{x} \cdot \mathbf{1} = 0$.

    (c) Suppose $\mathbf{b} \cdot \mathbf{1} \neq 0$ in part (b). What equation does $\mathbf{x} = P^T\mathbf{y}$ satisfy?

3.7. Write down the graph Laplacian matrices associated with the rectangular digraphs in Exercise 1.5. *Remark*: These matrices can be identified (modulo a suitable rescaling) with the matrices arising from the standard finite difference numerical discretization of the Laplace operator, cf. [180], which explains the original motivation for the term "graph Laplacian".

3.8. ◇ In Python, implement the power method (9.25) for computing a Fiedler vector using the value $\lambda = 2\max\{d_1, \ldots, d_m\}$ from Lemma 9.28. Test the method on some simple graphs and compare your result to eigenvalue solvers in **numpy** and **scipy**.

9.4   Binary Spectral Clustering

Python Notebook: Binary Spectral Clustering (.ipynb) 

The k-means clustering algorithm discussed in Section 7.5 works well for clusters that are roughly spherical, e.g., blob data. When a cluster has a more complicated geometry, a single cluster center may not be a good representative, and (Euclidean) distance to the center may not be a good indication of which cluster a data point belongs to. We show an example of this on the two moons and circles data sets in Figure 9.13. These data sets have two clusters with nonconvex shapes for which there are no good choices of cluster centers based on Euclidean distance. In this case, 2-means clustering performs poorly. In this section we will develop a class of clustering algorithms that exploit the graph structure of the data. Let $\mathcal{G}$ be a connected weighted graph with nodes $\mathcal{N}$ and weight matrix W. In the simplest version, we seek to cluster the nodes $\mathcal{N}$ of $\mathcal{G}$ into two groups in a manner that respects the graph structure. We will discuss how to handle the case of more than 2 clusters in Section 9.7.2.

Since we expect edges to connect *similar* data points, it is sensible to seek a partition of the nodes into two complementary subsets, $\mathcal{A} \subset \mathcal{N}$ and $\mathcal{A}^c = \mathcal{N} \setminus \mathcal{A}$, so that there are as few edges connecting a node in $\mathcal{A}$ to a node in $\mathcal{A}^c$ as possible. A natural way to do this is to minimize the *graph cut energy*

$$\mathrm{cut}(\mathcal{A}) = \sum_{i \in \mathcal{A}} \sum_{j \in \mathcal{A}^c} w_{ij} \tag{9.29}$$

over subsets $\mathcal{A} \subset \mathcal{N}$ of the nodes. The graph cut energy equals the sum of the edge weights corresponding to edges that would need to be *cut* in order to partition the graph into $\mathcal{A}$ and $\mathcal{A}^c$. While minimizing the graph cut energy seems reasonable, it turns out in practice that it can give poor clusterings, since it is often minimized by selecting $\mathcal{A}$ to contain a single outlying vertex, say the one with smallest degree, or even by setting $\mathcal{A} = \varnothing$ if this trivial case is not explicitly avoided in the optimization.

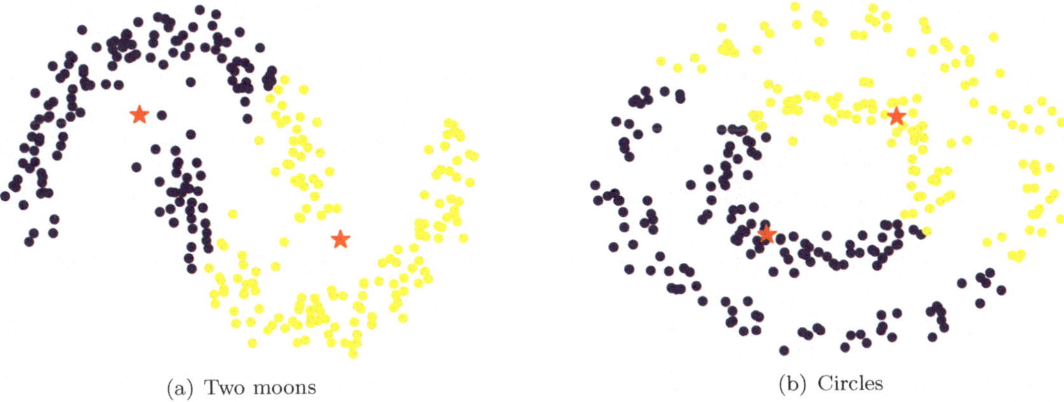

(a) Two moons   (b) Circles

Figure 9.13: 2-means clustering on the two moons and circles data sets fails to uncover the true clusters. The red stars indicate the positions of the two means.

In order to encourage the clusters $\mathcal{A}$ and $\mathcal{A}^c$ to have similar sizes, we can normalize the graph cut energy (9.29) in various ways. One approach is the *average cut energy*

$$\text{cut}_{\text{avg}}(\mathcal{A}) = \frac{\text{cut}(\mathcal{A})}{\#\mathcal{A}\ \#\mathcal{A}^c}. \tag{9.30}$$

Dividing by the product of $\#\mathcal{A}$ and $\#\mathcal{A}^c$ helps ensure that neither $\mathcal{A}$ nor $\mathcal{A}^c$ is too small. Indeed, if $\#\mathcal{A} = 0$ or $\#\mathcal{A}^c = 0$ then the average cut energy is infinity, while the product $\#\mathcal{A}\ \#\mathcal{A}^c$ is largest when the clustering is balanced, i.e., when $\#\mathcal{A} = \#\mathcal{A}^c$; see Exercise 4.1. Another approach is to minimize the *normalized cut energy*

$$\text{cut}_{\text{norm}}(\mathcal{A}) = \frac{\text{cut}(\mathcal{A})}{|\mathcal{A}|\,|\mathcal{A}^c|}, \tag{9.31}$$

where

$$|\mathcal{A}| = \sum_{i \in \mathcal{A}} d_i = \sum_{i \in \mathcal{A}} \sum_{j=1}^{m} w_{ij} \tag{9.32}$$

equals the sum of the degrees of the nodes in $\mathcal{A}$, i.e., its *total degree*. The degree serves as a measure of the importance of a vertex, so it is reasonable to incorporate it into a measure of the size of a cluster. The difference here is how we measure the size of $\mathcal{A}$: normalized cuts count the total degree whereas average cuts simply count the number of nodes.

Minimizing either the average cut (9.30) or normalized cut (9.31) energies gives very good clusterings, but it turns out to be a very hard problem computationally, especially when the graph has a large number of nodes. In fact, they are both NP hard problems [213]. When encountering such intractable computational problems, a well-paved road to follow is to find some way to relax or approximate it by one that is similar, but easier to solve computationally. Here the relaxed version is based on the connection between the average cut energy and the graph Laplacian. A similar result holds for the normalized cut energy; see Exercise 4.2.

Lemma 9.30. *Let $\mathcal{A} \subset \mathcal{N}$. Set $b = \#\mathcal{A}/\#\mathcal{A}^c$, $\mathbf{u} = \mathbf{1}_{\mathcal{A}} - b\,\mathbf{1}_{\mathcal{A}^c}$. Then $\mathbf{u}\cdot\mathbf{1} = 0$, and*

$$\text{cut}_{\text{avg}}(\mathcal{A}) = \frac{1}{m}\frac{\mathbf{u}^T L \mathbf{u}}{\|\mathbf{u}\|^2}, \tag{9.33}$$

Proof. For simplicity we write $k = \#\mathcal{A}$, so $m - k = \#\mathcal{A}^c$, and hence

$$b = \frac{k}{m-k}, \qquad (1+b)^2 = \frac{m^2}{(m-k)^2}.$$

The choice of b was made so that $\mathbf{u} \cdot \mathbf{1} = k - b(m-k) = 0$. Since $u_i = 1$ if $i \in \mathcal{A}$ and 0 if $i \in \mathcal{A}^c$, using (9.17),

$$\mathbf{u}^T L \mathbf{u} = \frac{1}{2} \sum_{i,j=1}^{m} w_{ij} (u_i - u_j)^2 = \sum_{i \in \mathcal{A}} \sum_{j \in \mathcal{A}^c} w_{ij} (1+b)^2$$
$$= (1+b)^2 \operatorname{cut}(\mathcal{A}) = \frac{m^2}{(m-k)^2} \operatorname{cut}(\mathcal{A}) = \frac{m^2 k}{m-k} \operatorname{cut}_{\mathrm{avg}}(\mathcal{A}). \tag{9.34}$$

Furthermore,

$$\|\mathbf{u}\|^2 = \sum_{i \in \mathcal{A}} 1^2 + \sum_{i \in \mathcal{A}^c} b^2 = \#\mathcal{A} + b^2 \#\mathcal{A}^c = k + \frac{m^2}{(m-k)^2}(m-k) = \frac{mk}{m-k}. \tag{9.35}$$

Dividing (9.34) by (9.35) establishes (9.33). ∎

Lemma 9.30 allows us to reformulate the average graph cut binary clustering problem as

$$\min \left\{ \frac{\mathbf{u}^T L \mathbf{u}}{\|\mathbf{u}\|^2} \,\middle|\, \mathbf{u} \in \mathbb{R}^m, \quad \mathbf{u} \cdot \mathbf{1} = 0 \right\},$$

where we restrict $\mathbf{u}$ to a binary vector with entries $u_i = 1$ or $-b$ with $b > 0$, i.e., of the form $\mathbf{u} = \mathbf{1}_{\mathcal{A}} - b\mathbf{1}_{\mathcal{A}^c}$ for some set $\mathcal{A} \subset \mathcal{N}$. The minimizer thus defines the minimizing set $\mathcal{A}$ for the average cut energy (9.33). Note that we can also identify $\mathcal{A}$ by the signs[10] of the elements of $\mathbf{u}^*$; that is,

$$C_1 = \{\, i \mid u_i^* > 0 \,\}, \qquad C_2 = \{\, i \mid u_i^* \leq 0 \,\} \tag{9.36}$$

are the two found clusters, and $\mathcal{A} = C_1$, $\mathcal{A}^c = C_2$, or vice-versa. While this reformulation is intriguing, since it involves the graph Laplacian, it remains exact and hence an NP hard problem.

In order to make the graph cut problem tractable, we relax the binary condition on the entries of $\mathbf{u}$, and minimize over $\mathbf{u} \in \mathbb{R}^m$, still subject to the mean zero condition $\mathbf{u} \cdot \mathbf{1} = 0$. This leads to the *binary spectral clustering* problem

$$\min \left\{ \frac{\mathbf{u}^T L \mathbf{u}}{\|\mathbf{u}\|^2} \,\middle|\, \mathbf{0} \neq \mathbf{u} \in \mathbb{R}^m, \ \mathbf{u} \cdot \mathbf{1} = 0 \right\} = \min \left\{ \mathbf{u}^T L \mathbf{u} \mid \|\mathbf{u}\| = 1, \ \mathbf{u} \cdot \mathbf{1} = 0 \right\}. \tag{9.37}$$

As above, after finding a minimizer $\mathbf{u}^*$ of (9.37), the induced clustering is obtained by the sign of the elements of $\mathbf{u}^*$ as in (9.36). However, we should note that it is possible to make other choices besides 0 for the threshold in (9.36). In particular, if we had some prior knowledge of the relative sizes of the two clusters, we would choose a threshold $\theta \in \mathbb{R}$ that resulted in clusters of the appropriate sizes.

As an immediate consequence of Theorem 9.25, we deduce that any Fiedler vector is a solution of the binary spectral clustering problem (9.37). Thus, our simple approximation of allowing $\mathbf{u} \in \mathbb{R}^m$ to be real-valued, instead of binary-valued, relaxed the graph-cut problem

---

[10]In floating point arithmetic, we almost never encounter $u_i = 0$, so the choice to put these points into C_2 is largely irrelevant. We could just as easily make any other choice, or assign such points to clusters at random.

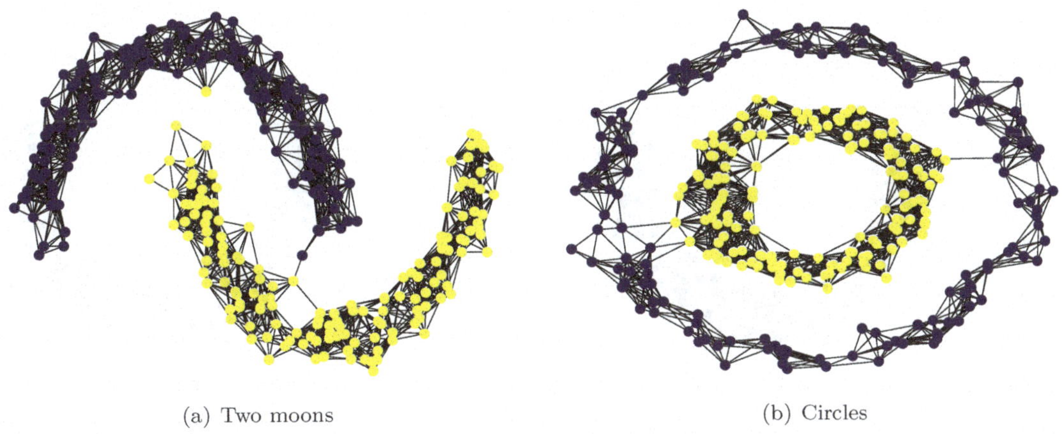

(a) Two moons   (b) Circles

Figure 9.14: Two side by side figures illustrating the similarity graph constructed over the two moons and circles data sets.

from an NP hard problem to the relatively simple computational problem of finding a Fiedler vector for the graph Laplacian. For real data sets, all eigenvalues of the graph Laplacian are generically simple, so the Fiedler vector $\mathbf{u}_F$ is often unique, up to a sign, which does not affect the clustering. Indeed, the negative Fiedler vector $-\mathbf{u}_F$ just switches the cluster labels, modulo the decision of where to place data points with value $u_i = 0$. This is also the reason for the name *spectral clustering* — we are using the spectrum (i.e., the eigenvalues and eigenvectors) of the graph Laplacian matrix for clustering.

> **Theorem 9.31.** *The Fiedler vector $\mathbf{u}_F = \mathbf{u}_2$ solves the binary spectral clustering problem (9.37).*

While spectral clustering is a tractable relaxation of graph cut clustering based on the average cut energy, it is important to note that this truly is an approximation, so the solution to spectral clustering will, in general, not agree with the solution of the graph-cut problem. The question of when the two methods do agree, or how closely they agree, is the subject of research at the intersection of graph theory and machine learning [59, 162]. In any event, spectral clustering often gives very good results for clustering nonlinear and complicated data sets, even if it is not exactly solving the motivational graph-cut problem.

Let us return to the two moons and circles clustering problems in Figure 9.13, which were not clustered correctly by k-means. We apply spectral clustering with Gaussian weights (9.7), and connectivity scale $\varepsilon = 0.1$. To make the graphs sparse, we set the weights to zero when $\|\mathbf{x}_i - \mathbf{x}_j\| \geq 0.25$. Figure 9.14 shows the edges in the similarity graphs for both data sets, colored by the ground truth cluster membership of each vertex. We can see that there are very few edges between the points in different clusters, so the graph structure nicely captures the desired clustering. Figure 9.15 shows the Fiedler vector colored with lowest values purple, and highest values yellow, along with the result of spectral clustering obtained by thresholding the Fiedler vector at 0. Spectral clustering succeeds in finding the correct clustering for the two moons and circles data set, which illustrates its ability to handle complicated cluster geometries.

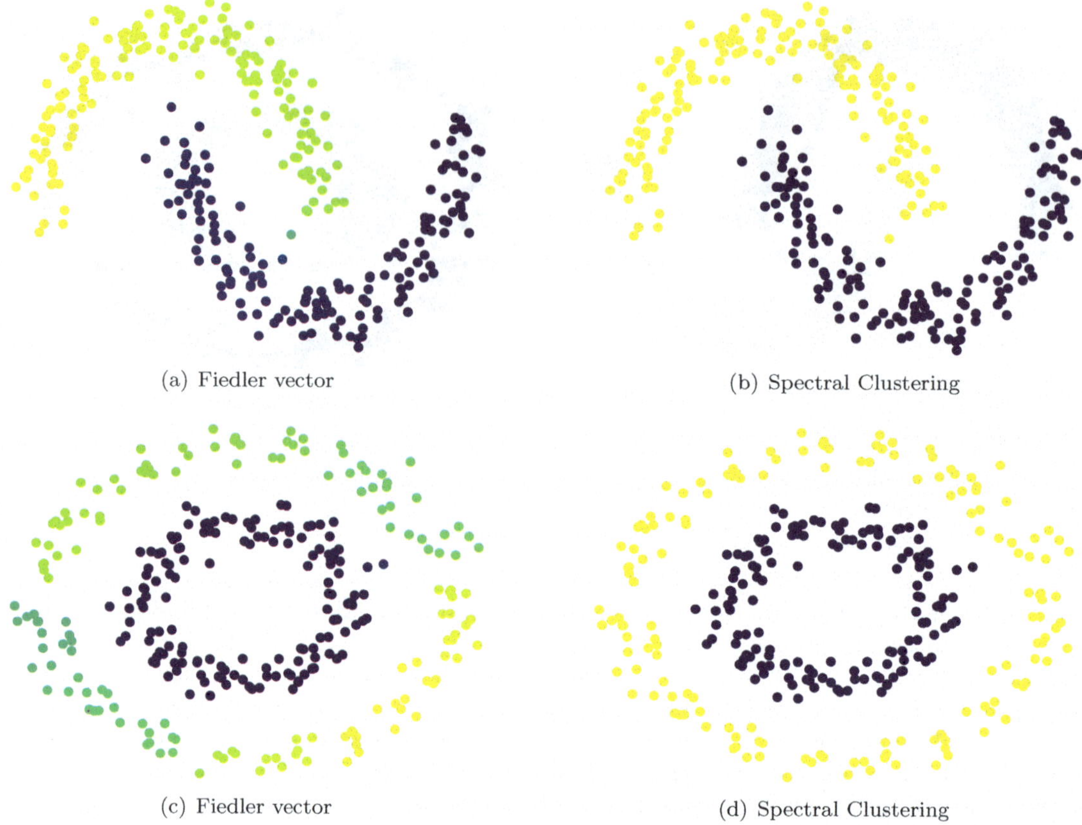

(a) Fiedler vector

(b) Spectral Clustering

(c) Fiedler vector

(d) Spectral Clustering

Figure 9.15: The Fiedler vector and spectral clustering on the two moons and circles data sets.

We next return to the problem of clustering MNIST digits from Section 7.5, to see if we can improve upon the results of k-means clustering. The choice of Gaussian weights (9.7) is not practical for the extensive data sets arising in real world applications for a couple of reasons. First, it produces a potentially intractably large dense weight matrix W, all of whose entries need to be stored. Second, there are difficulties using the same connectivity scale ε over the whole graph, since some areas of the graph may be very dense and a smaller value of ε would be appropriate, while other areas may be more sparse and require a larger ε to ensure they are well-connected to neighboring points. We show in Table 9.16 the results of binary spectral clustering of all pairs of MNIST digits using a $k = 10$ symmetrized nearest neighbor weighting, as described in Section 9.1.1. We generally see a noticeable improvement over the corresponding results for the k-means algorithm given in Table 7.31. However, there are still some pairs of digits that are difficult to separate, such as $(4, 9)$ and $(7, 9)$.

Finally, we consider Zachary's karate club graph that was introduced in Section 9.1.1. Here, the graph structure is given a priori, and does not need to be constructed. A direct application of spectral clustering breaks the $m = 34$ individuals — the nodes in the graph — into two groups of size 19 and 15. Only two individuals are placed in the wrong group; individuals 3 and 9. Inspecting the graph in Figure 9.5(a) we see that both 3 and 9 have more edges to the opposite group than to their own, so it would be natural to misclassify these individuals. However, it turns out that if we change the threshold of the Fiedler vector

Digit	1	2	3	4	5	6	7	8	9
0	99.8	98.9	99.5	99.8	99.5	98.7	99.7	99.2	99.3
1		97.0	99.3	99.1	99.4	99.7	98.8	99.1	99.6
2			98.3	99.5	99.1	99.5	98.0	98.6	99.3
3				99.6	82.3	99.6	99.0	91.8	97.9
4					99.6	99.3	98.9	98.9	53.4
5						97.9	99.8	90.0	98.3
6							99.8	99.0	99.7
7								99.1	70.9
8									97.0
9									

Table 9.16: Accuracy for binary spectral clustering of pairs of MNIST digits. The results are generally an improvement over 2-means clustering (see Table 7.31), but we still see some pairs of digits, such (4,9) and (7,9) are hard to separate.

slightly, to $\theta = 0.007$, then spectral clustering predicts the correct groups for *all* of the participants, which the reader can see in the accompanying Python notebook. We chose this threshold by hand with knowledge of the ground truth groups, but it could also be determined automatically provided one knew, before hand, that the karate club split into two equal sized groups with 17 members each.

Remark. We described in this section the *binary* spectral clustering algorithm that clusters data into two groups. We postpone the development of spectral clustering for more than 2 clusters until Section 9.7.2. ▲

9.4.1 Community Detection

Community detection refers to a class of graph-based algorithms that aim to find tightly knit subsets by looking for clusters of nodes that contain a relatively large number of edges [175]. This perspective strongly contrasts with the graph cut problems described previously, in which we sought clusters of nodes that could easily be *cut out* of the graph by removing only a few edges. Graph cuts do not factor into account how well-connected the found clusters are, while community detection does not directly account for how many edges must be cut to separate the communities in the graph. Nevertheless, the approaches are very similar from a mathematical perspective.

Many community detection algorithms are based on the concept of *modularity*, which is related to the graph cut energies described previously. Let $\mathcal{G}$ be a weighted graph with m nodes and e edges. At a high level, the *modularity* of a subset $\mathcal{A} \subset \mathcal{G}$ is the difference between the number of edges between pairs of points $i, j \in \mathcal{A}$, and the number of edges we would expect to find in $\mathcal{A}$ if the edges were placed at random. Subsets $\mathcal{A}$ with high (positive) modularity have more edges than expected, and correspond to well-connected subsets of the graph.

To derive a formula for modularity, we restrict ourselves for the moment to unweighted graphs. Let A denote the adjacency matrix, with entries $a_{ij} \in \{0,1\}$ counting the number of edges between nodes i and j, which, because the graph is simple, is either 0 or 1. Let $\mathbf{d}$ be the degree vector, which in this case is integer-valued, whose entries count the number of edges connected to the corresponding node. The total number of edges e in the graph satisfies $2e = \mathbf{d} \cdot \mathbf{1}$, since each edge is counted twice when summing the degree vector. Fix two nodes $i, j \in \mathcal{N}$. Let us suppose that all e edges are placed at random in such a way that the degrees of all the nodes are correct. We wish to compute the expected number of edges between nodes i and j in this situation. The analysis turns out to be simpler if we allow multiple edges and loops when selecting edges at random. Thus, for each of the d_i random edges attached to node i, the probability of the edge terminating at node j is $d_j/(2e)$ — indeed, each edge terminates at two nodes, so we are choosing from among $2e$ possible locations[11], and d_j of them correspond to node j. Since there are d_i edges ending at node i, the expected number of edges between nodes i and j is $d_i d_j/(2e)$, and so the quantity $a_{ij} - d_i d_j/(2e)$ represents the difference between the actual number of edges between nodes i and j and the expected number of edges. Summing these over a subset $\mathcal{A} \subset \mathcal{N}$ yields its *modularity*

$$\mathrm{mod}\,(\mathcal{A}) = \sum_{i,j \in \mathcal{A}} \left(a_{ij} - \frac{d_i d_j}{2e} \right). \tag{9.38}$$

The goal is to find subsets $\mathcal{A} \subset \mathcal{N}$ with large positive modularity.

Before proceeding, let us generalize modularity to weighted graphs. Let $W = W^T$ be a symmetric $m \times m$ weight matrix for the graph $\mathcal{G}$. We replace the adjacency matrix A by the weight matrix W, so a_{ij} becomes w_{ij} above. We also recall that in the unweighted case we have $2e = \mathbf{d} \cdot \mathbf{1} = \mathbf{d}^T \mathbf{1}$. Making these substitutions in (9.38) leads us to define

$$\mathrm{mod}\,(\mathcal{A}) = \sum_{i,j \in \mathcal{A}} \left(w_{ij} - \frac{d_i d_j}{\mathbf{d}^T \mathbf{1}} \right). \tag{9.39}$$

In this section, we will focus on splitting a weighted graph $\mathcal{G}$ into two communities $\mathcal{A}$ and $\mathcal{A}^c$ in a way that maximizes the sum of their modularities. Thus, we define the *modularity energy*

$$\mathrm{E}_{\mathrm{mod}}(\mathcal{A}) = \mathrm{mod}\,(\mathcal{A}) + \mathrm{mod}\,(\mathcal{A}^c). \tag{9.40}$$

Community detection based on modularity optimization corresponds to the problem of finding a subset $\mathcal{A} \subset \mathcal{N}$ that maximizes the modularity energy. As with the average and ratio graph cut problems described previously, maximizing modularity is a challenging computational problem, but it can be relaxed to a spectral modularity problem that is more tractable. Doing so requires defining the modularity matrix.

Definition 9.32. Let $\mathcal{G}$ be a weighted graph with weight matrix W and degree vector $\mathbf{d}$. The associated *modularity matrix* is the $m \times m$ matrix M defined by

$$M = W - \frac{\mathbf{d}\mathbf{d}^T}{\mathbf{d}^T \mathbf{1}}. \tag{9.41}$$

We note that the entries

$$m_{ij} = w_{ij} - \frac{d_i d_j}{d_1 + \cdots + d_m}, \tag{9.42}$$

of the modularity matrix M are exactly the quantities that appear in the definition (9.39) of modularity. The following lemma connects the modularity matrix to the modularity energy.

Lemma 9.33. *Let $\mathcal{A} \subset \mathcal{N}$ and set $\mathbf{u} = \mathbf{1}_{\mathcal{A}} - \mathbf{1}_{\mathcal{A}^c}$. Then $\|\mathbf{u}\|^2 = m$ and*

$$\mathrm{E}_{\mathrm{mod}}(\mathcal{A}) = \tfrac{1}{2} \mathbf{u}^T M \mathbf{u}. \tag{9.43}$$

---

[11]Actually, there are $2e - 1$ possible locations, since the edge cannot connect a node to itself, but this difference is negligible, especially for large graphs.

Proof. Note that the quantity $\frac{1}{2}(u_i u_j + 1)$ is one when i and j both belong to $\mathcal{A}$, or both belong to $\mathcal{A}^c$, and is zero otherwise. Therefore, we can write

$$\mathrm{E}_{\mathrm{mod}}(\mathcal{A}) = \frac{1}{2} \sum_{i,j=1}^{m} \left(w_{ij} - \frac{d_i d_j}{\mathbf{d}^T \mathbf{1}} \right) (u_i u_j + 1) = \tfrac{1}{2} (\mathbf{u}^T M \mathbf{u} + \mathbf{1}^T M \mathbf{1}).$$

Furthermore, as with the graph Laplacian, the ones vector $\mathbf{1} \in \mathbb{R}^m$ belongs to the kernel of the modularity matrix, since

$$M\mathbf{1} = W\mathbf{1} - \frac{\mathbf{d}\mathbf{d}^T \mathbf{1}}{\mathbf{d}^T \mathbf{1}} = \mathbf{d} - \mathbf{d} = \mathbf{0}.$$

Substituting this into the preceding equation verifies (9.43). $\blacksquare$

By Lemma 9.33, the modularity of the entire graph and of its complement, vanish:

$$\mathrm{E}_{\mathrm{mod}}(\mathcal{N}) = \mathrm{E}_{\mathrm{mod}}(\varnothing) = 0. \tag{9.44}$$

Thus, if there are subsets $\mathcal{A} \subset \mathcal{N}$ with positive modularity, then the modularity energy is not minimized by the trivial communities, in contrast to the case of graph cuts. It also follows from Lemma 9.33 that community detection based on maximizing modularity is equivalent to the optimization problem

$$\max \left\{ \mathbf{u}^T M \mathbf{u} \mid \mathbf{u} \in \{-1, 1\}^m \right\}. \tag{9.45}$$

As usual, the binary constraint, requiring the entries of $\mathbf{u}$ to be either $+1$ or -1, makes the modularity optimization problem computationally challenging. And, as before, we will relax the constraint by allow $\mathbf{u} \in \mathbb{R}^m$ to be any unit vector, $\|\mathbf{u}\| = 1$, instead of the condition $\|\mathbf{u}\|^2 = m$ used in Lemma 9.33, for simplicity. This yields the *spectral modularity community detection* problem

$$\max \left\{ \mathbf{u}^T M \mathbf{u} \mid \|\mathbf{u}\| = 1 \right\}. \tag{9.46}$$

According to (5.50), the solution is the top eigenvector $\mathbf{u}^*$ of the modularity matrix M, and we use the signs of the the entries of $\mathbf{u}^*$ to determine the communities, as we did in (9.36) for spectral clustering. In the case that the normalized ones vector $\mathbf{u} = 1/\sqrt{m}$ is the top eigenvector, modularity optimization returns a single community — the entire graph. This case is discussed further below after we study the modularity matrix M more closely.

Spectral modularity shares some similarities with spectral clustering, but also has notable differences that are illuminated by comparing the modularity matrix M with the graph Laplacian. At first glance, the modularity matrix (9.41) appears similar to the negative of the graph Laplacian, which can be written as

$$-L = W - D = W - \mathrm{diag}\,\mathbf{d}.$$

The difference is that the modularity matrix uses the rank one matrix $\mathbf{d}\mathbf{d}^T/(\mathbf{d}^T \mathbf{1})$ in place of the diagonal degree matrix D. As we saw in the proof of Lemma 9.33, the ones vector $\mathbf{1}$ belongs to the kernel of the modularity matrix M, as it does for the graph Laplacian matrix. However, the similarities between the two matrices do not go much further. In particular, the kernel of the modularity matrix *does not* indicate the connected components in the graph.

Indeed, if

$$M\mathbf{u} = \mathbf{0} \quad \text{then} \quad W\mathbf{u} = c\,\mathbf{d} = c\,W\mathbf{1}, \quad \text{where} \quad c = \frac{\mathbf{d} \cdot \mathbf{u}}{\mathbf{d} \cdot \mathbf{1}},$$

and so $\mathbf{u} - c\mathbf{1} \in \ker W$. Thus

$$\ker M = \{\, c\mathbf{1} + \mathbf{z} \mid c \in \mathbb{R}, \ \mathbf{z} \in \ker W \,\} = \operatorname{span}\{\mathbf{1}\} + \ker W$$

is the subspace spanned by the ones vector and all vectors in the kernel of the weight matrix. For many graphs — in particular, for all of the examples below — the weight matrix is nonsingular, and so the trivial case $\ker M = \operatorname{span}\{\mathbf{1}\}$ is not an uncommon occurrence, even for graphs that are not connected. In fact, even in the case of unweighted graphs, the kernel of the adjacency matrix is not simple to characterize; we refer to [211] for more details.

The other key point of departure from the graph Laplacian is positive definiteness. Since the modularity matrix appears similar to the negative of the graph Laplacian, one might expect M to be negative semidefinite. However, this is not necessarily the case.

Example 9.34. Consider the graph from Example 9.24 with 4 nodes and two connected components $\{1,2\}$ and $\{3,4\}$. The degree vector is $\mathbf{d} = \mathbf{1}$ and $\mathbf{d}^T\mathbf{1} = 4$. Therefore the modularity matrix is

$$M = W - \frac{\mathbf{1}\mathbf{1}^T}{4} = \frac{1}{4}\begin{pmatrix} -1 & 3 & -1 & -1 \\ 3 & -1 & -1 & -1 \\ -1 & -1 & -1 & 3 \\ -1 & -1 & 3 & -1 \end{pmatrix}.$$

The ones vector $\mathbf{1}$ is clearly a null eigenvector. By inspection, we can also see that $(1,1,-1,-1)^T$ is an eigenvector for eigenvalue $\lambda = 1$, and $(1,-1,0,0)^T$ and $(0,0,1,-1)^T$ are independent eigenvectors for eigenvalue $\lambda = -1$. Thus, M is neither positive nor negative semidefinite. On the other hand, the signs of the entries of the largest positive eigenvector $(1,1,-1-1)^T$ indicates the two connected components of the graph. ▲

In fact, the lack of semidefiniteness of the modularity matrix is a *feature* of modularity optimization for community detection. We are seeking communities with large *positive* modularity, indicating groups with more edges than expected. When no such groups exist, the modularity is negative for *all* subsets of the graph, and the modularity matrix M is *negative semidefinite*. We can state this result as a theorem, whose proof follows directly from Lemma 9.33, equation (9.44), and the optimization principle for eigenvectors of symmetric matrices, Theorem 5.29.

Theorem 9.35. *The modularity matrix M is negative semidefinite if and only if*

$$\max_{\mathcal{A} \subset \mathcal{N}} \mathrm{E}_{\mathrm{mod}}(\mathcal{A}) = \mathrm{E}_{\mathrm{mod}}(\mathcal{N}) = 0.$$

Thus, a key difference between modularity and spectral clustering is that modularity optimization can decide not to identify any groups, which occurs exactly when the modularity matrix M is negative semidefinite.

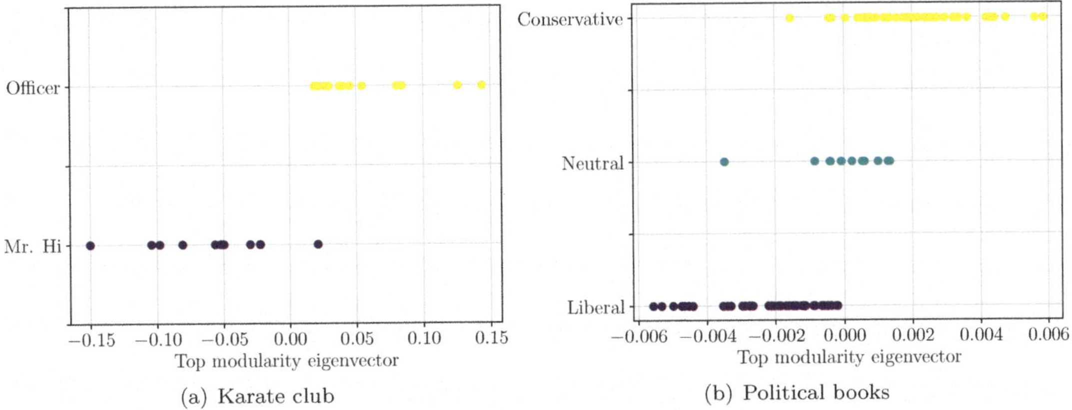

(a) Karate club   (b) Political books

Figure 9.17: The top modularity vector for Zachary's karate club graph, and Krebs' political books graph. The values of the top modularity eigenvector are displayed along the x-axis, while the class labels are displayed along the y-axis, and also displayed by the color of each point.

Example 9.36. Let $\mathcal{G}$ be the complete graph on m nodes, for which we have $W = \mathbf{1}\mathbf{1}^T - I$, $\mathbf{d} = (m-1)\mathbf{1}$, and $\mathbf{d}^T\mathbf{1} = m(m-1)$. Then the modularity matrix (9.41) is

$$M = \mathbf{1}\mathbf{1}^T - I - \frac{(m-1)^2\,\mathbf{1}\mathbf{1}^T}{m(m-1)} = \frac{1}{m}\mathbf{1}\mathbf{1}^T - I = -P,$$

where P is the orthogonal projection matrix onto the subspace $\mathbf{1}^\perp$, the orthogonal complement to the ones vector, i.e., the space of mean zero vectors. According to Exercise 6.3, every orthogonal projection matrix is positive semidefinite, so M is negative semidefinite. Theorem 9.35 implies that the maximum modularity is 0, and so modularity optimization does not find any significant communities. Indeed, all nodes are identical in the complete graph, and so there is no community structure to detect.

Since P projects onto an $m-1$ dimensional subspace, the zero eigenvalue of M is simple, and the other $m-1$ eigenvalues of M are $\lambda = -1$. Since there is a spectral gap between the trivial zero eigenvalue and the remainder of the spectrum, this suggests that even moderate perturbations of the complete graph will result in a negative semidefinite modularity matrix, and no detectable community structure. The perturbation would have to be sufficiently large so that one of the $\lambda = -1$ eigenvalues becomes positive before modularity optimization detects community structure. ▲

We now turn to some experiments. For Zachary's karate club graph, community detection by spectral modularity splits the members very well into two groups, misclassifying only one individual; see Figure 9.17(a). The modularity $\mathrm{E}_{\mathrm{mod}}(\mathcal{A})$ for the detected communities is positive, and equal to $0.74\,e$, where e is the number of edges. For Krebs' political books graph, we show the top modularity vector in Figure 9.17(b), where we can see that thresholding at zero classifies correctly all liberal books, and all but three of the conservative books. The books identified as neutral are roughly evenly split between the two groups, indicating that they do not strongly belong to either group. The modularity is $\mathrm{E}_{\mathrm{mod}}(\mathcal{A}) = 0.89\,e$. In both cases, the detected communities had positive modularity, indicating that the found communities are significant.

Both spectral modularity optimization and spectral clustering have variations that can find more than two clusters or communities on graphs. One approach is to recursively split the groups, which is described in detail for community detection in [175]. The other standard approach is to consider the top k eigenvectors of the modularity matrix or graph Laplacian, which we postpone until Section 9.7.2.

Exercises

4.1. ♡ Show that the quantity $(m - k)k$ is maximized over integers $0 \leq k \leq m$ by setting $k = \frac{1}{2}m$ when m is even, and $k = \frac{1}{2}(m - 1)$ or $k = \frac{1}{2}(m + 1)$ when m is odd.

4.2. ◇ Given a graph $\mathcal{G}$ with node set $\mathcal{N}$, let $\mathcal{A} \subset \mathcal{N}$, $b = |\mathcal{A}|/|\mathcal{A}^c|$, and $\mathbf{u} = \mathbf{1}_{\mathcal{A}} - b\,\mathbf{1}_{\mathcal{A}^c}$. Let D be the diagonal degree matrix. (a) Show that $\mathbf{u}^T D \mathbf{1} = 0$, and

$$\text{cut}_{\text{norm}}(\mathcal{A}) = \frac{1}{|\mathcal{N}|}\frac{\mathbf{u}^T L \mathbf{u}}{\mathbf{u}^T D \mathbf{u}}. \tag{9.47}$$

Thus, the version of spectral clustering based on normalized cuts corresponds to

$$\min\left\{\frac{\mathbf{u}^T L \mathbf{u}}{\mathbf{u}^T D \mathbf{u}} \;\middle|\; \mathbf{u} \in \mathbb{R}^m, \;\; \mathbf{u}^T D \mathbf{1} = 0\right\}. \tag{9.48}$$

(b) Show that the solution of (9.48) is the eigenvector of $D^{-1}L$ corresponding to the second smallest eigenvalue $\lambda_2(D^{-1}L)$. (c) Show that the solution of (9.48) is the eigenvector of $D^{-1}W$ corresponding to the second *largest* eigenvalue $\lambda_{m-1}(D^{-1}W)$.

4.3. ◇ Consider power method $\mathbf{u}_{k+1} = D^{-1}W\mathbf{u}_k/\|D^{-1}W\mathbf{u}_k\|$, initialized from $\mathbf{u}_0$ satisfying $\mathbf{u}_0^T D \mathbf{1} = 0$, to perform normalized cut spectral clustering as outlined in Exercise 4.2. Write a Python program that implements the method, and try your program on some of the examples from this section. *Note*: You can assume $\lambda_{m-1}(D^{-1}W) > |\lambda_1(D^{-1}W)|$ so that a spectral shift is not needed.

4.4. ♡ Suppose we allow loops in a graph, by permitting one or more diagonal entries of the weight matrix to be positive: $w_{ii} > 0$. Recall from Exercise 3.3 that the Laplacian matrix is unchanged by loops. Is this also true for the modularity matrix?

4.5. ◇ Let $\mathcal{G}$ be a graph with graph Laplacian matrix L and modularity matrix M.

(a) Show that $\lambda_{min}(M) \geq \lambda_{min}(W) - \dfrac{\|\mathbf{d}\|^2}{\mathbf{d}\cdot\mathbf{1}}$.

(b) Use part (a) to show that $\lambda_{min}(M) \geq -\left(\|W\|_F + \dfrac{\|\mathbf{d}\|^2}{\mathbf{d}\cdot\mathbf{1}}\right)$, where $\|W\|_F$ is the Frobenius norm of the weight matrix.

(c) Show that $\lambda_{max}(M) \leq \lambda_{max}(W) \leq \|W\|_F$.

4.6. Let $\mathcal{G}$ be a graph with weight matrix W, graph Laplacian matrix L, and modularity matrix M. Assume that all the nodes in $\mathcal{G}$ have the same degree, and so the degree vector is constant $\mathbf{d} = k\mathbf{1}$ for $k > 0$. This holds, for example, in a k-nearest neighbor graph. (a) Show that M and L are simultaneously complete, meaning that there exists an orthonormal basis $\mathbf{u}_1, \ldots, \mathbf{u}_m$ of $\mathbb{R}^m$ for which each $\mathbf{u}_i$ is an eigenvector of both M and L. *Hint*: Choose $\mathbf{u}_i$ to be the eigenvectors of L and show that they are also eigenvectors of M. What are the eigenvalues? (b) Assume that $\lambda_{max}(M) > 0$. Show that $\mathbf{u} \in \mathbb{R}^m$ is a top eigenvector of the modularity matrix M, with eigenvalue $\lambda_{max}(M)$, if and only if $\mathbf{u}$ is also a Fiedler vector for the graph $\mathcal{G}$.

4.7. ♡   Implement spectral modularity optimization for community detection in Python via the power method and detect communities in Zachary's karate club graph and Krebs' political books graph to reproduce the results from this section. Try other graphs in the `graphlearning` package as well. *Hint*: You will have to use a spectral shift of the modularity matrix, as we did for finding the Fiedler vector with the power iteration; see Exercise 4.5(a).

9.5   Distances on Graphs

Python Notebook: Graph Distances (.ipynb)

Let $\mathcal{G}$ be a connected weighted digraph with m nodes, e edges, and weight matrix W. Since the graph is directed, W need not be symmetric. For an edge from i to j, the associated weight w_{ij} gives us an indication of how similar the nodes are; larger values of w_{ij} indicate more similarity, and smaller values less. When there is no edge from i to j, the corresponding weight $w_{ij} = 0$. However, this *does not* necessarily imply that i and j are dissimilar or, in some sense, *nearby* in the graph — it merely indicates that the existing edges do not directly give us any information about their similarity. There could be an intermediate node k that is a neighbor of both, so that i and j are *2-hop neighbors* of each other — that is, there is a path with two edges that connects them. In this case, we may consider the nodes i and j as reasonably close, especially compared to other pairs of nodes that, say, require more intermediate hops.

In this section we study the problem of comparing the similarity, or equivalently distance, between nodes i and j in a digraph when they are not neighbors. We will do this here by measuring how quickly one can move from node i to node j following a *path* in the graph. Since we are assuming $\mathcal{G}$ is connected, there will, in general, be many paths connecting any two nodes, so we will choose the *shortest path*. In network science or navigation applications, this can be interpreted as the path with least travel time for a packet routed through a network, or a car driven between two cities, where the edges of the graph represent roads. In Zachary's karate club graph, the shortest path is the one with the fewest number of social connection hops between two members. Shortest path distances on graphs allow us to compare *every* pair of nodes in the graph, and assign a distance between them that respects the graph structure. We will see applications of this to graph visualization later in Section 9.5.3.

To be precisely mathematically, we will consider walks in the graph, so that we do not have to bother with the restriction that vertices and/or edges are distinct. We recall that a *walk* in a graph is simply an ordered list of edges $\varepsilon_1, \varepsilon_2, \ldots, \varepsilon_k \in I_n$ connecting nodes $m_1, m_2, \ldots, m_{k+1} \in I_m$ — that is, edge ε_i connects node m_i to node m_{i+1} in the prescribed direction. We will describe a walk by the vector $\mathbf{p} = (\varepsilon_1, \varepsilon_2, \ldots, \varepsilon_k)^T$ containing the sequence of edges belonging to the walk. Let $\mathcal{P}$ denote the set of all walks in the graph $\mathcal{G}$, and $\mathcal{P}_{ij} \subset \mathcal{P}$ those that begin at node i and end at node j.

Definition 9.37. The *length*, or *travel time*, of a walk $\mathbf{p} \in \mathcal{P}$, denoted by $\tau(\mathbf{p})$, is defined as

$$\tau(\mathbf{p}) = \sum_{(i,j) \in \mathbf{P}} w_{ij}^{-1}. \tag{9.49}$$

Since we treat the weight $w_{ij} \geq 0$ as a measure of similarity between data points, the distance, or travel time, along the edge from i to j is defined to be its reciprocal: w_{ij}^{-1}. This also has the convenient interpretation that when $w_{ij} = 0$, which indicates no edge exists from i to j, we have infinite travel time between them.

Remark 9.38. Suppose the nodes correspond to data points $\mathbf{x}_1, \ldots, \mathbf{x}_m \in \mathbb{R}^n$, and we choose weights that are inversely proportional to the interpoint distances, that is

$$w_{ij} = \begin{cases} \|\mathbf{x}_i - \mathbf{x}_j\|^{-1}, & \|\mathbf{x}_i - \mathbf{x}_j\| \leq \varepsilon \\ 0, & \text{otherwise}, \end{cases}$$

for some chosen length scale $\varepsilon > 0$ on which the data points are connected. Then the length of a path coincides with its Euclidean length:

$$\tau(\mathbf{p}) = \sum_{(i,j)\in\mathbf{p}} \|\mathbf{x}_i - \mathbf{x}_j\|.$$

In this situation, paths through the point cloud are restricted to hop a distance at most ε at each step. $\blacktriangle$

The travel times can be conveniently collected into a matrix.

Definition 9.39. The *distance matrix* T of a weighted digraph $\mathcal{G}$ is the $m \times m$ matrix with zeros on the diagonal, $t_{ii} = 0$, and off-diagonal entries

$$t_{ij} = \min_{\mathbf{p}\in\mathcal{P}_{ij}} \tau(\mathbf{p}), \qquad i \neq j. \tag{9.50}$$

A *minimizing walk* $\mathbf{p}$ is called an *optimal path* from i to j.

The value of t_{ij} is the length of the shortest walk that connects node i to node j. This walk must be a *path*, since repeating any node along the way can only make the walk longer. We can also interpret t_{ij} as the travel time from node i to node j, assuming one travels along the shortest path. We let $\mathbf{t}_i^T \in \mathbb{R}^m$ denote the i-th row of T, whose entries indicate the distances from node i to every other node in the graph. For a digraph, the distance matrix T need not be symmetric, since paths must traverse edges in the correct direction. On the other hand, if we have a weighted graph, so that $W = W^T$, then the distance matrix is also symmetric:[12] $T = T^T$.

Figure 9.18 shows examples of shortest paths in k-nearest neighbor graphs on the two moons and circles data sets. In each figure we show shortest paths between two different pairs of nodes in red. Notice that nodes corresponding to data points that are nearby in Euclidean distance may in fact be very far apart in terms of the graph distance, since the shortest path is required to navigate the graph. In Figure 9.19 we show the distance vectors $\mathbf{t}_i$ to a single node i, shown as a red star, which indicate the length of the shortest path from each node back to the red star. In this case, the edges are not directed, so the distance matrix is symmetric.

The distance matrix satisfies the following triangle inequality, the proof of which we leave to the reader in Exercise 5.1.

Lemma 9.40. *For all nodes* $i, j, k \in \mathcal{N}$, *the graph distances satisfy* $t_{ik} \leq t_{ij} + t_{jk}$.

---

[12]We apologize for the unfortunate clash of notation when writing the transpose of the distance matrix T.

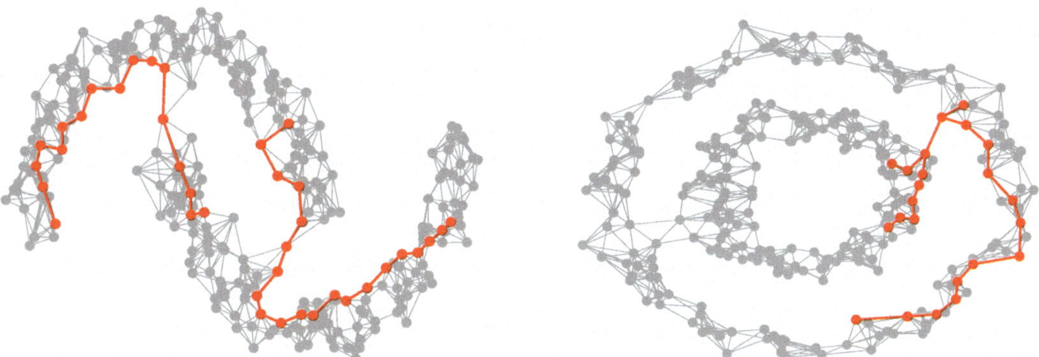

Figure 9.18: Examples of paths between pairs of points on the two moons and circles data sets. Each figure shows two shortest paths in red. The graphs are unweighted symmetrized k-nearest neighbor graphs with $k = 6$. Notice that points that are close in Euclidean distance may be far apart in the graph distance.

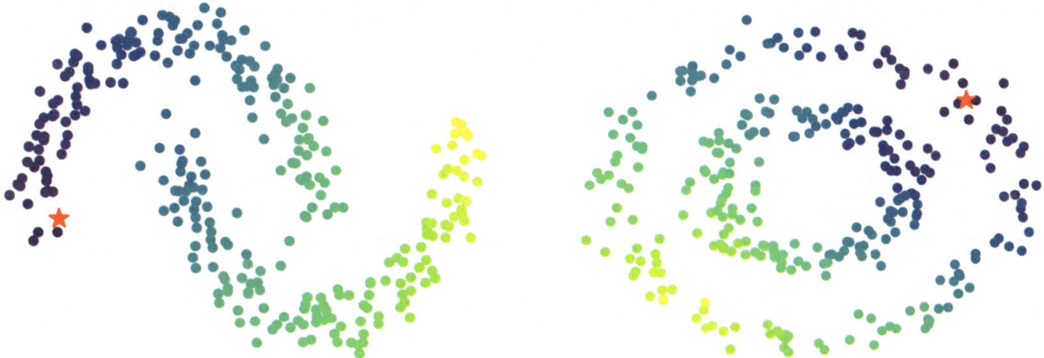

Figure 9.19: Examples of the distance vectors based on the red star node.

For a weighted graph, the distance matrix is symmetric, so $t_{ij} = t_{ji} > 0$ for $i \neq j$; moreover $t_{ii} = 0$. Therefore, it satisfies the axioms in Section 2.7.4 and hence defines a distance function on the nodes in the graph. Hence, the shortest path distance on a weighted graph endows the graph with the structure of a *metric space*, as per Definition 2.38. Applications of this fact with appear in Section 9.5.3.

9.5.1 Computing the Shortest Path Distance

We now turn to the problem of efficiently computing the distance matrix T. We will proceed in slightly more generality at first. For a subset $\mathcal{D} \subset \mathcal{N}$ of the graph nodes, we define the distance vector $\mathbf{t}_{\mathcal{D}}$ with entries

$$t_{\mathcal{D},i} = \min_{j \in \mathcal{D}} t_{ji}. \tag{9.51}$$

That is, the i-th element of $\mathbf{t}_{\mathcal{D}}$ gives the shortest path distance from the closest node $j \in \mathcal{D}$ to the node i. In particular, $t_{\mathcal{D},i} = 0$ if $i \in \mathcal{D}$. When $\mathcal{D} = \{i\}$ contains a single node, $\mathbf{t}_{\mathcal{D}} = \mathbf{t}_i$ is the corresponding row of the distance matrix.

We will focus on the computation of the distance vector $\mathbf{t}_{\mathcal{D}}$ for a fixed set of nodes $\mathcal{D} \subset \mathcal{N}$, which we will do directly, *without* first computing the distance matrix T. The main idea is encapsulated in the following lemma, which is called a *dynamic programming principle* — the reason for this name will be explained below.

Lemma 9.41. *Let $\mathcal{D} \subset \mathcal{N}$ with complement $\mathcal{D}^c = \mathcal{N} \setminus \mathcal{D}$. Then the entries $t_i = t_{\mathcal{D},i}$ of the distance vector $\mathbf{t}_{\mathcal{D}}$ satisfy*

$$t_i = \begin{cases} \min_j \left(t_j + w_{ji}^{-1} \right), & i \in \mathcal{D}^c, \\ 0, & i \in \mathcal{D}. \end{cases} \tag{9.52}$$

Moreover, $\mathbf{t}_{\mathcal{D}}$ is the unique solution to (9.52).

Remark 9.42. The minimum in (9.52) is over all nodes, so $j = 1, \ldots, m$, with the understanding that $w_{ji}^{-1} = \infty$ if $w_{ji} = 0$. Thus, the minimum will be attained at a j for which $w_{ji} > 0$ and hence there is an edge from j to i. $\blacktriangle$

Proof. Let k be a node that minimizes the right hand side of (9.52). Then $w_{ki} > 0$ since t_i is finite. Let $\mathbf{p}$ be a path from a node in $\mathcal{D}$ to node k for which $t_k = \tau(\mathbf{p})$. Let $\mathbf{q}$ be the walk obtained by appending the edge $\varepsilon = (k, i)$ to $\mathbf{p}$. Then

$$t_i \leq \tau(\mathbf{q}) = \tau(\mathbf{p}) + w_{ki}^{-1} = t_k + w_{ki}^{-1} = \min_j \left(t_j + w_{ji}^{-1} \right).$$

To show the opposite inequality, let $\mathbf{q}$ be a path connecting a node in $\mathcal{D}$ to node i for which $\tau(\mathbf{q}) = t_i$. Let $\mathbf{p}$ denote the path obtained by removing the last edge $\varepsilon = (k, i)$ from $\mathbf{q}$, which thereby connects a node in $\mathcal{D}$ to k. Then, $t_k \leq \tau(\mathbf{p})$, and hence

$$t_i = \tau(\mathbf{q}) = \tau(\mathbf{p}) + w_{ki}^{-1} \geq t_k + w_{ki}^{-1} \geq \min_j \left(t_j + w_{ji}^{-1} \right).$$

To prove uniqueness, suppose $\widetilde{\mathbf{t}} = \left(\widetilde{t}_1, \widetilde{t}_2, \ldots, \widetilde{t}_m \right)^T$ also satisfies (9.52). If $\widetilde{\mathbf{t}} \neq \mathbf{t}_{\mathcal{D}}$, there exists $i \in \mathcal{D}^c$ such that $\widetilde{t}_i \neq t_i$; without loss of generality, $\widetilde{t}_i > t_i$. (Otherwise, just reverse the roles of $\widetilde{\mathbf{t}}$ and $\mathbf{t}_{\mathcal{D}}$.) Then there exists $\lambda > 1$ such that $\widetilde{t}_i > \lambda t_i$. Now let i be the index for which $\widetilde{t}_i - \lambda t_i$ is largest, which need not be the same index i selected before. Since $\widetilde{t}_i - \lambda t_i > 0$, we know that $i \in \mathcal{D}^c$ and

$$\widetilde{t}_i - \lambda t_i \geq \widetilde{t}_j - \lambda t_j \qquad \text{for all} \qquad j = 1, \ldots, m.$$

Rearranging this and adding w_{ji}^{-1} to both sides produces

$$\widetilde{t}_j + w_{ji}^{-1} - \widetilde{t}_i \leq \lambda (t_j + w_{ji}^{-1} - t_i) + (1 - \lambda) w_{ji}^{-1}.$$

Since both $\widetilde{\mathbf{t}}$ and $\mathbf{t}_{\mathcal{D}}$ satisfy (9.52), the minima, over all $j = 1, \ldots, m$, of the left hand side and the quantity in parentheses on the right hand side are both 0. However, since $1 - \lambda < 0$ and $w_{ji} > 0$, the minimum of the final term is negative, which is a contradiction. $\blacksquare$

The term *dynamic programming* generally refers to mathematical algorithms that break a problem down into a series of sub-problems that are easier to solve, often in a recursive fashion. Here, the sub-problems given by Lemma 9.41 are encapsulated in (9.52), and are recursive, since each sub-problem depends on the values of t_j that appear in other sub-problems. Computation of shortest paths through networks is one of the seminal problems to which dynamic programming has been applied; for a survey we refer the reader to [22].

Lemma 9.41 allows us to break up the problem of finding a shortest path into many small localized optimization problems (9.52) — one for each node in the graph. The dynamic programming principle (9.52) indicates an iterative approach for computing the distance vector $\mathbf{t}_{\mathcal{D}}$. Namely, given an approximation $\mathbf{s}_k$ for $\mathbf{t}_{\mathcal{D}}$ with entries $s_{k,i}$, we construct a (hopefully) better approximation $\mathbf{s}_{k+1}$ by defining its entries to be

$$s_{k+1,i} = \begin{cases} \min\limits_{j}\left(s_{k,j} + w_{ji}^{-1}\right), & i \in \mathcal{D}^c, \\ 0, & i \in \mathcal{D}. \end{cases} \tag{9.53}$$

Better means that $s_{k+1,i} \leq s_{k,i}$ for all $i = 1, \ldots, m$, since our goal is to minimize path lengths. As we demonstrate below, the iterates must converge in finitely many steps, where convergence means $\mathbf{s}_{k+1} = \mathbf{s}_k$, at which point further iteration produces nothing new. Moreover, once they converge, the final iterate necessarily equals the distance vector, i.e., $\mathbf{t}_{\mathcal{D}} = \mathbf{s}_{k+1} = \mathbf{s}_k$.

The one tricky point is how to properly start the iterations. Initially, we may lack any information to determine an approximation to any of the t_i, except for those nodes that lie in $\mathcal{D}$, for which $t_i = 0$. Rather than try to guess values of t_i for $i \in \mathcal{D}^c$, we employ the following clever device. Let us take our initial "approximation" $\mathbf{s}_0$ to be the vector whose entries $s_{0,i}$ equal ∞ if $i \in \mathcal{D}^c$ and 0 if $i \in \mathcal{D}$; formally, $\mathbf{s}_0 = \infty \mathbf{1}_{\mathcal{D}^c}$. Recall that we have already introduced infinity when dealing with w_{ij}^{-1} when there is no edge between nodes i and j. In the algorithm, we will only use the properties that, whenever $a \in \mathbb{R}$ is a real number, $a < \infty$, and $a + \infty = \infty = \infty + \infty$. (In practice, many programming languages, including Python, allow treatment of ∞ in this same way.) This choice of $\mathbf{s}_0$ allows us to start the dynamic programming iteration (9.53).

Now, observe that the i-th entry $s_{1,i}$ of the first iterate $\mathbf{s}_1$ equals 0 if $i \in \mathcal{D}$, equals the minimal distance w_{ji}^{-1} from a neighboring node $j \in \mathcal{D}$ to node i, or equals ∞ if i is not a neighbor of any node in $\mathcal{D}$. Indeed, if $i \in \mathcal{D}^c$, the only way that $s_{1,i}$, as defined by (9.53), is not equal to ∞ is if there is at least one j such that $s_{0,j} = 0$, and hence $j \in \mathcal{D}$, and, further, there is at least one edge connecting j to i, so $w_{ji} > 0$. Thus, its finite nonzero entries are exactly the minimal path distances from the nodes in $\mathcal{D}^c$ that are connected to $\mathcal{D}$ by a path with a single edge. The process then continues. By a similar reasoning, the i-th entry $s_{2,i}$ of the following iterate $\mathbf{s}_2$ equals 0 if $i \in \mathcal{D}$, equals the minimal distance along any path with at most 2 edges connecting node i to a node in $\mathcal{D}$, or equals ∞ if i is not connected to any node in $\mathcal{D}$ by a path with at most 2 edges. The entries of the k-th iterate are similarly characterized by the minimal lengths of paths with at most k edges connecting each node to $\mathcal{D}$. Note that this characterization implies that the entries of the iterates decrease monotonically: $s_{k+1,i} \leq s_{k,i}$ since the former minimizes over a larger collection of paths. Moreover, they are bounded from below, $s_{k,i} \geq s_{\mathcal{D},i}$, since the latter minimizes distances over all possible paths.

Finally, to prove convergence, let k^* denote the maximal number of edges in any path connecting a node $i \in \mathcal{D}^c$ to a node in $\mathcal{D}$. Since the graph is connected, $k^* \leq m - \#\mathcal{D}$, which is the number of edges in a path (if such exists) that goes through every node in $\mathcal{D}^c$ before arriving in $\mathcal{D}$. By the above characterization, the nonzero entries of $\mathbf{s}_{k^*}$ must equal the minimal distances along all possible paths connecting the corresponding nodes to $\mathcal{D}$, and hence $\mathbf{s}_{k^*} = \mathbf{s}_{k^*+1} = \mathbf{t}_{\mathcal{D}}$. This implies that the iterates must have converged for some $k \leq k^*$, and thereby completes the justification of the dynamic programming algorithm. In practice, convergence can take place considerably faster than the indicated bound, $m - \#\mathcal{D}$.

We show in Figure 9.20 an experiment with the dynamic programming iterations on the two moons data set, where the method converges in 38 steps. The red points indicate nodes where the iterates still take the value ∞. Notice how the solution propagates from the initial node, denoted by the red star, through the entire graph. As noted above, this propagation is

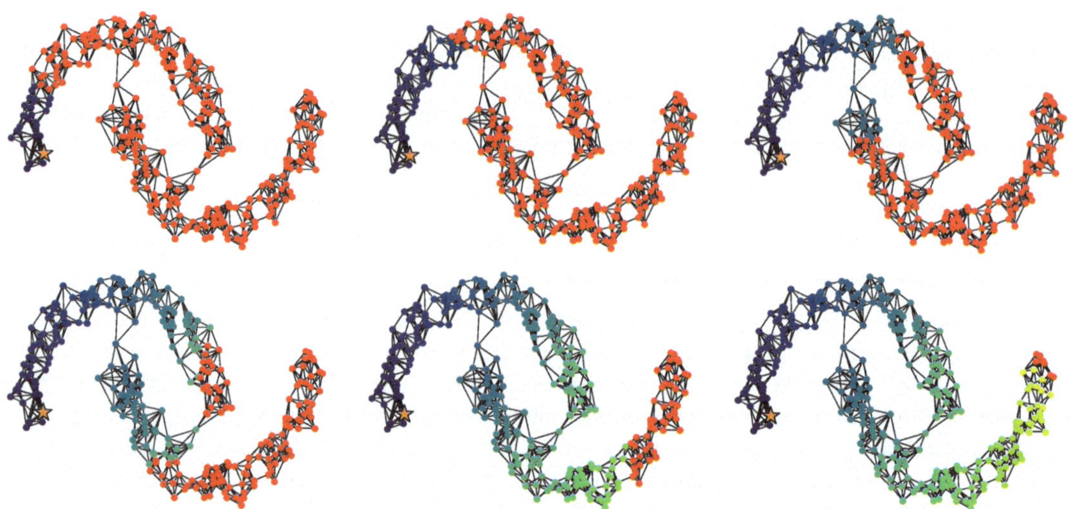

Figure 9.20: Illustration the dynamic programming iterations (9.53) for computing the distance vector $\mathbf{t}_i$ to the i-th node, shown as the orange star in the figure. The current iterate $\mathbf{s}_k$ is ∞ at the red nodes, while the other colors indicate the finalized values of the distance vector. The 6 figures show the iterations $k = 5, 10, 15, 20, 28, 34$ from left to right and top to bottom. The iterations converged on step $k = 35$.

essentially along shortest paths in the graph. Here, $m = 300$ and $\#\mathcal{D} = 1$. Convergence took 38 steps — much less than $m - \#\mathcal{D} = 299$.

Remark 9.43. (*Computational complexity*) If the graph $\mathcal{G}$ has e edges, then each iteration of the dynamic programming principle (9.53) takes $\mathrm{O}(e)$ operations. Thus, the computational complexity for computing $\mathbf{t}_\mathcal{D}$ is at most $\mathrm{O}\big(e(m - \#\mathcal{D})\big)$. The complexity of computing the entire distance matrix T in this manner is thus $\mathrm{O}(m^2 e)$. There are various simple tricks for accelerating convergence. One such method is to use a so-called Gauss–Seidel iteration[13], where instead of updating all of the entries of $\mathbf{s}_{k+1}$ at the same time using (9.53), we update them sequentially from $i = 1, \ldots, m$, and use the updated value at node i when updating node $i + 1$. This is no more computationally intensive than (9.53), but allows information to propagate faster on the graph, and hence convergence often takes place more quickly. However, proving the convergence is faster is challenging, as it depends heavily on the order in which the vertices are updated — often this is done at random, requiring probabilistic techniques, and there are other computational issues, such as the ability to parallelize code, which may be impacted. We refer the reader to Exercise 5.9. The Gauss–Seidel version of the iteration is closely related to the celebrated Floyd–Warshall algorithm, which was published by Robert Floyd in 1962 [75], but had been discovered independently by Bernard Roy in 1959 [200] and by Stephen Warshall in 1962 [244] in slightly different settings. The Floyd–Warshall algorithm can compute the entire distance matrix T in $\mathrm{O}(m^3)$ computations.   ▲

Dijkstra's Algorithm

If we organize the computations in a more efficient manner, we can obtain a much faster algorithm known as *Dijkstra's algorithm* for computing shortest path distances on graphs,

---

[13]The Gauss–Seidel scheme is a basic iterative method used to solve systems of linear algebraic equations. Details can be found in [181].

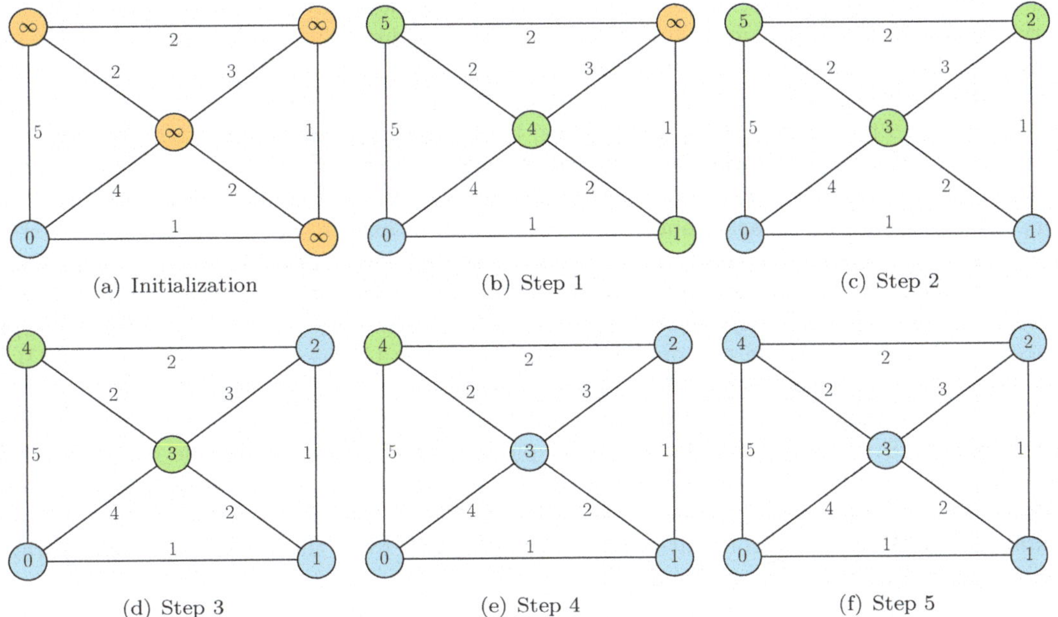

(a) Initialization

(b) Step 1

(c) Step 2

(d) Step 3

(e) Step 4

(f) Step 5

Figure 9.21: Example of Dijkstra's algorithm on a toy graph with 5 vertices to compute the shortest path graph distance to the bottom left node.

which was invented by computer scientist Edsger W. Dijkstra in 1959 [60]. The key insight of Dijkstra is to visit the nodes in the *correct order*, from closest to furthest in the graph distance; that way, each node will be visited exactly once. This is not a trivial matter to accomplish, since the distances are unknown beforehand — they are exactly the quantities we wish to compute! Below, we briefly explain how Dijkstra's algorithm works, and refer to [228] for more details.

Dijstra's algorithm splits the nodes in the graph into three sets: *finalized nodes*, *active nodes*, and *unvisited nodes*. Initially, the finalized nodes are exactly those belonging to $\mathcal{D}$, and their distances are *finalized* to be exactly zero, as they should be. The rest of the nodes are designated *unvisited*, and they each get assigned a distance of ∞. We then start the algorithm by designating all unvisited nodes that are adjacent to a node in $\mathcal{D}$ as *active nodes*, and removing them from the unvisited set. Each of these active nodes gets an initial guess for its shortest path distance, which is just the smallest value of w_{ij}^{-1} among all edges that connect the node to $\mathcal{D}$. This can be considered as the first step of the algorithm. The main part of Dijkstra's algorithm iterates the following steps until all nodes have been finalized: (1) the active node i with the smallest guess t_i for its distance has this distance finalized and is moved to the finalized set; (2) all unvisited graph neighbors j of node i are designated as active nodes, and their initial distances are set to $t_i + w_{ij}^{-1}$, which would be the distance to j if the optimal path went through node i; and then (3) all active nodes j that are neighbors of node i have their distances updated to $t_i + w_{ij}^{-1}$, *provided* this quantity is smaller than their current distance t_j.

Figure 9.21 gives an example of Dijkstra's algorithm applied to a toy graph with 5 nodes, where the set $\mathcal{D}$ contains just the bottom left node. The travel times w_{ij}^{-1} across each edge are depicted in the figures. The blue nodes are finalized, the green nodes are active, and the orange nodes are unvisited. The numerical value on each node is either the best current guess of the distance to that node, or the finalized distance value when it is blue. Notice how the shortest path to the top left node travels around the outside of the graph, and hence its value of 4 is not correct until the 3rd step of Dijkstra's algorithm.

Dijkstra's algorithm works because the order in which the nodes are finalized coincides with their distance to the source nodes in $\mathcal{D}$. Thus, these distances can be finalized without considering paths that travel through the unvisited nodes, which necessarily have longer length. We do not give a full proof of correctness and instead refer to [228]. Dijkstra's algorithm visits every node outside of $\mathcal{D}$ exactly once, so it terminates in exactly $m - \#\mathcal{D}$ steps. Each step requires a bit of work to determine which active node has the smallest distance. Using a data structure called a *heap*, this active node can be retrieved in $\mathrm{O}\big(\log(m - \#D)\big)$ computational steps [228]. Furthermore, Dijkstra's algorithm visits every edge at most once in a digraph (twice in a graph), thus the computational complexity is $\mathrm{O}\big(e + (m - \#\mathcal{D})\log(m - \#\mathcal{D})\big)$. Note that this is significantly faster than the dynamic programming iteration (9.53).

Note that to compute the entire distance matrix T using Dijkstra's algorithm would require $\mathrm{O}(me + m^2\log m)$ computational time. For sparse graphs, where each node has at most k edges, we have $e \le km$ and so the complexity is $\mathrm{O}(km^2\log m)$. If k is small, such as when working with similarity graphs like a k-nearest neighbor graphs, then the computational complexity of Dijkstra's algorithm is not much different from the $\mathrm{O}(m^2)$ memory storage required for storing the $m \times m$ distance matrix T.

Remark 9.44. Both Dijkstra's algorithm and the Gauss–Seidel version of dynamic programming iteration have inspired variants, called fast marching [212] and fast sweeping [263], respectively, for solving a partial differential equation called the *eikonal equation*, which is closely related to the computation of distance functions on subsets of Euclidean space. ▲

9.5.2 Computing Shortest Paths via Dynamic Programming

It is important to point out that the dynamic programming iteration provides an algorithm for computing the distance vector $\mathbf{t}_{\mathcal{D}}$, and hence the entire distance matrix T, *without* explicitly computing *any* optimal paths between *any* pairs of nodes in the graph. This is quite remarkable, since the distance matrix T is *defined* by the problem of finding optimal paths, and we have, up until now, done nothing of the sort. It turns out that once we have computed the distance vector $\mathbf{t}_{\mathcal{D}}$, it is straightforward to recover the optimal paths from *any* node j to the set $\mathcal{D}$ through a procedure also called *dynamic programming*. This procedure traces the optimal path *backwards* from a node j to the set $\mathcal{D}$, and proceeds as follows: The first node in the path is $i_0 = j$. Then, given i_k, the next node i_{k+1} in the path is chosen to be any node j that minimizes the right hand side of the dynamic programming principle (9.52). That is, setting $\mathbf{t} = \mathbf{t}_{\mathcal{D}}$, the path generated by dynamic programming satisfies

$$t_{i_k} = t_{i_{k+1}} + w^{-1}_{i_{k+1},i_k}. \tag{9.54}$$

Notice we are traversing the edges backwards in the path, since w_{i_{k+1},i_k} is the edge weight from i_{k+1} to i_k, and not the other way around. We note that dynamic programming is guaranteed to stop in a finite number of steps p, since by (9.54) we have $t_{i_{k+1}} < t_{i_k}$, and so we can never visit a node twice. In particular, dynamic programming stops when $i_p \in \mathcal{D}$ for some $p \ge 1$, and generates a path $\mathbf{p}$ with nodes $i_p, i_{p-1}, \ldots, i_1, i_0 = j$ that goes from $\mathcal{D}$ to j and is guaranteed to be optimal.

Lemma 9.45. *The path* $\mathbf{p}$ *constructed by dynamic programming* (9.54) *starting at node* j *is optimal, that is* $\tau(\mathbf{p}) = t_j$.

Proof. We rewrite (9.54) to read

$$t_{i_k} - t_{i_{k+1}} = w^{-1}_{i_{k+1}, i_k}$$

and sum from $k = 0$ to $k = p - 1$. The left hand side telescopes and we obtain

$$t_{i_0} - t_{i_p} = \sum_{k=0}^{p-1} w^{-1}_{i_{k+1}, i_k} = \tau(\mathbf{p}).$$

Since $i_0 = j$ and $i_p \in \mathcal{D}$, so $t_{i_p} = 0$, we have $t_j = \tau(\mathbf{p})$, which completes the proof. ∎

Remark 9.46. Equation (9.52) is the key identity involved in dynamic programming, which justifies the choice of its name. Dynamic programming is, in general, used to great effect in optimal control theory; see [13]. The shortest path problem is, in fact, a very special case, where the control is the choice of the *next* node in the path, and the objective is to find the shortest (i.e., optimal) path. ▲

Remark 9.47. It also is possible to view dynamic programming as gradient descent on the distance vector $\mathbf{t}$. This is easiest to see when the graph is unweighted, so $W = A$ is an adjacency matrix, i.e., $w_{ij} = 1$ whenever there is an edge from i to j, and $w_{ij} = 0$ otherwise. In this case, given i_k, we chose i_{k+1} to minimize t_j over j with $w_{j,i_k} > 0$; that is, dynamic programming chooses the neighbor of i_k with the smallest distance to $\mathcal{D}$, or rather, the neighbor that is closest to $\mathcal{D}$. This is analogous to how gradient descent moves in the direction that decreases an objective function most rapidly. ▲

9.5.3 ISOMAP and Metric Multidimensional Scaling

Let $\mathcal{G}$ be a weighted graph, so $W = W^T$ is symmetric, and let T be the associated *symmetric* distance matrix. The metric space structure induced by T finds immediate applications to graph embeddings and visualizations. In particular, we can embed the graph into Euclidean space by seeking an *isometric embedding* of the nodes in the graph, which is a collection of vectors $\mathbf{x}_1, \ldots, \mathbf{x}_m \in \mathbb{R}^k$, with $k \geq 1$, such that

$$t_{ij} = \| \mathbf{x}_i - \mathbf{x}_j \|. \tag{9.55}$$

This is exactly the classical multidimensional scaling (MDS) problem studied in Section 8.5. By Theorem 8.22 we know that an isometric embedding exists if and only if the centered matrix $H = -\frac{1}{2} J (T \circ T) J$ is positive semidefinite, where $T \circ T$ is the Hadamard product of T with itself, which has entries t_{ij}^2, and J is the $m \times m$ centering matrix defined in (7.5). Furthermore, by Corollary 8.23, when an isometric embedding exists, we can easily construct the embedding points in $\mathbb{R}^k$, where $k = \operatorname{rank} H$, through the spectral decomposition of H. To embed the points into a lower dimensional space, say $\mathbb{R}^2$ for visualization, the classical MDS algorithm simply restricts the spectral decomposition to the top 2 eigenvectors, in order to construct an embedding with the least amount of distortion in (9.55).

Now, it is natural to wonder whether the centered matrix H is always positive semidefinite when T defines a metric, that is, the triangle inequality in Lemma 9.40 holds. Certainly whenever (9.55) holds, the matrix T must satisfy the triangle inequality; see Exercise 5.2. The question is whether the converse statement is true; does the triangle inequality imply the existence of an isometric embedding, or equivalently, can any finite metric space be embedded into Euclidean space?

Clearly any $m = 2$ point metric space can be trivially embedded in $\mathbb{R}$, and it is possible to show (see Exercise 5.10) that any $m = 3$ point metric space can be embedded in $\mathbb{R}^2$ as the

(a) Two Moons   (b) Circles

Figure 9.22: ISOMAP embeddings for the two moons and circles data sets. Both embeddings are linearly separable, while the original point clouds were not.

vertices of a triangle with appropriate side lengths. However, it turns out that as soon as we get to an $m = 4$ point metric space, we can no longer expect to find exact embeddings into *any* Euclidean space $\mathbb{R}^n$. Hence, the centered matrix H may fail to be positive semidefinite.

Example 9.48. Consider the $m = 4$ point metric space with distances $t_{12} = t_{13} = t_{14} = 1$ and $t_{ij} = 2$ for $i, j \geq 2$ and $i \neq j$. Thus, the first point is distance 1 from the remaining three points, which are equidistant from each other. The reader should check that these distances satisfy the triangle inequality, and hence define a metric (see Exercise 5.7). We claim that there does not exist $\mathbf{x}_1, \mathbf{x}_2, \mathbf{x}_3, \mathbf{x}_4 \in \mathbb{R}^k$ for *any* $k \geq 1$ such that $\|\mathbf{x}_i - \mathbf{x}_j\| = t_{ij}$. Indeed, suppose such an embedding exists. Without loss of generality we may take $\mathbf{x}_1 = \mathbf{0}$. Then since $1 = \|\mathbf{x}_i - \mathbf{x}_1\| = \|\mathbf{x}_i\|$ for $i = 2, 3, 4$, the remaining three points lie on the unit sphere $S_1 = \{\|\mathbf{x}\| = 1\}$, and form an equilateral triangle with side lengths 2, since $\|\mathbf{x}_i - \mathbf{x}_j\| = 2$ for $i, j \geq 2$ and $i \neq j$. However, according to Exercise 5.8, the largest equilateral triangle that can be inscribed in the unit sphere in *any* dimension has side lengths $\sqrt{3} < 2$, which is a contradiction. $\blacktriangle$

Since we cannot always expect to find an isometric embedding of the graph with zero distortion, we resort to the classical MDS algorithm developed in Section 8.5, which finds the embedding into $\mathbb{R}^k$, for visualization $k = 2$ or $k = 3$, with the least amount of distortion. We briefly recall here that the embedding vectors $\mathbf{x}_i^T$ are the rows of the data matrix

$$X = P_k (\Lambda_k^+)^{1/2},$$

where the columns of $P_k = (\mathbf{p}_1 \cdots \mathbf{p}_k)$ are the top k eigenvectors of H, and the diagonal entries of $\Lambda_k^+ = \mathrm{diag}(\lambda_1^+, \ldots, \lambda_k^+)$ are the positive parts of the top k eigenvalues, where we sort the eigenvalues, and eigenvectors, of H from largest to smallest. When the distance matrix T is obtained from a nearest neighbor or ε-ball graph over a high dimensional point cloud with inverse distance weights, as described in Remark 9.38, so that path lengths are exactly the Euclidean lengths of path, this algorithm is known as ISOMAP, which stands for finding an isometric mapping [229]. In the more general setting of embedding metric spaces, this process is known as *metric multidimensional scaling*, or metric MDS, while when the process is applied to pairwise similarities that do not necessarily define a metric, as we did in Section 8.5, it is called *non-metric MDS*.

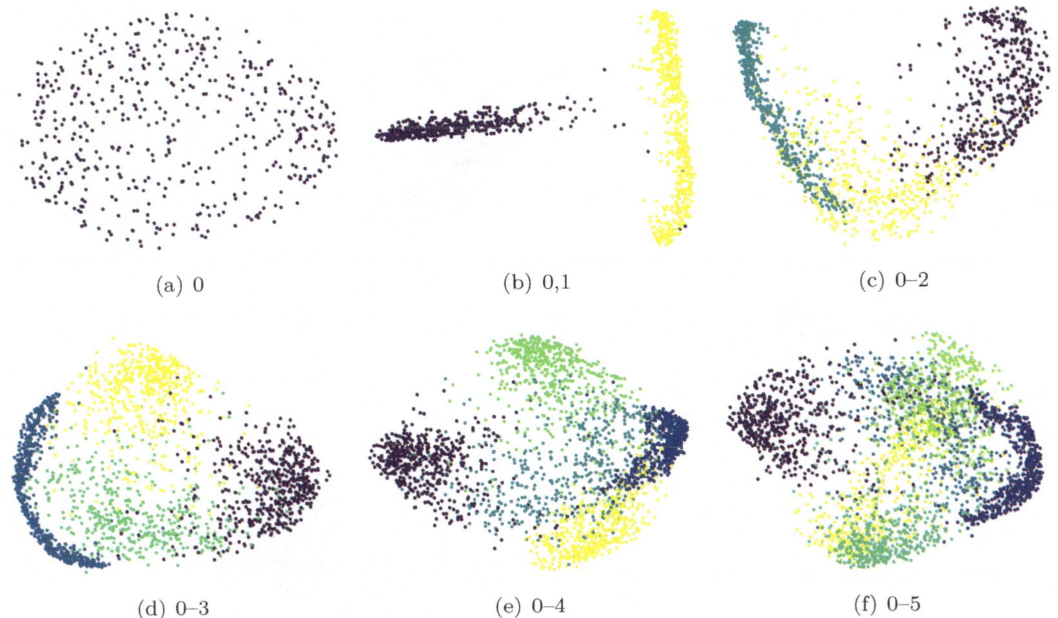

(a) 0   (b) 0,1   (c) 0–2

(d) 0–3   (e) 0–4   (f) 0–5

Figure 9.23: Six plots of ISOMAP embeddings of the MNIST data set. The plots are colored by the underlying digit label, starting with just the zeros, and incrementally adding digits up to 5.

We proceed to show some experimental results. In Figure 9.22 we show the ISOMAP algorithm applied to the two moons and circles data sets. Here, we used a graph with inverse distance weights and connected points within distance $\varepsilon = 0.25$ by edges. We see that in both cases, the ISOMAP algorithm produces an embedding that linearly separates the two clusters, so that a classification algorithm like linear SVM, as developed in Section 7.3, would work well. The edges that are drawn in the figure correspond to the edges in the original graph constructed on each data set. It is interesting to note that in the two circles case, the circular structure of the outer circle is preserved, while the inner circle is flattened.

In Figure 9.23 we show the ISOMAP embeddings of various subsets of MNIST digits. Here, we used a symmetrized k-nearest neighbor graph with $k = 10$ and inverse distance weights. We restricted to a random subset of 5000 MNIST digits to make the computations tractable. We started with just the zero digit, and added one digit at a time up to digit 5. The digits are fairly well-clustered in the embedding up until adding the 5th digit, at which point we start to see significant overlap between many digits.

Finally, we turn to some applications of metric MDS. In Figure 9.24 we show the metric MDS embeddings of Zachary's karate club graph, Krebs' political books graph, and the PubMed graph. All three graphs are unweighted, so the length of a path is simply the number of edges. For the karate club and political books graphs, the figure shows the edges in the graph, and we can see that, generally speaking, adjacent nodes are embedded at nearby points. For the PubMed graph, we do not display the edges, as the figure becomes too cluttered and is difficult to interpret. The metric MDS embedding for the karate graph separates the two groups well. For the political books graph, the liberal and conservative books are well separated, which correspond to the purple and yellow nodes, respectively, while the neutral books are more evenly scattered. For PubMed, two of the classes are well separated, but there is substantial overlap between the yellow and purple classes.

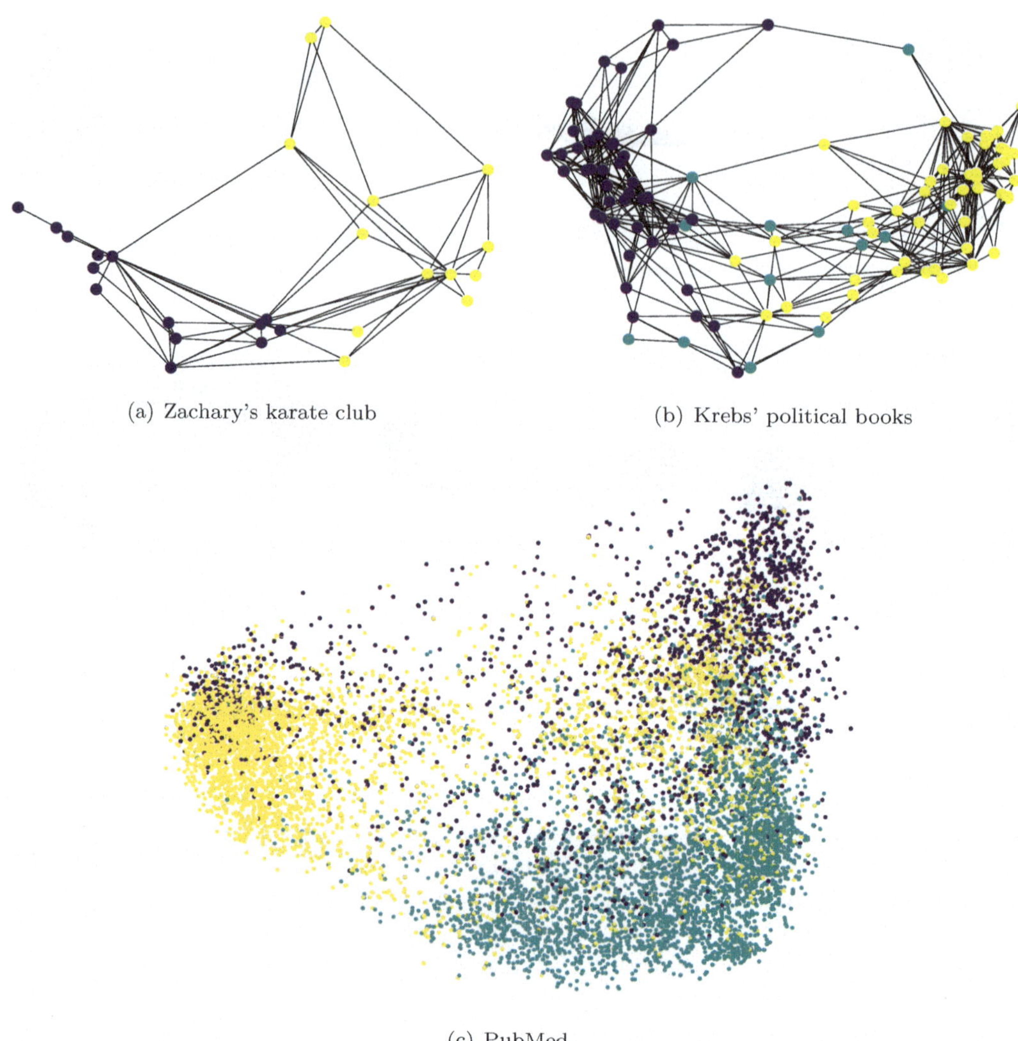

(a) Zachary's karate club

(b) Krebs' political books

(c) PubMed

Figure 9.24: Metric MDS for (a) Zachary's karate club graph, (b) Krebs' political books graph, and (c) the PubMed graph. For Krebs' political books, the purple nodes are liberal books while the yellow are conservative and the green are bipartisan.

Exercises

5.1. Prove Lemma 9.40.

5.2. ♡ Let $\mathbf{x}_1, \ldots, \mathbf{x}_m \in \mathbb{R}^n$, and let T be the $m \times m$ matrix with entries $t_{ij} = \|\mathbf{x}_i - \mathbf{x}_j\|$. Show that T satisfies the triangle inequality $t_{ik} \le t_{ij} + t_{jk}$ for all $1 \le i, j, k \le m$.

5.3. ♡ Show that the dynamic programming principle (9.52) can be reformulated as

$$t_i = 0 \quad \text{for} \quad i \in \mathcal{D}, \quad \text{and} \quad \max_j w_{ji}(t_i - t_j) = 1 \quad \text{for} \quad i \in \mathcal{D}^c. \tag{9.56}$$

5.4. ♡ Let $\mathcal{G} = (\mathcal{N}, \mathcal{E})$ be a disconnected graph and $\mathcal{D} \subset \mathcal{N}$. Suppose there is a connected component $\mathcal{H} \subset \mathcal{N}$ for which $\mathcal{H} \cap \mathcal{D} = \varnothing$. Show that there is no solution $\mathbf{t} \in \mathbb{R}^m$ to the dynamic programming principle (9.52). *Hint:* Suppose a solution $\mathbf{t}$ exists, and consider the node i that minimizes t_i over $i \in \mathcal{H}$. It is easier to work with the reformulation (9.56).

5.5. ♡ Let $\mathcal{G} = (\mathcal{N}, \mathcal{E})$ be a connected graph with m nodes, and let $\mathcal{D} \subset \mathcal{N}$. Assume that $\mathbf{u}, \mathbf{v} \in \mathbb{R}^m$ satisfy the reformulated dynamic programming principle (9.56) with inequalities

$$\max_{j} w_{ji}(u_i - u_j) \leq 1 \leq \max_{j} w_{ji}(v_i - v_j)$$

for all $i \in \mathcal{D}^c$, and suppose that $u_i \leq 0 \leq v_i$ for $i \in \mathcal{D}$. Show that $\mathbf{u} \leq \mathbf{t}_\mathcal{D} \leq \mathbf{v}$. *Hint:* Use a similar argument to the proof of Lemma 9.41.

5.6. ◇ Let $\mathcal{G}$ be a graph with m nodes and weight matrix W. Let $\widetilde{\mathcal{G}}$ be a graph with m nodes and weight matrix $\widetilde{W}$ satisfying

$$(1 - \varepsilon)\, w_{ij} \leq \widetilde{w}_{ij} \leq (1 + \varepsilon)\, w_{ij}$$

for all i, j and some $0 \leq \varepsilon < 1$. Let T be the distance matrix for $\mathcal{G}$ and $\widetilde{T}$ be the distance matrix for $\widetilde{\mathcal{G}}$. (a) Show that

$$(1 + \varepsilon)^{-1} t_{ij} \leq \widetilde{t}_{ij} \leq (1 - \varepsilon)^{-1} t_{ij}.$$

This shows that the distance matrix has some degree of robustness under perturbations of the graph. (b) Show that part (a) is not true for additive perturbations. In particular, give an example where $w_{ij} - \frac{1}{2} \leq \widetilde{w}_{ij} \leq w_{ij} + \frac{1}{2}$, but the distance matrices T and $\widetilde{T}$ differ by far more than $\frac{1}{2}$. *Hint:* Find an example of an unweighted graph with m nodes for which adding a single edge to the graph changes one of the distances t_{ij} by $m - 3$.

5.7. Verify that the distances t_{ij} defined over $m = 4$ points in Example 9.48 satisfy the triangle inequality.

5.8. ◇ Consider three points $\mathbf{x}_1, \mathbf{x}_2, \mathbf{x}_3 \in \mathbb{R}^n$ that form an equilateral triangle with side length $h > 0$ centered at the origin, so $\|\mathbf{x}_i - \mathbf{x}_j\| = h$ for $i \neq j$, and $\mathbf{x}_1 + \mathbf{x}_2 + \mathbf{x}_3 = \mathbf{0}$. Show that $\|\mathbf{x}_i\| = h / \sqrt{3}$ for $i = 1, 2, 3$. Thus, the largest equilateral triangle that can be inscribed in the unit sphere has side lengths $\sqrt{3}$.

5.9. ◇ The Python notebook from this section will be helpful when solving the following multipart exercise. (a) Run the dynamic programming iterations in Python on the PubMed graph to find the distance function to a particular node. How many iterations does it take to converge? (b) Modify the code to use the Gauss-Seidel method. Does Gauss-Seidel converge faster? If so, by how much? (c) In the Gauss-Seidel method, iterate over the nodes of the graph in order of the value of the distance vector, starting from smallest to largest. You can compute the distance vector using **graphlearning** for this, and use **numpy.argsort**. Can you get Gauss-Seidel to converge in one iteration?

5.10. Show that (a) any two-point metric space can be isometrically embedded into $\mathbb{R}$;
(b) any thee-point metric space can be isometrically embedded into $\mathbb{R}^2$.

9.6   Diffusion on Graphs and Digraphs

Many applications of graphs in data science, machine learning, engineering, and the natural sciences involve processes that propagate matter (or, equivalently, information, population, heat, etc.), from each node to its neighbors through a process called *diffusion*. In population dynamics, species diffuse spatially and interact with other species and their environment; in physics, heat diffuses between atoms in a material; in chemistry chemical concentrations diffuse (as well as react). There is an extensive field within probability studying random walks, where a probability distribution diffuses on the graph. Diffusion also turns out to have very important applications in visualizing, simplifying, and interpreting data, which we will explore. In this section, we introduce and study the mathematics of diffusion on graphs and digraphs. While most of the analysis is restricted to graphs, since they possess symmetric positive semidefinite Laplacian matrices, the basic set-up carries through for digraphs. A particularly important application of the latter is Google's PageRank algorithm, in which the digraph is based on connections between webpages on the internet.

We begin by formulating the diffusion process on a general digraph. Let $\mathcal{G}$ be a weighted digraph with m nodes, weight matrix W, which need not be symmetric, and diagonal degree matrix D. We assume that every node has at least one outgoing edge, so that D is invertible. Let $\mathbf{x} \in \mathbb{R}^m$ be a vector whose components are nonnegative, $x_i \geq 0$, that we think of as representing the amount of some material located at node i on the graph. In the applications described above, the interpretation of x_i could be the population of a particular species at location i, the temperature of a material at location i, the concentration of a chemical solvent, or the probability that a random walker can be found at node i. To fix notation, for the rest of the section we will refer to x_i as the amount of *mass* located at node i. The *total mass* is thus $\mathbf{x} \cdot \mathbf{1} = x_1 + \cdots + x_m$.

Diffusion proceeds by repeatedly redistributing the mass from each node to its neighbors on the graph, with the amount of mass sent from node i to node j being proportional to the edge weight w_{ij}. In particular, node i sends $w_{ij} d_i^{-1} x_i$ amount of mass to node j. Note that

$$\sum_{j=1}^{m} w_{ij} d_i^{-1} x_i = d_i d_i^{-1} x_i = x_i,$$

and hence node i is simply redistributing *all* of its mass to its neighbors. In the random walk interpretation of diffusion, the random walker moves from node i to node j with probability $d_i^{-1} w_{ij}$, and so the probability mass of the random walker is being distributed to its neighbors.

Let $\mathbf{y} \in \mathbb{R}^m$ denote the mass distribution after diffusing $\mathbf{x}$ for one step. Then the mass y_j at node j is the sum of the mass sent from each of its neighbors; that is,

$$y_j = \sum_{i=1}^{n} w_{ij} d_i^{-1} x_i, \qquad j = 1, \ldots, m. \tag{9.57}$$

We can write this system of equations in vector form as

$$\mathbf{y} = W^T D^{-1} \mathbf{x} = P \mathbf{x}, \tag{9.58}$$

where the matrix $P = W^T D^{-1}$ is a *transition matrix*, since its columns sum to one, cf. (5.80). Indeed,

$$\mathbf{1}^T P = \mathbf{1}^T W^T D^{-1} = \mathbf{d}^T D^{-1} = \mathbf{1}^T.$$

This directly implies *conservation of mass*, that is, if $\mathbf{y} = P\mathbf{x}$, then

$$\mathbf{1} \cdot \mathbf{y} = \mathbf{1}^T P \mathbf{x} = \mathbf{1}^T \mathbf{x} = \mathbf{1} \cdot \mathbf{x},$$

and so mass is neither created nor destroyed. This should be expected, since we derived the equation from a physical process of mass spreading around the graph. To make a connection to the graph Laplacian, notice we can write (9.58) as

$$\mathbf{y} = \mathbf{x} - (I - W^T D^{-1})\mathbf{x} = \mathbf{x} - (D - W)^T D^{-1}\mathbf{x} = \mathbf{x} - L^T D^{-1}\mathbf{x}.$$

This leads to the following definition.

Definition 9.49. The *random walk graph Laplacian matrix* is defined by

$$L_{\mathrm{rw}} = I - D^{-1}W = D^{-1}L, \tag{9.59}$$

where $L = D - W$ is the graph Laplacian matrix.

We can now rewrite the diffusion step (9.58) as

$$\mathbf{y} = \mathbf{x} - L_{\mathrm{rw}}^T \mathbf{x}.$$

If we start with initial mass distribution $\mathbf{x}_0$ and run the diffusion process indefinitely, we obtain a sequence of mass distributions $\mathbf{x}_0, \mathbf{x}_1, \mathbf{x}_2, \ldots$ satisfying

$$\mathbf{x}_{k+1} = \mathbf{x}_k - L_{\mathrm{rw}}^T \mathbf{x}_k. \tag{9.60}$$

The iterative system (9.60) constitutes one type of *diffusion* on a graph. The conservation of mass argument above shows that $\mathbf{x}_k \cdot \mathbf{1} = \mathbf{x}_0 \cdot \mathbf{1}$ for all $k \geq 1$, so the amount of mass is unchanged during diffusion — it simply spreads around as the process evolves.

The random walk graph Laplacian shares some similarities with the graph Laplacian. We note that since the degree matrix D is invertible, the graph Laplacian L and random walk Laplacian L_{rw} have the same kernel, so $\ker L = \ker L_{\mathrm{rw}}$. In particular, we always have $L_{\mathrm{rw}}\mathbf{1} = \mathbf{0}$. However, as we show below, their other eigenvectors and eigenvalues are in general different. The term random walk graph Laplacian in Definition 9.49 originates from the probabilistic interpretation of diffusion, where $\mathbf{x}_k$ is the probability mass distribution for a random walker after taking k steps of a random walk on the graph. To make this connection more clear, let us suppose that $\mathbf{x}_0$ is a probability vector, so its entries are nonnegative and sum to one — that is $\mathbf{x}_0 \cdot \mathbf{1} = 1$. By conservation of mass and positivity of all components of the matrix P, it follows that each $\mathbf{x}_k$ is also a probability vector. In other words, the diffusion process (9.60) is a Markov process; see equation (5.78) and the subsequent discussion.

We define the k-th *expectation* of a vector $\mathbf{v} \in \mathbb{R}^m$ with respect to the random walk by

$$\mathbb{E}_k(\mathbf{v}) = \mathbf{x}_k \cdot \mathbf{v} = \sum_{i=1}^{m} x_{k,i} v_i,$$

which is simply the weighted average of the components v_i of $\mathbf{v}$ against the probability distribution $\mathbf{x}_k$ of the random walker. Then the expected change in $\mathbf{v}$ from step k to $k+1$ is given by

$$\mathbb{E}_{k+1}(\mathbf{v}) = \mathbb{E}_k(\mathbf{v}) - \mathbb{E}_k(L_{\mathrm{rw}}\mathbf{v}). \tag{9.61}$$

Indeed, we simply take expectations on both sides of (9.60), i.e., dot both sides with $\mathbf{v}$, and use that

$$(L_{\mathrm{rw}}^T \mathbf{x}_k) \cdot \mathbf{v} = (L_{\mathrm{rw}}^T \mathbf{x}_k)^T \mathbf{v} = \mathbf{x}_k^T L_{\mathrm{rw}} \mathbf{v} = \mathbf{x}_k \cdot (L_{\mathrm{rw}}\mathbf{v}) = \mathbb{E}_k(L_{\mathrm{rw}}\mathbf{v}).$$

This means that the random walk graph Laplacian L_{rw} is the *generator* of the random walk, or diffusion, on the graph, hence the name. In fact, the diffusion process (9.60) can be equivalently defined by asking that the condition (9.61) holds for all $\mathbf{v} \in \mathbb{R}^m$.

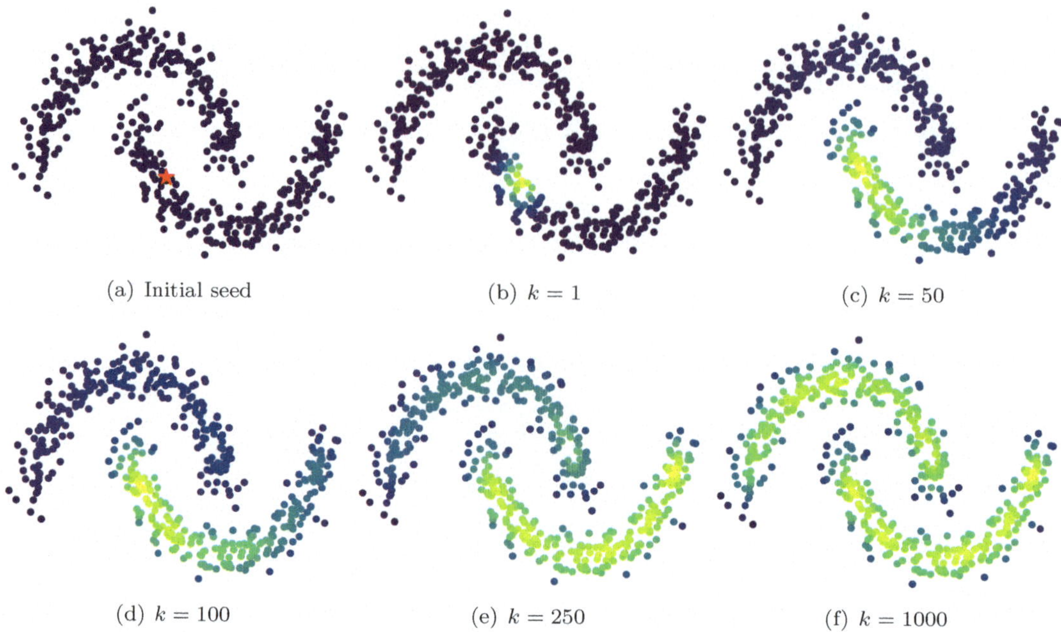

(a) Initial seed   (b) $k = 1$ (c) $k = 50$

(d) $k = 100$ (e) $k = 250$ (f) $k = 1000$

Figure 9.25: Example of diffusion of a unit of mass in the two moons data set concentrated on the red star in (a) over 1000 steps of diffusion. The mass spreads first over the lower cluster that the seed point belongs to, and then eventually to both clusters.

We show in Figure 9.25 an example of graph diffusion on the two moons data set starting with all of the mass concentrated on a single seed node, i.e., $\mathbf{x}_0 = \mathbf{e}_i$ for some i, which is shown as a red star in Figure 9.25(a). We see that the diffusion initial spreads quickly in the cluster that the seed node belongs to, and then over longer time scales spreads to the other cluster. In this example, we used a geometric graph with weights $w_{ij} = \exp\!\left(-\|\mathbf{x}_i - \mathbf{x}_j\|^2 / (2\varepsilon^2)\right)$ for the particular value $\varepsilon = 0.15$.

While we derived the diffusion equation (9.60) under the assumption that the components of $\mathbf{x}_0$, and hence those of $\mathbf{x}_k$, are nonnegative, the diffusion equation makes sense for any initial vector $\mathbf{x}_0 \in \mathbb{R}^m$, and so we drop the positivity assumption from now on.

A closely related process is governed by the *adjoint diffusion equation*

$$\mathbf{z}_{k+1} = \mathbf{z}_k - L_{\mathrm{rw}}\,\mathbf{z}_k, \tag{9.62}$$

where we have simply replaced L_{rw}^T by L_{rw} in (9.60). In fact, if $\mathbf{x}_k$ satisfies the diffusion equation (9.60) and we set $\mathbf{z}_k = D^{-1}\mathbf{x}_k$, then, provided $W = W^T$ is symmetric, the vectors $\mathbf{z}_k$ solve the adjoint equation (9.62) (see according to Exercise 6.1). They are also related by the following lemma.

Lemma 9.50. *Suppose that $\mathbf{x}_k$ satisfy the diffusion equation (9.60) and $\mathbf{z}_k$ satisfy the adjoint diffusion equation (9.62). Then $\mathbf{x}_k \cdot \mathbf{z}_0 = \mathbf{x}_0 \cdot \mathbf{z}_k$ for all $k \geq 0$.*

Proof. Since we can write (9.60) as $\mathbf{x}_{k+1} = (I - L_{\mathrm{rw}}^T)\,\mathbf{x}_k$,

$$\mathbf{x}_k = \left(I - L_{\mathrm{rw}}^T\right)^k \mathbf{x}_0 = \left[\left(I - L_{\mathrm{rw}}\right)^k\right]^T \mathbf{x}_0. \tag{9.63}$$

Likewise we can write $\mathbf{z}_k = (I - L_{\mathrm{rw}})^k \mathbf{z}_0$, and therefore

$$\mathbf{x}_k \cdot \mathbf{z}_0 = \mathbf{x}_k^T \mathbf{z}_0 = \mathbf{x}_0^T (I - L_{\mathrm{rw}})^k \mathbf{z}_0 = \mathbf{x}_0 \cdot \mathbf{z}_k. \qquad \blacksquare$$

While the adjoint diffusion equation (9.62) is directly related to the original diffusion equation (9.60), it has important differences. First, it does not in general conserve mass. For example, let us suppose that $\mathbf{z}_0 = \mathbf{e}_i$, so we start with mass concentrated on the i-th node. Then Lemma 9.50 with $\mathbf{x}_0 = \mathbf{1}$ says that the total mass at the k-th step is

$$\mathbf{z}_k \cdot \mathbf{1} = \mathbf{x}_k \cdot \mathbf{z}_0 = \mathbf{x}_k \cdot \mathbf{e}_i = x_{k,i},$$

while $\mathbf{z}_0 \cdot \mathbf{1} = 1$. So, if (9.62) were to preserve mass, then, starting with $\mathbf{x}_0 = \mathbf{1}$, we would need to have $\mathbf{x}_k = \mathbf{1}$ for all $k \geq 0$, which does not hold in general; see Exercise 6.4.

Instead, it turns out the adjoint diffusion equation (9.62) satisfies the following *maximum principle*.

Proof. We note that

$$\mathbf{z}_{k+1} = \mathbf{z}_k - L_{\mathrm{rw}} \mathbf{z}_k = D^{-1} W \mathbf{z}_k,$$

and so

$$z_{k+1,i} = \frac{1}{d_i} \sum_{j=1}^m w_{ij} z_{k,j} \leq \frac{1}{d_i} \left(\max_j z_{k,j} \right) \sum_{j=1}^m w_{ij} = \max_j z_{k,j}.$$

By induction we obtain $z_{k,i} \leq \max z_{0,j}$. The other inequality $z_{k,i} \geq \min z_{0,j}$ is proved by a similar argument. $\blacksquare$

Remark. The maximum principle shows that the components of $\mathbf{z}_k$ cannot exceed the maximum or minimum values of the initial distribution $\mathbf{z}_0$. The same result is not true for the diffusion equation (9.60) satisfied by $\mathbf{x}_k$. For example, consider the graph with $m = 3$ nodes and weight matrix

$$W = \begin{pmatrix} 0 & 1 & 0 \\ 1 & 0 & 1 \\ 0 & 1 & 0 \end{pmatrix}. \qquad (9.64)$$

This is simply a graph where nodes 1 and 2 are connected by an edge, as are nodes 2 and 3. If $x_0 = (1, 0, 1)^T$, so we initially have unit mass at nodes 1 and 3, but no mass at node 2, then after one step of diffusion we have $\mathbf{x}_1 = (0, 2, 0)^T$, which violates the maximum principle. $\blacktriangle$

The reader should note that our discussion has not assumed symmetry of the weight matrix, and all diffusion equations thus far are posed on digraphs. This observation will be important later in Section 9.6.1, when we introduce the PageRank algorithm.

For the rest of this section, we work with a weighted graph $\mathcal{G}$ with symmetric weight matrix $W = W^T$ — that is, the graph is no longer directed. Unless the degree matrix D is a multiple of the identity, which means that all nodes in the graph have the same degree, then the random walk graph Laplacian matrix $L_{\mathrm{rw}} = D^{-1} L$ is not symmetric. However, according to Proposition 4.19, it is *self-adjoint* with respect to the degree inner product $\langle \mathbf{x}, \mathbf{y} \rangle_D = \mathbf{x}^T D \mathbf{y}$. Nevertheless, it is sometimes convenient to work with a symmetric version of the graph Laplacian, which we now introduce.

Definition 9.52. The *symmetric normalized graph Laplacian matrix* is defined by

$$L_{\mathrm{sym}} = I - D^{-1/2} W D^{-1/2} = D^{-1/2} L D^{-1/2}, \tag{9.65}$$

where L is the graph Laplacian matrix.

Both L_{sym} and L_{rw} are closely related to the graph Laplacian L, and we can thus immediately characterize their kernels as $\ker L_{\mathrm{rw}} = \ker L$ and

$$\ker L_{\mathrm{sym}} = \left\{ D^{1/2} \mathbf{x} \;\middle|\; \mathbf{x} \in \ker L \right\}.$$

Thus, by Theorem 9.22, nullity L_{sym} and nullity L_{rw} both equal the number of connected components in the underlying graph, as they do for the graph Laplacian L. The reader should be cautioned, however, that, aside from their kernels, there is no simple relationship between the other eigenspaces of L_{sym} and L. Notice also that Proposition 9.16 implies

$$\mathbf{x}^T L_{\mathrm{sym}} \mathbf{x} = (D^{-1/2}\mathbf{x})^T L D^{-1/2} \mathbf{x} = \frac{1}{2} \sum_{i,j=1}^{m} w_{ij} \left(\frac{x_i}{\sqrt{d_i}} - \frac{x_j}{\sqrt{d_j}} \right)^2. \tag{9.66}$$

It follows that L_{sym} is positive semidefinite.

We now turn to the long time behavior of graph diffusions, that is, we wish to know what happens as the number of steps $k \to \infty$. This requires studying the spectral decomposition of L_{sym} and L_{rw}. Throughout the rest of this section, let $\mathbf{q}_1, \ldots, \mathbf{q}_m \in \mathbb{R}^m$ be orthonormal eigenvectors of L_{sym} with corresponding eigenvalues $0 \le \lambda_1 \le \lambda_2 \le \cdots \le \lambda_m$ arranged in increasing order. The matrix L_{rw} is self-adjoint, so it also has an orthonormal — in the degree inner product $\langle \cdot, \cdot \rangle_D$ — basis of eigenvectors. It turns out that L_{rw} and L_{sym} have the same eigenvalues and their eigenvectors are simply related.

Lemma 9.53. *The vector $\mathbf{p}_i = D^{-1/2}\mathbf{q}_i$ is an eigenvector of L_{rw} with eigenvalue λ_i. Moreover, $\mathbf{p}_1, \ldots, \mathbf{p}_m$ are orthonormal with respect to the degree inner product.*

The proof of Lemma 9.53 is deferred to Exercise 6.2. Since L_{sym} and L_{rw} have the same eigenvalues and L_{rw} is positive definite, it follows that the random walk graph Laplacian L_{rw} is self adjoint and positive definite. Therefore, by Theorem 5.35 we have the spectral decomposition

$$L_{\mathrm{rw}} = \sum_{i=1}^{m} \lambda_i \mathbf{p}_i \mathbf{p}_i^T D, \tag{9.67}$$

which we can use to explicitly write the solution of the diffusion equation (9.60). The result is simplest to state for the adjoint diffusion equation (9.62); we leave the other cases to Exercises 6.5 and 6.6.

Theorem 9.54. *Let $\mathbf{z}_k$ satisfy the adjoint diffusion equation (9.62). Then*

$$\mathbf{z}_k = \sum_{i=1}^{m} (1 - \lambda_i)^k \langle \mathbf{p}_i, \mathbf{z}_0 \rangle_D \, \mathbf{p}_i. \tag{9.68}$$

Proof. As in (9.67), we write out the spectral decomposition

$$I - L_{\text{rw}} = \sum_{i=1}^{m} (1 - \lambda_i)\, \mathbf{p}_i\, \mathbf{p}_i^T D.$$

Recalling the formula for $\mathbf{z}_k$ from the proof of Lemma 9.50, and taking powers as in (5.33), produces

$$\mathbf{z}_k = (I - L_{\text{rw}})^k \mathbf{z}_0 = \sum_{i=1}^{m} (1 - \lambda_i)^k\, \mathbf{p}_i\, (\mathbf{p}_i^T D \mathbf{z}_0). \qquad \blacksquare$$

We can see by Theorem 9.54 that the asymptotic behavior, as $k \to \infty$, of graph diffusion depends upon the sizes of the eigenvalues λ_i.

Lemma 9.55. *The eigenvalues of L_{rw} and L_{sym} lie in the interval $[0, 2]$.*

Proof. Since L_{sym} is positive semidefinite, its eigenvalues are all nonnegative. Since $W^T D^{-1}$ is a transition matrix, i.e., its columns sum to one, it follows from Proposition 4.57 that

$$\| D^{-1} W \|_\infty = \| W^T D^{-1} \|_1 = 1. \qquad (9.69)$$

Now, let $\mathbf{q} \neq \mathbf{0}$ be an eigenvector of L_{rw} with eigenvalue $\lambda \geq 0$. Then we have

$$\lambda \mathbf{q} = L_{\text{rw}} \mathbf{q} = \mathbf{q} - D^{-1} W \mathbf{q},$$

and so $(1 - \lambda)\mathbf{q} = D^{-1} W \mathbf{q}$. Taking the ∞ norm on both sides and using (9.69) yields

$$|1 - \lambda|\, \| \mathbf{q} \|_\infty = \| D^{-1} W \mathbf{q} \|_\infty \leq \| \mathbf{q} \|_\infty.$$

Therefore $|1 - \lambda| \leq 1$, which implies that $0 \leq \lambda \leq 2$. $\qquad \blacksquare$

Remark 9.56. If the weight matrix W is positive semidefinite, then we can furthermore say that the eigenvalues of L_{rw} and L_{sym} belong to the interval $[0, 1]$. Indeed, in this case, the matrix $D^{-1/2} W D^{-1/2}$ is also positive semidefinite, and so the eigenvalues of the symmetric normalized graph Laplacian matrix $L_{\text{sym}} = I - D^{-1/2} W D^{-1/2}$ can be no greater than one. There are some applications where weight matrices are positive semidefinite. For example, in Section 7.6 we showed that the weight matrix with radial basis function kernel function entries $w_{ij} = \exp\left[-\| \mathbf{x}_i - \mathbf{x}_j \|^2 / (2\varepsilon^2) \right]$ is positive semidefinite. $\qquad \blacktriangle$

The eigenvalue bounds established in Lemma 9.55 guarantee that $|1 - \lambda_i| \leq 1$ for all i, and so the terms $(1 - \lambda_i)^k$ in the solution formula (9.68) remain bounded as $k \to \infty$. In fact, for any i with $|1 - \lambda_i| < 1$ we have

$$\lim_{k \to \infty} (1 - \lambda_i)^k = 0.$$

The only terms that will survive when $k \to \infty$ are eigenvalues λ_i for which $|1 - \lambda_i| = 1$, that is $\lambda_i = 0$ or $\lambda_i = 2$. The case $\lambda_i = 0$ corresponds to the kernel of L_{rw}, and is convergent, since $(1 - \lambda_i)^k = 1$ for all k. The case of $\lambda_i = 2$ is more problematic, since then we have $(1 - \lambda_i)^k = (-1)^k$ which oscillates and does not converge as $k \to \infty$. In order to avoid this problematic case, we make the following definition.

Definition 9.57. A graph $\mathcal{G}$ is said to be *aperiodic* if $\lambda_m < 2$.

Suppose the graph fails to be aperiodic, so $\lambda_m = 2$, and consider the initial condition $\mathbf{z}_0 = \mathbf{p}_m$ for the adjoint diffusion equation (9.62). Then $\langle \mathbf{p}_i, \mathbf{z}_0 \rangle_D = \langle \mathbf{p}_i, \mathbf{p}_n \rangle_D = 0$ if $i < m$ and $\langle \mathbf{p}_i, \mathbf{z}_0 \rangle_D = 1$ if $i = m$. Thus, by Theorem 9.54, $\mathbf{z}_k = (-1)^k \mathbf{p}_m = (-1)^k \mathbf{z}_0$. This indicates a *periodic* structure in the graph which prevents the diffusion process from converging to a steady state as the number of diffusion steps k tends to infinity, which justifies the name.

Example 9.58. It is easy to construct an example of a graph that fails to be aperiodic. Consider the canonical two node graph with weight matrix $W = \begin{pmatrix} 0 & 1 \\ 1 & 0 \end{pmatrix}$. Then all three graph Laplacians are given by $L = \begin{pmatrix} 1 & -1 \\ -1 & 1 \end{pmatrix}$, whose eigenvectors are $\begin{pmatrix} 1 \\ 1 \end{pmatrix}$ and $\begin{pmatrix} 1 \\ -1 \end{pmatrix}$ with respective eigenvalues $\lambda_1 = 0$ and $\lambda_2 = 2$. The periodic structure in this graph is as follows: if we initialize diffusion with all the mass on node 1, then it will all move to node 2 in the first step, and then back to node 1 in the second step, and continue to oscillate back and forth between the two nodes in all future steps. In this way, the mass does not spread out, and instead just jumps back and forth between the two nodes. ▲

Aperiodicity of a graph is guaranteed by the regularity of the associated transition matrix $P = W^T D^{-1}$, meaning, as per Definition 5.61, that some positive power P^k has all strictly positive entries. This means that there is a non-zero probability of getting from one node to any other node (including the same node) in exactly k steps which, by Exercise 5.14, is also the case for any $\ell \geq k$. Combining the Perron–Frobenius Theorem 5.62 and Lemma 9.55, regularity of the transition matrix implies that its eigenvalues satisfy $-1 < 1 - \lambda_j \leq 1$ and hence $0 \leq \lambda_j < 2$, which says that the graph is aperiodic.

Theorem 9.59. *If $\mathcal{G}$ is a weighted graph with regular transition matrix $P = W^T D^{-1}$, then $\mathcal{G}$ is aperiodic.*

The long time behavior of diffusion on an aperiodic graph is stable and convergent. The convergence rate is determined by the following quantity.

Definition 9.60. The *spectral gap* of a graph $\mathcal{G}$, denoted by λ, is the value of the smallest nonzero eigenvalue of L_{rw}.

We can now prove our main result on convergence of diffusion on aperiodic graphs.

Theorem 9.61. *Let $\mathbf{z}_k$ be the solution of the adjoint diffusion equation (9.62). Let*

$$\mathbf{z}_\infty = \sum_{\mathcal{H}} \frac{\langle \mathbf{1}_{\mathcal{H}}, \mathbf{z}_0 \rangle_D}{\| \mathbf{1}_{\mathcal{H}} \|_D^2} \mathbf{1}_{\mathcal{H}}, \tag{9.70}$$

where the sum is over the connected components $\mathcal{H}$ of $\mathcal{G}$, be the orthogonal projection of the initial condition $\mathbf{z}_0$ onto $\ker L_{\mathrm{rw}}$. Then, for $k \geq 0$,

$$\| \mathbf{z}_k - \mathbf{z}_\infty \|_D \leq \tau^k \| \mathbf{z}_0 \|_D, \qquad \text{where} \qquad \tau = \max\{ |1 - \lambda|, |1 - \lambda_m| \}. \tag{9.71}$$

In particular, if $\mathcal{G}$ is aperiodic, then $\tau < 1$ and hence

$$\lim_{k \to \infty} \mathbf{z}_k = \mathbf{z}_\infty. \tag{9.72}$$

Proof. As above, let $\mathbf{p}_1, \ldots, \mathbf{p}_m$ be the orthonormal eigenvector basis of L_{rw}, where the corresponding eigenvalues are in increasing order: $0 = \lambda_1 \leq \lambda_2 \leq \cdots \leq \lambda_m$. If $\mathcal{G}$ has j connected components $\mathcal{H}_1, \ldots, \mathcal{H}_j$, then nullity $L_{\mathrm{rw}} = j$, and we can take the j null eigenvectors to be the corresponding normalized indicator vectors:

$$\mathbf{p}_i = \frac{\mathbf{1}_{\mathcal{H}_i}}{\|\mathbf{1}_{\mathcal{H}_i}\|_D}, \qquad i = 1, \ldots, j. \tag{9.73}$$

Moreover, the spectral gap is $\lambda = \lambda_{j+1}$, and $|1 - \lambda_i| \leq \tau$ for all $i \geq j+1$. Therefore,

$$\mathbf{z}_k = \sum_{i=1}^{m} (1-\lambda_i)^k \langle \mathbf{p}_i, \mathbf{z}_0 \rangle_D \, \mathbf{p}_i = \sum_{i=1}^{j} \langle \mathbf{p}_i, \mathbf{z}_0 \rangle_D \, \mathbf{p}_i + \sum_{i=j+1}^{m} (1-\lambda_i)^k \langle \mathbf{p}_i, \mathbf{z}_0 \rangle_D \, \mathbf{p}_i.$$

The first summation is, in view of (9.73), the vector $\mathbf{z}_\infty$ given in (9.70). By the preceding inequality and orthonormality of the basis eigenvectors, the norm of the second summation can be estimated by

$$\|\mathbf{z}_k - \mathbf{z}_\infty\|_D^2 = \left\| \sum_{i=j+1}^{m} (1-\lambda_i)^k \langle \mathbf{p}_i, \mathbf{z}_0 \rangle_D \, \mathbf{p}_i \right\|_D^2$$
$$= \sum_{i=j+1}^{m} (1-\lambda_i)^{2k} \langle \mathbf{p}_i, \mathbf{z}_0 \rangle_D^2 \leq \tau^{2k} \sum_{i=j+1}^{m} \langle \mathbf{p}_i, \mathbf{z}_0 \rangle_D^2 \leq \tau^{2k} \|\mathbf{z}_0\|_D^2. \qquad \blacksquare$$

Remark. Suppose $\mathcal{G}$ is aperiodic. The results of Theorem 9.61 can be easily restated for the solution $\mathbf{x}_k = D\mathbf{z}_k$ of the diffusion equation (9.60). Equation (9.72) becomes

$$\lim_{k \to \infty} \mathbf{x}_k = \mathbf{x}_\infty = D \sum_{\mathcal{H}} \frac{\mathbf{x}_0 \cdot \mathbf{1}_{\mathcal{H}}}{\|\mathbf{1}_{\mathcal{H}}\|_D^2} \, \mathbf{1}_{\mathcal{H}}. \tag{9.74}$$

Thus, diffusion on an aperiodic graph converges to a steady state mass distribution $\mathbf{x}_\infty$ that belongs kernel of L_{rw}^T, that is, belongs to the subspace spanned by the rescaled indicator vectors $D\mathbf{1}_{\mathcal{H}}$ of the connected components of $\mathcal{G}$. Furthermore, the limiting distribution $\mathbf{x}_\infty$ has no mass on a component $\mathcal{H}$ when, at the initial time we have $\mathbf{x}_0 \cdot \mathbf{1}_{\mathcal{H}} = 0$. In the case that $\mathbf{x}_0$ has nonnegative components, this implies that $\mathbf{x}_0$ is identically zero over the component $\mathcal{H}$, that is, it has no initial mass within $\mathcal{H}$, and the diffusion is unable to spread any mass to $\mathcal{H}$. When $\mathbf{x}_0$ is allowed to have negative values, then $\mathbf{x}_0 \cdot \mathbf{1}_{\mathcal{H}} = 0$ is a mean-zero condition on $\mathcal{H}$, and the positive and negative mass on this component will asymptotically cancel out during diffusion as $k \to \infty$. $\blacktriangle$

Finally, we note that if the graph is connected, so that there is only one connected component $\mathcal{H} = \mathcal{G}$ with indicator vector $\mathbf{1}$, then the limiting mass distribution is given by

$$\mathbf{x}_\infty = \frac{\mathbf{x}_0 \cdot \mathbf{1}}{\mathbf{d} \cdot \mathbf{1}} \, \mathbf{d}, \tag{9.75}$$

which is simply proportional to the degree vector $\mathbf{d}$ and has the same mass as the initial condition $\mathbf{x}_0$ (the initial mass is the numerator). This explains Figure 9.25, where the limiting distribution after $k = 1000$ steps is not constant over the graph, but instead is higher in dense regions where the degree is larger. In Figure 9.26, we show the same diffusion example but we display the difference $\mathbf{x}_k - \mathbf{x}_\infty$ between the current mass distribution and the limiting one. This centering counteracts the limiting behavior and displays, in the limit, the first nontrivial eigenvector $\mathbf{p}_2$. This is useful for clustering, as in the binary spectral clustering discussed in Section 9.4, though the reader should be careful to note that $\mathbf{p}_2$ is not the Fiedler vector. (Why not?)

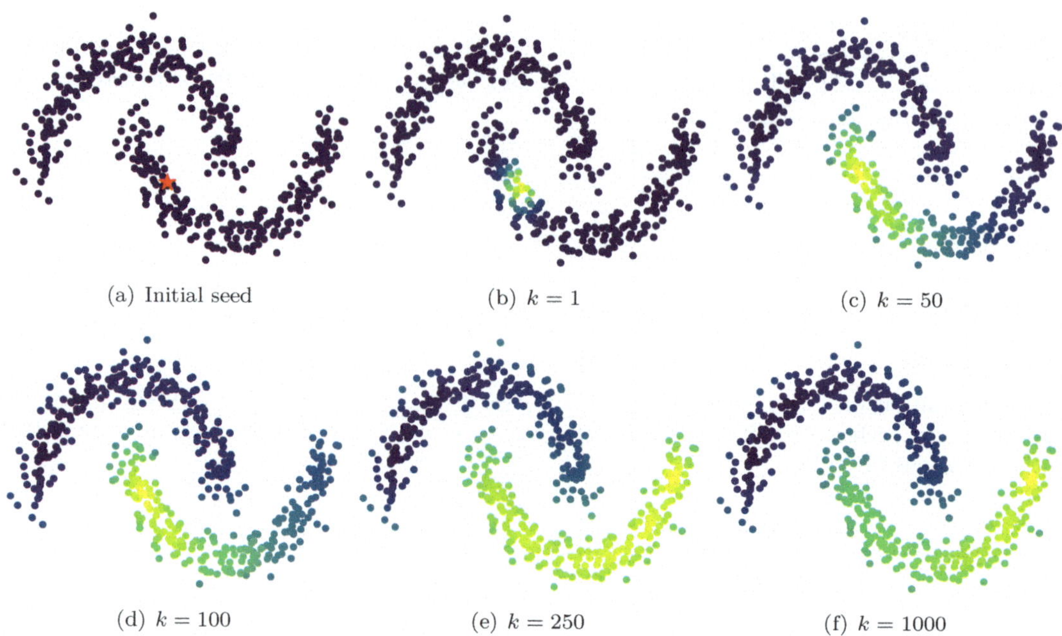

(a) Initial seed   (b) $k = 1$ (c) $k = 50$

(d) $k = 100$ (e) $k = 250$ (f) $k = 1000$

Figure 9.26: Example of centered diffusion of a unit of mass concentrated on the red point in (a) over 1000 steps of diffusion. The plot shows $\mathbf{x}_k - \mathbf{x}_\infty$, illustrating how the centering counteracts the limiting behavior of the diffusion at large time scales.

9.6.1   Jump Diffusion and PageRank

Python Notebook: PageRank (.ipynb)

There are many other mechanisms that can be introduced into the diffusion process, yielding a vast family of general types of diffusion equations on graphs and digraphs that have important and interesting practical applications. One possibility is to include a *source term*, which adds or removes mass at each node at a specified rate, which may depend the iteration k; see Exercise 6.8. Another possibility is the *lazy random walk*, which is a diffusion process where each node *keeps* some of its mass at each iteration; details can be found in Exercise 6.12.

In this section, we introduce the *jump diffusion process*, which augments the ordinary diffusion process by additionally sending a fraction of the mass at a node to *all* other nodes in the digraph, regardless of whether or not they are connected by edges. Jump diffusion has a regularizing effect that makes the diffusion less sensitive to connectivity and periodic structures. One of the most famous applications of jump diffusions on digraphs is the PageRank algorithm, which was used by Google to rank internet search results until around 2006. While PageRank originally gained popularity through its ability to rank Google's search results, it has found a wide range of applications in many other fields, including biology (e.g., GeneRank), chemistry, ecology, neuroscience, physics, sports, and computer systems [86]. It has even been employed in modern graph neural network architectures [82], which are the subject of Section 10.4.

We begin by discussing jump diffusion in general, and then proceed to applications to PageRank. Let $\mathcal{G}$ be a weighted digraph, with weight matrix W, which need not be symmetric. Let $\alpha \in [0,1]$ and let $\mathbf{v} \in \mathbb{R}^m$ be a fixed *probability vector*. In a *jump diffusion process*, instead of distributing all of the mass x_i at node i to its graph neighbors, a fraction of the mass *jumps* and is distributed across the *entire* graph. The parameter α controls the fraction of mass αx_i at node i that is distributed via diffusion, and the fraction $(1-\alpha)x_i$ that jumps. The vector $\mathbf{v}$ is called the *jump distribution* and controls how the mass that jumps is distributed among the nodes. To state it mathematically, the amount of mass that node i sends to node j is

$$\left[\alpha w_{ij} d_i^{-1} + (1-\alpha)v_j\right] x_i. \tag{9.76}$$

Note that some of the mass that *jumps* will arrive back at node i provided that $v_i > 0$. The first term involving α corresponds to diffusion, while the second term involving $1 - \alpha$ corresponds to the jump. Summing the expression in (9.76) over $j = 1, \ldots, m$ yields exactly x_i, which means that all of the mass at node i is distributed, as in the case of diffusion; see Exercise 6.9.

If we sum the expression (9.76) over i, we obtain the amount of mass that arrives at node j after one step of diffusion. In particular, if $\mathbf{x}$ is the initial distribution and $\mathbf{y}$ is the mass distribution after one step of this jump diffusion process, then

$$y_j = \alpha \sum_{i=1}^m w_{ij} d_i^{-1} x_i + (1-\alpha)v_j \sum_{i=1}^m x_i. \tag{9.77}$$

Assuming that $\mathbf{x}$ is a probability vector, so $\mathbf{x} \cdot \mathbf{1} = 1$, we can write (9.77) in vectorial form

$$\mathbf{y} = \alpha P \mathbf{x} + (1-\alpha)\mathbf{v}. \tag{9.78}$$

Recall from (9.58) that $P = W^T D^{-1}$ is a transition matrix, and so $\|P\|_1 = 1$; it also preserves mass, so $(P\mathbf{x}) \cdot \mathbf{1} = \mathbf{x} \cdot \mathbf{1} = 1$.

Now, since $\mathbf{v}$ is a probability vector, so $\mathbf{v} \cdot \mathbf{1} = 1$, we can dot both sides of (9.78) with the ones vector to obtain

$$\mathbf{y} \cdot \mathbf{1} = \alpha(P\mathbf{x}) \cdot \mathbf{1} + (1-\alpha)\mathbf{v} \cdot \mathbf{1} = \alpha \mathbf{x} \cdot \mathbf{1} + (1-\alpha)\mathbf{v} \cdot \mathbf{1} = \alpha + (1-\alpha) = 1.$$

Moreover, since all terms on the right hand side of (9.77) are nonnegative, we conclude that $\mathbf{y}$ is a probability vector. Hence, if we start from an initial probability vector $\mathbf{x}_0 \in \mathbb{R}^m$ and iterate the jump diffusion propagation, with

$$\mathbf{x}_{k+1} = \alpha P \mathbf{x}_k + (1-\alpha)\mathbf{v}, \tag{9.79}$$

then each $\mathbf{x}_k$ remains a probability vector.

The jump diffusion process (9.79) can also be viewed from the random walk perspective by modifying the dynamics to include a *teleportation* step, where the random walker jumps to any other node in the graph at random according to the probability vector $\mathbf{v}$, often called the *teleportation distribution*. Instead of a random walk, this process is called a *random surf*, or *random surfing* [86]. At each step of random surfing, we flip a coin that comes up heads with probability α and tails with probability $1 - \alpha$. If the coin comes up heads, then the surfer takes a random walk step to one of the neighbors of the current node. If the coin comes up tails, then the random surfer teleports to another node in the graph according to the probability vector $\mathbf{v}$; that is, the probability of teleporting to node j is exactly v_j. The parameter α is often called the *teleportation parameter*. The probability distributions $\mathbf{x}_k$ of the random surfer after k steps exactly solve the jump diffusion iteration (9.79). From the random surfing perspective, we can see how the teleportation step provides a regularization effect provided $\alpha < 1$; it ensures that the random walker cannot get trapped in a periodic loop, or in any particular connected component of the graph, for too long.

We turn to analyzing the dynamics of jump diffusion. In the case that the weight matrix is symmetric, $W = W^T$, we can employ the spectral decomposition of the graph Laplacian, as we did for ordinary diffusion; we defer this analysis to Exercise 6.11. Here, we focus on the asymptotics of jump diffusion in the case of a digraph with non-symmetric weight matrix W, where, we cannot use spectral theory since L_{rw} is not self-adjoint. Instead, we directly study the iterative process (9.79). To do this, we first note that if the jump diffusion were convergent, then the limiting probability vector $\mathbf{x}_\infty = \lim_{k \to \infty} \mathbf{x}_k$ must clearly satisfy the limit equation

$$\mathbf{x}_\infty = \alpha P \mathbf{x}_\infty + (1 - \alpha)\mathbf{v}, \tag{9.80}$$

which we obtain simply by taking limits on both sides of (9.79). It turns out that when $\alpha < 1$ the solution of (9.80) exists and is unique.

Lemma 9.62. *If $0 \le \alpha < 1$ then, for every $\mathbf{v} \in \mathbb{R}^m$, there exists a unique vector $\mathbf{x}_\infty \in \mathbb{R}^m$ satisfying (9.80). Furthermore, $\mathbf{x}_\infty \cdot \mathbf{1} = \mathbf{v} \cdot \mathbf{1}$.*

Proof. We rewrite (9.80) as

$$(\mathrm{I} - \alpha P)\mathbf{x}_\infty = (1 - \alpha)\mathbf{v},$$

which is uniquely solvable if and only if the matrix $\mathrm{I} - \alpha P$ is nonsingular. To show this, suppose $\mathbf{z} \in \ker(\mathrm{I} - \alpha P)$, so that $\mathbf{z} - \alpha P \mathbf{z} = \mathbf{0}$, or $\alpha P \mathbf{z} = \mathbf{z}$. According to Exercise 5.15, $\|P\|_1 = 1$, and hence

$$\|\mathbf{z}\|_1 = \|\alpha P \mathbf{z}\|_1 = \alpha \|P \mathbf{z}\|_1 \le \alpha \|P\|_1 \|\mathbf{z}\|_1 = \alpha \|\mathbf{z}\|_1.$$

Since $\alpha < 1$, this requires $\mathbf{z} = \mathbf{0}$, which establishes the claim.

Finally, dotting both sides of (9.80) with the ones vector yields

$$\mathbf{x}_\infty \cdot \mathbf{1} = \alpha P \mathbf{x}_\infty \cdot \mathbf{1} + (1 - \alpha)\mathbf{v} \cdot \mathbf{1} = \alpha \mathbf{x}_\infty \cdot \mathbf{1} + (1 - \alpha)\mathbf{v} \cdot \mathbf{1},$$

from which we obtain $\mathbf{x}_\infty \cdot \mathbf{1} = \mathbf{v} \cdot \mathbf{1}$. $\blacksquare$

We can now establish convergence of the jump diffusion as $k \to \infty$.

Theorem 9.63. *Let $0 \le \alpha < 1$ and $\mathbf{v} \in \mathbb{R}^m$. Let $\mathbf{x}_k$ satisfy the jump diffusion (9.79), and let $\mathbf{x}_\infty$ be the unique solution of (9.80). Then*

$$\|\mathbf{x}_k - \mathbf{x}_\infty\|_1 \le \alpha^k \|\mathbf{x}_0 - \mathbf{x}_\infty\|_1. \tag{9.81}$$

Proof. By (9.79) and (9.80),

$$\mathbf{x}_{k+1} - \mathbf{x}_\infty = \alpha P \mathbf{x}_k + (1 - \alpha)\mathbf{v} - \left[\alpha P \mathbf{x}_\infty + (1 - \alpha)\mathbf{v}\right] = \alpha P(\mathbf{x}_k - \mathbf{x}_\infty).$$

Taking 1 norms on both sides and using that $\|P\|_1 = 1$ produces

$$\|\mathbf{x}_{k+1} - \mathbf{x}_\infty\|_1 = \|\alpha P(\mathbf{x}_k - \mathbf{x}_\infty)\|_1 = \alpha \|P(\mathbf{x}_k - \mathbf{x}_\infty)\|_1 \le \alpha \|\mathbf{x}_k - \mathbf{x}_\infty\|_1.$$

Iterating this inequality yields (9.81). $\blacksquare$

Remark. Notice that, in contrast to Theorem 9.61, the convergence of the jump diffusion process does not require the graph to be aperiodic when $\alpha < 1$, and the limit $\mathbf{x}_\infty$ is not sensitive to the initialization $\mathbf{x}_0$, even when the graph is disconnected. The effect of the jump part of the diffusion, that is, the redistribution of mass according to $\mathbf{v}$, regularizes the diffusion process so that periodic structures and disconnected components in the graph do not cause a problem for convergence. Furthermore, the convergence rate α is independent of the spectral gap, and is a user-specified parameter. ▲

We already noted that, if $\mathbf{x}_0$ is a probability vector, then each iterate $\mathbf{x}_k$ of the jump diffusion process is a probability vector, and hence the limiting vector $\mathbf{x}_\infty$ must also be a probability vector. This implies:

Corollary 9.64. *Let $0 \leq \alpha < 1$, and let $\mathbf{x}_0$ and $\mathbf{v}$ be probability vectors in $\mathbb{R}^m$. Then the solution $\mathbf{x}_\infty$ of (9.80) is also a probability vector.*

PageRank is a graph-based algorithm that ranks websites based on their link structure. The idea behind the PageRank algorithm is to randomly surf the internet for a very long time, using the digraph structure coming from hyperlinks between websites. The websites that are visited more often get higher ranks than those that are visited less often, and the *PageRank* of a site, used to rank the site in search results, is the limit as $k \to \infty$ of the average amount of time spent on that particular site. The PageRank of a website is a robust notion of how well-connected and important a website is that takes into account the graph structure, and not simply how many hyperlinks the site has. The *PageRank vector* is the vector containing the PageRank scores for each website.

Since the internet is a complicated real world digraph that may have disconnected components or periodic structures, it is important to use the random surfing framework with $\alpha < 1$ so that the PageRank vector is well-defined, does not depend on where the random surfer is initially placed, and does not require a connected graph. After a very long random surf, or as $k \to \infty$, the entries of the PageRank vector indicate the asymptotic probabilities of finding the random surfer at the different nodes [63]. Thus, the PageRank vector is the limit of $\mathbf{x}_k$ of the jump diffusion process as $k \to \infty$, or, alternatively, the solution $\mathbf{x}$ of the limiting equation (9.80), which is often called the *PageRank equation*. Let us mention that, in the application to Google search, PageRank is only one part of the search engine; other data base retrieval techniques are used to find webpages that are similar to a given search query, and PageRank is used to decide which are important enough to show to the user and in what order they should appear.

Remark 9.65. In practice, one standard way to compute the PageRank vector, i.e., the solution of the PageRank problem (9.80), is via the jump diffusion iteration (9.79), which is often called the *PageRank iteration* and converges at the linear rate $\alpha < 1$ by Theorem 9.63. The PageRank iteration can also be interpreted as the power method for computing the largest eigenvector of a matrix. Indeed, since $\mathbf{x}_k \cdot \mathbf{1} = 1$ we can rewrite (9.79) as

$$\mathbf{x}_{k+1} = \alpha P \mathbf{x}_k + (1-\alpha)\mathbf{v}\,\mathbf{1}^T \mathbf{x}_k = P_\alpha \mathbf{x}_k, \qquad \text{where} \qquad P_\alpha = \alpha P + (1-\alpha)\mathbf{v}\,\mathbf{1}^T.$$

This is exactly the matrix power method (5.66), except that the normalization step is performed in the 1 norm, that is $\mathbf{x}_{k+1} = P_\alpha \mathbf{x}_k / \|P_\alpha \mathbf{x}_k\|_1$, since $\|P_\alpha \mathbf{x}_k\|_1 = 1$. ▲

There are two important parameters in the PageRank algorithm; the teleportation parameter α and the teleportation distribution $\mathbf{v}$. If α is chosen too small, then the teleportation step dominates, which does not convey any useful information. Indeed, by Exercise 6.10 we have $\|\mathbf{x}_\infty - \mathbf{v}\|_1 \leq 2\alpha$. Thus, when α is small, we simply get $\mathbf{x}_\infty \approx \mathbf{v}$, so the PageRank

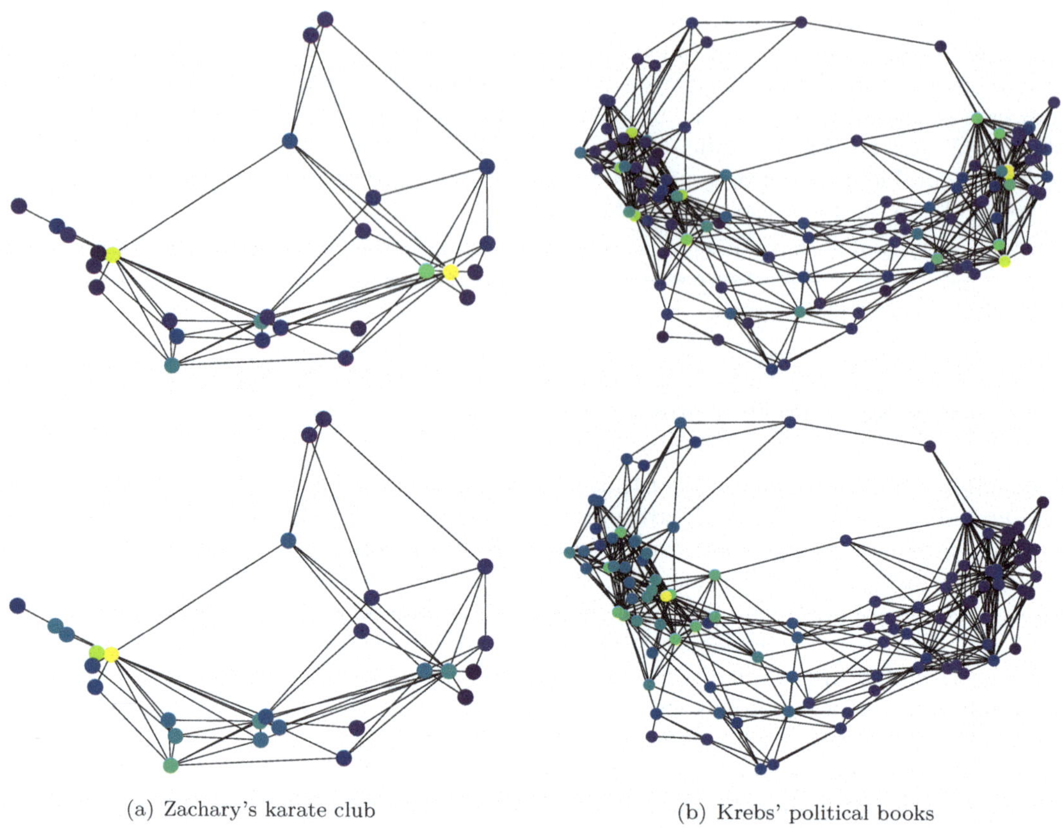

(a) Zachary's karate club   (b) Krebs' political books

Figure 9.27: Illustration of the PageRank vector on Zachary's karate club graph and Krebs' political books graph. On the top is the usual PageRank vector with uniform teleportation $\mathbf{v} = m^{-1}\mathbf{1}$, while on the bottom we show localized PageRank, relative to the yellow node. The color of each node represents the value of the PageRank vector, with yellow being the highest rank, and purple being the smallest.

vector mimics the teleportation distribution. On the other hand, if α is too close to 1, then the convergence rate α of the PageRank iteration is very poor and it may take many iterations to converge. A commonly used value is $\alpha = 0.85$, cf. [86].

There are many interesting choices for the teleportation distribution. The standard choice is the uniform distribution $\mathbf{v} = \mathbf{1}/m$, in which case the random surfer teleports uniformly at random to another site on the internet. Another typical choice is $\mathbf{v} = \mathbf{e}_i$, so that the random surfer always teleports back to the i-th node. This is called *localized PageRank* or *personalized PageRank*, and is used to rank all pages relative to page i. In particular, the localized PageRank vector for node i gives a notion of *similarity* between node i and all other nodes in the graph, where nodes with larger values are closer to node i. There are, of course, a continuum of possible teleportation distributions lying between the uniform random distribution and localized PageRank.

We now turn to some numerical experiments. In Figure 9.27 we show the PageRank vectors on Zachary's karate club graph and Krebs' political books graph. On the top we show the standard PageRank vector with uniform teleportation distribution, while on the bottom we show the localized PageRank vector, for ranking nodes relative to the yellow node

Figure 9.28: An example of using personalized PageRank for image retrieval. In each row the image on the left is the query image, and the following 19 images are the top retrieved images using personalized PageRank.

in the figure. For localized PageRank we can see how the nearby nodes in the graph structure are ranked higher, which indicates they are more similar to the yellow seed node. In all experiments we took $\alpha = 0.85$.

We now turn to an example of using localized PageRank for image retrieval, which takes a query image and tries to find similar images in a data set. We consider the MNIST data set of handwritten digits, and construct a k-nearest neighbor graph over the data set. To retrieve images similar to image j, we use localized PageRank with $\mathbf{v} = \mathbf{e}_j$ in order to rank all images based on their similarity to image j. We took one image from each of the 10 MNIST digits and ran personalized PageRank to retrieve the top 19 similar images. Figure 9.28 shows the results. The image on the left is the query image, and the following 19 images are the retrieved images. We note that most digits are from the same class, and are written in a very similar way to the query digit. There are a handful retrieved digits that are from different classes than the query; for example, a 2 and a 7 in the 3's row, an 8 and a 9 in the 4's row, and a 3 in the 5's row.

We also mention that there are other methods for ranking on graphs that use different techniques for regularizing the diffusion process. One such example is the LeaderRank algorithm [155], which adds an additional "ground" node to the graph which has bidirectional edges to *all* nodes in the graph. This makes any graph connected, so that the diffusion process is convergent. We explore LeaderRank more in Exercise 6.13.

Exercises

Note: Assume that the weight matrix $W = W^T$ is symmetric, unless otherwise specified.

6.1. ♡ Let $\mathbf{x}_k$ satisfy the diffusion equation (9.60) and set $\mathbf{z}_k = D^{-1}\mathbf{x}_k$. Show that $\mathbf{z}_k$ solves the adjoint equation (9.62).

6.2. Prove Lemma 9.53.

6.3. ♡ The symmetric normalized graph Laplacian (9.65) defines another type of diffusion on graphs, which is given by

$$\mathbf{y}_{k+1} = \mathbf{y}_k - L_{\text{sym}}\,\mathbf{y}_k. \tag{9.82}$$

Let $\mathbf{x}_k$ satisfy the diffusion equation (9.60) and set $\mathbf{y}_k = D^{-1/2}\mathbf{x}_k$. Show that $\mathbf{y}_k$ solves the symmetric diffusion equation (9.82).

6.4. ♡ Let $\mathbf{x}_k$ satisfy the diffusion equation (9.60). Give an example of a graph where $\mathbf{x}_0 = \mathbf{1}$, but $\mathbf{x}_k \neq \mathbf{1}$ when $k \geq 1$.

6.5. ♡ Show that the solution $\mathbf{x}_k$ of the diffusion equation (9.60) is given by

$$\mathbf{x}_k = D \sum_{i=1}^{m} (1 - \lambda_i)^k\,(\mathbf{x}_0 \cdot \mathbf{p}_i)\,\mathbf{p}_i, \tag{9.83}$$

where $\mathbf{p}_1, \ldots, \mathbf{p}_m$ are the orthonormal eigenvector basis of L_{rw} with eigenvalues $\lambda_1, \ldots, \lambda_m$.

6.6. Show that the solution $\mathbf{y}_k$ of the symmetric normalized graph Laplacian diffusion equation (9.82) is given by

$$\mathbf{y}_k = \sum_{i=1}^{m} (1 - \lambda_i)^k\,(\mathbf{y}_0 \cdot \mathbf{q}_i)\,\mathbf{q}_i, \tag{9.84}$$

where $\mathbf{q}_1, \ldots, \mathbf{q}_m$ are the orthonormal eigenvector basis of L_{sym}.

6.7. ♡ State and prove Theorem 9.61 for $\mathbf{x}_k$ and $\mathbf{y}_k$, the respective solutions of the diffusion equations (9.60) and (9.82).

6.8. ◇ Consider diffusion with a source term $\mathbf{f} \in \mathbb{R}^m$:

$$\mathbf{x}_{k+1} = \mathbf{x}_k - L_{\text{rw}}^T \mathbf{x}_k + \mathbf{f}. \tag{9.85}$$

The interpretation is that we are adding or removing mass, based on the sign of $\mathbf{f}$, at each iteration. Assume the graph is connected throughout this exercise.
 (a) Show that

$$\mathbf{x}_{k+1} \cdot \mathbf{1} = \mathbf{x}_k \cdot \mathbf{1} + \mathbf{f} \cdot \mathbf{1}.$$

For the rest of the exercise assume that $\mathbf{f} \cdot \mathbf{1} = 0$, so that (9.85) conserves mass.
 (b) Show that the mean zero condition $\mathbf{f} \cdot \mathbf{1} = 0$ implies that $\mathbf{f} \in \operatorname{img} L_{\text{rw}}^T$, and so there exists $\mathbf{x}^\star \in \mathbb{R}^m$ such that $L_{\text{rw}}^T \mathbf{x}^\star = \mathbf{f}$.
 (c) Set $\mathbf{u}_k = \mathbf{x}_k - \mathbf{x}^\star$ and show that $\mathbf{u}_k$ satisfies the diffusion equation (9.60), that is

$$\mathbf{u}_{k+1} = \mathbf{u}_k - L_{\text{rw}}^T \mathbf{u}_k.$$

 (d) Suppose that $\mathbf{x}_0 = \mathbf{0}$. Use Exercise 6.5 to solve for $\mathbf{u}_k$, and then show that

$$\mathbf{x}_k = \mathbf{x}^\star - D \sum_{i=1}^{m} (1 - \lambda_i)^k\,(\mathbf{x}^\star \cdot \mathbf{p}_i)\,\mathbf{p}_i,$$

where $\mathbf{p}_i, \lambda_i$ are the eigenvectors and eigenvalues of the random walk graph Laplacian L_{rw}.
 (e) Assume the graph is connected and aperiodic. Show that if $\mathbf{x}^\star \cdot \mathbf{1} = 0$, then $\mathbf{x}_k \to \mathbf{x}^\star$ as $k \to \infty$. What happens if $\mathbf{x}^\star \cdot \mathbf{1} \neq 0$?

6.9. Show that the sum of (9.76) over $j = 1, \ldots, m$ is exactly x_i. Do not assume $W = W^T$.

6.10. $\heartsuit$ Let $\mathbf{x}_\infty$ solve (9.80) with W not necessarily symmetric. Let $\mathbf{v}$ be a probability vector. Show that $\|\mathbf{x}_\infty - \mathbf{v}\|_1 \leq 2\alpha$.

6.11. Prove Theorem 9.63 using the spectral decomposition of L_{rw}, as we did in Theorem 9.61. Use the fact that $W = W^T$ is symmetric.

6.12. $\diamondsuit$ Suppose we modify the diffusion process on the graph so that instead of node i sending all of its mass to its neighbors, we fix some $0 < \alpha < 1$ and x_i sends αx_i mass to its neighbors, and keeps $(1 - \alpha) x_i$ mass for itself.

(a) Show that the corresponding diffusion equation for this process is

$$\mathbf{x}_{k+1} = \mathbf{x}_k - \alpha L_{\mathrm{rw}}^T \mathbf{x}_k.$$

(b) Use the same ideas as in the proof of Theorem 9.54 and Exercise 6.5 to show that

$$\mathbf{x}_k = D \sum_{i=1}^{m} (1 - \alpha \lambda_i)^k (\mathbf{x}_0 \cdot \mathbf{p}_i) \mathbf{p}_i.$$

where $\mathbf{p}_1, \ldots, \mathbf{p}_m$ and $\lambda_1, \ldots, \lambda_m$ are the eigenvectors and eigenvalues of L_{rw}.

(c) Show that if $0 < \alpha < 1$, then

$$\lim_{k \to \infty} \mathbf{x}_k = \sum_{\mathcal{H}} \frac{\mathbf{x}_0 \cdot \mathbf{1}_{\mathcal{H}}}{\mathbf{1}_{\mathcal{H}} \cdot (D \mathbf{1}_{\mathcal{H}})} D \mathbf{1}_{\mathcal{H}}.$$

Note that this holds even if the graph is *not* aperiodic.

(d) What is the rate of convergence in part (c)?

6.13. The *LeaderRank* algorithm, [155], ranks nodes in a weighted digraph by appending an extra node, called the "ground node" to the graph, and adding bi-directional edges between all nodes and the ground node. Then the LeaderRank score is obtained from the limiting distribution $\mathbf{x}_\infty$ of diffusion on the graph, i.e., PageRank with $\alpha = 1$. (a) Explain why adding the ground node ensures the graph is connected. (b) Implement LeaderRank in Python and compare it to PageRank on some of the graphs used in this section.

6.14. $\diamondsuit$ Write Python code to run personalized PageRank on the MNIST data set with teleportation distribution $\mathbf{v} = \mathbf{1}_{\mathcal{H}}$, where $\mathcal{H}$ is one of the MNIST digit classes (say, all the zeros). Plot some of the highest and lowest ranked digits within $\mathcal{H}$. What do these represent? Can you explain your findings?

6.15. $\heartsuit$ Let $\mathcal{G}$ be a connected weighted graph. Show that there exists $C > 0$, depending only on $\mathcal{G}$, such that

$$\sum_{i=1}^{m} u_i^2 \leq \frac{C}{2} \sum_{i,j=1}^{m} w_{ij}(u_i - u_j)^2, \tag{9.86}$$

for all $\mathbf{u} \in \mathbb{R}^m$ satisfying $\mathbf{u} \cdot \mathbf{1} = 0$. The result is known as a *Poincaré inequality*.

6.16. Does the Poincaré inequality (9.86) hold without assuming that $\mathbf{u} \cdot \mathbf{1} = 0$?

9.7   Diffusion Maps and Spectral Embeddings

Python Notebook: Spectral Embeddings and Clustering (.ipynb)

We now turn to using diffusion to measure the *similarity* between different nodes in a digraph. In fact, we already saw a version of this when we used personalized PageRank for image retrieval — that is, finding image similar to a query image — in Section 9.6.1. As we will see, using diffusion on graphs to measure similarity between nodes is useful for uncovering graph structures, such as well-separated clusters, and has close connections to the eigenvectors of the various normalization of the graph Laplacian. We describe in this section *diffusion maps* and *diffusion distances*, as introduced in [45, 171]. This will naturally lead us to various types of spectral embeddings, which use the top eigenvectors of the graph Laplacian to embed the graph nodes into Euclidean space, where the Euclidean geometry simplifies the data analysis steps. We will then see applications to clustering in the *multicluster spectral clustering* algorithm [176, 239] in Section 9.7.2.

9.7.1   Diffusion Distance

Let $\mathcal{G}$ be a connected weighted digraph. The diffusion distance gives a way to measure the distance between two nodes i and j that takes into account similarities in graph structure at the two nodes at different scales. The idea is to compare how mass concentrated at each node diffuses through the graph. In particular, we fix a number of diffusion steps k, and diffuse a unit of mass starting at each node for k steps, according to the diffusion equation (9.60). This gives two different mass distributions over the graph, $P^k \mathbf{e}_i$ for node i and $P^k \mathbf{e}_j$ for node j, where $P = \mathrm{I} - L_{\mathrm{rw}}^T = W^T D^{-1}$, cf. (9.59). The diffusion distance between the two nodes measures how far apart these distributions are after a fixed number of steps.

Mathematically, let us fix an inner product $\langle \mathbf{x}, \mathbf{y} \rangle_C = \mathbf{x}^T C \mathbf{y}$, where C is a symmetric positive definite matrix. Distance will be measured using the associated norm $\|\mathbf{x}\|_C = \sqrt{\mathbf{x}^T C \mathbf{x}}$, and, for the moment, we leave the choice of C open. After some analysis, natural choices will appear.

Definition 9.66. The k-step *diffusion distance* $d_{ij}^{(k)}$ between nodes i and j is defined by

$$d_{ij}^{(k)} = \| P^k (\mathbf{e}_i - \mathbf{e}_j) \|_C. \tag{9.87}$$

To see what the diffusion distance is measuring, let us write out the diffusion explicitly in a few special cases. Clearly $d_{ij}^{(0)} = \| \mathbf{e}_i - \mathbf{e}_j \|_C$ is trivial. Note that

$$(P\mathbf{e}_i)_j = p_{ji}, \qquad (P^2\mathbf{e}_i)_j = \sum_{\ell=1}^{m} p_{j\ell} p_{\ell i}, \qquad \text{where} \qquad p_{ij} = w_{ji} d_j^{-1}$$

are the entries of P. In the second equation above, we note that the j-th component of $P^2 \mathbf{e}_i$ is a sum over all *paths* from i to j with two edges, since $p_{j\ell} p_{\ell i} > 0$ exactly when there is an edge from i to ℓ and from ℓ to j. In general, we have

$$(P^k \mathbf{e}_i)_j = \sum_{\ell_1=1}^{m} \cdots \sum_{\ell_{k-1}=1}^{m} p_{j\ell_{k-1}} p_{\ell_{k-1}\ell_{k-2}} \cdots p_{\ell_1 i},$$

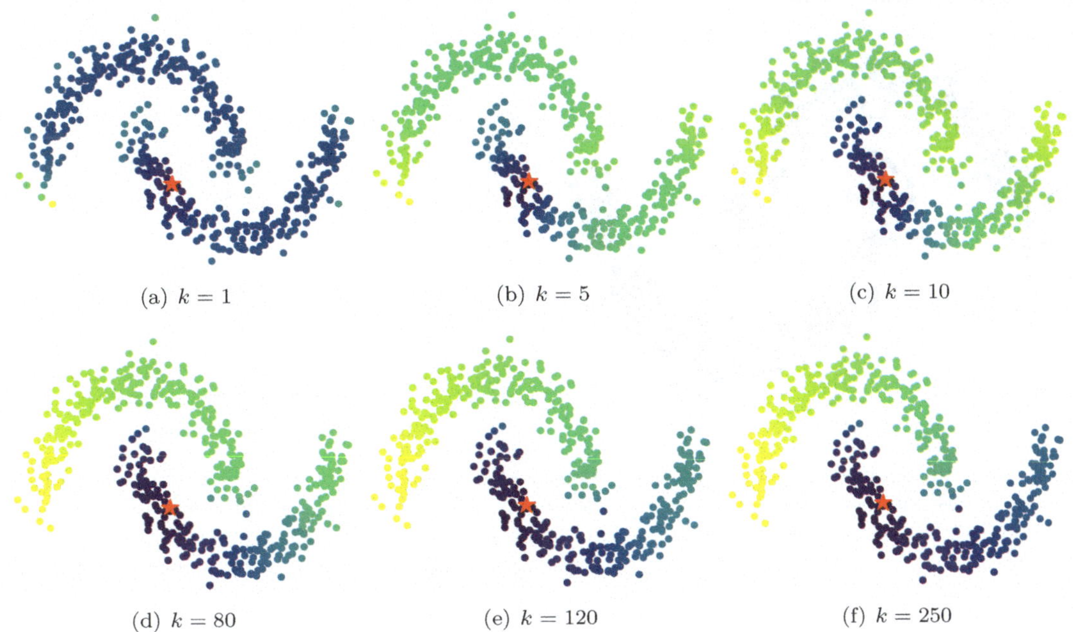

(a) $k = 1$ (b) $k = 5$ (c) $k = 10$

(d) $k = 80$ (e) $k = 120$ (f) $k = 250$

Figure 9.29: Example of the diffusion distance to the red star at different time scales.

which is a sum over all paths from i to j with k edges. Thus, in contrast to the graph distances we discussed in Section 9.5 that use information from only the shortest path between two nodes, diffusion distances make use of information from *all* paths between two nodes with a fixed number of edges k, which makes them less sensitive to noise or corruption in the edge information, and, in some cases, better able to detect underlying structure in the graph.

In Figure 9.29, we plot the diffusion distance to a point on the two moons graph as k is increased. For small values, the mass has not diffused enough and the distance does not capture the cluster structure at all. For moderate to large values of k, the diffusion distance to points in the same cluster as the seed point is small, but notably larger to points in the other cluster. Thus, the diffusion distance naturally captures the cluster structure in this data set, provided the number of steps is chosen appropriately.

When working with diffusion distances, a good choice of the number of steps k is important. If k is too small, the diffusion has not had enough time to progress, and if k is too large, fine details in the graph may be lumped together. In Figure 9.30 we show the behavior of a certain ratio of diffusion distances versus the number of steps. The ratio displayed is the diffusion distance between the two orange stars in the figure divided by the diffusion distance from the green star to the closest orange one. The orange stars are in the same cluster and we would like them to be close together in the diffusion distance, while the green star is in the other cluster and we'd like it to be far away. Thus, we want this ratio to be less than one. In Figure 9.30(b) we show the diffusion distance ratio over the first 400 diffusion steps. The ratio is less than one initially for about the first 10 steps, though this is a coincidence and may not happen in general. The important point is that if we diffuse for long enough — here about 120 steps — then the ratio drops below one and stays there. Thus, a diffusion distance with $k \geq 120$ is able to well separate the two moons, which we can also observe in Figure 9.29. In contrast, the shortest path distance depicted in Figure 9.31 is unable to separate the moons, since it is sensitive to the small number of stray edges connecting points in the opposing moons.

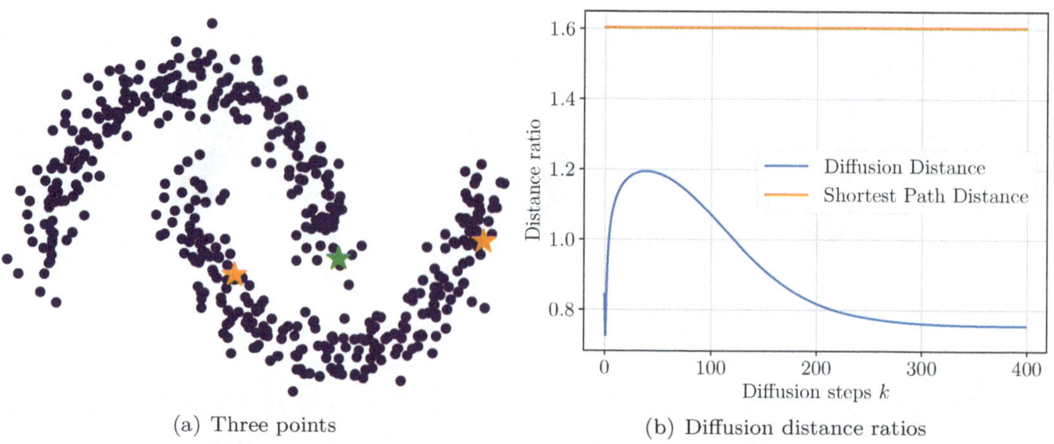

(a) Three points

(b) Diffusion distance ratios

Figure 9.30: Illustration of how the diffusion distance between points in different clusters is strongly affected by the number of diffusion steps k. When the diffusion distance ratio is less than one, the two moons are well separated, in the sense that the orange stars are closer together than the green is to either orange. In contrast, the shortest path distance is unable to completely separate the moons.

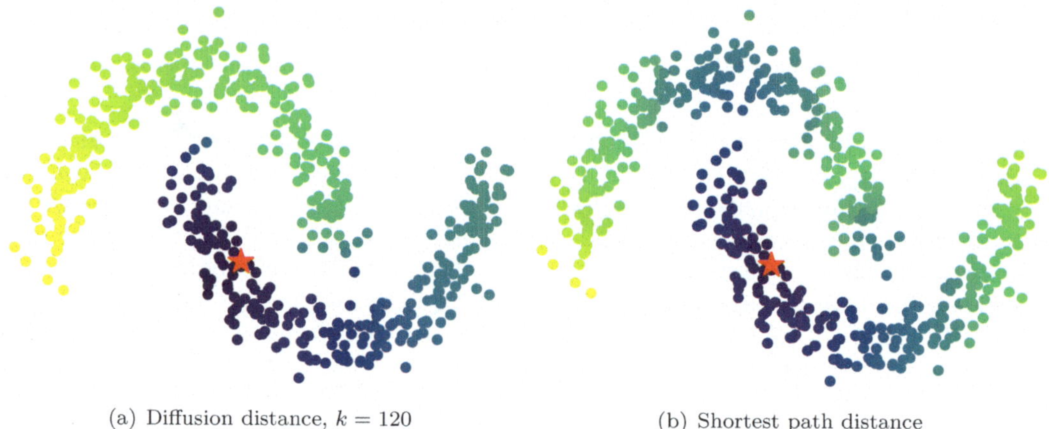

(a) Diffusion distance, $k = 120$

(b) Shortest path distance

Figure 9.31: Comparison of diffusion distance with shortest path distance, showing how diffusion distance is better able to separate the two moons.

A natural question is whether diffusion distance defines a *metric* on the graph. Distances between nodes are clearly symmetric, $d_{ij}^{(k)} = d_{ji}^{(k)}$, and satisfy the triangle inequality

$$d_{i\ell}^{(k)} \le d_{ij}^{(k)} + d_{j\ell}^{(k)} \qquad \text{for all} \qquad i, j, \ell, \tag{9.88}$$

which the reader is asked to justify in Exercise 7.1. The final requirement is that $d_{ij}^{(k)} = 0$ if and only if $i = j$.

Theorem 9.67. *If the weight matrix W is nonsingular then the diffusion distances $d_{ij}^{(k)}$ define a metric on the nodes of the graph for all $k \ge 1$.*

Proof. We only need to show that $d_{ij}^{(k)} = 0$ if and only if $i = j$. Clearly $d_{ii}^{(k)} = 0$. Since W

is nonsingular, so is W^T, $P = W^T D^{-1}$ and P^k. Thus, if $d_{ij}^{(k)} = 0$ then $P^k(\mathbf{e}_i - \mathbf{e}_j) = 0$, so $\mathbf{e}_i = \mathbf{e}_j$, which implies $i = j$. $\blacksquare$

Remark 9.68. In some situations, the weight matrix W may be singular, and Theorem 9.67 may fail to hold. In such case the diffusion distance defines a *pseudometric*, meaning that we may have $d_{ij}^{(k)} = 0$ when $i \neq j$. As an example, consider the three node graph with rank 2 weight matrix given in (9.64), whose kernel spanned by the vector $\mathbf{v} = (1, 0, -1)^T$. Here, diffusion starting at node 1 is indistinguishable from diffusion starting from node 3 after one step; indeed, in either case all the mass moves to node 2 in the first step — that is, if $\mathbf{x}_0 = \mathbf{e}_1$ or $\mathbf{x}_0 = \mathbf{e}_3$, then $\mathbf{x}_1 = \mathbf{e}_2$ — and the diffusions then proceed identically from that point on. Therefore $d_{13}^{(k)} = 0$ for all $k \geq 2$. $\blacktriangle$

Following our treatment of shortest path distances in Section 9.5, let us look for an isometric embedding of the graph nodes into Euclidean space that respects the diffusion distance between nodes. That is, we seek vectors $\mathbf{x}_1, \ldots, \mathbf{x}_m \in \mathbb{R}^n$ for some $n \geq 1$ such that

$$d_{ij}^{(k)} = \| \mathbf{x}_i - \mathbf{x}_j \| \qquad \text{for all} \qquad 1 \leq i, j \leq m.$$

In contrast to the general setting of finite metric spaces, which cannot always be embedded exactly, it turns out that the diffusion distance, for a particular choice of inner product, can *always* be isometrically embedded into Euclidean space.

The embedding utilizes the spectral decomposition of the random walk graph Laplacian, so for the rest of the section we work with a symmetric weight matrix $W = W^T$. Let $\mathbf{p}_1, \ldots, \mathbf{p}_m$ and $\lambda_1, \ldots, \lambda_m$ be the eigenvectors and eigenvalues of L_{rw}, with the eigenvalues sorted from smallest to largest. Set $\Lambda = \mathrm{diag}\,(\lambda_1, \ldots, \lambda_m)$ and $Q = (\mathbf{p}_1 \cdots \mathbf{p}_m)$. We can assume that the eigenvectors are orthonormal with respect to the degree inner product $\langle \mathbf{x}, \mathbf{y} \rangle_D = \mathbf{x}^T D \mathbf{y}$, so that $Q^T D Q = I$. The following result explains how to construct the embedding.

Theorem 9.69. *Let $C \in \mathcal{M}_{m \times m}$ be symmetric, positive definite. Set*

$$B = Q^T D C D Q \quad and \quad X = Q\,(1 - \Lambda)^k = (\mathbf{x}_1, \ldots, \mathbf{x}_m)^T, \tag{9.89}$$

so $\mathbf{x}_i^T$ is the i-th row of X. Then B is symmetric positive definite; let $\| \mathbf{x} \|_B = \sqrt{\mathbf{x}^T B \mathbf{x}}$ be the associated norm. Then the corresponding diffusion distance can be expressed as

$$d_{ij}^{(k)} = \| \mathbf{x}_i - \mathbf{x}_j \|_B. \tag{9.90}$$

Proof. By the spectral decomposition of the random walk graph Laplacian (9.67) we have

$$P^k = (I - L_{\mathrm{rw}}^T)^k = D \sum_{i=1}^m (1 - \lambda_i)^k \mathbf{p}_i \mathbf{p}_i^T = D Q (1 - \Lambda)^k Q^T.$$

Therefore, using the definitions (9.89) of B and X,

$$\begin{aligned}
(d_{ij}^{(k)})^2 &= \| D Q (1 - \Lambda)^k Q^T (\mathbf{e}_i - \mathbf{e}_j) \|_C^2 \\
&= (\mathbf{e}_i - \mathbf{e}_j)^T Q (1 - \Lambda)^k Q^T D C D Q (1 - \Lambda)^k Q^T (\mathbf{e}_i - \mathbf{e}_j) \\
&= \| (1 - \Lambda)^k Q^T (\mathbf{e}_i - \mathbf{e}_j) \|_B^2 = \| X^T \mathbf{e}_i - X^T \mathbf{e}_j \|_B^2 = \| \mathbf{x}_i - \mathbf{x}_j \|_B^2.
\end{aligned}$$

$\blacksquare$

Theorem 9.69 almost gives us an isometric embedding, except that the norm used on the right hand side of (9.90) is *not* the Euclidean norm, and is instead the norm induced by the matrix B. We now make a choice of the inner product matrix C defining the diffusion distance in Definition 9.66. If we set $C = D^{-1}$, then

$$B = Q^T D C D Q = Q^T D D^{-1} D Q = Q^T D Q = \mathrm{I}$$

is the identity matrix! In this case, Theorem 9.69 gives an isometric embedding of the graph with the diffusion distance metric $d_{ij}^{(k)}$ into $\mathbb{R}^m$ with the Euclidean distance. For this reason, the choice $C = D^{-1}$ is the standard one for diffusion distances; see [45, 171]. From now on, we fix $C = D^{-1}$.

Remark 9.70. It may initially seem strange that we ended up choosing the inner product defined by $C = D^{-1}$, while the random walk graph Laplacian is self-adjoint in the inner product defined by D. The reason for this is that the eigenvectors of P are *not* the same as those of L_{rw}. In fact, the eigenvectors of P are $D\mathbf{p}_i$; see the proof of Theorem 9.72. If we were to work with the eigenvectors of P instead of L_{rw}, the statement of Theorem 9.69 would be somewhat simpler; see Exercise 7.4.   ▲

Now, in practice, the embedded points $\mathbf{x}_i \in \mathbb{R}^m$ may still belong to a very high dimensional space, since we may be working very large graphs and m is the number of nodes. Fortunately, we can obtain a low distortion embedding simply by truncating the eigenvectors. For an integer $1 \le d \le m$ we define $\Lambda_d = \mathrm{diag}\,(\lambda_1, \ldots, \lambda_d)$ and $Q_d = (\mathbf{p}_1 \, \cdots \, \mathbf{p}_d)$. This serves to motivate the following definition.

> **Definition 9.71.** For any integers $d, k \ge 1$, we define the *diffusion map embedding matrix*
> $$X_d^{(k)} = Q_d (\mathrm{I} - \Lambda_d)^k \in \mathcal{M}_{m \times d}. \tag{9.91}$$

The rows of $X_d^{(k)}$ are vectors in $\mathbb{R}^d$, so the diffusion map embedding matrix (9.91) produces an embedding of the nodes into $\mathbb{R}^d$. We are free to choose d as we like, and for visualization purposes we may select $d = 2$ or $d = 3$. While we are thinking of k as the number of steps of diffusion, which is an integer, in (9.91) we may take $k > 0$ to be any positive real number, and can think of it as a *diffusion time*.

Recall from Remark 9.56 that many common similarity weight matrices are positive semidefinite. In this case, we can easily bound the distortion of the embedding.

> **Theorem 9.72.** *Assume that W is positive semidefinite. Let $1 \le d \le m - 1$, $k \ge 1$, and write $X_d^{(k)} = (\mathbf{x}_1, \ldots, \mathbf{x}_m)^T$, where $\mathbf{x}_i \in \mathbb{R}^d$. Then*
>
> $$\|\mathbf{x}_i - \mathbf{x}_j\| \le d_{ij}^{(k)} \le \|\mathbf{x}_i - \mathbf{x}_j\| + \alpha_{ij}(1 - \lambda_{d+1})^k, \quad \text{for all} \quad 1 \le i, j \le m, \tag{9.92}$$
>
> *where the constant $\alpha_{ij} = \sqrt{d_i^{-1} + d_j^{-1}}$ depends only on the degree vector $\mathbf{d}$.*

Proof. Since W is positive semidefinite, by Remark 9.56 we have that $\lambda_i \in [0, 1]$ for all i, and so

$$(1 - \lambda_1)^k \ge (1 - \lambda_2)^k \ge \cdots \ge (1 - \lambda_m)^k \ge 0. \tag{9.93}$$

(a) Two Moons   (b) Circles

Figure 9.32: Toy examples of diffusion map embeddings on the two moons and circles data sets. In both cases, the data sets are linearly separable in the embedded space.

Let us now define the spectrally truncated matrix

$$P_d = D \sum_{i=1}^{d} (1 - \lambda_i) \mathbf{p}_i \mathbf{p}_i^T, \qquad \text{for which} \qquad P_d^k = D \sum_{i=1}^{d} (1 - \lambda_i)^k \mathbf{p}_i \mathbf{p}_i^T.$$

By an argument similar to the proof of Theorem 9.69,

$$\| P_d^k (\mathbf{e}_i - \mathbf{e}_j) \|_{D^{-1}} = \| \mathbf{x}_i - \mathbf{x}_j \|.$$

We now write P_d^k in a slightly different way. Define $\mathbf{u}_i = D\mathbf{p}_i$ and note that $\mathbf{u}_1, \ldots, \mathbf{u}_m$ are orthonormal for the inner product $\langle \mathbf{x}, \mathbf{y} \rangle_{D^{-1}}$, and, in fact, are exactly the eigenvectors of P; see Exercise 7.2. Thus, we can write

$$P_d^k = \sum_{i=1}^{d} (1 - \lambda_i)^k \mathbf{u}_i \mathbf{u}_i^T D^{-1}, \quad \text{so} \quad P_d^k \mathbf{v} = \sum_{i=1}^{d} (1 - \lambda_i)^k \langle \mathbf{u}_i, \mathbf{v} \rangle_{D^{-1}} \mathbf{u}_i, \quad \text{for any} \quad \mathbf{v} \in \mathbb{R}^n.$$

We now use (9.93) to compute, for $\mathbf{v} = \mathbf{e}_i - \mathbf{e}_j$, that

$$\left(d_{ij}^{(k)} \right)^2 = \| P^k \mathbf{v} \|_{D^{-1}}^2 = \sum_{\ell=1}^{m} (1 - \lambda_\ell)^{2k} \langle \mathbf{u}_\ell, \mathbf{v} \rangle_{D^{-1}}^2$$

$$= \sum_{\ell=1}^{d} (1 - \lambda_\ell)^{2k} \langle \mathbf{u}_\ell, \mathbf{v} \rangle_{D^{-1}}^2 + \sum_{\ell=d+1}^{m} (1 - \lambda_\ell)^{2k} \langle \mathbf{u}_\ell, \mathbf{v} \rangle_{D^{-1}}^2$$

$$= \| P_d^k \mathbf{v} \|_{D^{-1}}^2 + \sum_{\ell=d+1}^{m} (1 - \lambda_\ell)^{2k} \langle \mathbf{u}_\ell, \mathbf{v} \rangle_{D^{-1}}^2 = \| \mathbf{x}_i - \mathbf{x}_j \|^2 + \sum_{\ell=d+1}^{m} (1 - \lambda_\ell)^{2k} \langle \mathbf{u}_\ell, \mathbf{v} \rangle_{D^{-1}}^2$$

$$\leq \| \mathbf{x}_i - \mathbf{x}_j \|^2 + (1 - \lambda_{d+1})^{2k} \sum_{\ell=1}^{m} \langle \mathbf{u}_\ell, \mathbf{v} \rangle_{D^{-1}}^2 = \| \mathbf{x}_i - \mathbf{x}_j \|^2 + (1 - \lambda_{d+1})^{2k} \| \mathbf{v} \|_{D^{-1}}^2.$$

Taking square roots on both sides and using that $\sqrt{a^2 + b^2} \leq a + b$ when $a, b \geq 0$, it follows that

$$\| \mathbf{x}_i - \mathbf{x}_j \| \leq d_{ij}^{(k)} \leq \| \mathbf{x}_i - \mathbf{x}_j \| + (1 - \lambda_{d+1})^k \| \mathbf{v} \|_{D^{-1}}.$$

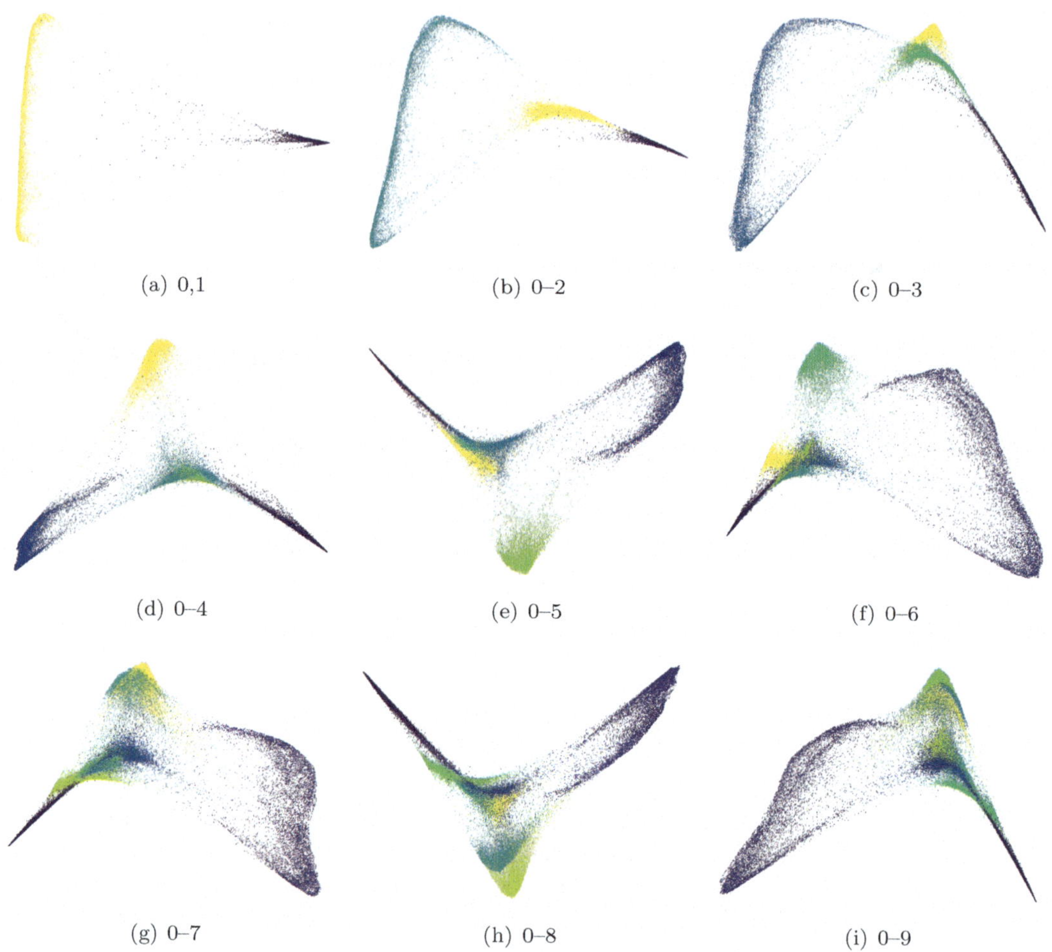

Figure 9.33: Diffusion map embeddings of MNIST digits with $k = 3$ and $d = 3$, where we omit the uninformative first component. Diffusion maps separates the digits well when there are a handful of digits in the embedding space, but we are still unable to separate all 10 MNIST digits.

To complete the proof, we compute

$$\|\mathbf{v}\|_{D^{-1}}^2 = \|\mathbf{e}_i - \mathbf{e}_j\|_{D^{-1}}^2 = \|\mathbf{e}_i\|_{D^{-1}}^2 + \|\mathbf{e}_j\|_{D^{-1}}^2 = d_i^{-1} + d_j^{-1}. \qquad \blacksquare$$

We now turn to some experiments with real data. When visualizing data with diffusion maps the first column of $X_d^{(k)}$ corresponds to the first eigenvector of the random walk Laplacian, which is trivial, and so is often omitted. Thus, when we show the diffusion maps embedding into the plane $\mathbb{R}^2$, we are actually computing $X_3^{(k)}$ and deleting the first column. Figure 9.32 shows the diffusion map embeddings of the two moons and circles data sets with diffusion time $k = 3$. Figure 9.33 shows the diffusion map embedding, also with $k = 3$, on the full MNIST data set containing 70000 digits. Here, the second and third eigenvalues λ_2, λ_3 are very similar, and so the value k of the diffusion time does not significantly affect the embedding. In this illustration, we used a symmetrized 10 nearest neighbor graph construction. We can see that the digits are very well separated when there are 4 or less present, but then we start to see some mixing in the embedded space.

Remark 9.73. The diffusion maps algorithm in [45] uses an additional normalization of the weight matrix that we have not discussed yet, and overview briefly here. Let $\alpha \geq 0$ and define the α–normalized weight and degree matrices

$$W_\alpha = D^{-\alpha} W D^{-\alpha}, \quad D_\alpha = \operatorname{diag}(d_{\alpha,1}, \ldots, d_{\alpha,m}), \quad \text{where} \quad d_{\alpha,i} = \sum_{j=1}^{m} w_{\alpha,ij}. \tag{9.94}$$

The α–*normalized graph Laplacian* is defined as the random walk graph Laplacian for the weight matrix W_α, that is

$$L_\alpha = I - D_\alpha^{-1} W_\alpha. \tag{9.95}$$

In the diffusion maps algorithm [45], the graph Laplacian L_α is used in place of L_{rw} in the definition of diffusion distance in (9.87). Since we worked with a general weight matrix W in this section, the only difference is that we would in general take the extra step of replacing W with W_α. The reason for introducing the parameter α is to allow one to control how sensitive the Laplacian is to the *density* of the data, which is often reflected in the degrees of each node in the graph. The choice $\alpha = 1$ is used in the standard diffusion maps framework, and removes the dependence on density entirely, revealing the geometric nature of the data; see [45] for further details. $\blacktriangle$

We mention here are several other related dimension reduction techniques, including local linear embedding [199] and Laplacian eigenmaps [20].

9.7.2 Spectral Clustering

We saw above how diffusion maps are able to embed a graph into a low dimensional Euclidean space while preserving the diffusion distance (at least approximately), thereby providing a simple low dimensional Euclidean representation of a graph that respects its structure. This low dimensional embedding is useful for many downstream machine learning tasks, such as clustering, semi-supervised learning, or even fully supervised learning. When the downstream task is clustering, this leads to the *multiclass spectral clustering* algorithm, which we describe in this section (recall the case of binary spectral clustering was introduced in Section 9.4).

Let $\mathcal{G}$ be a weighted graph, which in applications of spectral clustering is often built over a point cloud. There are several varieties of spectral clustering, but they all work on this same principle: project the graph onto the first k eigenvectors of a graph Laplacian, and then apply k-means clustering. While one can certainly use the diffusion map embedding matrices from Definition 9.71, it is more common to simply use the eigenvectors directly, and ignore the eigenvalues and the diffusion time parameter. That is, to partition into k clusters, we define the *spectral embedding* of our graph into $\mathbb{R}^k$ by the $m \times k$ data matrix

$$X = (\mathbf{p}_1 \; \cdots \; \mathbf{p}_k) \in \mathcal{M}_{m \times k},$$

where $\mathbf{p}_1, \ldots, \mathbf{p}_k$ are the first k eigenvectors of a graph Laplacian matrix. The *rows* $\mathbf{x}_1, \ldots, \mathbf{x}_m$ of $X = (\mathbf{x}_1 \; \cdots \; \mathbf{x}_m)^T$ are the embedded points $\mathbf{x}_i \in \mathbb{R}^k$ upon which we run the k-means clustering algorithm. The coordinates of each data point $\mathbf{x}_i$ are simply the values of the first k eigenvectors $\mathbf{p}_1, \ldots, \mathbf{p}_k$ when evaluated on the i-th node in the graph.

Spectral clustering methods are distinguished by which normalization of the graph Laplacian is used. The method proposed by Shi and Malik [213] uses the eigenvectors of the random walk graph Laplacian L_{rw}, while the method proposed by Ng, Jordan, and Weiss [176] uses the symmetric normalized graph Laplacian. In fact, in [176] an additional step is taken to project each $\mathbf{x}_i$ to the unit Euclidean sphere, so the embedded points are defined as $\widehat{\mathbf{x}}_i = \mathbf{x}_i / \|\mathbf{x}_i\|$.

This can improve the separation of clusters in the k-means algorithm, since much of the overlap between clusters occurs near the origin, where the eigenvectors are changing sign. It is certainly also possible to use the eigenvectors of the graph Laplacian directly, but normalizations tend to produce better results when $k > 2$. We note that this essentially coincides with the binary spectral clustering algorithm introduced in Section 9.4 when $k = 2$, since the first eigenvector is trivial, while the second is the Fiedler vector (if one is using the unnormalized or random walk Laplacian), the only difference being that here we use 2-means clustering instead of thresholding the Fiedler vector by its sign.

Now, there is a very good reason to use the top k eigenvectors, where k is also the desired number of clusters. In the case where the graph is truly composed of k disconnected components, i.e., the clusters, then, by Theorem 9.22, the first k eigenvectors of the graph Laplacian exactly indicate these k components. Thus, in this ideal disconnected case, spectral clustering will perfectly recover the k disconnected components. Using fewer than k eigenvectors would fail to recover all components, while using more would lead to the possibility of splitting a connected component into more than one cluster. When the graph is connected, but there are relatively few edges between nodes in different clusters, then we expect the first k eigenvectors to be a perturbation of the ideal disconnected case, and to therefore reflect an approximate cluster structure. A proof of such a perturbation result is well beyond the scope of this book; see, e.g., [109, 140, 231, 240]. We also mention that this is only a brief introduction to spectral clustering; we refer the reader to [239] for more in-depth survey. Let us now proceed with applications.

Figure 9.34 shows the result of spectral clustering on the MNIST data set using $k = 12$ clusters. We used a symmetrized 10-nearest neighbor graph constructed with the Euclidean distance between images, and the spectral clustering method of Ng, Jordan and Weiss [176]. We can see that some digits, like the 1's, are split into two clusters depending on how the digit is written, while other found clusters contain mixes of two or more digits. In particular, note the two clusters containing fours and nines. We chose $k = 12$ since we expected some digits to split into multiple clusters. The clustering purity, which was introduced in Section 7.5, over all 12 clusters is $p = 0.80$, while 8 of the cluster purities are above 0.9 and 5 are above 0.95.

Another common application of spectral clustering is image segmentation, which was the motivation for the algorithm of Shi and Malik [213]. Image segmentation refers to the process of partitioning an image in a meaningful way, such as separating objects of interest from the background, and is one of the most fundamental tasks in computer vision. It is closely related to data clustering, with the clusters corresponding to objects and the background in the image.

To apply spectral clustering to image segmentation, as in [213] we build a graph over the pixels. The graph is a variant of our geometric similarity graph construction, where the weights depend on both the spatial proximity of pixels, and the similarity in their colors or features. One choice is a weight of the form

$$w_{ij} = \exp\left(-\frac{\|\mathbf{x}_i - \mathbf{x}_j\|^2}{2\varepsilon_X^2} - \frac{\|\mathbf{y}_i - \mathbf{y}_j\|^2}{2\varepsilon_Y^2}\right), \tag{9.96}$$

where $\mathbf{x}_i$ is the spatial location and $\mathbf{y}_i$ the color of pixel i, respectively. We can truncate the graph by setting $w_{ij} = 0$ when $\|\mathbf{x}_i - \mathbf{x}_j\| \geq r$ in order to create a sparse graph for computations. The values of the parameters ε_X and ε_Y are tuned by the user. We can also replace the pixel values at $\mathbf{y}_i$ with other image features, like texture descriptors for texture segmentation.

We show an example of image segmentation via spectral clustering in Figure 9.35. There

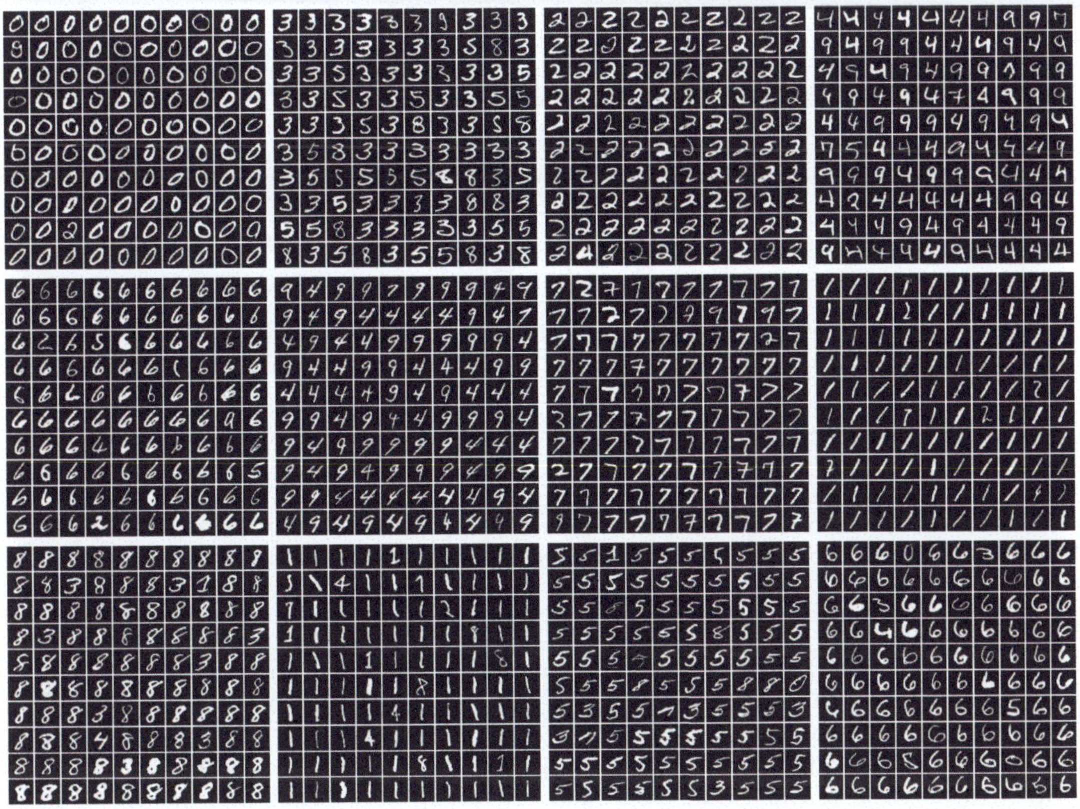

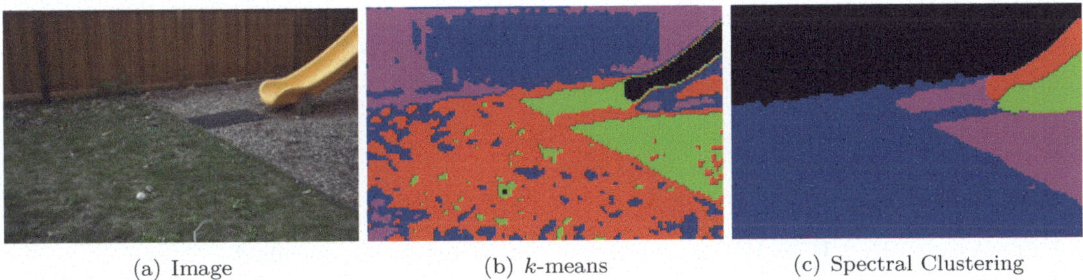

Figure 9.34: The 12 clusters found by spectral clustering on the MNIST data set. We can see some digits appear to have multiple clusters (e.g., the ones digit, depending on whether it is written with a slant or not), while the fours and nines are mixed into two clusters.

(a) Image   (b) k-means   (c) Spectral Clustering

Figure 9.35: Comparison of k-means ($k = 5$) and spectral clustering with five classes applied to an image containing a slide, grass, fence, and woodchip area.

are roughly five clusters in the image: the grass, the fence, the slide, the woodchip area, and the black mat. Spectral clustering is able to identify four of these clusters quite well, especially compared to k-means clustering shown in Figure 9.35(b), which is applied to the pixel values directly and doesn't take into account that the clusters should be grouped together spatially.

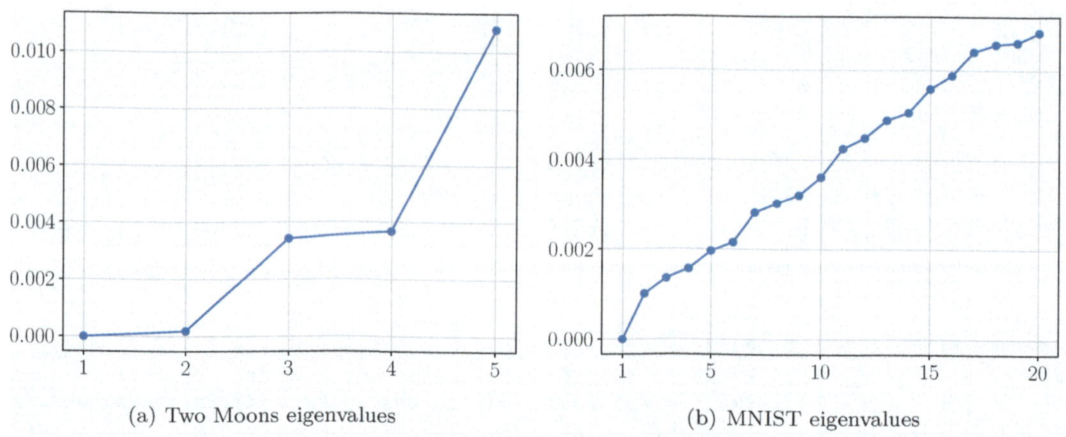

(a) Two Moons eigenvalues   (b) MNIST eigenvalues

Figure 9.36: The low lying eigenvalues for the symmetric normalized graph Laplacian on the two moons and MNIST data sets. We can see a clear spectral gap after the 2 clusters on two moons. However, with MNIST it is more difficult to deduce from the eigenvalues how many clusters may be present.

As was the case for the Fiedler value, the first k eigenvalues of the graph Laplacian can give us some indication of how well clustered the data set is. Indeed, if these eigenvalues all vanish, then the graph has at least k disconnected components, which we interpret as clusters. When the graph is connected but has k well-defined clusters with very few edges between them, then we expect the eigenvalues $\lambda_2, \ldots, \lambda_k$ to be small, and the size of these eigenvalues can give us some information about how well clustered the graph is, and how to choose the number of clusters k. In particular, we can try to look for a k such that the first k eigenvalues are very small, and then there is a large jump to λ_{k+1}. This would indicate that there are k very well-separated clusters in the data set, but not $k+1$ clusters. This idea works well with synthetic data sets, but its performance on real-world data is often of varying quality. In Figure 9.36 we show the first few eigenvalues for the two moons data set and the MNIST data set. For two moons we clearly see a spectral jump after the first 2 eigenvalues, indicating there are two clusters — the two moons. On MNIST, we do not see a very clear signal, and just observe a gradual increase in the eigenvalues.

Exercises

Note: Assume that $W = W^T$ is symmetric, unless otherwise specified.

7.1. ♡ Prove that the diffusion distance satisfies the triangle inequality (9.88).

7.2. Define $\mathbf{u}_i$ and λ_i as in the proof of Theorem 9.72. Show that (a) $\mathbf{u}_1, \ldots, \mathbf{u}_m$ are orthonormal in the inner product $\langle \mathbf{x}, \mathbf{y} \rangle_{D^{-1}} = \mathbf{x}^T D^{-1} \mathbf{y}$, and (b) $P\mathbf{u}_i = (1 - \lambda_i)\mathbf{u}_i$ for all i.

7.3. Suppose that $W = W^T$ is a symmetric and singular weight matrix, so the diffusion distance $d_{ij}^{(k)}$ may not be a metric; see Theorem 9.67. Suppose that we add self loops with weight $\tau > 0$ to all nodes in the graph, that is, we replace W with $W + \tau$. For which values of τ is the corresponding diffusion distance is a metric?

7.4. ♡ Let S be a self-adjoint matrix for the inner product $\langle \mathbf{x}, \mathbf{y} \rangle_C = \mathbf{x}^T C \mathbf{y}$, where C is symmetric positive definite, and let $S = Q \Lambda Q^T C$ be its spectral decomposition. Follow the proof of Theorem 9.69 to show that $\| S^k (\mathbf{e}_i - \mathbf{e}_j) \|_C = \| \mathbf{x}_i - \mathbf{x}_j \|$, where $\mathbf{x}_1, \ldots, \mathbf{x}_m \in \mathbb{R}^m$ are the rows of the matrix $X = C Q \Lambda^k$.

7.5. ◇ Write Python code to perform multiclass spectral clustering via the orthogonal iteration method described in Section 5.6. Implement both Shi and Malik [213] and Ng, Jordan and Weiss methods [176]. Test your method on the MNIST data set, or subsets thereof. You may use the k-means clustering code from the `sklearn` package. For the orthogonal iteration, you may find the notebook from Exercise 1.8 in Chapter 8 helpful.

7.6. ♡ Use Remark 8.12 to show that the first k eigenvectors $\mathbf{q}_1, \ldots, \mathbf{q}_k$ of the graph Laplacian L solve the optimization problem

$$\min \sum_{i=1}^{k} \mathbf{u}_i^T L \mathbf{u}_i = \operatorname{tr} (U^T L U), \tag{9.97}$$

where the minimum is over all $m \times k$ matrices $U = (\, \mathbf{u}_1 \ \ldots \ \mathbf{u}_k \,)$ with orthonormal columns. What is the minimum value of this optimization problem?

7.7. Formulate and prove Exercise 7.6 for the symmetric normalized and random walk graph Laplacians.

7.8. ◇ Let $\mathcal{G}$ be an unweighted graph and let $\mathcal{D}_1, \ldots, \mathcal{D}_k \subset \mathcal{N}$ be a partition of the nodes $\mathcal{N}$ in the graph $\mathcal{G}$ into k disjoint subsets. Assume that each set $\mathcal{D}_i$ has at most $p_i \geq 0$ edges that connect nodes in $\mathcal{D}_i$ to nodes in the complement $\mathcal{D}_i^c = \mathcal{N} \setminus \mathcal{D}_i$ (if $p_i = 0$ then $\mathcal{D}_i$ is disconnected from the rest of the graph). Let $\lambda_1, \ldots, \lambda_k$ denote the first k eigenvalues of the graph Laplacian matrix L. Use Exercise 7.6 to show that

$$\lambda_1 + \cdots + \lambda_k \leq \frac{p_1}{\#\mathcal{D}_1} + \cdots + \frac{p_k}{\#\mathcal{D}_k}. \tag{9.98}$$

9.8   t-SNE Embedding

Python Notebook: t-SNE (.ipynb)

We have already seen several methods for embedding data and graphs into low dimensional spaces, such as $\mathbb{R}^2$ or $\mathbb{R}^3$, for visualization purposes. We introduced PCA, LDA, and MDS in Chapter 8, the ISOMAP algorithm in Section 9.5.3, and diffusion maps in Section 9.7. Each of these methods struggled in various ways. We refer to, for example, Figure 9.33 showing the diffusion map embeddings of MNIST digits, which are not able to show the cluster structure — that is, there is significant overlap between many of the digits. In this section, we discuss a graph-based embedding and visualization technique called the *t-distributed stochastic neighbor embedding* (t-SNE).[14] The t-SNE embedding is used to visualize high dimensional data, and was originally proposed by van der Maaten and Hinton in 2008 [235]. It is now widely used in the natural sciences, e.g., gene analysis and math biology, for visualizing data and discovering cluster structure [130, 147], and has inspired new and improved variants, such as the uniform manifold approximation (UMAP) algorithm [160].

As we saw in Section 9.5.3, it is generally impossible to isometrically embed even small data sets into Euclidean space, much less a low dimensional space like $\mathbb{R}^2$. Indeed, in Example 9.48 we gave an example of a 4 point metric space that cannot be isometrically embedded in *any* Euclidean space. All of the methods we have considered thus far, like PCA, LDA, MDS, ISOMAP, and diffusion maps, address this by looking for low distortion embeddings. For example, given pairwise distances t_{ij} between data points, ISOMAP, or metric MDS, looks for embedded points $\mathbf{z}_i$ such that $t_{ij} \approx \|\mathbf{z}_i - \mathbf{z}_j\|$ for all i and j. When embedding into a low dimensional space like $\mathbb{R}^2$ for visualization, we expect to have a great deal of distortion, and the issue with such approaches is that they spread the distortion equally over *all* pairwise distances, yielding an embedding where most distances are poorly represented.

The t-SNE embedding algorithm seeks to *prioritize* which pairwise distances should be preserved, and which are allowed to be distorted. Since we cannot faithfully reconstruct all distances, it is better to intelligently select where to allow distortion. The key idea behind t-SNE is that the *local* structure should be prioritized over global structure. That is, we should prioritize faithfully preserving small distances t_{ij}, while allowing distortions of large distances. This gives the embedding a sufficient degree of flexibility to enable it to accurately preserve local structures in data, such as nearest neighbor information.

Let us now describe the t-SNE method mathematically. The starting point is a weighted graph with $m \times m$ weight matrix W, not necessarily symmetric. An example is the k-nearest neighbor weight matrix constructed at the end of Section 9.1, but the t-SNE method uses a specific graph construction based on the notion of *perplexity*, which we discuss later in this section. Given a weight matrix W, we construct a matrix

$$P = \frac{1}{2m}\left(D^{-1}W + W^T D^{-1}\right) \tag{9.99}$$

by normalizing and symmetrizing W using the associated diagonal degree matrix D. Note that $P = P^T$ is symmetric, with zero diagonal entries $p_{ii} = 0$, since $w_{ii} = 0$, and nonnegative off-diagonal entries: $p_{ij} = p_{ji} \geq 0$. Moreover, the sum of all the entries in P is one, i.e., $\mathbf{1}^T P \mathbf{1} = \sum_{i,j} p_{ij} = 1$; see Exercise 8.1. We will call P a *probability matrix*.[15]

The t-SNE embedding aims to find embedded points $\mathbf{z}_1, \mathbf{z}_2, \ldots, \mathbf{z}_m \in \mathbb{R}^d$, where usually $d = 2$ or 3, so that the similarity between $\mathbf{z}_i$ and $\mathbf{z}_j$ matches p_{ij} as closely as possible. The similarity matrix for the $\mathbf{z}_i$, denoted by Q, is the $m \times m$ matrix with entries

$$q_{ij} = \begin{cases} \dfrac{1}{\gamma\left(1 + \|\mathbf{z}_i - \mathbf{z}_j\|^2\right)}, & i \neq j \\ 0 & i = j, \end{cases} \quad \text{where} \quad \gamma = \sum_{\substack{i,j=1 \\ i \neq j}}^{m} \frac{1}{1 + \|\mathbf{z}_i - \mathbf{z}_j\|^2}. \tag{9.100}$$

The factor γ is included so that the entries of the matrix Q sum to 1, and hence Q is also a probability matrix.

The t-SNE algorithm chooses the points $\mathbf{z}_i$ so as to minimize the distance between the probability distributions P and Q. Specifically, it minimizes the *Kullback–Leibler divergence*

---

[14]t-SNE is pronounced like "Disney".
[15]*Warning*: in some texts, the term "probability matrix" refers to a transition matrix satisfying (5.79).

between P and Q, which is given by

$$E(\mathbf{z}_1, \mathbf{z}_2, \ldots, \mathbf{z}_m) = D(P \,\|\, Q) = \sum_{i,j=1}^{m} p_{ij} \log\left(\frac{p_{ij}}{q_{ij}}\right) = \sum_{i,j=1}^{m} \left(p_{ij} \log p_{ij} - p_{ij} \log q_{ij}\right), \quad (9.101)$$

and was introduced in Section 6.7. In particular, $D(P \,\|\, Q) \geq 0$ and equals zero when $P = Q$; recall Exercise 7.7 from Chapter 6. We also note that since $\lim_{x \to 0^+} x \log x = 0$, we interpret the term in the sum as zero whenever $p_{ij} = 0$, regardless of the value of q_{ij}. While the Kullback–Leibler divergence is a convex function of P and Q — see Example 6.43 — it is not a convex function of the embedding points $\mathbf{z}_i$.

Let us say a few words about the choice of the Kullback–Leibler divergence, in place of some other distance between probability distributions. In fact, since $D(P \,\|\, Q)$ is not symmetric in P and Q it is even natural to wonder whether the symmetrized Kullback–Leibler divergence $D(P \,\|\, Q) + D(Q \,\|\, P)$ would be more suitable. The choice of the non-symmetric Kullback–Leibler is actually quite intentional in t-SNE, and it is exactly what allows the method to preserve local structure, while allowing distortions at larger scales. Indeed, notice that if $p_{ij} > 0$ is large, then the (i, j) term in the Kullback–Leibler divergence measures the difference between p_{ij} and q_{ij}, and is only zero when $p_{ij} = q_{ij}$.[16] This places a strong emphasis on ensuring that q_{ij} matches p_{ij} well when nodes i and j are similar, and hence encourages *local* structure to be preserved in the embedding. In contrast, the Kullback–Leibler divergence places little to no penalty on discrepancy between p_{ij} and q_{ij} when $p_{ij} = 0$ or $p_{ij} \ll 1$ — that is, when nodes i and j are not similar. This allows large distortions of dissimilar pairs of points. This preference for preserving local structure is what allows the t-SNE embedding to give meaningful and useful results, even though two- or three-dimensional space cannot capture all of the details and nuances of high dimensional data.

The t-SNE algorithm uses gradient descent to minimize (9.101), starting from a random initial configuration. The only place the $\mathbf{z}_i$ appear is through Q, so we may as well simplify the energy by dropping the terms only involving P, producing

$$\widehat{E}(\mathbf{z}_1, \ldots, \mathbf{z}_m) = -\sum_{i,j=1}^{m} p_{ij} \log q_{ij}, \quad (9.102)$$

at least for the purposes of computing ∇E. By the definition of Q in (9.100), we can split this into two terms

$$\widehat{E}(\mathbf{z}_1, \ldots, \mathbf{z}_m) = \sum_{i,j=1}^{m} p_{ij} \log\left(1 + \|\mathbf{z}_i - \mathbf{z}_j\|^2\right) + \log\left(\sum_{\substack{i,j=1 \\ i \neq j}}^{m} \left(1 + \|\mathbf{z}_i - \mathbf{z}_j\|^2\right)^{-1}\right), \quad (9.103)$$

where we used that $\sum_{i,j} p_{ij} = 1$. The gradient of E has a component in the direction of each $\mathbf{z}_i$, which we denote by $\nabla_{\mathbf{z}_i} E$. To compute them, we begin by noting that

$$\nabla_{\mathbf{z}_\ell} \|\mathbf{z}_i - \mathbf{z}_j\|^2 = \begin{cases} 2(\mathbf{z}_i - \mathbf{z}_j), & \ell = i, \\ 2(\mathbf{z}_j - \mathbf{z}_i), & \ell = j, \\ 0, & \text{otherwise.} \end{cases} \quad (9.104)$$

---

[16] Of course, setting $q_{ij} > p_{ij}$ would make this term negative, but since the Kullback–Leibler divergence is nonnegative, another term in the sum would necessarily increase.

Therefore, using the chain rule,[17]

$$
\nabla_{\mathbf{z}_\ell} \sum_{i,j=1}^{m} p_{ij} \log\left(1 + \|\mathbf{z}_i - \mathbf{z}_j\|^2\right) = \sum_{i,j=1}^{m} p_{ij} \nabla_{\mathbf{z}_\ell} \log\left(1 + \|\mathbf{z}_i - \mathbf{z}_j\|^2\right)
$$

$$
= \sum_{i,j=1}^{m} p_{ij}\left(1 + \|\mathbf{z}_i - \mathbf{z}_j\|^2\right)^{-1} \nabla_{\mathbf{z}_\ell} \|\mathbf{z}_i - \mathbf{z}_j\|^2 \tag{9.105}
$$

$$
= 2\sum_{j=1}^{m} p_{\ell j}\left(1 + \|\mathbf{z}_\ell - \mathbf{z}_j\|^2\right)^{-1}(\mathbf{z}_\ell - \mathbf{z}_j) - 2\sum_{i=1}^{m} p_{i\ell}\left(1 + \|\mathbf{z}_i - \mathbf{z}_\ell\|^2\right)^{-1}(\mathbf{z}_i - \mathbf{z}_\ell)
$$

$$
= 4\sum_{j=1}^{m} p_{\ell j}\left(1 + \|\mathbf{z}_\ell - \mathbf{z}_j\|^2\right))^{-1}(\mathbf{z}_\ell - \mathbf{z}_j) = 4\gamma \sum_{j=1}^{m} p_{\ell j} q_{\ell j}(\mathbf{z}_\ell - \mathbf{z}_j),
$$

where the final expression follows from (9.100). This determines the gradient of the first term in (9.103). By a similar computation — see Exercise 8.2 — the gradient of the second term is found to be

$$
\nabla_{\mathbf{z}_\ell} \log\left(\sum_{\substack{i,j=1 \\ i \neq j}}^{m} \left(1 + \|\mathbf{z}_i - \mathbf{z}_j\|^2\right)^{-1} \right) = -4\gamma \sum_{j=1}^{m} q_{\ell j}^2 (\mathbf{z}_\ell - \mathbf{z}_j). \tag{9.106}
$$

Combining (9.105) and (9.106), and replacing ℓ with i, we find

$$
\nabla_{\mathbf{z}_i} E = 4\gamma \sum_{j=1}^{m} p_{ij} q_{ij}(\mathbf{z}_i - \mathbf{z}_j) - 4\gamma \sum_{j=1}^{m} q_{ij}^2 (\mathbf{z}_i - \mathbf{z}_j). \tag{9.107}
$$

The first term in the gradient is an attraction term; moving in the negative of this direction will move points towards their neighbors, pushing the points $\mathbf{z}_i$ and $\mathbf{z}_j$ together when their weights p_{ij} and q_{ij} are similar and large. The second term is a repulsion term that attempts to spread apart nearby points, and does not depend on the similarity p_{ij}. We also note that the terms in the gradient can be combined to simplify its form as

$$
\nabla_{\mathbf{z}_i} E = 4\gamma \sum_{j=1}^{m} (p_{ij} - q_{ij}) q_{ij}(\mathbf{z}_i - \mathbf{z}_j), \tag{9.108}
$$

but the separation in terms of attraction and repulsion terms is convenient.

The t-SNE energy (9.101) is minimized by an application of gradient descent

$$
\mathbf{z}_i^{(k+1)} = \mathbf{z}_i^{(k)} - \alpha \nabla_{\mathbf{z}_i} E(\mathbf{z}_1^{(k)}, \mathbf{z}_2^{(k)}, \ldots, \mathbf{z}_m^{(k)}), \qquad i = 1, \ldots, m, \tag{9.109}
$$

where $\alpha > 0$ is the time step which, for simplicity, is assumed to be fixed. Since the energy E is highly nonconvex as a function of the embedding points $\mathbf{z}_i$, there may be many different local and global minimizers, and there are no guarantees that gradient descent will converge to a minimizer. While the complexity of computing the gradient, as written in (9.107), is $O(m^2)$, which is computationally intractable for very large data sets, there are fast tree-based approaches to approximating the gradient that can reduce this to $O(m \log m)$ and are

---

[17]Keep in mind that ℓ is fixed while i, j are summation indices.

widely used in practice [234].[18] It is also common to preprocess the data with another linear dimension reduction technique, like PCA, before the t-SNE algorithm is applied in order to reduce the computational burden and remove noise.

Nevertheless, for moderate or large numbers of data points m, gradient descent is very slow to converge, and can sometimes produce poor embeddings and visualizations (compare Figures 9.37 and 9.38 below). To address this, it was proposed in [235] to enhance the attractive forces at the beginning of gradient descent, which is called *early exaggeration*. The gradient $\nabla_{\mathbf{z}_i} E$ is replaced by

$$\widehat{\nabla}_{\mathbf{z}_i} E = 4\gamma\beta \sum_{j=1}^{m} p_{ij} q_{ij} (\mathbf{z}_i - \mathbf{z}_j) - 4\gamma \sum_{j=1}^{m} q_{ij}^2 (\mathbf{z}_i - \mathbf{z}_j), \qquad (9.110)$$

where $\beta > 1$ is the early exaggeration amplification factor, often chosen around $\beta = 10$. The early exaggeration process strongly favors attraction forces, and quickly begins to form clusters in the embedded space. After some number of early exaggeration steps, usually a few hundred, the method switches to ordinary gradient descent, which restores the repulsive forces and spreads out the clusters more evenly. Aside from leading to faster convergence, early exaggeration also produces qualitatively better visualizations, presumably by finding a better local or global minimizer. This process is still poorly understood, especially since the early exaggeration period is *not* gradient descent, and the energy E may not be decreasing during this time.

Remark 9.74. The choice of time step α in t-SNE is crucial to ensure fast convergence. Notice that due to the heavy tails in the definition of q_{ij} in (9.100), the entries q_{ij} will not vary drastically in size. Since Q is a probability matrix, so the q_{ij} sum to one, it is reasonable to assume that $q_{ij} = O(1/m^2)$. A similar argument can be made for the p_{ij}, although for a sparse graph the largest values of p_{ij} will be significantly larger than $O(1/m^2)$ (and many will be zero).[19] This suggests that the norm of the gradient $\nabla_{\mathbf{z}_i} E$ depends on m and decreases substantially for large data sets, where m is large. To compensate for this, it is common to choose very large time steps α in t-SNE; a common choice is $\alpha = m$. ▲

Up to now, we have not discussed how to construct the weight matrix W for t-SNE, and have assumed it is given — recall P is constructed from W in (9.99). It turns out this is important for achieving good results. The version in [235] uses Gaussian weights

$$w_{ij} = \exp\left(-\frac{\|\mathbf{x}_i - \mathbf{x}_j\|^2}{2\,\varepsilon_i^2}\right).$$

The value of each ε_i is tuned to a specified perplexity level s, usually in the range 5 to 50. The *perplexity* of the ith row of w_{ij} is defined as

$$\text{perplexity} = 2^{h_i}, \qquad \text{where} \qquad h_i = -\sum_{j=1}^{m} p_j \log p_j \qquad (9.111)$$

is the *entropy* of the probability vector $\mathbf{p}$ with entries $p_j = w_{ij}/d_i$. The value of ε_i is determined so that the perplexity 2^{h_i} equals the desired user-specified value s, that is we

---

[18]Since p_{ij} appears in the attraction term, using a sparse graph can greatly accelerate the computation of the attraction portion of the gradient. However, the repulsion portion involves the term q_{ij}^2, which is not sparse, but can be approximated by tree-based techniques.

[19]For example, for a k-nearest neighbor graph, km entries of the matrix P would be nonzero, and since the entries sum to one, the nonzero values would be $O(1/(km))$ in size.

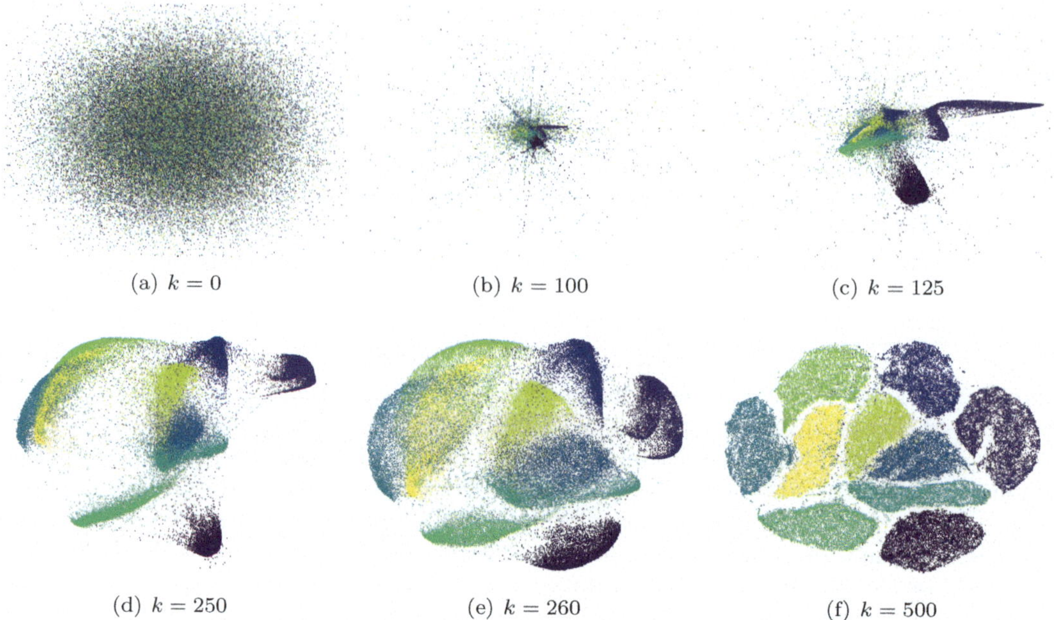

(a) $k = 0$ (b) $k = 100$ (c) $k = 125$

(d) $k = 250$ (e) $k = 260$ (f) $k = 500$

Figure 9.37: Evolution of the t-SNE embedding over the iterations of gradient descent. The result after $k = 250$ iterations in (d) is the end of the early exaggeration phase. Results do not change significantly after 500 iterations.

solve $s = 2^{h_i}$ by choosing ε_i so that

$$\log_2 s = -\sum_{j=1}^{m} p_j \log p_j \qquad \text{where} \qquad p_j = \frac{\exp\left(-\dfrac{\|\mathbf{x}_i - \mathbf{x}_j\|^2}{2\,\varepsilon_i^2}\right)}{\displaystyle\sum_{\ell=1}^{m} \exp\left(-\dfrac{\|\mathbf{x}_i - \mathbf{x}_\ell\|^2}{2\,\varepsilon_i^2}\right)}.$$

Distributions with large entropy h_i are more spread out: a uniform distribution has high entropy, whereas a concentrated distribution like $\mathbf{p} = \mathbf{e}_i$ has low entropy; see Exercise 8.3.

Since the entropy h_i is monotonically increasing with ε_i, we can use a bisection search to find ε_i; see Exercises 8.4, 8.5. Thus, by specifying the same perplexity for all rows of W, we are asking that the distributions of weights in each row are similarly spread out. Larger values of ε_i spread out the weights more, and are roughly equivalent to using a larger neighborhood in the graph construction, so the perplexity graph shares some similarities with k-nearest neighbor graphs, but is not equivalent. As a rule of thumb, one can think of the perplexity value s as the equivalent of the number k of neighbors in a k-nearest neighbor graph.

We now turn to some numerical experiments. We show in Figure 9.37 the evolution of the t-SNE embedding over iterations of gradient descent on the Kullback–Leibler divergence. We used a time step of $\alpha = m = 70000$ and 250 iterations of early exaggeration with $\beta = 12$.[20] In particular, the plot in Figure 9.37(d) shows the result after the end of the early exaggeration phase. We see all the digits are nicely clustered in the embedding space after only 500 iterations. In contrast, we show in Figure 9.38 the results *without* the use of early exaggeration. The clusters in the embedding still have high purity, but many digits are split into several different clusters in the visualization.

---

[20]During early exaggeration, we reduced the time step to $\alpha = m/\beta$ for stability. More sophisticated time step adaptations are common in t-SNE; see, e.g., [234].

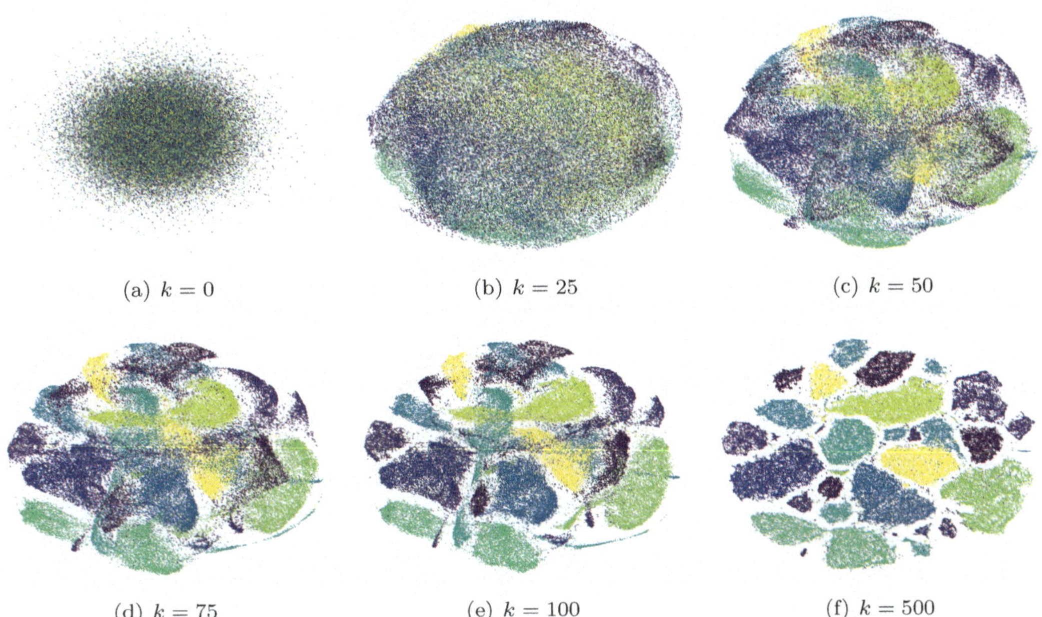

(a) $k = 0$ (b) $k = 25$ (c) $k = 50$

(d) $k = 75$ (e) $k = 100$ (f) $k = 500$

Figure 9.38: Evolution of the t-SNE embedding over the iterations of gradient descent *without* early exaggeration. The results do not change significantly after 500 iterations.

To give an idea of how t-SNE works on toy data sets, where we know the geometry and can easily visualize the ground truth, we show in Figure 9.39 the results of applying t-SNE to the two moons and circles data sets, as well as a data set consisting of $m = 1000$ points sampled uniformly along the standard parabola $y = x^2$ for $-1 \leq x \leq 1$. In the figure we show three different values for perplexity, 5, 30, and 50 from left to right. Larger perplexities encourage the embedding to preserve more global structure, since the similarity weight matrix W has a larger scale of interactions. It is interesting to note that very few geometric properties of the data sets are preserved in the embeddings. It is only with perplexity of 50 that the parabola is embedded in a way that somewhat resembles a parabola, although even then it more closely resembles a circle, with circular artifacts introduced at each end. Finally, we also show in Figure 9.40 three different t-SNE embeddings of the two circles data set, obtained from different initializations of gradient descent. In only one case is the outer circle embedded as a single cluster. In the other two embeddings, the outer circle is broken into 3 and 4 clusters, respectively. This reflects the t-SNE algorithm's preference for local structure over global, which allows it to sometimes hallucinate clusters that are not present in the original data set. A rigorous understanding of the ability of t-SNE to faithfully represent clusters in data is currently lacking.

Remark 9.75. The name t-distributed stochastic neighbor embedding (t-SNE) was derived from two aspects; the choice of weights q_{ij} in the embedded space is a Student's t-distribution with parameter $\nu = 1$, cf. [131], and the construction of the weight matrix W is a similarity graph encoding the neighborhood information in the high dimensional space. Another method named stochastic neighbor embedding (SNE) was proposed earlier in 2002, [106], the main difference being that SNE uses Gaussian weights in the definition of q_{ij} in (9.100); see Exercise 8.8. At an intuitive level, the very fast decay of the Gaussian weights for q_{ij} in the SNE method made it difficult to optimize the energy E and produced less impressive embeddings.

(a) Two Moons

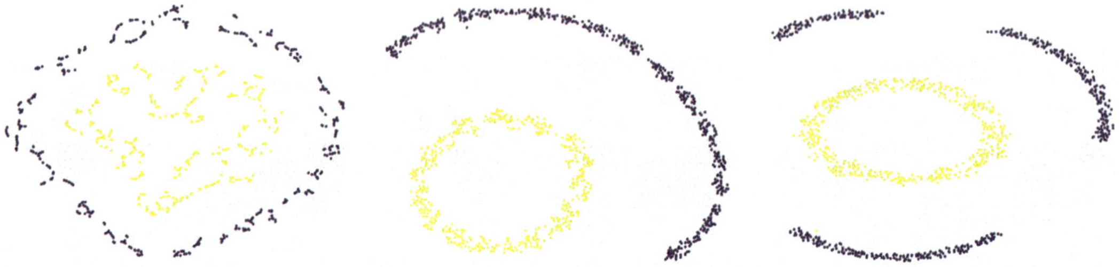

(b) Circles

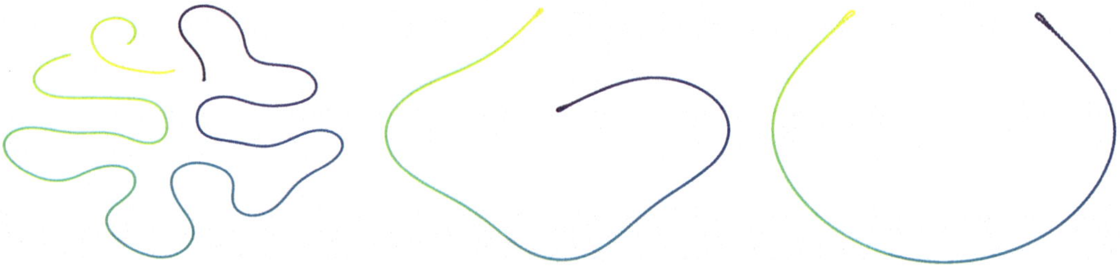

(c) Parabola

Figure 9.39: t-SNE embeddings of the two moons and circles data sets as well as a parabola embedded in $\mathbb{R}^{10}$. The perplexity values from left to right are 5, 30, and 50.

Figure 9.40: Different t-SNE embeddings of the circles data set with perplexity 30.

Switching to the heavier tailed t-distribution — that is, the inverse squared distance weights in (9.100) — regularizes the energy landscape, making the t-SNE algorithm easier to optimize, yielding better visualizations. Precise mathematical statements about the difference between SNE and t-SNE are still lacking and is an active area of research; see [148, 149, 221]. We also mention a more recent and related algorithm known as uniform manifold approximation (UMAP) [160]; see Exercise 8.10.

Exercises

8.1. ◇ Show that the entries of the matrix P defined in (9.99) satisfy $\sum_{i,j} p_{ij} = 1$.

8.2. ♡ Show that equation (9.106) holds.

8.3. ◇ Let $\mathbf{p} \in \mathbb{R}^m$ be a probability vector, and let $H(\mathbf{p}) = -\sum_{i=1}^{m} p_i \log p_i$ be its entropy, where we interpret $0\log 0 = 0$ by continuity. (a) Show that $H(\mathbf{p})$ is a concave function of $\mathbf{p}$. (b) Show that $H(\mathbf{p}) \geq 0$ for all probability vectors $\mathbf{p}$. (c) Show that $H(\mathbf{e}_i) = 0$ for any $i = 1,\ldots,m$. (d) Show that $H(\mathbf{p}) = 0$ if and only if $\mathbf{p} = \mathbf{e}_i$ for some i. (e) Let $\mathbf{q} = 1/m$. Show that $H(\mathbf{q}) = \log m$. (f) Show that $H(\mathbf{p}) \leq \log m$. *Hint:* Use Jensen's inequality (6.90) and the concavity of $\log x$.

8.4. ♡ Consider a probability vector $\mathbf{p}_\lambda = (\lambda, 1-\lambda)^T \in \mathbb{R}^2$, where $0 < \lambda < 1$, and let $h(\lambda)$ be its entropy. (a) Write down a formula for $h(\lambda)$. (b) Show that $h(\lambda)$ is increasing for $0 < \lambda < \frac{1}{2}$ and decreasing for $\frac{1}{2} < \lambda < 1$.

8.5. ◇ Let $H(\mathbf{p})$ be the entropy of a vector $\mathbf{p}$, defined in Exercise 8.3. (a) Suppose that $p_i > p_j$ for some indices i, j. Show that $H(\mathbf{p} + \lambda(\mathbf{e}_i - \mathbf{e}_j))$ is an increasing function of λ for $0 < \lambda < (p_i - p_j)/2$. That is, entropy increases when we redistribute mass from a heavy index i to a lighter index j. (b) Let $\mathbf{p}, \mathbf{q} \in \mathbb{R}^m$ be two probability vectors with ordered entries, so $p_1 \geq \cdots \geq p_m$ and $q_1 \geq \cdots \geq q_m$. Suppose there exists $1 \leq j \leq m - 1$ such that $p_i \geq q_i$ for $i \leq j$ and $p_i \leq q_i$ for $i > j$, meaning that $\mathbf{p}$ is more *concentrated* than $\mathbf{q}$. Use part (a) to explain why $H(\mathbf{p}) \leq H(\mathbf{q})$. (c) Use part (b) to show that the entropy h_i is an increasing function of ε_i in the perplexity graph construction in (9.111).

8.6. ◇ Write a Python program to experiment with the choice of weight matrix for the t-SNE embedding. Try using a k-nearest neighbor graph or ε-ball graph instead of the perplexity graph construction. Can you get similar results on some of the toy synthetic data sets? The Python notebook at the beginning of this section will help you get started.

8.7. ♡ Suppose we generalize the definition of Q in (9.100) to read

$$q_{ij} = \begin{cases} \dfrac{1}{\gamma} f\big(\|\mathbf{z}_i - \mathbf{z}_j\|^2 \big), & i \neq j \\ 0 & i = j, \end{cases} \qquad \text{where} \qquad \gamma = \sum_{\substack{i,j=1 \\ i \neq j}}^{m} f\big(\|\mathbf{z}_i - \mathbf{z}_j\|^2 \big), \qquad (9.112)$$

where $f: \mathbb{R} \to \mathbb{R}$. Show that the gradient of the corresponding Kullback–Leibler divergence (9.101) is given by

$$\nabla_{\mathbf{z}_i} E = \frac{4}{\gamma} \sum_{\substack{i,j=1 \\ i \neq j}} \left(\frac{p_{ij}}{q_{ij}} - 1 \right) g\big(\|\mathbf{z}_i - \mathbf{z}_j\|^2 \big) (\mathbf{z}_i - \mathbf{z}_j), \qquad (9.113)$$

where $g(t) = -f'(t)$. Verify that this gives the correct t-SNE gradient when $f(t) = 1/(1+t)$.

8.8. The original SNE algorithm [106] uses Gaussian weights in the embedding space, which are given by (9.112) with $f(t) = e^{-t}$. Use Exercise 8.7 to show that the gradient of the corresponding SNE energy is

$$\nabla_{\mathbf{z}_i} E = 4 \sum_{j=1}^{m} (p_{ij} - q_{ij})(\mathbf{z}_i - \mathbf{z}_j),$$

8.9. ♡ Implement the SNE algorithm in Python and compare the results to t-SNE on small toy data sets. The Python notebook from this section will be helpful.

8.10. ♡ This exercise explores the uniform manifold approximation (UMAP) algorithm [160], which is a variant of t-SNE that can often give better visualizations. The algorithm starts with a weight matrix W constructed over the high dimensional data for which the maximum entry in every row is exactly one; that is $\max_j w_{ij} = 1$ for all i.[21]

(a) The UMAP algorithm uses the symmetrization $p_{ij} = w_{ij} + w_{ji} - w_{ij}w_{ji}$, which in particular is not normalized to be a probability matrix. Show that $0 \le p_{ij} \le 1$.

(b) For the embedded data points, UMAP uses the weights

$$q_{ij} = \frac{1}{1 + a\,\|\,\mathbf{z}_i - \mathbf{z}_j\,\|^{2b}}, \tag{9.114}$$

for parameters a, b, which are also not normalized. The loss function for UMAP is defined by

$$E(\mathbf{z}_1, \ldots, \mathbf{z}_m) = \sum_{i=1}^{m} \sum_{\substack{j=1 \\ j \ne i}}^{m} \left[p_{ij} \log\left(\frac{p_{ij}}{q_{ij}} \right) + (1 - p_{ij}) \log\left(\frac{1 - p_{ij}}{1 - q_{ij}} \right) \right], \tag{9.115}$$

which is referred to as *fuzzy cross-entropy*. Take $a = b = 1$, and show that

$$\nabla_{\mathbf{z}_i} E = 4 \sum_{j=1}^{m} (1 - q_{ij})^{-1} q_{ij}(p_{ij} - q_{ij})(\mathbf{z}_i - \mathbf{z}_j). \tag{9.116}$$

8.11. Implement gradient descent on the UMAP energy from Exercise 8.10 in Python and compare the results to t-SNE on small toy data sets. The Python notebook from this section will be helpful. You will need to choose a very small time step α and you will initially get strange results. Try initializing UMAP from the result of t-SNE. You may also need to clip the values of q_{ij} so that $q_{ij} \le 1 - \delta$ for a $\delta > 0$.[22]

9.9 Graph-based Semi-supervised Learning

Python Notebook: Graph-based SSL (.ipynb)

We now turn to the problem of semi-supervised learning, which was briefly introduced in Section 7.1.6. In semi-supervised learning, we only have access to a small number[23] of labeled examples, and to compensate for this, we assume we can obtain a large amount of *unlabeled data*; that is, data points $\mathbf{x}_i$ without the associated labels $\mathbf{y}_i$. The goal is to use both labeled and unlabeled data to train a classifier, so that we can obtain better results than if only the labeled data is used, i.e., the fully supervised case. A toy example is given in Figure 9.41. In

---

[21]The matrix W is constructed using a similar perplexity construction as in t-SNE, though there are minor differences.

[22]The UMAP method uses a stochastic approximation of gradient descent that involves sampling only some of the edges at each iteration. It turns out that this sampling procedure introduces a bias and as a result the UMAP method actually minimizes a different loss function; we refer to [51] for details.

[23]There must, at the very least, be one labeled node; otherwise we are in the unsupervised context.

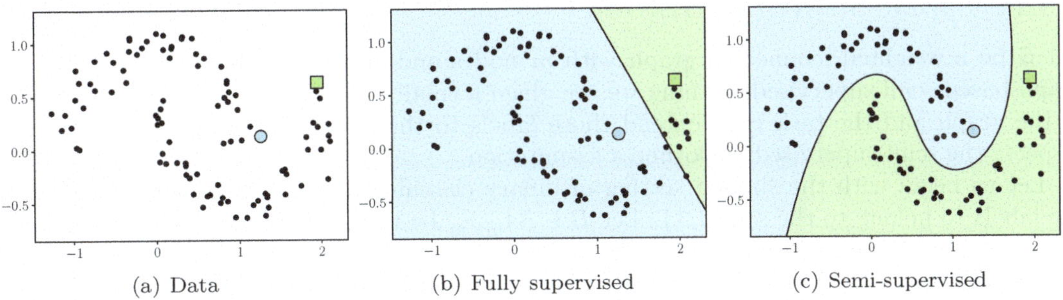

(a) Data   (b) Fully supervised   (c) Semi-supervised

Figure 9.41: A toy example comparing fully and semi-supervised learning. In (a) we show a data set with 2 labeled examples — the blue circle and green square — along with 98 unlabeled data points — the black dots. In (b) we show the decision regions from training a fully supervised classification algorithm, while in (c) we show the decision regions for a semi-supervised learning algorithm, which uses the unlabeled data to inform the decision boundary.

Figure 9.41(a), we show a data set with 2 labeled examples, the green square and blue circle, and 98 unlabeled data points, the black dots. In Figure 9.41(b), we show the result of fully supervised classification, using a linear SVM classifier as in Section 7.3, which only makes use of the two labeled data points and ignores the unlabeled data. In Figure 9.41(c) we show the result of a semi-supervised learning algorithm, which uses the unlabeled data to guide the placement of the decision boundary; in this case placing it so that it does not cut through either of the moons, while still classifying both labeled data points correctly.[24]

The main question is how do we effectively utilize the unlabeled data? One common and widely used approach is to impose the *semi-supervised smoothness assumption*, which stipulates that the label predictions should vary smoothly through dense regions of the unlabeled data. This assumption is violated in Figure 9.41(b), where the predicted labels change abruptly in the lower moon, while it is satisfied in Figure 9.41(c), since the labels change in the void region between the moons.

We recall from Section 7.1.6 that semi-supervised learning can be *transductive*, where we make predictions only at the unlabeled data points, or *inductive*, where we learn a general rule for classifying any new data point. The classifier shown in Figure 9.41(c) is inductive. We will focus in this section on the transductive setting, though once we have label predictions for all the unlabeled data points, we can easily train a fully supervised algorithm to obtain an inductive classifier, as we did in Figure 9.41(c).

One common way to encode the unlabeled data for semi-supervised learning is through the construction of a similarity graph, which offers a convenient way to represent relationships between high dimensional data points. In other contexts, the data for semi-supervised learning may already have an *intrinsic* graph structure, like in the PubMed or political books graphs described earlier. Either way, this leads to a class of algorithms known as *graph-based semi-supervised learning*. While the methods we introduce here are initially motivated by classification, they apply equally well to graph-based regression, where the task is predict a continuous real-valued quantity.

---

[24]In more detail, we applied a graph-based semi-supervised learning algorithm, as described in this section, to classify the unlabeled data points, and then we used these predicted labels as training data for a fully supervised SVM classifier with radial basis function kernel, as described in Section 7.6.

9.9.1 Laplacian Regularization

Let $\mathcal{G}$ be a weighted, connected graph with m nodes and symmetric weight matrix W. In graph-based semi-supervised learning we are given a small number $m_l < m$ of labeled nodes in the graph and the task is to extend those labels to the rest of the graph in a way that respects the semi-supervised smoothness assumption.

Let us begin with the simpler setting of binary classification, in which case we can take the labels to belong to the set $\{0,1\}$. Let $B \in \mathcal{M}_{m \times m}$ be the diagonal matrix that indicates the locations of the labeled data points, so that $b_{ii} = 1$ if i is a labeled data point and $b_{ii} = 0$ otherwise. Let $\mathbf{y} \in \{0,1\}^m$ be the vector containing the known labels in their correct positions. Thus, if $b_{ii} = 1$, so node i is labeled, then $y_i \in \{0,1\}$ is the corresponding binary label. The values of y_i at unlabeled data points, where $b_{ii} = 0$ are irrelevant; we take $y_i = 0$ at these points to be concrete.

The semi-supervised smoothness assumption can enforced by using *Laplacian regularized* graph-based learning, which minimizes a function of the form

$$E(\mathbf{u}) = \| B(\mathbf{u} - \mathbf{y}) \|^2 + \lambda \, \mathbf{u}^T L \mathbf{u}, \tag{9.117}$$

where $L = D - W$ is the graph Laplacian and $\lambda > 0$ is a parameter. The function E has two terms, the second is the Dirichlet energy, which can be written in the form (9.17). As in the case of binary spectral clustering in Section 9.4, minimizing the Dirichlet energy encourages the labels u_i to vary smoothly over the graph, thereby enforcing the semi-supervised smoothness assumption. The first term in (9.117) is a data fidelity term that encourages $\mathbf{u}$ to agree with the given labels at the labeled nodes, where the matrix B serves to restrict this term to only depend on the labeled nodes. The parameter λ controls the tradeoff between the two terms. By taking $\lambda \to 0$ we can enforce the hard label constraint that $B\mathbf{u} = B\mathbf{y}$, i.e., $u_i = y_i$ at every labeled node i. However, we can take a more moderate value for λ when we expect to see noise in the labels, noting that larger values of λ encourage more smoothness in the minimizer $\mathbf{u}$. Laplacian regularization was originally pioneered for semi-supervised learning in [270], and has been widely used in the field ever since.

The Laplacian regularized graph-based learning problem (9.117) is a special case of ridge regression introduced in Section 7.2. Indeed, we can rewrite the energy (9.117) as

$$E(\mathbf{u}) = \| B\mathbf{u} - \mathbf{b} \|^2 + \lambda \| L^{1/2} \mathbf{u} \|^2, \tag{9.118}$$

where $\mathbf{b} = B\mathbf{y}$. In this case, by (7.41) — see Exercise 2.3 in Chapter 7 — provided that $B + \lambda L$ is nonsingular, the minimizer is given by

$$\mathbf{u} = (B + \lambda L)^{-1} B\mathbf{y}. \tag{9.119}$$

Indeed, this equation is valid for all positive λ.

Lemma 9.76. *If $\lambda > 0$, then $B + \lambda L$ is nonsingular.*

Proof. Since both B and L are symmetric positive semidefinite matrices, so is $B + \lambda L$. If $\mathbf{x} \in \ker(B + \lambda L)$, then

$$\mathbf{0} = \mathbf{x}^T (B + \lambda L)\mathbf{x} = \mathbf{x}^T B\mathbf{x} + \lambda \mathbf{x}^T L\mathbf{x}.$$

By positivity, both terms must be zero, and hence $\mathbf{x}$ belongs to both $\ker B$ and $\ker L$. The latter condition requires $\mathbf{x} = c\mathbf{1}$ for some $c \in \mathbb{R}$. On the other hand, the former condition $B\mathbf{x} = \mathbf{0}$ requires $x_i = 0$ at all the labeled nodes, and hence $\mathbf{x} = \mathbf{0}$. $\blacksquare$

Of course, we should not actually invert the matrix $B + \lambda L$, and instead solve the linear system

$$(B + \lambda L)\mathbf{u} = B\mathbf{y} \tag{9.120}$$

by use of the QR factorization, as in Section 4.7, or Gaussian elimination [181], at least for small to moderate size graphs. For very large sparse graphs, a more efficient approach is to use preconditioned conjugate gradients, see Section 6.5, or a spectrally truncated approximation; see Exercise 9.8. The latter is especially useful when solving the equation repeatedly for various $\mathbf{y}$.

As was the case with binary spectral clustering, the minimizer of (9.117) is not a binary vector, and instead a real-valued vector $\mathbf{u} \in \mathbb{R}^m$, so we make classification predictions by thresholding at 0.5, so $c_1 = \{\, i \,|\, u_i \leq 0.5 \,\}$ is one class and $c_2 = \{\, i \,|\, u_i > 0.5 \,\}$ is the other. It turns out that the values u_i are always between zero and one. Hence, we can view u_i as the *probability* that node i belongs to one class or the other.

> **Proposition 9.77.** *Let $\mathbf{u}$ be the minimizer of (9.117), which is the solution of (9.120). Then $0 \leq u_i \leq 1$ for all $i = 1, \ldots, m$.*

Proof. Given $\mathbf{u} \in \mathbb{R}^m$, let $\mathbf{v} \in [0,1]^m$ have entries

$$v_i = \min\{\, \max\{0, u_i\}, 1 \,\} = \begin{cases} 0 & u_i \leq 0, \\ u_i & 0 \leq u_i \leq 1, \\ 1 & u_i \geq 1. \end{cases}$$

That is, $\mathbf{v}$ is obtained by clipping the values of u_i to the interval $[0,1]$. We claim that $E(\mathbf{v}) \leq E(\mathbf{u})$, which then shows that the minimizer of (9.117) belongs to $[0,1]^m$. To see this, since the labels y_i are either zero or one, we clearly have $\| B(\mathbf{v} - \mathbf{y}) \| \leq \| B(\mathbf{u} - \mathbf{y}) \|$, since clipping u_i to the interval $[0,1]$ can only make the values closer to the true labels. Similarly, $|\, v_i - v_j \,| \leq |\, u_i - u_j \,|$ for all i, j, and therefore, by the formula (9.17) for the Dirichlet energy, $\mathbf{v}^T L \mathbf{v} \leq \mathbf{u}^T L \mathbf{u}$. $\blacksquare$

In Figure 9.42 we plot an application of graph-based semi-supervised learning on the two moons data set subject to differing amounts of noise. We used 5 labeled examples in each moon, chosen uniformly at random, and small value for λ to ensure the known labels are attained with high fidelity. We see very good classification results, even when there is a high degree of noise.

The extension to the multiclass setting can be done with the one-vs-rest approach introduced in Section 7.3. Suppose we have c different classes. In the one-vs-rest approach we solve c binary classification problems

$$(B + \lambda L)\mathbf{u}_i = B\mathbf{y}_i, \tag{9.121}$$

where B encodes the locations of all the labels, and $\mathbf{y}_i$ is the label vector taking the value $y_j = 1$ when j is a labeled nodes in class i, and $y_j = 0$ at all other nodes. We solve these c linear systems, for $i = 1, \ldots, c$, and then the classification decision for node j is the index i maximizing $\mathbf{u}_i \cdot \mathbf{e}_j$. To put this into matrix form, let $U = (\, \mathbf{u}_1 \ldots \mathbf{u}_c \,)$ be the $m \times c$ matrix containing the solutions $\mathbf{u}_i$ of (9.121) as columns. Let $Y = (\, \mathbf{y}_1 \ldots \mathbf{y}_c \,)$ be the corresponding matrix with the label vectors $\mathbf{y}_i$ as columns; its rows exactly encode the labels as one-hot vectors, i.e., $\mathbf{e}_i$ for the i-th class. Then we can express all c equations in (9.121) at once in the matrix equation

$$(B + \lambda L)U = BY. \tag{9.122}$$

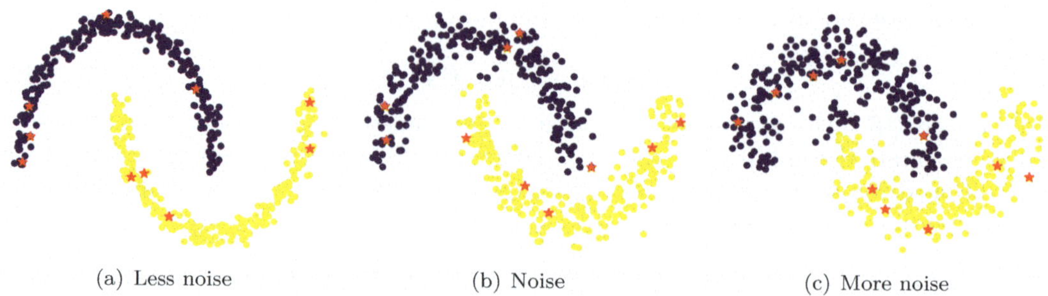

(a) Less noise   (b) Noise   (c) More noise

Figure 9.42: Example of graph-based semi-supervised learning on the two moons data set with different amounts of noise on the moons. The red stars indicate the randomly chosen locations of the labeled data points. The classifications are nearly perfect in all cases, except that increasing the noise results in some bleeding over of the upper moon into the lower one.

From the optimization perspective, each $\mathbf{u}_i$ minimizes the energy

$$E_i(\mathbf{u}) = \| B (\mathbf{u} - \mathbf{y}_i) \|^2 + \lambda \mathbf{u}^T L \mathbf{u}, \tag{9.123}$$

and so the matrix $U = (\, \mathbf{u}_1 \, \ldots \, \mathbf{u}_c \,)$ minimizes the *vectorized energy*

$$E(U) = \sum_{i=1}^{c} E_i(\mathbf{u}_i). \tag{9.124}$$

In fact, according to Exercise 9.1, we can write the vectorized energy in the form

$$E(U) = \| B (U - Y) \|_F^2 + \lambda \, \mathrm{tr}\, (U^T L U). \tag{9.125}$$

The matrix norm for the first term is the Frobenius norm (4.85) on the space of $m \times c$ matrices, and the second term is the Frobenius inner product (4.84) of U with LU. We note that the trace term above can be expressed as

$$\mathrm{tr}\, (U^T L U) = \frac{1}{2} \sum_{i,j=1}^{m} w_{ij} \| \mathbf{v}_i - \mathbf{v}_j \|^2, \tag{9.126}$$

where $\mathbf{v}_i^T$ is the i-th row of U; this is simply a *vectorized* version of the Dirichlet energy (9.17). We leave the verification of (9.126) to Exercise 9.2. We treat U as the matrix of label vectors $\mathbf{v}_i$ that we wish to learn for each node i in the graph. Thus, the multiclass version of Laplacian regularized learning has the same interpretation as the binary version; we wish to learn a label function that varies as smoothly as possible over the graph, and agrees with the given labeled nodes.

We now illustrate the method with the classification of MNIST digits. Here, we use a $k = 10$ nearest neighbor graph based on Euclidean distance between pixel values, as well a similar graph built over the diffusion map embedding of the MNIST data set into dimension $d = 50$ with $k = 64$ diffusion steps, as described in Section 9.7. We used the hard-constraint version of Laplacian regularized learning, called *Laplace learning*, discussed below in Section 9.9.2, which takes $\lambda \to 0$. We experimented with different label rates, randomly choosing 1, 5, 10, and 100 labeled data points per class, equivalent to 10, 50, 100, and 1000 labels. For each label rate, we ran 100 trials randomizing which images are labeled. Table 9.43 shows the results of the experiment. We can see the method gives good accuracy even with only a

# Labels	10	50	100	1000
Graph Nearest Neighbor	55.1 (5.5)	76.4 (2.5)	81.5 (1.5)	89.8 (0.3)
Laplace Learning	16.5 (6.9)	61.3 (10.6)	85.1 (3.8)	94.4 (0.2)
Graph Nearest Neighbor (DiffMap)	66.5 (7.2)	84.9 (2.6)	88.4 (1.4)	92.3 (0.3)
Laplace Learning (DiffMap)	65.7 (8.1)	87.3 (2.3)	90.4 (1.4)	94.0 (0.1)

Table 9.43: Laplace learning on MNIST with 1, 5, 10 and 100 labels per class. The average (standard deviation) classification accuracy over 100 trials is shown.

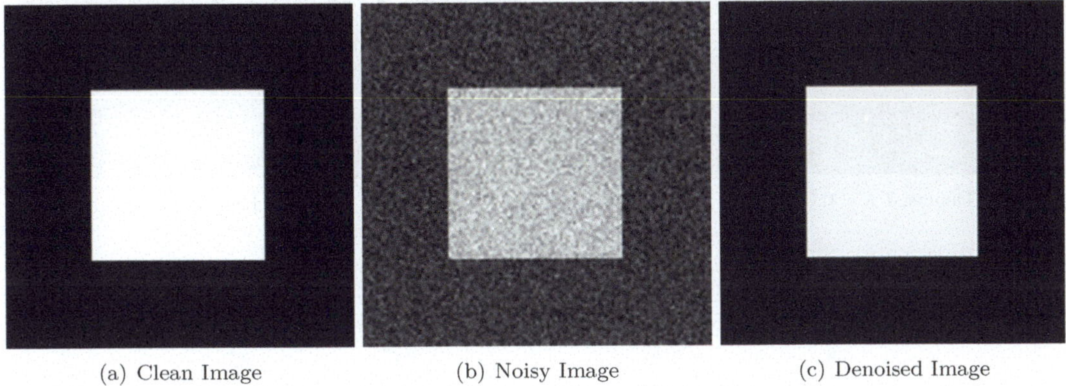

(a) Clean Image   (b) Noisy Image   (c) Denoised Image

Figure 9.44: Example of image denoising using graph-based regression by minimizing the denoising energy (9.127). Nearly all of the noise is perfectly removed from the image in this toy example.

handful of training examples, showing the power of semi-supervised learning. We also see that utilizing the diffusion map embedding increases the accuracy, especially when fewer labeled examples are available. We also compare these results with a nearest neighbor classifier, using the shortest path graph distance introduced in Section 9.5. In almost all cases, Laplace learning performs better. Note that to obtain the same result of 94% accuracy with SVM — see Section 7.3 — we needed around 14000 labeled examples, while we needed about 7000 examples for a k-nearest neighbor classifier — see Section 7.4. The fact that we can match this performance with only 1000 labeled examples illustrates the power of semi-supervised learning.

While we have motivated these methods by classification, they are equally well-suited for regression problems. An important application is to image denoising. In this setup, all of the nodes in the graph have labels, but they are treated as noisy labels and the goal is to remove the noise. In this case $B = I$ is the identity matrix, and the image denoising problem has the vectorized energy

$$E(U) = \|U - Y\|^2 + \lambda \operatorname{tr}\left(U^T L U\right). \tag{9.127}$$

Here, $Y \in \mathcal{M}_{m \times c}$ is a matrix containing all m pixels in an image with c color channels, and the minimizer $U \in \mathcal{M}_{m \times k}$ is the denoised image. We choose the weight matrix W for the graph construction in the same way as was done for image segmentation with spectral clustering in Section 9.7.2; see (9.96).

In Figure 9.44 we show a simple example of denoising a piecewise constant image with a white square in the foreground. We show (a) the clean image, (b) the noisy image Y with

(a) Clean Image　　　　　　(b) Noisy Image　　　　　　(c) Denoised $\lambda = 0.1$

(d) Denoised $\lambda = 0.2$　　　　　　(e) Denoised $\lambda = 1$　　　　　　(f) Denoised $\lambda = 10$

Figure 9.45: Example of image denoising using graph-based regression by minimizing the denoising energy (9.127) with different values of λ. Larger values of λ result in more aggressive smoothing of the image and removal of noise, texture, and possibly other important details.

additive Gaussian noise, and (c) the denoised image U. In this case we obtain a nearly perfect denoising. This is due to the fact that the weights (9.96) are defined so that only pixels with similar intensities are connected by edges, so there is very little diffusion occurring between the white and black region. We also used a very large value $\lambda = 100$ so there is a high degree of smoothing in the reconstruction.

In Figure 9.45 we show a more realistic example on the same image used for spectral clustering in Section 9.7.2. Here, the original is a color image with three channels, and again we imposed additive Gaussian noise. In Figure 9.45 we show the results of denoising with different values of the fidelity parameter λ. When λ is larger, the denoising is more aggressive and smooths the image more, while for smaller values of λ the denoising more closely resembles the noisy image.

9.9.2　Label Propagation and Hard Constraints

While the scalar energy E in (9.117), or the components E_i of the vector energy (9.125), can be efficiently minimized with the conjugate gradient method, it is instructive to consider minimizing it via a simple gradient descent algorithm. This has an elegant interpretation of propagating labels on graphs, and is often the method of choice in practice for solving the equations. In fact, label propagation was one of the main ideas underlying the work in [269, 270].

For simplicity, we restrict ourselves to the binary classification setting in this subsection. Let us define the weighted inner product $\langle \mathbf{x}, \mathbf{y} \rangle_D = \mathbf{x}^T D \mathbf{y}$ on $\mathbb{R}^m$, where D is the diagonal degree matrix of the graph. This has the effect that each node in $\mathcal{G}$ is weighted by its degree, so the higher the degree the "more important" the node. The (ordinary) gradient of the quadratic energy (9.117) is

$$\nabla E(\mathbf{u}) = 2B(\mathbf{u} - \mathbf{y}) + 2\lambda L\mathbf{u}. \tag{9.128}$$

As in Section 6.3, the gradient with respect to the inner product $\langle \mathbf{x}, \mathbf{y} \rangle_D$ is

$$\nabla_D E(\mathbf{u}) = D^{-1}\nabla E(\mathbf{u}) = D^{-1}\nabla E(\mathbf{u}) = 2D^{-1}B(\mathbf{u} - \mathbf{y}) + 2\lambda L_{\mathrm{rw}}\mathbf{u}, \tag{9.129}$$

where L_{rw} is the random walk graph Laplacian (9.59). Gradient descent with time step of $\alpha = 1/(2\lambda)$ thus corresponds to

$$\mathbf{u}_{k+1} = \mathbf{u}_k - \frac{1}{2\lambda}\nabla_D E(\mathbf{u}_k) = D^{-1}W\mathbf{u}_k + \lambda^{-1}D^{-1}B(\mathbf{y} - \mathbf{u}_k). \tag{9.130}$$

This is exactly a graph diffusion equation, as introduced in Section 9.6. The first term $D^{-1}W\mathbf{u}_k$ is the average of the labels over each nodes' neighbors in the graph, while the second term acts to push the labels towards their correct values at the labeled data points. We can run (9.130) until convergence to *train* the graph-based semi-supervised learning algorithm.

An important setting is the one with hard label constraints, where we send $\lambda \to 0$ in (9.117). In this case, minimizing the graph-based learning energy (9.117) is equivalent to the problem

$$\min\left\{\mathbf{u}^T L\mathbf{u} \mid \mathbf{u} \in \mathbb{R}^m, \ B\mathbf{u} = B\mathbf{y}\right\}. \tag{9.131}$$

We can solve the constrained problem with projected gradient descent, where at each iteration we take a step of gradient descent on the Dirichlet energy $\mathbf{u}^T L\mathbf{u}$ at the unlabeled nodes, while fixing the label condition $B\mathbf{u} = B\mathbf{y}$ at the labeled nodes. This can be conveniently written in the form

$$\mathbf{u}_{k+1} = (I - B)D^{-1}W\mathbf{u}_k + B\mathbf{y}. \tag{9.132}$$

The equation (9.132) is called *label propagation* in the literature, and is one of the fundamental methods in graph-based semi-supervised learning [269, 270]. Sometimes label propagation can be stopped early to give better results and prevent a type of overfitting or oversmoothing at very low label rates; see Exercise 9.3.

Remark 9.78. There is, of course, a vectorized version of label propagation for the multiclass setting given in matrix form by

$$U_{k+1} = (I - B)D^{-1}W U_k + BY. \tag{9.133}$$

The reader should compare this to the matrix form of the graph-based learning energy given in (9.125). ▲

While label propagation (9.132) has a nice intuitive interpretation via averaging labels with their neighbors on a graph until convergence, it is not necessarily the most efficient way to solve the constrained problem (9.131). In fact, we can easily convert (9.131) into an unconstrained problem that can be solved with any sparse iterative solver. Let m_u be the number of unlabeled data points, and let $i_1, \ldots, i_{m_u}$ denote the indices of the unlabeled nodes. We let

$$Q = \left(\mathbf{e}_{i_1}\, \mathbf{e}_{i_2}\, \cdots\, \mathbf{e}_{i_{m_u}}\right)$$

be the $m \times m_u$ matrix whose columns are the one-hot vectors indicating the positions of the unlabeled data points. Note that the columns of Q are orthogonal, so $Q^T Q = I$. Then any $\mathbf{u} \in \mathbb{R}^m$ satisfying the constraint $B\mathbf{u} = B\mathbf{y}$ can be written in the form

$$\mathbf{u} = Q\mathbf{v} + B\mathbf{y} \qquad \text{for some} \qquad \mathbf{v} \in \mathbb{R}^{m_l}.$$

The vector $\mathbf{v}$ represents the unknown labels at the unlabeled data points, which are the only degrees of freedom in the constrained problem (9.131). We can now write the Dirichlet energy as

$$\mathbf{u}^T L\mathbf{u} = \mathbf{v}^T Q^T L Q\mathbf{v} + 2\mathbf{v}^T Q^T LB\mathbf{y} + \mathbf{y}^T BLB\mathbf{y},$$

and minimize the right hand side above over $\mathbf{v} \in \mathbb{R}^{m_u}$. The matrix $M := Q^T L Q$ is the $m_u \times m_u$ sub-matrix of L corresponding to the unlabeled data points. We claim that M is positive definite, and hence invertible, so the unique minimizer is given by

$$\mathbf{u} = M^{-1} Q^T L B \mathbf{y} = (Q^T L Q)^{-1} Q^T L B \mathbf{y}. \tag{9.134}$$

Now that we have reduced the constrained problem (9.131) to solving the positive definite linear system (9.134), we can use any technique we like to solve (9.134).

To prove the claim, it is clear that M is positive semidefinite, so we just need to show that its kernel is trivial. Suppose that $Q^T L Q \mathbf{v} = 0$. Then $\mathbf{v}^T Q^T L Q \mathbf{v} = 0$, and upon setting $\mathbf{u} = Q \mathbf{v}$ we have $\mathbf{u}^T L \mathbf{u} = 0$. Since the graph is connected we have that $\mathbf{u}$ has all equal entries: $\mathbf{u} = c\mathbf{1}$. Moreover, $u_i = 0$ for any labeled node i, so we must have $\mathbf{u} = \mathbf{0}$. Thus, $\mathbf{v} = Q^T \mathbf{u} = \mathbf{0}$, which establishes the claim.

Let us mention that since the seminal work in [270], graph-based techniques have proliferated in semi-supervised learning [19, 21, 23, 25, 37, 97, 139, 165, 166, 214, 243, 266]. Recent work has utilized graph neural networks, which we will discuss in Section 10.4.

Exercises

9.1. ◇ Show that (9.125) holds.

9.2. ◇ Show that (9.126) holds.

9.3. Write a Python program to run the binary label propagation algorithm (9.132) or the vectorized version (9.133) on either a toy synthetic data set or MNIST. Plot the accuracy over each iteration k and at convergence as $k \to \infty$. Can you achieve a higher accuracy by "stopping early"? The Python notebook at the start of this section has some code to get you started.

9.4. Modify binary label propagation from Exercise 9.3 so that every p iterations the current iterate $\mathbf{u}_k$ is thresholded at 0.5 to labels $\mathbf{u}_k \in \{0,1\}^m$. The parameter p is a positive integer that you can experiment with (it is a hyperparameter). Implement this in Python for binary classification of pairs MNIST digits. Can you obtain better results than with label propagation for a good choice of p?[25]

9.5. ♡ Implement the following binary semi-supervised learning algorithm in Python. Let $\mathbf{y} \in \mathbb{R}^m$ be the label vector encoding the given labels, with $y_i = +1$ if i is in one class, $y_i = -1$ if i is in the other class, and $y_i = 0$ otherwise. Assume the number of labels in each class is the same. Start from the zero vector $\mathbf{u}_0 = \mathbf{0}$, and run the iterations

$$\mathbf{u}_{k+1} = D^{-1} W \mathbf{u}_k + D^{-1} \mathbf{y} \tag{9.135}$$

until convergence. This is essentially the power iteration for spectral clustering that we say previously in Exercise 4.3, except we are adding the labels as a *source term*. After convergence, threshold the vector $\mathbf{u}_k$ at zero to produce the two classes. Try your algorithm on some simple binary classification problems, like two moons, circles, and pairs of MNIST digits.

9.6. ♡ Under the assumptions of Exercise 9.5, show that $\lim_{k \to \infty} \mathbf{u}_k = \mathbf{u}$, where $\mathbf{u}$ is the unique solution of $L\mathbf{u} = \mathbf{y}$ that satisfies $\langle \mathbf{u}, \mathbf{1} \rangle_D = 0$.

---

[25] The alternation of diffusion and thresholding is used in the MBO methods; see [25] for details.

9.7. ♡ Under the assumptions of Exercises 9.5, 9.6,[26] show that $\mathbf{u}$ is the unique minimizer of $E(\mathbf{u}) = \frac{1}{2}\mathbf{x}^T L \mathbf{x} - \mathbf{y}^T \mathbf{u}$ satisfying $\langle \mathbf{u}, \mathbf{1} \rangle_D = 0$. *Hint*: Use Theorem 9.61.

9.8. ♡ For very large sparse graphs, direct methods struggle to solve the Laplacian regularized learning problem (9.120). Implement a spectrally truncated solver, as outlined in Exercise 3.7 in Chapter 5, and apply it to graph-based semi-supervised binary classification on the MNIST data set.

9.9. ◇ Repeat Exercise 9.8, except use the conjugate gradient method to solve the linear system.

9.10. Let $\mathcal{G}$ be a connected weighted graph and let $\mathcal{D} \subset \mathcal{N}$. Show that there exists $C > 0$ depending only on $\mathcal{G}$ and $\mathcal{D}$ such that

$$\sum_{i=1}^m u_i^2 \le \frac{C}{2} \sum_{i,j=1}^m w_{ij}\,(u_i - u_j)^2,$$

holds for all $\mathbf{u} \in \mathbb{R}^m$ such that $u_i = 0$ for all $i \in \mathcal{D}$. This is another version of the *Poincaré inequality* introduced in Exercise 6.15. *Hint*: The argument at the end of this section showing that M is invertible will be useful.

9.11. Show that the Poincaré inequalities in Exercises 6.15, 9.10 hold in the p norm; that is, for $p \ge 1$ there exists $C > 0$ depending only on p and $\mathcal{G}$ (and $\mathcal{D}$ for Exercise 9.10) such that

$$\sum_{i=1}^m |u_i|^p \le \frac{C}{2} \sum_{i,j=1}^m w_{ij}\,|u_i - u_j|^p,$$

Hint: Use Exercise 7.19 in Chapter 6.

9.12. ♡ Let $\mathcal{G}$ be an unweighted graph with adjacency matrix A. Let $k \ge 0$ such that for each pair of adjacent nodes i, j — so that $a_{ij} = 1$ — there are at least k other nodes that are adjacent to both i and j; that is the number of indices $\ell \ne i, j$ for which $a_{i\ell} = 1 = a_{j\ell}$ is at least k. Show that

$$|u_i - u_j| \le \sqrt{\frac{2\,\mathbf{u}^T L \mathbf{u}}{k+2}} \tag{9.136}$$

holds for all $\mathbf{u} \in \mathbb{R}^m$ and all adjacent nodes i, j. *Note*: The inequality (9.136) allows us to estimate the difference in the label predictions between two adjacent nodes in terms of the size of the regularization term $\mathbf{u}^T L \mathbf{u}$.

9.10 The Discrete Fourier Transform

Python Notebook: Discrete Fourier Transform (.ipynb)

In this section, we introduce the discrete Fourier transform (DFT), which decomposes a signal into its trigonometric constituents, and is of immense importance not just in machine

---

[26]The method outlined in Exercises 9.5 and 9.6 is called *Poisson learning*; we refer to [37] for more details.

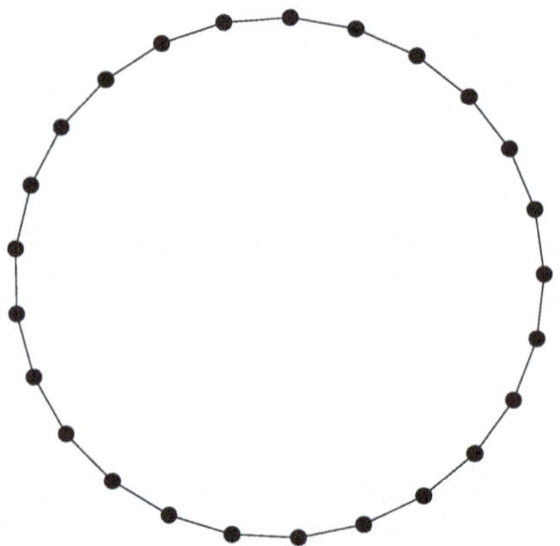

Figure 9.46: The cyclic graph with $m = 25$ nodes

learning but also signal processing, numerical solution of differential equations, and many other fields. We then develop the fast Fourier transform (FFT), which is a fast method for computing the DFT that makes large scale applications computationally feasible. We will first see how the DFT naturally arises from the eigendecomposition of the graph Laplacian associated with a circular graph. This connection serves to motivate generalizations of the DFT, based on convolution, to other graphs of importance in machine learning.

The starting point is the circular *cyclic graph* on m vertices. Each node has degree 2, and its two edges connect it to its two immediate neighbors; an example is depicted in Figure 9.46. The adjacency matrix for such a graph is the $m \times m$ symmetric matrix

$$W_m = \begin{pmatrix} 0 & 1 & 0 & \cdots & 0 & 1 \\ 1 & 0 & 1 & \cdots & 0 & 0 \\ 0 & 1 & 0 & \cdots & 0 & 0 \\ \vdots & \vdots & \vdots & \ddots & \vdots & \vdots \\ 0 & 0 & 0 & \cdots & 0 & 1 \\ 1 & 0 & 0 & \cdots & 1 & 0 \end{pmatrix}, \tag{9.137}$$

which, since the graph is taken to be unweighted, is also its weight matrix. The matrix W_m has 1's along the sub- and super-diagonals, as well as in the $(1, m)$ and $(m, 1)$ slots; all other entries are 0. (It is a type of circulant matrix; see Section 9.10.5.)

We are interested in determining the eigenvalues and eigenvectors of the graph Laplacian matrix associated with this graph. Since each node has degree $d_i = 2$, the degree matrix is $D = 2\,I$, and so the graph Laplacian matrix is $L = 2\,I - W_m$. Since D is a multiple of the identity, all powers of D commute with W_m, and so the other normalizations (9.59), (9.65) of the graph Laplacian are the same, up to a factor of $\frac{1}{2}$, that is

$$L_{\mathrm{rw}} = L_{\mathrm{sym}} = \tfrac{1}{2} L = I - \tfrac{1}{2} W_m, \tag{9.138}$$

Thus, the eigenvectors for all three normalizations of the graph Laplacian are the same, and coincide with the eigenvectors of the weight matrix W_m, so we will focus on the latter matrix.

If $\mathbf{x} \in \mathbb{R}^m$ is an eigenvector of W_m, then it satisfies $W_m \mathbf{x} = \lambda \mathbf{x}$ for some λ, which amounts to

$$x_{j-1} + x_{j+1} = \lambda x_j \tag{9.139}$$

for all $j = 2, \ldots, m - 1$. The cases of $j = 1$ and $j = m$ also follow the same formula if we identify $x_0 = x_m$ and $x_{m+1} = x_1$. It turns out that we can find solutions of the eigenvector equation (9.139) via some basic trigonometric identities. Recall the addition formulae

$$\cos\big((j-1)\theta\big) + \cos\big((j+1)\theta\big) = 2\cos\theta\,\cos(j\theta),$$
$$\sin\big((j-1)\theta\big) + \sin\big(((j+1)\theta\big) = 2\cos\theta\,\sin(j\theta). \tag{9.140}$$

This suggests that the vector $\mathbf{x}$ with entries $x_j = \cos(j\theta)$ is an eigenvector of W_m for any θ, with eigenvalue $\lambda = 2\cos\theta$, as is the vector $\mathbf{y}$ with $y_j = \sin(j\theta)$. However, we have to ensure the identifications $x_0 = x_m$ and $x_{m+1} = x_1$ hold, which requires that $\cos(0) = 1 = \cos(m\theta)$, and hence $\theta = 2\pi k/m$ for some integer k.

This suggests defining, for any integer k, the following vectors in $\mathbb{R}^m$:

$$\mathbf{x}_k = \left(1,\quad \cos\frac{2k\pi}{m},\quad \cos\frac{4k\pi}{m},\quad \cos\frac{6k\pi}{m},\quad \ldots,\quad \cos\frac{2(m-1)k\pi}{m}\right)^T,$$

$$\mathbf{y}_k = \left(0,\quad \sin\frac{2k\pi}{m},\quad \sin\frac{4k\pi}{m},\quad \sin\frac{6k\pi}{m},\quad \ldots,\quad \sin\frac{2(m-1)k\pi}{m}\right)^T. \tag{9.141}$$

We note that

$$\mathbf{x}_{k+m} = \mathbf{x}_k,\quad \mathbf{x}_{-k} = \mathbf{x}_k,\quad \mathbf{x}_0 = \mathbf{1},\quad \mathbf{y}_{k+m} = \mathbf{y}_k,\quad \mathbf{y}_{-k} = -\mathbf{y}_k,\quad \mathbf{y}_0 = \mathbf{0}. \tag{9.142}$$

Consequently, in the following discussion, we will view m as fixed and the index k to be computed modulo m. The trigonometric identities (9.140) can then be recast in vectorial form as

$$W_m\mathbf{x}_k = \lambda_k\mathbf{x}_k,\qquad W_m\mathbf{y}_k = \lambda_k\mathbf{y}_k,\qquad \text{where}\qquad \lambda_k = 2\cos\frac{2k\pi}{m}. \tag{9.143}$$

In other words, the vectors $\mathbf{x}_k, \mathbf{y}_k$, when nonzero, are eigenvectors of the weight matrix W_m with corresponding eigenvalue λ_k. Now, we know that an $m \times m$ matrix can have at most m linearly independent eigenvectors; but this is confirmed by the identities (9.142). Thus, we can explicitly write down the eigenvalues and eigenvectors of W_m. The precise statement depends upon whether m is even or odd.

Lemma 9.79. *Let $m = 2n$ or $m = 2n+1$ for $n \in \mathbb{N}$. Then the weight matrix W_m has $n+1$ distinct eigenvalues $\lambda_0, \ldots, \lambda_{n+1}$ as given by formula (9.143). The corresponding eigenspaces are spanned by the pair of vectors $\mathbf{x}_k, \mathbf{y}_k$ given in (9.141), where $\mathbf{y}_0 = \mathbf{0}$ and, if $m = 2n$ is even, $\mathbf{y}_n = \mathbf{0}$. Thus, the eigenspaces associated with $\lambda_0 = 1$ and, in the even case, $\lambda_n = 2$ are one-dimensional, while all the others have dimension 2. The indicated eigenvectors (i.e., excluding any zero vectors) form an orthogonal basis of $\mathbb{R}^m$ under the dot product.*

Proof. Orthogonality of eigenvectors belonging to different eigenvalues follows from the symmetry of W using Theorem 5.29. However, to establish orthogonality of $\mathbf{x}_k$ and $\mathbf{y}_k$, one must verify directly that $\mathbf{x}_k \cdot \mathbf{y}_k = 0$, which also follows from some basic trigonometric identities. Details are left to the reader in Exercise 10.3. ∎

We note further that, as a consequence of similar trigonometric identities, the eigenvectors have Euclidean norms

$$\|\mathbf{x}_0\| = \|\mathbf{x}_n\| = \sqrt{m},\qquad \text{when } m = 2n \text{ is even, while}$$

$$\|\mathbf{x}_k\| = \|\mathbf{y}_k\| = \sqrt{\frac{m}{2}},\qquad \text{for all}\quad 1 \le k < \tfrac{1}{2}m. \tag{9.144}$$

Again, the proofs are relegated to Exercise 10.3. As usual, dividing the nonzero vectors $\mathbf{x}_k, \mathbf{y}_k$ by their norms produces an orthonormal basis of $\mathbb{R}^m$.

Since the graph Laplacian (9.15) is simply $L = 2\,I - W_m$, using Exercise 1.7 in Chapter 5, the eigenvalues of L are readily seen to be

$$\mu_k = 2 - 2\cos\frac{2k\pi}{m}, \qquad k = 0, \ldots, \left\lfloor \frac{m}{2} \right\rfloor, \tag{9.145}$$

and by (9.138), the random walk Laplacian L_{rw} and symmetric normalized Laplacian L_{sym} have eigenvalues $\frac{1}{2}\mu_k$ and the same eigenspaces.

The *discrete Fourier transform* of a vector $\mathbf{v} \in \mathbb{R}^m$ is identified with the coordinates of $\mathbf{v}$ with respect to the orthogonal eigenvector basis of the circular graph Laplacian. Explicitly, if $m = 2n$ is even,

$$\mathbf{v} = a_0\,\mathbf{x}_0 + \sum_{k=1}^{n-1} \left(a_k\,\mathbf{x}_k + b_k\,\mathbf{y}_k \right) + a_n\,\mathbf{x}_n, \tag{9.146}$$

where, by orthogonality, and the eigenvector norm formulas (9.144),

$$a_0 = \frac{1}{m}\,\mathbf{v}\cdot\mathbf{x}_0, \quad a_k = \frac{2}{m}\,\mathbf{v}\cdot\mathbf{x}_k, \quad a_n = \frac{1}{m}\,\mathbf{v}\cdot\mathbf{x}_m, \quad b_k = \frac{2}{m}\,\mathbf{v}\cdot\mathbf{y}_k, \quad k = 1,\ldots,n-1, \tag{9.147}$$

and the discrete Fourier transform of $\mathbf{v} \in \mathbb{R}^m$ is the vector

$$\mathbf{w} = \left(a_0, \ldots, a_n, b_1, \ldots, b_{n-1} \right)^T \in \mathbb{R}^m. \tag{9.148}$$

On the other hand, if $m = 2n + 1$ is odd,

$$\mathbf{v} = a_0\,\mathbf{x}_0 + \sum_{k=1}^{n} \left(a_k\,\mathbf{x}_k + b_k\,\mathbf{y}_k \right), \tag{9.149}$$

where

$$a_0 = \frac{1}{m}\,\mathbf{v}\cdot\mathbf{x}_0, \qquad a_k = \frac{2}{m}\,\mathbf{v}\cdot\mathbf{x}_k, \qquad b_k = \frac{2}{m}\,\mathbf{v}\cdot\mathbf{y}_k, \qquad k = 1,\ldots,n, \tag{9.150}$$

and the discrete Fourier transform of $\mathbf{v} \in \mathbb{R}^m$ is the vector

$$\mathbf{w} = \left(a_0, \ldots, a_n, b_1, \ldots, b_n \right)^T \in \mathbb{R}^m. \tag{9.151}$$

In signal processing, one interprets the entries of a vector $\mathbf{v} \in \mathbb{R}^m$ as the sample values of a function $f \colon [0, 2\pi] \to \mathbb{R}$ at the m evenly spaced data points $x_j = 2j\pi/m$, for $j = 0, \ldots, m-1$, so that

$$\mathbf{v} = \left(f(0),\ f\!\left(\frac{2\pi}{m} \right),\ f\!\left(\frac{4\pi}{m} \right),\ \ldots,\ f\!\left(\frac{2(m-1)\pi}{m} \right) \right)^T. \tag{9.152}$$

In particular, the basis vectors $\mathbf{x}_k, \mathbf{y}_k$ in (9.141) can be identified as the sample vectors for the basic trigonometric functions $\cos kx$, $\sin kx$. Thus, the reconstruction formulas (9.146), (9.149) can be interpreted as decomposing the sampled function $f(x)$ into a linear combination of basic trigonometric functions:

$$f(x) \ \sim\ q_m(x) = \begin{cases} a_0 + \displaystyle\sum_{k=1}^{n-1} \left(a_k \cos kx + b_k \sin kx \right) + a_n \cos nx, & m = 2n, \\[2ex] a_0 + \displaystyle\sum_{k=1}^{n} \left(a_k \cos kx + b_k \sin kx \right), & m = 2n+1. \end{cases} \tag{9.153}$$

The symbol $\sim$ in (9.153) means that $f(x)$ agrees with the trigonometric polynomial on the right hand side at the sample points:

$$f\left(\frac{2j\pi}{m}\right) = q_m\left(\frac{2j\pi}{m}\right), \qquad j = 0, \ldots, m-1. \tag{9.154}$$

If $f(2\pi) = f(0)$, the same occurs at the right hand endpoint: $f(2\pi) = q_m(2\pi) = q_m(0)$, because the trigonometric polynomial $q_m(x)$ is 2π periodic. We defer showing an example until after we have explained the simpler complex reformulation of these equations.

9.10.1 Complexification

Computations involving trigonometric functions are almost always simplified by use of complex exponentials, and this is certainly the case here. The complex reformulation of the DFT is based on *Euler's formula*

$$e^{ix} = \cos x + i \sin x, \tag{9.155}$$

relating the complex exponential with the real sine and cosine functions. Here $i = \sqrt{-1}$ is the imaginary unit. Vice versa, since

$$e^{-ix} = \cos x - i \sin x,$$

one can combine these two identities to express the basic trigonometric functions in terms of complex exponentials:

$$\cos x = \frac{e^{ix} + e^{-ix}}{2}, \qquad \sin x = \frac{e^{ix} - e^{-ix}}{2i}. \tag{9.156}$$

Note: Although we promised to only use real numbers and vectors in this book, this is the one place where venturing into the complex realm becomes quite useful. (Another, that we encountered earlier, is when real matrices possess complex eigenvalues.) Thus, before going further, let us review the basics of complex arithmetic and vectors.

A *complex number* is an expression of the form $z = x + iy$, where $x, y \in \mathbb{R}$ are real. The set of all complex numbers (scalars) is denoted by $\mathbb{C}$. We call $x = \text{Re } z$ the *real part* and $y = \text{Im } z$ the *imaginary part* of $z = x + iy$. *Note*: The imaginary part is the real number y, *not* iy. A real number x is merely a complex number with zero imaginary part, so $x = x + i0$ and $\text{Im } x = 0$, and so we may regard $\mathbb{R} \subset \mathbb{C}$. Complex addition and multiplication are based on simple adaptations of the rules of real arithmetic to include the identity $i^2 = -1$, and so

$$\begin{aligned}(x + iy) + (u + iv) &= (x + u) + i(y + v), \\ (x + iy)(u + iv) &= (xu - yv) + i(xv + yu).\end{aligned} \tag{9.157}$$

Complex numbers enjoy all the usual laws of real addition and multiplication, *including commutativity*: $zw = wz$.

We can identify a complex number $x + iy$ with a vector $(x, y)^T \in \mathbb{R}^2$ in the real plane. (Indeed, this is how they are stored on a computer.) For this reason, $\mathbb{C}$ is sometimes referred to as the *complex plane*. Complex addition (9.157) corresponds to vector addition; however, complex multiplication does not have a readily identifiable vector counterpart.

Definition 9.80. The *complex conjugate* of $z = x + iy$ is $\bar{z} = x - iy$, whereby $\text{Re } \bar{z} = \text{Re } z$, while $\text{Im } \bar{z} = -\text{Im } z$.

Conjugating twice brings you back to where you started: $\overline{\overline{z}} = z$. Moreover, complex conjugation is compatible with complex arithmetic:

$$\overline{z + w} = \overline{z} + \overline{w}, \qquad \overline{z\,w} = \overline{z}\,\overline{w}.$$

In particular, $\overline{z} = z$ if and only if z is real. Note that

$$\operatorname{Re} z = \frac{z + \overline{z}}{2}, \qquad \operatorname{Im} z = \frac{z - \overline{z}}{2\,\mathrm{i}}. \tag{9.158}$$

Observe that the product of any complex number and its conjugate,

$$z\,\overline{z} = (x + \mathrm{i}\,y)\,(x - \mathrm{i}\,y) = x^2 + y^2, \tag{9.159}$$

is real and nonnegative. Its square root is known as the *modulus* or *norm* of the complex number $z = x + \mathrm{i}\,y$, and written

$$|z| = \sqrt{z\,\overline{z}} = \sqrt{x^2 + y^2}\,. \tag{9.160}$$

Thus, $|z| \geq 0$, with $|z| = 0$ if and only if $z = 0$. The modulus $|z|$ generalizes the absolute value of a real number, and coincides with the standard Euclidean norm in the $x\,y$–plane, which implies the validity of the triangle inequality

$$|z + w| \leq |z| + |w|. \tag{9.161}$$

Moreover, the modulus respects complex multiplication:

$$|z\,w| = |z|\,|w|, \qquad \text{and hence} \qquad |z^k| = |z|^k. \tag{9.162}$$

An important consequence of the latter identity is that the powers of a complex number converge to 0, so $z^k \to 0$ as $k \to \infty$, if and only if its modulus $|z| < 1$. In view of Euler's formula (9.155), we can write any complex number in polar coordinate form

$$x + \mathrm{i}\,y = z = r\,e^{\mathrm{i}\theta} = r\cos\theta + \mathrm{i}\,(r\sin\theta) \qquad \text{where} \qquad r = |z|. \tag{9.163}$$

The set of all column vectors $\mathbf{z} = (\,z_1, z_2, \ldots, z_m\,)^T$ with m complex entries $z_1, \ldots, z_m \in \mathbb{C}$ is the m-dimensional complex vector space $\mathbb{C}^m$. Vector addition and scalar multiplication are defined in the obvious manner, and satisfy all the arithmetic properties listed in Section 1.1. We can write any complex vector $\mathbf{z} = \mathbf{x} + \mathrm{i}\,\mathbf{y} \in \mathbb{C}^m$ as a linear combination of two real vectors: $\mathbf{x} = \operatorname{Re}\mathbf{z}$, $\mathbf{y} = \operatorname{Im}\mathbf{z} \in \mathbb{R}^m$, called its *real* and *imaginary parts*. Its *complex conjugate* $\overline{\mathbf{z}} = \mathbf{x} - \mathrm{i}\,\mathbf{y}$ is obtained by taking the complex conjugates of its individual entries. Thus, for example, if

$$\mathbf{z} = \begin{pmatrix} 1 + 2\,\mathrm{i} \\ -3 \\ 5\,\mathrm{i} \end{pmatrix} = \begin{pmatrix} 1 \\ -3 \\ 0 \end{pmatrix} + \mathrm{i} \begin{pmatrix} 2 \\ 0 \\ 5 \end{pmatrix}, \qquad \text{then} \qquad \operatorname{Re}\mathbf{z} = \begin{pmatrix} 1 \\ -3 \\ 0 \end{pmatrix}, \quad \operatorname{Im}\mathbf{z} = \begin{pmatrix} 2 \\ 0 \\ 5 \end{pmatrix},$$

and so its complex conjugate is $\overline{\mathbf{z}} = \begin{pmatrix} 1 - 2\,\mathrm{i} \\ -3 \\ -5\,\mathrm{i} \end{pmatrix} = \begin{pmatrix} 1 \\ -3 \\ 0 \end{pmatrix} - \mathrm{i} \begin{pmatrix} 2 \\ 0 \\ 5 \end{pmatrix}$. In particular, a vector

$\mathbf{z} \in \mathbb{R}^m \subset \mathbb{C}^m$ is real if and only if $\mathbf{z} = \overline{\mathbf{z}}$. Note that we can identify $\mathbb{C}^m$ with the real vector space $\mathbb{R}^{2m}$ of twice the dimension by identifying

$$\mathbf{z} = (\,z_1, z_2, \ldots, z_n\,)^T = (\,x_1 + \mathrm{i}\,y_1, \ldots, x_m + \mathrm{i}\,y_m\,)^T \simeq \begin{pmatrix} \mathbf{x} \\ \mathbf{y} \end{pmatrix} = (\,x_1, \ldots, x_m, y_1, \ldots, y_m\,)^T.$$

Keep in mind that this identification does not respect multiplication by complex scalars.

Most of the vector space concepts we developed in the real domain, including span, linear independence, basis, and dimension, can be straightforwardly extended to the complex regime. The one exception is the concept of an inner product, which requires a little thought. In analysis, the primary applications of inner products and norms rely on the associated inequalities: Cauchy–Schwarz and triangle. But there is no natural ordering of the complex numbers, and so one *cannot* assign a meaning to a complex inequality like $z < w$. Inequalities make sense only in the real domain, and so the norm of a complex vector should still be a positive and real. With this in mind, the naïve idea of simply summing the squares of the entries of a complex vector will *not* define a norm on $\mathbb{C}^m$, since the result will typically be complex. Moreover, some nonzero complex vectors, e.g., $(1, i)^T \in \mathbb{C}^2$, would then have zero "norm".

The correct definition is modeled on the formula (9.160) for the modulus or norm of a single complex number z, in which we view the quantity inside the square root as the inner product of z with itself. The vector version of this construction is named after the nineteenth-century French mathematician Charles Hermite, and called the *Hermitian dot product* on $\mathbb{C}^m$. It has the explicit formula

$$\mathbf{z} \cdot \mathbf{w} = \mathbf{z}^T \overline{\mathbf{w}} = z_1 \overline{w}_1 + z_2 \overline{w}_2 + \cdots + z_m \overline{w}_m, \quad \text{for} \quad \mathbf{z} = \begin{pmatrix} z_1 \\ z_2 \\ \vdots \\ z_m \end{pmatrix}, \quad \mathbf{w} = \begin{pmatrix} w_1 \\ w_2 \\ \vdots \\ w_m \end{pmatrix}. \tag{9.164}$$

Pay attention to the fact that we must apply complex conjugation to all the entries of the second vector. For example, if $\mathbf{z} = \begin{pmatrix} 1 + i \\ 3 + 2i \end{pmatrix}$, $\mathbf{w} = \begin{pmatrix} 1 + 2i \\ i \end{pmatrix}$, then

$$\mathbf{z} \cdot \mathbf{w} = (1 + i)(1 - 2i) + (3 + 2i)(-i) = 5 - 4i,$$
$$\mathbf{w} \cdot \mathbf{z} = (1 + 2i)(1 - i) + i(3 - 2i) = 5 + 4i.$$

We conclude that the Hermitian dot product is *not* symmetric. Indeed, reversing the order of the vectors conjugates their dot product:

$$\mathbf{w} \cdot \mathbf{z} = \overline{\mathbf{z} \cdot \mathbf{w}}. \tag{9.165}$$

This is an unexpected complication, but it does have the desired effect that the induced norm, namely

$$0 \le \|\mathbf{z}\| = \sqrt{\mathbf{z} \cdot \mathbf{z}} = \sqrt{\mathbf{z}^T \overline{\mathbf{z}}}$$
$$= \sqrt{|z_1|^2 + \cdots + |z_m|^2} = \sqrt{x_1^2 + \cdots + x_m^2 + y_1^2 + \cdots + y_m^2}, \tag{9.166}$$

where $\mathbf{z}_j = \mathbf{x}_j + i\mathbf{y}_j$, is strictly positive for all $\mathbf{0} \ne \mathbf{z} \in \mathbb{C}^m$. Thus, the Hermitian norm on $\mathbb{C}^m$ coincides with the ordinary Euclidean norm on $\mathbb{R}^{2m}$. For example, if

$$\mathbf{z} = \begin{pmatrix} 1 + 3i \\ -2i \\ -5 \end{pmatrix}, \quad \text{then} \quad \|\mathbf{z}\| = \sqrt{|1 + 3i|^2 + |-2i|^2 + |-5|^2} = \sqrt{39}.$$

The Hermitian dot product is well behaved under complex vector addition:

$$(\mathbf{z} + \mathbf{u}) \cdot \mathbf{w} = \mathbf{z} \cdot \mathbf{w} + \mathbf{u} \cdot \mathbf{w}, \quad \mathbf{z} \cdot (\mathbf{w} + \mathbf{v}) = \mathbf{z} \cdot \mathbf{w} + \mathbf{z} \cdot \mathbf{v}. \tag{9.167}$$

However, while complex scalar multiples can be extracted from the first vector without alteration, when they multiply the second vector, they emerge as complex conjugates:

$$(c\,\mathbf{z}) \cdot \mathbf{w} = c\,(\mathbf{z} \cdot \mathbf{w}), \qquad \mathbf{z} \cdot (c\,\mathbf{w}) = \overline{c}\,(\mathbf{z} \cdot \mathbf{w}), \qquad c \in \mathbb{C}. \tag{9.168}$$

Thus, the Hermitian dot product is not bilinear in the strict sense, but satisfies something that, for lack of a better name, is known as *sesquilinearity*.

9.10.2 Roots of Unity

When formulating the complex version of the discrete Fourier transform, the crux of the matter is the properties of the remarkable complex numbers

$$\zeta_m = e^{2\pi i/m} = \cos\frac{2\pi}{m} + i\sin\frac{2\pi}{m}, \qquad \text{where} \qquad m = 1, 2, 3, \ldots. \tag{9.169}$$

Particular cases include

$$\zeta_2 = -1, \qquad \zeta_3 = -\frac{1}{2} + \frac{\sqrt{3}}{2}\,i, \qquad \zeta_4 = i, \qquad \text{and} \qquad \zeta_8 = \frac{\sqrt{2}}{2} + \frac{\sqrt{2}}{2}\,i. \tag{9.170}$$

The m-th power of ζ_m is

$$\zeta_m^m = \left(e^{2\pi i/m}\right)^m = e^{2\pi i} = 1,$$

which follows from Euler's formula (9.155). Thus, ζ_m is one of the complex *m-th roots of unity*: $\zeta_m = \sqrt[m]{1}$. There are, in fact, m distinct complex m-th roots of 1, including 1 itself, namely the powers of ζ_m:

$$\zeta_m^k = e^{2k\pi i/m} = \cos\frac{2k\pi}{m} + i\sin\frac{2k\pi}{m}, \qquad k = 0, \ldots, m-1, \tag{9.171}$$

since $(\zeta_m^k)^m = (\zeta_m^m)^k = 1^k = 1$. Since it generates all the others, ζ_m is known as a *primitive m-th root of unity*. Geometrically, the m-th roots (9.171) are the vertices of a regular unit m-gon inscribed in the unit circle $|z| = 1$. In other words, they are the vertices of a circular graph! The primitive root ζ_m is the first vertex we encounter as we go around the m-gon in a counterclockwise direction, starting at 1. Continuing around, the other roots appear in their natural order $\zeta_m^2, \zeta_m^3, \ldots, \zeta_m^{m-1}$, cycling back to $\zeta_m^m = 1$. The complex conjugate of ζ_m is the "last" m-th root:

$$e^{-2\pi i/m} = \overline{\zeta}_m = \frac{1}{\zeta_m} = \zeta_m^{m-1} = e^{2(m-1)\pi i/m}. \tag{9.172}$$

More generally, we have

$$\zeta_m^{m+k} = \zeta_m^k, \qquad \overline{\zeta_m^k} = \zeta_m^{-k}, \qquad \zeta_m^0 = 1, \tag{9.173}$$

and, if $m = 2n$ is even,

$$\zeta_m^n = \zeta_m^{m/2} = -1. \tag{9.174}$$

The complex numbers (9.171) are a complete set of roots of the polynomial $z^m - 1$, which can therefore be factored:

$$z^m - 1 = (z - 1)(z - \zeta_m)(z - \zeta_m^2) \cdots (z - \zeta_m^{m-1}).$$

On the other hand, elementary algebra provides us with the real factorization

$$z^m - 1 = (z - 1)(1 + z + z^2 + \cdots + z^{m-1}).$$

Comparing the two, we conclude that

$$1 + z + z^2 + \cdots + z^{m-1} = (z - \zeta_m)(z - \zeta_m^2) \cdots (z - \zeta_m^{m-1}).$$

Substituting $z = \zeta_m^k$ into both sides of this identity, we deduce the useful formula

$$1 + \zeta_m^k + \zeta_m^{2k} + \cdots + \zeta_m^{(m-1)k} = \begin{cases} m, & k = 0, \\ 0, & 0 < k < m, \end{cases} \tag{9.175}$$

the case $k = 0$ being calculated directly. Using the first formula in (9.173), this formula can easily be extended to general integers k; the sum is equal to m if m evenly divides k, which we write as $k \equiv 0 \mod m$, and is 0 otherwise.

9.10.3 The Complex Discrete Fourier Transform

Now we can discuss the complex version of the discrete Fourier transform. Recalling the vectors (9.141), for any $k \in \mathbb{Z}$, we define the complex vectors

$$\begin{aligned}
\mathbf{z}_k &= \frac{\mathbf{x}_k + i\,\mathbf{y}_k}{\sqrt{m}} \\
&= \frac{1}{\sqrt{m}} \left(1, \cos \frac{2k\pi}{m} + i \sin \frac{2k\pi}{m}, \ldots, \cos \frac{2(m-1)k\pi}{m} + i \sin \frac{2(m-1)k\pi}{m} \right)^T \tag{9.176} \\
&= \frac{1}{\sqrt{m}} \left(1, e^{2k\pi i/m}, e^{4k\pi i/m}, \ldots, e^{2(m-1)k\pi i/m} \right)^T = \frac{1}{\sqrt{m}} \left(1, \zeta_m^k, \zeta_m^{2k}, \ldots, \zeta_m^{(m-1)k} \right)^T .
\end{aligned}$$

The factor of $1/\sqrt{m}$ is included to simplify the subsequent formulas. Since $\mathbf{z}_k$ is a linear combination of two eigenvectors belonging to the same eigenspace of the graph Laplacian, they are also eigenvectors:

$$L\mathbf{z}_k = \mu_k \mathbf{z}_k, \tag{9.177}$$

where the eigenvalues are given by formula (9.145), which is valid for all k. Note that, as in (9.142),

$$\mathbf{z}_{k+m} = \mathbf{z}_k, \qquad \overline{\mathbf{z}}_k = \mathbf{z}_{-k}, \qquad \mathbf{z}_0 = \mathbf{1}. \tag{9.178}$$

Furthermore, as a consequence of formula (9.175) and its extension to general k, we find

$$\mathbf{z}_k \cdot \mathbf{z}_\ell = \mathbf{z}_k^T \overline{\mathbf{z}}_\ell = \frac{1}{m} \sum_{j=0}^{m-1} \zeta_m^{jk} \overline{\zeta}_m^{j\ell} = \frac{1}{m} \sum_{j=0}^{m-1} \zeta_m^{j(k-\ell)} = \begin{cases} 1, & k \equiv \ell \mod m, \\ 0, & \text{otherwise.} \end{cases} \tag{9.179}$$

In particular, $\| \mathbf{z}_k \| = 1$ for all k, from which we conclude that the vectors $\mathbf{z}_0, \ldots, \mathbf{z}_{m-1}$ form an orthonormal basis of $\mathbb{C}^m$ under the Hermitian dot product.

The complex version of the discrete Fourier transform decomposes a general vector[27] $\mathbf{f} = \left(f_0, f_1, \ldots, f_{m-1} \right)^T \in \mathbb{C}^m$ into a linear combination of the complex eigenvectors:

$$\mathbf{f} = c_0 \mathbf{z}_0 + c_1 \mathbf{z}_1 + \cdots + c_{m-1} \mathbf{z}_{m-1}. \tag{9.180}$$

Orthonormality of the basis vectors implies that, unlike the real version, the coefficients are all given by the same formula:[28]

$$c_k = \mathbf{f} \cdot \mathbf{z}_k = \frac{1}{\sqrt{m}} \sum_{j=0}^{m-1} f_j \, \zeta_m^{-jk}. \tag{9.181}$$

---

[27] It is convenient to label the entries of vectors starting at 0 instead of 1 here.

[28] Keep in mind that when computing the Hermitian dot product, one takes complex conjugates of the entries of the second vector.

Thus the DFT of a vector $\mathbf{f} \in \mathbb{C}^m$ is the coefficient vector $\mathbf{c} = \left(c_0, c_1, \ldots, c_{m-1} \right)^T \in \mathbb{C}^m$. On the other hand, given the Fourier coefficients $c_0, \ldots, c_{m-1}$, the reconstruction formula (9.180) implies that the corresponding vector $\mathbf{f} \in \mathbb{C}^m$ has components

$$f_k = \frac{1}{\sqrt{m}} \sum_{j=0}^{m-1} c_j \, \zeta_m^{jk}, \qquad k = 0, \ldots, m-1. \tag{9.182}$$

The formula (9.181) mapping a vector $\mathbf{f} = \left(f_0, f_1, \ldots, f_{m-1} \right)^T \in \mathbb{C}^m$ to its coefficient vector $\mathbf{c} = \left(c_0, c_1, \ldots, c_{m-1} \right)^T \in \mathbb{C}^m$ is the complex *discrete Fourier transform* (DFT). The reconstruction formula (9.180) mapping $\mathbf{c}$ back to $\mathbf{f}$ is the *inverse discrete Fourier transform* (IDFT). Note that the only difference between the direct and inverse discrete Fourier transform formulas (9.181) and (9.182) is a minus sign!

Observe that the function mapping a vector to its discrete Fourier coefficients is linear. In other words, if $\mathbf{f}, \mathbf{g} \in \mathbb{C}^m$ map to their respective coefficient vectors $\mathbf{c}, \mathbf{d} \in \mathbb{C}^m$ and $a, b \in \mathbb{C}$ are any scalars, then the Fourier coefficients of the linear combination $a\,\mathbf{f} + b\,\mathbf{g}$ are the components of the vector $a\,\mathbf{c} + b\,\mathbf{d}$. Thus, according to Theorem 3.33 and Proposition 3.35 (which both work as stated in the complex domain), the discrete Fourier transform and its inverse can be realized by multiplying the vectors by an $m \times m$ matrix. Let us first look at the inverse DFT. Using the matrix multiplication formula (3.21), we see that the reconstruction formula (9.180) is equivalent to the vector equation

$$\mathbf{f} = F_m \, \mathbf{c}, \qquad \text{and hence} \qquad \mathbf{c} = F_m^{-1} \, \mathbf{f}, \tag{9.183}$$

where the $m \times m$ *Fourier matrix* F_m has columns $\mathbf{z}_0, \ldots, \mathbf{z}_{m-1}$. Explicitly,

$$F_m = \frac{G_m}{\sqrt{m}}, \quad \text{where} \quad G_m = \begin{pmatrix} 1 & 1 & 1 & 1 & \cdots & 1 \\ 1 & \zeta_m & \zeta_m^2 & \zeta_m^3 & \cdots & \zeta_m^{m-1} \\ 1 & \zeta_m^2 & \zeta_m^4 & \zeta_m^6 & \cdots & \zeta_m^{2(m-1)} \\ 1 & \zeta_m^3 & \zeta_m^6 & \zeta_m^9 & \cdots & \zeta_m^{3(m-1)} \\ \vdots & \vdots & \vdots & \vdots & \ddots & \vdots \\ 1 & \zeta_m^{m-1} & \zeta_m^{2(m-1)} & \zeta_m^{3(m-1)} & \cdots & \zeta_m^{(m-1)^2} \end{pmatrix}, \tag{9.184}$$

is an $m \times m$ complex matrix, all of whose entries are m-th roots of unity. Observe that F_m is a symmetric complex matrix whose columns form an orthonormal basis of $\mathbb{C}^m$ under the Hermitian dot product. However, since it has complex entries, it is *not* an orthogonal matrix; the proper mathematical term is *unitary matrix*. The orthonormality conditions (9.179) imply that the rows of the inverse matrix F_m^{-1}, which governs the direct DFT (9.181) are the complex conjugates of the transposes of the columns of F_m. Thus,

$$F_m^{-1} = F_m^{\dagger} = \frac{G_m^{\dagger}}{\sqrt{m}} \quad \text{where} \quad G_m^{\dagger} = \begin{pmatrix} 1 & 1 & 1 & \cdots & 1 \\ 1 & \zeta_m^{-1} & \zeta_m^{-2} & \cdots & \zeta_m^{-(m-1)} \\ 1 & \zeta_m^{-2} & \zeta_m^{-4} & \cdots & \zeta_m^{-2(m-1)} \\ \vdots & \vdots & \vdots & \ddots & \vdots \\ 1 & \zeta_m^{-(m-1)} & \zeta_m^{-2(m-1)} & \cdots & \zeta_m^{-(m-1)^2} \end{pmatrix}. \tag{9.185}$$

Here the dagger notation is used to denote the *Hermitian adjoint* of a matrix, which coincides with the transpose of its complex conjugate, the latter applied entrywise:

$$A^{\dagger} = \overline{A}^{\,T}, \tag{9.186}$$

The Hermitian adjoint is defined so that the complex form of the adjoint equation (4.19) holds:

$$\mathbf{z} \cdot (A^\dagger \mathbf{w}) = (A\mathbf{z}) \cdot \mathbf{w} \qquad \text{for all} \qquad \mathbf{z}, \mathbf{w} \in \mathbb{C}^m. \tag{9.187}$$

Note finally that $G_m^{-1} = G_m^\dagger/m$.

Since

$$F_m^\dagger F_m = F_m F_m^\dagger = I,$$

we also have

$$\mathbf{z} \cdot \mathbf{w} = (F_m^\dagger F_m \mathbf{z}) \cdot \mathbf{w} = (F_m \mathbf{z}) \cdot (F_m \mathbf{w}), \tag{9.188}$$

a formula known as *Parseval's identity*. In other words, the Hermitian dot product between vectors equals that between their Fourier coefficients. In particular, Parseval's identity implies that the DFT preserves norms: $\| \mathbf{z} \| = \| F_m \mathbf{z} \|$ for all $\mathbf{z} \in \mathbb{C}^m$.

9.10.4 Sampling, Trigonometric Interpolation, and Aliasing

As noted above, one can interpret a vector $\mathbf{f} \in \mathbb{C}^m$ as the sample values of a, in this case, complex-valued function $f(x)$ on the interval $[0, 2\pi]$, so that $f_j = f(2j\pi/m)$ for $j = 0, \ldots, m-1$. Of course, $f(x)$ could be real-valued, in which case $\mathbf{f} \in \mathbb{R}^m \subset \mathbb{C}^m$. The vector $\mathbf{z}_k$ given in (9.176) is the sample vectors of the rescaled complex exponential function $f_k(x) = e^{ikx}/\sqrt{m}$. Thus, the reconstruction formula (9.180) expresses the sample values of f as a linear combination of the first m sampled exponentials, so

$$f(x) \sim p_m(x) = \frac{c_0 + c_1 e^{ix} + c_2 e^{2ix} + \cdots + c_{m-1} e^{(m-1)ix}}{\sqrt{m}} = \frac{1}{\sqrt{m}} \sum_{k=0}^{m-1} c_k e^{ikx}. \tag{9.189}$$

As above, the symbol $\sim$ in (9.189) means that the function $f(x)$ and the sum $p_m(x)$ agree on the sample points:

$$f\left(\frac{2j\pi}{m}\right) = p_m\left(\frac{2j\pi}{m}\right), \qquad j = 0, \ldots, m-1, \tag{9.190}$$

which, in view of the 2π periodicity of the complex exponentials, also holds when $j = m$ provided f is 2π periodic, so $f(2\pi) = f(0) = p_m(0) = p_m(2\pi)$. Therefore, $p_m(x)$ can be viewed as an *interpolating trigonometric polynomial* for the sample data $f_0, \ldots, f_{m-1}$.

Remark. If $f(x)$ is real, then $p_m(x)$ is also real on the sample points, but the interpolating trigonometric polynomial (9.189) may very well be complex-valued in between. To avoid this unsatisfying state of affairs, we will usually discard its imaginary component, and regard the real part of $p_m(x)$ as "the" interpolating trigonometric polynomial. On the other hand, sticking with a purely real construction unnecessarily complicates the underlying mathematical analysis, and so it is extremely convenient to retain the complex exponential form of the discrete Fourier sum.   ▲

Sampling cannot distinguish between functions that have the same values at all of the sample points — from the sampler's point of view they are identical. For example, the periodic complex exponential function

$$f(x) = e^{imx} = \cos mx + i \sin mx$$

has sample values

$$f_j = f\left(\frac{2j\pi}{m}\right) = \exp\left(im\,\frac{2j\pi}{m}\right) = e^{2j\pi i} = 1 \qquad \text{for all} \qquad j = 0, \ldots, m-1,$$

and hence is indistinguishable from the constant function $g(x) \equiv 1$ — both produce the *same* sample vector $(1, 1, \ldots, 1)^T$. This has the important implication that sampling at m equally spaced sample points *cannot* detect periodic signals of frequency m. More generally, the two complex exponential signals

$$e^{\mathrm{i}(k+m)x} \qquad \text{and} \qquad e^{\mathrm{i}kx}$$

are also indistinguishable when sampled. In particular, exponentials $e^{-\mathrm{i}kx}$ of "negative" frequency can all be converted into positive versions, namely $e^{\mathrm{i}(m-k)x}$, by the same sampling argument. For example,

$$e^{-\mathrm{i}x} = \cos x - \mathrm{i}\sin x \qquad \text{and} \qquad e^{(m-1)\mathrm{i}x} = \cos(m-1)x + \mathrm{i}\sin(m-1)x$$

have identical values on the sample points. However, elsewhere they are quite different; the former is slowly varying, while the latter represents a high-frequency oscillation.

Figure 9.47 shows the Fourier modes $\mathbf{x}_k$ for $m = 8$ and $k = 0, \ldots, 8$. We see in the figures how the frequencies $k \geq 5$ are *aliased* back to lower frequencies, by the identity $\mathbf{x}_{m-k} = \mathbf{x}_k$. The highest frequency discrete mode is in fact the middle eigenvector $\mathbf{x}_4$, which is in fact the eigenvector with largest eigenvalue μ_k. The figure for the $\mathbf{y}_k$ modes would look similar, except shifted slightly.

This effect is commonly referred to as *aliasing*.[29] If you view a moving particle under a stroboscopic light that flashes only eight times, you would be unable to determine which of the two graphs the particle was following. Aliasing is the cause of a well-known artifact in movies: spoked wheels can appear to be rotating backwards when our brain interprets the discretization of the high-frequency forward motion imposed by the frames of the film as an equivalently discretized low-frequency motion in reverse, [24].

To mitigate the effect of high frequency aliasing, we endeavor to utilize the lowest frequencies possible in our formulas. This is accomplished by replacing (9.180) by the low frequency alternative[30]

$$\mathbf{f} = \begin{cases} c_{-n}\mathbf{z}_{-n} + \cdots + c_0\mathbf{z}_0 + \cdots + c_n\mathbf{z}_n, & m = 2n+1, \\[2mm] c_{1-n}\mathbf{z}_{1-n} + \cdots + c_0\mathbf{z}_0 + \cdots + c_n\mathbf{z}_n, & m = 2n. \end{cases} \tag{9.191}$$

It is important to emphasize that, in view of (9.178) and the corresponding formulae (9.181) for the Fourier coefficients, (9.191) is the exact same formula as (9.180). However, the alternative reconstruction formula (9.191) produces a different interpolation formula:

$$f(x) \sim \widehat{p}_m(x)$$
$$= \begin{cases} \dfrac{c_{n+1}e^{-\mathrm{i}nx} + \cdots + c_{2n}e^{-\mathrm{i}x} + c_0 + c_1 e^{\mathrm{i}x} + \cdots + c_n e^{\mathrm{i}nx}}{\sqrt{m}}, & m = 2n+1, \\[4mm] \dfrac{c_{n+1}e^{-\mathrm{i}(n-1)x} + \cdots + c_{2n-1}e^{-\mathrm{i}x} + c_0 + c_1 e^{\mathrm{i}x} + \cdots + c_n e^{\mathrm{i}nx}}{\sqrt{m}}, & m = 2n. \end{cases}$$
$$\tag{9.192}$$

In all cases, the coefficients are given by the DFT formula (9.181); also, bear in mind that we can identify $c_k = c_{k-m}$. Although $\widehat{p}_m(x)$ is a different trigonometric function, it has the same values (9.190) as $p_m(x)$ and $f(x)$ at the interpolation points. Formula (9.192) is known

---

[29] In computer graphics, [206], the term "aliasing" is used in a much broader sense that covers a variety of artifacts introduced by discretization — particularly, the jagged appearance of lines and smooth curves on a digital monitor.

[30] One can use either $\mathbf{z}_n$ or $\mathbf{z}_{-n}$ in the even order case as they have the same frequency.

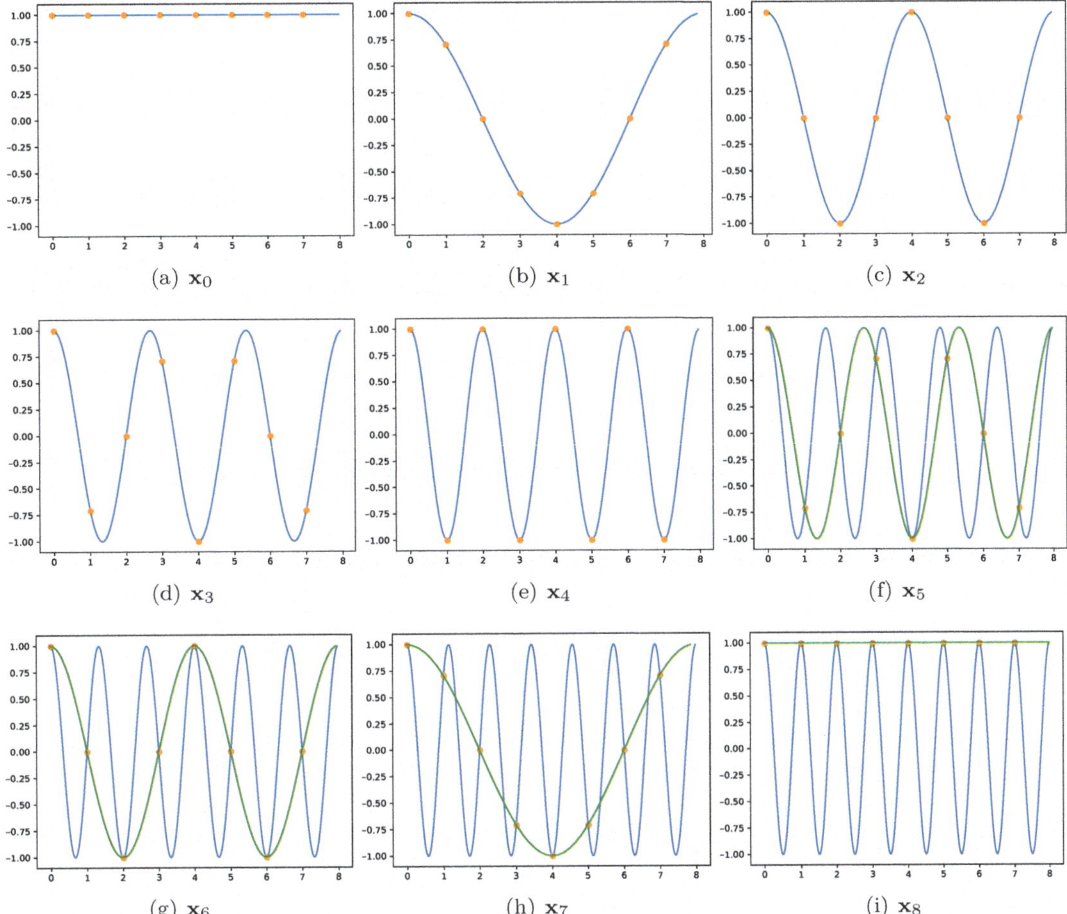

Figure 9.47: Real parts $\mathbf{x}_k$ of the discrete Fourier modes for $m = 8$. The orange dots show the vector $\mathbf{x}_k$ for $k = 0, 1, \ldots, 8$, which is sampled from the blue curves. For $k = 5, 6, 7, 8$, the sampled frequency is aliased to the lower frequency modes $k = 0, 1, 2, 3$, respectively, which is equivalent to sampling the green curve, since $\mathbf{x}_{m-k} = \mathbf{x}_k$. The discrete Fourier mode with the highest frequency is $k = 4$.

as the *low frequency* trigonometric interpolant, while formula (9.189) will be called the *high frequency* version. This is because the trigonometric frequencies appearing in the real and imaginary parts of (9.192) are lower overall than those appearing in (9.189). While the high frequency version is easier to use in mathematical manipulations, since it does not require treating even and odd m differently, the low frequency version is better for interpolation away from the sample points and avoiding aliasing artifacts.

Example 9.81. If $m = 4$, then $\zeta_4 = i$. The corresponding rescaled sampled exponential vectors

$$\mathbf{z}_0 = \frac{1}{2}\begin{pmatrix} 1 \\ 1 \\ 1 \\ 1 \end{pmatrix}, \qquad \mathbf{z}_1 = \frac{1}{2}\begin{pmatrix} 1 \\ i \\ -1 \\ -i \end{pmatrix}, \qquad \mathbf{z}_2 = \frac{1}{2}\begin{pmatrix} 1 \\ -1 \\ 1 \\ -1 \end{pmatrix}, \qquad \mathbf{z}_3 = \frac{1}{2}\begin{pmatrix} 1 \\ -i \\ -1 \\ i \end{pmatrix},$$

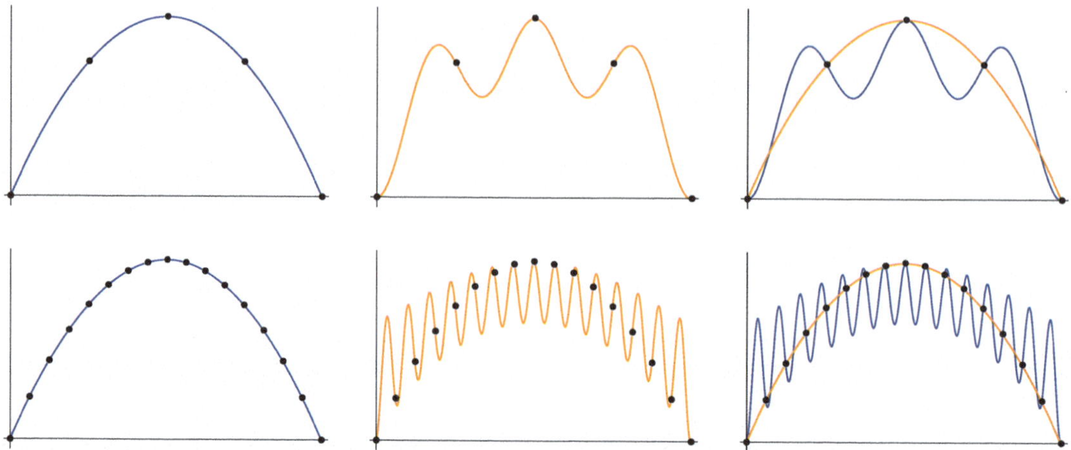

Figure 9.48: The Discrete Fourier Representation of $2\pi x - x^2$.

form an orthonormal basis of $\mathbb{C}^4$ with respect to the Hermitian dot product

$$\mathbf{v} \cdot \mathbf{w} = v_0 \,\overline{w_0} + v_1 \,\overline{w_1} + v_2 \,\overline{w_2} + v_3 \,\overline{w_3}, \qquad \text{where} \qquad \mathbf{v} = \begin{pmatrix} v_0 \\ v_1 \\ v_2 \\ v_3 \end{pmatrix}, \qquad \mathbf{w} = \begin{pmatrix} w_0 \\ w_1 \\ w_2 \\ w_3 \end{pmatrix}.$$

Given the sampled function values

$$f_0 = f(0), \qquad f_1 = f\left(\tfrac{1}{2}\pi\right), \qquad f_2 = f(\pi), \qquad f_3 = f\left(\tfrac{3}{2}\pi\right),$$

we construct the discrete Fourier representation

$$\mathbf{f} = c_0\, \mathbf{z}_0 + c_1\, \mathbf{z}_1 + c_2\, \mathbf{z}_2 + c_3\, \mathbf{z}_3, \tag{9.193}$$

where

$$\begin{aligned}
c_0 &= \mathbf{f} \cdot \mathbf{z}_0 = \tfrac{1}{2}(f_0 + f_1 + f_2 + f_3), & c_1 &= \mathbf{f} \cdot \mathbf{z}_1 = \tfrac{1}{2}(f_0 - \mathrm{i}\, f_1 - f_2 + \mathrm{i}\, f_3), \\
c_2 &= \mathbf{f} \cdot \mathbf{z}_2 = \tfrac{1}{2}(f_0 - f_1 + f_2 - f_3), & c_3 &= \mathbf{f} \cdot \mathbf{z}_3 = \tfrac{1}{2}(f_0 + \mathrm{i}\, f_1 - f_2 - \mathrm{i}\, f_3).
\end{aligned}$$

We interpret this decomposition as the complex exponential interpolant

$$f(x) \ \sim \ p_4(x) = \tfrac{1}{2}\left(c_0 + c_1\, e^{\mathrm{i}\,x} + c_2\, e^{2\,\mathrm{i}\,x} + c_3\, e^{3\,\mathrm{i}\,x} \right)$$

that agrees with $f(x)$ on the 4 sample points.

For instance, if

$$f(x) = 2\pi x - x^2, \tag{9.194}$$

then

$$f_0 = 0, \qquad f_1 = 7.4022, \qquad f_2 = 9.8696, \qquad f_3 = 7.4022,$$

and hence

$$c_0 = 12.3370, \qquad c_1 = -4.9348, \qquad c_2 = -2.4674, \qquad c_3 = -4.9348.$$

Therefore, the interpolating trigonometric polynomial is given by the real part of

$$p_4(x) = 6.1685 - 2.4674\, e^{\mathrm{i}\,x} - 1.2337\, e^{2\,\mathrm{i}\,x} - 2.4674\, e^{3\,\mathrm{i}\,x}, \tag{9.195}$$

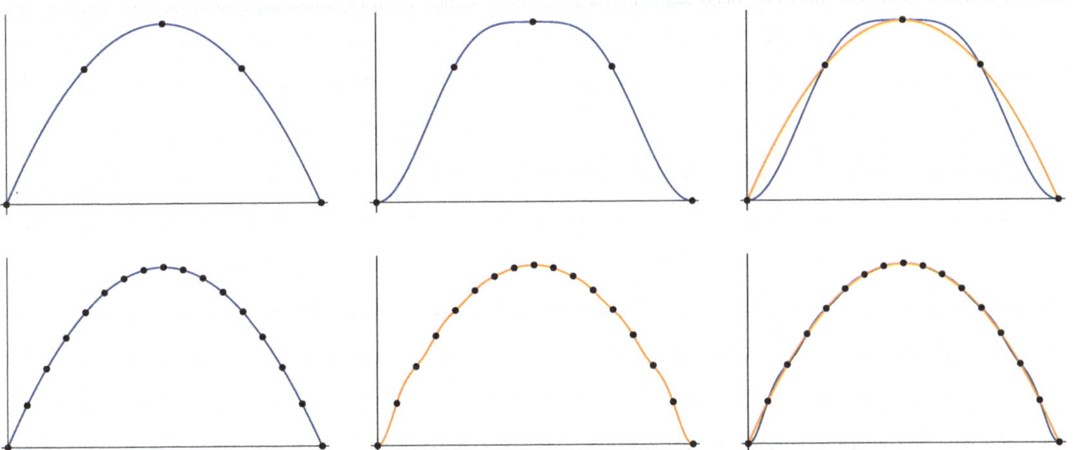

Figure 9.49: The Low–Frequency Discrete Fourier Representation of $2\pi x - x^2$.

namely,

$$\mathrm{Re}\, p_4(x) = 6.1685 - 2.4674 \cos x - 1.2337 \cos 2x - 2.4674 \cos 3x. \tag{9.196}$$

In Figure 9.48, we compare the function, with the interpolation points indicated, and its discrete Fourier representations (9.196) for both $m = 4$ in the first row, and $m = 16$ points in the second. The resulting graphs point out a significant difficulty with the discrete Fourier transform as developed so far. While the trigonometric polynomials do indeed correctly match the sampled function values, their pronounced oscillatory behavior makes them completely unsuitable for interpolation away from the sample points.

In fact, for interpolating purposes, we should replace (9.195) by the equivalent low-frequency interpolant

$$\widehat{p}_4(x) = -2.4674\, e^{-\mathrm{i}\,x} + 6.1685 - 2.4674\, e^{\mathrm{i}\,x} - 1.2337\, e^{2\,\mathrm{i}\,x}, \tag{9.197}$$

with real part

$$\mathrm{Re}\, \widehat{p}_4(x) = 6.1685 - 4.9348 \cos x - 1.2337 \cos 2x.$$

Graphs of the $m = 4$ and 16 low-frequency trigonometric interpolants can be seen in Figure 9.49. Thus, by utilizing only the lowest-frequency exponentials, we successfully suppress the aliasing artifacts, resulting in a quite reasonable trigonometric interpolant to the given function on the entire interval. ▲

9.10.5 Convolution and the DFT

In this section, we investigate the relationship between the discrete Fourier transform and the operation of convolution, which is of importance in machine learning and mathematical analysis, as well as signal and image processing, particularly when dealing with noisy signals and images.

Throughout this section, we will continue to index vectors in $\mathbb{R}^m$ or $\mathbb{C}^m$ starting with $i = 0$, so $\mathbf{x} = \left(x_0, x_1, \ldots, x_{m-1} \right)^T$. The indices k are always to be computed modulo m, so that $x_i = x_j$ whenever $i \equiv j \bmod m$. We begin by introducing a certain class of matrices.

Definition 9.82. The *circulant matrix* $C = C_{\mathbf{x}}$ associated with a vector $\mathbf{x} \in \mathbb{C}^m$ has entries $c_{ij} = x_{i-j}$ for $i, j = 0, \ldots, m - 1$.

Explicitly, a circulant matrix has the form

$$
C_{\mathbf{x}} = \begin{pmatrix}
x_0 & x_{m-1} & x_{m-2} & x_{m-3} & \cdots & x_1 \\
x_1 & x_0 & x_{m-1} & x_{m-2} & \cdots & x_2 \\
x_2 & x_1 & x_0 & x_{m-1} & \cdots & x_3 \\
x_3 & x_2 & x_1 & x_0 & \cdots & x_4 \\
\vdots & \vdots & \vdots & \vdots & \ddots & \vdots \\
x_{m-1} & x_{m-2} & x_{m-3} & x_{m-4} & \cdots & x_0
\end{pmatrix},
\tag{9.198}
$$

so that its entries are the same along each periodic diagonal and its first column is just $\mathbf{x} = C_{\mathbf{x}} \mathbf{e}_0$. Examples include the circular graph weight matrix (9.137), which is the circulant matrix associated with the vector $\mathbf{e}_1 + \mathbf{e}_{m-1} = \mathbf{e}_1 + \mathbf{e}_{-1}$ (recalling the periodicity convention on indices) and the corresponding graph Laplacian (9.138), which is the circulant matrix associated with the vector $2\mathbf{e}_0 - \mathbf{e}_1 - \mathbf{e}_{-1}$.

Of particular importance are the *cyclic translation matrices* $T_j = C_{\mathbf{e}_j}$, which are the circulant matrices associated with the elementary basis vectors $\mathbf{e}_j$ for $j = 0, \ldots, m - 1$. For example, $T_0 = I$ is the identity matrix, while

$$
T_1 = \begin{pmatrix}
0 & 0 & 0 & \cdots & 0 & 0 & 1 \\
1 & 0 & 0 & \cdots & 0 & 0 & 0 \\
0 & 1 & 0 & \cdots & 0 & 0 & 0 \\
0 & 0 & 1 & \cdots & 0 & 0 & 0 \\
\vdots & \vdots & \vdots & \ddots & \vdots & \vdots & \vdots \\
0 & 0 & 0 & \cdots & 1 & 0 & 0 \\
0 & 0 & 0 & \cdots & 0 & 1 & 0
\end{pmatrix}.
\tag{9.199}
$$

Observe that the i-th entry of $\mathbf{y} = T_j \mathbf{x}$ is simply $y_i = x_{i-j}$. The translation T_j simply shifts the entries in $\mathbf{x}$ down by j slots, looping around the ends using periodicity. Translation of the standard basis vectors satisfies the important property

$$
T_j \mathbf{e}_k = \mathbf{e}_{j+k}.
\tag{9.200}
$$

If one regards vectors as sample points of a function $f(x)$, the linear maps defined by cyclic translations can be viewed as the discrete analogs of translations in $x \mapsto x - 2j\pi/m$.

Translation matrices satisfy what is known as the *semi-group property*, which means that they mutually commute:

$$
T_j T_k = T_{j+k} = T_k T_j \qquad \text{whereby} \qquad T_j = T_1^j.
\tag{9.201}
$$

Since circulant matrices depend linearly on the vector $\mathbf{x}$, writing $\mathbf{x} = x_0 \mathbf{e}_0 + \cdots + x_{m-1} \mathbf{e}_{m-1} \in \mathbb{C}^m$, we see that we can write any circulant matrix as a linear combination of translation matrices, and hence, by (9.201), as a polynomial in the translation matrix (9.199):

$$
C_{\mathbf{x}} = x_0 T_0 + x_1 T_1 + \cdots + x_{m-1} T_{m-1} = x_{m-1} T_1^{m-1} + x_{m-2} T_1^{m-2} + \cdots + x_1 T_1 + x_0 I. \tag{9.202}
$$

Thus, a circulant matrix represents a generalized cyclic translation. As a consequence of (9.201), circulant matrices mutually commute: $C_{\mathbf{x}} C_{\mathbf{y}} = C_{\mathbf{y}} C_{\mathbf{x}}$ for all $\mathbf{x}, \mathbf{y} \in \mathbb{C}^m$. In fact, they are uniquely characterized by this property.

Proposition 9.83. *An $m \times m$ matrix A commutes with all translation matrices, so*

$$AT_j = T_j A, \qquad j = 0, \ldots, m-1, \tag{9.203}$$

if and only if it is a circulant matrix: $A = C_{\mathbf{x}}$ *for some* $\mathbf{x} \in \mathbb{C}^m$.

Proof. In view of (9.201), it suffices to prove that if A commutes with T_1 then it is a circulant matrix. The (i,j) entry of AT_1 is $a_{i,j+1}$, while the (i,j) entry of T_1A is $a_{i-1,j}$, which must be equal for any i, j, as always computed modulo m. Letting $\mathbf{x} = A\mathbf{e}_0$ be the first column of A, we use the preceding observation to deduce that $a_{i,1} = a_{i-1,0} = x_{i-1}$ and further, by induction, $a_{i,j} = a_{i-j,0} = x_{i-j}$, which proves that $A = C_{\mathbf{x}}$ is a circulant matrix. $\blacksquare$

Remark. The commutation property (9.203) is also known as *translation equivariance*, and plays an important role in applications to image and signal processing. In particular, since they are circulant matrices, the weight matrix and graph Laplacians associated with a circular graph are translation equivariant. $\blacktriangle$

We now connect these matrices with the operation of vector convolution. Interpreting vectors as sample values of functions, their convolution can be viewed as the discrete analog of the convolution integral of the corresponding functions, cf. [67]. As always in this section, indices are computed modulo m.

Definition 9.84. The *convolution* of two vectors $\mathbf{x}, \mathbf{y} \in \mathbb{C}^m$ is the vector

$$\mathbf{z} = \mathbf{x} * \mathbf{y} := C_{\mathbf{x}}\mathbf{y}, \tag{9.204}$$

whose entries are given by

$$z_k = \sum_{j=0}^{m-1} x_{k-j}\, y_j = \sum_{j=0}^{m-1} x_j\, y_{k-j}, \qquad k = 0, \ldots, m-1. \tag{9.205}$$

Note that if $\mathbf{x}, \mathbf{y} \in \mathbb{R}^m$ are real, so is their convolution $\mathbf{x} * \mathbf{y} \in \mathbb{R}^m$. Convolution satisfies the following basic properties for all vectors $\mathbf{x}, \mathbf{y}, \mathbf{z} \in \mathbb{C}^m$, and scalars $c, d \in \mathbb{C}$; the proofs are left to the reader as Exercise 10.5.

(a) *Distributivity (Bilinearity)*:

$$(c\mathbf{x} + d\mathbf{y}) * \mathbf{z} = c\,(\mathbf{x} * \mathbf{z}) + d\,(\mathbf{y} * \mathbf{z}), \qquad \mathbf{x} * (c\mathbf{y} + d\mathbf{z}) = c\,(\mathbf{x} * \mathbf{y}) + d\,(\mathbf{x} * \mathbf{z}). \tag{9.206}$$

(b) *Commutativity (Symmetry)*: $\quad \mathbf{x} * \mathbf{y} = \mathbf{y} * \mathbf{x}.$ $\tag{9.207}$

(c) *Associativity*: $\quad \mathbf{x} * (\mathbf{y} * \mathbf{z}) = (\mathbf{x} * \mathbf{y}) * \mathbf{z}.$ $\tag{9.208}$

Furthermore, in view of (9.202), the convolution operation is a generalized cyclic translation. Indeed,

$$\mathbf{x} * \mathbf{y} = C_{\mathbf{x}}\mathbf{y}, \qquad \text{where} \qquad C_{\mathbf{x}} = \sum_{j=0}^{m-1} x_j\, T_j. \tag{9.209}$$

As a consequence of Proposition 9.83, every translation equivariant linear map is given by convolution.

Proposition 9.85. *An $m \times m$ matrix A is translation equivariant if and only if $A\mathbf{x} = \mathbf{w} * \mathbf{x}$ for all $\mathbf{x}$, where $\mathbf{w} = A\mathbf{e}_0$.*

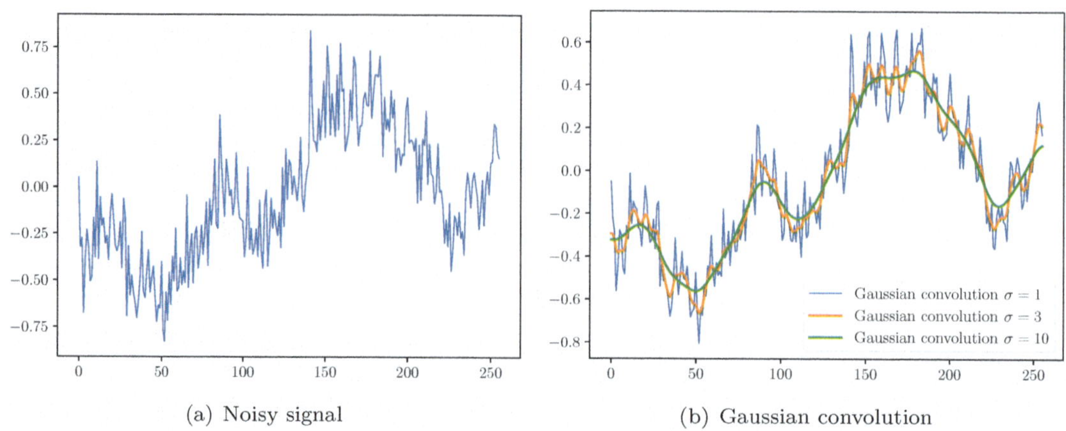

(a) Noisy signal   (b) Gaussian convolution

Figure 9.50: Example of the convolution of a noisy signal with a Gaussian filter.

Given $\mathbf{w}$, the convolution operation $\mathbf{x} \mapsto \mathbf{w} * \mathbf{x}$ can be thought of as averaging a signal $\mathbf{x}$ against shifted and reflected copies of $\mathbf{w}$. If $\mathbf{w}$ is a probability vector, and so its entries are nonnegative and sum to 1, then the i-th entry of $\mathbf{w} * \mathbf{x}$ can be interpreted as a weighted average of the signal values x_j centered at $j = i$. A common application of convolution in signal processing is to smooth out a noisy signal in order to remove noise. In Figure 9.50 we show the convolution of a noisy signal $\mathbf{x}$ with the normalized Gaussian $\mathbf{w} = \widehat{\mathbf{w}}/\|\widehat{\mathbf{w}}\|_1$ where $\widehat{w}_k = e^{-k^2/\sigma^2}$ for some $\sigma > 0$, and so $\mathbf{w}$ is a probability vector. We performed the *even extension* of $\mathbf{x}$ and $\mathbf{w}$ before convolving to avoid boundary effects in the convolution.[31] Notice that in Figure 9.50, as the variance σ^2 of the Gaussian increases, so does the degree of smoothing and noise removal.

The fact that convolution commutes with the graph Laplacian indicates that it is well behaved under the discrete Fourier transform, which relies on its eigenvectors. This is indeed the case, as the following theorem demonstrates. Recall the definition of the Hadamard product in Exercise 1.17. In particular, the Hadamard product of two vectors $\mathbf{x}, \mathbf{y} \in \mathbb{C}^m$ is the vector $\mathbf{z} = \mathbf{x} \circ \mathbf{y}$ whose entries $z_i = x_i y_i$ are obtained by entrywise multiplication. At the level of signals, if $\mathbf{x}$ represents the sampled values of the function $f(x)$ and $\mathbf{y}$ represents the sampled values of the function $g(x)$, then $\mathbf{z} = \mathbf{x} \circ \mathbf{y}$ represents the sampled values of the product function $f(x)\, g(x)$.

Theorem 9.86. *If* $\mathbf{z}, \mathbf{w} \in \mathbb{C}^m$, *then*

$$F_m(\mathbf{z} * \mathbf{w}) = \sqrt{m}\, F_m \mathbf{z} \circ F_m \mathbf{w}, \qquad F_m(\mathbf{z} \circ \mathbf{w}) = \sqrt{m}\, F_m \mathbf{z} * F_m \mathbf{w},$$
$$F_m^\dagger(\mathbf{z} * \mathbf{w}) = \sqrt{m}\, F_m^\dagger \mathbf{z} \circ F_m^\dagger \mathbf{w}, \qquad F_m^\dagger(\mathbf{z} \circ \mathbf{w}) = \sqrt{m}\, F_m^\dagger \mathbf{z} * F_m^\dagger \mathbf{w}. \tag{9.210}$$

In other words, up to a multiplicative factor, the DFT (IDFT) of the convolution of two vectors is the pointwise product of their DFT's (IDFT's) and vice versa.

---

[31] The even extension is obtained by mirroring the signal to one of length $2m$, so that the periodic extension inherent in the discrete Fourier transform does not introduce a discontinuity when the signal wraps around.

Proof. We prove the first identity, leaving the other three to the reader. Let $\mathbf{a} = F_m(\mathbf{z} * \mathbf{w})$, $\mathbf{c} = F_m \mathbf{z}$, $\mathbf{d} = F_m \mathbf{w}$. Then, in view of (9.181), (9.204),

$$a_k = \frac{1}{\sqrt{m}} \sum_{j=0}^{m-1} \zeta_m^{jk} \left(\sum_{i=0}^{m-1} x_{j-i} y_i \right) = \frac{1}{\sqrt{m}} \sum_{i=0}^{m-1} \left(\sum_{j=0}^{m-1} \zeta_m^{(j-i)k} x_{j-i} \right) \zeta_m^{ik} y_i$$

$$= \frac{1}{\sqrt{m}} \left(\sum_{\ell=0}^{m-1} \zeta_m^{\ell k} x_\ell \right) \left(\sum_{i=0}^{m-1} \zeta_m^{ik} y_i \right) = \sqrt{m}\, c_k\, d_k,$$

where passing from the first to the second line used the periodicity of x_j and $\zeta_m^{\ell k}$. Thus, $\mathbf{a} = \sqrt{m}\, \mathbf{c} \circ \mathbf{d}$, which establishes the first equation in (9.210). ∎

The convolution formulas (9.210) simplify slightly if one uses the unnormalized DFT and IDFT matrices, as in (9.184), (9.185):

$$\begin{aligned} G_m(\mathbf{z} * \mathbf{w}) &= G_m \mathbf{z} \circ G_m \mathbf{w}, & G_m(\mathbf{z} \circ \mathbf{w}) &= G_m \mathbf{z} * G_m \mathbf{w}, \\ G_m^\dagger(\mathbf{z} * \mathbf{w}) &= G_m^\dagger \mathbf{z} \circ G_m^\dagger \mathbf{w}, & G_m^\dagger(\mathbf{z} \circ \mathbf{w}) &= G_m^\dagger \mathbf{z} * G_m^\dagger \mathbf{w}. \end{aligned} \tag{9.211}$$

Theorem 9.86 is a fundamental property of the DFT and its inverse — that convolution is converted into multiplication and vice versa. As an application, we can write the convolution of two vectors $\mathbf{x}, \mathbf{y} \in \mathbb{R}^m$ entirely in terms of the (unnormalized) DFT and inverse DFT as follows:

$$\mathbf{x} * \mathbf{y} = G_m(G_m^\dagger \mathbf{x} \circ G_m^\dagger \mathbf{y}). \tag{9.212}$$

The FFT, to be described below, provides a fast method for computing the unnormalized DFT and inverse DFT, which will make (9.212) an extremely efficient method for computing the convolution of two vectors.

Signal Denoising

As an example, we recall the image denoising example from Section 9.9, which required solving the linear system (9.120) with $B = I$. When the image has one channel, and so is represented by a vector, the system has the form

$$(I + \lambda L)\mathbf{u} = \mathbf{y}, \tag{9.213}$$

where $\mathbf{y} \in \mathbb{R}^m$ is the noisy image and $\mathbf{u} \in \mathbb{R}^m$ is the denoised image. Here, $\lambda > 0$ is a parameter that controls the amount of denoising. Larger values of λ lead to more smoothing of the signal or image, and vice-versa for smaller values of λ. We consider here the case of denoising a periodic signal in one dimension, in which case Theorem 9.84 shows that the denoised signal $\mathbf{u}$ solving (9.213) can be written as

$$\mathbf{u} = \mathbf{w}_\lambda * \mathbf{y}, \qquad \text{where} \qquad (1 + \lambda L)\mathbf{w}_\lambda = \mathbf{e}_0. \tag{9.214}$$

Thus, the denoised signal is obtained by convolving, or locally averaging, the noisy signal against the convolutional kernel function $\mathbf{w}_\lambda$.

Figure 9.51 shows the result of denoising a noisy signal using (9.214) for different values of λ. We can see that larger values of λ lead to higher degrees of noise removal and smoothing. We also plot the convolutional kernel function $\mathbf{w}_\lambda$ in Figure 9.52(a), noting that when λ is small, it is concentrated near the origin, which means the denoising is implemented by averaging over small localized neighborhoods. These spread out as λ increases, so the denoising

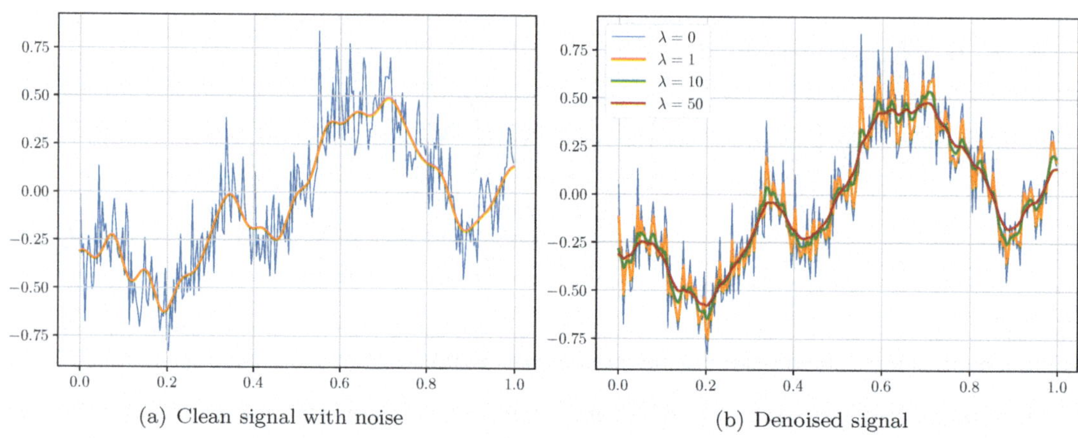

(a) Clean signal with noise

(b) Denoised signal

Figure 9.51: Example of signal denoising with the graph Laplacian.

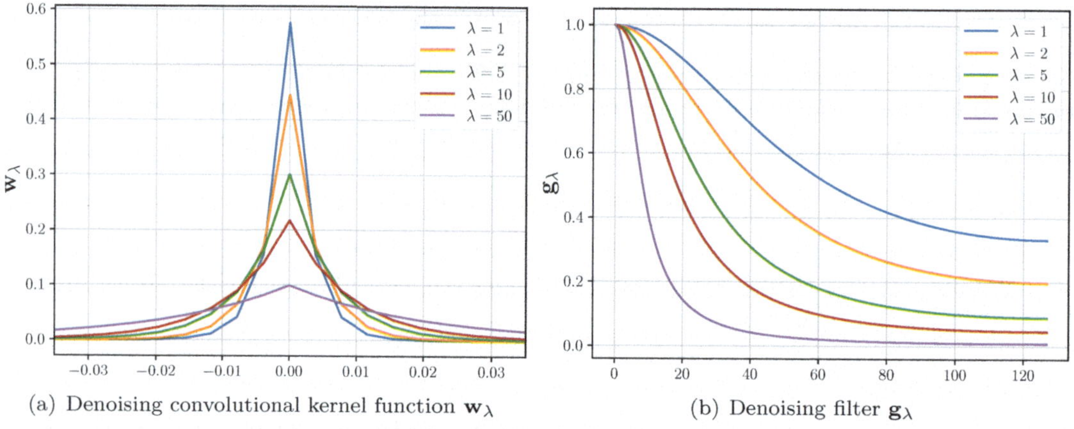

(a) Denoising convolutional kernel function $\mathbf{w}_\lambda$

(b) Denoising filter $\mathbf{g}_\lambda$

Figure 9.52: Depiction of denoising convolutional kernel functions and filters.

effectively averages over progressively larger neighborhoods, resulting in a higher degree of smoothing.

We can give a frequency domain interpretation of the denoising by using Theorem 9.86 to obtain

$$F_m^\dagger \mathbf{u} = \mathbf{g}_\lambda \circ F_m^\dagger \mathbf{y}, \qquad \text{where} \qquad \mathbf{g}_\lambda = \sqrt{m}\, F_m^\dagger \mathbf{w}_\lambda. \tag{9.215}$$

Thus, denoising simply modifies the Fourier modes by entrywise multiplication by $\mathbf{g}_\lambda$, which is the DFT of the convolutional kernel function $\mathbf{w}_\lambda$. In fact, in practice, we compute the denoised signal $\mathbf{u}$ with the fast Fourier transform, discussed below in Section 9.10.6, by applying the IDFT matrix F_m to both sides of (9.215):

$$\mathbf{u} = F_m(\mathbf{g}_\lambda \circ F_m^\dagger \mathbf{y}). \tag{9.216}$$

Using the fast Fourier transform, the denoised signal $\mathbf{u}$ can be computed in $\mathrm{O}(m \log m)$ operations, compared to the direct convolution formula (9.214) that can take, in the worst case scenario, $\mathrm{O}(m^2)$ operations.

It turns out we can also easily compute $\mathbf{g}_\lambda$. We simply take the dot product with $\mathbf{z}_k$ on both sides of (9.214) to obtain

$$\mathbf{z}_k \cdot \mathbf{e}_0 = \mathbf{z}_k \cdot (I + \lambda L)\mathbf{w}_\lambda = (I + \lambda L)\mathbf{z}_k \cdot \mathbf{w}_\lambda = (1 + \lambda \mu_k)\mathbf{z}_k \cdot \mathbf{w}_\lambda,$$

where we used symmetry of L and its eigenvector equation (9.177), where the eigenvalue μ_k is given by (9.145). Solving for $\mathbf{z}_k \cdot \mathbf{w}_\lambda$ and using that $\mathbf{z}_k \cdot \mathbf{e}_0 = 1/\sqrt{m}$ produces

$$(\mathbf{g}_\lambda)_k = \sqrt{m}\,(F_m^\dagger \mathbf{w}_\lambda)_k = \sqrt{m}\,\mathbf{z}_k \cdot \mathbf{w}_\lambda = \frac{1}{1 + \lambda \mu_k} = \frac{1}{1 + 2\lambda - 2\lambda \cos(2\pi k/m)}.$$

We show a plot of $\mathbf{g}_\lambda$ in Figure 9.52(b), which shows that the denoising essentially attenuates high frequencies, while allowing low frequencies to pass through unchanged. Increasing λ shifts the cutoff between low and high frequencies. The DFT $\mathbf{g}_\lambda$ of the convolutional kernel function is often called a *filter* in signal processing, since it acts to filter out certain frequencies.

Graph Convolutions

Finally, let us indicate how the preceding constructions can be extended to more general graphs. This will be of use in our applications to graph convolutional neural networks in Section 10.4. As in the preceding circular graph case, operations that commute with the graph Laplacian play a special role in spectral graph theory, and provide the generalization of the DFT and convolution.

Let $\mathcal{G}$ be a symmetric connected graph with weight matrix W, diagonal degree matrix D, and graph Laplacian $L = D - W$. Let $Q = (q_1 \ \cdots \ q_m)$ be the $m \times m$ orthogonal matrix whose columns are the eigenvector basis of the graph Laplacian. (Here we will concentrate on the real eigenvectors, although in certain cases their complex versions may facilitate computations.) The following definition provides a natural generalization of the DFT to an arbitrary graph.

Definition 9.87. The *graph Fourier transform* of a vector $\mathbf{x} \in \mathbb{R}^m$ is the vector $Q^T \mathbf{x}$.

Thus, the graph Fourier transform is simply the change of coordinates from $\mathbf{x}$ into the graph Laplacian eigenvector basis $\mathbf{q}_1, \ldots, \mathbf{q}_m$. We can also define a graph Fourier transform using the eigenvectors for the random walk graph Laplacian L_{rw} or the symmetric normalized graph Laplacian L_{sym}; see (9.59), (9.65). In the case of the circular graph they are all the same, but in general there are differences.

Given $\mathbf{x} \in \mathbb{R}^m$ and an $m \times m$ matrix A, it turns out that the graph Fourier transform of $A\mathbf{x}$ has a special form when A commutes with the graph Laplacian.

Lemma 9.88. *Suppose that the graph Laplacian L has distinct eigenvalues. Assume $A \in \mathcal{M}_{m \times m}$ commutes with L, so that $LA = AL$. Let $\mathbf{g} = (\gamma_1, \gamma_2, \ldots, \gamma_m)^T \in \mathbb{R}^m$ be the vector containing the eigenvalues γ_j of A. Then $\gamma_j = \mathbf{q}_j^T A \mathbf{q}_j$, and*

$$A\mathbf{x} = Q\big(\mathbf{g} \circ (Q^T\mathbf{x})\big) \qquad \text{for all} \qquad \mathbf{x} \in \mathbb{R}^m. \tag{9.217}$$

Proof. As a consequence of Theorem 5.19, A is complete and has the same eigenvectors as L, so that $A\mathbf{q}_j = \gamma_j \mathbf{q}_j$ for $j = 1, \ldots, m$; the formula for γ_j follows from their orthonormality. This implies that we can diagonalize $A = Q\Gamma Q^T$ using the same eigenvectors, where $\Gamma = \operatorname{diag}\mathbf{g}$ is its diagonal eigenvalue matrix. We conclude that $A\mathbf{x} = Q\Gamma Q^T\mathbf{x} = Q(\mathbf{g} \circ (Q^T\mathbf{x}))$. $\blacksquare$

Lemma 9.88 says that $A\mathbf{x}$ can be expressed as an entrywise product in the graph Fourier basis of a vector $\mathbf{g}$ containing the eigenvalues of A with the graph Fourier transform of $\mathbf{x}$. This occurs because L and A commute, so they share the same eigenvectors and can be *simultaneously* diagonalized by Q. The expression (9.217) in Lemma 9.88 is the generalization of the convolution formula (9.212) for the DFT in the case of the circular graph.[32] Thus, it is natural to view matrices A that commute with the graph Laplacian L as a generalization of the notion of convolution from the circular graph to general weighted graphs. We will explore this further in Section 10.4 in the context of graph convolutional neural networks.

9.10.6   The Fast Fourier Transform

If we were to compute the DFT or IDFT as written in (9.184) and (9.185), multiplying an $m \times m$ complex matrix by a vector requires $O(m^2)$ operations. This is too slow for most important applications, such as to audio or image processing, where the number of a samples or pixels m can be quite large. It turns out that many of the $O(m^2)$ operations are redundant, and, if the computation is done intelligently, it can be reduced to $O(m \log m)$ operations. The resulting algorithm is called the fast Fourier transform (FFT), whose computational efficiency is perhaps the dominant factor in the widespread applicability of Fourier methods in problems like signal processing, image and audio compression, solving partial differential equations, and many others. The FFT, credited to the American mathematicians James Cooley and John Tukey [46] (although intimations can be found in earlier papers, including an unpublished 1805 work by Gauss on the orbits of asteroids that even predates Fourier!), is widely viewed as one of the top 10 algorithms of 20th Century [44, 61, 195].

Throughout this section, we will assume, for simplicity, that $m = 2^p$ for some positive integer p (see Exercise 10.13 for the FFT for m that are not powers of 2). Since we treat all vectors as periodic, any $\mathbf{z} \in \mathbb{C}^{m/2}$ can also be treated as a vector $\mathbf{z} \in \mathbb{C}^m$ by periodic extension. Thus, with a slight abuse of notation, we will add and subtract vectors the number of whose entries differ by a multiple of two. It will be computationally simpler to work with the unnormalized version of the DFT, i.e., we use the matrix G_m in (9.184) and avoid the extra operation of dividing by $\sqrt{m}$ during the course of the computation, saving it for the final step, if necessary. Moreover, if $m = 2^p$ is a power of 2, then dividing by $\sqrt{m} = 2^{p/2}$ is almost trivial on an electronic computer, being just a bit shift in fixed point, or a change of exponent in floating point. It is important to note that $G_1 = F_1 = 1$.

The main idea behind the FFT is to recursively break up the computation into smaller half-size problems. This is done by subsampling the vector $\mathbf{z} \in \mathbb{C}^m$ into its even and odd parts, and applying the DFT or IDFT to these two components separately.

Definition 9.89. Given a vector $\mathbf{z} \in \mathbb{C}^{2n}$, we define its even and odd parts as the vectors $\mathbf{z}_e \in \mathbb{C}^n$ and $\mathbf{z}_o \in \mathbb{C}^n$ with respective entries

$$z_{e,k} = z_{2k} \qquad \text{and} \qquad z_{o,k} = z_{2k+1}, \qquad \text{for} \qquad k = 0, \dots, n-1.$$

The key technical observation is stated as follows.

---

[32] To make the formulas look more similar, we can write $\mathbf{g} = Q^T \mathbf{y}$, or we can also compare to (9.216).

Lemma 9.90. *Let* $m = 2n$. *Then, for each* $\mathbf{z} \in \mathbb{C}^m = \mathbb{C}^{2n}$,

$$G_m \mathbf{z} = G_{2n} \mathbf{z} = G_n \mathbf{z}_e + \mathbf{w} \circ (G_n \mathbf{z}_o),$$
$$G_m^\dagger \mathbf{z} = G_{2n}^\dagger \mathbf{z} = G_n^\dagger \mathbf{z}_e + \overline{\mathbf{w}} \circ (G_n^\dagger \mathbf{z}_o),$$

(9.218)

where the complex conjugate vectors $\mathbf{w}, \overline{\mathbf{w}} \in \mathbb{C}^n$ *have entries*

$$w_k = \zeta_m^k = e^{2k\pi i/m} = e^{k\pi i/n},$$
$$\overline{w}_k = \zeta_m^{-k} = e^{-2k\pi i/m} = e^{-k\pi i/n}, \qquad k = 0, \dots, n-1.$$

(9.219)

Proof. We prove the first formula, leaving the second as Exercise 10.10. Observe that if $k = 2j$ is even, then

$$\zeta_m^k = \zeta_m^{2j} = e^{4j\pi i/m} = e^{2j\pi i/n} = \zeta_n^j.$$

On the other hand, if $k = 2j + 1$ is odd, then

$$\zeta_m^k = \zeta_m^{(2j+1)} = e^{2(2j+1)\pi i/m} = e^{2j\pi i/n} e^{\pi i/n} = e^{\pi i/n} \zeta_n^j.$$

Thus, splitting the sum defining the ℓ-th component of the unnormalized IDFT of $\mathbf{z}$ into its even and odd parts yields

$$(G_m \mathbf{z})_\ell = \sum_{k=0}^{m-1} z_k \zeta_m^{k\ell} = \sum_{j=0}^{n-1} z_{2j} \zeta_m^{2j\ell} + \sum_{j=0}^{n-1} z_{2j+1} \zeta_m^{(2j+1)\ell}$$

$$= \sum_{j=0}^{n-1} z_{2j} \zeta_n^{j\ell} + e^{\ell\pi i/n} \sum_{j=0}^{n-1} z_{2j+1} \zeta_n^{j\ell} = (G_n \mathbf{z}_e)_\ell + w_\ell (G_n \mathbf{z}_o)_\ell. \qquad \blacksquare$$

At a high level, the FFT works by using Lemma 9.90 to recursively split the problem up into smaller subproblems. Each split reduces the number of entries in the signals by half, and after $\log_2 m$ such splits, the problem is reduced to computing the DFT of a signal with a single entry, which requires no computation since $G_1 = 1$. Each time we recombine the results from subproblems using (9.218) we incur a cost of $O(m)$ operations, yielding an overall complexity of $O(m \log_2 m)$. Notice how the assumption that $m = 2^p$ is used to cleanly subdivide the problem all the way down to $m = 1$ length signals.

The description above is just a summary of the main ideas. Below, we show Python code for the FFT algorithm using a recursive implementation. The algorithm checks if the input signal has length $m = 1$, and if so, it returns f, since $G_1^\dagger$ is the identity (as noted above). If $m \geq 2$ then the algorithm splits the signal into its even and odd parts, takes their DFTs recursively, and then recombines the resulting DFTs using (9.218). We also point out that steps with `hstack` are extending $G_n^\dagger \mathbf{z}_e$ and $G_n^\dagger \mathbf{z}_o$ from $\mathbb{C}^n$ to $\mathbb{C}^m = \mathbb{C}^{2n}$ by periodicity, and that in the last line `1j` $= \sqrt{-1}$ in Python.

We now carefully analyze the computational complexity of the FFT. In order to do so, it is important to consider how one chooses to count operations. Here, we will count an operation of addition, subtraction, multiplication or division[33] of *real* numbers as a single operation.

$$(a + ib) + (c + id) = (a + b) + i(c + d)$$

---

[33] In some treatments, multiplication or division are treated differently than addition and subtraction, since the latter tend to be performed faster on a typical computer chip.

Algorithm The Fast Fourier Transform (FFT) in Python

```python
import numpy as np

def fft(z):
    m = len(z)
    k = np.arange(m)
    if m == 1:
        return z
    else:
        Gze = fft(z[::2])
        Gzo = fft(z[1::2])
        Gze = np.hstack((Gze,Gze))
        Gzo = np.hstack((Gzo,Gzo))
        return Gze + np.exp(-2*np.pi*1j*k/m)*Gzo
```

---

requires two real additions — adding $a + b$ and $c + d$). The complex number $a + \mathrm{i}\,b$ is stored
in memory as a pair (a, b), with the understanding that a is the real part and b the imaginary
part of the complex number. So complex addition is equivalent to real vector addition, and is
simply stored as $(a + b, c + d)$. There is no need to treat the complex unit i: multiplication of
i by $c + d$ is not done by the computer (how could it be?), and so this does not count as an
operation. The result of this discussion is that addition or subtraction of complex numbers
takes two real operations.

Multiplication of complex numbers takes more than two real operations, since

$$(a + \mathrm{i}\,b) + (c + \mathrm{i}\,d) = (ac - bd) + \mathrm{i}\,(ad + bc).$$

So multiplying two complex numbers requires computing $ac - bd$ and $ad + bc$, which takes 6 real
operations — 4 multiplications and 2 additions. (See Exercise 10.2 for an alternative method
that requires 3 multiplications and 5 additions, which is used if the former is significantly
slower than the latter.) Note it is also possible to multiply numbers expressed in polar
coordinates (9.163) with fewer operations, since

$$r_1 e^{\mathrm{i}\theta_1} r_2 e^{\mathrm{i}\theta_2} = r_1 r_2 e^{\mathrm{i}(\theta_1 + \theta_2)}$$

requires only 2 real operations — one multiplication and one addition. However, adding or
subtracting in polar coordinates is more expensive, since one must convert back to Cartesian
coordinates $a + \mathrm{i}\,b$ first via Euler's formula. Thus, we will conduct our analysis assuming
the complex numbers are stored in Cartesian format, requiring 2 operations per addition or
subtraction and 6 operations for a multiplication.

We define

$$a_m = \text{Number of real operations used to compute the FFT on } \mathbb{C}^m,$$

noting that this count will be the same for both the DFT and the IDFT. In particular, $a_1 = 0$
since $G_1 = 1$ is the identity. We can use Lemma 9.90 to obtain a recursive formula for a_m.
Indeed, using (9.218) to compute $G_{2n}\mathbf{z}$ requires computing $G_n \mathbf{z}_e$ and $G_n \mathbf{z}_o$, both of which
take a_n operations. Then we need to perform the complex addition and multiplication in
(9.218), which takes $2 + 6 = 8$ operations. This has to be done $m = 2n$ times, yielding
$8m = 16n$ real operations, and hence

$$a_{2n} = 2a_n + 16n. \tag{9.220}$$

Notice we did not allocate any computation time to compute the root of unity $e^{k\pi\,\mathrm{i}/n}$. This quantity is independent of the vector $\mathbf{z}$ and can be computed once, offline, and the values stored in memory. A straightforward induction — see Exercise 10.11 — shows that when m is a power of 2, then the solution to the recursive formula with $a_1 = 0$ is

$$a_m = 8\,m\,p = 8\,m\log_2 m, \qquad \text{when} \qquad m = 2^p. \tag{9.221}$$

We conclude that the FFT takes at most $\mathrm{O}(m\log_2 m)$ operations to execute, which is an order of magnitude improvement over the naïve $\mathrm{O}(m^2)$ implementation of DFT and IDFT based on (9.181) and (9.182).

Nevertheless, it turns out this complexity can be further improved by removing some redundant computations. Referring back to (9.219), since

$$w_{\ell+n} = e^{(\ell+n)\,\pi\,\mathrm{i}/n} = e^{\ell\pi\,\mathrm{i}/n}\,e^{\pi\,\mathrm{i}} = -\,e^{\ell\pi\,\mathrm{i}/n} = -\,w_\ell,$$

and both $G_n\mathbf{z}_e$ and $G_n\mathbf{z}_o$ are n periodic, we have

$$
\begin{aligned}
(G_m\mathbf{z})_\ell &= (G_n\mathbf{z}_e)_\ell + w_\ell\,(G_n\mathbf{z}_o)_\ell, \\
(G_m\mathbf{z})_{\ell+n} &= (G_n\mathbf{z}_e)_\ell - w_\ell\,(G_n\mathbf{z}_o)_\ell,
\end{aligned}
\qquad \ell = 0,\ldots,n-1.
$$

Thus there are now only n complex multiplications to perform, since they are common between the two equations above. We still need $m = 2n$ complex additions, so the number of real operations is reduced from $16n$, as above, to $6n + 4n = 10n$ real operations, and the recursion (9.220) becomes $a_{2n} = 2\,a_n + 10n$. Thus, by Exercise 10.11, the total number of operations is reduced to $a_m = 5\,m\log_2 m$ when $m = 2^p$.

The computational complexity can be further improved to $4\,m\log_2 m$ by considering a 3-way split, where the odd terms are further split in half before the recursion. Exercise 10.12 explores this algorithm, called the split-radix FFT [257], which enjoyed its role as the fastest FFT for power-of-two n for quite some time, until the relatively recent work [120] made some modifications to the algorithm and improved the complexity to $\frac{34}{9}\,m\log_2 m$.

It is also important to note that the power-of-two assumption is usually not restrictive in practice, since one normally has a choice over the length of the signal; in signal or image processing the DFT is applied to signal or image blocks of a user specified size, which can be chosen as a power of 2. There are extensions of the FFT described in this section to m that are not a power of two; we defer this to Exercise 10.13.

Exercises

10.1. ♡ Find a formula for the the reciprocal $1/z$ of a nonzero complex number $z \neq 0$ in terms of its real and imaginary parts. Then write out the formula for complex division z/w. *Hint:* Use the formula $z\,\overline{z} - |z|^2$.

10.2. ◇ Show how to use the algebraic identity $ad + bc = (a+b)(c+d) - ac - bd$ for $a,b,c,d \in \mathbb{R}$ to perform complex multiplication using 3 real multiplications and 5 real additions/subtractions.

10.3. ♡ Use the complex orthogonality formulas (9.179) to prove that the real vectors $\mathbf{x}_k, \mathbf{y}_k$ in (9.141) are mutually orthogonal, and, moreover, the Euclidean norm formulas (9.144) hold. *Remark:* A purely real proof, based on trigonometric identities, is doable, but more challenging.

10.4. Use the symmetries $\mathbf{x}_{n-k} = \mathbf{x}_k$ and $\mathbf{y}_{n-k} = -\mathbf{y}_k$ to simplify (9.146), (9.149), or, equivalently, (9.153).

10.5. Let $\mathbf{x}, \mathbf{y}, \mathbf{z} \in \mathbb{R}^m$. Prove the convolution identities (9.206), (9.207), and (9.208).

10.6. Prove that all $m \times m$ circulant matrices have the *same* eigenvectors, namely $\mathbf{z}_0, \ldots, \mathbf{z}_{m-1}$. What are the eigenvalues?

10.7. ♡ Prove the unnormalized DFT convolution formulas in (9.211).

10.8. Use the fast Fourier transform to find the discrete Fourier coefficients for the following functions using the indicated number of sample points. Carefully indicate each step in your analysis.

(a)♡ $\dfrac{x}{\pi}$, $n = 4$; (b)◇ $\sin x$, $n = 8$; (c)♡ $|x - \pi|$, $n = 8$; (d) $\operatorname{sign}(x - \pi)$, $n = 16$.

10.9. Use the inverse fast Fourier transform to reassemble the sampled function data corresponding to the following discrete Fourier coefficients. Carefully indicate each step.

(a)♡ $c_0 = c_2 = 1$, $c_1 = c_3 = -1$, (b)◇ $c_0 = c_1 = c_4 = 2$, $c_2 = c_6 = 0$, $c_3 = c_5 = c_7 = -1$.

10.10. ◇ Prove the second FFT formula in (9.218).

10.11. ◇ Let $b \in \mathbb{R}$ be fixed. Use induction to show that when $m = 2^p$, the solution to the recursive equation $a_{2n} = 2a_n + 2bn$ with $a_1 = 0$ is $a_m = bmp = bm \log_2 m$.

10.12. (Split-radix FFT) Assume $m \geq 4$ is a power of 2. For $\mathbf{z} \in \mathbb{C}^m$ define $\mathbf{z}_e \in \mathbb{C}^{m/2}$ and $\mathbf{z}_p, \mathbf{z}_q \in \mathbb{C}^{m/4}$ by $z_{e,k} = z_{2k}$, $z_{p,k} = z_{4k+1}$, $z_{q,k} = z_{4k+3}$.
 (a) Show that

$$G_m^\dagger \mathbf{z} = G_{m/2}^\dagger \mathbf{z}_e + \mathbf{w} \circ (G_{m/4}^\dagger \mathbf{z}_p) + \mathbf{u} \circ (G_{m/4}^\dagger \mathbf{z}_q), \tag{9.222}$$

where $\mathbf{u}, \mathbf{w} \in \mathbb{C}^{m/4}$ are given by $u_k = e^{-6k\pi i/m}$ and $w_k = e^{-2k\pi i/m}$.

 (b) The FFT algorithm based on the 3-way split in (9.222) is called the split-radix FFT. Show that $G_m^\dagger \mathbf{z}$ can be computed via (9.222) in $6m$ real operations. *Hint:* You'll need to break up $\{0, \ldots, m-1\}$ into 4 equal size intervals, instead of 2 as was done above. Note also that multiplications with ± 1 or $\pm i$ do not count, since they amount to negation of real and/or imaginary parts, which can be absorbed into the next operation by changing it from addition to subtraction or vice versa.

 (c) Show that the number of real operations taken by the split-radix FFT, denoted again as a_m, satisfies the recursion $a_m = a_{m/2} + 2a_{m/4} + 6m$. Explain why $a_1 = 0$ and $a_2 = 4$. Use this to show that[34] $a_m \leq 4m \log_2 m$. *Hint:* Define $b_m = a_m - 4m \log_2 m$ and show that b_m satisfies $b_m = b_{m/2} + 2b_{m/4}$, with $b_1 = 0$ and $b_2 = -4$. Use this to argue that $b_m \leq 0$ for all m that are powers of 2.

10.13. In this exercise you'll generalize Lemma 9.90 to some values of m that are not powers of two. Assume that $m = pq$ is composite, with factors $p, q \geq 2$. For $k = 0, \ldots, q-1$ define $\mathbf{z}^{(k)}, \mathbf{w}^{(k)} \in \mathbb{C}^p$ by $z_j^{(k)} = z_{jq+k}$ and $w_j^{(k)} = e^{-2kj\pi i/m}$ for $j = 0, \ldots, p-1$. Show that

$$G_m^\dagger \mathbf{z} = \sum_{k=0}^{q-1} \mathbf{w}^{(k)} \circ (G_p^\dagger \mathbf{z}^{(k)}).$$

Explain how to use this identity to formulate an FFT algorithm for some values of m that are not powers of two.

---

[34] If one is more careful about redundant computations — there are additional multiplications with ± 1 or $\pm i$ that can be skipped — then the complexity of the split-radix FFT algorithm is actually $4m \log_2 m - 6m + 8$ real operations [257].

10.14. ♡ Let $\mathcal{G}$ and $\widetilde{\mathcal{G}}$ be two weighted graphs with m and $\widetilde{m}$ nodes, and with weight matrices W and $\widetilde{W}$, respectively. The *Cartesian product graph* $\mathcal{G} \times \widetilde{\mathcal{G}}$, whose nodes are indexed by pairs (i, j) of nodes in each graph, so $1 \le i \le m$ and $1 \le j \le \widetilde{m}$, and whose weight matrix is denoted by $W \times \widetilde{W}$, with edge weights

$$(W \times \widetilde{W})_{(i,j),(k,\ell)} = \begin{cases} w_{ik}, & \text{if } j = \ell \\ \widetilde{w}_{j\ell}, & \text{if } i = k \\ 0, & \text{otherwise.} \end{cases}$$

Note that $W \times \widetilde{W}$ is an $m\widetilde{m} \times m\widetilde{m}$ matrix, and the nodal set for $\mathcal{G} \times \widetilde{\mathcal{G}}$ is the Cartesian product of the nodal sets of each graph. Let L and $\widetilde{L}$ denote the graph Laplacian matrices for $\mathcal{G}$ and $\widetilde{\mathcal{G}}$, respectively. For $\mathbf{u} \in \mathbb{R}^m$ and $\widetilde{\mathbf{u}} \in \mathbb{R}^{\widetilde{m}}$, let $\mathbf{u} \times \widetilde{\mathbf{u}} \in \mathcal{M}_{m \times \widetilde{m}}$ be the matrix with (i, j) entry $u_i \widetilde{u}_j$ which we identify in the usual manner with a vector in $\mathbb{R}^{m\widetilde{m}}$. Let $L \times \widetilde{L}$ denote the $m\widetilde{m} \times m\widetilde{m}$ graph Laplacian matrix on $\mathcal{G} \times \widetilde{\mathcal{G}}$.

(a) Show that if $\mathbf{u}$ is an eigenvector of L with eigenvalue λ, and $\widetilde{\mathbf{u}}$ is an eigenvector of $\widetilde{L}$ with eigenvalue $\widetilde{\lambda}$, then $\mathbf{u} \times \widetilde{\mathbf{u}}$ is an eigenvector of $L \times \widetilde{L}$ with eigenvalue $\lambda + \widetilde{\lambda}$.

(b) Do the eigenvectors obtained this way form a basis for $\mathbb{R}^{m\widetilde{m}}$?

10.15. Let $\mathcal{G}^d$ be the d-fold Cartesian product, defined in Exercise 10.14, of the cyclic graph on m nodes. The nodes of $\mathcal{G}^d$ can be identified with integer vectors $k = (k_1, \dots, k_d) \in \mathbb{Z}^d$ modulo m in each entry. Use the complex version of Exercise 10.14 to show that the eigenvectors of the graph Laplacian on $\mathcal{G}^d$ are given by the vectors $\mathbf{z}_k$, $k \in \mathbb{Z}^d$, whose j-th entry for $j = (j_1, \dots, j_d) \in \mathbb{Z}^d$ is given by

$$\mathbf{z}_{k,j} = \frac{\zeta_m^{k \cdot j}}{m^{d/2}} = \frac{e^{2\pi i k \cdot j / m}}{m^{d/2}},$$

where $k \cdot j = k_1 j_1 + \cdots + k_d j_d$. This is the d-dimensional discrete Fourier transform.

10.16. ♡ Write a Python notebook to extend the signal denoising example with the DFT to two dimensional signals (i.e., images) and use it to denoise natural images corrupted by noise. The Python notebook from this section will be helpful to start out.

Chapter 10

Neural Networks and Deep Learning

The past few years have witnessed an exponential growth in the power of artificial intelligence over a broad range of tasks, from computer vision to natural language processing. This wave is being propelled by *deep learning*, which refers to machine learning algorithms that use artificial multilayer neural networks, which were very loosely inspired by biological neural networks.

While deep learning has reached impressive heights relatively recently, the concept of an artificial neural network, and much of the mathematical and computer science techniques required to train one, date back to the mid 20th century. The *artificial neuron* was invented by Warren McCulloch and Walter Pitts in 1943 [159]. In 1957 Frank Rosenblatt invented the *perceptron*, which used artificial neurons to mimic the structure of the brain and showed the ability to learn [197]. The ideas behind *backpropagation*, also called *reverse mode automatic differentiation*, for training neural networks first appeared in 1970 [150], while the version that is standard today was developed by Geoffrey Hinton in 1985 [203]. Some of the main ideas behind convolutional neural networks, which are used for computer vision tasks, were initially developed by Kunihiko Fukushima in 1980, while the modern version of the convolutional neural network is due to Yann LeCun, who pioneered them for handwritten digit recognition starting in 1989 [137, 138]. Yoshua Bengio and LeCun expanded the applications of convolutional neural networks shortly thereafter [136]. Recurrent neural networks for time-series analysis also have a long history dating back to the network introduced by John Hopfield in 1982 [113] and long short term memory (LSTM) recurrent neural networks in 1997 [107].

Throughout much of their long history, artificial neural networks were not widely used in machine learning and did not yield the impressive results that we are witnessing today, which was in large part due to a lack of sufficient computational resources and data. The situation dramatically changed in 2012 when a deep convolutional neural network named *AlexNet*, designed by Alex Krizhevsky in collaboration with Ilya Sutskever and Geoffrey Hinton, won the ImageNet competition [132]. ImageNet is a highly challenging image classification data set with around 14 natural million images from around 20,000 different classes.[1] AlexNet achieved a top-5 accuracy[2] of 84.7%, which was 10.8% better than the second place entry,

---

[1] To give an idea of the difficulty of properly classifying ImageNet images, even for humans, the ImageNet data set has 120 classes that simply correspond to different breeds of dogs!

[2] The ImageNet problem was so challenging that accuracy was initially reported as top-k accuracy, in which

and a top-1 accuracy of 63.3%. Since then there have been dramatic improvements, and, as of 2025, the best top-1 accuracy[3] is 91.0% [260], which is widely regarded as superhuman performance [204].

The ImageNet breakthrough opened the floodgates to a rapidly growing list of applications of deep learning, which we only briefly overview here. Deep reinforcement learning has mastered the game of Go and handily defeated the world's best professional Go players, all without learning from even a single game of Go between human players [215]. Generative deep learning [90, 218] is now able to produce nearly any image requested by the user. Large language models, like ChatGPT, which is powered by the new transformer architecture that was developed in 2017 [237], can recall information, have conversations, and edit books and computer code more quickly and efficiently than most humans. Deep learning is also revolutionizing the fields of computational science and engineering; see, e.g., [191]. This recent explosion in deep learning was driven by the confluent arrival of massive data and massive computational resources, and is sustained by new algorithmic breakthroughs. For their pioneering work in developing the field of deep learning, Yoshua Bengio, Yann LeCun, and Geoffrey Hinton were awarded the 2018 Turing Award, which is the equivalent of the Nobel prize for computing. Moreover, in 2024, Hinton and Hopfield were jointly awarded the Nobel Prize in Physics "for foundational discoveries and inventions that enable machine learning with artificial neural networks".

This chapter contains a introduction to artificial neural networks and deep learning. We cover fully connected networks, including results on backpropagation, automatic differentiation, and universal approximation theory, and include brief overview of convolutional neural networks, recurrent neural networks, graph neural networks and transformers. For a more in depth view of deep learning we refer the reader to [89].

Throughout this chapter, we will make use of the Python package `pytorch` for deep learning computations. The notebook below gives an introduction to `pytorch`.

Python Notebook: Intro to PyTorch (.ipynb)

10.1   Fully Connected Networks

Python Notebook: Fully Connected Neural Networks (.ipynb)

Artificial neural networks, which we will call *neural networks* from now on, are parameterized functions that are constructed using two types of basic building blocks: affine functions and simple nonlinearities. These are combined to form an *artificial neuron*, which is defined as a function $F: \mathbb{R}^n \to \mathbb{R}$ of the particular form

$$F(\mathbf{x}) = \sigma(\mathbf{w} \cdot \mathbf{x} + b), \tag{10.1}$$

---

an algorithm predicts k possible classes for an image and is counted as correct if any of the k predictions are correct. Top-1 accuracy is the usual notion of accuracy.

[3]Top-5 accuracy is often no longer reported.

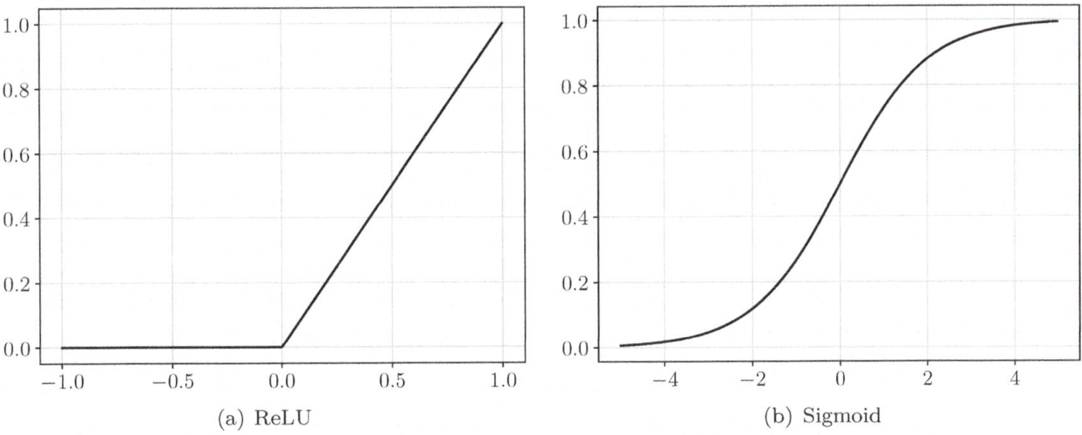

(a) ReLU   (b) Sigmoid

Figure 10.1: Plots of the ReLU and Sigmoid activation functions. Both activation functions have the behavior that they give zero, or close to zero, responses when the input is below a certain threshold, and give positive responses above.

obtained by composing a real-valued affine function $\mathbf{x} \mapsto \mathbf{w} \cdot \mathbf{x} + b$, with $\mathbf{w} \in \mathbb{R}^n$ called the *weight* and $b \in \mathbb{R}$ the *bias*, with a nonlinear scalar activation function $\sigma \colon \mathbb{R} \to \mathbb{R}$. The key property of an activation function is that the response is zero, or close to zero, when the input lies below a certain threshold, but is positive — i.e., *activated* — when above the threshold. The preceding model for a (mathematical) neuron is very loosely based on the behavior of neurons in human and animal brains, which are activated only when their input rises to a certain threshold. However, the analogy should not be taken any further than this.

Two common choices for the activation function are the *rectified linear unit* (ReLU)

$$\sigma(t) = \max\{t, 0\} = t_+, \tag{10.2}$$

and the *sigmoid* activation function

$$\sigma(t) = \frac{1}{1 + e^{-t}}, \tag{10.3}$$

though many others have been used in practice. Figure 10.1 plots both activation functions. One advantage of the sigmoid activation is that it is continuously differentiable — in fact, analytic. On the other hand, the ReLU activation is only Lipschitz continuous — see Definition 6.55 — and not differentiable at $t = 0$, while it has the advantage of being 1-*homogeneous*, meaning $\sigma(at) = a\,\sigma(t)$ for $a > 0$, and also (non-strictly) convex. In applications where a continuously differentiable activation function is necessary, we can also use the soft-plus function defined in (7.66), which is a convex and continuously differentiable approximation of the ReLU activation.

The single artificial neuron (10.1) can serve as a linear binary classifier when σ is the sigmoid activation, provided we use class labels of 0 and 1. This is similar to how we used linear functions for classification in the context of support vector machines (SVM) in Section 7.3. If we have more than two classes, we can consider a collection of artificial neurons, which we can write concisely in matrix notation as the function $F \colon \mathbb{R}^n \to \mathbb{R}^p$ given by

$$F(\mathbf{x}) = \sigma(W\mathbf{x} + \mathbf{b}), \tag{10.4}$$

where $W \in \mathcal{M}_{n \times p}$ is called a *weight matrix*, $\mathbf{b} \in \mathbb{R}^p$ is a *vector of biases*, and the activation function σ is applied componentwise to vectors. If we set $\mathbf{z} = F(\mathbf{x})$ then $z_i = \sigma(\mathbf{w}_i \cdot \mathbf{x} + b_i)$

is simply the output of the i-th neuron, where $\mathbf{w}_i^T$ denotes the i-th row of W. Thus, (10.4) is simply a compact way of writing the outputs of p artificial neurons all acting on the same input $\mathbf{x} \in \mathbb{R}^n$. Recalling our discussion of one-vs-rest and one-vs-all for multiclass classification in Section 7.3, we could obtain a multiclass classifier by, say, taking the class prediction to be the maximal component of F.

The function F in (10.4) is called a *one layer* neural network. Such one layer networks can be useful for linear classification, but in general, the class of functions that can be approximated with a single layer is quite restrictive. To obtain a more expressive family of functions, we can compose functions of the form (10.4) to obtain a *feedforward neural network*, which is also called a *multilayer perceptron*, or a *fully connected neural network*. Mathematically, a fully connected neural network with L layers is a function $F \colon \mathbb{R}^n \to \mathbb{R}^p$ of the form

$$F = F_L \circ F_{L-1} \circ \cdots \circ F_2 \circ F_1, \tag{10.5}$$

where the k-th layer $F_k \colon \mathbb{R}^{n_{k-1}} \to \mathbb{R}^{n_k}$ has the form (10.4), that is

$$F_k(\mathbf{x}) = \sigma_k(W_k \mathbf{x} + \mathbf{b}_k), \tag{10.6}$$

where $W_k \in \mathcal{M}_{n_k \times n_{k-1}}$ and $\mathbf{b}_k \in \mathbb{R}^{n_k}$ are the parameters, while σ_k is an activation function, which can vary from layer to layer. In particular, $n_0 = n$, and $n_L = p$. We call n_k the *width*, W_k the *weight matrix*, and $\mathbf{b}_k$ the *bias vector* of the k-th layer. The weight matrices $W_1, \ldots, W_L$ and bias vectors $\mathbf{b}_1, \ldots, \mathbf{b}_L$ are the tunable parameters in the model, while the activation functions are fixed in advance. When convenient, we will write a neural network as $F = F(\mathbf{x}; \mathbf{w})$, where $\mathbf{w}$ is the vector that contains the values of all the tunable parameters in the neural network.

It will often be the case that the final activation function σ_L is taken to be the identity, $\sigma_L(t) = t$, in which case we will say that the network has $L - 1$ *hidden layers*. For example, a one hidden layer neural network has the form

$$F(\mathbf{x}) = W_2 \sigma(W_1 \mathbf{x} + \mathbf{b}_1) + \mathbf{b}_2. \tag{10.7}$$

If the output dimension is $p = 1$, we can write a scalar-valued one hidden layer neural network $F \colon \mathbb{R}^n \to \mathbb{R}$ as

$$F(\mathbf{x}) = \sum_{i=1}^{n_1} a_i \sigma(\mathbf{w}_i \cdot \mathbf{x} + b_i) + b \tag{10.8}$$

where we have relabeled the parameters W_2 and $\mathbf{b}_2$ as a_i and b, respectively. Proceeding to the next level, a two hidden layer neural network has the form

$$F(\mathbf{x}) = W_3 \sigma_2\big(W_2 \sigma_1(W_1 \mathbf{x} + \mathbf{b}_1) + \mathbf{b}_2\big) + \mathbf{b}_3, \tag{10.9}$$

and similarly for the scalar-valued case.

We can think of a fully connected neural network as a network of interconnected neurons that are organized into L layers, so that each neuron in the k-th layer takes as its inputs all the outputs of neurons in the preceding, $(k - 1)$-st layer. Thus, the directed graph representing a neural network is an acyclic digraph; see Exercise 2.9 in Chapter 9. Figure 10.2 sketches an example of a six hidden layer fully connected neural network. The blue nodes are the input, the orange nodes are the hidden layers, while the green nodes are the output. The illustrated neural network corresponds to a map $F \colon \mathbb{R}^3 \to \mathbb{R}^2$, and the hidden layer dimensions are $n_1 = 5$, $n_2 = 10$, $n_3 = 2$, $n_4 = 4$, $n_5 = 10$, and $n_6 = 7$.

Although we can allow the activation functions of the hidden layers to be different, in practice we will often take them to be the same: $\sigma_1 = \sigma_2 = \cdots = \sigma_{L-1} = \sigma$. In order to

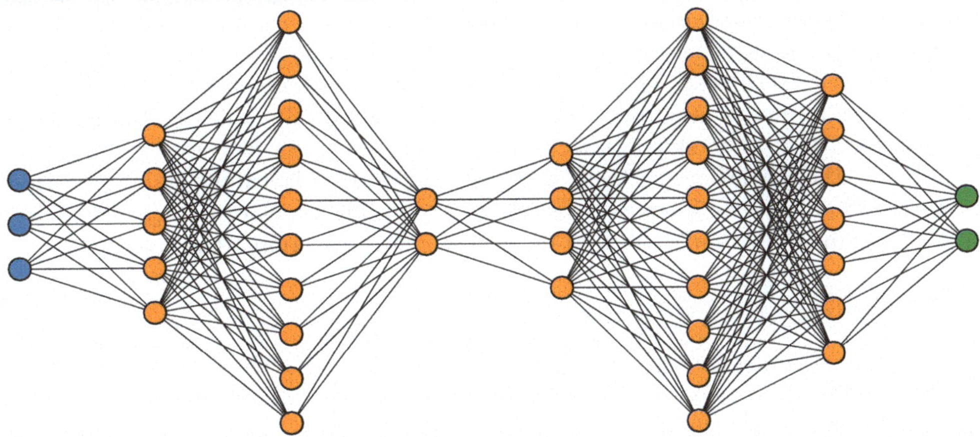

Figure 10.2: An example of a fully connected neural network with six hidden layers. The lines represent the interconnections between neurons, which go from a layer to its successor.

obtain an interesting and expressive class of functions, we must place some restrictions on the choice of activation function. For instance, if σ is the identity, then F is an affine function, being a composition of affine functions, cf. (3.64). If σ is a polynomial of degree q, then F is a polynomial of degree at most qL. As we will see later, in Section 10.6, both of these choices are not satisfactory, since they do not produce a sufficiently expressive class of functions that can approximate the general nonlinear relationships that arise in data.

Let $H_k \colon \mathbb{R}^n \to \mathbb{R}^{n_k}$ be the function that determines the output of the k-th layer of our neural network, that is,

$$H_k = F_k \circ F_{k-1} \circ \cdots \circ F_2 \circ F_1, \qquad k = 1, \ldots, L. \tag{10.10}$$

We clearly have the relationships

$$H_1 = F_1, \qquad H_k = F_k \circ H_{k-1}, \qquad F = H_L = F_L \circ \cdots \circ F_k \circ H_{k-1}, \qquad k \geq 2. \tag{10.11}$$

For a given input $\mathbf{x} \in \mathbb{R}^n$, let $\mathbf{z}_k = H_k(\mathbf{x})$ denote the output of the k-th layer. It follows that

$$\mathbf{z}_k = F_k(\mathbf{z}_{k-1}) = \sigma_k \left(W_k \mathbf{z}_{k-1} + \mathbf{b}_k \right), \qquad k = 1, \ldots, L, \tag{10.12}$$

where, in particular, $\mathbf{z}_0 = \mathbf{x}$ and $\mathbf{z}_L = F(\mathbf{x})$. We can also view this in a slightly different manner by defining the *preactivations*

$$\mathbf{p}_k = W_k \mathbf{z}_{k-1} + \mathbf{b}_k, \tag{10.13}$$

which are the values immediately preceding the application of activation function of the k-th layer. Then

$$\mathbf{z}_k = \sigma_k(\mathbf{p}_k), \qquad \text{and so} \qquad \mathbf{p}_k = W_k \sigma_{k-1}(\mathbf{p}_{k-1}) + \mathbf{b}_k. \tag{10.14}$$

Notice the similarity to affine iteration (5.71), which occurs when $\sigma_k(t) = t$ for all k.

There is a wide variety of other neural network architectures that have been proposed in the last decade, many of which can perform substantially better than the fully connected neural networks described above, especially in the context of very deep neural networks. The *residual neural network* (ResNet) architecture [100] uses the basic building blocks

$$F_k(\mathbf{x}) = \mathbf{x} + V_k \sigma(W_k \mathbf{x} + \mathbf{b}_k). \tag{10.15}$$

The idea is to learn the *residual* $F_k(\mathbf{x}) - \mathbf{x}$, hence the name. Note that ResNet adds a *skip connection* between one layer and the next — it allows the value of the input $\mathbf{x}$ to be passed directly, i.e., skipped, to the next layer. This can alleviate vanishing gradient problems — see Example 10.1 — and allows very deep networks to be more easily trained. Taking this idea further, DenseNet [115] adds skip connections between *all* layers in a feedforward fashion. There are many other variations of these architectures in deep learning. In later sections in this chapter, we'll discuss architectures designed for specific types of data, such as images (Section 10.3), graphs (Section 10.4), and natural language (Section 10.5).

10.1.1   Training and Optimization

A neural network is trained for a particular task by adjusting all of the parameters it depends on in order to achieve a desired behavior, or at least come reasonably close. In order to do this, the output $F(\mathbf{x}, \mathbf{w})$ of the neural network is fed into a loss function $\ell\colon \mathbb{R}^p \times \mathbb{R}^p \to \mathbb{R}$, and the total loss

$$\mathcal{L}(\mathbf{w}) = \frac{1}{m} \sum_{i=1}^{m} \ell(F(\mathbf{x}_i; \mathbf{w}), \mathbf{y}_i) \tag{10.16}$$

is computed based on the fully supervised training data $(\mathbf{x}_1, \mathbf{y}_1), \ldots, (\mathbf{x}_m, \mathbf{y}_m) \in \mathbb{R}^n \times \mathbb{R}^p$; see (7.24) and the ensuing discussion for details. The loss function (10.16) measures the performance of the network for the given learning task, and neural networks are trained by minimizing the total loss (10.16) over the choices of all the tunable parameters $\mathbf{w}$.

The loss is normally minimized by the application of some version of gradient[4] descent, that is

$$\mathbf{w}_{k+1} = \mathbf{w}_k - \alpha \nabla \mathcal{L}(\mathbf{w}_k), \tag{10.17}$$

where $\alpha > 0$ is the time step, also called the *learning rate*, and the gradient of the loss is given by

$$\nabla \mathcal{L}(\mathbf{w}) = \frac{1}{m} \sum_{i=1}^{m} \nabla_{\mathbf{w}} \ell(F(\mathbf{x}_i; \mathbf{w}), \mathbf{y}_i).$$

Each step of gradient descent is supposed to improve the performance of the network for the task at hand, which is interpreted as *learning*. The loss function $\mathcal{L}$ is generally *not* a convex function of the parameters $\mathbf{w}$, so there is no guarantee that gradient descent will converge to a minimizer. In general, it may be challenging, if not impossible, to find global minimizers of the loss $\mathcal{L}$, and we must often be satisfied with approximate local minimizers. We discuss this further in Section 11.7.

For modern machine learning problems with very large training sets, meaning that m is very large, it is sometimes impractical to compute the full gradient $\nabla \mathcal{L}$ since this involves *all* of the training data, which may not even fit into computer memory. *Stochastic gradient descent* (SGD) alleviates this concern by computing the gradient over a random subset of the training data of a fixed size m_b called a *mini-batch*. That is, letting $I_b \subset \{1, 2, \ldots, m\}$ denote the indices of the m_b points in the mini-batch, SGD uses the approximate gradient

$$\widetilde{\nabla} \mathcal{L}(\mathbf{w}) = \frac{1}{m_b} \sum_{i \in I_b} \nabla_{\mathbf{w}} \ell(F(\mathbf{x}_i; \mathbf{w}), \mathbf{y}_i). \tag{10.18}$$

The mini-batch changes, at random, at each iteration of SGD. In practice, this is done by splitting the data set, at random, into m/m_b mini-batches each of size m_b.[5] Then one *epoch* of

---

[4] Throughout this chapter, gradients are computed with respect to the dot product, and so ∇ is the usual Euclidean gradient (6.27).

[5] If m_b does not divide evenly into m, then one batch will have fewer than m_b points.

training refers to taking a step of gradient descent on each of the m/m_b mini-batches; that is, passing over each data point exactly one time. Training then proceeds over a certain number of epochs, or until the loss is sufficiently small. The term *full batch* gradient descent refers to ordinary gradient descent using the whole training set, so $m = m_b$. We study the theoretical properties of SGD in Section 11.5. In particular, we will see that SGD with a fixed learning rate does not converge, and we have to consider a schedule for decreasing the learning rate to obtain convergence, even for convex functions.

Various other optimization techniques, in addition to SGD, are used to improve the rate of convergence of gradient descent, and to prevent issues such as *vanishing gradients* and *overfitting*.

Example 10.1. It is entirely possible for the gradient of the loss to vanish, that is $\nabla \mathcal{L}(\mathbf{w}) = 0$ at some early point during training, far before the optimization procedure has reached a minimizer. For example, consider the one hidden layer network (10.8) with ReLU activation function $\sigma(t) = t_+$. If $\mathbf{w}_i \cdot \mathbf{x}_j + b_i < 0$ for all neurons $i = 1, \ldots, n_1$ and all training data points $j = 1, \ldots, m$, then $F(x) = b$ and $\nabla_{a_i} \mathcal{L} = \nabla_{\mathbf{w}_i} \mathcal{L} = \nabla_{b_i} \mathcal{L} = \mathbf{0}$ for all $i = 1, \ldots, n_1$. That is, all the gradients vanish and hence the weights $a_i, \mathbf{w}_i, b_i$ will not change during training. The possibility of vanishing gradients increases rapidly with the depth of the network. ▲

One way to address the issue of vanishing gradients is to use batch normalization [116], which is a modification of the neural network so that the preactivations $\mathbf{z}_k$ of any batch normalized layers are standardized (over the mini-batch) to have mean zero and unit variance. This ensures that the preactivations are centered around the *active* part of the activation function, which can help to prevent vanishing gradients and accelerate the convergence of training deep neural networks. Other ways to address vanishing gradients involve modifying the architecture of the network, such as using a ResNet architecture, as described above.

Neural networks are often employed with a very large number of parameters and have the potential to overfit the training data. As discussed in Chapter 7, we can add explicit penalties on the parameters $\mathbf{w}$ to attempt to regularize the model and prevent overfitting; however, this strategy has not been overly successful in deep learning. A common way to preclude overfitting in deep learning is to employ various types of implicit regularization during training. One example of this is dropout regularization [220], whereby during training, neurons passing through a dropout layer are dropped — their values set to zero — with some probability $0 \le p < 1$. This occurs every time an input $\mathbf{x}$ is passed through the network, with different random combinations of neurons being dropped each time. At a high level, this discourages the network from relying too much on any small collection of neurons, which may lead to overfitting.

Other optimization tools that are employed in training deep neural networks are *momentum* and *acceleration*, as well as *adaptive preconditioning*. Deep learning optimizers that include adaptive preconditioning and/or momentum include Adadelta [262], Adagrad [65], Adam [127], and RMSProp [201], among many others. We refer to [201] for an overview. Even the choice of (random) initialization of the weights and biases (i.e., the parameters) in the network has an effect on the convergence rate [135]. The brief, and by no means complete, basket of optimization tools described here are extremely important in practice, and can result in orders of magnitude faster convergence of the training process. In Chapter 11 we study basic optimization methods that include momentum and acceleration, including the heavy ball method and Nesterov acceleration, along with stochastic gradient descent, and show how they can accelerate convergence of gradient-based optimization in certain situations. We recall that the role of preconditioning was discussed in Chapter 6. An analysis of optimization methods like Adam is outside the scope of this book.

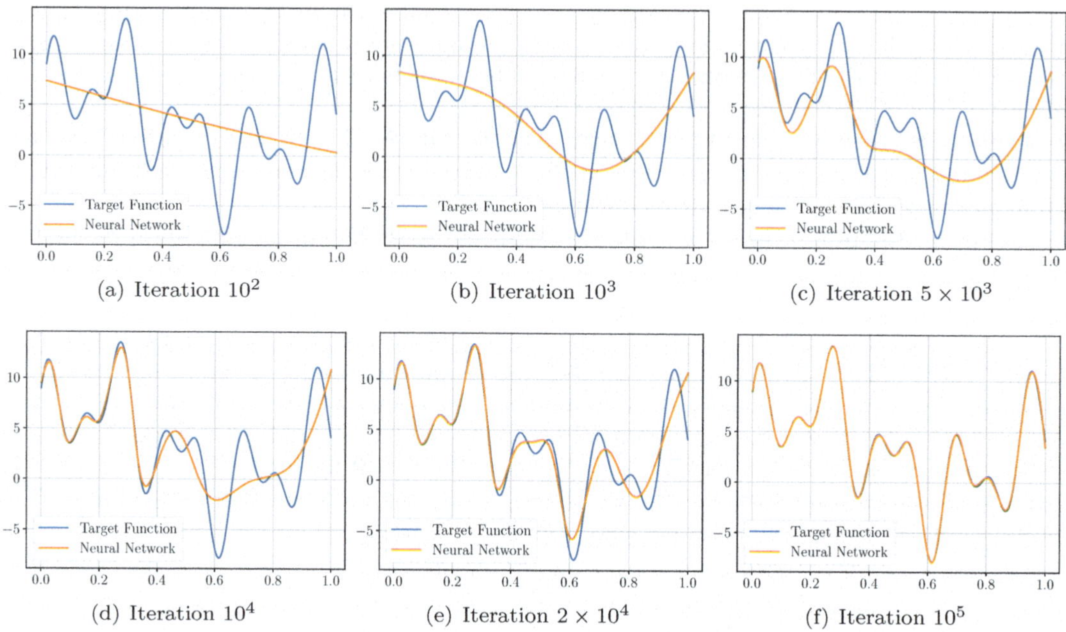

(a) Iteration 10^2 (b) Iteration 10^3 (c) Iteration 5×10^3

(d) Iteration 10^4 (e) Iteration 2×10^4 (f) Iteration 10^5

Figure 10.3: Plots of the intermediate stages of training a neural network to approximate the function given by the blue curve.

10.1.2   Approximation of Functions

A key property of neural networks is their ability to approximate very general functions. We begin here with a simple illustrative example. Consider training a one hidden layer neural network to approximate a given function $G(x)$ on the interval $[0, 1]$. We take evenly spaced points $0 = x_1 \leq x_2 \leq \cdots \leq x_m = 1$, and use the mean squared error loss function

$$\mathcal{L}(\mathbf{w}) = \frac{1}{m} \sum_{i=1}^{m} \big[F(x_i; \mathbf{w}) - G(x_i) \big]^2.$$

We choose $m = 1000$ points, and a one hidden layer neural network with 1000 hidden nodes and a sigmoid activation function. For the function G, we used a linear combination of trigonometric functions depicted as the blue curve in Figure 10.3. To train the neural network, we ran gradient descent using the Adam optimizer for 10^5 iterations. (For details, we refer the reader to the Python notebook for this section.)

Figure 10.3 shows the intermediate steps of gradient descent and the final result. We see that the neural network initially fits the low frequency parts of the function, and only later in training is it able to capture the small scale details that correspond to higher frequencies. This property is referred to as the *frequency principle* [196, 250, 251] in neural networks and significant research has been recently devoted to it.[6]

Figure 10.4 shows the mean squared error loss over all 10^5 training iterations. For comparison, we also experimented with plain gradient descent with the largest stable learning rate.

---

[6] The frequency principle has been cited as one explanation for why neural networks do not easily overfit. Basically, overfitting requires fitting the spurious parts of the data that may appear as noise or high frequency artifacts, which a neural network cannot typically do during the early phases of training.

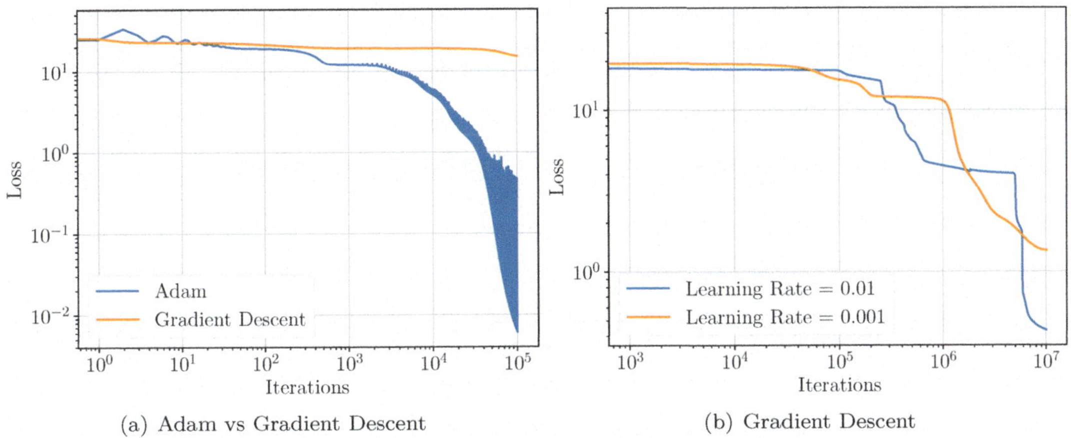

(a) Adam vs Gradient Descent   (b) Gradient Descent

Figure 10.4: Plots of the loss over training iterations for function approximation with neural networks. We show the results for both the Adam optimizer and plain gradient descent.

In Figure 10.4(a) we can see that the Adam optimizer reaches a loss of 10^{-2} after 10^5 iterations, while gradient descent still has a loss above 10. In Figure 10.4(b) we see that gradient descent must be run for at least 10^7 iterations before the loss can be reduced to below one. Thus, in this example, the Adam optimizer is at least 100 times faster than gradient descent.

With that being said, the 10^5 iterations of Adam optimization took about one minute to perform on a powerful graphics processing unit (GPU). This is highly inefficient for one-dimensional function approximation, which can be carried out with orthogonal polynomials or trigonometric functions (e.g., Fourier series) much more efficiently. This is simply to say that neural networks are not efficient at function approximation in low dimensions. Their power comes from their ability to both scale to high dimensions, where other techniques are not feasible due to the "curse of dimensionality", and automatically adapt to the data, which are properties we explore in the rest of this chapter.

10.1.3 Classification

We now turn to describing how to use fully connected neural networks for classification problems. As discussed in Chapter 7, for a problem with c classes, the output of the neural network $F(\mathbf{x})$ will have c components. In other words, the number of nodes in the last layer is $n_L = c$. Recall that our label vectors are taken to be the one-hot vectors $\mathbf{e}_1, \ldots, \mathbf{e}_c \in \mathbb{R}^c$, i.e., the standard basis vectors, whereby $\mathbf{e}_i$ represents the i-th class. The classification of the input $\mathbf{x}$ is taken to be the index of the largest component of the output $F(\mathbf{x})$. In particular, we do not require that $F(\mathbf{x})$ exactly fit the one-hot label vectors, and just need its largest component to be correct, which is a much easier task.

Let $\mathbf{x}_1, \ldots, \mathbf{x}_m \in \mathbb{R}^n$ denote the training data points, and $\mathbf{y}_1, \ldots, \mathbf{y}_m \in \mathbb{R}^c$ the corresponding one-hot label vectors, so that $\mathbf{y}_i = \mathbf{e}_j$ when $\mathbf{x}_i$ belongs to class j. Then a natural loss function would be

$$\ell\big(F(\mathbf{x}_i), \mathbf{y}_i\big) = -\mathbf{y}_i \cdot F(\mathbf{x}_i), \tag{10.19}$$

which is then used to assemble the total loss function (10.16). The minus sign means that minimizing the loss will maximize the dot product $F(\mathbf{x}_i) \cdot \mathbf{y}_i$, which is precisely the component of $F(\mathbf{x}_i)$ corresponding to the class that $\mathbf{x}_i$ belongs to. Unfortunately, this does not work since the loss can simply be arbitrarily decreased by multiplying F by a large positive or negative constant C, depending on the sign of the total loss, which is not productive for learning.

Thus, we need some way to constrain the loss. One natural option is to divide by the norm of the output of the network, that is set

$$\ell\big(F(\mathbf{x}_i), \mathbf{y}_i\big) = -\frac{F(\mathbf{x}_i)}{\|F(\mathbf{x}_i)\|} \cdot \mathbf{y}_i. \tag{10.20}$$

It turns out that this works reasonably well in practice, although it is not commonly used; see Exercise 1.3.

A more common approach is to normalize the outputs of the network so that they form a probability vector. This can be done by introducing the *soft-max function*

$$\mathbf{p}_i := \frac{\exp F(\mathbf{x}_i)}{\|\exp F(\mathbf{x}_i)\|_1}, \qquad i = 1, \ldots, m, \tag{10.21}$$

where the exponential function $\exp(t) = e^t$ is applied componentwise to $F(\mathbf{x}_i)$ and, because all entries of $\exp F(\mathbf{x}_i)$ are positive, its 1 norm is just the sum of its entries. The vector $\mathbf{p}_i$ is now a probability vector — its entries are nonnegative and sum to $1 = p_1 + \cdots + p_n = \|\mathbf{p}_i\|_1$. We then use the negative log likelihood loss

$$\begin{aligned}
\ell\big(F(\mathbf{x}_i), \mathbf{y}_i\big) &= -\mathbf{y}_i \cdot \log \mathbf{p}_i \\
&= -\mathbf{y}_i \cdot \log \left[\frac{\exp F(\mathbf{x}_i)}{\|\exp F(\mathbf{x}_i)\|_1} \right] = -\mathbf{y}_i \cdot F(\mathbf{x}_i) + \log \|\exp F(\mathbf{x}_i)\|_1, \tag{10.22}
\end{aligned}$$

where in the middle expression, the natural logarithm $\log t$ is also applied componentwise. Thus, the composition of soft-max and the logarithm produces a loss that is not so far removed from our original idea. Indeed, minimizing the first term in (10.22) is exactly what we proposed in (10.19), while minimizing the second term serves to to ensure that the loss is bounded below and cannot be minimized simply by scaling F.

Remark 10.2. The computation of the final logarithmic term in (10.22) is subject to numerical instabilities, in particular floating point overflow, when the entries of F become large. Fortunately, quantities like this can be computed in a numerically stable way with the following trick. We simply note that for any $a > 0$ we have

$$\log \|\exp F(\mathbf{x}_i)\|_1 = \log \|e^a \exp[F(\mathbf{x}_i) - a]\|_1 = a + \log \|\exp[F(\mathbf{x}_i) - a]\|_1.$$

If we set $a = \max_{1 \leq j \leq c} F_j(\mathbf{x}_i)$, where $F(\mathbf{x}) = (F_1(\mathbf{x}), \ldots, F_c(\mathbf{x}))$, then we can write

$$\log \|\exp F(\mathbf{x}_i)\|_1 = \max_{1 \leq j \leq c} F_j(\mathbf{x}_i) + \log \left\| \exp \left[F(\mathbf{x}_i) - \max_j F_j(\mathbf{x}_i) \right] \right\|_1.$$

The final term will not suffer from overflow since it is the exponential of a number that is less than one.   ▲

In Figure 10.5 we show the results of training a one hidden layer neural network on some toy binary classification problems in two dimensions. On the left we used a network with only a single neuron, so it is restricted to a linear decision boundary, similar to SVM; see Section 7.3. However, in contrast to SVM, the neural network is clearly not finding a maximum margin classifier. The other examples in Figure 10.5 use networks with one hidden layer containing 64 nodes. While the decision boundaries for the two moons and circles examples in Figure 10.5 appear to be simple explanations of the data that should generalize well, the neural network

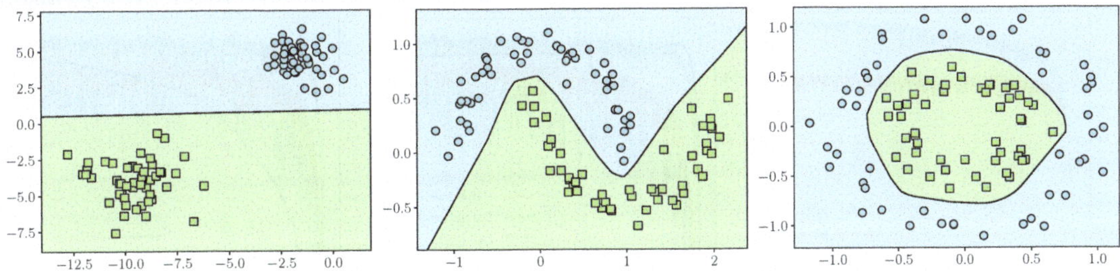

Figure 10.5: Toy examples of binary classification using neural networks

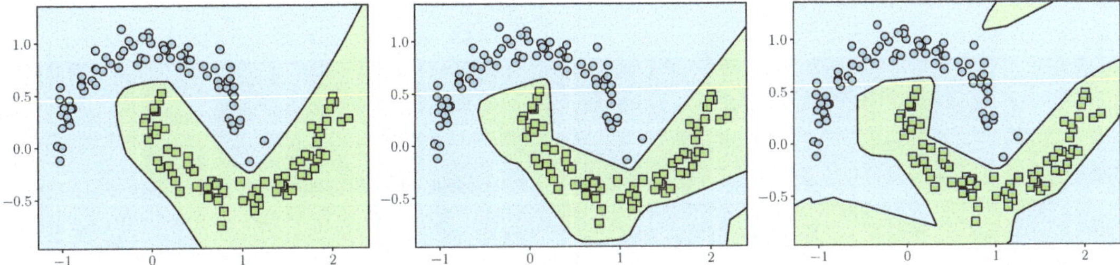

Figure 10.6: Examples of other classification functions obtained by the same neural network as in Figure 10.5.

is capable of producing far stranger results. In Figure 10.6 we show decision boundaries for the same neural network trained in a different way[7] that encourages overfitting. Observe that even a relatively small network can exhibit overfitting, although this does not occur when it is sensibly trained.

Next, we consider the classification of MNIST digits. We used a one hidden layer neural network with 64 hidden nodes and ReLU activation, and ran the Adam optimizer in addition to SGD with mini-batch size $m_b = 480$, using 40 epochs of training, which is 5,000 steps of gradient descent given that the training data has $m = 60,000$ images. Figure 10.7(a) shows the training and testing accuracy during training. We see both increase quickly, with testing accuracy leveling off around 97%, and training accuracy tending towards 100%, indicating a small amount of overfitting. We also experimented with different batch sizes, including $m_b = 30, m_b = 60$, and full batch $m_b = 60,000$. In Figure 10.7(b) we see that the small batch sizes of 30 and 60 produce slightly worse results, which is due to their poor quality approximation of the gradient. The batch size of 480 produces results similar to full-batch, and in fact exhibits slightly less overfitting in the long run. However, there is a key difference in computation time, since each iteration of SGD is far less expensive than full batch gradient descent. Figure 10.7(c) shows the accuracy plotted against CPU time[8] in seconds. Here, we see that SGD with $m_b = 480$ performs the best in terms of producing an accurate model in the least amount of time, reaching 97% in about one second, while full batch gradient descent does not reach 97% until after more than 10 seconds of training.

Finally, we point out that it is often difficult to interpret how a neural network is making its predictions. Even in this simple case of a one hidden layer network with 64 neurons in the hidden layer, we are unable to fully understand the resulting predictions. The best we can do

---

[7]We added a handful of corrupted labels at opportune locations.
[8]We did not use a GPU for this experiment.

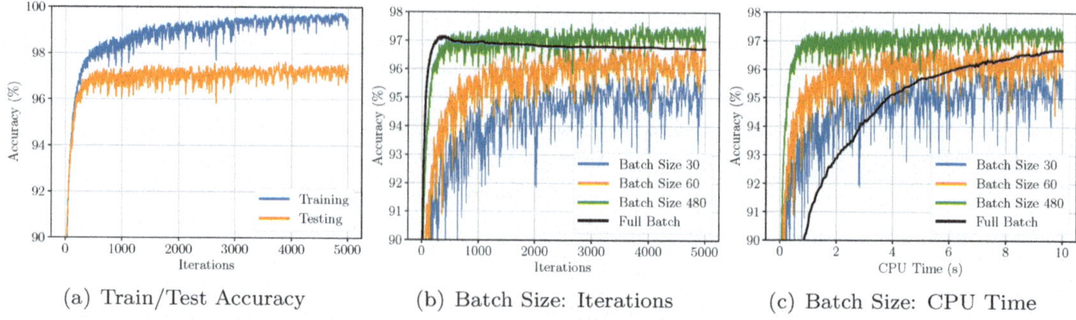

(a) Train/Test Accuracy   (b) Batch Size: Iterations   (c) Batch Size: CPU Time

Figure 10.7: Training and testing accuracy for training a neural network to classify MNIST digits.

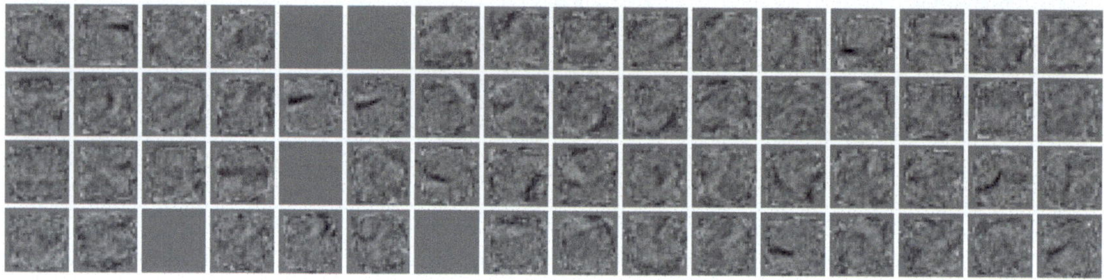

Figure 10.8: Plots of the weight vectors in the 64 hidden neurons in the MNIST classification example.

is examine the weights in the network, or examine the inputs that maximally activate parts of the network. In Figure 10.8 we plot the weight vectors associated with the 64 hidden neurons. The first layer of the network takes the dot products of an input MNIST image with each of these 64 images, and then adds a bias and applies the ReLU activation. Aside from some short black lines in a few of the images, it is unclear what these hidden neurons are looking for in the input image. This is unfortunately the case for many applications of deep learning, and the field of interpretable deep learning is a currently active area of research [39].

10.1.4   Connection to Kernel Methods

In this section we outline some connections between deep learning and kernel methods, which were introduced in Section 7.6. Since the last layer of a neural network normally has no activation function, we can write an L layer neural network in the form

$$F(\mathbf{x}; \mathbf{w}, W, \mathbf{b}) = W\,\phi(\mathbf{x}; \mathbf{w}) + \mathbf{b}.$$

Here $\phi(\mathbf{x}; \mathbf{w})$ is an $L - 1$ layer neural network with weights $\mathbf{w}$, mapping $\mathbf{x} \in \mathbb{R}^n$ to $\phi(\mathbf{x}; \mathbf{w}) \in \mathbb{R}^{n_L - 1}$, while $W \in \mathcal{M}_{n_L \times n_{L-1}}$ is the weight matrix and $\mathbf{b} \in \mathbb{R}^{n_L}$ the bias vector for the final layer. If we set $\mathbf{z}_i = \phi(\mathbf{x}_i; \mathbf{w})$ for $i = 1, \ldots, m$, then the loss function (10.16) has the form

$$\mathcal{L}(\mathbf{w}, W, \mathbf{b}) = \frac{1}{m} \sum_{i=1}^{m} \ell(W\mathbf{z}_i + \mathbf{b}, \mathbf{y}_i),$$

where $\mathbf{z}_i$ implicitly depends on the weights $\mathbf{w}$. In particular, if the loss is the least squared error $\ell(\mathbf{v}, \mathbf{y}) = \|\mathbf{v} - \mathbf{y}\|^2$, then this resembles a linear regression problem

$$\mathcal{L}(\mathbf{w}, W, \mathbf{b}) = \frac{1}{m} \sum_{i=1}^{m} \| W\mathbf{z}_i + \mathbf{b} - \mathbf{y}_i \|^2.$$

The difference with linear regression is that the $\mathbf{z}_i$ are not fixed data points, but rather they are the outputs of the first $L - 1$ layers of the neural network, which depend on the weights $\mathbf{w}$ that are being optimized over. Now, if a deep neural network is able to be trained to well-classify a data set, then, since the output is a linear function of the transformed data points $\mathbf{z}_i$, these data points must be essentially *linearly separable*. That is, the map $\phi(\mathbf{x}; \mathbf{w})$, which in this case is an $L - 1$ layer neural network, must be able to linearly separate the classes in the data, and so it is essentially a good *feature map* in the context of kernel methods.

Recall from Section 7.6 that we motivated kernel methods with the problem of choosing a feature map ϕ that can linearly separate the data, and then applying a machine learning method to the features $\mathbf{z}_i = \phi(\mathbf{x}_i)$. Since it is hard to choose ϕ by hand, we then used the *kernel trick*, whereby we replaced the choice of ϕ with the selection of a kernel function $\mathcal{K}$. In contrast, deep learning does not use the kernel trick, and instead can be interpreted as a method for learning a good feature map ϕ from the data itself.

Once we have trained a deep neural network to produce a feature map $\phi(\mathbf{x}; \mathbf{w})$, we can compute the associated kernel function

$$\mathcal{K}_\phi(\mathbf{x}, \mathbf{y}; \mathbf{w}) = \phi(\mathbf{x}; \mathbf{w}) \cdot \phi(\mathbf{y}; \mathbf{w}). \tag{10.23}$$

We maintain the dependence on the weights $\mathbf{w}$ in the first $L - 1$ layers to emphasize that the kernel function is learned by training the network, and is not selected by hand, e.g., a radial basis function kernel, as we did in Section 7.6.

It can be instructive to compute the kernel functions for some simple neural networks. Consider, for example, an $L = 2$ layer neural network with one hidden layer and scalar output, which is the function $F : \mathbb{R}^n \to \mathbb{R}$ given in (10.8). In this case $\phi(\mathbf{x}; \mathbf{w}) \in \mathbb{R}^{n_1}$ is the output of the first layer of the network, which is the vector

$$\phi(\mathbf{x}; \mathbf{w}) = \left(\sigma(\mathbf{w}_1 \cdot \mathbf{x} + b_1), \ \ldots \ , \sigma(\mathbf{w}_{n_1} \cdot \mathbf{x} + b_{n_1}) \right)^T \in \mathbb{R}^{n_1},$$

where $\mathbf{w} = (\mathbf{w}_1, b_1, \ldots, \mathbf{w}_{n_1}, b_{n_1})$ denotes all of the parameters in the first layer. The kernel function associated with this neural network, (10.23), is then given by

$$\mathcal{K}_\phi(\mathbf{x}, \mathbf{y}; \mathbf{w}) = \sum_{i=1}^{n_1} \sigma(\mathbf{w}_i \cdot \mathbf{x} + b_i)\, \sigma(\mathbf{w}_i \cdot \mathbf{y} + b_i).$$

Analyzing the kernel function associated to a neural network is one avenue towards understanding why deep learning works well. In some special cases these kernel functions are well-understood in the infinite width limit, where $n_1 \to \infty$, [40]. Kernel functions of this form also show up in the analysis of gradient descent for optimizing the loss function in deep learning, which we explore further in Section 11.7.

Exercises

1.1. ♡ Explain why we can, without loss of generality, assume that all but the last activation function in a neural network can be taken to be non-affine.

1.2.♡ Consider a *bias-free* ResNet, as in (10.15) but where all the bias vectors $\mathbf{b}_k$ are removed, i.e., we set $\mathbf{b}_k = 0$. Explain why any ResNet with input $\mathbf{x} \in \mathbb{R}^n$ has a bias-free version that gives the same outputs by taking $\begin{pmatrix} \mathbf{x} \\ 1 \end{pmatrix} \in \mathbb{R}^{n+1}$ as input.

1.3.♡ Modify the Python notebooks from this section to use the loss (10.20) for classification instead of the softmax and negative log likelihood loss. Try different norms for the $\| F(\mathbf{x}_i) \|$ term. In particular, try the p norm with different values for p; in particular, try $p > 2$ (for example, $p = 5$). Are you able to get similar accuracy to the softmax composed with negative log likelihood loss?

1.4.◇ Based on the expression (10.22) for the softmax/negative log likelihood loss, it is natural to consider a loss of the form

$$\ell\big(F(\mathbf{x}_i), \mathbf{y}_i\big) = -\mathbf{y}_i \cdot F(\mathbf{x}_i) + \| F(\mathbf{x}_i) \|.$$

Modify the Python notebook from this section to use this loss and try different norms. Are you able to get similar results?

1.5.◇ Modify the Python notebook from this section to use the 1 loss instead of the 2 loss for function approximation.

1.6. Modify the Python notebook from this section to train a neural network for classification on another test data set. For example, you can try FashionMNIST or Sign Language MNIST from the `graphlearning` package, or a test data set from `sklearn`.

1.7. Modify the Python notebook from this section to use far less than the usual 60000 MNIST images for training. Are you able to get the neural network to significantly overfit?

1.8. Modify the Python notebook from this section to change the architecture of the neural network, but adding more layers and more hidden nodes. Does the accuracy on MNIST change significantly?

10.2   Backpropagation and Automatic Differentiation

In order to train a neural network (10.5), we need to compute the gradients of the loss function (10.16) with respect to all the weights and biases. That is, given an input $\mathbf{x} \in \mathbb{R}^n$ to the network, and a desired output $\mathbf{y} \in \mathbb{R}^p$, we need to compute the gradients

$$\nabla_{W_k} \ell\big(F(\mathbf{x}), \mathbf{y}\big), \qquad \nabla_{\mathbf{b}_k} \ell\big(F(\mathbf{x}), \mathbf{y}\big), \qquad \text{for} \qquad k = 1, \ldots, L. \qquad (10.24)$$

Here, the second quantity is the usual gradient (6.27), while the first denotes the $n_k \times n_{k-1}$ matrix G whose entries are the partial derivatives $g_{ij} = \partial\ell(F(\mathbf{x}), \mathbf{y})/\partial w_{k,ij}$, where $w_{k,ij}$ denotes the (i, j) entry of W_k.

Since a neural network is formed by an L-fold composition of functions, the gradients (10.24) can in principle be computed by employing the usual chain rule. However, a naïve application to each weight and bias separately would be very inefficient, as it would involve redundant operations between the layers. It turns out there are relationships between gradients of each layer, also following from the chain rule, that can be used to efficiently compute *all* of the required gradients in a single backward sweep of the network. This is known as *backpropagation*, and is a special case of a more general class of techniques called *automatic differentiation*.

We begin this section by developing the backpropagation[9] equations for fully connected neural networks. The following formulas provide a recursive methods of computing the required gradients.

Theorem 10.3 (Backpropagation). *Let $\mathbf{x} \in \mathbb{R}^n$, $\mathbf{y} \in \mathbb{R}^c$, and suppose we are at layer $1 \leq k \leq L$. Let $\mathbf{p}_k$ denote the preactivation (10.13), and let $\mathbf{z}_k = \sigma_k(\mathbf{p}_k)$ be the output (10.12). Define $\mathbf{v}_k = \nabla_{\mathbf{z}_k} \ell(F(\mathbf{x}), \mathbf{y})$ and the diagonal matrix*

$$S_k = \operatorname{diag} \sigma_k'(\mathbf{p}_k) \in \mathcal{M}_{n_k \times n_k}, \tag{10.25}$$

where σ_k' is applied componentwise. Then

$$\nabla_{\mathbf{b}_k} \ell(F(\mathbf{x}), \mathbf{y}) = S_k \mathbf{v}_k, \quad \nabla_{W_k} \ell(F(\mathbf{x}), \mathbf{y}) = S_k \mathbf{v}_k \mathbf{z}_{k-1}^T, \quad \mathbf{v}_{k-1} = W_k^T S_k \mathbf{v}_k. \tag{10.26}$$

Proof. Fixing $\mathbf{y}$, let us write $G_L(\mathbf{z}) = \ell(\mathbf{z}, \mathbf{y})$ and, more generally

$$G_k = G_L \circ F_L \circ \cdots \circ F_{k+1}.$$

Observe that $\ell(F(\mathbf{x}), \mathbf{y}) = G_k(\mathbf{z}_k)$ for any $k = 0, \ldots, L$. Since $G_k \circ F_k = G_{k-1}$, and, further, $\mathbf{z}_k = F_k(\mathbf{z}_{k-1})$, the chain rule (6.77) yields

$$\mathbf{v}_{k-1} = \nabla G_{k-1}(\mathbf{z}_{k-1}) = \nabla G_k\big(F_k(\mathbf{z}_{k-1})\big) = \mathbf{D}F_k(\mathbf{z}_{k-1})^T \nabla G_k(\mathbf{z}_k) = \mathbf{D}F_k(\mathbf{z}_{k-1})^T \mathbf{v}_k.$$

Hence, to establish the third formula in (10.26), we just need to compute the Jacobian of the k-th layer of the network, which, according to Exercise 2.1, is given by

$$\mathbf{D}F_k(\mathbf{z}_{k-1}) = \mathbf{D}\big[\sigma\left(W_k \mathbf{z}_{k-1} + \mathbf{b}_k\right)\big] = S_k W_k.$$

To establish the first and second formulas, note that by (10.12),

$$\ell(F(\mathbf{x}), \mathbf{y}) = G_k(\mathbf{z}_k) = G_k\big(\sigma_k(W_k \mathbf{z}_{k-1} + \mathbf{b}_k)\big).$$

Thus, the chain rule (6.77) and Exercise 2.1 produce

$$\nabla_{\mathbf{b}_k} \ell(F(\mathbf{x}), \mathbf{y}) = S_k \nabla G_k(\mathbf{z}_k) = S_k \mathbf{v}_k,$$

while Exercise 2.2 yields

$$\nabla_{W_k} \ell(F(\mathbf{x}), \mathbf{y}) = S_k \nabla G_k(\mathbf{z}_k) \mathbf{z}_{k-1}^T = S_k \mathbf{v}_k \mathbf{z}_{k-1}^T. \qquad \blacksquare$$

The formulas in Theorem 10.3 suggest an algorithm, known as *backpropagation*, for efficiently computing the gradients (10.24) via recursion. The important quantity to compute is $\mathbf{v}_k = \nabla_{\mathbf{z}_k} \ell(F(\mathbf{x}), \mathbf{y})$, since both $\nabla_{\mathbf{b}_k} \ell(F(\mathbf{x}), \mathbf{y})$ and $\nabla_{W_k} \ell(F(\mathbf{x}), \mathbf{y})$ can be expressed in terms of $\mathbf{v}_k$ using (10.26). This is done with the recursion formula $\mathbf{v}_{k-1} = W_k^T S_k \mathbf{v}_k$ in (10.26), starting with $\mathbf{v}_L$ and iterating backwards through the network to compute $\mathbf{v}_{L-1}$, $\mathbf{v}_{L-2}$, and so on, through $\mathbf{v}_1$, hence the name backpropagation. This requires saving the values of the preactivations $\mathbf{p}_k$ from the forward propagation of the network, so that we can form the matrix S_k in (10.25).

The starting point for backpropagation, $\mathbf{v}_L$, depends on the choice of loss function ℓ, since $\mathbf{z}_L = F(\mathbf{x})$ and $\mathbf{v}_L = \nabla_{\mathbf{z}_L} \ell(F(\mathbf{x}), \mathbf{y})$. For concreteness, suppose we take the 2 loss

$$\ell(\mathbf{z}, \mathbf{y}) = \tfrac{1}{2} \|\mathbf{z} - \mathbf{y}\|_2^2. \tag{10.27}$$

---

[9]Commonly written without a space between the words.

Then we have

$$\nabla_{\mathbf{z}_L}\ell(\mathbf{z}_L,\mathbf{y}) = \mathbf{z}_L - \mathbf{y}, \qquad \text{and so} \qquad \mathbf{v}_L = \nabla_{\mathbf{z}_L}\ell(F(\mathbf{x}),\mathbf{y}) = F(\mathbf{x}) - \mathbf{y}.$$

To write out the formulas for $\mathbf{v}_k$ more explicitly, let $A_k = W_k^T S_k$ and note that from the recursion formula $\mathbf{v}_{k-1} = A_k \mathbf{v}_k$ we obtain

$$\mathbf{v}_k = A_{k+1}A_{k+2}\cdots A_L\mathbf{v}_L = A_{k+1}A_{k+2}\cdots A_L(F(\mathbf{x})-\mathbf{y}). \tag{10.28}$$

Here, we can see how the computations of $\mathbf{v}_k$ and $\mathbf{v}_j$ for $k \neq j$ involve many of the same operations. The idea behind backpropagation is that by computing $\mathbf{v}_k$ with the recursion $\mathbf{v}_{k-1} = A_k\mathbf{v}_k$ instead of (10.28), we can dispense of these wasteful repeated computations. It is important to point out that this recursion is necessarily backwards; since A_k is not invertible — it is usually not even square! — we cannot express $\mathbf{v}_k$ in terms of $\mathbf{v}_{k-1}$. Nevertheless, we can certainly compute $\mathbf{v}_k$ via *forward propagation*, whereby we perform the computation on the right hand side of (10.28) as written, from left to right. However, this would require multiplying matrices at each iteration, which is more expensive than matrix/vector multiplication, and we would have to perform the computations separately for each k (i.e., we cannot share computations as is done in backpropagation). We postpone a detailed discussion of computational complexity and forward propagation until later in this section.

We can think of the process of backpropagation as propagating the *error* $\nabla_{\mathbf{z}_L}\ell(F(\mathbf{x}),\mathbf{y})$ backwards through the network to see how all the weights and biases must change in order to decrease the error. In practice, we compute these gradients not for a single input $\mathbf{x}$, but for a mini-batch $\mathbf{x}_1,\ldots,\mathbf{x}_{m_b}$ of inputs. The computations are the same for each $\mathbf{x}_i$ and can easily be done in parallel.

Example 10.4. Consider the one hidden layer neural network (10.7), namely,

$$F(\mathbf{x}) = W_2\,\sigma(W_1\mathbf{x}+\mathbf{b}_1)+\mathbf{b}_2,$$

along with the 2 loss (10.27), so that $\mathbf{v}_2 = \nabla_{\mathbf{z}_2}\ell(F(\mathbf{x}),\mathbf{y}) = F(\mathbf{x}) - \mathbf{y}$. Here,

$$\mathbf{z}_0 = \mathbf{x}, \quad \mathbf{z}_1 = \sigma(W_1\mathbf{x}+\mathbf{b}_1), \quad \mathbf{z}_2 = F(\mathbf{x}), \quad S_1 = \mathrm{diag}(\sigma'(W_1\mathbf{x}+\mathbf{b}_1)), \quad S_2 = I.$$

Thus, by the backpropagation Theorem 10.3,

$$\nabla_{\mathbf{b}_2}\ell(F(\mathbf{x}),\mathbf{y}) = \mathbf{v}_2 = F(\mathbf{x}) - \mathbf{y},$$
$$\nabla_{W_2}\ell(F(\mathbf{x}),\mathbf{y}) = \mathbf{v}_2\mathbf{z}_1^T = [F(\mathbf{x})-\mathbf{y}]\,\sigma(W_1\mathbf{x}+\mathbf{b}_1)^T,$$

and, moreover, $\mathbf{v}_1 = W_2^T\mathbf{v}_2 = W_2^T[F(\mathbf{x})-\mathbf{y}]$, and

$$\nabla_{\mathbf{b}_1}\ell(F(\mathbf{x}),\mathbf{y}) = S_1\mathbf{v}_1 = \mathrm{diag}(\sigma'(W_1\mathbf{x}+\mathbf{b}_1))W_2^T[F(\mathbf{x})-\mathbf{y}],$$
$$\nabla_{W_1}\ell(F(\mathbf{x}),\mathbf{y}) = \mathrm{diag}(\sigma'(W_1\mathbf{x}+\mathbf{b}_1))W_2^T[F(\mathbf{x})-\mathbf{y}]\mathbf{x}^T.$$

Notice that, as expected, all of these gradients vanish when $F(\mathbf{x}) = \mathbf{y}$.

If we use the soft-max with negative log likelihood loss (10.22), i.e.,

$$\ell(\mathbf{z},\mathbf{y}) = -\mathbf{y}\cdot\mathbf{z} + \log\|\exp\mathbf{z}\|_1,$$

as is common in classification, then

$$\mathbf{v}_2 = \nabla_{\mathbf{z}_2}\ell(F(\mathbf{x}),\mathbf{y}) = \frac{\exp F(\mathbf{x})}{\|\exp F(\mathbf{x})\|_1} - \mathbf{y} = \mathbf{p} - \mathbf{y}, \tag{10.29}$$

where $\mathbf{p}$ is the output of the soft-max function applied to the output of the network $F(\mathbf{x})$. The equations above would all be the same, except that we would replace $F(\mathbf{x}) - \mathbf{y}$ with the expression on the right hand side of (10.29).   ▲

Backpropagation is a special case of an algorithmic differentiation technique called *automatic differentiation*, and the backpropagation result in Theorem 10.3 is called *reverse mode automatic differentiation*. It is useful to state a more general result, in order that the backpropagation equations do not need to be re-derived when making changes to the neural network architecture. To do this, we consider a general L-fold composition

$$F = F_L \circ \cdots \circ F_1, \tag{10.30}$$

where $F_k : \mathbb{R}^{n_{k-1}} \times \mathbb{R}^{m_k} \to \mathbb{R}^{n_k}$ is a parameterized function $F_k(\mathbf{z}; \mathbf{w})$, with $\mathbf{z} \in \mathbb{R}^{n_{k-1}}$ the input and $\mathbf{w} \in \mathbb{R}^{m_k}$ the parameters controlling its behavior. In the context of neural networks, $\mathbf{w}$ encodes all of the weights and biases in the k-th layer. Note that for the purpose of composition in (10.30), we treat F_k as a function of $\mathbf{z}$, mapping from $\mathbb{R}^{n_{k-1}}$ to $\mathbb{R}^{n_k}$. As before, we write $\mathbf{z}_0 = \mathbf{x}$ for the input, and view (10.30) as an iteration

$$\mathbf{z}_k = F_k(\mathbf{z}_{k-1}; \mathbf{w}_k), \tag{10.31}$$

where $\mathbf{w}_k$ denotes the weights for the k-th layer, and hence $\mathbf{z}_L = F(\mathbf{x})$. As usual, the output $F(\mathbf{x})$ is fed into a loss function $\ell : \mathbb{R}^{n_k} \to \mathbb{R}$, and we are interested in the gradients $\nabla_{\mathbf{w}_k} \ell(F(\mathbf{x}))$.

The following result summarizes the main ideas behind reverse mode automatic differentiation. The proof is very similar to Theorem 10.3, and so we leave it to the reader as Exercise 2.4. We use the notation $\mathbf{D}_\mathbf{z} F_k$ and $\mathbf{D}_\mathbf{w} F_k$ for the Jacobian matrices of F_k with respect to the input $\mathbf{z}$ and the weights $\mathbf{w}$, respectively.

Theorem 10.5 (Reverse Mode Automatic Differentiation). *Let F be defined by* (10.30), $\mathbf{z}_k$ *by* (10.31), *and set* $\mathbf{v}_k = \nabla_{\mathbf{z}_k} \ell(F(\mathbf{x}))$. *Then, for* $k = 1, \ldots, L$,

$$v_{k-1} = \mathbf{D}_\mathbf{z} F_k(\mathbf{z}_{k-1}; \mathbf{w}_k)^T \mathbf{v}_k, \qquad \nabla_{\mathbf{w}_k} \ell(F(\mathbf{x})) = \mathbf{D}_\mathbf{w} F_k(\mathbf{z}_{k-1}; \mathbf{w}_k)^T \mathbf{v}_k. \tag{10.32}$$

It is important to note that automatic differentiation is *not* the same as symbolic differentiation. The latter produces a formula for the gradient, through symbolic applications of the chain rule and other rules for differentiation. The symbolic gradient can be applied to any input to compute the gradient, without updating the symbolic expression. In contrast, automatic differentiation computes gradients for a given input $\mathbf{x}$ and weights $\mathbf{w}_k$ by multiplying the Jacobian matrices that arise from the chain rule. To compute the gradient for a different input or choices of weights $\mathbf{w}_k$ requires recomputing everything from scratch, i.e., computing (10.32) for $k = L, \ldots, 1$.

Remark 10.6. We remark briefly on the computational complexity of reverse mode automatic differentiation. For simplicity, let us assume that $n_k = n$ and $m_k = m$ for all k. Then $\mathbf{D}_z F_k \in \mathcal{M}_{n \times n}$ and each step of backpropagation (10.32) takes $\mathrm{O}(n^2)$ operations. Since $\mathbf{D}_\mathbf{w} F_k \in \mathcal{M}_{n \times m}$, computing $\nabla_{\mathbf{w}_k} \ell(F(\mathbf{x}))$ requires an extra $\mathrm{O}(mn)$ operations. Since there are L steps, the total complexity is $\mathrm{O}(n^2 L + mnL)$. This is similar in complexity to what one would expect from simply running a forward pass to compute the value $F(\mathbf{x})$, so we cannot in general hope to do any better. ▲

The Python packages `pytorch` and `tensorflow` use reverse mode automatic differentiation to compute the gradients of neural networks for training. In both packages, the sequence of operations in the composition (10.30) is tracked implicitly, so the compositional form (10.30) is not directly provided by the user. The user need only invoke backpropagation, when desired, and specify with respect to which variables, i.e., which weights $\mathbf{w}_k$, the gradients are to be computed.

There are many algorithmic issues with implementing reverse mode differentiation in such a general and implicit manner that that are not covered by Theorem 10.5. In particular, there may be many different computational paths that contribute to the loss. As a simple example, we could have two functions F and G of the form (10.30), with each possibly having a different numbers of layers L, but with some shared parameters $\mathbf{w}_k$ between the two functions. If the loss ℓ depends on both $F(\mathbf{x})$ and $G(\mathbf{x})$, then the gradient $\nabla_{\mathbf{z}_k}\ell$ will have contributions from the reverse mode differentiation of both F and G. In general, the situation can be quite complicated, and is handled algorithmically by the construction of an acyclic digraph[10] that records the computational dependencies. Every computation is a node of the graph, and backpropagation flows backwards along this computational graph, using the reverse differentiation equation (10.32) at each node to additively *update* the gradients with respect to any parameters involved in the computation. A given parameter's gradients may be updated multiple times during backpropagation. We refer the reader to [16, 182] for details.

As we hinted at above, automatic differentiation is not restricted to operating in the reverse direction, i.e., backpropagation. We can also propagate forwards, called *forward mode automatic differentiation*. However, the gradients that are computed in one sweep of either method are fundamentally different. While reverse mode automatic differentiation computes the gradients of a given output — in this case the loss $\ell(F(\mathbf{x}))$ — with respect to *all* parameters and inputs that appear throughout the network, forward mode automatic differentiation computes the gradients of the output of *every* layer in the network with respect to a given, fixed input or parameter. For simplicity, we state the result below for the gradients with respect to the input $\mathbf{x}$ of the network, while similar results hold for the gradient with respect to any of the parameters.

> **Theorem 10.7** (Forward Mode Automatic Differentiation). *Let F be defined by* (10.30), $\mathbf{z}_k$ *by* (10.31), *and set* $H_k = F_k \circ \cdots \circ F_1$. *For* $k = 1, \ldots, L$,
>
> $$\mathbf{D}H_k(\mathbf{x}) = \mathbf{D}F_k(\mathbf{z}_{k-1})\,\mathbf{D}H_{k-1}(\mathbf{x}). \qquad (10.33)$$

Proof. This easily follows by applying the chain rule (6.76) to $H_k = F_k \circ H_{k-1}$. ∎

Remark 10.8. As in Remark 10.6, we can analyze the computational complexity of forward mode automatic differentiation in the setting where $\mathbf{x} \in \mathbb{R}^m$, and $n_k = n$ for all $k \geq 1$. To iterate (10.33) for L steps, the resulting computational complexity is $\mathrm{O}(n^2 m L)$. This computes all of the Jacobians $\mathbf{D}H_k(\mathbf{x})$ for $k = 1, \ldots, L$, similar to the way in which reverse mode automatic differentiation computes all the gradients $\nabla_{\mathbf{w}_k}\ell(F(\mathbf{x}))$ for $k = 1, \ldots, L$. If we want to take the directional derivative in a certain direction $\mathbf{v} \in \mathbb{R}^m$, we can multiply both sides of (10.33) by $\mathbf{v}$ to obtain the iterative formula

$$\mathbf{D}H_k(\mathbf{x})\,\mathbf{v} = \mathbf{D}F_k(\mathbf{z}_{k-1})\big[\mathbf{D}H_{k-1}(\mathbf{x})\,\mathbf{v}\big].$$

This requires only $\mathrm{O}(n^2)$ operations per step, and hence $\mathrm{O}(n^2 L)$ operations in total, similar to reverse mode automatic differentiation.

However, the Jacobians $\mathbf{D}H_k(\mathbf{x})$, or the directional derivatives $\mathbf{D}H_k(\mathbf{x})\,\mathbf{v}$, are generally not useful in training neural networks. Instead we require gradients with respect to the weights $\mathbf{w}_k$ appearing in each layer. We can formulate similar versions of Theorem 10.7 where the Jacobians are taken with respect to any of the weights. If $\mathbf{w}_k \in \mathbb{R}^m$, then as

---

[10]See Exercise 2.9 in Chapter 9.

above the complexity is $O(n^2 m (L - k))$ to compute the gradient with respect to a single $\mathbf{w}_k$. To compute all the gradients would require a separate forward pass of the form (10.33) for each weight vector $\mathbf{w}_k$, starting at layer k. Thus, the complexity of forward propagation is $O(n^2 m L^2)$, which is significantly worse than reverse mode automatic differentiation. Hence, when training neural networks, reverse mode automatic differentiation, or backpropagation, is widely preferred and is the method of choice in practice. ▲

Finally, we remark that while both forward and reverse mode automatic differentiation can be used to compute the gradient $\nabla_{\mathbf{x}} \ell(F(\mathbf{x}))$ of the loss with respect to the input $\mathbf{x}$, reverse mode is more efficient, since the output is scalar, so reverse mode tracks gradients, while forward mode tracks Jacobians. These formulas are used, for instance, in generating adversarial examples in deep learning, which are small, usually imperceptible, changes to the input $\mathbf{x}$ that result in a noticeable change in the classification prediction [91, 157]. A classic example is adding imperceptible (to the human eye) noise to an image to change its classification from the correct class label to an incorrect one [6]. One way to generate adversarial examples is to take several steps of gradient ascent on the loss $\ell(F(\mathbf{x}))$ with respect to the input $\mathbf{x}$.

Exercises

2.1. ♡ Show that the Jacobian of the function $F(\mathbf{x}) = \sigma(W\mathbf{x} + \mathbf{b})$ equals $\mathbf{D}F(\mathbf{x}) = DW$, where $D = \operatorname{diag} \sigma'(W\mathbf{x} + \mathbf{b})$.

2.2. ◇ Let $F(W) = G\big(\sigma(W\mathbf{x} + \mathbf{b})\big)$, where G is a scalar-valued function. Let $\nabla_W F$ denote the matrix whose (i, j) entry is the partial derivative of F with respect to w_{ij}. Show that $\nabla_W F(W) = D \nabla G(\mathbf{z}) \mathbf{x}^T$, where $D = \operatorname{diag} \sigma'(W\mathbf{x} + \mathbf{b})$ and $\mathbf{z} = \sigma(W\mathbf{x} + \mathbf{b})$.

2.3. Given the ResNet layers (10.15), follow an argument similar to the proof of Theorem 10.3 to establish the ResNet backpropagation equations

$$
\begin{aligned}
\mathbf{v}_k &= \nabla_{\mathbf{z}_k} \ell(F(\mathbf{x}), \mathbf{y}), \\
\mathbf{v}_{k-1} &= (I + W_k^T S_k V_k^T) \mathbf{v}_k,
\end{aligned}
\quad \text{and} \quad
\begin{aligned}
\nabla_{\mathbf{b}_k} \ell(F(\mathbf{x}), \mathbf{y}) &= S_k V_k \mathbf{v}_k, \\
\nabla_{V_k} \ell(F(\mathbf{x}), \mathbf{y}) &= \mathbf{v}_k \sigma(\mathbf{p}_k)^T \\
\nabla_{W_k} \ell(F(\mathbf{x}), \mathbf{y}) &= V_k S_k \mathbf{v}_k \mathbf{z}_{k-1}^T.
\end{aligned}
$$

2.4. ♡ Prove Theorem 10.5. *Hint*: Mimic the proof of Theorem 10.3.

2.5. ♡ Implement a one hidden layer neural network in Python using only the **numpy** package. Use the formulas from Example 10.4 to compute the gradients of the network and implement plain gradient descent for training. Try your network on some of the basic classification examples from Section 10.1, including MNIST.

10.3 Convolutional Neural Networks

Python Notebook: Convolutional Neural Networks (.ipynb)

One of the challenges in image processing and computer vision is that the same object may be viewed in many different ways, each of which can produce an image that looks vastly different at the pixel level. For instance, we may move the camera position or angle, zoom in or out, change lighting conditions, etc. The human visual system is good at identifying these different views as images of the same object, and we would like our computer vision algorithms to enjoy the same properties. That is, we would like an image to be classified correctly regardless of how it is viewed, or rather, that the classification decision is insensitive to transformations that only change the view of the object.

Convolutional neural networks (CNNs) are powerful machine learning tools for image processing and computer vision that encode translation invariance (or rather, equivariance) into the neural network architecture directly. Invariance under other transformations, such as scaling, rotation, contrast changes, blurring, etc, is usually enforced through another mechanism called *data augmentation*, which we discuss at the end of this section.

In Section 9.10.5 we showed that the convolution operation is the unique linear transformation that is translation equivariant. While that section focused on one-dimensional signals in the periodic setting, the results apply equally well to images (2 or 3 dimensional signals), videos, or even to signals in higher dimensions. Thus, when the input to the neural network is an image, it is natural to replace the linear transformation $W\mathbf{x}$ in each layer of a neural network with a convolution of the image represented by $\mathbf{x}$. This ensures that each layer of the network enjoys translation equivariance, at least for pixels that are not near the boundary of the image.

In convolutional neural networks, the closely related *cross correlation* operation is used as an alternative to convolution. To define this mathematically, as in the remarks after (3.2), we will think of an $m \times n$ image I as a function

$$I \colon \mathbb{N}_m \times \mathbb{N}_n \longrightarrow \mathbb{R}^d,$$

where $\mathbb{N}_m = \{1, 2, \ldots, m\}$ and d is the number of (color) channels, e.g., for a grayscale image $d = 1$. For $\mathbf{x} = (i, j) \in \mathbb{N}_m \times \mathbb{N}_n$, the value $I(\mathbf{x}) = I(i, j) \in \mathbb{R}^d$ contains the color information of the pixel at location $\mathbf{x} = (i, j)$ in the image. For a color image with $d = 3$ channels, the three values of $I(\mathbf{x}) \in \mathbb{R}^3$ are often interpreted as the amount of red, green, and blue in the given pixel, although other color spaces are also used in practice.

A *convolutional kernel* W of width w is a function

$$W \colon \mathbb{Z}_w \times \mathbb{Z}_w \longrightarrow \mathbb{R}^d, \quad \text{where} \quad \mathbb{Z}_w = \{-w \ldots, -1, 0, 1, \ldots, w\}.$$

The convolutional kernel W can certainly be identified with a matrix of size $(2w+1) \times (2w+1)$, though it is less notationally convenient for us. The *cross correlation* $I * W$ of an image I with a kernel function W is defined as the grayscale image

$$(I * W)(\mathbf{x}) = \sum_{\mathbf{y} \in \mathbb{Z}_w^2} I(\mathbf{x} + \mathbf{y}) \cdot W(\mathbf{y}), \tag{10.34}$$

whose size we will discuss below. Essentially we are taking the dot product of W with the image locally, centered at the point $\mathbf{x}$. The closely related *convolution* operation is defined by the very similar looking expression

$$\sum_{\mathbf{y} \in \mathbb{Z}_w^2} I(\mathbf{x} - \mathbf{y}) \cdot W(\mathbf{y}),$$

the only difference being the $-\mathbf{y}$ in convolution. Since the convolutional kernel W is learned, we could use either convolution or cross correlation in practice, though most implementations of CNNs use the latter.

(a) Original Image   (b) Horizontal edges

(c) Vertical edges   (d) Laplacian (edge detector)

Figure 10.9: An example of convolutions of an image with different convolutional filters, which in this case are able to detect edges in various directions.

We note that the sum in (10.34) is not defined for all $\mathbf{x} \in \mathbb{N}_m \times \mathbb{N}_n$, since $\mathbf{x} + \mathbf{y}$ may lie outside the domain of the image. There are several ways to handle this in practice. One option is to extend the image I outside of its domain by either padding with additional pixels, often zeros, or by reflecting the image about the boundary. Another option, which is used by default in packages like `pytorch`, is to simply reduce the size of the convolved image so that extensions and/or padding are not needed. In this case, the convolved image $I * W$ has size $(m - 2w) \times (m - 2w)$, since we must remove w pixels from each side of the image.

Convolution, or cross correlation, is one of the most basic image processing operations, and convolutional kernels can identify edges or other types of local geometric shapes in an image. Figure 10.9 gives an example of the cross correlation of a test image with three different convolutional kernels, of width $w = 1$, namely

$$W_H = \begin{pmatrix} 1 & 2 & 1 \\ 0 & 0 & 0 \\ -1 & -2 & -1 \end{pmatrix}, \qquad W_V = \begin{pmatrix} -1 & 0 & 1 \\ -2 & 0 & 2 \\ -1 & 0 & 1 \end{pmatrix}, \qquad W_L = \begin{pmatrix} -1 & -1 & -1 \\ -1 & 8 & -1 \\ -1 & -1 & -1 \end{pmatrix},$$

which, respectively, detect horizontal edges, vertical edges, and all edges. The last filter W_L can be identified with the result of applying a graph Laplacian to the pixel values in the image, where the graph is constructed by connecting each pixel to its 8 nearest neighbors. In the figure, we applied the convolutional filters to each color channel separately, so we interpret $d = 1$ in this setting.

In CNNs, the convolutional kernel width w is often much smaller than the size of the image, which introduces the other important property of convolutional neural networks: *spatial locality*. By taking the width w to be small (often $w = 1, 2$), the network is forced to learn localized features, such as edges or small shapes, in the early layers, which agrees with a rough understanding of how the human visual system works [178]. The translation equivariance and locality restrictions imposed on convolutional neural networks drastically reduce the number of parameters involved.

Example 10.9. A single 3×3 convolutional filter applied to a 28×28 pixel MNIST image has only 9 parameters, but it produces a 26×26 image as an output, so it can be compared to a fully connected neural network with $26^2 = 676$ neurons, which has $676 \times 784 \approx 500{,}000$ parameters (not counting the biases). Thus, the translation equivariance and filter locality drastically reduce the number of parameters in a convolutional layer. This property helps ensure that the convolutional layers learn features that generalize well to new data. ▲

In a CNN, the cross correlation operation replaces the linear mapping in the fully connected neuron (10.4). The output of the convolution $I * W$ is an image as well, and each layer can have any number of convolutional kernels, so the output of a given layer is simply an image with a possibly large number of channels. Just like with fully connected networks, CNNs use biases and activation functions, normally ReLU, after each layer, and can stack many layers deep.

CNNs also make use of a technique called *pooling*, which occurs between some of the layers. Pooling is a form of subsampling that reduces the resolution of the image by aggregating features locally. Common pooling strategies are max-pooling and average-pooling. Max-pooling by 2 in each direction corresponds to splitting the image into 2×2 pixel blocks, and replacing each block with a single pixel taking the maximum pixel value over the block. Average-pooling uses the average value instead. Pooling introduces invariance to small shifts and allows deeper layers to detect large scale features within the image while keeping the convolutional kernels small.

The output of the convolutional and pooling layers in a CNN is an image of reduced size, but with many channels; usually many more channels than were present in the original image. This image is then flattened into a long vector which is called the *convolutional features* of the image. The convolutional features are used for training the classifier. Usually the classifier is a small fully connected neural network, though in some cases it can just be a linear function. As we described in Section 10.1.3, for a classification task, the output of the network is usually fed into a soft-max function and negative log likelihood loss, and then the entire network — both the convolutional and fully connected layers — is trained using gradient descent, or some variant thereof, on the loss function.

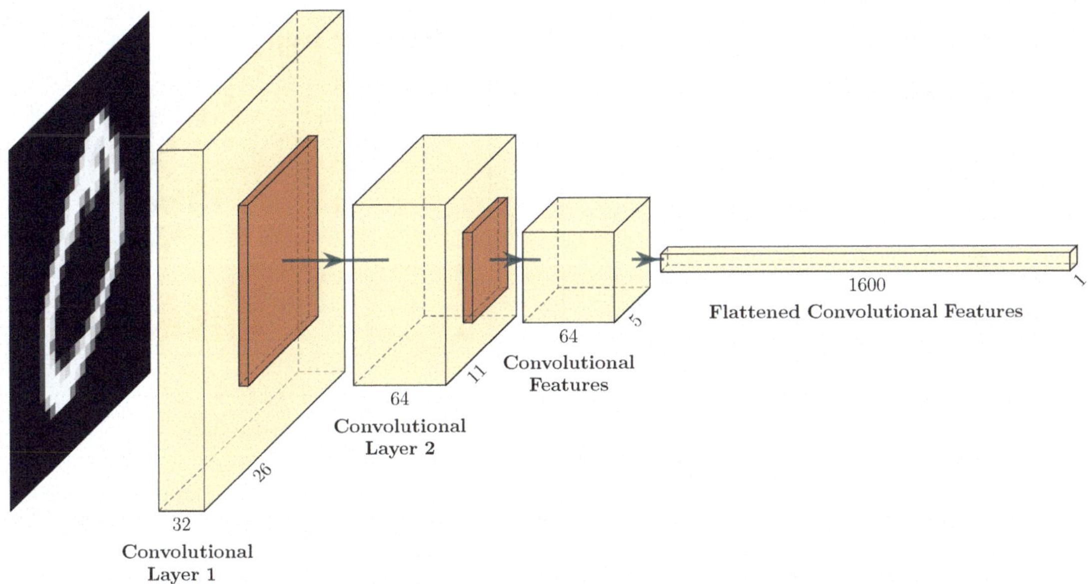

Figure 10.10: Example of a convolutional neural network architecture for MNIST digit classification. The flattened convolutional features are then fed into a fully connected neural network for to complete the classification.

Figure 10.10 gives a visual illustration of a simple 2-layer CNN that we will use for classifying MNIST digits, which are grayscale images of size 28×28. The first convolutional layer has 32 convolutional filters of size 3×3, so the output is of size $26 \times 26 \times 32$, and is fed into a max-pool on 2×2 blocks, reducing the size to $13 \times 13 \times 32$. This is then fed into a second convolutional layer with 64 convolutional filters of size $3 \times 3 \times 32$, resulting in an image of size $11 \times 11 \times 64$. This is then fed into another max-pool layer leading to a $5 \times 5 \times 64$ image as the output of the convolutional part of the network.[11] These convolutional features are then flattened into a vector of length $64 \times 5 \times 5 = 1600$ which is fed into a fully connected neural network with one hidden layer containing 1024 neurons, and outputs a vector in $\mathbb{R}^{10}$ since there are 10 classes in MNIST.

We used stochastic gradient descent (SGD) with batch size 64 and the Adadelta optimizer, though we could have equally well used the Adam optimizer (see Exercise 3.1). We trained for 20 epochs with a learning rate of $\alpha = 1$ and obtained a testing accuracy of between 99.3% and 99.4% at the end of training, depending on the random initialization of the weights, while accuracy was above 98% after the first epoch.[12] Figure 10.12 shows all of the misclassified test images. Figure 10.11(a) shows the accuracy during each mini-batch optimization step in training, where we note that one epoch corresponds to roughly 1000 iterations.

In order to understand more about how our CNN classifies images, we show in Figure 10.13 the outputs of the first and second convolutional layers *after* the ReLU activation, but before the max-pooling, for a given input image of a zero. We can see that the second layer is detecting edges in different directions at specific locations in the image, which is information that is undoubtedly useful for classifying MNIST digits.

---

[11] Note that when the size of the image is odd, 2×2 max-pooling omits the last column and row of the image.

[12] The best known result is 99.87% [35]. The other .13% are generally understood as incorrectly labeled, so there is not much chance of improving the result.

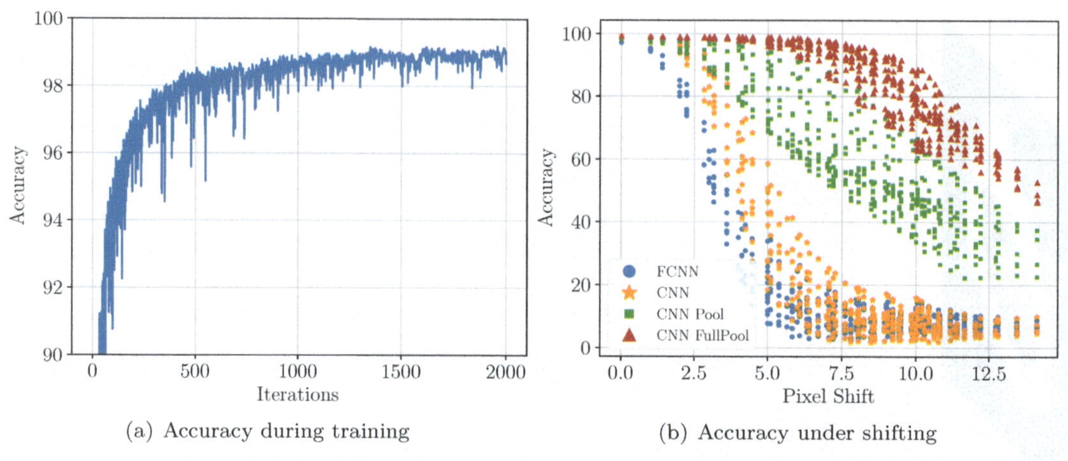

(a) Accuracy during training   (b) Accuracy under shifting

Figure 10.11: Plots of (a) the accuracy during training, where each iteration is a step of stochastic gradient descent, and (b), the accuracy on translated copies of the MNIST digits. In (b) we also compare to fully connected neural networks (FCNN) from Section 10.1.

Figure 10.12: All of the digits that were misclassified by our CNN trained for MNIST digit classification. There are a few consistent patterns; all of the 2's were misclassified as 7's, the last four 4's were misclassified as 9's, and all of the 5's except for the last (for which a 6 was predicted) and third from last (for which a 0 was predicted) were misclassified as 3's.

Translation Invariance

In Figure 10.11(b) we show the testing accuracy on shifted copies of the MNIST testing set, with shifts up to 10 pixels in both the vertical and horizontal directions; see Figure 10.14 for an example of various shifts of an MNIST digit. We considered shifts that were large enough to crop off parts of the MNIST digits, which can make classification challenging. We see in Figure 10.11(b) that our CNN is slightly better than the fully connected network from Section 10.1.3, but still suffers from poor accuracy after even small shifts of 2 or 3 pixels. This is

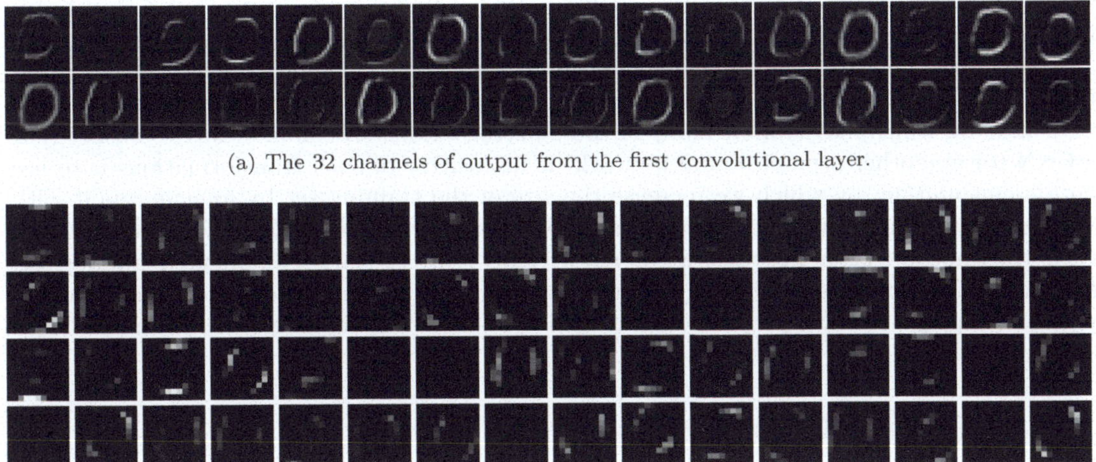

(a) The 32 channels of output from the first convolutional layer.

(b) The 64 channels of output from the second layer.

Figure 10.13: The outputs of the first and second convolutional layers, after the ReLU activation, but before max pooling, for an input image of the digit 0. Notice the channels in the second layer appear to be detecting edges in various directions.

Figure 10.14: Shifted, or translated, copies of an MNIST digit.

due to our use of very small max-pooling sizes and our choice to flatten the 5×5 images into convolutional features, which allows the network to use the precise locations of features for classification.

We modified the network in a few ways to show how additional translation invariance can be introduced. First, we added a 5×5 average pooling layer after the convolutional layers, so that the network cannot use the precision locations of features. This results in a length 64 convolutional features vector, and is called "CNN Pool" in Figure 10.11(b), which yields much better robustness to translations. Second, we considered an architecture that is completely translation invariant by removing the max-pooling layers, and padding the images by zeros so that they are not reduced in size by the convolutions. We increased the filter size to 7×7 to account for the lack of max-pooling layers, and increased the number of filters to 64 in the first layer and 128 in the second. As before, we added an average pooling of size 28 after the convolutional layers, so we have a length 128 convolutional feature vector. This network, called "CNN FullPool" in the figure, is completely translation invariant, with the exception of when translation clips parts of the depicted digit. This final network is the most robust with respect to shifts, with the accuracy decreasing by a non-negligible amount only after translation by 5 pixels or more, in which case parts of the digits begin to be clipped off.

Data Augmentation

As we described at the beginning of this section, there are many other transformations, aside from translations, that we would often like an image classification model to be insensitive

to, such as rotations or image resizing. It may also be desirable to ensure that a classifier is
insensitive to other perturbations of an image, such as small shifts in the color of an image,
adding noise or blurring, and reflecting or cropping images. We should expect a classifier that
is insensitive to the types of image perturbations seen practice to generalize better to new,
unseen, data, which may exhibit many of these perturbations. A common way to ensure that
a CNN (or any other type of neural network) is insensitive to such transformations is to use
data augmentation, by which we increase the size of the training set by *augmenting* it with
various perturbations and transformations of the images from the training set, and assigning
them the same labels. In practice, we do not actually store the larger augmented training set,
and instead, given a choice of image augmentations (e.g., horizontal reflection and Gaussian
noise), we apply a randomly chosen augmentation to each training image every time it is used
during training.

10.3.1   Transfer Learning

We can view the convolutional layers of a CNN as a *feature map* that maps an image into the
convolutional feature space, in which the classes are better separated, since the convolutional
features are more meaningful than pixel values for comparing images. In the case of our
simple CNN for MNIST, the convolutional layers provide a mapping $\phi \colon \mathbb{R}^{784} \to \mathbb{R}^{1600}$, so the
MNIST images are mapped into a *higher* dimensional feature space, in analogy with kernel
methods discussed in Section 7.6. The classification is then performed by a standard fully
connected neural network operating on the convolutional features of the image.

   We generally need very large image data sets to effectively train CNNs. On the other
hand, many practical real-world problems have only small amounts of labeled data. Since the
convolutional layers have very few parameters and are restricted to learn localized features,
it is often the case that the convolutional layers *generalize well* to other data sets that they
were not trained on. In other words, the feature map learned from one image classification
problem can be useful in another context. This is often used in practice with a very large image
classification problem like ImageNet, as was mentioned in the introduction to this chapter.
The convolutional features learned on ImageNet are often very useful in a wide variety of
image classification problems, and can be effectively *transferred* to other problems where very
little data is available. This is called *transfer learning* and it has proved to be very effective
in image classification.

   Transfer learning can be done in a couple different ways. The simplest is to simply use
a pre-trained network, say on ImageNet, as the starting point for training on a new image
classification problem. The weights in the fully connected part of the CNN are reinitialized,
and then we train for a small number of iterations, often called fine-tuning. It is also common
to freeze the convolutional part of the pre-trained network, and only train the fully connected
part, which is specific to the image classification task. This is more efficient, since the weights
in the convolutional filters do not need to be updated during training, and their gradients do
not need to be tracked with backpropagation. It can also be less likely to overfit, since the
model has fewer parameters.

   As a simple example of transfer learning, we consider transferring the convolutional features from our CNN for MNIST to another data set called FashionMNIST [249]. Fashion-
MNIST is a drop-in replacement for MNIST, except here the classes are different items of
clothing with pictures taken from a fashion catalog. Figure 10.15 shows an example of some of
the FashionMNIST images. Classification of the FashionMNIST dataset is more challenging
than MNIST, but is still a toy problem for modern deep learning. We employed transfer
learning in the second way described above, by fixing the convolutional part of the network
and only training the fully connected layers for classification. We achieved 86% accuracy after

Figure 10.15: Example of some images from the FashionMNIST dataset. Each image is a 28×28 pixel image of an item of clothing from a fashion catalog.

one epoch of training and 87% after 2 epochs. Training for much longer, the same 20 epochs as we did for MNIST, results in 89.62% accuracy. This is in contrast to training the neural network from scratch on FashionMNIST, which results in 90.89% accuracy. Transfer learning is also widely used in large language models, which are discussed in Section 10.5.

Exercises

3.1. ♡ Modify the Python notebook from this section to use the Adam optimizer instead of Adadelta and see if you can obtain similar results. You may have to adjust the learning rate to ensure the loss decreases during training.

3.2. ♡ How many parameters are in the convolutional part and the fully connected part of the CNN used for MNIST digit recognition in this section? Count just the weight matrices and ignore the biases, for simplicity.

3.3. Train the same CNN from this section on the FashionMNIST data set.

3.4. ♡ Modify the Python notebook from this this section so that the CNN can be trained for fully supervised classification on the CIFAR-10 data set, which has color images of size $32 \times 32 \times 3$, and train the network in a similar way as we did for MNIST (in particular, keep the size of the network roughly the same). What accuracy do you get?

3.5. ◊ Let $n \geq 3$ and let $\mathcal{N} = \{\mathbf{x} = (i,j) \,|\, 1 \leq i,j \leq n\}$ be an $n \times n$ grid representing the locations of pixels in an $n \times n$ image. Consider the unweighted graph $\mathcal{G}$ with vertex set $\mathcal{N}$ and edges between any two vertices $\mathbf{x}, \mathbf{y} \in \mathcal{N}$ for which $\|\mathbf{x} - \mathbf{y}\|_\infty \leq 1$. Show that the application of the graph Laplacian $L = D - A$ to a vector $\mathbf{u} \in \mathbb{R}^{n \times n}$ is equivalent to applying the Laplacian edge detector convolutional filter W_L depicted in Figure 10.9(d) at all interior pixels where $2 \leq i,j \leq n-1$.

3.6. Modify the code from this section to train a CNN to classify MNIST digits using a far smaller subset of the data set for training. Try around $m = 100$ labeled examples. Do you see overfitting? How does changing the architecture, in particular decreasing the number of parameters, affect the testing accuracy?

10.4   Graph Convolutional Neural Networks

Python Notebook: Intro to GCNs (.ipynb)

When a data set is endowed with the structure of a graph, weighted or unweighted, this additional information can be utilized to enhance the performance of neural networks. *Graph neural networks* refers to a wide and growing field concerned with how to effectively utilize graph structures, like the adjacency or weight matrix, within a neural network architecture. Graph neural networks are applied to many different types of machine learning problems, including graph classification, semi-supervised learning, as described in Section 9.9, graph embeddings and visualizations, and graph generation. Graph classification and generation find important applications in drug discovery and molecular property prediction [27, 118].

This section is focused on *graph convolutional neural networks* (GCN). Let $\mathcal{G}$ be a weighted graph with m nodes and symmetric weight matrix $A \in \mathcal{M}_{m \times m}$. We slightly abuse notation here by using the letter A for the graph weight matrix, which is usually the adjacency matrix, since we will reserve W for the neural network weight matrices. Let $\mathbf{x}_1, \ldots, \mathbf{x}_m \in \mathbb{R}^n$ be associated node feature vectors. In general, the weight matrix and node features may not be directly related, i.e., the weight matrix is generally not just a k-nearest neighbor graph constructed over the node features. As an example, we recall the PubMed graph introduced in Section 9.1.1 where the nodes are academic papers and the graph represents citations between papers, while the node feature $\mathbf{x}_i$ is a summary encoding of the abstract of paper i. The goal is to utilize both the node features vectors $\mathbf{x}_i$ and the graph structure in a neural network architecture. The reader should note that this setting is different from Section 9.9 where we used the graph structure alone, and did not incorporate any additional node information.

As an application we will focus on semi-supervised learning, in order to extend the ideas introduced in Section 9.9 to the neural network setting. However, the applications of graph convolutional neural networks are broader, as discussed above. Recall that in semi-supervised learning, we usually have access to a limited amount of training data. Thus, we cannot expect a model with a large number of parameters to be successful. The convolution operation on images used in Section 10.3 exploits translation equivariance of natural images to construct a neural network layer — the convolutional layer — that commutes with translations, extracts local features from an image, and has very few parameters compared to fully connected layers, thus generalizing well for image data. It is natural to seek an analogous convolution operation on graphs that has these main properties: is compatible with the graph structure, captures local graph information, and has relatively few parameters.

Since image convolution averages information from neighboring pixels, a very natural extension to graphs is to average information from neighboring nodes, as we did in label propagation in Section 9.9.2. This corresponds to the *aggregation operation*

$$\mathbf{z}_i = \frac{1}{d_i} \sum_{j=1}^{m} a_{ij} \mathbf{x}_j, \qquad \text{where} \qquad d_i = \sum_{j=1}^{m} a_{ij} \tag{10.35}$$

is the degree of node i, cf. Definition 9.7. The degrees are the entries of the degree vector $\mathbf{d} = (d_1, \ldots, d_m)^T \in \mathbb{R}^m$, and $D = \operatorname{diag} \mathbf{d}$ is the corresponding $m \times m$ diagonal degree matrix. If we stack all the feature vectors together into an $m \times n$ data matrix X — see (7.1) — whose i-th row is $\mathbf{x}_i^T$, then we can write (10.35) as

$$Z = D^{-1} A X, \tag{10.36}$$

where Z is the corresponding data matrix for $\mathbf{z}_1, \ldots, \mathbf{z}_m$. We also note that the linear part of a fully connected neural network layer, as in (10.4), applied to the node features directly can be written as $Z = XW$, where W is an $n \times k$ weight matrix mapping features vectors from $\mathbb{R}^n$ to $\mathbb{R}^k$, whose weights are all tunable parameters. In this section, we omit the bias $\mathbf{b}$ for simplicity, noting that X can be augmented with a column of ones to implicitly include a bias; see Exercise 1.2. Thus, graph operations like the aggregation in (10.36) operate on the *left* of X, since they act on the graph structure, while operations directly on the feature vectors themselves operate on the *right*.

A natural way to combine the aggregation (10.36) with a neural network structure is to do both operations at the same time, followed by an activation function. This leads to our first graph convolutional layer

$$Z = \sigma(D^{-1}AXW), \tag{10.37}$$

with tunable weight matrix W. The layer (10.37) is a basic form of what is called a *graph convolutional layer*, and intuitively it is simply applying a linear function to the node feature vectors — with the same matrix W for all nodes — followed by averaging, or diffusing, the results over the graph.[13] The graph convolutional layer (10.37) is a special case of what are called *message passing neural networks* [248, 265], since the aggregation step can be viewed as each node sending messages to its graph neighbors. Note that X is an $m \times n$ matrix and Z is an $m \times k$ matrix; the length of the feature vectors may change, but the first dimension remains the same since the graph has m nodes.

Remark 10.10. It is very common to add *self-loops* to the adjacency/weight matrix A, so we replace A by $A + I$ and update D accordingly in (10.37). This is called the *renormalization trick* in [128] and it improves results by allowing the local averaging in (10.35) to average with the self node $\mathbf{x}_i$, instead of relying solely on neighbors. Indeed, the renormalization trick is equivalent to changing (10.35) to

$$\mathbf{z}_i = \frac{1}{d_i + 1}\left(\mathbf{x}_i + \sum_{j=1}^{m} a_{ij}\mathbf{x}_j\right). \tag{10.38}$$

We also point out that for a graph with self loops and no edges, so $A = I = D$, the convolutional layer (10.37) reduces to a fully connected layer (10.4) without bias. We will work with the graph weight matrix A throughout this section, and assume the renormalization trick has already been applied, if desired. ▲

10.4.1 Convolution on Graphs

We now take some time to study convolution on graphs in more detail. Our goal is to provide a rigorous justification for calling $D^{-1}A$ a *graph convolution* and explore what other types of convolution operations we can define on graphs. Here we will work with the symmetric normalized graph Laplacian $L_{\mathrm{sym}} = I - D^{-1/2}AD^{-1/2}$, as defined in (9.65). Let $L_{\mathrm{sym}} = Q\Lambda Q^T$ be its spectral decomposition (5.31), so Q is the matrix whose columns are its orthonormal eigenvectors, and Λ the corresponding diagonal eigenvalue matrix.

Motivated by the connection between the discrete Fourier transform (DFT) and the circular graph, Definition 9.87 proposes that the graph-based DFT $\mathbf{c}$ of a data vector $\mathbf{x}$ is obtained by multiplication with the transpose of its graph Laplacian eigenvector matrix: $\mathbf{c} = Q^T\mathbf{x}$. The entries of $\mathbf{c}$ are the coordinates of $\mathbf{x}$ in the orthonormal eigenvector basis, i.e., $c_i = \mathbf{q}_i \cdot \mathbf{x} = \mathbf{q}_i^T\mathbf{x}$.

---

[13]Of course, by associativity of matrix multiplication, we can view the graph convolutional layer in the opposite way, as averaging over the graph followed by applying the matrix W.

Recall further the formula (9.212) that expresses the convolution of two data vectors in terms of the Hadamard product of their DFT's. Motivated by this connection between the DFT and convolution, we make the following definition.

Definition 10.11. Given a graph $\mathcal{G}$ with graph Laplacian eigenvector matrix Q, the *spectral graph convolution* of $\mathbf{x}, \mathbf{y} \in \mathbb{R}^m$ is defined by

$$\mathbf{x} *_{\mathcal{G}} \mathbf{y} = Q\left(Q^T\mathbf{x} \circ Q^T\mathbf{y}\right). \tag{10.39}$$

To simplify the convolution equation (10.39), we parameterize by the weight vector $\mathbf{w} = Q^T\mathbf{y}$ instead of $\mathbf{y}$, yielding the spectral graph convolution formula

$$\mathbf{x} *_{\mathcal{G}} \mathbf{y} = Q\left(\mathbf{w} \circ Q^T\mathbf{x}\right) = Q\left(\operatorname{diag} \mathbf{w}\right) Q^T\mathbf{x}. \tag{10.40}$$

We recall from Lemma 9.88, see (9.217), that we derived this same expression for $A\mathbf{x}$, where A is any matrix that commutes with the graph Laplacian (and $\mathbf{w}$ depended on A). Thus, we may think of graph convolutions, as per Definition 10.11, as precisely the class of linear operations on graphs that commute with the graph Laplacian — in this case the symmetric normalized Laplacian L_{sym}.

The spectral convolution can be used in a graph neural network architecture, but there are some disadvantages. It requires computing all of the eigenvectors of the graph Laplacian, which may be computationally hard for large graphs, the number of parameters m in the parameter vector $\mathbf{w}$ is still very large, and the spectral convolution is not *localized* on the graph in any way. One approach to overcoming some of these issues is to use only the first k eigenvectors of L_{sym} in the definition of the convolution, that is, we use the truncated spectral convolution

$$\mathbf{x} *_{\mathcal{G}_k} \mathbf{y} = Q_k\left(\mathbf{w} \circ Q_k^T\mathbf{x}\right), \tag{10.41}$$

where the columns of Q_k are the first k eigenvectors[14] of L_{sym}, and $\mathbf{w} \in \mathbb{R}^k$, which can have far fewer than m parameters. While this helps with computational issues, it still requires computing eigenvectors, which may be undesirable, and we have not addressed the locality issue. Indeed, any notion of convolution that is based on the spectrum of the graph Laplacian is inherently nonlocal and uses information from the entire graph.

It turns out we can express the spectral graph convolution in a way that can be made as local as we desire, by use of the following result.

Lemma 10.12. *Assume the eigenvalues of L_{sym} are distinct. Let $\mathbf{w} \in \mathbb{R}^m$, and set $\mathbf{y} = Q\mathbf{w}$. Then there exists a polynomial*

$$p(x) = c_0 + c_1 x + \cdots + c_{m-1} x^{m-1}$$

of degree $\leq m - 1$, depending on the eigenvalues of L_{sym} and $\mathbf{w}$, such that

$$\mathbf{x} *_{\mathcal{G}} \mathbf{y} = p(L_{\text{sym}})\mathbf{x} = \sum_{j=0}^{m-1} c_j L_{\text{sym}}^j \mathbf{x}. \tag{10.42}$$

Proof. Since the eigenvalues of L_{sym} are distinct, and there are m of them, according to Exercise 2.11, there exists an interpolating polynomial p of degree $\leq m-1$ such that $p(\lambda_i) = w_i$

---

[14]As in Chapter 9, we order the eigenvalues from smallest to largest.

for all $i = 1, \ldots, m$, and hence $p(\Lambda) = \operatorname{diag} \mathbf{w}$. Recalling that $L_{\mathrm{sym}}^j = Q \Lambda^j Q^T$, we have

$$p(L_{\mathrm{sym}}) = \sum_{j=0}^{m-1} c_j L_{\mathrm{sym}}^j = Q \left(\sum_{j=0}^{m-1} c_j \Lambda^j \right) Q^T = Q\, p(\Lambda)\, Q^T = Q\, (\operatorname{diag} \mathbf{w})\, Q^T.$$

It follows that

$$\sum_{j=0}^{m-1} c_j L_{\mathrm{sym}}^j \mathbf{x} = Q\, (\operatorname{diag} \mathbf{w})\, Q^T \mathbf{x} = Q\, (\mathbf{w} \circ Q^T \mathbf{x}) = \mathbf{x} *_{\mathcal{G}} \mathbf{y}. \qquad \blacksquare$$

Remark 10.13. Lemma 10.12 allows us to write the spectral graph convolution in a form that depends only on powers of the graph Laplacian. This is quite natural in light of the discussion in Section 9.10.5, and in particular Lemma 9.88, which showed how the convolution operation can be interpreted as the family of linear operations that *commute* with the graph Laplacian, a property that is clearly satisfied by its powers as well as any linear combination thereof, i.e., polynomials $p(L_{\mathrm{sym}})$. $\blacktriangle$

By Lemma 10.12, we can compute convolutions on graphs without computing eigenvectors of L_{sym}. The coefficients c_j of the polynomial p may depend in a complicated way on $\mathbf{w}$ and L_{sym}, but we may now just treat them as learnable parameters and forget about $\mathbf{w}$. Since we take powers of L_{sym} up to order $m - 1$, the operation is still not localized on the graph, and involves a large number of parameters m. However, this can be easily addressed by truncating the polynomial to lower degree.

> **Definition 10.14.** Let $1 \leq k \leq m - 1$. The k-th order *graph convolution* of $\mathbf{x} \in \mathbb{R}^m$ with the convolutional kernel $\mathbf{w} \in \mathbb{R}^{k+1}$ is defined by
>
> $$\mathbf{x} *_k \mathbf{w} = \sum_{j=0}^{k} w_{j+1} L_{\mathrm{sym}}^j \mathbf{x}. \qquad (10.43)$$

The k-th order convolution has only $k + 1$ parameters $\mathbf{w} \in \mathbb{R}^{k+1}$, where the user has the freedom to choose k. Observe that, since the nonzero non-diagonal entries of the graph Laplacian correspond to edges, it is *localized* on the graph to include information only from nodes that are reachable within k hops — that is, nodes that are reachable by a path containing at most k edges.

Example 10.15. Consider the first order, so $k = 1$, graph convolution with $\mathbf{w} = (1, -1)^T$. Then

$$\mathbf{x} *_k \mathbf{w} = w_1 \mathbf{x} + w_2 L_{\mathrm{sym}} \mathbf{x} = \mathbf{x} - \mathbf{x} + D^{-1/2} A D^{-1/2} \mathbf{x} = D^{-1/2} A D^{-1/2} \mathbf{x}.$$

Thus, the aggregation operation $D^{-1/2} A D^{-1/2} \mathbf{x}$, which is the type of graph convolution used in one of the seminal works [128], is a first order convolution. Note that if we had defined the spectral graph convolution using L_{rw} instead of L_{sym}, we would have obtained the convolution operation $D^{-1} A \mathbf{x}$ used at the beginning of this section in (10.36). $\blacktriangle$

Example 10.16. Referring back to Section 9.6, the symmetric graph diffusion equation $\mathbf{u}_{k+1} = \mathbf{u}_k - \alpha L_{\mathrm{sym}} \mathbf{u}_k$, applied for k steps with $\mathbf{u}_0 = \mathbf{x}$, can be viewed as a k-th order graph convolution since

$$\mathbf{u}_k = (I - \alpha L_{\mathrm{sym}})^k \mathbf{x} = p(L_{\mathrm{sym}}) \mathbf{x},$$

where $p(x) = (1 - \alpha x)^k$ is a k-th order polynomial. $\blacktriangle$

While the localized convolution in Definition 10.14 seems to solve all of our problems, there is a numerical issue with (10.43). We recall from Section 9.6 that the eigenvalues of L_{sym} lie in the interval $[0, 2]$, so taking powers of L_{sym} is numerically unstable. Furthermore, the most important eigenvalues for learning large scale and important graph structures are near 0, and will be heavily attenuated for larger powers. In contrast, the large eigenvalues near 2, which correspond to high frequency information, will be amplified, which may be undesirable.[15] We recall from Section 9.6 that diffusion on graphs corresponds to powers of the matrix $I - L_{\text{sym}}$, whose eigenvalues lie in $[-1, 1]$, and so taking powers is numerically stable. Furthermore, the eigenvalues near 1 are the most important for understanding the large scale structural properties of the graph. Thus, it is better for numerical stability and interpretability to parametrize the k-th order polynomial in terms of diffusion propagators, as

$$\mathbf{x} *_k \mathbf{w} = \sum_{j=0}^{k} w_{j+1}(I - L_{\text{sym}})^j \mathbf{x}. \tag{10.44}$$

This approach was taken in [83] to improve the original spectral convolutional method in [55].

The method introduced in [55] addressed this numerical issue through the use of an alternative polynomial approximation, namely

$$g_{\mathbf{w}}(\Lambda) = \sum_{j=0}^{k} w_{j+1} T_j(\widetilde{\Lambda}), \qquad \text{where} \qquad \widetilde{\Lambda} = \frac{2}{\lambda_{max}(L_{\text{sym}})}\Lambda - I, \tag{10.45}$$

and T_j is the j-th order *Chebyshev polynomial* (see also Section 11.3), which is defined by the recursive formula

$$T_0(x) = 1, \qquad T_1(x) = x, \qquad T_j(x) = 2x\, T_{j-1}(x) - T_{j-2}(x), \qquad j \geq 2. \tag{10.46}$$

The Chebyshev polynomials form an orthogonal basis for real-valued functions on the interval $[-1, 1]$ and serve as an alternative to the standard monomial, [181]. Notice the role of $\widetilde{\Lambda}$ is to map the eigenvalues to the interval $[-1, 1]$.

10.4.2 Numerical Experiments

We now turn to some numerical experiments. We use a two hidden layer GCN

$$F(X) = \sigma\big(D^{-1}A\,\sigma(D^{-1}AXW_1)\,W_2\big)W_3, \tag{10.47}$$

based on (10.36), and consider semi-supervised node classification problems with c classes. The dimensions of the tunable weight matrices W_1 and W_2 are $n \times n_h$ and $n_h \times n_h$, where n_h denotes the number of *hidden nodes* in the network. The dimensions of the final weight matrix W_3 are $n_h \times c$, so that $Z = F(X)$ is an $m \times c$ matrix, whose i-th row is the output of the network corresponding to the i-th node in the graph, which we write as $\mathbf{z}_i^T$, where $\mathbf{z}_i \in \mathbb{R}^c$. The components of $\mathbf{z}_i$ are scores for each class, and the label prediction is given by the class with the largest score. As in Section 10.1.3, the loss function is the composition of a soft-max with the negative log likelihood loss. The difference here compared to fully supervised learning is that the loss applies only to the *training* data on the graph, which is often a small subset of the entire graph. The loss is minimized by gradient descent using the Adam optimizer.

---

[15] Recent work on *heterophyllic graphs*, where there are many edges between nodes from different classes, makes use of the highest frequency eigenvectors, and oftentimes the spectral form of convolution; see [145, 254, 264, 268] for details.

Algorithm:	MLP	Laplace	Laplace with diffusion maps	GCN
Accuracy:	65.8 (2.5)	46.9 (10.1)	71.7 (2.7)	76.1 (2.1)

Table 10.16: Accuracy of various methods for semi-supervised node classification on the PubMed graph using 58 labeled examples. The accuracies are averaged over 100 trials randomizing the training set, and the standard deviations are shown in parentheses.

We applied GCN for semi-supervised node classification on the PubMed data set with 58 labeled examples, which is the same label rate of 0.3% used in [128]. Recall that the PubMed data set has $m = 19,717$ nodes. We used $n_h = 50$ hidden nodes in order to keep the model's complexity small, and trained with the Adam optimizer with a learning rate of 0.01 for 100 epochs. We used early stopping with a validation set of size 500 to select the best model over the 100 epochs. The training takes under one second to complete on a computer with a single GPU.

We show in Table 10.16 the accuracy of the GCN compared to a fully connected neural network, denoted MLP for multilayer perceptron, with the same number of hidden nodes and layers, as well as against Laplace learning with and without the diffusion maps preprocessing as discussed in Section 9.9. The GCN method outperforms the other methods, none of which combine the node features *and* graph structure together — Laplace learning uses only the graph structure, and MLP operates only on the node features. We note that the results reported in [128] are slightly higher, at 79%, using the same label rate and a similar architecture. The authors in [128] use additional regularization techniques, including dropout, and perform hyperparameter optimization using a validation set.

A GCN can also be used to construct node embeddings of graphs. We can do this by training the GCN on a semi-supervised node classification task, and then taking the output of an intermediate layer as the node embedding. Here, we consider the output

$$Z = D^{-1}A\,\sigma(D^{-1}AXW_1)\,W_2.$$

of the second layer, before the activation. The matrix Z is $m \times n_h$, and, for each node i in the graph, gives the node embedding $\mathbf{z}_i \in \mathbb{R}^{n_h}$ for the GCN (10.47).

In Figure 10.17 we show the GCN embeddings of the karate club, political books, and PubMed graphs introduced in Section 9.1.1. In each case, the embeddings are reduced to two-dimensions (from R^{n_h}) via PCA for visualization. For karate club and political books, we labeled one example per class, shown by the red dots in the figure. The PubMed embedding is from one trial of the semi-supervised learning experiment conducted above where the accuracy is roughly 78%. We observe that the node embeddings separate the different classes well. The karate club and political books graphs do not have feature vectors attached to the nodes, and we set the feature matrix $X = I_m$ to be the $m \times m$ identity matrix, assigning each node a unique one hot vector.

Remark 10.17. It is important to note that the dimensions of the learned weight matrices W_1 and W_2 in (10.47) depend only on the dimension of the node features n and the chosen number of hidden nodes n_h. In particular, their size is independent of the number of nodes in the graph, as well as any other properties of the graph. This means that the graph convolutional neural network (10.47) can be *transferred* to any other graph that has node features of the same dimensions. Of course, the network may not perform well on other graphs, unless the node features and graph structures have a sufficient amount of similarity to the graph the network was trained on. ▲

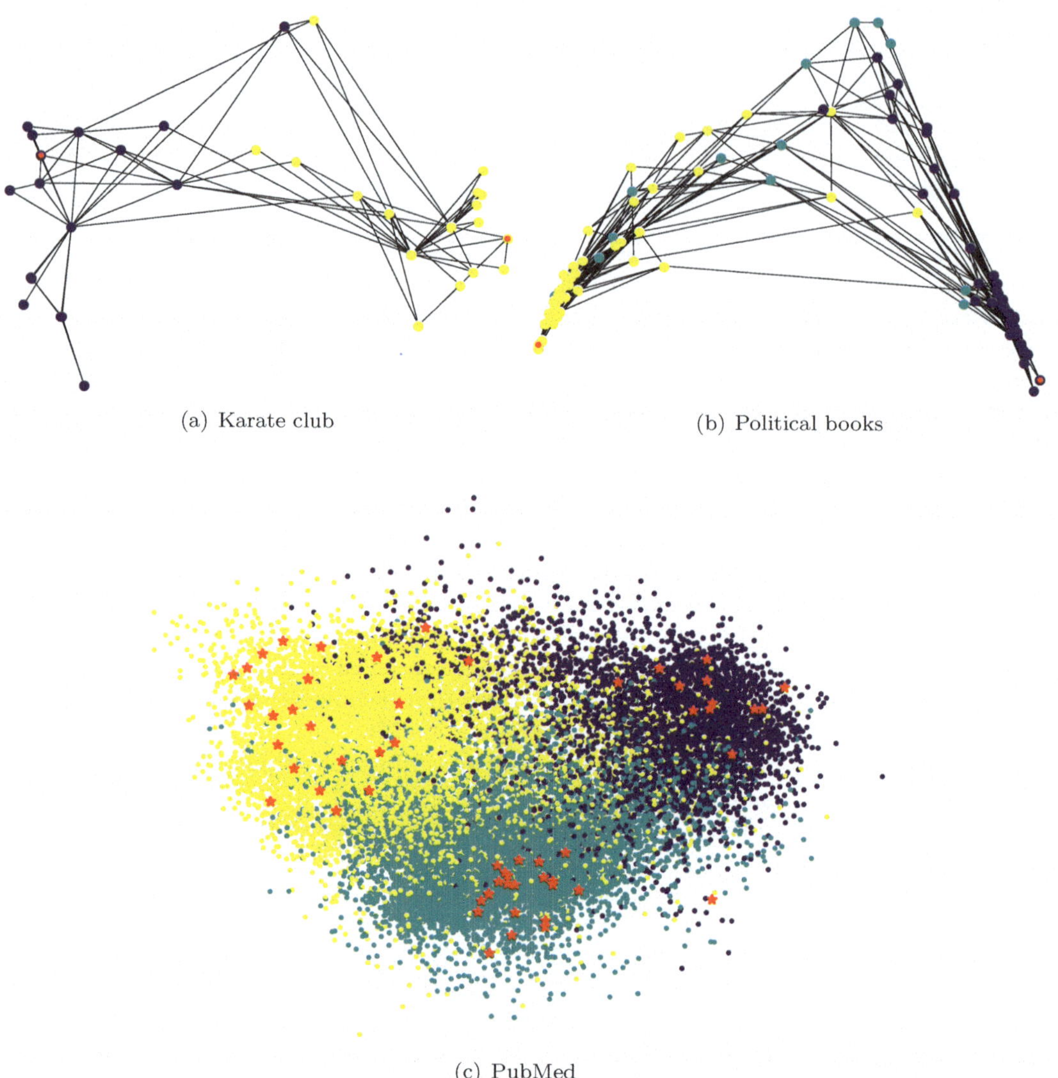

(a) Karate club

(b) Political books

(c) PubMed

Figure 10.17: Node embeddings obtained by the GCN (10.47) on the karate club graph, political books graph, and PubMed. The labeled nodes are indicated by red dots in the karate club and political books graphs, and by red stars for PubMed. There are too many edges in PubMed to display in an informative way.

We also mention that in other applications, like graph classification where a single prediction is made over the whole graph, we need a notion of *graph pooling*, analogous to the pooling operation used in convolutional neural networks in Section 10.3. Common approaches include using techniques like spectral clustering, either binary or multiclass — see Sections 9.4 and 9.7.2 — to hierarchically decompose the graph into groups, and run the pooling operation by reversing this decomposition. There is also a much wider class of graph neural networks that use aggregation techniques that are learned and vary over the graph; see [248, 265] for surveys.

Exercises

4.1. ♡ What conditions could you place on $\mathbf{w}$ in Lemma 10.12 so that the result (10.42) remains true even when the eigenvalues of L_{sym} are not distinct?

4.2. ♡ Show that the spectral graph convolution in Definition 10.11 can be written as

$$\mathbf{x} *_{\mathcal{G}} \mathbf{y} = \sum_{i=1}^{m} (\mathbf{q}_i \cdot \mathbf{x})(\mathbf{q}_i \cdot \mathbf{y}) \mathbf{q}_i. \tag{10.48}$$

4.3. ◇ Does the spectral graph convolution depend on the *choices* made in diagonalizing L_{sym}? When the eigenvalues are distinct, we can choose $\mathbf{q}_i$ or $-\mathbf{q}_i$, while when the eigenvalues have higher multiplicity, there are more choices that can be made.

4.4. Use `pytorch` to experiment with GCNs for semisupervised learning on PubMed. Start with the notebook from this section.
   (a) Try a deeper GCN, with more than 5 layers. How does performance change?
   (b) Try replacing $D^{-1}A$ with $D^{-1/2}AD^{-1/2}$ in the definition of convolution.
   (c) Experiment without the renormalization trick (i.e., do not add self-loops).

4.5. ♡ Implement a diffusion GCN, which uses diffusion (10.44) as the convolution in each layer. This has tunable parameters w_j that need to be learned. Write your code where the degree of the convolution k can be arbitrarily chosen. You may want to use `nn.Parameter` to define the weights w_j in the neural network. For simplicity, use the same convolution in each layer. Try this on the PubMed data set, and compare to the Python notebook from this section.

4.6. Try using a semi-supervised GCN for node embeddings on the MNIST data set. How do the embeddings compare to other techniques, like PCA, MDS, t-SNE, etc.?

10.5   Transformers and Large Language Models

We now turn to a brief study of the applications of neural networks to *natural language processing*, which is concerned with enabling machine learning to understand, interpret, and generate human language. Applications include language translation, chatbots, auto-correct, text summarization, and many more. The main topic of this section is the *transformer* neural network architecture, which is used by modern large language models, and powers chatbots like OpenAI's ChatGPT [2].

In order to apply machine learning, language models use *tokenization* to convert words into tokens, which are represented by integers $1, \ldots, M$, where M is the number of tokens. Oftentimes a whole word is a token, while some larger words can be broken up into several tokens. The coloring of the phrase below shows how it would be tokenized by GPT-4:

Large language models use tokenization to represent words.

Breaking longer words into tokens allows language models to understand composite words, like "overpythonized", even if they do not appear in any dictionary. Code from programming languages as well as mathematics is also tokenized in a similar way. After the text is tokenized, the tokens, which are just integers, are prescribed by vector embeddings in $\mathbb{R}^n$. The token embeddings are organized into an $M \times n$ matrix E whose i-th row $\mathbf{x}_i^T$ represents the vector embedding of the i-th token. The transformer model uses *learned* token embeddings, so the matrix E is initialized at random and its values are free parameters that are learned during training, along with the rest of the model that is described below. We can view this initial vector embedding of the tokens as a standard linear neural network layer applied to the one-hot vector representation of the tokens; see Exercise 5.1. Throughout this section, we will work with the vector token embeddings $\mathbf{x}_1, \mathbf{x}_2, \mathbf{x}_3, \cdots \in \mathbb{R}^n$.

Large language models are built upon pretrained generative machine learning models that learn the statistical properties of the time series so that they can *generate* new samples $\tilde{\mathbf{x}}_1, \tilde{\mathbf{x}}_2, \tilde{\mathbf{x}}_3, \ldots$ that have the same statistical properties and are ideally indistinguishable from the training data $\mathbf{x}_1, \mathbf{x}_2, \mathbf{x}_3, \ldots$. To do this, *generative language models* are trained on the task of *predicting* the next token in the sequence. For example, given the phrase "The quick brown fox jumped over the", the model would attempt to predict the next word "lazy". Mathematically, we are given tokens $\mathbf{x}_1, \ldots, \mathbf{x}_m$ and the goal is to predict the next token $\mathbf{x}_{m+1}$. In practice, we cannot use the entire history of m tokens, and instead, at time $t \in \mathbb{N}$ we use the previous m tokens $\mathbf{x}_{t-m}, \mathbf{x}_{t-m+1}, \ldots, \mathbf{x}_{t-1}$ to predict $\mathbf{x}_t$. The previous m tokens are called the *context* while m is the *context size*. Generative large language models are trained over massive text data sets that contain much of the content on the internet, as well as various textbooks, until they are able to accurately guess the next word in any phrase. The language model is essentially trained to be a very powerful version of auto-correct.

To be precise, language models do not actually predict tokens (i.e., words or sub-words), but rather they predict *probabilities*. That is, the prediction of the next token is a probability vector $\mathbf{p} \in \mathbb{R}^M$ over a dictionary of M possible tokens. The model is trained by minimizing a loss that compares $\mathbf{p}$ to the probability vector concentrated on the true next word, namely $\mathbf{e}_i$ if the next word is the i-th word in our dictionary. As with classification, the output of the network $\mathbf{p}$ is usually compared to $\mathbf{e}_i$ using the negative log likelihood loss. When a pretrained large language model is applied to *generate* new text, the tokens that are produced are sampled at random from the predicted probability vectors, so the output of the model is random and may be different every time it is sampled. Furthermore, language models are *autoregressive*, meaning that when they generate new tokens, the context $\mathbf{x}_{t-m}, \mathbf{x}_{t-m+1}, \ldots, \mathbf{x}_{t-1}$ used to generate $\mathbf{x}_t$ consists of tokens that were previously generated by the model.

Recurrent neural networks (RNNs) are one of the oldest approaches to natural language processing [107, 113]. A very simple (memoryless) RNN has the form

$$\mathbf{p}_t = F(\mathbf{x}_{t-m}, \mathbf{x}_{t-m+1}, \ldots, \mathbf{x}_{t-1}), \tag{10.49}$$

where F is parameterized as a feedforward neural network, and $\mathbf{p}_t$ is the prediction output from the network, which in the example above is the probability vector predicting the next token $\mathbf{x}_t$. A drawback of memoryless RNNs is that the context size m cannot be too large, so the model cannot learn long range context dependencies, which are crucial for understanding human language. This can be addressed to a degree by adding a *hidden state* to the RNN, which allows it to have a longer term memory. An RNN with hidden state has the form

$$(\mathbf{p}_t, \mathbf{h}_t) = F(\mathbf{x}_{t-m}, \mathbf{x}_{t-m+1}, \ldots, \mathbf{x}_{t-1}, \mathbf{h}_{t-1}), \tag{10.50}$$

where $\mathbf{h}_t$ is the hidden state, which is propagated from one step to the next. Long Short Term Memory (LSTM) recurrent neural networks introduced in [107] are one of the most commonly used types of RNNs with hidden states and memory. In theory, LSTMs can model very long distance relationships in language, but in practice this feature is limited since it is challenging to know precisely what information to keep in the hidden state for future recall. It is also challenging to scale up the training of LSTMs to massive data sets, since the recursive structure of (10.50) renders it difficult to train in parallel.

The recent striking advances in large language models are due to the *transformer* architecture, introduced in [237]. This architecture dispenses with the recursive structure of the LSTM, so at a high level has the form (10.49), where the context size m can be extremely large. For example, the current versions of some large language models have a context size of up to $m = 128 \times 10^3$ tokens, which may be expected to increase even further in the near future. By dispensing with the recursive architecture, transformers can be trained in parallel, allowing them to be scaled up to massive data sets. The large context size m in transformers necessitates that we construct the function F in (10.49) in an intelligent way, and not just with a vanilla fully connected neural network. The transformer does this with an *attention mechanism* and *positional encoding*. Neither idea is new to the transformer architecture; in fact, the attention mechanism was used previously in the context of long short term memory (LSTM) recurrent neural networks [41,93]. Combining these ideas, and only these ideas, into a flexible and scalable neural network architecture is the transformer's key advance. Furthermore, as we will see below, since the positional information is encoded into the tokens themselves, and not explicitly part of the function F, the transformer architecture also works for *variable context sizes*; that is, the m in (10.49) does not need to be fixed during training, and models trained on one context size can generalize to others.

10.5.1 The Attention Mechanism

The crucial idea underlying transformers is the *attention mechanism*, which allows the model to pay attention to certain pairs of words in the context window. Consider the task of completing the sentence below:

> I walked through a field on a clear and sunny day. The color of the sky was

Most people would say "blue" is a good guess for the next word. In order to make this prediction correctly, there are certain words that are important to *pay attention to* in the context window above. Perhaps the most important are "color" and "sky", although this may be insufficient, since the sky can also be black at night and sometimes gray or red due to certain weather conditions. Paying attention in addition to the words "clear", "sunny", and "day" gives more context and suggests that "blue" is more likely than "red" or "gray". Other words like "I", "walked", and to some extent "field", are less important and do not deserve as much attention. Of course, the ordering of the words is also extremely important; if the phrase were instead

> I walked through a field on a clear and sunny day. The sky was the color of

then the prediction would be substantially different, perhaps "faded denim". For the moment we will ignore the ordering of tokens and focus on the attention mechanism between pairs of tokens.

Assume our context window has m tokens represented by vectors $\mathbf{x}_1, \ldots, \mathbf{x}_m \in \mathbb{R}^n$. The self-attention layer in a transformer produces a new set of vectors[16] $\mathbf{z}_1, \ldots, \mathbf{z}_m \in \mathbb{R}^n$. The transformed vectors are constructed through an attention mechanism that learns to pay attention to certain pairs of words. The *self-attention* between tokens i and j is defined by

$$a_{ij} = \exp\left(\beta\, \mathbf{x}_i^T W \mathbf{x}_j\right), \tag{10.51}$$

where $W \in \mathcal{M}_{n \times n}$ is a learnable weight matrix, and $\beta \in \mathbb{R}$ is a positive scale factor that we'll specify later. Self-attention allows the model to *learn*, through the matrix W, which pairs

---

[16]In general, the dimension n of the $\mathbf{z}_i$ may be different, but we keep it the same for simplicity.

of tokens are important to *pay attention* to. We may think of $\mathbf{x}_i^T W \mathbf{x}_j$ as an inner product between $\mathbf{x}_i$ and $\mathbf{x}_j$, except that in general the matrix W is neither symmetric, nor positive definite.

If we assemble the attentions a_{ij}, which are, by construction, nonnegative, into a matrix $A \in \mathcal{M}_{m \times m}$, then we essentially have the structure of a *complete weighted graph* over the context window, where the nodes in the graph are the m context tokens $\mathbf{x}_1, \ldots, \mathbf{x}_m$, and the edge weights a_{ij} are the self-attentions. The transformer layer is then equivalent to the graph convolutional layer we introduced in Section 10.4. That is, the tokens $\mathbf{z}_1, \ldots, \mathbf{z}_m$ are computed via

$$\mathbf{z}_i = \frac{1}{d_i} \sum_{j=1}^{m} a_{ij} V^T \mathbf{x}_j, \qquad d_i = \sum_{j=1}^{m} a_{ij}, \tag{10.52}$$

where $V \in \mathcal{M}_{n \times n}$ is another tunable weight matrix. That is, $\mathbf{z}_i$ is the weighted average of the linearly transformed tokens $V \mathbf{x}_j$, weighted by the attention scores. As in Section 10.4, we can write this in matrix form, by defining the diagonal degree matrix $D = \operatorname{diag}(d_1, \ldots, d_m)$, and assembling the tokens $\mathbf{x}_i$ and $\mathbf{z}_i$ into the rows of data matrices X and Z — see (7.1). Then (10.52) is equivalent to

$$Z = D^{-1} A X V, \tag{10.53}$$

The self-attention layer (10.53), along with the positional encoding described below, are the fundamental components of the transformer architecture. The rest of the layer contains routine operations; the self-attention is followed by a residual connection, layer normalization, and a fully connected neural network layer. It is important to note that the graph structure, i.e., the matrices D and A, are computed from the token embeddings $\mathbf{x}_i$ themselves, and are not *fixed* like they were in the context of graph neural networks in Section 10.4. We also mention that in some other references, e.g., [237], the exponential is omitted in the definition of attention in (10.51), and is introduced later in the definition of the transformer layer via the application of a soft-max; we leave the equivalence of this definition to Exercise 5.2.

Now, the matrix W is learned through a low rank factorization as $W = Q^T K$, where Q is called the *queries matrix*[17] and K is the *keys matrix*. The terminology comes from database retrieval problems, and in this context the matrix V is often called the *values matrix*. The dimensions of Q and K are both $n_k \times n$, so the rank of W is at most n_k, which is usually chosen to be much smaller than n. The scaling factor β in the definition of attention in (10.51) is then chosen as $\beta = 1/\sqrt{n_k}$. Thus, with this factorization of W and choice of β, we can write the attention from (10.51) as

$$a_{ij} = \exp\left(\frac{1}{\sqrt{n_k}} (Q \mathbf{x}_i)^T K \mathbf{x}_j \right). \tag{10.54}$$

With this choice of β the attention is called the *scaled dot product attention*; the scaling is chosen so that the values inside the exponential do not get too large, which can cause numerical instabilities. Transformers also use *multihead attention*, where multiple transformer layers (10.53) are used in parallel with different learned weight matrices Q, K, V, allowing each head to pay attention to different language patterns. The outputs of the heads are concatenated and projected back to $\mathbb{R}^n$ with another learnable weight matrix.

Remark 10.18. It's very important to note that the dimensions of the tunable weight matrices Q, K, V depend on the token embedding dimension n, and the chosen rank n_k. In particular, they do *not* depend on the context size m. Thus, the context size does not have to be fixed in training, and can even vary during inference.   ▲

---

[17] *Warning*: The queries matrix Q is not an orthogonal matrix.

Remark 10.19. In order to prevent the model from looking ahead to future tokens, self-attention is often *masked* by setting $a_{ij} = 0$ for $j > i$, so that in the transformer layer (10.53), the computation of $\mathbf{z}_i$ uses only the tokens $\mathbf{x}_1, \ldots, \mathbf{x}_i$ from the past, and does not make use of any tokens $\mathbf{x}_{i+1}, \ldots, \mathbf{x}_m$ from the future part of the sequence.   ▲

Remark 10.20. There are other types of attention that are used in transformers, depending on the natural language processing application. In the context of chatbots or language translation, there is also a user input that is separate from the tokens $\mathbf{x}_1, \ldots, \mathbf{x}_m$ that are being produced by the model. For chatbots, the user input is the question posed by the human, while in language translation the input is the word or phrase to be translated into another language. Let $\mathbf{y}_1, \ldots, \mathbf{y}_p$ be the input tokens. In this context, transformers use encoder-decoder attention, where attention is paid between the current outputs $\mathbf{x}_i$ of the model (i.e., the decoder) and the fixed inputs $\mathbf{y}_i$ (i.e., the encoder). In this case, the keys and values come from the encoded input symbols $\mathbf{y}_i$, so the attention layer becomes

$$a_{ij} = \exp\left(\frac{1}{\sqrt{n_k}} (Q\mathbf{x}_i)^T K\mathbf{y}_j \right) \qquad \text{and} \qquad Z = D^{-1} AYV. \qquad ▲$$

10.5.2   Positional Encoding

Note that in the formulation of self-attention described above, we have omitted the *ordering* of the tokens, which is clearly very important for natural language processing tasks. In order to utilize ordering information, transformers use *positional encoding*, which is a way to encode the position of a token into the token vector $\mathbf{x}_i$ itself, in a manner that allows the model to easily utilize *relative* positional information about the tokens.

The original transformer architecture [237] uses additive positional encoding. If the tokens $\mathbf{x}_1, \ldots, \mathbf{x}_m \in \mathbb{R}^n$ are ordered, then we define positional vectors $\mathbf{p}_1, \ldots, \mathbf{p}_m \in \mathbb{R}^n$, and simply add the two. That is, we replace each token vector $\mathbf{x}_j$ with

$$\widetilde{\mathbf{x}}_j = \mathbf{x}_j + \mathbf{p}_j$$

before the tokens make their way into the first transformer layer. The main question is how to choose the positional encoding vectors. One possibility is to treat them as free parameters that can be learned during training of the model, in the same way that the vector embeddings of tokens are learned. This works well and is often done in practice. However, there are also simple ways to hard-code the positional vectors so that the model has fewer parameters to learn. Below we describe the original positional encoding framework from [237], as well as some modifications that have been made more recently.

It is useful to first examine the properties we'd like the positional encoding vectors to satisfy. They should clearly be distinct, so that $\mathbf{p}_j = \mathbf{p}_k$ if and only if $j = k$. However, this is not sufficient. We would also like the model to be able to deduce the *relative ordering* of tokens easily with a linear transformation. That is, we'd like there to be a linear relationship of the form

$$\mathbf{p}_k = A_{k-\ell} \mathbf{p}_\ell \tag{10.55}$$

for any indices k, ℓ, where the matrix $A_{k-\ell} \in \mathbb{R}^{n \times n}$ depends only on the relative positional difference $k - \ell$. Thus, while we are encoding the absolute positions of the tokens, we'd really prefer for the model to be able to deduce relative positions easily, so that it can generalize to variable context sizes.

For some motivation, consider the binary representation of numbers. The columns of the matrix below contain all of the 4 bit binary numbers:

$$
\begin{pmatrix}
0 & 1 & 0 & 1 & 0 & 1 & 0 & 1 & 0 & 1 & 0 & 1 & 0 & 1 & 0 & 1 \\
0 & 0 & 1 & 1 & 0 & 0 & 1 & 1 & 0 & 0 & 1 & 1 & 0 & 0 & 1 & 1 \\
0 & 0 & 0 & 0 & 1 & 1 & 1 & 1 & 0 & 0 & 0 & 0 & 1 & 1 & 1 & 1 \\
0 & 0 & 0 & 0 & 0 & 0 & 0 & 0 & 1 & 1 & 1 & 1 & 1 & 1 & 1 & 1
\end{pmatrix}
\tag{10.56}
$$

Notice each bit oscillates at a different frequency. The first bit oscillates with the highest frequency, switching from 0 to 1 at every column, while each subsequent bit oscillates at half the frequency. In fact, if we replace 0 by -1 throughout, then we can write the j-th (modified) bit of the integer k as

$$
b_{jk} = - \operatorname{sign}\left[\sin\left(\frac{k\pi}{2^{j-1}} \right) \right],
\tag{10.57}
$$

provided we interpret $\operatorname{sign}\left[\sin(2n\pi) \right] = +1$ and $\operatorname{sign}\left[\sin((2n+1)\pi) \right] = -1$ when $n \in \mathbb{Z}$; see Exercise 5.4. Thus, sinusoidal waves with different frequencies — here following the geometric progression 2^{j-1} — can be used to encode positional information.

Thus motivated, let us take the positional vectors to be different complex exponentials.[18] Assume that n is an even number. Let $w_1, \ldots, w_{n/2}$ with $0 < w_j \le 2\pi$ be different frequencies, to be selected below. Define complex positional vectors $\mathbf{q}_1, \ldots, \mathbf{q}_m \in \mathbb{C}^{n/2}$ with entries

$$
q_{k,j} = e^{k w_j \, \mathrm{i}} = \cos(k w_j) + \mathrm{i}\,\sin(k w_j), \qquad j = 1, \ldots, \frac{n}{2}.
\tag{10.58}
$$

We choose complex vectors with $\frac{1}{2}n$ entries since $\mathbb{C}^{n/2} \simeq \mathbb{R}^n$. Now it is straightforward to arrange that $\mathbf{q}_1, \ldots, \mathbf{q}_m$ be distinct.

Lemma 10.21. *If at least one frequency satisfies $0 < w_j < 2\pi/(m-1)$, then the positional vectors $\mathbf{q}_1, \ldots, \mathbf{q}_m \in \mathbb{C}^{n/2}$ defined by (10.58) are distinct, i.e., $\mathbf{q}_k = \mathbf{q}_\ell$ if and only if $k = \ell$.*

Proof. We note that $\mathbf{q}_k = \mathbf{q}_\ell$ if and only if $e^{k w_j \mathrm{i}} = e^{\ell w_j \mathrm{i}}$ for all j, which is equivalent to $e^{(k-\ell) w_j \mathrm{i}} = 1$. Thus, $(k - \ell) w_j$ must be an integer multiple of 2π which, by the assumed inequality on one of the w_j and the fact that $|k - \ell| \le m - 1$, implies $k = \ell$. ∎

Lemma 10.21 shows that the positional vectors are distinct provided at least one of the coordinates oscillates at a sufficiently low frequency $w_j < 2\pi/(m-1)$. This is not surprising; indeed, this frequency is so low that it does not repeat itself over $k = 1, \ldots, m$. If we only cared about having distinct vectors, then we would only need this one low frequency and could ignore the others. However, there are other considerations at play, in particular, the relative ordering property (10.55). The complex positional vectors certainly enjoy this property, since

$$
q_{k,j} = e^{(k+\ell-\ell) w_j \mathrm{i}} = e^{(k-\ell) w_j \mathrm{i}} \, e^{\ell w_j \mathrm{i}} = q_{k-\ell,j}\, q_{\ell,j},
$$

or, equivalently, their Hadamard products satisfy

$$
\mathbf{q}_k = \mathbf{q}_{k-\ell} \circ \mathbf{q}_\ell.
$$

In terms of the real and imaginary parts of the positional vectors, these amount to the trigonometric identities

$$
\begin{pmatrix} \cos(k w_j) \\ \sin(k w_j) \end{pmatrix} = A_{k-\ell,j} \begin{pmatrix} \cos(\ell w_j) \\ \sin(\ell w_j) \end{pmatrix}, \qquad \text{where} \qquad A_{p,j} = \begin{pmatrix} \cos(p w_j) & -\sin(p w_j) \\ \sin(p w_j) & \cos(p w_j) \end{pmatrix}
$$

---

[18] Complex exponentials and the complex vector space $\mathbb{C}^m$ were introduced in Section 9.10.1.

is the rotation matrix that rotates vectors in $\mathbb{R}^2$ by pw_j radians; see Example 4.36. Thus, $A_{k-\ell,j}$ is a rotation matrix that depends only on the relative positional difference $k - \ell$. If $w_j < 2\pi/(m-1)$, the matrices $A_{k-\ell,j}$ are distinct, so it also appears the relative positional property requires only one frequency!

However, in practice it would be difficult for the model to learn the matrices $A_{p,j}$ for a single low frequency w_j, as the following example illustrates.

Example 10.22. Suppose that $m = 1000$ and fix $w_j = \pi/1000$. Then the matrices for relative positions of 1 and 10 are given by

$$A_{1,j} \approx \begin{pmatrix} 0.999995 & -0.003141 \\ 0.003141 & 0.999995 \end{pmatrix}, \qquad A_{10,j} \approx \begin{pmatrix} 0.999506 & -0.031410 \\ 0.031410 & 0.999506 \end{pmatrix}. \qquad (10.59)$$

Thus, $A_{1,j}$ and $A_{10,j}$ differ only in the 4th decimal place along the diagonal, and the 2nd decimal place off-diagonal, which may be difficult for the model to detect (and context size is normally much larger than $m = 1000$). In other words, when measured at low frequencies, small positional shifts are hard to distinguish. A relative positional difference of 1 token or 10 tokens can be extremely important for understanding language; indeed, a 1 position difference can be a sub-token within the same word, while 10 tokens can be several words apart.

Thus, in order to measure smaller relative differences, we need to consider higher frequency sinusoids that change more quickly on small scales. For example, if we had chosen instead $w_j = \pi/10$, the corresponding matrices

$$A_{1,j} \approx \begin{pmatrix} 0.95 & -0.31 \\ 0.31 & 0.95 \end{pmatrix}, \qquad A_{10,j} = \begin{pmatrix} -1 & 0 \\ 0 & -1 \end{pmatrix}.$$

are clearly more easily distinguished than those in (10.59), and so we expect the model to have far less difficulty in learning them when the relative positional differences of 1 and 10 are important. ▲

Example 10.22 shows that we need a wide variety of frequencies in the positional encoding in order to allow the model to easily compare *any* relative positional differences, no matter what scale they occur on. The lowest frequency should be less than $2\pi/(m-1)$ to satisfy Lemma 10.21, while the highest frequency w_1 should be on the order of 2π so that it changes sign every step. The example of binary numbers in (10.56) motivates a geometric progression of the form $w_j = b^j$, and a choice of $b = (1/N)^{1/n}$ satisfies our requirements for a sufficiently large $N \gg m$. Hence, our complex positional encoding vectors are given by

$$q_{k,j} = e^{k N^{-j/n} \mathrm{i}} = \cos\left(\frac{k}{N^{j/n}}\right) + \mathrm{i}\sin\left(\frac{k}{N^{j/n}}\right), \qquad j = 1, \ldots, \frac{n}{2}.$$

In practice, the real and imaginary parts are concatenated into a vector twice as long, so we define the real positional vectors $\mathbf{p}_1, \ldots, \mathbf{p}_m \in \mathbb{R}^n$ by

$$p_{k,2j-1} = \cos\left(\frac{k}{N^{j/n}}\right), \qquad p_{k,2j} = \sin\left(\frac{k}{N^{j/n}}\right), \qquad j = 1, \ldots, \frac{n}{2}.$$

By fixing an N much larger than the context size m, we ensure that the positional encoding is not dependent on the context size, similar to the attention mechanism. Thus, all aspects of the transformer model can generalize to different context sizes.

Now, we can think of the coordinates of the positional vectors $\mathbf{p}_1, \ldots, \mathbf{p}_k \in \mathbb{R}^n$ as *clocks* that run at different speeds. The first coordinates are clocks that run the fastest, while the last coordinates run the slowest. Depending on the scale of positional differences to be measured,

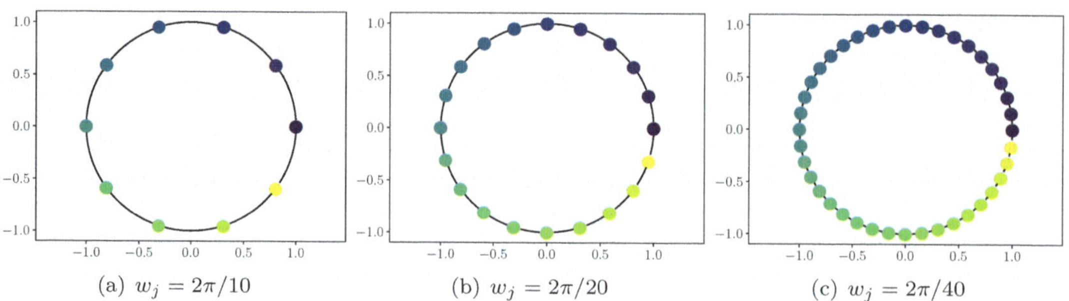

(a) $w_j = 2\pi/10$ (b) $w_j = 2\pi/20$ (c) $w_j = 2\pi/40$

Figure 10.18: Illustration of how the coordinates of the positional encoding vectors $\mathbf{p}_k$ traverse the unit circle at different speeds. The images are colored by $k = 1, \ldots, 2\pi/w_j$, and relative positional differences are measured through the angles between points. The coordinates in (a) are good at measuring positional differences less than 10, while the coordinates in (c) are useful for measuring larger positional differences, but are not as sensitive to small positional differences.

the model can choose which clocks to use by selecting coordinates of the positional vectors. We can visualize pairs of these clocks, namely $(p_{k,2j-1}, p_{k,2j})$ for j fixed and $k = 1, \ldots, 2\pi/w_j$, as simply traversing the unit circle in $\mathbb{R}^2$ at different speeds. Figure 10.18 shows three such clocks that run slower from left to right. The clock in Figure 10.18(a) is good at measuring differences of at most 10 positions via the angle between pairs of points, but for differences larger than 10, the periodicity of the clock renders it less useful, and should be replaced by clocks with lower frequency, as in (b) and (c).

Remark 10.23. It may seem more natural *a priori* to concatenate the token vector $\mathbf{x}_i$ with the positional vector $\mathbf{p}_i$, i.e., $\widetilde{\mathbf{x}}_i = (\mathbf{x}_i, \mathbf{p}_i) \in \mathbb{R}^{2n}$, since adding them would seem to contaminate the token and position information. One possible explanation as to why additive positional encoding works well is that the token embeddings $\mathbf{x}_i \in \mathbb{R}^n$ are learned, so it is possible for the model to, say, keep the positional information and token information in *orthogonal subspaces*, as they would be if they were concatenated. If this is the case, then additive positional encoding is simply a more efficient means of concatenating the positional and token information, since true concatenation would require doubling the dimension, which may not be desirable.

We can also examine how the additive positional vectors affect the attention mechanism (10.51). Omitting the exponential, the self-attention of the positional encoded vectors $\widetilde{\mathbf{x}}_i = \mathbf{x}_i + \mathbf{p}_i$ satisfies

$$\widetilde{\mathbf{x}}_i^T W \widetilde{\mathbf{x}}_j = \mathbf{x}_i^T W \mathbf{x}_j + \mathbf{p}_i^T W \mathbf{x}_j + \mathbf{x}_i^T W \mathbf{p}_j + \mathbf{p}_i^T W \mathbf{p}_j. \tag{10.60}$$

Thus, we have four attention terms; the first term is the attention without any positional information, the last one is the attention of *only* the positional information, while the middle two are attention terms that mix the positional and token information. Some recent works [125] have suggested that the mixed terms are not helpful, and that different matrices W should be used for the token and position attention, yielding alternative positional attentions of the form

$$a_{ij} = \exp\left(\beta_1 \mathbf{x}_i^T W_1 \mathbf{x}_j + \beta_2 \mathbf{p}_i^T W_2 \mathbf{p}_j \right),$$

where β_1, β_2 are scaling parameters, and W_1, W_2 are learnable weight matrices.   ▲

```
KING RICHARD III:                          RICHARD III:
O Buckingham, now do I play the touch,     So thou hast stay'd the will.
To try if thou be current gold indeed
Young Edward lives: think now what I would say.   DUKE OF YORK:
                                           Thou art continued to the king is friend.
BUCKINGHAM:
Say on, my loving lord.                    BUCKINGHAM:
                                           Stanley, be it within the head, and I will not have king.
KING RICHARD III:                          If it had rather lived my will be faired to do it.
Why, Buckingham, I say, I would be king,
                                           KING RICHARD III:
BUCKINGHAM:                                Well, and so I beg, I can go for me;
Why, so you are, my thrice renowned liege.    Tear it in my parting as my mind.

KING RICHARD III:                          GLOUCESTER:
Ha! am I king? 'tis so: but Edward lives.    Would you speak to choose your hands, my troth,
                                           And that he should make me to say?
BUCKINGHAM:
True, noble prince.                        KING EDWARD IV:
                                           Well, my gracious lord, bear thee in diademned.
```

Figure 10.19: An excerpt from the source material — the collected works of William Shakespeare — and a sample generated by a small character-level transformer model trained from scratch on the source material. Can you distinguish the true source material from the generated sample?

10.5.3 Experiments with Character-based Models

We now turn to some numerical experiments. We used a small transformer model called nanoGPT that can be easily trained from scratch on a laptop.[19] The model is trained on individual characters (i.e., letters, numbers, punctuation, symbols, and newline characters) instead of tokens, using a context size of 256 characters. It is a 6-layer transformer with 6 attention heads in each layer. On a single GPU it can train from scratch on a small data set in a few minutes.

The model was trained on the collected works of William Shakespeare. After training, the model is able to generate random text that looks, at a statistical level, like it belongs in a play written by Shakespeare. An excerpt from one of Shakespeare's plays (Richard III) is shown in Figure 10.19 along with a sample generated by the trained transformer. Since the model is relatively small and works only on a character level, it is not difficult to distinguish the source material from the transformer generated sample, though we encourage the reader to try this themselves. While this is already impressive, larger models that work on tokens instead of characters produce far better results.

Remark 10.24. Currently, as of 2024, one of the largest transformer based models is GPT-4, which was developed and trained by OpenAI [2]. GPT stands for *generative pretrained transformer*. It is pretrained in a similar manner as described in this section, except that the size of the training set and model are massive.[20] The training set is not disclosed, but is believed to contain a refined version of the internet, as well as a large collection of textbooks. The model was trained for about 100 days on 25,000 GPUs for a cost that likely exceeded $60 million USD. The carbon footprint of training large language models such as GPT-4 is immense, though it may pale in comparison to the footprint of inference, i.e., using the model to continuously answer vast numbers of queries; we refer to [183] for a discussion of the environmental impact of deep learning.

After pretraining, the GPT model is fine-tuned for specific tasks, such as being a chatbot (i.e., ChatGPT) or to work as a LaTeX or Python code editor, email composer, etc. The fine-tuning uses reinforcement learning with human feedback in order to shape the responses based

---

[19]The code was developed by Andrej Karpathy; see https://github.com/karpathy/nanoGPT.

[20]The model has over 1 trillion parameters, and the training set has around 13 trillion tokens.

on what is expected and informative, while attempting to avoid bias, offense, or providing information that can cause harm or break any laws.   ▲

Remark 10.25. We briefly mention that transformers have also recently been used in computer vision tasks, and in some cases achieved performance better than CNNs [62, 126]. In image processing, the tokens are patches, or blocks, of pixels in the image, and the transformer architecture allows for attention between *any* two image patches, and is thus, unlike convolutions, not *localized*.   ▲

Exercises

5.1. ♡ Explain how the initial vector embedding of tokens into $\mathbb{R}^n$ can be viewed as a linear function, i.e., matrix multiplication, of the $M \times n$ token embedding matrix E with a matrix containing the one-hot encoding $\mathbf{e}_i$ of each token in the context window.

5.2. Let $P \in \mathcal{M}_{m \times m}$ be the matrix with entries by $p_{ij} = \beta \mathbf{x}_i^T W \mathbf{x}_j$. Explain how the matrix $D^{-1} A$ in (10.53) can be obtained by applying the soft-max function to the rows of P.

5.3. ♡ Let π be a permutation of the integers $1, \ldots, m$. (a) Show that if $\pi(i) = i$, then the output $\mathbf{z}_i$ of the attention layer (10.52) is unchanged by permuting the tokens with π, that is

$$\mathbf{z}_i = \frac{1}{d_{\pi(i)}} \sum_{j=1}^{m} a_{\pi(i)\,\pi(j)} V \mathbf{x}_{\pi(j)}.$$

(b) Show that the same is true for masked self-attention described in Remark 10.19 provided $\pi(i) = i$ and $\pi(j) \leq i$ for all $j \leq i$.

5.4. ◇ Show that the j-th binary bit can be expressed as (10.57).

5.5. ◇ Let $V \subset \mathbb{R}^n$ be a subspace and $V^\perp$ its orthogonal complement. Suppose that the token embeddings $\mathbf{x}_i \in V$ for all i, and the positional vectors $\mathbf{p}_i \in V^\perp$ for all i. Let

$$W = W_1 W_1^T + W_2 W_2^T$$

where $W_i \in \mathcal{M}_{n \times k_i}$ with $\operatorname{img} W_1 \subset V$ and $\operatorname{img} W_2 \subset V^\perp$. Show that attention with positional information given in (10.60) reduces to

$$\widetilde{\mathbf{x}}_i^T W \widetilde{\mathbf{x}}_j = \mathbf{x}_i^T W_1 W_1^T \mathbf{x}_j + \mathbf{p}_i^T W_2 W_2^T \mathbf{p}_j.$$

5.6. Experiment with the `https://github.com/karpathy/nanoGPT` to train a character-based transformer model from scratch. You can find text files of publicly available books here: `https://www.gutenberg.org/`.

10.6   Universal Approximation

One reason for the general effectiveness of neural networks is their ability, given a sufficient number of neurons, to approximate any continuous function to arbitrary accuracy. This property is known as *universal approximation*. There are of course many other mathematical approaches to function approximation that have been well-developed and have a long history in mathematics, dating back to Newton, and including polynomial interpolation, rational function (Padé) interpolation, splines, Fourier series, and many others [11, 163, 189]. These traditional methods work well, and have rigorous guarantees when approximating functions in relatively few dimensions, but they all suffer from the *curse of dimensionality*. That is, the number of required parameters scales exponentially with dimension, rendering the methods inapplicable to the high dimensional data that is commonplace in machine learning applications. For example, even the MNIST dataset, which is in essence a toy example, lives in 784 dimensions!

Example 10.26. Consider a degree d multivariate polynomial $p(x)$ in n dimensions, which includes a coefficient for every monomial of degree d or less. For example, when $n = 2$ and $d = 2$ we have

$$p(x) = c_1 x_1^2 + c_2 x_1 x_2 + c_3 x_2^2 + c_4 x_1 + c_5 x_2 + c_6,$$

which has 6 coefficients. In general, the number of coefficients is the binomial coefficient $\binom{n+d}{d}$, which grows extremely rapidly with n and d. For example, suppose our data is $n = 100$ dimensional. Then a degree $d = 10$ polynomial would have $\binom{110}{10} \approx 5 \times 10^{13}$ coefficients. This is about 50 times larger than the number of parameters in the largest deep learning models that exist today, such as ChatGPT4. If we scaled up to $n = 1000$ dimensions, we'd need $\binom{1010}{10} \approx 3 \times 10^{23}$ coefficients! (When n is a billion, the number of coefficients exceeds the number of atoms in the universe.) ▲

Neural networks offer a flexible method of function approximation that does not, *a priori*, suffer from the curse of dimensionality. By this, we mean that a neural network representing a function $F \colon \mathbb{R}^n \to \mathbb{R}$ can be defined with as few as $O(n)$ parameters, or with many more if desired. In practice, it is common to *overparametrize* a neural network, which means using more parameters than the number of training examples, making them easier to train. On the other hand, overparametrized models still have far fewer parameters than the numbers appearing in Example 10.26. The number of parameters one chooses in a neural network has less to do with the extrinsic dimension n and more to do with the complexity of the underlying function to be approximated, making them easy to apply in various settings. If the underlying function is very complicated, a neural network may still suffer from the curse of dimensionality, in that the number of parameters required to achieve good approximation accuracy may scale poorly with dimension. However, real world data often exhibits some sort of low dimensional structure, and neural networks appear to be good at detecting and exploiting this. Obtaining a clear mathematical statement of this is still an active area of research.

In this section, we give a brief overview of universal approximation using neural networks and other types of functions, such as polynomials. We mainly focus on approximation of scalar functions, for simplicity of exposition, though we do include some results for multivariate functions approximation in Section 10.6.7.

Let us fix some notation. We will consider scalar functions $u \colon \mathbb{R} \to \mathbb{R}$ defined on all of $\mathbb{R}$. Let $\mathrm{Lip}(\mathbb{R})$ denote the space of Lipschitz continuous scalar functions, as defined in Section 6.8. Throughout this section we also use the notation $\lceil x \rceil$ to denote the smallest integer greater than $x \in \mathbb{R}$, and $\lfloor x \rfloor$ to denote the largest integer smaller than x.

Definition 10.27. A set of scalar functions $\mathcal{F}$ satisfies the *universal approximation property* if for any $u \in \mathrm{Lip}(\mathbb{R})$, any interval $[a, b]$, and any $\varepsilon > 0$, there exists $f \in \mathcal{F}$ such that

$$| f(x) - u(x) | < \varepsilon \qquad \text{for all} \qquad a \leq x \leq b. \tag{10.61}$$

In other words, the universal approximation property requires that we can uniformly approximate any Lipschitz continuous function on any bounded interval by a function in our prescribed set.

We present several examples of classes of functions that possess the universal approximation property, starting with the classical analytical cases of polynomials and trigonometric polynomials, then piecewise affine functions that arise in classical numerical approximation and interpolation theory, and concluding with several classes of neural networks. In all cases, the function classes $\mathcal{F}$ we consider possess an affine invariance property.

Definition 10.28. We say a class of functions $\mathcal{F}$ on $\mathbb{R}$ is *affine invariant* if for each $f \in \mathcal{F}$ and $c, d \in \mathbb{R}$, the function $g(x) = f(cx + d)$ belongs to $\mathcal{F}$.

For an affine invariant class of functions $\mathcal{F}$, it is sufficient to prove the universal approximation property in Definition 10.27 on a single interval, which we take to be the unit interval $[0, 1]$; see Exercise 6.2. For the rest of this section, we restrict our attention to the unit interval.

Remark 10.29. All the spaces of functions $\mathcal{F}$ that are considered here are *vector spaces* of functions. This means that the sum $f(x) = f_1(x) + f_2(x)$ of any two functions $f_1, f_2 \in \mathcal{F}$ in the space, is also in the space, $f \in \mathcal{F}$, as is any constant multiple of such a function, $g(x) = c f_1(x)$, so $g \in \mathcal{F}$ for any $c \in \mathbb{R}$. As with $\mathbb{R}^n$, this immediately implies that any linear combination of elements of the space also lies in it:

$$c_1 f_1(x) + \cdots + c_k f_k(x) \in \mathcal{F} \qquad \text{whenever} \qquad f_1, \ldots, f_k \in \mathcal{F}, \quad c_1, \ldots, c_k \in \mathbb{R}.$$

Thus, for example, since the sum of two polynomials is a polynomial, as is any constant multiple of a polynomial, as is any linear combination of polynomials, the space of polynomials $\mathcal{P}$ is a vector space. We refer the reader to [181] for an introduction to general vector spaces and subspaces, which are modeled on $\mathbb{R}^n$ and its subspaces. ▲

Remark 10.30. In the Definition 10.27 of universal approximation, it is often required that any *continuous* function u can be approximated, while we have added the stronger condition that u be Lipschitz continuous. We chose to concentrate on this slightly restrictive class of functions since the proofs are somewhat simpler, almost all functions that show up in machine learning are at least of this nature, and we have already studied them in detail in Section 6.8. We also mention that, while we have restricted our study to scalar functions $u \colon \mathbb{R} \to \mathbb{R}$ for simplicity and clarity of presentation, all of the results have natural extensions to multivariate functions $u \colon \mathbb{R}^n \to \mathbb{R}$, some of which we discuss in remarks throughout the section and in Section 10.6.7. ▲

10.6.1   Polynomials

We begin with polynomial functions. The fact that they satisfy the universal approximation property is the celebrated Stone–Weierstrass theorem, [202].

Theorem 10.31. *The space of all polynomial functions $\mathcal{P}$ satisfies the universal approximation property.*

Proof. We outline the key points in a short proof, which was in fact the original proof used by Weierstrass. Given $u \in \mathrm{Lip}(\mathbb{R})$, and $t > 0$, define

$$u_t(x) = \int_{-\infty}^{\infty} u(x - y)\,\Phi_t(y)\,dy, \qquad \text{where} \qquad \Phi_t(x) = \frac{e^{-x^2/(4t)}}{2\sqrt{\pi t}}. \tag{10.62}$$

Remark. For those familiar with partial differential equations, cf. [180], $u_t(x)$ is the solution of the heat equation with the initial condition $u_0(x) = u(x)$ at $t = 0$. ▲

It can be shown that $\lim_{t \to 0^+} u_t(x) = u(x)$ for all $x \in \mathbb{R}$, and that the convergence is uniform on $[0, 1]$; that is

$$\lim_{t \to 0^+} \max_{0 \le x \le 1} |\, u_t(x) - u(x)\,| = 0.$$

Indeed, using the fact — see Exercise 6.7 — that $\int_{-\infty}^{\infty} \Phi_t(y)\,dy = 1$, we have

$$u_t(x) - u(x) = \int_{-\infty}^{\infty} \big[\, u(x - y) - u(x)\,\big]\Phi_t(y)\,dy.$$

Taking absolute values on both sides and using that u is Lipschitz continuous we have

$$|\, u_t(x) - u(x)\,| \le \int_{-\infty}^{\infty} |\, u(x - y) - u(x)\,|\,\Phi_t(y)\,dy \le \mathrm{Lip}(u) \int_{-\infty}^{\infty} |\, y\,|\,\Phi_t(y)\,dy.$$

The term on the right hand side is independent of x and goes to zero as $t \to 0$ — again, see Exercise 6.7 — which establishes the claim.

Thus, it suffices to show that u_t can be approximated by polynomials for any $t > 0$. For this, we make a change of variables to write

$$u_t(x) = \int_{-\infty}^{\infty} u(y)\,\Phi_t(x - y)\,dy = \frac{1}{2\sqrt{\pi t}} \int_{-\infty}^{\infty} u(y)\,e^{-(x-y)^2/(4t)}\,dy.$$

Expanding the exponential in a Taylor series yields

$$u_t(x) = \sum_{k=0}^{\infty} p_k(x), \qquad \text{where} \qquad p_k(x) = \frac{1}{2\sqrt{\pi t}} \int_{-\infty}^{\infty} u(y)\,\frac{(x - y)^{2k}}{k!\,2^{2k}\,t^k}\,dy.$$

Using the binomial formula to expand $(x - y)^{2k}$, we easily see that, for fixed t, each $p_k(x)$ is a polynomial of degree k. Thus, truncating the Taylor expansion implies that u_t can be uniformly approximated by polynomials. ∎

10.6.2 Trigonometric Polynomials

A second important class is the set $\mathcal{T}$ of *trigonometric polynomials*, that is, polynomial functions of $\cos x$ and $\sin x$. Use of trigonometric identities shows that the functions in $\mathcal{T}$ can all be uniquely written in the form

$$p(x) = \frac{a_0}{2} + \sum_{k=1}^{N} \big[\, a_k \cos(c_k\,x) + b_k \sin(c_k\,x)\,\big], \tag{10.63}$$

where $N \in \mathbb{N}_0$ is its *degree*, $a_0, \ldots, a_N$, $b_1, \ldots, b_N$, and $c_1, \ldots, c_N$ are real constants, and the factor of $\frac{1}{2}$ in the constant term is introduced for convenience. The trigonometric polynomials form a subspace $\mathcal{T} \subset \mathrm{Lip}(\mathbb{R})$, which forms the foundation of the theory of Fourier series, [67, 180], of overwhelming importance in mathematical analysis, partial differential equations, quantum mechanics, optics, signal processing, and many other applications, as well as areas in pure mathematics, e.g., number theory.

Theorem 10.32. *The space of all trigonometric polynomials $\mathcal{T} \subset \mathrm{Lip}(\mathbb{R})$ satisfies the universal approximation property.*

Proof. As noted above, we only need to check the universal approximation property on one specific interval $[0, 1]$. On this interval, we set $c_k = k$ in (10.63) and use the trigonometric polynomial

$$p(x) = \frac{a_0}{2} + \sum_{k=1}^{N} \left[a_k \cos(k\,x) + b_k \sin(k\,x) \right], \qquad (10.64)$$

which is 2π-periodic; that is $p(x + 2\pi) = p(x)$ for all $x \in \mathbb{R}$. These trigonometric polynomials can clearly only approximate 2π-periodic functions, but this is sufficient for us, since we need only check the universal approximation property on the interval $[0, 1]$.

Given $u \in \mathrm{Lip}(\mathbb{R})$, we let $\widetilde{u} \in \mathrm{Lip}(\mathbb{R})$ be any 2π-periodic function that agrees with u on $[0, 1]$; the existence of such a $\widetilde{u}$ follows from Exercise 6.4. Its degree N trigonometric polynomial approximant has the form (10.64) whose coefficients are given by the Fourier formulae

$$a_k = \frac{1}{\pi} \int_{-\pi}^{\pi} \widetilde{u}(x) \cos k\,x \, dx, \qquad k = 0, 1, 2, 3, \ldots,$$

$$b_k = \frac{1}{\pi} \int_{-\pi}^{\pi} \widetilde{u}(x) \sin k\,x \, dx, \qquad k = 1, 2, 3, \ldots. \qquad (10.65)$$

We remark that the Fourier series for $\widetilde{u}$ is the limiting trigonometric series obtained by using the formulas (10.65) for the coefficients and letting $N \to \infty$. Since the Fourier series of a Lipschitz continuous function converges uniformly, [67], we have constructed a sequence of trigonometric polynomials that converge uniformly to $\widetilde{u}$ on the interval $[0, 2\pi]$, and hence to u on the unit interval $[0, 1]$. $\blacksquare$

10.6.3   Piecewise Affine Functions

The next example, which forms the basis of the universal approximation property of ReLU neural networks, is the set $\mathcal{A} = \mathcal{A}(\mathbb{R})$ consisting of all continuous *piecewise affine functions* (also often called piecewise linear) on $\mathbb{R}$. In other words, the graph of a function in $\mathcal{A}$ consists of a finite number of straight lines that are connected at a finite number of nodes[21] $x_i \in \mathbb{R}$. It is not hard to see that $\mathcal{A}$ forms a vector space, meaning that the sum of two piecewise affine functions is piecewise affine (even when they are based on different nodes) as is any constant multiple of a piecewise affine function; see Exercise 6.1. The space $\mathcal{A}$ is also *affine invariant*, as per Definition 10.28.

We will focus our attention on piecewise affine functions with uniformly spaced nodes, and prove that they already satisfy the universal approximation property, which implies the following theorem.

---

[21]These are not to be confused with the nodes in a graph.

Theorem 10.33. *The space of all piecewise affine functions $\mathcal{A}$ satisfies the universal approximation property.*

Given $N \in \mathbb{N}$, let $h = 1/N$ be the *step size*. Let $x_j = j\,h,\ j = 0,\dots,N$, denote $N+1$ equally spaced *nodes*, so that $x_{j+1} - x_j = h$ for $0 \le j < n$, and $0 = x_0 < x_1 < x_2 < \cdots < x_N = 1$. A piecewise affine function $g \colon \mathbb{R} \to \mathbb{R}$ based on the given points takes the form

$$g(x) = a_j + b_j\,x \qquad \text{for} \qquad x_j < x < x_{j+1}, \qquad j = 0,\dots,N-1, \tag{10.66}$$

for constants $a_0,\dots,a_{N-1}, b_0,\dots,b_{N-1} \in \mathbb{R}$, and is *continuous* if and only if

$$g(x_j^-) = a_{j-1} + b_{j-1}x_j = a_j + b_j x_j = g(x_j^+), \qquad j = 1,\dots,N-1. \tag{10.67}$$

Given a continuous function $u \colon \mathbb{R} \to \mathbb{R}$, its N-th order *piecewise affine interpolant* is the unique continuous piecewise affine function $g \in \mathcal{A}$ that agrees with u at the data points: $g(x_j) = u(x_j),\ j = 0,\dots,N$. We leave it as an exercise for the reader to determine the formulas for the coefficients a_j, b_j for such an interpolating function. (Or see (10.70) below.)

The proof of Theorem 10.33 is contained in the following lemma, which also gives a quantitative bound on the approximation error.

Lemma 10.34. *Let $u \in \mathrm{Lip}\,[0,1]$. Let $g_N(x)$ denote the piecewise affine interpolant to $u(x)$ based at the nodes $x_j = j\,h,\ j = 0,\dots,N$, where $h = 1/N$. Then*

$$|\,u(x) - g_N(x)\,| \le h\,\mathrm{Lip}\,u \qquad \textit{for all} \qquad 0 \le x \le 1. \tag{10.68}$$

Proof. Let $x \in [0,1]$. Then there exists a grid point $x_j = j\,h$ such that $|x - x_j| \le \tfrac{1}{2}h$. Using the interpolant property $u(x_j) = g_N(x_j)$ and the basic Lipschitz inequality (6.113), we compute

$$|\,u(x) - g_N(x)\,| = |\,u(x) - u(x_j) + g_N(x_j) - g_N(x)\,| \le |\,u(x) - u(x_j)\,| + |\,g_N(x_j) - g_N(x)\,|$$

$$\le \mathrm{Lip}(u)\,|\,x - x_j\,| + \mathrm{Lip}(g_N)\,|\,x - x_j\,| \le \frac{\mathrm{Lip}(u) + \mathrm{Lip}(g_N)}{2}\,h.$$

The proof is completed by noting — see Exercise 6.8 — that $\mathrm{Lip}(g_N) \le \mathrm{Lip}(u)$. ∎

Later we will require the following strengthened piecewise affine interpolation estimate for functions whose derivatives are Lipschitz.

Lemma 10.35. *Suppose that $u, u' \in \mathrm{Lip}\,[0,1]$, and let $g_N(x)$ denote the piecewise affine interpolant to $u(x)$ based at the nodes $x_j = j\,h,\ j = 0,\dots,N$, where $h = 1/N$. Then*

$$|\,u(x) - g_N(x)\,| \le h^2\,\mathrm{Lip}(u') \qquad \textit{for all} \qquad 0 \le x \le 1. \tag{10.69}$$

Proof. Since u' is Lipschitz continuous, we can apply the mean value theorem, whereby

$$\frac{u(x_{j+1}) - u(x_j)}{x_{j+1} - x_j} = u'(y_j) \qquad \text{for some} \qquad x_j \le y_j \le x_{j+1}.$$

Thus, since

$$
g_N(x) = u(x_j) + \frac{u(x_{j+1}) - u(x_j)}{x_{j+1} - x_j}\,(x - x_j) = u(x_j) + u'(y_j)\,(x - x_j) \tag{10.70}
$$

for $x_j \le x \le x_{j+1}$, we have, for such x,

$$
\begin{aligned}
|\,u(x) - g_N(x)\,| &= |\,u(x) - u(x_j) - u'(y_j)\,(x - x_j)\,| \\
&= |\,[u(x) - u(y_j) - u'(y_j)\,(x - y_j)] - [u(x_j) - u(y_j) - u'(y_j)\,(x_j - y_j)]\,| \\
&\le |\,u(x) - u(y_j) - u'(y_j)\,(x - y_j)\,| + |\,u(x_j) - u(y_j) - u'(y_j)\,(x_j - y_j)\,|.
\end{aligned}
$$

Applying the Taylor estimate in Proposition 6.61 to each of the final summands, and noting that both $|\,x - y_j\,|$ and $|\,x_j - y_j\,|$ are bounded by h, establishes (10.69). ∎

Remark. Piecewise affine interpolation plays a fundamental role in numerical analysis, particularly the finite element method, [180, 225]. One can generalize to interpolation by piecewise polynomial functions, which have the ability to better approximate the function and one of more of its derivatives. The most important case are the piecewise cubic *splines*, which have become ubiquitous in numerical analysis, geometric modeling, design and manufacturing, computer graphics, animation, gaming, and beyond, [73, 210].   ▲

10.6.4   Two Layer Neural Networks

We next show that a large class of 2-layer neural networks also has the universal approximation property. Recall that a 2-layer neural network with N hidden units and activation function $\sigma : \mathbb{R} \to \mathbb{R}$ has the form

$$
f_N(x) = \sum_{i=1}^{N} a_i\,\sigma(w_i x + b_i), \tag{10.71}
$$

where the tunable parameters are $a_1, \ldots, a_N, w_1, \ldots, w_N, b_1, \ldots, b_N \in \mathbb{R}$. Let $\mathcal{N}_\sigma$ denote the set of all such functions, which is easily seen to form a vector space. Moreover, $\mathcal{N}_\sigma$ is also clearly affine invariant.

> **Theorem 10.36.** *Assume σ is infinitely differentiable. Then the space of 2-layer neural networks $\mathcal{N}_\sigma$ satisfies the universal approximation property if and only if σ is not a polynomial.*

Proof. If σ is a polynomial, say of degree k, then any 2-layer neural network is also a polynomial of degree $\le k$. However, polynomials of bounded degree cannot uniformly approximate all continuous functions — in fact, even a degree $k + 1$ polynomial cannot be approximated. On the other hand, if σ is not a polynomial, we will show that we can uniformly approximate *any* polynomial by such a 2-layer neural network. The theorem then follows from the Stone–Weierstrass Theorem 10.31; see Exercise 6.3.

Let us fix $w, b \in \mathbb{R}$, and for $h > 0$ consider the finite difference function

$$
g_h(x) = \frac{\sigma\big((w + h)x + b\big) - \sigma(wx + b)}{h},
$$

which is clearly of the form (10.71), i.e., $g_h \in \mathcal{N}_\sigma$. Moreover,

$$
\lim_{h \to 0} g_h(x) = \frac{d}{dw}\,\sigma(wx + b) = x\,\sigma'(wx + b).
$$

Equation (6.123) implies that, for each fixed $w, b, x \in \mathbb{R}$ we have

$$|g_h(x) - x\,\sigma'(wx + b)| \leq C(x)\,h, \qquad \text{where} \qquad C(x) = \frac{x^2}{2} \max_{w \leq v \leq w+h} \sigma''(vx + b). \qquad (10.72)$$

Moreover, equation (10.72) holds uniformly for $x \in [0, 1]$ by replacing $C(x)$ with $C = \max_{0 \leq x \leq 1} C(x)$. Therefore, given $\varepsilon > 0$ we can choose h sufficiently small so that $Ch \leq \varepsilon$ to conclude that the function $x\,\sigma'(wx + b)$ can be approximated by a 2 layer neural network for $x \in [0, 1]$ and for any fixed choices of $w, b \in \mathbb{R}$. In particular, setting $w = 0$, we find that $x\,\sigma'(b)$ can be so approximated for any choice of b. Choosing b so that $\sigma'(b) \neq 0$, which is possible since σ is not a constant polynomial, and dividing by $\sigma'(b)$, we see, thanks to the vector space property of $\mathcal{N}_\sigma$, that we can uniformly approximate the function x.

The next step is to iterate this process. Since we can approximate the function $x\,\sigma'(wx+b)$ for any w and b, we can approximate its finite difference

$$g_h(x) = \frac{x\,\sigma'\big((w + h)\,x + b\big) - x\,\sigma'(wx + b)}{h},$$

which satisfies

$$\lim_{h \to 0} g_h(x) = \frac{d^2}{dw^2}\,\sigma(wx + b) = x^2\,\sigma''(wx + b).$$

By a similar argument as above, this limit is attained uniformly for $x \in [0, 1]$, and so a 2 layer neural network can approximate the function $x^2\,\sigma''(wx + b)$ for any choice of w and b. We again choose $w = 0$ and b so that $\sigma''(b) \neq 0$, which is possible since σ is not a linear polynomial. Note that this choice of b may be different from the one chosen in the first step of the proof. Dividing by $\sigma''(b)$ we find that a 2 layer neural network can approximate the function x^2. Clearly, we can continue this process by induction to find that we can approximate each monomial x^n, and thus all polynomials, on the unit interval $[0, 1]$.   ■

Since Theorem 10.36 requires the activation function σ be infinitely differentiable, the result does not apply directly to the ReLU activation, though it does apply to its smooth approximation soft-plus (7.66). Nevertheless, the theorem can be extended to nonsmooth activations with some additional technical details and assumptions, but this is outside the scope of this book; we refer the reader to [58]. In Section 10.6.5, we provide quantitative approximation results for ReLU networks that not only establish universal approximation, but also give estimates on how many parameters are needed to obtain a desired approximation accuracy.

Remark 10.37. The proof of Theorem 10.36 extends naturally to 2-layer neural networks with inputs in $\mathbb{R}^n$, which have the form

$$f_N(\mathbf{x}) = \sum_{i=1}^{N} a_i\,\sigma(\mathbf{w}_i \cdot \mathbf{x} + b_i),$$

where the weights $\mathbf{w}_i \in \mathbb{R}^n$ are vectors, and, as before, $a_i, b_i \in \mathbb{R}$ for $i = 1, \ldots, N$. In this case, we can take partial derivatives of the function $\sigma(\mathbf{w} \cdot \mathbf{x} + b)$ in the components of $\mathbf{w}$, to find that all multivariate polynomials belong to $\mathcal{N}_\sigma$, and then use the multivariate version of Stone–Weierstrass Theorem 10.31.

An alternative way of showing that neural networks can approximate functions in dimensions $n \geq 2$ is to use the Kolmogorov–Arnold representation theorem, cf. [154], which shows that any continuous function $f : \Omega \to \mathbb{R}$ on a compact set $\Omega \subset \mathbb{R}^n$ can be written in the form

$$f(\mathbf{x}) = \sum_{i=1}^{2n} h_i \left(\sum_{j=1}^{n} g_j(x_j) \right),$$

for continuous scalar functions $h_1, \ldots, h_{2n}, g_1, \ldots, g_n : \mathbb{R} \to \mathbb{R}$. The scalar functions h_i and g_i can be approximated arbitrarily well on compact domains by one-dimensional neural networks, and so we can approximate $f(\mathbf{x})$ on the n-dimensional unit *hypercube* $[0,1]^n$ using $3n$ neural networks with inputs in $\mathbb{R}$. ▲

10.6.5 Two Layer ReLU Networks

We now turn to studying the special case of ReLU activation functions, which are commonly used in practice. In this case, a 2-layer ReLU neural network with M hidden nodes has the form

$$f_M(x) = \sum_{i=1}^{M} a_i \sigma(w_i x + b_i) = \sum_{i=1}^{M} a_i (w_i x + b_i)_+, \tag{10.73}$$

where $\sigma(t) = t_+ = \max\{t, 0\}$ is the ReLU activation function. These networks are not covered by our previous Theorem 10.36 since the ReLU activation function is not continuously differentiable and, moreover, $\sigma''(t) = 0$ for all $t \neq 0$. On the other hand, as we will show directly, the universal approximation property continues to hold. Moreover, under certain conditions on the function to be approximated, we can establish quantitative error estimates, which allow us to estimate how many neurons are required to achieve a desired approximation accuracy.

Let $\mathcal{R}$ denote the space of all 2-layer ReLU neural network functions (10.73). Note that, as above, $\mathcal{R}$ is a vector space, since any linear combination of 2-layer ReLU neural network functions is again a 2-layer ReLU neural network function. Moreover, $\mathcal{R}$ is also clearly affine invariant.

Theorem 10.38. *The space $\mathcal{R}$ of all 2-layer ReLU neural network functions satisfies the universal approximation property.*

The theorem is an immediate consequence of the piecewise affine universal approximation Theorem 10.33 because every piecewise affine function is a 2-layer ReLU network. Specifically:

Theorem 10.39. *Every piecewise affine function based on $N+1$ equally spaced nodes $x_j = jh$, where $h = 1/N$ and $j = 0, \ldots, N$, is a 2-layer ReLU network with $M \leq N+3$ hidden nodes.*

Proof. Let us first consider the piecewise affine *hat function*

$$\theta(x) = (1 - |x|)_+ = (x+1)_+ - 2x_+ + (x-1)_+ = \begin{cases} 1 - x, & 0 \leq x \leq 1, \\ x + 1, & -1 \leq x \leq 0, \\ 0, & \text{otherwise}, \end{cases} \tag{10.74}$$

which is depicted in Figure 10.20(a). In particular,

$$\theta(k) = \begin{cases} 1, & k = 0, \\ 0, & 0 \neq k \in \mathbb{Z} \end{cases}. \tag{10.75}$$

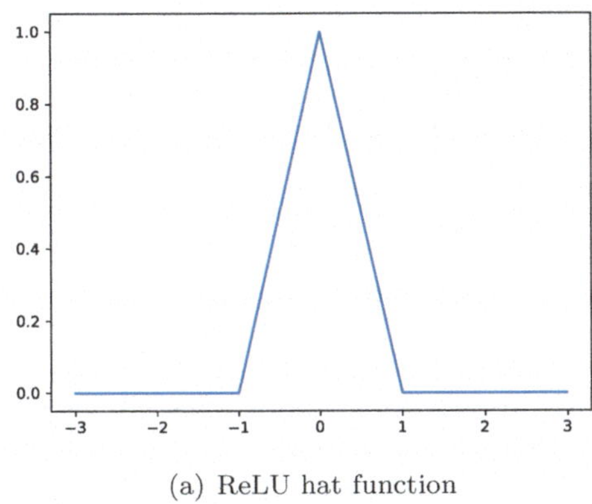

(a) ReLU hat function

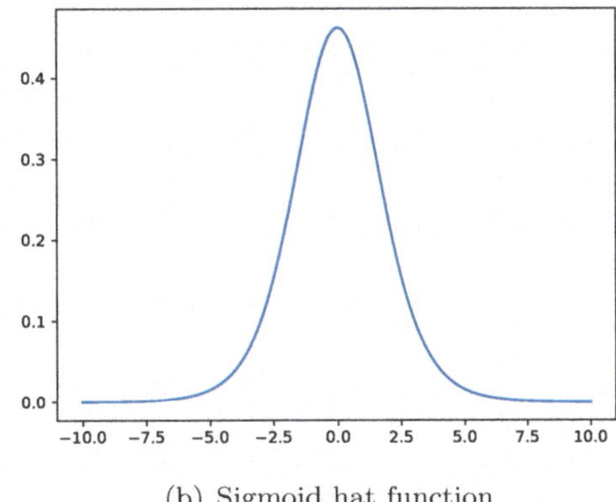

(b) Sigmoid hat function

Figure 10.20: Depiction of the ReLU hat function and sigmoid bump function constructed from 2-layer neural networks with ReLU and sigmoid activations, respectively.

From the second formula, the hat function is a 2-layer ReLU neural network with 3 hidden nodes.

Given our equally spaced nodes, we define the scaled and translated hat functions

$$\theta_i(x) = \theta\left(\frac{x - x_i}{h}\right) = \frac{(x - x_{i-1})_+ - 2(x - x_i)_+ + (x - x_{i+1})_+}{h}, \qquad i = 0, \ldots, N, \quad (10.76)$$

which are also 2-layer ReLU neural networks with 3 hidden nodes. In the definition of θ_0 and θ_N we set $x_{-1} = -h$ and $x_{N+1} = (N+1)h$, respectively; that is, the formula $x_j = jh$ is taken for all $j = -1, \ldots, N+1$. As a consequence of (10.75), these have the interpolation values

$$\theta_i(x_j) = \begin{cases} 1, & i = j, \\ 0, & i \neq j. \end{cases} \qquad (10.77)$$

Hence, hat functions play a role analogous to the standard basis vectors in the subspace of piecewise affine functions based on the given nodes.

Given a continuous function $u \colon [0, 1] \to \mathbb{R}$, let $g_N(x)$ denote its piecewise affine interpolant based at these points, so that $g_N(x_i) = u(x_i)$ for $i = 0, \ldots, N$. We claim that

$$g_N(x) = \sum_{i=0}^{N} u(x_i)\, \theta_i(x) = \sum_{i=0}^{N} u(x_i)\, \theta\left(\frac{x - x_i}{h}\right), \qquad 0 \leq x \leq 1. \qquad (10.78)$$

Indeed, the right hand side is a linear combination of piecewise affine functions, and hence itself piecewise affine. Moreover, at $x = x_i$, by (10.77), it has value $g_N(x_i) = u(x_i)$. Since a piecewise affine function is uniquely determined by its values at the nodes, this establishes (10.78).

Now, each summand in (10.78) is a 2-layer ReLU neural network, and hence so is g_N. While formula (10.78) seems to indicate we need $3N$ hidden nodes to compute g_N, namely 3 for each hat function, in fact at most $N+3$ are needed because, by substituting the formulas (10.76), the resulting function is a linear combination of the $N+3$ ReLU functions $(x - x_i)_+$ for $i = -1, \ldots, N+1$. ■

The key to the proof of Theorem 10.39 is the construction of the hat function $\theta(x)$ in (10.74) in terms of ReLU neurons. We can construct hat functions for other activation functions in a similar way. For example, the sigmoid hat function requires only 2 hidden nodes, namely $g(x) = \sigma(x+1) - \sigma(x-1)$, since the sigmoid activation, as graphed in Figure 10.1, is bounded and does not grow linearly like ReLU. Figure 10.20(b) shows the sigmoid hat function. It is possible to extend Theorem 10.39 to other activations by using their hat functions, but there are additional technical details when the activation is not ReLU.

Remark 10.40. The above proof shows that the hat functions θ_i defined in (10.76) satisfy

$$\sum_{i=0}^{N} \theta_i(x) = \sum_{i=0}^{N} \theta\left(\frac{x - x_i}{h}\right) = 1 \qquad \text{for all} \qquad 0 \le x \le 1. \tag{10.79}$$

Indeed, the left hand side is simply the piecewise affine interpolant for the constant function $u(x) \equiv 1$, which is equal to 1 everywhere in $[0, 1]$. A collection of nonnegative functions that sum to one on an interval is called a *partition of unity*, a concept of importance in analysis and differential geometry, [1]. It will also be important in Section 10.6.6 when we extend these results to higher dimensions. ▲

We can also combine the representation (10.78) with the estimate in Lemma 10.34 to establish bounds on the number of nodes required for an approximating ReLU neural network.

Theorem 10.41. *Let $u \in \mathrm{Lip}\,[0, 1]$. Then for any $\varepsilon > 0$, there exists a 2-layer ReLU neural network $f_M(x)$ of the form (10.73) with*

$$M \le \mathrm{Lip}(u)\,\varepsilon^{-1} + 4 \tag{10.80}$$

hidden nodes such that

$$|f_M(x) - u(x)| \le \varepsilon \qquad \text{for all} \qquad 0 \le x \le 1. \tag{10.81}$$

Proof. Let $g_N(x)$ be the piecewise affine interpolant based on N equally spaced nodes, which, as noted in Theorem 10.41, is a ReLU neural network with $M \le N + 3$ hidden nodes. It follows from (10.68) that

$$|g_N(x) - u(x)| \le \mathrm{Lip}(u)\,h = \frac{\mathrm{Lip}(u)}{N} \le \varepsilon,$$

provided $N \ge \mathrm{Lip}(u)\,\varepsilon^{-1}$. We choose N to be the least integer satisfying this inequality, and so $N \le \mathrm{Lip}(u)\,\varepsilon^{-1} + 1$. Combining this with $N \ge M - 3$ yields (10.80). ∎

If we strengthen our requirement on the function u so that its derivative u' is Lipschitz, then we can reduce the required number of hidden nodes in the network. Using Lemma 10.35, a similar reasoning produces the following improved version of Theorem 10.41, whose proof we leave to Exercise 6.11.

Theorem 10.42. *Assume that $u, u' \in \mathrm{Lip}(\mathbb{R})$. Then for any $\varepsilon > 0$, there exists a 2-layer ReLU neural network $f_M(x)$ of the form (10.73) with $M \le \sqrt{\mathrm{Lip}(u')}\,\varepsilon^{-1/2} + 4$ hidden nodes such that*

$$|f_N(x) - u(x)| \le \varepsilon \qquad \text{for all} \qquad 0 \le x \le 1. \tag{10.82}$$

10.6.6 Deep ReLU Networks

While 2-layer neural networks are universal approximators, deeper networks can offer more efficient ways to parameterize certain classes of functions. In fact, in some settings, the complexity of a ReLU network grows *exponentially* with its depth. A simple example of this is the computation of the maximum

$$s_n = \max\{x_1, \ldots, x_n\}.$$

of a list of numbers. We can compute the maximum (or minimum) of $n = 2$ numbers with a 2-layer neural network with 3 hidden nodes, since

$$\max\{x_1, x_2\} = \max\{x_1 - x_2, 0\} + x_2 = (x_1 - x_2)_+ + x_{2+} - (-x_2)_+, \tag{10.83}$$

where we used that $a = a_+ - (-a)_+$ for any $a \in \mathbb{R}$. Similarly, to compute the maximum of 4 numbers, we can compose the maximum operation twice

$$\max\{x_1, x_2, x_3, x_4\} = \max\{\max\{x_1, x_2\}, \max\{x_3, x_4\}\}.$$

This requires a 3-layer neural network with 9 hidden nodes. Iterating this argument establishes the following result.

> **Lemma 10.43.** *The maximum of $n = 2^k$ numbers can be computed with a $(k+1)$-layer ReLU neural network with $N = 3(n-1)$ hidden nodes.*

Proof. Given we can compute the maximum of n numbers, we compute the maximum of $2n$ numbers by the recursion

$$\max\{x_1, \ldots, x_{2n}\} = \max\{\max\{x_1, \ldots, x_n\}, \max\{x_{n+1}, \ldots, x_{2n}\}\}, \tag{10.84}$$

and we use the neural network in (10.83) to compute the outer max on the right hand side. Thus, doubling the number of terms in the maximum requires one additional layer and 3 additional hidden nodes, which shows that we need $k + 1$ layers for $n = 2^k$ numbers. The number N_k of hidden nodes required to compute the max of $n = 2^k$ numbers satisfies $N_0 = 0$, $N_1 = 3$, $N_2 = 9$, and, by (10.84), the recurrence $N_{k+1} = 2N_k + 3$. An easy induction, cf. Exercise 6.13 shows that $N_k = 3(2^k - 1)$ for all $k \geq 0$. $\blacksquare$

Remark 10.44. To compute the maximum of n numbers $x_1, \ldots, x_n$ where n is not a power of 2, we can simply repeat the last coordinate x_n a sufficient number of times and reduce to the max over 2^k hidden nodes and $k + 1$ layers, where $k = \lceil \log_2 n \rceil$. Therefore, we can compute the maximum of n numbers with a ReLU neural network with $\lceil \log_2 n \rceil + 1$ layers and at most $3(2^k - 1) \leq 6n$ hidden nodes. ▲

For the rest of this section, we will switch to using big O notation (see the Preface for details) to count hidden nodes and layers, in order to simplify the presentation. Then Lemma 10.43 and Remark 10.44 show that the maximum of n numbers can be computed with a ReLU network with $O(\log n)$ layers and $O(n)$ hidden nodes. We also use the following terminology.

Definition 10.45. Let $\varepsilon > 0$. A scalar function $u\colon \mathbb{R} \to \mathbb{R}$ can be *ε-approximated* on an interval $[a, b]$ with a neural network with a specified depth and number of hidden nodes if there exists a neural network $f\colon \mathbb{R} \to \mathbb{R}$ of the specified size such that

$$|u(x) - f(x)| \leq \varepsilon \qquad \text{for all} \qquad a \leq x \leq b.$$

When the interval is omitted, it is implied that u can be so approximated on any bounded interval $[a, b]$.

Throughout the remainder of this section, we will always assume $0 < \varepsilon \leq \frac{1}{2}$, so that expressions like $\log \varepsilon^{-1}$ are well-defined and strictly positive. Using the terminology of Definition 10.45, Theorem 10.41 shows that any Lipschitz function $u\colon \mathbb{R} \to \mathbb{R}$ can be ε-approximated on $[0, 1]$ with a 2-Layer neural network with $O(\varepsilon^{-1})$ hidden nodes. By the affine invariance of neural networks, this also holds for any bounded interval $[a, b]$. In contrast, Theorem 10.42 showed that a function with Lipschitz derivative can be ε-approximated with by a 2-layer ReLU network with $O(\varepsilon^{-1/2})$ hidden nodes.

We now show that the number of hidden nodes can be substantially reduced for certain types of functions. In order to do this, we need to produce other types of exponential complexity in deep ReLU networks. As we saw above, a natural way to do this is to compose an operation recursively. In this case, we define the hat function

$$g(x) = \begin{cases} 2x, & 0 \leq x \leq \frac{1}{2} \\ 2 - 2x, & \frac{1}{2} \leq x \leq 1, \end{cases} \tag{10.85}$$

which is similar to that in (10.74), and hence can also be constructed with a 2-layer ReLU network with 3 hidden nodes. Let

$$g_k = \underbrace{g \circ g \circ \cdots \circ g}_{k \text{ times}}$$

denote its k-fold composition, which can be thus implemented with a ReLU network with $O(k)$ layers and hidden nodes. According to Exercise 6.5, g_k is a piecewise affine *sawtooth function* with 2^{k-1} teeth, satisfying

$$g_k\left(\frac{j}{2^k}\right) = \begin{cases} 1, & j \text{ odd}, \\ 0, & j \text{ even}. \end{cases} \tag{10.86}$$

Figure 10.21 shows g_1, g_2 and g_3. Notice that g_k has *exponentially* many pieces, namely $O(2^k)$, with respect to the size of the network $O(k)$. If we were to use the method in Theorem 10.41 to construct g_k with a 2-Layer ReLU network, we would need $O(2^k)$ hidden nodes.

We now show that we can use the functions $g_1, \ldots, g_k$ to efficiently construct piecewise affine interpolants of certain functions. We start with the quadratic function $u(x) = x^2$, so that $\mathrm{Lip}(u') = 2$. Let $f_m(x)$ be the piecewise affine interpolant of $u(x)$ with 2^m pieces, so that

$$f_m(2^{-m}k) = 2^{-2m}k^2, \qquad k = 0, \ldots, 2^m. \tag{10.87}$$

The functions f_0, f_1, f_2 and f_3 are depicted in Figure 10.21. By Lemma 10.35, with $h = 2^{-m}$,

$$|f_m(x) - x^2| \leq 2h^2 = 2^{1-2m} \qquad \text{for} \qquad x \in [0, 1]. \tag{10.88}$$

The following lemma shows how we can construct the piecewise affine interpolant f_m from the sawtooth functions g_k.

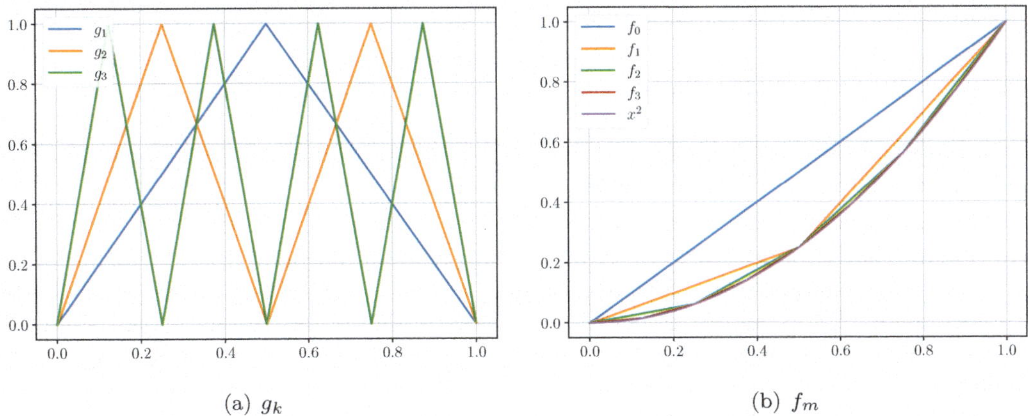

(a) g_k (b) f_m

Figure 10.21: A depiction of the functions g_k and f_m used to approximate the parabola x^2.

Lemma 10.46. *For any $0 \le x \le 1$ and $m \ge 1$,*

$$f_m(x) = x - \sum_{k=1}^{m} \frac{g_k(x)}{2^{2k}}. \tag{10.89}$$

Proof. The proof proceeds by induction on m. The case $m = 0$ is trivial. Observe that

$$f_{m+1}(x) = f_m(x) - 2^{-2m-2} g_{m+1}(x)$$

is piecewise affine on the nodes $2^{-m-1}j$ for $j = 0, \ldots, 2^{m+1}$, and the aim is to prove that it interpolates x^2. Fixing $0 \le k < 2^m$, note that $g_{m+1}(2^{-m}k) = 0$, and hence

$$f_{m+1}\big(2^{-m-1}(2k)\big) = f_m(2^{-m}k) = 2^{-2m}k^2 = \big(2^{-m-1}(2k)\big)^2. \tag{10.90}$$

On the other hand, because $f_m(x)$ is affine for $2^{-m}k \le x \le 2^{-m}(k+1)$, and $2^{-m-1}(2k+1)$ is the midpoint of this interval,

$$f_m\big(2^{-m-1}(2k+1)\big) = \tfrac{1}{2}\big[\, f_m(2^{-m}k) + f_m(2^{-m}(k+1))\,\big]$$
$$= \tfrac{1}{2}\big[\, 2^{-2m}k^2 + 2^{-2m}(k+1)^2 \,\big] = 2^{-2m-1}(2k^2 + 2k + 1).$$

Thus,

$$f_{m+1}\big(2^{-m-1}(2k+1)\big) = f_m\big(2^{-m-1}(2k+1)\big) - 2^{-2m-2} g_{m+1}\big(2^{-m-1}(2k+1)\big)$$
$$= 2^{-2m-1}(2k^2 + 2k + 1) - 2^{-2m-2} = \big(2^{-m-1}(2k+1)\big)^2. \tag{10.91}$$

Equations (10.90), (10.91) show that f_{m+1} has the correct interpolation values at all the nodes, as in (10.87) with m replaced by $m + 1$, which completes the proof. ∎

Lemma 10.46 shows that the function f_m can be expressed as a linear combination of $g_1, \ldots, g_m$, all of which can be computed with one ReLU network with $O(m)$ layers and hidden nodes. The intermediate g_i are computed at the intermediate layers, and the summation in (10.89) can be computed along the way as well; see Exercise 6.14. Thus, we can implement f_m

with a ReLU network with $O(m)$ layers and hidden nodes. Recalling the approximation error in (10.88), we see that a ReLU network with $O(m)$ nodes can approximate x^2 for $x \in [0,1]$ to an accuracy of 2^{1-2m}. If we desire an accuracy ε, we need to choose $m = \frac{1}{2}(1 + \lceil \log_2(\varepsilon^{-1}) \rceil)$ in order that $2^{1-2m} \leq \varepsilon$. This discussion is summarized in the following result.

Proposition 10.47. *The function x^2 can be ε-approximated on $[0,1]$ by a ReLU network with $O(\log \varepsilon^{-1})$ layers and hidden nodes*

We remark that Proposition 10.47 is much sharper than Theorem 10.41, which required $O(\varepsilon^{-1/2})$ nodes to guarantee the same approximation error. However, a key difference is that in Proposition 10.47, the number of layers must tend to infinity, though only logarithmically, as the approximation error ε is taken to zero, while Theorem 10.41 uses a simple 2-layer ReLU network.

We can extend Proposition 10.47 to approximate x^2 on $[-1,1]$ with the same size network by adding an initial layer that computes the absolute value $|x|$; see Exercise 6.12. We can also extend the approximation to any bounded interval $[-a,a]$ for $a > 0$ by modifying the initial absolute value layer to compute $|x/a|$, which is done by simply adjusting the weights in the first layer, and then adding a final linear layer that multiplies by a^2. Thus, the function x^2 can also be ε-approximated on $[-a,a]$, and hence on any interval, by a ReLU network with $O(\log \varepsilon^{-1})$ layers.

Now that we can approximate x^2, we can use the identity

$$xy = \left(\frac{x+y}{2}\right)^2 - \left(\frac{x-y}{2}\right)^2 \tag{10.92}$$

to approximate the product of two numbers. From here, we can approximate polynomials, and then general continuous functions (by Theorem 10.31).

Remark 10.48. Before proceeding, it is worthwhile to note that for $x, y \in [-1,1]$, the ReLU neural network approximations of x^2 and xy that we have constructed thus far also belong to the interval $[-1,1]$. This is clear for x^2, since the ReLU approximation is exactly its affine interpolant. For the product xy, note that when $x, y \in [-1,1]$, so $|x|, |y| \leq 1$, we have $(x \pm y)/2 \in [-1,1]$ and so the squares of these numbers lie in the interval $[0,1]$. Their difference, which is used in (10.92), thus lies in the interval $[-1,1]$. ▲

We now show that we can iterate the identity (10.92) to efficiently approximate the product of n numbers with a deep ReLU network.

Lemma 10.49. *The product $x_1 x_2 \cdots x_n$ of $n = 2^k$ numbers $x_i \in [-1,1]$ can be $n\varepsilon$-approximated with a ReLU network with $O(k \log \varepsilon^{-1})$ layers and $O(n \log \varepsilon^{-1})$ hidden nodes.*

Proof. The proof proceeds by induction on k. By the identity (10.92) and Proposition 10.47, we can ε-approximate the product of $n = 2$ numbers $x_1, x_2 \in [-1,1]$ with a ReLU network with $O(\log \varepsilon^{-1})$ layers and hidden nodes. The inductive step is similar to Lemma 10.43: we compute the product recursively as follows

$$x_1 \cdots x_{2n} = a_n b_n, \qquad \text{where} \qquad a_n = x_1 \cdots x_n, \qquad b_n = x_{n+1} \cdots x_{2n}, \tag{10.93}$$

are computed recursively, and their product $a_n b_n$ is computed with the ε-approximate network from the base case.

The errors will accumulate over this recursion. Let e_k be the error for computing the 2^k-fold product. We have $e_0 = 0$ and $e_1 = \varepsilon$. Let $\tilde{a}_n, \tilde{b}_n$ denote the outputs of the network approximating a_n and b_n, which, thanks to Remark 10.48, satisfy

$$|\tilde{a}_n| \le 1, \qquad |\tilde{b}_n| \le 1, \qquad |\tilde{a}_n - a_n| \le e_k, \qquad |\tilde{b}_n - b_n| \le e_k.$$

It follows that

$$|\tilde{a}_n \tilde{b}_n - a_n b_n| = |\tilde{a}_n(\tilde{b}_n - b_n) + b_n(\tilde{a}_n - a_n)| \le |\tilde{b}_n - b_n| + |\tilde{a}_n - a_n| \le e_k + e_k = 2e_k.$$

Multiplying $\tilde{a}_n$ and $\tilde{b}_n$ using the network in the base case yields an additional ε error, so the error satisfies the recursion $e_{k+1} = 2e_k + \varepsilon$. As similar argument as in Lemma 10.43 yields $e_k = (2^k - 1)\varepsilon \le 2^k \varepsilon = n\varepsilon$.

We now count the number of layers and nodes are in our network, denoted L_k and N_k, respectively. Each recursive step in (10.93) adds $\mathrm{O}(\log \varepsilon^{-1})$ layers, so $L_k = \mathrm{O}(k \log \varepsilon^{-1})$. The base case network has at most $N_1 = C \log \varepsilon^{-1}$ layers and nodes for some $C > 0$. The recursion (10.93) implies $N_{k+1} = 2N_k + C \log \varepsilon^{-1}$, and hence, as before,

$$N_k = C \log(\varepsilon^{-1})(2^k - 1) = \mathrm{O}(n \log \varepsilon^{-1}). \qquad \blacksquare$$

Remark 10.50. Let us fix $\delta > 0$ and set $\varepsilon = \delta/n$. Then Lemma 10.49 shows we can δ-approximate the $n = 2^k$-fold product of numbers in $[-1, 1]$ with a ReLU network with $\mathrm{O}(k \log(n\delta^{-1}))$ layers and $\mathrm{O}(n \log(n\delta^{-1}))$ hidden nodes. As in Remark 10.44 we can extend the result to n that are not powers of 2 by padding with ones until the next power of two. This simply replaces n by at most $2n$ in the formulas above, and the factor of 2 can be absorbed into the big O notation. $\blacktriangle$

Now that we can efficiently compute products of numbers with ReLU networks, we can also compute powers x^k, and thus polynomials – see Exercise 6.15 — and any functions that are sufficiently well-approximated by polynomials, e.g., real analytic functions. We refer the interested reader to [256], where many of the results above, and more, are established.

10.6.7 Approximating Multivariate Functions

We conclude this section with some results for approximating Lipschitz functions in more than one variable with deep ReLU networks. These results crucially rely on approximating products of numbers via Lemma 10.49. The key idea is to construct a type of hat function, like θ in Theorem 10.39, for which shifted and scaled versions form a *partition of unity* in $\mathbb{R}^n$; recall Remark 10.40. It turns out that the higher dimensional version of the hat function can be taken as the coordinatewise product of the one dimensional hat functions $\theta(x_i)$. That is, we define $\Phi \colon \mathbb{R}^n \to \mathbb{R}$ by

$$\Phi(\mathbf{x}) = \prod_{i=1}^{n} \theta(x_i) = \prod_{i=1}^{n} \left((1 - |x_i|)_+\right). \tag{10.94}$$

Now, it is not simple to *exactly* construct Φ with a ReLU network, unless $n = 1$, but since Φ is an n-fold product of the hat functions θ, each of which can be exactly represented with 2-layer ReLU networks with 3 hidden nodes, it follows from Lemma 10.49 that we can ε-approximate Φ with a ReLU network with $\mathrm{O}\left((\log n) \log(n\varepsilon^{-1})\right)$ layers and $\mathrm{O}\left(n \log(n\varepsilon^{-1})\right)$ hidden nodes.

We first show that Φ satisfies the partition of unity property, which allows us to approximate any Lipschitz continuous function with a linear combination of shifted and scaled

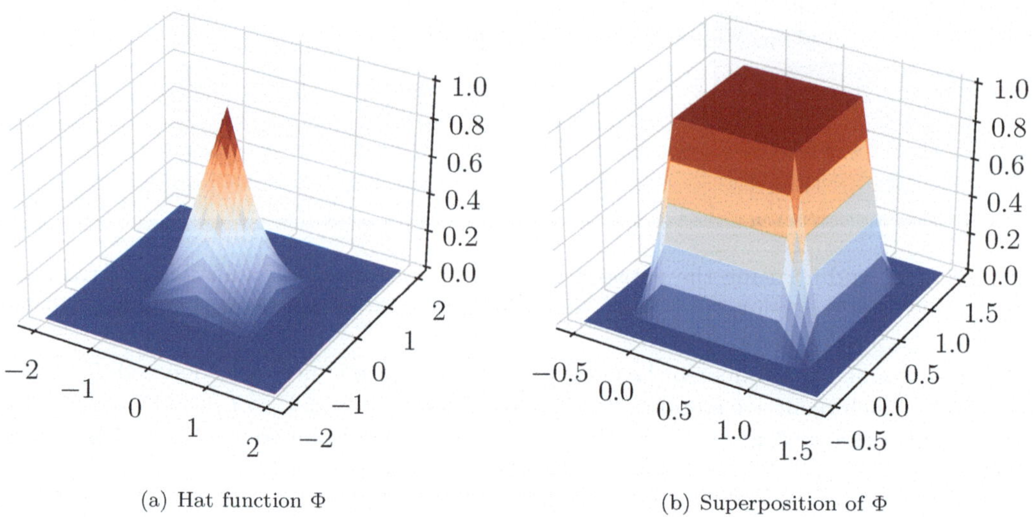

(a) Hat function Φ (b) Superposition of Φ

Figure 10.22: (a) The two-dimensional hat function Φ defined in (10.94), and (b) the superposition of such hat functions, illustrating that they form a partition of unity.

versions of the hat function; see Theorem 10.51 below. To state this result, we let m be a positive integer and set $h = 1/m$. Let

$$G_m = \left\{ \mathbf{x}_I := (i_1 h, \ \ldots \ , i_n h) \,\middle|\, 0 \le i_j \le m \right\} \subset [0,1]^n$$

be the uniform grid with internode spacing h on the n-dimensional unit hypercube $[0,1]^n$. Then, by (10.94) and the one-dimensional partition of unity formula (10.79),

$$\sum_I \Phi\left(\frac{\mathbf{x} - \mathbf{x}_I}{h}\right) = \sum_{i_1=0}^{m} \cdots \sum_{i_n=0}^{m} \left[\prod_{j=1}^{n} \theta\left(\frac{x_j - i_j h}{h}\right) \right]$$

$$= \prod_{j=1}^{n} \left[\sum_{i_j=0}^{m} \theta\left(\frac{x_j - i_j h}{h}\right) \right] = 1 \qquad (10.95)$$

for all $\mathbf{x} \in [0,1]^n$. We note that the initial summation is over all $I = (i_1, \ldots, i_n)$ in the definition of G_m above. See Figure 10.22 for a depiction of the two-dimensional hat function Φ along with the partition of unity property.

We now verify that this partition of unity allows us to approximate Lipschitz functions.

Theorem 10.51. *Let $U \colon [0,1]^n \to \mathbb{R}$ be Lipschitz continuous and define*

$$F(\mathbf{x}) = \sum_I U(\mathbf{x}_I)\, \Phi\left(\frac{\mathbf{x} - \mathbf{x}_I}{h}\right). \qquad (10.96)$$

Then

$$|U(\mathbf{x}) - F(\mathbf{x})| \le \sqrt{n}\,\mathrm{Lip}(U)\,h \qquad \text{for all} \qquad \mathbf{x} \in [0,1]^n. \qquad (10.97)$$

Proof. By the partition of unity (10.95), we can write

$$U(\mathbf{x}) = \sum_I U(\mathbf{x}) \, \Phi\left(\frac{\mathbf{x} - \mathbf{x}_I}{h}\right) \qquad \text{for all} \qquad \mathbf{x} \in [0,1]^n.$$

Subtracting this from F yields

$$|F(\mathbf{x}) - U(\mathbf{x})| = \left| \sum_I \left[U(\mathbf{x}_I) - U(\mathbf{x}) \right] \Phi\left(\frac{\mathbf{x} - \mathbf{x}_I}{h}\right) \right|$$

$$\leq \sum_I |U(\mathbf{x}_I) - U(\mathbf{x})| \, \Phi\left(\frac{\mathbf{x} - \mathbf{x}_I}{h}\right) \leq \mathrm{Lip}(u) \sum_I \|\mathbf{x} - \mathbf{x}_I\| \, \Phi\left(\frac{\mathbf{x} - \mathbf{x}_I}{h}\right).$$

We now claim that

$$\|\mathbf{x}\| \, \Phi\left(\frac{\mathbf{x}}{h}\right) \leq \sqrt{n} \, h \, \Phi\left(\frac{\mathbf{x}}{h}\right) \qquad \text{for all} \qquad \mathbf{x} \in \mathbb{R}^n. \tag{10.98}$$

Indeed, if $\mathbf{x}/h \in [-1,1]^n$ then $\|\mathbf{x}\| \leq h\sqrt{n}$, while if $\mathbf{x}/h \notin [-1,1]^n$, then $\Phi(\mathbf{x}/h) = 0$, and hence the inequality holds for all $\mathbf{x}$. We now use (10.98) along with (10.95) to obtain

$$|F(\mathbf{x}) - U(\mathbf{x})| \leq \sqrt{n} \, \mathrm{Lip}(u) \, h \sum_I \Phi\left(\frac{\mathbf{x} - \mathbf{x}_I}{h}\right) = \sqrt{n} \, \mathrm{Lip}(u) \, h. \qquad \blacksquare$$

Remark 10.52. If we use Theorem 10.51 to construct a ReLU neural network approximating a Lipschitz function $U \colon [0,1]^n \to \mathbb{R}$, by approximating each shifted/scaled hat function $\Phi\big((\mathbf{x} - \mathbf{x}_I)/h\big)$ and combining them via (10.96), the number of hidden nodes required grows exponentially with the dimension n, while the number of layers remains $O(n \log(n\varepsilon^{-1}))$. Indeed, even if we could exactly represent Φ with a ReLU network with $O(1)$ layers and hidden nodes — we in fact need more layers and nodes as described above — the issue is that we simply need to construct too many hat functions. We need one hat function for every $\mathbf{x}_I \in G_m$; that is we need $(m+1)^n$ hat functions. Since the error in Theorem 10.51 is $O(h)$, we would need to set $h = \varepsilon$ (or possibly smaller) to achieve an ε-approximation of u. Thus $m = \varepsilon^{-1}$ and so we need $O(\varepsilon^{-n})$ hat functions, which requires at least $O(\varepsilon^{-n})$ hidden nodes. This is the same *curse of dimensionality* we encountered in Example 10.26 in the context of approximating high dimensional functions with multivariate polynomials. However, in the case of neural networks, the curse of dimensionality does not prohibit us from using the methods on high dimensional data with far fewer than $O(\varepsilon^{-n})$ parameters; it simply suggests that we may not be able to represent all functions in this case, and so the question is whether the functions of importance in machine learning can be so represented. ▲

While neural networks suffer from the curse of dimensionality when approximating general Lipschitz functions in high dimensions, there are classes of high dimensional functions with *lower complexity* that can be efficiently approximated. We may expect problems with real data to exhibit some type of low dimensional structure, and neural networks turn out to be an expressive and adaptable set of functions that are able to exploit this. For example, if a function $F \colon \mathbb{R}^n \to \mathbb{R}$ has a linear low dimensional structure, it is the only the lower intrinsic dimension that matters; see Exercise 6.17. The low dimensional structure need not be linear and can instead be represented by a smooth embedded manifold; see [39]. Another example is the class of *Barron functions*, which have a low dimensional structure that allows their approximation with 2-layer neural networks whose width avoids the curse of dimensionality; see e.g., [14, 15].

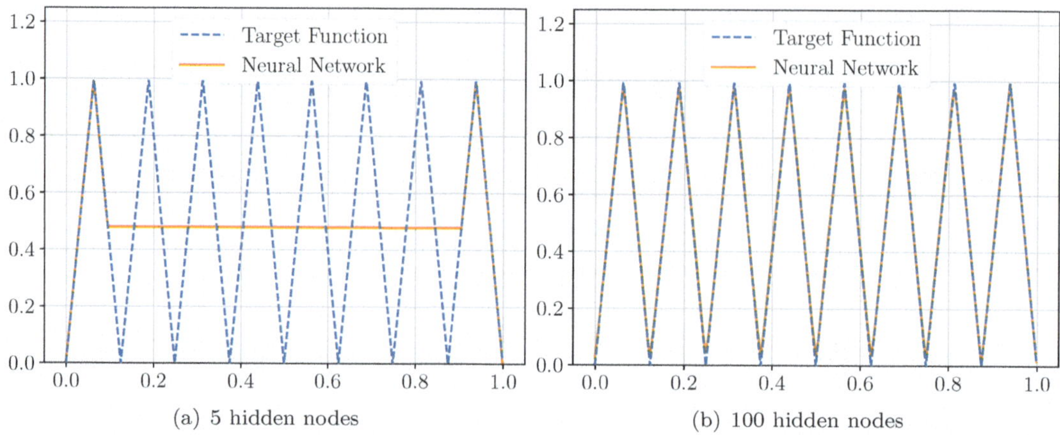

(a) 5 hidden nodes   (b) 100 hidden nodes

Figure 10.23: 6-layer ReLU networks approximating the sawtooth function g_4 with (a) 5 hidden nodes per layer and (b) 100 hidden nodes per layer. In (a) we show the result after 10^5 epochs of gradient descent on the least squares loss function, while in (b) we show the result after convergence at 5000 steps. Even though both networks are able to exactly represent g_4, the optimization is unable to find a good solution without *overparametrization*.

It is also important to mention that the universal approximation theorems in these sections generally do not explain why deep learning works well in practice, since the networks we constructed by hand may not be similar to those that are encountered in practice when the weights and biases are learned by training with stochastic gradient descent on a loss function. In fact, a network with the theoretically minimal number of parameters required to approximate a specific function is often more difficult to train than one with more parameters. As an example, we show in Figure 10.23 the result of training two 6-layer ReLU networks to approximate the sawtooth function g_4 via gradient descent on the least squares loss function using the Adam optimizer. In Figure 10.23(a) we used 5 hidden nodes per layer, while in Figure 10.23(b) we used 100. In theory, we can exactly represent g_4 with a 5-layer network with 3 hidden node per layer, so both networks are sufficiently large. However, in practice, the smaller network is difficult to train and has not converged even after 10^5 iterations of gradient descent. On the other hand, the larger overparametrized network in Figure 10.23(b) converges to the sawtooth function in less than 5000 iterations of gradient descent. The difference in the performance of these two networks has to do with the dynamics of gradient descent for training the networks, and cannot be understood through universal approximation results. In some sense, overparametrized networks have so many extraneous degrees of freedom that they are less likely to become trapped in spurious local minimizers during optimization.

Exercises

6.1. ♡ Prove that the space of piecewise affine functions $\mathcal{A}$ forms a vector space. Explain in detail why it is also affine invariant.

6.2. Explain why, if a class of functions $\mathcal{F}$ is affine invariant, then the universal approximation property (10.61) needs only be verified for the unit interval $[0, 1]$.

6.3. Justify the statement that if one can uniformly approximate any polynomial on any interval by functions in a class $\mathcal{F}$, then the class satisfies the universal approximation property.

6.4. ♡ Prove that if $u \in \mathrm{Lip}(\mathbb{R})$, there exists a 2π-periodic function $\tilde{u} \in \mathrm{Lip}(\mathbb{R})$ such that $u(x) = \tilde{u}(x)$ for all $0 \le x \le 1$.

6.5. ♡ Prove that, for each $k \ge 1$, the function (10.85) is continuous, piecewise affine, and satisfies (10.86).

6.6. ♡ (a) Use polar coordinates to prove that, for any $a > 0$,

$$\iint_{\mathbb{R}^2} e^{-a(x^2+y^2)}\,dx\,dy = \frac{\pi}{a}. \tag{10.99}$$

(b) Explain why

$$\int_{-\infty}^{\infty} e^{-ax^2}\,dx = \sqrt{\frac{\pi}{a}}. \tag{10.100}$$

6.7. ◊ Let $\Phi_t(x)$ be the function defined in (10.62).

(a) Use Exercise 6.6 to show that $\displaystyle\int_{-\infty}^{\infty} \Phi_t(y)\,dy = 1$.

(b) Show that $\displaystyle\int_{0}^{\infty} y\,\Phi_t(y)\,dy = \sqrt{\frac{t}{\pi}}$. Conclude that $\displaystyle\int_{-\infty}^{\infty} |\,y\,|\,\Phi_t(y)\,dy = 2\sqrt{\frac{t}{\pi}}$.

6.8. ♡ Let $u \in \mathrm{Lip}(\mathbb{R})$, and let g be a piecewise affine interpolant of u on an interval $[a,b]$. Prove that $\mathrm{Lip}_{[a,b]}(g) \le \mathrm{Lip}_{[a,b]}(u)$.

6.9. ♡ Suppose $u\colon [a,b] \to [c,d]$. Let g be a piecewise affine interpolant of u on $[a,b]$. Prove that $g\colon [a,b] \to [c,d]$.

6.10. ♡ Write down the hat functions corresponding to unequally spaced nodes, and prove that they still form a partition of unity.

6.11. ◊ Prove Theorem 10.42.

6.12. Show that (i) the identity function x, and (ii) the absolute value function $|\,x\,|$, can both be computed with a 2-layer ReLU network with 2 nodes.

6.13. ◊ Prove that $N_k = 3(2^k - 1)$ in Lemma 10.43.

6.14. Carefully write down and sketch a diagram of the ReLU network used to compute f_m.

6.15. Using big O notation, determine how many layers and hidden nodes are needed to approximate a degree k polynomial on the unit interval $[0,1]$ using a ReLU network. Does your answer change if you know how to factor the polynomial? *Hint:* Use Lemma 10.49.

6.16. How does your answer to Exercise 6.15 change if we consider a multivariate polynomial on the d-dimensional unit hypercube $[0,1]^d$? *Hint:* Try $d = 2$ first.

6.17. Let $\mathbf{v} \in \mathbb{R}^n$ be a unit vector $\|\mathbf{v}\| = 1$ and define the orthogonal projection matrix $P = \mathbf{v}\mathbf{v}^T$. Assume $F\colon \mathbb{R}^n \to \mathbb{R}$ is Lipschitz continuous and satisfies $F(\mathbf{x}) = F(P\mathbf{x})$ for all $\mathbf{x} \in \mathbb{R}^n$. Show that F can be ε-approximated on $[0,1]^n$ with a ReLU network with 2 layers and $O(\varepsilon^{-1}\sqrt{n})$ hidden nodes, and is thus not subject to the curse of dimensionality. *Hint:* Use Theorem 10.41 and also show that the diameter of the unit cube $[0,1]^n$ is $\sqrt{n}$.

Chapter 11

Advanced Optimization

In this final chapter, we return to our study of optimization that we began in Chapter 6. Our focus here is on more advanced optimization techniques that have recently found important applications in the context of deep learning. In particular, we study momentum-based gradient descent, including the heavy ball method, Krylov subspace methods, conjugate gradients, and Nesterov acceleration, as well stochastic gradient descent. We establish results that show how each of these methods improves the gradient descent convergence results obtained in Chapter 6. In the final section, we present a unifying analysis based on the continuum limit of our iterative techniques that leads to a differential equation interpretation of optimization.

Since the methods and proofs are more sophisticated than those in Chapter 6, we will work with the dot product and the ordinary (Euclidean) gradient, leaving extensions to general inner products to the exercises. Thus, throughout this chapter $\|\mathbf{x}\| = \sqrt{\mathbf{x} \cdot \mathbf{x}} = \sqrt{\mathbf{x}^T \mathbf{x}}$ denotes the Euclidean norm on $\mathbb{R}^n$, and $\nabla F(\mathbf{x})$ denotes the Euclidean gradient, as defined in (6.27), of a differentiable function $F \colon \mathbb{R}^n \to \mathbb{R}$. All functions F are assumed to admit at least one global minimizer, which we denote by $\mathbf{x}^\star \in \mathbb{R}^n$, and we denote the minimal value by $F^\star = F(\mathbf{x}^\star)$. Before proceeding with this chapter, the reader should ensure that they are familiar with the notion of Lipschitz continuity from Section 6.8, the theory of convex and strongly convex functions from Section 6.7, and the basic convergence results we established for gradient descent in Section 6.9.

11.1 Linear Convergence of Gradient Descent

We return to our study of gradient descent

$$\mathbf{x}_{k+1} = \mathbf{x}_k - \alpha_k \nabla F(\mathbf{x}_k).$$

(11.1)

In much of Chapter 6, we assumed the objective function $F \colon \mathbb{R}^n \to \mathbb{R}$ to be strongly convex, in which case, according to Theorem 6.68, $\mathbf{x}_k$ converges at a linear rate to the global minimizer $\mathbf{x}^\star$. In this section, we seek to relax the strong convexity assumption on F while retaining the linear convergence rate.

We recall from Theorem 6.52 that any μ-strongly convex function F satisfies the Polyak–Lojasiewicz inequality (6.106), which we reproduce here for convenience:

$$F(\mathbf{x}) - F^\star \le \frac{1}{2\mu} \|\nabla F(\mathbf{x})\|^2 \qquad \text{for all} \qquad \mathbf{x} \in \mathbb{R}^n.$$

(11.2)

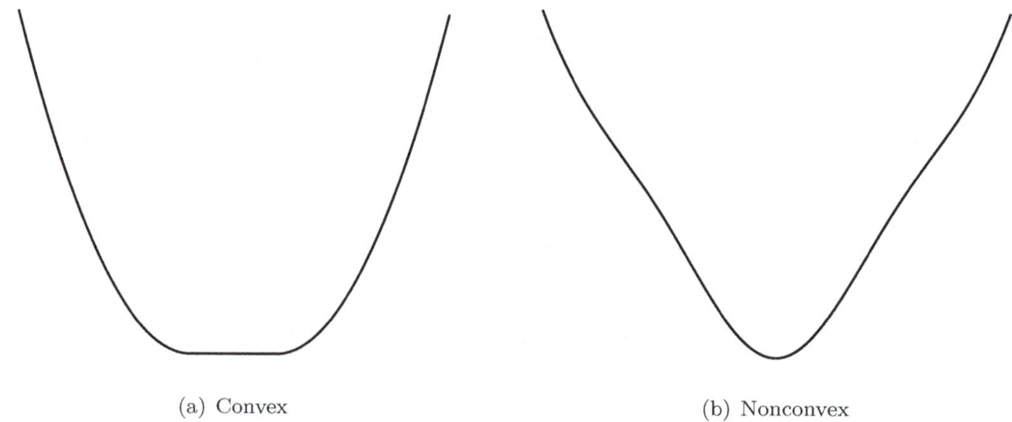

(a) Convex   (b) Nonconvex

Figure 11.1: Examples of μ-PL functions that are *not* strongly convex. In (a) the function, which is given in Example 11.4, is convex, but is simply too flat at the minimum to be strongly convex. In (b) the function, which is given in Example 11.5, is nonconvex away from the minimizer.

As noted above $F^\star = F(\mathbf{x}^\star)$ denotes the value of F at a (and hence any) global minimizer. On the other hand, not all functions satisfying the inequality (11.2) are strongly convex; this serves to inspire the following definition.

Definition 11.1. Let $\mu > 0$. We say that F is μ-*PL* if F admits a global minimizer $\mathbf{x}^\star$ and satisfies the Polyak–Lojasiewicz inequality (11.2).

The condition that F is μ-PL is weaker than strong convexity. In fact, some convex functions that fail to be strongly convex are still μ-PL, and there are even *nonconvex* functions that are μ-PL. Before presenting some examples, let us first restate our linear convergence result, Theorem 6.68, in terms of μ-PL functions. We leave the proof to the reader in Exercise 1.3.

Theorem 11.2. *Assume F is μ-PL, and that ∇F is Lipschitz continuous. Let $\mathbf{x}_k$, for $k \geq 0$, be the iterations of the gradient descent algorithm (11.1) with fixed time step $\alpha_k = \alpha$. If $0 < \alpha \leq \mathrm{Lip}(\nabla F)^{-1}$, then, for any integer $k \geq 0$,*

$$F(\mathbf{x}_k) - F^\star \leq (1 - \alpha\mu)^k \left(F(\mathbf{x}_0) - F^\star\right). \tag{11.3}$$

Remark 11.3. As we did in (6.49), we can use the inequality $1 - x \leq e^{-x}$ to rewrite (11.3) as

$$F(\mathbf{x}_k) - F^\star \leq e^{-\alpha\mu k} \left(F(\mathbf{x}_0) - F^\star\right), \tag{11.4}$$

which illustrates how a linear convergence rate depends exponentially on the index k.   ▲

Example 11.4. Consider the function

$$F(x) = \begin{cases} \frac{1}{2} x^2, & x < 0 \\ 0, & 0 \leq x \leq 1 \\ \frac{1}{2} (x-1)^2, & x > 1, \end{cases} \tag{11.5}$$

which is depicted in Figure 11.1(a). The function F is convex, but not μ-strongly convex for any $\mu > 0$ since it is flat on the interval $[0, 1]$; see Exercise 1.4. Nevertheless, the global minimum is $F^\star = 0$. We claim that

$$F(x) \le \tfrac{1}{2} F'(x)^2, \qquad (11.6)$$

which implies that the μ-PL inequality (11.2) holds for $\mu = 1$, so F is 1-PL. The proof of (11.6) is relegated to Exercise 1.1. ▲

Example 11.5. Consider the scalar function

$$F(x) = x^2 + \frac{\pi}{3} \sin^2 x,$$

depicted in Figure 11.1(b). We claim that F is *not* convex, but is μ-PL with $\mu = \frac{2}{7}$. To see this, note that F has a unique global minimizer at $x^\star = 0$ and minimum value $F^\star = 0$. Furthermore,

$$F'(x) = 2x + \frac{2\pi}{3} \sin x \cos x = 2x + \frac{\pi}{3} \sin(2x),$$

$$F''(x) = 2 + \frac{2\pi}{3} \cos(2x), \quad \text{and, in particular,} \quad F''\left(\frac{\pi}{2}\right) = 2 - \frac{2\pi}{3} < 0,$$

which implies that F is not convex. To see that F is μ-PL, we compute

$$F'(x)^2 = 4x^2 + \frac{4\pi}{3} x \sin(2x) + \frac{\pi^2}{9} \sin^2(2x).$$

We split the proof into two cases. When $|x| \le \tfrac{1}{2}\pi$, we have $x \sin(2x) \ge 0$ and so $F'(x)^2 \ge 4x^2$. Since $\sin^2 x \le x^2$,

$$F(x) \le \left(1 + \frac{\pi}{3}\right) x^2 \le 4x^2 \le F'(x)^2.$$

Therefore, the PL-inequality (11.2) holds with $\mu = \tfrac{1}{2}$ — and hence for any $\mu \le \tfrac{1}{2}$ — provided $|x| \le \tfrac{1}{2}\pi$. On the other hand, when $|x| > \tfrac{1}{2}\pi$, we have $x^2 \ge \tfrac{1}{2}\pi|x|$. Therefore,

$$F'(x)^2 \ge 4x^2 - \frac{4\pi}{3}|x| \ge \frac{4}{3} x^2, \qquad \text{and so} \qquad F(x) \le \frac{7}{3} x^2 \le \frac{7}{4} F'(x)^2.$$

This implies that the PL-inequality (11.2) holds, with $\mu = \frac{2}{7}$, for all x. ▲

Example 11.6. Consider the quadratic function

$$F(\mathbf{x}) = \tfrac{1}{2} \mathbf{x}^T H \mathbf{x} - \mathbf{x}^T \mathbf{b} + c, \qquad \mathbf{x} \in \mathbb{R}^n, \qquad (11.7)$$

where the coefficient matrix $H \ge 0$ is assumed to be symmetric positive *semidefinite*, so that F is not strongly convex. Let $k = \operatorname{rank} H < n$, and let us denote the eigenvalues of H by

$$\lambda_1 \ge \lambda_2 \ge \cdots \ge \lambda_k > 0 = \lambda_{k+1} = \cdots = \lambda_n,$$

with corresponding orthonormal eigenvectors $\mathbf{p}_1, \ldots, \mathbf{p}_n \in \mathbb{R}^n$.

We assume that $\mathbf{b} \in \operatorname{img} H = (\ker H)^\perp$ so that F admits global minimizers, and we recall from Theorem 6.9 that every such minimizer belongs to the affine solution space:

$$\mathbf{x}^\star \in S = \{\mathbf{y} \in \mathbb{R}^n \mid H\mathbf{y} = \mathbf{b}\}. \qquad (11.8)$$

Note that minimizers are not unique since $\mathbf{x}^\star + \mathbf{z} \in S$ whenever $\mathbf{z} \in \ker H$.

Now, since H is not positive definite, Example 6.49 tells us that F is not strongly convex. However, we claim that F is μ-PL where $\mu = \lambda_k$ is the smallest positive eigenvalue of H. To see this, note that $\nabla F(\mathbf{x}) = H\mathbf{x} - \mathbf{b}$, so the PL-inequality (11.2) reads

$$F(\mathbf{x}) - F^\star \le \frac{1}{2\mu} \| H\mathbf{x} - \mathbf{b} \|^2.$$

Now, if $\mathbf{x}^\star$ is any minimizer of F, so $H\mathbf{x}^\star = \mathbf{b}$, we can complete the square and write

$$F(\mathbf{x}) = \tfrac{1}{2} (\mathbf{x} - \mathbf{x}^\star)^T H (\mathbf{x} - \mathbf{x}^\star) + F^\star. \tag{11.9}$$

Therefore, by the Cauchy-Schwarz inequality,

$$F(\mathbf{x}) - F^\star \le \tfrac{1}{2} \| \mathbf{x} - \mathbf{x}^\star \| \, \| H(\mathbf{x} - \mathbf{x}^\star) \| = \tfrac{1}{2} \| \mathbf{x} - \mathbf{x}^\star \| \, \| H\mathbf{x} - \mathbf{b} \|. \tag{11.10}$$

To deal with the term $\| \mathbf{x} - \mathbf{x}^\star \|$, we choose $\mathbf{x}^\star$ to be the orthogonal projection of $\mathbf{x}$ onto the affine solution space (11.8), which is the closest minimizer of F to $\mathbf{x}$ in the Euclidean norm. This choice of $\mathbf{x}^\star$ satisfies $\mathbf{x} - \mathbf{x}^\star \in (\ker H)^\perp$, and so

$$\mathbf{x} - \mathbf{x}^\star = \sum_{i=1}^{k} c_i \, \mathbf{p}_i, \qquad H\mathbf{x} - \mathbf{b} = H(\mathbf{x} - \mathbf{x}^\star) = \sum_{i=1}^{k} c_i \lambda_i \, \mathbf{p}_i. \tag{11.11}$$

Therefore,

$$\| H\mathbf{x} - H\mathbf{x}^\star \|^2 = \| H\mathbf{x} - \mathbf{b} \|^2 = \sum_{i=1}^{k} c_i^2 \lambda_i^2 \ge \lambda_k^2 \sum_{i=1}^{k} c_i^2 = \lambda_k^2 \| \mathbf{x} - \mathbf{x}^\star \|^2.$$

Inserting this into (11.10) yields

$$F(\mathbf{x}) - F^\star \le \frac{1}{2} \| \mathbf{x} - \mathbf{x}^\star \| \, \| H\mathbf{x} - \mathbf{b} \| \le \frac{1}{2\lambda_k} \| H\mathbf{x} - \mathbf{b} \|^2,$$

which establishes the claim.   ▲

Even though none of the functions in Examples 11.4, 11.5, and 11.6 are strongly convex, they are μ-PL, and so Theorem 11.2 guarantees that gradient descent converges at a linear rate. Of course, not all convex functions are μ-PL; see Exercise 1.7. It is important to note, however, that Examples 11.4 and 11.6 show that the minimizers of μ-PL functions need not be unique. Thus, we cannot hope to prove a convergence rate of $\mathbf{x}_k \to \mathbf{x}^\star$, as we did in Remark 6.69, unless we can identify which $\mathbf{x}^\star$ the gradient descent iterations converge to. In some cases, we can make this identification, as shown in the theorem below.

Theorem 11.7. *Suppose that H is symmetric positive semidefinite, and let $\mathbf{b} \in \operatorname{img} H$. Let $\mu > 0$ denote the smallest nonzero eigenvalue of H, and let $0 < \alpha \le 1/\lambda_{max}(H)$. Let $\mathbf{x}_k$, for $k \ge 0$, be the iterations of the gradient descent algorithm (11.1) with fixed time step $\alpha_k = \alpha$ applied to the quadratic function (11.7), and let $\mathbf{x}^\star$ denote the orthogonal projection of $\mathbf{x}_0$ onto the affine solution space defined in (11.8). Then*

$$\frac{\mu}{2} \| \mathbf{x} - \mathbf{x}^\star \|^2 \le (1 - \alpha\mu)^k \left(F(\mathbf{x}_0) - F^\star \right). \tag{11.12}$$

Proof. Replacing $\mathbf{x}$ by $\mathbf{x}_k$ in (11.11) and using the orthonormality of the eigenvectors, we deduce

$$F(\mathbf{x}_k) - F^\star = \frac{1}{2}(\mathbf{x}_k - \mathbf{x}^\star)^T H (\mathbf{x}_k - \mathbf{x}^\star) = \frac{1}{2}\left(\sum_{i=1}^{k} c_i \mathbf{P}_i\right)^T \left(\sum_{j=1}^{k} c_j \lambda_j \mathbf{P}_j\right)$$

$$= \sum_{i=1}^{k} c_i^2 \lambda_i \geq \frac{\lambda_k}{2}\sum_{i=1}^{k} c_i^2 = \frac{\mu}{2}\|\mathbf{x} - \mathbf{x}^\star\|^2.$$

The proof is completed by invoking Theorem 11.2. $\blacksquare$

Remark 11.8. We finally remark that we can define μ-PL functions with respect to a general inner product $\langle \mathbf{x}, \mathbf{y}\rangle_C = \mathbf{x}^T C \mathbf{y}$, where C is symmetric positive definite. Indeed, Theorem 6.52 holds in this general setting. In this case we have $\nabla_C F(\mathbf{x}) = C^{-1}\nabla F(\mathbf{x})$ and (11.2) becomes

$$F(\mathbf{x}) - F^\star \leq \frac{1}{2\mu}\|\nabla_C F(\mathbf{x})\|_C^2 = \frac{1}{2\mu}\nabla F(\mathbf{x})^T C^{-1}\nabla F(\mathbf{x}) = \frac{1}{2\mu}\|\nabla F(\mathbf{x})\|_{C^{-1}}^2.$$

Exercises 1.5, 2.2, 4.2, and 5.7 below work with this general definition. $\blacktriangle$

Exercises

1.1. ♡ Prove the inequality (11.6).

1.2. ◇ Show that if F is μ-PL and ∇F is Lipschitz, then $\mu \leq \mathrm{Lip}(\nabla F)$.

1.3. ♡ Prove Theorem 11.2.
  Hint: The proof is nearly identical to Theorem 6.68. Use Lemma 6.64 and Exercise 1.2.

1.4. ♡ Show that the function F defined in (11.5) is not strongly convex.

1.5. ◇ Let F be the quadratic function defined in Example 11.6, with H symmetric positive definite. Let C be a symmetric positive definite matrix that is simultaneously complete with H — that is, the eigenvectors of H are also eigenvectors of C. Let μ_C be the μ-PL constant of F with respect to the inner product defined by $\langle \mathbf{x}, \mathbf{y}\rangle = \mathbf{x}^T C \mathbf{y}$, as described in Remark 11.8. Show that μ_C is the smallest nonzero eigenvalue of $C^{-1}H$. What choice of C makes $\mu_C = 1$? *Hint*: H is not invertible, so you cannot choose $C = H$.

1.6. Let x_k denote the iterations of gradient descent on the function F defined in (11.5). Which minimizer does x_k converge to, and how does it depend on the initial condition? State and prove a result similar to Theorem 11.7 showing a convergence rate for x_k towards a minimizer of F.

1.7. Let $1 < p \in \mathbb{R}$. Show that the scalar function $F(x) = |x|^p/p$ satisfies the modified PL inequality $F(x) - F^\star \leq |F'(x)|^{p/(p-1)}/p$.

1.8. Consider the general PL inequality

$$[F(\mathbf{x}) - F^\star]^\theta \leq \frac{1}{2\mu}\|\nabla F(\mathbf{x})\|^2 \qquad \text{for all} \qquad \mathbf{x} \in \mathbb{R}^n, \tag{11.13}$$

where $\theta > 0$. (The standard PL inequality (11.2) takes $\theta = 1$.) In the following parts, assume that F satisfies (11.13) for $\mu > 0$ and $\theta > 0$, ∇F is Lipschitz continuous, and $0 < \alpha \leq \mathrm{Lip}(\nabla F)^{-1}$. Let $\mathbf{x}_k$, for $k \geq 0$, be the iterations of the gradient descent algorithm (11.1) with fixed time step $\alpha_k = \alpha$. Also assume that $F(\mathbf{x}_0) - F^\star \leq 1$ and $\alpha \mu \leq 1$.

  (a) If $0 < \theta \leq 1$, show that $F(\mathbf{x}_k) - F^\star \leq (1 - \alpha \mu)^k \left[F(\mathbf{x}_0) - F^\star \right]$.

  (b) If $\theta > 1$, show that $F(\mathbf{x}_k) - F^\star \leq \left(\dfrac{1}{(\theta - 1)\alpha \mu k + 1} \right)^{1/(\theta - 1)}$.

Hint: For both parts, follow the proof of Theorem 6.68. For part (b), the results established in Example 6.67 will be useful.

11.2   The Heavy Ball Method

Even in the strongly convex setting, or, more generally, when the PL-inequality (11.2) holds, where gradient descent converges at a linear rate, the convergence may slow down considerably as the iterates approach the minimizer. This is due to the fact that the gradient ∇F vanishes at the minimizer, and hence is small nearby. Indeed, in the basic iterative equation (11.1), the change from $\mathbf{x}_k$ to $\mathbf{x}_{k+1}$ is $\alpha_k \nabla F(\mathbf{x}_k)$, which, for fixed time step α_k, becomes vanishingly small near the minimizer. To compensate, one can try to take larger time steps, but this often leads to the iterates bouncing back and forth across the energy landscape, thereby limiting progress towards the minimizer, or even leading to instability and nonconvergence. Figures 11.2(b) and 11.2(c) show examples of the bouncing effect of gradient descent for large time steps. In the figure, we are minimizing the quadratic function

$$F(x_1, x_2) = (x_1 + x_2)^2 + 8(x_1 - x_2)^2,$$

which is strongly convex with a global minimum at $x_1 = x_2 = 0$. These effects are the consequence of placing too much trust in the gradient direction, which in general does *not* point directly towards the minimizer.

   The convergence of gradient descent can be accelerated by utilizing *momentum*. In momentum methods, the negative gradient direction $-\nabla F(\mathbf{x}_k)$ is viewed as a *force* that causes *acceleration*, but does not instantaneously change the descent direction. One can imagine rolling a ball with some positive mass down the energy landscape defined by F; the presence of friction and other dissipative forces cause the ball to slow down. On the other hand, momentum can build up speed over time, provided the descent directions are similar over many steps, leading to faster convergence near the minimizer. When the descent directions change rapidly at each step, as in Figures 11.2(b) and 11.2(c), momentum acts to average out the directions over time and reduces the amount of bouncing and backtracking, leading again to faster convergence. In fact, it may even be possible for momentum to propel the method through spurious local minimizers, though it is challenging to study this rigorously.

   One of the oldest momentum based methods is the *heavy ball method* of Polyak [187], which iterates

$$\mathbf{x}_{k+1} = \mathbf{x}_k - \alpha \nabla F(\mathbf{x}_k) + \beta (\mathbf{x}_k - \mathbf{x}_{k-1}), \tag{11.14}$$

where α is the (fixed) time step, and $\beta > 0$ is a parameter. We note that the heavy ball method uses the previous direction moved, namely $\mathbf{x}_k - \mathbf{x}_{k-1}$, in combination with the negative gradient direction $-\nabla F(\mathbf{x}_k)$ to decide on a new descent direction. Since the heavy ball method uses two previous iterates, $\mathbf{x}_k$ and $\mathbf{x}_{k-1}$, we need two initial conditions $\mathbf{x}_0$ and $\mathbf{x}_1$. They are normally taken to be equal $\mathbf{x}_1 = \mathbf{x}_0$, or we can take $\mathbf{x}_1$ as the initial step of ordinary

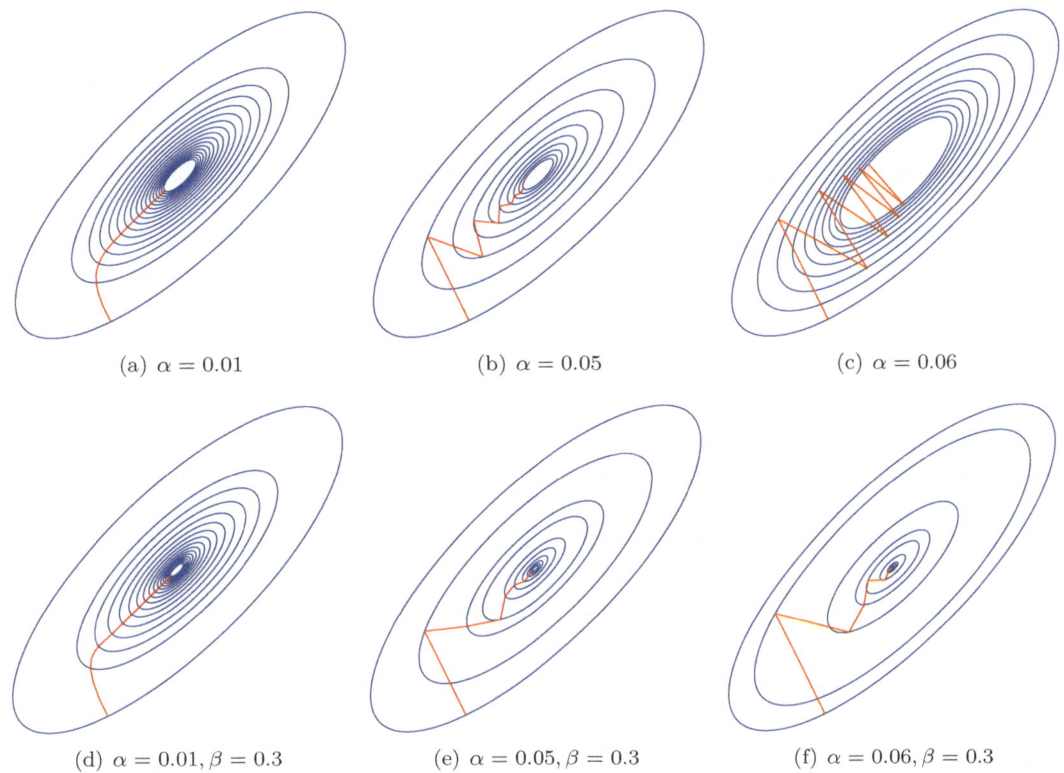

(a) $\alpha = 0.01$ (b) $\alpha = 0.05$ (c) $\alpha = 0.06$

(d) $\alpha = 0.01, \beta = 0.3$ (e) $\alpha = 0.05, \beta = 0.3$ (f) $\alpha = 0.06, \beta = 0.3$

Figure 11.2: Gradient descent (top row) and the heavy ball method (bottom row) applied to a quadratic function with different choices of time steps α and momentum parameter β. In (a) and (d) we took 50 steps, while in the other cases we took 10 steps. For gradient descent with larger time steps the iterations bounce back and forth, limiting progress towards the minimizer. For the heavy ball method, the bouncing effect is reduced and the method makes faster progress towards the minimizer, even in the absence of bouncing.

gradient descent starting at $\mathbf{x}_0$. Figures 11.2(e) and 11.2(f) show how momentum mitigates the bouncing effect seen in Figures 11.2(b) and 11.2(c) at large time steps. Even in the absence of bouncing, the heavy ball method makes faster progress toward the minimizer, as shown in Figures 11.2(a) and 11.2(d). We will see in Section 11.6 how the iterative system (11.14) is connected to the equations of motion for a rolling ball with friction.

Example 11.9. Consider the heavy ball method (11.14) in dimension $n = 1$. At, or nearby, a minimizer, where $F'(x_k) = 0$, it is close to the linear iterative equation

$$x_{k+1} = (1 + \beta)\, x_k - \beta\, x_{k-1}. \tag{11.15}$$

Setting $\mathbf{c}_k = (x_k, x_{k-1})^T \in \mathbb{R}^2$, we can write this as a vectorial first order iterative system[1]

$$\mathbf{c}_{k+1} = B\,\mathbf{c}_k, \qquad \text{where} \qquad B = \begin{pmatrix} 1+\beta & -\beta \\ 1 & 0 \end{pmatrix}, \qquad \text{and so} \qquad \mathbf{c}_k = B^{k-1}\,\mathbf{c}_1.$$

The matrix B has eigenvalues $1, \beta$, and thus, when $\beta > 1$, the iteration is *unstable*, with $\|\mathbf{c}_k\| \to \infty$ exponentially fast as $k \to \infty$ for most choices of initial condition $\mathbf{c}_1$. ▲

---

[1] This is analogous to the usual method of writing s second order ordinary differential equation as a first order system; see [30, 181].

Example 11.9 shows that we should choose the parameter $\beta \in [0,1]$. Values of $\beta \in [-1,0]$ would also be stable, as in Example 11.9, but this would amount to moving a small amount in the opposite direction from the previous step, which would slow down convergence, and is therefore not useful. For small positive values of β, the heavy ball method behaves in a similar manner as gradient descent. It turns out — see Theorem 11.10 below — that we need to take β close to 1 in order to see accelerated convergence.

Let us analyze the convergence of the heavy ball method in the case of the quadratic function (11.7) with symmetric positive definite coefficient matrix H. In this case, $\nabla F(\mathbf{x}) = H\mathbf{x} - \mathbf{b}$, and so the unique minimizer of F, denoted $\mathbf{x}^\star$, is also the unique solution of the linear system $H\mathbf{x}^\star = \mathbf{b}$. The heavy ball method in this setting becomes

$$\mathbf{x}_{k+1} = \mathbf{x}_k - \alpha\,(H\mathbf{x}_k - \mathbf{b}) + \beta\,(\mathbf{x}_k - \mathbf{x}_{k-1}). \tag{11.16}$$

The following result establishes a convergence rate in this simplified setting.

Theorem 11.10. *Let H be symmetric positive definite. Let $\mathbf{x}^\star$ be the unique solution of the linear system $H\mathbf{x}^\star = \mathbf{b}$. Let $\mathbf{x}_k$ for $k \geq 2$ satisfy the heavy ball iteration (11.16), and let $\mathbf{x}_1 = \mathbf{x}_0$. Assume that*

$$0 < \alpha \leq \frac{1}{\lambda_{max}(H)} \qquad and \qquad \left(1 - \sqrt{\alpha\,\lambda_{min}(H)}\,\right)^2 \leq \beta < 1. \tag{11.17}$$

Then for every $\varepsilon > 0$, there exists $N \geq 1$ such that

$$\|\mathbf{x}_k - \mathbf{x}^\star\| \leq \sqrt{2}\,\left(\sqrt{\beta} + \varepsilon\,\right)^k \|\mathbf{x}_0 - \mathbf{x}^\star\| \qquad for\ all \qquad k \geq N. \tag{11.18}$$

In particular, by choosing ε sufficiently small, we conclude that $\mathbf{x}_k \to \mathbf{x}^\star$ as $k \to \infty$.

Proof. Let $\mathbf{y}_k = \mathbf{x}_k - \mathbf{x}^\star$, in terms of which the heavy ball system (11.16) becomes

$$\mathbf{y}_{k+1} = (1 + \beta)\,\mathbf{y}_k - \alpha\,H\mathbf{y}_k - \beta\,\mathbf{y}_{k-1}.$$

We rewrite this second order iterative system as a first order system in $\mathbb{R}^{2n}$:

$$\mathbf{w}_{k+1} = M\,\mathbf{w}_k, \quad \text{where} \quad \mathbf{w}_k = \begin{pmatrix} \mathbf{y}_k \\ \mathbf{y}_{k-1} \end{pmatrix} \in \mathbb{R}^{2n}, \qquad M = \begin{pmatrix} (1+\beta)\,\mathrm{I} - \alpha H & -\beta\,\mathrm{I} \\ \mathrm{I} & \mathrm{O} \end{pmatrix},$$

which has size $2n \times 2n$. Thus, $\mathbf{w}_k = M^k\mathbf{w}_0$, and convergence of the iterates is determined by the spectral radius of M.

To determine the eigenvalues μ of M, first note that, provided $\beta \neq 0$, the matrix M has trivial kernel (why?), and hence 0 is not an eigenvalue. Suppose $M\mathbf{w} = \mu\mathbf{w}$. Writing

$$\mathbf{w} = \begin{pmatrix} \mathbf{y} \\ \mathbf{z} \end{pmatrix}, \quad \text{this requires} \quad (1 + \beta)\,\mathbf{y} - \alpha\,H\mathbf{y} - \beta\,\mathbf{z} = \mu\,\mathbf{y}, \qquad \mathbf{y} = \mu\,\mathbf{z}.$$

Since $\mu \neq 0$, this implies

$$H\mathbf{y} = \frac{1 + \beta - \mu - \beta/\mu}{\alpha}\,\mathbf{y}.$$

Thus, $\mathbf{y}$ must be an eigenvector of H whose eigenvalue λ is given by the fraction on the right hand side of the last equation. Equating the fraction to λ yields

$$\mu^2 - (1 + \beta - \alpha\,\lambda)\,\mu + \beta = 0. \tag{11.19}$$

Our assumption (11.15) implies that the discriminant of this quadratic equation is negative, and hence it has two complex conjugate roots, namely,

$$\mu_{\pm} = \frac{1 + \beta - \alpha\lambda \pm i\sqrt{4\beta - (1 + \beta - \alpha\lambda)^2}}{2},$$

which, by the preceding argument, are both (complex) eigenvalues of M. Moreover, a straight-forward calculation shows that they have the same modulus, $|\mu_{\pm}| = \sqrt{\beta}$, which is independent of the magnitude of the originating eigenvalue λ. Thus the matrix M has spectral radius $\rho(M) = \sqrt{\beta}$ and hence is convergent provided $|\beta| < 1$.

We now use the estimate (5.24) to deduce that, given $\varepsilon > 0$, there exists a positive integer N such that, for all $k \geq N$,

$$\|\mathbf{y}_k\| \leq \|\mathbf{w}_k\| = \|M^k\mathbf{w}_0\| \leq \left(\rho(M) + \varepsilon\right)^k \|\mathbf{w}_0\| = (\sqrt{\beta} + \varepsilon)^k \|\mathbf{w}_0\|.$$

Finally, we note that since we assumed $\mathbf{x}_1 = \mathbf{x}_0$, and hence $\mathbf{y}_1 = \mathbf{y}_0$,

$$\|\mathbf{w}_0\|^2 = \|\mathbf{y}_1\|^2 + \|\mathbf{y}_0\|^2 = 2\|\mathbf{y}_0\|^2.$$

Substituting this back into the preceding inequality and then replacing $\mathbf{y}_k = \mathbf{x}_k - \mathbf{x}^\star$ and $\mathbf{y}_0 = \mathbf{x}_0 - \mathbf{x}^\star$ completes the demonstration of (11.18).   ∎

Remark 11.11. Ignoring the ε term, which can be made arbitrarily small, Theorem 11.10 shows that heavy ball method for solving $H\mathbf{x} = \mathbf{b}$ converges at the linear rate $\sqrt{\beta}$ provided α and β satisfy the condition in equation (11.17). In particular, if $\beta \geq 1$ then the heavy ball method is nonconvergent. This also suggests the optimal choices $\alpha = 1/\lambda_{max}(H)$ and $\beta = \left(1 - \kappa^{-1/2}\right)^2$, where $\kappa = \kappa(H) = \lambda_{max}(H)/\lambda_{min}(H)$ is the condition number of H. In this case, the rate (11.18) becomes

$$\|\mathbf{x}_k - \mathbf{x}^\star\| \leq \sqrt{2}\left(1 - \kappa^{-1/2} + \varepsilon\right)^k \|\mathbf{x}_0 - \mathbf{x}^\star\|.$$

So, with the optimal choices of parameters α and β, and ignoring ε, the heavy ball method converges at the linear rate $1 - \kappa^{-1/2}$. We recall that gradient descent in the same setting converges at the rate $1 - \kappa^{-1}$; see (6.48). Thus, while both methods converge at a linear rate, the heavy ball rate is significantly faster, especially when the matrix H is poorly conditioned, with large condition number κ.

We also mention that the proof of Theorem 11.10 can be improved by allowing the time step α to be slightly larger, selecting

$$\alpha = \frac{4}{\left(\sqrt{\lambda_{min}(H)} + \sqrt{\lambda_{max}(H)}\right)^2}, \qquad \beta = \left(\frac{\sqrt{\kappa} - 1}{\sqrt{\kappa} + 1}\right)^2,$$

which yields the convergence rate

$$\sqrt{\beta} = \frac{\sqrt{\kappa} - 1}{\sqrt{\kappa} + 1} = \frac{1 - \kappa^{-1/2}}{1 + \kappa^{-1/2}}.$$

The proof of this rate is more tedious; we refer the reader to [187] for details. We note that when H is very poorly conditioned, so $\kappa \gg 1$, we can use the Taylor expansion $1/(1+x) \approx 1-x$ to obtain

$$\sqrt{\beta} \approx \left(1 - \kappa^{-1/2}\right)^2 = 1 - 2\kappa^{-1/2} + \kappa^{-1}. \tag{11.20}$$

Up to first order, so neglecting the smaller κ^{-1} term, the optimal rate differs from Theorem 11.10 simply by a factor of 2.   ▲

If we examine the proof of Theorem 11.10, we see that when H is positive semidefinite, and hence has a zero eigenvalue $\lambda = 0$ for some j, then the roots of the quadratic equation determining the eigenvalues of M are $\mu = \beta$ and 1. Thus, the proof would seem to fail in this case. However, the method is still convergent, similar to the case of gradient descent in Theorem 11.7. A very similar argument as in Theorem 11.10 shows that the residual $\|H\mathbf{x}_k - \mathbf{b}\|$ converges to zero at the same rate in this case; see Exercise 2.3.

The analysis of the heavy ball method for general convex functions F is far more complicated, and outside the scope of this book; we refer the reader to, for example, [84, 143]. Even for strongly convex functions, the method may sometimes fail to converge and get trapped in limit cycles [143].

Exercises

2.1. ♡ Suppose $\beta > 1$ in (11.15). For which initial conditions x_0, x_1 do the iterates *not* become unbounded as $k \to \infty$?

2.2. ◇ Consider the heavy ball method (11.14) where the gradient ∇F is defined with respect to an inner product $\langle \cdot, \cdot \rangle$ on $\mathbb{R}^n$. Formulate and prove a version of Theorem 11.10 in this setting.

2.3. ♡ Suppose that H is only positive semidefinite in Theorem 11.10. Assume $\mathbf{b} \in \operatorname{img} H$ so that there exists $\mathbf{x}^\star$ such that $H\mathbf{x}^\star = \mathbf{b}$. Let UU^T be the orthogonal projection matrix onto $\operatorname{img} H$. Show that for any choice of $\mathbf{x}^\star$ we have

$$\|UU^T(\mathbf{x}_k - \mathbf{x}^\star)\| \le \sqrt{2}\left(\sqrt{\beta} + \varepsilon\right)^k \|UU^T(\mathbf{x}_0 - \mathbf{x}^\star)\| \tag{11.21}$$

using all the same conditions as Theorem 11.10, except that $\lambda_{min}(H)$ is replaced by the smallest positive eigenvalue in (11.17). Can you prove anything about the convergence of $\mathbf{x}_k$ to $\mathbf{x}^\star$?

2.4. Implement the heavy ball method (11.16) in `numpy` in Python for a random symmetric positive definite matrix H. Try using the optimal parameters from Theorem 11.10 as well as those from Remark 11.11. Plot the error on a loglog plot and compare to gradient descent with optimal time step α. Find choices of parameters α, β for which the heavy ball method does not converge, and ones for which it converges but is slower than gradient descent.

2.5. Try your code from Exercise 2.4 for the heavy ball method on a matrix H that is positive semidefinite. You can construct a random semidefinite matrix by setting $H = A^T A$ for a random matrix $A \in \mathcal{M}_{k \times n}$ with $k < n$.

11.3   Krylov Subspace Methods and Conjugate Gradients

In recent years, for dealing with large sparse linear systems, such as those arising in machine learning and the numerical solution of partial differential equations, semidirect iterative methods based on Krylov subspaces have become quite popular. The original ideas were introduced in the 1930's by the Russian naval engineer Alexei Krylov, who was in search of an efficient and reliable method for numerically computing eigenvalues. In recent years, Krylov methods have been well developed in a wide variety of directions, [88, 181, 205, 236].

The starting point is an $n \times n$ matrix H, here assumed to be symmetric and positive definite. When dealing with large matrices, Krylov subspace methods are particularly effective when H is also sparse, meaning many of its entries are 0, but they can also be usefully applied to more general matrices.

Definition 11.12. Given an $n \times n$ real matrix H, the *Krylov subspace* of *order* $k \geq 1$ generated by a nonzero vector $\mathbf{0} \neq \mathbf{b} \in \mathbb{R}^n$ is the subspace $V_k \subset \mathbb{R}^n$ spanned by the vectors $\mathbf{b}, H\mathbf{b}, H^2\mathbf{b}, \ldots, H^{k-1}\mathbf{b}$. We also set $V_0 = \{\mathbf{0}\}$ by convention.

For example, if $\mathbf{b}$ is an eigenvector of H, so $H\mathbf{b} = \lambda\mathbf{b}$, then $V_2 = V_1$ is the one-dimensional eigenspace spanned by $\mathbf{b}$; conversely, if V_2 is one-dimensional, then $\mathbf{b}$ is necessarily an eigenvector, and hence $V_k = V_1$ for all $k \geq 1$. More generally, if $V_{j+1} = V_j$ for some $j \geq 0$, then $V_k = V_j$ for all $k \geq j$; the proof is left to the reader as Exercise 3.1.

An iterative method is called a *Krylov subspace method* if each iterate, starting with $\mathbf{x}_0 = \mathbf{0}$, belongs to a Krylov subspace: $\mathbf{x}_k \in V_k \subset \mathbb{R}^n$ for some $\mathbf{b}$. Many of the iterative methods considered in this text are, in fact, Krylov methods. An evident example is the power method for computing eigenvalues, which consists of successively multiplying a randomly chosen vector by the matrix H, possibly followed by dividing by the norm of the resulting vector; see Section 5.6.1 for details.

Ordinary gradient descent (11.1) for minimizing the quadratic function

$$F(\mathbf{x}) = \frac{1}{2}\mathbf{x}^T H\mathbf{x} - \mathbf{b} \cdot \mathbf{x} + c,$$

starting with $\mathbf{x}_0 = \mathbf{0}$, is a Krylov subspace method. The first few iterates are given by

$$\mathbf{x}_1 = \mathbf{x}_0 - \alpha(H\mathbf{x}_0 - \mathbf{b}) = \alpha\mathbf{b}, \qquad \mathbf{x}_2 = \mathbf{x}_1 - \alpha(H\mathbf{x}_1 - \mathbf{b}) = 2\alpha\mathbf{b} - \alpha^2 H\mathbf{b},$$

and

$$\mathbf{x}_3 = \mathbf{x}_2 - \alpha(H\mathbf{x}_2 - \mathbf{b}) = 3\alpha\mathbf{b} - 3\alpha^2 H\mathbf{b} + \alpha^3 H^2\mathbf{b}.$$

The same is true for the heavy ball method. In fact, if we initialize the heavy ball method from $\mathbf{x}_0 = \mathbf{x}_1 = \mathbf{0}$, then $\mathbf{x}_k \in V_{k-1}$, but if we initialize $\mathbf{x}_0 = \mathbf{0}$ and $\mathbf{x}_1 = \alpha\mathbf{b}$ with a step of gradient descent, then $\mathbf{x}_k \in V_k$.

We claim that the conjugate gradient method is also a Krylov subspace method. In fact, as we show below, it is, in a certain sense, the optimal Krylov method for minimizing a quadratic function.

We assume that the conjugate gradient algorithm has not terminated at step k, meaning that the k-th residual $\mathbf{r}_k = \mathbf{b} - H\mathbf{x}_k \neq \mathbf{0}$. (Otherwise we have solved the equation and nothing more needs to be done.) Referring back to (6.70), (6.71), (6.72), let us show that $\mathbf{x}_k$ and $\mathbf{v}_k$ belong to the Krylov subspace V_k and, moreover, $\mathbf{r}_k \in V_{k+1}$. This is easily proved by induction. First, $\mathbf{v}_0 = \mathbf{x}_0 = \mathbf{0}$ and $\mathbf{r}_1 = \mathbf{b} \in V_1$. As for the induction step, since $\mathbf{v}_{k+1}$ is a linear combination of $\mathbf{r}_k$ and $\mathbf{v}_k$, it belongs to V_{k+1}. Similarly, $\mathbf{x}_{k+1}$ is a linear combination of $\mathbf{x}_k$ and $\mathbf{v}_{k+1}$, so it also belongs to V_{k+1}, and hence is a linear combination of the vectors $\mathbf{b}, H\mathbf{b}, \ldots, H^k\mathbf{b}$. Therefore $H\mathbf{x}_{k+1}$ is a linear combination of $H\mathbf{b}, H^2\mathbf{b}, \ldots, H^{k+1}\mathbf{b}$ and hence belongs to V_{k+2}, as does $r_{k+1} = \mathbf{b} - H\mathbf{x}_{k+1}$, which completes the induction step.

We claim that $\mathbf{r}_k$ is orthogonal to the entire Krylov subspace V_k under the dot product. To prove this, we first note that $\mathbf{v}_1, \ldots, \mathbf{v}_k$ form a basis for V_k because they are H-orthogonal and nonzero: $\mathbf{0} \neq \mathbf{v}_j \in V_j \subset V_k$ for $j = 1, \ldots, k$; see Theorem 2.18. We again work by induction, the case $k = 0$ being trivial since $V_0 = \{\mathbf{0}\}$. To establish the induction step, we are given that the k-th residual $\mathbf{0} \neq \mathbf{r}_k \in V_{k+1}$ is orthogonal to V_k, and hence $\mathbf{v}_1, \ldots, \mathbf{v}_k, \mathbf{r}_k$

forms a basis for V_{k+1}. Moreover, since $\mathbf{r}_{k+1}$ is a linear combination of $\mathbf{r}_k$ and $H\mathbf{v}_{k+1}$, it too is orthogonal to V_k; moreover, by construction, it is also orthogonal to $\mathbf{r}_k$, and hence to the entire Krylov subspace V_{k+1}, which completes the induction step.

Since the residual $\mathbf{r}_k = \mathbf{b} - H\mathbf{x}_k$ is orthogonal to V_k, Corollary 6.13 immediately implies the following result, which says that among all Krylov subspace methods for minimizing a quadratic function, conjugate gradients is optimal in the sense that it minimizes the function the most at each step.

> **Theorem 11.13.** *Starting with $\mathbf{x}_0 = \mathbf{0}$, the conjugate gradient iterate $\mathbf{x}_k$ is the unique minimizer of the quadratic function $F(\mathbf{x})$ over all vectors $\mathbf{x}$ belonging to the Krylov subspace V_k.*

Next, let us investigate the convergence of Krylov subspace methods, focusing on conjugate gradients, for minimizing a quadratic function. Let $\mathbf{x}^\star = H^{-1}\mathbf{b}$ be the unique minimizer of $F(\mathbf{x})$. Using the identity in (6.15), given any $\mathbf{x} \in \mathbb{R}^n$,

$$0 \leq F(\mathbf{x}) - F(\mathbf{x}^\star) = \tfrac{1}{2}(\mathbf{x} - \mathbf{x}^\star)^T H (\mathbf{x} - \mathbf{x}^\star) \tag{11.22}$$

Let

$$\lambda_1 \geq \lambda_2 \geq \cdots \geq \lambda_n > 0$$

be the eigenvalues of the coefficient matrix H, ordered, as usual from largest to smallest. According to the minimization principle in (5.50), the right hand side of (11.22) can be bounded from below by

$$F(\mathbf{x}) - F(\mathbf{x}^\star) \geq \frac{\lambda_n}{2}\,\|\mathbf{x} - \mathbf{x}^\star\|^2, \qquad \text{for all} \qquad \mathbf{x} \in \mathbb{R}^n, \tag{11.23}$$

where λ_n is the smallest eigenvalue of H. Let us next establish an upper bound, assuming that $\mathbf{x}$ belongs to a Krylov subspace.

> **Lemma 11.14.** *Suppose*
>
> $$\mathbf{x} = a_0 \mathbf{b} + a_1 H\mathbf{b} + \cdots + a_{k-1} H^{k-1}\mathbf{b} \in V_k. \tag{11.24}$$
>
> *Then,*
>
> $$F(\mathbf{x}) - F(\mathbf{x}^\star) \leq \frac{1}{2}\left(\mathbf{x}^{\star T} H \mathbf{x}^\star\right) \max_{1 \leq i \leq n} q(\lambda_i)^2, \tag{11.25}$$
>
> *where*
>
> $$q(\lambda) = 1 - a_0\lambda - a_1\lambda^2 - \cdots - a_{k-1}\lambda^k. \tag{11.26}$$

Proof. Let $\mathbf{v}_1, \ldots, \mathbf{v}_n$ be the orthonormal eigenvectors of H, and write $\mathbf{x}^\star = \sum_{i=1}^{n} c_i \mathbf{v}_i$. Since $\mathbf{b} = H\mathbf{x}^\star$, we can write

$$\mathbf{x} = \sum_{j=0}^{k-1} a_j H^j H\mathbf{x}^\star = \sum_{j=0}^{k-1} a_j H^{j+1} \sum_{i=1}^{n} c_i \mathbf{v}_i = \sum_{i=1}^{n} c_i \sum_{j=0}^{k-1} a_j \lambda_i^{j+1} \mathbf{v}_i.$$

Thus,

$$\mathbf{x} - \mathbf{x}^\star = \sum_{i=1}^{n} c_i \left(\sum_{j=0}^{k-1} a_j \lambda_i^{j+1} - 1\right) \mathbf{v}_i = -\sum_{i=1}^{n} c_i\, q(\lambda_i)\, \mathbf{v}_i.$$

In view of the identity in (11.22) and the orthonormality of the eigenvectors, it follows that

$$F(\mathbf{x}) - F(\mathbf{x}^\star) = \tfrac{1}{2}(\mathbf{x} - \mathbf{x}^\star)^T H (\mathbf{x} - \mathbf{x}^\star) = \frac{1}{2}\sum_{i=1}^{n} c_i^2 \lambda_i\, q(\lambda_i)^2.$$

We now make the estimate

$$F(\mathbf{x}) - F(\mathbf{x}^\star) \le \frac{1}{2}\left(\sum_{i=1}^{n} c_i^2 \lambda_i\right) \max_{1\le i\le n} q(\lambda_i)^2 = \tfrac{1}{2}(\mathbf{x}^{\star T} H \mathbf{x}^\star) \max_{1\le i\le n} q(\lambda_i)^2. \qquad \blacksquare$$

There are two issues confronting the application of the estimate (11.25) to a Krylov subspace iterate $\mathbf{x}_k \in V_k$. First of all, since the vectors $H^j\mathbf{b}$ do not form an orthogonal basis of the Krylov subspace, it may not be so easy to determine the coefficients $\mathbf{a}_j$ of the vector $\mathbf{x}_k$ or, equivalently, the polynomial $q(\lambda)$. Second, even if we know $q(\lambda)$, we need to know, or at least approximate, its values on all the eigenvalues of H on order to get a handle on the upper bound.

Fortunately, in the case of the conjugate gradient algorithm, the minimization property of the iterates in Theorem 11.13 enable us to overcome these difficulties. The first remark is that since we can bound

$$F(\mathbf{x}_k) - F(\mathbf{x}^\star) \le F(\mathbf{x}) - F(\mathbf{x}^\star)$$

for *any* $\mathbf{x} \in V_k$, and we can bound the right hand side using (11.25) for $\mathbf{x}$ instead of $\mathbf{x}_k$, we can take the coefficients in (11.24) to be anything we like. Thus,

$$F(\mathbf{x}_k) - F(\mathbf{x}^\star) \le \tfrac{1}{2}(\mathbf{x}^{\star T} H \mathbf{x}^\star) \max_{1\le i\le n} q(\lambda_i)^2, \qquad (11.27)$$

where $q(\lambda)$ can be *any* polynomial of degree k whatsoever, provided only that $q(0) = 1$. The name of the game is then to choose a particular polynomial that is small when evaluated on the eigenvalues of H.

For this purpose, we recall the k-th order *Chebyshev polynomial*, which we encountered in recursive form in (10.46). They are, in fact, defined by the equation[2]

$$T_k(x) = \begin{cases} \cos(k\arccos x), & |x| \le 1, \\[2mm] \dfrac{\left(x + \sqrt{x^2 - 1}\right)^k + \left(x - \sqrt{x^2 - 1}\right)^k}{2}, & |x| \ge 1. \end{cases} \qquad (11.28)$$

They clearly satisfy

$$|T_k(x)| \le 1, \qquad |x| \le 1. \qquad (11.29)$$

On the other hand, because $T_k(x)$ is a polynomial of degree k, it grows quickly, as $O(x^k)$, outside of the interval $[-1, 1]$. Further details, including the fact that they satisfy the recursive equation (10.46), are relegated to Exercise 3.2.

By our ordering convention, the eigenvalues of H lie in the interval $0 < \lambda_n \le \lambda_j \le \lambda_1$. Thus, in order to apply the Chebyshev polynomial T_k to select q, we need to shift and scale the Chebyshev polynomials in order to adapt them to the range of the eigenvalues $[\lambda_n, \lambda_1]$. Specifically, let us assume $\lambda_1 > \lambda_n$.[3] Set

$$P_k(\lambda) = T_k\left(\frac{\lambda_1 + \lambda_n - 2\lambda}{\lambda_1 - \lambda_n}\right), \qquad Q_k(\lambda) = \frac{P_k(\lambda)}{P_k(0)}. \qquad (11.30)$$

---

[2] The fact that these two formulas define the same polynomial is a part of Exercise 3.2.

[3] If they are equal, then H has all equal eigenvalues, and is hence a multiple of the identity matrix. In this case, finding the minimizer of the corresponding quadratic function is trivial and need not concern us.

If $P_k(0) \neq 0$, then $Q_k(\lambda)$ is a polynomial of degree k and satisfies $Q_k(0) = 1$. Moreover, since

$$-1 \leq \frac{\lambda_1 + \lambda_n - 2\lambda}{\lambda_1 - \lambda_n} \leq 1 \qquad \text{whenever} \qquad \lambda_1 \leq \lambda \leq \lambda_n,$$

we have

$$|Q_k(\lambda)| \leq \frac{1}{|P_k(0)|} \qquad \text{when} \qquad \lambda_1 \leq \lambda \leq \lambda_n.$$

Using $q = Q_k$ in (11.27), we conclude that the conjugate gradient iterates satisfy

$$F(\mathbf{x}_k) - F(\mathbf{x}^\star) \leq \frac{\mathbf{x}^{\star T} H \mathbf{x}^\star}{2\, P_k(0)^2}. \tag{11.31}$$

It thus remains to determine the value of $P_k(0)$. According to (11.30),

$$P_k(0) = T_k\left(\frac{\lambda_1 + \lambda_n}{\lambda_1 - \lambda_n}\right) = T_k\left(\frac{\kappa + 1}{\kappa - 1}\right), \qquad \text{where} \qquad \kappa = \kappa(H) = \frac{\lambda_1}{\lambda_n}$$

is the condition number of H; see Remark 5.83. Since $\kappa > 1$, we have $(\kappa+1)/(\kappa-1) > 1$, and hence we must use the second definition in (11.28) to evaluate $P_k(0)$. A short calculation, left to the reader as part of Exercise 3.2, demonstrates that

$$P_k(0) = T_k\left(\frac{\kappa+1}{\kappa-1}\right) = \frac{1}{2}\left[\left(\frac{\sqrt{\kappa}+1}{\sqrt{\kappa}-1}\right)^k + \left(\frac{\sqrt{\kappa}-1}{\sqrt{\kappa}+1}\right)^k\right] \geq \frac{1}{2}\left(\frac{\sqrt{\kappa}+1}{\sqrt{\kappa}-1}\right)^k.$$

Using the latter estimate in (11.31) combined with (11.23) completes the proof of the following.

Theorem 11.15. *Given a quadratic function $F(\mathbf{x}) = \frac{1}{2}\mathbf{x}^T H \mathbf{x} - \mathbf{x}^T \mathbf{b} + c$, with H symmetric positive definite, let $\mathbf{x}^\star = H^{-1}\mathbf{b}$ be the unique minimizer. Let $\mathbf{x}_k$ be the conjugate gradient iterates, starting with $\mathbf{x}_0 = \mathbf{0}$. Then*

$$\frac{\lambda_{min}(H)}{2}\|\mathbf{x}_k - \mathbf{x}^\star\|^2 \leq F(\mathbf{x}_k) - F(\mathbf{x}^\star) \leq 2\left(\mathbf{x}^{\star T} H \mathbf{x}^\star\right)\left(\frac{\sqrt{\kappa}-1}{\sqrt{\kappa}+1}\right)^{2k}, \tag{11.32}$$

where $\kappa = \kappa(H)$ is the condition number of H.

Notice that the convergence rate in Theorem 11.15 is the same as the asymptotic rate for the heavy ball method with optimal choices of parameters given in Remark 11.11. It is quite remarkable that such a simple two step linear iteration like the heavy ball method can produce an optimal iterative technique. However, the reader should be cautioned that in the heavy ball method, the choices of α and β are crucial; for poor choices the method may not converge, or may converge more slowly than gradient descent; see Exercise 2.4.

In Figure 11.4 we show a comparison of the heavy ball method, the conjugate gradient method, and gradient descent for minimizing a quadratic function F with $H \in \mathcal{M}_{n \times n}$ with $n = 100$. For future reference, we also show Nesterov's accelerated gradient descent method described in Section 11.4. We used the optimal choices of α and β from Remark 11.11 for the heavy ball method, and set $\alpha = 1/\lambda_{max}(H)$ for gradient descent. We set $H = A^T A + I$, where A is a Gaussian random matrix with zero mean and variance 1. We see that both heavy ball and conjugate gradient converge at a significantly faster rate than gradient descent, while the rates for heavy ball and conjugate gradient are similar for a period of time. The conjugate

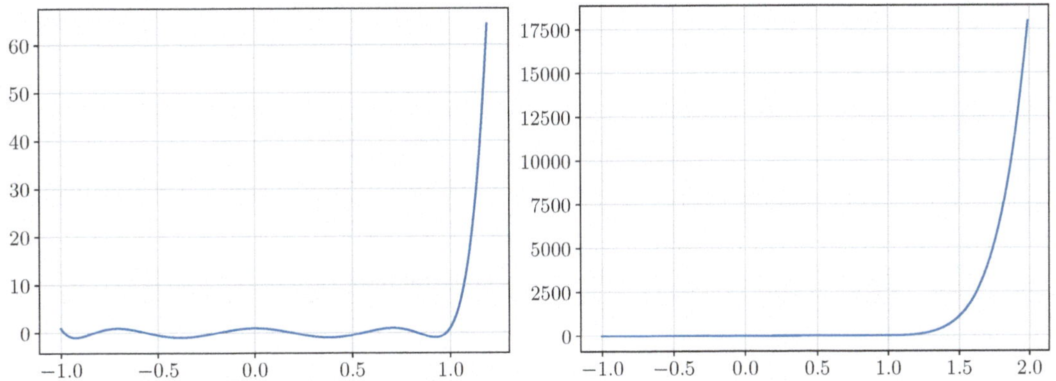

Figure 11.3: Plots of the Chebyshev polynomial T_8 on two different scales.

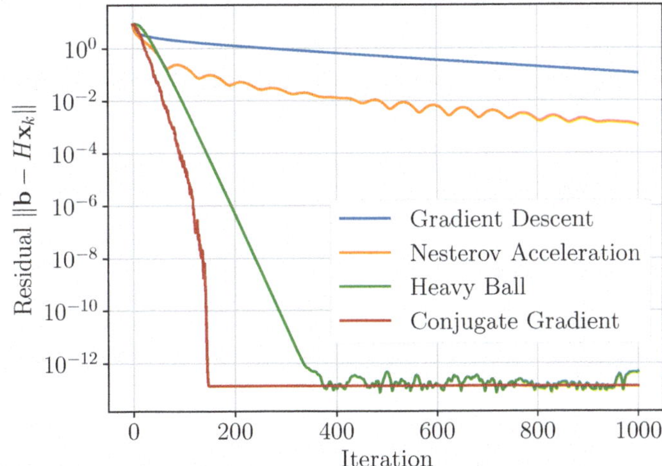

Figure 11.4: Comparison of the heavy ball method, conjugate gradient, gradient descent, and Nesterov acceleration (see Section 11.4) for minimizing a quadratic function with $H \in \mathcal{M}_{n \times n}$ and $n = 100$.

gradient method, being an optimal algorithm, converges more quickly to machine precision in practice. It also has the advantage of having no parameters to tune, like the α and β in the heavy ball method. In fact, it is possible to write the conjugate gradient method as a two step recursion for $\mathbf{x}_k$, like the heavy ball method (11.16), except that the time step α and momentum parameter β depend on k; see Exercise 3.7.

It is also interesting to note that the conjugate gradient method does not converge to the exact solution after $n = 100$ steps, which is should according to the discussion in Section 6.5. This is due to accumulating floating point roundoff errors, which cause the vectors $\mathbf{v}_1, \mathbf{v}_2, \ldots$ to fail to be H orthogonal. This is similar to how the Gram-Schmidt algorithm can suffer from loss of orthogonality for ill-conditioned matrices. Reorthogonalization can be used to improve the conjugate gradient method [129]. We also mention that the conjugate gradient method and gradient descent are both *descent methods*, which are guaranteed to decrease the objective F at every step (up to roundoff errors for conjugate gradient), while the heavy ball method does not have the descent property. Indeed, we can see in Figure 11.4 that the heavy ball method increases the residual considerably at the start.

Exercises

3.1. ♡　Prove by induction that if the Krylov subspaces satsify $V_{j+1} = V_j$ for some $j \geq 0$, then $V_k = V_j$ for all $k \geq j$.

3.2. ♡　Let $T_k(x)$ denote the k-th order Chebyshev polynomial (11.28).

(a) Show that both formulas in (11.28) satisfy the recursive formula (10.46). *Hint*: Use the trigonometric identity $\cos(x+y) + \cos(x-y) = 2\cos x \cos y$.

(b) Explain why both formulas define the same polynomial of degree k.

(c) Write down expressions for the Chebyshev polynomials $T_3(x)$ and $T_4(x)$.

(d) For $\kappa > 1$, show that $\quad T_k\left(\dfrac{\kappa+1}{\kappa-1}\right) = \dfrac{1}{2}\left[\left(\dfrac{\sqrt{\kappa}+1}{\sqrt{\kappa}-1}\right)^k + \left(\dfrac{\sqrt{\kappa}-1}{\sqrt{\kappa}+1}\right)^k\right].$

3.3.　Implement the conjugate gradient method in **numpy** in Python and compare against the heavy ball method and gradient descent, with optimal choices of parameters for each. As in Exercise 2.4, use a random symmetric positive definite matrix H and examine the error on a loglog plot.

3.4.　Try your code from Exercise 3.3 for the conjugate gradient method on a matrix H that is positive semidefinite.

3.5. ♡　Choose a polynomial $q(\lambda)$ in the proof of Theorem 11.15 to show that if H has at exactly k distinct eigenvalues then $\mathbf{x}_k = \mathbf{x}^\star$.

3.6. ◇　Suppose that $\mathbf{x}^\star = c_1 \mathbf{v}_1 + \cdots + c_n \mathbf{v}_n$ and that at most k of the coefficients c_i are nonzero. Modify the proof of Theorem 11.15 to show that $\mathbf{x}_k = \mathbf{x}^\star$.

3.7.　Suppose that $\mathbf{x}_0 = \mathbf{0}$, $\mathbf{x}_1 = \alpha_1 \mathbf{b}$. Show that the subsequent iterates of the conjugate gradient method satisfy

$$\mathbf{x}_{k+1} = \mathbf{x}_k - \alpha_{k+1}(H\mathbf{x}_k - \mathbf{b}) + \frac{\alpha_{k+1}\beta_k}{\alpha_k}(\mathbf{x}_k - \mathbf{x}_{k-1}), \qquad k \geq 1.$$

Can you find formulas for α_{k+1}, α_k, and β_k that depend only on $\mathbf{x}_k$ and $\mathbf{x}_{k-1}$? (Refer to Section 6.5 for the explicit conjugate gradient iteration).

11.4　Nesterov's Accelerated Gradient Descent

Nesterov's accelerated gradient descent [172] is a momentum-based descent method that overcomes the convergence issues appearing in the heavy ball method, and, when F is convex, is provably convergent with an optimal rate. There are several versions; the one we will study has the form

$$\mathbf{x}_{k+1} = \mathbf{x}_k + \beta_k(\mathbf{x}_k - \mathbf{x}_{k-1}) - \alpha\nabla F\big(\mathbf{x}_k + \beta_k(\mathbf{x}_k - \mathbf{x}_{k-1})\big). \tag{11.33}$$

The key differences between this and the heavy ball method (11.14) are that the momentum parameter β_k is allowed to vary with the iteration index, and it uses a *look-ahead* inside ∇F, whereby the gradient is evaluated *after* the momentum step. Intuitively, this allows for a correction when the momentum step moves in the wrong direction.

It is common to write *Nesterov's method* as a two step procedure in the form

$$\mathbf{y}_k = \mathbf{x}_k + \beta_k(\mathbf{x}_k - \mathbf{x}_{k-1}), \qquad \mathbf{x}_{k+1} = \mathbf{y}_k - \alpha\nabla F(\mathbf{y}_k), \qquad k \geq 1, \tag{11.34}$$

starting from an initial guess $\mathbf{x}_0 = \mathbf{x}_1$. The momentum parameter β_k can be chosen in various ways. The original method in [172] uses

$$\beta_k = \frac{\lambda_{k-1} - 1}{\lambda_k}, \quad \text{where} \quad \lambda_0 = 0, \quad \text{and} \quad \lambda_k = \frac{1 + \sqrt{1 + 4\lambda_{k-1}^2}}{2}, \quad \text{for} \quad k \geq 1. \quad (11.35)$$

This choice of λ_k may seem opaque at first, so we make a few important observations. First, note that $\lambda_1 = 1$, which means that $\beta_1 = -1$, so $\mathbf{y}_1 = \mathbf{x}_0$. Further, $\beta_2 = 0$, and so $\mathbf{y}_2 = \mathbf{x}_2$. We also note that the choice of λ_k in (11.35) was made so that the following holds:

$$\lambda_{k-1}^2 = \lambda_k^2 - \lambda_k = \lambda_k(\lambda_k - 1). \quad (11.36)$$

Indeed, (11.36) is a quadratic equation for λ_k, and the formula in (11.35) is simply the quadratic formula for its largest root. In fact, the reader can note that in the proof of Theorem 11.18 below, formula (11.36) is used directly, and not (11.35).

We also note that λ_k grows linearly in k, and in fact we have the following result.

Proposition 11.16. *For all $k \geq 1$ we have*

$$\frac{k}{2} \leq \lambda_k \leq \frac{k}{2} + \frac{\log(k-1) + 3}{4}. \quad (11.37)$$

Proof. According to (11.35), $\lambda_k \geq \frac{1}{2} + \lambda_{k-1}$. Since $\lambda_0 = 0$, we conclude by induction that $\lambda_k \geq \frac{1}{2}k$ for all $k \geq 0$. Now, the upper bound is clearly true for $k = 1$, since $\lambda_1 = 1$, so we can assume $k \geq 2$ from here on. We now use the inequality

$$\sqrt{x+1} \leq \sqrt{x} + \frac{1}{2\sqrt{x}}, \quad \text{for all} \quad x > 0,$$

which holds because the right hand side is the tangent line to the *concave* function $f(t) = \sqrt{x+t}$ centered at the point $t = 0$ and evaluated at $t = 1$; see Theorem 6.39. Applying this with $x = 4\lambda_{k-1}^2$ yields

$$\lambda_k \leq \frac{1}{2} + \frac{1}{2}\left(2\lambda_{k-1} + \frac{1}{4\lambda_{k-1}}\right) \leq \frac{1}{2} + \lambda_{k-1} + \frac{1}{4(k-1)},$$

where we used the lower bound in the last inequality. Replacing k with j and summing both sides of the preceding formula from $j = 2$ to $j = k$ produces

$$\lambda_k - \lambda_1 \leq \frac{k-1}{2} + \frac{1}{4}\sum_{j=2}^{k}\frac{1}{j-1} \leq \frac{k-1}{2} + \frac{1}{4} + \frac{1}{4}\int_2^k \frac{1}{x-1}\,dx = \frac{k}{2} + \frac{\log(k-1) - 1}{4}.$$

Since $\lambda_1 = 1$, the proof is complete.   ∎

Remark 11.17. Proposition 11.16 says that $\lambda_k \approx \frac{1}{2}k$ as $k \to \infty$. Thus, the momentum coefficient β_k is asymptotic to

$$\beta_k = \frac{\lambda_{k-1} - 1}{\lambda_k} \approx \frac{\frac{1}{2}(k-1) - 1}{\frac{1}{2}k} = 1 - \frac{3}{k} \longrightarrow 1 \quad \text{as} \quad k \longrightarrow \infty. \quad (11.38)$$

This is sensible for general convex optimization, where the objective function may be nearly flat around the minimizer. In this case, the Hessian $\nabla^2 F(\mathbf{x})$, which is H in Theorem 11.10,

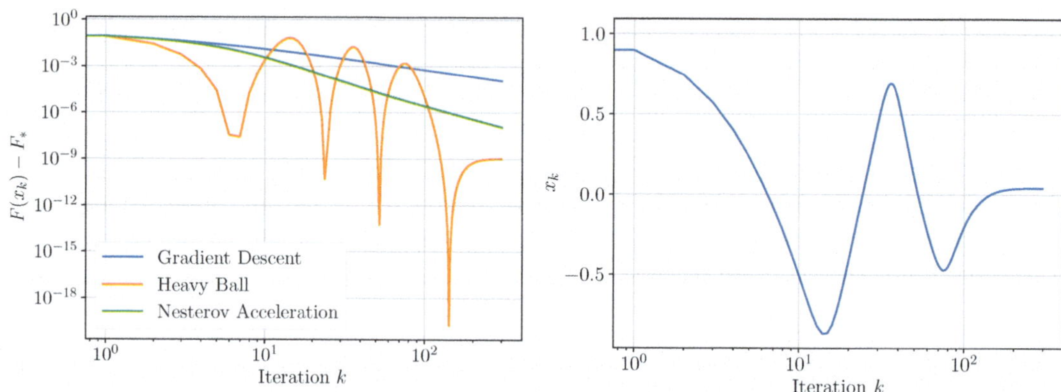

Figure 11.5: In (a) we show a comparison of gradient descent, the heavy ball method, and Nesterov acceleration for minimizing the function $F(x) = \frac{1}{6}x^6$, which is convex, but not strongly. In (b) we show the iterates x_k of the heavy ball method. The "ball" is rolling too quickly and oscillates rapidly about the global minimizer, before getting stuck slightly above the minimizer.

may be semidefinite at or near the minimizer, and so the momentum parameter β must be very close to 1 in order to satisfy the convergence requirement (11.17).

We also mention that it is common in Nesterov acceleration to replace the momentum update step in (11.34) with the asymptotic equivalent

$$\mathbf{y}_k = \mathbf{x}_k + \left(1 - \frac{3}{k}\right)(\mathbf{x}_k - \mathbf{x}_{k-1}), \tag{11.39}$$

at least for $k \geq 3$, since the method is simpler to implement and interpret. ▲

We now turn to establishing a convergence rate for Nesterov's accelerated gradient descent method. The proof is rather technical, and so is placed at the end of this section. We also note that a simpler proof of a related result is given in the continuum setting in Section 11.6.

Theorem 11.18. *Assume F is convex and ∇F is Lipschitz. Let $\mathbf{x}^\star$ be any minimizer of F. Let $\mathbf{x}_k$ and $\mathbf{y}_k$ satisfy Nesterov's accelerated gradient descent iteration (11.34), and assume that $0 < \alpha \leq \mathrm{Lip}(\nabla F)^{-1}$. Then for $k \geq 2$,*

$$F(\mathbf{x}_k) - F(\mathbf{x}^\star) \leq \frac{2\|\mathbf{x}_0 - \mathbf{x}^\star\|^2}{\alpha(k-1)^2}. \tag{11.40}$$

Theorem 11.18 shows that Nesterov's accelerated gradient descent method converges at a rate of $\mathrm{O}(k^{-2})$ for convex functions, which is a clear improvement over the $\mathrm{O}(k^{-1})$ rate for gradient descent in Theorem 6.65. It turns out the $\mathrm{O}(k^{-2})$ rate is optimal for minimizing convex functions with first order gradient-based methods [173]. While the version of Nesterov's accelerated gradient descent that we introduced is designed for general convex functions, it also often works well to give accelerated convergence rates for strongly convex functions. In addition, there are various modifications of the method that can be made to improve its convergence rate in the strongly convex case; we refer the reader to [173].

Figure 11.5(a) shows a comparison of gradient descent, the heavy ball method, and Nesterov's accelerated gradient descent for minimizing the function $F(x) = \frac{1}{6}x^6$, which is convex, but not strongly convex. We chose $\alpha = 0.1$ in all methods, and $\beta = 0.96$ for the heavy ball method, which we obtained by trial and error. Using a line of best fit near the end of the plot, gradient descent converges at a rate of $O(k^{-1.48})$, while Nesterov acceleration converges at a rate of $O(k^{-2.87})$. Both are are better than the theoretical $O(k^{-1})$ and $O(k^{-2})$ rates, respectively, which allow for even flatter functions, like $F(x) = x^p/p$ for $p \gg 0$.

We also mention that the heavy ball method seems to be rather unstable and oscillates rapidly. It also does not appear to converge, leveling off at an error of about 10^{-9}. In Figure 11.5(b) we plot the iterates x_k of the heavy ball method, where we can see that the "ball" is oscillating around the minimizer. Every downward spike in Figure 11.5(a) corresponds to a zero crossing of the iterates x_k, whereby the iterate overshoots the minimum and begins to climb up the other side of the energy well. Figure 11.4 in Section 11.2 shows the performance of Nesterov acceleration on a strongly convex quadratic function, where it performs better than gradient descent, but not as well as the heavy ball method or the conjugate gradient method, both of which are designed for positive definite quadratic functions.

Proof of Theorem 11.18. By (11.34),

$$\nabla F(\mathbf{y}_k) = -\frac{1}{\alpha}(\mathbf{x}_{k+1} - \mathbf{y}_k).$$

Combining this with Lemma 6.64 produces

$$F(\mathbf{x}_{k+1}) \le F(\mathbf{y}_k) - \frac{1}{2\alpha}\|\mathbf{x}_{k+1} - \mathbf{y}_k\|^2. \tag{11.41}$$

Since F is convex, we can use (6.91) and (11.34) again to obtain, for any $\mathbf{x} \in \mathbb{R}^n$,

$$F(\mathbf{x}) \ge F(\mathbf{y}_k) + \nabla F(\mathbf{y}_k) \cdot (\mathbf{x} - \mathbf{y}_k) = F(\mathbf{y}_k) - \frac{1}{\alpha}(\mathbf{x}_{k+1} - \mathbf{y}_k) \cdot (\mathbf{x} - \mathbf{y}_k),$$

which can be rearranged to read

$$F(\mathbf{y}_k) \le F(\mathbf{x}) - \frac{1}{\alpha}(\mathbf{x}_{k+1} - \mathbf{y}_k) \cdot (\mathbf{y}_k - \mathbf{x}).$$

Inserting this into (11.41), we have

$$F(\mathbf{x}_{k+1}) - F(\mathbf{x}) \le -\frac{1}{2\alpha}\left[\|\mathbf{x}_{k+1} - \mathbf{y}_k\|^2 + 2(\mathbf{x}_{k+1} - \mathbf{y}_k) \cdot (\mathbf{y}_k - \mathbf{x})\right]. \tag{11.42}$$

We now use (11.42) with $\mathbf{x} = \mathbf{x}_k$, and multiply both sides by $\lambda_k - 1$:

$$(\lambda_k - 1)\left[F(\mathbf{x}_{k+1}) - F(\mathbf{x}_k)\right] \le -\frac{\lambda_k - 1}{2\alpha}\left[\|\mathbf{x}_{k+1} - \mathbf{y}_k\|^2 + 2(\mathbf{x}_{k+1} - \mathbf{y}_k) \cdot (\mathbf{y}_k - \mathbf{x}_k)\right]. \tag{11.43}$$

We also set $\mathbf{x} = \mathbf{x}^\star$ in (11.42) to obtain

$$F(\mathbf{x}_{k+1}) - F(\mathbf{x}^\star) \le -\frac{1}{2\alpha}\left[\|\mathbf{x}_{k+1} - \mathbf{y}_k\|^2 + 2(\mathbf{x}_{k+1} - \mathbf{y}_k) \cdot (\mathbf{y}_k - \mathbf{x}^\star)\right]. \tag{11.44}$$

We now carefully add (11.43) and (11.44). Adding the left hand sides produces

$$(\lambda_k - 1)\left[F(\mathbf{x}_{k+1}) - F(\mathbf{x}_k)\right] + F(\mathbf{x}_{k+1}) - F(\mathbf{x}^\star)$$
$$= \lambda_k F(\mathbf{x}_{k+1}) - (\lambda_k - 1)F(\mathbf{x}_k) - F(\mathbf{x}^\star) = \lambda_k r_{k+1} - (\lambda_k - 1)r_k,$$

where $r_k = F(\mathbf{x}_k) - F(\mathbf{x}^\star)$. Proceeding to add the right hand sides of (11.43) and (11.44) as well, we obtain

$$\lambda_k r_{k+1} - (\lambda_k - 1) r_k \leq -\frac{\lambda_k}{2\alpha} \|\mathbf{x}_{k+1} - \mathbf{y}_k\|^2 - \frac{1}{\alpha}\left[(\mathbf{x}_{k+1} - \mathbf{y}_k) \cdot (\lambda_k \mathbf{y}_k - (\lambda_k - 1)\mathbf{x}_k - \mathbf{x}^\star) \right].$$

We now multiply by λ_k on both sides, use (11.36), and complete the square:

$$\begin{aligned}
\lambda_k^2 r_{k+1} - \lambda_{k-1}^2 r_k &= \lambda_k^2 r_{k+1} - \lambda_k(\lambda_k - 1) r_k \\
&\leq -\frac{1}{2\alpha}\left[\|\lambda_k(\mathbf{x}_{k+1} - \mathbf{y}_k)\|^2 + 2\lambda_k(\mathbf{x}_{k+1} - \mathbf{y}_k) \cdot (\lambda_k \mathbf{y}_k - (\lambda_k - 1)\mathbf{x}_k - \mathbf{x}^\star) \right] \\
&= -\frac{1}{2\alpha}\left[\|(\lambda_k - 1)(\mathbf{x}_{k+1} - \mathbf{x}_k) + \mathbf{x}_{k+1} - \mathbf{x}^\star\|^2 - \|\lambda_k \mathbf{y}_k - (\lambda_k - 1)\mathbf{x}_k - \mathbf{x}^\star\|^2 \right] \\
&= -\frac{1}{2\alpha}\left[\|\lambda_{k+1}\mathbf{y}_{k+1} - (\lambda_{k+1} - 1)\mathbf{x}_{k+1} - \mathbf{x}^\star\|^2 - \|\lambda_k \mathbf{y}_k - (\lambda_k - 1)\mathbf{x}_k - \mathbf{x}^\star\|^2 \right],
\end{aligned}$$

where we used (11.34) to simplify in the last line. We now sum both sides of the preceding inequality from $k = 1$ to $j - 1$. Both sums telescope, and so we find

$$\lambda_{j-1}^2 r_j - \lambda_0^2 r_1 \leq -\frac{1}{2\alpha}\left[\|\lambda_j \mathbf{y}_j - (\lambda_j - 1)\mathbf{x}_j - \mathbf{x}^\star\|^2 - \|\lambda_1 \mathbf{y}_1 - (\lambda_1 - 1)\mathbf{x}_1 - \mathbf{x}^\star\|^2 \right].$$

We now replace j with k, and use that $\lambda_0 = 0$ and $\lambda_1 = 1$ to obtain

$$\lambda_{k-1}^2\left[F(\mathbf{x}_k) - F(\mathbf{x}^\star) \right] = \lambda_{k-1}^2 r_k \leq \frac{1}{2\alpha} \|\mathbf{y}_1 - \mathbf{x}^\star\|^2.$$

The proof is completed by dividing by λ_{k-1}^2 and then recalling $\lambda_{k-1} \geq \frac{1}{2}(k - 1)$ from Proposition 11.16. ∎

Exercises

4.1. Show that Nesterov's accelerated gradient descent (11.33) for minimizing the quadratic function $F(\mathbf{x}) = \frac{1}{2}\mathbf{x}^T H \mathbf{x} - \mathbf{b} \cdot \mathbf{x} + c$ is given by

$$\mathbf{x}_{k+1} = \mathbf{x}_k - \alpha(H\mathbf{x}_k - b) + \beta_k(I - \alpha H)(\mathbf{x}_k - \mathbf{x}_{k-1}).$$

4.2. ♡ Consider Nesterov's accelerated gradient descent (11.33) where the gradient ∇F is defined with respect to an inner product $\langle \cdot, \cdot \rangle$ on $\mathbb{R}^n$. Formulate and prove a version of Theorem 11.18 in this setting.

4.3. ♡ Implement Nesterov's accelerated gradient descent on a convex, but not strongly convex, function F in **numpy** in Python. Compare it with ordinary gradient descent and with the heavy ball method.

4.4. ◇ Implement Nesterov's acceleration with the alternative update (11.39) in **numpy** in Python, and compare to the original version you implemented in Exercise 4.3.

4.5. Consider Nesterov's accelerated gradient descent applied to a quadratic function, as in Exercise 4.1. Mimic the proof of Theorem 11.10 to obtain an asymptotic linear convergence rate. How does your rate depend on the condition number of H?

4.6. Consider Nesterov's gradient descent without the *look ahead* feature:

$$\mathbf{x}_{k+1} = \mathbf{x}_k + \beta_k(\mathbf{x}_k - \mathbf{x}_{k-1}) - \alpha\nabla F(\mathbf{x}_k),$$

which is essentially the heavy ball method with time varying momentum parameter β_k. Do you expect this method to be convergent? *Hint*: Consider the method for quadratic functions, and do Exercise 4.5 first, or just examine the proof of Theorem 11.10.

11.5 Stochastic Gradient Descent

Python Notebook: SGD (.ipynb)

In machine learning, we often encounter objective functions of the form

$$F(\mathbf{x}) = \frac{1}{m}\sum_{i=1}^{m} F_i(\mathbf{x}), \tag{11.45}$$

where m is very large. For example, when training neural networks, see (10.16), $\mathbf{x}$ represents the trainable weights in the network and F_i is the per-item loss ℓ applied to a single training example, as a function of the parameters in the network. When the objective function has the form (11.45) and m is very large, it can sometimes be computationally burdensome to evaluate the full gradient $\nabla F(\mathbf{x})$ in order to take steps of gradient descent (11.1). In order to accelerate convergence, we can try to approximate the gradient of F by the gradient of any of the F_i, yielding the *stochastic gradient descent* (SGD) algorithm, which was introduced earlier in (10.18). Here it takes the form

$$\mathbf{x}_{k+1} = \mathbf{x}_k - \alpha_k\nabla F_{i_k}(\mathbf{x}_k), \tag{11.46}$$

where the index $1 \le i_k \le m$ is chosen uniformly at random at each step. Note we have also allowed the learning rate α_k to vary with k as well; we shall see the importance of this later on.

Remark 11.19. The version of SGD described above has a *batch size* of 1. In general, as in (10.18), we may choose a batch size $b \ge 1$ and use the *mini-batch SGD* optimization method

$$\mathbf{x}_{k+1} = \mathbf{x}_k - \frac{\alpha_k}{b}\sum_{j\in I_k} \nabla F_j(\mathbf{x}_k),$$

where $I_k \subset \{1,\ldots,m\}$ is a set of b indices indicating the mini-batch and is randomly chosen at each iteration. While we choose to study $b = 1$ in this section, to simplify the exposition, all results extend, with minor adjustments, to mini-batch SGD. ▲

The analysis in this section requires some basic facts from probability, and in particular, properties of conditional expectation, which we briefly review at a high level. In k steps of the SGD algorithm, we make k random choices when we select the indices $i_0, i_1, \ldots, i_{k-1}$. In the language of probability theory, the choices of indices $i_0, i_1, \ldots$ are a sequence of independent random variables uniformly distributed on $\{1,\ldots,m\}$. The *expectation* $\mathbb{E}$ of any quantity that depends on these random choices is simply the average over all possible choices, treating each as equally likely. That is, if $G = G(i_0)$ depends only on the first choice then

$$\mathbb{E}\,G = \frac{1}{m} \sum_{i_0=1}^{m} G(i_0).$$

If G depends on more than one choice, e.g., $G = G(i_0, \ldots, i_k)$, then we average over all m^{k+1} possible choices, and so the expectation is

$$\mathbb{E}\,G = \frac{1}{m^{k+1}} \sum_{i_0=1}^{m} \cdots \sum_{i_k=1}^{m} G(i_0, \ldots, i_k). \tag{11.47}$$

An example of such a function is the iterate $\mathbf{x}_{k+1}$ of SGD, which is a random variable that depends on the choices $i_0, \ldots, i_k$ made in the first $k+1$ steps of the algorithm. Hence, we could explicitly write

$$\mathbf{x}_{k+1} = \mathbf{x}_{k+1}(i_0, \ldots, i_k),$$

but this extra notation becomes burdensome, so we will not explicitly keep track of this dependence.

We define the *conditional expectation* of G, conditioned on $i_0, \ldots, i_{k-1}$, which we denote as $\mathbb{E}_k G$, by

$$\mathbb{E}_k\, G(i_0, \ldots, i_{k-1}) = \frac{1}{m} \sum_{i_k=1}^{m} G(i_0, \ldots, i_{k-1}, i_k). \tag{11.48}$$

The conditional expectation averages over just the most recent choice of i_k, leaving all the other previous choices fixed. Thus, $\mathbb{E}_k G$ is a random variable depending on $i_0, \ldots, i_{k-1}$, but not on i_k. In particular, if G does not depend on i_k, then $\mathbb{E}_k G = G$. Therefore, by (11.47) and (11.48),

$$\mathbb{E}\,\mathbb{E}_k\, G = \frac{1}{m^k} \sum_{i_0=1}^{m} \cdots \sum_{i_{k-1}=1}^{m} \mathbb{E}_k\, G(i_0, \ldots, i_{k-1}) = \frac{1}{m^{k+1}} \sum_{i_0=1}^{m} \cdots \sum_{i_{k-1}=1}^{m} \sum_{i_k=1}^{m} G(i_0, \ldots, i_k).$$

Since the right hand side is $\mathbb{E}\,G$, we have

$$\mathbb{E}\,\mathbb{E}_k\, G = \mathbb{E}\,G, \tag{11.49}$$

which is known as the *law of iterated expectations*. It is a key result that we will use in the analysis below. Let us also note that expectations are *linear* operators. That is,

$$\mathbb{E}\,(G + H) = \mathbb{E}\,G + \mathbb{E}\,H, \qquad \mathbb{E}(\lambda G) = \lambda \mathbb{E}\,G,$$

for any $\lambda \in \mathbb{R}$. These linearity properties also hold for the conditional expectations $\mathbb{E}_k$.

In the context of SGD, the conditional expectation of a gradient term is given by

$$\mathbb{E}_k \nabla F_{i_k}(\mathbf{x}_k) = \frac{1}{m} \sum_{i_k=1}^{m} \nabla F_{i_k}(\mathbf{x}_k) = \nabla F(\mathbf{x}_k). \tag{11.50}$$

This means that ∇F_{i_k} is an *unbiased* estimator of ∇F. Let us write

$$\xi_k = \nabla F_{i_k}(\mathbf{x}_k) - \nabla F(\mathbf{x}_k), \tag{11.51}$$

so that SGD (11.46) has the form

$$\mathbf{x}_{k+1} = \mathbf{x}_k - \alpha_k \nabla F(\mathbf{x}_k) + \alpha_k \xi_k. \tag{11.52}$$

Thus, we can view SGD as a noisy version of gradient descent, that has been corrupted by the noise vector ξ_k.

Since $\nabla F(\mathbf{x}_k)$ does not depend on i_k — it depends only on $i_0, \ldots, i_{k-1}$ — we have $\mathbb{E}_k \nabla F(\mathbf{x}_k) = \nabla F(\mathbf{x}_k)$. Thus, by (11.50) we have

$$\mathbb{E}_k \xi_k = \mathbb{E}_k \big[\nabla F_{i_k}(\mathbf{x}_k) - \nabla F(\mathbf{x}_k) \big] = \mathbb{E}_k \nabla F_{i_k}(\mathbf{x}_k) - \nabla F(\mathbf{x}_k) = \mathbf{0},$$

and so the noise vector ξ_k has mean zero. However, its variance, namely $\mathbb{E}_k \| \xi_k \|^2$, may be large, even near a critical point where $\nabla F(\mathbf{x}) = \mathbf{0}$. Indeed, we compute

$$\mathbb{E}_k \| \xi_k \|^2 = \mathbb{E}_k \| \nabla F_{i_k}(\mathbf{x}_k) - \nabla F(\mathbf{x}_k) \|^2 = \frac{1}{m} \sum_{i=1}^m \| \nabla F_i(\mathbf{x}_k) - \nabla F(\mathbf{x}_k) \|^2$$

$$= \frac{1}{m} \sum_{i=1}^m \| \nabla F_i(\mathbf{x}_k) \|^2 - 2 \nabla F(\mathbf{x}_k) \cdot \left[\frac{1}{m} \sum_{i=1}^m \nabla F_i(\mathbf{x}_k) \right] + \| \nabla F(\mathbf{x}_k) \|^2 \tag{11.53}$$

$$= \frac{1}{m} \sum_{i=1}^m \| \nabla F_i(\mathbf{x}_k) \|^2 - \| \nabla F(\mathbf{x}_k) \|^2,$$

where we used (11.50) in the last line. Since the summation term is $O(1)$, even near critical points, we cannot expect the noise vector ξ_k to have small variance. A reasonable assumption to make is that there exists $\sigma \geq 0$ so that

$$\mathbb{E}_k \| \xi_k \|^2 \leq \sigma^2 \qquad \text{for all} \qquad k \geq 0. \tag{11.54}$$

By (11.53) this is equivalent to assuming that

$$\frac{1}{m} \sum_{i=1}^m \| \nabla F_i(\mathbf{x}) \|^2 \leq \| \nabla F(\mathbf{x}) \| + \sigma^2 \qquad \text{for all} \qquad \mathbf{x} \in \mathbb{R}^n. \tag{11.55}$$

We can also write this in terms of conditional expectation, since

$$\mathbb{E}_k \| \nabla F_{i_k}(\mathbf{x}_k) \|^2 = \frac{1}{m} \sum_{i=1}^m \| \nabla F_i(\mathbf{x}_k) \|^2 \leq \| \nabla F(\mathbf{x}_k) \|^2 + \sigma^2. \tag{11.56}$$

A similar bound that will be useful, and follows from (11.54), is

$$\mathbb{E}_k \big[\xi_k \cdot \nabla F_{i_k}(\mathbf{x}_k) \big] \leq \sigma^2. \tag{11.57}$$

We relegate the proof of this to Exercise 5.1.

Recall that our analysis of gradient descent was based on the descent Lemma 6.64 that provided guarantees on how much the objective is decreased at each iteration. This is our natural starting place with SGD.

Lemma 11.20. *Assume ∇F is Lipschitz continuous, and let $L = \mathrm{Lip}(\nabla F)$. Suppose that there exists $\sigma > 0$ so that (11.54) holds. Let $\mathbf{x}_k$ satisfy the SGD iteration (11.46) and let $0 < \alpha_k \leq L^{-1}$. Then for any $k \geq 1$ we have*

$$\mathbb{E}_k F(\mathbf{x}_{k+1}) \leq F(\mathbf{x}_k) - \frac{\alpha_k}{2} \| \nabla F(\mathbf{x}_k) \|^2 + \frac{L \alpha_k^2}{2} \sigma^2. \tag{11.58}$$

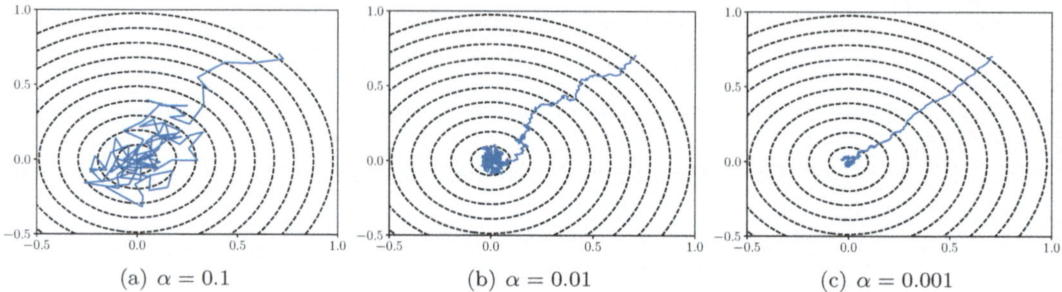

(a) $\alpha = 0.1$ (b) $\alpha = 0.01$ (c) $\alpha = 0.001$

Figure 11.6: Examples of SGD with different time steps α for minimizing the quadratic function (11.61). Smaller time steps are able to more accurately identify the minimizer.

Proof. The proof follows Lemma 6.64 closely, with modifications to account for stochasticity. Since ∇F is Lipschitz and $F(\mathbf{x}_{k+1}) = F(\mathbf{x}_k - \alpha_k \nabla F_{i_k}(\mathbf{x}_k))$ we have

$$F(\mathbf{x}_{k+1}) \le F(\mathbf{x}_k) - \alpha_k \nabla F(\mathbf{x}_k) \cdot \nabla F_{i_k}(\mathbf{x}_k) + \frac{L\alpha_k^2}{2} \|\nabla F_{i_k}(\mathbf{x}_k)\|^2. \tag{11.59}$$

Taking the conditional expectation $\mathbb{E}_k$ on both sides yields

$$\mathbb{E}_k F(\mathbf{x}_{k+1}) \le F(\mathbf{x}_k) - \alpha_k \nabla F(\mathbf{x}_k) \cdot \mathbb{E}_k \nabla F_{i_k}(\mathbf{x}_k) + \frac{L\alpha_k^2}{2} \mathbb{E}_k \|\nabla F_{i_k}(\mathbf{x}_k)\|^2,$$

where we used that $\mathbb{E}_k F(\mathbf{x}_k) = F(\mathbf{x}_k)$, since $\mathbf{x}_k$ does not depend on i_k. Using (11.50) and (11.56) we obtain

$$\mathbb{E}_k F(\mathbf{x}_{k+1}) \le F(\mathbf{x}_k) - \alpha_k \|\nabla F(\mathbf{x}_k)\|^2 + \frac{L\alpha_k^2}{2} \|\nabla F(\mathbf{x}_k)\|^2 + \frac{L\alpha_k^2 \sigma^2}{2}.$$

Since $\alpha_k \le 1/L$, we have $L\alpha_k^2 \le \alpha_k$. Inserting this intro the preceding inequality completes the proof. ∎

By Lemma 11.20, the variance σ^2 of the noise vector is ξ_k appears in the descent inequality (11.58) through the additional final term. In order to guarantee descent, at least on average, we need this term to be smaller than the descent term $\frac{1}{2}\alpha_k \|\nabla F(\mathbf{x}_k)\|^2$, which requires

$$\alpha_k \le \frac{\|\nabla F(\mathbf{x}_k)\|^2}{\mathrm{Lip}(\nabla F)\,\sigma^2}. \tag{11.60}$$

Since $\|\nabla F(\mathbf{x}_k)\|^2$ is small near any minimizer, or critical point for that matter, and in fact $\nabla F(\mathbf{x}_k) \to 0$ if SGD converges, we clearly need our time step α_k to be adaptive, and to choose it in such a way that $\alpha_k \to 0$ as $k \to \infty$. As we will see below, this can be done in a way that ensures convergence of SGD, in expectation, but this will cause the gradient descent convergence rate to be slower in comparison with the noise-free case, where $\sigma = 0$.

Example 11.21. As an example, we consider the function $F \colon \mathbb{R}^2 \to \mathbb{R}$ given by

$$F(\mathbf{x}) = \frac{1}{m} \sum_{i=1}^{m} \left(\tfrac{1}{2}\|\mathbf{x}\|^2 + \mathbf{v}_i \cdot \mathbf{x} \right). \tag{11.61}$$

If we choose the vectors $\mathbf{v}_i$ so that $\mathbf{v}_1 + \cdots + \mathbf{v}_m = \mathbf{0}$, then $F(\mathbf{x}) = \frac{1}{2}\|\mathbf{x}\|^2$. A simple way to do this is to set $\mathbf{v}_i = \left(\cos(2\pi i/m), \sin(2\pi i/m) \right)^T$, provided m is even. This function has the form (11.45) with $F_i(\mathbf{x}) = \frac{1}{2}\|\mathbf{x}\|^2 + \mathbf{v}_i \cdot \mathbf{x}$ and $\nabla F_i(\mathbf{x}) = \mathbf{x} + \mathbf{v}_i$. In this case the noise vector $\xi_{i_k} = \mathbf{v}_{i_k}$, with $\|\xi_{i_k}\| = 1$, so $\sigma = 1$. Figure 11.6 shows the paths taken by SGD starting from $\left(\sqrt{2}, \sqrt{2}\right)^T$ with $m = 16$ and different choices of fixed time step α. In each experiment we took $k = 10\,L\alpha^{-1}$ steps. We can see that large time steps lead to fast initial progress toward the minimizer, but at a certain point the noise overwhelms the gradient, and (11.60) fails to hold, which causes SGD to wander around unable to make further progress towards the minimum. Small time steps behave more like gradient descent, and appear to converge in a more direct fashion, at the expense of the higher computational cost of taking many more steps. As we will see below, a compromise is to use adaptive time steps that vary with the SGD iteration, sending $\alpha_k \to 0$ as $k \to \infty$. ▲

We now turn to our first convergence result for SGD. The convergence rate will depend on how we vary the time steps α_k through the quantity

$$\varepsilon_k = \frac{\alpha_0^2 + \alpha_1^2 + \cdots + \alpha_{k-1}^2}{\alpha_0 + \alpha_1 + \cdots + \alpha_{k-1}} = \frac{\alpha_0^2 + \cdots + \alpha_{k-1}^2}{k\,\overline{\alpha}_k}, \tag{11.62}$$

where

$$\overline{\alpha}_k = \frac{\alpha_0 + \alpha_1 + \cdots + \alpha_{k-1}}{k} \tag{11.63}$$

is the average time step over the first k steps.

Theorem 11.22. *Assume ∇F is Lipschitz continuous, and let $L = \mathrm{Lip}(\nabla F)$. Suppose that there exists $\sigma > 0$ so that (11.54) holds. Let $\mathbf{x}^\star$ be any minimizer of F. Let $\mathbf{x}_k$ satisfy the SGD iteration (11.46) and let $0 < \alpha_k \leq L^{-1}$. Then, for any $k \geq 1$, the following estimate holds:*

$$\min_{0 \leq j \leq k-1} \mathbb{E}\|\nabla F(\mathbf{x}_j)\|^2 \leq \frac{2}{k\,\overline{\alpha}_k}\left[F(\mathbf{x}_0) - F(\mathbf{x}^\star)\right] + L\sigma^2\varepsilon_k. \tag{11.64}$$

Moreover, if F is convex then

$$\min_{1 \leq j \leq k}\left[\mathbb{E}F(\mathbf{x}_j) - F(\mathbf{x}^\star)\right] \leq \frac{\|\mathbf{x}_0 - \mathbf{x}^\star\|^2}{2k\,\overline{\alpha}_k} + \sigma^2\varepsilon_k. \tag{11.65}$$

Proof. We first prove (11.64). Taking the expectation $\mathbb{E}$ on both sides of (11.58), using the law of iterated expectation (11.49), and rearranging terms yields

$$\alpha_k\mathbb{E}\|\nabla F(\mathbf{x}_k)\|^2 \leq 2\left[\mathbb{E}F(\mathbf{x}_k) - \mathbb{E}F(\mathbf{x}_{k+1})\right] + L\alpha_k^2\sigma^2.$$

Summing both sides from 0 to $k - 1$, the right hand side telescopes, and so

$$\sum_{j=0}^{k-1} \alpha_j\mathbb{E}\|\nabla F(\mathbf{x}_j)\|^2 \leq 2\left[F(\mathbf{x}_0) - \mathbb{E}F(\mathbf{x}_k)\right] + L\sigma^2\sum_{j=0}^{k-1}\alpha_j^2. \tag{11.66}$$

On the other hand,

$$\sum_{j=0}^{k-1}\alpha_j\mathbb{E}\|\nabla F(\mathbf{x}_j)\|^2 \geq \left(\min_{0 \leq j \leq k-1}\mathbb{E}\|\nabla F(\mathbf{x}_j)\|^2\right)k\,\overline{\alpha}_k.$$

The proof of (11.64) is completed by using this lower bound on the left hand side of (11.66), rearranging terms, and then noting that $\mathbb{E}F(\mathbf{x}_k) \geq F^\star$.

We now prove (11.65). We subtract $F^\star$ from both sides of (11.59) and then apply the convexity inequality (6.92) along with the assumption $\alpha_k \leq 1/L$, followed by the definition (11.51) of ξ_k, and then replacing the first two occurrences of $\nabla F_{i_k}(\mathbf{x}_k)$ using (11.46). We thereby obtain

$$
\begin{aligned}
F(\mathbf{x}_{k+1}) - F(\mathbf{x}^\star) &\leq F(\mathbf{x}_k) - F(\mathbf{x}^\star) - \alpha_k \nabla F(\mathbf{x}_k) \cdot \nabla F_{i_k}(\mathbf{x}_k) + \frac{L\alpha_k^2}{2}\|\nabla F_{i_k}(\mathbf{x}_k)\|^2 \\
&\leq \nabla F(\mathbf{x}_k) \cdot (\mathbf{x}_k - \mathbf{x}^\star) - \alpha_k \nabla F(\mathbf{x}_k) \cdot \nabla F_{i_k}(\mathbf{x}_k) + \frac{\alpha_k}{2}\|\nabla F_{i_k}(\mathbf{x}_k)\|^2 \\
&\leq \nabla F_{i_k}(\mathbf{x}_k) \cdot (\mathbf{x}_k - \mathbf{x}^\star) - \frac{\alpha_k}{2}\|\nabla F_{i_k}(\mathbf{x}_k)\|^2 - \xi_k \cdot \big[\mathbf{x}_k - \mathbf{x}^\star - \alpha_k \nabla F_{i_k}(\mathbf{x}_k)\big] \\
&= \frac{1}{\alpha_k}(\mathbf{x}_k - \mathbf{x}_{k+1}) \cdot (\mathbf{x}_k - \mathbf{x}^\star) - \frac{1}{2\alpha_k}\|\mathbf{x}_k - \mathbf{x}_{k+1}\|^2 - \xi_k \cdot \big[\mathbf{x}_k - \mathbf{x}^\star - \alpha_k \nabla F_{i_k}(\mathbf{x}_k)\big] \\
&= \frac{1}{2\alpha_k}\big[\|\mathbf{x}_k - \mathbf{x}^\star\|^2 - \|\mathbf{x}_{k+1} - \mathbf{x}^\star\|^2\big] - \xi_k \cdot \big[\mathbf{x}_k - \mathbf{x}^\star - \alpha_k \nabla F_{i_k}(\mathbf{x}_k)\big].
\end{aligned}
$$

We now take the conditional expectation $\mathbb{E}_k$ of both sides, using that $\mathbf{x}_k$ is independent of i_k, while $\mathbb{E}_k \xi_k = 0$:

$$
\begin{aligned}
\mathbb{E}_k F(\mathbf{x}_{k+1}) - F(\mathbf{x}^\star) &\leq \frac{1}{2\alpha_k}\big[\|\mathbf{x}_k - \mathbf{x}^\star\|^2 - \mathbb{E}_k\|\mathbf{x}_{k+1} - \mathbf{x}^\star\|^2\big] + \alpha_k \mathbb{E}_k\big[\xi_k \cdot \nabla F_{i_k}(\mathbf{x}_k)\big] \\
&\leq \frac{1}{2\alpha_k}\big[\|\mathbf{x}_k - \mathbf{x}^\star\|^2 - \mathbb{E}_k\|\mathbf{x}_{k+1} - \mathbf{x}^\star\|^2\big] + \alpha_k \sigma^2,
\end{aligned}
$$

where the final inequality follows from (11.57). We now take the expectation $\mathbb{E}$ of both sides, and multiply by α_k to find that

$$
\alpha_k\big[\mathbb{E}F(\mathbf{x}_{k+1}) - F(\mathbf{x}^\star)\big] \leq \frac{1}{2}\big[\mathbb{E}\|\mathbf{x}_k - \mathbf{x}^\star\|^2 - \mathbb{E}\|\mathbf{x}_{k+1} - \mathbf{x}^\star\|^2\big] + \alpha_k^2 \sigma^2.
$$

Summing from 0 to $k-1$, and using that the first term on the right hand side telescopes, produces

$$
\sum_{j=0}^{k-1} \alpha_j\big[\mathbb{E}F(\mathbf{x}_{j+1}) - F(\mathbf{x}^\star)\big] \leq \frac{1}{2}\|\mathbf{x}_0 - \mathbf{x}^\star\|^2 + \sigma^2 \sum_{j=0}^{k-1} \alpha_j^2.
$$

We now use the lower bound

$$
\sum_{j=0}^{k-1} \alpha_j\big[\mathbb{E}F(\mathbf{x}_{j+1}) - F(\mathbf{x}^\star)\big] \geq \min_{1 \leq j \leq k}\big[\mathbb{E}F(\mathbf{x}_j) - F(\mathbf{x}^\star)\big]\, k\overline{\alpha}_k,
$$

and divide both sides by $k\overline{\alpha}_k$ to complete the proof. ∎

If $\sigma = 0$, then the convergence rates in Theorem 11.22 are essentially the same as those in Theorems 6.65 and 6.66, except that we have allowed the time steps α_k to depend on k. In the noise free $\sigma = 0$ case, we can set $\alpha_k = \alpha$ so that $\overline{\alpha}_k = \alpha$ and obtain the familiar $\mathrm{O}(k^{-1})$ gradient descent convergence rate.

In the noisy case, when $\sigma > 0$, Theorem 11.22 shows that the convergence rate for SGD depends on the additional quantity ε_k defined in (11.62), which depends on the choice of time steps α_k. In order to obtain convergence of SGD, we must choose the time steps α_k so that

$$
\varepsilon_k = \frac{\alpha_0^2 + \cdots + \alpha_{k-1}^2}{\alpha_0 + \cdots + \alpha_{k-1}} \longrightarrow 0 \quad \text{and} \quad k\overline{\alpha}_k = \alpha_0 + \cdots + \alpha_{k-1} \longrightarrow \infty \quad \text{as} \quad k \longrightarrow \infty,
$$

both as quickly as possible. The first important observation to make is that, as expected from our discussion above in Example 11.21, the choice of constant time step $\alpha_k = \alpha$ is *not convergent*. Indeed, this implies $\varepsilon_k = k\alpha^2/(k\alpha) = \alpha$, and so ε_k does not converge to zero as $k \to \infty$. Hence, we need to decrease the time steps α_k as $k \to \infty$, and our goal is to do so in a way that optimizes the convergence rate.

A natural way to choose the time steps is to balance the two error terms in Theorem 11.22. Indeed, if either term is much smaller than the other, then it stands to reason that we could modify our choice of α_k to decrease the larger term slightly to get a better rate. Equating the coefficients amounts to asking that

$$\varepsilon_k = \frac{1}{k\,\overline{\alpha}_k}, \qquad \text{which requires} \qquad \alpha_0^2 + \cdots + \alpha_{k-1}^2 = 1. \tag{11.67}$$

or at least that this sum is close to O(1). There are, of course, many choices of time steps α_k that would satisfy (11.67), but we would like the largest possible time steps, so that the error terms decrease as quickly as possible. Any choice of $\alpha_k = O(k^{-p})$ for $p > \frac{1}{2}$ will satisfy (11.67) — see Exercise 5.5 — and so it is natural to take the limiting case of $p = \frac{1}{2}$, and so $\alpha_k = O(k^{-1/2})$. In this case $\varepsilon_k = O(k^{-1/2})$, up to logarithmic factors, as we verify in Proposition 11.23 below. In fact, the choice $\alpha_k = O(k^{-p})$ for any $0 < p \leq 1$ results in $\varepsilon_k \to 0$ and $k\overline{\alpha}_k \to \infty$, which implies convergence of SGD, while the choice of $p = \frac{1}{2}$ gives the fastest rate (see Exercise 5.6).

It turns out that $k^{-1/2}$ is a fundamental limit for how quickly ε_k can converge to zero. Indeed, by the Cauchy-Schwarz inequality,

$$k\,\overline{\alpha}_k = \sum_{j=0}^{k-1} \alpha_j \leq \sqrt{\sum_{j=0}^{k-1} 1^2} \sqrt{\sum_{j=0}^{k-1} \alpha_j^2} = \sqrt{k} \sqrt{\sum_{j=0}^{k-1} \alpha_j^2}.$$

Therefore,

$$\varepsilon_k = \frac{\alpha_0^2 + \cdots + \alpha_{k-1}^2}{k\,\overline{\alpha}_k} \geq \frac{\sqrt{\alpha_0^2 + \cdots + \alpha_{k-1}^2}}{\sqrt{k}} \geq \frac{\alpha_0}{\sqrt{k}}.$$

Hence, the rate for SGD in Theorem 11.22 cannot be faster than $O\left(k^{-1/2}\right)$, and our argument above suggests that a choice like $\alpha_k = O(k^{-1/2})$ yields this rate. This is readily confirmed by the following result.

Proposition 11.23. *Let ε_k be defined by (11.62), and set $\alpha_k = \alpha/\sqrt{k+1}$ for $k \geq 0$. Then*

$$k\,\overline{\alpha}_k \geq 2\alpha\sqrt{k+1} \qquad \text{and} \qquad \varepsilon_k \leq \frac{\alpha\,(\log k + 1)}{2\sqrt{k+1}}.$$

Proof. We first note that

$$\sum_{j=0}^{k-1} \alpha_j^2 = \alpha^2 \sum_{j=0}^{k-1} \frac{1}{j+1} \leq \alpha^2 + \int_0^{k-1} \frac{1}{x+1}\, dx = \alpha^2(\log k + 1).$$

On the other hand,

$$k\,\overline{\alpha}_k = \sum_{j=0}^{k-1} \alpha_j = \alpha \sum_{j=0}^{k-1} \frac{1}{\sqrt{j+1}} \geq \alpha \int_0^k \frac{1}{\sqrt{x+1}}\, dx = 2\alpha\sqrt{k+1}.$$

The proof follows upon dividing these two inequalities and simplifying. ∎

Proposition 11.23 shows that the best convergence rate we can obtain for SGD in the general setting, without further assumptions on F, is, up to logarithmic factors, the sublinear rate $O(k^{-1/2})$, which is obtained with the choice of time steps $\alpha_k = O(k^{-1/2})$. Compare this with the marginally better $O(k^{-1})$ rate that we obtained in the noiseless setting in Theorem 6.65.

Now, let us suppose that F is μ-PL, which includes μ-strongly convex functions, and see what kind of improvement we can obtain. The first step is to modify the descent Lemma 11.20 in this setting; see Exercise 5.2.

Lemma 11.24. *Assume F is μ-PL, ∇F is Lipschitz, and assume there exists $\sigma > 0$ so that (11.54) holds. Let $\mathbf{x}^\star$ be any minimizer of F, and let $F^\star = F(\mathbf{x}^\star)$. Let $\mathbf{x}_k$ satisfy the SGD iteration (11.46) and let $0 < \alpha_k \le \mathrm{Lip}(\nabla F)^{-1}$. Then, for any $k \ge 1$,*

$$\mathbb{E}_k F(\mathbf{x}_{k+1}) - F^\star \le (1 - \alpha_k \mu)\big(F(\mathbf{x}_k) - F^\star\big) + \frac{1}{2} \mathrm{Lip}(\nabla F)\alpha_k^2 \sigma^2. \tag{11.68}$$

Again, we need a restriction on the time step α_k in order to ensure F is strictly decreasing, on average, with each iteration of SGD. It is instructional to consider the simple quadratic function $F(\mathbf{x}) = \|\mathbf{x} - \mathbf{x}^\star\|^2$ in (11.68), the general μ-PL case being quite similar. In this case, SGD is guaranteed to decrease F at the k-th step provided

$$\|\mathbf{x}_k - \mathbf{x}^\star\|^2 = F(\mathbf{x}_k) - F^\star \le \frac{\mathrm{Lip}(\nabla F)\alpha_k^2 \sigma^2}{2(1 - \alpha_k \mu)}.$$

Thus, if we fix the time step, $\alpha_k = \alpha$, then we decrease F only until $\|\mathbf{x}_k - \mathbf{x}^\star\| = O(\alpha)$. This gives an additional explanation for the results in Figure 11.6, which illustrates how SGD gets stuck once the distance $\|\mathbf{x}_k - \mathbf{x}^\star\|$ to the minimizer is proportional to the time step α, and emphasizes why we need to send $\alpha_k \to 0$, even when F is strongly convex.

To see what convergence rate we may expect when F is μ-PL, let us take expectations on both sides of (11.68):

$$\mathbb{E}\, F(\mathbf{x}_{k+1}) - F^\star \le (1 - \alpha_k \mu)\big[\mathbb{E}\, F(\mathbf{x}_k) - F^\star\big] + \frac{\mathrm{Lip}(\nabla F)\alpha_k^2 \sigma^2}{2}. \tag{11.69}$$

To simplify the analysis, we set

$$\tau_k = \alpha_k \mu, \qquad \delta_k = \frac{2\mu^2 \big[\mathbb{E}\, F(\mathbf{x}_k) - F^\star\big]}{\mathrm{Lip}(\nabla F)\sigma^2}, \tag{11.70}$$

so that (11.69) can be written as

$$\delta_{k+1} \le (1 - \tau_k)\delta_k + \tau_k^2. \tag{11.71}$$

Now, the question is how to choose τ_k so that $\delta_k \to 0$ as quickly as possible. A natural approach is to minimize the right hand side of (11.71), which amounts to setting $\tau_k = \frac{1}{2}\delta_k$. With this choice, (11.71) becomes

$$\delta_{k+1} \le \delta_k - \tfrac{1}{4}\delta_k^2, \tag{11.72}$$

and this is fastest we are able to decrease δ_k at each iteration. This iteration is essentially the same as (6.138) from Example 6.67 with $p = 3$, except that here we have an inequality. By the same argument that produced (6.140), we find

$$\delta_k \le \frac{4}{k + 4\delta_0^{-1}}, \tag{11.73}$$

and this estimate is tight in the sense that if (11.72) were to hold with equality, then we can prove a lower bound for δ_k in a similar form as (11.73); see (6.142).

The argument above shows that the fastest rate we can expect from SGD is the slow $\mathrm{O}(k^{-1})$ sublinear convergence rate, even when F is μ-PL. We can easily obtain this rate, up to logarithmic factors, with a time step $\alpha_k = \mathrm{O}(k^{-1})$.

Theorem 11.25. *Assume F is μ-PL, ∇F is Lipschitz, and assume there exists $\sigma > 0$ so that $\mathbb{E}_k \|\xi_k\|^2 \leq \sigma^2$ for all $k \geq 0$. Let $\mathbf{x}_k$ satisfy the SGD iteration (11.46) with $\alpha_k = 1/(\mu k + \mathrm{Lip}(\nabla F))$. Then, for all $k \geq 1$,*

$$\mathbb{E}\, F(\mathbf{x}_k) - F^\star \leq \frac{\beta}{k+\beta}\left[F(\mathbf{x}_0) - F^\star + \frac{\mathrm{Lip}(\nabla F)\,\sigma^2\,(1 + \log(k+\beta))}{2\,\mu^2\,\beta} \right], \qquad (11.74)$$

where $\beta = \mu^{-1}\,\mathrm{Lip}(\nabla F) - 1 \geq 0$.

Proof. First, the fact that $\beta \geq 0$ follows from Exercise 1.2. The choice of time step clearly satisfies $\alpha_k \leq \mathrm{Lip}(\nabla F)^{-1}$, so that Lemma 11.24 applies. We also have

$$\tau_k = \alpha_k\,\mu = \frac{1}{k+\beta+1}.$$

Substituting this into (11.71) yields

$$\delta_k \leq (1 - \tau_{k-1})\delta_{k-1} + \tau_{k-1}^2 = \left(\frac{k+\beta-1}{k+\beta}\right)\delta_{k-1} + \frac{1}{(k+\beta)^2}, \qquad k \geq 1.$$

Iterating this inequality yields

$$\delta_k \leq \left(\frac{\beta}{k+\beta}\right)\delta_0 + \frac{1}{k+\beta}\sum_{j=1}^{k}\frac{1}{j+\beta}. \qquad (11.75)$$

We now estimate the sum

$$\sum_{j=1}^{k}\frac{1}{j+\beta} \leq \frac{1}{\beta+1} + \int_1^k \frac{1}{x+\beta}\,dx = \frac{1}{\beta+1} + \log(k+\beta) - \log(\beta+1) \leq 1 + \log(k+\beta).$$

Substituting this into (11.75) yields

$$\delta_k \leq \left(\frac{\beta}{k+\beta}\right)\delta_0 + \frac{1 + \log(k+\beta)}{k+\beta}.$$

Recalling (11.70) and making some algebraic simplifications completes the proof. ∎

Theorem 11.25 shows that when F is μ-PL, we can only improve the $\mathrm{O}(k^{-1/2})$ convergence rate of SGD to the sublinear rate $\mathrm{O}(k^{-1})$, up to logarithmic factors. On the other hand, this is considerably slower than the linear convergence rates we obtained for gradient descent in Theorem 6.68 for strongly convex functions, and Theorem 11.2 for μ-PL functions. Even so, the method is extremely useful in practice for two main reasons. First, each step of SGD is significantly faster and requires less memory than full batch gradient descent steps, so we can afford to take many more steps of SGD, while it may not even be computationally tractable to perform gradient descent. The other reason is that it is oftentimes not necessary to run SGD until convergence; in practice we may be satisfied once the loss decreases beyond a certain threshold. As the result below shows, the convergence rate of SGD is actually much faster in the initial stages — essentially linear convergence — before the noise becomes comparable in size to the gradient. And this initial phase of SGD is more important for applications than the long time convergence rates.

Lemma 11.26. *Assume F is μ-PL, ∇F is Lipschitz, and assume there exists $\sigma > 0$ so that the inequality (11.54) holds. Let $\mathbf{x}_k$ satisfy the SGD iteration (11.46) with fixed time step $\alpha \le \mathrm{Lip}(\nabla F)^{-1}$. Then for any $k \ge 1$ we have*

$$\mathbb{E}\, F(\mathbf{x}_k) - F^\star \le (1 - \alpha\mu)^k \big[F(\mathbf{x}_0) - F^\star \big] + \frac{L\sigma^2\alpha}{2\mu}.$$

Proof. We start with (11.69) with $\alpha_k = \alpha$. Let us write $L = \mathrm{Lip}(\nabla F)$ and subtract $\frac{L\sigma^2\alpha}{2\mu}$ from both sides to obtain

$$\mathbb{E}\, F(\mathbf{x}_{k+1}) - F^\star - \frac{L\sigma^2\alpha}{2\mu} \le (1 - \alpha\mu)\big[\mathbb{E}\, F(\mathbf{x}_k) - F^\star\big] + \frac{L\alpha^2\sigma^2}{2} - \frac{L\sigma^2\alpha}{2\mu}.$$

which simplifies to

$$\mathbb{E}\, F(\mathbf{x}_{k+1}) - F^\star - \frac{L\sigma^2\alpha}{2\mu} \le (1 - \alpha\mu)\big[\mathbb{E}\, F(\mathbf{x}_k) - F^\star - \frac{L\sigma^2\alpha}{2\mu}\big].$$

Iterating this completes the proof. ∎

Lemma 11.26 shows that if we are initially sufficiently far away from any minimizer $\mathbf{x}^*$, so that $F(\mathbf{x}_0) - F^\star$ is much larger than $L\sigma^2\alpha/(2\mu)$, then SGD with a fixed time step exhibits roughly linear convergence, as expected. Thus, unless we happen to be very close to a minimizer, we expect relatively rapid improvement in the objective function early on, while, over the longer term, the slower sublinear convergence rates kick in, provided we adapt the time step as discussed above.

Exercises

5.1. ♡ Show that (11.54) implies (11.57).

5.2. ◇ Prove Lemma 11.24.

5.3. ◇ Let $\alpha_k = \alpha/(k+1)^p$ where $0 < p < 1$. Show that

$$k\,\overline{\alpha}_k \ge \frac{\alpha}{1-p}\big[(k+1)^{1-p} - 1\big].$$

What happens when $p = 1$?

5.4. ♡ Let $\alpha_k = \alpha/(k+1)^p$ for $0 < p < \frac{1}{2}$. Show that

$$\sum_{j=0}^{k-1} \alpha_j^2 \le \frac{\alpha^2\,(k^{1-2p} - 2p)}{1 - 2p}.$$

What happens when $p = \frac{1}{2}$?

5.5. Let $\alpha_k = \alpha/(k+1)^p$ for $p > \frac{1}{2}$. Show that

$$\sum_{j=0}^{k-1} \alpha_j^2 \leq \frac{2p\alpha^2}{2p-1}.$$

5.6. ♡ Let $\alpha_k = \alpha/(k+1)^p$ where $0 < p \leq 1$. Use the previous three exercises and Theorem 11.22 to establish convergence rates for SGD with this choice of time step (use big O notation for simplicity). For which values of p is the method convergent? What happens when $p > 1$?

5.7. Consider SGD (11.46) where the gradient ∇F is defined with respect to a general inner product on $\mathbb{R}^n$, as in (6.36). Formulate and prove versions of Theorems 11.22 and 11.25 in this setting.

5.8. Modify the code in the Python notebook from this section to use a time varying time step α_k. Try the choices $\alpha_k = \alpha/(k+1)^p$ from this section. Which choice gives the fastest convergence rate?

11.6 Continuum Analysis of Optimization

In general, by the *continuum limit* of a discrete iterative system, we mean the equations — typically differential equations — that govern the limiting behavior as the time step becomes vanishingly small. Many of the first order optimization algorithms we have covered have continuum limits that correspond to ordinary differential equation (ODEs), which can oftentimes be easier to analyze and thereby give more insight into the algorithms.

For example, let us consider the basic gradient descent algorithm (11.1) for an objective function $F \colon \mathbb{R}^n \to \mathbb{R}$ with fixed time step $\alpha_k = \alpha$, which upon rearranging, reads

$$\frac{\mathbf{x}_{k+1} - \mathbf{x}_k}{\alpha} = -\nabla F(\mathbf{x}_k). \tag{11.76}$$

Let us imagine that the iterates $\mathbf{x}_k \in \mathbb{R}^n$ are obtained by sampling a smooth curve $\mathbf{x}(t)$, which is a function $\mathbf{x} \colon \mathbb{R} \to \mathbb{R}^n$, at multiples $t_k = \alpha k$ of the time step α; that is, $\mathbf{x}_k = \mathbf{x}(\alpha k)$. Then the left hand side of (11.76) is a finite difference approximation, [32], to the derivative $\mathbf{x}'(t)$ at the time $t = \alpha k$; indeed,

$$\mathbf{x}'(t) = \lim_{\alpha \to 0} \frac{\mathbf{x}(t+\alpha) - \mathbf{x}(t)}{\alpha} = \lim_{\alpha \to 0} \frac{\mathbf{x}_{k+1} - \mathbf{x}_k}{\alpha}. \tag{11.77}$$

Hence, by passing to the limit as $\alpha \to 0$, we obtain the first order ordinary differential equation

$$\mathbf{x}'(t) = -\nabla F(\mathbf{x}(t)) \qquad \text{with initial condition} \qquad \mathbf{x}(0) = \mathbf{x}_0. \tag{11.78}$$

The initial value problem (11.78) can be interpreted as the continuum limit of gradient descent, since the time t is now allowed to take any value in the continuum of real numbers $\mathbb{R}$. The first order ordinary differential equation simply says that the *velocity* $\mathbf{x}'(t)$ of a particle following the trajectory $\mathbf{x}(t)$ should be equal to the negative gradient $-\nabla F$ of the objective function, so that the particle moves in the direction of steepest descent at all times.

Throughout this section, we assume that ∇F is Lipschitz continuous, which, according to the basic existence and uniqueness results concerning initial value problems for ordinary differential equations ensures that the initial value problem (11.78) has a unique solution $\mathbf{x}(t)$

that is continuously differentiable in time [5, 30]. Incidentally, the Lipschitz constant $\text{Lip}(\nabla F)$ does not appear in any of our results because we are not discretizing in time.

The sublinear and linear convergence rates for gradient descent are, in fact, simpler to prove in the continuum setting, since we can make use of the power of calculus. The basic convergence result below is the continuum counterpart to Theorem 6.66.

Theorem 11.27. *Assume that F is convex and let $\mathbf{x}^\star$ be any minimizer. Let $\mathbf{x}(t)$ satisfy the gradient descent initial value problem (11.78). Then*

$$F\big(\mathbf{x}(t)\big) - F(\mathbf{x}^\star) \le \frac{\|\mathbf{x}_0 - \mathbf{x}^\star\|^2}{2\,t} \qquad \textit{for any} \qquad t > 0. \tag{11.79}$$

Proof. We show that the *energy*

$$E(t) = t\left[F\big(\mathbf{x}(t)\big) - F(\mathbf{x}^\star)\right] + \tfrac{1}{2}\|\mathbf{x}(t) - \mathbf{x}^\star\|^2$$

is decreasing with t, so $E(t) \ge E(s)$ whenever $t \le s$. It follows that

$$t\left[F\big(\mathbf{x}(t)\big) - F(\mathbf{x}^\star)\right] \le E(t) \le E(0) = \tfrac{1}{2}\|\mathbf{x}_0 - \mathbf{x}^\star\|^2.$$

Rearranging this inequality produces (11.79).

To prove that $E(t)$ is decreasing, it suffices to show that its derivative $E'(t) \le 0$. Using the chain rule, we compute

$$\begin{aligned}
E'(t) &= F\big(\mathbf{x}(t)\big) - F(\mathbf{x}^\star) + t\,\nabla F\big(\mathbf{x}(t)\big) \cdot \mathbf{x}'(t) + (\mathbf{x}(t) - \mathbf{x}^\star) \cdot \mathbf{x}'(t) \\
&= F\big(\mathbf{x}(t)\big) - F(\mathbf{x}^\star) - (\mathbf{x}(t) - \mathbf{x}^\star) \cdot \nabla F\big(\mathbf{x}(t)\big) - t\,\|\nabla F\big(\mathbf{x}(t)\big)\|^2,
\end{aligned}$$

where we used (11.78) to replace $\mathbf{x}'(t)$ in the last line. We now apply the convexity inequality (6.92) with $\mathbf{x} = \mathbf{x}(t)$ to obtain

$$E'(t) \le -t\,\|\nabla F\big(\mathbf{x}(t)\big)\|^2 \le 0. \qquad \blacksquare$$

We can similarly obtain a linear convergence rate, analogous to Theorem 11.2, when F is μ-PL, as per Definition 11.1.

Theorem 11.28. *Assume F is μ-PL and and let $\mathbf{x}^\star$ be any minimizer. Let $\mathbf{x}(t)$ satisfy the gradient descent initial value problem (11.78). Then*

$$F\big(\mathbf{x}(t)\big) - F(\mathbf{x}^\star) \le \left[F(\mathbf{x}_0) - F(\mathbf{x}^\star)\right]e^{-2\mu t} \qquad \textit{for any} \qquad t > 0. \tag{11.80}$$

Proof. In this case, we define the energy

$$E(t) = e^{2\mu t}\left[F\big(\mathbf{x}(t)\big) - F(\mathbf{x}^\star)\right],$$

and compute its derivative

$$\begin{aligned}
E'(t) &= 2\mu e^{2\mu t}\left[F\big(\mathbf{x}(t)\big) - F(\mathbf{x}^\star)\right] + e^{2\mu t}\,\nabla F\big(\mathbf{x}(t)\big) \cdot \mathbf{x}'(t) \\
&= e^{2\mu t}\left(2\mu\left[F\big(\mathbf{x}(t)\big) - F(\mathbf{x}^\star)\right] - \|\nabla F\big(\mathbf{x}(t)\big)\|^2\right) \le 0,
\end{aligned}$$

where the last step follows from the PL inequality (11.2). Therefore $E(t)$ is decreasing with t, and so

$$e^{2\mu t}\left[F\big(\mathbf{x}(t)\big) - F(\mathbf{x}^\star)\right] = E(t) \le E(0) = F(\mathbf{x}_0) - F(\mathbf{x}^\star),$$

which, upon dividing by $e^{2\mu t}$, establishes (11.80). $\qquad \blacksquare$

The reader should compare the exponential rate $e^{-2\mu t}$ from (11.80) with the rate $e^{-\alpha\mu k}$ from (11.4) in the discrete setting. The rates are nearly the same, when we identify $t = \alpha k$, except for the factor 2 improvement in the continuum rate. The reason for the difference is that in the proofs of Theorems 6.68 and 11.2 we used the upper bound $\alpha \leq \mathrm{Lip}(\nabla F)^{-1}$, which is convenient but slightly loose; see Exercise 9.4 from Chapter 6.

We now turn to the continuum versions of accelerated gradient descent methods, such as the heavy ball and Nesterov methods. The situation is slightly different in that they are second order iterative schemes, and hence we expect their continuum limits to be second order ordinary differential equations. To see why, we rearrange the heavy ball iteration (11.14) to read

$$\frac{\mathbf{x}_{k+1} - 2\mathbf{x}_k + \mathbf{x}_{k-1}}{\alpha} + (1 - \beta_k)\frac{\mathbf{x}_k - \mathbf{x}_{k-1}}{\alpha} = -\nabla F(\mathbf{x}_k), \tag{11.81}$$

where we allow β_k to change with k so our analysis applies to Nesterov's method as well. We recognize that the second term on the right hand side involves the finite difference approximation (11.77) to the derivative $\mathbf{x}'(t)$. The first term is similarly related to the standard *centered difference* approximation to the second derivative, [32]:

$$\mathbf{x}''(t) = \lim_{h \to 0} \frac{\mathbf{x}(t + h) - 2\mathbf{x}(t) + \mathbf{x}(t - h)}{h^2}, \tag{11.82}$$

which is readily established by Taylor expanding the first and third terms in the numerator. The centered difference can be identified with the first term in (11.81) provided we set $\alpha = h^2$, so $h = \sqrt{\alpha}$. That is, we view the iterates $\mathbf{x}_k$ of the heavy ball method as sample values $\mathbf{x}_k = \mathbf{x}(k\sqrt{\alpha})$ of a smooth curve $\mathbf{x}(t)$. Then, by (11.77) and (11.82),

$$\mathbf{x}'(k\sqrt{\alpha}) \approx \frac{\mathbf{x}_k - \mathbf{x}_{k-1}}{\sqrt{\alpha}}, \qquad \mathbf{x}''(k\sqrt{\alpha}) \approx \frac{\mathbf{x}_{k+1} - 2\mathbf{x}_k + \mathbf{x}_{k-1}}{\alpha}.$$

Therefore, the limiting ordinary differential equation corresponding to such momentum descent methods is

$$\mathbf{x}''(t) + \frac{1 - \beta_k}{\sqrt{\alpha}}\, \mathbf{x}'(t) = -\nabla F(\mathbf{x}(t)), \tag{11.83}$$

where β_k may depend on $t = \alpha k$. While it may seem odd that the time step α appears in the continuum ordinary differential equation, it turns out that it drops out when β_k takes on reasonable values. Recall from Theorem 11.10 that the optimal momentum parameter for minimizing the quadratic function (11.7) with the heavy ball method is of the form

$$\beta_k = \left(1 - \sqrt{\alpha \lambda_{min}(H)}\right)^2 = 1 - 2\sqrt{\alpha \lambda_{min}(H)} + \alpha \lambda_{min}(H) \approx 1 - 2\sqrt{\alpha \lambda_{min}(H)}, \tag{11.84}$$

when α is small. In this case, $\beta_k = \beta$ is independent of k, and we have

$$\frac{1 - \beta}{\sqrt{\alpha}} \approx 2\sqrt{\lambda_{min}(H)},$$

with the approximation becoming exact as $\alpha \to 0$.

We can make a similar argument for Nesterov's accelerated gradient descent, the only difference being that $\beta_k \approx 1 - 3/k$ by (11.38), and so

$$\frac{1 - \beta_k}{\sqrt{\alpha}} \approx \frac{3}{k\sqrt{\alpha}} = \frac{3}{t}.$$

Thus, both the heavy ball method and Nesterov's accelerated gradient descent have continuum versions governed by *second* order ordinary differential equations of the form

$$\mathbf{x}''(t) + a(t)\,\mathbf{x}'(t) = -\nabla F\big(\mathbf{x}(t)\big) \tag{11.85}$$

for particular choices of the *friction coefficient* $a(t)$. In particular, the heavy ball method corresponds to a constant friction $a(t) = a > 0$, while Nesterov acceleration corresponds to the asymptotically *vanishing* friction $a(t) = 3/t$. Since the ordinary differential equation (11.85) is second order, it requires two initial conditions. In practice, we often set $\mathbf{x}_1 = \mathbf{x}_0$ in accelerated methods, which corresponds to the initial conditions $\mathbf{x}(0) = \mathbf{x}_0$, $\mathbf{x}'(0) = \mathbf{0}$ (there are some subtleties in the initial conditions for Nesterov acceleration due to the singular friction term $a(t) = 3/t$, which are discussed later in this section).

A few remarks are in order.

Remark 11.29. The second order ordinary differential equation (11.85) can be viewed as the Newton equations governing the motion of a particle with unit mass $m = 1$ and position $\mathbf{x}(t) \in \mathbb{R}^n$. The initial term $\mathbf{x}''(t)$ represents mass times acceleration, which is balanced by a conservative external force $-\nabla F\big(\mathbf{x}(t)\big)$ with potential $F(\mathbf{x})$, while being slowed by a time-varying frictional force proportional to its velocity $\mathbf{x}'(t)$, with constant of proportionality $a(t)$. This is the reason for the use of terms like *acceleration*, or *heavy ball method*; we can view the optimization methods precisely as discretizations of the equations of motion for a ball rolling down the energy landscape defined by the objective function subject to the forces of gravity and friction. ▲

Remark 11.30. It is also interesting to note that Nesterov's choice of the momentum parameter $\beta_k = 1 - r/k$ for $r \geq 0$ is the *only* choice for which the friction coefficient $(1 - \beta_k)/\sqrt{\alpha}$ in (11.83) depends solely on the continuum time variable $t = k\sqrt{\alpha}$, provided we insist that β_k depends only on k. See Exercise 6.5 for a justification. ▲

Let us next address the convergence of the heavy ball ordinary differential equation in the particular case of solving a linear system $H\mathbf{x} = \mathbf{b}$. This result is the continuum version of Theorem 11.10.

Theorem 11.31. *Let H be a symmetric positive definite matrix, let $\mathbf{b} \in \mathbb{R}^n$, and let $\mathbf{x}^\star$ be the solution of $H\mathbf{x}^\star = \mathbf{b}$. Let $a > 0$ be constant, and let $\mathbf{x}(t)$ be the solution of the heavy ball initial value problem*

$$\mathbf{x}''(t) + a\,\mathbf{x}'(t) = \mathbf{b} - H\mathbf{x}(t), \qquad \mathbf{x}(0) = \mathbf{x}_0, \quad \mathbf{x}'(0) = \mathbf{0}. \tag{11.86}$$

If $a \leq 2\sqrt{\lambda_{min}(H)}$, then

$$\|\mathbf{x}(t) - \mathbf{x}^\star\| \leq \big(1 + \tfrac{1}{2}at\big)\, e^{-at/2}\, \|\mathbf{x}_0 - \mathbf{x}^\star\| \qquad \textit{for all} \qquad t > 0. \tag{11.87}$$

Proof. The proof is, at a high level, similar to that of Theorem 11.10. As before, since $H\mathbf{x}^\star = \mathbf{b}$, the function $\mathbf{y}(t) = \mathbf{x}(t) - \mathbf{x}^\star$ satisfies the corresponding homogeneous ordinary differential equation

$$\mathbf{y}''(t) + a\,\mathbf{y}'(t) + H\mathbf{y}(t) = \mathbf{0}. \tag{11.88}$$

Let us write the solution $\mathbf{y}(t)$ as a time-varying linear combination of the orthonormal eigenvector basis $\mathbf{v}_1, \mathbf{v}_2, \ldots, \mathbf{v}_n$ of H, with corresponding eigenvalues $\lambda_1 \geq \cdots \geq \lambda_n > 0$, so

$$\mathbf{y}(t) = \sum_{i=1}^{n} c_i(t)\,\mathbf{v}_i, \qquad \text{where} \qquad c_i(t) = \mathbf{y}(t) \cdot \mathbf{v}_i.$$

Then

$$\mathbf{y}'(t) = \sum_{i=1}^{n} c_i'(t)\,\mathbf{v}_i, \qquad \mathbf{y}''(t) = \sum_{i=1}^{n} c_i''(t)\,\mathbf{v}_i, \qquad H\,\mathbf{y}(t) = \sum_{i=1}^{n} c_i(t)\,\lambda_i\mathbf{v}_i.$$

Substituting these identities into (11.88) and using the linear independence of the eigenvectors, we deduce that the coefficients must satisfy the following decoupled scalar second order linear ordinary differential equations

$$c_i''(t) + a\,c_i'(t) + \lambda_i\,c_i(t) = 0, \qquad i = 1, \ldots, n.$$

The initial condition $\mathbf{x}(0) = \mathbf{x}_0$ implies that $c_i(0) = \mathbf{y}(0) \cdot \mathbf{v}_i = (\mathbf{x}_0 - \mathbf{x}^\star) \cdot \mathbf{v}_i$, and the initial condition $\mathbf{x}'(0) = \mathbf{0}$ implies that $c_i'(0) = 0$.

By assumption we have $(a/2)^2 \leq \lambda_i$, and so, according to Exercise 6.1,

$$|\,c_i(t)\,| \leq \left(1 + \frac{at}{2}\right) e^{-at/2}\,|\,c_i(0)\,|.$$

Therefore

$$\|\mathbf{x}(t) - \mathbf{x}^\star\|^2 = \|\mathbf{y}(t)\|^2 = \sum_{i=1}^{n} c_i(t)^2 \leq \left(1 + \frac{at}{2}\right)^2 e^{-at} \sum_{i=1}^{n} c_i(0)^2$$

$$= \left(1 + \frac{at}{2}\right)^2 e^{-at}\,\|\mathbf{y}(0)\|^2 = \left(1 + \frac{at}{2}\right)^2 e^{-at}\,\|\mathbf{x}_0 - \mathbf{x}^\star\|^2. \qquad \blacksquare$$

Remark 11.32. The optimal choice for the friction coefficient is clearly the largest possible choice of a satisfying the assumptions of Theorem 11.31, which is $a = 2\sqrt{\mu}$, where $\mu = \lambda_{min}(H)$. In this case, the convergence rate is $e^{-\sqrt{\mu}\,t}$. We compare this to the convergence rate of $e^{-2\mu t}$ of gradient descent in the continuum given in Theorem 11.28. In order to fairly compare, we can assume, by possibly rescaling H, that $\lambda_{max}(H) = 1$. Then $\mu = \lambda_{min}(H) = \kappa(H)^{-1}$, and we see the same phenomenon in the continuum as we did for the discrete counterparts, whereby the heavy ball method gives an order of magnitude improvement for its convergence rate. $\blacktriangle$

Figure 11.7(a) compares the heavy ball ordinary differential equation (11.86) to the gradient descent ordinary differential equation (11.78) for minimizing a strongly convex quadratic function in $n = 100$ variables; the function is the same as the one used in Figure 11.4. We use the optimal choice of friction coefficient $a = a^* = 2\sqrt{\lambda_{min}(H)}$ as well as an *overdamped*[4] choice of $a = 5a^*$ and an *underdamped* case $a = a^*/5$. We can see in Figure 11.7(a) that convergence is fastest for the optimal choice a^*, while the underdamped choice produces oscillations that slow the convergence, and the overdamped case behaves similarly to gradient descent.

We now turn to the convergence rate for the continuous time Nesterov ordinary differential equation

$$\mathbf{x}''(t) + \frac{3}{t}\,\mathbf{x}'(t) = -\nabla F\big(\mathbf{x}(t)\big). \tag{11.89}$$

---

[4]See [30] for a discussion of overdamped, critically damped, and underdamped second order ordinary differential equations.

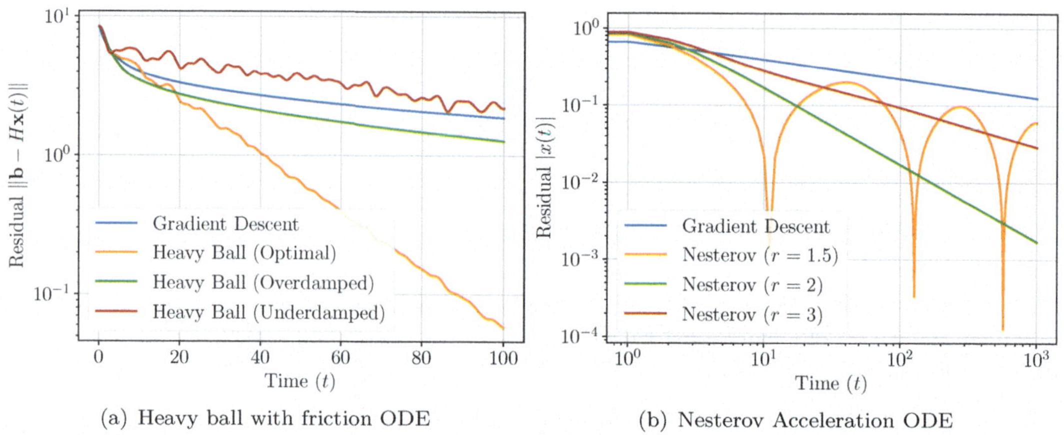

(a) Heavy ball with friction ODE   (b) Nesterov Acceleration ODE

Figure 11.7: Simulations of the heavy ball ODE (11.86) and the Nesterov ODE (11.89) with different choices of parameters. In (a) we are minimizing a strongly convex quadratic function, while in (b) we are minimizing a convex, but not strongly, function. In (b) the friction coefficient is $a(t) = r/t$, with the choice $r = 3$ standard in Nesterov's method.

However, before stating the result, it is worth saying a few words about the existence and uniqueness of solutions to this equation. The reader who is not familiar with differential equations can skip this discussion and proceed to Theorem 11.34. The existence and uniqueness of a solution $\mathbf{x}(t)$ to (11.89) satisfying an initial condition at $t = 0$ is somewhat complicated by the presence of the friction term $3/t$, which is singular at $t = 0$ and fails to be Lipschitz continuous. It turns out the equation is still well-posed (i.e., it has a unique solution given any initial condition $\mathbf{x}(0)$). To see this, multiply by t^3 on both sides of (11.89) to find that

$$\frac{d}{dt}\big[t^3\mathbf{x}'(t)\big] = t^3\mathbf{x}''(t) + 3t^2\mathbf{x}'(t) = -t^3\nabla F(\mathbf{x}(t)), \quad t > 0.$$

We now integrate both sides from 0 to t and divide by t^3 to obtain the first order differential equation

$$\mathbf{x}'(t) = -\frac{1}{t^3}\int_0^t s^3\nabla F(\mathbf{x}(s))\,ds. \tag{11.90}$$

Standard arguments, meaning Picard iteration [5, 30], can be applied to establish existence and uniqueness of a solution $\mathbf{x}(t)$, provided ∇F is Lipschitz continuous, which is our standing assumption. It is then possible to verify that any solution of (11.90) is twice continuously differentiable and satisfies (11.89). However, since (11.90) is a *first order* differential equation, we can only specify one initial condition, i.e., $\mathbf{x}(0)$. It turns out that any solution of (11.90) automatically satisfies $\mathbf{x}'(0) = \mathbf{0}$ and $\mathbf{x}''(0) = -\frac{1}{4}\nabla F(\mathbf{x}(0))$; see Exercise 6.6.

Remark 11.33. The reformulation (11.90) gives us another way to view acceleration; the velocity $\mathbf{x}'(t)$ in Nesterov's method is the temporal average of the previous (negative) gradient directions. However, it is not quite an average, since the weighting function s^3 integrates over the interval $[0, t]$ to $\frac{1}{4}t^4$ but we are dividing by t^3. When t is large, that is when $t \geq 4$ so that $\frac{1}{4}t^4 \geq t^3$, this has the effect of magnifying the average gradient by the factor $\frac{1}{4}t$, which is a way to interpret the accelerated convergence of Nesterov's method.

It is also interesting to note that when $t < 4$ we are underweighting the average, and for small $t > 0$ we can approximate (11.90) as

$$\mathbf{x}'(t) \approx -\frac{t}{4}\,\nabla F(\mathbf{x}(t)). \tag{11.91}$$

Thus, for short time, when $t \approx 0$, Nesterov acceleration looks like an extremely slow version of gradient descent. We can perform a very similar analysis for the heavy ball method; see Exercise 6.7. $\blacktriangle$

We now show that the continuous version of Nesterov's method converges at a rate that is $O(t^{-2})$.

> **Theorem 11.34.** *Let F be convex and let $\mathbf{x}^\star \in \mathbb{R}^n$ be any minimizer. Let $\mathbf{x}(t)$ satisfy the continuous time Nesterov equation (11.89) with $\mathbf{x}(0) = \mathbf{x}_0$. Then*
>
> $$F\big(\mathbf{x}(t)\big) - F(\mathbf{x}^\star) \le \frac{2\,\|\,\mathbf{x}_0 - \mathbf{x}^\star\,\|^2}{t^2} \qquad \text{for all} \qquad t > 0. \tag{11.92}$$

Proof. Define the energy

$$E(t) = t^2\left[\,F\big(\mathbf{x}(t)\big) - F(\mathbf{x}^\star)\,\right] + 2\,\big\|\,\mathbf{x}(t) + \tfrac{1}{2}\,t\,\mathbf{x}'(t) - \mathbf{x}^\star\,\big\|^2.$$

Differentiation yields

$$\begin{aligned}
E'(t) &= 2t\left[\,F\big(\mathbf{x}(t)\big) - F(\mathbf{x}^\star)\,\right] + t^2\,\nabla F\big(\mathbf{x}(t)\big)\cdot\mathbf{x}'(t)\\
&\qquad + 4\left[\,\mathbf{x}(t) + \tfrac{1}{2}\,t\,\mathbf{x}'(t) - \mathbf{x}^\star\,\right]\cdot\left[\,\tfrac{3}{2}\,\mathbf{x}'(t) + \tfrac{1}{2}\,t\,\mathbf{x}''(t)\,\right]\\
&= 2t\left[\,F\big(\mathbf{x}(t)\big) - F(\mathbf{x}^\star) - \nabla F\big(\mathbf{x}(t)\big)\cdot(\mathbf{x}(t) - \mathbf{x}^\star)\,\right] \le 0,
\end{aligned}$$

where we used the ordinary differential equation (11.89) to replace $\mathbf{x}''(t)$. The final inequality follows from the convexity of F, cf. (6.92), and the assumption that $t > 0$. We conclude that $E(t)$ is a decreasing function of t, and hence

$$t^2\left[\,F\big(\mathbf{x}(t)\big) - F(\mathbf{x}^\star)\,\right] \le E(t) \le E(0) = 2\,\|\,\mathbf{x}_0 - \mathbf{x}^\star\,\|^2.$$

Dividing by t^2 completes the proof. $\blacksquare$

It is quite remarkable how simple the proof of Theorem 11.34, in the continuum setting, is compared to the proof of discrete $O(k^{-2})$ convergence rate in Theorem 11.18. We also note that there is nothing particularly special about the factor 3 in the Nesterov ordinary differential equation (11.89); as the reader can verify in Exercise 6.4, we can replace the friction term by r/t for certain choices of r. In Figure 11.7(b) we compare the continuous time Nesterov ordinary differential equation (11.89) with different values of r to gradient descent (11.78) for minimizing the non-strongly convex function $F(x) = \frac{1}{6}x^6$ of a single variable $x \in \mathbb{R}$, which is the same function appearing in Figure 11.5(a). For all choices of r shown, the convergence rate is faster than gradient descent. In this case, we see failure when $r < 2$ — although see Exercise 6.4 for the general case.

For more details on the continuous ordinary differential equation interpretation of Nesterov acceleration, we refer the reader to [226, 246]. The continuous interpretation of heavy ball with friction was the original inspiration for Polyak's heavy ball method, and is discussed in his seminal paper [187]. We also mention the papers [8, 9], which study the ordinary differential equation governing the motion of a heavy ball with friction in various settings.

Exercises

6.1. ♡ (*a*) Solve the initial value problem

$$c''(t) + 2\,b\,c'(t) + \lambda\,c(t) = 0, \qquad c(0) = c_0, \qquad c'(0) = 0,$$

where $b > 0$, $\lambda > 0$, and $c_0 \in \mathbb{R}$.

    (*b*) Show that $|c(t)| \le |c_0|\,(1 + bt)\,e^{-bt}$ whenever $b^2 \le \lambda$. *Hint*: Use $\sin x \le x$.

    (*c*) What can you say about the solution if $b^2 > \lambda$?

6.2. ◇ In the proof of Theorem 11.31, suppose that $c_i(0) = (\mathbf{x}_0 - \mathbf{x}^\star)\cdot\mathbf{v}_i = 0$ for $i = k+1, \dots, n$. Show that you can choose a larger value for the friction coefficient a in this setting and prove a faster convergence rate. What is the optimal value for the friction coefficient a, and the optimal rate?

6.3. Formulate and prove a version of Theorem 11.31 where H is only positive semidefinite.

6.4. For what values of $r > 0$ does the Nesterov ordinary differential equation

$$\mathbf{x}''(t) + \frac{r}{t}\,\mathbf{x}'(t) = -\,\nabla F\big(\mathbf{x}(t)\big)$$

exhibit the $O(t^2)$ convergence rate from Theorem 11.34? Modify the proof of Theorem 11.34 to work in this case.

6.5. ♡ Define $\beta_k = 1 - \sqrt{\alpha}\,F\big(k\sqrt{\alpha}\big)$. Prove that β_k is independent of α if and only if $F(t) = r/t$ for some $r \in \mathbb{R}$. *Hint*: Differentiate the expression for β_k with respect to α.

6.6. ♡ Let $\mathbf{x}(t)$ be twice continuously differentiable and satisfy (11.90). Assume that ∇F is Lipschitz continuous. Show that $\mathbf{x}'(0) = \mathbf{0}$ and $\mathbf{x}''(0) = -\frac{1}{4}\,\nabla F(\mathbf{x}(0))$.
Hint: For $\mathbf{x}''$, use (11.91).

6.7. Let $\mathbf{x}(t)$ be twice continuously differentiable and satisfy the heavy ball equation

$$\mathbf{x}''(t) + a\,\mathbf{x}'(t) = -\,\nabla F(\mathbf{x}(t)), \qquad t > 0. \tag{11.93}$$

Show that

$$\mathbf{x}'(t) = e^{-at}\,\mathbf{x}'(0) - \int_0^t e^{a\,(s-t)}\,\nabla F(\mathbf{x}(s))\,ds. \tag{11.94}$$

11.7   Optimizing Neural Networks

When training a neural network, the loss is often a convex function of the output of the neural network. However, optimization is taken over the parameters, which often appear in a nonconvex manner. For example, suppose we have a neural network with input $\mathbf{x} \in \mathbb{R}^n$, parameters $\mathbf{w} \in \mathbb{R}^d$, scalar-valued output $F(\mathbf{x}; \mathbf{w}) \in \mathbb{R}$, and we use the squared error loss function

$$\mathcal{L}(\mathbf{w}) = \frac{1}{m} \sum_{i=1}^{m} \big(F(\mathbf{x}_i; \mathbf{w}) - y_i\big)^2 \tag{11.95}$$

over training data $\mathbf{x}_1, \dots, \mathbf{x}_m \in \mathbb{R}^n$ and associated scalar labels $y_1, \dots, y_m \in \mathbb{R}$. Even though (11.95) is a convex function of the outputs $F(\mathbf{x}_i; \mathbf{w})$, the optimization is performed over the weights $\mathbf{w}$, which renders the problem nonconvex, even for a neural network with 1 hidden layer. Thus, the results in this chapter and Chapter 6, which are concerned with global convergence of optimization towards a global minimizer, do not directly apply.[5]

However, it is possible to exploit the convexity of the per-item loss function, including the squared loss (11.95), to deduce some partial results. We work in general from here on. Let $\ell \colon \mathbb{R} \times \mathbb{R} \to \mathbb{R}$ be the per-item loss function. For example, the least squares loss corresponds to $\ell(z, y) = \frac{1}{2}(z - y)^2$. We denote by ℓ_z the partial derivative of ℓ with respect to z; for the least squares loss $\ell_z(z, y) = z - y$. The general loss function we seek to minimize has the form

$$\mathcal{L}(\mathbf{w}) = \frac{1}{m} \sum_{i=1}^{m} \ell\big(F(\mathbf{x}_i; \mathbf{w}), y_i\big).$$

Its gradient with respect to $\mathbf{w}$ is given by

$$\nabla \mathcal{L}(\mathbf{w}) = \frac{1}{m} \sum_{i=1}^{m} \ell_z\big(F(\mathbf{x}_i; \mathbf{w}), y_i\big) \, \nabla_{\mathbf{w}} F(\mathbf{x}_i; \mathbf{w}). \tag{11.96}$$

The continuum version of gradient descent, i.e., (11.78), is given by

$$\mathbf{w}'(t) = -\nabla \mathcal{L}\big(\mathbf{w}(t)\big). \tag{11.97}$$

At the level of the weights in the network, the optimization landscape is still highly nonconvex.

The key observation is to switch perspectives and consider the dynamics of the *output* of the neural network applied to the training data over time, instead of focusing on the weights. We denote the output by $\mathbf{z}(t) = F\big(X; \mathbf{w}(t)\big) \in \mathbb{R}^m$, where $X = (\mathbf{x}_1, \ldots, \mathbf{x}_m)^T \in \mathcal{M}_{m \times n}$ is the data matrix — see (7.1) — whose rows are the training data points, and $F(X; \mathbf{w}) = (F(\mathbf{x}_1; \mathbf{w}), \ldots, F(\mathbf{x}_m; \mathbf{w}))^T \in \mathbb{R}^m$. That is, we simply have $z_i(t) = F\big(\mathbf{x}_i; \mathbf{w}(t)\big)$. Now, applying the chain rule and using (11.97),

$$z_i'(t) = \nabla_{\mathbf{w}} F\big(\mathbf{x}_i; \mathbf{w}(t)\big) \cdot \mathbf{w}'(t) = -\nabla_{\mathbf{w}} F\big(\mathbf{x}_i; \mathbf{w}(t)\big) \cdot \nabla \mathcal{L}\big(\mathbf{w}(t)\big)$$

$$= -\frac{1}{m} \sum_{j=1}^{m} \nabla_{\mathbf{w}} F\big(\mathbf{x}_i; \mathbf{w}(t)\big) \cdot \nabla_{\mathbf{w}} F\big(\mathbf{x}_j; \mathbf{w}(t)\big) \, \ell_z\big(z_j(t), y_j\big). \tag{11.98}$$

The resulting formula appears to have the form of matrix multiplication. Indeed, let us define the $m \times m$ matrix $K = K(\mathbf{w})$ with entries

$$K_{ij} = \nabla_{\mathbf{w}} F(\mathbf{x}_i; \mathbf{w}) \cdot \nabla_{\mathbf{w}} F(\mathbf{x}_j; \mathbf{w}). \tag{11.99}$$

This has the form of a kernel matrix, as per Definition 7.16, corresponding to the kernel function

$$\mathcal{K}_\phi(\mathbf{x}, \mathbf{y}) = \phi(\mathbf{x}) \cdot \phi(\mathbf{y}), \qquad \text{where} \qquad \phi(\mathbf{x}) = \nabla_{\mathbf{w}} F(\mathbf{x}; \mathbf{w}) \tag{11.100}$$

can be viewed as the feature map. For notational simplicity, we will write $K(t) = K\big(\mathbf{w}(t)\big)$. We also define the vectorized loss

$$\ell(\mathbf{z}, \mathbf{y}) = \frac{1}{m} \sum_{i=1}^{m} \ell(z_i, y_i), \qquad \text{for} \quad \mathbf{z}, \mathbf{y} \in \mathbb{R}^m,$$

whose gradient is given by

$$\nabla_{\mathbf{z}} \ell(\mathbf{z}, \mathbf{y}) = \frac{1}{m} \big(\ell_z(z_1, y_1), \ldots, \ell_z(z_m, y_m) \big)^T.$$

---

[5] The local sublinear convergence result in Theorem 6.65 does apply, although it does not yield convergence to a global minimizer.

Using this notation, (11.98) becomes

$$\mathbf{z}'(t) = - K(t)\, \nabla_{\mathbf{z}} \ell(\mathbf{z}(t), \mathbf{y}), \qquad\qquad (11.101)$$

where $\mathbf{y} = (y_1, \ldots, y_m)^T \in \mathbb{R}^m$ is the label vector.

Being a kernel matrix, and hence a Gram matrix, $K(t)$ is, for each $t \in \mathbb{R}$, automatically symmetric positive semidefinite. If it is positive definite, then $\langle \mathbf{x}, \mathbf{y} \rangle_{C(t)} = \mathbf{x}^T C(t)\, \mathbf{y}$, where $C(t) = K(t)^{-1}$, defines an inner product on $\mathbb{R}^m$. According to (6.36), the gradient of a function $G \colon \mathbb{R}^m \to \mathbb{R}$ with respect to this inner product is given by

$$\nabla_{C(t)} G(\mathbf{z}) = C(t)^{-1} \nabla G(\mathbf{z}) = K(t)\, \nabla G(\mathbf{z}).$$

Thus, (11.101) can be interpreted as preconditioned gradient descent

$$\mathbf{z}'(t) = - \nabla_{C(t)} \ell(\mathbf{z}(t), \mathbf{y}), \qquad\qquad (11.102)$$

where the preconditioning matrix $C(t)$ is changing in time, and given by the inverse of the kernel matrix $K(t)$. Noting that $\ell(\mathbf{z}(t), \mathbf{y}) = \mathcal{L}(\mathbf{w}(t))$, we see that (11.102) is still minimizing the same loss, it is just that we are switching perspectives and considering the dynamics of $\mathbf{z}(t)$, the output of the neural network, instead of the weights $\mathbf{w}(t)$. This makes the analysis tractable because the loss $\ell(\mathbf{z}, \mathbf{y})$ can be assumed to be *convex* as a function of $\mathbf{z}$, while $\mathcal{L}(\mathbf{w})$ is certainly not convex in $\mathbf{w}$.

Now, provided the kernel matrix remains well-conditioned, we can prove convergence of $\mathbf{z}(t)$ to a global minimizer of the vectorized loss $\ell(\mathbf{z}, \mathbf{y})$. Let

$$\ell^{\star} = \min_{\mathbf{z} \in \mathbb{R}^m} \ell(\mathbf{z}, \mathbf{y}), \qquad \mathcal{L}^{\star} = \min_{\mathbf{w} \in \mathbb{R}^m} \mathcal{L}(\mathbf{w}). \qquad\qquad (11.103)$$

Since $\ell(\mathbf{z}, \mathbf{y}) = \mathcal{L}(\mathbf{w})$ when $\mathbf{z} = F(X; \mathbf{w})$, we clearly have that $\ell^{\star} \leq \mathcal{L}^{\star}$. We now come to our main result in this section.

Theorem 11.35. *Assume that, for each label vector $\mathbf{y}$, the function $\mathbf{z} \mapsto \ell(\mathbf{z}, \mathbf{y})$ is μ-PL. Let $\mathbf{w}(t)$ satisfy (11.97) with $\mathbf{w}(0) = \mathbf{w}_0$, and set $\mathbf{z}(t) = F(X; \mathbf{w}(t))$. Assume there exists $\lambda > 0$ and $T > 0$ such that, for all $0 \leq t \leq T$,*

$$\mathbf{v}(t)^T K(t)\, \mathbf{v}(t) \geq \lambda \, \| \mathbf{v}(t) \|^2 \qquad where \qquad \mathbf{v}(t) = \nabla_{\mathbf{z}} \ell(\mathbf{z}(t), \mathbf{y}). \qquad (11.104)$$

Then

$$\mathcal{L}(\mathbf{w}(t)) - \mathcal{L}^{\star} \leq \ell(\mathbf{z}(t), \mathbf{y}) - \ell^{\star} \leq \big(\ell(\mathbf{z}_0, \mathbf{y}) - \ell^{\star} \big)\, e^{-2\mu\lambda t}, \qquad 0 \leq t \leq T. \quad (11.105)$$

Proof. The first inequality in (11.105) is trivial, since $\mathcal{L}(\mathbf{w}(t)) = \ell(\mathbf{z}(t), \mathbf{y})$ and $\mathcal{L}^{\star} \geq \ell^{\star}$. To prove the second, we define the energy

$$E(t) = e^{2\mu\lambda t} \big[\ell(\mathbf{z}(t), \mathbf{y}) - \ell^{\star} \big],$$

and compute, using (11.102) and (11.104), its derivative:

$$
\begin{aligned}
E'(t) &= 2\mu\lambda e^{2\mu\lambda t} \big[\ell(\mathbf{z}(t), \mathbf{y}) - \ell^{\star} \big] + e^{2\mu\lambda t} \nabla_{\mathbf{z}} \ell(\mathbf{z}(t), \mathbf{y}) \cdot \mathbf{z}'(t) \\
&= e^{2\mu\lambda t} \big[2\mu\lambda \big(\ell(\mathbf{z}(t), \mathbf{y}) - \ell^{\star} \big) - \nabla_{\mathbf{z}} \ell(\mathbf{z}(t), \mathbf{y})^T K(t)\, \nabla_{\mathbf{z}} \ell(\mathbf{z}(t), \mathbf{y}) \big] \\
&\leq \lambda e^{2\mu t} \big[2\mu \big(\ell(\mathbf{z}(t), \mathbf{y}) - \ell^{\star} \big) - \| \nabla_{\mathbf{z}} \ell(\mathbf{z}(t), \mathbf{y}) \|^2 \big] \leq 0,
\end{aligned}
$$

where the last step follows from the PL inequality (11.2). Therefore $E(t)$ is a decreasing function of t, and the proof is completed similarly to Theorem 11.28. $\blacksquare$

Theorem 11.35 shows that gradient descent with respect to the weights $\mathbf{w}$ on the loss function $\mathcal{L}(\mathbf{w})$ converges to a global minimizer that exactly fits the data, i.e., achieves the optimal loss $\ell^\star$. This would appear to be a very powerful result, especially since we have used nothing about the particular form of $F(\mathbf{x}; \mathbf{w})$; everything above works for *any* parameterized function. The catch is that assumption (11.104) is a rather strong condition; it requires more justification and necessitates specifying the type of function $F(\mathbf{x}; \mathbf{w})$ that we are working with. Note that we can simplify this condition by assuming instead that

$$\lambda_{min}(K(t)) \geq \lambda \qquad \text{for all} \qquad 0 \leq t \leq T, \tag{11.106}$$

which, since $K(t)$ is symmetric positive definite, is equivalent to the inequality

$$\mathbf{v}^T K(t)\mathbf{v} \geq \lambda \|\mathbf{v}\|^2 \qquad \text{for all} \qquad \mathbf{v} \in \mathbb{R}^m \quad \text{and} \quad 0 \leq t \leq T,$$

not just the directions $\mathbf{v}$ that align with the gradient of the loss. However, (11.106) is an even stronger condition that is limiting even for linear regression, as the following example shows.

Example 11.36. Consider the problem of linear regression introduced in Section 7.2, in which $F(\mathbf{x}; \mathbf{w}) = \mathbf{w} \cdot \mathbf{x}$ is a linear function and $\ell(\mathbf{z}, \mathbf{y}) = \|\mathbf{z} - \mathbf{y}\|^2 / (2m)$. Then $\nabla_\mathbf{w} F(\mathbf{x}; \mathbf{w}) = \mathbf{x}$ is independent of $\mathbf{w}$, and so the entries of the kernel matrix $K = K(\mathbf{w}) \in \mathcal{M}_{m \times m}$ are given by

$$k_{ij} = \nabla_\mathbf{w} F(\mathbf{x}_i; \mathbf{w}) \cdot \nabla_\mathbf{w} F(\mathbf{x}_j; \mathbf{w}) = \mathbf{x}_i \cdot \mathbf{x}_j.$$

Thus, we can write the kernel matrix as the Gram matrix associated with the data vectors: $K = XX^T$, where $X = (\, \mathbf{x}_1 \, \ldots \, \mathbf{x}_m \,)^T$ is the data matrix. We also have

$$\nabla_\mathbf{z} \ell(\mathbf{z}, \mathbf{y}) = \frac{1}{m}(\mathbf{z} - \mathbf{y}), \qquad \text{and so} \qquad \nabla_\mathbf{z} \ell\big(F(X; \mathbf{w}), \mathbf{y}\big) = \frac{1}{m}(X\mathbf{w} - \mathbf{y})$$

Since $\operatorname{rank} K = \operatorname{rank} X$ and $X \in \mathcal{M}_{m \times n}$, we have $\operatorname{rank} K = m$, i.e., (11.106) holds, if and only if $\mathbf{x}_1, \ldots, \mathbf{x}_m$ are linearly independent, which is possible only when $n \geq m$, i.e., there are more dimensions than data points. This is a very strong assumption for linear regression, which is often used in the setting where $m \gg n$. Notice, as well, that K does not depend on the given labels $\mathbf{y}$.

In the general setting, we turn to the weaker condition (11.104), which does involve the labels $\mathbf{y}$. We assume that we can fit the labels exactly, so $\mathbf{y} \in \operatorname{img} X$ and so there exists $\mathbf{w}^\star \in \mathbb{R}^n$ such that $X\mathbf{w}^\star = \mathbf{y}$. We claim that, under these conditions, (11.104) holds with $\lambda = \sigma_{min}(X)^2$. To see this, we compute, using Exercise 7.1, that

$$\nabla_\mathbf{z} \ell(F(X; \mathbf{w}), \mathbf{y})^T K \nabla_\mathbf{z} \ell(F(X; \mathbf{w}), \mathbf{y}) = \frac{1}{m^2}(X\mathbf{w} - \mathbf{y})^T XX^T(X\mathbf{w} - \mathbf{y})$$

$$= \frac{1}{m^2}\|X^T X(\mathbf{w} - \mathbf{w}^\star)\|^2 \geq \frac{\sigma_{min}(X)^2}{m^2}\|X(\mathbf{w} - \mathbf{w}^\star)\|^2$$

$$= \sigma_{min}(X)^2 \|\nabla_\mathbf{z} \ell(F(X; \mathbf{w}), \mathbf{y})\|^2. \qquad \blacktriangle$$

As Example 11.36 illustrates, the condition (11.104) as well as the stronger condition (11.106) are in some sense connected to the ability of the functions $F(\mathbf{x}; \mathbf{w})$ to exactly represent the given data, and to do so through gradient descent based optimization. The main difference between (11.104) and (11.106) is that the latter does not depend on the labels $\mathbf{y}$. Thus, the condition (11.106) is basically asking that the functions $F(\mathbf{x}; \mathbf{w})$ can approximate *any* labels $\mathbf{y}$ with an appropriate choice of $\mathbf{w}$, and that gradient descent converges linearly to a solution. In the case of linear regression in Example 11.36, this means that we should, for any $\mathbf{y} \in \mathbb{R}^m$, be able to solve $X\mathbf{w} = \mathbf{y}$, though not necessarily uniquely. This is only true when rank $X = m$ which requires $n \geq m$ and the rows of X are linearly independent. On the other hand, the weaker condition (11.104) in Theorem 11.35 involves the given labels $\mathbf{y}$ and can be thought of as measuring our ability to fit only these labels.

When $F(\mathbf{x}; \mathbf{w})$ is a neural network, the kernel matrix K is more complicated. For simplicity, let us consider the two layer network

$$F(\mathbf{x}; \mathbf{w}) = \frac{1}{\sqrt{N}} \sum_{k=1}^{N} \sigma(\mathbf{w}_k \cdot \mathbf{x}), \tag{11.107}$$

with N hidden nodes, activation function σ, and weight vector[6] $\mathbf{w} = (\mathbf{w}_1^T, \ldots, \mathbf{w}_N^T)^T \in \mathbb{R}^{Nn}$ that is simply the concatenation of each neuron's weight vector. Then,

$$\nabla_{\mathbf{w}} F(\mathbf{x}; \mathbf{w}) = \frac{1}{\sqrt{N}} \left(\sigma'(\mathbf{w}_1 \cdot \mathbf{x})\,\mathbf{x}^T, \; \ldots \;, \sigma'(\mathbf{w}_N \cdot \mathbf{x})\,\mathbf{x}^T \right)^T \in \mathbb{R}^{Nn},$$

and so the kernel matrix $K(\mathbf{w})$ has entries

$$K(\mathbf{w})_{ij} = \nabla_{\mathbf{w}} F(\mathbf{x}_i; \mathbf{w}) \cdot \nabla_{\mathbf{w}} F(\mathbf{x}_j; \mathbf{w}) = \frac{\mathbf{x}_i \cdot \mathbf{x}_j}{N} \sum_{k=1}^{N} \sigma'(\mathbf{w}_k \cdot \mathbf{x}_i)\, \sigma'(\mathbf{w}_k \cdot \mathbf{x}_j). \tag{11.108}$$

It is generally difficult to study kernel matrices like this directly, without imposing further conditions on the weights $\mathbf{w}_k$. A natural simplifying assumption is that the weight vectors $\mathbf{w}_1, \ldots, \mathbf{w}_N$ are independent and identically distributed random variables on $\mathbb{R}^n$ with probability density $\rho(\mathbf{w})$. This is how the weights in a neural network are typically initialized, usually with the Gaussian distribution, and so this is valid at least initially during optimization. In this case we can consider the expectations of the entries of the kernel matrix, which we denote by

$$K_{ij}^{\infty} = \mathbb{E}\, K_{ij}(\mathbf{w}) = (\mathbf{x}_i \cdot \mathbf{x}_j) \int_{\mathbb{R}^n} \sigma'(\mathbf{w} \cdot \mathbf{x}_i)\, \sigma'(\mathbf{w} \cdot \mathbf{x}_j)\, \rho(\mathbf{w})\, d\mathbf{w}.$$

In fact, by the law of large numbers [193] the kernel matrix $K(\mathbf{w})$ converges to K^{∞} as the width of the network $N \to \infty$, which justifies its name. Since the weights no longer appear in K^{∞}, the limiting matrix is simpler to study, and so very wide neural networks have attracted much attention, at least in this setting. Notice that the choice of $1/\sqrt{N}$ normalization factor in (11.107) was selected precisely so that a factor of $1/N$ appeared in the kernel matrix $K(\mathbf{w})$, and we could view it as an average to which we can apply the law of large numbers.

For some choices of activation function σ, the limiting kernel matrix K^{∞} can be explicitly computed. One example is the ReLU activation $\sigma(x) = x_+ = \max\{x, 0\}$, when ρ is a standard normal distribution. Here, it turns out that K^{∞} is the so-called arc-cosine kernel matrix, with entries

$$K_{ij}^{\infty} = (\mathbf{x}_i \cdot \mathbf{x}_j) \left[\frac{1}{2} - \frac{1}{2\pi} \arccos\left(\frac{\mathbf{x}_i \cdot \mathbf{x}_j}{\|\mathbf{x}_i\|\,\|\mathbf{x}_j\|} \right) \right].$$

A proof of this fact, as well as identification of the limiting kernel matrices for many other activations, can be found in [40].

Since we can identify the kernel matrix K^{∞} in the infinite-width limit, it is possible to study its spectrum and establish convergence results for shallow and wide neural networks,

---

[6]Note: $\mathbf{w}$ is *not* to be regarded as a matrix.

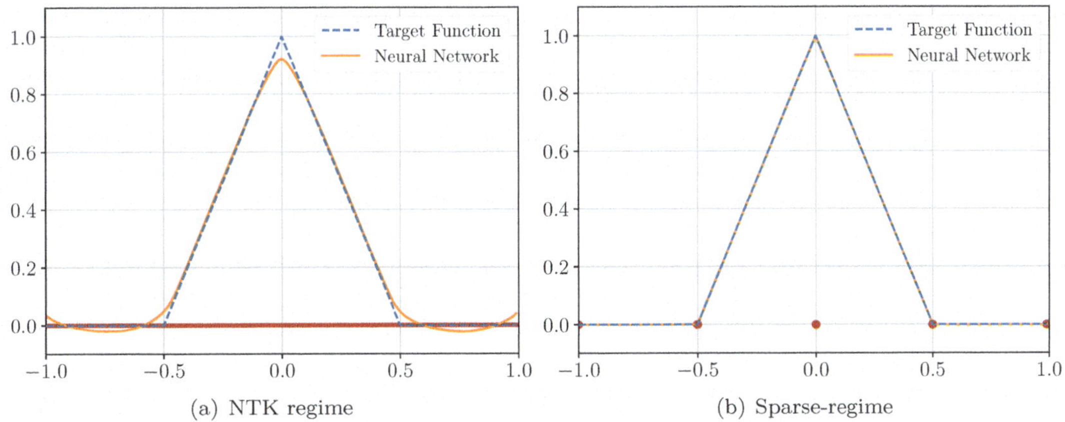

(a) NTK regime

(b) Sparse-regime

Figure 11.8: Comparison of (a) the NTK regime where the number of neurons $N \to \infty$ (here, $N = 10^4$) and (b) the sparse regime where N is small (here, $N = 100$). The red dots indicate the locations of the zero crossings of each neuron.

like Theorem 11.35, except that the condition (11.104) is no longer needed. The main idea is that when N is very large, the neural network can fit a function with only very small changes from the initial randomized weights $\mathbf{w}_0$. Thus, one can make the approximations $\mathbf{w}(t) \approx \mathbf{w}(0) = \mathbf{w}_0$ and $K(t) \approx K(0) \approx K_{ij}^\infty$, and then the analysis of the limiting kernel matrix K_{ij}^∞ is sufficient to prove a result similar to Theorem 11.35. We refer the reader to [117] for more details on this analysis, and note that K^∞ is often called the *neural tangent kernel matrix*, and training in this setting is called the NTK regime.

Since in the very wide NTK regime the weights do not change much during training, $\mathbf{w}(t) \approx \mathbf{w}_0$, the neural network is well-approximated by its linearization around the initial weights

$$F(\mathbf{x}; \mathbf{w}) \approx F(\mathbf{x}; \mathbf{w}_0) + \nabla_{\mathbf{w}} F(\mathbf{x}; \mathbf{w}_0) \cdot (\mathbf{w} - \mathbf{w}_0) = F(\mathbf{x}; \mathbf{w}_0) + \phi(\mathbf{x}) \cdot (\mathbf{w} - \mathbf{w}_0), \qquad (11.109)$$

where we recall the feature map ϕ is defined in (11.100). Thus, training an infinite-width neural network is actually somewhat similar to using a *linear* classifier on the feature data $\mathbf{z}_i = \phi(\mathbf{x}_i)$, which makes another connection between deep learning and kernel methods. In fact, we can even go further and *replace* the neural network $F(\mathbf{x}; \mathbf{w})$ by its linearization, so that we can use fast linear techniques, e.g., linear regression. These approaches are known as *random feature methods*, and were introduced in [190]. The term *random feature* refers to the feature map (11.100), which in this setting is the output of the first layer of a neural network with random choices for the weights.

While the analysis of very wide neural network has received much attention, it is important to point out that the results of training such networks are significantly different than those obtained by training narrow networks. As an experiment, we trained a 2-layer ReLU neural network of the form

$$F(\mathbf{x}; \mathbf{w}) = \frac{1}{\sqrt{N}} \sum_{k=1}^{N} a_k (\mathbf{w}_k \cdot \mathbf{x} + b_k)_+ + c,$$

to approximate the one dimensional hat function (10.74) that we utilized in some of the proofs of universal approximation in Section 10.6. Recall this function can be exactly expressed with a 2-layer ReLU network with 3 hidden nodes. We initialized all parameters in the network as independent standard normal random variables. In Figure 11.8 we show the results of training in the very wide NTK regime, where $N = 10^4$, as well as in the sparse regime, where $N = 100$. We used plain gradient descent on the least squares loss, using 200 equally spaced training points on the interval $[-1, 1]$, and trained both for 5×10^4 iterations with a time step of $\alpha = 0.1$ in the NTK regime and $\alpha = 1$ in the sparse example. The red dots show the locations of the zero crossings of the ReLU neurons; that is, we plot the points $-b_k/w_k$ for $k = 1, \ldots, N$ on the x-axis. For the sparse network, the activations coalesce around the corners of the hat function, indicating that the network is learning something quite similar to the optimal 3-node network. On the other hand, the very wide NTK regime exhibits a quite different behavior. The neurons do not concentrate on the corners, and instead they remain spread out over the entire interval $[-1, 1]$, and the neural network appears to be a smoothed out approximation of the hat function.[7] Thus, even though both networks are theoretically able to learn the optimal network with 3 hidden nodes, only the sparse network is able to do this in practice, given the dynamics of gradient descent. Recent research has focused on understanding phenomena like this in the training of neural networks; we refer to [247] for more details.

Another related approach is the *mean-field limit* of the training dynamics of neural networks, originally introduced in [161], whereby we adjust the scaling factor in the neural network to be $1/N$:

$$F(\mathbf{x}; \mathbf{w}) = \frac{1}{N} \sum_{k=1}^{N} a_k \sigma(\mathbf{w}_k \cdot \mathbf{x} + b_k) + c,$$

and we treat the neural network itself with the law of large numbers in the infinite width limit $N \to \infty$. In this case, the parameters of the neural network are better approximated by a mathematical object called a *measure*, and the training dynamics are substantially different than the NTK regime. We refer the reader to [161] for additional details.

Exercises

7.1. ♡ Given $\mathrm{O} \neq X \in \mathcal{M}_{m \times n}$, show that $\|X^T X \mathbf{u}\| \geq \sigma_{min}(X)\|X\mathbf{u}\|$ for all $\mathbf{u} \in \mathbb{R}^n$.

7.2. ♡ Suppose F has the form

$$F(\mathbf{x}; \mathbf{w}) = \sum_{k=1}^{m} w_k \Phi_k(\mathbf{x}), \tag{11.110}$$

where each $\Phi_k \colon \mathbb{R}^n \to \mathbb{R}$ is a scalar-valued function. Assume that the functions Φ_k can *distinguish* the training points, in the sense that $\Phi_k(\mathbf{x}_\ell) = 0$ for $k \neq \ell$ and $\Phi_k(\mathbf{x}_k) = 1$. Show that the corresponding kernel matrix K defined in (11.99) is the identity matrix.

7.3. ◇ Suppose F has the form (11.110), and, for any $y_1, \ldots, y_m \in \mathbb{R}$, there exists a choice of weights $\mathbf{w} \in \mathbb{R}^m$ such that $F(\mathbf{x}_i; \mathbf{w}) = y_i$ for all i. Show that the corresponding kernel matrix K defined in (11.99) is positive definite. Can you characterize $\lambda_{min}(K)$?

7.4. Here, we weaken the assumptions in Exercise 7.3. Suppose F has the form (11.110) and choose $y_1, \ldots, y_m$ so that there exists $\mathbf{w} \in \mathbb{R}^m$ such that $F(\mathbf{x}; \mathbf{w}_i) = y_i$ for all i (but do not assume this is true for any choice of y_i). Assume the loss $\ell(\mathbf{z}, \mathbf{y}) = \frac{1}{2}\|\mathbf{z} - \mathbf{y}\|^2$. Show that while the kernel matrix K may be singular, the value of λ in Theorem 11.35 is positive.

---

[7]This is a manifestation of the *frequency bias* in training very wide neural networks that we introduced in Chapter 10; see also [196, 250, 251].

7.5. $\diamond$ Suppose F is a kernel function of the form

$$F(\mathbf{x}; \mathbf{w}) = \sum_{k=1}^{m} w_k\, \Phi\left(\frac{\mathbf{x} - \mathbf{x}_k}{\varepsilon}\right),$$

where $\Phi : \mathbb{R}^n \to \mathbb{R}$ is a scalar-valued function satisfying $\Phi(\mathbf{0}) = 1$ and $\Phi(\mathbf{x}) = 0$ for $\|\mathbf{x}\| \geq 1$. Assume all the training data points $\mathbf{x}_1, \dots, \mathbf{x}_m \in \mathbb{R}^n$ are distinct. Show that the corresponding kernel matrix K defined in (11.99) is the identity matrix for $\varepsilon > 0$ sufficiently small. How small must we take ε? Do you think this is a good model that will generalize to new data points? *Hint*: Use Exercise 7.2, and show that the distinguishing property holds for $\varepsilon > 0$ sufficiently small.

Bibliography

Note: The numbers in brackets refer to the page(s) where each reference is cited.

[1] R. Abraham, J. E. Marsden, and T. Ratiu. *Manifolds, Tensor Analysis, and Applications*. Springer Science & Business Media, 2012. {xiii, 47, 49, 374, 536}

[2] J. Achiam, S. Adler, S. Agarwal, L. Ahmad, I. Akkaya, F. L. Aleman, D. Almeida, J. Altenschmidt, S. Altman, S. Anadkat, et al. Gpt-4 technical report. *arXiv:2303.08774*, 2023. {517, 525}

[3] D. Aloise, A. Deshpande, P. Hansen, and P. Popat. NP-hardness of Euclidean sum-of-squares clustering. *Machine Learning*, 75:245–248, 2009. {289}

[4] T. M. Apostol. *Calculus*. John Wiley & Sons, 1969, 1991. {xii, 76, 183, 186, 193, 210}

[5] V. I. Arnold. *Ordinary Differential Equations*. Springer Science & Business Media, 1992. {578, 582}

[6] D. Arthur and S. Vassilvitskii. k-means++: the advantages of careful seeding. In *Proceedings of the Eighteenth Annual Acm-Siam Symposium on Discrete Algorithms*, pages 1027–1035, 2007. {291, 294, 501}

[7] S. Arya, D. M. Mount, N. S. Netanyahu, R. Silverman, and A. Y. Wu. An optimal algorithm for approximate nearest neighbor searching fixed dimensions. *J. ACM*, 45(6):891–923, 1998. {286}

[8] H. Attouch and F. Alvarez. The heavy ball with friction dynamical system for convex constrained minimization problems. In *Optimization: Proceedings of the 9th Belgian-French-German Conference on Optimization Namur, September 7–11, 1998*, pages 25–35. Springer, 2000. {583}

[9] H. Attouch, X. Goudou, and P. Redont. The heavy ball with friction method, I. The continuous dynamical system: global exploration of the local minima of a real-valued function by asymptotic analysis of a dissipative dynamical system. *Communications in Contemporary Mathematics*, 2(01):1–34, 2000. {583}

[10] A. Bagirov, N. Karmitsa, and M. M. Mäkelä. *Introduction to Nonsmooth Optimization: Theory, Practice and Software*, volume 12. Springer, 2014. {270}

[11] G. A. Baker Jr and P. Graves-Morris. Padé Approximants, volume 59 of. *Encyclopedia of Mathematics and its Applications*, 22, 1996. {527}

[12] J. B. Barbour. *Absolute or Relative Motion?: A Study from a Machian Point of View of the Discovery and the Structure of Dynamical Theories*. Cambridge University Press, 1989. {3}

[13] M. Bardi, I. C. Dolcetta, et al. *Optimal Control and Viscosity Solutions of Hamilton-Jacobi-Bellman Equations*, volume 12. Springer, 1997. {405}

[14] A. R. Barron. Neural net approximation. In *Proc. 7th Yale Workshop on Adaptive and Learning Systems*, volume 1, pages 69–72, 1992. {543}

[15] A. R. Barron. Approximation and estimation bounds for artificial neural networks. *Machine Learning*, 14:115–133, 1994. {543}

[16] A. G. Baydin, B. A. Pearlmutter, A. A. Radul, and J. M. Siskind. Automatic differentiation in machine learning: a survey. *J. Marchine Learning Research*, 18:1–43, 2018. {500}

[17] A. Beck and M. Teboulle. A fast iterative shrinkage-thresholding algorithm for linear inverse problems. *SIAM J. Imaging Sciences*, 2(1):183–202, 2009. {270}

[18] E. Behrends. *Introduction to Markov Chains*. Springer, 2000. {156, 158}

[19] M. Belkin, I. Matveeva, and P. Niyogi. Regularization and semi-supervised learning on large graphs. In *Learning Theory: 17th Annual Conference on Learning Theory, COLT 2004, Banff, Canada, July 1-4, 2004. Proceedings 17*, pages 624–638. Springer, 2004. {454}

[20] M. Belkin and P. Niyogi. Laplacian eigenmaps for dimensionality reduction and data representation. *Neural Computation*, 15(6):1373–1396, 2003. {433}

[21] M. Belkin and P. Niyogi. Semi-supervised learning on Riemannian manifolds. *Machine Learning*, 56:209–239, 2004. {454}

[22] R. E. Bellman and S. E. Dreyfus. *Applied Dynamic Programming*. Princeton University Press, 2015. {400}

[23] Y. Bengio, O. Delalleau, and N. Le Roux. Label Propagation and Quadratic Criterion. In *Semi-Supervised Learning*, pages 193–216. MIT Press, January 2006. {454}

[24] M. Bertalmío. *Image Processing for Cinema*. CRC Press, 2014. {102, 466}

[25] A. L. Bertozzi and A. Flenner. Diffuse interface models on graphs for classification of high dimensional data. *SIAM Review*, 58(2):293–328, 2016. {454}

[26] C. M. Bishop. *Pattern Recognition and Machine Learning*. springer, 2006. {344}

[27] P. Bongini, M. Bianchini, and F. Scarselli. Molecular generative graph neural networks for drug discovery. *Neurocomputing*, 450:242–252, 2021. {365, 510}

[28] I. Borg and P. J. Groenen. *Modern Multidimensional Scaling: Theory and Applications*. Springer Science & Business Media, 2005. {354, 355}

[29] L. Bottou and Y. Bengio. Convergence properties of the k-means algorithms. *Advances in Neural Information Processing Systems*, 7, 1994. {290, 294}

[30] W. E. Boyce, R. C. DiPrima, and D. B. Meade. *Elementary Differential Equations and Boundary Value Problems*. John Wiley & Sons, 8th edition, 2004. {xiii, 553, 578, 581, 582}

[31] S. P. Boyd and L. Vandenberghe. *Convex Optimization*. Cambridge University Press, 2004. {225, 242}

[32] B. Bradie. *A Friendly Introduction to Numerical Analysis.* Prentice–Hall, 2006. {122, 577, 579}

[33] L. Bronsard and R. V. Kohn. Motion by mean curvature as the singular limit of Ginzburg-Landau dynamics. *J. Differential Equations*, 90(2):211–237, 1991. {243}

[34] S. R. Buss. *3D Computer Graphics: A Mathematical Introduction with OpenGL.* Cambridge University Press, 2003. {103}

[35] A. Byerly, T. Kalganova, and I. Dear. No routing needed between capsules. *Neurocomputing*, 463:545–553, 2021. {248, 505}

[36] J. Calder, R. Coil, J. A. Melton, P. J. Olver, G. Tostevin, and K. Yezzi-Woodley. Use and misuse of machine learning in anthropology. *IEEE BITS: the Information Theory Magazine*, 2(1):102–115, 2022. {259}

[37] J. Calder, B. Cook, M. Thorpe, and D. Slepcev. Poisson learning: Graph based semi-supervised learning at very low label rates. In *International Conference on Machine Learning*, pages 1306–1316. PMLR, 2020. {454, 455}

[38] A. Cauchy. Compte rendu des s eances de lacademie des sciences. *Comptes Rendus Hebd. Seances Acad. Sci*, 21(25):536–538, 1847. {181}

[39] C. Chen, O. Li, D. Tao, A. Barnett, C. Rudin, and J. K. Su. This looks like that: deep learning for interpretable image recognition. *Advances in Neural Information Processing Systems*, 32, 2019. {494, 543}

[40] Y. Cho and L. Saul. Kernel methods for deep learning. *Advances in Neural Information Processing Systems*, 22, 2009. {495, 588}

[41] J. K. Chorowski, D. Bahdanau, D. Serdyuk, K. Cho, and Y. Bengio. Attention-based models for speech recognition. *Advances in Neural Information Processing Systems*, 28, 2015. {519}

[42] F. R. Chung. *Spectral Graph Theory.* American Mathematical Soc., 1997. {357}

[43] K. W. Church. Word2Vec. *Natural Language Engineering*, 23(1):155–162, 2017. {255}

[44] B. A. Cipra. The best of the 20th century: Editors name top 10 algorithms. *SIAM News*, 33(4):1–2, 2000. {476}

[45] R. R. Coifman and S. Lafon. Diffusion maps. *Applied and Computational Harmonic Analysis*, 21(1):5–30, 2006. {426, 430, 433}

[46] J. W. Cooley and J. W. Tukey. An algorithm for the machine calculation of complex Fourier series. *Mathematics of Computation*, 19(90):297–301, 1965. {476}

[47] T. H. Cormen, C. E. Leiserson, R. L. Rivest, and C. Stein. *Introduction to algorithms.* MIT Press, 2022. {287}

[48] R. Courant. Variational methods for the solution of problems of equilibrium and vibrations. *Bull. Amer. Math. Soc.*, 49:1–23, 1943. {181}

[49] T. A. Cover and J. A. Thomas. *Elements of Information Theory.* Wiley-Interscience, 2006. {257}

[50] H. B. Curry. The method of steepest descent for non-linear minimization problems. *Quarterly Applied Mathematics*, 2(3):258–261, 1944. {181}

[51] S. Damrich and F. A. Hamprecht. On UMAP's true loss function. *Advances in Neural Information Processing Systems*, 34:5798–5809, 2021. {446}

[52] G. B. Dantzig and M. N. Thapa. *Linear Programming: Introduction*, volume 1. Springer, 1997. {185}

[53] G. B. Dantzig and M. N. Thapa. *Linear Programming: Theory and Extensions*, volume 2. Springer, 2003. {185}

[54] P. de Fermat. Maxima et minima (Latin). 1636–1642. In P. Tannery and C. Henry, editors, *Oeuvres de Fermat*, volume 1, pages 133–179, 1891. {181}

[55] M. Defferrard, X. Bresson, and P. Vandergheynst. Convolutional neural networks on graphs with fast localized spectral filtering. *Advances in Neural Information Processing Systems*, 29, 2016. {514}

[56] J. W. Demmel. *Applied Numerical Linear Algebra*. SIAM, 1997. {xii, 39, 40, 64, 209}

[57] N. Deng, Y. Tian, and C. Zhang. *Support Vector Machines: Optimization Based Theory, Algorithms, and Extensions*. CRC Press, 2012. {307}

[58] R. DeVore, B. Hanin, and G. Petrova. Neural network approximation. *Acta Numerica*, 30:327–444, 2021. {533}

[59] I. S. Dhillon, Y. Guan, and B. Kulis. *A Unified View of Kernel k-Means, Spectral Clustering and Graph Cuts*. Citeseer, 2004. {389}

[60] E. Dijkstra. A note on two problems in connexion with graphs. *Numerische Mathematik*, 1(1):269–271, 1959. {403}

[61] J. Dongarra and F. Sullivan. Guest editors introduction to the top 10 algorithms. *Computing in Science & Engineering*, 2(01):22–23, 2000. {476}

[62] A. Dosovitskiy, L. Beyer, A. Kolesnikov, D. Weissenborn, X. Zhai, T. Unterthiner, M. Dehghani, M. Minderer, G. Heigold, S. Gelly, et al. An image is worth 16x16 words: Transformers for image recognition at scale. In *International Conference on Learning Representations*, 2021. {526}

[63] D. Down, S. P. Meyn, and R. L. Tweedie. Exponential and uniform ergodicity of Markov processes. *Annals of Probability*, 23(4):1671–1691, 1995. {421}

[64] D.-Z. Du, P. M. Pardalos, and W. Wu. History of optimization. In C. A. Floudas and P. M. Pardalos, editors, *Encyclopedia of Optimization*, pages 1538–1542, Boston, MA, 2009. Springer US. {181}

[65] J. Duchi, E. Hazan, and Y. Singer. Adaptive subgradient methods for online learning and stochastic optimization. *J. Machine Learning Research*, 12(7), 2011. {489}

[66] R. Durrett. *Essentials of Stochastic Processes*. Springer, 1999. {156}

[67] H. Dym and H. P. McKean. *Fourier Series and Integrals*. Academic Press, 1972. {471, 530}

[68] C. Eckart and G. Young. The approximation of one matrix by another of lower rank. *Psychometrika*, 1(3):211–218, 1936. {328}

[69] B. Efron, T. Hastie, I. Johnstone, and R. Tibshirani. Least angle regression. *Annals of Statistics*, 32(2):407–499, 2004. {248, 253}

[70] C. K. Enders. *Applied Missing Data Analysis*. Guilford Publications, 2022. {250}

[71] L. C. Evans. *Partial Differential Equations*, volume 19. American Mathematical Society, 2nd edition, 2010. {230}

[72] K. Fan. Minimax theorems. *Proceedings of the National Academy of Sciences*, 39(1):42–47, 1953. {306}

[73] G. E. Farin. *Curves and Surfaces for CAGD: A Practical Guide*. Morgan Kaufmann, 2002. {532}

[74] B. Fine and G. Rosenberger. *The Fundamental Theorem of Algebra*. Springer Science & Business Media, 1997. {125}

[75] R. W. Floyd. Algorithm 97: shortest path. *Communications ACM*, 5(6):345, 1962. {402}

[76] S. Fortunato. Community detection in graphs. *Physics Reports*, 486(3-5):75–174, 2010. {365}

[77] S. Foucart and H. Rauhut. *An Invitation to Compressive Sensing*. Springer, 2013. {250}

[78] J. G. Francis. The QR transformation: A unitary analogue to the LR transformation — Part 1. *Computer Journal*, 4(3):265–271, 1961. {171}

[79] J. G. Francis. The QR transformation — Part 2. *Computer Journal*, 4(4):332–345, 1962. {171}

[80] J. H. Friedman, J. L. Bentley, and R. A. Finkel. An algorithm for finding best matches in logarithmic expected time. *ACM Transactions Mathematical Software*, 3(3):209–226, 1977. {286}

[81] G. Gan, C. Ma, and J. Wu. *Data Clustering: Theory, Algorithms, and Applications*. SIAM, 2020. {295}

[82] J. Gasteiger, A. Bojchevski, and S. Günnemann. Predict then propagate: Graph neural networks meet personalized PageRank. In *International Conference on Learning Representations*, 2018. {418}

[83] J. Gasteiger, S. Weißenberger, and S. Günnemann. Diffusion improves graph learning. *Advances in Neural Information Processing Systems*, 32, 2019. {514}

[84] E. Ghadimi, H. R. Feyzmahdavian, and M. Johansson. Global convergence of the heavy-ball method for convex optimization. In *2015 European Control Conference (ECC)*, pages 310–315. IEEE, 2015. {556}

[85] R. Gilmore. *Lie Groups, Physics, and Geometry: An Introduction for Physicists, Engineers and Chemists*. Cambridge University Press, 2008. {102}

[86] D. F. Gleich. PageRank beyond the Web. *SIAM Review*, 57(3):321–363, 2015. {418, 419, 422}

[87] H. Goldstein. *Classical Mechanics*. Addison–Wesley, 2nd edition, 1980. {102}

[88] G. H. Golub and C. F. Van Loan. *Matrix Computations*. Johns Hopkins University Press, 3rd edition, 2013. {xii, 39, 40, 161, 176, 556}

[89] I. Goodfellow, Y. Bengio, and A. Courville. *Deep Learning*. MIT Press, 2016. `http://www.deeplearningbook.org`. {484}

[90] I. Goodfellow, J. Pouget-Abadie, M. Mirza, B. Xu, D. Warde-Farley, S. Ozair, A. Courville, and Y. Bengio. Generative adversarial networks. *Communications ACM*, 63(11):139–144, 2020. {484}

[91] I. J. Goodfellow, J. Shlens, and C. Szegedy. Explaining and harnessing adversarial examples. In *International Conference on Learning Representations*, 2015. {501}

[92] R. L. Graham, D. E. Knuth, and O. Patashnik. *Concrete Mathematics: A Foundation for Computer Science*. Addison-Wesley Professional, 1994. {xxiii, 296}

[93] A. Graves. Generating sequences with recurrent neural networks. *arXiv:1308.0850*, 2013. {519}

[94] M. Greenacre, P. J. Groenen, T. Hastie, A. I. d'Enza, A. Markos, and E. Tuzhilina. Principal component analysis. *Nature Reviews Methods Primers*, 2(1):100, 2022. {311}

[95] J. Hadamard. *Mémoire sur le Problème d'Analyse Relatif à l'Équilibre des Plaques Élastiques Encastrées*, volume 33. Imprimerie Nationale, 1908. {181}

[96] W. W. Hager and H. Zhang. A survey of nonlinear conjugate gradient methods. *Pacific J. Optimization*, 2(1):35–58, 2006. {209}

[97] J. Ham, D. Lee, and L. Saul. Semisupervised alignment of manifolds. In *International Workshop on Artificial Intelligence and Statistics*, pages 120–127. PMLR, 2005. {454}

[98] S. Han and M. Boutin. The hidden structure of image datasets. In *2015 IEEE International Conference on Image Processing (ICIP)*, pages 1095–1099. IEEE, 2015. {298}

[99] E. M. Harrell. Double wells. *Communications Mathematical Physics*, 75:239–261, 1980. {243}

[100] K. He, X. Zhang, S. Ren, and J. Sun. Deep residual learning for image recognition. In *Proceedings of the IEEE Conference on Computer Vision and Pattern Recognition*, pages 770–778, 2016. {487}

[101] T. L. Heath. *A History of Greek Mathematics*, volume 1. Cambridge University Press, 2013. {241}

[102] J. M. Hendrickx and A. Olshevsky. Matrix p-norms are NP-hard to approximate if $p \neq 1, 2, \infty$. *SIAM J. on Matrix Analysis and Applications*, 31(5):2802–2812, 2010. {118}

[103] M. R. Hestenes and E. Stiefel. Methods of conjugate gradients for solving linear systems. *J. Research of the National Bureau of Standards*, 49(6):409–436, 1952. {206}

[104] N. J. Higham. Stable iterations for the matrix square root. *Numerical Algorithms*, 15:227–242, 1997. {140, 241}

[105] N. J. Higham. *Accuracy and Stability of Numerical Algorithms*. SIAM, 2nd edition, 2002. {xii, 39, 40, 64, 161, 240}

[106] G. E. Hinton and S. Roweis. Stochastic neighbor embedding. *Advances in Neural Information Processing Systems*, 15, 2002. {443, 445}

[107] S. Hochreiter and J. Schmidhuber. Long short-term memory. *Neural Computation*, 9(8):1735–1780, 1997. {483, 518}

[108] R. W. Hoerl. Ridge regression: a historical context. *Technometrics*, 62(4):420–425, 2020. {263}

[109] F. Hoffmann, B. Hosseini, A. A. Oberai, and A. M. Stuart. Spectral analysis of weighted Laplacians arising in data clustering. *Applied and Computational Harmonic Analysis*, 56:189–249, 2022. {434}

[110] R. V. Hogg, E. A. Tanis, and D. L. Zimmerman. *Probability and Statistical Inference*. Macmillan New York, 9th edition, 1977. {22, 251, 252}

[111] L. Hooi-Tong. On a class of directed graphs–with an application to traffic-flow problems. *Operations Research*, 18(1):87–94, 1970. {362}

[112] S. Hoory, N. Linial, and A. Wigderson. Expander graphs and their applications. *Bulletin of the American Mathematical Society*, 43(4):439–561, 2006. {384}

[113] J. J. Hopfield. Neural networks and physical systems with emergent collective computational abilities. *Proceedings National Academy of Sciences*, 79(8):2554–2558, 1982. {483, 518}

[114] H. Hotelling. Analysis of a complex of statistical variables into principal components. *J. Educational Psychology*, 24(6):417, 1933. {311}

[115] G. Huang, Z. Liu, L. Van Der Maaten, and K. Q. Weinberger. Densely connected convolutional networks. In *Proceedings of the IEEE Conference on Computer Vision and Pattern Recognition*, pages 4700–4708, 2017. {488}

[116] S. Ioffe and C. Szegedy. Batch normalization: Accelerating deep network training by reducing internal covariate shift. In *International Conference on Machine Learning*, pages 448–456. PMLR, 2015. {489}

[117] A. Jacot, F. Gabriel, and C. Hongler. Neural tangent kernel: Convergence and generalization in neural networks. *Advances in Neural Information Processing Systems*, 31, 2018. {589}

[118] D. Jiang, Z. Wu, C.-Y. Hsieh, G. Chen, B. Liao, Z. Wang, C. Shen, D. Cao, J. Wu, and T. Hou. Could graph neural networks learn better molecular representation for drug discovery? A comparison study of descriptor-based and graph-based models. *J. Cheminformatics*, 13(1):1–23, 2021. {365, 510}

[119] X. Jin, M. Zhao, T. W. Chow, and M. Pecht. Motor bearing fault diagnosis using trace ratio linear discriminant analysis. *IEEE Transactions on Industrial Electronics*, 61(5):2441–2451, 2013. {345}

[120] S. G. Johnson and M. Frigo. A modified split-radix FFT with fewer arithmetic operations. *IEEE Transactions on Signal Processing*, 55(1):111–119, 2006. {479}

[121] I. T. Jolliffe. *Principal Component Analysis for Special Types of Data.* Springer, 2nd edition, 2002. {311}

[122] I. T. Jolliffe and J. Cadima. Principal component analysis: a review and recent developments. *Philosophical Transactions Royal Society A: Mathematical, Physical and Engineering Sciences*, 374(2065):20150202, 2016. {311}

[123] M. M. Julian. *Foundations of Crystallography with Computer Applications.* CRC Press, 2014. {102}

[124] L. V. Kantorovich and G. P. Akilov. *Functional Analysis.* Elsevier, 2016. {303}

[125] G. Ke, D. He, and T.-Y. Liu. Rethinking positional encoding in language pre-training. In *International Conference on Learning Representations*, 2020. {524}

[126] S. Khan, M. Naseer, M. Hayat, S. W. Zamir, F. S. Khan, and M. Shah. Transformers in vision: A survey. *ACM Computing Surveys*, 54(10s):1–41, 2022. {526}

[127] D. P. Kingma and J. Ba. Adam: A method for stochastic optimization. In *International Conference on Learning Representations*, 2015. {489}

[128] T. N. Kipf and M. Welling. Semi-supervised classification with graph convolutional networks. In *International Conference on Learning Representations*, 2017. {511, 513, 515}

[129] A. V. Knyazev. Toward the optimal preconditioned eigensolver: Locally optimal block preconditioned conjugate gradient method. *SIAM J. Scientific Computing*, 23(2):517–541, 2001. {208, 561}

[130] D. Kobak and P. Berens. The art of using t-SNE for single-cell transcriptomics. *Nature Communications*, 10(1):5416, 2019. {438}

[131] S. Kotz and S. Nadarajah. *Multivariate t-distributions and their Applications.* Cambridge University Press, 2004. {443}

[132] A. Krizhevsky, I. Sutskever, and G. E. Hinton. Imagenet classification with deep convolutional neural networks. *Advances in Neural Information Processing Systems*, 25, 2012. {483}

[133] V. N. Kublanovskaya. On some algorithms for the solution of the complete eigenvalue problem. *USSR Computational Mathematics and Mathematical Physics*, 1(3):637–657, 1962. {171}

[134] S. Kullback. *Information Theory and Statistics.* Courier Corporation, 1997. {50}

[135] S. K. Kumar. On weight initialization in deep neural networks. *arXiv:1704.08863*, 2017. {489}

[136] Y. LeCun and Y. Bengio. Convolutional networks for images, speech, and time series. *The Handbook of Brain Theory and Neural Networks*, 3361(10):1995, 1995. {483}

[137] Y. LeCun, B. Boser, J. Denker, D. Henderson, R. Howard, W. Hubbard, and L. Jackel. Handwritten digit recognition with a back-propagation network. *Advances in Neural Information Processing Systems*, 2, 1989. {483}

[138] Y. LeCun, L. Bottou, Y. Bengio, and P. Haffner. Gradient-based learning applied to document recognition. *Proceedings IEEE*, 86(11):2278–2324, 1998. {483}

[139] W.-Y. Lee, L.-C. Hsieh, G.-L. Wu, and W. Hsu. Graph-based semi-supervised learning with multi-modality propagation for large-scale image datasets. *J. Visual Communication and Image Representation*, 24(3):295–302, 2013. {454}

[140] J. Lei and A. Rinaldo. Consistency of spectral clustering in stochastic block models. *Annals of Statistics*, 43(1):215 – 237, 2015. {434}

[141] G. Lerman and T. Maunu. Fast, robust and non-convex subspace recovery. *Information and Inference*, 7(2):277–336, 2018. {330, 332}

[142] G. Lerman and T. Maunu. An overview of robust subspace recovery. *Proceedings IEEE*, 106(8):1380–1410, 2018. {330}

[143] L. Lessard, B. Recht, and A. Packard. Analysis and design of optimization algorithms via integral quadratic constraints. *SIAM J. Optimization*, 26(1):57–95, 2016. {556}

[144] C.-K. Li and W. So. Isometries of ℓ_p–norm. *American Mathematical Monthly*, 101(5):452–453, 1994. {100}

[145] X. Li, R. Zhu, Y. Cheng, C. Shan, S. Luo, D. Li, and W. Qian. Finding global homophily in graph neural networks when meeting heterophily. In *International Conference on Machine Learning*, pages 13242–13256. PMLR, 2022. {514}

[146] Z. Li, F. Nie, X. Chang, and Y. Yang. Beyond trace ratio: weighted harmonic mean of trace ratios for multiclass discriminant analysis. *IEEE Transactions on Knowledge and Data Engineering*, 29(10):2100–2110, 2017. {345}

[147] G. C. Linderman, M. Rachh, J. G. Hoskins, S. Steinerberger, and Y. Kluger. Fast interpolation-based t-SNE for improved visualization of single-cell RNA-seq data. *Nature Methods*, 16(3):243–245, 2019. {261, 438}

[148] G. C. Linderman and S. Steinerberger. Clustering with t-SNE, provably. *SIAM J. Mathematics of Data Science*, 1(2):313–332, 2019. {444}

[149] G. C. Linderman and S. Steinerberger. Dimensionality reduction via dynamical systems: the case of t-SNE. *SIAM Review*, 64(1):153–178, 2022. {444}

[150] S. Linnainmaa. The representation of the cumulative rounding error of an algorithm as a Taylor expansion of the local rounding errors. *M.S. thesis, Dept. Comput. Sci., Univ. Helsinki*, pages 6–7, 1970. {483}

[151] T. Liu, A. Moore, K. Yang, and A. Gray. An investigation of practical approximate nearest neighbor algorithms. *Advances in Neural Information Processing Systems*, 17, 2004. {286}

[152] S. Lloyd. Least squares quantization in PCM. *IEEE Transactions Information Theory*, 28(2):129–137, 1982. {289}

[153] S. Lojasiewicz. A topological property of real analytic subsets. *Coll. du CNRS, Les équations aux dérivées partielles*, 117(87-89):2, 1963. {224}

[154] G. Lorentz. *Approximation of Functions*. Holt, Rinehart and Winston, 1966. {533}

[155] L. Lü, Y.-C. Zhang, C. H. Yeung, and T. Zhou. Leaders in social networks, the delicious case. *PloS One*, 6(6):e21202, 2011. {423, 425}

[156] D. J. MacKay. *Information Theory, Inference and Learning Algorithms*. Cambridge University Press, 2003. {50}

[157] A. Madry, A. Makelov, L. Schmidt, D. Tsipras, and A. Vladu. Towards deep learning models resistant to adversarial attacks. In *International Conference on Learning Representations*, 2017. {501}

[158] J. E. Marsden and A. J. Tromba. *Vector Calculus*. W.H. Freeman, 6th edition, 2012. {xii, 76, 146, 193, 210}

[159] W. S. McCulloch and W. Pitts. A logical calculus of the ideas immanent in nervous activity. *Bulletin Mathematical Biophysics*, 5:115–133, 1943. {483}

[160] L. McInnes, J. Healy, N. Saul, and L. Großberger. UMAP: Uniform manifold approximation and projection. *Journal of Open Source Software*, 3(29):861, 2018. {438, 444, 446}

[161] S. Mei, A. Montanari, and P.-M. Nguyen. A mean field view of the landscape of two-layer neural networks. *Proc. National Academy of Sciences*, 115(33):E7665–E7671, 2018. {590}

[162] M. Meilua and J. Shi. A random walks view of spectral segmentation. In *International Workshop on Artificial Intelligence and Statistics*, pages 203–208. PMLR, 2001. {389}

[163] G. Meinardus. *Approximation of Functions: Theory and Numerical Methods*. Springer Science & Business Media, 2012. {527}

[164] J. Mercer. Functions of positive and negative type, and their connection the theory of integral equations. *Philosophical Transactions Royal Society of London. Series A*, 209:415–446, 1909. {303}

[165] E. Merkurjev, A. Bertozzi, X. Yan, and K. Lerman. Modified Cheeger and ratio cut methods using the Ginzburg–Landau functional for classification of high-dimensional data. *Inverse Problems*, 33(7):074003, 2017. {454}

[166] E. Merkurjev, A. L. Bertozzi, and F. Chung. A semi-supervised heat kernel pagerank MBO algorithm for data classification. *Communications in Mathematical Sciences*, 16(5):1241–1265, 2018. {454}

[167] A. Messiah. *Quantum Mechanics*. John Wiley & Sons, 1976. {102, 138}

[168] L. Mirsky. Symmetric gauge functions and unitarily invariant norms. *Quarterly J. Mathematics*, 11(1):50–59, 1960. {328, 332}

[169] C. W. Misner, K. S. Thorne, and J. A. Wheeler. *Gravitation*. Macmillan, 1973. {84}

[170] M. Muja and D. G. Lowe. Fast approximate nearest neighbors with automatic algorithm configuration. *VISAPP (1)*, 2(331-340):2, 2009. {370}

[171] B. Nadler, S. Lafon, I. Kevrekidis, and R. Coifman. Diffusion maps, spectral clustering and eigenfunctions of Fokker-Planck operators. *Advances in Neural Information Processing Systems*, 18, 2005. {426, 430}

[172] Y. Nesterov. A method of solving a convex programming problem with convergence rate O $\left(1/k^2\right)$. *Doklady Akademii Nauk*, 269:543–547, 1983. {562, 563}

[173] Y. Nesterov. *Introductory Lectures on Convex Optimization: A Basic Course*, volume 87. Springer Science & Business Media, 2013. {564}

[174] Y. Nesterov and B. T. Polyak. Cubic regularization of Newton method and its global performance. *Mathematical Programming*, 108(1):177–205, 2006. {240, 242}

[175] M. E. Newman. Modularity and community structure in networks. *Proceedings National Academy of Sciences*, 103(23):8577–8582, 2006. {365, 391, 395}

[176] A. Y. Ng, M. I. Jordan, and Y. Weiss. On spectral clustering: analysis and an algorithm. *Advances in Neural Information Processing Systems*, 2:849–856, 2002. {426, 433, 434, 437}

[177] J. Nocedal and S. J. Wright. Quadratic programming. *Numerical Optimization*, pages 448–492, 2006. {308}

[178] B. A. Olshausen and D. J. Field. Emergence of simple-cell receptive field properties by learning a sparse code for natural images. *Nature*, 381(6583):607–609, 1996. {504}

[179] P. J. Olver. *Classical Invariant Theory*. Cambridge University Press, 1999. {210}

[180] P. J. Olver. *Introduction to Partial Differential Equations*. Springer, 2014. {93, 146, 210, 378, 386, 529, 530, 532}

[181] P. J. Olver and C. Shakiban. *Applied Linear Algebra*. Springer, 2nd edition, 2006. {xi, xiv, xxi, xxiii, 4, 15, 63–65, 74, 88, 102, 108, 109, 112, 114, 121, 123, 125, 128–130, 132–134, 138, 154, 161, 164, 168, 170, 177, 189, 209, 271, 272, 315, 380, 402, 449, 514, 528, 553, 556}

[182] A. Paszke, S. Gross, S. Chintala, G. Chanan, E. Yang, Z. DeVito, Z. Lin, A. Desmaison, L. Antiga, and A. Lerer. Automatic differentiation in pytorch, 2017. {500}

[183] D. Patterson, J. Gonzalez, U. Hölzle, Q. Le, C. Liang, L.-M. Munguia, D. Rothchild, D. R. So, M. Texier, and J. Dean. The carbon footprint of machine learning training will plateau, then shrink. *Computer*, 55(7):18–28, 2022. {525}

[184] K. Pearson. On lines and planes of closest fit to systems of points in space. *The London, Edinburgh, and Dublin Philosophical Magazine and Journal of Science*, 2(11):559–572, 1901. {311}

[185] H.-O. Peitgen and P. H. Richter. *The Beauty of Fractals: Images of Complex Dynamical Systems*. Springer Science & Business Media, 1986. {242}

[186] B. T. Polyak. Gradient methods for minimizing functionals. *Zhurnal Vychislitel'noi Matematiki i Matematicheskoi Fiziki*, 3(4):643–653, 1963. {224}

[187] B. T. Polyak. Some methods of speeding up the convergence of iteration methods. *USSR Computational Mathematics and Mathematical Physics*, 4(5):1–17, 1964. {552, 555, 583}

[188] K. T. Poole and H. Rosenthal. A spatial model for legislative roll call analysis. *American J. Political Science*, pages 357–384, 1985. {355}

[189] M. J. D. Powell. *Approximation Theory and Methods.* Cambridge University Press, 1981. {527}

[190] A. Rahimi and B. Recht. Random features for large-scale kernel machines. *Advances in Neural Information Processing Systems,* 20, 2007. {589}

[191] M. Raissi, P. Perdikaris, and G. E. Karniadakis. Physics-informed neural networks: A deep learning framework for solving forward and inverse problems involving nonlinear partial differential equations. *J. Computational Physics,* 378:686–707, 2019. {484}

[192] M. Reed and B. Simon. *Methods of Modern Mathematical Physics.* Academic Press, 1972. {93, 138, 149}

[193] A. Rényi. *Probability Theory.* Courier Corporation, 2007. {588}

[194] R. Rifkin, M. Pontil, and A. Verri. A note on support vector machine degeneracy. In *International Conference on Algorithmic Learning Theory,* pages 252–263. Springer, 1999. {277}

[195] D. N. Rockmore. The FFT: an algorithm the whole family can use. *Computing in Science & Engineering,* 2(1):60–64, 2000. {476}

[196] B. Ronen, D. Jacobs, Y. Kasten, and S. Kritchman. The convergence rate of neural networks for learned functions of different frequencies. *Advances in Neural Information Processing Systems,* 32, 2019. {490, 590}

[197] F. Rosenblatt. The perceptron: a probabilistic model for information storage and organization in the brain. *Psychological Review,* 65(6):386, 1958. {483}

[198] R. M. Roth. Introduction to Coding Theory. *IET Communications,* 47(18-19):4, 2006. {50}

[199] S. T. Roweis and L. K. Saul. Nonlinear dimensionality reduction by locally linear embedding. *Science,* 290(5500):2323–2326, 2000. {433}

[200] B. Roy. Transitivité et connexité. *CR Acad. Sci. Paris,* 249(216-218):182, 1959. {402}

[201] S. Ruder. An overview of gradient descent optimization algorithms. *arXiv:1609.04747,* 2016. {489}

[202] W. Rudin. *Principles of Mathematical Analysis.* McGraw–Hill, 3rd edition, 1976. {xiii, 47, 49, 528}

[203] D. E. Rumelhart, G. E. Hinton, and R. J. Williams. Learning internal representations by error propagation. *Technical Report, Institute for Cognitive Science, University of California, San Diego,* 1985. {483}

[204] O. Russakovsky, J. Deng, H. Su, J. Krause, S. Satheesh, S. Ma, Z. Huang, A. Karpathy, A. Khosla, M. Bernstein, et al. Imagenet large scale visual recognition challenge. *International J. Computer Vision,* 115:211–252, 2015. {484}

[205] Y. Saad. *Numerical Methods for Large Eigenvalue Problems.* SIAM, 2011. {161, 315, 556}

[206] D. Salomon. *Computer Graphics and Geometric Modeling.* Springer Science & Business Media, 2012. {103, 466}

[207] M. J. Schervish and M. H. DeGroot. *Probability and Statistics*. Pearson Education, 3rd edition, 2014. {251}

[208] E. Schmidt. Zur Theorie der linearen und nichtlinearen Integralgleichungen. *Mathematische Annalen*, 63(4):433–476, 1907. {328}

[209] I. J. Schoenberg. Remarks to Maurice Fréchet's article "Sur la définition axiomatique d'une classe d'espaces vectoriels distanciés applicables vectoriellement sur l'espace de Hilbert". *Annals of Mathematics*, 36:724–732, 1935. {349}

[210] L. Schumaker. *Spline Functions: Basic Theory*. Cambridge University Press, 2007. {532}

[211] I. Sciriha. A characterization of singular graphs. *Electronic J. Linear Algebra*, 16:451–462, 2007. {394}

[212] J. A. Sethian. A fast marching level set method for monotonically advancing fronts. *Proceedings National Academy of Sciences*, 93(4):1591–1595, 1996. {404}

[213] J. Shi and J. Malik. Normalized cuts and image segmentation. *IEEE Transactions on Pattern Analysis and Machine Intelligence*, 22(8):888–905, 2000. {387, 433, 434, 437}

[214] Z. Shi, S. Osher, and W. Zhu. Weighted nonlocal laplacian on interpolation from sparse data. *J. Scientific Computing*, 73:1164–1177, 2017. {454}

[215] D. Silver, J. Schrittwieser, K. Simonyan, I. Antonoglou, A. Huang, A. Guez, T. Hubert, L. Baker, M. Lai, A. Bolton, et al. Mastering the game of go without human knowledge. *Nature*, 550(7676):354–359, 2017. {484}

[216] M. Sion. On general minimax theorems. *Pacific J. Math.*, pages 171–176, 1958. {306}

[217] L. Sirovich and M. Kirby. Low-dimensional procedure for the characterization of human faces. *J. Optical Society of America A*, 4(3):519–524, 1987. {318}

[218] J. Sohl-Dickstein, E. Weiss, N. Maheswaranathan, and S. Ganguli. Deep unsupervised learning using nonequilibrium thermodynamics. In *International Conference on Machine Learning*, pages 2256–2265. PMLR, 2015. {484}

[219] D. Spielman. Spectral graph theory. In *Combinatorial Scientific Computing*, chapter 18, pages 495–524. CRC Press, Boca Raton, FL, 2012. {357}

[220] N. Srivastava, G. Hinton, A. Krizhevsky, I. Sutskever, and R. Salakhutdinov. Dropout: a simple way to prevent neural networks from overfitting. *J. Machine Learning Research*, 15(1):1929–1958, 2014. {489}

[221] S. Steinerberger and Y. Zhang. t-SNE, forceful colorings, and mean field limits. *Research in the Mathematical Sciences*, 9(3):42, 2022. {444}

[222] G. W. Stewart. *Matrix Algorithms: Volume 1: Basic Decompositions*. SIAM, 1998. {112}

[223] Z. Strakovs and P. Tichý. On error estimation in the conjugate gradient method and why it works in finite precision computations. *ETNA. Electronic Transactions on Numerical Analysis*, 13:56–80, 2002. {208}

[224] G. Strang. *Linear Algebra and Its Applications*. Harcourt, Brace, Jovanovich, 3rd edition, 1988. {63, 74, 102, 108, 112, 121, 125}

[225] G. Strang and G. Fix. *An Analysis of the Finite Element Method*. SIAM, 2008. {93, 122, 532}

[226] W. Su, S. Boyd, and E. J. Candes. A differential equation for modeling Nesterov's accelerated gradient method: Theory and insights. *J. of Machine Learning Research*, 17(153):1–43, 2016. {583}

[227] R. A. Tapia, J. E. Dennis Jr, and J. P. Schäfermeyer. Inverse, shifted inverse, and Rayleigh quotient iteration as Newton's method. *Siam Review*, 60(1):3–55, 2018. {122}

[228] R. E. Tarjan. *Data Structures and Network Algorithms*. SIAM, 1983. {403, 404}

[229] J. B. Tenenbaum, V. d. Silva, and J. C. Langford. A global geometric framework for nonlinear dimensionality reduction. *Science*, 290(5500):2319–2323, 2000. {406}

[230] S. A. Teukolsky, B. P. Flannery, W. Press, and W. Vetterling. *Numerical Recipes in C: The Art of Scientific Computing*. Cambridge University Press, 2nd edition, 1995. {xii, 39, 176, 209, 240}

[231] N. G. Trillos and D. Slepvcev. A variational approach to the consistency of spectral clustering. *Applied and Computational Harmonic Analysis*, 45(2):239–281, 2018. {434}

[232] M. A. Turk and A. P. Pentland. Face recognition using eigenfaces. In *Proceedings. 1991 IEEE Computer Society Conference on Computer Vision and Pattern Recognition*, pages 586–587. IEEE Computer Society, 1991. {318}

[233] B. E. Usevitch. A tutorial on modern lossy wavelet image compression: foundations of JPEG 2000. *IEEE Signal Processing Magazine*, 18(5):22–35, 2001. {338}

[234] L. Van Der Maaten. Accelerating t-SNE using tree-based algorithms. *J. Machine Learning Research*, 15(1):3221–3245, 2014. {441, 442}

[235] L. Van der Maaten and G. Hinton. Visualizing data using t-SNE. *J. Machine Learning Research*, 9(11), 2008. {438, 441}

[236] H. A. Van der Vorst. *Iterative Krylov Methods for Large Linear Systems*. Cambridge University Press, 2003. {556}

[237] A. Vaswani, N. Shazeer, N. Parmar, J. Uszkoreit, L. Jones, A. N. Gomez, K. Kaiser, and I. Polosukhin. Attention is all you need. *Advances in Neural Information Processing Systems*, 30, 2017. {484, 519–521}

[238] N. Vaswani, T. Bouwmans, S. Javed, and P. Narayanamurthy. Robust subspace learning: Robust PCA, robust subspace tracking, and robust subspace recovery. *IEEE Signal Processing Magazine*, 35(4):32–55, 2018. {330}

[239] U. Von Luxburg. A tutorial on spectral clustering. *Statistics and Computing*, 17:395–416, 2007. {426, 434}

[240] U. Von Luxburg, M. Belkin, and O. Bousquet. Consistency of spectral clustering. *Annals of Statistics*, pages 555–586, 2008. {434}

[241] R. E. Walpole, R. H. Myers, S. L. Myers, and K. Ye. *Probability and Statistics for Engineers and Scientists*. Macmillan New York, 9th edition, 1993. {22, 251}

[242] G. G. Walter and X. Shen. *Wavelets and Other Orthogonal Systems*. CRC Press, 2000. {15}

[243] B. Wang, Z. Tu, and J. K. Tsotsos. Dynamic label propagation for semi-supervised multi-class multi-label classification. In *Proceedings of the IEEE International Conference on Computer Vision*, pages 425–432, 2013. {454}

[244] S. Warshall. A theorem on Boolean matrices. *J. ACM*, 9(1):11–12, 1962. {402}

[245] D. S. Watkins. *Fundamentals of Matrix Computations*. John Wiley & Sons, 2004. {xii, 161, 176, 209}

[246] A. Wibisono, A. C. Wilson, and M. I. Jordan. A variational perspective on accelerated methods in optimization. *Proc. National Academy of Sciences*, 113(47):E7351–E7358, 2016. {583}

[247] F. Williams, M. Trager, D. Panozzo, C. Silva, D. Zorin, and J. Bruna. Gradient dynamics of shallow univariate relu networks. *Advances in Neural Information Processing Systems*, 32, 2019. {590}

[248] Z. Wu, S. Pan, F. Chen, G. Long, C. Zhang, and S. Y. Philip. A comprehensive survey on graph neural networks. *IEEE Transactions on Neural Networks and Learning Systems*, 32(1):4–24, 2020. {511, 516}

[249] H. Xiao, K. Rasul, and R. Vollgraf. Fashion-mnist: a novel image dataset for benchmarking machine learning algorithms. *arXiv:1708.07747*, 2017. {508}

[250] Z.-Q. J. Xu, Y. Zhang, T. Luo, Y. Xiao, and Z. Ma. Frequency principle: Fourier analysis sheds light on deep neural networks. *Communications in Computational Physics*, 28:1746–1767, 2020. {490, 590}

[251] Z.-Q. J. Xu, Y. Zhang, and Y. Xiao. Training behavior of deep neural network in frequency domain. In *Neural Information Processing: 26th International Conference, ICONIP 2019, Sydney, NSW, Australia, December 12–15, 2019, Proceedings, Part I 26*, pages 264–274. Springer, 2019. {490, 590}

[252] I. M. Yaglom. *Felix Klein and Sophus Lie: Evolution of the Idea of Symmetry in the Nineteenth Century*. Birkhäuser, 1988. {102}

[253] P. B. Yale. *Geometry and Symmetry*. Holden–Day, 1968. {103}

[254] Y. Yan, M. Hashemi, K. Swersky, Y. Yang, and D. Koutra. Two sides of the same coin: Heterophily and oversmoothing in graph convolutional neural networks. In *2022 IEEE International Conference on Data Mining (ICDM)*, pages 1287–1292. IEEE, 2022. {514}

[255] Z. Yang, W. Cohen, and R. Salakhudinov. Revisiting semi-supervised learning with graph embeddings. In *International Conference on Machine Learning*, pages 40–48. PMLR, 2016. {365}

[256] D. Yarotsky. Error bounds for approximations with deep ReLU networks. *Neural Networks*, 94:103–114, 2017. {541}

[257] R. Yavne. An economical method for calculating the discrete Fourier transform. In *Proceedings of the December 9-11, 1968, Fall Joint Computer Conference, part I*, pages 115–125, 1968. {479, 480}

[258] K. Yezzi-Woodley, A. Terwilliger, J. Li, E. Chen, M. Tappen, J. Calder, and P. J. Olver. Using machine learning on new feature sets extracted from 3D models of broken animal bones to classify fragments according to break agent. *J. Human Evolution*, 2024. {259}

[259] G. Young and A. S. Householder. Discussion of a set of points in terms of their mutual distances. *Psychometrika*, 3(1):19–22, 1938. {349}

[260] J. Yu, Z. Wang, V. Vasudevan, L. Yeung, M. Seyedhosseini, and Y. Wu. CoCa: Contrastive captioners are image-text foundation models. *Transactions on Machine Learning Research*, 2022. {484}

[261] W. W. Zachary. An information flow model for conflict and fission in small groups. *J. Anthropological Research*, 33(4):452–473, 1977. {364}

[262] M. D. Zeiler. Adadelta: An adaptive learning rate method. *arXiv:1212.5701*, 2012. {489}

[263] H. Zhao. A fast sweeping method for eikonal equations. *Mathematics of Computation*, 74(250):603–627, 2005. {404}

[264] X. Zheng, Y. Liu, S. Pan, M. Zhang, D. Jin, and P. S. Yu. Graph neural networks for graphs with heterophily: A survey. *arXiv:2202.07082*, 2022. {514}

[265] J. Zhou, G. Cui, S. Hu, Z. Zhang, C. Yang, Z. Liu, L. Wang, C. Li, and M. Sun. Graph neural networks: A review of methods and applications. *AI open*, 1:57–81, 2020. {511, 516}

[266] X. Zhou and M. Belkin. Semi-supervised learning by higher order regularization. In *Proceedings of the Fourteenth International Conference on Artificial Intelligence and Statistics*, pages 892–900. JMLR Workshop and Conference Proceedings, 2011. {454}

[267] J. Zhu, S. Rosset, R. Tibshirani, and T. Hastie. 1-norm support vector machines. *Advances in Neural Information Processing Systems*, 16, 2003. {276}

[268] J. Zhu, Y. Yan, L. Zhao, M. Heimann, L. Akoglu, and D. Koutra. Beyond homophily in graph neural networks: Current limitations and effective designs. *Advances in Neural Information Processing Systems*, 33:7793–7804, 2020. {514}

[269] X. Zhu and Z. Ghahramani. Learning from labeled and unlabeled data with label propagation. In *Technical Report CMU-CALD-02-107*. Carnegie Mellon University, 2002. {452, 453}

[270] X. Zhu, Z. Ghahramani, and J. D. Lafferty. Semi-supervised learning using Gaussian fields and harmonic functions. In *Proceedings of the 20th International Conference on Machine Learning*, pages 912–919, 2003. {448, 452–454}

Index

The manufacturer's authorised representative in the EU is Springer
Nature Customer Service Centre GmbH, Europaplatz 3, 69115 Heidelberg,
Germany. If you have any concerns regarding our products, please
contact ProductSafety@springernature.com

Printed and bound by CPI Group (UK) Ltd, Croydon, CR0 4YY
26/06/2026
02148989-0001